Lewin's Essential GENES

Third Edition

Jones & Bartlett Learning Titles in Biological Science

Marine Environmental Biology and Conservation
Daniel W. Beckman

Plant Structure: A Colour Guide
Brian G. Bowes, James D. Mauseth

Lewin's CELLS, Second Edition
Lynne Cassimeris
George Plopper
Vishwanath R. Lingappa

Environmental Science, Ninth Edition
Daniel D. Chiras

Human Biology, Eighth Edition
Daniel D. Chiras

Plants, Genes, and Crop Biotechnology
Maarten J. Chrispeels
David E. Sadava

Plant Biochemistry
Florence Gleason

Restoration Ecology
Sigurdur Greipsson

Evolution: Principles and Processes
Brian K. Hall

Strickberger's Evolution, Fourth Edition
Brian K. Hall
Benedikt Hallgrímsson

Essential Genetics: A Genomics Perspective, Fifth Edition
Daniel L. Hartl

Genetics: Analysis of Genes and Genomes, Eighth Edition
Daniel L. Hartl
Elizabeth W. Jones

Genetics of Populations, Fourth Edition
Philip W. Hedrick

Lewin's Essential GENES, Third Edition
Jocelyn E. Krebs
Elliott S. Goldstein
Stephen T. Kilpatrick

Lewin's GENES X
Jocelyn E. Krebs
Elliott S. Goldstein
Stephen T. Kilpatrick

Tropical Forests
Bernard A. Marcus

Botany: An Introduction to Plant Biology, Fourth Edition
James D. Mauseth

Plants & People
James D. Mauseth

Environmental Science: Systems and Solutions, Fourth Edition
Michael McKinney
Robert Schoch
Logan Yonavjak

RNA Interference and Model Organisms
Joanna A. Miller

An Introduction to Marine Mammal Biology and Conservation
E.C.M. Parsons

Alcamo's Fundamentals of Microbiology: Body Systems Edition, Second Edition
Jeffery C. Pommerville

Principles of Cell Biology
George Plopper

An Introduction to Zoology: Investigating the Animal World
Joseph T. Springer
Dennis Holley

Introduction to the Biology of Marine Life, Tenth Edition
John Morrissey
James L. Sumich

Exploring Bioinformatics: A Project-Based Approach
Caroline St. Clair
Jonathan E. Visick

The Ecology of Agroecosystems
John H. Vandermeer

Mammalogy, Fifth Edition
Terry A. Vaughan
James M. Ryan
Nicholas J. Czaplewski

Lewin's Essential GENES

Third Edition

Jocelyn E. Krebs
University of Alaska Anchorage

Elliott S. Goldstein
Arizona State University

Stephen T. Kilpatrick
University of Pittsburgh at Johnstown

JONES & BARTLETT
LEARNING

World Headquarters
Jones & Bartlett Learning
5 Wall Street
Burlington, MA 01803
978-443-5000
info@jblearning.com
www.jblearning.com

Jones & Bartlett Learning books and products are available through most bookstores and online booksellers. To contact Jones & Bartlett Learning directly, call 800-832-0034, fax 978-443-8000, or visit our website, www.jblearning.com.

Substantial discounts on bulk quantities of Jones & Bartlett Learning publications are available to corporations, professional associations, and other qualified organizations. For details and specific discount information, contact the special sales department at Jones & Bartlett Learning via the above contact information or send an email to specialsales@jblearning.com.

Production Credits
Chief Executive Officer: Ty Field
President: James Homer
SVP, Editor-in-Chief: Michael Johnson
SVP, Chief Technology Officer: Dean Fossella
SVP, Chief Marketing Officer: Alison M. Pendergast
Publisher: Cathleen Sether
Senior Acquisitions Editor: Erin O'Connor
Senior Associate Editor: Megan R. Turner
Editorial Assistant: Rachel Isaacs
Production Manager: Louis C. Bruno, Jr.
Senior Marketing Manager: Andrea DeFronzo
V.P., Manufacturing and Inventory Control: Therese Connell
Production Services and Composition: Aptara®, Inc.
Illustrations: Imagineering Media Services, Inc., Shepherd, Inc., and Aptara®, Inc.
Cover Design: Kristin E. Parker
Text Design: Anne Spencer
Permissions and Photo Research Supervisor: Anna Genoese
Cover Image: Courtesy of Leonid Mirny and Maxim Imakaev, MIT
Printing and Binding: Courier Kendallville
Cover Printing: Courier Kendallville

About the cover: The fractal globule model of 3D genome architecture in a human cell. The chromatin fiber is formed by DNA and nucleosomes. Nearby regions of the fiber are indicated using similar colors and are close in 3D. The model is a result of computer simulations and analysis of Hi-C data. For more details see Lieberman-Aiden E, van Berkum NL, et al, "Comprehensive mapping of long-range interactions reveals folding principles of the human genome" *Science* 2009; 326(5950):289–293.

To order this book, use ISBN 978-1-4496-4479-6

Library of Congress Cataloging-in-Publication Data
Krebs, Jocelyn E.
Lewin's essential genes / Jocelyn E. Krebs, Elliott S. Goldstein, Stephen T. Kilpatrick. — 3rd ed.
p. ; cm.
Essential genes
Condensed ed. of: Genes X / Benjamin Lewin. c2011.
Includes bibliographical references and index.
ISBN 978-1-4496-1265-8
1. Genetics. 2. Genes. I. Lewin, Benjamin. II. Goldstein, Elliott S. III. Kilpatrick, Stephen T. IV. Lewin, Benjamin. Genes X. V. Title. VI. Title: Essential genes.
[DNLM: 1. Genes. 2. DNA—genetics. 3. Genetic Processes. 4. Genome. 5. Proteins—genetics. 6. RNA—genetics. QU 470]
QH430.L4 2012
576.5—dc23 2011028260

6048

Printed in the United States of America
16 15 14 13 12 10 9 8 7 6 5 4 3 2 1

Dedication

To Benjamin Lewin, for setting the bar high.

To my mother, Ellen Baker, for raising me with a love of science; to the memory of my stepfather Barry Kiefer, for convincing me science would stay fun; and to my partner Susannah Morgan, for always pretending my biology jokes are funny; and to my son Rhys, who is far too young to enjoy this text. Finally, I would like to dedicate this edition to the memory of my mentor, Dr. Marietta Dunaway, a great inspiration who set my feet of the exciting path of chromatin biology.

Jocelyn Krebs

To my family: my wife, Suzanne, whose patience, understanding, and confidence in me are amazing; my children, Andy, Hyla, and Gary, who have taught me so much about using the computer; and my grandchildren, Seth and Elena, whose smiles and giggles inspire me. And to the memory of my mentor and dear friend, Lee A. Snyder, whose professionalism, guidance, and insight demonstrated the skills necessary to be a scientist and teacher. I have tried to live up to his expectations. This is for you, Doc.

Elliott Goldstein

To my wife, Lori; my parents, David and Sandra;
and my children, Jennifer, Andrew, and Sarah.

Stephen Kilpatrick

Brief Contents

Contents

PART II. DNA REPLICATION AND RECOMBINATION 269

Chapter 11. Replication Is Connected to the Cell Cycle 270

Chapter 12. The Replicon: Initiation of Replication 291

Chapter 13. DNA Replication 308

Chapter 30. Regulatory RNA 780

Preface

Of the diverse ways to study the living world, molecular biology has been most remarkable in the speed and breadth of its expansion. New data are acquired daily, and new insights into well-studied processes come on a scale measured in weeks or months rather than years. It's difficult to believe that the first complete organismal genome sequence was obtained less than 20 years ago, and, yet, routine individual whole-genome sequencing is just around the corner. The structure and function of genes and genomes and their associated cellular processes are sometimes elegantly and deceptively simple but frequently amazingly complex, and no single book can do justice to the realities and diversities of natural genetic systems. The purpose of this book is to provide a clear and concise overview of the field for the undergraduate student; it may also be appropriate for some medical school courses in the subject. Compared to the full edition, there is a redirected focus on essential topics and (in some areas) more background and introductory material.

Much of the revision and reorganization of this edition follows that of *Lewin's GENES X,* but there are many updates and features that are new to this book. Most notably, there are two new chapters in this edition: Chapter 3, Methods in Molecular Biology and Genetic Engineering, provides an introduction to the concepts and practice of laboratory techniques in molecular biology early on in the book, and Chapter 8, Genome Evolution, combines, expands, and updates material that had been scattered among various chapters in previous editions, and introduces a number of topics new to this book. This edition is generally reorganized for a more logical flow of topics, and many chapters and sections within chapters have been renamed to better indicate their contents. In particular, discussion of chromatin organization and nucleosome structure now precedes the discussion of eukaryotic transcription, because chromosome organization is critical to all DNA transactions in the cell, and current research in the field of transcriptional regulation is heavily biased toward the study of the role of chromatin in this process. The discussion of transcriptional activation and chromatin remodeling has accordingly been combined into one chapter (Chapter 28). Two chapters on transposons and retroposons have been combined into one (Chapter 17). In addition, some chapters have been revised to contain extensive new material. The original introductory chapter on messenger RNA has been entirely rewritten to cover more advanced topics (Chapter 22, mRNA Stability and Localization), and the regulatory RNA chapter has been dramatically expanded to include material on RNAi pathways (Chapter 30, Regulatory RNA). Many new figures are included in this book, some reflecting new developments in the field, particularly in the topics of chromatin structure and function, epigenetics, and regulation by noncoding and microRNAs in eukaryotes.

This book is organized into four parts. **Part I, Genes and Chromosomes,** comprises Chapters 1 through 10. Chapters 1 and 2 serve as an introduction to the structure and function of DNA and contain basic coverage of DNA replication and gene expression. Chapter 3 provides information on molecular laboratory techniques. Chapter 4 introduces the interrupted structures of eukaryotic genes, and Chapters 5 through 8 discuss genome structure and evolution. Chapter 9 discusses the structure of viral, prokaryotic, and eukaryotic chromosomes, while Chapter 10 focuses on the more detailed structure of eukaryotic chromatin.

Part II, DNA Replication and Recombination, comprises Chapters 11 through 18. Chapters 11 to 14 provide detailed discussions of DNA replication in plasmids, viruses, and prokaryotic and eukaryotic cells. Chapters 15 through 18 cover recombination and its roles in DNA repair and the human immune system, with Chapter 16 discussing DNA repair pathways in detail and Chapter 17 focusing on different types of transposable elements.

Part III, Gene Expression, includes Chapters 19 through 25. Chapters 19 and 20 provide more in-depth coverage of bacterial and eukaryotic transcription.

Chapters 21 through 23 are concerned with RNA, discussing messenger RNA, RNA stability and localization, RNA processing, and the catalytic roles of RNA. Chapters 24 and 25 discuss translation and the genetic code.

Part IV, Gene Regulation, comprises Chapters 26 through 30. In Chapter 26, the regulation of bacterial gene expression via operons is discussed. Chapter 27 covers the regulation of expression of genes during phage development as they infect bacterial cells. Chapters 28 and 29 cover eukaryotic gene regulation, including epigenetic modifications. Finally, Chapter 30 covers RNA-based control of gene expression in prokaryotes and eukaryotes.

For instructors who prefer to order topics with the essentials of DNA replication and gene expression followed by more advanced topics, the following chapter sequence is suggested:

Introduction: Chapters 1–2
Gene Structure and Genome Structure: Chapters 4–7
DNA Replication: Chapters 11–14
Transcription: Chapters 19–22
Translation: Chapters 24–25
Regulation of Gene Expression: Chapters 9–10 and 26–30

Other chapters can be covered at the instructor's discretion.

To the Instructor

This edition contains many pedagogical components to help the instructor engage students in the topic. Each chapter section concludes with Concept and Reasoning Checks: one or two questions for review, conceptual synthesis, hypothesizing, or application of the information. Each chapter includes a set of End-of-Chapter Questions with answers to half of the questions provided to the students; the other questions could be used as homework assignments or quizzes. There are additional instructional tools available on the Instructor's Media CD and accompanying website (see below).

To the Student

There are a number of features in the book to help you learn as you read. Each section is summarized with a bulleted list of Key Concepts. Key Terms are highlighted in boldface in the text and defined in the margin for easy reference and are compiled into the Glossary at the end of the book. Each chapter includes a set of End-of-Chapter Questions intended for self-assessment. Most chapters contain at least one feature box with additional background material or more in-depth details on an issue relevant to the chapter's focus. Boxes fall into one of four categories: Essential Ideas, Historical Perspectives, Methods and Techniques, and Medical Applications. In many cases these represent areas of ongoing research in the field. Finally, each chapter concludes with suggested Further Reading, a brief list of current reviews and pivotal papers to supplement and reinforce the chapter content.

Supplements

Jones & Bartlett Learning offers an impressive variety of traditional and interactive multimedia supplements to assist instructors and aid students in mastering molecular biology. Additional information and review copies of any of the following items are available through your Jones & Bartlett Learning sales representative or by going to http://www.jblearning.com.

For the Student

Companion Website

Jones & Bartlett Learning and Brent Nielsen of Brigham Young University have developed an interactive companion website dedicated exclusively to this title. Students will find a variety of study aids and resources at go.jblearning.com/essgenes3e, all designed to explore the concepts of molecular biology in more depth and to help students master the material in the book. A variety of activities are available to help students review class material, such as an interactive summary, web-based learning exercises, study quizzes, a searchable glossary, and links to animations, videos, and podcasts, all to help students master important terms and concepts.

For the Instructor

Instructor's Media CD

The Instructor's Media CD contains a suite of files to help professors teach their courses. The materials are cross-platform for Windows and Macintosh systems. All the files on the CD are ready for online courses using the WebCT or Blackboard formats.

- The **PowerPoint® Image Bank** provides the illustrations, photographs, and tables (to which Jones & Bartlett Learning holds the copyright or has permission to reprint digitally) inserted into PowerPoint slides. With the Microsoft PowerPoint program, you can quickly and easily copy individual image slides into your existing lecture slides.

- The **PowerPoint Lecture Outline** presentation package provides images and lecture notes, created by author Stephen T. Kilpatrick, for each chapter of *Lewin's Essential Genes, Third Edition.* A PowerPoint viewer is provided on the CD. Instructors with the Microsoft PowerPoint software can customize the outlines, figures, and order of presentation.

Answers to End-of-Chapter Questions
Each chapter of *Lewin's Essential Genes, Third Edition* includes a set of End-of-Chapter Questions with answers to half of the questions provided to the students; the other questions may be used as homework assignments or quizzes. An answer guide for the second half of these questions is available as an instructor download.

Test Bank
An electronic Test Bank, also created by Stephen T. Kilpatrick, is provided as a text file with over 700 questions in a variety of formats and is available as an instructor download.

Acknowledgements

The authors would like to thank the editorial, production, marketing, and sales teams at Jones & Bartlett Learning for their guidance in the preparation of this book. They have been exemplary in all aspects of this project. Megan Turner, Cathy Sether, Molly Steinbach, Anna Genoese, and Lou Bruno deserve special mention. Megan, Cathy, and Molly have provided guidance, leadership, and assistance from the original concept and throughout the production process. Anna and Lou have been especially helpful as the content has been shaped into an actual book, and we thank them for their patience with us.

This book was written and edited using Power XEditor (PXE) developed by Aptara, Inc. We thank Kelly Ricci of Aptara for interpreting our directions and intentions into an accurate and visually appealing book. We also thank the members of Aptara's technical support team for assistance as we made the transition to an online authoring and editing system.

Thanks for the creation of many of the End-of-Chapter Questions go to Brent Nielsen. We also thank the authors of the special topics boxes found throughout the text:

Loree Burns
Jamie Kass, New York Academy of Sciences
Brent Nielsen, Brigham Young University
Teri Shors, University of Wisconsin, Oshkosh
Esther Siegfried, Penn State–Altoona.

Finally, we would like to express our gratitude to the reviewers, whose feedback helped to shape the text in many ways:

Salem Al-Maloul, Hashemite University
James Botsford, New Mexico State University
David Bourgaize, Whittier College
John Boyle, University of Mississippi
Mary Connell, Appalachian State University
Robert Dotson, Tulane University
Julia Frugoli, Clemson University
Daniel Herman, University of Wisconsin, Eau Claire
Stan Ivey, Delaware State University
Christi Magrath, Troy University
Mitch McVey, Tufts University
Hao Nguyen, California State University, Sacramento
Stacy Darling Novak, University of La Verne
Eva Sapi, University of New Haven
Ben Stark, Illinois Institute of Technology
Takashi Ueda, Florida Gulf Coast University
Ramakrishna Wusirika, Michigan Technological University
Anastasia Zimmerman, College of Charleston

Jocelyn E. Krebs
Elliott S. Goldstein
Stephen T. Kilpatrick

About the Authors

Benjamin Lewin founded the journal *Cell* in 1974 and was Editor until 1999. He founded the Cell Press journals *Neuron, Immunity,* and *Molecular Cell.* In 2000, he founded Virtual Text, which was acquired by Jones & Bartlett Publishers in 2005. He is also the author of *GENES* and *CELLS.*

Jocelyn E. Krebs received a B.A. in Biology from Bard College, Annandale-on-Hudson, NY, and a Ph.D. in Molecular and Cell Biology from the University of California, Berkeley. For her Ph.D. dissertation, she studied the roles of DNA topology and insulator elements in transcriptional regulation in the laboratory of Dr. Marietta Dunaway. She performed her postdoctoral training as an American Cancer Society Fellow at the University of Massachusetts Medical School in the laboratory of Dr. Craig Peterson, where she focused on the roles of histone acetylation and chromatin remodeling in transcription. In 2000, Dr. Krebs joined the faculty in the Department of Biological Sciences at the University of Alaska Anchorage, where she is now a full professor. She also serves as the Director of Alaska INBRE (IDeA Networks of Biomedical Research Excellence). She directs a research group studying chromatin structure and function in transcription and DNA repair in the yeast *Saccharomyces cerevisiae* and the role of chromatin remodeling in embryonic development in the frog *Xenopus*. She teaches courses in molecular biology for undergraduates, graduate students, and first-year medical students. She also teaches a Molecular Biology of Cancer course and has taught Genetics and Introductory Biology. She lives in Eagle River, AK, with her partner, their son, and a house full of dogs and cats. Her non-work passions include hiking, camping, and snowshoeing.

Elliott S. Goldstein earned his B.S. in Biology from the University of Hartford (Connecticut) and his Ph.D. in Genetics from the University of Minnesota, Department of Genetics and Cell Biology. Following this, he was awarded an N.I.H. Postdoctoral Fellowship to work with Dr. Sheldon Penman at the Massachusetts Institute of Technology. After leaving Boston, he joined the faculty at Arizona State University in Tempe, where he is an associate professor in the Cellular, Molecular, and Biosciences program in the School of Life Sciences and in the Honors Disciplinary Program. His research interests are in the area of molecular and developmental genetics of early embryogenesis in *Drosophila melanogaster*. In recent years, he has focused on the *Drosophila* counterparts of the human proto-oncogenes *jun* and *fos*. His primary teaching responsibilities are in the undergraduate General Genetics course as well as the graduate level Molecular Genetics course. Dr. Goldstein lives in Tempe with his wife, his high school sweetheart. They have three children and two grandchildren. He is a bookworm who loves reading as well as underwater photography. His pictures can be found at http://www.public.asu.edu/~elliotg/.

Stephen T. Kilpatrick received a B.S. in Biology from Eastern College (now Eastern University) in St. Davids, PA, and a Ph.D. from the Program in Ecology and Evolutionary Biology at Brown University. His dissertation research was an investigation of the population genetics of interactions between the mitochondrial and nuclear genomes of *Drosophila melanogaster*. Since 1995, Dr. Kilpatrick has taught at the University of

Pittsburgh–Johnstown in Johnstown, PA. His regular teaching duties include undergraduate courses in nonmajors biology, introductory majors biology, genetics for nursing students, and advanced undergraduate courses in genetics, evolution, molecular genetics, and biostatistics. He has also supervised a number of undergraduate research projects in evolutionary genetics. Dr. Kilpatrick's major professional focus has been in biology education. He has participated in the development and authoring of ancillary materials for several introductory biology, genetics, and molecular genetics texts as well as writing articles for educational reference publications. For his classes at Pitt-Johnstown, Dr. Kilpatrick has developed many active learning exercises in introductory biology, genetics, and evolution. Dr. Kilpatrick resides in Johnstown, PA, with his family. Outside of scientific interests, he enjoys music, literature, and theater and occasionally performs in local community theater groups. He can be contacted at kilpatri@pitt.edu.

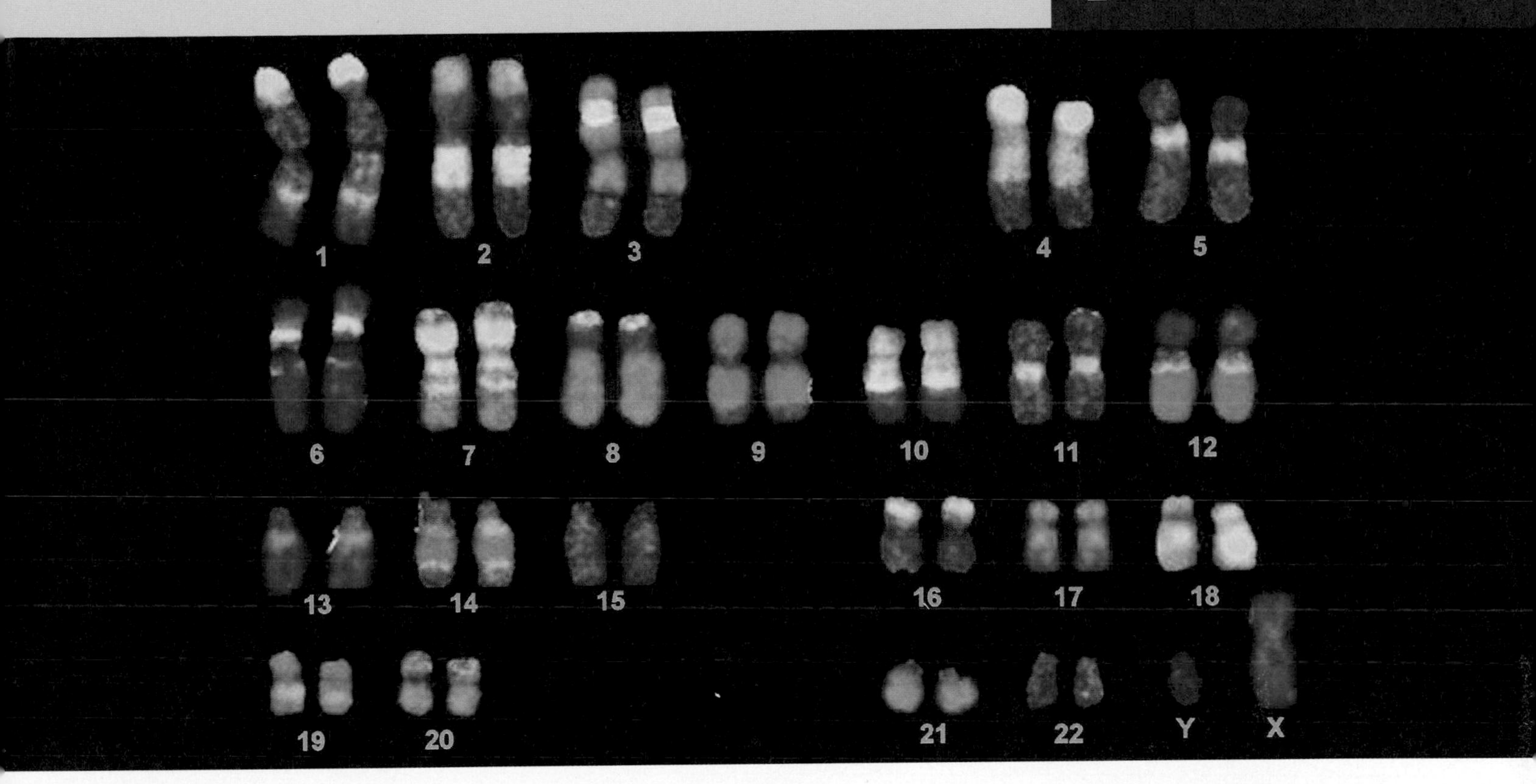

The human male karyotype, or complete set of chromosomes, shown using chromosome painting. The "pseudocolors" are created by hybridizing fluorescent molecular probes to specific chromosomal regions. Photo courtesy of Steven M. Carr, Memorial University. Adapted from a photograph by Genetix Ltd. Used with permission.

Genes and Chromosomes

1

A strand of DNA in blue and red. DNA is the genetic material of eukaryotic cells, bacteria, and many viruses.

Genes Are DNA

CHAPTER OUTLINE

1.1 Introduction

The hereditary basis of every living organism is its **genome**, a long sequence of DNA that provides the complete set of hereditary information carried by the organism's cells. The genome includes chromosomal DNA as well as DNA in plasmids and (in eukaryotes) organellar DNA as found in mitochondria and chloroplasts. We use the term *information* because the genome does not itself perform an active role in the development of the organism. It is the sequence of the individual subunits, or bases, of the DNA that determines development. By a complex series of interactions, the DNA sequence encodes all the RNAs and proteins of the organism that are to be produced at the appropriate time and place. Proteins serve a diverse series of roles in the development and functioning of an organism: they can form part of the structure of the organism, they have the capacity to build the structures, they perform the metabolic reactions necessary for life, and they participate in regulation as transcription factors, receptors, key players in signal transduction pathways, and other molecules. RNAs encoded by the genome, but that do not themselves encode proteins, also function in the expression of genes during development.

▸ **genome** The complete set of sequences in the genetic material of an organism. It includes the sequence of each chromosome plus any DNA in organelles.

Physically, the genome may be divided into a number of different DNA molecules, or **chromosomes**. The ultimate definition of a genome is the sequence of the DNA of each chromosome. Functionally, the genome is divided into genes. Each gene is a sequence of DNA that encodes a single type of RNA or polypeptide (although it may encode multiple versions of its product). Each of the discrete chromosomes comprising the genome may contain a large number of genes. Genomes for living organisms may contain as few as ~500 genes (for a mycoplasma, a type of bacterium) to ~20,000 to 25,000 for a human being and as many as 50,000 in some plants.

▸ **chromosome** A discrete unit of the genome carrying many genes. Each consists of a very long molecule of duplex DNA and (in eukaryotes) an approximately equal mass of proteins. It is visible as a morphological entity only during cell division.

In this chapter, we explore the gene in terms of its basic molecular construction. **FIGURE 1.1** summarizes the stages in the transition from the historical concept of the gene to the modern definition of the genome.

The first definition of the gene as a functional molecular unit followed from the discovery that individual genes are responsible for the production of specific proteins. The chemical differences between the DNA of the gene and its protein product led to the suggestion that a gene encodes a protein. This in turn led to the discovery of the complex apparatus by which the DNA sequence of a gene determines the amino acid sequence of a polypeptide.

Understanding the process by which a gene is expressed allows us to make a more rigorous definition of its nature. **FIGURE 1.2** shows the basic theme of this book. A gene is a sequence of DNA that directly produces a single strand of another nucleic acid, RNA, with a sequence that is identical to one of the two polynucleotide strands of DNA. In many cases, the RNA is in turn used to direct production of a polypeptide, whereas in other cases (such as rRNA, tRNA, and many other genes, the RNA transcribed from the gene is the functional end product. Thus a gene is a sequence of DNA that encodes an RNA, and in protein-coding (or **structural**) genes, the RNA in turn encodes a polypeptide.

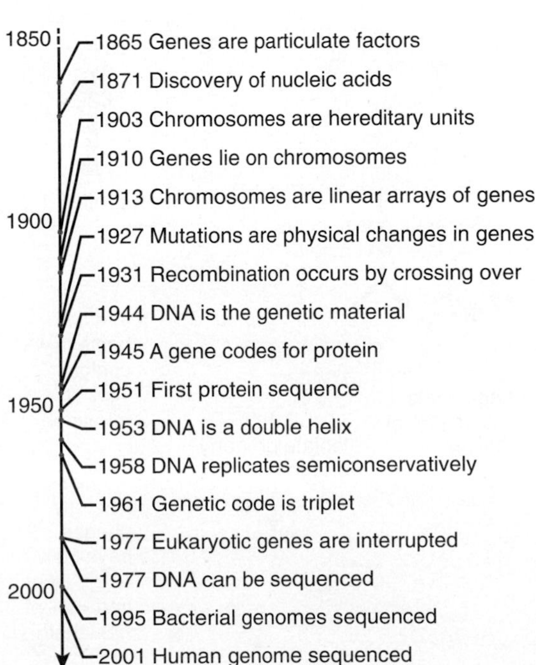

FIGURE 1.1 A brief history of genetics.

▸ **structural gene** A gene that encodes any RNA or polypeptide product other than a regulator.

From the demonstration that a gene consists of DNA and that a chromosome consists of a long stretch

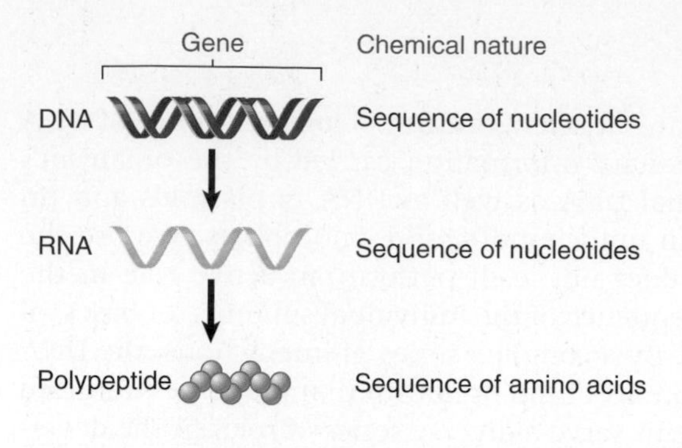

FIGURE 1.2 A gene encodes an RNA, which may encode a polypeptide.

of DNA representing many genes, we will move to the overall organization of the genome. In *Chapter 4, The Interrupted Gene,* we take up in more detail the organization of the gene and its representation in proteins. In *Chapter 5, The Content of the Genome,* we consider the total number of genes, and in *Chapter 7, Clusters and Repeats,* we discuss other components of the genome and the maintenance of its organization.

CONCEPT AND REASONING CHECK

Why is it accurate to say that a genome has information for the development of an organism but does not directly participate in development?

1.2 DNA Is the Genetic Material of Bacteria, Viruses, and Eukaryotic Cells

▸ **transformation** In bacteria, it is the acquisition of new genetic material by incorporation of added DNA.

The idea that the genetic material is DNA has its roots in the discovery of **transformation** by Frederick Griffith in 1928 (see the accompanying box, *Historical Perspectives: Determining That DNA Is the Genetic Material*). Purification of the transforming principle from bacterial cells in 1944 by Avery, MacLeod, and McCarty showed that it is deoxyribonucleic acid (DNA).

Having shown that DNA is the genetic material of bacteria, the next step was to demonstrate that DNA is the genetic material in a quite different system. Phage T2 is a bacteriophage virus (or phage) that infects the bacterium *Escherichia coli.* When phage particles are added to bacteria, they attach to the outside surface, some material enters the cell, and then ~20 minutes later each cell bursts open, or lyses, to release a large number of progeny phage.

FIGURE 1.3 illustrates the results of an experiment in 1952 by Alfred Hershey and Martha Chase in which bacteria were infected with T2 phages that had been radioactively labeled either in their DNA component (with ^{32}P) or in their protein component (with ^{35}S). The infected bacteria were agitated in a blender, and two fractions were separated by centrifugation. One fraction contained the empty phage "ghosts" that were released from the surface of the bacteria, and the other consisted of the infected bacteria themselves. Previously, it had been shown that phage replication occurs intracellularly, so that the genetic material of the phage would have to enter the cell during infection.

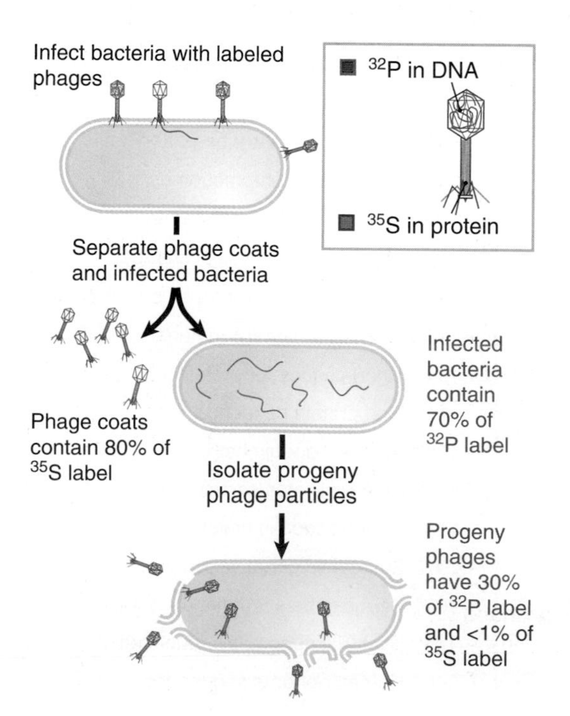

FIGURE 1.3 The genetic material of phage T2 is DNA.

Most of the ^{32}P label was present in the fraction containing infected bacteria. The progeny phage particles produced by the infection contained ~30% of the original ^{32}P label. The progeny received less than 1% of the protein contained in the original phage population. The phage ghosts consist of protein and therefore carried the ^{35}S radioactive label. This experiment directly showed that only the DNA of the parent phages enters the bacteria and becomes part of the progeny phages, which is exactly the pattern expected of genetic material.

A phage reproduces by commandeering the machinery of an infected host cell to manufacture more copies of itself. The phage possesses genetic material with properties analogous to those of cellular genomes: its traits are faithfully expressed and are subject to the same rules that govern inheritance of cellular traits. The case of T2 reinforces the general conclusion that DNA is genetic material of the genome of a cell or a virus.

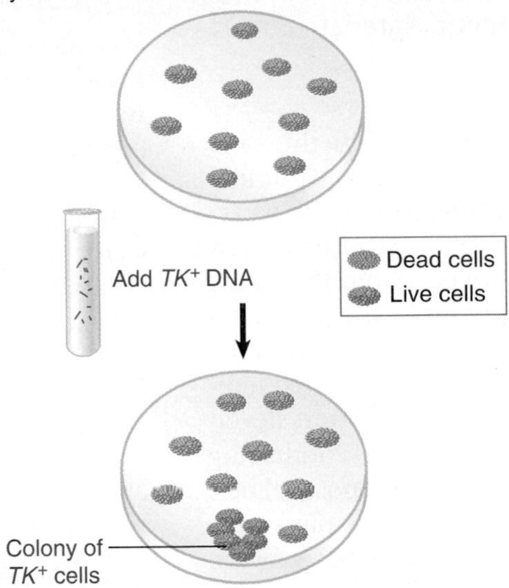

FIGURE 1.4 Eukaryotic cells can acquire a new phenotype as the result of transfection by added DNA.

When DNA is added to eukaryotic cells growing in culture, it enters the cells, and in some of them this results in the production of new proteins. When an isolated gene is used, its incorporation leads to the production of a particular protein, as depicted in **FIGURE 1.4**. Although for historical reasons these experiments are described as **transfection** when performed with animal cells, they are a direct counterpart to bacterial transformation. The DNA that is introduced into the recipient cell becomes part of its genome and is inherited with it, and expression of the new DNA results in a new trait. At first, these experiments were successful only with individual cells growing in culture, but in later experiments DNA was introduced into mouse eggs by microinjection and became a stable part of the genome of the mouse. Such experiments show directly that DNA is the genetic material in eukaryotes and that it can be transferred between different species and remain functional.

▸ **transfection** In eukaryotic cells, the acquisition of new genetic material by incorporation of added DNA.

The genetic material of all known organisms and many viruses is DNA. However, some viruses use RNA as the genetic material. Therefore, the general nature of the genetic material is that it is always nucleic acid; specifically, it is DNA, except in the RNA viruses.

KEY CONCEPTS

- Bacterial transformation provided the first support that DNA is the genetic material of bacteria. Genetic properties can be transferred from one bacterial strain to another by extracting DNA from the first strain and adding it to the second strain.
- Phage infection showed that DNA is the genetic material of viruses. When the DNA and protein components of bacteriophages are labeled with different radioactive isotopes, only the DNA is transmitted to the progeny phages produced by infecting bacteria.
- DNA can be used to introduce new genetic traits into animal cells or whole animals.
- In some viruses, the genetic material is RNA.

CONCEPT AND REASONING CHECK

If Hershey and Chase had observed that nearly none of the DNA but a substantial fraction of the protein of phage T2 enters *E. coli* cells during infection, what would they have concluded?

HISTORICAL PERSPECTIVES

Determining That DNA Is the Genetic Material

Pneumonia was a leading cause of death in the early part of the 20th century, and much effort was put into understanding the structure and function of the bacteria known to cause it: the pneumococci. Scientists knew that several pneumococcal types existed and that these types could be distinguished by the molecules (capsular polysaccharides) displayed on their surface, but they did not know whether the different types represented individual and stable strains of bacteria or different developmental stages of a single strain. While conducting experiments to distinguish between these possibilities, Frederick Griffith set in motion a series of experiments that ultimately led to the identification of DNA as the genetic material of all cells.

Griffith studied S forms and R forms of pneumococcal bacteria (so-called for the smooth and rough appearance of the bacteria when grown in laboratory culture; see *Section 1.2, DNA Is the Genetic Material of Bacteria, Viruses, and Eukaryotic Cells*). The R form, which was generated in the laboratory from the S form, produced no capsular polysaccharides and did not cause pneumonia when injected into mice. The S form was lethal to mice but could be inactivated by exposing the bacteria to high heat before injection. Surprisingly, Griffith found that when the R form was injected into mice alongside heat-killed S bacteria, the mice contracted pneumonia and died (**FIGURE B1.1**). What was more, live S-type bacteria—virulent and coated with capsular proteins—could be isolated from the dead mice. R-form bacteria had clearly been transformed into a stable S form, and the transformation process required some factor from the heat-killed S sample. Griffith suspected this factor, the so-called transforming principle, was a protein . . . and he was not alone.

Proteins are by far the most abundant macromolecules in living cells. Their essential components, the amino acids, come in twenty varieties and can be joined in

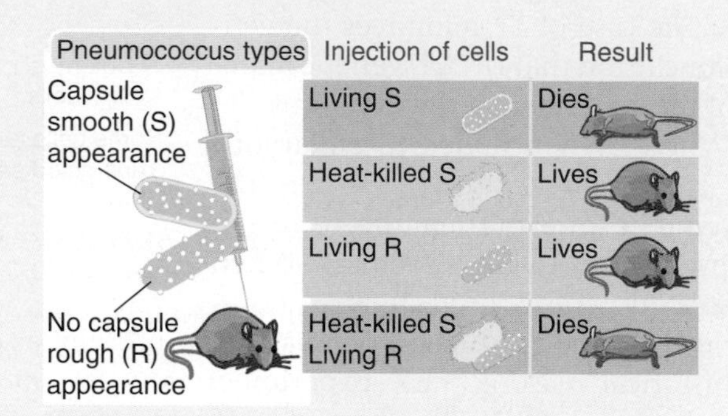

FIGURE B1.1 Neither heat-killed S-type nor live R-type bacteria can kill mice, but simultaneous injection of both can kill mice just as effectively as the live S-type.

1.3 Polynucleotide Chains Have Nitrogenous Bases Linked to a Sugar-Phosphate Backbone

- **purine** A double-ringed nitrogenous base, such as adenine or guanine.
- **pyrimidine** A single-ringed nitrogenous base, such as cytosine, thymine, or uracil.
- **nucleoside** A molecule consisting of a purine or pyrimidine base linked to the 1′ carbon of a pentose sugar.
- **nucleotide** A molecule consisting of a purine or pyrimidine base linked to the 1′ carbon of a pentose sugar and a phosphate group linked to either the 5′ or 3′ carbon of the sugar.
- **polynucleotide** A chain of nucleotides, such as DNA or RNA.

The essential component of nucleic acids (DNA and RNA) is the nucleotide, which has three parts:

- a nitrogenous base,
- a sugar, and
- one or more phosphates.

The nitrogenous base is a **purine** or **pyrimidine** ring. The base is linked to the 1′ ("one prime") carbon on a pentose sugar by a glycosidic bond from the N_1 of pyrimidines or the N_9 of purines. The pentose sugar linked to a nitrogenous base is called a **nucleoside**. Nucleic acids are named for the type of sugar: DNA has 2′-deoxyribose, whereas RNA has 2′-ribose. The difference is that the sugar in RNA has a hydroxyl (–OH) group on the 2′ carbon of the pentose ring. The sugar can be linked by its 5′ or 3′ carbon to a phosphate group. A nucleoside linked to a phosphate is a **nucleotide**.

A **polynucleotide** is a long chain of nucleotides. **FIGURE 1.5** shows that the backbone of the polynucleotide chain consists of an alternating series of pentose (sugar) and phosphate residues. The chain is formed by linking the 5′ carbon of one pentose ring to the 3′ carbon of the next pentose ring via a phosphate group, so the sugar-phosphate backbone is said to consist of 5′–3′ phosphodiester linkages. The

seemingly endless combinations to result in molecules with a tremendous variety of size, shape, and function. By contrast, DNA is a relatively minor component of cells and was known at the time to comprise only four nucleotide types. These nucleotides were thought to be arranged in a specific, repetitive pattern, which contributed to the notion that deoxyribonucleic acids were molecules of "monotonous uniformity" that were unlikely to have any biological specificity. And, so, it was universally assumed that the transforming principle was a protein.

In 1944, Oswald Avery, Colin MacLeod, and Maclyn McCarty published the first report on the chemical nature of the transforming principle, and their results were startling. The main component of their active sample of purified transforming principle was not protein at all; it was DNA. To show that the transforming activity of this sample was due to the DNA and not to some minor but active contaminant, the authors treated it with enzymes that degrade protein, RNA, and carbohydrate; in all cases the sample retained the ability to stably transform R bacteria into S bacteria. The authors concluded that the transforming principle "consists principally, if not solely, of a highly polymerized, viscous form of deoxyribonucleic acid."

Avery and his colleagues strengthened their claim that DNA was the transforming (or genetic) material by later showing that DNase I, an enzyme that specifically degrades DNA, completely eliminated the transforming activity of their purified sample, but the scientific community remained skeptical. It was not until 1952, when A. D. Hershey and Martha Chase studied the independent roles of protein and DNA in bacteriophage T2 infection, that the notion of DNA as the genetic material finally gained favor.

Bacteriophage T2 is a virus that infects bacterial cells in discrete steps: the phage particle attaches itself to the bacterial cell wall, phage material is injected into the cell, the host cell is induced to produce new phage particles, and, finally, the bacterial cell bursts, releasing hundreds of new phage particles. To follow the movement of phage protein and phage DNA during this process, Hershey and Chase radioactively labeled each macromolecule separately: the protein with a radioactive form of sulfur and the DNA with a radioactive form of phosphorous. In their experiment, the majority of the phosphorous label entered the bacterial cell during infection, whereas the majority of the sulfur label stayed outside the cell. The authors went on to show that the new phage particles produced by the infected bacterial cell contained a large percentage of the labeled DNA but virtually none of the labeled protein (see Figure 1.3). These results indicated that it was DNA from the infecting phage, not protein, that entered the bacterial cell and induced genetic changes.

nitrogenous bases "stick out" from the backbone.

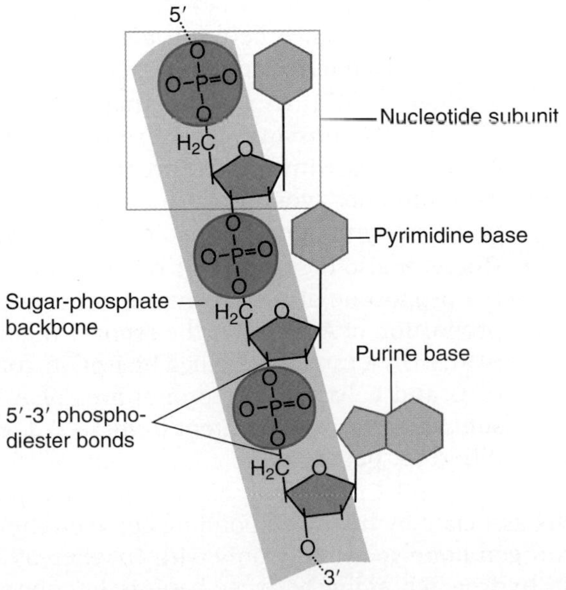

Each nucleic acid contains four types of nitrogenous base. The same two purines, adenine (A) and guanine (G), are present in both DNA and RNA. The two pyrimidines in DNA are cytosine (C) and thymine (T); in RNA uracil (U) is found instead of thymine. The only difference between uracil and thymine is the presence of a methyl group at position C_5.

The terminal nucleotide at one end of the chain has a free 5′ phosphate group, whereas the terminal nucleotide at the other end has a free 3′ hydroxyl group. It is conventional to write nucleic acid sequences in the 5′ to 3′ direction—that is, from the 5′ terminus at the left to the 3′ terminus at the right.

FIGURE 1.5 A polynucleotide chain consists of a series of 5′–3′ sugar-phosphate links that form a backbone from which the bases protrude.

KEY CONCEPTS

- A nucleoside consists of a purine or pyrimidine base linked to the 1′ carbon of a pentose sugar.
- The difference between DNA and RNA is in the group at the 2′ position of the sugar. DNA has a deoxyribose sugar (2′–H); RNA has a ribose sugar (2′–OH).
- A nucleotide consists of a nucleoside linked to a phosphate group on either the 5′ or 3′ carbon of the (deoxy)ribose.
- Successive (deoxy)ribose residues of a polynucleotide chain are joined by a phosphate group between the 3′ carbon of one sugar and the 5′ carbon of the next sugar.
- One end of the chain (conventionally written on the left) has a free 5′ end and the other end of the chain has a free 3′ end.
- DNA contains the four bases adenine, guanine, cytosine, and thymine; RNA normally has uracil instead of thymine.

CONCEPT AND REASONING CHECK

List the structural differences between DNA and RNA nucleotides.

1.4 DNA Is a Double Helix

By the 1950s, the observation by Erwin Chargaff that the nitrogenous bases are present in different amounts in the genomes of different species led to the concept that the sequence of bases is the form in which genetic information is carried. Given this concept, there were two remaining challenges: working out the structure of DNA and explaining how a sequence of bases in DNA could determine the sequence of amino acids in a protein.

Three pieces of evidence contributed to the construction of the double-helix model for DNA by James Watson and Francis Crick in 1953:

- X-ray diffraction data collected by Rosalind Franklin and Maurice Wilkins showed that the B-form of DNA (found in aqueous solution) is a regular helix, making a complete turn every 34 Å (3.4 nm), with a diameter of ~20 Å (2 nm). Since the distance between adjacent nucleotides is 3.4 Å (0.34 nm), there must be 10 nucleotides per turn.
- The density of DNA suggests that the helix must contain two polynucleotide chains. The constant diameter of the helix can be explained if the bases in each chain face inward and are restricted so that a purine is always paired with a pyrimidine, avoiding partnerships of purine-purine (which would be too wide) or pyrimidine-pyrimidine (which would be too narrow).
- Chargaff also observed that regardless of the absolute amounts of each base, the proportion of G is always the same as the proportion of C in DNA, and the proportion of A is always the same as that of T. Consequently, the composition of any DNA can be described by its G-C content, or the sum of the proportions of G and C bases. (The proportions of A and T bases can be determined by subtracting the G-C content from 1.) G-C content ranges from 0.26 to 0.74 for different species.

Watson and Crick proposed that the two polynucleotide chains in the double helix associate by hydrogen bonding between the nitrogenous bases. Normally, G can hydrogen bond specifically only with C, whereas A can bond specifically only with T. This hydrogen bonding between bases is described as *base pairing*, and the paired bases (G forming three hydrogen bonds with C, or A forming two hydrogen bonds with T) are said to be **complementary**. Base pairing occurs because of the complementary shapes of the complementary bases at the interfaces of where they pair, along with

▸ **complementary** Base pairs that match up in the pairing reactions in double helical nucleic acids (A with T in DNA or with U in RNA, and C with G).

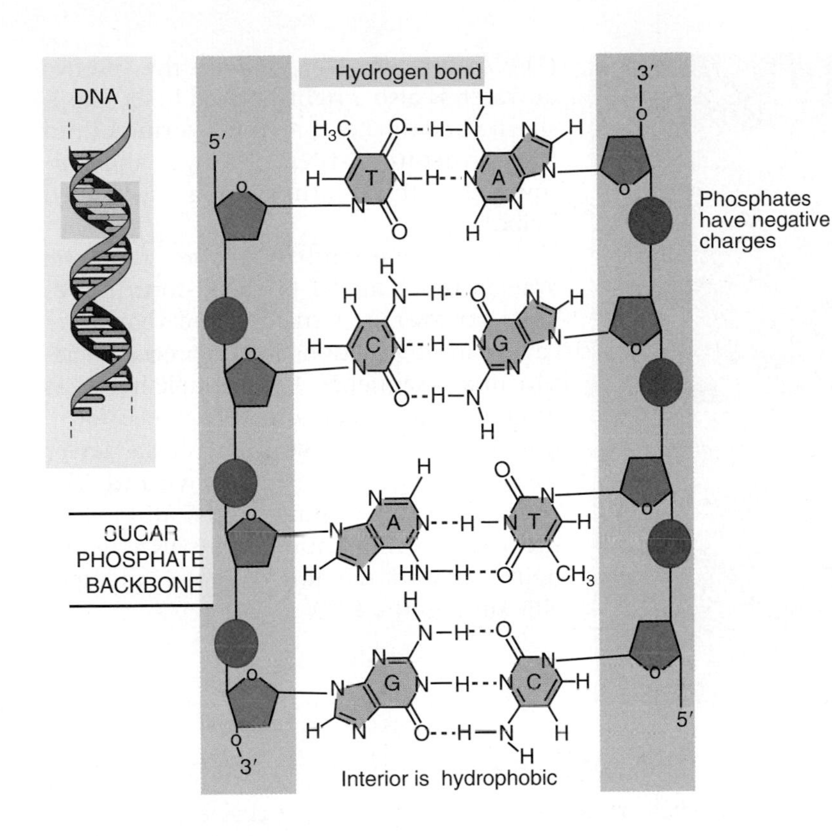

FIGURE 1.6 The DNA double helix maintains a constant width because purines always face pyrimidines in the complementary A-T and G-C base pairs. The sequence in the figure is T-A, C-G, A-T, G-C.

the location of just the right functional groups in just the right geometry along those interfaces so that hydrogen bonds can form.

The Watson-Crick model has the two polynucleotide chains running in opposite directions, so they are said to be **antiparallel**, as illustrated in **FIGURE 1.6**. Looking in one direction along the helix, one strand runs in the 5′ to 3′ direction, whereas its complement runs 3′ to 5′. (Double-stranded DNA is often called *duplex DNA*.)

▸ **antiparallel** Strands of the DNA double helix organized in opposite orientation, so that the 5′ end of one strand is aligned with the 3′ end of the other strand.

The sugar-phosphate backbones are on the outside of the double helix and carry negative charges on the phosphate groups. When DNA is in solution *in vitro*, the charges are neutralized by the binding of metal ions, typically sodium (Na^+). In the cell, positively charged proteins provide some of the neutralizing force. These proteins play important roles in determining the organization of DNA in the cell.

The base pairs (often abbreviated as bp) are on the inside of the double helix. They are flat and lie perpendicular to the axis of the helix. Using the analogy of the double helix as a spiral staircase, the base pairs form the steps, as illustrated schematically in **FIGURE 1.7**. Proceeding up the helix, bases are stacked above one another like a pile of plates.

Each base pair is rotated ~36° around the axis of the helix relative to the next base pair, so ~10 base pairs (in solution, 10.4 bp) make a complete turn of 360°. The twisting of the two strands around one another forms a double helix with a **minor groove** that is ~12 Å (1.2 nm) across and a **major groove** that is ~22 Å (2.2 nm) across, as can be seen from the scale model of **FIGURE 1.8**. In B-DNA, the double helix is said to be "right-handed"; the turns run clockwise as viewed along the helical axis.

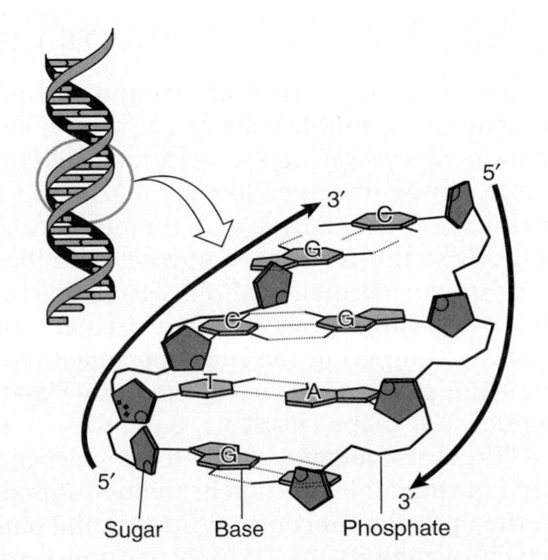

FIGURE 1.7 Flat base pairs lie perpendicular to the sugar-phosphate backbone.

▸ **minor groove** A fissure running the length of the DNA double helix that is 12 Å across.

▸ **major groove** A fissure running the length of the DNA double helix that is 22 Å across.

FIGURE 1.8 The two strands of DNA form a double helix. Photo © Photodisc.

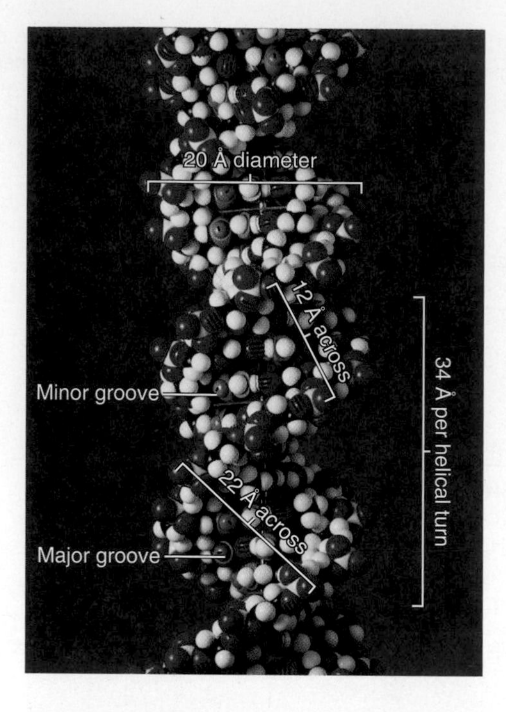

(The A-form of DNA, found in the absence of water, is also a right-handed helix that is shorter and thicker than the B-form. A third DNA structure, Z-DNA, is longer and narrower than the B-form and is a left-handed helix.)

It is important to realize that the Watson-Crick model of the B-form represents an average structure and that there can be local variations in the precise structure due to sequence. If the double helix has more base pairs per turn than the B-form, it is said to be **overwound**; if it has fewer base pairs per turn it is **underwound**. The degree of local winding can be affected by the overall conformation of the DNA double helix or by the binding of proteins to specific sites on the DNA.

▸ **overwound** B-form DNA that has more than 10.4 base pairs per turn of the helix.

▸ **underwound** B-form DNA that has fewer than 10.4 base pairs per turn of the helix.

KEY CONCEPTS

- The B-form of DNA is a double helix consisting of two polynucleotide chains that run antiparallel.
- The nitrogenous bases of each chain are flat purine or pyrimidine rings that face inward and pair with one another by hydrogen bonding to form only A-T or G-C pairs.
- The diameter of the double helix is 20 Å, and there is a complete turn every 34 Å, with ten base pairs per turn (~10.4 bp per turn in solution).
- The double helix has a major (wide) groove and a minor (narrow) groove.

CONCEPT AND REASONING CHECKS

1. Summarize the evidence that Watson and Crick used in proposing their model of the B-form of DNA.
2. If the G-C content of a DNA duplex is 0.44, what are the proportions of the four bases?

1.5 Supercoiling Affects the Structure of DNA

The two strands of DNA are wound around each other to form the double helical structure; the double helix can also wind around itself to change the overall conformation, or *topology*, of the DNA molecule in space. This is called **supercoiling**. The effect can be imagined like a rubber band twisted around itself. Supercoiling creates tension in the DNA and, therefore, can occur only if the DNA has no free ends (otherwise the free ends can rotate to relieve the tension) or in linear DNA if it is anchored to a protein scaffold, as in eukaryotic chromosomes. The simplest example of a DNA with no free ends is a circular molecule. The effect of supercoiling can be seen by comparing the nonsupercoiled circular DNA lying flat (**FIGURE 1.9**, center) with the supercoiled circular molecule that forms a twisted (and, therefore, more condensed) shape (**FIGURE 1.9**, bottom).

▸ **supercoiling** The coiling of a closed duplex DNA in space so that it crosses over its own axis.

The consequences of supercoiling depend on whether the DNA is twisted around itself in the same direction as the two strands within the double helix (clockwise) or in the opposite direction. Twisting in the same direction produces *positive supercoiling*, which overwinds the DNA so that there are fewer base pairs per turn than in the

B-form. Twisting in the opposite direction produces *negative supercoiling*, or underwinding, so there are more base pairs per turn than in the B-form. Both types of supercoiling of the double helix in space are tensions in the DNA (which is why DNA molecules with no supercoiling are said to be "relaxed"). Negative supercoiling can be thought of as creating tension in the DNA that is relieved by the unwinding of the double helix. The effect of severe negative supercoiling is to generate a region in which the two strands of DNA have separated (technically, zero base pairs per turn).

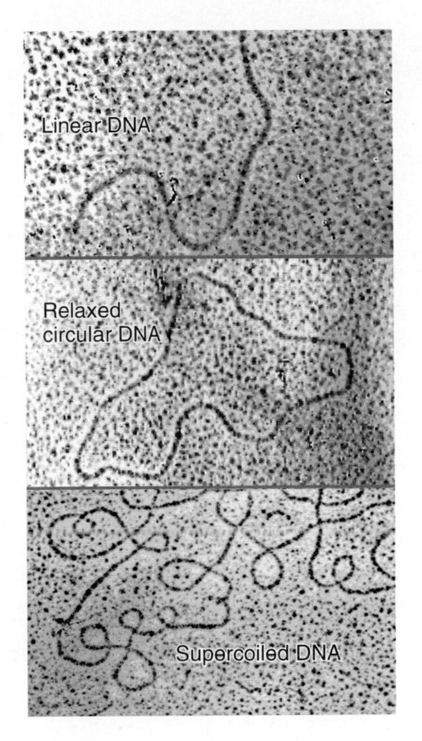

FIGURE 1.9 Linear DNA is extended (top); a circular DNA remains extended if it is relaxed (nonsupercoiled) (center); but a supercoiled DNA has a twisted and condensed form (bottom). Photos courtesy of Nirupam Roy Choudhury, International Centre for Genetic Engineering and Biotechnology (ICGEB).

Topological manipulation of DNA is a central aspect of all its functional activities (recombination, replication, and transcription) as well as of the organization of its higher-order structure. All synthetic activities involving double-stranded DNA require the strands to separate. The strands do not simply lie side by side, though; they are intertwined. Their separation therefore requires the strands to rotate about each other in space. Some possibilities for the unwinding reaction are illustrated in **FIGURE 1.10**.

Unwinding a short linear DNA presents no problems, as the DNA ends are free to spin around the axis of the double helix to relieve any tension. However, DNA in a typical chromosome is not only extremely long but is also coated with proteins that serve to anchor the DNA at numerous points. Therefore, even a linear eukaryotic chromosome does not functionally possess free ends.

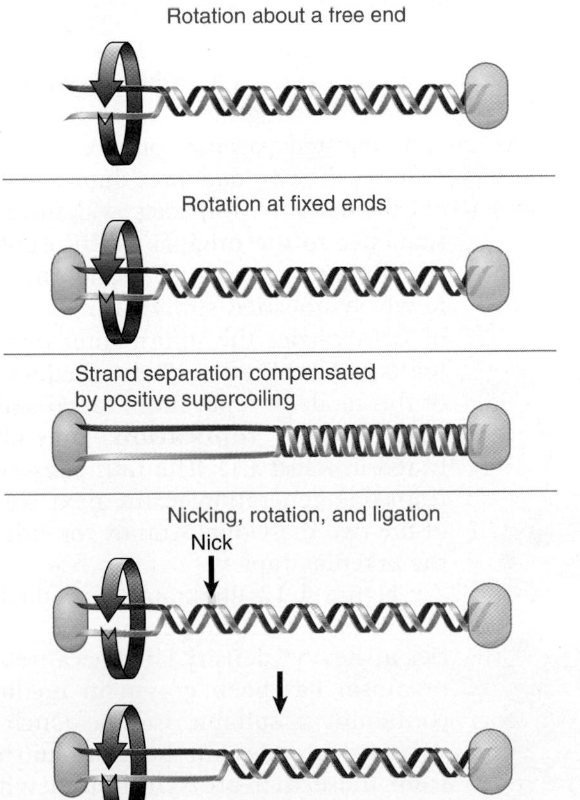

FIGURE 1.10 Separation of the strands of a DNA double helix can be achieved in several ways.

Consider the effects of separating the two strands in a molecule whose ends are not free to rotate. When two intertwined strands are pulled apart from one end, the result is to *increase* their winding about each other farther along the molecule, resulting in positive supercoiling elsewhere in the molecule to balance the underwinding generated in the single-stranded region. The problem can be overcome by introducing a transient nick in one strand. An internal free end allows the nicked strand to rotate about the intact strand, after which the nick can be sealed. Each repetition of the nicking and sealing reaction releases one superhelical turn. The topoisomerase enzymes that perform these reactions to control supercoiling in the cell will be discussed in *Section 15.7, Topoisomerases Relax or Introduce Supercoils in DNA*.

KEY CONCEPTS

- Supercoiling occurs only in "closed" DNA with no free ends.
- Closed DNA is either circular DNA or linear DNA in which the ends are anchored so that they are not free to rotate.

CONCEPT AND REASONING CHECK

Why does negative supercoiling facilitate unwinding of DNA but positive supercoiling inhibit unwinding?

1.6 DNA Replication Is Semiconservative

It is crucial that DNA be reproduced accurately. The two polynucleotide strands are joined only by hydrogen bonds, so they are able to separate without the breakage of covalent bonds. The specificity of base pairing suggests that both of the separated parental strands could act as template strands for the synthesis of complementary daughter strands. **FIGURE 1.11** shows the principle that a new daughter strand is assembled from each parental strand. The sequence of the daughter strand is determined by the parental strand: an A in the parental strand causes a T to be placed in the daughter strand, a parental G directs incorporation of a daughter C, and so on.

The top part of Figure 1.11 shows an unreplicated parental duplex with the original two parental strands. The lower part shows the two daughter duplexes produced by complementary base pairing. Each of the daughter duplexes is identical in sequence to the original parent duplex, containing one parental strand and one newly synthesized strand. The structure of DNA carries the information needed for its own replication. The consequences of this mode of replication, called **semiconservative replication**, are illustrated in **FIGURE 1.12**. The unit conserved from one generation to the next is one of the two individual strands comprising the parental duplex.

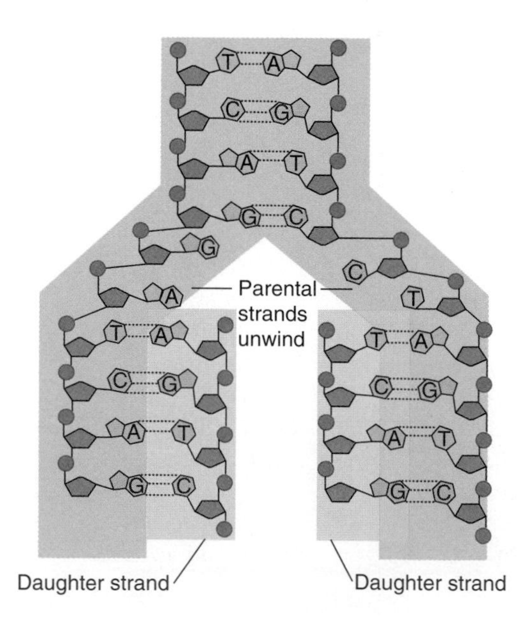

FIGURE 1.11 Base pairing provides the mechanism for replicating DNA.

▸ **semiconservative replication** DNA replication accomplished by separation of the strands of a parental duplex, each strand then acting as a template for synthesis of a complementary strand.

Figure 1.12 illustrates a prediction of this model. If the parental DNA carries a "heavy" density label because the organism has been grown in medium containing a suitable isotope (such as ^{15}N), its strands can be distinguished from those that are synthesized when the organism is transferred to a medium containing "light" isotopes. The parental DNA is a duplex of two "heavy" strands (red). After one generation of growth in "light" medium, the duplex DNA is *hybrid* in density—it consists of one heavy parental strand (red) and one light daughter strand (blue). After a second generation, the two strands of each hybrid duplex have separated. Each strand gains a light partner, so that now one half of the duplex DNA remains hybrid and the other half is entirely light (both strands are blue).

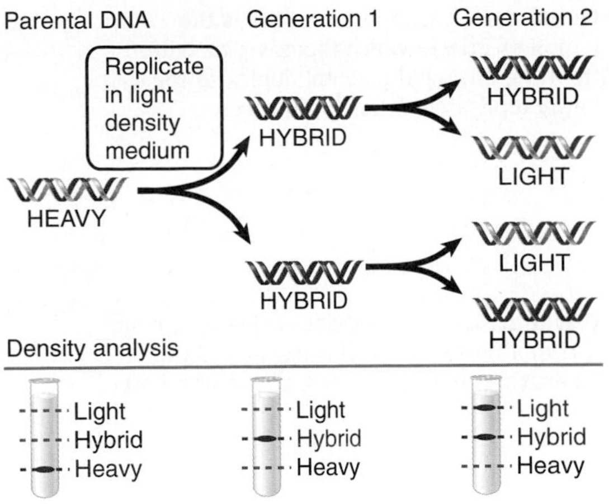

FIGURE 1.12 Replication of DNA is semiconservative.

The individual strands of these duplexes are entirely heavy or entirely light. This pattern was confirmed experimentally by Matthew Meselson and Franklin Stahl in 1958. Meselson and Stahl followed the semiconservative replication of DNA through three generations of growth of *E. coli*. When DNA was extracted from bacteria and separated in a density gradient by centrifugation, the DNA formed bands corresponding to its density—heavy for parental, hybrid for the first generation, and half hybrid bands and half light bands in the second generation, indicating that a single parental strand is retained in the daughter molecule. (See the box in *Chapter 12* entitled *Historical Perspectives: The Meselson-Stahl Experiment* for more detail on this experiment.)

KEY CONCEPTS

- The Meselson-Stahl experiment used "heavy" isotope labeling to show that the single polynucleotide strand is the unit of DNA that is conserved during replication.
- Each strand of a DNA duplex acts as a template for synthesis of a daughter strand.
- The sequences of the daughter strands are determined by complementary base pairing with the separated parental strands.

CONCEPT AND REASONING CHECK

What is the expected result of an experiment similar to that of Meselson and Stahl, but beginning with "light" DNA and culturing cells in a "heavy" medium?

1.7 Polymerases Act on Separated DNA Strands at the Replication Fork

Replication requires the two strands of the parental duplex to undergo separation, or **denaturation**. The disruption of the duplex, however, is only transient and is reversed, or undergoes **renaturation**, as the daughter duplex is formed. Only a small stretch of the duplex DNA is denatured at any moment during replication.

The helical structure of a molecule of DNA during replication is illustrated in **FIGURE 1.13**. The unreplicated region consists of the parental duplex, opening into the replicated region where the two daughter duplexes have formed. The duplex is disrupted at the junction between the two regions, which is called the **replication fork**. Replication involves movement of the replication fork along the parental DNA, so that there is continuous denaturation of the parental strands and formation of daughter duplexes.

- **denaturation** A molecule's conversion from the physiological conformation to some other (inactive) conformation. In DNA, this involves the separation of the two strands due to breaking of hydrogen bonds between bases.
- **renaturation** The reassociation of denatured complementary single strands of a DNA double helix.
- **replication fork** The point at which strands of parental duplex DNA are separated so that replication can proceed. A complex of proteins, including DNA polymerase, is found there.

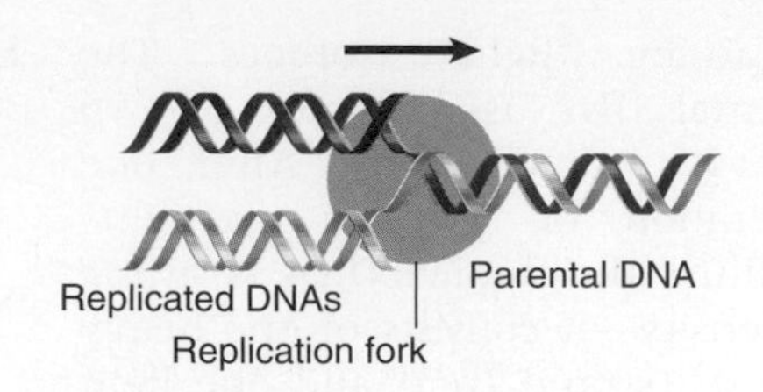

FIGURE 1.13 The replication fork is the region of DNA in which there is a transition from the unwound parental duplex to the newly replicated daughter duplexes.

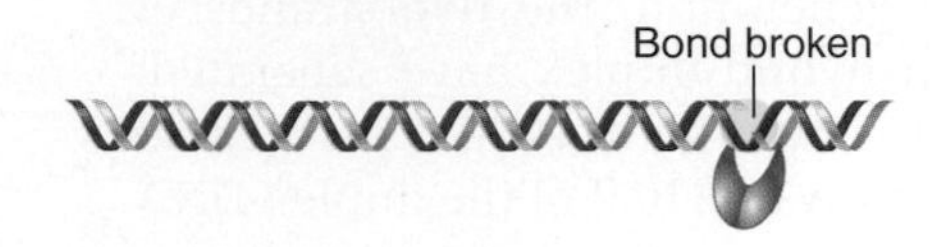

FIGURE 1.14 An endonuclease cleaves a bond within a nucleic acid. This example shows an enzyme that attacks one strand of a DNA duplex.

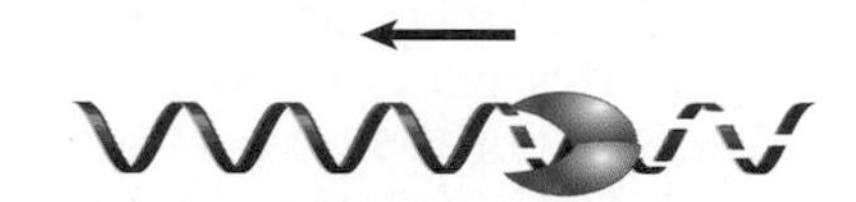

FIGURE 1.15 An exonuclease removes bases one at a time by cleaving the last bond in a polynucleotide chain.

- **DNA polymerase** An enzyme that synthesizes a daughter strand(s) of DNA (under direction from a DNA template). Any particular enzyme may be involved in repair or replication (or both).
- **DNase** An enzyme that degrades DNA.
- **RNase** An enzyme that degrades RNA.
- **exonuclease** An enzyme that cleaves nucleotides one at a time from the end of a polynucleotide chain; it may be specific for either the 5′ or 3′ end of DNA or RNA.
- **endonuclease** An enzyme that cleaves bonds within a nucleic acid chain; it may be specific for RNA or for single-stranded or double-stranded DNA.

The synthesis of DNA is aided by specific enzymes, **DNA polymerases**, that recognize the template strand and catalyze the addition of nucleotide subunits to the polynucleotide chain that is being synthesized. They are accompanied in DNA replication by ancillary enzymes such as helicases that unwind the DNA duplex, a primase that synthesizes an RNA primer required by DNA polymerase, and DNA ligase that connects discontinuous DNA fragments. Degradation of nucleic acids also requires specific enzymes: deoxyribonucleases (**DNases**) degrade DNA, and ribonucleases (**RNases**) degrade RNA. The nucleases (see *Chapter 3, Methods in Molecular Biology and Genetic Engineering*) fall into the general classes of **exonucleases** and **endonucleases**:

- Endonucleases break individual phosphodiester linkages within RNA or DNA molecules, generating discrete fragments. Some DNases cleave both strands of a duplex DNA at the target site, whereas others cleave only one of the two strands. Endonucleases are involved in cutting reactions, as shown in **FIGURE 1.14**.
- Exonucleases remove nucleotide residues one at a time from the end of the molecule, generating mononucleotides. They always function on a single nucleic acid strand, and each exonuclease proceeds in a specific direction, that is, starting either at a 5′ or at a 3′ end and proceeding toward the other end. They are involved in trimming reactions, as shown in **FIGURE 1.15**.

KEY CONCEPTS

- Replication of DNA is undertaken by a complex of enzymes that separate the parental strands and synthesize the daughter strands.
- The replication fork is the point at which the parental strands are separated.
- The enzymes that synthesize DNA are called DNA polymerases.
- Nucleases are enzymes that degrade nucleic acids; they include DNases and RNases and can be categorized as endonucleases or exonucleases.

CONCEPT AND REASONING CHECK

What are the functions of DNA polymerases, DNases, and RNases in living cells?

1.8 Genetic Information Can Be Provided by DNA or RNA

The **central dogma** is the dominant paradigm of molecular biology. Structural genes exist as sequences of nucleic acid, but they function by being expressed in the form of polypeptides. Replication makes possible the inheritance of genetic information, while transcription and translation are responsible for its expression to another form.

> ▸ **central dogma** Information cannot be transferred from polypeptide to polypeptide or polypeptide to nucleic acid, but can be transferred between nucleic acids and from nucleic acid to polypeptide.

> ▸ **RNA polymerase** An enzyme that synthesizes RNA using a DNA template (formally described as DNA-dependent RNA polymerases).

The central dogma includes several observations about the processes of replication, transcription, and translation (see Figure 1.29):

- Transcription of DNA by a DNA-dependent **RNA polymerase** generates RNA molecules. Messenger RNAs (mRNAs) are translated to polypeptides. Other types of RNA, such as rRNAs and tRNAs, are functional themselves and are not translated.
- A genetic system may involve either DNA or RNA as the genetic material. Cells use only DNA. Some viruses use RNA, and replication of viral RNA by an RNA-dependent RNA polymerase occurs in the infected cell.
- The expression of cellular genetic information is usually unidirectional. Transcription of DNA generates RNA molecules; the exception is the reverse transcription of retroviral RNA to DNA that occurs when retroviruses infect cells (see below). Generally polypeptides cannot be retrieved for use as genetic information; translation of RNA into polypeptide is always irreversible.

These mechanisms are equally effective for the cellular genetic information of prokaryotes or eukaryotes and for the information carried by viruses. The genomes of all living organisms consist of duplex DNA. Viruses have genomes that consist of DNA or RNA, and there are examples of each type that are double-stranded (dsDNA or dsRNA) or single-stranded (ssDNA or ssRNA). Details of the mechanism used to replicate the nucleic acid vary among viruses, but the principle of replication via synthesis of complementary strands remains the same, as illustrated in **FIGURE 1.16**.

The restriction of a unidirectional transfer of information from DNA to RNA in cells is not absolute. It is broken by the retroviruses, which have genomes consisting of a single-stranded RNA molecule. During the retroviral cycle of infection, the RNA is converted into a single-stranded DNA by the process of **reverse transcription**, which is accomplished by the enzyme *reverse transcriptase*, an RNA-dependent DNA polymerase. The resulting ssDNA is in turn converted into dsDNA. This duplex DNA becomes part of the genome of the host cell and is inherited like any other gene. So reverse transcription allows a sequence of RNA to be retrieved and used as DNA in a cell.

> ▸ **reverse transcription** Synthesis of DNA from a template of RNA by the enzyme reverse transcriptase.

The existence of RNA replication and reverse transcription establishes the general principle that information in the form of either type of nucleic acid sequence can be converted into the other type. In the usual course of events, however, the cell relies on the processes of DNA replication, transcription, and translation. But on rare occasions (possibly mediated by an RNA virus), information from a cellular RNA is converted into DNA and inserted into the genome. Although retroviral reverse transcription is not necessary for the regular operations of the cell, it becomes a mechanism

Double-stranded template

Old strand

New strands

Old strand

Replication generates two daughter duplexes each containing one parental strand and one newly synthesized strand

Single-stranded template

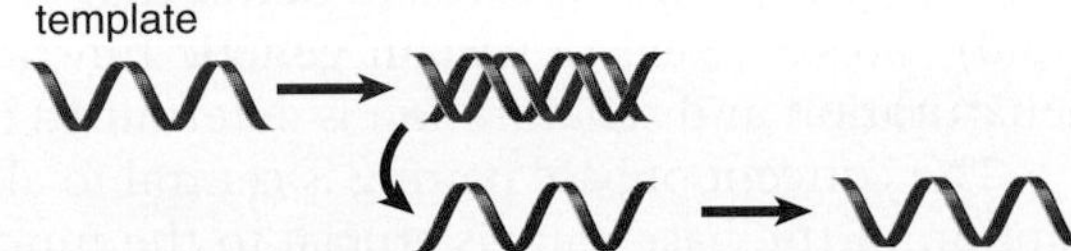

Single parental strand is used to synthesize complementary strand

Complementary strand is used to synthesize copy of parental strand

FIGURE 1.16 Double-stranded and single-stranded nucleic acids both replicate by synthesis of complementary strands governed by the rules of base pairing.

FIGURE 1.17 The size of the genome varies over an enormous range.

Genome	Gene Number	Base Pairs
Organisms		
Plants	<50,000	$<10^{11}$
Mammals	30,000	$\sim 3 \times 10^{9}$
Worms	14,000	$\sim 10^{8}$
Flies	12,000	1.6×10^{8}
Fungi	6,000	1.3×10^{7}
Bacteria	2–4,000	$<10^{7}$
Mycoplasma	500	$<10^{6}$
dsDNA Viruses		
Vaccinia	<300	187,000
Papova (SV40)	~6	5,226
Phage T4	~200	165,000
ssDNA Viruses		
Parvovirus	5	5,000
Phage fX174	11	5,387
dsRNA Viruses		
Reovirus	22	23,000
ssRNA Viruses		
Coronavirus	7	20,000
Influenza	12	13,500
TMV	4	6,400
Phage MS2	4	3,569
STNV	1	1,300
Viroids		
PSTV RNA	0	359

of potential importance when we consider the evolution of the genome (see *Chapter 8, Genome Evolution*).

The same principles for the perpetuation of genetic information apply to the massive genomes of plants or amphibians as well as the tiny genomes of mycoplasma and the even smaller genomes of DNA or RNA viruses. **FIGURE 1.17** presents some examples that illustrate the range of genome types and sizes. The reasons for such variation in genome size and gene number will be explored in *Chapters 5* and *6*.

Among the various living organisms, with genomes varying in size over a 100,000-fold range, a common principle prevails: the DNA encodes all of the polypeptides that the cell(s) of the organism must synthesize, and the polypeptides in turn (directly or indirectly) provide the functions needed for survival. A similar principle describes the function of the genetic information of viruses, whether DNA or RNA: the nucleic acid codes for the polypeptide(s) needed to package the genome and for any other functions in addition to those provided by the host cell that are needed to reproduce the virus. (The smallest virus—the satellite tobacco necrosis virus [STNV]—cannot replicate independently. It requires the presence of a "helper" virus—the tobacco necrosis virus [TNV], which is itself a normally infectious virus.)

KEY CONCEPTS

- Cellular genes are DNA, but viruses may have genomes of RNA.
- DNA is converted into RNA by transcription, and RNA may be converted into DNA by reverse transcription.
- The translation of RNA into protein is unidirectional.

CONCEPT AND REASONING CHECK

What types of enzymes would be necessary to replicate ssDNA, ssRNA, dsDNA, and dsRNA genomes to produce exact copies of the same type of nucleic acid?

1.9 Nucleic Acids Hybridize by Base Pairing

A crucial property of the double helix is the capacity to separate the two strands without disrupting the covalent bonds that form the polynucleotides and at the (very rapid) rates needed to sustain genetic functions. The specificity of the processes of denaturation and renaturation is determined by complementary base pairing.

The concept of base pairing is central to all processes involving nucleic acids. Disruption of the base pairs is crucial to the function of a double-stranded nucleic acid, whereas the ability to form base pairs is essential for the activity of a single-stranded nucleic acid. **FIGURE 1.18** shows that base pairing enables complementary single-stranded nucleic acids to form a duplex.

- An intramolecular duplex region can form by base pairing between two complementary sequences that are part of a single-stranded nucleic acid.
- A single-stranded nucleic acid may base pair with an independent, complementary single-stranded nucleic acid to form an intermolecular duplex.

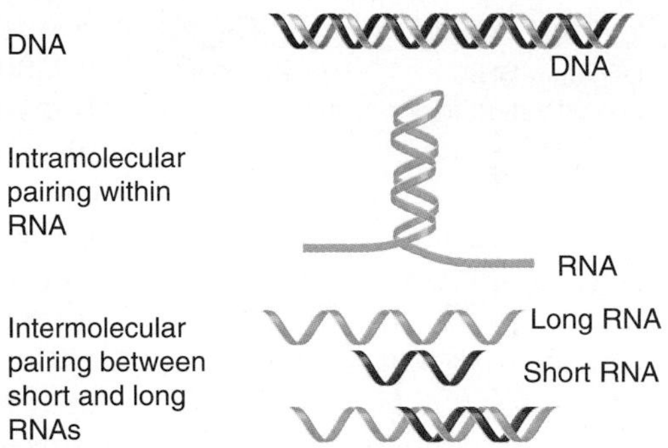

DNA

DNA

Intramolecular pairing within RNA

RNA

Intermolecular pairing between short and long RNAs

Long RNA

Short RNA

FIGURE 1.18 Base pairing occurs in duplex DNA and also in intra- and intermolecular interactions in single-stranded RNA (or DNA).

Formation of duplex regions from single-stranded nucleic acids is most prevalent in RNA but is also important for single-stranded viral DNA genomes. Base pairing between independent complementary single strands is not restricted to DNA–DNA or RNA–RNA but can also occur between DNA and RNA.

The lack of covalent bonds between complementary strands makes it possible to manipulate DNA *in vitro*. The hydrogen bonds that stabilize the double helix are disrupted by heating or low salt concentration. The two strands of a double helix separate entirely when all the hydrogen bonds between them are broken.

Denaturation of DNA occurs over a narrow temperature range and results in striking changes in many of its physical properties. The midpoint of the temperature range over which the strands of DNA separate is called the **melting temperature** (T_m) and depends on the G-C content of the duplex. Because each G-C base pair has three hydrogen bonds, it is more stable than an A-T base pair, which has only two hydrogen bonds. The more G-C base pairs in a DNA, the greater the energy that is needed to separate the two strands. In solution under physiological conditions, a DNA that is 40% G-C (a value typical of mammalian genomes) denatures with a T_m of about 87°C, so duplex DNA is stable at the temperature of the cell.

▸ **melting temperature** The midpoint of the temperature range over which the strands of DNA separate.

The denaturation of DNA is reversible under appropriate conditions. Renaturation depends on specific base pairing between the complementary strands. **FIGURE 1.19** shows that the reaction takes place in two stages. First, single strands of DNA in the solution encounter one another by chance; if their sequences are complementary, the two strands base pair to generate a short double-stranded region. This region of base pairing then extends along the molecule, much like a zipper, to form a lengthy duplex.

Complete renaturation restores the properties of the original double helix. The property of renaturation applies to any two complementary nucleic acid sequences. This is sometimes called **annealing**, but the reaction is more generally called **hybridization** whenever nucleic acids from different sources are involved, as in the case when DNA hybridizes to RNA. The ability of two nucleic acids to hybridize constitutes a precise test for their complementarity because only complementary sequences can form a duplex (although under certain conditions imperfect matches can be tolerated).

▸ **annealing** The renaturation of a duplex structure from single strands that were obtained by denaturing duplex DNA.

▸ **hybridization** The pairing of complementary RNA and DNA strands to give an RNA-DNA hybrid.

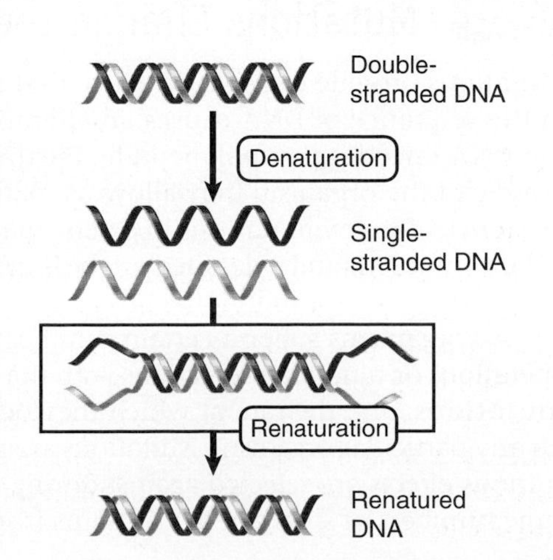

FIGURE 1.19 Denatured single strands of DNA can renature to give the duplex form.

The principle of the hybridization reaction is to combine two single-stranded nucleic acids in solution and then to measure the amount of double-stranded material that forms. **FIGURE 1.20** illustrates a procedure in which a DNA preparation is denatured and the single strands are attached to a filter. Then a second denatured DNA (or RNA) preparation is added. The filter is treated so that the second preparation can attach to it only if it is able to base pair with the DNA that

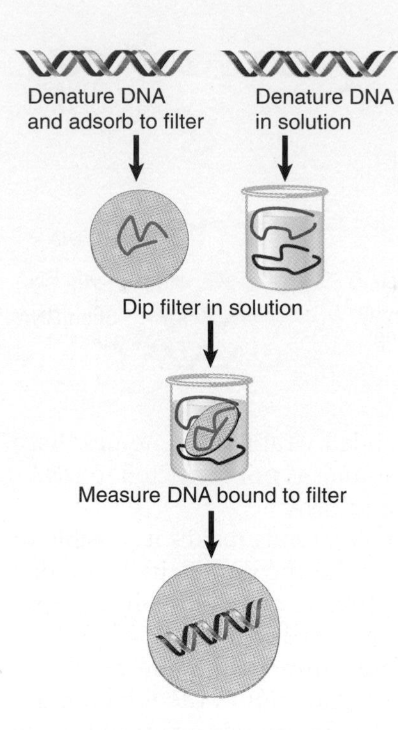

FIGURE 1.20 Filter hybridization establishes whether a solution of denatured DNA (or RNA) contains sequences complementary to the strands immobilized on the filter.

was originally attached. Usually the second preparation is labeled so that the hybridization reaction can be measured as the amount of label retained by the filter. Alternatively, hybridization in solution can be measured as the change in UV-absorbance of a nucleic acid solution at 260 nm as detected via spectrophotometry. As DNA denatures to single strands with increasing temperature, UV-absorbance of the DNA solution increases; UV-absorbance consequently decreases as ssDNA hybridizes to complementary DNA or RNA with decreasing temperature.

Two sequences need not be perfectly complementary to hybridize. If they are similar but not identical, an imperfect duplex is formed in which base pairing is interrupted at positions where the two single strands are not complementary.

KEY CONCEPTS

- Heating causes the two strands of a DNA duplex to separate.
- T_m is the midpoint of the temperature range for denaturation.
- Complementary single strands can renature when the temperature is reduced.
- Denaturation and renaturation/hybridization can occur with DNA–DNA, DNA–RNA, or RNA–RNA combinations and can be intermolecular or intramolecular.
- The ability of two single-stranded nucleic acids to hybridize is a measure of their complementarity.

CONCEPT AND REASONING CHECK

Describe how measures of hybridization between DNA from different species can be used to estimate the evolutionary relationships between those species.

1.10 Mutations Change the Sequence of DNA

Mutations provide decisive evidence that DNA is the genetic material. When a change in the sequence of DNA causes an alteration in a polypeptide, we may conclude that the DNA encodes that polypeptide. Furthermore, a corresponding change in the phenotype of the organism may allow us to identify the function of that polypeptide. The existence of many mutations in a gene may allow many variant forms of a polypeptide to be compared, and a detailed analysis can be used to identify regions of the polypeptide responsible for individual enzymatic or other functions.

All organisms suffer a certain number of mutations as the result of normal cellular operations or random interactions with the environment. These are called **spontaneous mutations**, and the rate at which they occur (the "background level") is characteristic for any particular organism. Mutations are rare events, and of course those that have deleterious effects are selected against during evolution. It is, therefore, difficult to observe large numbers of spontaneous mutants from natural populations.

▸ **spontaneous mutations** Mutations that occur in the absence of any added reagent to increase the mutation rate as the result of errors in replication (or other events involved in the reproduction of DNA) or by random changes to the chemical structure of bases.

The occurrence of mutations can be increased by treatment with certain compounds. These are called **mutagens**, and the changes they cause are called **induced mutations**. Most mutagens either modify a particular base of DNA or become incorporated into the nucleic acid. The potency of a mutagen is judged by how much it increases the rate of mutation above background. By using mutagens, it becomes possible to induce many changes in any gene.

Mutation rates can be measured at several levels of resolution: mutation across the whole genome (as the rate per genome per generation), mutation in a gene (as the rate per locus per generation), or mutation at a specific nucleotide site (as the rate per base pair per generation). These rates correspondingly decrease as smaller units are observed.

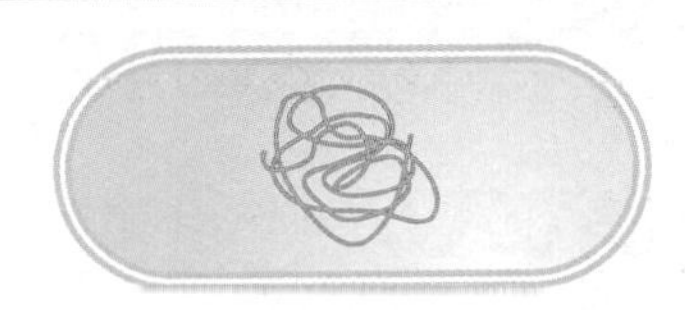

FIGURE 1.21 A base pair is mutated at a rate of 10^{-9}–10^{-10} per generation, a gene of 1000 bp is mutated at ~10^{-6} per generation, and a bacterial genome is mutated at 3×10^{-3} per generation.

▸ **mutagens** Substances that increase the rate of mutation by inducing changes in DNA sequence, directly or indirectly.

▸ **induced mutations** Mutations that result from the action of a mutagen. The mutagen may act directly on the bases in DNA, or it may act indirectly to trigger a pathway that leads to a change in DNA sequence.

Spontaneous mutations that inactivate gene function occur in bacteriophages and bacteria at a relatively constant rate of 3 to 4×10^{-3} per genome per generation. Given the large variation in genome sizes between bacteriophages and bacteria, this corresponds to great differences in the mutation rate per base pair. This suggests that the overall rate of mutation has been subject to selective forces that have balanced the deleterious effects of most mutations against the advantageous effects of some mutations. This conclusion is strengthened by the observation that an archaean that lives under harsh conditions of high temperature and acidity (which are expected to damage DNA) does not show an elevated mutation rate but in fact has an overall mutation rate just below the average range. **FIGURE 1.21** shows that in bacteria, the mutation rate corresponds to ~10^{-6} events per locus per generation or to an average rate of change per base pair of 10^{-9} to 10^{-10} per generation. The rate at individual base pairs varies very widely, over a 10,000-fold range. We have no accurate measurement of the rate of mutation in eukaryotes, although usually it is thought to be somewhat similar to that of bacteria on a per locus per generation basis. One reason that mutation rates vary across species is that the activity and efficacy of DNA repair systems also vary. DNA repair systems will be discussed in *Chapter 16, Repair Systems*.

KEY CONCEPTS

- All mutations are changes in the sequence of DNA.
- Mutations may occur spontaneously or may be induced by mutagens.

CONCEPT AND REASONING CHECK

What are the advantages of maintaining a nonzero mutation rate?

1.11 Mutations May Affect Single Base Pairs or Longer Sequences

Any base pair of DNA can be mutated. A **point mutation** changes only a single base pair and can be caused by either of two types of event:

- Chemical modification of DNA directly changes one base into a different base.
- An error during the replication of DNA causes the wrong base to be inserted into a polynucleotide.

▸ **point mutation** A change in the sequence of DNA involving a single base pair.

FIGURE 1.22 Mutations can be induced by chemical modification of a base.

- **transition** A mutation in which one pyrimidine is replaced by the other or in which one purine is replaced by the other.

- **transversion** A mutation in which a purine is replaced by a pyrimidine, or vice versa.

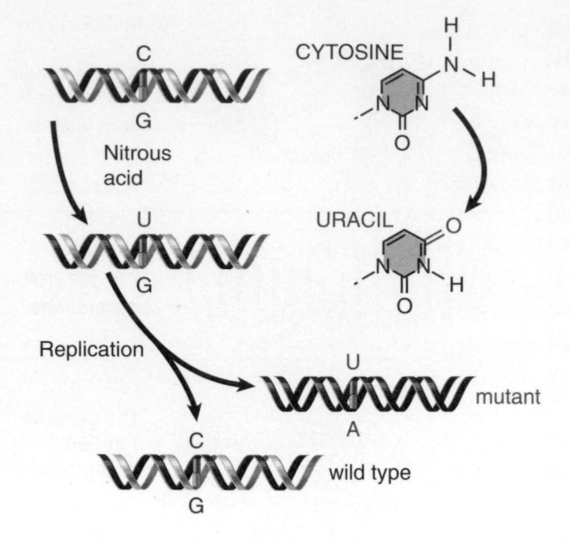

FIGURE 1.23 Mutations can be induced by the incorporation of base analogs into DNA.

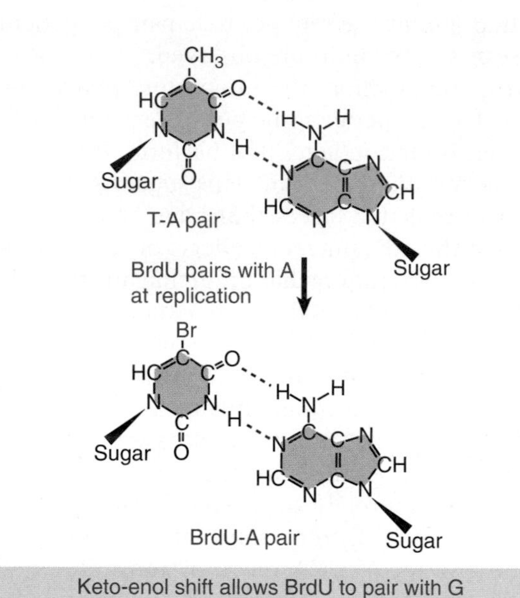

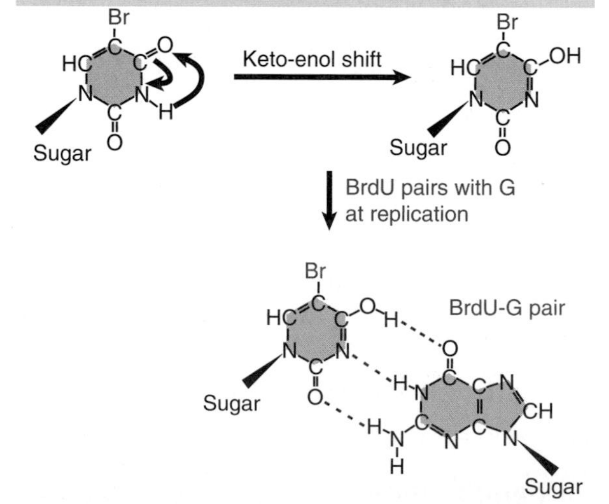

Point mutations can be divided into two types, depending on the nature of the base substitution:

- The most common class is the **transition**, resulting from the substitution of one pyrimidine by the other or of one purine by the other. This replaces a G-C pair with an A-T pair, or vice versa.
- The less common class is the **transversion**, in which a purine is replaced by a pyrimidine, or vice versa, so that (for example) an A-T pair becomes a T-A or C-G pair.

As shown in **FIGURE 1.22**, the mutagen nitrous acid performs an oxidative deamination that converts cytosine into uracil, resulting in a transition. In the replication cycle following the transition, the U pairs with an A instead of the G with which the original C would have paired. So the C-G pair is replaced by a T-A pair when the A pairs with the T in the next replication cycle. (Nitrous acid can also deaminate adenine, causing the reverse transition from A-T to G-C.)

Transitions are also caused by base mispairing, when noncomplementary bases pair instead of the usual Watson-Crick pairs. Base mispairing usually occurs as an aberration resulting from the incorporation into DNA of an abnormal base that has flexible pairing properties. **FIGURE 1.23** shows the example of the mutagen bromouracil (BrdU), an analog of thymine that contains a bromine atom in place of thymine's methyl group and can be incorporated into DNA in place of thymine. However, BrdU has flexible pairing properties because the presence of the bromine atom allows a *tautomeric shift* from a keto (=O) form to an enol (–OH) form. The enol form of BrdU can pair with guanine, which after replication leads to substitution of the original A-T pair by a G-C pair. Tautomeric shifts can also occur when a proton shifts position within a normal base and results in an anomalous, but more stable, base pairing. For example, while the common keto form of guanine pairs stably with cytosine, the rare enol form of guanine pairs stably with thymine.

Transversions are rarer than transitions, since they require a temporary purine-purine or pyrimidine-pyrimidine pairing that would alter the diameter of the DNA duplex. However, one cause of transversions is a proton shift followed by a 180° rotation of the base around the glycosidic bond. For example, a rotated synadenine (produced by a proton shift in adenine) will pair stably with a normal adenine.

Point mutations were thought for a long time to be the principal means of change in individual genes. However, we now know that insertions and deletions ("indels") of short sequences are quite frequent. Some mutagens, such as intercalating agents, can cause the insertion or deletion of a single base pair. An intercalating agent will insert itself between two adjacent base pairs in a DNA duplex, so that the duplex is distorted and a DNA polymerase can either skip or add a base during DNA replication. If this occurs in the coding sequence of a gene, a frameshift mutation will result (see *Section 2.7, The Genetic Code Is Triplet*). Often, the insertions are the result of transposable elements, which are sequences of DNA with the ability to move from one site to another (see *Chapter 17, Transposable Elements and Retroviruses*). An insertion within a coding region usually abolishes the activity of the gene. However, both insertions and deletions of short sequences can occur by other mechanisms—for example, those involving errors during replication or recombination. In addition, mutagens belonging to a class called acridines introduce very small insertions and deletions.

KEY CONCEPTS

- A point mutation changes a single base pair.
- Point mutations can be caused by the chemical conversion of one base into another or by errors that occur during replication.
- A transition replaces a G-C base pair with an A-T base pair, or vice versa.
- A transversion replaces a purine with a pyrimidine, such as changing A-T to T-A.
- Insertions can result from the movement of transposable elements.

CONCEPT AND REASONING CHECK

Why are transitions more common than tranversions? Consider how the DNA repair mechanisms might recognize errors and the effects of these mutations on DNA structure in your answer.

1.12 The Effects of Mutations Can Be Reversed

FIGURE 1.24 shows that the possibility of reversion mutations, or **revertants**, is an important characteristic that distinguishes point mutations and insertions from deletions. Mutations that inactivate a gene are called **forward mutations**. Their effects are reversed by **back mutations**, which are of two types: true reversions and second-site reversions.

- A point mutation can revert either by a true reversion or a second-site reversion.
- An insertion can revert by deletion of the inserted sequence.
- A deletion of a sequence cannot revert in the absence of some mechanism to restore the lost sequence.

An exact reversal of the original mutation is called a **true reversion**. For example, if an A-T pair was replaced by a G-C pair in the original mutation, another mutation to restore the A-T pair will exactly regenerate the original sequence. The exact removal of a transposable element following its insertion is another example of a true reversion.

The second type of back mutation, **second-site reversion**, may occur elsewhere in the gene, and its effects compensate for the first mutation. For example, one amino acid change in a protein may abolish its function, but a second alteration may

▸ **revertants** Reversions of a mutant cell or organism to the wild-type phenotype.

▸ **forward mutation** A mutation that inactivates a functional gene.

▸ **back mutation** A mutation that reverses the effect of a mutation that had inactivated a gene; thus it restores the original sequence or function of the gene product.

▸ **true reversion** A mutation that restores the original sequence of the DNA.

▸ **second-site reversion** A second mutation suppressing the effect of a first mutation.

FIGURE 1.24 Point mutations and insertions can revert, but deletions cannot revert.

ATCGGACTTACCGGTTA
TAGCCTGAATGGCCAAT
Point mutation
ATCGGACTCACCGGTTA
TAGCCTGAGTGGCCAAT
Reversion
ATCGGACTTACCGGTTA
TAGCCTGAATGGCCAAT

ATCGGACTTACCGGTTA
TAGCCTGAATGGCCAAT
Insertion
ATCGGACTTXXXXXACCGGTTA
TAGCCTGAAYYYYYTGGCCAAT
Reversion by deletion
ATCGGACTTACCGGTTA
TAGCCTGAATGGCCAAT

ATCGGACTTACCGGTTA
TAGCCTGAATGGCCAAT
Deletion
ATCGGACGGTTA
TAGCCTGCCAAT
No reversion possible

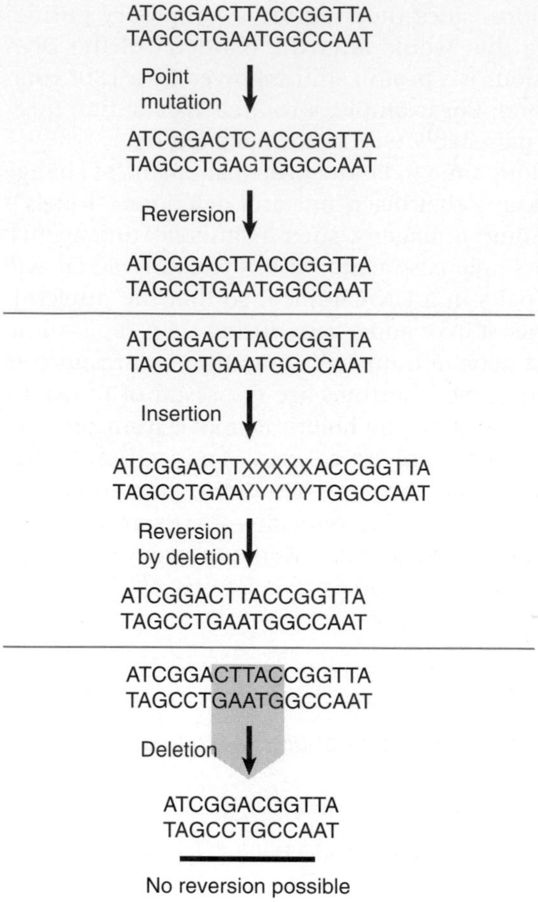

compensate for the first and restore protein activity.

A forward mutation results from any change that alters the function of a gene product, whereas a back mutation must restore the original function to the altered gene product. Therefore, the possibilities for back mutations are much more restricted than those for forward mutations. The rate of back mutations is correspondingly lower than that of forward mutations, typically by a factor of ~10.

Mutations in other genes can also occur to circumvent the effects of mutation in the original gene. This is called a **suppression mutation**. A locus in which a mutation suppresses the effect of a mutation in another locus is called a *suppressor*. For example, a point mutation may cause an amino acid substitution in a polypeptide, whereas a second mutation in a tRNA gene may cause it to recognize the mutated codon and, as a result, insert the original amino acid during translation. (Note that this suppresses the original mutation but causes errors during translation of other mRNAs.)

▸ **suppression mutation** A second event eliminates the effects of a mutation without reversing the original change in DNA.

KEY CONCEPTS

- Forward mutations alter the function of a gene, and back mutations (or revertants) reverse their effects.
- Insertions can revert by deletion of the inserted material, but deletions cannot revert.
- Suppression occurs when a mutation in a second gene bypasses the effect of mutation in the first gene.

CONCEPT AND REASONING CHECK

Transposable elements were originally identified because the rate of reversion of mutations caused by them is much higher than the reversion rate for point mutations. Explain why the reversion rate is so high.

1.13 Mutations Are Concentrated at Hotspots

So far we have presented mutations in terms of individual changes in the sequence of DNA that influence the activity of the DNA in which they occur. When we consider mutations in terms of the alteration of function of the gene, most genes within a species show more or less similar rates of mutation relative to their size. This suggests that the gene can be regarded as a target for mutation and that damage to any part of it can alter its function. As a result, susceptibility to mutation is roughly proportional to the size of the gene. But are all base pairs in a gene equally susceptible, or are some more likely to be mutated than others?

What happens when we isolate a large number of independent mutations in the same gene? Each is the result of an individual mutational event. Most mutations will occur at different sites, but some will occur at the same position. Two independently isolated mutations at the same site may constitute exactly the same change in DNA (in which case the same mutation has happened more than once), or they may constitute different changes (three different point mutations are possible at each base pair).

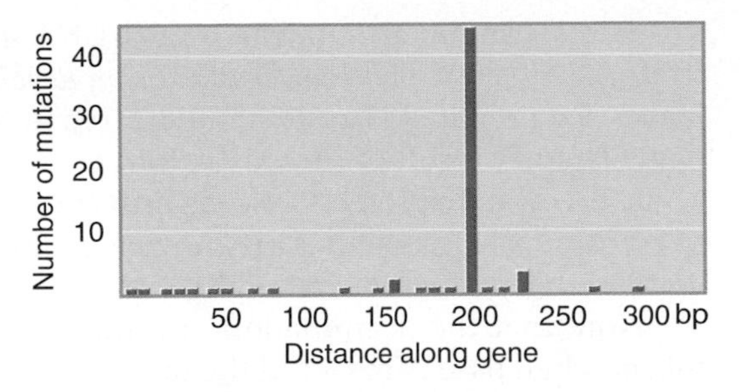

FIGURE 1.25 Spontaneous mutations occur throughout the *lacI* gene of *E. coli* but are concentrated at a hotspot.

The histogram of **FIGURE 1.25** shows the frequency with which mutations are found at each base pair in the *lacI* gene of *E. coli*. The statistical probability that more than one mutation occurs at a particular site is given by random-hit kinetics (as seen in the Poisson distribution). Some sites will gain one, two, or three mutations, whereas others will not gain any. Some sites gain far more than the number of mutations expected from a random distribution; they may have 10× or even 100× more mutations than predicted by random hits. These sites are called **hotspots**. Spontaneous mutations may occur at hotspots, and different mutagens may have different hotspots.

▸ **hotspots** A site in the genome at which the frequency of mutation (or recombination) is very much increased, usually by at least an order of magnitude relative to neighboring sites.

A major cause of spontaneous mutation is the presence of an unusual base in the DNA. In addition to the four standard bases of DNA, *modified bases* are sometimes found. The name reflects their origin; they are produced by chemical modification of one of the four standard bases. The most common modified base is 5-methylcytosine, which is generated when a methylase enzyme adds a methyl group to cytosine residues at specific sites in the DNA. Sites containing 5-methylcytosine are hotspots for spontaneous point mutation in *E. coli*. In each case, the mutation is a G-C to A-T transition. The hotspots are not found in mutant strains of *E. coli* that cannot methylate cytosine.

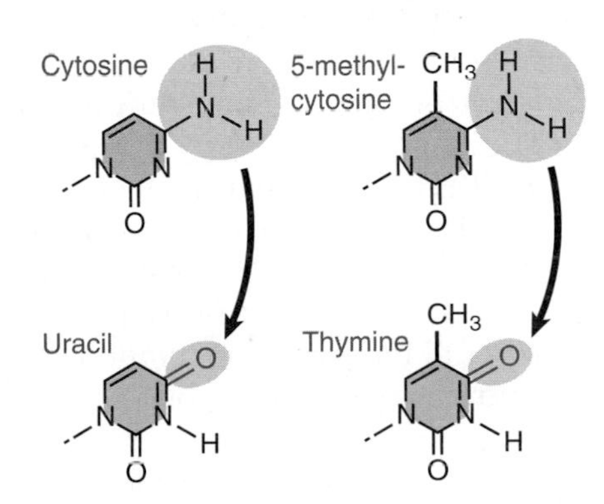

FIGURE 1.26 Deamination of cytosine produces uracil, whereas deamination of 5-methylcytosine produces thymine.

The reason for the existence of these hotspots is that cytosine bases suffer a higher frequency of spontaneous deamination. In this reaction, the amino group is replaced by a keto group. Recall that deamination of cytosine generates uracil (see Figure 1.22). **FIGURE 1.26** compares this reaction with the deamination of 5-methylcytosine where deamination generates thymine. The effect is to generate the mismatched base pairs G-U and G-T, respectively.

FIGURE 1.27 shows that the consequences of deamination are different for 5-methylcytosine and cytosine. Deaminating the (rare) 5-methylcytosine causes a mutation, whereas deaminating cytosine does not have this effect. This happens because the DNA repair systems are much more effective in recognizing G-U than G-T and always correct the U (which normally should be present only in RNA, not in DNA).

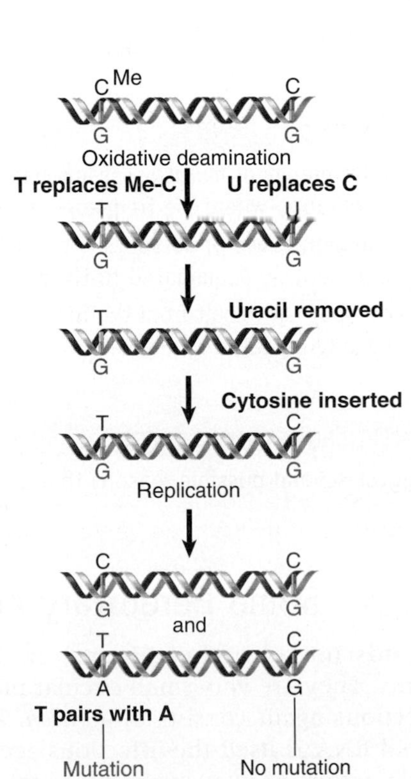

FIGURE 1.27 The deamination of 5-methylcytosine produces thymine (by C-G to T-A transitions), whereas the deamination of cytosine produces uracil (which usually is removed and then replaced by cytosine).

E. coli contain an enzyme, uracil-DNA-glycosidase, that removes uracil residues from DNA (see *Section 16.4, Base Excision Repair Systems Require Glycosylases*). This action leaves an unpaired G residue, and a repair system then inserts a complementary C base. The net result of these reactions is to restore the original sequence of the DNA. Thus, this system protects DNA against the consequences of spontaneous deamination of cytosine. (This system is not, however, efficient enough to prevent the effects of the increased deamination caused by nitrous acid; see Figure 1.22.)

Note that the deamination of 5-methylcytosine creates thymine and results in a mismatched base pair, G-T. If the mismatch is not corrected before the next replication cycle, a mutation results; the bases in the mispaired G-T separate, and then they pair with the correct complements to produce the original G-C and the mutant A-T.

Deamination of 5-methylcytosine is the most common cause of mismatched G-T pairs in DNA. Repair systems that act on G-T mismatches have a bias toward replacing the T with a C (rather than the alternative of replacing the G with an A), which helps to reduce the rate of mutation (see *Section 16.6, Controlling the Direction of Mismatch Repair*). However, these systems are not as effective as those that remove U from G-U mismatches. As a result, deamination of 5-methylcytosine leads to mutation much more often than does deamination of cytosine.

5-methylcytosine also creates hotspots in eukaryotic DNA. It is common at CpG dinucleotides that are concentrated in regions called *CpG islands* (see *Section 29.6, CpG Islands Are Subject to Methylation*). Although 5-methylcytosine accounts for ~1% of the bases in human DNA, sites containing the modified base account for ~30% of all point mutations.

The importance of repair systems in reducing the rate of mutation is emphasized by the effects of eliminating the mouse enzyme MBD4, a glycosylase that can remove T (or U) from mismatches with G. The result is to increase the mutation rate at CpG sites by a factor of 3. (The reason the effect is not greater is that MBD4 is only one of several systems that act on G-T mismatches; probably the elimination of all the systems would increase the mutation rate much more.)

Another type of hotspot, though not often found in coding regions, is the "slippery sequence"—a homopolymer run, or region where a very short sequence (one or a few nucleotides) is repeated many times in tandem. During replication, a DNA polymerase may skip one repeat or replicate the same repeat twice, leading to a decrease or increase in repeat number.

KEY CONCEPTS

- The frequency of mutation at any particular base pair is statistically equivalent, except for hotspots, where the frequency is increased by at least an order of magnitude.
- A common cause of hotspots is the modified base 5-methylcytosine, which is spontaneously deaminated to thymine.
- A hotspot can result from the high frequency of change in copy number of a short tandemly repeated sequence.

CONCEPT AND REASONING CHECK

Suggest several possible reasons that a particular base pair can be a mutational hotspot.

1.14 Some Hereditary Agents Are Extremely Small

▸ **viroid** A small infectious nucleic acid that does not have a protein coat.

Viroids (or subviral pathogens) are infectious agents that cause diseases in higher plants. They are very small circular molecules of RNA. Unlike viruses—for which the infectious agent consists of a *virion*, a genome encapsulated in a protein coat—the viroid RNA is itself the infectious agent. The viroid consists solely of the RNA molecule, which is extensively folded by imperfect base pairing, forming a characteristic

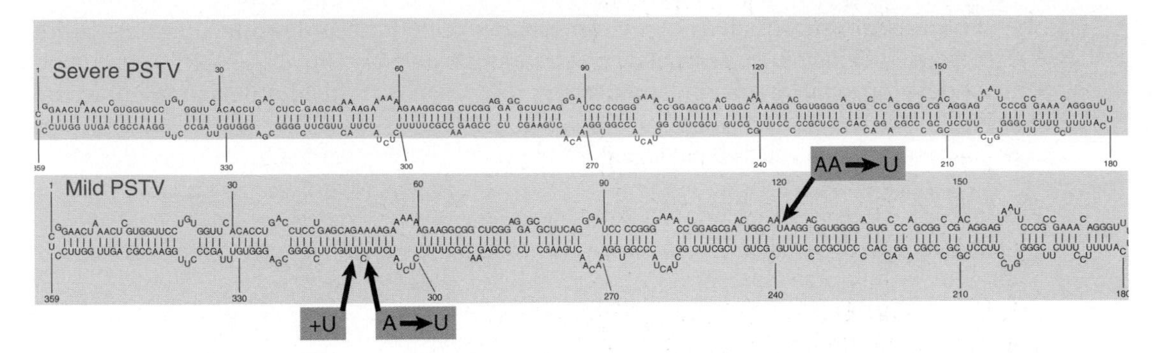

FIGURE 1.28 PSTV RNA is a circular molecule that forms an extensive double-stranded structure, interrupted by many interior loops. The severe and mild forms of PSTV have RNAs that differ at three sites.

rod as shown in **FIGURE 1.28**. Mutations that interfere with the structure of this rod reduce the infectivity of the viroid.

A viroid RNA consists of a single molecule that is replicated autonomously and accurately in infected cells. Viroids are categorized into several groups. A given viroid is assigned to a group according to sequence similarity with other members of the group. For example, four viroids in the PSTV (potato spindle tuber viroid) group have 70%–83% sequence similarity with PSTV. Different isolates of a particular viroid strain vary from one another in sequence, which may result in phenotypic differences among infected cells. For example, the "mild" and "severe" strains of PSTV differ by three nucleotide substitutions.

Viroids are similar to viruses in having heritable nucleic acid genomes but differ from viruses in both structure and function. Viroid RNA does not appear to be translated into polypeptide, so it cannot itself encode the functions needed for its survival. This situation poses two as yet unanswered questions: How does viroid RNA replicate, and how does it affect the phenotype of the infected plant cell?

Replication must be carried out by enzymes of the host cell. The heritability of the viroid sequence indicates that viroid RNA is the template for replication. Thus, the modern version of the central dogma, as presented in **FIGURE 1.29**, includes replication of RNA as a component in some systems.

Viroids are presumably pathogenic because they interfere with normal cellular processes. They might do this in a relatively random way; for example, they may take control of an essential enzyme for their own replication or interfere with the production of necessary cellular RNAs. Alternatively, they might behave as abnormal regulatory molecules, with particular effects upon the expression of individual genes.

▸ **prion** A proteinaceous infectious agent that behaves as an inheritable trait, although it contains no nucleic acid. One example is PrP^{Sc}, the agent of scrapie in sheep and bovine spongiform encephalopathy.

An even more unusual agent is the cause of scrapie, a degenerative neurological disease of sheep and goats. The disease is similar to the human diseases of kuru and Creutzfeldt-Jakob syndrome, which affect brain function. The infectious agent of scrapie does not contain nucleic acid. This extraordinary agent is called a **prion** (proteinaceous infectious agent). It is a 28-kD hydrophobic glycoprotein, PrP. PrP is encoded by a cellular gene (conserved among the mammals) that is expressed in normal brain cells. The protein exists in two forms: the version found in normal brain cells is called PrP^{c} and is entirely degraded by proteases during normal protein turnover. The version found in infected brains is called PrP^{sc} and is extremely resistant to degradation by proteases. PrP^{c} is converted to PrP^{sc} by a conformational change that confers protease resistance.

As the infectious agent of scrapie, PrP^{sc} must in some way modify

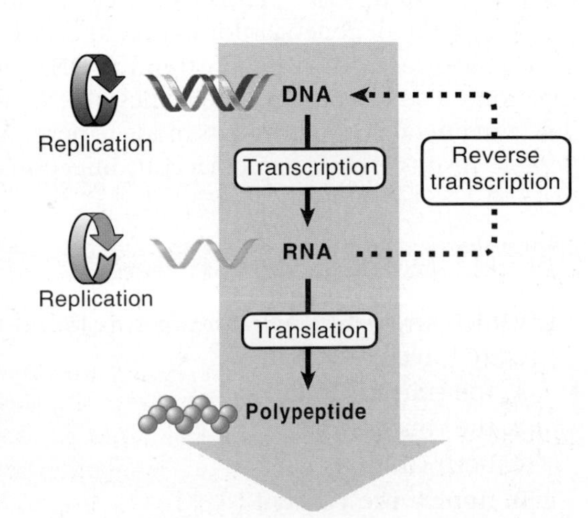

FIGURE 1.29 The central dogma states that information in nucleic acid can be perpetuated or transferred, but the transfer of information into a polypeptide (protein) is irreversible.

the synthesis of its normal cellular counterpart so that it becomes infectious instead of harmless (see *Section 29.9, Prions Cause Diseases in Mammals*). Mice that lack a PrP gene cannot develop scrapie, which demonstrates that PrP is essential for development of the disease.

KEY CONCEPT

- Some very small hereditary agents do not encode polypeptide but consist of RNA or protein with heritable properties.

CONCEPT AND REASONING CHECK

How would you distinguish whether a newly discovered infectious agent is an organism, a virus, a viroid, or a prion?

1.15 Summary

Two classic experiments provided strong evidence that DNA is the genetic material of bacteria, eukaryotic cells, and many viruses. DNA isolated from one strain of *Pneumococcus* bacteria can confer properties of that strain upon another strain. In addition, DNA is the only component that is inherited by progeny phages from parental phages. DNA can be used to transfect new properties into eukaryotic cells.

DNA is a double helix consisting of antiparallel strands in which the nucleotide units are linked by 5′ to 3′ phosphodiester bonds. The backbone is on the exterior; purine and pyrimidine bases are stacked in the interior in pairs in which A is complementary to T and G is complementary to C. In semiconservative replication, the two strands separate and daughter strands are assembled by complementary base pairing. Complementary base pairing is also used to transcribe an RNA from one strand of a DNA duplex.

A mutation consists of a change in the sequence of A-T and G-C base pairs in DNA. A mutation in a coding sequence may change the sequence of amino acids in the corresponding polypeptide. A point mutation changes only the amino acid represented by the codon in which the mutation occurs. Point mutations may be reverted by back mutation of the original mutation. Insertions may revert by loss of the inserted material, but deletions cannot revert. Mutations may also be suppressed indirectly when a mutation in a different gene counters the original defect.

The natural incidence of mutations is increased by mutagens. Mutations may be concentrated at hotspots. A type of hotspot responsible for some point mutations is caused by deamination of the modified base 5-methylcytosine. Forward mutations occur at a rate of $\sim 10^{-6}$ per locus per generation; back mutations are rarer.

Although all genetic information in cells is carried by DNA, viruses have genomes of double-stranded or single-stranded DNA or RNA. Viroids are subviral pathogens that consist solely of small molecules of RNA with no protective packaging. The RNA does not encode protein and its mode of perpetuation and of pathogenesis is unknown. Scrapie results from a proteinaceous infectious agent, or prion.

CHAPTER QUESTIONS

1. Which strain of *S. pneumoniae* was found to be avirulent (nonlethal) when mice were infected with it?
   A. the smooth strain
   B. the rough strain
   C. both strains
   D. none were virulent

2. What isotope can be used to specifically label protein?
   A. ^{14}C
   B. ^{3}H
   C. ^{32}P
   D. ^{35}S
3. The difference between RNA and DNA is the presence of a:
   A. 2′-PO_4 group on the ribose sugar in RNA.
   B. 3′-PO_4 group on the ribose sugar in RNA.
   C. 2′-OH group on the ribose sugar in RNA.
   D. 3′-OH group on the ribose sugar in RNA.
4. Which pair of scientists determined that DNA replication is semiconservative?
   A. Meselson and Stahl
   B. Watson and Crick
   C. Okazaki and Okazaki
   D. Griffith and Avery
5. In the density labeling experiment to study DNA replication, parental DNA in the cell was labeled with a high density isotope and, after one or more generations of growth, subjected to density gradient centrifugation. Which of the following populations was not detected after the second generation?
   A. the light density population
   B. the hybrid density population
   C. the heavy density population
   D. both the light and heavy density populations
6. Which class of infectious agent is inserted into the host genome as a double-stranded DNA segment?
   A. retroviruses
   B. double-stranded RNA viruses
   C. double-stranded DNA viruses
   D. viroids
7. The mutation rate in bacteria is about:
   A. 10^{-6} per locus per generation.
   B. 10^{-7} per locus per generation.
   C. 10^{-8} per locus per generation.
   D. 10^{-9} per locus per generation.
8. About 30% of human point mutations are associated with which of the following modified bases?
   A. 5-methylguanine
   B. 5-methyladenine
   C. 5-methylthymine
   D. 5-methylcytosine
9. The presence of the modified base in the previous question often leads to:
   A. transitions.
   B. transversions.
   C. deletions.
   D. insertions.
10. The reversal of an original base pair that was changed from A-T to G-C, then back to A-T, is an example of a:
    A. true reversion.
    B. second-site reversion.
    C. forward mutation.
    D. suppression.

KEY TERMS

annealing
antiparallel
back mutation
central dogma
chromosome
complementary
denaturation
DNA polymerase
DNase
endonuclease
exonuclease
forward mutation
genome
hotspots
hybridization
induced mutations
major groove
melting temperature
minor groove
mutagens
nucleoside
nucleotide
overwound
point mutation
polynucleotide
prion
purine
pyrimidine
renaturation
replication fork
reverse transcription
revertants
RNA polymerase
RNase
second-site reversion
semiconservative replication
spontaneous mutations
structural gene
supercoiling
suppression mutation
transfection
transformation
transforming principle
transition
transversion
true reversion
underwound
viroid

FURTHER READING

Holmes, F. (2001). *Meselson, Stahl, and the Replication of DNA: A History of the Most Beautiful Experiment in Biology*. Yale University Press, New Haven, CT. An account of Meselson and Stahl's scientific partnership with a unique look into the daily business of "doing science."

Maki, H. (2002). Origins of spontaneous mutations: specificity and directionality of base-substitution, frameshift, and sequence-substitution mutageneses. *Annu. Rev. Genet.* **36**, 279–303. A review of causes and effects of spontaneous mutations and mutational hotspots.

Prusiner, S. B. (1998). Prions. *Proc. Natl. Acad. Sci. USA* **95**, 13363–13383. An edited version of Prusiner's Nobel lecture, including an overview of the biology of prions and an account of their discovery.

Watson, J. D. (1981). *The Double Helix: A Personal Account of the Discovery of the Structure of DNA* (Norton Critical Editions). W. W. Norton, New York, NY. Watson's 1968 best-selling personal account of the discovery of the double helix along with reprints of original publications and additional commentary.

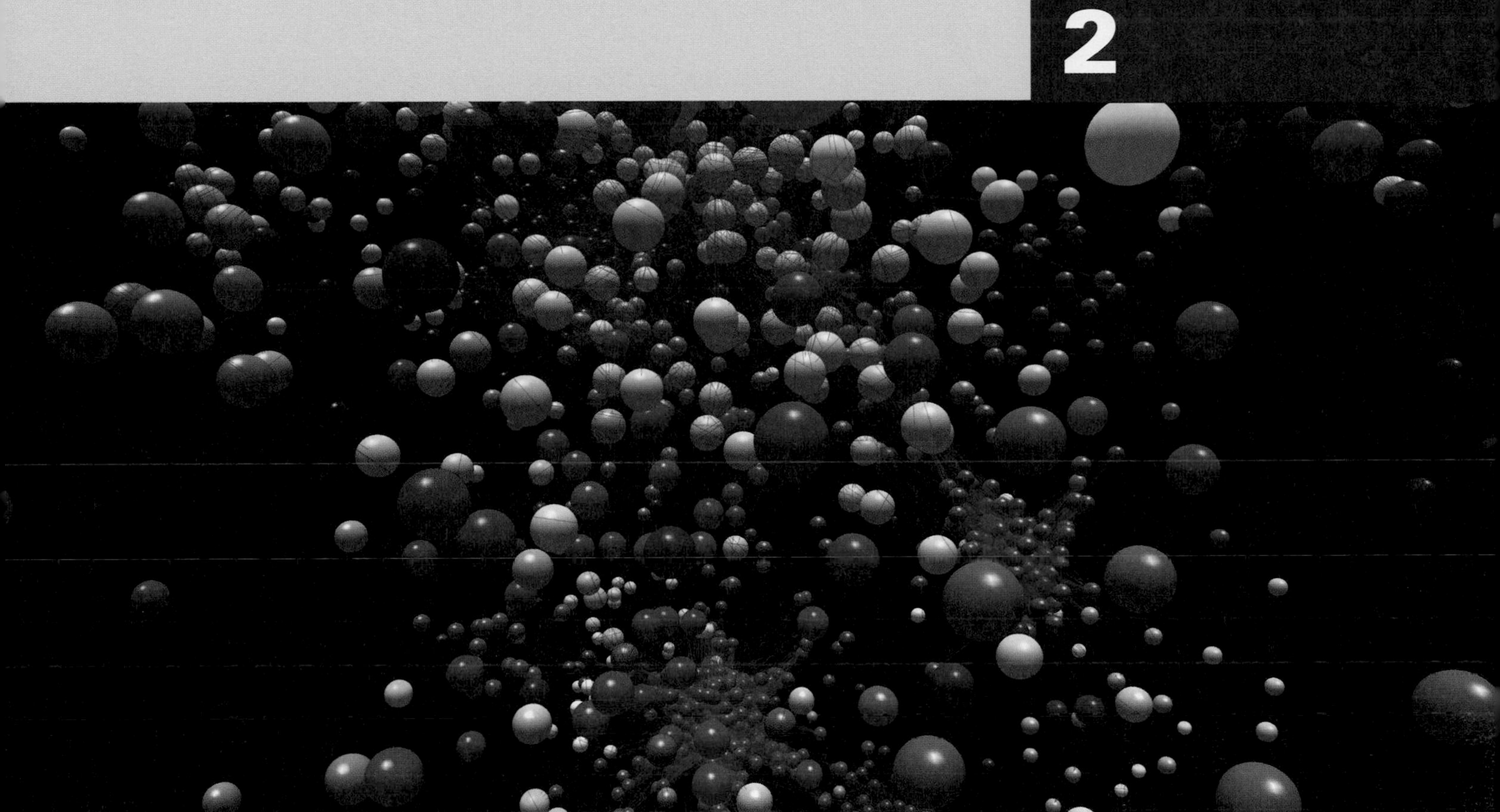

A visual representation of gene expression analysis that allows researchers to see patterns in a network of expressed genes. Each dot represents a gene, and links between the dots occur where there are similar patterns of expression between the genes. © Anton Enright, The Sanger Institute/Wellcome Images.

Genes Encode RNAs and Polypeptides

CHAPTER OUTLINE

2.1 Introduction

The gene is the functional unit of heredity. Each gene is a DNA sequence that functions by encoding a discrete product, which may be a polypeptide or an RNA. The basic pattern of inheritance of a gene was proposed by Mendel more than a century ago. Summarized in his two major principles of *segregation* and *independent assortment*, the gene was recognized as a "particulate factor" that passes unchanged from parent to progeny. A gene may exist in alternative forms, called **alleles**.

▸ **allele** One of several alternative forms of a gene occupying a given locus on a chromosome.

In diploid organisms, which have two sets of chromosomes, one copy of each chromosome is inherited from each parent. This is the same pattern of inheritance that is displayed by genes. One of the two copies of each gene is the paternal allele (inherited from the father); the other is the maternal allele (inherited from the mother). The shared pattern of inheritance of genes and chromosomes led to the discovery that chromosomes in fact carry the genes.

Each chromosome consists of a linear array of genes, and each gene resides at a particular location on the chromosome. The location is more formally called a genetic **locus**. The alleles of a gene are the different forms that are found at its locus. Although generally there are up to two alleles per locus in an individual, a population may have many alleles.

▸ **locus** The position on a chromosome at which the gene for a particular trait resides. It may be occupied by any one of the alleles for the gene.

▸ **linkage** The tendency of genes to be inherited together as a result of their location on the same chromosome. This is measured by percent recombination between loci.

The key to understanding the organization of genes into chromosomes was the discovery of genetic **linkage**—the tendency for genes on the same chromosome to remain together in the progeny instead of assorting independently, as predicted by Mendel's principle. Once the unit of *recombination* (reassortment) was introduced as a measure of linkage, the construction of genetic maps became possible.

The resolution of the linkage map of a multicellular eukaryote is restricted by the small number of progeny that can be obtained from each mating. Recombination occurs so infrequently between nearby points that it is rarely observed between different variable sites in the same gene. As a result, classical linkage maps of eukaryotes can place the genes in order but cannot resolve the locations of variable sites within a gene. By using a microbial system in which a very large number of progeny can be obtained from each genetic cross, researchers could demonstrate that recombination occurs within genes and that it follows the same rules as those for recombination between genes.

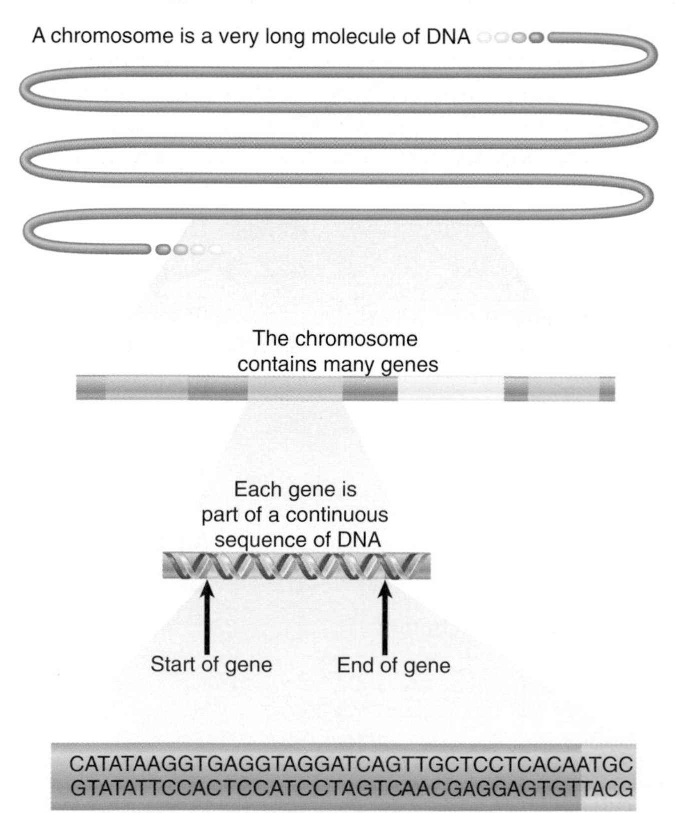

FIGURE 2.1 Each chromosome has a single long molecule of DNA, within which are the sequences of individual genes.

Variable nucleotide sites among alleles of a gene can be mapped into a linear order, showing that the gene itself has the same linear construction as the array of genes on a chromosome. In other words, the genetic map is linear within, as well as between, loci as an unbroken sequence of nucleotides. This conclusion leads naturally to the modern view summarized in **FIGURE 2.1** that the genetic material of a chromosome consists of an uninterrupted length of DNA representing many genes.

CONCEPT AND REASONING CHECK

Early work measuring recombination frequencies between genes led to the establishment of "linkage groups," sets of genes that do not assort independently. How does this support the concept that genes are carried on chromosomes?

2.2 Most Genes Encode Polypeptides

Sir Archibald Garrod was the first person to make a connection between a metabolic defect and a heritable factor, suggesting that gene action has chemical effects in cells. In studying a patient with alkaptonuria (a disease characterized by brownish-black urine) in 1902 he determined both that the disorder resulted from an enzymatic deficiency and that its pattern of inheritance in the patient's family was consistent with an autosomal recessive allele. Garrod later identified several other "inborn errors of metabolism" with autosomal recessive inheritance.

The work of George Beadle and Edward Tatum in the 1940s was the first systematic attempt to associate genes with enzymes. Beadle and Tatum showed that each step in a metabolic pathway is catalyzed by a single enzyme and can be blocked by mutation in a single gene. This led to the **one gene : one enzyme hypothesis**. A mutation in a gene alters the activity of the protein enzyme it encodes.

A modification in the hypothesis is needed to apply to proteins that consist of more than one polypeptide subunit. If the subunits are all the same, the protein is a **homomultimer**, and is encoded by a single gene. If the subunits are different, the protein is a **heteromultimer**, and each different subunit is encoded by a different gene. Stated as a more general rule applicable to any heteromultimeric protein, the one gene : one enzyme hypothesis becomes more precisely expressed as **one gene : one polypeptide hypothesis**. (Even this modification is not completely descriptive of the relationship between genes and proteins, as many genes encode alternate versions of a polypeptide; see *Section 21.11, Alternative Splicing Is a Rule, Rather Than an Exception, in Multicellular Eukaryotes.*)

Identifying the biochemical effects of a particular mutation can be a protracted task. The mutation responsible for creating Mendel's wrinkled-pea phenotype was identified only in 1990 as an alteration that inactivates the gene for a starch debranching enzyme!

It is important to remember that a gene does not directly generate a polypeptide. As shown in Figure 1.2, a gene encodes an RNA, which may in turn encode a polypeptide. Most genes are structural genes that encode polypeptides, but some genes encode RNAs that are not translated to polypeptides. These RNAs may be structural components of the protein synthesis machinery or may have roles in regulating gene expression. The basic principle is that *the gene is a sequence of DNA that specifies the sequence of an independent product.* The process of gene expression may terminate in a product that is either RNA or polypeptide.

A mutation in a coding region is generally a random event with regard to the structure and function of the gene (but see *Section 1.13, Mutations Are Concentrated at Hotspots*); mutations can have little or no effect (as in the case of neutral mutations), or they can damage or even abolish gene function. Most mutations that affect gene function are recessive; they result in an absence of function because the mutant allele does not produce its usual polypeptide. **FIGURE 2.2** illustrates

- **one gene : one enzyme hypothesis** Beadle and Tatum's hypothesis that a gene is responsible for the production of a single enzyme.
- **homomultimer** A molecular complex (such as a protein) in which the subunits are identical.
- **heteromultimer** A molecular complex (such as a protein) composed of different subunits.
- **one gene : one polypeptide hypothesis** A modified version of the not generally correct one gene : one enzyme hypothesis; the hypothesis that a gene is responsible for the production of a single polypeptide.

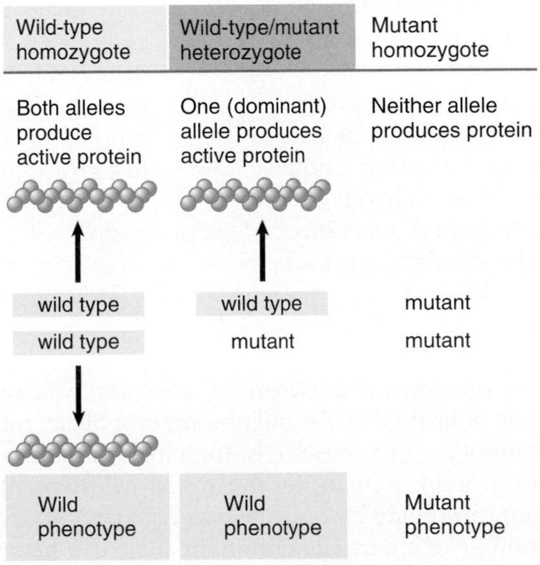

FIGURE 2.2 Genes encode proteins; dominance is explained by the properties of mutant proteins. A recessive allele does not contribute to the phenotype because it produces no protein (or protein that is nonfunctional).

HISTORICAL PERSPECTIVES

One Gene : One Enzyme—George W. Beadle and Edward L. Tatum, 1941

Genetic Control of Biochemical Reactions in *Neurospora*

How do genes control metabolic processes? The suggestion that genes are responsible for the production of enzymes was made very early in the history of genetics, most notably by the British physician Archibald Garrod in his 1908 book *Inborn Errors of Metabolism*. But the precise relationship between genes and enzymes was still uncertain. Perhaps each enzyme was controlled by more than one gene, or perhaps each gene contributed to the control of several enzymes. The classic experiments of George Beadle and Edward Tatum showed that the relationship is often remarkably simple: one gene encodes one enzyme. The pioneering experiments united genetics and biochemistry and for the "one gene, one enzyme" concept, Beadle and Tatum were awarded a Nobel Prize in 1958 (Joshua Lederberg shared the prize for his contributions to microbial genetics). Because we now know that some enzymes contain polypeptide chains encoded by two (or occasionally more) different genes, a more accurate statement of the principle is "one gene, one polypeptide." Beadle and Tatum's experiments also demonstrate the importance of choosing the right organism. *Neurospora* had been introduced as a genetic model organism only a few years earlier, and Beadle and Tatum realized that they could take advantage of the ability of this organism to grow on a simple medium composed of known substances.

> "From the standpoint of physiological genetics the development and functioning of an organism consist essentially of an integrated system of chemical reactions controlled in some manner by genes. . . . In investigating the roles of genes, the physiological geneticist usually attempts to determine the physiological and biochemical bases of already known hereditary traits. . . . There are, however, a number of limitations inherent in this approach. Perhaps the most serious of these is that the investigator must in general confine himself to the study of nonlethal heritable characters. Such characters are likely to involve more or less non-essential so-called "terminal" reactions. . . . A second difficulty . . . is that the standard approach to the problem implies the use of characters with visible manifestations. Many such characters involve morphological variations, and these are likely to be based on systems of biochemical reactions so complex as to make analysis exceedingly difficult. Considerations such as those just outlined have led us to investigate the general problem of the genetic control of development and metabolic reactions by reversing the ordinary procedure and, instead of attempting to work out the chemical bases of known genetic characters, to set out to determine if and how genes control known biochemical reactions. The ascomycete *Neurospora* offers many advantages for such an approach and is well suited to genetic studies. Accordingly, our program has been built around this organism. The procedure is based on the assumption that x-ray treatment will induce mutations in genes concerned with the control of known specific chemical reactions. If the organism must be able to carry out a certain chemical reaction to survive on a given medium, a mutant unable to do this will obviously be lethal on this medium. Such a mutant can be maintained and studied, however, if it will grow on a medium to which has been added the essential product of the genetically blocked reaction. . . . Among approximately 2000 . . . strains [derived from single cells after x-ray treatment], three mutants have been found that grow essentially normally on the complete medium and scarcely at all on the minimal medium. . . . One of these strains . . . proved to be unable to synthesize vitamin B6 (pyridoxine). A second strain . . . turned out to be unable to synthesize vitamin B1 (thiamine). . . . A third strain . . . has been found to be unable to synthesize para-aminobenzoic acid. . . . [These] preliminary results . . . appear to us to indicate that [this] approach . . . may offer considerable promise as a method of learning more about how genes regulate development and function. For example, it should be possible, by finding a number of mutants unable to carry out a particular step in a given synthesis, to determine whether only one gene is ordinarily concerned with the immediate regulation of a given specific chemical reaction."

Source: G. W. Beadle and E. L. Tatum. (1941). *Proc. Natl. Acad. Sci. USA* **27**, pp. 499–506.

the relationship between recessive and wild-type alleles. When a heterozygote contains one wild-type allele and one mutant allele, the wild-type allele is able to direct production of the enzyme and is, therefore, dominant. (This assumes that an adequate amount of protein is made by the single wild-type allele. When this is not true, the smaller amount made by one allele as compared to two alleles results in the intermediate phenotype of a partially dominant allele in a heterozygote.)

KEY CONCEPTS

- The one gene : one polypeptide hypothesis summarizes the basis of modern genetics : that a typical gene is a stretch of DNA encoding a single polypeptide chain.
- Genes can also encode RNA products that are not translated into polypeptides.
- Mutations can damage gene function.

CONCEPT AND REASONING CHECK

Propose a situation in which a mutant allele is dominant in a heterozygote with a wild-type allele.

2.3 Mutations in the Same Gene Cannot Complement

How do we determine whether two mutations that cause a similar phenotype have occurred in the same gene? If they map to positions that are very close together (i.e., they recombine very rarely), they may be alleles. However, in the absence of information about their relative positions, they could also represent mutations in two different genes whose proteins are involved in the same function. The **complementation test** is used to determine whether two recessive mutations are alleles of the same gene or in different genes. The test consists of generating a heterozygote for the two mutations (by mating parents homozygous for each mutation) and observing its phenotype.

▸ **complementation test** A test that determines whether two mutations are alleles of the same gene. It is accomplished by crossing two different recessive mutations that have the same phenotype and determining whether the wild-type phenotype can be produced. If so, the mutations are said to complement each other and are probably not mutations in the same gene.

If the mutations are alleles of the same gene, the parental genotypes can be represented as:

$$\frac{m_1}{m_1} \text{ and } \frac{m_2}{m_2}$$

The first parent provides an m_1 mutant allele and the second parent provides an m_2 allele, so that the heterozygote progeny have the genotype:

$$\frac{m_1}{m_2}$$

No wild-type allele is present, so the heterozygotes have mutant phenotypes.

If the mutations lie in different linked genes, the parental genotypes can be represented as:

$$\frac{m_1\ +}{m_1\ +} \text{ and } \frac{+\ m_2}{+\ m_2}$$

Each chromosome has one wild-type allele at one locus (represented by the plus sign, +) and one mutant allele at the other locus. Then the heterozygote progeny have the genotype:

$$\frac{m_1\ +}{+\ m_2}$$

in which the two parents between them have provided a wild-type allele from each gene. The heterozygotes have wild-type phenotypes, and the two genes are said to complement.

The complementation test is shown in more detail in **FIGURE 2.3**. The basic test consists of the comparison shown in the top part of the figure. If two mutations are alleles of the same gene, we see a difference in the phenotypes of the *trans* configuration (both mutations are not in the same allele) and the *cis* configuration (both mutations are in the same allele). The *trans* configuration is mutant, because each allele has a different mutation. However, the *cis* configuration is wild-type, because one allele has two mutations and the other allele has no mutations. The lower part of the figure shows that if the two mutations are in different genes, we always see a wild-type phenotype. There is always one wild type and one mutant allele of each gene in both the

FIGURE 2.3 The cistron is defined by the complementation test. Genes are represented by DNA helices; red stars identify sites of mutation.

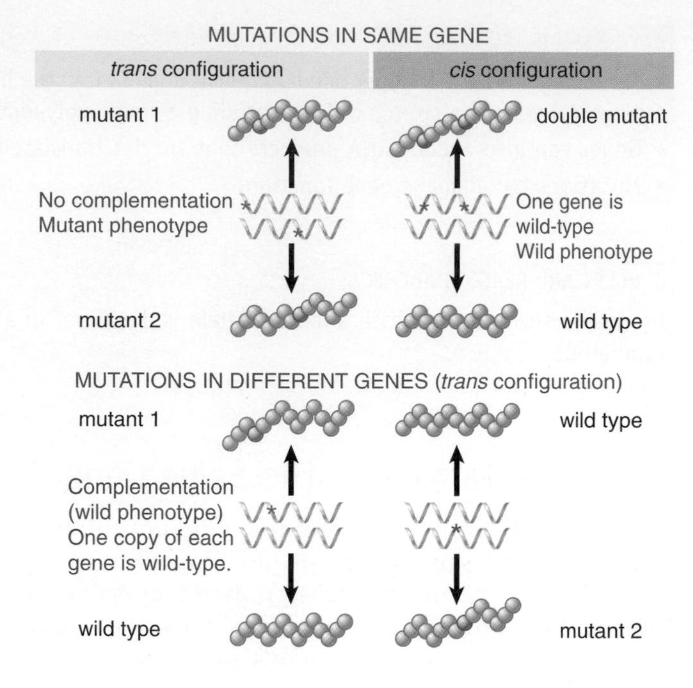

cis and *trans* configurations. "Failure to complement" means that the two mutations occurred in the same gene. The term **cistron** is used to describe the gene as defined by the complementation test and (like "gene") describes a stretch of DNA that functions as a unit to produce an RNA or polypeptide product. The properties of the gene with regard to complementation are explained by the fact that its product is a single molecule that behaves as a functional unit.

▸ **cistron** The genetic unit defined by the complementation test; it is equivalent to a gene and includes all noncomplementing alleles.

KEY CONCEPTS

- A mutation in a gene affects only the polypeptide encoded by the mutant copy of the gene and does not affect the polypeptide encoded by any other allele.
- Failure of two mutations to complement (produce wild-type phenotype when they are present in *trans* configuration in a heterozygote) means that they are alleles of the same gene.

CONCEPT AND REASONING CHECK

Would the complementation test work for mutations that are dominant? Why or why not?

2.4 Mutations May Cause Loss of Function or Gain of Function

The various possible effects of mutation in a gene are summarized in **FIGURE 2.4**.

In principle, when a gene has been identified, insight into its function can be gained by generating a mutant organism that entirely lacks the gene. A mutation that completely eliminates gene function—usually because the gene has been deleted—is called a **null mutation**. If the gene is essential to the organism's survival, a null mutation is lethal.

▸ **null mutation** A mutation that completely eliminates the function of a gene.

To determine how a gene affects the phenotype, it is necessary to characterize the effect of a null mutation. When a mutation fails to affect the phenotype, it is possible that it is a "leaky" mutation—enough active product is made to fulfill its function, even though the activity is quantitatively reduced or qualitatively different from the wild type. However, if a null mutant fails to affect a phenotype, we may safely conclude that the gene function is not essential.

Null mutations, or other mutations that impede gene function (but do not necessarily abolish it entirely), are called **loss-of-function mutations**. A loss-of-function mutation is generally recessive (as in the example of Figure 2.2). Sometimes a mutation has the opposite effect and causes a protein to acquire a new function; such a change is called a **gain-of-function mutation**. A gain-of-function mutation is generally dominant.

Not all mutations in genes lead to a detectable change in the phenotype. Mutations without apparent phenotypic effect are called **silent mutations**. They fall into two categories: (1) base changes in a gene that do not cause any change in the amino acid in the resulting polypeptide; and (2) base changes in a gene that change the amino acid, but the replacement in the polypeptide does not affect its activity. Silent mutations in the second category are called **neutral substitutions**.

Wild-type gene codes for protein

Silent mutation does not affect protein

Point mutation may damage function

Null mutation makes no protein

Point mutation may create new function

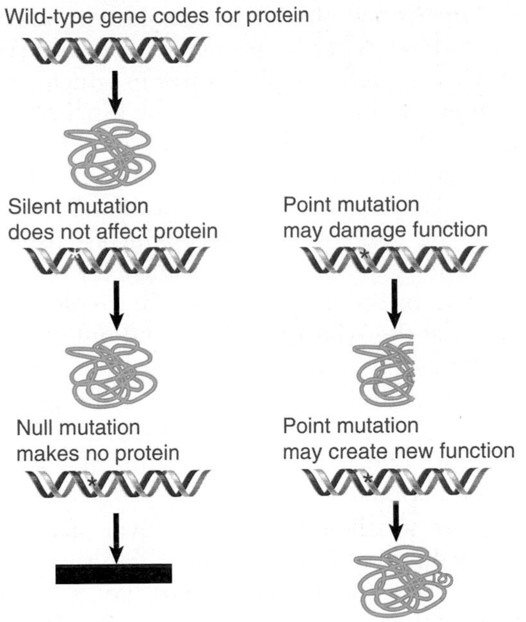

FIGURE 2.4 Mutations that do not affect protein sequence or function are silent. Mutations that abolish all protein activity are null. Point mutations that cause loss of function are recessive; those that cause gain of function are dominant.

- **loss-of-function mutation** A mutation that eliminates or reduces the activity of a gene. It is often, but not always, recessive.
- **gain-of-function mutation** A mutation that causes an increase in the normal gene activity. It sometimes represents acquisition of certain abnormal properties. It is often, but not always, dominant.
- **silent mutation** A mutation that does not change the sequence of a polypeptide because it produces synonymous codons.
- **neutral substitutions** Mutations that cause changes in amino acids of the protein product but that do not affect the protein's activity.

KEY CONCEPTS

- Recessive mutations are due to loss of function by the polypeptide product.
- Dominant mutations result from a gain of function.
- Testing whether a gene is essential requires a null mutation (one that completely eliminates its function).
- Silent mutations have no phenotypic effect, either because the base change does not change the sequence or amount of polypeptide or because the change in polypeptide sequence has no effect.
- "Leaky" mutations do affect the function of the gene product but are not shown in the phenotype because sufficient activity remains.

CONCEPT AND REASONING CHECK

Explain why loss-of-function mutations are generally recessive and gain-of-function mutations are generally dominant.

2.5 A Locus May Have Many Alleles

If a recessive mutation is produced by every change in a gene that prevents the production of an active protein, there should be a large potential number of such mutations for any one gene. Many amino acid replacements may change the structure of the protein sufficiently to impede its function.

Different variants of the same gene are called *multiple alleles*, and their existence makes it possible to generate heterozygotes with two mutant alleles. The relationships between these multiple alleles can take various forms.

In the simplest case, a wild-type allele encodes a polypeptide product that is functional, whereas mutant allele(s) encode polypeptides that are nonfunctional. However, there are often cases in which a series of mutant alleles have different phenotypes. For example, wild-type function of the *white* locus of *Drosophila melanogaster* is required for development of the normal red color of the eye. The locus is named for the effect of null mutations, which, in homozygotes, cause the fly to have white eyes.

In Drosophila, the name of the wild-type allele is indicated by a "plus" superscript after the name of the locus, which is usually named after a mutant phenotype. For example, w^+ is the wild-type allele for red eye color in *D. melanogaster*, and the locus is named for the mutant "white" eye color phenotype. Sometimes + is used by itself to describe the wild-type allele, and only the mutant alleles are indicated by the name of the locus.

An entirely defective form of the gene (or absence of phenotype) may be indicated by a "minus" superscript. To distinguish among a variety of mutant alleles with different effects, other superscripts may be introduced, such as w^i (ivory eye color) or w^a (apricot eye color).

The w^+ allele is dominant over any other allele in heterozygotes, and there are many different mutant alleles for this locus. **FIGURE 2.5** shows a small sample. Although some alleles have no eye color (i.e., a white eye), many alleles produce some color. Each of these mutant alleles must, therefore, represent a different mutation of the gene, many of which do not eliminate its function entirely but leave a residual activity that produces a characteristic phenotype. These alleles are named for the color of the eye in a homozygote. Most *w* mutations affect the quantity of pigment in the eye. The examples in the figure are arranged in roughly declining amount of color, but others, such as w^{sp}, affect the pattern in which pigment is deposited.

Allele	Phenotype of homozygote
w^+	red eye (wild type)
w^{bl}	blood
w^{ch}	cherry
w^{bf}	buff
w^h	honey
w^a	apricot
w^e	eosin
w^i	ivory
w^z	zeste (lemon-yellow)
w^{sp}	mottled, color varies
w^1	white (no color)

FIGURE 2.5 The *w* locus in *Drosophila melanogaster* has an extensive series of alleles whose phenotypes extend from wild-type (red) color to complete lack of pigment.

When multiple alleles exist, an organism may be a heterozygote that carries two different mutant alleles. The phenotype of such a heterozygote depends on the nature of the residual activity of each allele. The relationship between two mutant alleles is, in principle, no different from that between wild-type and mutant alleles: one allele may be dominant, there may be partial dominance, or there may be codominance.

There is not necessarily a unique wild-type allele for any particular locus. Control of the *ABO* human blood group system provides an example. Lack of function is represented by the null, or *O*, allele. However, the functional alleles *A* and *B* are codominant with one another and dominant to the *O* allele. The basis for this relationship is illustrated in **FIGURE 2.6**.

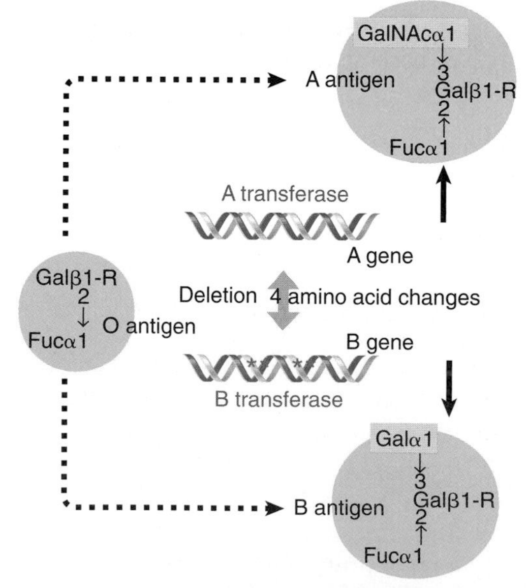

Phenotype	Genotype	Activity
O	OO	None
A	AO or AA	N-Ac-gal transferase
B	BO or BB	Gal transferase
AB	AB	GalN-Ac-Gal-transferase

FIGURE 2.6 The *ABO* human blood group locus encodes a galactosyltransferase whose specificity determines the blood group.

The O antigen is generated in all individuals and consists of a particular carbohydrate group that is added to proteins. The *ABO* locus encodes a galactosyltransferase enzyme that puts a additional sugar group on the O antigen. The specificity of this enzyme determines the blood group. The *A* allele produces an enzyme that uses the cofactor UDP-N-acetylgalactose, forming the A antigen. The *B* allele produces an enzyme that uses the cofactor UDP-galactose, forming the B antigen. The A and B versions of the transferase enzyme differ in four amino acids that presumably affect its recognition of the type of cofactor. The *O* allele has a small deletion that eliminates the activity of the transferase, so no modification of the O antigen occurs.

This explains why *A* and *B* alleles are dominant in the *AO* and *BO* heterozygotes: the corresponding transferase activity forms the A or B antigen. The *A* and *B* alleles are codominant in *AB* heterozygotes, because both transferase activities are expressed. The *OO* homozygote is a null that has neither activity and therefore lacks both antigens.

Neither *A* nor *B* alleles can be regarded as uniquely wild-type because they represent alternative activities rather than loss or gain of function. A situation such as this, in which there are multiple functional alleles in a population, is described as a **polymorphism** (see *Section 3.3, Individual Genomes Show Extensive Variation*)

polymorphism The simultaneous occurrence in the population of alleles showing variations at a given position.

KEY CONCEPTS

- The existence of multiple alleles allows the possibility of heterozygotes representing any pairwise combination of alleles.
- A locus may have a polymorphic distribution of alleles with no individual allele that can be considered to be the sole wild type.

CONCEPT AND REASONING CHECK

Explain why an individual fly with a new recessive mutation of the *white* locus will have a wild-type phenotype.

2.6 Recombination Occurs by Physical Exchange of DNA

The term **genetic recombination** describes the generation of new combinations of alleles at each generation in diploid organisms. Generally, each chromosome has a *homologous* partner with which it pairs during meiosis. The two homologous copies of each chromosome may have different alleles at some loci. By the exchange of corresponding segments between the homologs, called *crossing over*, recombinant chromosomes that are different from the parental chromosomes can be generated.

genetic recombination A process by which separate DNA molecules are joined into a single molecule due to such processes as crossing over or transposition.

Recombination results from a physical exchange of chromosomal material. For example, recombination may result from the crossing over that occurs during meiosis (the specialized division that produces haploid germ cells), specifically during Prophase I. Meiosis starts with a cell that has duplicated its chromosomes, so that it has four copies of each chromatid. (A *chromatid* is one of two identical copies of a chromosome following its duplication; since there are two homologs per diploid cell, there are four chromatids of each type of chromosome.) Early in meiosis, all four copies are closely associated (synapsed) in a structure called a *bivalent* and, later, a *tetrad*. At this point, pairwise exchanges of material between two nonsister (of the four total) chromatids may occur.

The point of synapsis between homologs is called a **chiasma** and is illustrated diagrammatically in **FIGURE 2.7**. A chiasma represents a site at which one strand in each of two nonsister chromatids in a tetrad has been broken and exchanged. If during the resolution of the chiasma the previously unbroken strands are also broken and exchanged, recombinant chromatids will be generated. Each recombinant chromatid consists of material derived from one chromatid on one side of the chiasma, with

chiasma A site at which two homologous chromosomes synapse during Prophase I of meiosis.

FIGURE 2.7 Chiasma formation at Prophase I of meiosis is responsible for generating recombinant chromosomes.

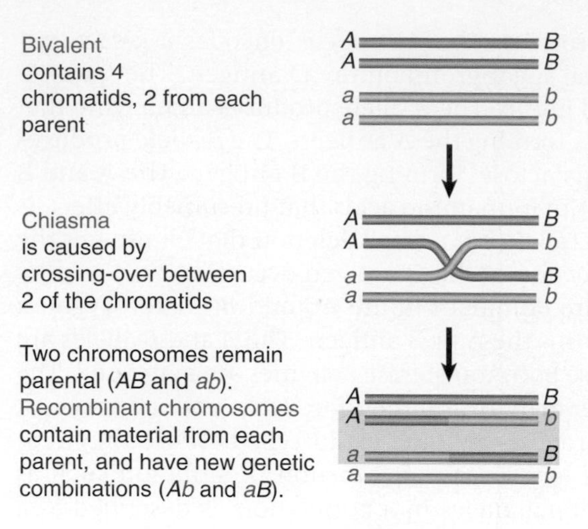

material from the other chromatid on the opposite side. The two recombinant chromatids have reciprocal structures. The event can be described as a "breakage and reunion." Because each individual crossing-over event involves only two of the four associated chromatids, a single recombination event can produce only 50% recombinants.

The complementarity of the two strands of DNA is essential for the recombination process. Each of the chromatids shown in Figure 2.7 consists of a very long duplex of DNA. For them to be broken and reconnected without any addition or loss of material requires a mechanism to recognize exactly corresponding positions; this mechanism is complementary base pairing.

▸ **heteroduplex DNA** DNA that is generated by base pairing between complementary single strands derived from the different parental duplex molecules; it occurs during crossing over.

Recombination results from a process in which the single strands in the region of the crossover exchange their partners, resulting in a branch that may migrate for some distance in either direction. This creates a stretch of **heteroduplex DNA**, in which the single strand of one duplex is paired with its complement from the other duplex. The mechanism, of course, involves other stages in which strands must be broken and religated, which we discuss in more detail in *Chapter 15, Homologous and Site-Specific Recombination,* but the crucial feature that makes precise recombination possible is the complementarity of DNA strands. A stretch of heteroduplex DNA forms in the recombination intermediate when a single strand crosses over from one duplex to the other. Each recombinant consists of one parental duplex DNA which is connected by a stretch of heteroduplex DNA to the other parental duplex. Each duplex DNA corresponds to one of the chromatids involved in recombination in Figure 2.7.

The formation of heteroduplex DNA requires the sequences of the two recombining duplexes to be close enough to allow pairing between the complementary strands. If there are no differences between the two parental genomes in this region, formation of heteroduplex DNA will be perfect. However, pairing can still occur even when there are small differences. In this case, the hybrid DNA has points of mismatch, at which a base in one strand is paired with a base in the other strand that is not complementary to it. The correction of such mismatches is another feature of genetic recombination (see *Chapter 16, Repair Systems*).

Over chromosomal distances, recombination events occur more or less at random, with a characteristic frequency. The probability that a crossover will occur within any specific region of the chromosome is more or less proportional to the length of the region, up to a saturation point. For example, a large human chromosome usually has three or four crossover events per meiosis, but a small chromosome may have only one on average.

FIGURE 2.8 compares three situations: two genes on different chromosomes, two genes that are far apart on the same chromosome, and two genes that are close together on the same chromosome. Genes on different chromosomes segregate independently according to Mendel's laws, resulting in the production of 50% "parental" types and 50% "recombinant" types during meiosis. When genes are sufficiently far apart on the same chromosome, the probability of at least one crossover in the region between them becomes so high that their association is the same as that of genes on different chromosomes, and they show 50% recombination.

What if genes are close together on the same chromosome? The probability of a crossover between them is reduced, and recombination occurs only in some proportion of meioses. For example, if it occurs in one-quarter of the meioses, the overall rate of recombination is 12.5% (because a single recombination event produces

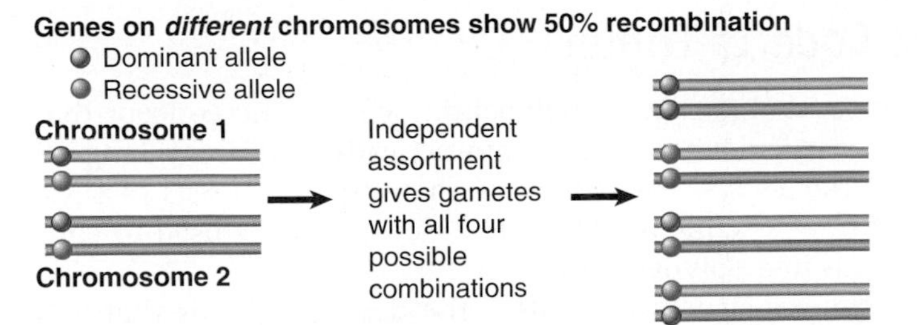

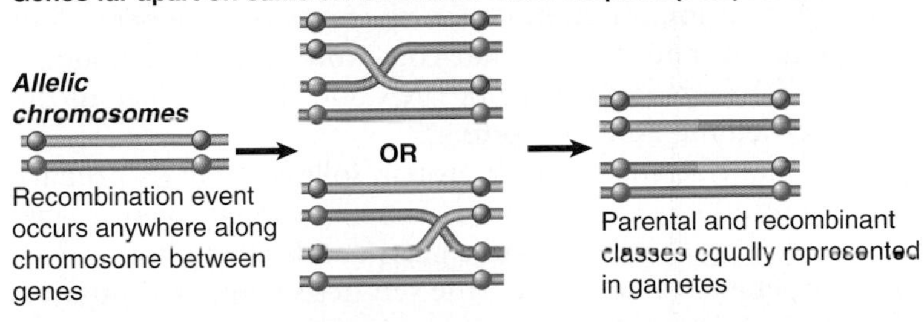

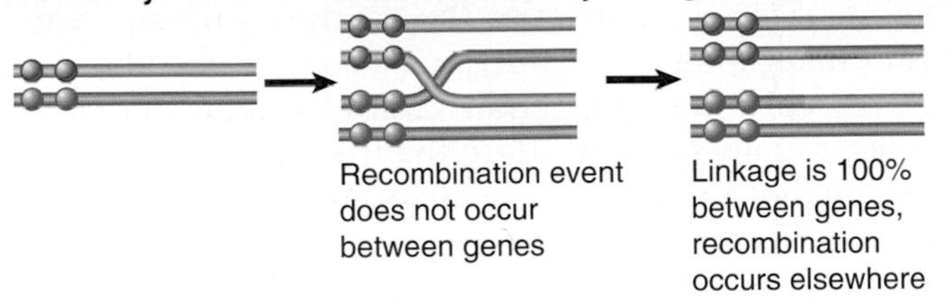

FIGURE 2.8 Genes on different chromosomes segregate independently so that all possible combinations of alleles are produced in equal proportions. Crossing over occurs so frequently between genes that are far apart on the same chromosome that they effectively segregate independently. But recombination is reduced when genes are closer together, and for adjacent genes it may hardly ever occur.

50% recombination, and this occurs in 25% of meioses). When genes are very close together, as shown in the bottom panel of Figure 2.8, recombination between them may never be observed in phenotypes of multicellular eukaryotes (because they produce few offspring).

This leads us to the concept that a chromosome is an array of many genes. Each protein-coding gene is an independent unit of expression and is represented in one or more polypeptide chains. The properties of a gene can be changed by mutation. The allelic combinations present on a chromosome can be changed by crossing over. We can now ask, "What is the relationship between the sequence of a gene and the sequence of the polypeptide chain it encodes?"

KEY CONCEPTS

- Recombination is the result of crossing over that occurs at a chiasma during meiosis and involves two of the four chromatids.
- Recombination occurs by a breakage and reunion that proceeds via an intermediate of heteroduplex DNA.
- The distance between genes on the same chromosome is determined by the frequency of recombination between them.
- Genes that are close together are tightly linked because recombination between them is rare.
- Genes that are far apart may not show linkage because recombination is so frequent as to produce the same result as for genes on different chromosomes.

CONCEPT AND REASONING CHECK

Explain why the maximum frequency of recombination that can be measured between two loci is 50%.

2.7 The Genetic Code Is Triplet

Each protein-coding gene encodes a particular polypeptide chain. The concept that each polypeptide consists of a particular sequence of amino acids dates from Sanger's characterization of insulin in the 1950s. The discovery that a gene consists of DNA presents us with the issue of how a sequence of nucleotides in DNA is used to construct a sequence of amino acids in a polypeptide.

A crucial feature of the general structure of DNA, the double helix, is that it is independent of the particular sequence of its component nucleotides. The sequence of nucleotides in a structural gene is important not because of its structure per se, but because it encodes the sequence of amino acids that constitutes the corresponding polypeptide. The relationship between a sequence of DNA and the sequence of the corresponding polypeptide is called the **genetic code**.

▸ **genetic code** The correspondence between triplets in DNA (or RNA) and amino acids in polypeptide.

The structure and/or enzymatic activity of each protein follows from its primary sequence of amino acids. By determining the sequence of amino acids, the gene is able to carry all the information needed to specify an active polypeptide chain. In this way, a single type of structure—the gene—is able to direct the synthesis of many thousands of polypeptide types in a cell.

Together the various proteins of a cell undertake the catalytic and structural activities that are responsible for establishing its phenotype. Of course, in addition to sequences that encode proteins, DNA also contains certain control sequences that are recognized by regulator molecules, usually proteins. Here the function of the DNA is determined by its sequence directly, not via any intermediary molecule. Both types of sequence—genes expressed as proteins and sequences recognized by proteins—constitute genetic information.

The coding region of a gene is deciphered by a complex apparatus that interprets the nucleic acid sequence. In any given region, it is usually the case that only one of the two strands of DNA encodes a functional RNA, so we write the genetic code as a sequence of bases (rather than base pairs). (Recent evidence suggests that both strands are transcribed in some regions, but it is not clear that both resulting transcripts have functional importance.)

A coding sequence is read in groups of three nucleotides, each group representing one amino acid. Each trinucleotide sequence is called a **codon**. A gene includes a series of codons that is read sequentially from a starting point at one end to a termination point at the other end. Written in the conventional 5′ to 3′ direction, the nucleotide sequence of the DNA strand that encodes a polypeptide corresponds to the amino acid sequence of the polypeptide written in the direction from N-terminus to C-terminus.

▸ **codon** A triplet of nucleotides that codes for an amino acid, or a termination signal.

A coding sequence is read in nonoverlapping triplets from a fixed starting point:

- *Nonoverlapping* implies that each codon consists of three nucleotides and that successive codons are represented by successive trinucleotides. An individual nucleotide is part of only one codon.
- The use of a *fixed starting point* means that assembly of a polypeptide must start at one end and proceed to the other, so that different parts of the coding sequence cannot be read independently.

The nature of the code predicts that two types of mutations will have different effects. If a particular sequence is read sequentially, such as:

UUU AAA GGG CCC (codons)
aa1 aa2 aa3 aa4 (amino acids)

then a nucleotide substitution will affect only one amino acid. For example, the substitution of an A by some other base (X) causes aa2 to be replaced by aa5:

UUU AAX GGG CCC
aa1 aa5 aa3 aa4

because only the second codon has been changed.

However, a mutation that inserts or deletes a single base will change the triplet sets for the entire subsequent sequence. A change of this sort is called a **frameshift**. An insertion might take the form:

UUU AAX AGG GCC C

aa1 aa5 aa6 aa7

Because the new sequence of triplets is completely different from the old one, the entire amino acid sequence of the polypeptide is altered downstream from the site of mutation, so the function of the protein is likely to be lost completely.

Frameshift mutations are induced by the **acridines**, compounds that bind to DNA and distort the structure of the double helix, causing additional bases to be incorporated or omitted during replication. Each mutagenic event in the presence of an acridine results in the addition or removal of a single base pair.

If an acridine mutant is produced by, say, addition of a nucleotide, it should revert to wild type by deletion of the nucleotide. However, reversion also can be caused by deletion of a different base at a site close to the first. Combinations of such mutations provided revealing evidence about the nature of the genetic code.

FIGURE 2.9 illustrates the properties of frameshift mutations. An insertion or deletion changes the entire polypeptide sequence following the site of mutation. However, the combination of the insertion of a single nucleotide and the deletion of a single nucleotide causes the code to be read incorrectly only between the two sites of mutation; reading in the original frame resumes after the second site.

In 1961, genetic analysis of acridine mutations in the *rII* region of the phage T6 showed that all the mutations could be classified into one of two sets, described as (+) and (–). Either type of mutation by itself causes a frameshift: the (+) type by virtue of a base addition and the (–) type by virtue of a base deletion. Double mutant combinations of the types (++) and (– –) continue to show mutant behavior. However, combinations of the types (+–) or (–+) suppress one another, so that one mutation is described as a frameshift suppressor of the other. (In the context of this work, "suppressor" is used in an unusual sense because the second mutation is in the same gene as the first; in fact, these are second-site reversions.)

These results show that the genetic code must be read as a sequence that is fixed by the starting point. Therefore, a single base addition and deletion compensate for each

▸ **frameshift** A mutation caused by deletions or insertions that are not a multiple of three base pairs. They change the frame in which triplets are translated into polypeptide.

▸ **acridines** Mutagens that act on DNA to cause the insertion or deletion of a single base pair. They were useful in defining the triplet nature of the genetic code.

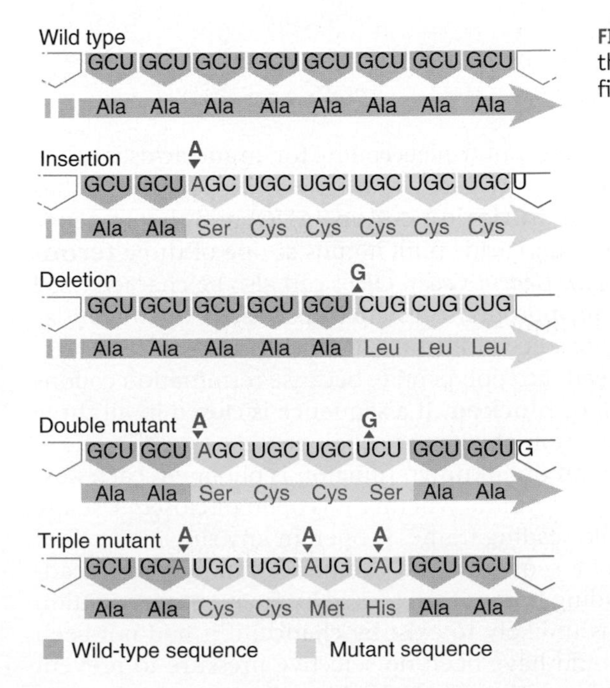

FIGURE 2.9 Frameshift mutations show that the genetic code is read in triplets from a fixed starting point.

other, whereas double additions or double deletions remain frameshift mutants. However, these observations do not suggest how many nucleotides make up each codon.

When triple mutants are constructed, only (+++) and (– – –) combinations show the wild-type phenotype, whereas other combinations remain mutant. If we take three single base additions or three deletions to correspond respectively to the addition or omission overall of a single amino acid, this implies that the code is read in triplets. An incorrect amino acid sequence is found between the two outside sites of mutation, and the sequence on either side remains wild type, as indicated in Figure 2.9.

KEY CONCEPTS

- The genetic code is read in nucleotide triplets called *codons*.
- The triplets are nonoverlapping and are read from a fixed starting point.
- Mutations that insert or delete individual bases cause a shift in the triplet sets after the site of mutation; these are frameshift mutations.
- Combinations of mutations that together insert or delete three bases (or multiples of three) insert or delete amino acids, but do not change the reading of the triplets beyond the last site of mutation.

CONCEPT AND REASONING CHECK

Consider mutagens that insert or delete two adjacent nucleotides. Using "+" to mean an insertion of two nucleotides and "–" to mean a deletion of two nucleotides, what combinations of insertions and deletions would suppress one another?

2.8 Every Coding Sequence Has Three Possible Reading Frames

If a coding sequence is read in nonoverlapping triplets, there are three possible ways of translating any nucleotide sequence into polypeptide, depending on the starting point. These are called **reading frames**. For the sequence

A C G A C G A C G A C G A C G A C G

the three possible reading frames are

ACG ACG ACG ACG ACG ACG

CGA CGA CGA CGA CGA CG

GAC GAC GAC GAC GAC G

A reading frame that consists exclusively of triplets coding for amino acids is called an **open reading frame,** or **ORF**. A sequence that is translated into polypeptide has a reading frame that starts with a special **initiation codon** (AUG) and then extends through a series of triplets coding for amino acids until it ends at one of three **termination codons** (see *Chapter 25, Using the Genetic Code*). ORFs can also be characterized by consensus promoter sequences at an appropriate distance upstream from the initiation codon, and (in eukaryotes) the presence of intron splicing junctions.

A reading frame that cannot be read into polypeptide because termination codons occur frequently is said to be **closed,** or **blocked**. If a sequence is closed in all three reading frames, it cannot have the function of coding for polypeptide.

When the sequence of a DNA region of unknown function is obtained, each possible reading frame can be analyzed to determine whether it is open or closed. Usually no more than one of the three possible reading frames is open in any single stretch of DNA. **FIGURE 2.10** shows an example of a sequence that can be read in only one reading frame because the alternative reading frames are blocked by frequent termination codons. A long open reading frame is unlikely to exist by chance; if it had not been translated into polypeptide, there would have been no selective pressure to prevent the accumulation of termination codons. Therefore, the identification of a lengthy

- **reading frame** One of three possible ways of reading a nucleotide sequence. Each divides the sequence into a series of successive triplets.
- **open reading frame (ORF)** A sequence of DNA consisting of triplets that can be translated into amino acids starting with an initiation codon and ending with a termination codon.
- **initiation codon** A special codon (usually AUG) used to start synthesis of a polypeptide.
- **termination codon** One of the three codons (UAA, UAG, UGA) that signal the termination of translation of a polypeptide. They are also known as stop codons.
- **closed (blocked) reading frame** A reading frame that cannot be translated into polypeptide because of the occurrence of termination codons.

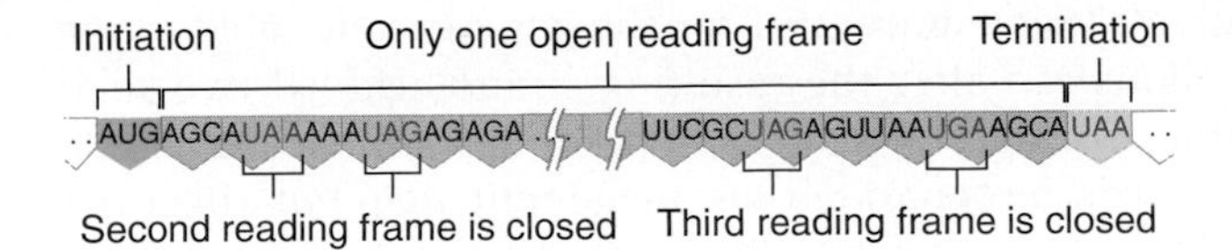

FIGURE 2.10 An open reading frame starts with AUG and continues in triplets to a termination codon. Closed reading frames may be interrupted frequently by termination codons.

open reading frame is taken to be *prima facie* evidence that the sequence is (or until recently has been) translated into a polypeptide in that frame. An ORF for which no polypeptide product has been identified is sometimes called an **unidentified reading frame (URF)**.

▸ **unidentified reading frame (URF)** An open reading frame with an as yet undetermined function.

KEY CONCEPT

- Usually only one of the three possible reading frames is translated, and the other two are closed by frequent termination signals.

CONCEPT AND REASONING CHECK

Suppose you identify a stretch of DNA in which only one of the three possible reading frames is closed. What would you hypothesize from this observation?

2.9 Bacterial Genes Are Colinear with Their Products

By comparing the nucleotide sequence of a gene with the amino acid sequence of its polypeptide product, we can determine whether the gene and the polypeptide are **colinear**—that is, whether the sequence of nucleotides in the gene exactly corresponds to the sequence of amino acids in the polypeptide. In bacteria and their viruses, genes and their products are colinear. Each gene is a continuous stretch of DNA with a coding region that is three times the number of amino acids in the polypeptide which it encodes (due to the triplet nature of the genetic code). In other words, if a polypeptide contains *N* amino acids, the gene encoding that polypeptide contains 3*N* nucleotides.

▸ **colinearity** The relationship that describes the 1:1 correspondence of a sequence of triplet nucleotides to a sequence of amino acids.

The correspondence of the bacterial gene and its product means that a physical map of DNA will exactly match an amino acid map of the polypeptide. How well do these maps match the recombination map?

The colinearity of gene and polypeptide was originally investigated in the tryptophan synthetase gene of *E. coli*. Genetic distance was measured as the percent recombination between mutations; amino acid distance was measured as the number of amino acids separating sites of replacement. **FIGURE 2.11** compares the two maps. The order of seven sites of mutation is the same as the order of the corresponding sites of amino acid replacement, and the recombination distances are roughly similar to the actual distances in the protein. The recombination map expands the distances between some mutations, but otherwise there is little distortion of the recombination map relative to the physical map.

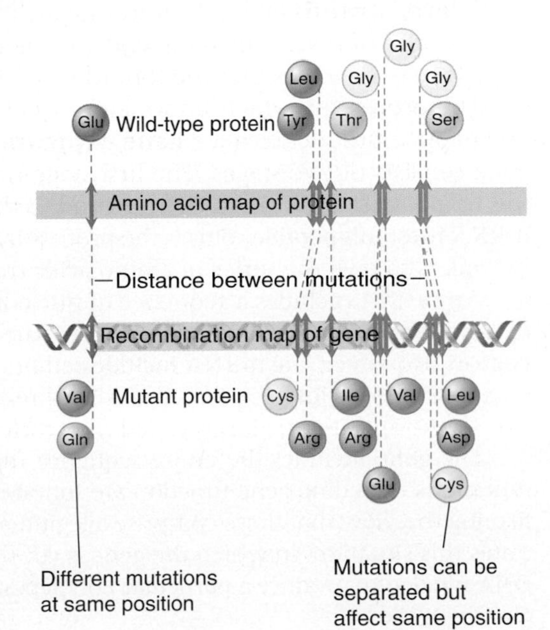

FIGURE 2.11 The recombination map of the tryptophan synthetase gene corresponds with the amino acid sequence of the polypeptide.

The recombination map leads to two further general points about the organization of the gene. Different mutations may cause a wild-type amino acid to be replaced with different alternatives. (See Figure 25.1 for the genetic code table to see how base changes in specific codons can alter the resulting amino acid.) If two such mutations cannot recombine, they must involve different point mutations at the same position in DNA. If the mutations can be separated on the genetic map but affect the same amino acid on the upper map (the connecting lines converge in the figure), they must involve point mutations at different positions in the same codon. This happens because the minimum size unit of genetic recombination (1 bp) is smaller than the unit coding for the amino acid (3 bp).

KEY CONCEPTS

- A bacterial gene consists of a continuous length of 3*N* nucleotides that encodes *N* amino acids.
- The gene is colinear with both its mRNA and polypeptide products.

CONCEPT AND REASONING CHECK

Although a recombination map and a physical map will show genetic markers in the same order, the distance between markers will usually be different when the two maps are compared. Why?

2.10 Several Processes Are Required to Express the Product of a Gene

- **messenger RNA (mRNA)** The intermediate that represents one strand of a gene encoding a polypeptide. Its coding region is related to the polypeptide sequence by the triplet genetic code.
- **antisense (template) strand** The DNA strand that is complementary to the sense strand and acts as the template for synthesis of mRNA.
- **coding (sense) strand** The DNA strand that has the same sequence as the mRNA and is related by the genetic code to the polypeptide sequence that it represents.
- **gene expression** The process by which the information in a sequence of DNA in a gene is used to produce an RNA or polypeptide, involving transcription and (for polypeptides) translation.
- **transcription** Synthesis of RNA from a DNA template.
- **translation** Synthesis of polypeptide from an mRNA template.
- **coding region** A part of a gene that encodes a polypeptide sequence.
- **leader (5′ UTR)** In mRNA, the untranslated sequence at the 5′ end that precedes the initiation codon.
- **trailer (3′ UTR)** An untranslated sequence at the 3′ end of an mRNA following the termination codon.

In comparing a gene and its polypeptide product, we are restricted to the sequence of DNA that lies between the points corresponding to the N-terminus and C-terminus of the polypeptide. However, a gene is not directly translated into polypeptide but is expressed via the production of a **messenger RNA** (abbreviated as **mRNA**), a nucleic acid intermediate actually used to synthesize a polypeptide (as we see in detail in *Chapter 24, Translation*).

Messenger RNA is synthesized by the same process of complementary base pairing used to replicate DNA, with the important difference that it corresponds to only one strand of the DNA double helix. **FIGURE 2.12** shows that the sequence of mRNA is complementary to the sequence of one strand of DNA (called the **antisense,** or **template, strand**) and is identical (apart from the replacement of T with U) to the other strand of DNA (called the **coding,** or **sense, strand**). The convention for writing DNA sequences is that the top strand is the coding strand and runs 5′ to 3′.

The process by which information from a gene is used to synthesize an RNA or polypeptide product is called **gene expression**. In bacteria, expression of a structural gene consists of two stages. The first stage is **transcription**, when an mRNA copy of the template strand of the DNA is produced. The second stage is **translation** of the mRNA into polypeptide. This is the process by which the sequence of an mRNA is read in triplets to give the series of amino acids that make the corresponding polypeptide.

An mRNA includes a sequence of nucleotides that corresponds with the sequence of amino acids in the polypeptide. This part of the nucleic acid is called the **coding region**. However, the mRNA includes additional sequences on either end that do not encode amino acids. The 5′ untranslated region is called the **leader,** or **5′ UTR**, and the 3′ untranslated region is called the **trailer,** or **3′ UTR**.

The gene includes the entire sequence represented in messenger RNA. Sometimes mutations impeding gene function are found in the additional, noncoding regions, confirming the view that these comprise a legitimate part of the genetic unit. **FIGURE 2.13** illustrates this situation, in which the gene is considered to comprise a continuous stretch of DNA needed to produce a particular polypeptide, including the leader, the coding region, and the trailer.

DNA consists of two base-paired strands

top strand

5′ ATGCCGTTAGACCGTTAGCGGACCTGAC

3′ TACGGCAATCTGGCAATCGCCTGGACTG

bottom strand

RNA synthesis

5′ AUGCCGUUAGACCGUUAGCGGACCUGAC 3

RNA has same sequence as DNA top strand; is complementary to DNA bottom strand

FIGURE 2.12 RNA is synthesized by using one strand of DNA as a template for complementary base pairing.

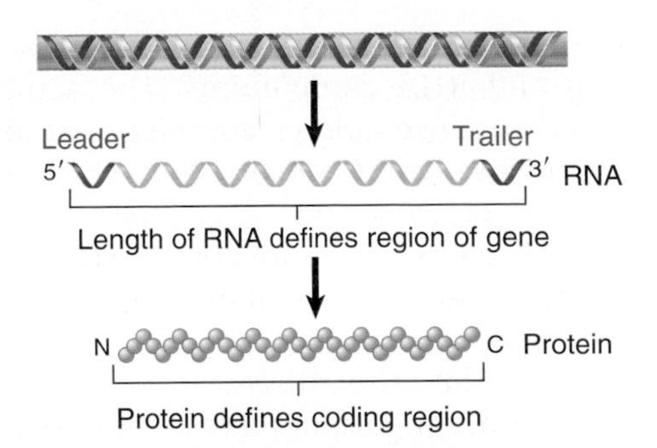

FIGURE 2.13 The gene is usually longer than the sequence encoding the polypeptide.

A bacterial cell has only a single compartment, so transcription and translation occur in the same place, as illustrated in **FIGURE 2.14**. In eukaryotes, transcription occurs in the nucleus, but the RNA product must be transported to the cytoplasm in order to be translated. This results in a spatial separation between transcription (in the nucleus) and translation (in the cytoplasm). The simplest eukaryotic genes are like bacterial genes; the transcript RNA is in fact the mRNA. However, for more complex genes, the primary transcript of the gene is a **pre-mRNA** that requires processing to generate the mature mRNA. The basic stages of gene expression in a eukaryote are outlined in **FIGURE 2.15**.

The most important stage in **RNA processing** is **splicing**. Many genes in eukaryotes (and a majority in multicellular eukaryotes) contain internal regions called **introns** that do not carry coding information for the polypeptide products encoded by those genes. The process of splicing removes introns from the pre-mRNA to generate an RNA that has a continuous open reading frame (see Figure 4.1). Other processing events that occur at this stage involve the modification of the 5′ and 3′ ends of the pre-mRNA (see Figure 21.1).

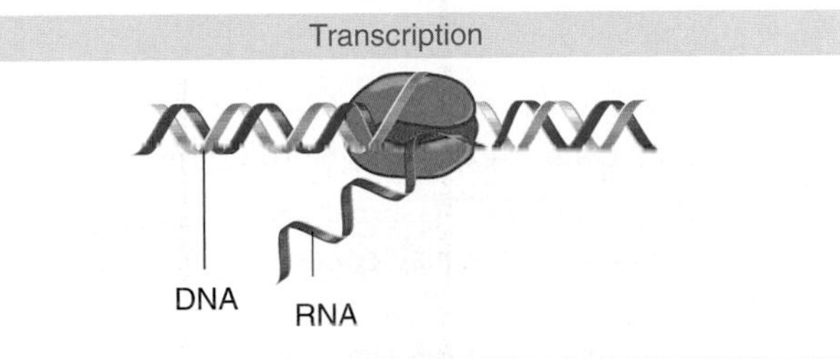

FIGURE 2.14 Transcription and translation take place in the same compartment in bacteria.

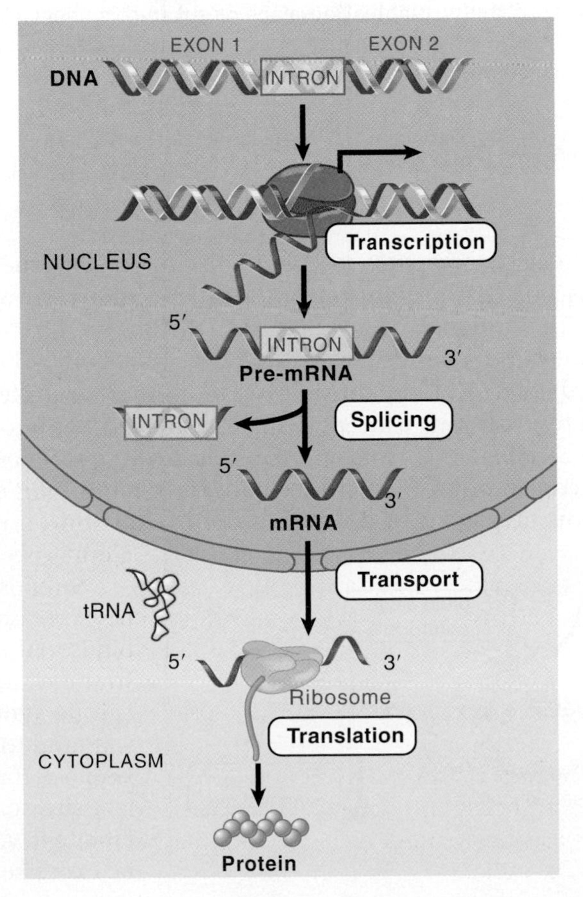

FIGURE 2.15 In eukaryotes, transcription occurs in the nucleus, and translation occurs in the cytoplasm.

- **pre-mRNA** The nuclear transcript that is processed by modification and splicing to produce a mature mRNA.
- **RNA processing** Modifications to RNA transcripts of eukaryotic genes. This may include alterations to the 3′ and 5′ ends and the removal of introns.
- **splicing** The process of excising introns from RNA and connecting the exons into a continuous mRNA.
- **intron** A segment of DNA that is transcribed but later removed from within the transcript by splicing together the sequences (exons) on either side of it.

- **ribosome** A large assembly of RNA and proteins that synthesizes polypeptides under direction from an mRNA template.
- **ribosomal RNAs (rRNAs)** A major component of the ribosome.
- **transfer RNA (tRNA)** The intermediate in protein synthesis that interprets the genetic code. Each molecule can be linked to an amino acid. It has an anticodon sequence that is complementary to a triplet codon representing the amino acid.

Translation is accomplished by a complex apparatus that includes both protein and RNA components. The actual "machine" that undertakes the process is the **ribosome**, a large complex that includes some large RNAs (**ribosomal RNAs**, abbreviated as **rRNAs**) and many small proteins. The process of recognizing which amino acid corresponds to a particular nucleotide triplet requires an intermediate **transfer RNA** (abbreviated as **tRNA**); there is at least one tRNA species for every amino acid. Many ancillary proteins are involved. We describe translation in *Chapter 24, Translation*, but note for now that the ribosomes are the large structures in that translate the mRNA.

It is an important point to note that the process of gene expression involves RNA not only as the essential substrate but also in providing components of the apparatus. The rRNA and tRNA components are encoded by genes and are generated by the process of transcription (like mRNA), but they are not translated to polypeptide.

KEY CONCEPTS

- A bacterial gene is expressed by transcription into mRNA and then by translation of the mRNA into polypeptide.
- In eukaryotes, a gene may contain introns that are not represented in the polypeptide product.
- Introns are removed from the RNA transcript by splicing to give an mRNA that is colinear with the polypeptide product.
- Each mRNA consists of an untranslated 5′ leader, a coding region, and an untranslated 3′ trailer.

CONCEPT AND REASONING CHECKS

1. Why is bacterial gene expression generally faster than eukaryotic gene expression?
2. Why do the recombination maps of eukaryotic genes usually not correspond to the amino acid maps of their products?

2.11 Proteins Are *trans*-Acting, but Sites on DNA Are *cis*-Acting

A crucial progression in the definition of the gene was the realization that all its parts must be present on one contiguous stretch of DNA. In genetic terminology, sites that are located on the same DNA are said to be in *cis*. Sites that are located on two different molecules of DNA are described as being in *trans*. So two mutations may be in *cis* (on the same DNA) or in *trans* (on different DNAs). The complementation test uses this concept to determine whether two mutations are in the same gene (see *Section 2.3, Mutations in the Same Gene Cannot Complement*). We may now extend the concept of the difference between *cis* and *trans* effects from defining the coding region of a gene to describing the interaction between a gene and its regulatory elements.

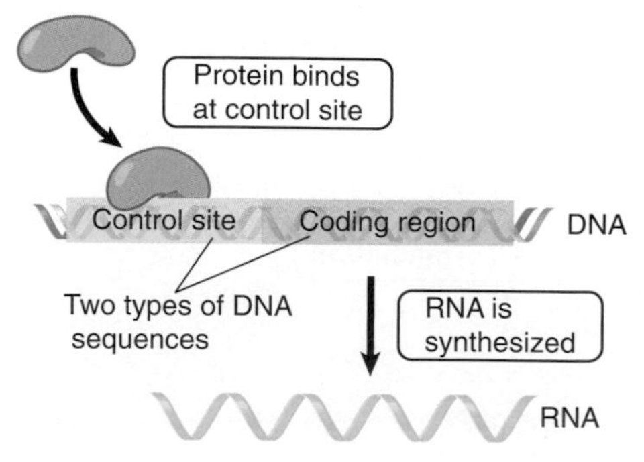

FIGURE 2.16 Control sites in DNA provide binding sites for proteins; coding regions are expressed via the synthesis of RNA.

Suppose that the ability of a gene to be expressed is controlled by a protein that binds to the DNA close to the coding region. In the example depicted in FIGURE 2.16, RNA can be synthesized only when the protein is bound to a control site on the DNA. Now suppose that a mutation occurs in the control site so that the protein can no longer bind to it. As a result, the gene can no longer be expressed.

So gene expression can be inactivated either by a mutation in a control site or by a mutation in a coding region. The mutations cannot be distinguished genetically because both have the property of acting only on the DNA sequence of the single allele in which they occur. They have identical properties in the complementation test, so a mutation in a control region is defined as comprising part of the gene in the same way as a mutation in the coding region.

FIGURE 2.17 shows that a change in the control site affects only the coding region to which it is connected; it does not affect the ability of the homologous allele to be expressed. A mutation that acts solely by affecting the properties of the contiguous sequence of DNA is called ***cis*-acting**.

FIGURE 2.17 A *cis*-acting site controls the adjacent DNA but does not influence the other allele.

We may contrast the behavior of the *cis*-acting mutation shown in Figure 2.17 with the result of a mutation in the gene encoding the regulatory protein. **FIGURE 2.18** shows that the absence of regulatory protein would prevent both alleles from being expressed. A mutation of this sort is said to be ***trans*-acting**.

Reversing the argument, if a mutation is *trans*-acting, we know that its effects must be exerted through some diffusible product (typically a protein) that acts on multiple targets within a cell. However, if a mutation is *cis*-acting, it must function via affecting directly the properties of the contiguous DNA, which means that it is not expressed in the form of RNA or protein.

- ***cis*-acting sequence** A site that affects the activity only of sequences on its own molecule of DNA (or RNA); this property usually implies that the site does not encode polypeptide.
- ***trans*-acting sequence** DNA sequence encoding a product that can function on any copy of its target DNA. This implies that it is a diffusible protein or RNA.

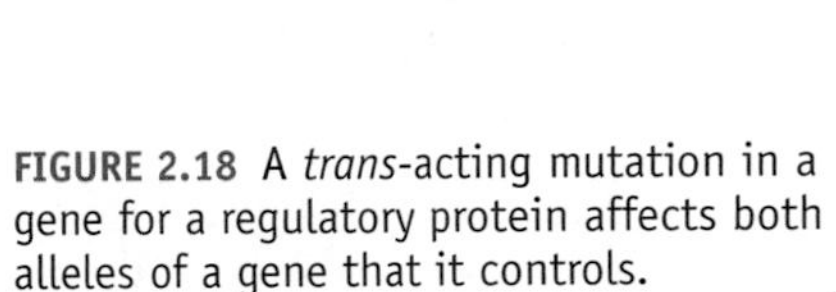

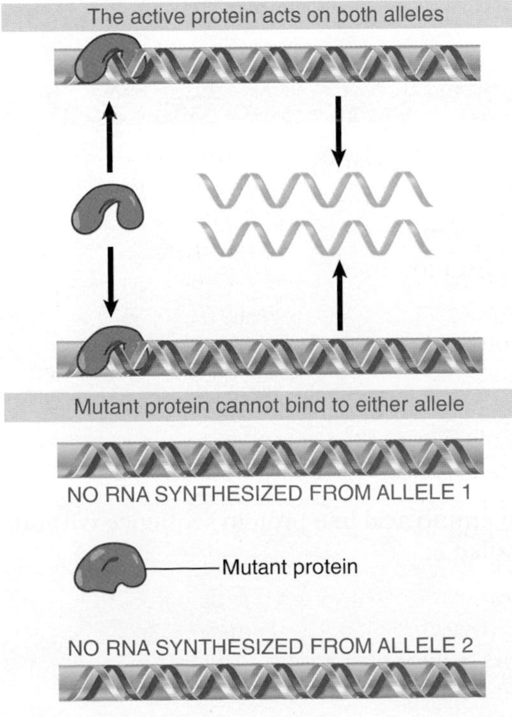

FIGURE 2.18 A *trans*-acting mutation in a gene for a regulatory protein affects both alleles of a gene that it controls.

KEY CONCEPTS

- All gene products (RNA or polypeptides) are *trans*-acting. They can act on any copy of a gene in the cell.
- *cis*-acting mutations identify sequences of DNA that are targets for recognition by *trans*-acting products. They are not expressed as RNA or polypeptide and affect only the contiguous stretch of DNA.

CONCEPT AND REASONING CHECK

Can a mutation in a control site alter the function of a gene's protein product without inactivating its expression? Explain.

2.12 Summary

A chromosome consists of an uninterrupted length of duplex DNA that contains many genes. Each gene is transcribed into an RNA product, which in turn is translated into a polypeptide sequence if it is a structural gene. An RNA or protein product of a gene is said to be *trans*-acting. A gene is defined as a unit of a single stretch of DNA by the complementation test. A site on DNA that regulates the activity of an adjacent gene is said to be *cis*-acting.

When a gene encodes a polypeptide, the relationship between the sequence of DNA and sequence of the polypeptide is given by the genetic code. Only one of the two strands of DNA codes for polypeptide. A codon consists of three nucleotides that represent a single amino acid. A coding sequence of DNA consists of a series of codons, read from a fixed starting point. Usually only one of the three possible reading frames can be translated into polypeptide.

A gene may have multiple alleles. Recessive alleles are caused by loss-of-function mutations that interfere with the function of the protein. A null allele has total loss of function. Dominant alleles are caused by gain-of-function mutations that create a new property in the protein.

CHAPTER QUESTIONS

1. Failure to complement means that two mutations are:
   A. part of the same genetic unit.
   B. in related genetic units on the same chromosome.
   C. in related genetic units on different chromosomes.
   D. in totally unrelated genetic units.
2. Most mutations that affect gene function are:
   A. dominant.
   B. recessive.
   C. codominant.
   D. corecessive.
3. A mutation in the DNA that changes an amino acid in a protein sequence without affecting the activity of the protein is called a:
   A. recessive mutation.
   B. neutral substitution.
   C. leaky mutation.
   D. null mutation.

4. The *w* gene for eye color in *Drosophila* is an example of:
   A. partial dominance.
   B. multiple alleles.
   C. codominance.
   D. a leaky mutation.
5. Which blood group is the null phenotype?
   A. A
   B. B
   C. AB
   D. O
6. Any given segment of a genome could have ______ possible reading frame(s) in a single strand of the DNA.
   A. 1
   B. 2
   C. 3
   D. 4
7. What types of mutations do acridines cause?
   A. frameshift mutations
   B. point mutations
   C. long deletions
   D. long insertions
8. The change in base sequence from AAA AGC TTC GAC to AAA GCT TCG ACC is an example of a:
   A. point mutation.
   B. insertion.
   C. frameshift mutation.
   D. deamination.
9. The convention for writing one strand of the DNA sequence of a protein-coding gene is:
   A. 5′ to 3′, with the sequence being the same as for the mRNA (except for T instead of U).
   B. 3′ to 5′, with the sequence being the same as for the mRNA (except for T instead of U).
   C. 5′ to 3′, with the sequence being opposite that for the mRNA (except for T instead of U).
   D. 3′ to 5′, with the sequence being opposite that for the mRNA (except for T instead of U).
10. What type of mutation would prevent both alleles of a gene from being expressed?
   A. dominant
   B. recessive
   C. *trans*-acting
   D. *cis*-acting

KEY TERMS

acridines
allele
antisense (template) strand
chiasma
***cis*-acting sequence**
cistron
closed (blocked) reading frame
coding region
coding (sense) strand
codon
colinearity
complementation test
frameshift
gain-of-function mutation
gene expression
genetic code
genetic recombination
heteroduplex DNA
heteromultimer
homomultimer

initiation codon
intron
leader (5′ UTR)
linkage
locus
loss-of-function mutation
messenger RNA (mRNA)
neutral substitutions
null mutation
one gene : one enzyme hypothesis
one gene : one polypeptide hypothesis
open reading frame (ORF)
polymorphism
pre-mRNA
reading frame
ribosomal RNAs (rRNAs)
ribosome
RNA processing
silent mutation
splicing
termination codon
trailer (3′ UTR)
trans-acting sequence
transcription
transfer RNA (tRNA)
translation
unidentified reading frame (URF)

FURTHER READING

Carter, C. W., Jr. (2008). Whence the genetic code?: thawing the 'frozen accident.' *Heredity* **100**, 339–340. A brief review of current hypotheses for the origin of the genetic code.

Ripley, L. S. (1990). Frameshift mutation: determinants of specificity. *Annu. Rev. Genet.* **24**, 189–213. A review of the mechanisms of frameshift mutations.

Yanofsky, C. (2001). Advancing our knowledge in biochemistry, genetics, and microbiology through studies on tryptophan metabolism. *Annu. Rev. Biochem.* **70**, 1–37. A personal account of Yanofsky's research, including establishing the colinearity of genes and their protein products and the regulation of *trp* expression via attenuation.

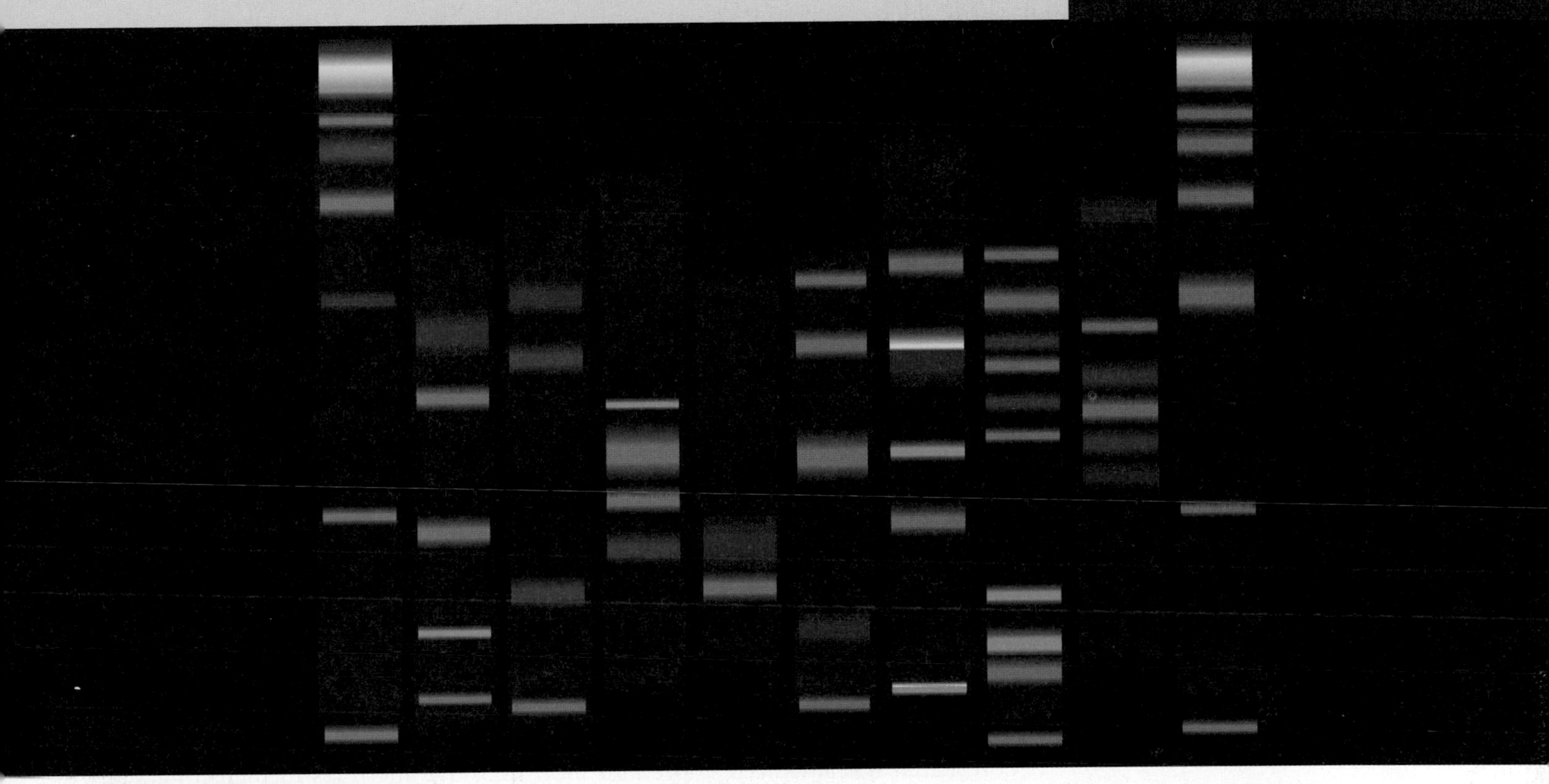

DNA fragments separated by gel electrophoresis. © Nicemonkey/Dreamstime.com

3 Methods in Molecular Biology and Genetic Engineering

CHAPTER OUTLINE

3.1 Introduction

Today, the field of molecular biology focuses on the mechanisms by which cellular processes are carried out by the various biological macromolecules in the cell, with a particular emphasis on the structure and function of genes and genomes. Molecular biology as a field, however, was originally born from the development of tools and methods that allow the direct manipulation of DNA both *in vitro* and *in vivo* in numerous organisms.

Two essential items in the molecular biologist's toolkit are **restriction endonucleases**, which allow DNA to be cut into precise pieces, and **cloning vectors**, such as plasmids or phages, used to "carry" inserted foreign DNA fragments for the purposes of producing more material or a protein product. The term *genetic engineering* was originally used to describe the range of manipulations of DNA that become possible with the ability to clone a gene by placing its DNA into another context in which it could be propagated. From this beginning, when recombinant DNA was used as a tool to analyze gene structure and expression, we moved to the ability to change the DNA content of bacteria and eukaryotic cells by directly introducing cloned DNA that could become part of the genome. Then, by changing the genetic content in conjunction with the ability to develop an animal from an embryonic cell, it became possible to generate multicellular eukaryotes with deletions or additions of specific genes that are inherited via the germline. We now use genetic engineering to describe a range of activities, including the manipulation of DNA, the introduction of changes into specific somatic cells within an animal or plant, and even changes in the germline itself.

- **restriction endonuclease** An enzyme that recognizes specific short sequences of DNA and cleaves the duplex (sometimes at the target site and sometimes elsewhere, depending on type).
- **cloning vectors** DNA (often derived from a plasmid or a bacteriophage genome) that can be used to propagate an incorporated DNA sequence in a host cell; vectors contain selectable markers and replication origins to allow identification and maintenance in the host.

As research has advanced, more and more sensitive methods for detecting and amplifying DNA have been developed. Now that we have entered the era of routine whole-genome sequencing, methods to assess the content, function, and expression of entire genomes have become commonplace. This chapter will discuss some of the most common methods used in molecular biology, ranging from the very first tools developed by molecular biologists to some of the most recently developed methods now in use.

The advances in molecular biology over the past 40 years have been largely catalyzed by technological breakthroughs in methods and procedures such as cloning, DNA sequencing, PCR, and others that have allowed us to ask questions previously unthinkable. One can only imagine the new technologies that lie over the horizon that will allow the students of today to ask questions not even dreamed of yet.

3.2 Nucleases

Nucleases are one of the most valuable tools in a molecular biology laboratory. One class of enzymes, the restriction endonucleases that we discuss below, was critical for the cloning revolution. **Nucleases** are enzymes that degrade nucleic acids, the opposite function of polymerases. They hydrolyze, or break, an ester bond in a phosphodiester linkage between adjacent nucleotides in a polynucleotide chain, as shown in **FIGURE 3.1**.

There is another, related class of enzymes that can hydrolyze an ester bond in a nucleotide chain (a monoesterase, usually called a **phosphatase**). The critical difference between a phosphatase and a nuclease is shown in Figure 3.1. A phosphatase can hydrolyze only a terminal ester bond linking a phosphate (or di- or triphosphate) to a terminal nucleotide at the 3′ or 5′ end, whereas a nuclease can hydrolyze an internal ester bond in a diester link between adjacent bases.

Phosphatases are important enzymes in the laboratory because they allow the removal of a terminal phosphate from a polynucleotide chain. This is often required for a subsequent step of connecting, or **ligating**, chains together. This also allows one to replace the phosphate with a radioactive ^{32}P molecule.

- **nuclease** An enzyme that can break a phosphodiester bond.
- **phosphatase** An enzyme that breaks a phosphomonoester bond, cleaving a terminal phosphate.
- **ligate** To covalently link two ends of nucleic acid chains; they may be two ends of one chain or two ends of different chains, either DNA or RNA.

We can divide nucleases into different groups based on a number of different features. First, we can distinguish between **endonucleases** and **exonucleases,** as shown in Figure 3.1. An endonuclease can hydrolyze internal bonds within a polynucleotide chain, whereas an exonuclease must start at the end of a chain and hydrolyze from that end position.

The specificity of nucleases ranges from none to extreme. Nucleases may be specific for DNA, as DNases, or RNA, as RNases, or even be specific for a DNA/RNA hybrid, as RNaseH (which cleaves the RNA strand of a hybrid RNA-DNA duplex). Nucleases may be specific for either single-strand nucleotide chains, duplex chains, or both.

When a nuclease, either endo- or exo-, hydrolyzes an ester bond in a phosphodiester linkage, it will have specificity for either of the two ester bonds, generating either 5′ nucleotides or 3′ nucleotides, as seen in Figure 3.1. An exonuclease may attack a polynucleotide chain from either the 5′ end and hydrolyze 5′ to 3′ or attack from the 3′ end and hydrolyze 3′ to 5′.

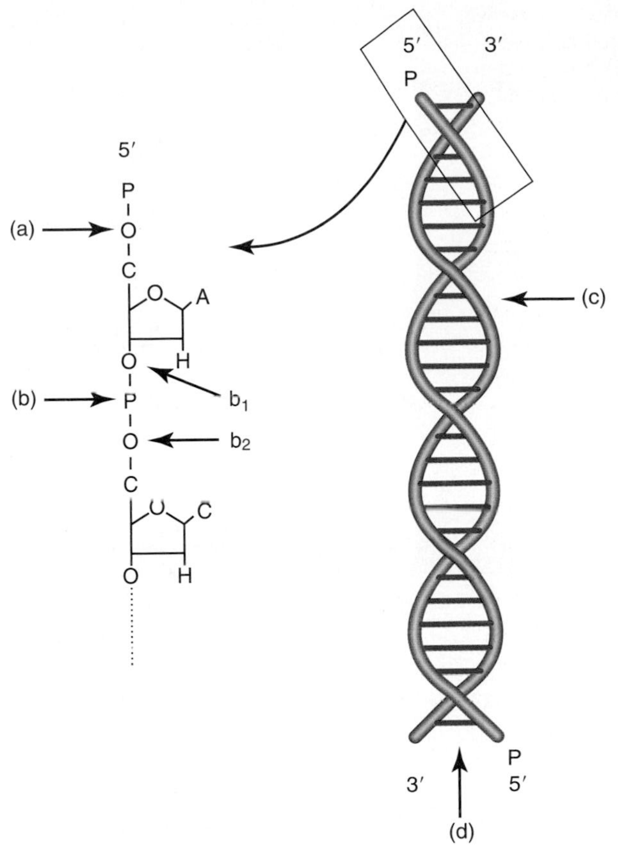

FIGURE 3.1 The target of a phosphatase is shown in (a), a terminal phosphomonoester bond. The target of a nuclease is shown in (b), the phosphodiester bond between two adjacent nucleotides. Note that the nuclease can cleave either the first ester bond from the 3′ end of the terminal nucleotide (b1) or the second ester bond from the 5′ end of the next nucleotide (b2). Nucleases can cleave internal bonds (c) as an endonuclease or start from an end and progress into the fragment (d) as an exonuclease.

- **endonuclease** An enzyme that cleaves bonds within a nucleic acid chain; it may be specific for RNA or for single-stranded or double-stranded DNA.
- **exonuclease** An enzyme that cleaves nucleotides one at a time from the end of a polynucleotide chain; it may be specific for either the 5′ or 3′ end of DNA or RNA.

Nucleases may have a sequence preference, such as pancreatic RNase A, which preferentially cuts after a pyrimidine, or T1 RNase, which cuts single-stranded RNA chains after a G. At the extreme end of sequence specificity lie the *restriction endonucleases,* usually called *restriction enzymes.* These are endonucleases from eubacteria and archaea that recognize a specific DNA sequence. Their name typically derives from the bacteria in which they were discovered. For example, EcoR1 is the first restriction enzyme from an *E. coli R* strain.

Broadly speaking, there are three different classes of restriction enzymes and several subclasses. In 1978, the Nobel Prize in Medicine was awarded to Daniel Nathans, Werner Arber, and Hamilton Smith for the discovery of restriction endonucleases and their application to problems in molecular genetics. It was this discovery that enabled scientists to develop the methods to clone DNA, as we will see in the next section. Thousands of restriction enzymes are known, many of which are now commercially available. Restriction enzymes have to do two things: (1) recognize a specific sequence and (2) cut, or restrict, that sequence.

The type II restriction enzymes (with several subgroups) are the most common. Type II enzymes are distinguished because the recognition site and cleavage site are the same. These sites range in length from 4 to 8 bp. The sites are typically *inversely palindromic,* that is, reading the same forward and backward on complementary strands, as shown in **FIGURE 3.2**. Restriction enzymes can cut the DNA in two different ways. The first and more common is a staggered cut, which leaves single-stranded overhangs, or "sticky ends." The overhang may be a 3′ or a 5′ overhang. The second way is a blunt double-stranded cut, which does not leave an overhang. An additional level of specificity determines whether or not the enzyme will cut DNA containing a methylated base. The degree of specificity in the site also varies. Most enzymes are

FIGURE 3.2 (a) A restriction endonuclease may cleave its recognition site and make a staggered cut, leaving a 5′ overhang or a 3′ overhang. (b) A restriction endonuclease may cleave its recognition site and make a blunt end cut.

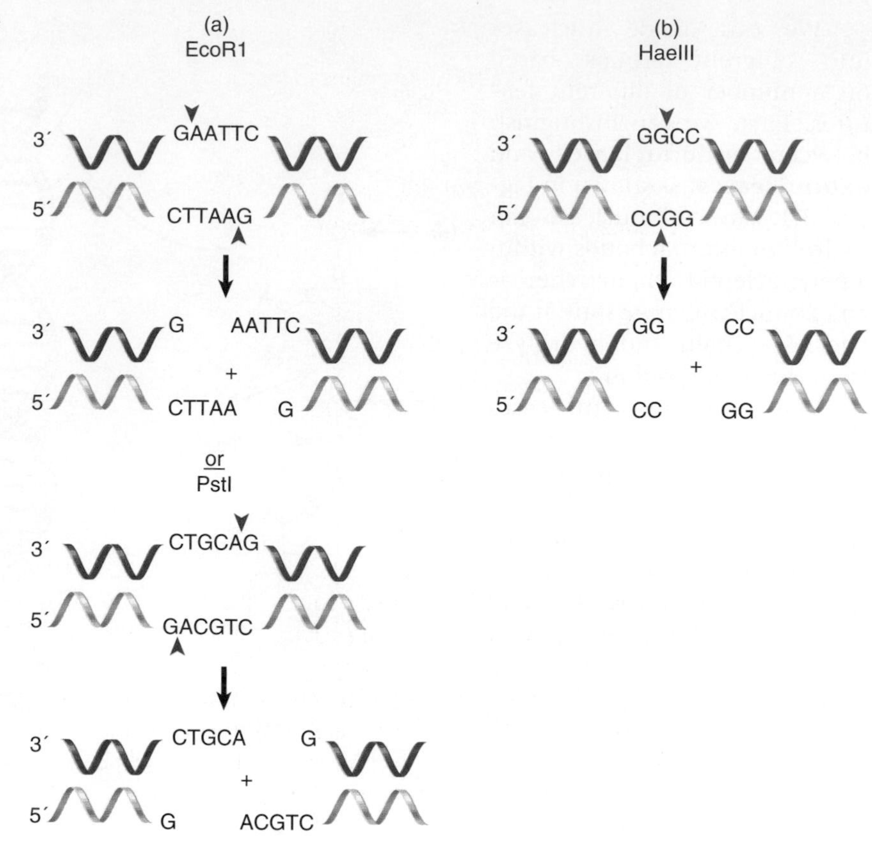

very specific, whereas some will allow multiple bases at one or two positions within the site.

Restriction enzymes from different bacteria may have the same recognition site but cut the DNA differently. One may make a blunt cut and the other may make a staggered cut, or one may leave a 3′ overhang but the second may leave a 5′ overhang. These different enzymes are called *isoschizomers*.

Types I and III enzymes differ from type II enzymes in that the recognition site and cleavage site are different and are usually not palindromes. With a type I enzyme, the cleavage site can be up to 1000 bp away from the recognition site. Type III enzymes have closer cleavage sites, usually 20 to 30 bp away.

A *restriction map* represents a linear sequence of the sites at which particular restriction enzymes find their targets. When a DNA molecule is cut with a suitable restriction enzyme, it is cleaved into distinct negatively charged fragments. These fragments can be separated on the basis of their size by gel electrophoresis (described later, in *Section 3.6, DNA Separation Techniques*; see Figure 3.14). By analyzing the restriction fragments of DNA, we can generate a map of the original molecule in the form shown in **FIGURE 3.3**. The map shows the positions at which particular restriction enzymes cut DNA. *So the DNA is divided into a series of regions of defined lengths that lie between sites recognized by the restriction enzymes*. A restriction map can be obtained for any sequence of DNA, irrespective of whether we have any knowledge of its function. If the sequence of the DNA is known, a restriction map can be generated *in silico* by simply searching for the recognition sites of known enzymes. Knowing the restriction map of a DNA sequence of interest is extremely valuable in DNA cloning, which is described in the next section.

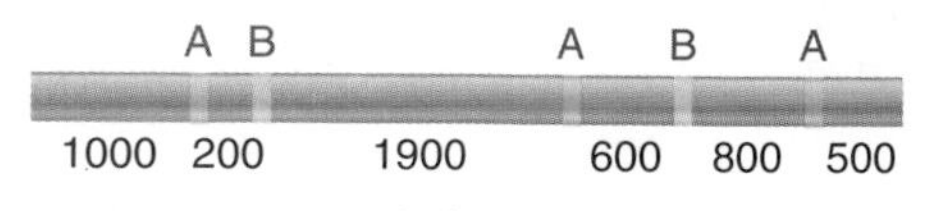

FIGURE 3.3 A restriction map is a linear sequence of sites separated by defined distances on DNA. The map identifies the three sites cleaved by enzyme A and the two sites cleaved by enzyme B. Thus A produces four fragments, which overlap those of B, and B produces three fragments, which overlap those of A; digestion with both A and B will result in six fragments.

KEY CONCEPTS

- Nucleases hydrolyze an ester bond within a phosphodiester bond.
- Phosphatases hydrolyze the ester bond in a phosphomonoester bond.
- Nucleases have a multiplicity of specificities.
- Restriction endonucleases can be used to cleave DNA into defined fragments.
- A map can be generated by different restriction enzymes.

CONCEPT AND REASONING CHECK

What would be the advantage of a naturally produced restriction endonuclease in a bacterial cell?

3.3 Cloning

Cloning has a very simple definition: to *clone* is to make identical copies, whether it is done by a copy machine for a piece of paper, cloning Dolly the sheep, or cloning DNA, which is what we will discuss here. Cloning can also be considered an amplification process, in which we currently have one copy and we want many identical copies. Cloning DNA typically involves **recombinant DNA**. This also has a very simple definition: a DNA molecule from two (or more) different sources.

In order to clone a fragment of DNA, a recombinant DNA molecule must be created and copied many times. There are two different DNAs needed: a *cloning* ***vector*** and an **insert**, or the molecule to be cloned. The two most popular classes of vectors are derived from plasmids and viruses, respectively.

Over the years, vectors have been specifically engineered for safety, selection ability, and high growth rate. "Safety" means that the vector will not integrate into a genome (unless engineered specifically for that purpose) and the recombinant vector will not autotransfer to another cell. (We discuss selection shortly.) In general, about a microgram of vector DNA will be ligated with about a microgram of the insert DNA that we wish to clone. Both the vector and insert should be restricted with the same restriction endonuclease to create compatible DNA ends. Let us now examine the details and the variables that will affect the process.

We will start with the insert, the DNA fragment to be amplified. The insert could come from one of many different sources, such as restricted genomic DNA, either size selected on an agarose gel or unselected; a larger fragment from another clone to be **subcloned** (meaning taking a smaller part of the larger fragment); a PCR fragment (see *Section 3.8, PCR and RT-PCR*); or even a DNA fragment synthesized *in vitro*. The size and the nature of the fragment ends must be known. Are the ends blunt or do they have overhanging single strands (recall *Section 3.2, Nucleases*); if so, what are their sequences? The answer to this question comes from how the fragments were created (what restriction enzyme(s) were used to cut the DNA or what PCR primers were used to amplify the DNA).

The vector is selected based on the answers to these questions. For this exercise, we will use a common type of plasmid cloning vector called a *blue/white selection vector*, as shown in **FIGURE 3.4**. This vector has been constructed with a number of important elements. It has an *ori*, or origin of replication (see *Chapter 14, Extrachromosomal Replication*), to allow plasmid replication, which will provide the actual amplification step in a bacterial cell. It contains a gene that codes for resistance to the antibiotic ampicillin, *amp^r^*, which will allow selection of bacteria that contain the vector. It also contains the *E. coli lacZ* gene (see *Chapter 26, The Operon*), which will allow selection of an insert DNA fragment in the vector.

The *lacZ* gene has been engineered to contain a **multiple cloning site**, or MCS. This is an oligonucleotide sequence with a series of different restriction endonuclease

- **cloning** Propagation of a DNA sequence by incorporating it into a hybrid construct that can be replicated in a host cell.
- **recombinant DNA** An artificial DNA molecule created by joining two (or more) DNA molecules from different sources.
- **vector** A plasmid or phage chromosome that is used to perpetuate a cloned DNA segment.
- **insert** A fragment of DNA that is to be cloned in a vector.
- **subclone** The process of breaking a cloned fragment into smaller fragments for further cloning.
- **multiple cloning site** An artificial DNA sequence in a cloning vector containing multiple restriction endonuclease sites for cloning.

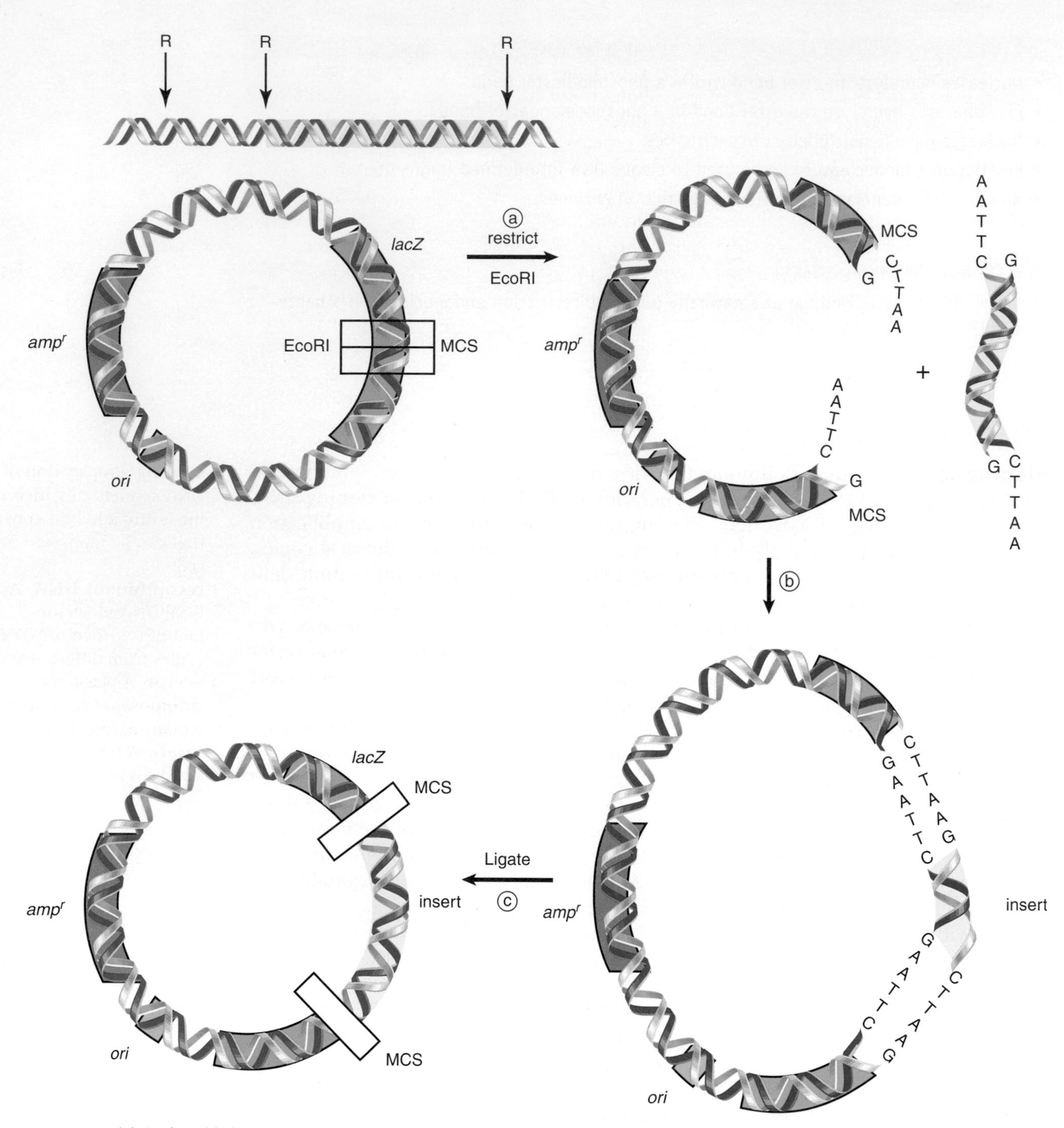

FIGURE 3.4 (a) A plasmid that contains three key sites (an origin of replication, *ori*; a gene for ampicillin resistance, *amp^r*; and *lacZ* with an MCS), together with the insert DNA to be cloned, is restricted with EcoR1. (b) Restricted insert fragments and vector will be combined and (c) ligated together. The final pool of this DNA will be transformed into *E. coli*.

recognition sites arranged in tandem in the same reading frame as the *lacZ* gene itself. This is the heart of blue/white selection. The *lacZ* gene codes for the β-galactosidase (β-gal) enzyme, which cleaves the galactoside bond in lactose. It will also cleave the galactoside bond in an artificial substrate called X-gal (5-bromo-4-chloro-3-indolyl-beta-D-galactopyranoside), which can be added to bacterial growth media and has a blue color when cleaved by the intact enzyme. *If a fragment of DNA is cloned (inserted) into the MCS, the lacZ gene will be disrupted, inactivating it, and the resulting β-gal will no longer be able to cleave X-gal, resulting in white bacterial colonies rather than blue colonies.* This is the blue/white selection mechanism.

Let us now begin the cloning experiment. Following along in Figure 3.4, both the vector and the insert are cut with the same restriction enzyme in order to generate compatible single-stranded sticky ends. The variables here are the abilities to select different enzymes that recognize different restriction sites as long as they generate the same overhang sequence. An enzyme that makes a blunt cut can also be used, although that will make the next step, ligation, less efficient. Two completely different ends with different overhangs can also be used if an exonuclease is used to trim the ends and produce blunt ends. (Continuing with the same reasoning, randomly sheared DNA can also be used if the ends are then blunted for ligation.) If we are forced to use a type I or type III restriction enzyme, the ends must also be blunted. An important alternative is to use two different restriction enzymes that leave different overhangs on each end. The advantages to this are that neither the vector nor the insert will self-circularize, and the orientation of how the insert goes into the vector can be controlled; this is called *directional cloning*. We will select the vector that has the appropriate restriction endonuclease sites.

The next step is to combine the two pools of DNA fragments, vector and insert, in order to connect or ligate them. A 5- or 10-to-1 molar ratio of insert to vector is usually used. Too much vector and vector-vector dimers will be produced. Too much insert and multiple inserts per vector will be produced. The size of the insert is important; if it is too large (over ~10 kb) an insert will not be efficiently cloned in a plasmid vector, which will necessitate using an alternative virus-based vector. Ligation is often performed overnight at low temperature to slow the ligation reaction down and generate fewer multimers.

The pool of randomly generated ligated DNA molecules is now used to "transform" *E. coli*. **Transformation** is the process by which DNA is introduced into a host cell. *E. coli* does not normally undergo physiological transformation. As a result, DNA must be forced into the cell. There are two common methods of transformation: washing the bacteria in a high salt wash of $CaCl_2$, or *electroporation*, in which an electric current is applied. Both methods create small pores or holes in the cell wall. Even with these methods, only a tiny fraction of bacterial cells will be transformed. The strain of *E. coli* is important. It should not have a restriction system or a modification system to methylate the incoming DNA. The strain should also be compatible with the blue/white system, which means that it should contain the α-complementing fragment of LacZ (the *lacZ* gene contained in most plasmids does not function without this fragment). DH5α is a commonly used strain.

▸ **transformation** The acquisition of new exogenous genetic material by a cell.

Transformation results in a pool of multiple types of bacteria, most of which are not wanted because they either contain vector with no insert or have not taken up any DNA at all. We must select the handful of bacteria that contain recombinant plasmid from the millions that do not. The transformed bacterial cells are plated on an agar plate containing both the antibiotic ampicillin and an artificial β-gal inducer called IPTG (Isopropyl Thiogalactoside). The ampicillin in the plate will kill the vast majority of bacterial cells, namely, all those that have not been transformed with the *amp*^{r} plasmid. The remaining bacteria can now grow and form visible colonies. As shown in **FIGURE 3.5**, two different types of colonies—blue ones that contain a vector without an insert (because β-gal cleaved X-gal into a blue compound) and white ones, that we want, for which the inactivated β-gal did not cleave X-gal and so remained colorless, are seen.

This is not quite the end of the story. False positive clones, such as those that were formed as vector-only dimers, must be identified and removed. In order to do so, plasmid DNA must be at least partly purified from each candidate colony and restricted and run on a gel to check for the insert size. Sequencing the fragment to be absolutely certain a random contaminant has not been cloned is also suggested. Sequencing is described in *Section 3.7, DNA Sequencing*.

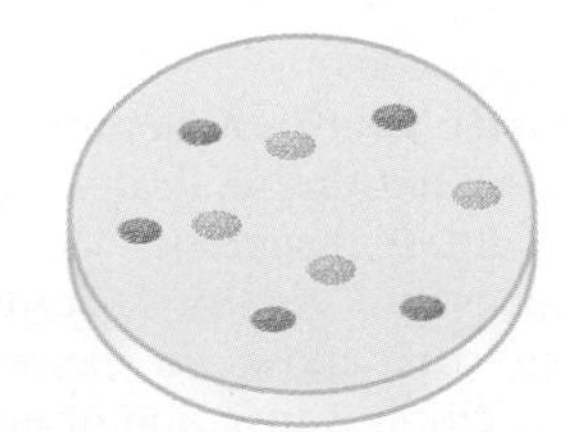

FIGURE 3.5 After transformation into *E. coli* of restricted and ligated vector plus insert DNA, the bacterial cells are plated onto agar plates containing ampicillin, IPTG, and the color indicator, X-gal. Overnight incubation at 37°C will yield both blue and white colonies. The white colonies will be used to prepare DNA for further analysis.

KEY CONCEPTS

- Cloning a fragment of DNA requires a specially engineered vector.
- Blue/white selection allows the identification of bacteria that contain the vector plasmid and vector plasmids that contain an insert.

CONCEPT AND REASONING CHECK

If you digest a vector and donor DNA with restriction enzymes that make blunt ends, do you have to use the same enzyme for both vector and donor DNA? Why or why not?

3.4 Cloning Vectors Can Be Specialized for Different Purposes

In the example in the previous section, we described the use of a vector that is designed simply for amplifying insert DNA, with inserts up to ~10 kb. It is often desirable to clone larger inserts, though, and sometimes the goal is not just to amplify the DNA, but also to express cloned genes in cells, investigate properties of a promoter, or create various fusion proteins (defined shortly). **FIGURE 3.6** summarizes the properties of the most common classes of cloning vectors. These include vectors based on bacteriophage genomes, which can be used in bacteria but have the disadvantage that only a limited amount of DNA can be packaged into the viral coat (although more than can be carried in a plasmid). The advantages of plasmids and phages are combined in the **cosmid**, which propagates like a plasmid but uses the packaging mechanism of phage lambda to deliver the DNA to the bacterial cells. Cosmids can carry inserts of up to 47 kb (the maximum length of DNA that can be packaged into the phage head).

FIGURE 3.6 Cloning vectors may be based on plasmids or phages or may mimic eukaryotic chromosomes.

Vector	Features	Isolation of DNA	DNA limit
Plasmid	High copy number	Physical	10 kb
Phage	Infects bacteria	Via phage packaging	20 kb
Cosmid	High copy number	Via phage packaging	48 kb
BAC	Based on F plasmid	Physical	300 kb
YAC	Origin + centromere + telomere	Physical	3000 kb

The vector used for cloning the largest possible DNA inserts is the **yeast artificial chromosome (YAC)**. A YAC has a yeast origin to support replication, a centromere to ensure proper segregation, and telomeres to afford stability. In effect, it is propagated just like a yeast chromosome. YACs have the largest capacity of any cloning vector, and can propagate with inserts measured in the Mb-length range.

An extremely useful class of vectors known as **shuttle vectors** can be used in more than one species of host cell. The example shown in **FIGURE 3.7** contains origins of replication and selectable markers for both *E. coli* and the yeast *S. cerevisiae*. It can replicate as a circular multicopy plasmid in *E. coli*. It has a yeast centromere and also has yeast telomeres adjacent to BamH1 restriction sites, so that cleavage with BamH1 generates a YAC that can be propagated in yeast.

Other vectors, such as **expression vectors**, may contain promoters to drive expression of genes. Any open reading frame can be inserted into the vector and expressed without further modification. These promoters can be continuously active or may be *inducible*, so that they are only expressed under specific conditions.

Alternatively, the goal may be to study the function of a cloned promoter of interest in order to understand the normal regulation of a gene. In this case, rather than using the actual gene, we can use an easily detected **reporter gene** under control of the promoter of interest.

The type of reporter gene that is most appropriate depends on whether we are interested in quantitating the efficiency of the promoter (and, for example, determining the effects of mutations in it or the activities of transcription factors that bind to it) or determining its tissue-specific pattern of expression. **FIGURE 3.8** summarizes a common system for assaying promoter activity. A cloning vector is created that has

- **cosmid** Cloning vector derived from a bacterial plasmid by incorporating the *cos* sites of phage lambda, which make the plasmid DNA a substrate for the lambda packaging system.
- **yeast artificial chromosome (YAC)** A cloning vector used to clone very large DNA fragments, up to 3000 kb in size, containing yeast telomeres, a centromere, and a replication origin so that it can propagate in yeast cells.
- **shuttle vector** A cloning vector that can replicate in two different species.
- **expression vector** A cloning vector that allows the expression, either translation or just transcription, of the insert.
- **reporter gene** A sequence that is attached to another gene, which codes for a peptide that is easily identified or measured.

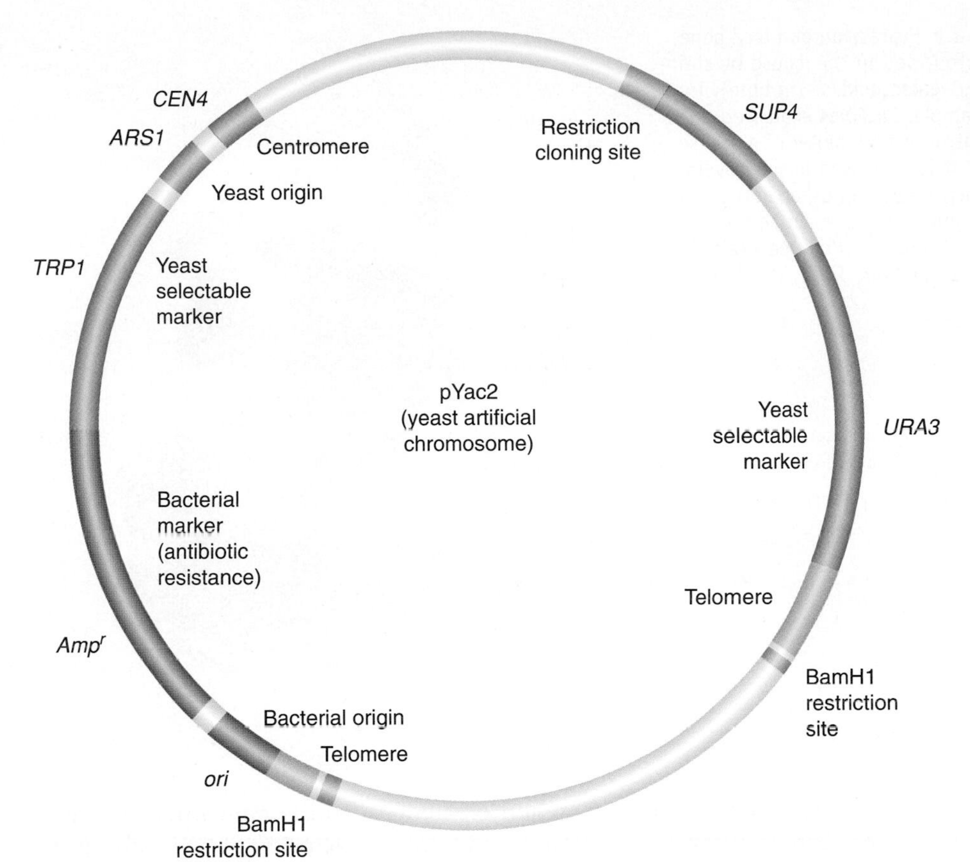

FIGURE 3.7 pYac2 is a cloning vector with features that allow replication and selection in both bacteria and yeast. Bacterial features (described in blue) include an origin of replication and antibiotic resistance gene. Yeast features (described in orange and yellow) include an origin, a centromere, two selectable markers, and telomeres.

a eukaryotic promoter linked to the coding region of *luciferase,* a gene that encodes the enzyme responsible for bioluminescence in the firefly. In general, a transcription termination signal is added to ensure the proper generation of the mRNA. The hybrid vector is introduced into target cells, and the cells are grown and subjected to any appropriate experimental treatments. The level of luciferase activity is measured by addition of its substrate luciferin. Luciferase activity results in light emission that can be measured at 562 nanometers (nm) and is directly proportional to the amount of enzyme that was made, which in turn depends upon the activity of the promoter.

Some very striking reporters are now available for visualizing gene expression. The *lacZ* gene, described in the blue-white selection strategy before, also serves as a very useful reporter gene. **FIGURE 3.9** shows what happens when the *lacZ* gene is placed under the control of a tissue-specific promoter. The tissues in which this promoter is normally active can be visualized by providing the X-gal substrate to stain the embryo.

One of the most popular reporters that can be used to visualize patterns of gene expression is GFP (green fluorescent protein), which is obtained from jellyfish. GFP

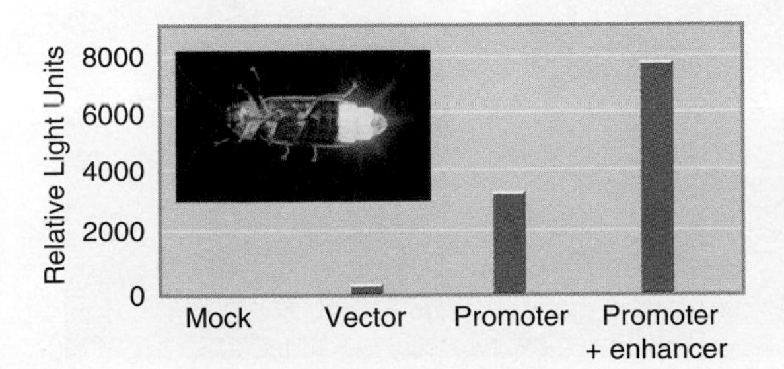

FIGURE 3.8 Luciferase (derived from fireflies such as the one shown here) is a popular reporter. The graph shows the results from mammalian cells transfected with a luciferase vector driven by a minimal promoter or the promoter plus a putative enhancer. The levels of luciferase activity correlate with the activities of the promoters. Photo © Cathy Keifer/Dreamstime.com.

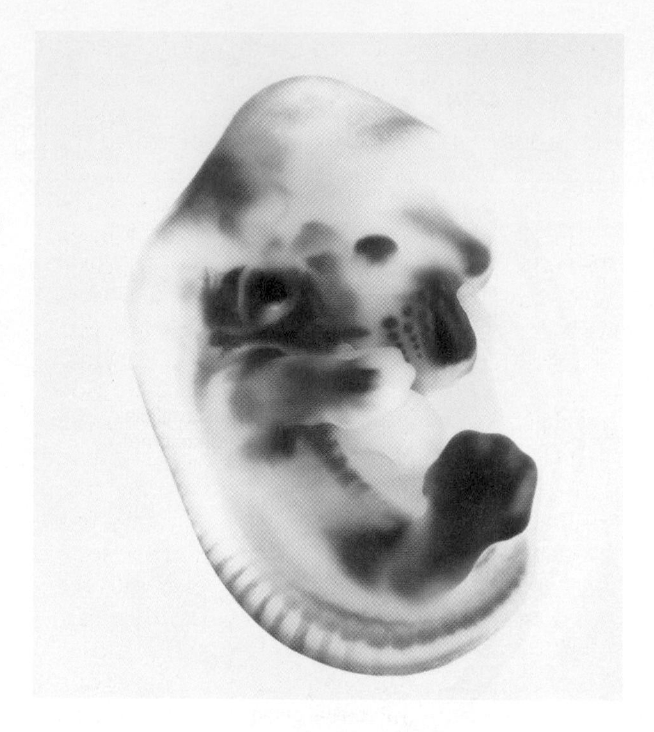

FIGURE 3.9 Expression of a *lacZ* gene can be followed in the mouse by staining for β-galactosidase (in blue). In this example, *lacZ* was expressed under the control of a promoter of a mouse gene that is expressed in the developing heart, limbs and other tissues. The corresponding tissues can be visualized by blue staining. © Philippe Psaila/ Photo Researchers, Inc.

is a naturally fluorescent protein that, when excited with one wavelength of light, emits fluorescence in another wavelength. In addition to the original GFP, numerous variants that fluoresce in different colors, such as yellow (YFP), cyan (CFP), and blue (BFP), have been developed. GFP and its variants can be used as reporter genes on their own, or they can be used to generate *fusion proteins*, in which a protein of interest is fused to GFP and can thus be visualized in living tissues, as is shown in the example in **FIGURE 3.10**.

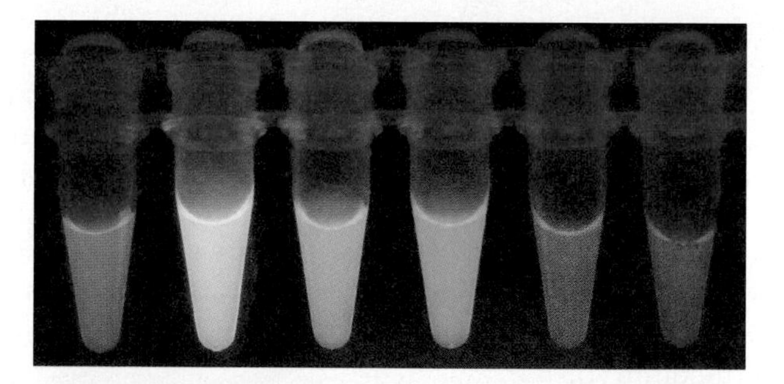

FIGURE 3.10 (a) Since the discovery of GFP, derivatives that fluoresce in different colors have been engineered. Photo courtesy of Joachim Goedhart, Molecular Cytology, SILS, University of Amsterdam. (b) A live transgenic mouse expressing human rhodopsin (a protein expressed in the retina of the eye) fused to GFP. Reprinted from *Vision Res.*, vol. 45, T. G. Wensel et al., Rhodopsin-EGFP knock-ins . . . , pp. 3445–3453. Copyright 2005, with permission from Elsevier (http://www.sciencedirect.com/science/journal/00426989). Photo courtesy of Theodore G. Wensel, Baylor College of Medicine.

Vectors are introduced into different species in a variety of different ways. Bacteria and simple eukaryotes like yeast can be transformed easily, using chemical treatments that permeabilize the cell membranes, as discussed in *Section 3.3, Cloning*. Many types of cells cannot be transformed so easily, though, and other methods must be used, as summarized in **FIGURE 3.11**. Some types of cloning vectors use natural methods of infection to pass the DNA into the cell, such as a viral vector that uses the viral infective process to enter the cell. *Liposomes* are small spheres made from artificial membranes, which can contain DNA or other biological materials. Liposomes can fuse with plasma membranes and release their contents into the cell. *Microinjection* uses a very fine needle to puncture the cell membrane. A solution containing DNA can be introduced into the cytoplasm, or directly into the nucleus in the case where the nucleus is large enough to be chosen as a target (such as an egg). The thick cell walls of plants are an impediment to many transfer methods, and the "gene gun" was invented as a means for overcoming this obstacle. A gene gun shoots very small particles into the cell by propelling them through the wall at high velocity. The particles can consist of gold or nanospheres coated with DNA. This method now has been adapted for use with a variety of species, including mammalian cells.

A viral vector introduces DNA by infection

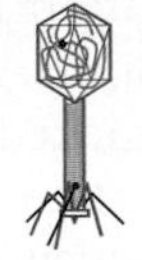

Liposomes may fuse with the membrane

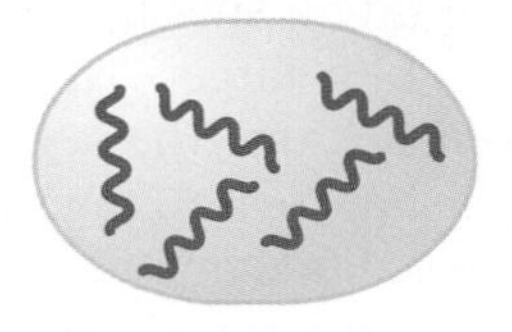

Microinjection introduces DNA directly into the cytoplasm or nucleus

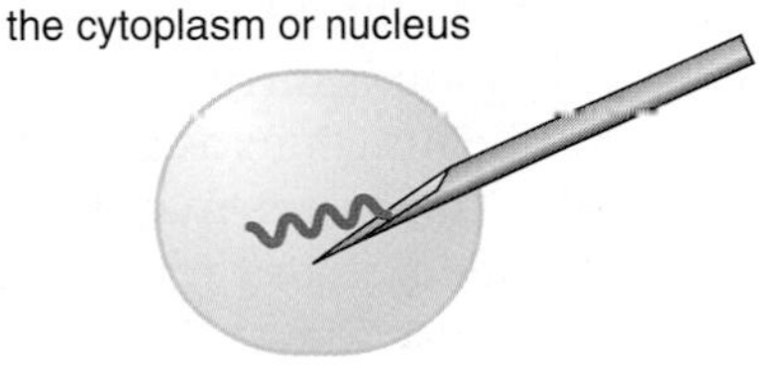

Nanospheres can be shot into the cell by a gene gun

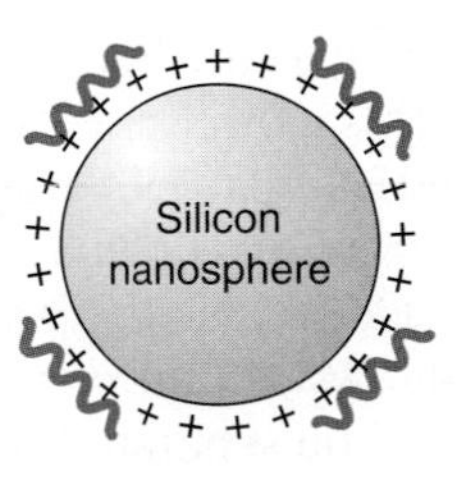

FIGURE 3.11 DNA can be released into target cells by methods that pass it across the membrane naturally, such as by means of a viral vector (in the same way as a viral infection) or by encapsulating it in a liposome (which fuses with the membrane). Alternatively, it can be passed manually, by microinjection, or by coating it on the exterior of nanoparticles that are shot into the cell by a gene gun that punctures the membrane at very high velocity.

KEY CONCEPTS

- Cloning vectors may be bacterial plasmids, phages, cosmids, or yeast artificial chromosomes.
- Shuttle vectors can be propagated in more than one type of host cell.
- Expression vectors contain promoters that allow transcription of any cloned gene.
- Reporter genes can be used to measure promoter activity or tissue-specific expression.
- Numerous methods exist to introduce DNA into different target cells.

CONCEPT AND REASONING CHECK

Why must a GFP fusion reporter be ligated in the same reading frame as the protein of interest?

3.5 Nucleic Acid Detection

There are a number of different ways to detect DNA and RNA. The classical method relies on ability of nucleic acids to absorb light with a wavelength of 260 nm. The amount of light absorbed is proportional to the amount of nucleic acid present. There is a slight difference in the amount of absorption by single-stranded as compared to double-stranded nucleic acids, but not DNA versus RNA. Protein contamination can affect the outcome, but because proteins absorb maximally at 280 nm, tables have been published of 260/280 ratios that allow quantitation of the amount of nucleic acid present.

DNA and RNA can be nonspecifically stained with ethidium bromide (EtBr) to make visualization more sensitive. EtBr is an organic tricyclic compound that binds strongly to double-stranded DNA (and RNA) by intercalating into the double helix between the stacked base pairs. It binds to DNA; as a result it is a strong mutagen, and care must be taken when using it. EtBr fluoresces when exposed to UV light, which increases the sensitivity. SYBR green is a safer alternate DNA stain.

We will focus here on the detection of *specific* sequences of nucleic acids. The ability to identify a specific sequence relies on hybridization of a **probe** with a known sequence to a target. The probe will detect and bind to a sequence to which it is **complementary**. The percent of match does not have to be perfect, but as the match percentage decreases, the stability of the nucleic acid hybrid decreases. G-C base pairs are more stable than A-T base pairs so that base composition (usually referred to as % G-C) is an important variable. The second set of variables that affects hybrid stability is extrinsic; it includes the buffer conditions (concentration and composition) and the temperature at which hybridization occurs. This is called the **stringency** under which the hybridization is carried out.

- **probe** A labeled nucleic acid used to identify a complementary sequence.
- **complementary** Base pairs that match up in the pairing reactions in double stranded nucleic acids (A with T in DNA or with U in RNA, and C with G).
- **stringency** A measure of the exactness of complementarity required between two nucleic acid strands to allow them to hybridize. Stringency is related to buffer ionic strength and reaction temperature.

The probe functions as a single-stranded molecule (if it is double-stranded, it must be melted). The target may be single-stranded or double-stranded. If the target is double-stranded, it also must be melted to single strands to begin the hybridization process. The reaction can take place in solution (for example, during sequencing or PCR; see *Sections 3.7, DNA Sequencing* and *3.8, PCR and RT-PCR*) or can be performed when the target has been bound to a membrane support such as a nitrocellulose filter (see *Section 3.9, Blotting Methods*). The target may be DNA (called a Southern blot) or RNA (called a Northern blot); the probe is usually DNA.

For this exercise, let's use a Southern blot from an experiment in which we have restricted a large DNA fragment into smaller fragments and subcloned the individual fragments (see *Section 3.2, Cloning*). Starting with the clones on the plate from Figure 3.5, we will isolate plasmid DNA from each white clone and restrict the DNA with the same restriction enzymes that we used to clone the fragments. The DNA fragments will be separated on an agarose gel and blotted onto nitrocellulose (see *Section 3.6, DNA Separation Techniques*).

In order to increase the sensitivity from the optical range, the probe must be labeled. We will begin with radiolabeling and then describe alternative labeling without radioactivity. For most reactions, ^{32}P is used, but ^{33}P (with a longer half-life but less penetrating ability) and ^{3}H (for special purposes described later) are also used. Probes can be radiolabeled in several different ways. One is *end labeling*, in which a strand of DNA (which has no 5′ phosphate) is labeled using a kinase and ^{32}P. Alternatively, a probe can be generated by *nick-translation* or *random priming* with ^{32}P-labeled nucleotides using the Klenow DNA polymerase fragment (see *Section 13.3, DNA Polymerases Have Various Nuclease Activities*) or during a PCR reaction (see *Section 3.8, PCR and RT-PCR*).

In performing nucleic acid hybridization studies, standard procedures are typically used that allow hybridization over a large range of G-C content. Hybridization experiments are performed in a standardized buffer called SSC (standard sodium citrate), which is usually prepared as a 20× concentrated stock solution. Hybridization is typically carried out within a standard temperature range of 45°C to 65°C, depending upon the required stringency.

The actual hybridization between a labeled probe and a target DNA bound to a membrane usually takes place in a closed (or sealed) container in a buffer that contains a set of molecules to reduce background hybridization of probe to the filter. Hybridization experiments typically are performed overnight to ensure maximum probe-to-target hybridization. The hybridization reaction is stochastic and depends upon the abundance of each different sequence. The more copies of a sequence, the greater the chance of a given probe molecule encountering its complementary sequence.

The next step is to wash the filter to remove all the probe that is not specifically bound to a complementary sequence of nucleic acid. Depending on the type of experiment, the stringency of the wash is usually set quite high to avoid spurious results. Higher-stringency conditions include higher temperature (closer to the melting temperature of the probe) and lower concentration of cations. (Lower salt concentrations result in less shielding of the negative phosphate groups of the DNA backbone, which in turn inhibits strand annealing.) In some experiments, however, where one is looking specifically for hybridization to targets with a lower percent match (such as finding a copy of species X DNA using a probe from species Y), hybridization would be performed at lower stringency.

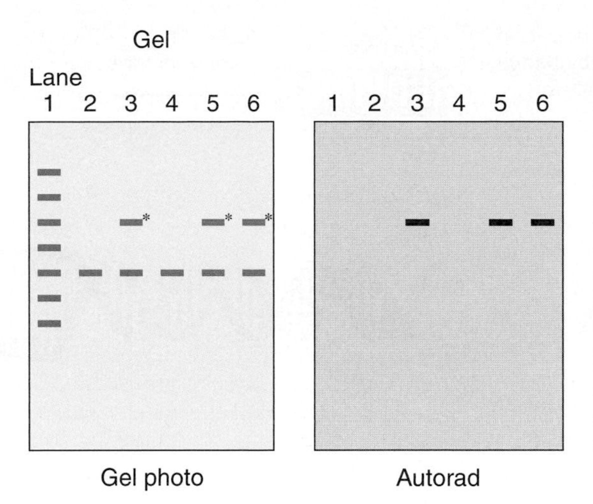

FIGURE 3.12 A cartoon of an autoradiogram of a gel prepared from the colonies described in Figure 3.5. The gel was blotted onto nitrocellulose and probed with a radioactive gene fragment. Lane 1 contains a set of standard DNA size markers. Lane 2 is the original vector cleaved with EcoR1. Lanes 3 to 6 each contain plasmid DNA from one of the white clones from Figure 3.4 that was restricted with EcoR1. A cartoon of the photograph of the gel is on the left; the radioactive bands are marked with an asterisk.

The last step is the identification of which target DNA band on the gel (and thus the filter) has been bound by the radiolabeled probe. The washed nitrocellulose filter is subjected to **autoradiography**. The dried filter will be placed against a sheet of X-ray film. To amplify the radioactive signal, intensifying screens can be used. These are special screens placed on either side of the filter/film pair that act to bounce the radiation back through the film. Alternatively, a *phosphorimaging* screen (a solid-state liquid scintillation device) can be used. This is more sensitive and faster than X-ray film but results in somewhat lower resolution. The length of time for autoradiography is empirical. An estimate of the total radioactivity can be made with a handheld radiation monitor. Sample results are seen in **FIGURE 3.12**. One band on the filter has blackened the X-ray film. The film can be aligned to the filter to determine which band corresponds to the probe.

▸ **autoradiography** A method of capturing an image of radioactive materials on film or nuclear emulsion.

A simple modification of the autoradiography procedure called *in situ* **hybridization** allows one to peer into a cell and determine the location, at a microscopic level, of specific nucleic acid sequences. We simply modify a few steps in the preceding process to perform the hybridization between our probe, usually labeled with ^3H, and complementary nucleic acids in an intact cell or tissue. The goal is to determine exactly where the target is located. The cell or tissue slice is mounted on a microscope slide. Following hybridization, a photographic emulsion instead of film is applied to the slide, covering it. The emulsion, when developed, is transparent to visible light so that it is possible to see the exact location in the cell where the grains in the emulsion blackened by the radioactivity are located. Development time can be weeks to months because ^3H has less energetic radiation and its longer half-life results in lower activity.

▸ **hybridization** The pairing of complementary nucleic acid strands from different sources to give a hybrid.

There are nonradioactive alternatives to the procedures described earlier that use either colorimetric or fluorescence labeling. Digoxygenin-labeled probe is a commonly used colorimetric procedure. Probe bound to target is localized with an antidigoxygenin antibody coupled to alkaline phosphatase, an enzyme that acts on a colorless substrate to develop color. The advantage is the time required to see the results. It is typically a single day, but sensitivity is usually less than with radioactivity. Fluorescence *in situ* hybridization, or FISH, is another very common nonradioactive procedure that uses a fluorescently labeled probe. This method is illustrated in **FIGURE 3.13**. Multiple fluorophores in different colors are available—about a dozen now—but ratios of different probe colors combinations can be used to create additional colors.

These procedures are more picturesque but less quantitative than traditional scintillation counting. At best, these procedures can be called semiquantitative. It is possible to use an optical scanner to quantitate the amount of signal produced on film, but care must be taken to ensure the time of exposure during the experiment is within a linear range.

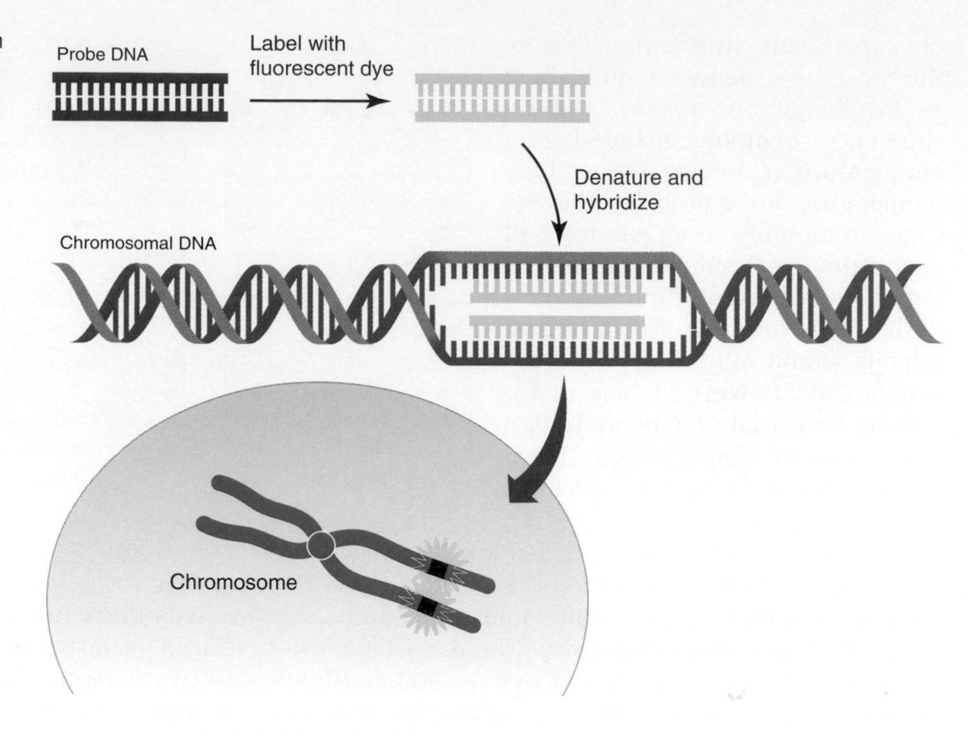

FIGURE 3.13 Fluorescence *in situ* hybridization (FISH). Adapted from an illustration by Darryl Leja, National Human Genome Research Institute (www.genome.gov).

KEY CONCEPT

- Hybridization of labeled nucleic acid to complementary sequences can identify specific nucleic acids in a mixture.

CONCEPT AND REASONING CHECK

Why is the % G-C content an important variable in hybridization reaction?

3.6 DNA Separation Techniques

With a few exceptions, the individual pieces of DNA (chromosomes) making up a living organism's genome are on the order of megabases in length, making them too physically large to be manipulated easily in the laboratory. Individual genes or chromosomal regions of interest by contrast are often quite small and readily manageable, on the order of hundreds or a few thousand base pairs in length. A necessary first step, therefore, in many experimental processes investigating a specific gene or region is to break the large original chromosomal DNA molecule down into smaller manageable pieces and then begin isolation and selection of the particular relevant fragment or fragments of interest. This breakage can be done by mechanical shearing of chromosomes in a process that produces breakages randomly to produce a uniform size distribution of assorted molecules. This approach is useful if a randomness in breakpoints is required, such as to create a library of short DNA molecules that "tile," or partially overlap, each other while together representing a much larger genomic region, such as an entire chromosome or genome. Alternatively, restriction endonucleases (see *Section 3.2, Nucleases*) may be employed to cut large DNA molecules into defined shorter segments in a way that is reproducible. This reproducibility is frequently useful, in that a DNA section of interest can be identified in part by its size. Consider a hypothetical gene *genX* on a bacterial chromosome, with the entire gene lying between two EcoRI sites spaced 2.3 kb apart. Digestion of the bacterial DNA with EcoRI will yield a range of small DNA molecules, but *genX* will always occur on the same 2.3 kb fragment. Depending on the size and complexity of the starting genome, there may be several other DNA segments of similar size produced, or in a simple enough system, this 2.3 kb size may be unique to the *genX* fragment. In this latter case, detection or visualization of a 2.3 kb fragment is enough to definitively identify the presence of *genX*. Many of the earliest laboratory techniques developed in working with DNA relate to separating and concentrating DNA molecules based on size expressly to take advantage of these concepts. An ability to separate DNA molecules based on size allows for taking a complex mixture of many fragment sizes and selecting a much smaller, less-complex subset of interest for further study.

The simplest method for separation and visualization of DNA molecules based on size is gel electrophoresis. In neutral agarose gel electrophoresis, the most basic type of gel, this is done by preparing a small slab of gel in an electrically conductive, mildly basic buffer. While similar to the gelatins used to make dessert dishes, this type of gel is made from agarose, a polysaccharide that is derived from seaweed and has very uniform molecular sizes. Preparation of agarose gels of a specific percentage of agarose by mass (usually in the range of 0.8% to 3%) creates, in effect, a molecular sieve, with a "mesh" pore size being determined by the percentage of agarose (higher percentages yielding smaller pores). The gel is poured in a molten state into a rectangular container, with discrete wells being formed near one end of the product. After cooling and solidifying, the slab is submerged in the same conductive, mildly alkaline, buffer, and samples of mixed DNA fragments are placed in the preformed wells. A DC electric current is then applied to the gel, with the positive charge being at the opposite end of the gel from the wells. The alkalinity of the solution ensures that the DNA molecules have a uniform negative charge from their backbone phosphates, and the DNA fragments begin to be drawn electrostatically toward the positive electrode. Shorter DNA fragments are able to move through the agarose pores with less resistance than longer fragments, and so over time the smallest DNA molecules move the furthest from the wells and the largest move the least. All fragments of a given size will move at about the same rate, effectively concentrating any population of equal-sized molecules into a discrete band at the same distance from the well. Addition of a DNA-binding fluorescent dye, such as ethidium bromide or SYBR green, to the gel stains these DNA bands such that they can be directly seen by eye when the gel is exposed to fluorescence-exciting light. In practice, a standard sample consisting of a set of DNA molecules of a known size is run in one of the wells, with sizes of bands in other wells estimated in comparison to the standard, as shown in **FIGURE 3.14**. DNA molecules of roughly 50 to 10,000 bp can be quickly separated, identified, and sized to within about 10% accuracy by this simple method, which remains a common laboratory technique. DNAs can be separated not only by size, but also

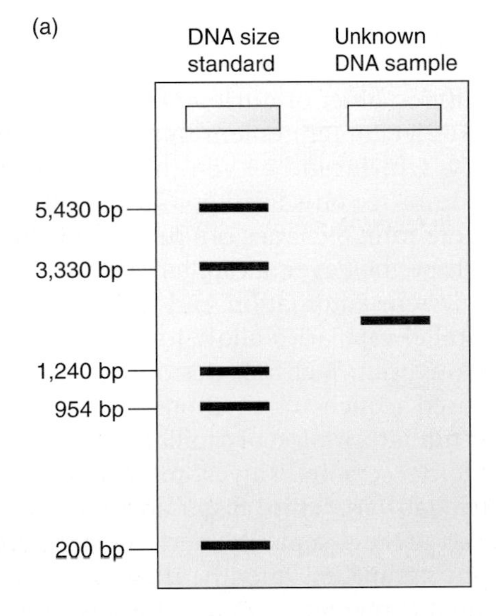

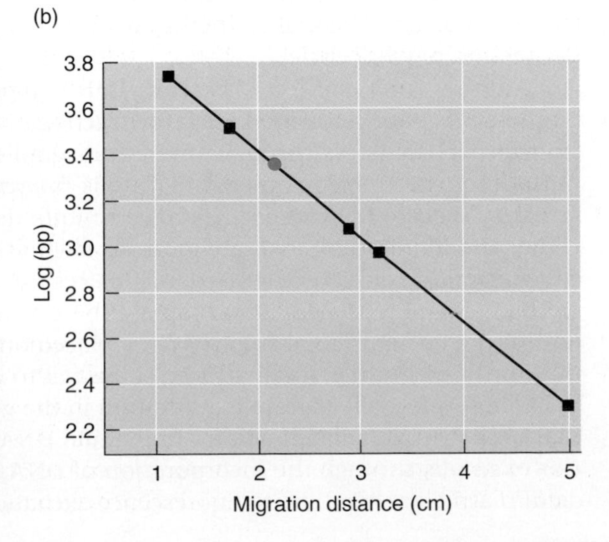

FIGURE 3.14 DNA sizes can be determined by gel electrophoresis. A DNA size standard and a DNA of unknown size are run in two lanes of a gel, depicted schematically. The migration of the DNAs of known size in the standard is graphed to create a standard curve (migration distance in cm vs. log bp). The point shown in green is for the DNA of unknown size. Adapted from an illustration by Michael Blaber, Florida State University.

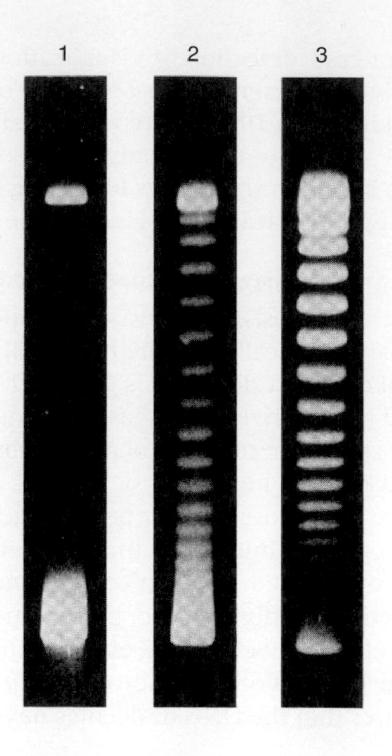

FIGURE 3.15 Supercoiled DNAs separated by agarose gel electrophoresis. Lane 1 contains untreated negatively supercoiled DNA (lower band). Lanes 2 and 3 contain the same DNA that was treated with a type 1 topoisomerase for 5 and 30 minutes, respectively. The topoisomerase makes a single strand break in the DNA and relaxes negative supercoils in single steps (one supercoil relaxed per strand broken and reformed). Reproduced from W. Keller, *Proc. Natl. Acad. Sci. USA* 72 (1975): 2550–2554. Photo courtesy of Walter Keller, University of Basel.

by shape. Supercoiled DNA, which is compact compared to relaxed or linear DNA, migrates more rapidly on a gel, and the more supercoiling, the faster the migration, as seen in **FIGURE 3.15**.

Variations on this method primarily relate to changing the gel matrix from agarose to other molecules such as synthetic polyacrylamides, which can have even more precisely controlled pore sizes. These can offer finer-size resolution of DNA molecules from roughly 10 to 1500 bp in size. Both resolution and sensitivity are further improved by making these types of gels as thin as possible, normally requiring they be formed between glass plates for mechanical strength. When chemical denaturants such as urea are added to the buffer system, the DNA molecules are forced to unfold (losing any secondary structures) and take on hydrodynamic properties related only to molecule length. This approach can clearly resolve DNA molecules differing in length by only a single nucleotide. Denaturing polyacrylamide electrophoresis is a key component of common DNA sequencing techniques whereby the separation and detection of a series of single nucleotide length difference DNA products allows for the reading of the underlying order of nucleotide bases (see *Section 3.7, DNA Sequencing*).

The next level of refinement to this technique is to place the gel matrix in a very fine capillary, which can be even thinner than a glass plate–supported gel and thus still further improve on sensitivity and resolution capacity. Unlike a glass-supported slab gel, where multiple lanes can be run side by side, a capillary can handle only one sample at a time; however, a capillary can be run clean of sample and reused, making it ideal for system automation and high-throughput applications. Instruments with multiple parallel capillaries allow for parallel analysis of multiple samples to further increase throughput. Technologies of this form mark the apparent apex of chain termination-based sequencing methods.

Further miniaturization of capillaries onto the surfaces of inert "chips" with etched-in microfluidic reservoirs, valves, pumps, and mixing chambers can be employed to create entire "lab-on-a-chip" disposable nucleic acid sample analysis cartridges. These cartridges can process, separate, perform size analysis, and quantitate DNA or RNA in a small input sample. Frequently, these devices are controlled and have data output processed by a computer, which in turn will manipulate the data output in order to present it as a traditional stained agarose or polyacrylamide gel—in effect, bringing the technology full circle.

Another method for separating DNA molecules from other contaminating biomolecules—or, in some cases, for fractionation of specific small DNA molecules from other DNAs—is through the use of gradients, as depicted in **FIGURE 3.16**. The most frequent implementation of this is *isopycnic banding,* which is based on the fact that specific DNA molecules have unique densities based on their G-C content. Under the influence of extreme g-forces, such as through ultracentrifugation, a high-concentration solution of a salt (such as cesium chloride) will form a stable density gradient from low density (near top of tube/center of rotor) to high density (near bottom of tube or outside of rotor). When placed on top of this gradient (or even mixed uniformly within the gradient) and subjected to continued centrifugation, individual DNA molecules will migrate to a position in the gradient where their density matches that of the surrounding medium. Individual DNA bands can then be either visualized (for example, through the incorporation of DNA-binding fluorescent dyes in the gradient matrix and exposure to fluorescence excitation), or recovered by careful puncture

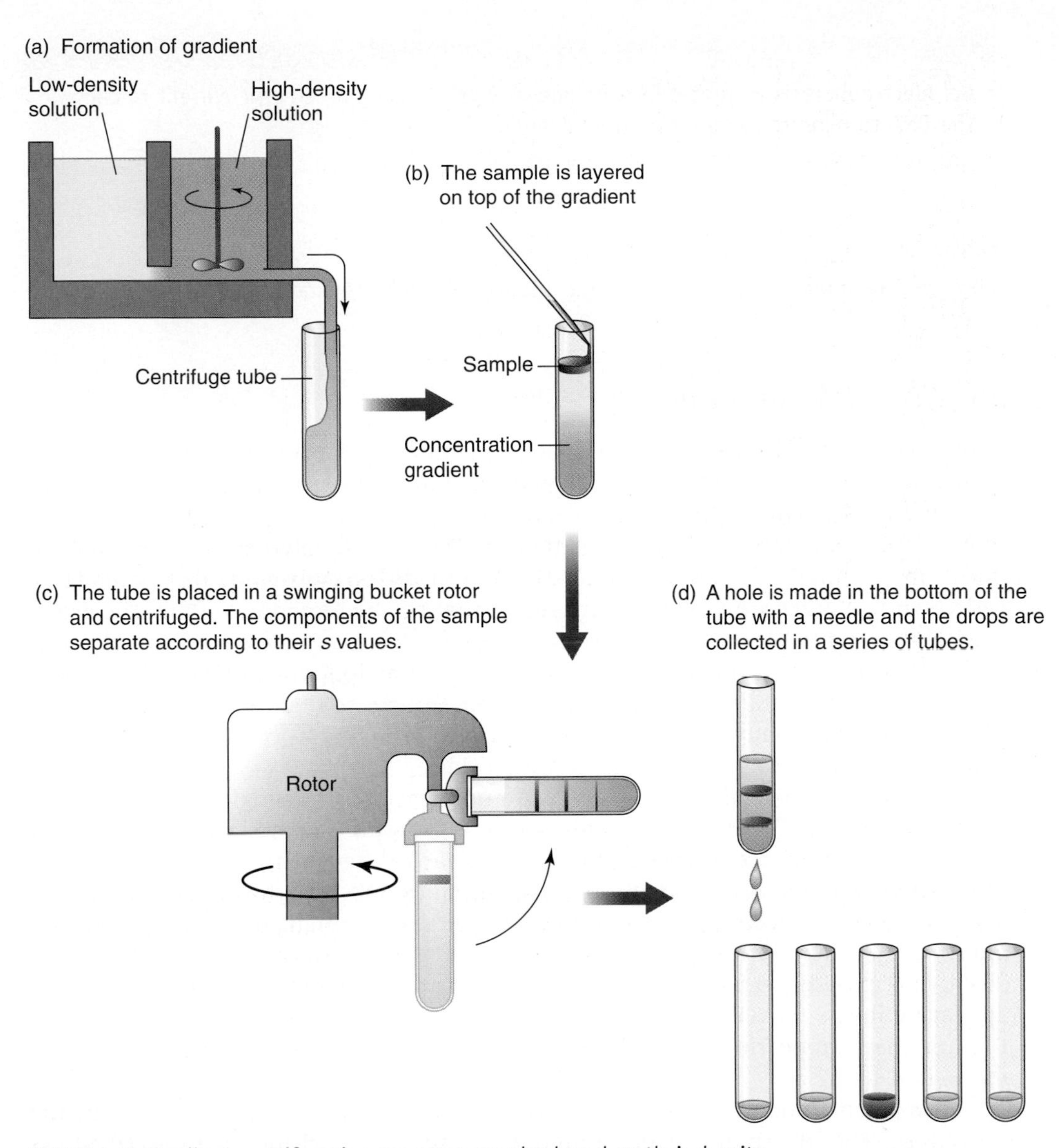

FIGURE 3.16 Gradient centrifugation separates samples based on their density.

of the centrifuge tube and fractional collection of the tube contents. This method can also be used to separate double-stranded from single-stranded molecules and RNA from DNA molecules, again based solely of density differences.

Choice of the gradient matrix material, its concentration, and the centrifugation conditions can influence the total density range separated by the process, with very narrow ranges being used to fractionate one particular type of DNA molecule from others, and wider ranges being used to separate DNAs in general from other biomolecules. Historically, one of the best-known uses of this technique was in the Meselson–Stahl experiment of 1958 (introduced in *Section 1.6, DNA Replication Is Semiconservative*), in which the stepwise density changes in the DNA genomes of bacteria shifted from growth in "heavy" nitrogen (^{15}N) to "regular" nitrogen (^{14}N) were observed. The method's capacity to differentially band DNA with pure ^{15}N, half ^{15}N/half ^{14}N, and pure ^{14}N conclusively demonstrated the semiconservative nature of DNA replication. Today, the method is most frequently employed as a large-scale preparative purification technique with wider density ranges to purify DNAs as a group away from proteins and RNAs.

KEY CONCEPTS

- Gel electrophoresis separates DNA fragments by size, using an electric current to cause the DNA to migrate toward a positive charge.
- DNA can also be isolated using density gradient centrifugation.

CONCEPT AND REASONING CHECK

Why are DNA fragments of different G-C contents able to be separated by CsCl gradients?

3.7 DNA Sequencing

▸ **primer** A short sequence that is paired with one strand of DNA and provides a 3′-OH end at which an DNA polymerase starts synthesis of a DNA chain.

▸ **dideoxynucleotide (ddNTP)** A chain-terminating nucleotide that lacks a 3′-OH group and, therefore, is not a substrate for DNA polymerization; used in DNA sequencing.

The most commonly used method of DNA sequencing hasn't changed much since Frederick Sanger and colleagues developed a technique in 1977 called *dideoxy sequencing*. This method requires many identical copies of the DNA, an oligonucleotide **primer** that is complementary to a short stretch of the DNA, DNA polymerase, deoxynucleotides (dNTPs: dATP, dCTP, dGTP, and dTTP), and **dideoxynucleotides (ddNTPs)**. Dideoxynucleotides are modified nucleotides that can be incorporated into the growing DNA strand but lack the 3′ hydroxyl group needed to attach the next nucleotide. Thus, their incorporation terminates the synthesis reaction. The ddNTPs are added at much lower concentrations than the normal nucleotides so that they are incorporated at a low rate, randomly, and often only after synthesis has proceeded normally for a strand length of up to several hundred nucleotides.

Originally, four separate reactions were necessary, with a single different ddNTP added to each one. The reason for this was that the strands were labeled with radioisotopes and could not be distinguished from each other on the basis of the label. Thus, the reactions were loaded into adjacent lanes on a denaturing acrylamide gel and separated by electrophoresis at a resolution that distinguished between strands differing by a length of one nucleotide. The gel was transferred to a solid support, dried, and exposed to a film. The results were read from top to bottom, with a band appearing in the ddATP lane indicating that the strand terminated with an adenine, the next band appearing in the ddTTP lane indicating that the next base was a thymine, and so on.

Two recent modifications have aided in the automation and scaling up of the procedure. The incorporation of a different fluorescent label for each ddNTP allows a single reaction to be run that is read as the strands are hit with a laser and pass by an optical sensor. The information as to which ddNTP terminated the fragment is fed directly into a computer. The second modification is the replacement of large slabs of polyacrylamide gels with very thin, long, glass capillary tubes filled with gel, as described previously in *Section 3.6, DNA Separation Techniques*. These tubes can dissipate heat more rapidly, allowing the electrophoresis to be run at a higher voltage, greatly reducing the time required for separation. A schematic illustrating this process is shown in **FIGURE 3.17**. These modifications, with their resulting automation and increased throughput, ushered in the era of whole-genome sequencing.

A number of "second-generation" sequencing technologies are currently coming into the market. These aim to eliminate the need for time-consuming gel separation and reliance on human labor. Sequencing by synthesis and sequencing through nanopores and from picolitre wells are some of the many new technologies.

Sequencing by synthesis relies on the detection and identification of each nucleotide as it is added to a growing strand. In one such application, the primer is tethered to a glass surface and the complementary DNA to be sequenced anneals to the primer. Sequencing proceeds by adding polymerase and fluorescently labeled nucleotides individually, washing away any unused dNTPs. After illuminating with a laser, the nucleotide that has been incorporated into the DNA strand can be detected. Other versions use nucleotides with reversible termination, so that only one nucleotide can

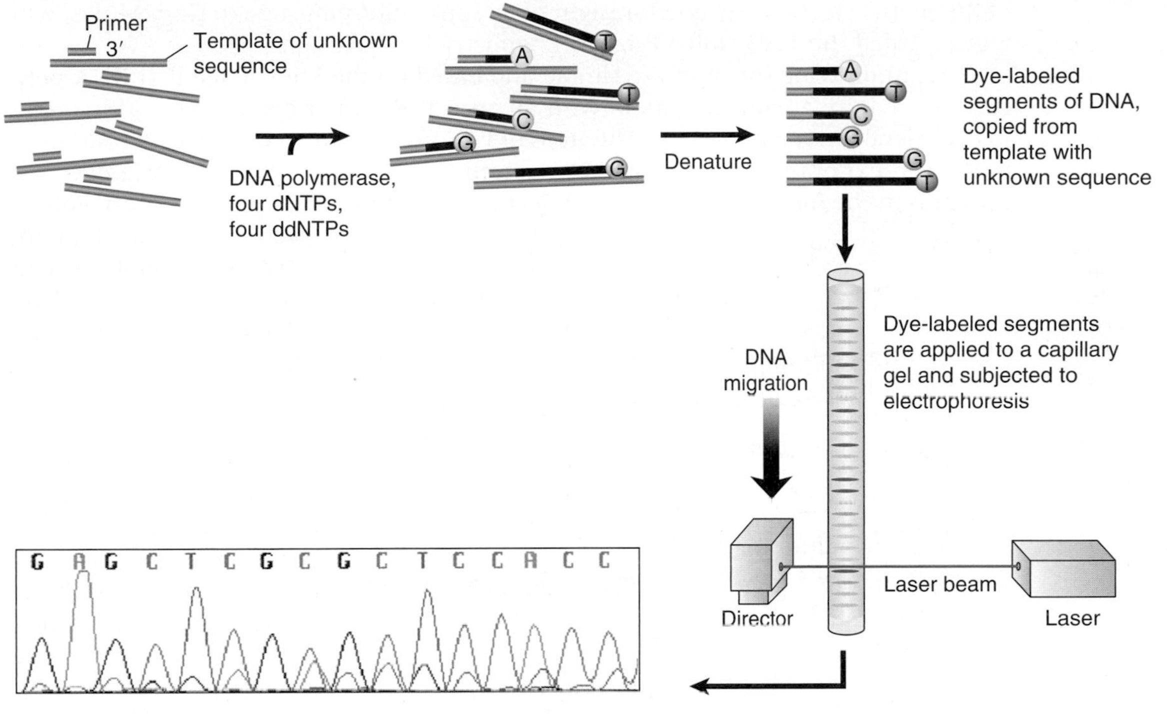

FIGURE 3.17 DideoxyNTP sequencing using fluorescent tags. Inset photo courtesy of Jan Kieleczawa.

be incorporated at a time even if there is a stretch of homopolymeric DNA (such as a run of adenines). Still another version, called *pyrosequencing,* detects the release of pyrophosphate from the newly added base. These technologies have the advantage that many parallel reactions can be run.

A completely different approach aims to detect individual nucleotides as a DNA sequence is run through a silicone nanopore. Tiny transistors are used to control a current passing through the pore. As a nucleotide passes through the pore, it disturbs the current in a manner unique to its chemical structure. This technology has the advantage of reading DNA by simply using electronics, with no chemistry or optical detection required. Nevertheless, there are still some kinks to work out of the process.

KEY CONCEPTS

- Chain-termination sequencing uses dideoxynucleotides to terminate DNA synthesis at particular nucleotides.
- Fluorescently tagged ddNTPs and capillary gel electrophoresis allow automated, high-throughput DNA sequencing.
- The next generation of sequencing techniques increases automation and decreases time and cost of sequencing.

CONCEPT AND REASONING CHECK

Why are dideoxynucleotides necessary for sequencing DNA?

3.8 PCR and RT-PCR

Few advances in the life sciences have had the broad-reaching and even paradigm-shifting impact of the **polymerase chain reaction** (**PCR**). Although evidence exists that the underlying core principles of the method were understood and in fact used in practice by a few isolated people prior to 1983, credit for independent conceptualization

▸ **polymerase chain reaction (PCR)** A process for the amplification of a defined nucleic acid section through repeated thermal cycles of denaturation, annealing, and polymerase extension.

of the mature technology and foresight of its applications must go to Kary Mullis, who was awarded the 1993 Nobel Prize in Chemistry for his insight.

The underlying concepts are simple and based on the knowledge that DNA polymerases require a template strand with an annealed primer containing a 3′ hydroxyl to commence strand extension. The steps of PCR are illustrated in **FIGURE 3.18**. Although in the context of normal cellular DNA replication (see *Chapter 13, DNA Replication*) this primer is in the form of a short RNA molecule provided by DNA primase, it can equally well be provided in the form of a short, single-stranded synthetic DNA oligonucleotide having a defined sequence complementary to the 3′ end of any known sequence of interest. Heating of the double-stranded target sequence of interest (known as the "template molecule," or just "template," for short) to near 100°C in appropriate buffer causes thermal denaturation as the template strands melt apart from each other (Figure 3.18a and b). Rapid cooling to the annealing temperature (or $\boldsymbol{T_m}$) of the primer/template pair and a vast molar excess of the short, kinetically active synthetic primer ensures that a primer molecule finds and appropriately anneals to its complementary target sequence more rapidly than the original opposing strand can do so (Figure 3.18c). If presented to a polymerase, this annealed primer presents a defined location from which to commence primer extension (Figure 3.18d). In general, this extension will occur until either the polymerase is forced off the template or it reaches the 5′ end of the template molecule and effectively runs out of template to copy.

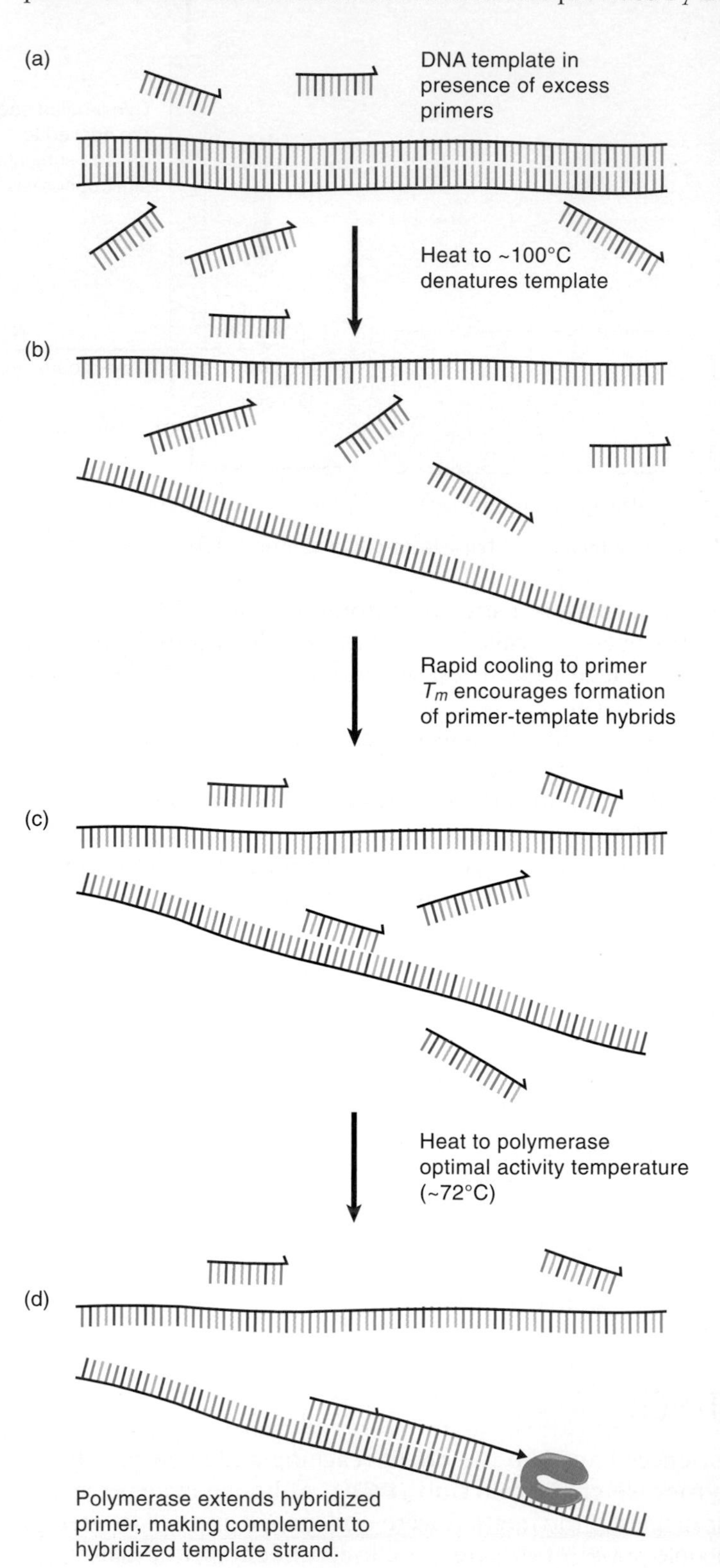

FIGURE 3.18 Denaturation (a) and rapid cooling (b) of a DNA template molecule in the presence of excess primer allows the primer to hybridize to any complementary sequence region of the template (c). This provides a substrate for polymerase action and primer extension (d), creating a complementary copy of one template strand downstream from the primer.

▸ $\boldsymbol{T_m}$ The theoretical melting temperature of a duplex nucleic acid segment into separate strands. T_m is dependent on parameters that include sequence composition, duplex length, and buffer ionic strength.

The ingenuity of PCR arises from simultaneously incorporating a nearby second primer of opposing polarity (that is, complementary to the opposite strand the first primer anneals to) and then subjecting the mixture of template, two primers (at high concentrations), thermostable DNA polymerase, and dNTP containing polymerase buffer to repeated cycles of thermal denaturation, annealing, and primer extension. Consider just the first cycle of the process: denaturation and annealing occur as described previously, but with both primers, creating the situation depicted in **FIGURE 3.19**. If polymerase extension is allowed to proceed for a short period of time (on

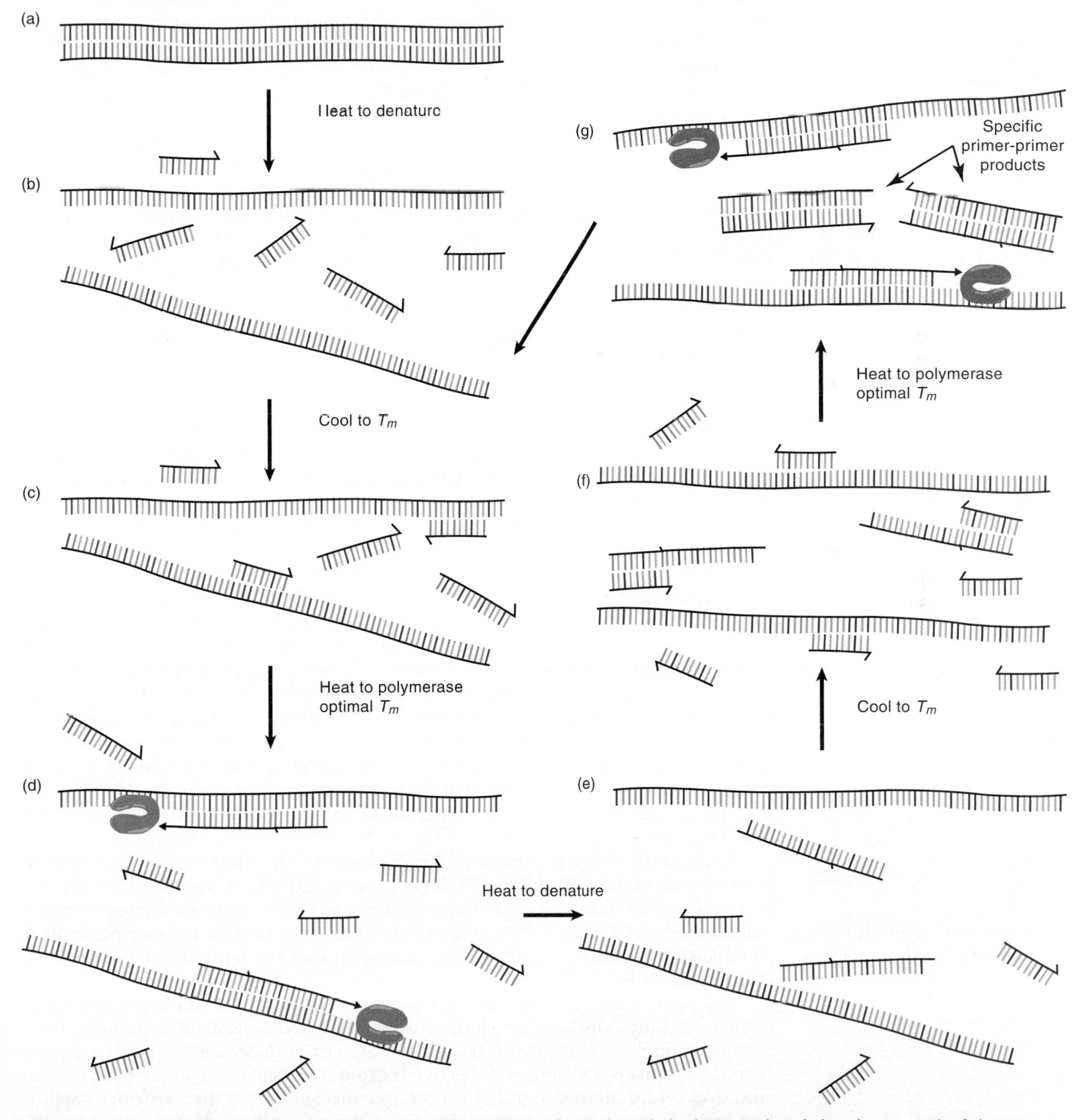

FIGURE 3.19 Thermally driven cycles of primer extension, where primers of opposite polarity have nearby priming sites on each of the two template strands, which leads to the exponential production of the short, primer-to-primer-defined sequence (the amplicon).

the order of 1 minute per 1000 bp), each of the primers will be extended out and past the location of the other, thus creating a new complementary annealing site for the opposing primer. Raising the temperature back to denaturation stops the primer elongation process and displaces the polymerases and newly created strands. As the system is cooled once more to the annealing temperature, each of the newly formed short, single DNA strands serves as an annealing site for its opposite polarity primer. In this second thermal cycle, extension of the primers proceeds only as far as the template exists—that is, the 5′ end of the opposing primer sequence. The process has now made both strands of the short, defined, precisely primer-to-primer DNA sequence. Repeating the thermal steps of denaturation, annealing, and primer extension lead to an exponential (2^N, where N is number of thermal cycles) increase in the number of this defined product, allowing for phenomenal levels of "sequence amplification." Close consideration of the process reveals that while this also creates uncertain length products from the extension of each primer off the original template molecule with each cycle, these products accrue in a linear fashion and are quickly vastly outnumbered by the primer-to-primer defined product (known as the **amplicon**). In fact, within 40 thermal cycles of an idealized PCR reaction, a single template DNA molecule generates approximately 10^{12} amplicons—more than enough to go from an invisible target to a clearly visible fluorescent dye stained product.

▸ **amplicon** The precise, primer-to-primer double-stranded nucleic acid product of a PCR or RT-PCR reaction.

Perhaps not surprisingly, there are many technical complexities underlying this deceptively simple description. Primer design must take into account issues such as DNA secondary structures, uniqueness of sequence, and similarity of T_m between primers. Use of a thermostable polymerase (that is, one that is not inactivated by the high temperatures used in the denaturation steps) is an essential concept identified by Mullis and coworkers. Within this constraint, though, different enzyme sources with differing properties (such as exonuclease activities for increased accuracy) can be exploited to meet individual application needs. Buffer composition (including agents such as DMSO to help reduce secondary structural barriers to effective amplification and inclusion of divalent cations such as Mg^{2+} at sufficient concentration to not be depleted by chelation to nucleotides) often needs some optimization for effective reactions. In general, the PCR process works best when the primers are within short distances of each other (100 to 500 bp), but well-optimized reactions have been successful at distances into the tens of kilobases. "Hot start" techniques—frequently through covalent modification of the polymerase—can be employed to ensure that no inappropriate primer annealing and extension can occur prior to the first denaturation step, thereby avoiding the production of incorrect products. Generally, somewhere around 40 thermal cycles marks an effective limit for a PCR reaction with good kinetics in the presence of appropriate template, as depletion of dNTPs into amplicons effectively occurs around this point and a "plateau phase" occurs wherein no more product is made. Conversely, if the appropriate template was not present in the reaction, proceeding beyond 40 cycles primarily increases the likelihood of production of rare, incorrect products.

Pairing PCR with a preliminary reverse transcription step (either random-primed or using one of the PCR primers to direct activity of the RNA-dependent DNA polymerase [reverse transcriptase]) allows for RNA templates to be converted to cDNA and then subject to regular PCR, in a variation known as **reverse transcription PCR** (**RT-PCR**). In general, the subsequent discussion uses the term *PCR* to refer to both PCR and RT-PCR.

▸ **reverse transcription PCR (RT-PCR)** A technique for the detection and quantification of expression of a gene by reverse transcription and amplification of RNAs.

Detection of PCR products can be done in a number of ways. Post-reaction "end-point techniques" include gel electrophoresis and DNA-specific dye staining. Long a staple of molecular biological techniques (described in *Section 3.6, DNA Separation Techniques*), this is a simple but effective technique to rapidly visualize both that an amplicon was produced and that it is of an expected size. If the particular application requires exact, to-the-nucleotide product sizing, capillary electrophoresis can be used instead. Hybridization of PCR products to microarrays or suspension bead arrays can be used to detect specific amplicons when more than one product sequence may come out of an assay. These in turn use a variety of methods for amplicon labeling, including chemiluminescence, fluorescence, and electrochemical techniques. Alternatively, **real-time PCR** methodologies employ some way of directly detecting the ongoing production of amplicons in the reaction vessel, most commonly through monitoring a direct or indirect fluorescence change linked to amplicon production by optical methods. These methods allow the reaction vessel to stay sealed throughout the process. In contrast to endpoint methods where final amplicon concentration bears little relationship to starting template concentration, real-time methods show good correlations between the thermocycle number at which clear signals are measurable (usually referred to as the **threshold cycle**, or **C_T**) and the starting template concentration. Thus, real-time methods are effective template quantification approaches. As a result, these methods are often referred to as **quantitative PCR** (**qPCR**) methods.

▸ **real-time PCR** A PCR technique with continuous monitoring of product formation as synthesis proceeds, usually through fluorometric methods.

▸ **threshold cycle (C_T)** The thermocycle number in a real-time PCR or RT-PCR reaction at which the product signal rises above a specified cutoff value to indicate amplicon production is occuring.

▸ **quantitative PCR (qPCR)** A PCR reaction used to amplify and simultaneously quantify an amplicon.

Conceptually, the simplest method for real-time PCR detection is based on the use of dyes that selectively bind and become fluorescent in the presence of double-stranded DNA, such as SYBR green. Production of a PCR product during thermocycling leads to an exponential increase in the amount of double-stranded product present at the annealing and extension thermal steps of each cycle. The real-time instrument monitors fluorescence in each reaction tube during these thermal steps of each cycle and calculates the change in fluorescence per cycle to generate a sigmoidal amplification curve. A cutoff threshold value placed approximately midrange in the exponential phase of this curve is used for calculating the C_T of each sample and can be used for quantitation if appropriate controls are present.

A potential issue with this approach is that the reporter dyes are not sequence specific, so any spurious products produced by the reaction can lead to false positive signals. In practice, this is usually controlled by performance of a melt-point analysis at the end of regular thermocycling. The reaction is cooled to the annealing temperature and then the temperature is slowly raised while fluorescence is constantly monitored. Specific amplicons will have a characteristic melt point at which fluorescence is lost, whereas nonspecific amplicons will demonstrate a broad range of melt points, giving a gradual loss in sample fluorescence.

A number of alternative approaches use probe-based fluorescence reporters, which avoid this potential nonspecific signal. Probe-based approaches work through the application of a process called **fluorescence resonant energy transfer** (**FRET**). In simple terms, FRET occurs when two fluorophores are in close proximity and the emission wavelength of one (the reporter) matches the excitation wavelength of the other (the quencher). Photons emitted at the reporter dye-emission wavelength are effectively captured by the nearby quencher dye and reemitted at the quencher-emission wavelength. In the simplest form of this approach, two short oligonucleotide probes with homology to adjoining sequences within the expected amplicon are included in the assay reaction; one probe carries the reporter dye and the other, the quencher. If specific PCR product is formed in the reaction, then at each annealing step these two probes can anneal to the single-stranded product and thereby place the reporter and quencher molecules close to each other. Illumination of the reaction with the excitation wavelength of the reporter dye will lead to FRET and fluorescence at the quencher dye's characteristic emission frequency. By contrast, if the homologous template for the probe molecules is not present (that is, the expected PCR product), the two dyes will not be colocalized and excitation of the reporter dye will lead to fluorescence at its emission frequency. This is illustrated in **FIGURE 3.20**. As with the DNA-binding dye approach, the real-time instrument monitors the quencher emission wavelength during each cycle and generates a similar sigmoidal amplification curve. Multiple alternate ways of exploiting FRET for this process exist, including 5′ fluorogenic nuclease assays, molecular beacons, and molecular scorpions. Although the details of these differ, the underlying concept is similar and all generate data in a similar fashion.

▸ **fluorescence resonant energy transfer (FRET)** A process whereby the emission from an excited fluorophore is captured and reemitted at a longer wavelength by a nearby second fluorophore whose excitation spectrum matches the emission frequency of the first fluorophore.

FIGURE 3.20 Fluorescence resonant energy transfer (FRET) occurs only when the reporter and quencher fluorophores are very close to each other, leading to the detection of light at the quencher emission frequency when the reporter is stimulated by light of its excitation frequency. If the reporter and quencher are not colocalized, stimulation of the reporter instead leads to detection of light at the reporter's emission frequency. By placing the reporter and quencher fluorophores on single-stranded nucleic acid probes complementary to the expected amplicon, different variations on this method can be designed such that the occurrence of FRET can be used to monitor the production of sequence-specific amplicons.

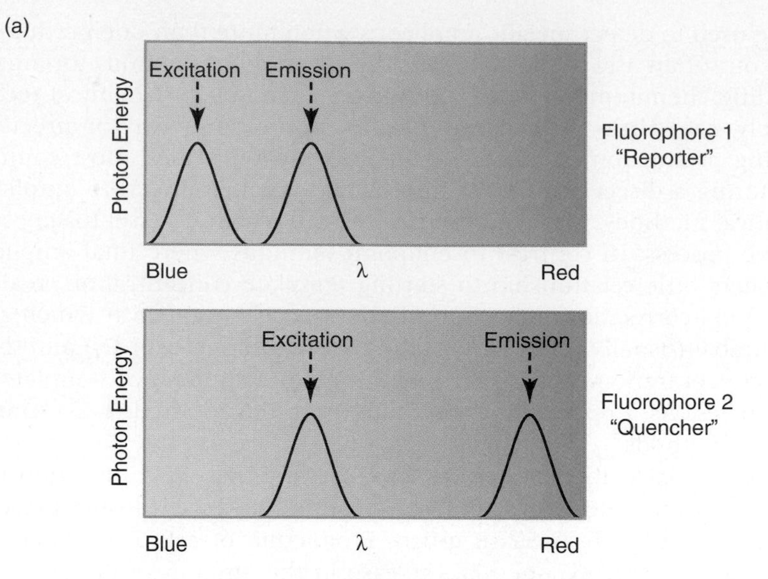

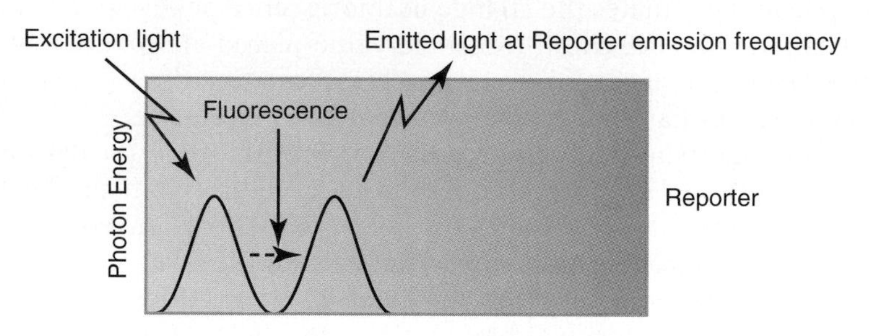

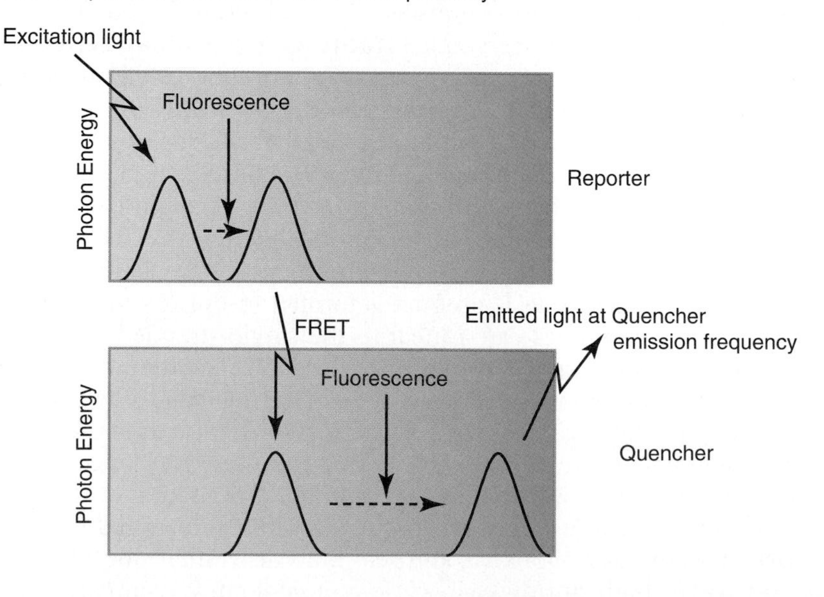

The applications of the PCR process are incredibly diverse. The simple appearance or nonappearance of an amplicon in a properly controlled reaction can be taken as evidence for the presence or absence, respectively, of the assay target template. This leads to medical applications such as the detection of infectious disease agents at sensitivities, specificities, and speeds much greater than alternate methods. Although the two primer sites must be of known sequence, the internal section may be any sequence of a general length; this fact leads directly to applications where a PCR product for a region known to vary between species (or even between individuals) can be produced and subject to sequence analysis to identify the species (or individual identity, in the latter case) of the sample template. Coupled with single molecule sensitivity, this has provided criminal forensics with tools powerful enough to identify individuals from residual DNA on crime scene traces as simple as cigarette butts, smudged fingerprints, or a single hair. Evolutionary biologists have made use of PCR to amplify DNA from well-preserved samples, such as insects in amber millions of years old, with subsequent sequencing and phylogenetic analysis yielding fascinating results on the continuity and evolution of life on Earth. Quantitative real-time approaches have applications in medicine (for example, monitoring viral loads in transplant patients), research (such as examining transcriptional activation of a specific target gene in a single cell), or environmental monitoring (for instance, water purification quality control).

In general, PCR reactions are run with carefully optimized T_m values that maximize sensitivity and amplification kinetics while ensuring that primers will anneal only to their exact hybridization matches. Lowering the T_m of a PCR reaction—in effect, relaxing the reaction stringency and allowing primers to anneal to not quite perfect hybridization partners—has useful applications as well, such as in searching a sample for an unknown sequence suspected to be similar to a known one. This technique has been successfully employed for the discovery of new virus species, when primers matching a similar virus species are employed. Similarly, during a PCR-directed cloning of a gene or region of interest, use of planned mismatches in the primer sequence and slightly lowered T_ms can be used to introduce wanted mutations in a process called *site-directed mutagenesis*. Differential detection of single nucleotide polymorphisms (SNPs; see *Section 5.3, Individual Genomes Show Extensive Variation*), which can be directly indicative of particular genotypes or serve as surrogate linked markers for nearby genetic targets of interest, can be done through design of PCR primers with a 3′ terminal nucleotide specific to the expected polymorphism. At the optimal T_m, this final crucial nucleotide can hybridize and provide a 3′ hydroxyl to the waiting polymerase only if the matching SNP occurs, in a process known by several names, including amplification refractory mutation selection (ARMS) or allele-specific PCR (ASPE).

The PCR process described thus far has been restricted to amplification of a single target per reaction, or "simplex" PCR. Although this is the most common application, it is possible to combine multiple, independent PCR reactions into a single reaction, allowing for an experiment to query a single minute specimen for the presence, absence, or possibly the amount of multiple unrelated sequences. This *multiplex PCR* is particularly useful in forensics applications and medical diagnostic situations but entails rapidly increasing levels of complexity in ensuring that multiple primer sets do not have unwanted interactions that lead to undesired false products. At best, multiplexing tends to result in loss of some sensitivity for each individual PCR due to effective competition between them for limited polymerase and nucleotides.

A final point of interest to many students with regard to PCR is its consideration from a philosophical perspective. In practice, performance of this now incredibly pervasive method requires the use of a thermostable polymerase, as previously indicated. These polymerases (of which there are a number of varieties) primarily derive from bacterial DNA polymerases originally identified in extremophiles living in boiling hot springs and deep-sea volcanic thermal vents. Few people would have been likely to suspect that studying deep-sea thermal vent microbes would be of such direct importance to so many other aspects of science, including ones with impact on their daily lives. These unexpected links between topics serve to highlight the importance of basic research on all manner of subjects; critical discoveries can come from the least expected avenues of research.

KEY CONCEPTS

- PCR permits the exponential amplification of a desired sequence, using primers that anneal to the sequence of interest.
- RT-PCR uses reverse transcriptase to convert RNA to DNA for use in a PCR reaction.
- Real-time, or quantitative, PCR detects the products of PCR amplification during their synthesis and is more sensitive and quantitative than conventional PCR.
- PCR depends on the use of thermostable DNA polymerases that can withstand multiple cycles of template denaturation.

CONCEPT AND REASONING CHECK

Why does PCR depend on a thermostable DNA polymerase to function?

3.9 Blotting Methods

After nucleic acids are separated by size in a gel matrix, they can be detected using dyes that are nonsequence specific, or specific sequences can be detected using a method generically referred to as *blotting*. Although slower and more involved than direct visualization by fluorescent dye staining, blotting techniques have two major advantages: they have a greatly increased sensitivity relative to dye staining, and they allow for the specific detection of defined sequences of interest among many similarly sized bands on a gel.

The method was first developed for application to DNA agarose gels and was briefly introduced in *Section 3.5, Nucleic Acid Detection*. In this form, the method is referred to as **Southern blotting** (after the method's inventor, Dr. Edwin Southern). A schematic of this process is shown in **FIGURE 3.21**. A regular agarose gel is made and run (and, if desired, stained), as described previously. Following this, the gel is soaked in alkali buffer to denature the DNA and then placed in contact with a sheet of porous membrane (commonly nitrocellulose or nylon). A buffer is drawn through the gel and then the membrane, either by capillary action (for instance, by wicking into a stack of dry paper towel) or by a gentle vacuum pressure. This slow flow of buffer, in turn, draws each nucleic acid band in the gel out of the gel matrix and onto the membrane surface. Nucleic acids bind to the membrane, which in many cases is positively charged to increase efficiency of DNA binding. This in effect creates a "contact print" of the order and position of all nucleic acid bands as resolved by size in the gel. To make the elution of large DNA molecules from the gel matrix more efficient, the gel is sometimes treated with a mild acid after electrophoresis but before transfer. This induces acid depurination and creates random strand breaks in the DNA within the gel, such that large molecules are broken into smaller subsections that elute more readily but remain in the same physical location as their original gel band. A common variation of this method uses a vacuum to draw the buffer through the blot rather than using capillary action.

▸ **Southern blotting** A process for the transfer of DNA bands separated by gel electrophoresis from the gel matrix to a solid support membrane for subsequent probing and detection.

Following transfer, the nucleic acids are fixed to the membrane either through drying or through exposure to UV light, which can create physical cross-links between the membrane and the nucleic acids (primarily pyrimidines). The blot is now ready for blocking, where it is immersed in a warmed, low-salt buffer containing materials that will bind to and block areas of the blot that may bind organic compounds nonspecifically. Following blocking, a probe molecule is introduced. The probe consists of a labeled (isotopically or chemically, such as through incorporation of biotinylated nucleotides) copy of the target sequence of interest, which has been heat denatured and rapidly cooled to place it in a single-stranded form. When this is added to the warmed buffer and allowed to incubate with the blocked membrane, the probe will attempt to hybridize to homologous sequences on the membrane surface. Following this hybridization step, the membrane is generally washed in warm buffer without probe or blocking agent to remove nonspecifically associated probe molecules and

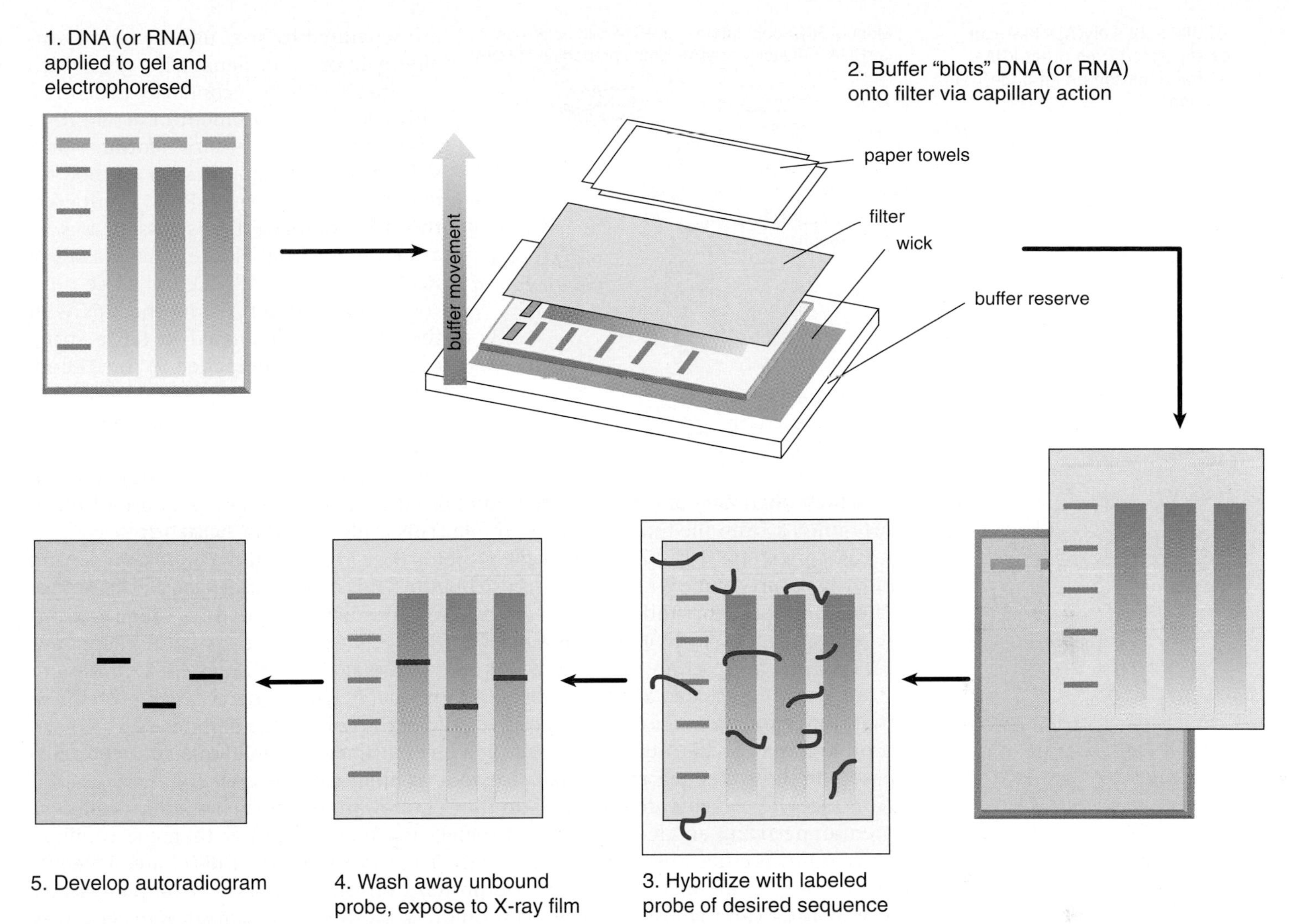

FIGURE 3.21 To perform a Southern blot, DNA digested with restriction enzymes is electrophoresed to separate fragments by size. Double-stranded DNA is denatured in an alkali solution either before or during blotting. The gel is placed on a wick (such as a sponge) in a container of transfer buffer and a membrane (nylon or nitrocellulose) is placed on top of the gel. Absorbent materials such as paper towels are placed on top. Buffer is drawn from the reservoir through the gel by capillary action, transferring the DNA to the membrane. The membrane is then incubated with a labeled probe (usually DNA). The unbound probe is washed away, and the bound probe is detected by autoradiography or phosphorimaging. In northern blotting, RNA, rather than DNA, is run on a gel.

then visualized; in the case of isotopically labeled probes, this can be done by simply exposing the membrane to a piece of film or a phosphor-imager screen. Decay of the label (usually ^{32}P or ^{35}S) leads to the production of an image in which any hybridized DNA bands become visible on the developed film or scanned phosphor screen. For chemically labeled probes, chemiluminescent or fluorescent detection strategies are used in an analogous manner.

A final benefit of the Southern blotting technique is that the observed band intensity is related to the amount of target on the membrane—in other words, it is a quantitative method. If a suitable standard (such as a dilution series of unlabeled probe sequence) is included in the gel, then comparison of this standard to target band intensities allows for determination of target quantity in the starting sample. This information can be useful for applications such as determining viral copy number in a host cell sample.

Numerous variations on the Southern blot approach exist, such as use of a denaturing gel matrix for an otherwise analogous process on RNA molecules (referred to as northern blotting). In this case, there is no initial digestion step, so intact RNAs

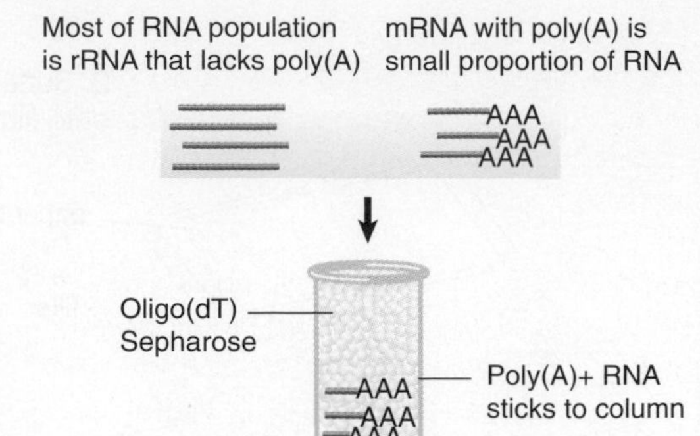

FIGURE 3.22 Poly(A)+ RNA can be separated from other RNAs by fractionation on an oligo(dT) column.

are separated by size, usually on a formaldehyde or other denaturing gel, which eliminates RNA secondary structures. This allows measurement of actual RNA sizes, and like Southern blotting, provides a similarly quantitative method for detection of any type of RNA. If mRNA is the target of interest, it is possible to separate mRNA from all the other classes of RNA in the cell. mRNA (and some noncoding RNA) differs from other RNAs in that it is polyadenylated (it has a string of adenine residues added to the 3′ end; see *Section 21.14, The 3′ Ends of mRNAs Are Generated by Cleavage and Polyadenylation*). Poly(A)+ mRNA can, therefore, be enriched by use of an oligo(dT) column, in which oligomers of oligo(dT) are immobilized on a solid support and used to capture mRNA from the total RNA in a sample. This is illustrated in **FIGURE 3.22**.

A conceptually similar process for proteins based on protein-separation gels and blotting to membrane is known as western blotting. This method is depicted in **FIGURE 3.23**. There are some key differences between the procedures for blotting proteins compared to nucleic acids. First, protein-separation gels typically contain the detergent SDS, which both serves to unfold the proteins so that they will migrate according to size rather than shape, and also provides a uniform negative charge to all proteins so that they will migrate toward the positive pole of the gel. (In the absence of SDS, each protein has a specific individual charge at a given pH; it is possible to separate proteins based on these charges, rather than size, in a technique called *isoelectric focusing*.)

Once the proteins are separated on the gel, they are transferred to a nitrocellulose membrane using an electric current to effect the transfer, rather than the capillary or vacuum methods used for nucleic acids. The most significant difference in western blotting is the method of detecting proteins on the membrane. Complementary base pairing can't be used to detect a protein, so westerns use *antibodies* to recognize the protein of interest. The antibody can either recognize the protein itself, if such

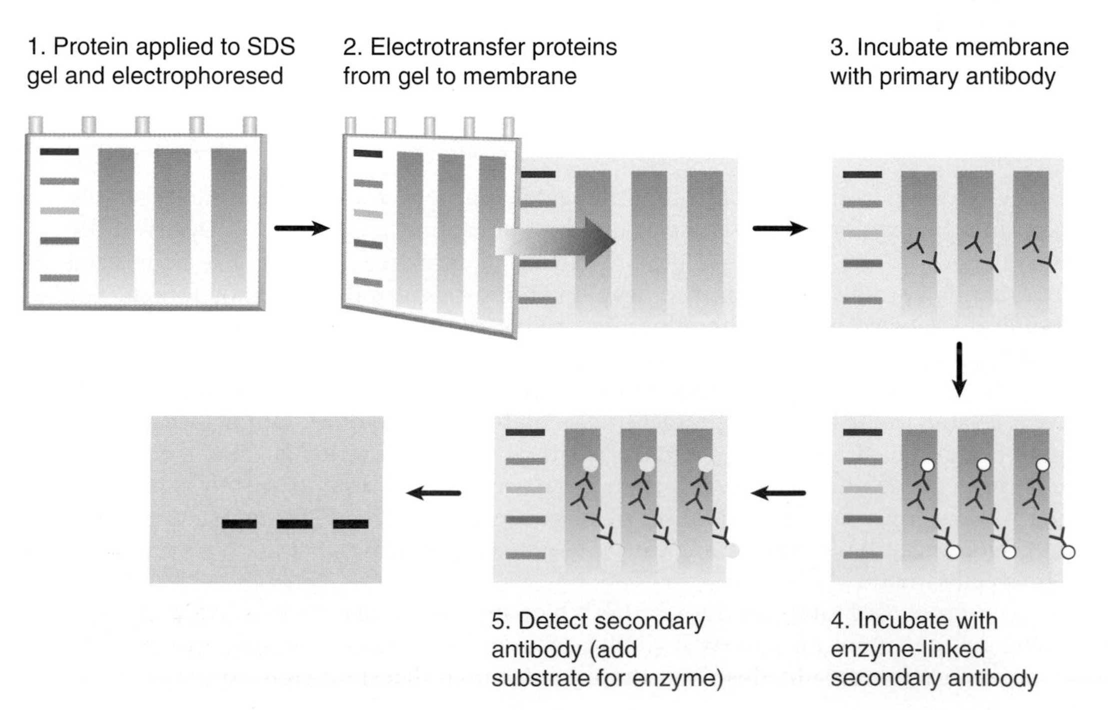

FIGURE 3.23 In a western blot, proteins are separated by size on an SDS gel, transferred to a nitrocellulose membrane, and detected using an antibody. The primary antibody detects the protein and the enzyme-linked secondary antibody detects the primary antibody. The secondary antibody is detected in this example via addition of a chemiluminescent substrate, which results in emission of light that can be detected on X-ray film.

an antibody is available, or can recognize an **epitope tag** that has been fused to the protein sequence. An epitope tag is a short peptide sequence that is recognized by a commercially available antibody; the DNA encoding the tag can be cloned in-frame to a gene of interest, resulting in a product containing the epitope (typically at the N- or C-terminus of the protein). Sequences for the most commonly used epitope tags (such as the HA, FLAG, and myc tags) are often available in expression vectors for ease of fusion (see *Section 3.4, Cloning Vectors Can Be Specialized for Different Purposes*).

▸ **epitope tag** A polypeptide that has been added to a protein that allows its identification by an antibody.

The antibody that recognizes the target on the membrane is known as the *primary antibody*. The final stage of western blotting is detection of the primary antibody with a *secondary antibody*, which is the antibody that can be visualized. Secondary antibodies are raised in a different species than the primary antibody used and recognize the constant region of the primary antibody. (For example, a "goat antirabbit" antibody will recognize a primary antibody raised in a rabbit; see *Chapter 18, Somatic Recombination and Hypermutation in the Immune System*, for a review of antibody structure.) The secondary antibody is typically linked to a moiety that allows its visualization—for example, a fluorescent dye or an enzyme such as alkaline phosphatase or horseradish peroxidase. These enzymes serve as visualization tools because they can convert added substrates to a colored product (*colorimetric detection*) or can release light as a reaction product (*chemiluminescent detection*). Use of primary and secondary antibodies (rather than linking a visualizer to the primary antibody) increases the sensitivity of western blotting. The result is semiquantitative detection of the protein of interest.

Continuing in the same vein, techniques used to identify interactions between DNA and proteins (through protein gel separation and blotting followed by probing with a DNA) are called southwestern blotting. When an RNA probe is used, the technique is called northwestern blotting.

KEY CONCEPTS

- Southern blotting involves the transfer of DNA from a gel to a membrane, followed by detection of specific sequences by hybridization with a labeled probe.
- Northern blotting is similar to Southern blotting but involves the transfer of RNA from a gel to a membrane.
- Western blotting entails separation of proteins on an SDS gel, transfer to a nitrocellulose membrane, and detection of proteins of interest using antibodies.

CONCEPT AND REASONING CHECK

Why does a Southern blot require blocking before a probe is added?

3.10 DNA Microarrays

A logical technical progression from Southern and northern blotting is the microarray. Instead of having the unknown sample on the membrane and the probe in solution, this effectively reverses the two. These originated in the form of "slot-blots" or "dot-blots," where a researcher would spot individual DNA sequences of interest directly onto a hybridization membrane, in an ordered pattern, with each spot consisting of a different, single-known sequence. Drying of the membrane immobilized these spots, creating a premade blotting array. In use, the researcher would then take a nucleic acid sample of interest, such as total cellular DNA, and fragment and randomly and uniformly label this DNA (originally with a radioisotopic label). This labeled mix of sample DNA could then be used, exactly as in a Southern blot, as a probe to hybridize to the premade blot. Labeled DNA sequences homologous to any of the array spots would hybridize and be retained in the known, fixed location of that spot and be visualized by autoradiography. By viewing the autoradiogram and knowing the physical location of each specific probe spot, the pattern of hybridized versus nonhybridized spots could be read out to indicate the presence or absence of each of the corresponding known sequences in the unknown sample.

Technological improvements to this approach followed rapidly through miniaturization of the size and physical density of the immobilized spots, going from membranes with 30 to 100 spots to glass microscope slides with up to 1000 spots. Today, silicon chip substrates have hundreds of thousands (and now up to a million or more) of individual spots in an area about the size of a postage stamp.

In order to visualize the distinct spots in such a high-density array, automated optical microscopy is used and fluorescence has replaced radiolabeling to allow for increased spatial resolution (higher spot density) as well as easier quantification of each hybridization signal. In parallel with the increased total number of spots per array, the length of each unique probe has generally become shorter, allowing for each spot in the array to be specific to a smaller target area—in effect, giving greater "resolution" on a molecular scale. Although the potential applications of microarrays are really limited only by the user's imagination, there are a number of particular applications where they have become standard tools.

The first of these is in gene expression profiling, where a total mRNA sample from a specimen of interest (such as tissue in a disease state or under a particular environmental challenge) is collected and converted en masse to cDNA by a random primed reverse transcription. A label is incorporated into the cDNA during its synthesis (either through use of labeled nucleotides or having the primers themselves with a label); this can either be a fluorophore ("direct labeling") or another hapten (such as biotin), which can at a later stage be exposed to a fluorophore conjugate that will bind the hapten (in the present example, streptavidin—phycoerythrin conjugate might be used) in what is called "indirect labeling." This labeled cDNA is then hybridized to an array where the immobilized spots consist of complementary strands to a number of known mRNAs from the target organism. Hybridization, washing, and visualization allow for the detection of those spots that have bound their complementary labeled cDNA and thus the readout of those genes being expressed in the original sample. This process is depicted in **FIGURE 3.24**. As with Southern blotting, the method is quantitative, meaning that the observed signal on each spot corresponds to the original level of its particular mRNA. Clever selection of the sequence of each of the immobilized spots, such as choosing short probe sequences that are complementary to particular alternate exons of a gene, can even allow the method to differentiate and quantitate the relative levels of alternate splicing products from a single gene. By comparison of the data from such experiments performed in parallel on experimental tissue and control tissue, an experiment can collect a snapshot of the total cellular "global" changes in gene expression patterns, often with useful insight to the state or condition of the experimental tissue.

A second major application is in genotyping. Analysis of the human genome (and other organisms) has led to the identification of large numbers of **single nucleotide polymorphisms (SNPs)**, which are single nucleotide substitutions at a specific genetic locus (see *Section 5.4, RFLPs and SNPs Can Be Used for Genetic Mapping*). Individual SNPs occur at known frequencies, which often differ between populations. The most straightforward examples are where the SNP creates a missense mutation within a gene of interest, such as one involved in metabolism of a drug. People carrying one allele of the SNP may clear a drug from circulation at a very different rate from those with an alternate allele, and thus determination of a patient's allele at this SNP may be an important consideration in choosing an appropriate drug dosage. An example of this that has come all the way from theory into everyday use is CYP450 SNP genotyping to determine appropriate dosage of the anticoagulant warfarin. Another is in SNP genotyping of the K-Ras oncogene in some types of cancer patients in order to determine whether EGFR-inhibitory drugs will be of therapeutic value. Other SNPs may be of no direct biological consequence but can become a valuable genetic marker if found to be closely associated to a particular allele of interest—that is, if in genetic terms it is closely linked. Hundreds of thousands of SNPs have been mapped in the human genome, and arrays that can be probed with a subject's DNA allow for the genotype at each of these to be simultaneously determined, with concurrent determination of what the linked genetic alleles are. In effect, this allows for much of the

▸ **single nucleotide polymorphism (SNP)** A polymorphism or variation in sequence between individuals caused by a change in a single nucleotide. This is responsible of much of the genetic variation between individuals.

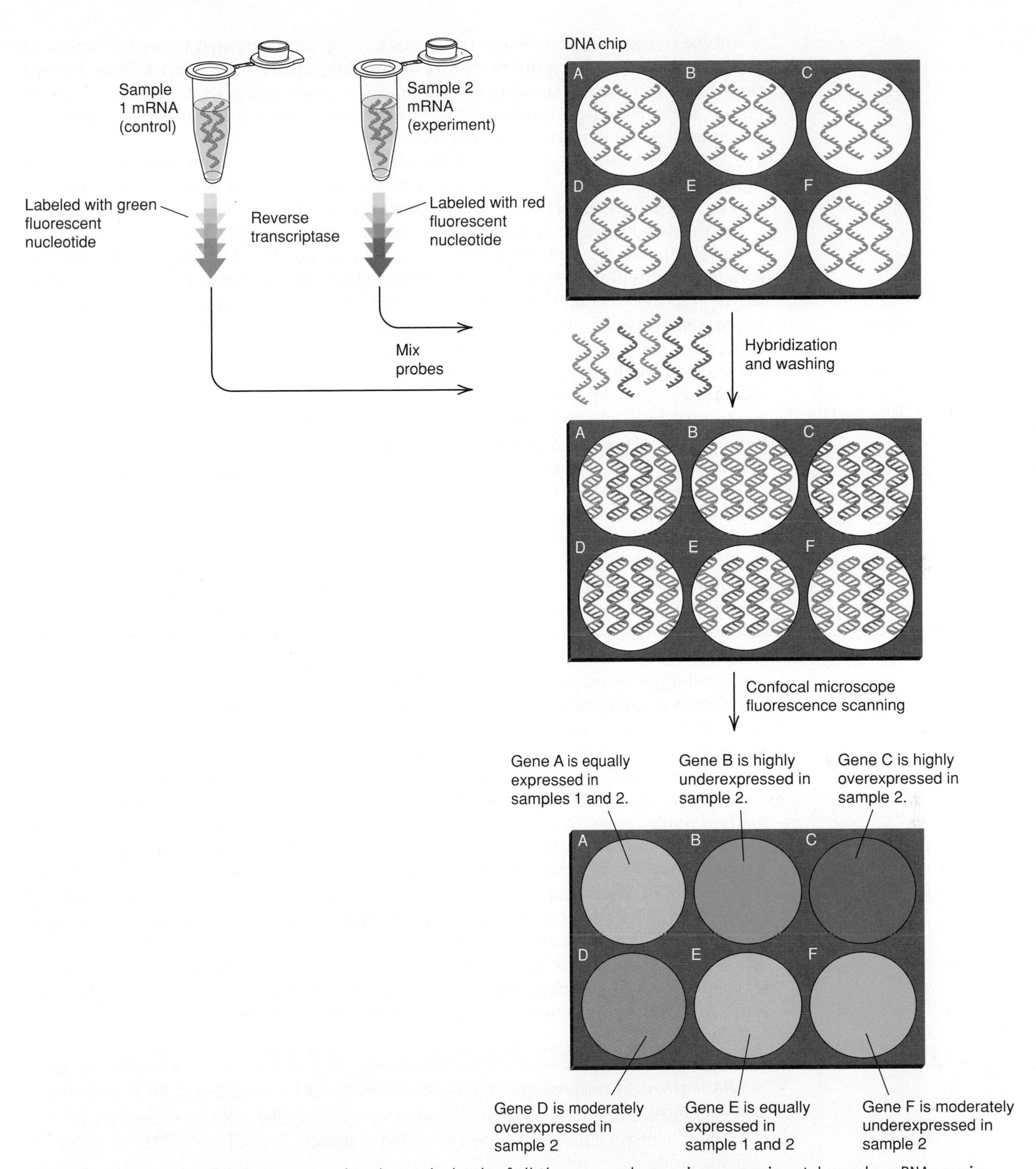

FIGURE 3.24 Gene expression arrays are used to detect the levels of all the expressed genes in an experimental sample. mRNAs are isolated from control and experimental cells or tissues and reverse transcribed in the presence of fluorescently labeled nucleotides (or primers), resulting in labeled cDNAs with different fluorophores (red and green strands) for each sample. Competitive hybridization of the red and green cDNAs to the microarray is proportional to the relative abundance of each mRNA in the two samples. The relative levels of red and green fluorescence are measured by microscopic scanning and displayed as a single color. Red or orange indicates increased expression in the red (experimental) sample, green or yellow-green indicates lower expression, and yellow indicates equal levels of expression in the control and experiment.

genotype of the subject to be inferred from a single experiment at vastly less time and expense than actually sequencing the entire subject genome. With a view toward the future, however, it should be noted that SNP genotyping, in the common case of linked alleles as opposed to direct missense mutation alleles, is indirect inference and has at least some potential for being inaccurate.

Sequencing, on the other hand, is definitive. If emerging sequencing technologies improve to the point of offering an entire human genome in 24 hours for a competitive cost to SNP genotyping, it may move to become the dominant approach for genotyping.

A third major application of DNA microarrays is *array comparative genome hybridization* (*array-CGH*). This is a technique that is augmenting, and in some cases replacing, cytogenetics for the detection and localization of chromosomal abnormalities that change the copy number of a given sequence—that is, deletions or duplications. In this technique, the array chip (known as a **tiling array**) is spotted with an organism's genomic sequences that together represent the entire genome; the higher the density of the array, the smaller genetic region each spot represents and thus the higher resolution the assay can provide. Two DNA samples (one from normal control tissue and one from the tissue of interest) are each randomly labeled with a different fluorophore, such that one sample, for example, is green and the other is red (similar to the mRNA labeling described earlier for the expression arrays). These two differentially labeled specimens are mixed at exactly equal ratios for total DNA and then hybridized to the chip. Regions of DNA that occur equally in the two samples will hybridize equally to their complementary array spots, giving a "mixed" color signal. By comparison, any DNA regions that occur more in one sample than the other will outcompete and thus show a stronger color on its complementary probe spot than the deficient sample will. Computer-assisted image analysis can read out and quantitate small color changes on each array spot and thus detect hemizygous loss of even very small regions in a test sample. The resolution and facility for automation provided by this technique compared to conventional cytogenetics is leading to its increasing adoption in diagnostic settings for the detection of chromosomal copy number changes associated with a range of hereditary diseases.

▸ **tiling array** An array of immobilized nucleic acid sequences, which together represent the entire genome of an organism. The shorter each array sequence is, the larger the total required number of spots is, but the greater the genetic resolution of the array.

Tiling arrays are also often used for chromatin immunoprecipitation (ChIP) studies, which can identify sequences interacting with a DNA-binding protein or complex on a genome-wide scale; this is described in *Section 3.11, Chromatin Immunoprecipitation*.

In addition to the chiplike solid phase arrays described, lower density arrays for focused applications (with up to a few hundred targets, as opposed to millions) can be made in microbead-based formats. In these approaches, each microscopic bead has a distinct optical signal or code, and its surface can be coated with the target DNA sequence. Different bead codes can be mixed and matched into a single sample of labeled sample DNA or cDNA, and then sorted, detected, and quantitated by optical and/or flow sorting methods. Although of much lower density than chip-type arrays, bead arrays can be modified and adapted much more readily to suit a particular focused biological question, and in practice show faster three-dimensional hybridization kinetics than chips, which effectively have two-dimensional kinetics.

KEY CONCEPTS

- DNA microarrays comprise known DNA sequences spotted or synthesized on a small chip.
- Genomewide transcription analysis is performed using labeled cDNA from experimental samples hybridized to a microarray containing sequences from all ORFs of the organism being used.
- SNP arrays permit genomewide genotyping of single nucleotide polymorphisms.
- Array comparative genome hybridization (array-CGH) allow the detection of copy number changes in any DNA sequence compared between two samples.

CONCEPT AND REASONING CHECK

How can microarray technology tell which genes are active in a given cell type?

3.11 Chromatin Immunoprecipitation

Most of the methods discussed thus far in this chapter are *in vitro* methods that allow the detection or manipulation of nucleic acids or proteins that have been isolated from cells (or produced synthetically). Many other powerful molecular techniques have been developed, though, that allow either direct visualization of the *in vivo* behavior of macromolecules (such as imaging of GFP fusions in live cells) or that allow researchers to take a "snapshot" of the *in vivo* localization or interactions of macromolecules at a particular condition or point in time.

Throughout this book, we will discuss numerous proteins that function by interacting directly with DNA, such as chromatin proteins, or the factors that perform replication, repair, and transcription. While much of our understanding of these processes is derived from *in vitro* reconstitution experiments, it is critical to map the dynamics of protein-DNA interactions in living cells in order to fully understand these complex functions. The powerful technique of **chromatin immunoprecipitation (ChIP)** was developed to capture such interactions. (*Chromatin* refers to the native state of eukaryotic DNA *in vivo*, in which it is packaged extensively with proteins; this is discussed in *Chapter 10, Chromatin*.) ChIP allows researchers to detect the presence of any protein of interest at a specific DNA sequence *in vivo*.

> **chromatin immunoprecipitation (ChIP)** A method for detecting *in vivo* protein DNA interactions that entails isolating proteins with an antibody and identifying DNA sequences that are associated with these proteins.

FIGURE 3.25 shows the process of chromatin immunoprecipitation. This method depends on the use of an antibody to detect the protein of interest. As was discussed earlier for western blots (see *Section 3.9, Blotting Methods*), this antibody can be against the protein itself or against an epitope-tagged target.

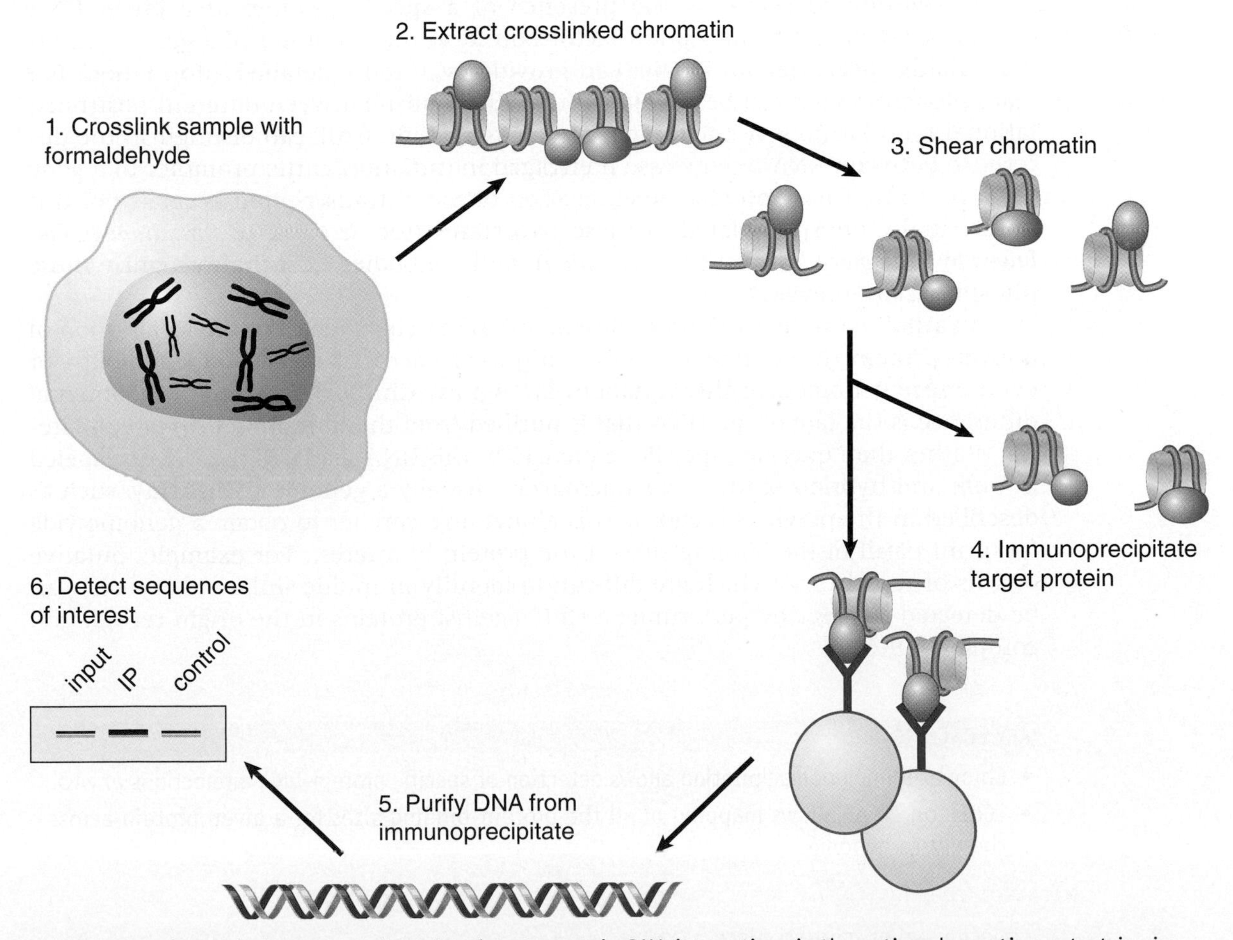

FIGURE 3.25 Chromatin immunoprecipitation detects protein-DNA interactions in the native chromatin context *in vivo*. Proteins and DNA are cross linked, chromatin is broken into small fragments, and an antibody is used to immunoprecipitate the protein of interest. Associated DNA is then purified and analyzed by either identifying specific sequences by PCR (as shown), or by labeling the DNA and applying to a tiling array to detect genomewide interactions.

The first step in ChIP is typically the cross linking of the cell (or tissue or organism) of interest by fixing it with formaldehyde. This serves two purposes: (1) it kills the cell and arrests all ongoing processes at the time of fixation, providing the snapshot of cellular activity; and (2) it covalently links any protein and DNA that are in very close proximity, thus preserving protein-DNA interactions through the subsequent analysis. ChIP can be performed on cells or tissues under different experimental conditions (such as different phases of the cell cycle, or after specific treatments) to look for changes in protein-DNA interactions under different conditions.

After cross linking, the chromatin is then isolated from the fixed material and cleaved into small chromatin fragments, usually 200 to 1000 bp each. This can be achieved by sonication, which uses high-intensity sound waves to nonspecifically shear the chromatin. Nucleases (either sequence specific or nonsequence specific) can be used to fragment the DNA. These small chromatin fragments are then incubated with the antibody against the protein target of interest. These antibodies can then be used to immunoprecipitate the protein by pulling the antibodies out of the solution using heavy beads coated with a protein (such as Protein A) that binds to the antibodies.

After washing away unbound material, the remaining material contains the protein of interest still cross linked to any DNA it was associated with *in vivo*. This is sometimes called a "guilt by association" assay because the DNA target is isolated only due to its interaction with the protein of interest. The final stages of ChIP entail reversal of the cross links so that the DNA can be purified, and detection of specific DNA sequences using PCR or blotting methods. Quantitative (real-time) PCR is usually the method of choice for detecting the DNA.

In addition to revealing the presence of a specific protein at a given DNA sequence (such as a transcription factor bound to the promoter of a gene of interest), highly specialized antibodies can provide even more detailed information. For example, antibodies can be developed that distinguish between different posttranslational modifications of the same protein. As a result, ChIP can distinguish the difference between RNA polymerase II engaged in initiation at the promoter of a gene from pol II that has entered the elongation phase of transcription because pol II is differentially phosphorylated in these two states (see *Section 20.8, Initiation Is Followed by Promoter Clearance and Elongation*), and antibodies exist that recognize these phosphorylation events.

A variation on the ChIP procedure allows researchers to query the localization of a given protein (or modified version of a protein) across large genomic regions—or even entire genomes. In this variation, known as "ChIP on chip," the fundamental difference is the fate of the DNA that is purified from the immunoprecipitated material. Rather than querying specific sequences in this DNA via PCR, the DNA is labeled in bulk and hybridized to a DNA microarray (usually a genome tiling array, such as described in the previous section). This allows a researcher to obtain a genomewide footprint of all of the binding sites of the protein of interest. For example, putative origins of replication (which are difficult to identify in multicellular eukaryotes) can be detected *en masse* by performing a ChIP against proteins in the origin recognition complex (ORC).

KEY CONCEPTS

- Chromatin immunoprecipitation allows detection of specific protein-DNA interactions *in vivo*.
- "ChIP on chip" allows mapping of all the protein-binding sites for a given protein across the entire genome.

CONCEPT AND REASONING CHECK

How does chromatin immunoprecipitation allow the development of a profile of all the binding sites for a given protein?

3.12 Gene Knockouts and Transgenics

An organism that gains new genetic information from the addition of foreign DNA is described as **transgenic**. For simple organisms, such as bacteria or yeast, it is easy to generate transgenics by transformation with DNA constructs containing sequences of interest. Transgenesis in multicellular organisms, however, can be much more challenging.

▸ **transgenic** An organism created by introducing DNA prepared in test tubes into the germline. The DNA may be inserted into the genome or exist in an extrachromosomal structure.

The approach of directly injecting DNA can be used with mouse eggs, as shown in **FIGURE 3.26**. Plasmids carrying the gene of interest are injected into the nucleus of the oocyte or into the pronucleus of the fertilized egg. The egg is implanted into a pseudopregnant mouse (a mouse that has mated with a vasectomized male to trigger a receptive state). After birth, the recipient mouse can be examined to see whether it has gained the foreign DNA and, if so, whether it is expressed. Typically, a minority (~15%) of the injected mice carry the transfected sequence. In general, multiple copies of the plasmid appear to have been integrated in a tandem array into a single chromosomal site. The number of copies varies from 1 to 150, and they are inherited by the progeny of the injected mouse. The levels of gene expression from *transgenes* introduced in this way is highly variable, both due to copy number and the site of integration. A gene may be highly expressed if it integrates within an active chromatin domain, but not if it integrates in or near a silenced region of the chromosome.

Transgenesis with novel or mutated genes can be used to study genes of interest in the whole animal. In addition, defective genes can be replaced by functional genes using transgenic techniques. One example is the cure of the defect in the *hypogonadal* mouse. The *hpg* mouse has a deletion that removes the distal part of the gene coding for the precursor to GnRH (gonadotropin-releasing hormone) and GnRH-associated peptide (GAP). As a result, the mouse is infertile. When an intact *hpg* gene is introduced into the mouse by transgenic techniques, it is expressed in the appropriate tissues. **FIGURE 3.27** summarizes experiments to introduce a transgene into a line of *hpg–/–* homozygous mutant mice. The resulting progeny are normal. This provides a striking demonstration that expression of a transgene under normal regulatory control can be indistinguishable from the behavior of the normal allele.

Although promising, there are impediments to using such techniques to cure human genetic defects. The transgene must be introduced into the germline of the *preceding* generation, the ability to express a transgene is not predictable, and an adequate level of expression of a transgene may be obtained in only a small minority of the transgenic individuals. In addition, the large number of transgenes that may

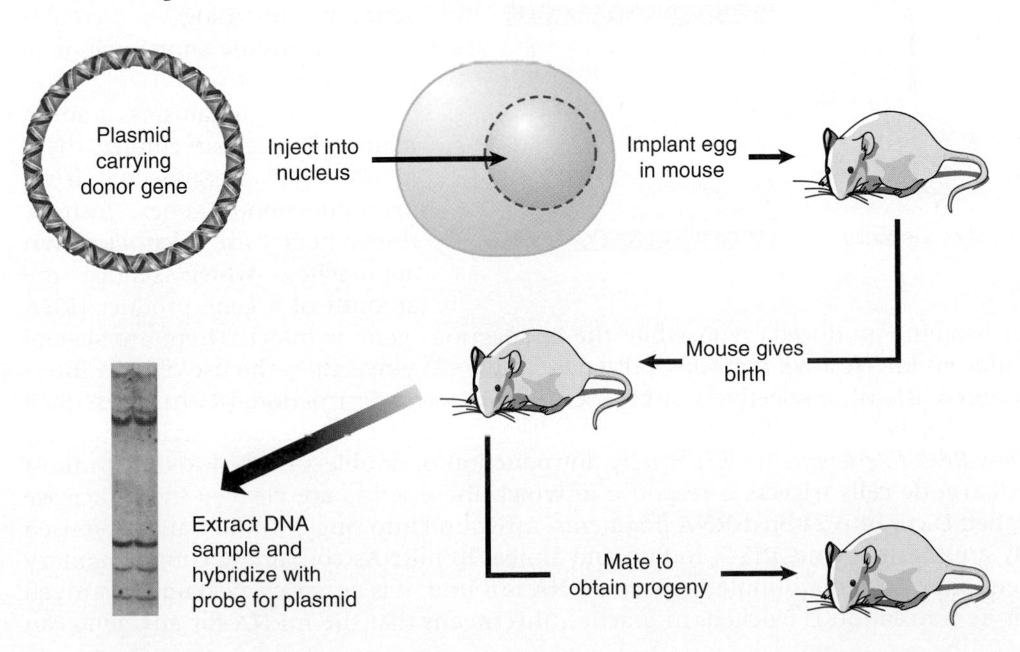

FIGURE 3.26 Transfection can introduce DNA directly into the germline of animals. Photo reproduced from P. Chambon, *Sci. Am.* 244 (1981): 60–71. Used with permission of Pierre Chambon, Institute of Genetics and Molecular and Cellular Biology, College of France.

FIGURE 3.27 Hypogonadism can be averted in the progeny of *hpg* mice by introducing a transgene that has the wild-type sequence.

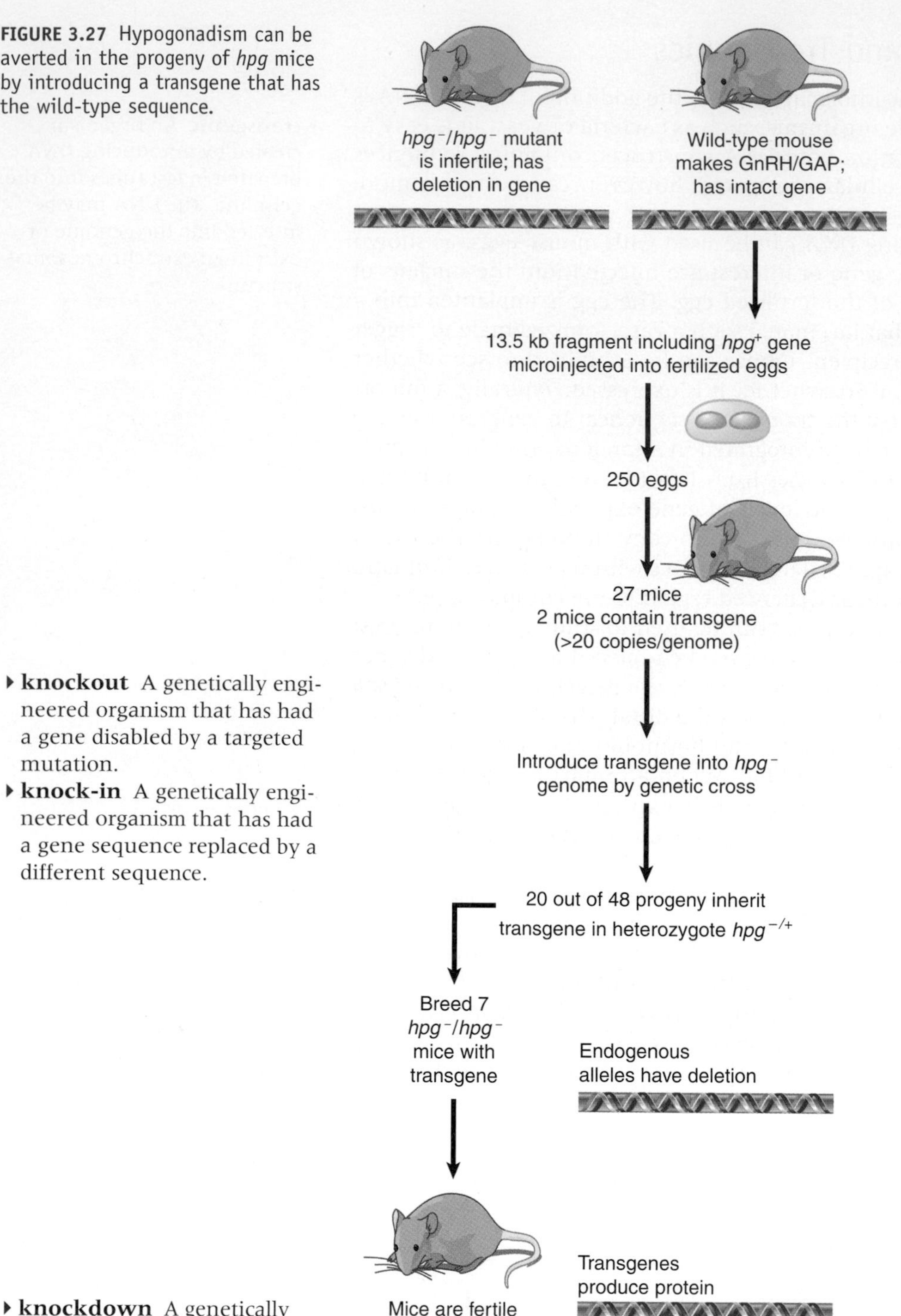

be introduced into the germline and their erratic expression could pose problems in cases in which overexpression of the transgene is harmful. In other cases, the transgene can integrate near an oncogene and activate it, promoting carcinogenesis.

A more versatile approach for studying the functions of genes is to eliminate the gene of interest. Transgenesis methods allow DNA to be *added* to cells or animals, but in order to understand the function of a gene, it is most useful to be able to *remove* the gene or its function and observe the resulting phenotype. The most powerful techniques for changing the genome use *gene targeting* to delete or replace genes by homologous recombination. Gene deletions are usually referred to as **knockouts**, whereas replacement of a gene with an alternative mutated version is called a **knock-in**.

- **knockout** A genetically engineered organism that has had a gene disabled by a targeted mutation.
- **knock-in** A genetically engineered organism that has had a gene sequence replaced by a different sequence.

In simple organisms such as yeast, this is again a very simple process in which DNA encoding a selectable marker flanked by short regions of homology to a target gene is transformed into the yeast. As little as 40 bp or so of homology will result in extremely efficient replacement of the target gene by the introduced marker gene, via homologous recombination using the short regions of homology.

In some organisms, and in mammalian cells in culture, there is no good method for deleting endogenous genes. Instead, researchers use **knockdown** approaches, which reduce the amount of a gene product (RNA or protein) produced, even while the endogenous gene is intact. There are several different knockdown methods, but one of the most powerful is the use of RNA interference (RNAi) to selectively target specific mRNAs for destruction. (RNAi is described in *Section 30.5, MicroRNAs Are Widespread Regulators in Eukaryotes*, and *Section 30.6, How Does RNA Interference Work?*) Briefly, introduction of double-stranded RNA into most eukaryotic cells triggers a response in which these RNAs are cleaved by a nuclease called Dicer into 21 bp dsRNA fragments, unwound into single strands, and then used by another enzyme, RISC, to find and anneal to mRNAs containing complementary sequence. When a complementary mRNA is found, it is either cleaved and destroyed or its translation is blocked. In practice, this means that the mRNA for any gene can

- **knockdown** A genetically engineered organism that has had a gene downregulated by introducing a silencing vector to reduce the expression (usually translation) of the gene.

be targeted for silencing by introduction of a dsRNA designed to anneal to the target of interest. The means of introducing the dsRNA depends on the species being targeted; in mammalian cells one method is transfection with DNA encoding a self-annealing RNA that forms a hairpin contain the targeting sequence.

In some multicellular organisms gene deletion is possible, but the process is more complicated than in organisms like yeast. In mammals, the target is usually the genome of an embryonic stem (ES) cell, which is then used to generate a mouse with the knockout. ES cells are derived from the mouse blastocyst (an early stage of development, which precedes implantation of the egg in the uterus). **FIGURE 3.28** illustrates the general approach.

ES cells are transfected with DNA in the usual way (most often by microinjection or electroporation). By using a donor that carries an additional sequence, such as a drug-resistance marker or some particular enzyme, it is possible to select ES cells that have obtained an integrated transgene carrying any particular donor trait. This results in a population of ES cells in which there is a high proportion carrying the marker.

These ES cells are then injected into a recipient blastocyst. The ability of the ES cells to participate in normal development of the blastocyst forms the basis of the technique. The blastocyst is implanted into a foster mother and in due course develops into a *chimeric* mouse. Some of the tissues of the chimeric mice are derived from the cells of the recipient blastocyst; other tissues are derived from the injected ES cells. The proportions of tissues in the adult mouse that are derived from cells in the recipient blastocyst and from injected ES cells vary widely in individual progeny; if a visible

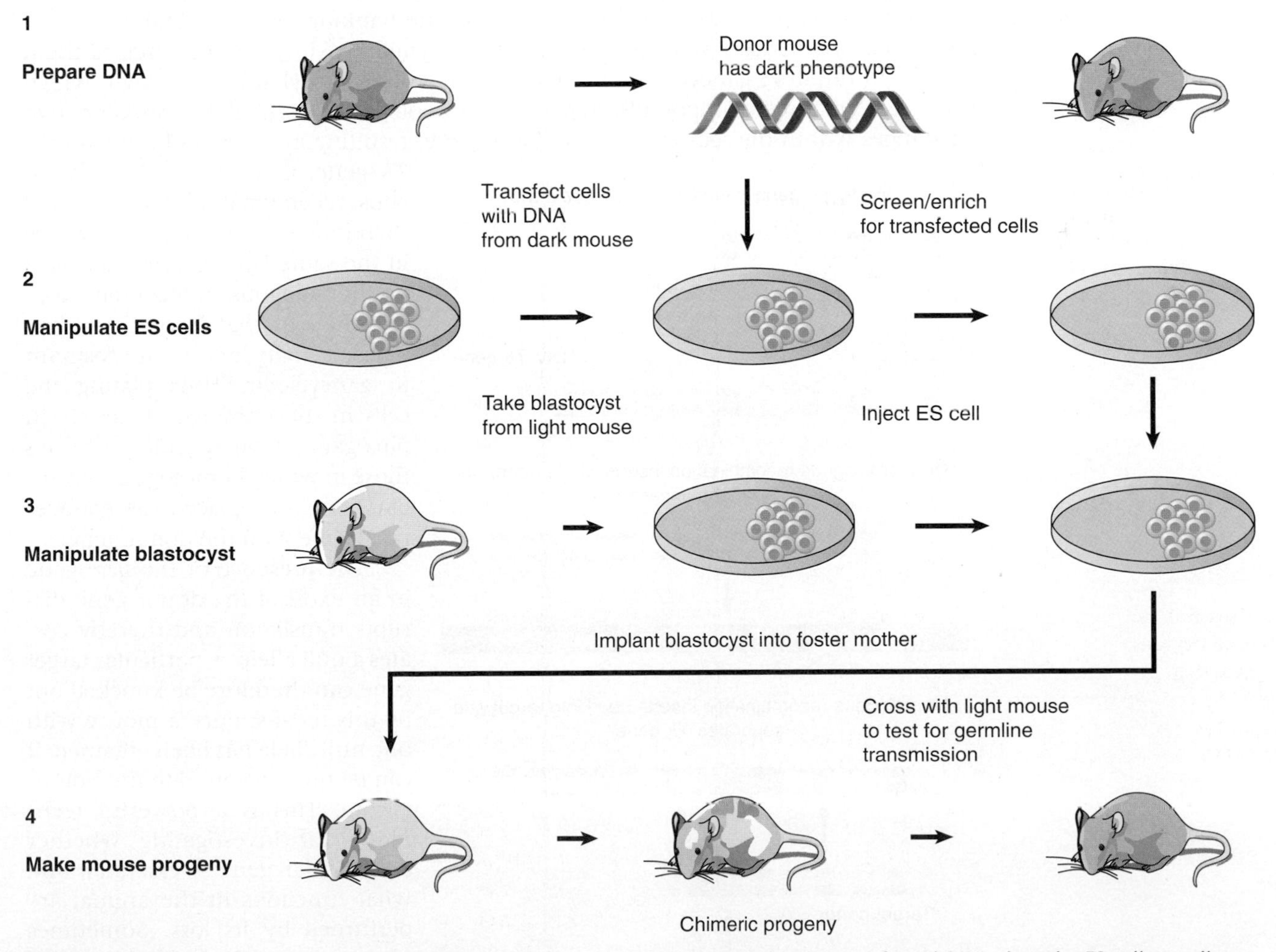

FIGURE 3.28 ES cells can be used to generate mouse chimeras, which breed true for the transfected DNA when the ES cell contributes to the germline.

marker (such as coat-color gene) is used, areas of tissue representing each type of cell can be seen.

To determine whether the ES cells contributed to the germline, the chimeric mouse is crossed with a mouse that lacks the donor trait. Any progeny that have the trait must be derived from germ cells that have descended from the injected ES cells. By this means, it is known that an entire mouse has been generated from an original ES cell!

When a donor DNA is introduced into the cell, it may insert into the genome by either nonhomologous or homologous recombination. Homologous recombination is relatively rare, probably representing <1% of all recombination events and thus occurring at a frequency of $\sim 10^{-7}$. By designing the donor DNA appropriately, though, we can use selective techniques to identify those cells in which homologous recombination has occurred.

FIGURE 3.29 illustrates the knockout technique that is used to disrupt endogenous genes. The basis for the technique is the design of a knockout construct with two different markers that are designed to allow nonhomologous and homologous recombination events in the ES cells to be distinguished. The donor DNA is homologous to a target gene but has two key modifications. First, the gene is inactivated by interrupting or replacing an exon with a gene encoding a selectable marker (most often the *neo*^{R} gene that confers resistance to the drug G418 is used). Second, a *counterselectable* marker (a gene that can be selected *against*) is added on one side of the gene—for example, the *TK* gene of the herpes virus.

When this knockout construct is introduced into an ES cell, homologous and nonhomologous recombinations will result in different outcomes. Nonhomologous recombination inserts the entire construct, including the flanking *TK* gene. These cells are resistant to neomycin, and they also express thymidine kinase, which makes them *sensitive* to the drug gancyclovir (thymidine kinase phosphorylates gancyclovir, which converts it to a toxic compound). In contrast, homologous recombination involves two exchanges within the sequence of the donor gene, resulting in the loss of the flanking *TK* gene. Cells in which homologous recombination has occurred therefore gain neomycin resistance in the same way as cells that have nonhomologous recombination, but they do *not* have thymidine kinase activity, and so are resistant to gancyclovir. Thus plating the cells in the presence of neomycin plus gancyclovir specifically selects those in which homologous recombination has replaced the endogenous gene with the donor gene.

The presence of the *neo*^{R} gene in an exon of the donor gene disrupts translation and thereby creates a null allele. A particular target gene can therefore be knocked out by this means; once a mouse with one null allele has been obtained, it can be bred to generate the homozygote. This is a powerful technique for investigating whether a particular gene is essential, and what functions in the animal are perturbed by its loss. Sometimes phenotypes can even be observed in the heterozygote.

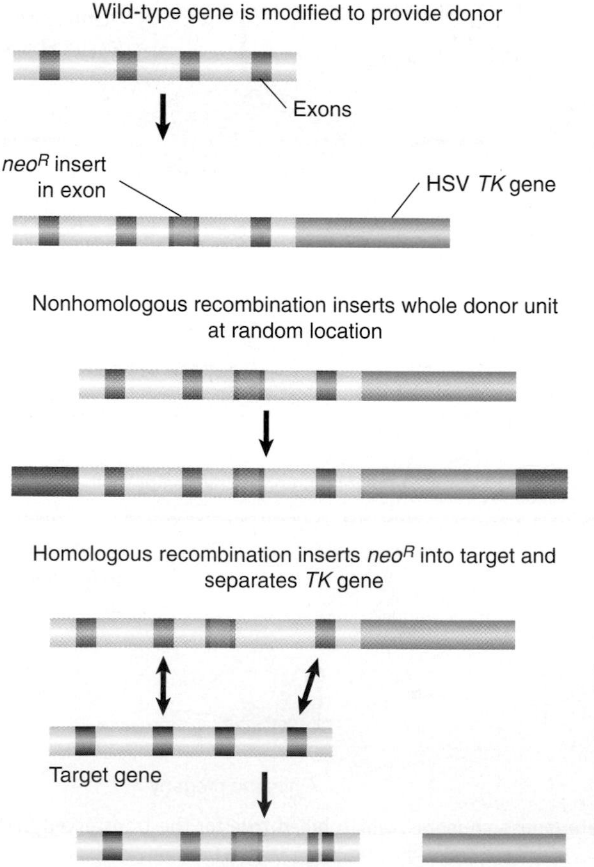

FIGURE 3.29 A transgene containing *neo*^{R} within an exon and *TK* downstream can be selected by resistance to G418 and loss of *TK* activity.

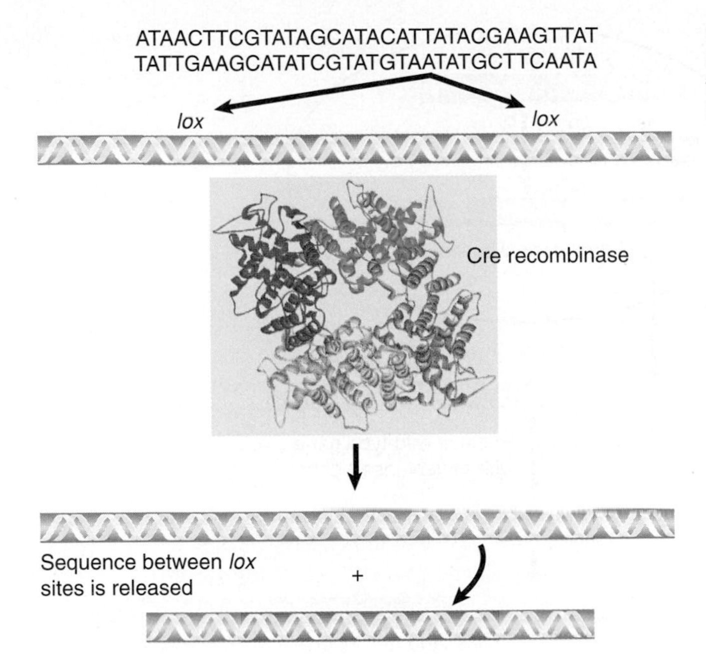

FIGURE 3.30 The Cre recombinase catalyzes a site-specific recombination between two identical *lox* sites, releasing the DNA between them. Structure from Protein Data Bank: 1OUQ. E. Ennifar et al., *Nucleic Acids Res.* 31 (2003): 5449–5460.

A major extension of ability to manipulate a target genome has been made possible by using the phage Cre/*lox* system to engineer site-specific recombination in a eukaryotic cell. The Cre enzyme catalyzes a site-specific recombination reaction between two *lox* sites, which are identical 34-bp sequences (see *Section 15.8, Site-Specific Recombination Resembles Topoisomerase Activity*). **FIGURE 3.30** shows that the consequence of the reaction is to excise the stretch of DNA between the two *lox* sites.

The great utility of the Cre/*lox* system is that it requires no additional components and works when the Cre enzyme is produced in any cell that has a pair of *lox* sites. **FIGURE 3.31** shows that we can control the reaction to make it work in a particular cell by placing the *cre* gene under the control of a regulated promoter. The procedure starts with two mice. One mouse has the *cre* gene, typically controlled by a promoter that can be turned on specifically in a certain cell or under certain conditions. The other mouse has a target sequence flanked by *lox* sites. When we cross the two mice, the progeny have both elements of the system; and the system can be turned on by controlling the promoter of the *cre* gene. This allows the sequence between the *lox* sites to be excised in a controlled way.

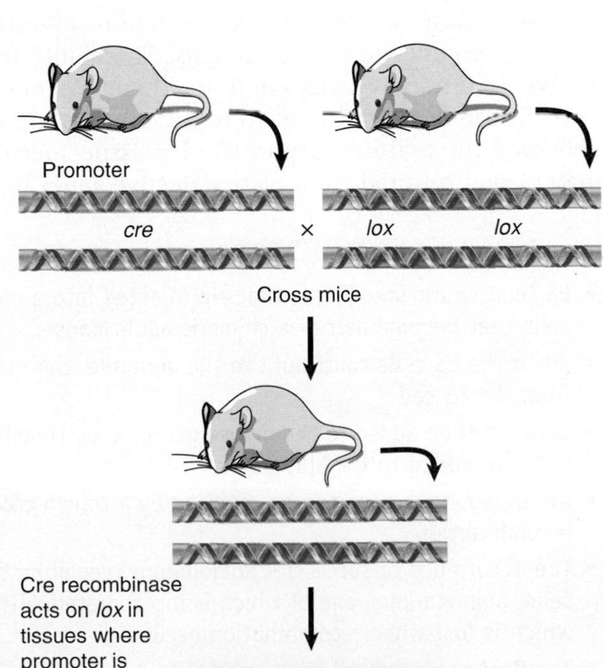

FIGURE 3.31 By placing the Cre recombinase under the control of a regulated promoter, it is possible to activate the excision system only in specific cells. One mouse is created that has a promoter-*cre* construct and another that has a target sequence flanked by *lox* sites. The mice are crossed to generate progeny that have both constructs. Then excision of the target sequence can be triggered by activating the promoter.

The Cre/*lox* system can be combined with the knockout technology to give us even more control over the genome. Inducible knockouts can be made by flanking the neo^R gene (or any other gene that is used similarly in a selective procedure) with *lox* sites. After the knockout has been made, the target gene can be reactivated by causing Cre to excise the neo^R gene in some particular circumstance (such as in a specific tissue).

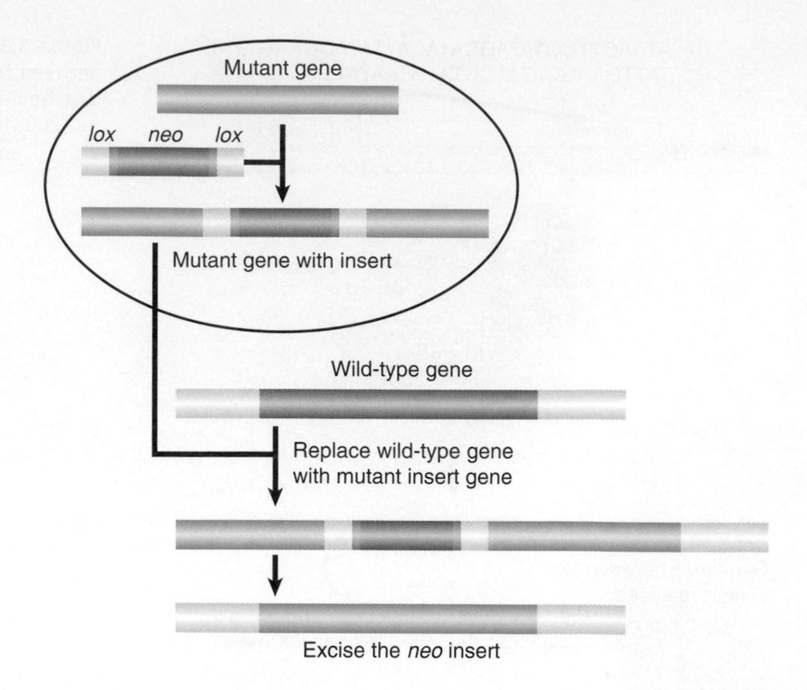

FIGURE 3.32 An endogenous gene is replaced in the same way as when a knockout is made (see Figure 3.29), but the neomycin gene is flanked by *lox* sites. After the gene replacement has been made using the selective procedure, the neomycin gene can be removed by expressing Cre recombinase, leaving an active insert.

FIGURE 3.32 shows a modification of this procedure that allows a knock-in to be created. Basically, we use a construct in which some mutant version of the target gene is used to replace the endogenous gene, replying on the usual selective procedures. Then, when the inserted gene is reactivated by excising the neo^R sequence, we have in effect replaced the original gene with a different version.

A useful variant of this method is to introduce a wild-type copy of the gene of interest in which the gene itself (or one of its exons) is flanked by *lox* sites. This results in a normal animal that can be crossed to a mouse containing *cre* under control of a tissue-specific or otherwise regulated promoter. The offspring of this cross are *conditional knockouts*, in which the function of the gene is lost only in cells that express the Cre recombinase. This is particularly useful for studying genes that are essential for embryonic development; genes in this class would be lethal in homozygous embryos and thus are very difficult to study.

With these techniques, we are able to investigate the functions and regulatory features of genes in whole animals. The ability to introduce DNA into the genome allows us to make changes in it, to add new genes that have had particular modifications introduced *in vitro*, or to inactivate existing genes. Thus it becomes possible to delineate the features responsible for tissue-specific gene expression. Ultimately, we may expect routinely to replace defective genes in the genome in a targeted manner.

KEY CONCEPTS

- ES (embryonic stem) cells that are injected into a mouse blastocyst generate descendant cells that become part of a chimeric adult mouse.
- When the ES cells contribute to the germline, the next generation of mice may be derived from the ES cell.
- Genes can be added to the mouse germline by transfecting them into ES cells before the cells are added to the blastocyst.
- An endogenous gene can be replaced by a transfected gene using homologous recombination.
- The occurrence of successful homologous recombination can be detected by using two selectable markers, one of which is incorporated with the integrated gene, the other of which is lost when recombination occurs.
- The Cre/*lox* system is widely used to make inducible knockouts and knock-ins.

CONCEPT AND REASONING CHECK

What is the advantage of being able to create a conditional knockout in a selected tissue or at a particular time in development?

3.13 Summary

DNA can be manipulated and propagated using the techniques of cloning. These include digestion by restriction endonucleases, which cut DNA at specific sequences, and insertion into cloning vectors, which permit DNA to be maintained and amplified in host cells such as bacteria. Cloning vectors can have specialized functions as well, such as allowing expression of the product of a gene of interest, or fusion of a promoter of interest to an easily assayed reporter gene.

DNA (and RNA) can be detected nonspecifically by the use of dyes that bind independently of sequence. Specific nucleic acid sequences can be detected using base complementarity. Specific primers can be used to detect and amplify particular DNA targets via PCR. RNA can be reverse transcribed into DNA to be used in PCR; this is known as reverse transcription (RT)-PCR. Labeled probes can be used to detect DNA or RNA on Southern or northern blots, respectively. Proteins are detected on western blots using antibodies.

DNA microarrays are solid supports (usually silicon chips or glass slides) on which DNA sequences corresponding to ORFs or complete genomic sequences are arrayed. Microarrays are used to detect gene expression, for SNP genotyping, and to detect changes in DNA copy number, as well as many other applications.

Protein-DNA interactions can be detected *in vivo* using chromatin immunoprecipitation. The DNA obtained in a chromatin immunoprecipitation experiment can be used as a probe on a genome tiling array to map all localization sites for a given protein in the genome.

New sequences of DNA may be introduced into a cultured cell by transfection or into an animal egg by microinjection. The foreign sequences may become integrated into the genome, often as large tandem arrays. The array appears to be inherited as a unit in a cultured cell. The sites of integration appear to be random. A transgenic animal arises when the integration event occurs into a genome that enters the germ cell lineage. Often a transgene responds to tissue and temporal regulation in a manner that resembles the endogenous gene. Under conditions that promote homologous recombination, an inactive sequence can be used to replace a functional gene, thus creating a knockout, or deletion, of the target locus. Extensions of this technique can be used to make conditional knockouts, where the activity of the gene can be turned on or off (such as by Cre-dependent recombination), and knock-ins, where a donor gene specifically replaces a target gene. Transgenic mice can be obtained by injecting recipient blastocysts with ES cells that carry transfected DNA. Knockdowns, mostly commonly achieved using RNA interference, can be used to eliminate gene products in cell types for which knockout technologies are not available.

CHAPTER QUESTIONS

1. Restriction enzymes that generate sticky ends:
   - **A.** make a single cut in the DNA backbone.
   - **B.** make two cuts in the DNA backbone in the same strand.
   - **C.** make two staggered cuts in the DNA backbone.
   - **D.** make two cuts in the DNA backbone directly across the double helix from each other.

2. Recombinant plasmids are usually introduced into bacterial cells by:
   A. gene gun.
   B. transformation of chemically treated cells.
   C. transfection using liposomes.
   D. microinjection.
3. The best vector for cloning very large (>1 Mb) fragments of DNA is:
   A. plasmid.
   B. bacteriophage.
   C. cosmid.
   D. YAC.
4. Reporter genes are:
   A. genes that produce products that are easy to detect and/or quantify.
   B. genes that are essential in all species.
   C. genes that never occur in nature.
   D. all of the above.
5. A knockout construct does not usually contain:
   A. regions homologous to an endogenous gene.
   B. a selectable marker.
   C. a marker for counterselection.
   D. a reporter gene.
6. Put the following events in the generation of a knockout mouse in the correct order:
   A. breed heterozygotes to generate homozygous null offspring.
   B. inject knockout construct into ES cells.
   C. breed chimeric offspring to obtain heterozygotes from ES-derived germ cells.
   D. implant blastocyst into foster mother and obtain offspring.
   E. make knockout construct to target gene of interest.
   F. introduce ES cells to blastocyst.
   G. select for ES cells that have integrated knockout construct via homologous recombination.

KEY TERMS

amplicon
autoradiography
chromatin immunoprecipitation (ChIP)
cloning
cloning vector
complementary
cosmid
dideoxynucleotide (ddNTP)
epitope tag
endonuclease
exonuclease
expression vector
fluorescence resonant energy transfer (FRET)
hybridization
insert
knockdown
knock-in
knockout
ligate
multiple cloning site
nuclease
phosphatase
polymerase chain reaction (PCR)
primer
probe
quantitative PCR (qPCR)
real time PCR
recombinant DNA
reporter gene
restriction endonuclease
reverse transcription PCR (RT-PCR)
shuttle vector
single nucleotide polymorphism (SNP)
Southern blotting
stringency
subclone
threshold cycle (C_T)
tiling array
T_m
transformation
transgene
vector
yeast artificial chromosome (YAC)

4

A self-splicing intron in an rRNA of the large ribosomal subunit. © Kenneth Eward/Photo Researchers, Inc.

The Interrupted Gene

CHAPTER OUTLINE

4.1 Introduction

The simplest form of a gene is a length of DNA that directly corresponds to its polypeptide product. Bacterial genes are almost always of this type, in which a continuous coding sequence of $3N$ base pairs encodes a polypeptide of N amino acids. In eukaryotes, however, a gene may include additional sequences that lie within the coding region and interrupt the sequence that encodes the polypeptide. These sequences are removed from the RNA product following transcription, generating an mRNA that includes a nucleotide sequence exactly corresponding to the polypeptide product according to the rules of the genetic code.

The sequences of DNA comprising an **interrupted gene** are divided into the two categories depicted in **FIGURE 4.1**.

- **Exons** are the sequences retained in the mature RNA product. By definition, a gene begins and ends with exons that correspond to the 5′ and 3′ ends of the RNA.
- **Introns** are the intervening sequences that are removed when the primary transcript is processed to give the mature RNA product.

The exon sequences are in the same order in the gene and in the RNA, but an interrupted gene is longer than its final RNA product because of the presence of the introns.

The processing of interrupted genes requires an additional step that is not needed for uninterrupted genes. The DNA of an interrupted gene is transcribed to an RNA copy (a transcript) that is exactly complementary to the original gene sequence. This RNA is only a precursor, though; it cannot yet be used to produce a polypeptide. First, the introns must be removed from the RNA to give a messenger RNA that consists only of the series of exons. This process is called **RNA splicing** (see *Section 2.10, Several Processes Are Required to Express the Product of a Gene*) and involves precisely deleting the introns from the primary transcript and then joining the ends of the RNA on either side to form a covalently intact molecule (see *Chapter 21, RNA Splicing and Processing*).

The original gene comprises the region in the genome between the points corresponding to the 5′ and 3′ terminal bases of the mature mRNA. We know that transcription starts at the DNA template corresponding to the 5′ end of the mRNA and usually extends beyond the complement to the 3′ end of the mature mRNA, which is generated by cleavage of the primary RNA transcript (see *Section 21.14, The 3′ Ends of mRNAs Are Generated by Cleavage and Polyadenylation*). The gene is also considered to include the regulatory regions on both sides of the gene that are required for initiating and (sometimes) terminating transcription.

▸ **interrupted gene** A gene in which the coding sequence is not continuous due to the presence of introns.

▸ **exon** Any segment of an interrupted gene that is represented in the mature RNA product.

▸ **intron** A segment of DNA that is transcribed but later removed from within the transcript by splicing together the sequences (exons) on either side of it.

▸ **RNA splicing** The process of excising introns from RNA and connecting the exons into a continuous mRNA.

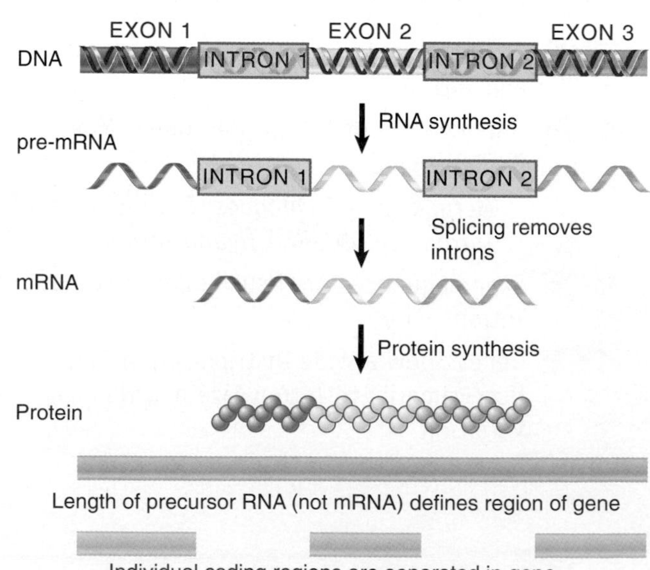

FIGURE 4.1 Interrupted genes are expressed via a precursor RNA. Introns are removed when the exons are spliced together. The mature mRNA has only the sequences of the exons.

CONCEPT AND REASONING CHECK

Why might it be difficult to express an intact human gene inserted into a bacterial cell?

4.2 An Interrupted Gene Consists of Exons and Introns

How does the existence of introns change our view of the gene? During splicing, the exons are always joined together in the same order they are found in the original DNA, so the correspondence between the gene and polypeptide sequences is maintained. **FIGURE 4.2** shows that the order of exons in the gene remains the same as the order of exons in the processed mRNA, but the distances between sites in the gene (as determined by recombination analysis) do not correspond to the distances between sites in the processed mRNA. The length of the gene is defined by the length of the primary RNA transcript instead of the length of the processed mRNA.

All the exons are present in the primary RNA transcript, and their splicing together occurs only as an intramolecular reaction. There is usually no joining of exons carried by different RNA transcripts, so the splicing mechanism excludes any splicing together of sequences from different alleles. (However, in a phenomenon known as *trans*-splicing, sequences from different mRNAs are ligated together into a single molecule for translation.) Mutations located in different exons of a gene cannot complement one another, so they continue to be defined as members of the same complementation group.

Mutations that directly affect the sequence of a polypeptide must occur in exons. What are the effects of mutations in the introns? The introns are not part of the processed messenger RNA, so mutations in them cannot directly affect the polypeptide sequence. However, they may affect the processing of the messenger RNA by inhibiting the splicing together of exons. A mutation of this sort acts only on the allele that carries it. As a result, it fails to complement any other mutation in that allele and is part of the same complementation group as the exons.

Mutations that affect splicing are usually deleterious. The majority are single-base substitutions at the junctions between introns and exons. They may cause an exon to be left out of the product, cause an intron to be included, or make splicing occur at a different site. The most common outcome is a termination codon that truncates the polypeptide sequence. About 15% of the point mutations that cause human diseases disrupt splicing.

Some eukaryotic genes are not interrupted, and, like prokaryotic genes, correspond directly with the polypeptide product. In yeast, most genes are uninterrupted. In multicellular eukaryotes, most genes are interrupted, and introns are usually much longer than exons, so that genes are considerably larger than their coding regions. (See Figure 4.6 for the example of the mammalian β-globin genes, in which exons can be ~37% of the total length of the gene.)

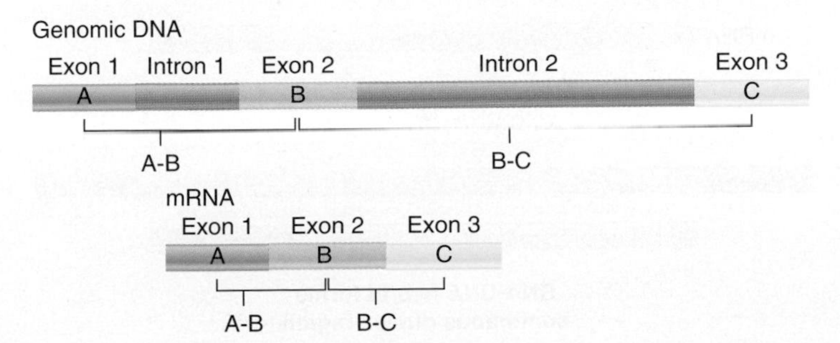

FIGURE 4.2 Exons remain in the same order in mRNA as in DNA, but distances along the gene do not correspond to distances along the mRNA or polypeptide products. The distance A-B in the gene is smaller than the distance B-C; but the distance A-B in the mRNA (and polypeptide) is greater than the distance B-C.

KEY CONCEPTS

- Introns are removed by the process of RNA splicing.
- Only mutations in exons can affect polypeptide sequence; however, mutations in introns can affect processing of the RNA and therefore prevent production of polypeptide.

CONCEPT AND REASONING CHECK

Describe the types of mutations that lead to abnormal splicing and their specific effects.

4.3 Organization of Interrupted Genes May Be Conserved

The characterization of eukaryotic genes was first made possible by the development of techniques for physically mapping DNA. When an mRNA is compared with the DNA sequence from which it was transcribed, the DNA sequence turns out to have extra regions that are not represented in the mRNA.

One technique for comparing mRNA with genomic DNA is to hybridize the mRNA with the complementary strand of the DNA. If the two sequences are colinear, a duplex is formed. **FIGURE 4.3** shows a typical result when an RNA transcribed from an uninterrupted gene is hybridized with a DNA that includes the gene. The sequences on either side of the gene are not represented in the RNA, but the DNA sequence of the gene hybridizes with the RNA to form a continuous duplex region.

Suppose we now perform the same experiment with RNA transcribed from an interrupted gene. The difference is that the sequences represented in the mRNA lie on either side of a sequence that is not in the mRNA. **FIGURE 4.4** shows that the RNA-DNA hybrid forms a duplex, but the unhybridized DNA sequence in the middle remains single stranded, forming a loop that extrudes from the duplex. The hybridizing regions correspond to the exons, and the extruded loop corresponds to the intron.

The structure of the mRNA-DNA hybrid can be visualized by electron microscopy. One of the very first examples of the visualization of an interrupted gene is

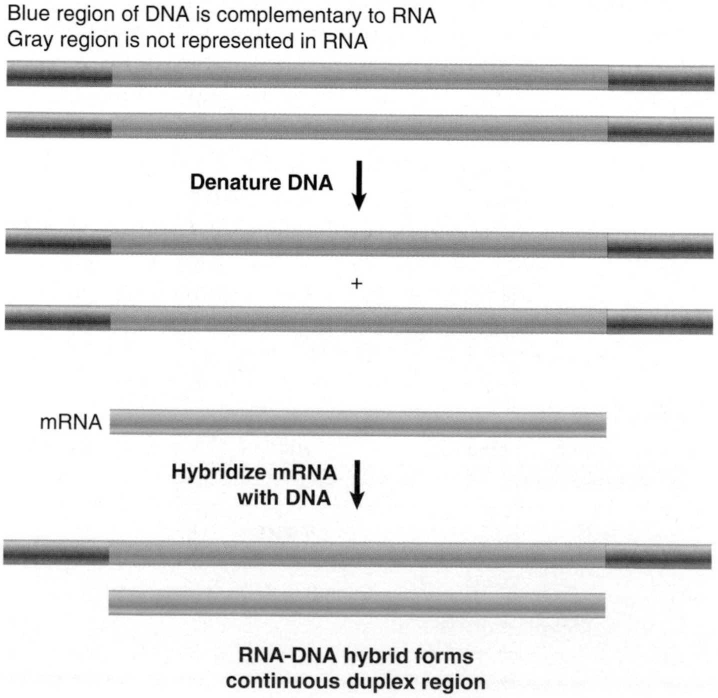

FIGURE 4.3 Hybridizing an mRNA from an uninterrupted gene with the DNA of the gene generates a duplex region corresponding to the gene.

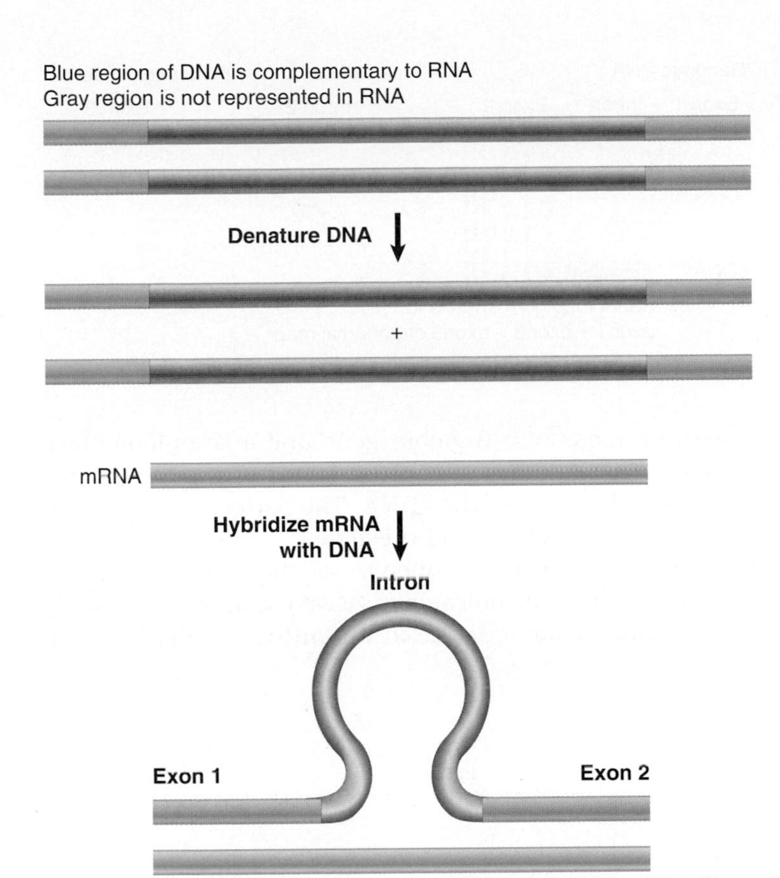

FIGURE 4.4 RNA hybridizing with the DNA from an interrupted gene produces a duplex corresponding to the exons, with an intron excluded as a single-stranded loop between the exons.

shown in **FIGURE 4.5**. On the right side of the figure, tracing the structure shows that three introns are located close to the very beginning of the gene.

When a gene is uninterrupted, the restriction map of its DNA corresponds exactly with the map of its mRNA. When a gene has introns, the maps of the gene and its mRNA are different except at each end, corresponding to the first and last exon. The gene map will have additional regions due to the presence of introns.

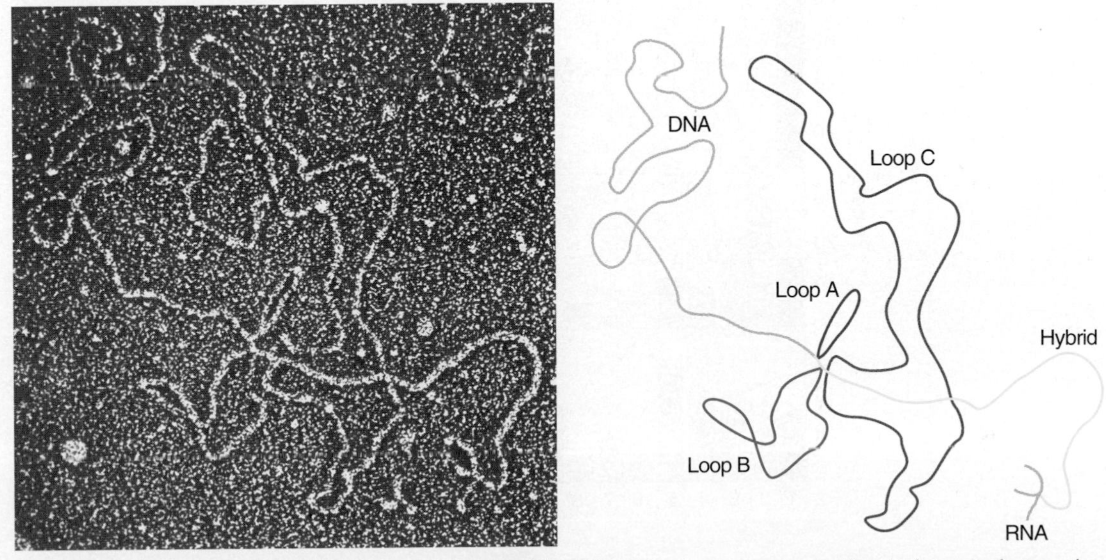

FIGURE 4.5 Hybridization between an adenovirus mRNA and its DNA identifies three loops corresponding to introns that are located at the beginning of the gene. Photo reproduced from S. M. Berget, C. Moore, and P. A. Sharp, *Proc. Natl. Acad. Sci. USA* 74 (1977): 3171–3175. Used with permission of Philip Sharp, Koch Institute for Integrative Cancer Research, Massachusetts Institute of Technology.

FIGURE 4.6 Comparison of the restriction maps of cDNA and genomic DNA for mouse β globin shows that the gene has two introns that are not present in the cDNA. The exons can be aligned exactly between cDNA and gene.

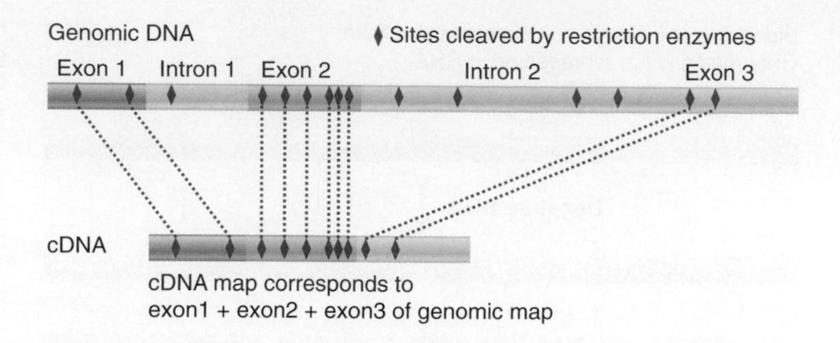

cDNA A single-stranded DNA complementary to an RNA and synthesized from it by reverse transcription *in vitro*.

FIGURE 4.6 compares the restriction maps of a β-globin gene and a complementary DNA (**cDNA**) copy of its mRNA. The gene has two introns, each of which contains a series of restriction sites that are absent from the cDNA. The pattern of restriction sites in the exons is the same in both the cDNA and the gene.

Ultimately, a comparison of the nucleotide sequences of the gene and mRNA sequences precisely identifies the introns. As indicated in FIGURE 4.7, an intron usually has no open reading frame. An intact reading frame in the mRNA results from the removal of the introns.

FIGURE 4.7 An intron is a sequence present in the gene but absent from the mRNA (here shown in terms of the cDNA sequence). The reading frame is indicated by the alternating open and shaded blocks; note that all three possible reading frames are closed by termination codons in the intron.

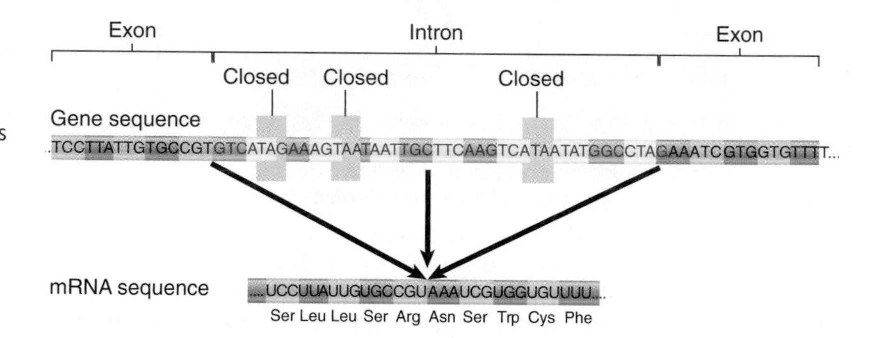

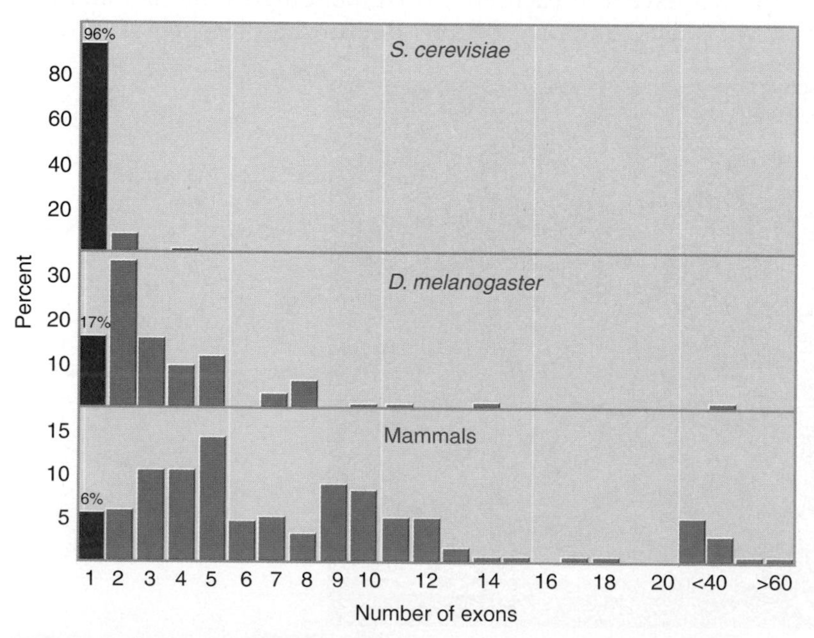

FIGURE 4.8 Most genes are uninterrupted in yeast, but most genes are interrupted in flies and mammals. (Uninterrupted genes have only one exon and are shown in the leftmost column in red.)

The structures of eukaryotic genes vary extensively. Some genes are uninterrupted, so that the gene sequence corresponds to the mRNA sequence. Most multicellular eukaryotic genes are interrupted, but among genes the introns vary enormously in both number and size. (See **FIGURE 4.8** for distributions of intron number variation among genes and **FIGURE 4.9** for distributions of intron-size variation.)

Genes encoding polypeptides, rRNA, or tRNA may all have introns. Introns are also found in mitochondrial genes of plants, fungi, protists, and one metazoan (a sea anemone), and in chloroplast genes. Genes with introns have been found in every class of eukaryotes, archaea, bacteria, and bacteriophages, although they are extremely rare in prokaryotic genomes.

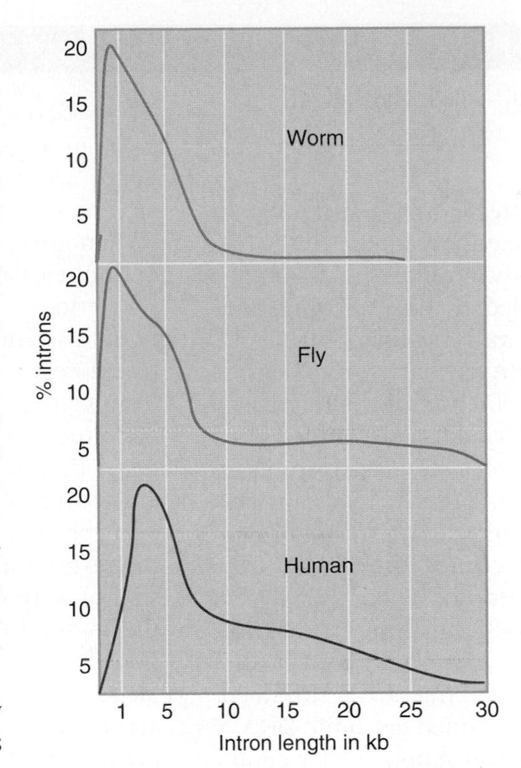

FIGURE 4.9 Introns range from very short to very long.

Some interrupted genes have only one or a few introns. The globin genes provide an extensively studied example (see *Section 4.8, Members of a Gene Family Have a Common Organization*). The two general classes of globin gene, α and β, share a common organization. They originated from an ancient gene duplication event and are described as **paralogous genes,** or **paralogs**. The consistent structure of mammalian globin genes is evident from the "generic" globin gene presented in **FIGURE 4.10**.

▸ **paralogous genes (paralogs)** Genes that share a common ancestry due to gene duplication.

Introns are found at homologous positions (relative to the coding sequence) in all known active globin genes, including those of mammals, birds, and frogs. Although intron lengths vary, the first intron is always fairly short and the second is usually longer. Most of the variation in the lengths of different globin genes results from length variation in the second intron. For example, the second intron in the mouse β-globin gene is 150 bp of the total 850 bp of the gene, whereas the homologous intron in the mouse major β-globin gene is 585 bp of the total 1382 bp. The difference in length of the genes is much greater than that of their mRNAs (β-globin mRNA = 585 bases; β-globin mRNA = 620 bases).

The globin genes are examples of a general phenomenon: genes that share a common ancestry have similar organizations with conservation of the positions (of at least some) of the introns. Variations in the lengths of the genes are primarily due to intron length variation.

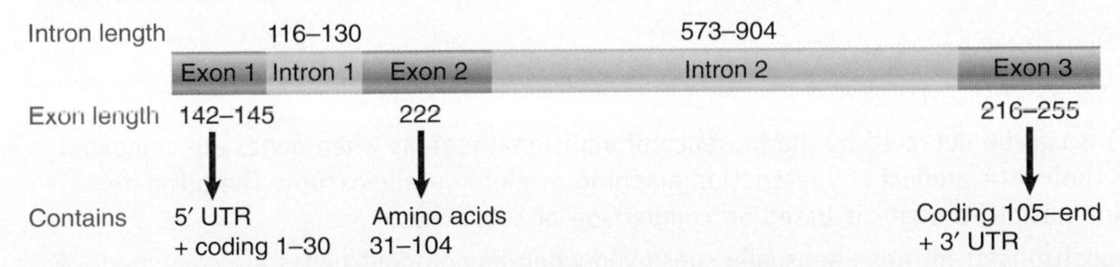

FIGURE 4.10 All functional globin genes have an interrupted structure with three exons. The lengths indicated in the figure apply to the mammalian β-globin genes.

METHODS AND TECHNIQUES

The Discovery of Introns by DNA-RNA Hybridization

The discovery of introns (intervening sequences) in eukaryotic genes was unexpectedly made in 1977. Two different experimental methods, both taking advantage of DNA-RNA hybridization, led to this startling finding. One avenue of research was the mapping of the adenovirus genome using RNA displacement loops to determine the position of mRNA transcripts relative to the viral genome. Adenovirus is a double-stranded virus that infects human epithelial cells, causing respiratory cold symptoms and stomach flu. This virus was an early model system for studying eukaryotic transcription.

In 1977, two independent groups, the labs of Richard Roberts at Cold Spring Harbor Laboratories and Phillip Sharp at the Massachusetts Institute of Technology, used the method of *R loop mapping* to determine the regions of the adenovirus genome that are transcribed. Using this technique, double-stranded genomic DNA is mixed with individual mRNA transcripts under conditions that promote DNA-RNA hybrids between complementary sequences. Since the mRNA will anneal to only one strand of the DNA, the other strand of DNA is displaced, resulting in a loop (R loop) of single-stranded DNA visible under the electron microscope (see Figure 4.4). When mapping the R loops formed between adenovirus DNA and viral mRNA transcripts expressed late in the infection cycle, it was observed that the 5′ and 3′ ends of the RNA did not anneal with the DNA. This was not surprising for the 3′ end of the mRNA transcript, which was known to be modified following transcription by the addition of a polyadenylate tail; however, the discontinuity of the 150 to 200 nucleotides at the 5′ end was unexpected. Hybridization of the 5′ terminal mRNA and discrete single-stranded fragments of genomic DNA demonstrated that this mRNA sequence is actually composed of sequences derived from three distinct and dispersed regions of the adenovirus genome (see Figure 4.5). This was the first evidence of the post-transcriptional process of mRNA splicing, and Drs. Roberts and Sharp were recognized for their discovery of "interrupted genes" in 1993 with a Nobel Prize in physiology or Medicine.

A second line of research, construction of physical maps of eukaryotic genes, revealed that interrupted genes are not unique to viral genomes. Prior to the advent of whole genome sequencing, physical maps of genomes were constructed by mapping restriction endonuclease cleavage sites in genomic DNA. Restriction endonucleases are bacterial enzymes that cleave double-stranded DNA in a site-specific fashion; different restriction endonucleases recognize and cleave distinct sequences (see *Section 3.2, Nucleases*). Whole-genomic DNA can be fragmented by cleavage with restriction endonucleases, and the resulting fragments are separated based on size through agarose gel electrophoresis. It is impossible to directly visualize the fragments for a specific gene because of the complexity and size of a typical eukaryotic genome; however, using the method of Southern blotting and DNA hybridization the fragments derived from a specific gene can be identified. The separated fragments of genomic DNA are denatured and transferred to a membrane in a process referred to as Southern blotting (named for its inventor Ed Southern; see *Section 3.9, Blotting Methods*). The membrane with the bound DNA is then immersed in a solution containing a specific radio-labeled DNA probe. The DNA probe will anneal only to the complementary sequences bound to the membrane, and when this membrane is washed to remove unbound probe, dried, and exposed to X-ray film, the fragments of DNA annealing to the probe are revealed. Typically, a physical map of a gene is constructed by using a cDNA (copied DNA from an mRNA transcript) to probe the genomic DNA. One of the first genes to be mapped in this fashion was the rabbit β-globin gene, performed by Jeffreys and Flavell in 1977. This led to the revelation that there is a sequence of 600 base pairs in the coding region of the genomic DNA that is not present in the cDNA. Another interrupted gene discovered at the same time is the chicken ovalbumin gene, which is also interrupted in the coding region. With time, it became clear that most eukaryotic genes are interrupted by intervening sequences that are transcribed and then removed by mRNA splicing. In 1978 Walter Gilbert proposed that these intervening sequences be referred to as *introns*.

KEY CONCEPTS

- Introns can be detected by the presence of additional regions when genes are compared with their RNA products by restriction mapping or electron microscopy. The ultimate determination, though, is based on comparison of sequences.
- The positions of introns are usually conserved when homologous genes are compared between different organisms. The lengths of the corresponding introns may vary greatly.

CONCEPT AND REASONING CHECK

Why would the genes of viruses with DNA genomes that infect eukaryotic cells be expected to have introns?

4.4 Exon Sequences Are Usually Conserved, but Introns Vary

Is a single-copy structural gene completely unique among other genes in a genome? The answer depends on how "completely unique" is defined. Considered as a whole, the gene is unique, but its exons may be related to those of other genes. As a general rule, when two genes are related, the relationship between their exons is closer than the relationship between their introns. In an extreme case, the exons of two different genes may encode the same polypeptide sequence while the introns are different. This situation can result from the duplication of a common ancestral gene followed by unique base substitutions in both copies, with substitutions restricted in the exons by the need to encode a functional polypeptide.

As we will see in *Chapter 8, Genome Evolution*, when we consider the evolution of the genome, exons can be considered basic building blocks that may be assembled in various combinations. It is possible for a gene to have some exons related to those of another gene, with the remaining exons unrelated. Usually, in such cases, the introns are not related at all. Such homologies between genes may result from duplication and translocation of individual exons.

The homology between two genes can be plotted in the form of a dot matrix comparison, as in **FIGURE 4.11**. A dot is placed in each position that is identical in both genes. The dots form a solid line on the diagonal of the matrix if the two sequences are completely identical. If they are not identical, the line is broken by gaps that lack homology and is displaced laterally or vertically by nucleotide deletions or insertions in one or the other sequence.

When the two mouse β-globin genes are compared in this way, a line of homology extends through the three exons and the small intron. The line disappears in the flanking UTRs and in the large intron. This is a typical pattern in related genes; the coding sequences and areas of introns adjacent to exons retain their similarity, but there is greater divergence in longer introns and the regions on either side of the coding sequence.

The overall degree of divergence between two homologous exons in related genes corresponds to the differences between the polypeptides. It is mostly a result of base substitutions. In the translated regions, changes in exon sequences are constrained by selection against mutations that alter or destroy the function of the polypeptide. Many of the preserved changes do not affect codon meanings because they change one codon into another for the same amino acid (i.e., they are synonymous substitutions). Similarly, there are higher rates of changes in nontranslated regions of the gene (specifically, those that are transcribed to the 5′ UTR [leader] and 3′ UTR [trailer] of the mRNA).

In homologous introns, the pattern of divergence involves both changes in length (due to deletions and insertions) and base substitutions. Introns evolve much more rapidly than exons. When a

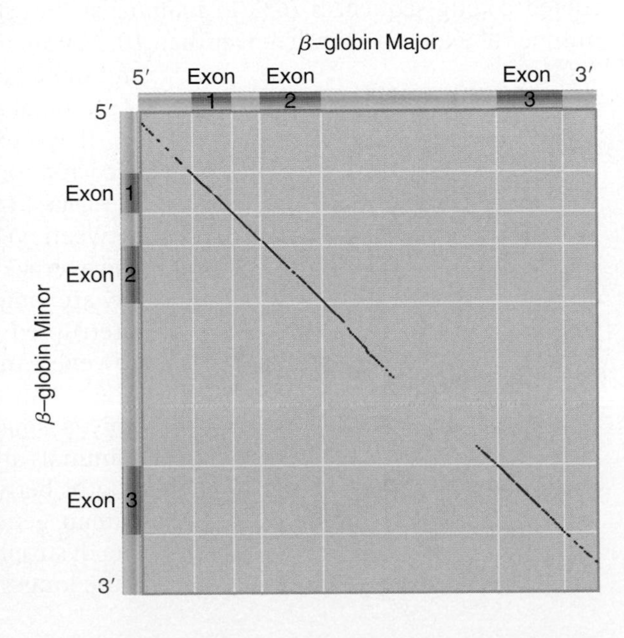

FIGURE 4.11 The sequences of the mouse β^{maj}- and β^{min}-globin genes are closely related in coding regions but differ in the flanking UTRs and the long intron. Data provided by Philip Leder, Harvard Medical School.

gene is compared among different species, there are instances where its exons are homologous but its introns have diverged so much that very little homology is retained. Although mutations in certain intron sequences (branch site, splicing junctions, and perhaps other sequences influencing splicing) will be subject to selection, most intron mutations are expected to be selectively neutral.

In general, mutations occur at the same rate in both exons and introns, but exon mutations are eliminated more effectively by selection. However, because of the low level of functional constraints, introns may more freely accumulate point substitutions and other changes. The empirical observation of faster evolution in introns implies that introns have fewer sequence-specific functions, but that does not necessarily mean that their presence is not required for normal gene function (see *Section 4.6, Some DNA Sequences Encode More Than One Polypeptide*).

KEY CONCEPTS

- Comparisons of related genes in different species show that the sequences of the corresponding exons are usually conserved but the sequences of the introns are much less similar.
- Introns evolve much more rapidly than exons because of the lack of selective pressure to produce a polypeptide with a useful sequence.

CONCEPT AND REASONING CHECK

In comparing two genes in different species, consider whether they are homologous given the following differences: (1) different exons; (2) the same exons, but completely different introns; (3) the same exons, and introns in the same positions but of different sizes.

4.5 Genes Show a Wide Distribution of Sizes Due Primarily to Intron Size and Number Variation

Figure 4.8 compares the organization of genes in a yeast, an insect, and mammals. In the yeast *Saccharomyces cerevisiae*, the majority of genes (~96%) are not interrupted, and those that have introns generally have three or fewer. There are virtually no *S. cerevisiae* genes with more than four exons.

In insects and mammals, the situation is reversed. Only a few genes have uninterrupted coding sequences (6% in mammals). Insect genes tend to have a fairly small number of exons—typically fewer than 10. Mammalian genes are split into more pieces, and some have more than 60 exons. About half of mammalian genes have more than 10 introns.

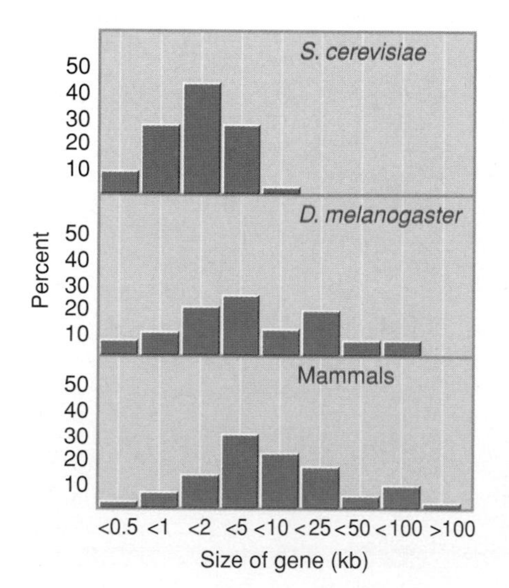

FIGURE 4.12 Yeast genes are short, but genes in flies and mammals have a dispersed distribution extending to very long sizes.

If we examine the effect of intron number variation on the total size of genes, we see in **FIGURE 4.12** that there is a striking difference between yeast and multicellular eukaryotes. The average yeast gene is 1.4 kb long, and very few are longer than 5 kb. The predominance of interrupted genes in multicellular eukaryotes, however, means that the gene can be much larger than the sum total of the exon lengths. Only a small percentage of genes in flies or mammals are shorter than 2 kb, and most have lengths between 5 kb and 100 kb. The average human gene is 27 kb long (see Figure 6.11). The dystrophin gene, with a length of 2000 kb, is the longest known human gene.

The switch from largely uninterrupted to largely interrupted genes seems to have occurred with the evolution of multicellular eukaryotes. In fungi other than *S. cerevisiae*, the majority of genes are interrupted, but they have a relatively small number of exons (<6) and are fairly short (<5 kb). In the fruit fly, gene sizes have a bimodal distribution—many are short but some are quite long. With this increase in the length of the gene due to the increased number of introns, the correlation between genome size and organism complexity becomes weak (see Figure 8.7).

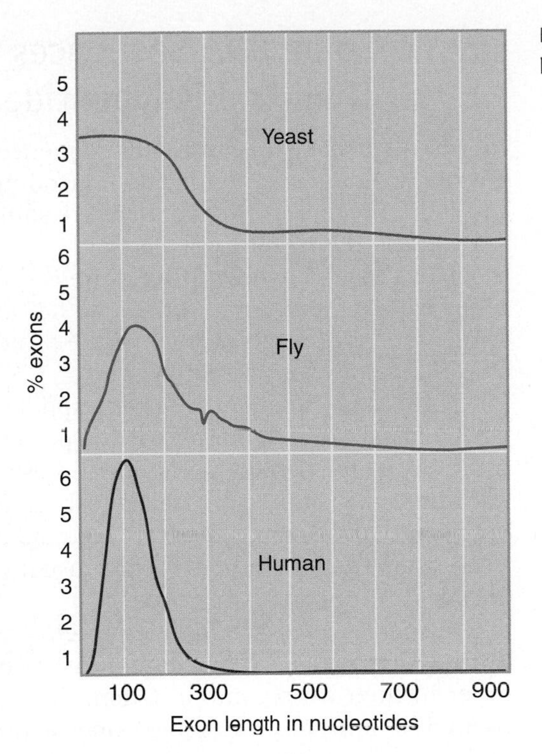

FIGURE 4.13 Exons encoding polypeptides are usually short.

Is there an evolutionary trend in intron length as well as intron number? Yes, organisms with larger genomes tend to have larger introns, whereas exon size does not tend to increase (**FIGURE 4.13**). In multicellular eukaryotes, the average exon codes for ~50 amino acids, and the general distribution is consistent with the hypothesis that genes have evolved by the gradual addition of exon units that code for short, functionally independent protein domains (see *Section 8.6, How Did Interrupted Genes Evolve?*). There is no significant difference in the average size of exons in different multicellular eukaryotes, although the size range is smaller in vertebrates for which there are few exons longer than 200 bp. In yeast, there are some longer exons that represent uninterrupted genes for which the coding sequence is intact.

Figure 4.9 shows that introns vary widely in size among multicellular eukaryotes. In worms and flies, the average intron is not much longer than the exons. There are no very long introns in worms, but flies contain many. In vertebrates, the size distribution is much wider, extending from approximately the same length as the exons (<200 bp) up to 60 kb in extreme cases. Some fish, such as fugu, have compressed genomes with shorter introns and intergenic spaces than mammals have.

Very long genes are the result of very long introns, not the result of encoding longer products. There is no correlation between total gene size and total exon size in multicellular eukaryotes, nor is there a good correlation between gene size and number of exons. The size of a gene is, therefore, determined primarily by the lengths of its individual introns. In mammals and insects, the average gene length is approximately five times that of the total length of its exons.

KEY CONCEPTS

- Most genes are uninterrupted in *S. cerevisiae* but are interrupted in most other eukaryotes.
- Exons are usually short, typically coding for <100 amino acids.
- Introns are short in unicellular eukaryotes but can be many kilobases in length in multicellular eukaryotes.
- The overall length of a gene is determined largely by its introns.

CONCEPT AND REASONING CHECK

Would you expect there to be a general evolutionary trend for increase or decrease in the size of introns? Why or why not?

4.6 Some DNA Sequences Encode More Than One Polypeptide

Most structural genes consist of a sequence of DNA that encodes a single polypeptide, although the gene may include noncoding regions at both ends and introns within the coding region. However, there are some cases in which a single sequence of DNA encodes more than one polypeptide.

▸ **overlapping gene** A gene in which part of the sequence is found within part of the sequence of another gene.

The simplest **overlapping gene** is one in which one gene is part of another. In other words, the first half (or second half) of a gene independently specifies a protein that is the first (or second) half of the protein specified by the full gene. This relationship is illustrated in **FIGURE 4.14**.

A more complex form of overlapping gene occurs when the same sequence of DNA encodes two nonhomologous proteins because it has more than one reading frame. Usually, a coding DNA sequence is read in only one of the three potential reading frames. In some viral and mitochondrial genes, however, there is some overlap between two adjacent genes that are read in different reading frames, as illustrated in **FIGURE 4.15**. The length of overlap is usually short, so that most of the DNA sequence encodes a unique protein sequence.

▸ **mature transcript** A modified RNA transcript. Modification may include the removal of intron sequences and alterations to the 5′ and 3′ ends.

▸ **alternative splicing** The production of different RNA products from a single product by changes in the usage of splicing junctions.

In many genes, alternative proteins result from switches in the process of connecting the exons. A single gene may generate a variety of mRNA products that differ in their exon content. Often, this is because there are pairs of exons that are treated as mutually exclusive—one or the other is included in the **mature transcript**, but not both. The alternative proteins have one part in common and one unique part. Such an example is presented in **FIGURE 4.16**. The 3′ half of the rat troponin T gene contains five exons, but only four are used to construct an individual mRNA. Three exons, W, X, and Z, are included in all mRNAs. However, in one **alternative splicing** pattern the α exon is included between X and Z, whereas in the other pattern it is replaced by the β exon. The α and β forms of troponin T therefore

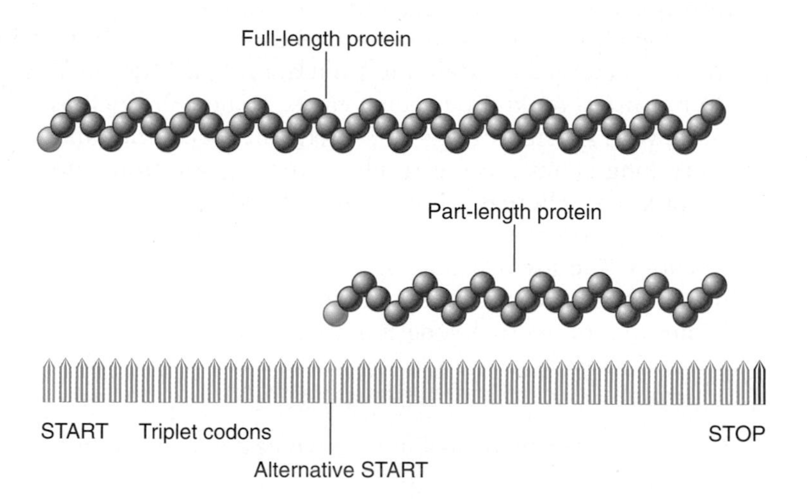

FIGURE 4.14 Two proteins can be generated from a single gene by starting (or terminating) expression at different points.

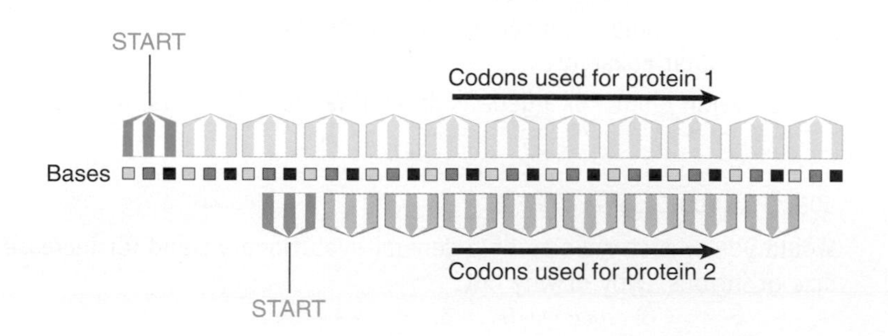

FIGURE 4.15 Two genes may overlap by reading the same DNA sequence in different frames.

differ in the sequence of the amino acids between W and Z, depending on which of the alternative exons (α or β) is used.

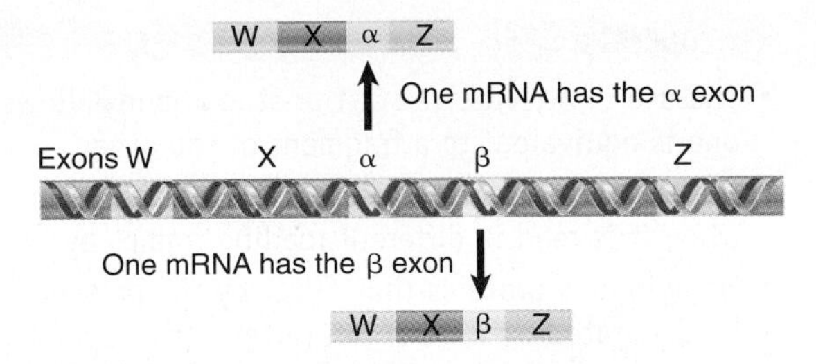

FIGURE 4.16 Alternative splicing generates the α and β variants of troponin T.

There are also cases of alternative splicing in which certain exons are optional; i.e., they may be included or spliced out, as in the example presented in FIGURE 4.17. There is a single **primary transcript** from the gene, but it can be spliced in either one of two ways. In the first, more standard way, the two introns are spliced out and the three exons are joined together. In the second way, the second exon is excluded along with the two introns as if a single large intron is spliced out. The two alternate proteins are the same at their ends, but one has an additional sequence in the middle. (Other types of combinations that are produced by alternative splicing are discussed in *Section 21.11, Alternative Splicing Is a Rule, Rather Than an Exception, in Multicellular Eukaryotes.*)

▸ **primary transcript** The original unmodified RNA product corresponding to the transcription unit of a gene.

Sometimes the two alternative splicing patterns operate simultaneously, with a certain proportion of the primary mRNA transcripts being spliced in each way. However, in some genes the splicing patterns are alternatives that are expressed under different conditions, e.g., one in one cell type and one in another cell type.

So, alternative (or differential) splicing can generate different proteins with related sequences from a single stretch of DNA. It is curious that the multicellular eukaryotic genome is often extremely large, with long genes that are often widely dispersed along a chromosome, but at the same time there may be multiple products from a single locus. Due to alternative splicing, there are ~15% more proteins than genes in flies and worms, but it is estimated that the majority of human genes are alternatively spliced (see *Section 6.5, The Human Genome Has Fewer Genes Than Originally Expected*).

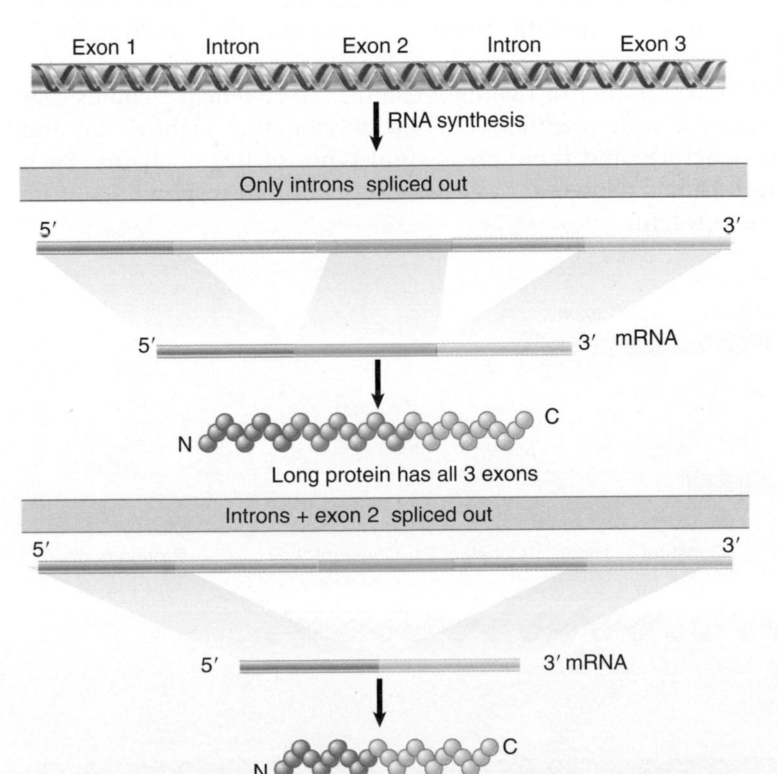

FIGURE 4.17 Alternative splicing uses the same pre-mRNA to generate mRNAs that have different combinations of exons.

KEY CONCEPTS

- The use of alternative start or stop codons allows two proteins to be generated where one is equivalent to a fragment of the other.
- Nonhomologous protein sequences can be produced from the same sequence of DNA when it is read in different reading frames by two (overlapping) genes.
- Homologous proteins that differ by the presence or absence of certain regions can be generated by differential (alternative) splicing when certain exons are included or excluded. This may take the form of including or excluding individual exons or of choosing between alternative exons.

CONCEPT AND REASONING CHECK

What are the advantages to having the same gene produce two or more related polypeptides? Why might different versions of the same polypeptide be necessary?

4.7 Some Exons Can Be Equated with Protein Functional Domains

The issue of the evolution of interrupted genes is more fully considered in *Section 8.6, How Did Interrupted Genes Evolve?* If proteins evolve by recombining parts of ancestral proteins that were originally separate, the accumulation of protein domains is likely to have occurred sequentially, with one exon added at a time. Are the current functions of these domains the same as their original functions? In other words, can we assign particular functions of current proteins to individual exons?

In some cases, there is a clear relationship between the structures of the gene and the protein. The example *par excellence* is provided by the immunoglobulin (antibody) proteins, which are encoded by genes in which every exon corresponds exactly to a known functional domain of the protein. **FIGURE 4.18** compares the structure of an immunoglobulin with its gene.

An immunoglobulin is a tetramer of two light chains and two heavy chains that covalently bond to generate a protein with several distinct domains. Light chains and heavy chains differ in structure, and there are several types of heavy chains. Each type of chain is produced from a gene that has a series of exons corresponding to the structural domains of the protein.

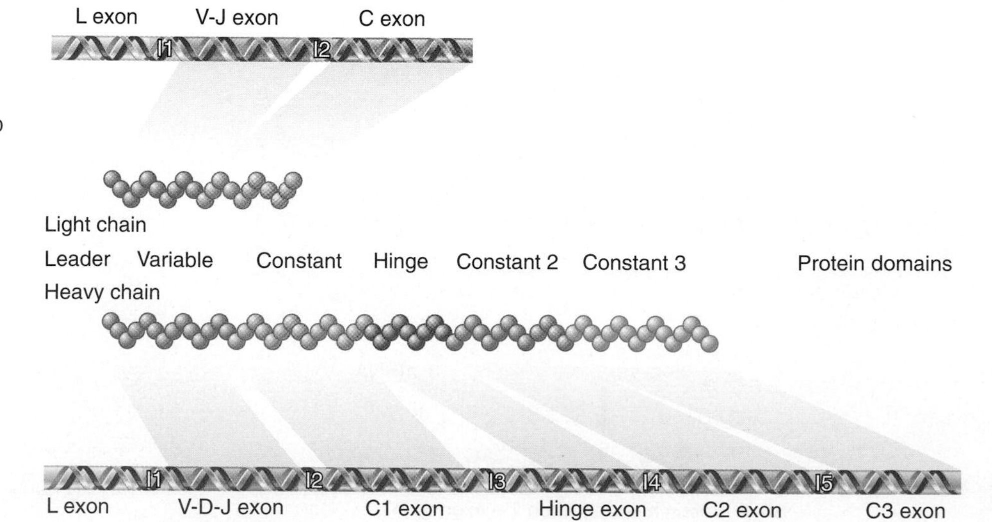

FIGURE 4.18 Immunoglobulin light chains and heavy chains are encoded by genes whose structures (in their expressed forms) correspond to the distinct domains in the protein. Each protein domain corresponds to an exon; introns are numbered I1 to I5.

In many instances, some of the exons of a gene can be identified with particular functions. In secretory proteins, such as insulin, the first exon often specifies the signal sequence involved in membrane secretion.

The view that exons are the functional building blocks of genes is supported by cases in which two genes may share some related exons but also have unique exons. **FIGURE 4.19** summarizes the relationship between the human LDL (plasma low density lipoprotein) receptor and other proteins. The LDL receptor gene has a series of exons related to the exons of the EGF (epidermal growth factor) precursor gene, and another series of exons related to those of the blood protein complement factor C9. Apparently, the LDL receptor gene evolved by the assembly of modules suitable for its various functions. These modules are also used in different combinations in other proteins.

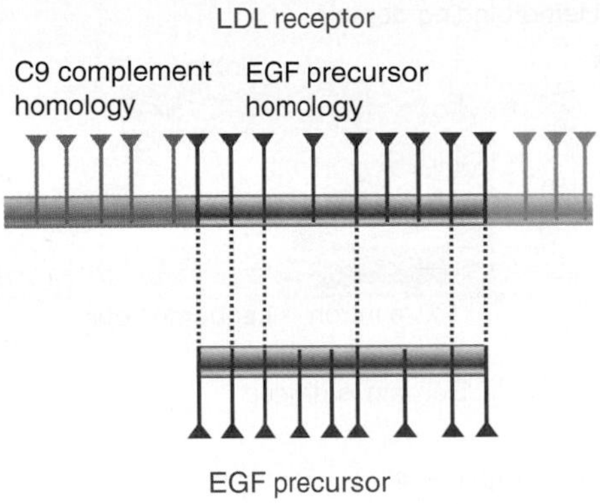

LDL receptor
C9 complement homology
EGF precursor homology
EGF precursor

FIGURE 4.19 The LDL receptor gene consists of 18 exons, some of which are related to EGF precursor exons and some of which are related to the C9 blood complement gene. Triangles mark the positions of introns.

Exons tend to be fairly small (see Figure 6.11), around the size of the smallest polypeptide that can assume a stable folded structure (~20 to 40 residues). It may be that proteins were originally assembled from rather small modules. Each individual module need not correspond to a current function; several modules could have combined to generate a new functional unit. Larger genes tend to have more exons, which is consistent with the view that proteins acquire multiple functions by successively adding appropriate modules.

KEY CONCEPTS

- Many exons can be equated with encoding polypeptide sequences that have particular functions.
- Related exons are found in different genes.

CONCEPT AND REASONING CHECK

Would you expect two membrane-bound enzymes that catalyze different reactions to have (1) completely different exons, (2) some similar exons and some different exons, or (3) all similar exons? Why?

4.8 Members of a Gene Family Have a Common Organization

Many genes in a multicellular eukaryotic genome are related to others in the same genome. A **gene family** is defined as a group of genes that encode related or identical products as a result of gene duplication events. After the first duplication event, the two copies are identical, but then they diverge as different mutations accumulate in them. Further duplications and divergences extend the family. The globin genes are an example of a family that can be divided into two subfamilies (α globin and β globin), but all its members have the same basic structure and function. In some cases, we can find genes that are more distantly related, but still can be recognized as having common ancestry. Such a group of gene families is called a **superfamily**.

A fascinating case of evolutionary conservation is presented by the α and β globins and two other proteins related to them. Myoglobin is a monomeric oxygen-binding protein in animals. Its amino acid sequence suggests a common (though ancient)

▸ **gene family** A set of genes within a genome that encode related or identical proteins or RNAs. The members originated from duplication of an ancestral gene followed by accumulation of changes in sequence between the copies. Most often the members are related but not identical.

▸ **superfamily** A set of genes all related by presumed descent from a common ancestor but now showing considerable variation.

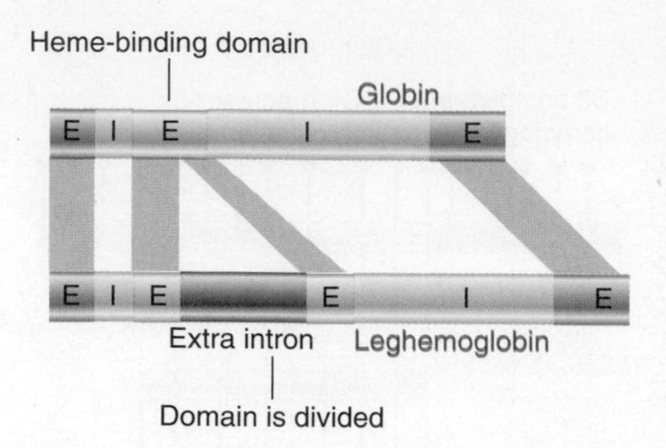

FIGURE 4.20 The exon structure of globin genes corresponds to protein function, but leghemoglobin has an extra intron in the central domain.

origin with globins. Leghemoglobins are oxygen-binding proteins found in legume plants; like myoglobin, they are monomeric and share a common origin with the other heme-binding proteins. Together, the globins, myoglobin, and leghemoglobins make up the globin superfamily—a set of gene families all descended from an ancient common ancestor.

Both α- and β-globin genes have three exons and two introns at conserved positions (see Figure 4.6). The central exon represents the heme-binding domain of the globin chain. There is a single myoglobin gene in the human genome and its structure is essentially the same as that of the globin genes. The three-exon structure therefore predates the common ancestor of the myoglobin and globin genes. Leghemoglobin genes contain three introns, the first and last of which are homologous to the two introns in the globin genes. This remarkable similarity suggests an exceedingly ancient origin for the interrupted structure of heme-binding proteins, as illustrated in **FIGURE 4.20**.

- **orthologous genes (orthologs)** Related genes in different species.
- **homologous genes (homologs)** Related genes in the same species, such as alleles on homologous chromosomes or multiple genes in the same genome sharing common ancestry.

Orthologous genes, or **orthologs**, are genes that are **homologous** (**homologs**) due to speciation; in other words, they are related genes in different species. Comparison of orthologs that differ in structure may provide information about their evolution. An example is insulin. Mammals and birds have only one gene for insulin, except for rodents, which have two. **FIGURE 4.21** illustrates the structures of these genes. We use the principle of parsimony in comparing the organization of orthologous genes by assuming that a common feature predates the evolutionary separation of the two species. In chickens, the single insulin gene has two introns; one of the two homologous rat genes has the same structure. The common structure implies that the ancestral insulin gene had two introns. However, since the second rat gene has only one intron, it must have evolved by a gene duplication in rodents that was followed by the precise removal of one intron from one of the homologs.

The organizations of some orthologs show extensive discrepancies between species. In these cases, there must have been deletion or insertion of introns during evolution. A well-characterized example is that of the actin genes. The common features of actin genes are a nontranslated leader of <100 bases, a coding region of ~1200 bases, and a trailer of ~200 bases. Most actin genes have introns, and their positions can be aligned with regard to the coding sequence (except for a single intron sometimes found in the leader). **FIGURE 4.22** shows that almost every actin gene is different in its pattern of intron positions. Among all the genes being compared, introns occur at 19 different sites. However, the range of intron number per gene is zero to six. How did this situation arise? If we suppose that the primordial actin gene had introns, and that all current actin genes are related to it by loss of introns, different introns have been lost in each evolutionary branch. Probably some introns have been lost entirely, so the primordial gene could well have had 20 introns or more. The alternative is to suppose that a process of intron insertion continued independently in the different lines of evolution.

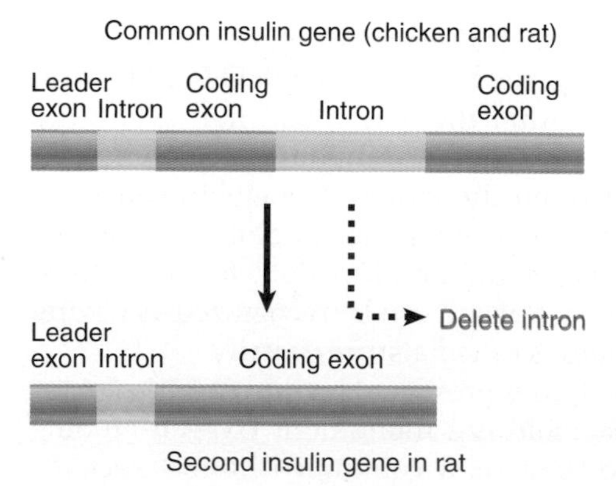

FIGURE 4.21 The rat insulin gene with one intron evolved by loss of an intron from an ancestor with two introns.

The relationship between individual exons and functional protein domains is somewhat erratic. In some cases there is a clear 1:1 relationship; in others no pattern can be discerned. One possibility is that the removal of introns has fused previously adjacent exons. This means that the intron must have been precisely removed, without changing the integrity of the coding region. An alternative is that some introns arose by

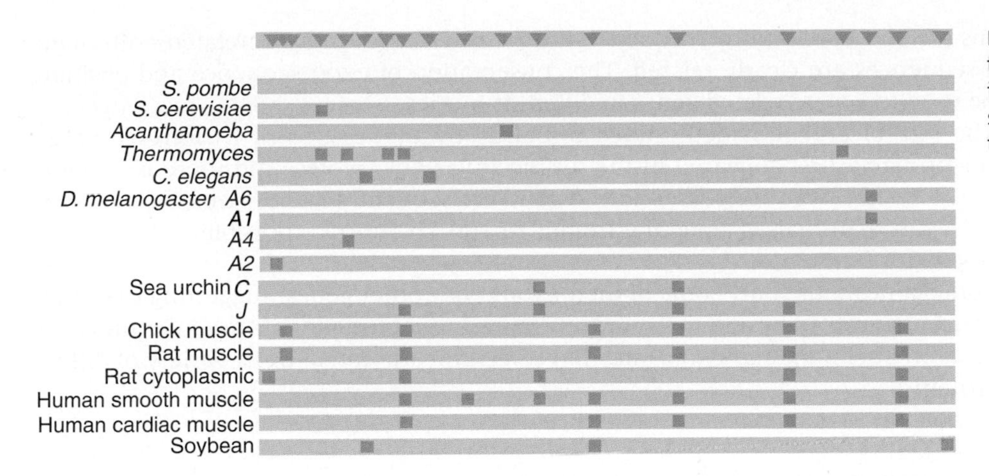

FIGURE 4.22 Actin genes vary widely in their organization. The sites of introns are indicated by dark boxes. The bar at the top summarizes all the intron positions among the different orthologs.

insertion into an exon encoding a single domain. Together with the variations that we see in exon placement in cases such as the actin genes, the conclusion is that intron positions can evolve.

The correspondence of at least some exons with protein domains and the presence of related exons in different proteins leave no doubt that the duplication and juxtaposition of exons have played important roles in evolution. It is possible that the number of ancestral exons—from which all proteins have been derived by duplication, variation, and recombination—could be relatively small, perhaps as little as a few thousand. The idea that exons are the building blocks of new genes is consistent with the "introns early" model for the origin of genes encoding proteins (see *Section 8.6, How Did Interrupted Genes Evolve?*).

KEY CONCEPTS

- A common feature in a set of genes is assumed to identify a property that preceded their separation in evolution.
- All globin genes have a common form of organization with three exons and two introns, suggesting that they are descended from a single ancestral gene.

CONCEPT AND REASONING CHECK

"Parsimony" is the least complex explanation for a given observation. For example, the explanation that modern actin genes evolved from an ancestral gene with many exons is more parsimonious than the explanation that modern actin genes gained introns independently. Evaluate the statement, "Selecting the most parsimonious explanation guarantees the choice of the correct explanation."

4.9 Summary

Virtually all eukaryotic genomes contain interrupted genes. The proportion of interrupted genes is low in some fungi, but few genes are uninterrupted in multicellular eukaryotes. Introns are found in all classes of eukaryotic genes. The structure of the interrupted gene is the same in all tissues: exons are spliced together in RNA in the same order as they are found in DNA, and the introns, which usually have no coding function, are removed from RNA by splicing.

Some genes are expressed by alternative splicing patterns, in which a particular sequence is removed as an intron in some situations, but retained as an exon in others. Often, when the organizations of orthologous genes are compared, the positions of introns are conserved. Intron sequences vary—and may even be unrelated—although exon sequences are clearly related. The conservation of exon sequence and position can be used to isolate related genes in different species.

The size of a gene is determined primarily by the lengths of its introns. Large introns probably first appeared in the multicellular eukaryotes, and there is an evolutionary trend toward increased intron (and consequently, gene) size. The range of gene sizes in mammals is generally from 1 to 100 kb, but it is possible to have even larger genes.

Some genes share only some of their exons with other genes, suggesting that they have been assembled by addition of exons representing functional "modular units" of the protein. Such modular exons may have been incorporated into a variety of different proteins.

CHAPTER QUESTIONS

1. Mutations that affect splicing:
   A. are usually inconsequential.
   B. are usually deleterious.
   C. are always deleterious.
   D. increase gene expression.
2. In which group of organisms are most genes interrupted (have introns)?
   A. bacteria
   B. yeast
   C. animals
   D. more than one of the above
3. The length of genes in a gene family often varies; this variation is most often determined by:
   A. length of the 5′ untranslated region.
   B. number and size of exons.
   C. number and size of introns.
   D. length of the 3′ untranslated region.
4. In general, for two genes that are related, which shows the closest relationship?
   A. exons
   B. introns
   C. both introns and exons
   D. the promoter region and first exon
5. What percentage of genes is uninterrupted in animals?
   A. <5%
   B. <10%
   C. <20%
   D. ~33%
6. In general, as genome size increases in different organisms:
   A. exons also increase in size.
   B. introns increase in size.
   C. both exons and introns increase in size.
   D. 5′ and 3′ untranslated regions increase in size.
7. Which group of organisms has the longest average intron size?
   A. bacteria
   B. flies
   C. worms
   D. mammals

8. The hypothesis that new genes are constructed by combining exons from existing genes is supported by the observation that genes with different functions:
   A. have all related introns.
   B. have some related exons.
   C. have all related exons.
   D. have some related introns.
9. Genes that encode related or identical proteins in an organism are part of:
   A. a gene family.
   B. a superfamily.
   C. homologous genes.
   D. orthologous genes.
10. α globin and β globin are:
   A. unrelated.
   B. orthologs of each other.
   C. paralogs of myoglobin.
   D. paralogs of each other.

KEY TERMS

alternative splicing
cDNA
exon
gene family
homologous genes (homologs)
interrupted gene
intron
mature transcript
orthologous genes (orthologs)
overlapping gene
paralogous genes (paralogs)
primary transcript
RNA splicing
superfamily

FURTHER READING

Black, D. L. (2003). Mechanisms of alternative pre-messenger RNA splicing. *Annu. Rev. Biochem.* **72,** 291–336. An in-depth review of specific examples of alternative splicing, including the *Drosophila* sex determination system.

Faustino, N. A., and Cooper, T. A. (2003). Pre-mRNA splicing and human disease. *Genes Dev.* **17,** 419–437. A review of the mechanisms by which problems with alternative splicing of pre-mRNA can result in human diseases.

Ponting, C. P., and Russell, R. R. (2002). The natural history of protein domains. *Annu. Rev. Biophys. Biomol.* **31,** 45–71. A discussion of the origin and evolution of protein domains, including the suggestion that domains be classified in a hierarchical taxonomic system.

Reddy, A. S. N. (2007). Alternative splicing of pre-messenger RNAs in plants in the genomic era. *Annu. Rev. Plant Biol.* **58,** 267–294. A review of unexpected recent discoveries of alternative splicing of plant genes.

Rodríguez-Trelles, F., Tarrío, R., and Ayala, F. J. (2006). Origins and evolution of spliceosomal introns. *Annu. Rev. Genet.* **40,** 47–76. A review of classic and newer hypotheses about the origins of introns.

5

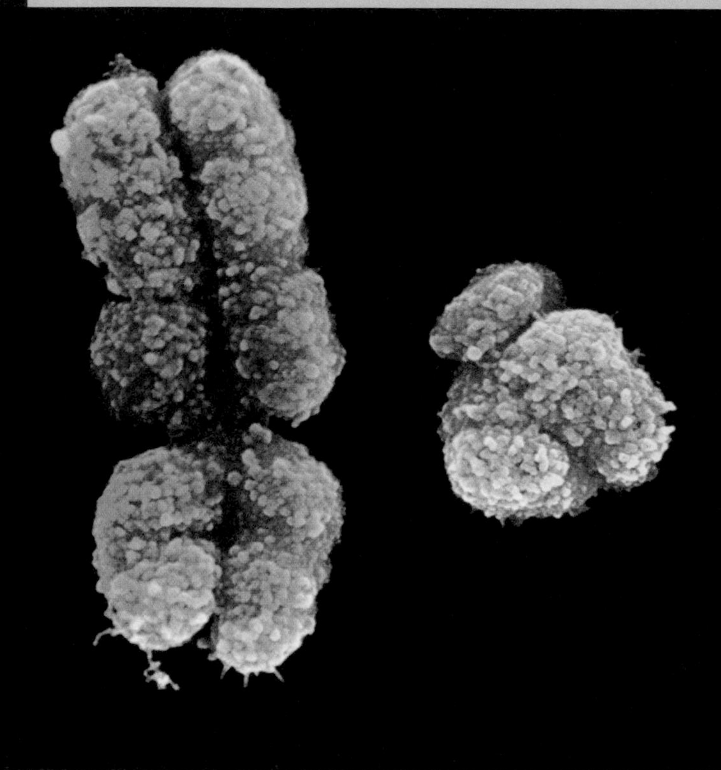

Scanning electron micrograph (SEM) of human X and Y chromosomes in metaphase (35,000×). Most of the content of the genome is found in chromosomal DNA, though some organelles contain their own genomes.

The Content of the Genome

CHAPTER OUTLINE

5.1 Introduction

One key question about any genome is how many genes it contains. However, an even more fundamental question is, "What is a gene?" Clearly, genes cannot solely be defined as a sequence of DNA that encodes a polypeptide because many genes encode multiple polypeptides, and many encode RNAs that serve other functions. Given the variety of RNA functions and the complexities of gene expression, it seems prudent to focus on the gene as a unit of transcription. However, large areas of chromosomes previously thought to be devoid of genes now appear to be extensively transcribed, so at present the definition of a "gene" is a moving target.

We can attempt to characterize both the total number of genes and the number of protein-coding genes at four levels, which correspond to successive stages in gene expression:

- The **genome** is the complete set of genes of an organism. Ultimately it is defined by the complete DNA sequence, although as a practical matter it may not be possible to identify every gene unequivocally solely on the basis of sequence.
- The **transcriptome** is the complete set of genes expressed under particular conditions. It is defined in terms of the set of RNA molecules present in a single cell type, a more complex assembly of cells, or a complete organism. Because some genes generate multiple mRNAs, the transcriptome is likely to be larger than the actual number of genes in the genome. The transcriptome includes noncoding RNAs (such as tRNAs, rRNAs, and microRNAs or miRNAs, described in *Section 30.5, MicroRNAs Are Widespread Regulators in Eukaryotes*) as well as mRNAs.
- The **proteome** is the complete set of polypeptides encoded by the whole genome or produced in any particular cell or tissue. It should correspond to the mRNAs in the transcriptome, although there can be differences of detail reflecting changes in the relative abundance or stabilities of mRNAs and proteins. There may also be posttranslational modifications to proteins that allow more than one protein to be produced from a single transcript (this is called *protein splicing*; see *Section 23.11, Protein Splicing Is Autocatalytic*).
- Proteins may function independently or as part of multiprotein or multimolecular complexes, such as holoenzymes and metabolic pathways where enzymes are clustered together. The RNA polymerase holoenzyme (see *Section 19.4, Bacterial RNA Polymerase Consists of the Core Enzyme and Sigma Factor*) and the spliceosome (see *Section 21.8, The Spliceosome Assembly Pathway*) are two examples. If we could identify all protein-protein interactions, we could define the total number of independent complexes of proteins. This is sometimes referred to as the **interactome**.

▸ **genome** The complete set of sequences in the genetic material of an organism. It includes the sequence of each chromosome plus any DNA in organelles.

▸ **transcriptome** The complete set of RNAs present in a cell, tissue, or organism. Its complexity is due mostly to mRNAs, but it also includes noncoding RNAs.

▸ **proteome** The complete set of proteins that is expressed by the entire genome. Sometimes the term is used to describe the complement of proteins expressed by a cell at any one time.

▸ **interactome** The complete set of protein complexes/protein-protein interactions present in a cell, tissue, or organism.

The maximum number of polypeptide-encoding genes in the genome can be identified directly by characterizing open reading frames. Large-scale analysis of this nature is complicated by the fact that interrupted genes may consist of many separated open reading frames. We do not necessarily have information about the functions of the protein products—or, indeed, proof that they are expressed at all—so this approach is restricted to defining the *potential* of the genome. However, it is presumed that any conserved open reading frame is likely to be expressed.

Another approach is to define the number of genes directly in terms of the transcriptome (by directly identifying all the RNAs) or proteome (by directly identifying all the polypeptides). This gives an assurance that we are dealing with *bona fide* genes that are expressed under known circumstances. It allows us to ask how many genes are expressed in a particular tissue or cell type, what variation exists in the relative levels of expression, and how many of the genes expressed in one particular cell are unique to that cell or are also expressed elsewhere. In addition, analysis of the transcriptome can reveal how many different mRNAs (e.g., mRNAs containing different combinations of exons) are generated from a particular gene.

Also, we may ask whether a particular gene is *essential*: what is the phenotypic effect of a null mutation in that gene? If a null mutation is lethal or the organism has a clear defect, we may conclude that the gene is essential or at least beneficial. However, the functions of some genes can be eliminated without apparent effect on the phenotype. Are these genes really dispensable, or does a selective disadvantage result from the absence of the gene, perhaps in other circumstances or over longer periods of time? In some cases, the absence of the functions of these genes could be offset by a redundant mechanism, such as a gene duplication, providing a backup for an essential function.

CONCEPT AND REASONING CHECK

Explain why, in animals, the transcriptome and proteome of a cell are generally smaller than its genome, whereas the transcriptome and proteome of the whole organism are generally larger than its genome.

5.2 Genomes Can Be Mapped at Several Levels of Resolution

Defining the contents of a genome essentially means making a map of the genetic loci found on the organism's chromosome(s). We can map loci and genomes at several levels of resolution:

▸ **linkage map** A map of the positions of loci or other genetic markers on a chromosome obtained by measuring recombination frequencies between markers.

▸ **restriction map** A linear array of restriction sites on DNA, determined by cutting the DNA with various restriction endonucleases or by scanning a known sequence for restriction sites.

- A **linkage map** presents the distance between loci in units based on recombination frequencies. It is limited by its dependence on the observation of recombination between variable markers that are either directly visible (such as phenotypic traits) or that can otherwise be visualized (such as by electrophoresis). For example, a linkage map can be constructed by measuring recombination frequencies between sites in genomic DNA that have sequence variations generating differences in the ability to be cut by certain restriction enzymes. Because such variable sites are common, such maps are easily prepared for any organism. Because recombination frequencies can be distorted relative to the physical distance between sites, a linkage map does not accurately represent physical distances along chromosomes.
- A **restriction map** is constructed by cutting DNA into fragments with restriction enzymes and measuring the physical distances, in terms of the length of DNA in base pairs (determined by migration on an electrophoretic gel) between the cut sites. A restriction map does not intrinsically identify sites of interest, such as a gene. For it to be compared to the linkage map, mutations have to be characterized in terms of their effects upon the restriction sites. Large changes in the genome can be recognized because they affect the sizes or numbers of restriction fragments. Point mutations are more difficult to detect because they may change only a single restriction site or they may lie between restriction sites and be undetectable.
- The ultimate genomic map is the DNA sequence of the genome. From the sequence, we can identify genes and the distances between them. By analyzing the protein-coding potential of a sequence of the DNA, we can hypothesize about its function. The basic assumption is that natural selection prevents the accumulation of deleterious mutations in sequences that encode functional products. Reversing the argument, we may assume that an intact coding sequence is likely to produce a functional polypeptide.

By comparing a wild-type DNA sequence with that of a mutant allele, we can determine the nature of a mutation and its exact location in the sequence. This provides a way to determine the relationship between the linkage map (based entirely on variable sites) and the physical map (based on, or even comprising, the sequence of DNA).

Similar techniques are used to identify and sequence genes and to map the genome, although there is, of course, a difference of scale. In each case, the approach is to characterize a series of overlapping fragments of DNA that can be connected into a continuous map. The crucial feature is that each segment is related to the next segment on the map by the overlap between them, so that we can be sure no segments are missing. This principle is applied both at the level of assembling large fragments into a map and in connecting the sequences that make up the fragments.

KEY CONCEPTS

- Linkage maps are based on the frequency of recombination between genetic markers; restriction maps are based on the physical distances between markers.
- Molecular characterization of mutations can be used to reconcile linkage maps with physical maps.

CONCEPT AND REASONING CHECK

If the same chromosome is sampled from different populations of the same organism, the physical map can be identical but the linkage map may be slightly different. Why?

5.3 Individual Genomes Show Extensive Variation

The original Mendelian view of the genome classified alleles as either wild type or mutant. Subsequently, we recognized the existence of multiple alleles for a gene in a population, each with a different effect on the phenotype. In some cases it may not even be appropriate to define any one allele as wild type.

The coexistence of multiple alleles at a locus in a population is called genetic **polymorphism**. Any site at which multiple alleles exist as stable components of the population is by definition polymorphic. A locus is usually defined as polymorphic if two or more alleles are present at a frequency of >1% in the population.

polymorphism The simultaneous occurrence in a population of alleles showing variations at a particular position.

Although not evident from the phenotype, the wild type may itself be polymorphic. Multiple versions of the wild-type allele may be distinguished by differences in sequence that do not affect their function and that, therefore, do not produce phenotypic variants. A population may have extensive polymorphism at the level of the genotype. Many different sequence variants may exist at a particular locus; some of them are evident because they affect the phenotype, but others are "hidden" because they have no visible effect. These mutant alleles are selectively neutral, with their fates mainly a result of random genetic drift (see *Chapter 8, Genome Evolution*).

So there may be a variety of changes at a locus, including those that change DNA sequence but do not change the sequence of the polypeptide product, those that change polypeptide sequence without changing its function, those that result in polypeptides with different functions, and those that result in altered polypeptides that are nonfunctional.

When alleles of the same locus are compared, a difference in a single nucleotide is called a **single nucleotide polymorphism (SNP)**. On average, one SNP occurs approximately every 1330 bases in the human genome. Defined by SNPs, every human being is unique. SNPs can be detected by various means, ranging from direct comparisons of sequences to mass spectroscopy or biochemical methods that produce differences based on sequence variations in a defined region.

single nucleotide polymorphism (SNP) A polymorphism (variation in sequence between individuals) caused by a change in a single nucleotide. This accounts for most of the genetic variation between individuals.

One aim of genetic mapping is to obtain a catalog of common variants. The observed frequency of SNPs per genome predicts that, in the human population as a whole (considering the genomes of all living human individuals), there should be >10 million SNPs that occur at a frequency of >1%. More than 6 million human SNPs have already been identified.

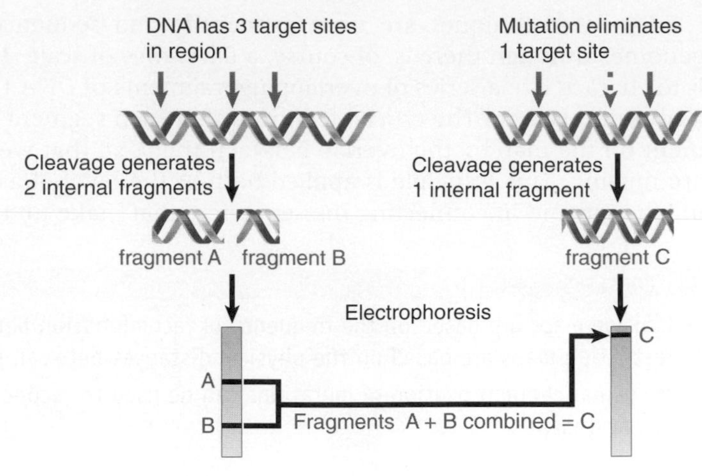

FIGURE 5.1 A point mutation that affects a restriction site is detected by a difference in restriction fragment lengths.

Some polymorphisms in the genome can be detected by comparing the restriction maps of different individuals. If a mutation occurs in the target site for a restriction enzyme, then the pattern of fragments produced by digestion with that enzyme will also be changed. **FIGURE 5.1** shows that when a target site is present in the genome of one individual and absent from that of another individual, the extra cleavage in the first genome will generate two fragments corresponding to the single fragment in the second genome. A difference in restriction maps between two individuals is called a **restriction fragment length polymorphism (RFLP)**, or "riflip." Basically, an RFLP is an SNP that is located in the target site for a restriction enzyme. It can be used as a genetic marker in exactly the same way as any other marker. Instead of detecting some visible feature of the phenotype, we directly assess the genotype as revealed by the restriction map. **FIGURE 5.2** shows a pedigree of an RFLP followed through three generations. It displays Mendelian segregation at the level of DNA marker fragments.

▸ **restriction fragment length polymorphism (RFLP)** Inherited differences in target sites for restriction enzymes (for example, caused by base changes in a target site) that result in differences in the lengths of the fragments produced by cleavage with the relevant restriction enzyme. They are used for genetic mapping to link the genome directly to a conventional genetic marker.

The restriction map is independent of gene function; as a result, an RFLP at this level can be detected whether or not the sequence change affects the phenotype. Probably very few of the RFLPs in a genome actually affect the phenotype. Most involve sequence changes that have no effect on the production of proteins (for example, because they lie between genes).

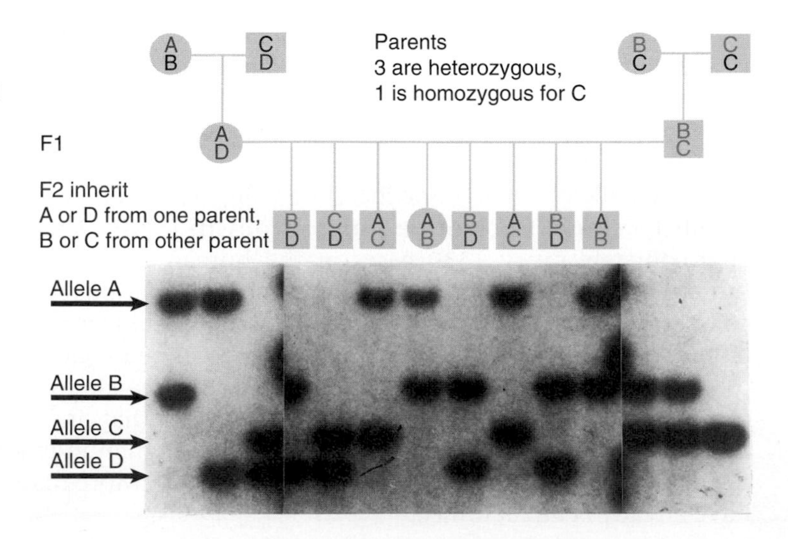

FIGURE 5.2 Restriction site polymorphisms are inherited according to Mendelian rules. As visualized by Southern blotting and probe hybridization, four alleles for a restriction marker are found in all possible pairwise combinations and segregate independently at each generation. Photo courtesy of Ray White, Ernest Gallo Clinic and Research Center, University of California, San Francisco.

KEY CONCEPTS

- Polymorphism may be detected at the phenotypic level when a sequence affects gene function, at the restriction fragment level when it affects a restriction enzyme target site, and at the sequence level by direct analysis of DNA.
- The alleles of a gene show extensive polymorphism at the sequence level, but many sequence changes do not affect function.

CONCEPT AND REASONING CHECK

Why might a mutation in the coding sequence of a gene result in a genetic, but not a phenotypic, polymorphism?

5.4 RFLPs and SNPs Can Be Used for Genetic Mapping

Recombination frequency can be measured between a restriction marker and a visible phenotypic marker, as illustrated in **FIGURE 5.3**. Thus a genetic map can include both genotypic and phenotypic markers.

Restriction markers are not limited to those genetic changes that affect the phenotype; as a result, they provide the basis for an extremely powerful technique for identifying genetic variants at the molecular level. A typical problem concerns a mutation with known effects on the phenotype, where the relevant genetic locus can be placed on a genetic map but for which we have no knowledge about the corresponding gene or its product. Many damaging or fatal human diseases fall into this category. For example, cystic fibrosis shows Mendelian inheritance, but the molecular nature of the mutant function was unknown until it could be identified as a result of characterizing the gene.

If restriction polymorphisms occur at random in the genome, there should be some near any particular target gene. We can identify such restriction markers by virtue of their close linkage to the gene responsible for the mutant phenotype. If we compare the restriction map of DNA from patients suffering from a disease with the DNA of healthy people, we may find that a particular restriction site is always present (or always absent) from the patients.

A hypothetical example is shown in **FIGURE 5.4**. This situation corresponds to finding 100% linkage between the restriction marker and the locus producing the phenotype. It would imply that the restriction marker lies so close to the mutant gene that it is never separated from it by crossing over; it may, in fact, be the same mutation.

The identification of such a marker has two important consequences:

- It may offer a diagnostic procedure for detecting the disease. Some of the human diseases that have a known inheritance pattern but are not well defined in molecular terms cannot be easily diagnosed. If a restriction marker is closely linked to the phenotype, then its presence can be used to diagnose the probability of carrying the disease allele.

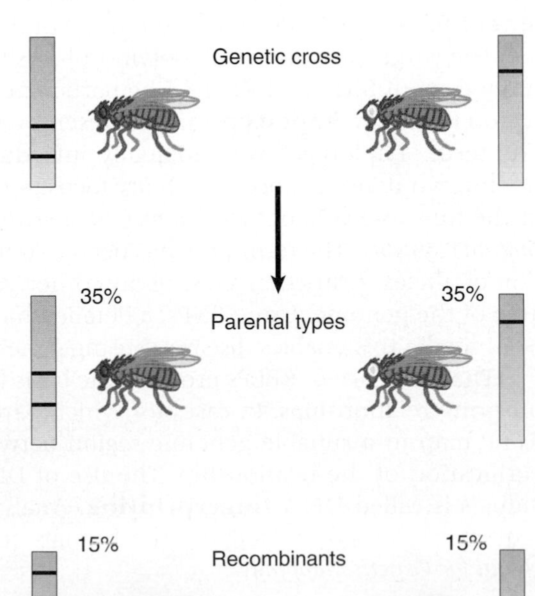

FIGURE 5.3 A restriction polymorphism can be used as a genetic marker to measure recombination distance from a phenotypic marker (such as eye color). The figure simplifies the situation by showing only the DNA bands corresponding to the allele of one genome in a diploid.

FIGURE 5.4 If a restriction marker is associated with a phenotypic characteristic, the restriction site must be located near the gene responsible for the phenotype. The mutation changing the band that is common in healthy people into the band that is common in patients is very closely linked to the disease gene.

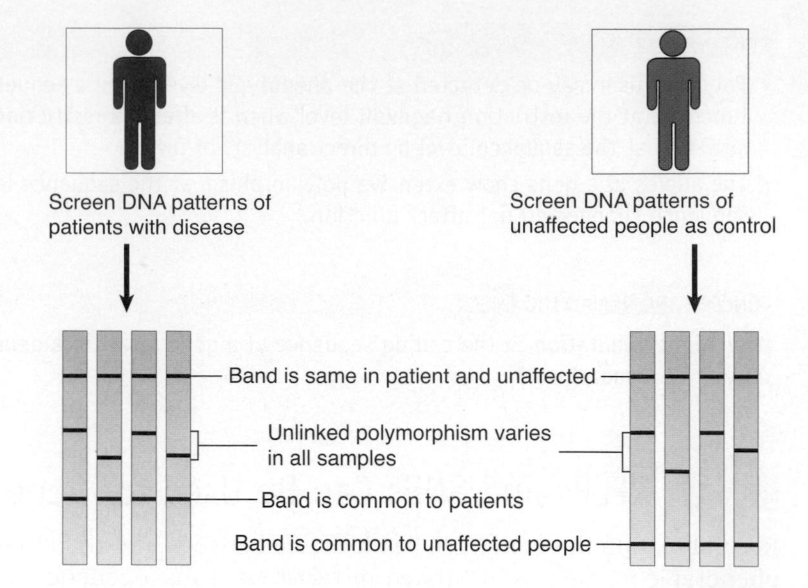

- It may lead to isolation of the gene. The restriction marker must lie relatively near the gene on the genetic map if the two loci rarely or never recombine. "Relatively near" in genetic terms can be a substantial distance in terms of base pairs of DNA; nonetheless, it provides a starting point from which one can then proceed to track down the gene itself.

The frequent occurrence of SNPs in the human genome makes them useful for genetic mapping. From the several million SNPs that have already been identified, there is on average an SNP every ~1 kb. This should allow rapid localization of new disease genes by locating them between the nearest SNPs.

On the same principle, RFLP mapping has been in use for some time. Once an RFLP has been assigned to a linkage group (i.e., a chromosome), it can be placed on the genetic map. RFLP mapping of both the human and mouse genomes has led to the construction of linkage maps for both. Any site with an unknown position can be tested for linkage to these sites and by this means can be rapidly placed on the map. There are fewer RFLPs than SNPs, so the resolution of the RFLP map is in principle more limited.

The large proportion of polymorphic sites means that every individual has a unique set of SNPs and RFLPs. The particular combination of sites found in a specific region is called a **haplotype** and represents a small portion of the complete genotype. The term "haplotype" was originally introduced to describe the genetic content of the human major histocompatibility locus, a region specifying proteins of importance in the immune system (see *Chapter 18, Somatic Recombination and Hypermutation in the Immune System*). The term now has been extended to describe the particular combination of alleles, restriction sites, or any other genetic markers present in some defined area of the genome. Using SNPs, a detailed haplotype map of the human genome has been made; this enables disease-causing genes to be mapped more easily.

The existence of RFLPs provides the basis for a technique to establish clear parent-offspring relationships. In cases for which parentage is in doubt, a comparison of the RFLP map in a suitable genomic region between potential parents and child allows verification of the relationship. The use of DNA restriction analysis to identify individuals is called **DNA fingerprinting**. Analysis of especially variable "minisatellite" sequences is used in mapping the human genome (see *Section 7.8, Minisatellites Are Useful for Genetic Mapping*).

KEY CONCEPT

- RFLPs and SNPs can be the basis for linkage maps and are useful for establishing parent-offspring relationships.

▸ **haplotype** The particular combination of alleles in a defined region of some chromosome—a small portion of the genotype. Originally used to describe combinations of major histocompatibility complex (MHC) alleles, it now may be used to describe particular combinations of RFLPs, SNPs, or other markers.

▸ **DNA fingerprinting** A technique for analyzing the differences between individuals in the fragments generated by using restriction enzymes to cut regions that contain short repeated sequences or by PCR. The lengths of the repeated regions are unique to every individual, and, as a result, the presence of a particular subset in any two individuals can be used to define their common inheritance (e.g., a parent-child relationship).

CONCEPT AND REASONING CHECK

Describe how the position of a disease allele can be narrowed down to a specific chromosomal region by linkage analysis using RFLPs from members of a large family in which the disease allele is segregating.

5.5 Eukaryotic Genomes Contain Both Nonrepetitive and Repetitive DNA Sequences

The general nature of the eukaryotic genome can be assessed by the kinetics of reassociation of denatured DNA. This technique was used extensively before large-scale DNA sequencing became possible.

Reassociation kinetics identifies two general types of genomic sequences:

- **Nonrepetitive DNA** consists of sequences that are unique: there is only one copy in a haploid genome.
- **Repetitive DNA** consists of sequences that are present in more than one copy in each genome.

Repetitive DNA often is divided into two general types:

- *Moderately repetitive DNA* consists of relatively short sequences that are repeated typically 10 to 1000× in the genome. The sequences are dispersed throughout the genome and are responsible for the high degree of secondary structure formation in pre-mRNA when inverted repeats in the introns pair to form duplex regions.
- *Highly repetitive DNA* consists of very short sequences (typically <100 bp) that are present many thousands of times in the genome, often organized as long regions of tandem repeats (see *Section 7.5, Satellite DNAs Often Lie in Heterochromatin*). Neither class is found in exons.

▸ **nonrepetitive DNA** DNA that is unique (present only once) in a genome.

▸ **repetitive DNA** DNA that is present in many (related or identical) copies in a genome.

The proportion of the genome occupied by nonrepetitive DNA varies widely among taxonomic groups. **FIGURE 5.5** summarizes the genome organization of some representative organisms. Prokaryotes contain nonrepetitive DNA almost exclusively. For unicellular eukaryotes, most of the DNA is nonrepetitive; <20% falls into one or more moderately repetitive components. In animal cells, up to half of the DNA often is represented by moderately and highly repetitive components. In plants and amphibians, the moderately and highly repetitive components may account for up to 80% of the genome, so that the nonrepetitive DNA is reduced to a small component.

A significant part of the moderately repetitive DNA consists of **transposons**, short sequences of DNA (up to ~5 kb) that have the ability to move to new locations in the genome and/or to make additional copies of themselves (see *Chapter 17, Transposable Elements and Retroviruses*). In some multicellular eukaryotic genomes they may even occupy more than half of the genome (see *Section 6.5, The Human Genome Has Fewer Genes Than Originally Expected*).

Transposons are sometimes viewed as **selfish DNA**, which is

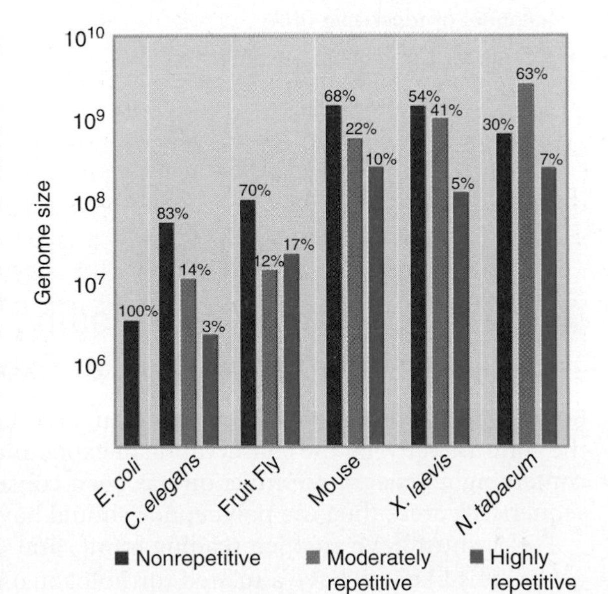

FIGURE 5.5 The proportions of different sequence components vary in eukaryotic genomes. The absolute content of nonrepetitive DNA increases with genome size but reaches a plateau at ~2 × 10^9 bp.

▸ **transposon** A DNA sequence able to insert itself (or a copy of itself) at a new location in the genome without having any sequence relationship with the target locus.

▸ **selfish DNA** DNA sequences that do not contribute to the phenotype of the organism but have self-perpetuation within the genome as their primary function.

defined as sequences that propagate themselves within a genome without contributing to the development and functioning of the organism. Transposons may cause genome rearrangements, which could confer selective advantages. It is fair to say, though, that we do not really understand why selective forces do not act against transposons becoming such a large proportion of the genome. It may be that they are selectively neutral as long as they do not interrupt or delete coding or regulatory regions. Many organisms actively suppress transposition, perhaps because in some cases deleterious chromosome breakages result (see Figure 17.7). Another term used to describe the apparent excess of DNA in some genomes is *junk DNA*, meaning genomic sequences without any apparent function, though this name may simply reflect our failure to understand the functions of many of these sequences. Of course, it is likely that there is a balance in the genome between the generation of new sequences and the elimination of unwanted sequences, and some proportion of DNA that apparently lacks function may be in the process of being eliminated.

The length of the nonrepetitive DNA component tends to increase with overall genome size as we proceed up to a total genome size of ~3×10^9 (characteristic of mammals). Further increases in genome size, however, generally reflect an increase in the amount and proportion of the repetitive components, so that it is rare for an organism to have a nonrepetitive DNA component less than 2×10^9. Therefore, the nonrepetitive DNA content of genomes is a better indication of the relative complexity of the organism. *E. coli* has 4.2×10^6 bp of nonrepetitive DNA, *C. elegans* has an order of magnitude more (6.6×10^7 bp), *D. melanogaster* has ~10^8 bp, and mammals yet another order of magnitude more at ~2×10^9 bp.

What type of DNA corresponds to polypeptide-encoding genes? Reassociation kinetics typically shows that mRNA is transcribed from nonrepetitive DNA. The amount of nonrepetitive DNA is, therefore, a better indication of the coding potential than is the size of the genome. (However, more detailed analysis based on genomic sequences shows that many exons have related sequences in other exons [see *Section 4.4, Exon Sequences Are Usually Conserved, but Introns Vary*]. Such exons evolve by a duplication to give copies that initially are identical but that then diverge in sequence during evolution.)

KEY CONCEPTS

- The kinetics of DNA reassociation after a genome has been denatured distinguish sequences by their frequency of repetition in the genome.
- Polypeptides are generally encoded by sequences in nonrepetitive DNA.
- Larger genomes within a taxonomic group do not contain more genes but have large amounts of repetitive DNA.
- A large part of moderately repetitive DNA may be made up of transposons.

CONCEPT AND REASONING CHECK

Explain how related species with no difference in complexity can have very different C-values.

5.6 Eukaryotic Protein-Coding Genes Can Be Identified by the Conservation of Exons

Some major approaches to identifying eukaryotic protein-coding genes are based on the contrast between the conservation of exons and the variation of introns. In a region containing a gene whose function has been conserved among a range of species, the sequence representing the polypeptide should have two distinctive properties:

- it must have an open reading frame, and
- it is likely to have a related (orthologous) sequence in other species.

These features can be used to identify functional genes.

Suppose we know by linkage analysis that a gene influencing a particular trait is located in a certain chromosomal region. If we lack knowledge about the nature of the gene product, how are we to identify the gene in a region that may be, for example, >1 Mb in size?

An approach that has proved successful with some genes of medical importance is to screen relatively short fragments from the region for the two properties expected of a conserved gene. First, we seek to identify sections that cross-hybridize with the genomes of other species, and then we examine these sections for open reading frames.

Sections of DNA that are conserved among species can be identified by performing a **zoo blot**. We use short fragments from the region as labeled probes to test for homologous DNA from a variety of species by Southern blotting (a technique for transferring DNA fragments from an electrophoretic gel to a filter membrane, followed by hybridization of a probe to detect the complementary or near-complementary sequence); see *Section 3.9, Blotting Methods*. If we find hybridizing fragments in several species related to that of the probe (which is usually prepared from human DNA), the probe becomes a candidate for an exon of the gene being sought.

▸ **zoo blot** The use of Southern blotting to test the ability of a DNA probe from one species to hybridize with the DNA from the genomes of a variety of other species.

The candidates are sequenced, and if they contain open reading frames, they are used to isolate surrounding genomic regions. If these appear to be part of an exon, they can then be used to identify the entire gene, to isolate the corresponding cDNA (DNA reverse transcribed from the mRNA) or mRNA itself, and, ultimately, to identify the protein. Alternatively, now that whole-genome sequencing has become more common, much of this analysis can be done by searching computer databases of complete genomes for homologs of the candidate gene of interest.

When a human disease is caused by a change in a known protein, the gene that is responsible can be identified because it encodes the protein, and its responsibility for the disease can be confirmed by showing that it has inactivating mutations in the DNA of patients but not in DNA of unaffected individuals. However, in many cases we do not know the cause of a disease at the molecular level, and it is necessary to identify the gene without any information about its protein product.

The basic criterion for identifying a gene involved in a human disease is to show that in every patient with the disease, the gene has a mutation that is not present in normal DNA. However, the extensive neutral polymorphism between individual genomes means that we may find many changes when we compare patient DNA with a standard reference DNA. Before the sequencing of the human genome, genetic linkage could be used to identify a region containing a disease gene, but the region could contain many candidate genes. For a very large gene, with introns spread over a long stretch of the genome, it was difficult to identify the critical mutations in patients. The availability of high-resolution SNP maps and of the genome sequence now makes it much easier to pinpoint a smaller region containing the gene in which sequences of standard reference and patient DNA can be directly compared.

An example of the process by which a disease gene can be tracked down is provided by the gene responsible for Duchenne muscular dystrophy (DMD), a degenerative disorder of muscle that is X-linked and affects 1 in 3500 human males. The steps in identifying the gene are summarized in **FIGURE 5.6**.

Linkage analysis localized the DMD locus to chromosomal region Xp21. Patients with the disease often have chromosomal rearrangements involving this region. By comparing the ability of DNA probes of this region to hybridize with DNA from patients with normal DNA, cloned fragments were obtained that correspond to the region that was rearranged or deleted in patients' DNA.

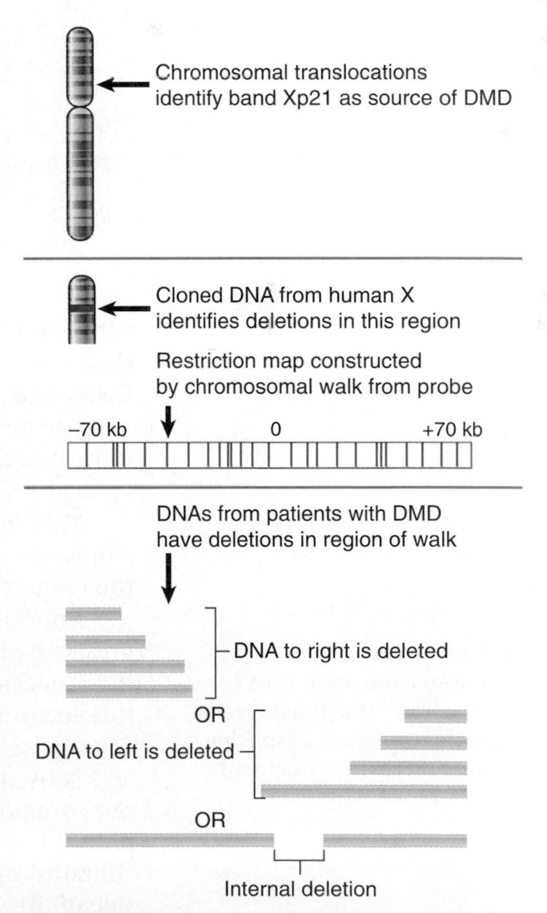

FIGURE 5.6 The gene involved in Duchenne muscular dystrophy was tracked down by chromosome mapping and "walking" to a region in which deletions can be identified with the occurrence of the disease.

FIGURE 5.7 The Duchenne muscular dystrophy gene was characterized by zoo blotting, cDNA hybridization, genomic hybridization, and identification of the protein.

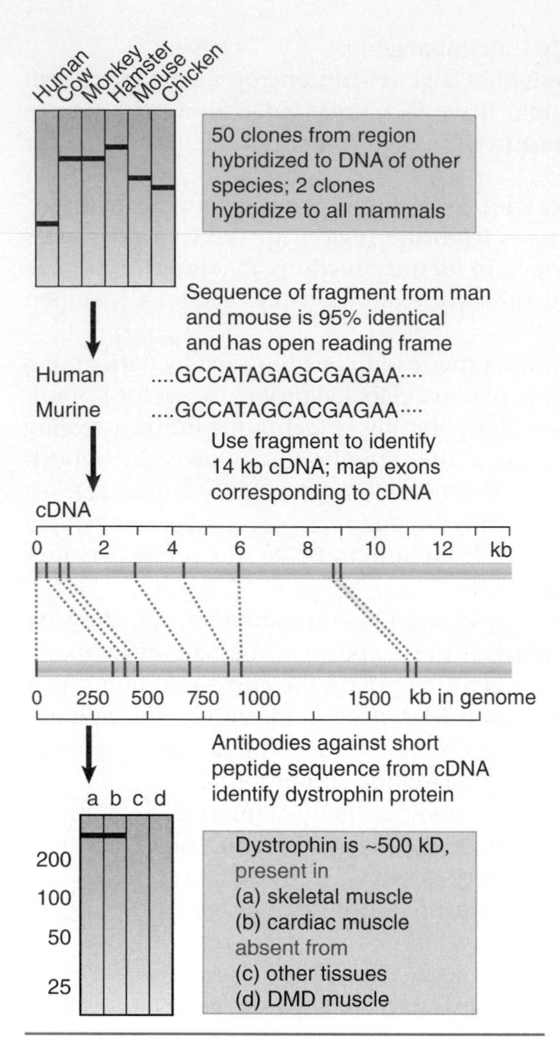

▸ **chromosomal walk** A technique for locating a gene by using the most closely linked markers as a probe for a genetic library.

Once some DNA in the general vicinity of the target gene has been obtained, it is possible to "walk" along the chromosome until the gene is reached. A **chromosomal walk** was used to construct a restriction map of the region on either side of the probe, which covered a stretch of >100 kb. Analysis of the DNA from a series of patients identified large deletions in this region that extended in either direction. The most telling deletion is one that is contained entirely within the region because this delineates a segment that must be important in gene function and indicates that the gene—or at least part of it—lies in this region.

After identifying the region of the gene, its exons and introns needed to be identified. For DMD, a zoo blot identified fragments that cross-hybridize with the mouse X chromosome and with other mammalian DNAs. As summarized in FIGURE 5.7, these were examined for open reading frames, and the sequences were typically found at exon-intron junctions. Fragments that met these criteria were used as probes to identify homologous sequences in a cDNA library prepared from muscle mRNA.

The cDNA corresponding to the gene identifies an unusually large (14 kb) mRNA. Hybridization back to the genome shows that the mRNA is encoded by >60 exons, which are spread over ~2000 kb of DNA. This makes DMD one of the longest identified genes.

The gene encodes a protein of ~500 kD called *dystrophin,* which is a component of muscle and is present in rather low amounts. All patients with the disease have deletions at this locus and lack (or have defective) dystrophin.

Muscle also has the distinction of having the largest known protein, titin, with almost 27,000 amino acids. The *titin* gene has the largest number of exons (178) and the longest single exon in the human genome (17 kb).

▸ **exon trapping** Inserting a genomic fragment into a vector whose function depends on the provision of splicing junctions by the fragment.

Another technique that allows genomic fragments to be rapidly scanned for the presence of exons is called **exon trapping**. FIGURE 5.8 shows that it starts with a vector that contains a strong promoter and has a single intron between two exons. When this vector is transfected into cells, its transcription generates large amounts of an RNA containing the sequences of the two exons. A restriction site lies within the intron and is used to insert genomic fragments from a region of interest. If a fragment does not contain an exon, there is no change in the splicing pattern, and the RNA contains only the same sequences as the parental vector. However, if the genomic fragment contains an exon flanked by two partial intron sequences, the splicing sites on either side of this exon are recognized and the sequence of the exon is inserted into the RNA between the two exons of the vector. This can be easily detected by reverse transcribing the cytoplasmic RNA into cDNA and using PCR (called *RT-PCR,* described in the next section and in *Section 3.8, PCR and RT-PCR*) to amplify the sequences between the two exons of the vector. The appearance of sequences from the genomic fragment

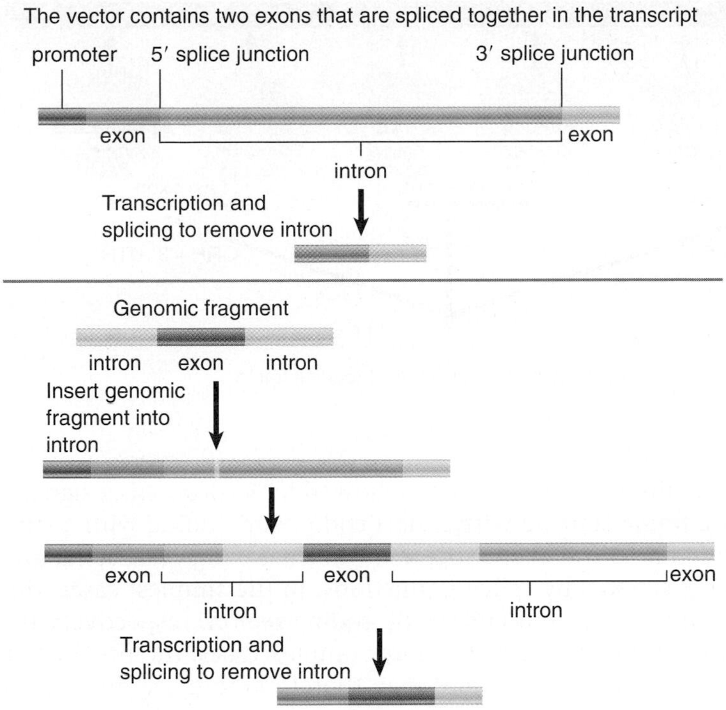

FIGURE 5.8 A special splicing vector is used for exon trapping. If an exon is present in the genomic fragment, its sequence will be recovered in the cytoplasmic RNA. If the genomic fragment consists solely of sequences from within an intron, though, splicing does not occur, and the mRNA is not exported to the cytoplasm.

in the amplified products indicates that an exon has been "trapped." In mammalian protein-coding genes, introns are usually large and exons are small, so there is a high probability that a random piece of genomic DNA will have the required structure of an exon surrounded by partial introns. In fact, exon trapping may mimic the events that have occurred naturally during evolution of genes (see *Section 8.6, How Did Interrupted Genes Evolve?*).

Ultimately, exons can be identified by the large-scale sequencing of cellular mRNAs that is now commonplace.

KEY CONCEPTS

- Conservation of exons can be used as the basis for identifying coding regions by identifying fragments whose sequences are present in multiple organisms.
- Human disease genes are identified by mapping and sequencing DNA of patients to find differences from normal DNA that are genetically linked to the disease.

CONCEPT AND REASONING CHECK

Probing for a particular exon may identify multiple sites in the genome. Why?

5.7 The Conservation of Genome Organization Helps to Identify Genes

Once we have determined the sequence of a genome, we still have to identify the genes within it. Coding sequences represent a very small fraction of the total genome. Potential exons can be identified as uninterrupted open reading frames flanked by appropriate sequences. What criteria need to be satisfied to identify a functional (intact) gene from a series of exons?

FIGURE 5.9 shows that a functional gene should consist of a series of exons for which the first exon immediately follows a promoter, the internal exons are flanked

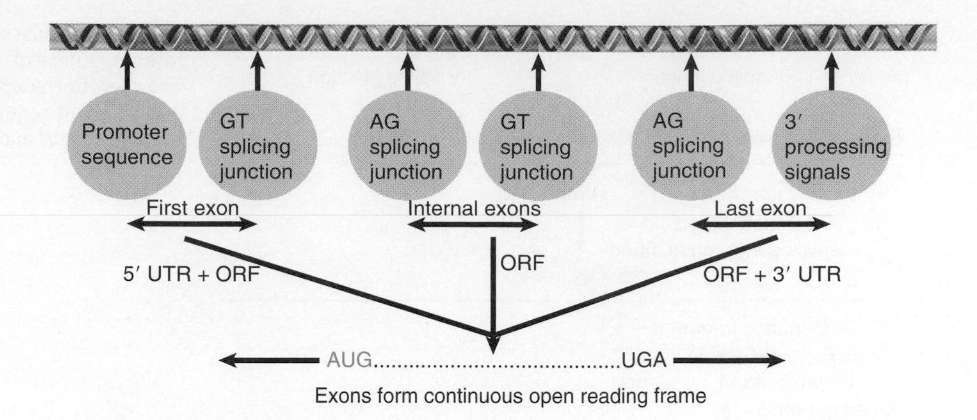

FIGURE 5.9 Exons of protein-coding genes are identified as coding sequences flanked by appropriate signals (with untranslated regions at both ends). The series of exons must generate an open reading frame with appropriate start and stop codons.

by appropriate splicing junctions, the last exon is followed by 3′ processing signals, and a single open reading frame starting with a start codon and ending with a stop codon can be deduced by joining the exons together. Internal exons can be identified as open reading frames flanked by splicing junctions. In the simplest cases, the first and last exons contain the start and end of the coding region, respectively (as well as the 5′ and 3′ untranslated regions). In more complex cases, the first or last exons may have only untranslated regions and may therefore be more difficult to identify.

The algorithms that are used to connect exons are not completely effective when the gene is very large and the exons may be separated by very large distances. For example, the initial analysis of the human genome mapped 170,000 exons into 32,000 genes. This is unlikely to be correct because it gives an average of 5.3 exons per gene, whereas the average of individual genes that have been fully characterized is 10.2. Either we have missed many exons, or they should be connected differently into a smaller number of genes in the whole genome sequence.

Even when the organization of a gene is correctly identified, there is the problem of distinguishing functional genes from pseudogenes. Many pseudogenes can be recognized by obvious defects in the form of multiple mutations that result in a nonfunctional coding sequence. Pseudogenes that have originated more recently have not accumulated so many mutations and thus may be more difficult to recognize. In an extreme example, the mouse has only one functional *Gapdh* gene (encoding glyceraldehyde phosphate dehydrogenase), but has ~400 homologous pseudogenes. Approximately 100 of these pseudogenes initially appeared to be functional in the mouse genome sequence, and individual examination was necessary to exclude them from the list of functional genes. Pseudogenes with relatively intact coding sequences but mutated transcription signals are more difficult to identify. (Some pseudogenes encode functional RNAs that play a role in gene regulation; see *Section 30.5, MicroRNAs Are Regulators in Eukaryotes*.)

How can suspected protein-coding genes be verified? If it can be shown that a DNA sequence is transcribed and processed into a translatable mRNA, it is assumed that it is functional. One technique for doing this is **reverse transcription polymerase chain reaction (RT-PCR),** in which RNA isolated from cells is reverse transcribed to DNA and subsequently amplified to many copies using the polymerase chain reaction (see *Section 3.8, PCR and RT-PCR*). The amplified DNA products can then be sequenced or otherwise analyzed to see if they have the appropriate structural features of a mature transcript. RT-PCR can also be used as a quantitative assessment of gene expression.

▸ **reverse transcription polymerase chain reaction (RT-PCR)** A technique for the detection and quantification of expression of a gene by reverse transcription and amplification of RNAs from a cell sample.

Confidence that a gene is functional can be increased by comparing regions of the genomes of different species. There has been extensive overall reorganization of sequences between the mouse and human genomes, as seen in the simple fact that there are 23 chromosomes in the human haploid genome and 20 chromosomes in the mouse haploid genome. However, at the local level the order of genes is generally the same: when pairs of human and mouse homologs are compared, the genes located on either side also tend to be homologs. This relationship is called **synteny**.

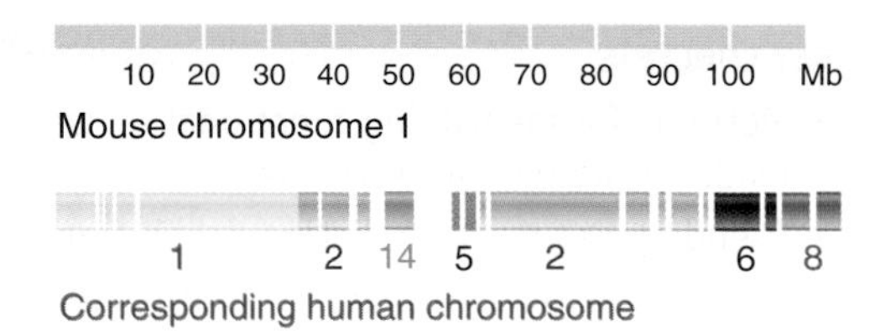

FIGURE 5.10 Mouse chromosome 1 has 21 segments of between 1 and 25 Mb in length that are syntenic with regions corresponding to parts of six human chromosomes.

▸ **synteny** A relationship between chromosomal regions of different species where homologous genes occur in the same order.

FIGURE 5.10 shows the relationship between mouse chromosome 1 and the human chromosomal set. Twenty-one segments in this mouse chromosome that have syntenic counterparts in human chromosomes have been identified. The extent of reshuffling that has occurred between the genomes is shown by the fact that the segments are spread among six different human chromosomes. The same types of relationships are found in all mouse chromosomes except for the X chromosome, which is syntenic only with the human X chromosome. This is explained by the fact that the X is a special case, subject to dosage compensation to adjust for the difference between the one copy of males and the two copies of females (see *Section 29.5, X Chromosomes Undergo Global Changes*). This restriction may apply selective pressure against the translocation of genes to and from the X chromosome.

Comparison of the mouse and human genome sequences shows that >90% of each genome lies in syntenic blocks that range widely in size from 300 kb to 65 Mb. There is a total of 342 syntenic segments, with an average length of 7 Mb (0.3% of the genome). Ninety-nine percent of mouse genes have a homolog in the human genome; for 96% that homolog is in a syntenic region.

Comparison of genomes provides interesting information about the evolution of species. The number of gene families in the mouse and human genomes is the same, and a major difference between the species is the differential expansion of particular families in the mouse genome. This is especially noticeable in genes that affect phenotypic features that are unique to the species. Of 25 gene families for which the size has been expanded in the mouse genome, 14 contain genes specifically involved in rodent reproduction, and 5 contain genes specific to the immune system.

A validation of the importance of the identification of syntenic blocks comes from pairwise comparisons of the genes within them. For example, a gene that is not in a syntenic location (that is, its context is different in the two species being compared) is twice as likely to be a pseudogene. Put another way, gene translocation away from the original locus tends to be associated with the formation of pseudogenes. The lack of a related gene in a syntenic position is, therefore, grounds for suspecting that an apparent gene may really be a pseudogene. Overall, >10% of the genes that are initially identified by analysis of the genome are likely to turn out to be pseudogenes.

As a general rule, comparisons between genomes add significantly to the effectiveness of gene prediction. When sequence features indicating active genes are conserved—for example, between human and mouse genomes—there is an increased probability that they identify active orthologs.

Identifying genes encoding RNAs other than mRNA is more difficult because we cannot use the criterion of the open reading frame. It is certainly true that the comparative genome analysis described above has increased the rigor of the analysis. For example, analysis of either the human or the mouse genome alone identifies ~500 genes encoding tRNAs, but comparison of their features suggests that fewer than 350 of these genes are in fact functional in each genome.

An active gene can be located through the use of an **expressed sequence tag (EST)**, a short portion of a transcribed sequence usually obtained from sequencing one or both ends of a cloned fragment from a cDNA library. An EST can confirm that a suspected gene is actually transcribed or help identify genes that influence particular disorders. Through the use of a physical mapping technique such as *in situ* hybridization (see *Section 7.5, Satellite DNAs Often Lie in Heterochromatin*), the chromosomal location of an EST can be determined.

▸ **expressed sequence tag (EST)** A short, sequenced fragment of a cDNA sequence that can be used to identify an actively expressed gene.

descended from a single population that lived in Africa ~200,000 years ago. This cannot be interpreted as evidence that there was only a single population at that time, however; there may have been many populations, and some or all of them may have contributed to modern human *nuclear* genetic variation.

KEY CONCEPTS

- Mitochondria and chloroplasts have genomes that show non-Mendelian inheritance. Typically they are maternally inherited.
- Organelle genomes may undergo somatic segregation in plants.
- Comparisons of human mitochondrial DNA suggest that it is descended from a single population that existed ~200,000 years ago in Africa.

CONCEPT AND REASONING CHECK

Why might the mutation rate of mitochondrial DNA in mammals be an order of magnitude higher than that of nuclear DNA? (*Hint:* What compounds are found in mitochondria and not in the nucleus?)

5.9 Organelle Genomes Are Circular DNAs That Encode Organelle Proteins

▸ **mtDNA** Mitochondrial DNA.

▸ **ctDNA (cpDNA)** Chloroplast DNA.

Most organelle genomes take the form of a single circular molecule of DNA of unique sequence (denoted **mtDNA** in the mitochondrion and **ctDNA** or **cpDNA** in the chloroplast). There are a few exceptions in unicellular eukaryotes for which mitochondrial DNA is a linear molecule.

Usually there are several copies of the genome in the individual organelle. There are also multiple organelles per cell; therefore, there are many organelle genomes per cell, so the organelle genome can be considered a repetitive sequence.

Chloroplast genomes are relatively large, usually ~140 kb in higher plants and <200 kb in unicellular eukaryotes. This is comparable to the size of a large bacteriophage genome, such as that of T4 at ~165 kb. There are multiple copies of the genome per organelle, typically 20 to 40 in a higher plant, and multiple copies of the organelle per cell, typically 20 to 40.

Mitochondrial genomes vary in total size by more than an order of magnitude. Animal cells have small mitochondrial genomes (approximately 16.6 kb in mammals). There are several hundred mitochondria per cell, and each mitochondrion has multiple copies of the DNA. The total amount of mitochondrial DNA relative to nuclear DNA is small; it is estimated to be <1%.

In yeast, the mitochondrial genome is much larger. In *Saccharomyces cerevisiae*, the exact size varies among different strains but averages ~80 kb. There are ~22 mitochondria per cell, which corresponds to ~4 genomes per organelle. In growing cells, the proportion of mitochondrial DNA can be as high as 18%.

Plants show an extremely wide range of variation in mitochondrial DNA size, with a minimum size of ~100 kb. The size of the genome makes it difficult to isolate, but restriction mapping in several plants suggests that the mitochondrial genome is usually a single sequence that is organized as a circle. Within this circle there are multiple copies of short homologous sequences. Recombination between these elements generates smaller, subgenomic circular molecules that coexist with the complete, "master" genome—a good example of the apparent complexity of plant mitochondrial DNAs.

With mitochondrial genomes sequenced from many organisms, we can now see some general patterns in the representation of functions in mitochondrial DNA. **FIGURE 5.13** summarizes the distribution of genes in mitochondrial genomes. The total number of protein-coding genes is rather small and does not correlate with the size of the genome. The 16.5 kb mammalian mitochondrial genomes encode 13 proteins, whereas the 60 to 80 kb yeast mitochondrial genomes encode as few as 8 proteins. The much larger plant mitochondrial genomes encode more proteins. Introns are found in most mitochondrial genomes, although not in the very small mammalian genomes.

The two major rRNAs are always encoded by the mitochondrial genome. The number of tRNAs encoded by the mitochondrial genome varies from none to the full complement (25 to 26 in mitochondria). This accounts for the variation in Figure 5.13.

Species	Size (kb)	Protein-coding genes	RNA-coding genes
Fungi	19–100	8–14	10–28
Protists	6–100	3–62	2–29
Plants	186–366	27–34	21–30
Animals	16–17	13	4–24

FIGURE 5.13 Mitochondrial genomes have genes encoding (mostly complex I–IV) proteins, rRNAs, and tRNAs.

The major part of the protein-coding activity is devoted to the components of the multisubunit assemblies of respiration complexes I–IV. Many ribosomal proteins are encoded in protist and plant mitochondrial genomes, but there are few or none in fungi and animal genomes. There are genes encoding proteins involved in cytoplasm-to-mitochondrion import in many protist mitochondrial genomes.

Animal mitochondrial DNA is extremely compact. There are extensive differences in the detailed gene organization found in different animal taxonomic groups, but the general principle of a small genome encoding a restricted number of functions is maintained. In mammalian mitochondria, the genome is extremely compact. There are no introns, some genes actually overlap, and almost every base pair can be assigned to a gene. With the exception of the **D-loop**, a region involved with the initiation of DNA replication, no more than 87 of the 16,569 bp of the human mitochondrial genome lie in intercistronic regions.

▸ **D-loop** A region of the animal mitochondrial DNA molecule that is variable in size and sequence and contains the origin of replication.

The complete nucleotide sequences of animal mitochondrial genomes show extensive homology in organization. The map of the human mitochondrial genome is summarized in **FIGURE 5.14**. There are 13 protein-coding regions. All the proteins are components of the electron transfer system of cellular respiration. These include cytochrome b, three subunits of cytochrome oxidase, one of the subunits of ATPase, and seven subunits (or associated proteins) of NADH dehydrogenase.

The fivefold discrepancy in size between the *S. cerevisiae* (84 kb) and mammalian (16 kb) mitochondrial genomes alone alerts us to the fact that there must be a great difference in their genetic organization in spite of their common function. The number of endogenously synthesized products concerned with mitochondrial enzymatic functions appears to be similar. Does the additional genetic material in yeast mitochondria encode other proteins, perhaps concerned with regulation, or is it unexpressed?

The map shown in **FIGURE 5.15** accounts for the major RNA and protein products of the yeast mitochondrion. The most notable feature is the dispersion of loci on the map.

The two largest loci are the interrupted genes *box* (encoding cytochrome b) and *oxi3* (encoding subunit 1 of cytochrome oxidase). Together these two genes are almost as long as the entire mitochondrial genome in mammals! Many of the long introns in these genes have open reading frames in register with the preceding exon (see *Section 23.5, Some Group I Introns Code for Endonucleases That Sponsor Mobility*). This adds several proteins, all synthesized in low amounts, to the complement of the yeast mitochondrion.

The remaining genes are uninterrupted. They correspond to the other two subunits of cytochrome oxidase encoded by the mitochondrion, to the subunit(s) of the ATPase, and (in the case of *var1*) to a mitochondrial ribosomal protein. The total number of yeast mitochondrial genes is unlikely to exceed ~25.

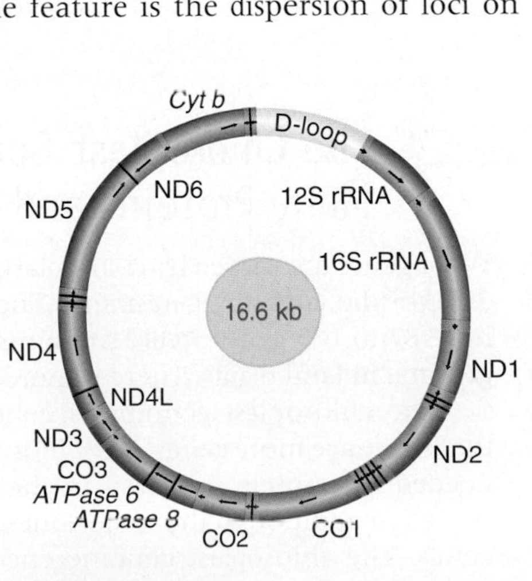

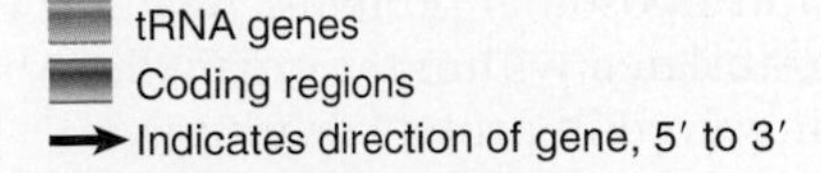

CO: cytochrome oxidase
ND: NADH dehydrogenase

FIGURE 5.14 Human mitochondrial DNA has 22 tRNA genes, 2 rRNA genes, and 13 protein-coding regions. Fourteen of the 15 protein-coding or rRNA-coding regions are transcribed in the same direction. Fourteen of the tRNA genes are expressed in the clockwise direction, and 8 are read counterclockwise.

FIGURE 5.15 The mitochondrial genome of *S. cerevisiae* contains both interrupted and uninterrupted protein-coding genes, rRNA genes, and tRNA genes (positions not indicated). Arrows indicate direction of transcription.

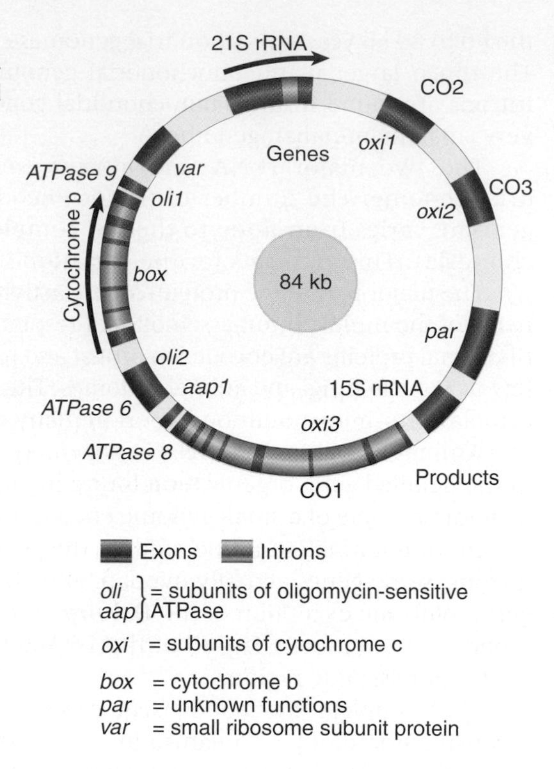

KEY CONCEPTS

- Organelle genomes are usually (but not always) circular molecules of DNA.
- Organelle genomes encode some of, but not all, the proteins used in the organelle.
- Animal cell mitochondrial DNA is extremely compact and typically encodes 13 proteins, 2 rRNAs, and 22 tRNAs.
- Yeast mitochondrial DNA is 5× longer than animal mtDNA because of the presence of long introns.

CONCEPT AND REASONING CHECK

What accounts for the variation in size of mitochondrial genomes among different eukaryotes?

5.10 The Chloroplast Genome Encodes Many Proteins and RNAs

Genes	Types
RNA coding	
16S rRNA	1
23S rRNA	1
4.5S rRNA	1
5S rRNA	1
tRNA	30–32
Gene Expression	
r-proteins	20–21
RNA polymerase	3
Others	2
Chloroplast functions	
Rubisco and thylakoids	31–32
NADH dehydrogenase	11
Total	105–113

FIGURE 5.16 The chloroplast genome in land plants encodes 4 rRNAs, 30 tRNAs, and ~60 proteins.

What genes are carried by chloroplasts? Chloroplast DNAs vary in length from ~120 to 217 kb (the largest in geranium). The sequenced chloroplast genomes (>100 in total) have 87 to 183 genes. **FIGURE 5.16** summarizes the functions encoded by the chloroplast genome in land plants. There is more variation in the chloroplast genomes of algae.

The chloroplast genome is generally similar to that of mitochondria, except that there are more genes. The chloroplast genome encodes all the rRNAs and tRNAs needed for protein synthesis in the chloroplast. The ribosome includes two small rRNAs in addition to the major ones. The tRNA set may include all of the necessary genes. The chloroplast genome encodes ~50 proteins, including RNA polymerase and ribosomal proteins. Again, the rule is that organelle genes are transcribed and translated within the organelle. About half of the chloroplast genes encode proteins involved in protein synthesis.

Introns in chloroplasts fall into two general classes. Those in tRNA genes are usually (although not inevitably) located in the anticodon loop, like the introns found in yeast nuclear tRNA genes (see *Section 21.13, tRNA Splicing Involves Cutting and Rejoining in Separate Reactions*). Those in protein-coding genes resemble the introns of mitochondrial genes (see *Chapter 23, Catalytic RNA*). This places the endosymbiotic event at a time in evolution before the separation of prokaryotes with uninterrupted genes.

The chloroplast is the site of photosynthesis. Many of its genes encode proteins of photosynthetic complexes located in the thylakoid membranes. The constitution of these complexes shows a different balance from that of mitochondrial complexes. Although some complexes are like mitochondrial complexes in that they have some subunits encoded by the organelle genome and some by the nuclear genome, other chloroplast complexes are encoded entirely by one genome. For example, the gene for the large subunit of ribulose bisphosphate carboxylase (RuBisCO, which catalyzes the carbon fixation reaction of the Calvin cycle), *rbcL*, is contained in the chloroplast genome; variation in this gene is frequently used as a basis for reconstructing plant phylogenies. However, the gene for the small rubisco subunit, *rbcS*, is usually carried in the nuclear genome. On the other hand, genes for photosystem protein complexes are found on the chloroplast genome, whereas those for the LHC (light-harvesting complex) proteins are nuclear encoded.

KEY CONCEPT

- Chloroplast genomes vary in size but are large enough to encode 50 to 100 proteins as well as the rRNAs and tRNAs.

CONCEPT AND REASONING CHECK

What advantage is there to chloroplast genomes encoding RNA polymerase, tRNAs, rRNAs, and ribosomal proteins?

5.11 Mitochondria and Chloroplasts Evolved by Endosymbiosis

How is it that an organelle evolved so that it contains genetic information for some of its functions, whereas the information for other functions is encoded in the nucleus? **FIGURE 5.17** shows the endosymbiotic hypothesis for mitochondrial evolution, in which primitive cells captured bacteria that provided the function of cellular respiration and over time evolved into mitochondria. At this point, the proto-organelle must have contained all the genes needed to direct its functions. A similar mechanism has been proposed for the origin of chloroplasts.

Sequence homologies suggest that mitochondria and chloroplasts evolved separately from lineages that are common with different eubacteria, with mitochondria sharing an origin with α-purple bacteria and chloroplasts sharing an origin with cyanobacteria. The closest known relative of mitochondria among the bacteria is *Rickettsia* (the causative agent of typhus), which is an obligate intracellular parasite that is probably descended from free-living bacteria. This reinforces the idea that mitochondria originated in an endosymbiotic event involving an ancestor that is also common to *Rickettsia*.

The endosymbiotic origin of the chloroplast is emphasized by the relationships between its genes and their counterparts in bacteria. The organization of the rRNA genes in particular is closely related to that of a cyanobacterium, which pins down more precisely the last common ancestor between chloroplasts and bacteria. Not surprisingly, cyanobacteria are photosynthetic.

Two changes must have occurred as the bacterium became integrated into the recipient cell and evolved into the mitochondrion (or chloroplast). The organelles have far fewer genes than an independent bacterium and have lost

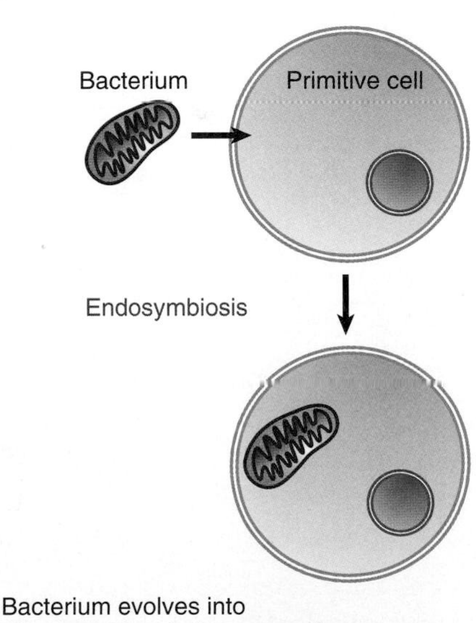

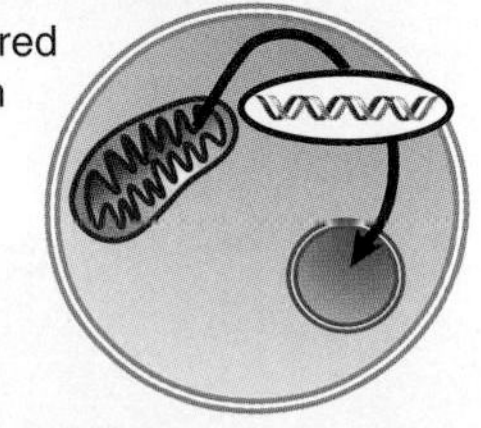

FIGURE 5.17 Mitochondria originated by an endosymbiotic event when a bacterium was captured by a eukaryotic cell.

many of the gene functions that are necessary for independent life (such as metabolic pathways). The majority of genes encoding organelle functions are in fact now located in the nucleus; thus these genes must have been transferred there from the organelle.

Transfer of DNA between an organelle and the nucleus has occurred over evolutionary history and still continues. The rate of transfer can be measured directly by introducing into an organelle a gene that can function only in the nucleus, because (for example) it contains a nuclear intron, or because the protein must function in the cytosol. In terms of providing the material for evolution, the transfer rates from organelle to nucleus are roughly equivalent to the rate of single gene mutation. DNA introduced into mitochondria is transferred to the nucleus at a rate of 2×10^{-5} per generation. Experiments to measure transfer in the reverse direction, from nucleus to mitochondrion, suggest that the rate is much lower, $<10^{-10}$. When a nuclear-specific antibiotic resistance gene is introduced into chloroplasts, its transfer to the nucleus and successful expression can be followed by screening seedlings for resistance to the antibiotic. This shows that transfer occurs at a rate of 1 in 16,000 seedlings, or 6×10^{-5} per generation.

Transfer of a gene from an organelle to the nucleus requires physical movement of the DNA, of course, but successful expression also requires changes in the coding sequence. Organelle proteins that are encoded by nuclear genes have special sequences that allow them to be imported into the organelle after they have been synthesized in the cytoplasm. These sequences are not required by proteins that are synthesized within the organelle. Perhaps the process of effective gene transfer occurred at a period when compartments were less rigidly defined, so that it was easier both for the DNA to be relocated and for the proteins to be incorporated into the organelle irrespective of the site of synthesis.

Phylogenetic analyses show that gene transfers have occurred independently in many different lineages. It appears that transfers of mitochondrial genes to the nucleus occurred only early in animal cell evolution, but it is possible that the process is still continuing in plant cells. The number of transfers can be large; there are >800 nuclear genes in *Arabidopsis* whose sequences are related to genes in the chloroplasts of other plants. These genes are candidates for evolution from genes that originated in the chloroplast.

KEY CONCEPTS

- Both mitochondria and chloroplasts are descended from bacterial ancestors.
- Most of the genes of the mitochondrial and chloroplast genomes have been transferred to the nucleus during the organelle's evolution.

CONCEPT AND REASONING CHECKS

1. What evidence is there that chloroplasts are descended from prokaryotic ancestors?
2. Design an experiment to measure the rate of transfer of genes from mitochondria to the nucleus in yeast cells.

5.12 Summary

The DNA sequences composing a eukaryotic genome can be classified into three groups:

- nonrepetitive sequences that are unique;
- moderately repetitive sequences that are dispersed and repeated a small number of times, with some copies not being identical; and
- highly repetitive sequences that are short and usually repeated as tandem arrays.

The proportions of these types of sequences are characteristic for each genome, although larger genomes tend to have a smaller proportion of nonrepetitive DNA.

Almost 50% of the human genome consists of repetitive sequences, the vast majority corresponding to transposon sequences. Most structural genes are located in nonrepetitive DNA. The amount of nonrepetitive DNA is a better reflection of the complexity of the organism than the total genome size; the greatest amount of nonrepetitive DNA in genomes is ~2×10^9 bp.

Non-Mendelian inheritance is explained by the presence of DNA in organelles in the cytoplasm. Mitochondria and chloroplasts are membrane-bounded systems in which some proteins are synthesized within the organelle, while others are imported. The organelle genome is usually a circular DNA that codes for all the RNAs and some of the proteins required by the organelle.

Mitochondrial genomes vary greatly in size from the small 16 kb mammalian genome to the 570 kb genome of higher plants. The larger genomes may encode additional functions. Chloroplast genomes range from ~120 to 217 kb. Those that have been sequenced have similar organizations and coding functions. In both mitochondria and chloroplasts, many of the major proteins contain some subunits synthesized in the organelle and some subunits imported from the cytosol. Transfers of DNA have occurred between chloroplasts or mitochondria and nuclear genomes.

CHAPTER QUESTIONS

1. Single nucleotide polymorphisms (SNPs) occur approximately how often in the human genome?
A. one per 560 bases
B. one per 950 bases
C. one per 1330 bases
D. one per 2040 bases

2. If a genetic cross between two flies, one with dark eye pigment and a single DNA fragment hybridizing with a probe and one lacking pigment and having two fragments hybridizing with the probe, results in progeny with 40% of one parental type 40% of the other parental type, and 10% of each of the recombinant phenotypes, what is the genetic map distance between the restriction marker and the eye-color marker?
A. 10%
B. 20%
C. 40%
D. 80%

3. Which of the following groups of organisms shows the least genome complexity?
A. insects
B. mollusks
C. fungi
D. algae

4. Which species shows the greatest percentage of repetitive DNA?
A. *C. elegans*
B. *X. laevis*
C. *D. melanogaster*
D. *H. sapiens*

5. Prokaryotic genomes generally contain:
A. nearly 100% nonrepetitive DNA.
B. mostly nonrepetitive DNA with about 20% moderately repetitive DNA.
C. mostly moderately or highly repetitive DNA with about 20% nonrepetitive DNA.
D. about equal amounts of moderately repetitive and nonrepetitive DNA.

6. Plant and amphibian genomes contain:
   A. nearly 100% nonrepetitive DNA.
   B. mostly nonrepetitive DNA with about 20% moderately repetitive DNA.
   C. mostly moderately or highly repetitive DNA with about 20% nonrepetitive DNA.
   D. about equal amounts of moderately repetitive and nonrepetitive DNA.
7. Duchenne muscular dystrophy (DMD) is caused by a ______________ in the gene for dystrophin:
   A. large deletion
   B. large insertion
   C. single base change
   D. small insertion
8. In most animals, mitochondrial DNA is inherited:
   A. equally from both parents.
   B. predominantly from the male parent.
   C. from the female parent only.
   D. from the male parent only.
9. Over the course of evolution, genes from the mitochondrial and chloroplast genomes have:
   A. been transferred to nuclear chromosomes.
   B. been transferred from one organelle to the other.
   C. undergone extensive recombination.
   D. evolved to other functions.
10. The closest known relative of mitochondria among bacteria is:
   A. cyanobacteria.
   B. *Rickettsia*.
   C. *Salmonella*.
   D. *Bacillus*.

KEY TERMS

ctDNA (cpDNA)
chromosomal walk
D-loop
DNA fingerprinting
exon trapping
expressed sequence tag (EST)
extranuclear genes
genome
haplotype
interactome
linkage map
maternal inheritance
mtDNA
non-Mendelian inheritance
nonrepetitive DNA
polymorphism
proteome
repetitive DNA
restriction fragment length polymorphism (RFLP)
restriction map
reverse transcription polymerase chain reaction (RT-PCR)
selfish DNA
single nucleotide polymorphism (SNP)
synteny
transcriptome
transposon
zoo blot

FURTHER READING

Altshuler, D., Brooks, L. D., Chakravarti, A., Collins, F. S., Daly, M. J., and Donnelly, P. (2005). A haplotype map of the human genome. *Nature* **437**, 1299–1320. A report on HapMap, the public database of SNPs (single nucleotide polymorphisms) that represents a large-scale attempt to document human genetic variation.

Gregory, T. R. (2001). Coincidence, coevolution, or causation? DNA content, cell size, and the C-value enigma. *Biol. Rev. Camb. Philos. Soc.* **76**, 65–101. A review of the C-value enigma (paradox) and the hypotheses to account for it, and a discussion of the positive correlation between C value and cell size.

Hinds, D. A., Stuve, L. L., Nilsen, G. B., Halperin, E., Eskin, E., Ballinger, D. G., Frazer, K. A., and Cox, D. R. (2005). Whole-genome patterns of common DNA variation in three human populations. *Science* **307**, 1072–1079. A report on more than 1.5 million SNPs from 71 human individuals descended from three different populations as a sample of human genetic variation.

Lang, B. F., Gray, M. W., and Burger, G. (1999). Mitochondrial genome evolution and the origin of eukaryotes. *Annu. Rev. Genet.* **33**, 351–397. Evidence suggests that mitochondria evolved from an endosymbiotic α-proteobacterium at about the same time that the nucleus evolved.

Sharp, A. J., Cheng, Z., and Eichler, E. E. (2006). Structural variation of the human genome. *Annu. Rev. Genom. Hum. G.* **7**, 407–442. A review of the structural rearrangements of the human genome and their effects on phenotypic variation.

Sugiura, M., Hirose, T., and Sugita, M. (1998). Evolution and mechanism of translation in chloroplasts. *Annu. Rev. Genet.* **32**, 437–459. Comparison of multiple chloroplast genomes reveals multiple mechanisms for translation within these organelles.

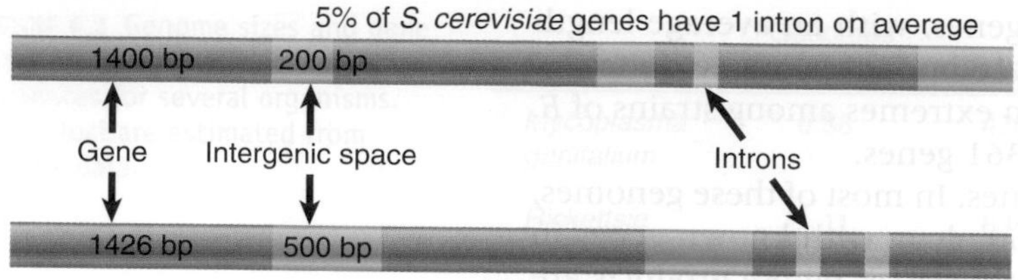

FIGURE 6.5 The *S. cerevisiae* genome of 13.5 Mb has 6000 genes, almost all uninterrupted. The *S. pombe* genome of 12.5 Mb has 5000 genes, almost half having introns. Gene sizes and spacing are fairly similar.

is high; organization is generally similar, although the spaces between genes are a bit shorter in *S. cerevisiae*. About half of the genes identified by sequence were either known previously or related to known genes. The remainder were previously unknown, which gives some indication of the number of new types of genes that may be discovered.

The identification of long reading frames on the basis of sequence is quite accurate. However, ORFs coding for less than 100 amino acids cannot be identified solely by sequence because of the high occurrence of false positives. Analysis of gene expression suggests that only ~300 of 600 such ORFs in *S. cerevisiae* are likely to be active genes.

A powerful way to validate gene structure is to compare sequences in closely related species: if a gene is active, it is likely to be conserved. Comparisons between the sequences of four closely related yeast species suggest that 503 of the genes originally identified in *S. cerevisiae* do not have counterparts in the other species and therefore should not be considered active genes. This reduces the total estimated gene number for *S. cerevisiae* to 5726.

The genome of *Caenorhabditis elegans* DNA varies between regions rich in genes and regions in which genes are more sparsely distributed. The total sequence contains ~18,500 genes. Only ~42% of the genes have suspected counterparts outside Nematoda.

The fly genome is larger than the worm genome, but there are fewer genes in some species (~14,000 in *D. melanogaster*) and more in others (e.g., ~23,000 in *D. persimilis*). The number of different transcripts is somewhat larger as the result of alternative splicing. We do not understand why *C. elegans*—arguably, a less complex organism—has 30% more genes than the fly, but it may be because *C. elegans* has a larger average number of genes per gene family than does *D. melanogaster*, so the numbers of *unique* genes of the two species are more similar. A comparison of 12 *Drosophila* genomes reveals that there can be a fairly large range of gene number among closely related species. In some cases, there are several thousand genes that are species specific. This emphasizes forcefully the lack of an exact relationship between gene number and complexity of the organism.

The plant *Arabidopsis thaliana* has a genome size intermediate between the worm and the fly but has a larger gene number (~25,000) than either. This again shows the lack of a clear relationship between complexity and genome size and also emphasizes the special quality of plants, which may have more genes (due to ancestral duplications) than animal cells. A majority of the *Arabidopsis* genome is found in duplicated segments, suggesting that there was an ancient doubling of the genome (to result in a tetraploid). Only 35% of *Arabidopsis* genes are present as single copies.

The genome of rice (*Oryza sativa*) is ~4× larger than that of *Arabidopsis*, but the number of genes is only ~25% larger, probably ~32,000. Repetitive DNA occupies 42% to 45% of the genome. More than 80% of the genes found in *Arabidopsis* are also found in rice. Of these common genes, ~8000 are found in *Arabidopsis* and rice but not in any of the bacterial or animal genomes that have been sequenced. This is probably the set of genes that codes for plant-specific functions, such as photosynthesis.

From the 12 sequenced *Drosophila* genomes, we can form an impression of how many genes are devoted to each type of function. **FIGURE 6.6** breaks down the functions into different categories. Among the genes that are identified, we find more than 3000 enzymes, ~900 transcription factors, and ~700 transporters and ion channels. About a quarter of the genes encode products of unknown function.

Polypeptide size increases from prokaryotes to eukaryotes. The archaean *M. jannaschi* and bacterium *E. coli* have average polypeptide lengths of 287 and 317

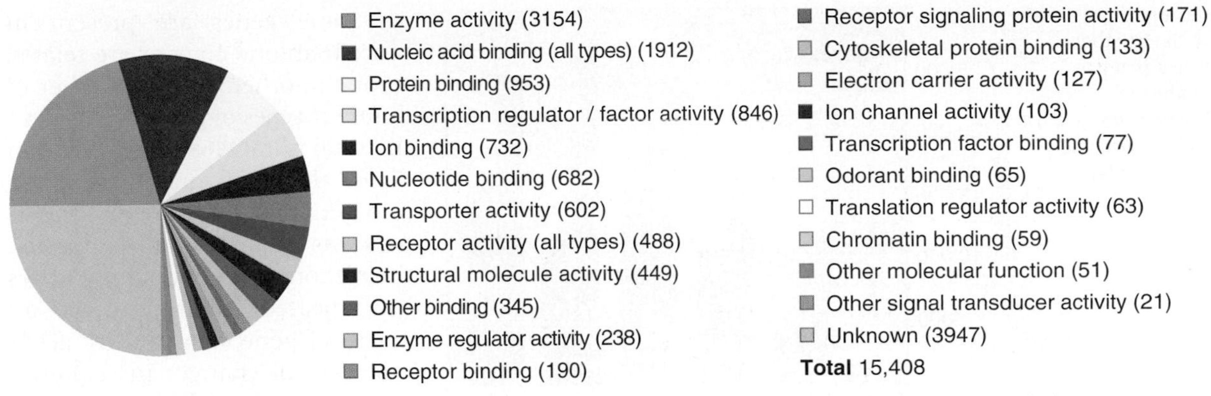

FIGURE 6.6 Functions of *Drosophila* genes based on comparative genomics of 12 species. The functions of about a quarter of the genes of *Drosophila* are unknown. Adapted from *Drosophila* 12 Genomes Consortium, "Evolution of genes and genomes on the *Drosophila* phylogeny," *Nature* **450** (2007): 203–218.

amino acids, respectively, whereas *S. cerevisiae* and *C. elegans* have average lengths of 484 and 442 amino acids, respectively. Large polypeptides (>500 amino acids) are rare in bacteria but comprise a significant component (~1/3) in eukaryotes. The increase in length is due to the addition of extra domains, with each domain typically constituting 100 to 300 amino acids. The increase in polypeptide size, however, is responsible for only a very small part of the increase in genome size.

Another insight into gene number is obtained by counting the number of expressed protein-coding genes. If we relied upon the estimates of the number of different mRNA species that can be counted in a cell, we would conclude that the average vertebrate cell expresses ~10,000 to 20,000 genes. The existence of significant overlaps between the mRNA populations in different cell types would suggest that the total expressed gene number for the organism should be within the same order of magnitude. The estimate for the total human gene number of 20,000 to 25,000 (see *Section 6.5, The Human Genome Has Fewer Genes Than Originally Expected*) would imply that a significant proportion of the total gene number is actually expressed in any given cell.

Eukaryotic genes are transcribed individually, with each gene producing a **monocistronic mRNA**. There is only one general exception to this rule: in the genome of *C. elegans*, ~15% of the genes are organized into **polycistronic mRNAs** (which are associated with the use of *trans*-splicing to allow expression of the downstream genes in these units; see *Section 21.12,* trans-*Splicing Reactions Use Small RNAs*).

- **monocistronic mRNA** mRNA that encodes a single polypeptide.
- **polycistronic mRNA** mRNA that includes coding regions representing more than one gene.

KEY CONCEPT

- There are 6000 genes in yeast; 18,500 in a worm; 13,600 in a fly; 25,000 in the small plant *Arabidopsis*; and probably 20,000 to 25,000 in mice and humans.

CONCEPT AND REASONING CHECK

Why is it that in multicellular eukaryotes the genome size is not a good indication of the number of genes?

6.4 How Many Different Types of Genes Are There?

Some genes are unique; others belong to families in which the other members are related (but not usually identical). The proportion of unique genes declines, and the proportion of genes in families increases, with increasing genome size.

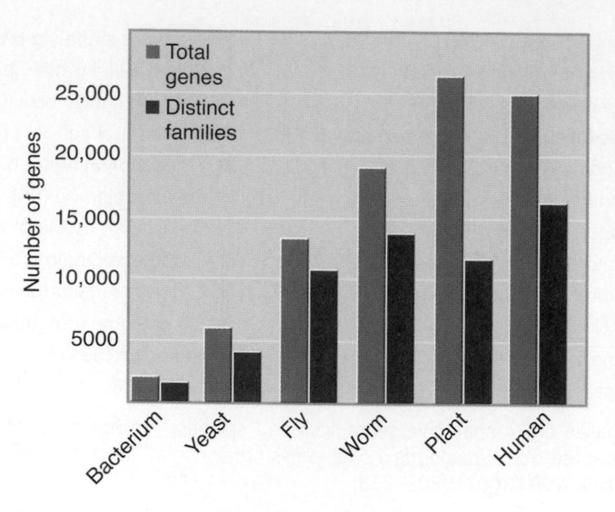

FIGURE 6.7 Many genes are duplicated, and as a result the number of different gene families is much less than the total number of genes. The histogram compares the total number of genes with the number of distinct gene families.

Some genes are present in more than one copy or are related to one another, so the number of different types of genes is less than the total number of genes. We can divide the total number of genes into sets that have related members, as defined by comparing their exons. (A gene family arises by repeated duplication of an ancestral gene followed by accumulation of changes in sequence among the copies. Most often the members of a family are similar but not identical.) The number of types of genes is calculated by adding the number of unique genes (for which there is no other related gene at all) to the numbers of families that have two or more members.

FIGURE 6.7 compares the total number of genes with the number of distinct families in each of six genomes. In bacteria, most genes are unique, so the number of distinct families is close to the total gene number. The situation is different even in the unicellular eukaryote *S. cerevisiae*, for which there is a significant proportion of repeated genes. The most striking effect is that the number of genes increases quite sharply in the multicellular eukaryotes, but the number of gene families does not change much.

FIGURE 6.8 shows that the proportion of unique genes drops sharply with increasing genome size. When genes are present in families, the number of members in a family is small in bacteria and unicellular eukaryotes but is large in multicellular eukaryotes. Much of the extra genome size of *Arabidopsis* is accounted for by families with more than four members.

If every gene is expressed, the total number of genes will account for the total number of polypeptides required to make the organism (the proteome). There are two conditions, however, that cause the size of the proteome to be different from the total gene number. First, genes can be duplicated, and as a result some of them encode the same polypeptide (although it may be expressed at a different time or in a different type of cell) and others may encode related polypeptides that also play the same role at different times or in different cell types. Second, the proteome can be larger than the number of genes because some genes can produce more than one polypeptide by means of alternative splicing.

What is the core proteome—the basic number of the different types of polypeptides in the organism? Although difficult to estimate because of the possibility of alternative splicing, a minimum estimate is given by the number of gene families, ranging from 1400 in the bacterium to ~4000 in the yeast and 11,000 to 14,000 for the fly and the worm.

What is the distribution of the proteome by type of protein? The 6000 proteins of the yeast proteome include 5000 soluble proteins and 1000 transmembrane proteins. About half of the proteins are cytoplasmic, a quarter are in the nucleus, and the remainder are split between the mitochondrion and the endoplasmic reticulum (ER)/Golgi system.

How many genes are common to all organisms (or to groups such as bacteria or multicellular eukaryotes), and how many are specific to lower-level taxonomic groups? **FIGURE 6.9** shows the comparison of fly genes to those of the

	Unique genes	Families with 2–4 members	Families with >4 members
H. influenzae	89%	10%	1%
S. cerevisiae	72%	19%	9%
D. melanogaster	72%	14%	14%
C. elegans	55%	20%	26%
A. thaliana	35%	24%	41%

FIGURE 6.8 The proportion of genes that are present in multiple copies increases with genome size in multicellular eukaryotes.

worm (another multicellular eukaryote) and yeast (a unicellular eukaryote). Genes that encode corresponding polypeptides in different organisms are called **orthologous genes,** or **orthologs** (see *Section 4.8, Members of a Gene Family Have a Common Organization*). Operationally, we usually consider that two genes in different organisms are orthologs if their sequences are similar over >80% of the length. By this criterion, ~20% of the fly genes have orthologs in both yeast and worm. These genes are probably required by all eukaryotes. The proportion increases to 30% when fly and worm are compared, probably representing the addition of gene functions that are common to multicellular eukaryotes. This still leaves a major proportion of genes as encoding proteins that are required specifically by either flies or worms, respectively.

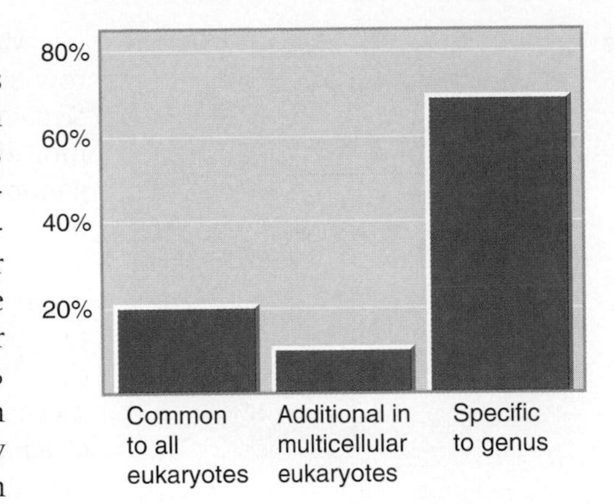

FIGURE 6.9 The fly genome can be divided into genes that are (probably) present in all eukaryotes, additional genes that are (probably) present in all multicellular eukaryotes, and genes that are more specific to subgroups of species that include flies.

▸ **orthologous genes (orthologs)** Related genes in different species.

A minimum estimate of the size of an organismal proteome can be deduced from the number and structures of genes, and a cellular or organismal proteome size can also be directly measured by analyzing the total polypeptide content of the cell or organism. By such approaches, some proteins have been identified that were not suspected on the basis of genome analysis; this has led to the identification of new genes. Several methods are used for large-scale analysis of proteins. Mass spectrometry can be used for separating and identifying proteins in a mixture obtained directly from cells or tissues. Hybrid proteins bearing tags can be obtained by expression of cDNAs made by linking the sequences of ORFs to appropriate expression vectors that incorporate the sequences for affinity tags. This allows array analysis to be used to analyze the products. These methods also can be effective in comparing the proteins of two tissues—for example, a tissue from a healthy individual and one from a patient with disease—to pinpoint the differences.

In addition to functional genes, there are also copies of genes that have become nonfunctional (identified as such by interruptions in their protein-coding sequences). These are called *pseudogenes* (see *Section 8.11, Pseudogenes Are Nonfunctional Gene Copies*). The number of pseudogenes can be large. In the mouse and human genomes, the number of pseudogenes is ~10% of the number of (potentially) active genes (see *Section 5.7, The Conservation of Genome Organization Helps to Identify Genes*). Some of these pseudogenes may serve some function by producing regulatory microRNAs; see *Chapter 30, Regulatory RNA*.

KEY CONCEPTS

- The sum of the number of unique genes and the number of gene families is an estimate of the number of types of genes.
- The minimum size of the proteome can be estimated from the number of types of genes.

CONCEPT AND REASONING CHECK

Why isn't the size of the proteome an accurate estimate of the number of types of genes?

6.5 The Human Genome Has Fewer Genes Than Originally Expected

The human genome was the first vertebrate genome to be sequenced. This massive task has revealed a wealth of information about the genetic makeup of our species and about the evolution of genomes in general. Our understanding is deepened further by the ability to compare the human genome sequence with other sequenced vertebrate genomes.

FIGURE 6.10 Genes occupy 25% of the human genome, but protein-coding sequences are only a tiny part of this fraction.

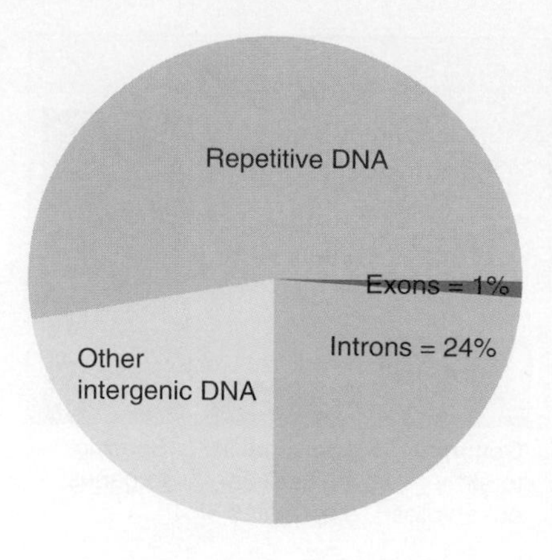

Mammal genomes generally fall into a narrow size range averaging about 3×10^9 bp (see *Section 8.7, Why Are Some Genomes So Large?*). The mouse genome is ~14% smaller than the human genome, probably because it has had a higher rate of deletion. The genomes contain similar gene families and genes, with most genes having an ortholog in the other genome but with differences in the number of members of a family, especially in those cases for which the functions are specific to the species (see *Section 5.7, The Conservation of Genome Organization Helps to Identify Genes*). Originally estimated to have ~30,000 genes, the mouse genome is now thought to have about the same number of genes as the human genome, 20,000 to 25,000. The 23,000 protein-coding genes are accompanied by ~3000 genes representing RNAs that do not encode proteins; these are generally small (aside from the ribosomal RNAs). Almost half of these genes encode transfer RNAs. In addition to the active genes, ~1200 pseudogenes have been identified.

The human (haploid) genome contains 22 autosomes plus the X and Y chromosomes. The chromosomes range in size from 45 to 279 Mb of DNA, making a total genome size of 3286 Mb (~3.3×10^9 bp). On the basis of chromosome structure, the genome can be divided into regions of euchromatin (containing many active genes) and heterochromatin, with a much lower density of active genes (see *Section 9.5, Chromatin Is Divided into Euchromatin and Heterochromatin*). The euchromatin comprises the majority of the genome, ~2.9×10^9 bp. The identified genome sequence represents ~90% of the euchromatin. In addition to providing information on the genetic content of the genome, the sequence also identifies features that may be of structural importance (see *Section 9.6, Chromosomes Have Banding Patterns*).

FIGURE 6.10 shows that a tiny proportion (~1%) of the human genome is accounted for by the exons that actually encode polypeptides. The introns that constitute the remaining sequences of protein-coding genes bring the total of DNA concerned with producing proteins to ~25%. As shown in **FIGURE 6.11**, the average human gene is 27 kb long, with nine exons that include a total coding sequence of 1340 bp. The average coding sequence is, therefore, only 5% of the length of an average protein-coding gene.

By any measure, the total human gene number of 20,000 to 25,000 is much less than was originally expected—most estimates before the genome was sequenced were ~100,000. It shows a relatively small increase over flies and worms (~13,600 and ~18,500, respectively), not to mention the plant *Arabidopsis thaliana* (~25,000; see Figure 6.2) and rice (32,000). We should not, however, be particularly surprised by the notion that it does not take a great number of additional genes to make a more complex organism. The difference in DNA sequences between the human and chimpanzee genomes is extremely small (there is >99% similarity), so it is clear that the functions and interactions between a similar set of genes can produce different results. The functions of specific groups of genes may be especially important because detailed comparisons of orthologous genes in humans and chimpanzees suggest that there has been rapid evolution of certain classes of genes, including some involved in early development, olfaction, and hearing—all functions that are relatively specialized in these species.

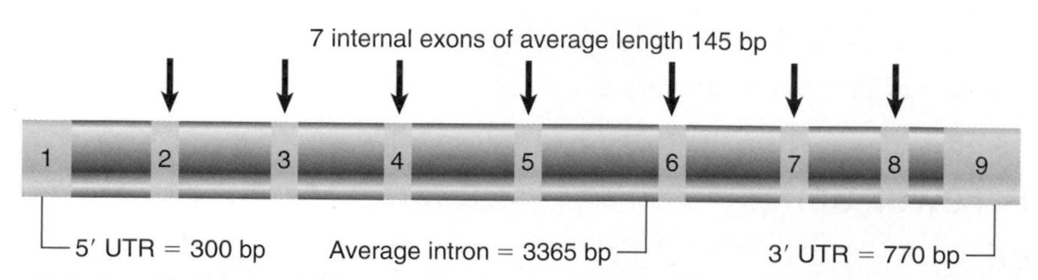

FIGURE 6.11 The average human gene is 27 kb long and has nine exons, usually comprising two longer exons at each end and seven internal exons. The UTRs in the terminal exons are the untranslated (noncoding) regions at each end of the gene. (This is based on the average. Some genes are extremely long, which makes the median length 14 kb with seven exons.)

The number of protein-coding genes is less than the number of potential polypeptides because of mechanisms such as alternative splicing, alternative promoter selection, and alternative poly(A) site selection that can result in several polypeptides from the same gene (see *Section 21.11, Alternative Splicing is a Rule Rather than an Exception in Higher Eukaryotic Cells*). The extent of alternative splicing is greater in humans than in flies or worms; it may affect as many as 60% of the genes, so the increase in size of the human proteome relative to that of the other eukaryotes may be larger than the increase in the number of genes. A sample of genes from two chromosomes suggests that the proportion of the alternative splices that actually result in changes in the polypeptide sequence may be as high as 80%. This could increase the size of the proteome to 50,000 to 60,000 members.

In terms of the diversity of the number of gene families, however, the discrepancy between humans and the other eukaryotes may not be so great. Many of the human genes belong to gene families. An analysis of ~25,000 genes identified 3500 unique genes and 10,300 gene pairs. As can be seen from Figure 6.7, this extrapolates to a number of gene families only slightly larger than that of worms or flies.

KEY CONCEPTS

- Only 1% of the human genome consists of exons.
- The exons comprise ~5% of each gene, so genes (exons plus introns) comprise ~25% of the genome.
- The human genome has 20,000 to 25,000 genes.
- ~60% of human genes are alternatively spliced.
- Up to 80% of the alternative splices change protein sequence, so the proteome has ~50,000 to 60,000 members.

CONCEPT AND REASONING CHECK

How was the original estimate of the number of human genes obtained? Why is the true number so much lower?

6.6 How Are Genes and Other Sequences Distributed in the Genome?

Are genes uniformly distributed in the genome? Some chromosomes are relatively gene poor and have >25% of their sequences as "deserts"—regions longer than 500 kb where there are no ORFs. Even the most gene-rich chromosomes have >10% of their sequences as deserts. So overall, ~20% of the human genome consists of deserts that have no protein-coding genes.

Repetitive sequences account for ~50% of the human genome, as seen in **FIGURE 6.12**. The repetitive sequences fall into five classes:

- Transposons (either active or inactive) account for the vast majority (45% of the genome). All transposons are found in multiple copies.
- Processed pseudogenes, ~3000 in all, account for ~0.1% of total DNA. (These are sequences that arise by insertion of a reverse transcribed DNA copy of an mRNA sequence into the genome; see *Section 8.11, Pseudogenes Are Nonfunctional Gene Copies.*)
- Simple sequence repeats (highly repetitive DNA such as [CA]) account for ~3%.
- Segmental duplications (blocks of 10 to 300 kb that have been duplicated into a new region) account for ~5%. Only a minority of these duplications are found on the same chromosome; in the other cases, the duplicates are on different chromosomes.
- Tandem repeats form blocks of one type of sequence (especially found at centromeres and telomeres).

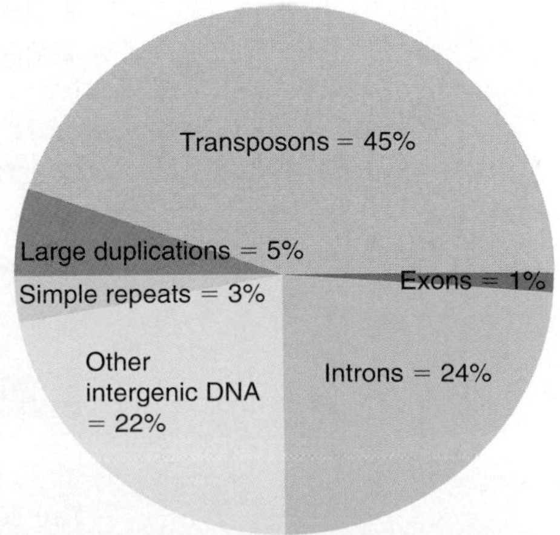

FIGURE 6.12 The largest component of the human genome consists of transposons. Other repetitive sequences include large duplications and simple repeats.

The sequence of the human genome emphasizes the importance of transposons. (Many transposons have the capacity to replicate themselves and insert into new locations. They may function exclusively as DNA elements or may have an active form that is RNA [see *Chapter 17, Transposable Elements and Retroviruses*]. Their distribution in the human genome is summarized in Figure 17.33.) Most of the transposons in the human genome are nonfunctional; very few are currently active. The high proportion of the genome occupied by these elements, however, indicates that they have played an active role in shaping the genome. One interesting feature is that some present genes originated as transposons and evolved into their present condition after losing the ability to transpose. At least 50 genes appear to have originated in this manner.

Segmental duplication at its simplest involves the tandem duplication of some region within a chromosome (typically because of an aberrant recombination event at meiosis; see *Section 7.2, Unequal Crossing Over Rearranges Gene Clusters*). In many cases, however, the duplicated regions are on different chromosomes, implying that either there was originally a tandem duplication followed by a translocation of one copy to a new site or that the duplication arose by some different mechanism altogether. The extreme case of a segmental duplication is when a whole genome is duplicated, in which case the diploid genome initially becomes tetraploid (see *Section 8.12, Genome Duplication Has Played a Role in Plant and Vertebrate Evolution*). As the duplicated copies evolve differences from one another, the genome may gradually become effectively a diploid again, although homologies between the diverged copies leave evidence of the event. This is especially common in plant genomes. The present state of analysis of the human genome identifies many individual duplicated regions, and there is evidence for a whole-genome duplication in the vertebrate lineage.

One curious feature of the human genome is the presence of sequences that do not appear to have coding functions but that nonetheless show an evolutionary conservation higher than the background level. As detected by comparison with other genomes (such as the mouse genome), these represent about 5% of the total genome. Are these sequences associated with protein-coding sequences in some functional way? Their density on chromosome 18 is the same as elsewhere in the genome, although chromosome 18 has a significantly lower concentration of protein-coding genes. This suggests indirectly that their function is not connected with structure or expression of protein-coding genes.

KEY CONCEPTS

- Repeated sequences (present in more than one copy) account for >50% of the human genome.
- The great bulk of repeated sequences consist of copies of nonfunctional transposons.
- There are many duplications of large chromosome regions.

CONCEPT AND REASONING CHECK

What mechanisms may result in tandem duplications of chromosomal segments? What mechanism may account for a displaced duplication found on a different chromosome?

6.7 The Y Chromosome Has Several Male-Specific Genes

The sequence of the human genome has significantly extended our understanding of the role of the sex chromosomes. It is generally thought that the X and Y chromosomes have descended from a common, very ancient autosome pair. Their evolution has involved a process in which the X chromosome has retained most of the original genes, whereas the Y chromosome has lost most of them.

The X chromosome is like the autosomes insofar as females have two copies and recombination can take place between them. The density of genes on the X chromosome is comparable to the density of genes on other chromosomes.

The Y chromosome is much smaller than the X chromosome and has many fewer genes. Its unique role results from the fact that only males have the Y chromosome, of which there is only one copy, so Y-linked loci are effectively haploid instead of diploid like all other human genes.

For many years, the Y chromosome was thought to carry almost no genes except for one or a few genes that determine maleness. The vast majority of the Y chromosome (>95% of its sequence) does not undergo crossing over with the X chromosome, which led to the view that it could not contain active genes because there would be no means to prevent the accumulation of deleterious mutations. This region is flanked by short pseudoautosomal regions that exchange frequently with the X chromosome during male meiosis. It was originally called the nonrecombining region but now has been renamed the *male-specific region*.

Detailed sequencing of the Y chromosome shows that the male-specific region contains three types of sequences, as illustrated in **FIGURE 6.13**.

- The *X-transposed sequences* consist of a total of 3.4 Mb, comprising some large blocks resulting from a transposition from band q21 in the X chromosome about 3 or 4 million years ago. This is specific to the human lineage. These sequences do not recombine with the X chromosome and have become largely inactive. They now contain only two active genes.
- The *X-degenerate segments* of the Y are sequences that have a common origin with the X chromosome (going back to the common autosome from which both X and Y have descended) and contain genes or pseudogenes related to X-linked genes. There are 14 active genes and 13 pseudogenes. The active genes have, in a sense, thus far defied the trend for genes to be eliminated from chromosomal regions that cannot recombine at meiosis.
- The *ampliconic segments* have a total length of 10.2 Mb and are internally repeated on the Y chromosome. There are eight large palindromic blocks. They include nine protein-coding gene families, with copy numbers per family ranging from 2 to 35. The name *amplicon* reflects the fact that the sequences have been internally amplified on the Y chromosome.

Totaling the genes in these three regions, the Y chromosome contains 156 transcription units, of which half represent protein-coding genes and half represent pseudogenes.

The presence of the active genes is explained by the fact that the existence of closely related gene copies in the ampliconic segments allows gene conversion

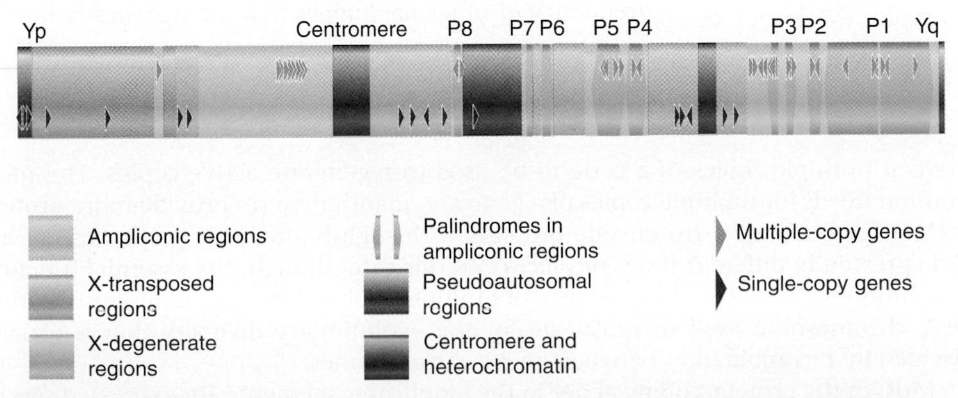

FIGURE 6.13 The Y chromosome consists of X-transposed regions, X-degenerate regions, and amplicons. The X-transposed and X-degenerate regions have 2 and 14 single-copy genes, respectively. The amplicons have 8 large palindromes (P1–P8), which contain 9 gene families. Each family contains at least 2 copies.

METHODS AND TECHNIQUES

Tracing Human History through the Y Chromosome

Because the Y chromosome does not undergo recombination along most of its length, genetic markers in the Y chromosome are completely linked and remain together as the chromosome is transmitted from generation to generation. The genetic relation between Y chromosomes can, therefore, be traced, because chromosomes that are closely related will share more alleles along their length than will more distantly related chromosomes. The set of alleles at two or more loci present in a particular chromosome is called a *haplotype*. For many genealogical studies of the Y chromosome, simple sequence repeat (SSR) polymorphisms are convenient because of their relatively high rate of mutation due to replication errors and the large number of alleles. The logic is that Y chromosomes with haplotypes that share alleles at each of 20 to 30 SSRs across the chromosome must have descended from the same ancestral Y chromosome in the very recent past. For haplotypes differing at a single locus the genetic relationship is less close, for those differing at two loci it is still less close, and so forth. This simple logic is the basis of tracing population history through Y-chromosome polymorphisms. Haplotypes that share many alleles have a more recent common ancestral Y chromosome than haplotypes that share fewer alleles. Furthermore, because the rate of SSR mutation can be estimated, the time at which the ancestral chromosome existed can be deduced. This reasoning forms the basis of the estimate that the most recent common ancestor of all extant human Y chromosomes existed 50,000 years ago. Such estimates are not highly precise, and there are many assumptions that must be made. Other studies using different markers yield estimates of 150,000 years. The reasons for the discrepancy are still unclear. Nevertheless, much can be learned about human population history through studies of the Y chromosome. For example, the legacy of Genghis Khan can be investigated by tracing the lineage of the Y chromosome.

At its maximum range, stretching from China to Russia through to the Middle East and then into Eastern Europe, the Mongol Empire of the 13th century comprised the largest land empire that history has known. The founder was born with the name Temujin around 1162. As a young man he organized a confederation of tribes, who around the year 1200 took to their small Mongolian ponies equipped with high wooden saddles and stirrups and armed with bow and arrow began to conquer their neighbors. Soon thereafter, Temujin adopted the name Genghis Khan, which means "universal ruler." He was often merciless, exterminating the men and boys of rebellious cities and kidnapping the women and girls. In answer to a question about the source of happiness, he is reputed to have said (as recorded by the chronicler, Rashid-ad-Din, in the *Jami al-tawarikh*), "The greatest happiness is to vanquish your enemies, to chase them before you, to rob them of their wealth, to see those dear to them bathed in tears, to clasp to your bosom their wives and daughters." Through their multiple wives, concubines, and innumerable unrecorded sexual conquests, Genghis Khan and his descendants were very prolific. His eldest son Tushi had 40 acknowledged sons, and his grandson Kublai Khan (under whom the Mongol Empire reached its greatest scope) had 22 acknowledged sons.

Although the legacy of Genghis Khan is well recorded in history, it was hardly expected that it would show up in studies of the Y chromosome. Genotyping studies of 32 markers along the Y chromosome of 2123 men sampled from throughout a large region of Asia, however, yielded the remarkable result shown in **FIGURE B6.1**. Each circle represents a population sample, with its area proportional to the sample size. The red sectors denote the relative frequency of a group of nearly identical Y-chromosome haplotypes, whereas the white sectors represent the relative frequency of other haplotypes that are genetically much

between multiple copies of a gene to be used to regenerate active copies. The most common needs for multiple copies of a gene are quantitative (to provide more protein product) or qualitative (to encode proteins with slightly different properties or that are expressed in different times or places). In this case, though, the essential function is evolutionary. In effect, the existence of multiple copies allows recombination within the Y chromosome itself to substitute for the evolutionary diversity that is usually provided by recombination between allelic chromosomes.

Most of the protein-coding genes in the ampliconic segments are expressed specifically in testis and are likely to be involved in male development. If there are ~60 such genes out of a total human gene set of ~25,000, then the genetic difference between male and female humans is ~0.2%.

more diverse. The most recent common ancestor of the closely related haplotypes is estimated to have existed 1000 ± 300 years ago. Furthermore, the geographical region in which the closely related haplotypes cluster is included largely within the Mongol Empire (shading). The sole exception is population 10, composed of the ethnic Hazara of Pakistan. This provides a clue to the origin of the closely related Y chromosomes, because the Hazara consider themselves to be of Mongol origin, and many claim to be direct male-line descendants of Genghis Khan. Whatever their origin, the closely related Y chromosomes are found in about 8% of the males throughout a large region of Asia (populations 1–16). Direct proof of the connection with Genghis Khan in principle could be obtained by determining the haplotype of the Y chromosome in material recovered from his grave. He died in 1227 from injuries sustained in a fall from a horse, and he is said to have been buried in secret at a site near his birthplace.

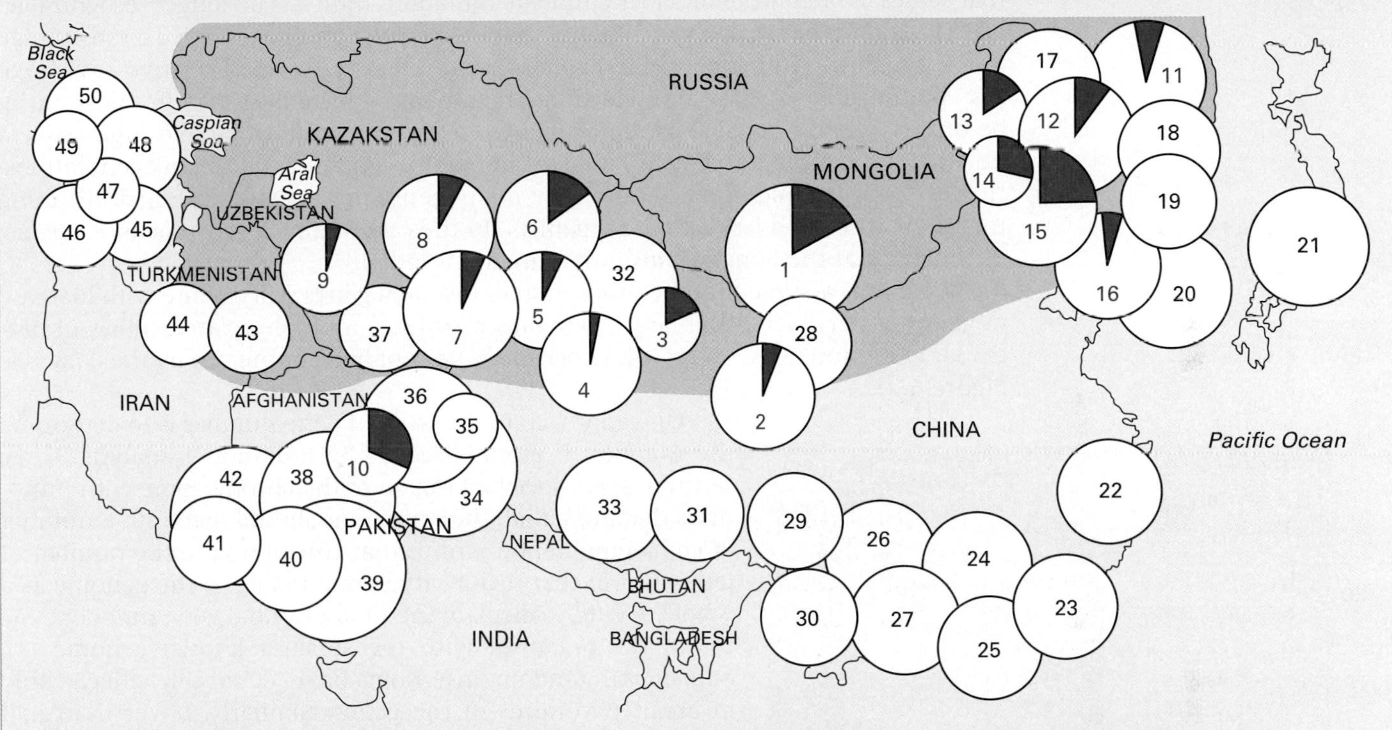

FIGURE B6.1 Distribution of Y-chromosome haplotypes (red), presumed to have descended from Genghis Khan or his close male relatives, among populations near and bordering the ancient Mongol Empire. The specific population groups are: (1) Mongolian, (2) Han [Gansu], (3) Chinese Kazak, (4) Han [Xinjiang], (5) Xibe, (6) Uyghur, (7) Kyrgyz, (8) Kazak, (9) Uzbek, (10) Hazara, (11) Hezhe, (12) Daur, (13) Ewenki, (14) Han [Inner Mongolian], (15) Inner Mongolian, (16) Manchu, (17) Oroqen, (18) Han [Heilongjiang], (19) Chinese Korean, (20) Korean, (21) Japanese, (22) Shezu, (23) Han [Guangdong], (24) Yaozu [Liannan], (25) Lizu, (26) Buyi, (27) Yaozu [Bama], (28) Huizu, (29) Han [Sichuan], (30) Hani, (31) Qiangzu, (32) Chinese Uyghur, (33) Tibetan, (34) Burusho, (35) Balti, (36) Kalash, (37) Tajik, (38) Baloch, (39) Parsi, (40) Makrani Negroid, (41) Makrani Baloch, (42) Brahui, (43) Turkmen, (44) Kurd, (45) Azeni, (46) Armenian, (47) Lezgi, (48) Georgian, (49) Ossetian, (50) Svan. [Adapted from T. Zerja,. *Am. J. Hum. Genet.* 72 (2003): 717–721.]

KEY CONCEPTS

- The Y chromosome has ~60 genes that are expressed specifically in the testis.
- The male-specific genes are present in multiple copies in repeated chromosomal segments.
- Gene conversion between multiple copies allows the active genes to be maintained during evolution.

CONCEPT AND REASONING CHECK

Why was it originally assumed that the human Y chromosome would have very few active genes?

6.8 How Many Genes Are Essential?

The force of natural selection ensures that functional genes are retained in the genome. Mutations occur at random, and a common mutational effect in an ORF will be to damage the protein product. An organism with a damaging mutation will be at a disadvantage in competition, and ultimately the mutation may be eliminated from a population. The frequency of a disadvantageous allele in the population is balanced, however, between the generation of new mutants and the elimination of the allele by selection. Reversing this argument, whenever we see an intact, expressed ORF in the genome, we assume that its product plays a useful role in the organism. Natural selection must have prevented mutations from accumulating in the gene. The ultimate fate of a gene that ceases to be functional is to accumulate mutations until it is no longer recognizable.

The maintenance of a gene implies that it does not confer a selective disadvantage to the organism. In the course of evolution, though, even a small relative advantage may be the subject of natural selection, and a phenotypic defect may not necessarily be immediately detectable as the result of a mutation. Also, in diploid organisms, a new recessive mutation may be "hidden" in heterozygous form for many generations. However, we should like to know how many genes are actually essential, meaning that their absence is lethal to the organism. In the case of diploid organisms, it means, of course, that the homozygous null mutation is lethal.

We might assume that the proportion of essential genes will decline with increase in genome size, given that larger genomes may have multiple related copies of particular gene functions. So far this expectation has not been borne out by the data (see Figure 6.2).

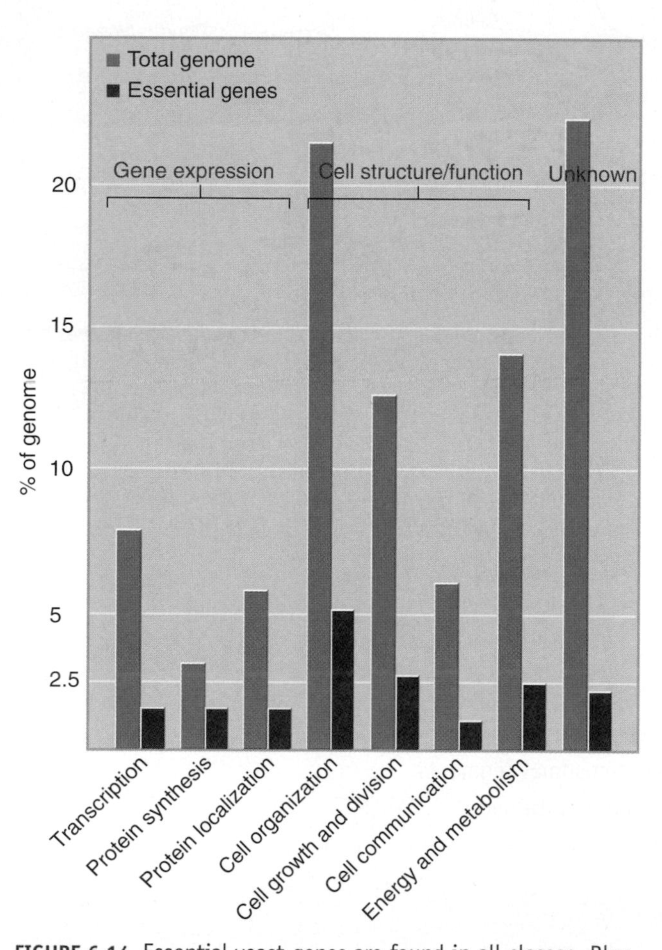

FIGURE 6.14 Essential yeast genes are found in all classes. Blue bars show total proportion of each class of genes, and red bars show those that are essential.

One approach to the issue of gene number is to determine the number of essential genes by mutational analysis. If we saturate some specified region of the chromosome with mutations that are lethal, the mutations should map into a number of complementation groups that correspond to the number of lethal loci in that region. By extrapolating to the genome as a whole, we may calculate the total essential gene number.

In the organism with the smallest known genome (*M. genitalium*), random insertions have detectable effects only in about two-thirds of the genes. Similarly, fewer than half of the genes of *E. coli* appear to be essential. The proportion is even lower in the yeast *S. cerevisiae*. When insertions were introduced at random into the genome in one early analysis, only 12% were lethal, and another 14% impeded growth. The majority (70%) of the insertions had no effect. A more systematic survey based on completely deleting each of 5916 genes (>96% of the identified genes) shows that only 18.7% are essential for growth on a rich medium (that is, when nutrients are fully provided). **FIGURE 6.14** shows that these include genes in all categories. The only notable concentration of defects is in genes encoding products involved in protein synthesis, where ~50% are essential. Of course, this approach underestimates the number of genes that are essential for the yeast to live in the wild when it is not so well provided with nutrients.

FIGURE 6.15 summarizes the results of a systematic analysis of the effects of loss of gene function in the worm *C. elegans*. The sequences of individual genes were predicted from the genome sequence, and by targeting an inhibitory RNA against these sequences (see *Section 30.3, Noncoding RNAs Can Be Used to Regulate Gene Expression*), a large collection of worms was made in which one (predicted) gene was prevented from functioning in each worm. Detectable effects on the phenotype were observed for only 10% of these knockdowns, suggesting that most genes do not play essential roles.

There is a greater proportion of essential genes (21%) among those worm genes that have counterparts in other eukaryotes, suggesting that highly conserved genes tend to have more basic functions. There is also an increased proportion of essential genes among those that are present in only one copy per haploid genome, compared with those where there are multiple copies of related or identical genes. This suggests that many of the multiple genes might be relatively recent duplications that can substitute for one another's functions.

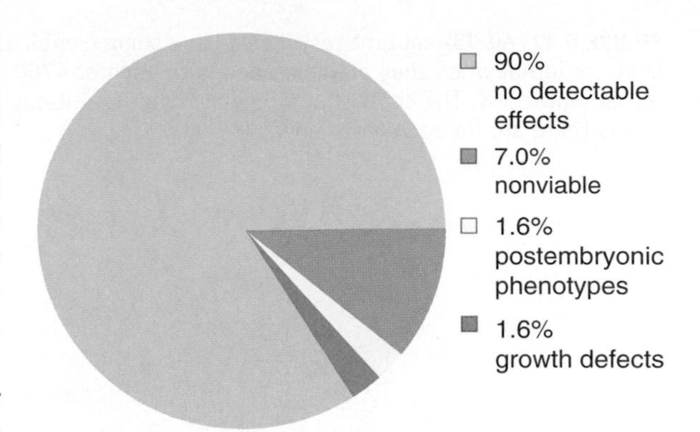

FIGURE 6.15 A systematic analysis of loss of function for 86% of worm genes shows that only 10% have detectable effects on the phenotype.

Extensive analyses of essential gene number in a multicellular eukaryote have been made in *Drosophila* through attempts to correlate visible aspects of chromosome structure with the number of functional genetic units. The notion that this might be possible originated from the presence of bands in the polytene chromosomes of *D. melanogaster*. (These chromosomes are found at certain developmental stages and represent an unusually extended physical form, in which a series of bands [more formally called chromomeres] are evident; see *Section 9.10, Polytene Chromosomes Form Bands*.) From the time of the early concept that the bands might represent a linear order of genes, there has been an attempt to correlate the organization of genes with the organization of bands. There are ~5000 bands in the *D. melanogaster* haploid set; they vary in size over an order of magnitude, but on average there is ~20 kb of DNA per band.

The basic approach is to saturate a chromosomal region with mutations. Usually the mutations are simply collected as lethals, without analyzing the cause of the lethality. *Any mutation that is lethal is taken to identify a locus that is essential for the organism. Sometimes mutations cause visible deleterious effects short of lethality, in which case we also define them as essential loci.* When the mutations are placed into complementation groups, the number can be compared with the number of bands in the region, or individual complementation groups may even be assigned to individual bands. The purpose of these experiments has been to determine whether there is a consistent relationship between bands and genes. For example, does every band contain a single gene?

Totaling the analyses that have been carried out over the past 40 years, the number of essential complementation groups is ~70% of the number of bands. It is an open question whether there is any functional significance to this relationship. Irrespective of the cause, the equivalence gives us a reasonable estimate for the essential gene number of ~3600. By any measure, the number of essential loci in *Drosophila* is significantly less than the total number of genes.

If the proportion of essential human genes is similar to that of other eukaryotes, we would predict a range of ~4000 to 8000 genes, in which mutations would be lethal or produce evidently damaging effects. At present, 1300 genes in which mutations cause evident defects have been identified. This is a substantial proportion of the expected total, especially in view of the fact that many lethal genes may act so early in development that we never see their effects. This sort of bias may also explain the results in **FIGURE 6.16**, which show that the majority of known genetic defects are due to point mutations (where there is more likely to be at least some residual function of the gene).

▸ **redundancy** The concept that two or more genes may fulfill the same function, so that no single one of them is essential.

How do we explain the persistence of genes whose deletion appears to have no effect? The most likely explanation is that the organism has alternative ways of fulfilling the same function. The simplest possibility is that there is **redundancy**, and that some genes are present in multiple copies. This is certainly true in some cases, in which multiple (related) genes must be knocked out in order to produce an effect. In a slightly more complex scenario, an organism might have two separate biochemical pathways capable of providing some activity. Inactivation of either pathway by itself would

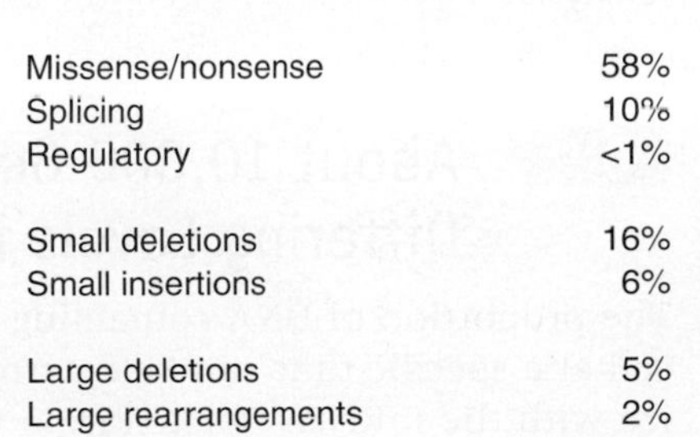

Missense/nonsense	58%
Splicing	10%
Regulatory	<1%
Small deletions	16%
Small insertions	6%
Large deletions	5%
Large rearrangements	2%

FIGURE 6.16 Most known genetic defects in human genes are due to point mutations. The majority directly affect the protein sequence. The remainder are due to insertions, deletions, or rearrangements of varying sizes.

FIGURE 6.17 All 132 mutant test genes have some combinations that are lethal when they are combined with each of 4700 nonlethal mutations. The chart shows how many lethal interacting genes there are for each test gene.

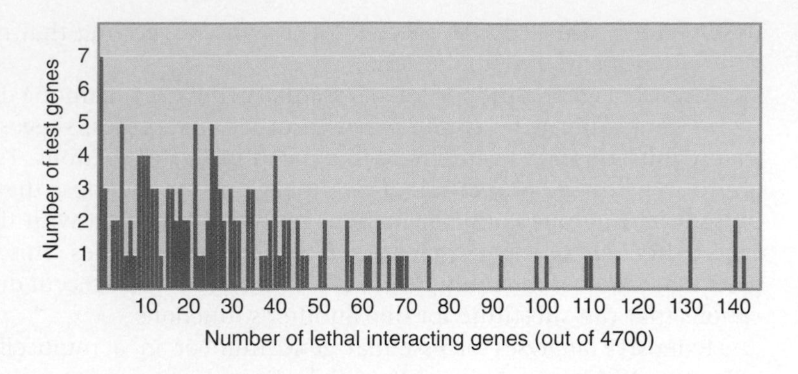

not be damaging, but the simultaneous occurrence of mutations in genes from both pathways would be deleterious.

Such situations can be tested by combining mutations. In this approach, deletions in two genes, neither of which is lethal by itself, are introduced into the same strain. If the double mutant dies, the strain is said to show **synthetic lethality**. This technique has been used to great effect with yeast, where the isolation of double mutants can be automated. The procedure is called **synthetic genetic array analysis (SGA)**. FIGURE 6.17 summarizes the results of an analysis in which an SGA screen was made for each of 132 viable deletions by testing whether it could survive in combination with any one of 4700 viable deletions. Every one of the tested genes had at least one partner with which the combination was lethal, and most of the tested genes had many such partners; the median is ~25 partners, and the greatest number is shown by one tested gene, which had 146 lethal partners. A small proportion (~10%) of the interacting mutant pairs encode polypeptides that interact physically.

- **synthetic lethality** An event that occurs when two mutations that are viable by themselves cause lethality when combined.
- **synthetic genetic array analysis (SGA)** An automated technique in budding yeast whereby a mutant is crossed to an array of approximately 5000 deletion mutants to determine if the mutations interact to cause a synthetic lethal phenotype.

This result goes some way toward explaining the apparent lack of effect of so many deletions. Natural selection will act against these deletions when they are found in lethal pairwise combinations. To some degree, the organism is protected against the damaging effects of mutations by built-in redundancy. However, there is a price in the form of accumulating the "genetic load" of mutations that are not deleterious in themselves but that may cause serious problems when combined with other such mutations in future generations. Presumably the loss of the individual genes in such circumstances produces a sufficient disadvantage to maintain the active gene during the course of evolution.

KEY CONCEPTS

- Not all genes are essential. In yeast and flies, deletions of <50% of the genes have detectable effects.
- When two or more genes are redundant, a mutation in any one of them may not have detectable effects.
- We do not fully understand the persistence of genes that are apparently dispensable in the genome.

CONCEPT AND REASONING CHECK

What are some of the problems in measuring the number of essential genes by mutational analysis?

6.9 About 10,000 Genes Are Expressed at Widely Differing Levels in a Eukaryotic Cell

The proportion of DNA containing protein-coding genes being expressed in a specific cell at a specific time can be determined by the amount of the DNA that can hybridize with the mRNAs isolated from that cell. Such a saturation analysis conducted for many cell types at various times typically identifies ~1% of the DNA being expressed as mRNA. From this we can calculate the number of protein-coding genes, so long as we know the average length of an mRNA. For a unicellular eukaryote such as yeast, the total number of expressed protein-coding genes is ~4000. For somatic tissues of multicellular eukaryotes, including both plants and vertebrates, the number usually is 10,000 to 15,000. (The only consistent exception to this type of value is presented by mammalian brain cells, for which much larger numbers of genes appear to be expressed, although the exact number is not certain.)

The average number of molecules of each mRNA per cell is called its **abundance**. It can be calculated quite simply if the total mass of a specific mRNA species in the cell is known. For example, in chick oviduct cells, the total mRNA can be accounted for as 100,000 copies of ovalbumin mRNA, 4000 copies of each of 7 or 8 other mRNAs, and only ~5 copies of each of the 13,000 remaining mRNAs.

▸ **abundance** The average number of mRNA molecules per cell.

We can divide the mRNA population into two general classes, according to their abundance:

- The oviduct is an extreme case, with so much of the mRNA represented by only one species, but most cells do contain a small number of RNAs present in many copies each. This **abundant mRNA** component typically consists of <100 different mRNAs present in 1000 to 10,000 copies per cell. It often corresponds to a major part of the mass, approaching 50% of the total mRNA.
- About half of the mass of the mRNA consists of a large number of sequences, of the order of 10,000, each represented by only a small number of copies in the mRNA—say, <10. This is the **scarce mRNA** (or **complex mRNA**) class.

▸ **abundant mRNA** Consists of a small number of individual species, each present in a large number of copies per cell.

▸ **scarce mRNA (complex mRNA)** mRNA that consists of a large number of individual mRNA species, each present in very few copies per cell. This accounts for most of the sequence complexity in RNA.

Many somatic tissues of multicellular eukaryotes have an expressed gene number in the range of 10,000 to 20,000. How much overlap is there between the genes expressed in different tissues? For example, the expressed gene number of chick liver is ~11,000 to 17,000, compared with the value for oviduct of ~13,000 to 15,000. How many of these two sets of genes are identical? How many are specific for each tissue? These questions are usually addressed by analyzing the transcriptome—the set of sequences represented in RNA.

We see immediately that there are likely to be substantial differences among the genes expressed in the abundant class. Ovalbumin, for example, is synthesized only in the oviduct and not at all in the liver. This means that 50% of the mass of mRNA in the oviduct is specific to that tissue.

The abundant mRNAs represent only a small proportion of the number of expressed genes, though. In terms of the total number of genes of the organism and of the number of changes in transcription that must be made between different cell types, we need to know the extent of overlap between the genes represented in the scarce mRNA classes of different cell phenotypes.

Comparisons between different tissues show that, for example, ~75% of the sequences expressed in liver and oviduct are the same. In other words, ~12,000 genes are expressed in both liver and oviduct, ~5000 additional genes are expressed only in liver, and ~3000 additional genes are expressed only in oviduct.

The scarce mRNAs overlap extensively. Between mouse liver and kidney, ~90% of the scarce mRNAs are identical, leaving a difference between the tissues of only 1000 to 2000 in terms of the number of expressed genes. The general result obtained in several comparisons of this sort is that only ~10% of the mRNA sequences of a cell are unique to it. The majority of sequences are common to many—perhaps even all—cell types.

This suggests that the common set of expressed gene functions, numbering perhaps ~10,000 in mammals, comprise functions that are needed in all cell types. Sometimes this type of function is referred to as a **housekeeping gene**, or **constitutive gene**. It contrasts with the activities represented by specialized functions (such as ovalbumin or globin) needed only for particular cell phenotypes. These are sometimes called **luxury genes**.

▸ **housekeeping gene (constitutive gene)** A gene that is (theoretically) expressed in all cells because it provides basic functions needed for sustenance of all cell types.

▸ **luxury gene** A gene encoding a specialized function, usually synthesized in large amounts in particular cell types.

▸ **crossover fixation** A possible consequence of unequal crossing over that allows a mutation in one member of a tandem cluster to spread through the whole cluster (or to be eliminated).

The **crossover fixation** model supposes that an entire cluster is subject to continual rearrangement by the mechanism of unequal crossing over. Such events can explain the concerted evolution of multiple genes if unequal crossing over causes all the copies to be regenerated physically from one copy.

Following the sort of event depicted in Figure 7.4, for example, the chromosome carrying a triple locus could suffer deletion of one of the genes. Of the two remaining genes, 1½ represent the sequence of one of the original copies; only ½ of the sequence of the other original copy has survived. Any mutation in the first region now exists in both genes and is subject to selection.

Tandem clustering provides frequent opportunities for "mispairing" of loci whose sequences are the same but which lie in different positions in their clusters. By continually expanding and contracting the number of units via unequal crossing over, it is possible for all the units in one cluster to be derived from a rather small proportion of those in an ancestral cluster. The variable lengths of the spacers are consistent with the idea that unequal crossing-over events take place in spacers that are internally mispaired. This can explain the homogeneity of the genes compared with the variability of the spacers. The genes are exposed to selection when individual repeating units are amplified within the cluster; however, the spacers are functionally irrelevant and can accumulate changes.

In a region of nonrepetitive DNA, recombination occurs between precisely matching points on the two homologous chromosomes, thus generating reciprocal recombinants. The basis for this precision is the ability of two duplex DNA sequences to align exactly. We know that unequal recombination can occur when there are multiple copies of genes whose exons are related, even though their flanking and intervening sequences may differ. This happens because of the mispairing between corresponding exons in nonallelic genes.

Imagine how much more frequently misalignment must occur in a tandem cluster of identical or nearly identical repeats. Except at the very ends of the cluster, the close relationship between successive repeats makes it impossible even to define the exactly corresponding repeats! This has two consequences: there is continual adjustment of the size of the cluster, and there is homogenization of the repeating unit.

Consider a sequence consisting of a repeating unit "ab" with ends "x" and "y." If we represent one chromosome in black and the other in red, the exact alignment between "allelic" sequences would be:

xababababababababababababababababy
xababababababababababababababababy

It is likely, however, that *any* sequence *ab* in one chromosome could pair with *any* sequence *ab* in the other chromosome. In a misalignment such as:

xababababababababababababababababy
      xababababababababababababababababy

the region of pairing is no less stable than in the perfectly aligned pair, although it is shorter. We do not know very much about how pairing is initiated prior to recombination, but very likely it starts between short corresponding regions and then spreads. If it starts within highly repetitive satellite DNA, it more likely than not involves repeating units that do not have exactly corresponding locations in their clusters.

Now suppose that a recombination event occurs within the unevenly paired region. The recombinants will have different numbers of repeating units. In one case, the cluster has become longer; in the other, it has become shorter,

xababababababababababababababababy
×
xababababababababababababababababy
↓
xababababababababababababababababababy
+
xababababababababababababababy

where "x" indicates the site of the crossover.

If this type of event is common, clusters of tandem repeats will undergo continual expansion and contraction. This can cause a particular repeating unit to spread through the cluster, as illustrated in **FIGURE 7.11**. Suppose that the cluster consists initially of a sequence *abcde*, where each letter represents a repeating unit. The different repeating units are similar enough to one another to mispair for recombination. Then by a series of unequal recombination events, the size of the repetitive region increases or decreases, and one unit spreads to replace all the others.

The crossover fixation model predicts that *any sequence of DNA that is not under selective pressure will be taken over by a series of identical tandem repeats generated in this way*. The critical assumption is that the process of crossover fixation is fairly rapid relative to mutation, so that new mutations either are eliminated (their repeats are lost) or come to take over the entire cluster. In the case of the rDNA cluster, of course, a further factor is imposed by selection for a functional transcribed sequence.

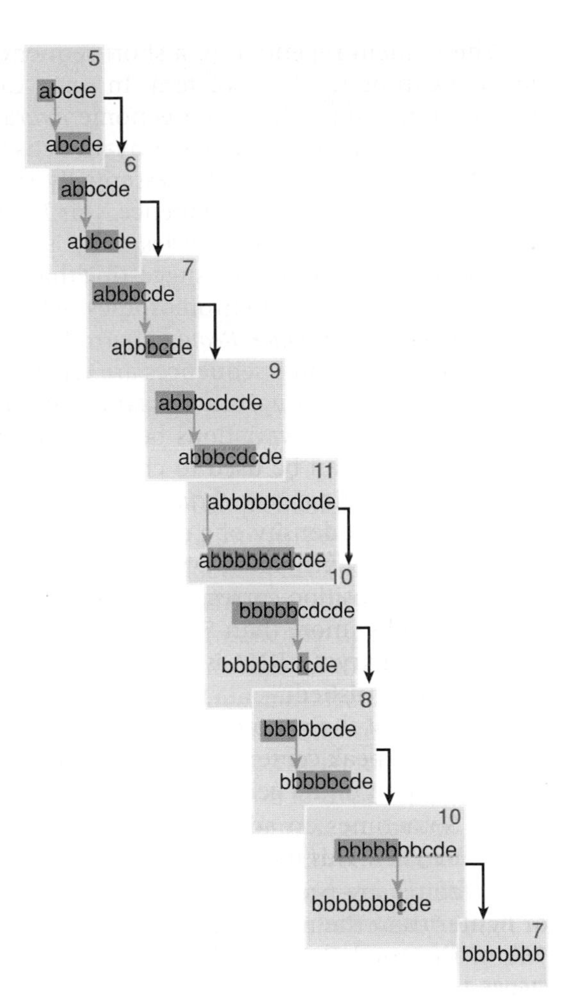

FIGURE 7.11 Unequal recombination allows one particular repeating unit to occupy the entire cluster. The numbers indicate the length of the repeating unit at each stage.

KEY CONCEPTS

- Unequal crossing over changes the size of a cluster of tandem repeats.
- Individual repeating units can be eliminated or can spread through the cluster.

CONCEPT AND REASONING CHECK

Describe two mechanisms by which concerted evolution of repeated sequences can take place.

7.5 Satellite DNAs Often Lie in Heterochromatin

Repetitive DNA is characterized by its relatively rapid rate of renaturation. The component that renatures most rapidly in a eukaryotic genome is called *highly repetitive* DNA and consists of very short sequences repeated many times in tandem in large clusters. As a result of its short repeating unit, it is sometimes described as **simple sequence DNA**. This type of component is present in almost all multicellular eukaryotic genomes, but its overall amount is extremely variable. In mammalian genomes it is typically less than 10%, but in (for example) *Drosophila virilis*, it amounts to ~50%. In addition to the large clusters in which this type of sequence was originally discovered, there are smaller clusters interspersed with nonrepetitive DNA. It typically consists of short sequences that are repeated in identical or related copies in the genome.

▸ **simple sequence DNA** Short repeating units of DNA sequence.

The main feature of these satellites is their very short repeating unit of only 7 bp. Similar satellites are found in other species. *D. melanogaster* has a variety of satellites, several of which have very short repeating units (5, 7, 10, or 12 bp). Comparable satellites are found in crustaceans.

The close sequence relationship found among the *D. virilis* satellites is not necessarily a feature of other genomes, for which the satellites may have unrelated sequences. *Each satellite has originated by a lateral amplification of a very short sequence.* This sequence may represent a variant of a previously existing satellite (as in *D. virilis*), or it could have some other origin.

Satellites are continually generated and lost from genomes. This makes it difficult to ascertain evolutionary relationships, because a current satellite could have evolved from some previous satellite that has since been lost. The important feature of these satellites is that *they represent very long stretches of DNA of very low sequence complexity, within which constancy of sequence can be maintained.*

One feature of many of these satellites is a pronounced asymmetry in the orientation of base pairs on the two strands. In the example of the *D. virilis* satellites shown in Figure 7.14, in each of the major satellites one of the strands is much richer in T and G bases. This increases its buoyant density, so that upon denaturation this heavy strand (H) can be separated from the complementary light strand (L). This can be useful in sequencing the satellite.

KEY CONCEPT

- The repeating units of arthropod satellite DNAs are only a few nucleotides long. Most of the copies of the sequence are identical.

CONCEPT AND REASONING CHECK

Why is it difficult to assess evolutionary relationships using satellite sequences?

7.7 Mammalian Satellites Consist of Hierarchical Repeats

In mammals, as typified by various rodent species, the sequences comprising each satellite show appreciable divergence between tandem repeats. Common short sequences can be recognized by their preponderance among the oligonucleotide fragments produced by chemical or enzymatic treatment. However, the predominant short sequence usually accounts for only a small minority of the copies. The other short sequences are related to the predominant sequence by a variety of substitutions, deletions, and insertions.

But a series of these variants of the short unit can constitute a longer repeating unit that is itself repeated in tandem with some variation. Thus mammalian satellite DNAs consist of a hierarchy of repeating units that can be detected by reassociation analyses or restriction digestion.

When any satellite DNA is digested with an enzyme that has a recognition site in its repeating unit, one fragment will be obtained for every repeating unit in which the site occurs. In fact, when the DNA of a eukaryotic genome is digested with a restriction enzyme, most of it gives a general smear, due to the random distribution of cleavage sites. But satellite DNA generates sharp bands, because a large number of fragments of identical or almost identical size are created by cleavage at restriction sites that lie a regular distance apart.

The satellite DNA of the mouse *M. musculus* is cleaved by the enzyme EcoRII into a series of bands, including a predominant monomeric fragment of 234 bp. This sequence must be repeated with few variations throughout the 60% to 70% of the

```
10        20        30        40        50        60        70     G  80        90        100       110
GGACCTGGAATATGGCGAGAAAACTGAAAATCACGGAAAATGAGAAATACACACTTTAGGACGTGAAATATGGCGAGAAAACTGAAAAAGGTGGAAAATTAGAAATGTCCACTGTA
                                                                                                     T

GGACGTGGAATATGGCAAGAAAACTGAAAATCATGGAAAATGAGAAACATCCACTTGACGACTTGAAAAATGACGAAATCACTAAAAAACGTGAAAAATGAGAAATGCACACTGAA
120       130       140       150       160       170       180       190       200       210       220       230
```

FIGURE 7.15 The repeating unit of mouse satellite DNA contains two half-repeats, which are aligned to show the identities (in blue).

satellite that is cleaved into the monomeric band. We may analyze this sequence in terms of its successively smaller constituent repeating units.

FIGURE 7.15 depicts the sequence in terms of two half-repeats. By writing the 234 bp sequence so that the first 117 bp are aligned with the second 117 bp, we see that the two halves are quite closely related. They differ at 22 positions, corresponding to 19% divergence. This means that the current 234 bp repeating unit must have been generated at some time in the past by duplicating a 117 bp repeating unit, after which differences accumulated between the duplicates.

```
   10        20        30        40        50
GGACCTGGAATATGGCGAGAAAACTGAAAATCACGGAAAATGAGAAATACACACTTTA

60        70      G 80        90        100       110
GGACGTGAAATATGGCGAGAAAACTGAAAAAGGTGGAAAATTAGAAATGTCCACTGTA
                                          T

120       130       140       150       160       170
GGACGTGGAATATGGCAAGAAAACTGAAAATCATGGAAAATGAGAAACATCCACTTGA

 180       190       200       210       220       230
CGACTTGAAAAATGACGAAATCACTAAAAAACGTGAAAAATGAGAAATGCACACTGAA
```

FIGURE 7.16 The alignment of quarter-repeats identifies homologies between the first and second half of each half-repeat. Positions that are the same in all four quarter repeats are shown in green. Identities that extend only through ¾ of the quarter repeats are in black, with the divergent sequences in red.

Within the 117 bp unit we can recognize two further subunits. Each of these is a quarter-repeat relative to the whole satellite. The four quarter-repeats are aligned in **FIGURE 7.16**. The upper two lines represent the first half-repeat of Figure 7.15; the lower two lines represent the second half-repeat. We see that the divergence between the four quarter-repeats has increased to 23 out of 58 positions, or 40%. The first three quarter-repeats are somewhat better related, and a large proportion of the divergence is due to changes in the fourth quarter-repeat.

Looking within the quarter-repeats, we find that each consists of two related subunits (eighth-repeats), shown as the α and β sequences in **FIGURE 7.17**. The α sequences all have an insertion of a C, and the β sequences all have an insertion of a trinucleotide relative to a common consensus sequence. This suggests that the quarter-repeat originated by the duplication of a sequence like the consensus sequence, after which changes occurred to generate the components we now see as α and β. Further changes then took place between tandemly repeated αβ sequences to generate the

```
α1          G G A C C T G G A A T A T G G C G A G A A         A A C T G A A
β1          A A T C A C G G A A A A T G A   G A A A T A C A C A C T T T A
α2          G G A C G T G A A A T A T G G C G A G A^G A       A A C T G A A
β2          A A A G G T G G A A A A T T^T A G A A A T G T C C A C T G T A
α3          G G A C G T G G A A T A T G G C A A G A A         A A C T G A A
β3          A A T C A T G G A A A A T G A   G A A A C A T C C A C T T G A
α4          C G A C T T G A A A A A T G A C G A A A T         C A C T A A A
β4          A A A C G T G A A A A A T G A   G A A A T G C A C A C T G A A
Consensus   A A A C G T G A A A A A T G A   G A A A T         C A C T G A A

Ancestral?  A A A C G T G A A A A A T G A   G A A A T G C A C A C T G A A
```

FIGURE 7.17 The alignment of eighth-repeats shows that each quarter-repeat consists of an α half and a β half. The consensus sequence gives the most common base at each position. The "ancestral" sequence shows a sequence very closely related to the consensus sequence, which could have been the predecessor to the α and β units. (The satellite sequence is continuous, so that for the purposes of deducing the consensus sequence we can treat it as a circular permutation, as indicated by joining the last GAA triplet to the first 6 bp.)

```
      G G A C C T
G G A A T A T G G C
G A G A A A A C T
G A A A A T C A C
G G A A A A T G A
G A A A T C A C T
T T A G G A C G T
G A A A T A T G G C
G A G A^G A A A C T
G A A A A A G G T
G G A A A A T^T T A
G A A A T* C A C T
G T A G G A C G T
G G A A T A T G G C
A A G A A A A C T
G A A A A T C A T
G G A A A A T G A
G A A A C* C A C T
T G A C G A C T T
G A A A A A T G A C
G A A A T C A C T
A A A A A A C G T
G A A A A A T G A
G A A A T* C A C T
G A A
```

$G_{20} A_{16} A_{21} A_{20} A_{12} A_{17} T_8 G_{11} A_5$

$T_7 C_5 A_8 C_9 T_{15}$

C_7

* indicates inserted triplet in β sequence
C in position 10 is extra base in α sequence

FIGURE 7.18 The existence of an overall consensus sequence is shown by writing the satellite sequence in terms of a 9 bp repeat.

individual quarter- and half-repeats that exist today. Among the one-eighth repeats, the present divergence is 19/31 = 61%.

The consensus sequence is analyzed directly in **FIGURE 7.18**, which demonstrates that the current satellite sequence can be treated as derivatives of a 9 bp sequence. We can recognize three variants of this sequence in the satellite, as indicated at the bottom of the figure. If in one of the repeats we take the next most frequent base at two positions instead of the most frequent, we obtain three closely related 9 bp sequences:

G A A A A A C G T
G A A A A A T G A
G A A A A A A C T

The origin of the satellite could well lie in an amplification of one of these three nonamers (9 bp units). The overall consensus sequence of the present satellite is GAAAAA$^{AG}_{TC}$T, which is effectively an amalgam of the three 9 bp repeats.

The average sequence of the monomeric fragment of the mouse satellite DNA explains its properties. The longest repeating unit of 234 bp is identified by the restriction cleavage. The unit of reassociation between single strands of denatured satellite DNA is probably the 117 bp half-repeat, because the 234 bp fragments can anneal both in register and in half-register (in the latter case, the first half-repeat of one strand renatures with the second half-repeat of the other).

So far, we have treated the present satellite as though it consisted of identical copies of the 234 bp repeating unit. Although this unit accounts for the majority of the satellite, variants of it also are present. Some of them are scattered at random throughout the satellite; others are clustered.

The existence of variants is implied by our description of the starting material for the sequence analysis as the "monomeric" fragment. When the satellite is digested by an enzyme that has one cleavage site in the 234 bp sequence, it also generates dimers, trimers, and tetramers relative to the 234 bp length. They arise when a repeating unit has lost the enzyme cleavage site as the result of mutation.

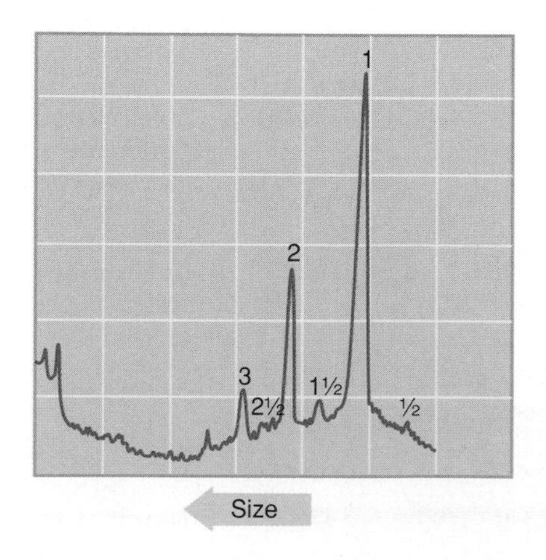

FIGURE 7.19 Digestion of mouse satellite DNA with the restriction enzyme EcoRII identifies a series of repeating units (1, 2, 3) that are multimers of 234 bp and also a minor series (½, 1½, 2½) that includes half-repeats (see previous text). The band at the far left is a fraction resistant to digestion.

The monomeric 234 bp unit is generated when two adjacent repeats each have the recognition site. A dimer occurs when one unit has lost the site, a trimer is generated when two adjacent units have lost the site, and so on. With some restriction enzymes, most of the satellite is cleaved into a member of this repeating series, as shown in the example of **FIGURE 7.19**. The declining number of dimers, trimers, and so forth shows that there is a random distribution of the repeats in which the enzyme's recognition site has been eliminated by mutation.

KEY CONCEPT

- Mouse satellite DNA has evolved by duplication and mutation of a short repeating unit to give a basic repeating unit of 234 bp in which the original half-, quarter-, and eighth-repeats can be recognized.

CONCEPT AND REASONING CHECK

Why is direct sequencing of satellite regions problematic?

7.8 Minisatellites Are Useful for Genetic Mapping

Sequences that resemble satellites (in that they consist of tandem repeats of a short unit) but that overall are much shorter—consisting of (for example) 5 to 50 repeats—are common in mammalian genomes. They were discovered by chance as fragments whose size is extremely variable in genomic libraries of human DNA. The variability is seen when a population contains fragments of many different sizes that represent the same genomic region; when individuals are examined, it turns out that there is extensive polymorphism and that many different alleles can be found.

Whether a repeat cluster is called a minisatellite or a microsatellite depends on both the length of the repeat unit and the number of repeats in the cluster. The name **microsatellite** is usually used when the length of the repeating unit is less than 10 bp; the number of repeats is smaller than that of minisatellites. The name *minisatellite* is used when the length of the repeating unit is ~10 to ~100 bp and there is a greater number of repeats. The terminology is not, however, precisely defined. These types of sequences are also called **variable number tandem repeat (VNTR)** regions. VNTRs used in human forensics are microsatellites that generally have fewer than 20 copies of a 2 to 6 bp repeat.

▸ **microsatellite** DNAs consisting of repetitions of extremely short (typically <10 bp) units.

▸ **variable number tandem repeat (VNTR)** Very short repeated sequences, including microsatellites and minisatellites.

The cause of the variation between individual genomes at microsatellites or minisatellites is that individual alleles have different numbers of the repeating unit. For example, one minisatellite has a repeat length of 64 bp and is found in the population with the following approximate distribution:

7% 18 repeats
11% 16 repeats
43% 14 repeats
36% 13 repeats
4% 10 repeats

The rate of genetic exchange at minisatellite sequences is high, ~10^{-4} per kb of DNA. (The frequency of exchanges per actual locus is assumed to be proportional to the length of the minisatellite.) This rate is ~10× greater than the rate of homologous recombination at meiosis for any random DNA sequence.

The high variability of minisatellites makes them especially useful for genomic mapping because there is a high probability that individuals will vary in their alleles at such a locus. An example of mapping by minisatellites is illustrated in **FIGURE 7.20**. This shows an extreme case in which two individuals both are heterozygous at a minisatellite locus, and, in fact, all four alleles are different. All progeny gain one allele from each parent in the usual way, and it is possible to unambiguously determine the source of every allele in the progeny. In the terminology of human genetics, the meioses described in this figure are highly informative because of the variation between alleles.

One family of minisatellites in the human genome shares a common "core" sequence. The core is a G-C-rich sequence of 10 to 15 bp, showing an asymmetry of purine/pyrimidine distribution on the two strands. Each individual minisatellite has

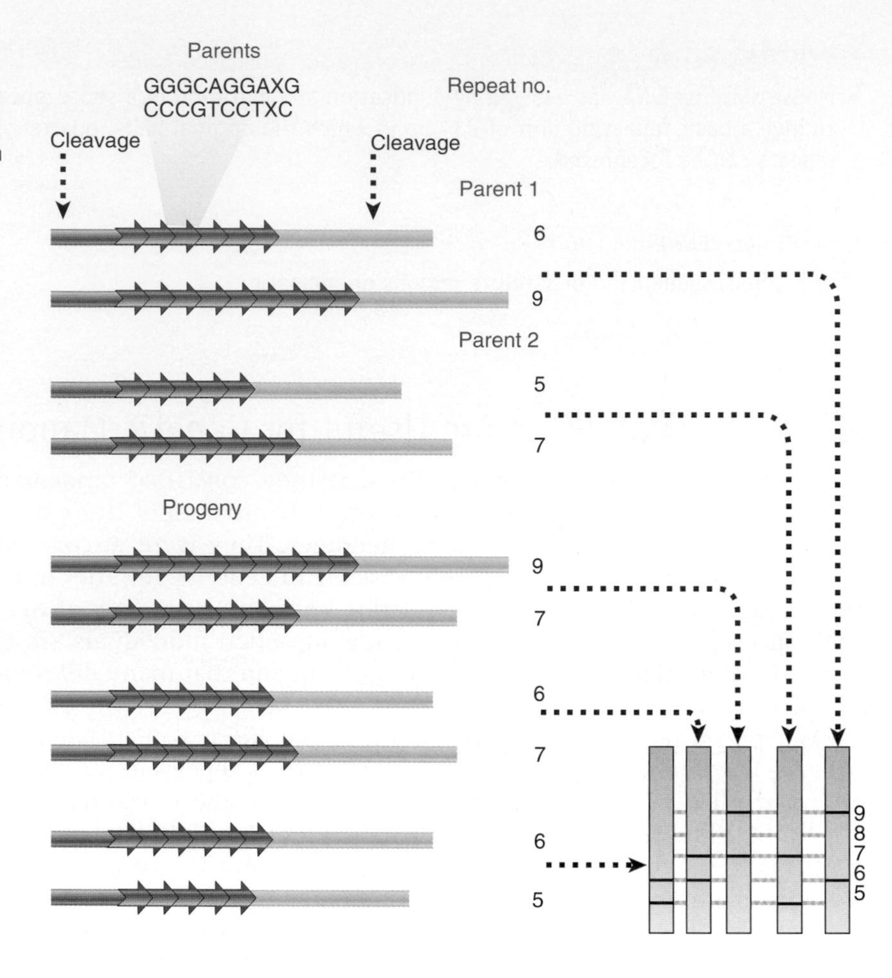

FIGURE 7.20 Alleles may differ in the number of repeats at a minisatellite locus, so that cleavage on either side generates restriction fragments that differ in length. By using a minisatellite with alleles that differ between parents, the pattern of inheritance can be followed.

a variant of the core sequence, but ~1000 minisatellites can be detected on Southern blot (see *Section 3.9, Blotting Methods*) by a probe consisting of the core sequence.

Consider the situation shown in Figure 7.20, but multiplied many times by the existence of many such sequences. The effect of the variation at individual loci is to create a unique pattern for every individual. This makes it possible to assign heredity unambiguously between parents and progeny by showing that 50% of the bands in any individual are inherited from a particular parent. This is the basis of the technique known as **DNA fingerprinting**.

▸ **DNA fingerprinting** Analysis of the differences between individuals of the fragments generated by using restriction enzymes to cleave regions that contain short repeated sequences or by PCR. The lengths of the repeated regions are unique to every individual, and, as a result, the presence of a particular subset in any two individuals can be used to define their common inheritance (e.g., a parent-child relationship).

Both microsatellites and minisatellites are unstable, although for different reasons. Microsatellites undergo intrastrand mispairing, when slippage during replication leads to expansion of the repeat, as shown in **FIGURE 7.21**. Systems that repair damage to DNA—in particular, those that recognize mismatched base pairs—are important in reversing such changes, as shown by a large increase in their frequency when repair genes are inactivated. Mutations in repair systems are an important contributory factor in the development of cancer; thus tumor cells often display variations in microsatellite sequences. Minisatellites undergo the same sort of unequal crossing over between repeats that we have discussed for repeating units (see Figure 7.3). One telling case is that increased variation is associated with a recombination hotspot. The recombination event is not usually associated with recombination between flanking markers but has a complex form in which the new mutant allele gains information from both the sister chromatid and the other (homologous) chromosome.

It is not clear at what repeating length the cause of the variation shifts from replication slippage to unequal crossing-over.

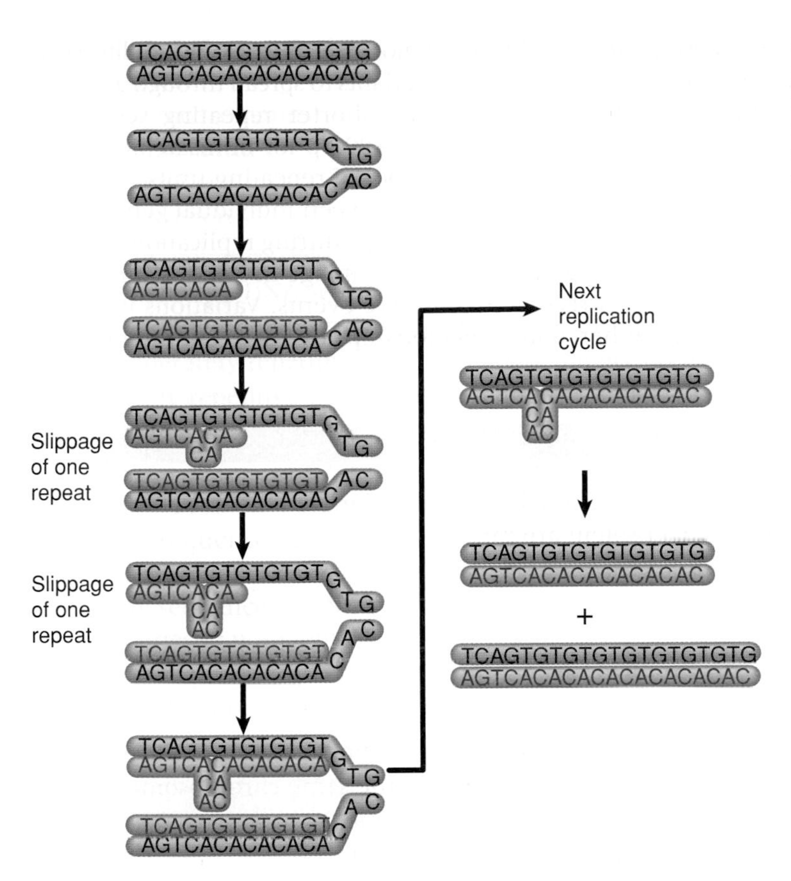

FIGURE 7.21 Replication slippage occurs when the daughter strand slips back one repeating unit in pairing with the template strand. Each slippage event adds one repeating unit to the daughter strand. The extra repeats are extruded as a single-strand loop. Replication of this daughter strand in the next cycle generates a duplex DNA with an increased number of repeats.

KEY CONCEPT

- The variation between microsatellites or minisatellites in individual genomes can be used to identify heredity unequivocally by showing that 50% of the bands in an individual are inherited from a particular parent.

CONCEPT AND REASONING CHECK

Why are VNTRs useful for DNA fingerprinting? Describe how the technique may be used to determine paternity.

7.9 Summary

Most genes belong to families, which are defined by the possession of related sequences in the exons of individual members. Families evolve by the duplication of a gene (or genes), followed by divergence between the copies. Some copies suffer inactivating mutations and become pseudogenes that no longer have any function.

A tandem cluster consists of many copies of a repeating unit that includes the transcribed sequence(s) and a nontranscribed spacer(s). rRNA gene clusters encode only a single rRNA precursor. Maintenance of active genes in clusters depends on mechanisms such as gene conversion or unequal crossing over, which cause mutations to spread through the cluster so that they become exposed to evolutionary forces, namely, selection.

Satellite DNA consists of very short sequences repeated many times in tandem. Its distinct centrifugation properties reflect its biased base composition. Satellite DNA is concentrated in centromeric heterochromatin, but its function (if any) is unknown. The individual repeating units of arthropod satellites are identical. Those of mammalian satellites are related and can be organized into a hierarchy reflecting the evolution of the satellite by the amplification and divergence of randomly chosen sequences.

mechanisms of damage (see *Section 8.14, There May Be Biases in Mutation, Gene Conversion, and Codon Usage*) and differences in the likelihood of repair of the damage.

For example, if one assumes that mutation from one base to any of the other three is equally probable, then *transversion mutations* (from a pyrimidine to a purine, or vice versa) would be twice as frequent as *transition mutations* (from one pyrimidine to another or one purine to another; see *Section 1.11, Mutations May Affect Single Base Pairs or Longer Sequences*). The observation is usually the opposite, though: transitions occur roughly twice as frequently as transversions. This may be because (1) spontaneous transitional errors occur more frequently than transversional errors; (2) transversional errors are more likely to be detected and corrected by DNA repair mechanisms; or (3) both of these are true. Given that transversional errors result in distortion of the DNA duplex as either pyrimidines or purines are paired together and that base-pair geometry is used as a fidelity mechanism (see *Section 13.4, DNA Polymerases Control the Fidelity of Replication*), it is less likely for a DNA polymerase to make a transversional error. The distortion also makes it easier for transversional errors to be detected by postreplication repair mechanisms.

If a mutation occurs in the coding region of a protein-coding gene, it can be characterized by its effect on the polypeptide product of the gene. A substitution mutation that does not change the amino acid sequence of the polypeptide product is a **synonymous mutation**; this is a specific type of *silent mutation*. (Silent mutations include those that occur in noncoding regions.) A **nonsynonymous mutation** in a coding region does alter the amino acid sequence of the polypeptide product, creating either a missense codon (for a different amino acid) or a nonsense (termination) codon. The effect of the mutation on the phenotype of the organism will influence the fate of the mutation in subsequent generations.

- **synonymous mutation** A mutation in a coding region that does not alter the amino acid sequence of the polypeptide product.
- **nonsynonymous mutation** A mutation in a coding region that alters the amino acid sequence of the polypeptide product.

Mutations in genes other than those encoding polypeptides and mutations in noncoding sequences may, of course, also be subject to selection. In noncoding regions, a mutational change may alter the regulation of a gene by directly changing a regulatory sequence or by changing the secondary structure of the DNA in such a way that some aspect of the gene's expression (such as transcription rate, RNA processing, or mRNA structure influencing translation rate) is affected. Many changes in noncoding regions, though, may be selectively **neutral mutations**, having no effect on the phenotype of the organism.

- **neutral mutation** A mutation that has no significant effect on evolutionary fitness and usually has no effect on the phenotype.
- **genetic drift** The chance fluctuation (without selective pressure) of the frequencies of alleles in a population.

If a mutation is selectively neutral or near neutral, then its fate is predictable only in terms of probability. The random changes in the frequency of a mutational variant in a population are called **genetic drift**; this is a type of "sampling error" in which, by chance, the offspring genotypes of a particular set of parents do not precisely match those predicted by Mendelian inheritance. In a very large population, the random effects of genetic drift tend to average out, so there is little change in the frequency of each variant. In a small population, however, these random changes can be quite rapid and large, so genetic drift can have a major effect on the genetic variation of the population. **FIGURE 8.1** shows a simulation comparing the random changes in allele frequency for seven populations of 10 individuals each with those of seven populations of 100 individuals each. Each population begins with two alleles, each with a frequency of 0.5. After 50 generations, most of the small populations have lost one or the other allele, whereas the large populations have retained both alleles (although their allele frequencies have randomly drifted from the original 0.5).

Genetic drift is a random process. The eventual fate of a particular variant is not strictly predictable, but the current frequency of the variant is a measure of the probability that it will eventually be *fixed* in the population. In other words, a new mutation (with a low frequency in a population) is very likely to be lost from the population by chance. If by chance it becomes more frequent, though, it has a greater probability of being retained in the population. Over the long term, a variant may either be lost from the population or fixed, replacing all other variants, but in the short term there may be randomly fluctuating variation for a given locus, particularly in smaller populations where **fixation** or loss occurs more quickly.

- **fixation** The process by which a new allele replaces the allele that was previously predominant in a population.

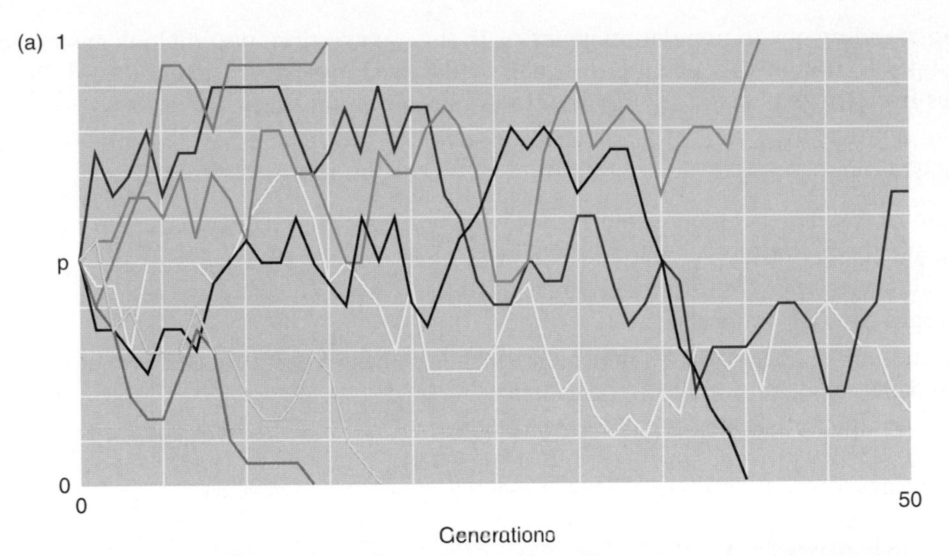

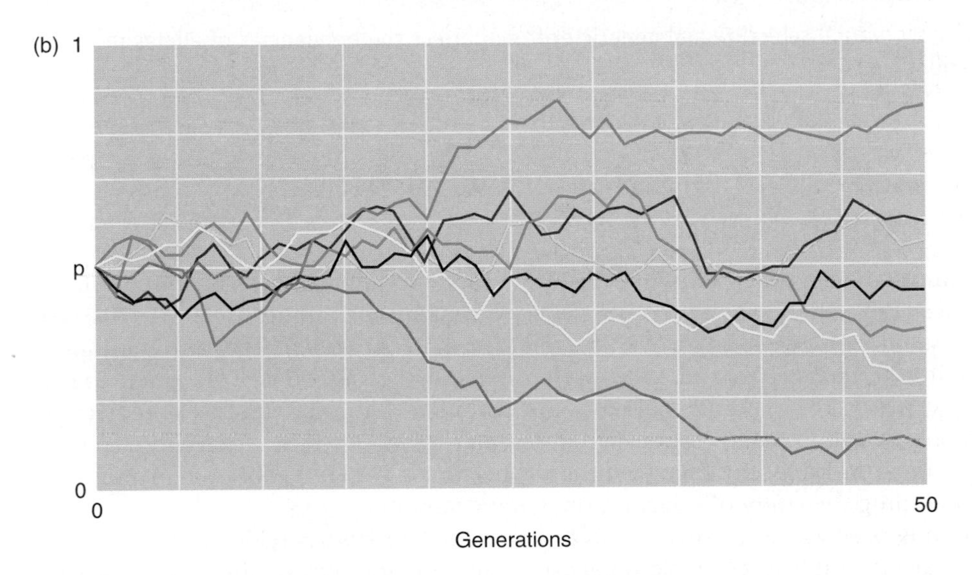

FIGURE 8.1 The fixation or loss of alleles by random genetic drift occurs more rapidly in populations of 10 (a) than in populations of 100 (b). *p* is the frequency of one of two alleles at a locus in the population. Data courtesy of Kent E. Holsinger, University of Connecticut (http://darwin.eeb.uconn.edu).

On the other hand, if a new mutation is not selectively neutral and does affect phenotype, natural selection will play a role in its increase or decrease in frequency in the population. The speed of its frequency change will partly depend on how much of an advantage or disadvantage the mutation confers to the organisms that carry it. It will also depend on whether it is dominant or recessive; in general, because dominant mutations are "exposed" to natural selection when they first appear, they are affected by selection more rapidly.

Mutations are random with regard to their effects, and thus the common result of a nonneutral mutation is for the phenotype to be negatively affected, so selection often acts primarily to eliminate new mutations (although this may be somewhat delayed in the likely event that the mutation is recessive). This is called *negative* (or *purifying*) *selection*. The overall result of negative selection is for there to be little variation within a population as new variants are generally eliminated. More rarely, a new mutation may be subject to *positive selection* if it happens to confer an advantageous phenotype. This type of selection will also tend to reduce variation within a population, as the new mutation eventually replaces the original sequence, but may result in greater variation *between* populations, provided they are isolated from one another, as different mutations occur in these different populations.

The question of how much observed genetic variation in a population or species (or the lack of such variation) is due to selection and how much is due to genetic drift is a long-standing one in population genetics. In the next section, we will look at some ways that selection on DNA sequences may be detected by testing for significant differences from the expectations of evolution of neutral mutations.

KEY CONCEPTS

- The probability of a mutation is influenced by the likelihood that the particular error will occur and the likelihood that it will be repaired.
- In small populations, the frequency of a mutation will change randomly and new mutations are likely to be eliminated by chance.
- The frequency of a neutral mutation largely depends on genetic drift, the strength of which depends on the size of the population.
- The frequency of a mutation that affects phenotype will be influenced by negative or positive selection.

CONCEPT AND REASONING CHECK

Explain how natural selection and genetic drift can affect the frequencies of alleles in populations.

8.3 Selection Can Be Detected by Measuring Sequence Variation

Many methods have been used over the years for analyzing selection on DNA sequences. With the development of DNA sequencing techniques in the 1970s (see *Chapter 3, Methods in Molecular Biology and Genetic Engineering*), the automation of sequencing in the 1990s, and the development of high-throughput sequencing in the 21st century, large numbers of partial or complete genome sequences are becoming available. Coupled with the polymerase chain reaction (PCR) to amplify specific genomic regions, DNA sequence analysis has become a valuable tool in many applications, including the study of selection on genetic variants.

There is now an abundance of DNA sequence data from a wide range of organisms in various publicly available databases. Homologous gene sequences have been obtained from many species as well as from different individuals of the same species. This allows for determination of genetic changes across species lineages as compared to changes within a species. These comparisons have led to the observation that some species (such as *Drosophila melanogaster*) have high levels of DNA sequence polymorphism among individuals, most likely as a result of neutral mutations and random genetic drift within populations. (Other species, such as humans, have moderate levels of polymorphism, and without further investigation the relative roles of genetic drift and selection in keeping these levels low is not immediately clear. This is one use for techniques to detect selection on sequences.) By conducting both interspecific (between species) and intraspecific (within species) DNA sequence analysis, the level of divergence due to species differences can be determined.

Some neutral mutations are synonymous mutations (see *Section 8.2, DNA Sequences Evolve by Mutation and a Sorting Mechanism*), but not all synonymous mutations are neutral. While this may at first seem unlikely, the concentrations of individual tRNAs that all encode a specific amino acid in a cell are not equal. Some cognate tRNAs (different tRNAs that carry the same amino acid) are more abundant than others, and a specific codon may lack sufficient tRNAs, whereas a different codon for the same amino acid may have a sufficient number. In the case of a codon that requires a rare tRNA in that organism, ribosomal frameshifting or other alterations in translation may occur (see *Section 25.13, Frameshifting Occurs at Slippery Sequences*). It also may be that a particular codon is necessary to maintain mRNA structure. Alternatively, there may be a nonsynonymous mutation to an amino acid with the same general characteristics, with little or no effect on the folding and activity of the polypeptide. In either case neutral sequence changes have little effect on the organism. A nonsynonymous mutation may result in an amino acid with different properties, however, such as a change from a polar to a nonpolar amino acid or from a hydrophobic amino acid to a hydrophilic one in a protein embedded in a phospholipid bilayer. Such changes are likely to have functional effects that are deleterious to the role of the polypeptide and thus to the organism. Depending on the location of the amino acid in the polypeptide, such a change may cause only slight disruption of protein folding and activity. Only in rare cases is an amino acid change advantageous; in this case the mutational change may become subjected to positive selection and ultimately lead to fixation of this variant in the population.

One common approach for determining selection is to use codon-based sequence information to study the evolutionary history of a gene. This can be done by counting the number of synonymous (K_s) and nonsynonymous (K_a) amino acid substitutions in orthologous genes (see *Section 6.4, How Many Different Types of Genes Are There?*), and determining the K_a/K_s ratio. This ratio is indicative of the selective constraints on the gene. A K_a/K_s ratio of 1 is expected for those genes that evolve neutrally, with amino acid sequence changes being neither favored nor disfavored. In this case the changes that occur do not usually affect the activity of the polypeptide, and this serves as a suitable control. A K_a/K_s ratio <1 is most commonly observed and indicates negative selection where amino acid replacements are disfavored because they affect the activity of the polypeptide. Thus there is selective pressure to retain the original functional amino acid at these sites in order to maintain proper protein function.

Positive selection, which is more rarely observed than negative selection, results in a K_a/K_s ratio that is >1. This indicates that the amino acid changes are advantageous and may become fixed in the population. One example of this is the antigenic proteins of some pathogens, such as viral coat proteins, which are under strong selection pressure to evade the immune response of the host. A second example is some reproductive proteins that are under *sexual selection* (selection on traits found in one sex). As a third example, the K_a/K_s ratios for the peptide-binding regions of mammalian MHC genes, the products of which function in immunological self-recognition by displaying both "self" and "nonself" antigens, are typically in the range of 2 to 10, indicating strong selection for new variants. This is expected since these proteins represent the cellular uniqueness of individual organisms. The detection of a positive K_a/K_s ratio may be rare in part because the average value of the ratio must be greater than 1 over a length of sequence. If a single substitution in a gene is being positively selected, but flanking regions are under negative selection, the average ratio across the sequence may actually be negative. In contrast, the K_a/K_s ratios for histone genes are typically much less than 1, suggesting strong purifying selection on these genes. Histones are DNA-binding proteins that make up the basic structure of chromatin (see *Chapter 10, Chromatin*) and alterations to their structures are likely to result in deleterious effects on chromosome integrity and gene expression.

In addition to the difficulty of detecting strong selection on a single substitution variant when K_a/K_s is averaged over a stretch of DNA, mutational hotspots may also affect this measure. There have been reports of unusually highly mutable regions of some protein-coding genes that encode a high proportion of polar amino acids; such a bias may influence the interpretation of the K_a/K_s ratio because a higher point mutation rate may be incorrectly interpreted as a higher substitution rate. Although codon based methods of detecting selection can be useful, their limitations must be taken into account.

Intraspecific DNA sequence analysis can be used to detect positive selection by comparing the nucleotide sequence between two alleles or two individuals of the same species. Nucleotide sequences evolve neutrally at a certain rate; variation in this rate at specific nucleotides affects the *heterozygosity* of a population (the proportion of heterozygotes at a locus). If a variant sequence is favored, the variant will increase in frequency and eventually become fixed in the population, and the site will show a reduction in nucleotide heterozygosity. Closely linked neutral variants may also become fixed, a phenomenon termed **genetic hitchhiking**. These regions are characterized by having a lower level of DNA sequence polymorphism. (It is important to remember, though, that reduced polymorphism can have other causes, such as negative selection or genetic drift.)

▸ **genetic hitchhiking** The change in frequency of an allele due to its association (functional association or physical linkage) with an allele at another locus that is also changing in frequency.

In practice it is more reliable to carry out both interspecific and intraspecific DNA sequence comparisons to detect deviations from neutral evolutionary expectations. By including sequence information from at least one closely related species, species-specific DNA polymorphisms can be distinguished from ancestral polymorphisms, and more accurate information regarding the link between the polymorphisms and between-species differences can be obtained. With this combined analysis, the degree of nonsynonymous changes between species can be determined. If evolution is primarily neutral, the ratio of nonsynonymous to synonymous changes *within* species is expected to be the same as the ratio *between* species. An excess of nonsynonymous changes may be evidence for positive selection on these amino acids, whereas a lower ratio may indicate that negative selection is conserving sequences.

One example is the comparison of 12 sequences of the *Adh* gene in *D. melanogaster* to each other and to *Adh* sequences from *D. simulans* and *D. yakuba*, as shown in **FIGURE 8.2**. A simple contingency chi-square test on these data shows that there are significantly more fixed nonsynonymous changes between species than similar polymorphisms in *D. melanogaster*. The high proportion of nonsynonymous differences among species suggests positive selection on *Adh* variants in these species, as does the lower proportion of such differences within one species, given that nonneutral variation would not be expected to persist for very long within a species.

Relative rate tests can also be used to detect the signature of selection. This involves (at a minimum) three related species, two that are closely related and one outgroup representative. The substitution rate is compared between the close relatives, and each is compared to the outgroup species to see if the substitution rates are similar. This removes the dependence of the analysis on time, as long as the phylogenetic relationship between the species is certain. If the rate of substitutions between relatives compared to the rate between these and the outgroup species is different, this may be an indication of selection on the sequence. For example, the protein lysozyme, which functions to digest bacterial cell walls and is a general antibiotic in many species, has evolved to be active at low pH in ruminating mammals, where it functions to digest dead bacteria in the gut. **FIGURE 8.3** shows that the number of amino acid (i.e., nonsynonymous) substitutions for lysozyme in the cow/deer (ruminant) lineage is higher than that of the nonruminant pig outgroup.

	Nonsynonymous	Synonymous
Fixed	7	17
Polymorphic	2	42

FIGURE 8.2 Nonsynonymous and synonymous variation in the *Adh* locus in *Drosophila melanogaster* ("polymorphic") and between *D. melanogaster*, *D. simulans*, and *D. yakuba* ("fixed"). Adapted from J. H. McDonald and M. Kreitman, *Nature* 351 (1991): 652–654.

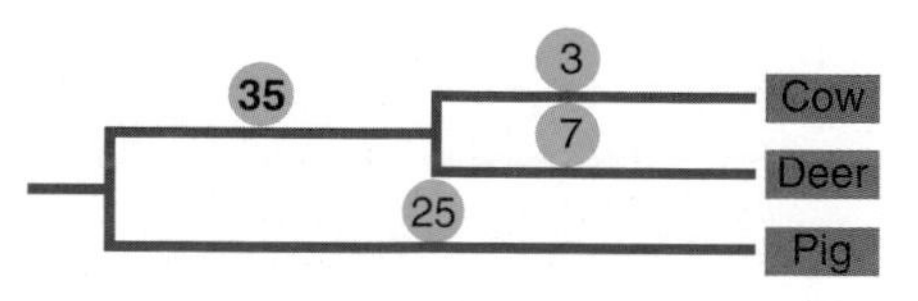

FIGURE 8.3 A higher number of nonsynonymous substitutions in lysozyme sequences in the cow/deer lineage as compared to the pig lineage is a result of adaptation of the protein for digestion in ruminant stomachs. Adapted from N. H. Barton et al., *Evolution*. Cold Spring Harbor Laboratory Press, 2007. Original figure appeared in J. H. Gillespie, *The Causes of Molecular Evolution*. Oxford University Press, 1994.

Another method for detecting selection utilizes estimates of polymorphism at specific genetic loci. For example, sequence analysis of the *Teosinte branched 1* (*tb1*) locus, an important gene in domesticated maize, has been used to characterize the nucleotide substitution rate in domesticated and native maize (teosinte) varieties, with an estimate of 2.9×10^{-8} to 3.3×10^{-8} base substitutions per year. **FIGURE 8.4** shows the ratio of a measure of nucleotide diversity (π) of the *tb1* region in domesticated maize to π in wild teosinte. For a neutrally evolving gene in these two species this ratio is ~0.75, but it is <0.1 in this region. The interpretation is that strong selection in domesticated maize has severely reduced variation for this gene.

As genomewide data on nucleotide diversity become available, regions of low diversity may indicate recent selection. Millions of single nucleotide polymorphisms (SNPs) are being characterized in humans, nonhuman animals, and plants, as well as in other species. One approach that has been applied to the human genome is to look for an association between an allele's frequency and

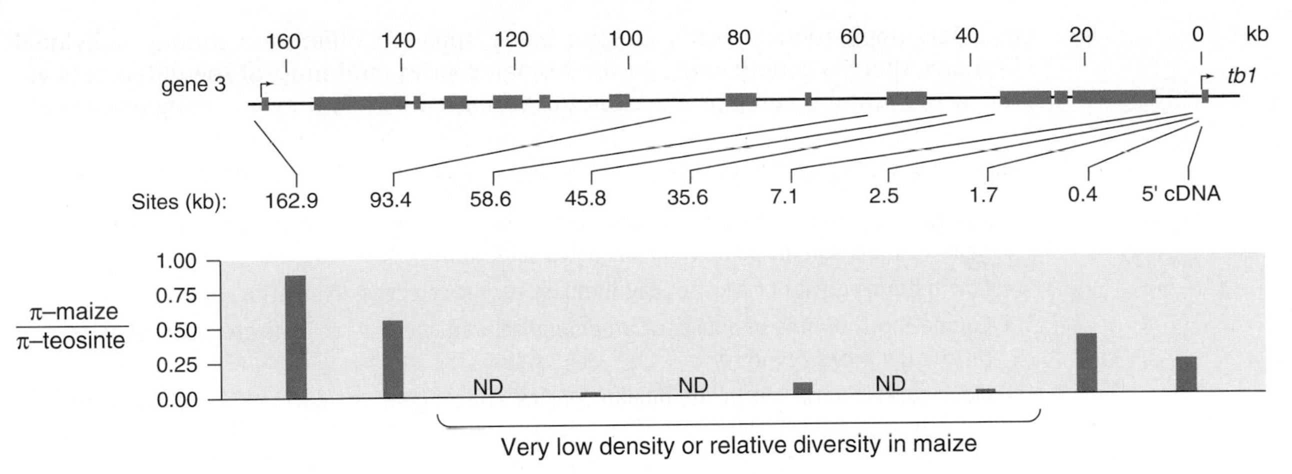

FIGURE 8.4 Nucleotide diversity (π) of the *tb1* region in domesticated maize is much lower than in wild teosinte, indicating strong selection on this locus in maize. Reproduced from R. M. Clark et al., *Proc. Natl. Acad. Sci. USA* 101 (2004): 700–707. © 2004 National Academy of Sciences, U.S.A. Courtesy of John F. Doebley, University of Wisconsin, Madison.

its **linkage disequilibrium** with other genetic markers surrounding it. Linkage disequilibrium is a measure of an association between an allele at one locus and an allele at a different locus. When a new mutation occurs on one chromosome, it initially has high linkage disequilibrium with alleles at other polymorphic loci on the same chromosome. In a large population, a neutral allele is expected to rise to fixation slowly, so recombination and mutation will break up associations between loci and linkage disequilibrium will decrease. On the other hand, an allele under positive selection will rise to fixation more quickly and linkage disequilibrium will be maintained. By sampling SNPs across the genome, a general background level of linkage disequilibrium that accounts for local variations in rates of recombination can be established, and any significantly higher measures of linkage disequilibrium can be detected. **FIGURE 8.5** shows the slowly decreasing linkage disequilibrium (measured by the increasing fraction of recombinant chromosomes) with increasing chromosomal distance from a variant of the *G6PD* locus that confers resistance to malaria in African human populations. This pattern suggests that this allele has been under strong recent selection—carrying along with it linked alleles at other loci—and that recombination has not yet had time to break up these interlocus associations.

▸ **linkage disequilibrium** A nonrandom association between alleles at different loci, whether due to linkage or some other cause, such as selection on a specific multilocus combination of alleles.

The availability of multiple complete human genome sequences and the ability to rapidly resequence specific regions of the genome in many individuals allows large-scale measurement of genetic variation in the human population. As described before, a local lack of genetic variation can indicate negative selection on that sequence, implying that the sequence is functional. If the analysis includes individuals from many populations, it can be determined whether individual variations are unique, shared by other members of a specific population, or found globally. Surprisingly, such studies show that the majority of *functional* variations in the human genome are *not* nonsynonymous changes in coding sequences, but are found in noncoding sequences such as introns or intergenic regions! In other words, protein variations account for only a small percentage of functional differences among humans. Presumably, the large percentage of functional variation in noncoding regions reflect differences in regulatory regions (see *Part 4, Gene Regulation*). Also, most of these variations are found in most or all sampled populations and are not limited to one

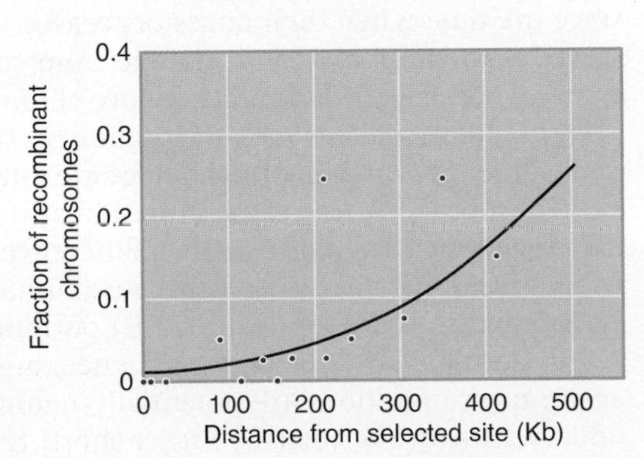

FIGURE 8.5 The fraction of recombinants between an allele of *G6PD* and alleles at nearby loci on a human chromosome remains low, suggesting that the allele has rapidly increased in frequency by positive selection. The allele confers resistance to malaria. Adapted from E. T. Wang et al., *Proc. Natl. Acad. Sci. USA* 103 (2006): 135–140.

or a few populations. Clearly, despite many apparent differences among individual humans, there is genetic unity to the human species, and most of the differences are not in the proteins being produced but in when and where they are being produced.

KEY CONCEPTS

- The ratio of nonsynonymous to synonymous substitutions in the evolutionary history of a gene is a measure of positive or negative selection.
- Low heterozygosity of a gene may indicate recent selective events.
- Comparing the rates of substitution among related species can indicate whether selection on the gene has occurred.
- Most genetic variation in the human species affects gene regulation and not variation in proteins.

CONCEPT AND REASONING CHECK

Explain why either positive or negative selection would result in different rates of synonymous and nonsynonymous substitutions in a gene.

8.4 A Constant Rate of Sequence Divergence Is a Molecular Clock

Most changes in gene sequences occur by mutations that accumulate slowly over time. Point mutations and small insertions and deletions occur by chance, probably with more or less equal probability in all regions of the genome. The exceptions to this are *hotspots*, where mutations occur much more frequently. Recall from *Section 8.2* that most nonsynonymous mutations are deleterious and will be eliminated by negative selection, whereas the rare advantageous substitution will spread through the population and eventually replace the original sequence (fixation). Neutral variants are expected to be lost or fixed in the population due to random genetic drift. What proportion of mutational changes in a protein-coding gene sequence are selectively neutral is a historically contentious issue.

The rate at which substitutions accumulate is a characteristic of each gene, presumably depending at least in part on its functional flexibility with regard to change. Within a species, a gene evolves by mutation, followed by fixation within the single population. Recall that when we study the genetic variation of a species, we see only the variants that have been maintained, whether by selection or genetic drift. When multiple variants are present they may be stable, or they may in fact be transient because they are in the process of being fixed (or lost).

When a single species separates into two new species, each of the resulting species now constitutes an independent evolutionary lineage. By comparing orthologous genes in two species, we see the differences that have accumulated between them since the time when their ancestors ceased to interbreed. Some genes are highly conserved, showing little or no change from species to species. This indicates that most changes are deleterious and therefore eliminated.

▸ **divergence** The corrected percent difference in nucleotide sequence between two related DNA sequences or in amino acid sequences between two polypeptides.

The difference between two genes is expressed as their **divergence**, the percent of positions at which the nucleotides are different, corrected for the possibility of convergent mutations (the same mutation at the same site in two separate lineages) and true revertants. There is usually a difference in the rate of evolution among the three codon positions within genes because mutations at the third base position often are synonymous, as are some at the first position.

In addition to the coding sequence, a gene contains untranslated regions. Here again, most mutations are potentially neutral, apart from their effects on either secondary structure or (usually rather short) regulatory signals.

Although synonymous mutations are expected to be neutral with regard to the polypeptide, they could affect gene expression via the sequence change in RNA (see *Section 8.2, DNA Sequences Evolve by Mutation and a Sorting Mechanism*). Another possibility is that a change in synonymous codons calls for a different tRNA to respond, influencing the efficiency of translation. Species generally show a **codon bias**; when there are multiple codons for the amino acid, one codon is found in protein-coding genes in a high percentage, whereas the remaining codons are found in low percentages. There is a corresponding percentage difference in the tRNA species that recognize these codons. Consequently, a change from a common to a rare synonymous codon may reduce the rate of translation due to a lower concentration of appropriate tRNAs. (Alternatively, there may be a nonadaptive explanation for codon bias; see *Section 8.14, There May Be Biases in Mutation, Gene Conversion, and Codon Usage*.)

▸ **codon bias** A higher usage of one codon in genes to encode amino acids for which there are several synonymous codons.

FIGURE 8.6 shows the divergence of three types of proteins (representing nonsynonymous changes in DNA) over time by comparing species for which there is paleontological evidence for the time of divergence. There are two striking features of these data. First, the three types of proteins evolve at different rates: fibrinopeptides evolve quickly, cytochrome *c* evolves slowly, and hemoglobin evolves at an intermediate rate. Second, for each protein type, the rate of evolution is approximately constant over millions of years. In other words, for a particular protein, the divergence between any pair of sequences is (more or less) proportional to the time since they separated. This provides a **molecular clock** that measures the accumulation of substitutions at an approximately constant rate during the evolution of a particular protein-coding gene.

▸ **molecular clock** An approximately constant rate of evolution that occurs in DNA sequences, such as by the genetic drift of neutral mutations.

There can also be molecular clocks for paralogous proteins diverging within a species lineage. To take the example of the human β- and δ-globin chains (see *Section 7.2, Unequal Crossing Over Rearranges Gene Clusters*, and *Section 8.10, Globin Clusters Arise by*

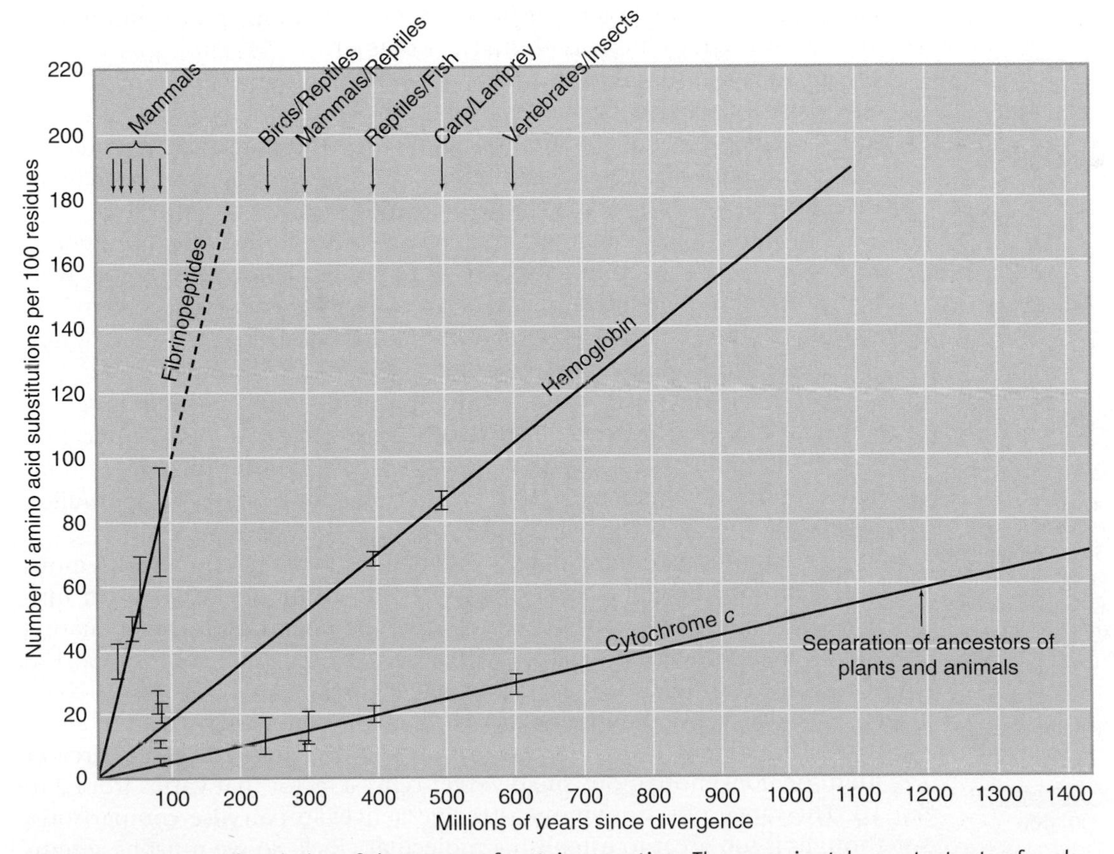

FIGURE 8.6 The rate of evolution of three types of proteins over time. The approximately constant rate of evolution of each protein type is a molecular clock. Reproduced with kind permission from Springer Science+Business Media: *J. Mol. Evol.*, The structure of cytochrome and the rates of molecular evolution, vol. 1, 1971, pp. 26–45, R. E. Dickerson, fig. 3. Courtesy of Richard Dickerson, University of California, Los Angeles.

Duplication and Divergence), there are 10 differences in 146 residues, a divergence of 6.9%. The DNA sequence has 31 changes in 441 residues. The nonsynonymous and synonymous changes are distributed very differently, however. There are 11 changes in the 330 nonsynonymous sites, but 20 changes in only 111 synonymous sites. This gives corrected rates of divergence of 3.7% in the nonsynonymous sites and 32% in the synonymous sites, an order of magnitude in difference.

The striking difference in the divergence of nonsynonymous and synonymous sites demonstrates the existence of much greater constraints on nucleotide changes that change polypeptide sequence compared to those that do not. Many fewer amino acid changes are neutral.

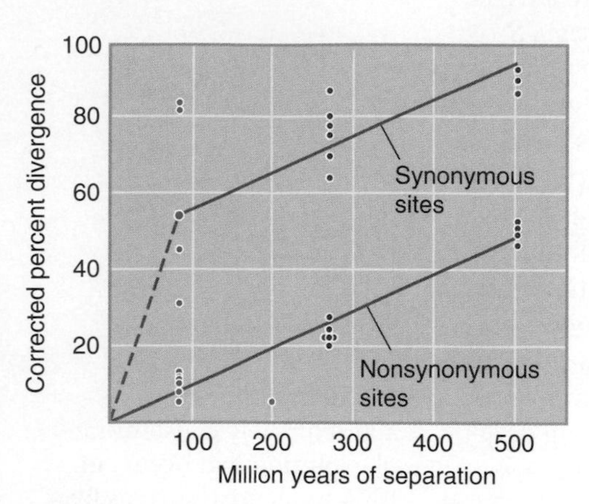

FIGURE 8.7 Divergence of DNA sequences depends on evolutionary separation. Each point on the graph represents a pairwise comparison.

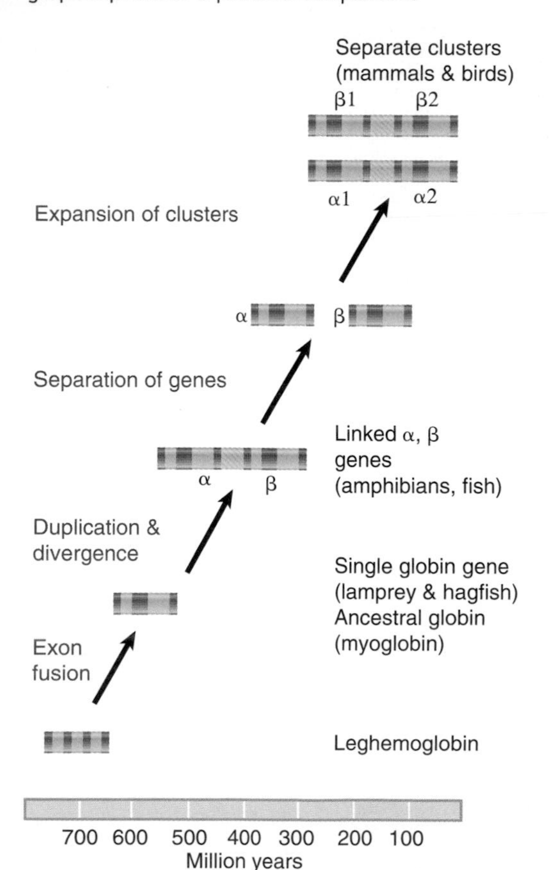

FIGURE 8.8 All globin genes have evolved by a series of duplications, transpositions, and mutations from a single ancestral gene.

Suppose we take the rate of synonymous substitutions to indicate the underlying rate of mutational fixation (assuming there is no selection at all at the synonymous sites). Then over the period since the β and δ genes diverged, there should have been changes at 32% of the 330 nonsynonymous sites, for a total of 105. All but 11 of them have been eliminated, which means that ~90% of the mutations were not maintained.

The rate of divergence can be measured as the percent difference per million years or as its reciprocal, the unit evolutionary period (UEP), the time in millions of years that it takes for 1% divergence to accrue. Once the rate of the molecular clock has been established by pairwise comparisons between species (remembering the practical difficulties in establishing the actual time since the existence of the common ancestor), it can be applied to paralogous genes within a species. From their divergence, we can calculate how much time has passed since the duplication that generated them.

By comparing the sequences of orthologous genes in different species, the rate of divergence at both nonsynonymous and synonymous sites can be determined, as plotted in **FIGURE 8.7**.

In pairwise comparisons, there is an average divergence of 10% in the nonsynonymous sites of either the α- or β-globin genes of mammal lineages that have been separated since the mammalian radiation occurred ~85 million years ago. This corresponds to a nonsynonymous divergence rate of 0.12% per million years.

The rate is approximately constant when the comparison is extended to genes that diverged in the more distant past. For example, the average nonsynonymous divergence between orthologous mammalian and chicken globin genes is 23%. Relative to a common ancestor at ~270 million years ago, this gives a rate of 0.09% per million years.

Going further back, we can compare the α- with the β-globin genes within a species. They have been diverging since the original duplication event 500 million years ago (see **FIGURE 8.8**). They have an average nonsynonymous divergence of ~50%, which gives a rate of 0.1% per million years.

The summary of these data in Figure 8.8 shows that nonsynonymous divergence in the globin genes has an average rate of ~0.096% per million years (for a UEP of 10.4). Considering the uncertainties in estimating the times at which the species diverged, the results lend good support to the idea that there is a constant molecular clock.

The data on synonymous site divergence are much less clear. In every case, it is evident that the synonymous site divergence is much greater than the nonsynonymous site divergence by a factor that varies from 2 to 10. The range of synonymous site divergences in pairwise comparisons, though, is too great to establish a molecular clock, so we must base temporal comparisons on the nonsynonymous sites.

From Figure 8.8, it is clear that the rate at synonymous sites is not constant over time. If we assume that there must be zero divergence at zero years of separation, we see that the rate of synonymous site divergence is much greater for the first ~100 million years of separation. One interpretation is that roughly half of the synonymous sites are rapidly (within 100 million years) saturated by mutations; this half behaves as neutral sites. The other half accumulates mutations more slowly, at a rate approximately the same as that of the nonsynonymous sites; this fraction represents sites that are synonymous with regard to the polypeptide, but that are under selective constraint for some other reason.

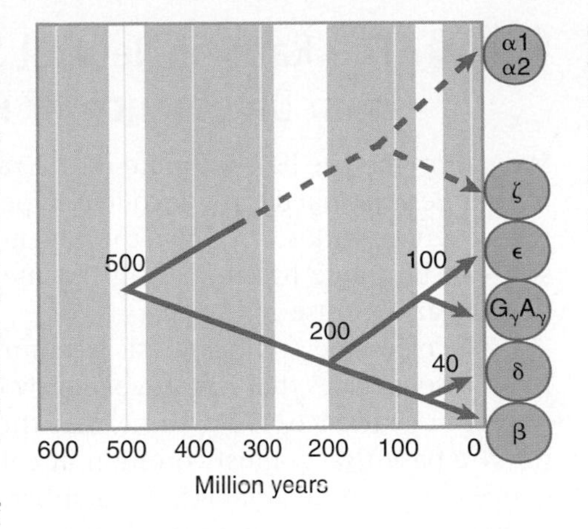

FIGURE 8.9 Nonsynonymous site divergences between pairs of β-globin genes allow the history of the human cluster to be reconstructed. This tree accounts for the separation of classes of globin genes.

Now we can reverse the calculation of divergence rates to estimate the times since paralogous genes were duplicated. The difference between the human β and δ genes is 3.7% for nonsynonymous sites. At a UEP of 10.4, these genes must have diverged 10.4 × 3.7 = 40 million years ago—about the time of the separation of the major primate lineages: New World monkeys, Old World monkeys, and great apes (including humans). All these taxonomic groups have both β and δ genes, which suggests that the gene divergence began just before this point in evolution.

Proceeding further back, the divergence between the nonsynonymous sites of γ and ε genes is 10%, which corresponds to a duplication event ~100 million years ago. The separation between embryonic and fetal globin genes therefore may have just preceded or accompanied the mammalian radiation.

An evolutionary tree for the human globin genes is presented in **FIGURE 8.9**. Paralogous groups that evolved before the mammalian radiation—such as the separation of β/δ from γ—should be found in all mammals. Paralogous groups that evolved afterward—such as the separation of β- and δ-globin genes—should be found in individual lineages of mammals.

In each species, there have been comparatively recent changes in the structures of the clusters. We know this because we see differences in gene number (one adult β-globin gene in humans, two in the mouse) or in type (most often concerning whether there are separate embryonic and fetal genes).

When sufficient data have been collected on the sequences of a particular gene or gene family, the analysis can be reversed, and comparisons between orthologous genes can be used to assess taxonomic relationships. If a molecular clock has been established, the time to common ancestry between the previously analyzed species and a species newly introduced to the analysis can be estimated.

KEY CONCEPTS

- The sequences of orthologous genes in different species vary at nonsynonymous sites (where mutations have caused amino acid substitutions) and synonymous sites (where mutation has not affected the amino acid sequence).
- Synonymous substitutions accumulate ~10× faster than nonsynonymous substitutions.
- The evolutionary divergence between two DNA sequences is measured by the corrected percent of positions at which the corresponding nucleotides differ.
- Substitutions may accumulate at a more or less constant rate after genes separate, so that the divergence between any pair of globin sequences is proportional to the time since they shared common ancestry.

CONCEPT AND REASONING CHECK

Explain why synonymous substitutions in coding regions occur at a much higher rate than nonsynonymous substitutions do.

8.5 The Rate of Neutral Substitution Can Be Measured from Divergence of Repeated Sequences

We can make the best estimate of the rate of substitution at neutral sites by examining sequences that do not encode polypeptide. (We use the term "neutral" here rather than "synonymous," because there is no coding potential.) An informative comparison can be made by comparing the members of a common repetitive family in the human and mouse genomes.

The principle of the analysis is summarized in **FIGURE 8.10**. We start with a family of related sequences that have evolved by duplication and substitution from an original ancestral sequence. We assume that the ancestral sequence can be deduced by taking the base that is most common at each position. Then we can calculate the divergence of each individual family member as the proportion of bases that differ from the deduced ancestral sequence. In this example, individual members vary from 0.13 to 0.18 divergence and the average is 0.16.

One family used for this analysis in the human and mouse genomes derives from a sequence that is thought to have ceased to be active at about the time of the common ancestor between humans and rodents (the LINEs family; see *Section 17.10, Retroelements Fall into Three Classes*). This means that it has been diverging under limited selective pressure for the same length of time in both species. Its average divergence in humans is ~0.17 substitutions per site, corresponding to a rate of 2.2×10^{-9} substitutions per base per year over the 75 million years since the separation. In the mouse genome, however, neutral substitutions have occurred at twice this rate, corresponding to 0.34 substitutions per site in the family, or a rate of 4.5×10^{-9}. Note, however, that if we calculated the rate per generation instead of per year, it would be greater in humans than in the mouse ($\sim 2.2 \times 10^{-8}$ as opposed to $\sim 10^{-9}$).

These figures probably underestimate the rate of substitution in the mouse; at the time of divergence the rates in both lineages would have been the same, and the difference must have arisen since then. The current rate of neutral substitution per year in the mouse is probably two to three times greater than the historical average. At first glance, these rates would seem to reflect the balance between the occurrence of mutations (which may be higher in species with higher metabolic rates, like the mouse) and the loss of them due to genetic drift, which is largely a function of population size, since genetic drift is a type of "sampling error" where allele frequencies fluctuate more widely in smaller populations. In addition to eliminating neutral alleles more quickly, smaller population sizes also allow faster fixation and loss of neutral alleles. Rodent species tend to have short generation times (allowing more opportunities for substitutions per year), but species with short generation times also tend to have larger population sizes; thus the effects of more substitutions per year but less fixation of neutral alleles would cancel each other out. The higher substitution rate in mice is probably due primarily to a higher mutation rate.

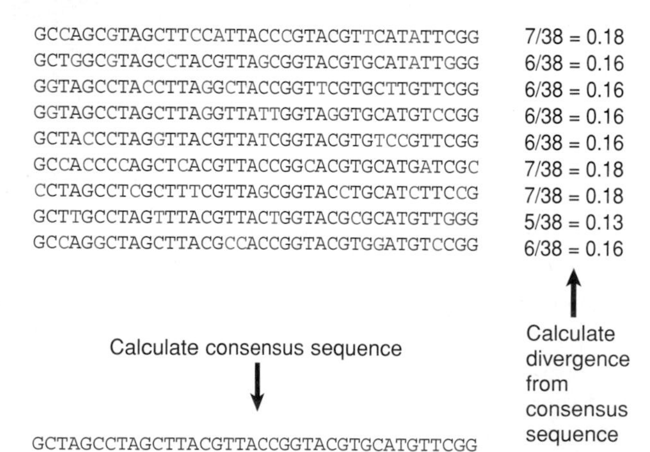

FIGURE 8.10 An ancestral consensus sequence for a family is calculated by taking the most common base at each position. The divergence of each existing current member of the family is calculated as the proportion of bases at which it differs from the ancestral sequence.

Comparing the mouse and human genomes allows us to assess whether syntenic (homologous) regions show signs of conservation or have differed at the rate predicted from accumulation of neutral substitutions. The proportion of sites that show signs of selection is ~5%. This is much higher than the proportion found in exons (~1%). This observation implies that the genome includes many more stretches whose sequence is important for functions other than coding for RNA. Known regulatory elements are likely to comprise only a small part of this proportion. This number also suggests that most (i.e., the rest) of the genome sequences do not have any function that depends on the exact sequence.

KEY CONCEPT

- The rate of substitution per year at neutral sites is greater in the mouse genome than in the human genome, probably because of a higher mutation rate.

CONCEPT AND REASONING CHECK

In comparing homologous sequences from related species, describe how both the ancestral sequence and the rates of evolution in the descendant lineages can be estimated.

8.6 How Did Interrupted Genes Evolve?

The structure of many eukaryotic genes suggests a concept of the eukaryotic genome as a sea of mostly unique DNA sequences in which exon "islands" separated by intron "shallows" are strung out in individual gene "archipelagoes." What was the original form of genes?

The **"introns early" model** is the proposal that introns have always been an integral part of the gene. Genes originated as interrupted structures, and those now without introns have lost them in the course of evolution.

The **"introns late" model** is the proposal that the ancestral protein-coding sequences were uninterrupted and that introns were subsequently inserted into them.

In simple terms, can the difference between eukaryotic and prokaryotic gene organizations be accounted for by the acquisition of introns in the eukaryotes or by the loss of introns from the prokaryotes?

One point in favor of the "introns early" model is that the mosaic structure of genes suggests an ancient combinatorial approach to the construction of genes to encode novel proteins, a hypothesis known as **exon shuffling**. Suppose that an early cell had a number of separate protein-coding sequences: it is likely to have evolved by reshuffling different polypeptide units to construct new proteins. Although we recognize the advantages of this mechanism for gene evolution, that does not necessarily mean that it was the primary reason for the *initial* evolution of the mosaic structure. Introns may have greatly assisted, but might not have been critical for, the recombination of protein-coding gene segments. Thus, a failure of support for the exon shuffling hypothesis would neither disprove the introns early hypothesis nor support the introns late hypothesis.

▸ **"introns early" model** The hypothesis that the earliest genes contained introns and some genes subsequently lost them.

▸ **"introns late" model** The hypothesis that the earliest genes did not contain introns and that introns were subsequently added to some genes.

▸ **exon shuffling** The hypothesis that genes have evolved by the recombination of various exons encoding functional protein domains.

If a protein-coding unit (now known as an exon) must be a continuous series of codons, every such reshuffling event would require a precise recombination of DNA to place separate protein-coding units in sequence and in the same reading frame (a ⅓ probability in any one random joining event). If, however, this combination doesn't produce a functional protein, the cell might be damaged because the original sequence of protein-coding units might have been lost.

The cell might survive, though, if some of the experimental recombination occurs in RNA transcripts, leaving the DNA intact. If a translocation event could place two protein-coding units in the same transcription unit, various RNA splicing experiments to combine the two proteins into a single polypeptide chain could be explored. If some combinations are not successful, the original protein-coding units remain available for further trials. In addition, this scenario does not require the two protein-coding units to be recombined precisely into a continuous coding sequence. There is evidence supporting this scenario: different genes have related exons, as if each gene had been assembled by a process of exon shuffling (see *Section 4.7, Some Exons Can Be Equated to Protein Functional Domains*).

FIGURE 8.11 illustrates the result of a translocation of a random sequence that includes an exon into a gene. In some organisms, exons are very small compared to introns, so it is likely that the exon will insert within an intron and be flanked by functional 5′ and 3′ splice junctions. Splicing junctions are recognized in sequential pairs, so the splicing mechanism should recognize the 5′ splicing junction of the original intron and the 3′ splicing junction of the introduced exon, instead of the 3′ splice junction of the original intron. Similarly, the 5′ splicing junction of the new exon and the 3′ splicing junction

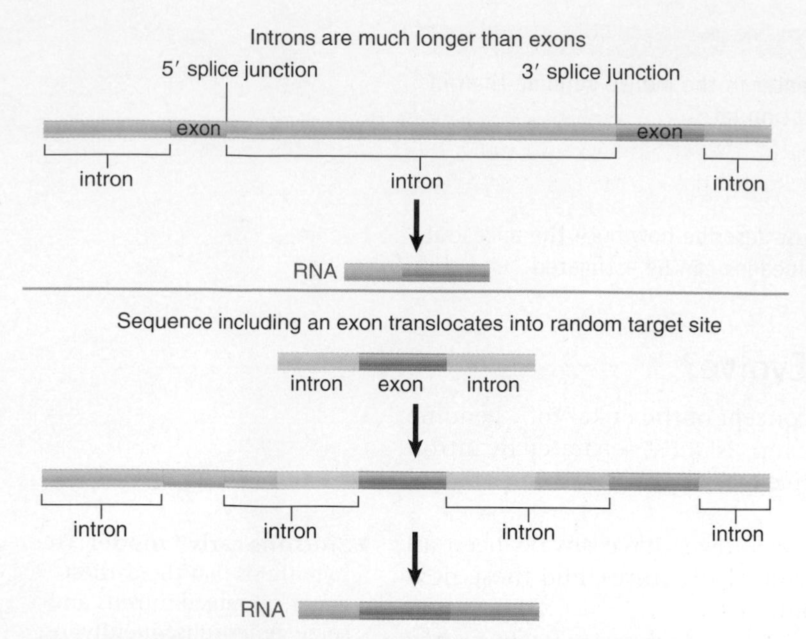

FIGURE 8.11 An exon surrounded by flanking sequences that is translocated into an intron may be spliced into the RNA product.

of the original intron may be recognized as a pair, so the new exon will remain between the original two exons in the mature RNA transcript. As long as the new exon is in the same reading frame as the original exons (a ⅓ probability at each end), a new, longer polypeptide will be produced. Exon shuffling events could have been responsible for generating new combinations of exons during evolution. (Note that the mechanism of this process is mimicked by the technique of *exon trapping* that is used to screen for functional exons [see Figure 5.8]).

Given that it is difficult to envision (1) the assembly of long chains of amino acids by some template-independent process and (2) that such assembled chains would be able to self-replicate, it is widely believed that the most successful early self-replicating molecules were nucleic acids—probably RNA. Indeed, RNA molecules can act both as coding templates and as catalysts (i.e., *ribozymes*; see *Chapter 23, Catalytic RNA*). It was probably by virtue of their catalytic activities that prototypic molecules in the early "RNA world" were able to self-replicate; the templating property would have emerged later.

Some species have alternative forms of rRNA and tRNA genes, both with and without introns. For tRNAs, which all have the same general conformation, it seems unlikely that the two regions of the gene evolved independently because the two regions base pair to fold the molecule into a functional shape. In this case, the intron must have been inserted into a continuous gene.

There is evidence that introns have been lost from some members of gene families. See *Section 4.8, Members of a Gene Family Have a Common Organization*, for examples from the insulin and actin gene families. In the case of the actin gene family, it is sometimes not clear whether the presence of an intron in a member of the family indicates the ancestral state or an insertion event. Overall, current evidence suggests that genes originally had sequences now called introns but can evolve with both the loss and gain of introns.

Organelle genomes show the evolutionary connections between prokaryotes and eukaryotes. There are many general similarities between mitochondria or chloroplasts and certain bacteria because those organelles originated by endosymbiosis, in which a bacterial cell dwelled within the cytoplasm of a eukaryotic prototype. Although there are similarities to bacterial genetic processes—such as protein and RNA synthesis—some organelle genes possess introns and therefore resemble eukaryotic nuclear genes. Introns are found in several chloroplast genes, including some that are homologous to *E. coli* genes. This suggests that the endosymbiotic event occurred before introns were lost from the prokaryotic lineage.

Mitochondrial genome comparisons are particularly striking. The genes of yeast and mammalian mitochondria encode virtually identical proteins, in spite of a considerable difference in gene organization. Vertebrate mitochondrial genomes are very small and extremely compact, whereas yeast mitochondrial genomes are larger and have some complex interrupted genes. Which is the ancestral form? Yeast mitochondrial introns (and certain other introns) can be mobile—they are independent sequences that can splice out of the RNA and insert DNA copies elsewhere—which suggests that they may have arisen by insertions into the genome (see *Section 23.5, Some Group I Introns Encode Endonucleases That Sponsor Mobility*, and *Section 23.6, Group II Introns May Encode Multifunction Proteins*). While most evidence supports "introns early," there is reason to believe that, in addition to the introduction of mobile elements, ongoing accommodations to various extrinsic and intrinsic (genomic) pressures might result, from time to time, in the emergence of new introns ("introns late").

KEY CONCEPTS

- A major evolutionary question is whether genes originated with introns or whether they were originally uninterrupted.
- Interrupted genes that correspond either to proteins or to independently functioning nonprotein-encoding RNAs probably originated in an interrupted form (the "introns early" hypothesis).
- A special class of introns is mobile and can insert themselves into genes.

CONCEPT AND REASONING CHECK

The "introns early" hypothesis is the proposal that cells that existed before eukaryotic cells evolved had genes with introns. Why might the loss of introns in the ancestral lineages of modern prokaryotes be advantageous, or at least not deleterious?

8.7 Why Are Some Genomes So Large?

The total amount of DNA in the (haploid) genome is a characteristic of each living species known as its **C-value**. There is enormous variation in the range of C-values, from $<10^6$ bp for a mycoplasma to $>10^{11}$ bp for some plants and amphibians.

▸ **C-value** The total amount of DNA in the genome (per haploid set of chromosomes).

FIGURE 8.12 summarizes the range of C-values found in different taxonomic groups. There is an increase in the minimum genome size found in each group as the complexity increases. Although C-values are greater in the multicellular eukaryotes, we do see some wide variations in the genome sizes within some groups.

Plotting the minimum amount of DNA required for a member of each group suggests in **FIGURE 8.13** that an increase in genome size is required for increased complexity in prokaryotes, fungi, and invertebrate animals.

Mycoplasma are the smallest prokaryotes and have genomes only ~3× the size of a large bacteriophage and smaller than those of some megaviruses. More typical bacterial genome sizes start at ~2×10^6 bp. Unicellular eukaryotes (whose lifestyles may resemble those of prokaryotes) also get by with genomes that are small, although they

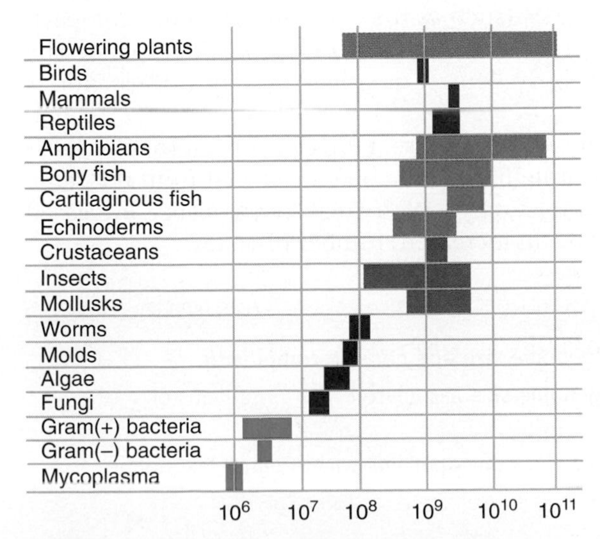

FIGURE 8.12 DNA content of the haploid genome increases with morphological complexity of lower eukaryotes but varies extensively within some groups of animals and plants. The range of DNA values within each group is indicated by the shaded area.

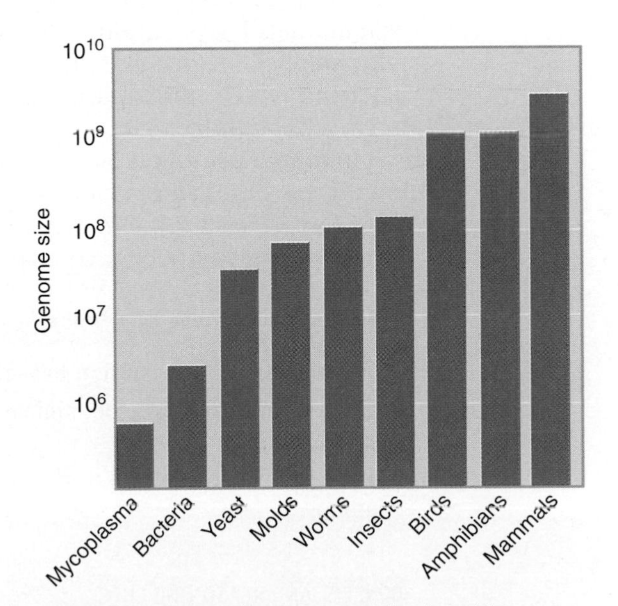

FIGURE 8.13 The minimum genome size found in each taxonomic group increases from prokaryotes to mammals.

Phylum	Species	Genome (bp)
Algae	*Pyrenomas salina*	6.6×10^5
Mycoplasma	*M. pneumoniae*	1.0×10^6
Bacterium	*E. coli*	4.2×10^6
Yeast	*S. cerevisiae*	1.3×10^7
Slime mold	*D. discoideum*	5.4×10^7
Nematode	*C. elegans*	8.0×10^7
Insect	*D. melanogaster*	1.8×10^8
Bird	*G. domesticus*	1.2×10^9
Amphibian	*X. laevis*	3.1×10^9
Mammal	*H. sapiens*	3.3×10^9

FIGURE 8.14 The genome sizes of some commonly studied organisms.

are larger than those of most bacteria. Being eukaryotic per se does not imply a vast increase in genome size; a yeast may have a genome size of ~1.3×10^7 bp, which is only about twice the size of an average bacterial genome.

A further twofold increase in genome size is adequate to support the slime mold *Dictyostelium discoideum*, which is able to live in either unicellular or multicellular modes. Another increase in complexity is necessary to produce fully multicellular organisms; the nematode worm *Caenorhabditis elegans* has a DNA content of 8×10^7 bp.

We also can see the steady increase in genome size with complexity in the listing in **FIGURE 8.14** of some of the most commonly studied organisms. It is necessary for insects, birds, amphibians, and mammals to have larger genomes than those of unicellular eukaryotes. After this point, though, there is no clear relationship between genome size and morphological complexity of the organism.

We know that eukaryotic genes are much larger than the sequences needed to encode polypeptides because exons may comprise only a small part of the total length of a gene. This explains why there is much more DNA than is needed to provide reading frames for all the proteins of the organism. Large parts of an interrupted gene may not encode polypeptide. In addition, in multicellular organisms there also may be significant lengths of DNA between genes, some of which functions in gene regulation. So it is not possible to deduce from the overall size of the genome anything about the number of genes or the complexity of the organism.

▸ **C-value paradox** The lack of relationship between the DNA content (C-value) of an organism and its coding potential.

The **C-value paradox** refers to the lack of correlation between genome size and genetic and morphological complexity (such as the number of different cell types). There are some extremely curious observations about relative genome size, such as that the toad *Xenopus* and humans have genomes of essentially the same size. In some taxonomic groups there are large variations in DNA content between organisms that do not vary much in complexity, as seen in Figure 8.12. (This is especially marked in insects, amphibians, and plants but does not occur in birds, reptiles, and mammals, which all show little variation within the group—with an ~2× range of genome sizes.) A cricket has a genome 11× the size of that of a fruit fly. In amphibians, the smallest genomes are $<10^9$ bp, whereas the largest are ~10^{11} bp. There is unlikely to be a large difference in the number of genes needed for the development of these amphibians. Some fish species have about the same number of genes as mammals have, but other fish genomes (such as that of fugu) are more compact, with smaller introns and shorter intergenic spaces. Still others are tetraploid. The extent to which this variation is selectively neutral or subject to natural selection is not yet fully understood.

In mammals, additional complexity is also a consequence of the alternative splicing of genes that allows two or more protein variants to be produced from the same gene (see *Chapter 21, RNA Splicing and Processing*). With such mechanisms, increased complexity need not be accompanied by an increased number of genes.

KEY CONCEPTS

- There is no clear correlation between genome size and genetic complexity.
- There is an increase in the minimum genome size associated with organisms of increasing complexity.
- There are wide variations in the genome sizes of organisms within many taxonomic groups.

CONCEPT AND REASONING CHECK

What are some advantages and disadvantages of a large genome? A small genome?

8.8 Morphological Complexity Evolves by Adding New Gene Functions

Comparison of the human genome sequence with sequences found in other species is revealing about the process of evolution. **FIGURE 8.15** shows an analysis of human genes according to the breadth of their distribution among all cellular organisms. Starting with the most generally distributed (top right corner of the figure), ~21% of genes are common to eukaryotes and prokaryotes. These tend to encode proteins that are essential for all living forms—typically basic metabolism, replication, transcription, and translation. Moving clockwise, another ~32% of genes are found in eukaryotes in general—for example, they may be found in yeast. These tend to encode proteins involved in functions that are general to eukaryotic cells but not to bacteria—for example, they may be concerned with specifying organelles or cytoskeletal components. Another ~24% of genes are generally found in animals. These include genes necessary for multicellularity and for development of different tissue types. Approximately 22% of genes are unique to vertebrates. These mostly encode proteins of the immune and nervous systems; they encode very few enzymes, consistent with the idea that enzymes have ancient origins and that metabolic pathways originated early in evolution. We see, therefore, that the evolution of more complex morphology and specialization requires the addition of groups of genes representing the necessary new functions.

One way to define essential proteins is to identify the proteins present in all proteomes. Comparing the human proteome in more detail with the proteomes of other organisms, 46% of the yeast proteome, 43% of the worm proteome, and 61% of the fly proteome is represented in the human proteome. A key group of ~1300 proteins is present in all four proteomes. The common proteins are basic "housekeeping" proteins required for essential functions, falling into the types summarized in **FIGURE 8.16**. The main functions are concerned with transcription and translation (35%), metabolism (22%), transport (12%), DNA replication and modification (10%), protein folding and degradation (8%), and cellular processes (6%), with the remaining 7% dedicated to various other functions.

One of the striking features of the human proteome is that it has many unique proteins compared with those of other eukaryotes but has relatively few unique protein domains (portions of proteins having a specific function). Most protein domains appear to be common to the animal kingdom. There are many unique protein architectures, however, defined as unique combinations of domains. **FIGURE 8.17** shows that the greatest proportion of unique proteins consists of transmembrane and extracellular proteins. In yeast, the vast majority of architectures are concerned with intracellular proteins. About twice as many intracellular architectures are found in flies (or nematodes), but there is a strikingly higher proportion of transmembrane and extracellular proteins, as might be expected from the additional functions required for the interactions between the cells of a multicellular organism. The additions in intracellular architectures required in a vertebrate (typified by the human genome)

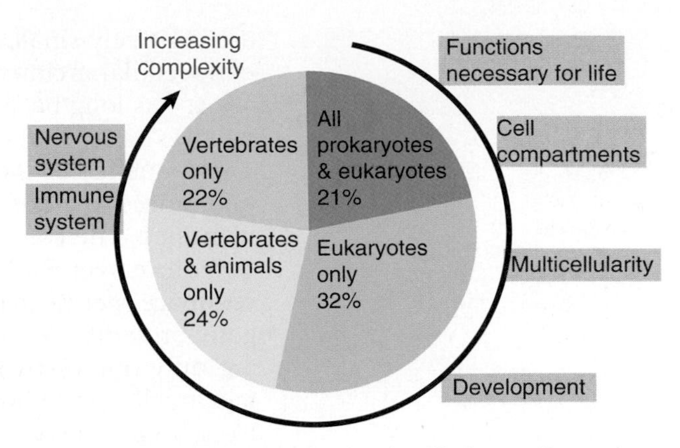

FIGURE 8.15 Human genes can be classified according to how widely their homologs are distributed in other species.

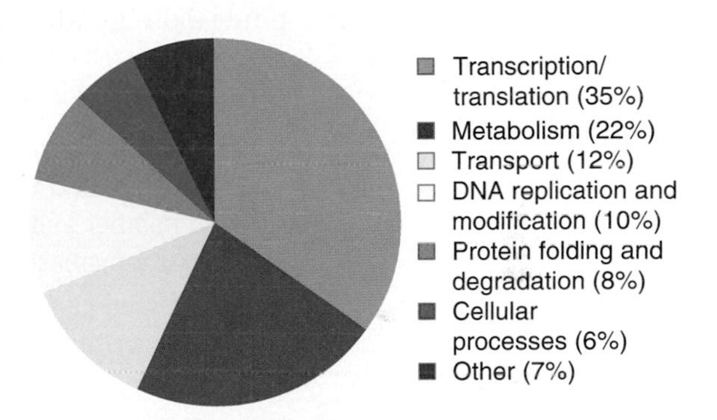

FIGURE 8.16 Common eukaryotic proteins are concerned with essential cellular functions.

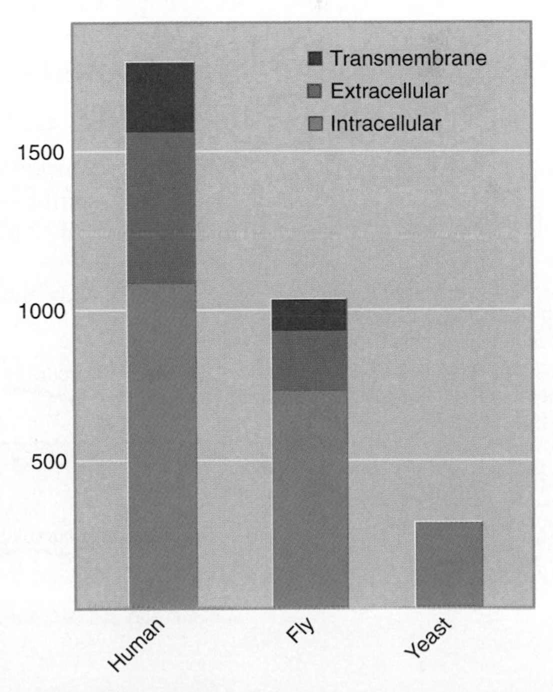

FIGURE 8.17 Increasing complexity in eukaryotes is accompanied by accumulation of new proteins for transmembrane and extracellular functions.

are relatively small, but there is, again, a higher proportion of transmembrane and extracellular architectures.

It has long been known that the genetic difference between humans and chimpanzees (our nearest relative) is very small, with ~99% identity between genomes. The sequence of the chimpanzee genome now allows us to investigate the 1% of differences in more detail to see whether features responsible for "humanity" can be identified. The comparison shows 35×10^6 nucleotide substitutions (1.2% sequence difference overall), 5×10^6 deletions or insertions (making ~1.5% of the euchromatic sequence specific to each species), and many chromosomal rearrangements. Homologous proteins are usually very similar: 29% are identical, and in most cases there are only one or two amino acid differences in the protein between the species. In fact, nucleotide substitutions occur less often in genes encoding polypeptides than are likely to be involved in specifically human traits, suggesting that protein evolution is not a major factor in human-chimpanzee differences. This leaves larger-scale changes in gene structure and/or changes in gene regulation as the major candidates. Some 25% of nucleotide substitutions occur in CpG dinucleotides (among which are many potential regulator sites).

KEY CONCEPTS

- In general, comparisons of eukaryotes to prokaryotes, multicellular to unicellular eukaryotes, and vertebrate to invertebrate animals show a positive correlation between gene number and morphological complexity as additional genes are needed with generally increased complexity.
- Most of the genes that are unique to vertebrates are concerned with the immune or nervous systems.

CONCEPT AND REASONING CHECK

Account for the observation that the human proteome has many unique proteins but few unique protein domains.

8.9 Gene Duplication Contributes to Genome Evolution

Exons act as modules for building genes that are tried out in the course of evolution in various combinations (see *Section 4.7, Some Exons Can Be Equated to Protein Functional Domains*). At one extreme, an individual exon from one gene may be copied and used in another gene. At the other extreme, an entire gene, including both exons and introns, may be duplicated. In such a case, mutations can accumulate in one copy without elimination by natural selection as long as the other copy is under selection to remain functional. The selectively neutral copy may then evolve to a new function, become expressed at a different time or in a different cell type from the first copy, or become a nonfunctional pseudogene.

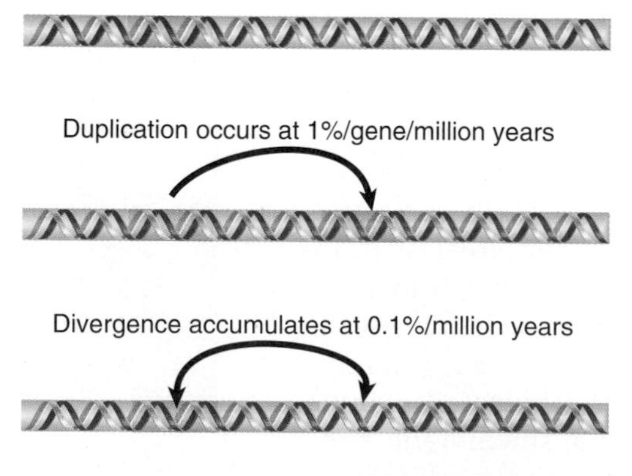

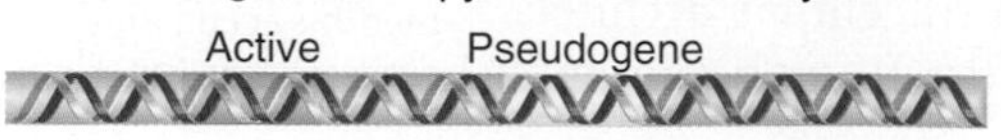

FIGURE 8.18 After a globin gene has been duplicated, differences may accumulate between the copies. The genes may acquire different functions or one of the copies may become inactive.

FIGURE 8.18 summarizes our present view of the rates at which these processes occur. There is a ~1% probability that a given gene will be included in a duplication in a period of one million years.

After the gene has duplicated, differences evolve as the result of the occurrence of different mutations in each copy. These accumulate at a rate of ~0.1% per million years (see *Section 8.4, A Constant Rate of Sequence Divergence Is a Molecular Clock*).

Unless the gene encodes a product that is required in high concentration in the cell, the organism is not likely to need to retain two identical copies of the gene. As differences evolve between the duplicated genes, one of two types of event is likely to occur:

- Both the gene copies remain necessary. This can happen either because the differences between them generate proteins with different functions, or because they are expressed specifically at different times or in different cell types.
- If this does not happen, one of the genes is likely to become a pseudogene because it will by chance gain a deleterious mutation, and there will be no purifying selection to eliminate this copy, so by genetic drift the mutant version may increase in frequency and fix in the species. Typically this takes ~4 million years for globin genes; in general, the time to fixation of a neutral mutant depends on the generation time and the effective population size, with genetic drift being a stronger force in smaller populations. In such a situation, it is purely a matter of chance which of the two copies becomes inactive. (This can contribute to incompatibility between different individuals, and ultimately to speciation, if different copies become inactive in different populations.)

  Analysis of the human genome sequence shows that ~5% of the genome comprises duplications of identifiable segments ranging in length from 10 to 300 kb. These duplications have arisen relatively recently; that is, there has not been sufficient time for divergence between them for their homology to become obscured. They include a proportional share (~6%) of the expressed exons, which shows that the duplications are occurring more or less irrespective of genetic content. The genes in these duplications may be especially interesting because of the implication that they have evolved recently and therefore could be important for recent evolutionary developments (such as the separation of the human lineage from that of other primates).

KEY CONCEPT

- Duplicated genes may diverge to generate different genes, or one copy may become an inactive pseudogene.

CONCEPT AND REASONING CHECK

A single duplication event is likely to increase the rate of additional duplication events involving those two gene copies. Why?

8.10 Globin Clusters Arise by Duplication and Divergence

The most common type of gene duplication generates a second copy of the gene close to the first copy. In some cases, the copies remain associated, and further duplication may generate a cluster of related genes. The best-characterized example of a gene cluster is that of the globin genes, which constitute an ancient gene family fulfilling a function that is central to animals: the transport of oxygen.

The major constituent of the vertebrate red blood cell is the globin tetramer, which is associated with its heme (iron-binding) group in the form of hemoglobin. Functional globin genes in all species have the same general structure: they are divided into three exons, as shown previously in Figure 4.4. We conclude that all globin genes have evolved from a single ancestral gene, and by tracing the history of individual globin genes within and between species we may learn about the mechanisms involved in the evolution of gene families.

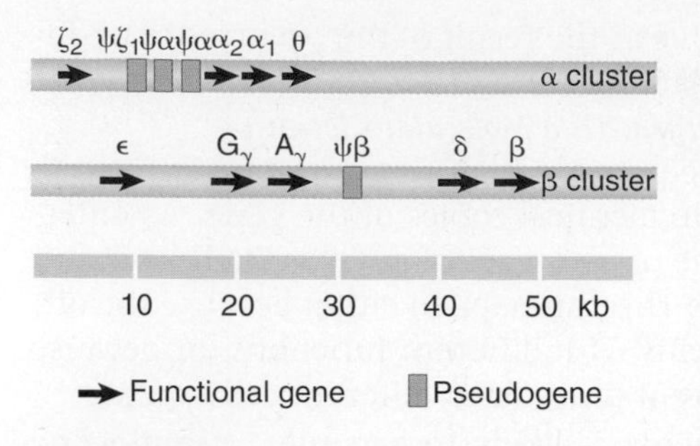

FIGURE 8.19 Each of the α-like and β-like globin gene families is organized into a single cluster, which includes functional genes and pseudogenes (ψ).

In red blood cells of adult mammals, the globin tetramer consists of two identical α chains and two identical β chains. Embryonic red blood cells contain hemoglobin tetramers that are different from the adult form. Each tetramer contains two identical α-like chains and two identical β-like chains, each of which is related to the adult polypeptide and is later replaced by it in the adult form of the protein. This is an example of developmental control, in which different genes are successively switched on and off to provide alternative products that fulfill the same function at different times.

The division of globin chains into α-like and β-like chains reflects the organization of the genes. Each type of globin is encoded by genes organized into a single cluster. The structures of the two clusters in the primate genome are illustrated in **FIGURE 8.19**. Pseudogenes are indicated by the symbol ψ.

Stretching over 50 kb, the β cluster contains five functional genes (ε, two γ, δ, and β) and one nonfunctional pseudogene (ψβ). The two γ genes differ in their coding sequence in only one amino acid: the G variant has glycine at position 136, whereas the A variant has alanine.

The more compact α cluster extends over 28 kb and includes one active ζ gene, one nonfunctional ζ pseudogene, two α genes, two nonfunctional α pseudogenes, and the θ gene of unknown function. The two α genes encode the same protein. Two (or more) identical genes present on the same chromosome are described as **nonallelic genes**.

▸ **nonallelic genes** Two (or more) copies of the same gene that are present at different locations in the genome (contrasted with alleles, which are copies of the same gene derived from different parents and present at the same location on the homologous chromosomes).

The details of the relationship between embryonic and adult hemoglobins vary with the species. The human pathway has three stages: embryonic, fetal, and adult. The distinction between embryonic and adult is common to mammals, but the number of preadult stages varies. In humans, ξ and α are the two α-like chains. ε, γ, δ, and β are the β-like chains. **FIGURE 8.20** shows how the chains are expressed at different stages of development. There is also tissue-specific expression associated with the developmental expression: embryonic hemoglobin genes are expressed in the yolk sac, fetal genes are expressed in the liver, and adult genes are expressed in bone marrow.

In the human pathway, ζ is the first α-like chain to be expressed, but it is soon replaced by α. In the β-pathway, ε and γ are expressed first, with δ and β replacing them later. In adults, the $\alpha_2\beta_2$ form provides 97% of the hemoglobin, $\alpha_2\delta_2$ provides ~2%, and ~1% is provided by persistence of the fetal form $\alpha_2\gamma_2$.

What is the significance of the differences between embryonic and adult globins? The embryonic and fetal forms have a higher affinity for oxygen, which is necessary in order to obtain oxygen from the mother's blood. This helps to explain why there is no direct equivalent (although there is temporal expression of globins) in, for example, the chicken, for which the embryonic stages occur outside the mother's body (that is, within the egg).

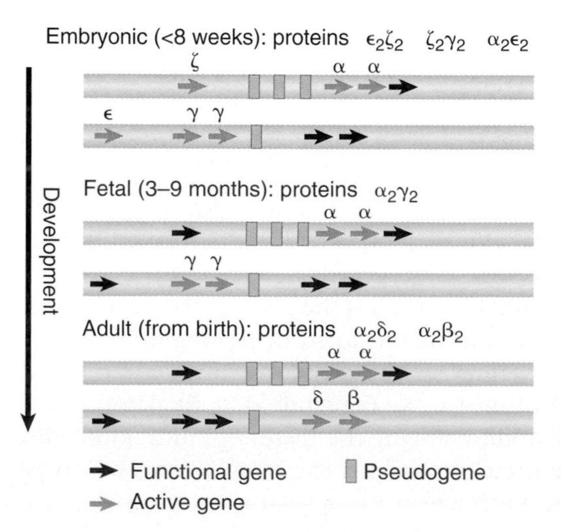

FIGURE 8.20 Different hemoglobin genes are expressed during embryonic, fetal, and adult periods of human development.

Functional genes are defined by their expression to RNA and ultimately by the polypeptides they encode. Pseudogenes are defined as such by their inability to produce functional polypeptides; the reasons for their inactivity vary, and the deficiencies may be in transcription or translation (or both). A similar general organization is found in other vertebrate globin gene clusters, but details of the types, numbers, and order of genes all vary, as illustrated in **FIGURE 8.21**. Each cluster contains both

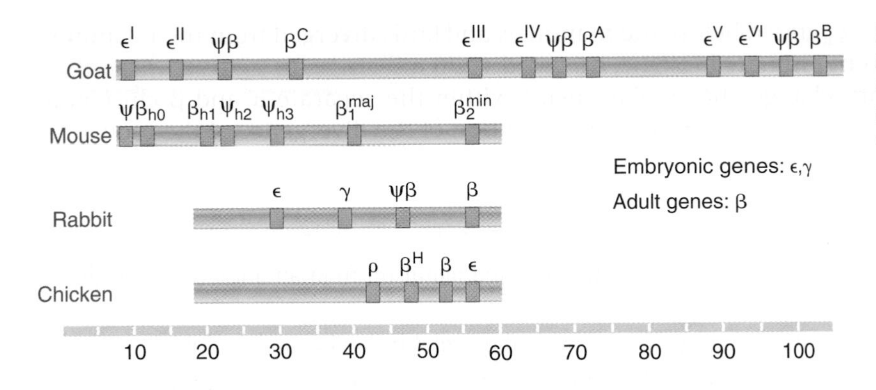

FIGURE 8.21 Clusters of β-globin genes and pseudogenes are found in vertebrates. Seven mouse genes include two early embryonic genes, one late embryonic gene, two adult genes, and two pseudogenes. Rabbits and chickens each have four genes.

embryonic and adult genes. The total lengths of the clusters vary widely. The longest known cluster is found in the goat genome, where a basic cluster of four genes has been duplicated twice. The distribution of active genes and pseudogenes differs in each case, illustrating the random nature of the evolution of one copy of a duplicated gene to a pseudogene.

The characterization of these gene clusters provides an important general point. There may be more members of a gene family, both functional and nonfunctional, than we would suspect on the basis of protein analysis. The extra functional genes may represent duplicates that encode identical polypeptides, or they may be related to—but different from—known proteins (and presumably expressed only briefly or in low amounts).

With regard to the question of how much DNA is needed to encode a particular function, we see that encoding the β-like globins requires a range of 20 to 120 kb in different mammals. This is much greater than we would expect just from scrutinizing the known β-globin proteins or from even considering the individual genes. Clusters of this type are not common, though; most genes are found as individual loci.

From the organization of globin genes in a variety of species, we should be able to trace the evolution of present globin gene clusters from a single ancestral globin gene. Our present view of the evolutionary history was pictured in Figure 8.8.

The leghemoglobin gene of plants, which is related to the globin genes, may provide some clues about the ancestral form, though of course the modern leghemoglobin gene has evolved for just as long as the animal globin genes. (Leghemoglobin is an oxygen carrier found in the nitrogen-fixing root nodules of legumes.) The furthest back that we can trace a true globin gene is to the sequence of the single chain of mammalian myoglobin, which diverged from the globin lineage ~800 million years ago in the ancestors of mammals. The myoglobin gene has the same organization as globin genes, so we may take the three-exon structure to represent that of their common ancestor.

Some members of the class Chondrichthyes (cartilaginous fish) have only a single type of globin chain, so they must have diverged from the lineage of other vertebrates before the ancestral globin gene was duplicated to give rise to the α and β variants. This appears to have occurred ~500 million years ago, during the evolution of the Osteichthyes (bony fish).

The next stage of globin evolution is represented by the state of the globin genes in the amphibian *Xenopus laevis*, which has two globin clusters. Each cluster, though, contains both α and β genes, of both larval and adult types. The cluster must, therefore, have evolved by duplication of a linked α-β pair, followed by divergence between the individual copies. Later the entire cluster was duplicated.

The amphibians separated from the reptilian/mammalian/avian line ~350 million years ago, so the separation of the α- and β-globin genes must have resulted from a transposition in the reptilian/mammalian/avian forerunner after this time. This probably occurred in the period of early tetrapod evolution. There are separate clusters for α and β globins in both birds and mammals; so the α and β genes must have

8.13 What Is The Role of Transposable Elements in Genome Evolution?

Transposable elements (TEs) are mobile genetic elements that can be integrated into the genome at multiple sites and (for some elements) also excised from an integration site. (See *Chapter 17, Transposable Elements and Retroviruses*, for an extensive discussion of the types and mechanisms of TEs.) The insertion of a TE at a new site in the genome is called **transposition**. One type of TE, the retrotransposon, transposes via an RNA intermediate; a new copy of the element is created by transcription, followed by reverse transcription to DNA and subsequent integration at a new site.

▸ **transposition** Movement of mobile genetic elements from one location in the genome to another.

Most TEs integrate at sequences that are random (at least with respect to their functions). As such, they are a major source of the problems associated with insertion mutations: frameshifts if inserted into coding regions and altered gene expression if inserted into regulatory regions. The number of copies of a given TE in a species' genome therefore depends on several factors: the rate of integration of the TE; its rate of excision (if any); selection on individuals with phenotypes altered by TE integration; and regulation of transposition.

TEs effectively act as intracellular parasites and, like other parasites, may need to strike an evolutionary balance between their own proliferation and the detrimental effects on the "host" organism. Studies on *Drosophila* TEs confirm that the mutational integration of TEs generally have deleterious, sometimes lethal, phenotypic effects. This suggests that negative selection plays an important role in the regulation of transposition; individuals with high levels of transposition are less likely to survive and reproduce. One might, however, expect that both TEs and their hosts may evolve mechanisms to limit transposition, and in fact both are observed. In one example of TE self-regulation, the *Drosophila* P element encodes a transposition repressor protein that is active in somatic tissue (see *Section 17.6, Transposition of P Elements Causes Hybrid Dysgenesis*). In addition, there are two major cellular mechanisms for transposition regulation:

- In an RNA interference-like mechanism (see *Section 30.6, How Does RNA Interference Work?*) involving piRNAs (see *Section 30.5, MicroRNAs Are Widespread Regulators in Eukaryotes*), the RNA intermediates of retrotransposons can be selectively degraded.
- In mammals, plants, and fungi, a DNA methyltransferase methylates cytosines within TEs, resulting in transcriptional silencing (see *Section 29.7, DNA Methylation Is Responsible for Imprinting*).

In any case, it is rare for TE proliferation to continue unchecked but rather to be limited by negative selection and/or regulation of transposition. Following introduction of a TE to a genome, though, the copy number may increase to many thousands or millions before some equilibrium is achieved, particularly if TEs are integrated into introns or intergenic DNA where phenotypic effects will be absent or minimal. As a result, genomes may contain a high proportion of moderately or highly repetitive sequences (see *Section 5.5, Eukaryotic Genomes Contain Both Nonrepetitive and Repetitive DNA Sequences*).

KEY CONCEPT

- Transposable elements tend to increase in copy number when introduced to a genome but are kept in check by negative selection and transposition regulation mechanisms.

CONCEPT AND REASONING CHECK

Why might it be advantageous for a transposable element to limit its own transposition?

8.14 There May Be Biases in Mutation, Gene Conversion, and Codon Usage

As discussed earlier in this chapter (see *Section 8.2, DNA Sequences Evolve by Mutation and a Sorting Mechanism*), the probability of a particular mutation is a function of the probability that a particular replication error or DNA-damaging event will occur and the probability that the error will be detected and repaired before the next DNA replication. To the extent that there is bias in these two events, there is bias in the types of mutations that occur (for example, a bias for transition mutations over transversion mutations despite the greater number of possible transversions).

Observations of the distributions of types of mutations over a taxonomically wide range of species (including prokaryotes and unicellular and multicellular eukaryotes), assessed by direct observation of mutational variants or by comparing sequence differences in pseudogenes, show a consistent pattern of a bias toward a high AT genomic content. The reasons for this are complex, and different mechanisms may be more or less important in different taxonomic groups, but there are two likely mechanisms. First, the common mutational source of spontaneous deamination of cytosine to uracil, or of 5-methylcytosine to thymine (see Figure 1.30), promotes the transition mutation of C-G to T-A. Uracil in DNA is more likely to be repaired than thymine (see *Section 1.13, Mutations Are Concentrated at Hotspots*), so methylated cytosines (often found in CG doublets) are not only mutation hotspots but specifically biased toward producing a T-A pair. Second, oxidation of guanine to 8-oxoguanine can result in a C-G to A-T transversion because 8-oxoguanine pairs more stably with adenine than with cytosine.

Despite this *mutational bias*, in analyses in which the expected equilibrium base composition is predicted from the observed rates of specific types of mutations, the observed AT content is generally lower than expected. This suggests that some mechanism or mechanisms are working to counteract the mutational bias toward AT. One possibility is that this is adaptive; a highly biased base composition limits the mutational possibilities and consequently limits evolutionary potential. As discussed next, though, there may be a nonadaptive explanation.

A second possible source of bias in genomic base composition is *gene conversion*, which occurs when heteroduplex DNA containing mismatched base pairs, often resulting from the resolution of a Holliday junction during recombination or double-strand break repair, is repaired using the mutated strand as template (see *Section 7.4, Crossover Fixation Could Maintain Identical Repeats*, and *Section 15.3, Double-Strand Breaks Initiate Recombination*). Interestingly, observations of gene conversion events in animals and fungi show a clear bias toward G-C, though the mechanism is unclear. In support of this observation, chromosomal regions of high recombinational activity show more mutations to G-C, and regions with low recombinational activity tend to be AT-rich. The observed rates of gene conversion per site tend to be of the same order of magnitude or higher than mutation rates; thus gene conversion bias alone may account for the lower-than-expected AT content being driven higher by mutational bias. *Gene conversion bias* may also be partly responsible for another universally observed bias in genome composition, *codon bias* (see *Section 8.4, A Constant Rate of Sequence Divergence Is a Molecular Clock*).

Due to the degeneracy of the genetic code, most of the amino acids found in polypeptides are represented by more than one codon in a genetic message. The alternate codons are not generally found in equal frequencies in genes, though; particularly in highly expressed genes, one codon of the two, four, or six that call for a particular amino acid is often used at a much higher frequency than the others. As discussed in *Section 8.4*, one explanation for this bias is that a particular codon may be more efficient at recruiting an abundant tRNA species, such that the rate or accuracy of translation is greater with higher usage of that codon. There may be additional adaptive consequences of particular exon sequences: some may contribute to splicing efficiency, form

9.2 Viral Genomes Are Packaged into Their Coats

From the perspective of packaging the individual sequence, there is an important difference between a cellular genome and a virus. The cellular genome is essentially indefinite in size; the number and location of individual sequences can be changed by duplication, deletion, and rearrangement. Thus it requires a generalized method for packaging its DNA, one that is insensitive to the total content or distribution of sequences. By contrast, two restrictions define the needs of a virus. The amount of nucleic acid to be packaged is predetermined by the size of the genome, and it must all fit within a coat assembled from a protein or proteins coded by the viral genes.

A virus particle is deceptively simple in its superficial appearance. The nucleic acid genome is contained within a **capsid**, which is a symmetrical or quasi-symmetrical structure assembled from one or only a few proteins. Attached to the capsid (or incorporated into it) are other structures; these structures are assembled from distinct proteins and are necessary for infection of the host cell.

▸ **capsid** The external protein coat of a virus particle.

The virus particle is tightly constructed. The internal volume of the capsid is rarely much greater than the volume of the nucleic acid it must hold. The difference is usually less than twofold, and often the internal volume is barely larger than the nucleic acid. There are two types of solution to the problem of how to construct a capsid that contains nucleic acid:

- The protein shell can be assembled around the nucleic acid, thereby condensing the DNA or RNA by protein–nucleic acid interactions during the process of assembly.
- The capsid can be constructed from its component(s) in the form of an empty shell, into which the nucleic acid must be inserted, being condensed as it enters.

The capsid is assembled around the genome for single-stranded RNA viruses. The principle of assembly is that *the position of the RNA within the capsid is determined directly by its binding to the proteins of the shell.* The best-characterized example is TMV (tobacco mosaic virus). Assembly starts at a duplex hairpin that lies within the sequence of the RNA genome. From this **nucleation center**, assembly proceeds bidirectionally along the RNA until it reaches the ends. The unit of the capsid is a two-layer disk, with each layer containing 17 identical protein subunits. The disk is a circular structure, which forms a helix as it interacts with the RNA. At the nucleation center, the RNA hairpin inserts into the central hole in the disk, and the disk changes conformation into a helical structure that surrounds the RNA. Additional disks are added, with each new disk pulling a new stretch of RNA into its central hole. The RNA becomes coiled in a helical array on the inside of the protein shell, as illustrated in **FIGURE 9.2**.

▸ **nucleation center** A duplex hairpin in TMV (tobacco mosaic virus) in which assembly of coat protein with RNA is initiated.

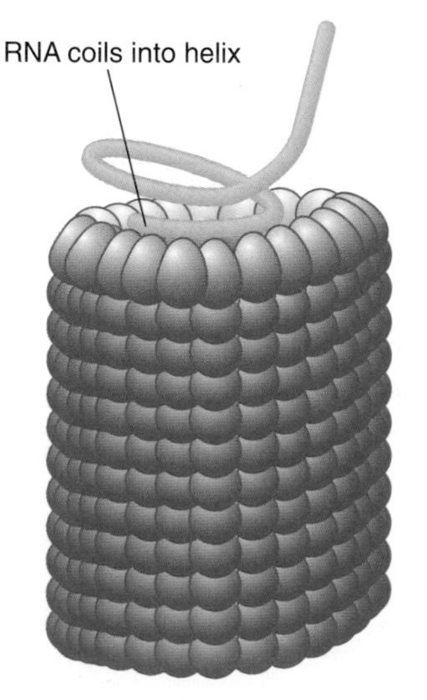

FIGURE 9.2 A helical path for TMV RNA is created by the stacking of protein subunits in the virus.

The spherical capsids of DNA viruses are assembled in a different way, as best characterized for the phages lambda and T4. In each case, an empty headshell is assembled from a small set of proteins. The duplex genome then is inserted into the head, accompanied by a structural change in the capsid.

FIGURE 9.3 summarizes the assembly of lambda. It starts with a small headshell that contains a protein "core." This is converted to an empty headshell of more distinct shape. At this point the DNA packaging begins, the headshell expands in size (though remaining the same shape), and finally the full head is sealed by the addition of the tail.

Inserting DNA into a phage head involves two types of reaction: *translocation* and *condensation*. Both are energetically unfavorable.

Translocation is an active process in which the DNA is driven into the head by an ATP-dependent mechanism. A

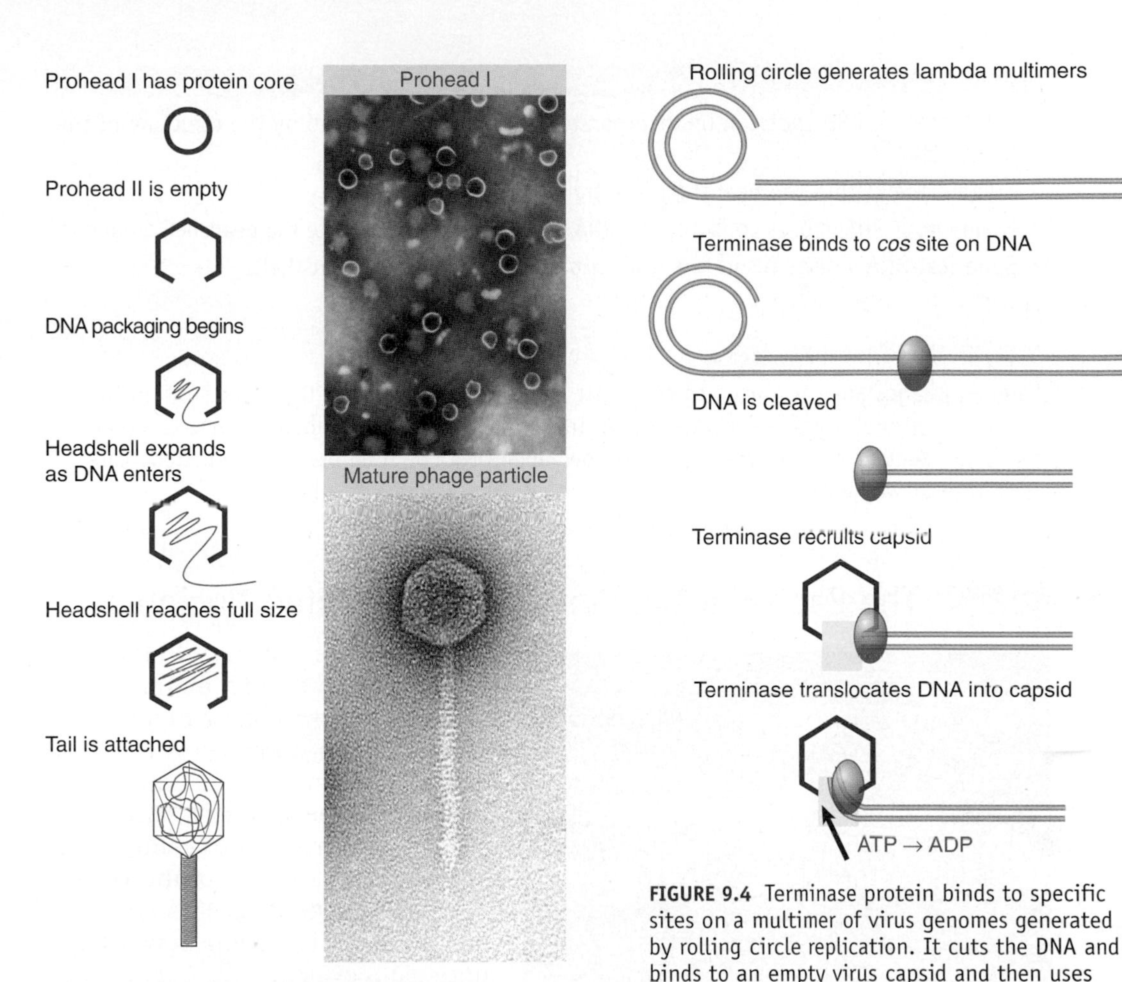

FIGURE 9.3 Maturation of phage lambda passes through several stages. The empty head changes shape and expands when it becomes filled with DNA. The electron micrographs show the particles at the start and the end of the maturation pathway. Top photo reproduced from D. Cue and M. Feiss, *Proc. Natl. Acad. Sci. USA* 90 (1993): 9290–9294. Copyright 1993 National Academy of Science, U.S.A. Photo courtesy of Michael G. Feiss, University of Iowa. Bottom photo courtesy of Robert Duda, University of Pittsburgh.

FIGURE 9.4 Terminase protein binds to specific sites on a multimer of virus genomes generated by rolling circle replication. It cuts the DNA and binds to an empty virus capsid and then uses energy from hydrolysis of ATP to insert the DNA into the capsid.

common mechanism is used for many viruses that replicate by a rolling circle mechanism to generate long tails that contain multimers of the viral genome. Phage lambda inserts its genome into the empty capsid using the **terminase** enzyme. **FIGURE 9.4** summarizes the process.

Little is known about the mechanism of condensation into an empty capsid, except that the capsid contains "internal proteins" as well as DNA. One possibility is that they provide some sort of scaffolding onto which the DNA condenses. (This would be a counterpart to the use of the proteins of the shell in the plant RNA viruses like TMV.)

How specific is the packaging? It cannot depend on particular sequences, because deletions, insertions, and substitutions all fail to interfere with the assembly process. The relationship between DNA and the headshell has been investigated directly by determining which regions of the DNA can be chemically crosslinked to the proteins of the capsid. The surprising answer is that all regions of the DNA are more or less equally susceptible. This probably means that when DNA is inserted into the head, it follows a general rule for condensing, but the pattern is not determined by particular sequences.

▸ **terminase** An enzyme that cleaves multimers of a viral genome and then uses hydrolysis of ATP to provide the energy to translocate the DNA into an empty viral capsid starting with the cleaved end.

KEY CONCEPTS

- The length of DNA that can be incorporated into a virus is limited by the structure of the headshell.
- Nucleic acid within the headshell is extremely condensed.
- Filamentous RNA viruses condense the RNA genome as they assemble the headshell around it.
- Spherical DNA viruses insert the DNA into a preassembled protein shell.

CONCEPT AND REASONING CHECK

Lambda phages are often used to construct *libraries* (collections of DNA representing the genomic material of different organisms), in which most of the lambda genome is replaced by library sequences. Is there a limit to how much DNA could be placed in a lambda carrier? Why or why not?

9.3 The Bacterial Genome Is a Supercoiled Nucleoid

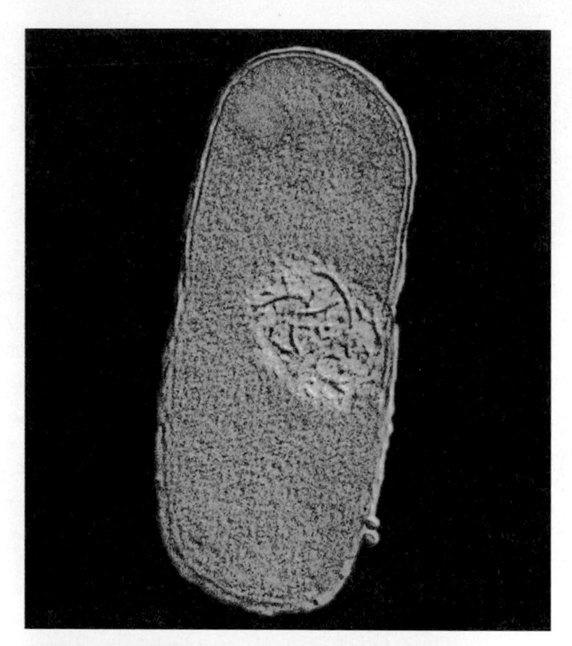

FIGURE 9.5 *E. coli* bacterium, colored transmission electron micrograph (TEM). This thin slice is cut through the bacterial cell lengthwise, showing the internal structure. Unlike eukaryotic organisms, bacteria do not have a membrane-bound cell nucleus. Instead, their genetic material is located in a confined area called the nucleoid. © Dr. Klaus Boller/Photo Researchers, Inc.

Although bacteria do not display structures with the distinct morphological features of eukaryotic chromosomes, their genomes nonetheless are organized into definite bodies. The genetic material can be seen as a fairly compact clump (or series of clumps) that occupies about a third of the volume of the cell. **FIGURE 9.5** displays a thin section through a bacterium in which this nucleoid is evident.

When *E. coli* cells are lysed, fibers are released in the form of loops attached to the broken envelope of the cell. As can be seen from **FIGURE 9.6**, the DNA of these loops is not found in the extended form of a free duplex but instead is compacted by association with proteins. It is not known which proteins are bound to the DNA to form the nucleoid.

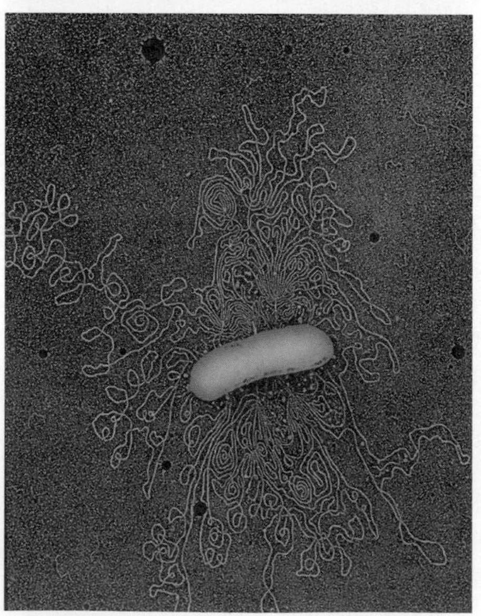

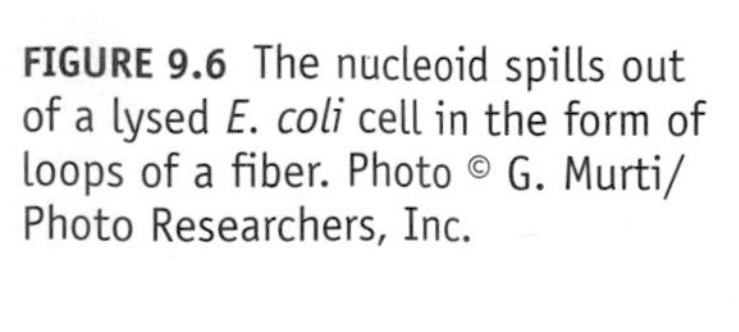

FIGURE 9.6 The nucleoid spills out of a lysed *E. coli* cell in the form of loops of a fiber. Photo © G. Murti/Photo Researchers, Inc.

The isolated nucleoid consists of ~80% DNA by mass. Eukaryotic chromosomes, in comparison, contain ~50% DNA by mass. The bacterial nucleoid can be unfolded by treatment with reagents that destroy RNA or protein, indicating that both RNA and protein are essential to maintain its structure.

The DNA of the bacterial nucleoid behaves as a closed duplex structure, as judged by its response to ethidium bromide. This small molecule intercalates between base pairs to generate positive superhelical turns in "closed" circular DNA molecules, that is, molecules in which both strands have covalent integrity. (In "open" circular molecules, which contain a nick in one strand, or with linear molecules, the DNA can rotate freely in response to the intercalation, thus relieving the tension.)

In a natural closed DNA that is negatively supercoiled, the intercalation of ethidium bromide first removes the negative supercoils and then introduces positive supercoils. The amount of ethidium bromide needed to achieve zero supercoiling is a measure of the original density of negative supercoils.

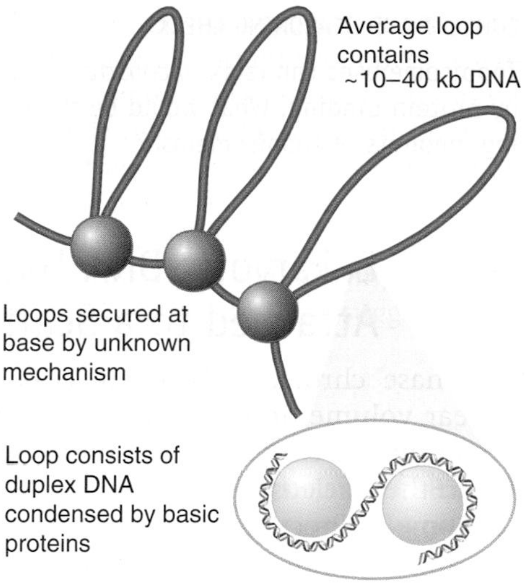

Some nicks occur in the compact nucleoid during its isolation; they can also be generated by limited treatment with DNase. This does not, however, abolish the ability of ethidium bromide to introduce positive supercoils. This capacity of the genome to retain its response to ethidium bromide in the face of nicking means that it must have many independent chromosomal **domains** and that *the supercoiling in each domain is not affected by events in the other domains.*

Average loop contains ~10–40 kb DNA

Loops secured at base by unknown mechanism

Loop consists of duplex DNA condensed by basic proteins

FIGURE 9.7 The bacterial genome consists of a large number of loops of duplex DNA (in the form of a fiber), each of which is secured at the base to form an independent structural domain.

‣ **domain** In reference to a chromosome, refers either to a discrete structural entity defined as a region within which supercoiling is independent of other regions or to an extensive region including an expressed gene that has heightened sensitivity to degradation by the enzyme DNase I. In a protein, it is a discrete continuous part of the amino acid sequence that can be equated with a particular function.

This autonomy suggests that the structure of the bacterial chromosome has the general organization depicted diagrammatically in **FIGURE 9.7**. Each domain consists of a loop of DNA, the ends of which are secured in some (unknown) way that does not allow rotational events to propagate from one domain to another. There are estimated to be ~400 domains of ~10 kb each in the *E. coli* genome.

The existence of separate domains could permit different degrees of supercoiling to be maintained in different regions of the genome. This could be relevant in considering the different susceptibilities of particular bacterial promoters to supercoiling (see *Section 19.14, Supercoiling Is an Important Feature of Transcription*).

As shown in **FIGURE 9.8**, supercoiling in the genome can take either of two forms:

- If a supercoiled DNA is free, its path is *unconstrained*, and negative supercoils generate a state of torsional tension that is transmitted freely along the DNA within a domain. This supercoiling can be relieved by unwinding the double helix, as described in *Section 1.5, Supercoiling Affects the Structure of DNA*. The DNA is in a dynamic equilibrium between the states of tension and unwinding.
- Supercoiling can be *constrained* if proteins are bound to the DNA to hold it in a particular three-dimensional configuration. In this case, the supercoils are represented by the path the DNA follows in its fixed association with the proteins. The energy of interaction between the proteins and the supercoiled DNA stabilizes the nucleic acid, so that no tension is transmitted along the molecule.

It is estimated that about half of the total supercoils in *E. coli* are unconstrained, and the total superhelical density averages one negative superhelical turn per 100 base pairs. There is likely to be variation about an average level, but it is clear that the level of superhelicity is sufficient to exert significant effects on DNA structure, for example, in assisting melting in particular regions such as origins or promoters.

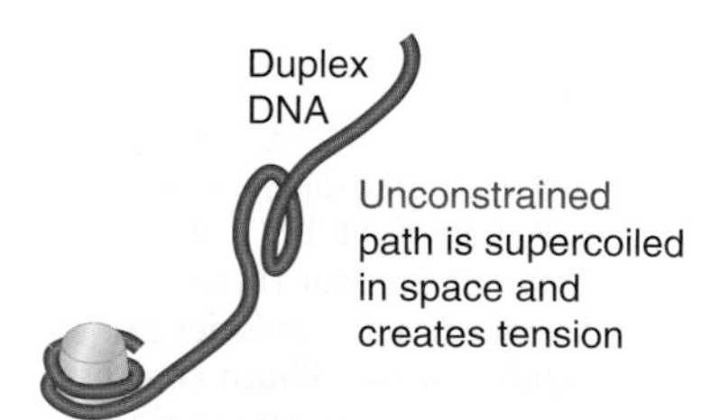

FIGURE 9.8 An unconstrained supercoil in the DNA path creates tension, but no tension is transmitted along DNA when a supercoil is restrained by protein binding.

KEY CONCEPTS

- The bacterial nucleoid is ~80% DNA by mass and can be unfolded by agents that act on RNA or protein.
- The proteins that are responsible for condensing the DNA have not been identified.
- The nucleoid has ~400 independent negatively supercoiled domains.
- The average density of supercoiling is ~1 supercoil/100 bp.

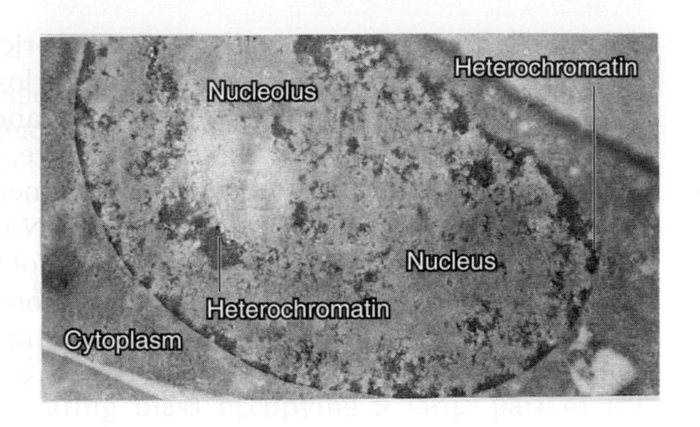

FIGURE 9.11 A thin section through a nucleus stained with feulgen shows heterochromatin as compact regions clustered near the nucleolus and nuclear membrane. Photo courtesy of Edmond Puvion, Centre National de la Recherche Scientifique.

amount of chromatin doubles. Chromatin is fibrillar, although the overall configuration of the fiber in space is hard to discern in detail. The fiber itself, however, is similar or identical to that of the mitotic chromosomes.

Chromatin can be divided into two types of material, which can be seen in the nuclear section of **FIGURE 9.11**:

- In most regions, the fibers are much less densely packed than in the mitotic chromosome. This material is called **euchromatin**. It has a relatively dispersed appearance, stains lightly with DNA-specific dyes, and occupies most of the nuclear region in Figure 9.11.
- Some regions of chromatin are very densely packed with fibers, displaying a density comparable to that of the chromosome at mitosis. This material is called **heterochromatin** and stains strongly with DNA-specific dyes. It is found at a number of locations within the chromatin, including the centromeres. It passes through the cell cycle with relatively little change in its degree of condensation. It forms a series of discrete clumps in Figure 9.11, but in some cell types the various heterochromatic regions aggregate into a densely staining **chromocenter**. The common form of heterochromatin that always remains heterochromatic is called *constitutive heterochromatin*. In contrast, there is another sort of heterochromatin, called *facultative heterochromatin*, in which regions of euchromatin are converted to a heterochromatic state.

▸ **euchromatin** The form of chromatin that comprises most of the genome in the interphase nucleus, which is less tightly coiled than heterochromatin, and contains most of the active or potentially active single copy genes.

▸ **heterochromatin** Regions of the genome that are highly condensed are less transcribed and are late-replicating. It is divided into two types: constitutive and facultative.

▸ **chromocenter** An aggregate of heterochromatin from different chromosomes.

The same fibers run continuously between euchromatin and heterochromatin, showing that these states represent different degrees of condensation of the genetic material. In the same way, euchromatic regions exist in different states of condensation during interphase and during mitosis. Thus the genetic material is organized in a manner that permits alternative states to be maintained side by side in chromatin and allows cyclical changes to occur in the packaging of euchromatin between interphase and mitosis. We discuss the molecular basis for these states in *Chapter 28, Eukaryotic Transcription Regulation*, and *Chapter 29, Epigenetic Effects Are Inherited*.

The structural condition of the genetic material is correlated with its activity. The common features of constitutive heterochromatin are as follows:

- It is permanently condensed.
- It replicates late in S phase and has a reduced frequency of genetic recombination.
- It often consists of multiple repeats of a few sequences of DNA that are not transcribed or are transcribed at very low levels. (Genes that reside in heterochromatic regions are generally less transcriptionally active than their euchromatic counterparts, but there are exceptions to this general rule.)
- The density of genes in this region is very much reduced compared with euchromatin, and genes that are translocated into or near it are often inactivated. The one dramatic exception to this is the ribosomal DNA in the nucleolus, which has the general compacted appearance and behavior of heterochromatin (such as late replication), yet is engaged in very active transcription.

Although most active genes are contained within euchromatin, only a subset of euchromatic genes are transcribed at any time. Thus location in euchromatin is *necessary* for gene expression (at least for RNA polymerase II) but is not *sufficient* for it.

KEY CONCEPTS

- Individual chromosomes can be seen only during mitosis.
- During interphase, the general mass of chromatin is in the form of euchromatin, which is less tightly packed than mitotic chromosomes.
- Regions of heterochromatin remain densely packed throughout interphase.

CONCEPT AND REASONING CHECK

Why are genes not transcribed in heterochromatin? What do you predict happens to transcription during mitosis?

9.6 Chromosomes Have Banding Patterns

Because of the diffuse state of interphase chromatin, it is difficult to determine the specificity of its organization. We can, however, ask whether the structure of the mitotic chromosome is ordered. Do particular sequences always lie at particular sites, or is the folding of the fiber into the overall structure a more random event?

At the level of the chromosome, each member of the complement has a different and reproducible ultrastructure. When mitotic chromosomes are subjected to proteolytic enzyme (trypsin) treatment and then stained with the chemical dye Giemsa, they generate distinct chromosome-specific patterns called **G-bands**. FIGURE 9.12 shows an example of the human set.

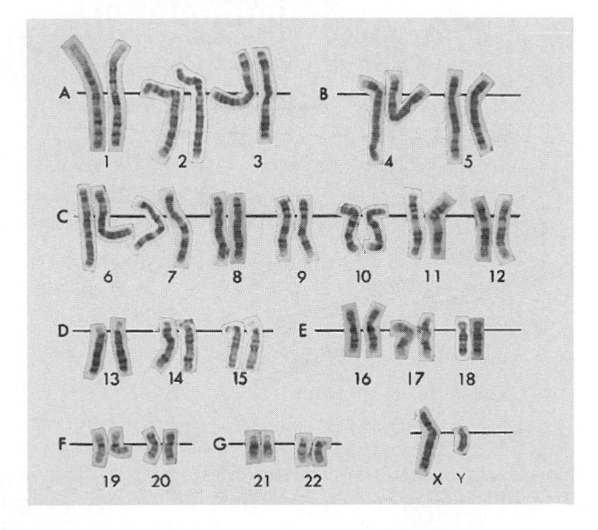

FIGURE 9.12 G-banding generates a characteristic lateral series of bands in each member of the chromosome set. Photo courtesy of Lisa Shaffer, Washington State University–Spokane.

▸ **G-bands** Bands generated on eukaryotic chromosomes by staining techniques that appear as a series of lateral striations. They are used for karyotyping (identifying chromosomes and chromosomal regions by the banding pattern).

Until the development of this technique, human chromosomes could be distinguished only by their overall size and the relative location of the centromere. G-banding allows each chromosome to be identified by its characteristic banding pattern. This pattern allows translocations from one chromosome to another to be identified by comparison with the original diploid set. FIGURE 9.13 shows a diagram of the bands of the human X chromosome. The bands are large structures, each ~10^7 bp of DNA, each of which could include many hundreds of genes. This figure also shows the nomenclature used to identify genetic positions on individual chromosomes. A given location is indicated by its position on the long (q) or short (p) arm and then by the region of the arm, band, and subband(s). For example, *CFTR*, the gene that is mutated in cystic fibrosis, is located at 7q31.2—in other words, on the long arm of chromosome 7, region 3, band 1, subband 2.

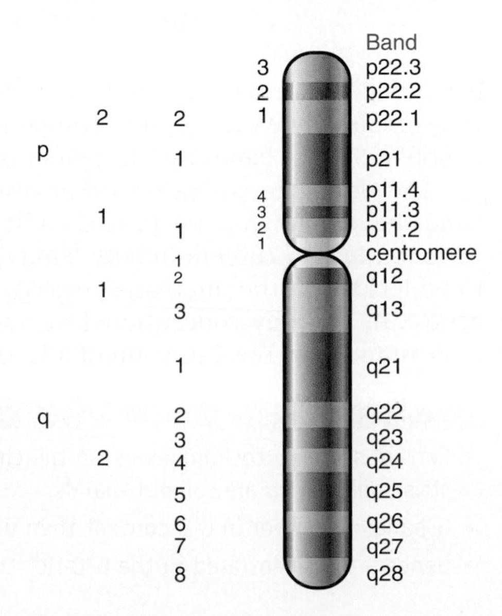

FIGURE 9.13 The human X chromosome can be divided into distinct regions by its banding pattern. The short arm is p and the long arm is q; each arm is divided into larger regions that are further subdivided. This map shows a low-resolution structure; at higher resolution, some bands are further subdivided into smaller bands and interbands—e.g., p21 is divided into p21.1, p21.2, and p21.3.

The banding technique is of enormous practical use, but the mechanism of banding remains a mystery. All that is certain is that the dye stains *untreated* chromosomes more or less uniformly. Thus the generation of

METHODS AND TECHNIQUES

FISH, Chromosome Painting, and Spectral Karyotyping

Classical staining techniques do not provide sufficient sensitivity to detect translocations, deletions, or insertions that involve small segments within a chromosome. However, investigators have taken advantage of lessons learned from molecular biology to devise a very sensitive technique for detecting even very small chromosomal changes in a sample fixed to a microscope slide.

This technique, called *fluorescent in situ hybridization (FISH)*, takes advantage of the fact that a DNA probe with an attached fluorescent dye will bind to a specific DNA sequence within a denatured chromosome (**FIGURE B9.1**). The fluorescent probes can be prepared by nick translation or by the polymerase chain reaction using a template corresponding to the sequence of interest. FISH has a wide variety of applications, which include detecting aneuploidy, identifying chromosomal aberrations, and locating genes and other DNA segments on

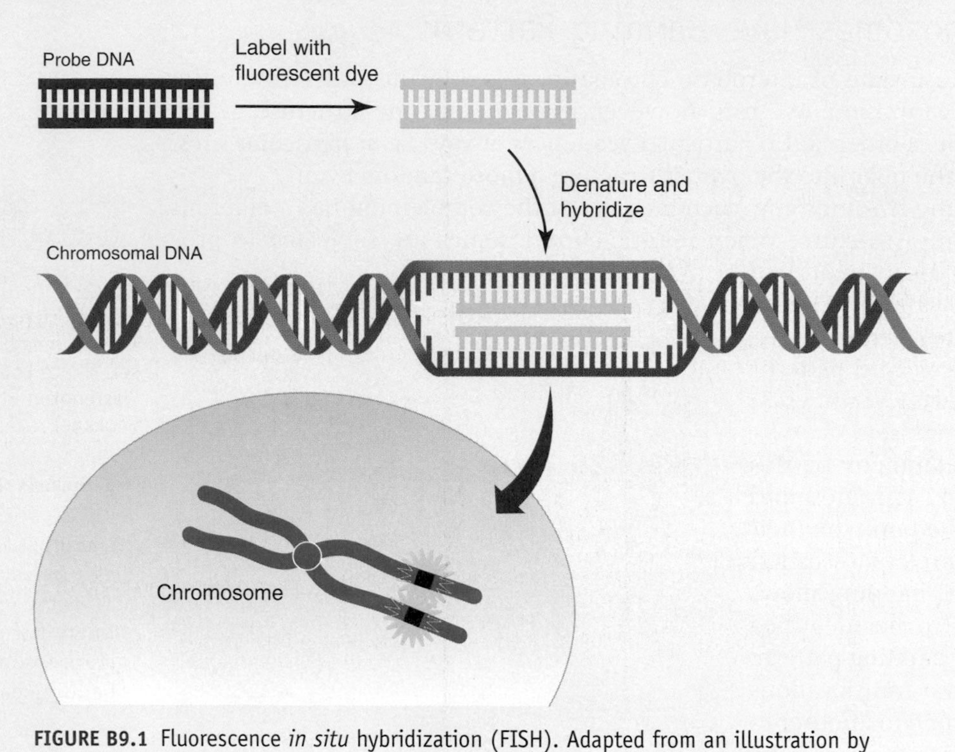

FIGURE B9.1 Fluorescence *in situ* hybridization (FISH). Adapted from an illustration by Darryl Leja, National Human Genome Research Institute (www.genome.gov).

bands depends on a variety of treatments that change the response of the chromosome (presumably by extracting the component that binds the stain from the nonbanded regions). Similar bands can be generated by an assortment of other treatments.

interbands The gene-rich regions of chromosomes that lie between the bands in Giemsa-stained chromosomes. More generally, interbands refer to any region between identified bands (G-bands, bands of polytene chromosomes, etc).

The only known features that distinguish bands from **interbands** is that the bands have an *average* lower G-C content than the interbands, although there is still variation in G-C content in both bands and interbands. There is a tendency for genes to be located in the interband regions. The human genome sequence indicates that genes are typically concentrated in regions of higher G-C content. We have yet to understand how the G-C content affects chromosome structure.

KEY CONCEPTS

- Certain staining techniques cause the chromosomes to have the appearance of a series of striations, which are called G-bands.
- G-bands are lower in G-C content than interbands.
- Genes are concentrated in the G-C-rich interbands.

a chromosome. It can also be used to locate DNA segments during interphase when the chromosome is not visible.

A variation of FISH, called *chromosome painting,* uses a fluorescent dye bound to DNA probes that bind all along a particular chromosome. The major shortcoming of FISH and chromosome painting is that they cannot be used to study all chromosomes at the same time because there are not enough fluorescent dyes with sufficient color differences to mark all 23 chromosomes in a unique color.

This problem was solved by labeling the painting probes for each chromosome with a different assortment of fluorescent dyes, a technique called *spectral karyotyping (SKY).* When the fluorescent probes hybridize to a chromosome, each kind of chromosome is labeled with a different assortment of fluorescent dye combinations. Stained chromosomes are then viewed through a series of filters, each of which transmits only light emitted by a single fluorescent dye. Alternatively, an interferometer determines the full spectrum of light emitted by the stained chromosome. In either case, a computer provides a composite picture that shows different chromosome pairs as if they were stained in different colors (**FIGURE B9.2**). Chromosome painting and spectral karyotyping are particularly useful for detecting chromosomal rearrangements (such as translocations).

(a)

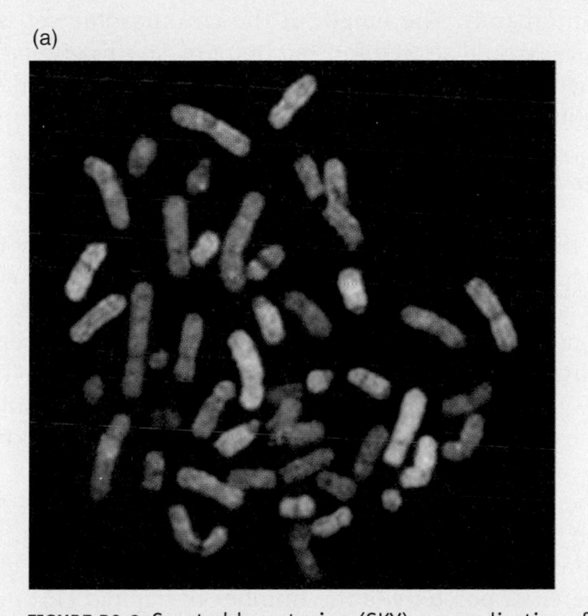

(b)

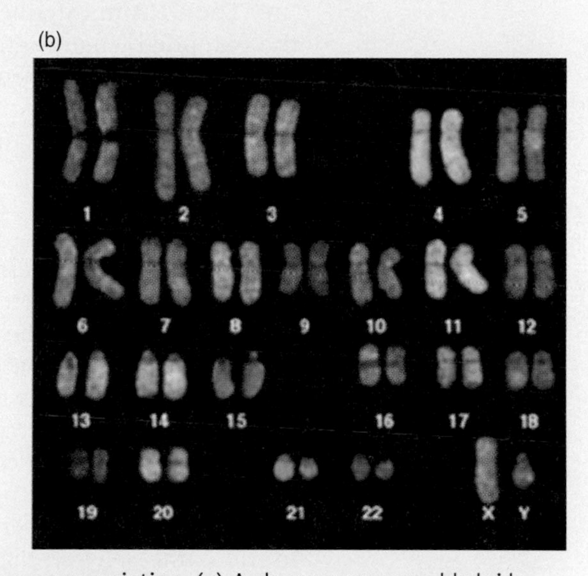

FIGURE B9.2 Spectral karyotyping (SKY), an application of chromosome painting. (a) A chromosome spread hybridized with SKY fluorescent probes. (b) The same labeled chromosomes as in (a), sorted by size to show the karyotype. Photos courtesy of Johannes Wienberg, Ludwig-Maximilians-University, and Thomas Ried, National Institutes of Health.

CONCEPT AND REASONING CHECK

How can G-banding be used to detect chromosomal deletions, inversions, and translocations?

9.7 Polytene Chromosomes Form Bands That Expand at Sites of Gene Expression

Much of our understanding of chromosome structure and the structural changes associated with transcription comes from the study of an interesting type of chromosome found in dipteran (two-winged) insects. The interphase nuclei of specific tissues of the larvae of dipteran flies contain chromosomes that are greatly enlarged relative to their usual condition. They possess both increased diameter and greater length. **FIGURE 9.14** shows an example of a chromosome set from the salivary gland of *D. melanogaster.* The members of this set are called **polytene chromosomes.**

▸ **polytene chromosomes** Chromosomes that are generated by successive replications of a chromosome set without separation of the replicas.

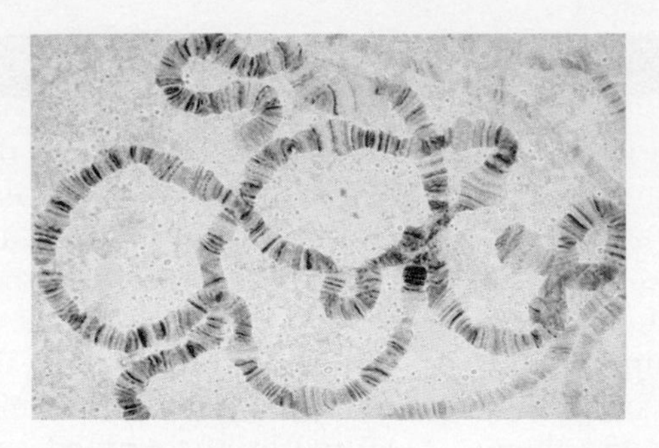

FIGURE 9.14 The polytene chromosomes of *D. melanogaster* form an alternating series of bands and interbands. Photo courtesy of José Bonner, Indiana University.

▸ **bands** Structures visible on polytene chromosomes as dense regions that contain the majority of DNA; they include active genes.

▸ **chromomeres** Densely staining granules visible in chromosomes under certain conditions, especially early in meiosis, when a chromosome may appear to consist of a series of chromomeres.

Each member of the polytene set consists of a visible series of **bands** (more properly, but rarely, described as **chromomeres**). The bands range in size from the largest, with a breadth of ~0.5 μm, to the smallest, at ~0.05 μm. (The smallest can be distinguished only under an electron microscope.) The bands contain most of the mass of DNA and stain intensely with appropriate reagents. The regions between them stain more lightly and are called *interbands* (like those found between G-bands, described previously). There are ~5000 bands in the *D. melanogaster* set.

The centromeres of all four chromosomes of *D. melanogaster* aggregate to form a chromocenter that consists largely of heterochromatin (in the male it includes the entire Y chromosome). Allowing for this, ~75% of the haploid DNA set is organized into alternating bands and interbands. The length of the chromosome set is ~2000 μm. The DNA in extended form would stretch for ~40,000 μm, so this gives a ratio of the extended length to the actual length of ~20. Typically, this ratio is about 1000 to 2000 for interphase chromatin and up to ~7000 for highly condensed mitotic chromosomes! This demonstrates vividly the extension of the genetic material of polytene chromosomes relative to the usual condensation states of interphase chromatin or mitotic chromosomes.

What is the structure of these giant chromosomes? Each is produced by the successive replications of a synapsed diploid pair of chromosomes. The replicas do not separate but instead remain attached to each other in their extended state. At the start of the process, each synapsed pair has a DNA content of 2C (where C represents the DNA content of the individual chromosome). This amount then doubles up to nine times, at its maximum, giving a content of 1024C. The number of doublings is different in the various tissues of the *D. melanogaster* larva. This process is known as *endoreduplication*.

Each chromosome can be visualized as a large number of parallel fibers running longitudinally that are tightly condensed in the bands and less so in the interbands. It is likely that each fiber represents a single (C) haploid chromosome. This gives rise to the name *polytene*: the degree of polyteny is the number of haploid chromosomes contained in the giant chromosome.

The banding pattern is characteristic for each strain of *Drosophila*. The constant number and linear arrangement of the bands was first noted in the 1930s, when it was realized that they form a *cytological map* of the chromosomes. Rearrangements—such as deletions, inversions, or duplications—result in alterations of the order of bands.

The linear array of bands can be equated with the linear array of genes. Thus genetic rearrangements, as seen in a linkage map, can be correlated with structural rearrangements of the cytological map. Ultimately, a particular mutation can be located in a particular band. The total number of genes in *D. melanogaster* exceeds the number of bands, so there are probably multiple genes in most or all bands.

▸ ***in situ* hybridization** Hybridization performed by denaturing the DNA of cells squashed on a microscope slide so that annealing is possible with an added single-stranded RNA or DNA; the added nucleic acid is fluorescently or radioactively labeled. *In situ* hybridization can also be performed in intact tissues.

The positions of particular genes on the cytological map can be determined directly by the technique of ***in situ* hybridization**. This method is described in the accompanying box, *Methods and Techniques: FISH, Chromosome Painting, and Spectral Karyotyping*. Although fluorescent probes are currently preferred, when the method was originally developed, a radioactive probe representing the gene of interest was used, and the position (or positions) of the corresponding genes was detected by autoradiography. An example is shown in **FIGURE 9.15**. Using *in situ* hybridization, it is possible to determine directly the band in which a particular sequence lies.

One of the intriguing features of the polytene chromosomes is that sites of active transcription can be visualized. Some of the bands pass transiently through an

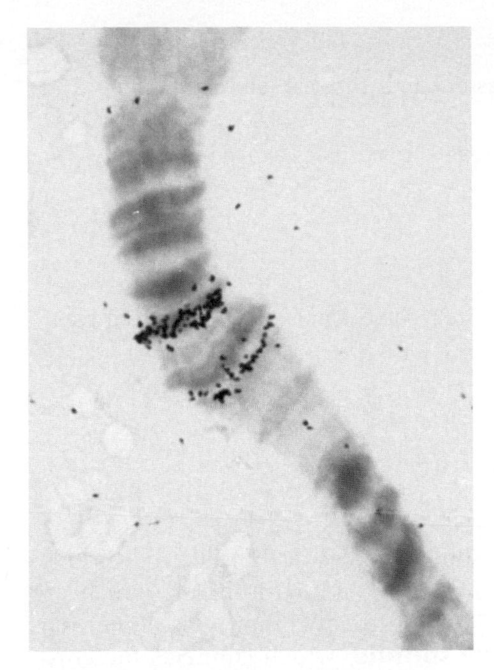

FIGURE 9.15 A magnified view of bands 87A and 87C shows their hybridization *in situ* with labeled RNA extracted from heat-shocked cells. Photo courtesy of José Bonner, Indiana University.

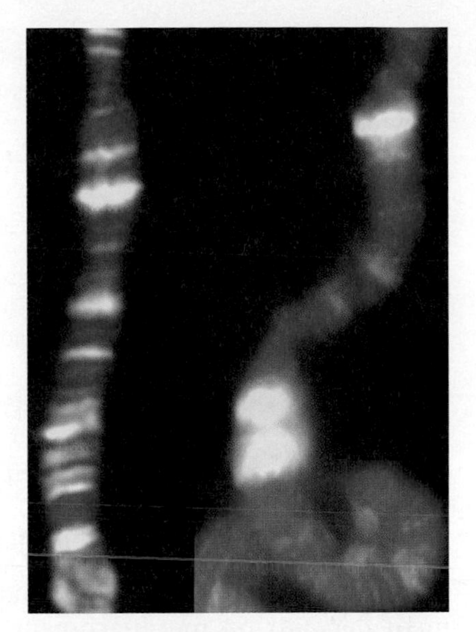

FIGURE 9.16 Heat-shock-induced puffing at major heat shock loci 87A and C. Displayed is a small segment of chromosome 3 before (left) and after (right) heat shock. Chromosomes are stained for DNA (blue) and for RNA polymerase II (yellow). Photo courtesy of Victor G. Corces, Emory University.

expanded state in which they appear like a **puff** on the chromosome, when chromosomal material is extruded from the axis. The dramatic puffing of a heat shock gene locus upon heat shock is shown in **FIGURE 9.16**. In this example, the correlation between puffing and transcription is illustrated by staining the chromosomes for RNA polymerase II, which is concentrated in the puffs.

▸ **puff** An expansion of a band of a polytene chromosome associated with the synthesis of RNA at some locus in the band.

What is the nature of the puff? It consists of a region in which the chromosome fibers unwind from their usual state of packing in the band. The fibers remain continuous with those in the chromosome axis. Puffs usually emanate from single bands, although when they are very large, the swelling may be so extensive as to obscure the underlying array of bands.

The pattern of puffs is related to gene expression. During larval development, puffs appear and regress in temporal and tissue-specific patterns. A characteristic pattern of puffs is found in each tissue at any given time. Many puffs are induced by the hormone ecdysone that controls *Drosophila* development. Puffs can also be induced by environmental changes, such as the puffs induced by temperature stress shown in Figure 9.16.

The puffs are *sites where RNA is being synthesized.* The accepted view of puffing has been that expansion of the band is a consequence of the need to relax its structure in order to synthesize RNA. Puffing has therefore been viewed as a consequence of transcription. A puff can be generated by a single active gene. The sites of puffing differ from ordinary bands in accumulating additional proteins, which include RNA polymerase II (as shown in Figure 9.16) and other proteins associated with transcription. These observations suggest that, in order to be transcribed, the genetic material is dispersed from its usual, more tightly packed state. The question to keep in mind is whether this dispersion at the gross level of the chromosome mimics the events that occur at the molecular level within the mass of ordinary interphase euchromatin.

KEY CONCEPTS

- Polytene chromosomes of dipterans have a series of bands that can be used as a cytological map.
- Bands that are sites of gene expression on polytene chromosomes expand to give "puffs."

CONCEPT AND REASONING CHECK

How do the bands on polytene chromosomes differ from the G-bands described in the previous section?

9.8 The Eukaryotic Chromosome Is a Segregation Device

During mitosis, the sister chromatids move to opposite poles of the cell. Their movement depends on the attachment of the chromosome to microtubules, which are connected at their other end to the poles. (The microtubules comprise a cellular filamentous system, reorganized at mitosis into the **spindle**, which connects the chromosomes to the poles of the cell.) The sites in the two regions where microtubule ends are organized—in the vicinity of the centrioles at the poles and at the chromosomes—are called **microtubule organizing centers (MTOCs)**.

‣ **spindle** A structure made up of microtubules that guides the movements of the chromosomes during mitosis.

‣ **microtubule organizing center (MTOC)** A region from which microtubules emanate. In animal cells the centrosome is the major microtubule organizing center.

‣ **centromere** A constricted region of a chromosome that includes the site of attachment (the kinetochore) to the mitotic or meiotic spindle. It may consist of unique DNA sequences or highly repetitive sequences and contains proteins not found anywhere else in the chromosome.

‣ **acentric fragment** A fragment of a chromosome (generated by breakage) that lacks a centromere and is lost at cell division.

FIGURE 9.17 illustrates the separation of sister chromatids as mitosis proceeds from metaphase to telophase. The region of the chromosome that is responsible for its segregation at mitosis and meiosis is called the **centromere**. The centromeric region on each sister chromatid is pulled by microtubules to the opposite pole. Opposing this motive force, "glue" proteins called *cohesins* hold the sister chromatids together. Initially the sister chromatids separate at their centromeres, and then they are released completely from one another during anaphase, when the cohesins are degraded. The centromere is pulled toward the pole during mitosis, and the attached chromosome appears to be "dragged along" behind it. The chromosome therefore provides a device for attaching a large number of genes to the apparatus for division. The centromere essentially acts as a handle for the chromosome, and its location typically appears as a constricted region connecting all four chromosome arms, as can be seen in the photo of Figure 9.10.

The centromere is essential for segregation, as shown by the behavior of chromosomes that have been broken. A single break generates one piece that retains the centromere, and another, an **acentric fragment**, that lacks it. The acentric fragment does not become attached to the mitotic spindle, and as a result it fails to be included in either of the daughter nuclei. When chromosome movement relies on discrete centromeres, there can be *only* one centromere per chromosome. When translocations generate chromosomes with more than one centromere, aberrant structures form at mitosis. This is because the two centromeres on the *same* sister chromatid can be pulled toward different poles, thus breaking the chromosome. In some species, though, the centromeres are *holocentric,* meaning they are diffuse and spread along

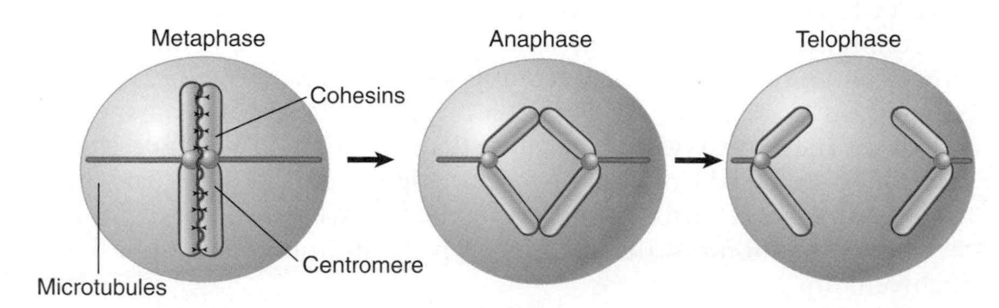

FIGURE 9.17 Chromosomes are pulled to the poles via microtubules that attach at the centromeres. The sister chromatids are held together until anaphase by glue proteins (cohesins). The centromere is shown here in the middle of the chromosome (metacentric), but it can be located anywhere along its length, including close to the end (acrocentric) and at the end (telocentric).

the entire length of the chromosome. Species with holocentric chromosomes still make spindle-fiber attachments for chromosome separation but do not require one and only one centromere per chromosome. Most of the molecular analysis of centromeres has been done on *point centromeres* (such as the very short centromeres found in budding yeast) or *regional centromeres* (the larger centromeres found in species such as *Drosophila*, mammals, and rice), each of which function only when there is a single centromere per chromosome.

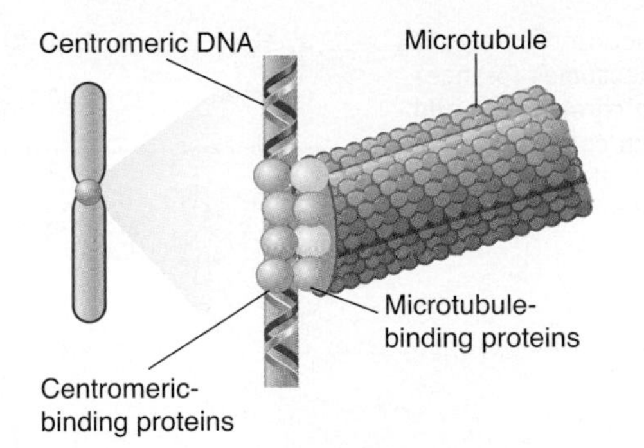

FIGURE 9.18 The centromere is identified by a DNA sequence that binds specific proteins. These proteins do not themselves bind to microtubules, but instead establish the site at which the microtubule-binding proteins in turn bind.

The regions flanking the centromere often are rich in repetitive satellite DNA sequences and frequently display a considerable amount of heterochromatin. The region of the chromosome at which the centromere forms is defined by DNA sequences. The centromeric DNA binds specific proteins that are responsible for establishing the structure that attaches the chromosome to the microtubules, which is called the **kinetochore**. The kinetochore is a darkly staining fibrous object of ~400 nm. The kinetochore provides the microtubule attachment point on a chromosome. **FIGURE 9.18** shows a generic example of the hierarchy of organization that connects centromeric DNA to the microtubules. Proteins bound to the centromeric DNA bind other proteins that bind to microtubules.

▸ **kinetochore** A small organelle associated with the surface of the centromere that attaches a chromosome to the microtubules of the mitotic spindle. Each mitotic chromosome contains two "sisters" that are positioned on opposite sides of its centromere and face in opposite directions.

KEY CONCEPTS

- A eukaryotic chromosome is held on the mitotic spindle by the attachment of microtubules to the kinetochore that forms in its centromeric region.
- Centromeres often have heterochromatin that is rich in satellite DNA sequences.
- Centromeres in higher eukaryotic chromosomes contain large amounts of repetitive DNA.
- The function of the repetitive DNA is not known.

CONCEPT AND REASONING CHECK

What would happen to a chromosome that, as a result of a translocation, contains two centromeres?

9.9 Regional Centromeres Contain a Centromeric Histone H3 Variant and Repetitive DNA

The region of the chromosome at which the centromere forms was originally thought to be defined by DNA sequences, yet recent studies in plants, animals and fungi have shown that centromeres are more likely to be specified epigenetically by chromatin structure. Centromere-specific histone H3 (CENP-A/CenH3; see *Section 10.5, Histone Variants Produce Alternative Nucleosomes*) appears to be a primary determinant in establishing functional centromeres and kinetochore assembly sites. This finding explains an old puzzle of why specific DNA sequences could not be identified as "the centromeric DNA" and why there is so much variation in centromere-associated DNA sequences among closely related species. **FIGURE. 9.19** shows a model for the epigenetic specification of centromeres, with the kinetochore connecting to clusters of CenH3-containing nucleosomes, which protrude from the bulk chromatin. New questions of centromere function include what determines or restricts the sites of CenH3 assembly and how chromosomes maintain one such region per chromosome.

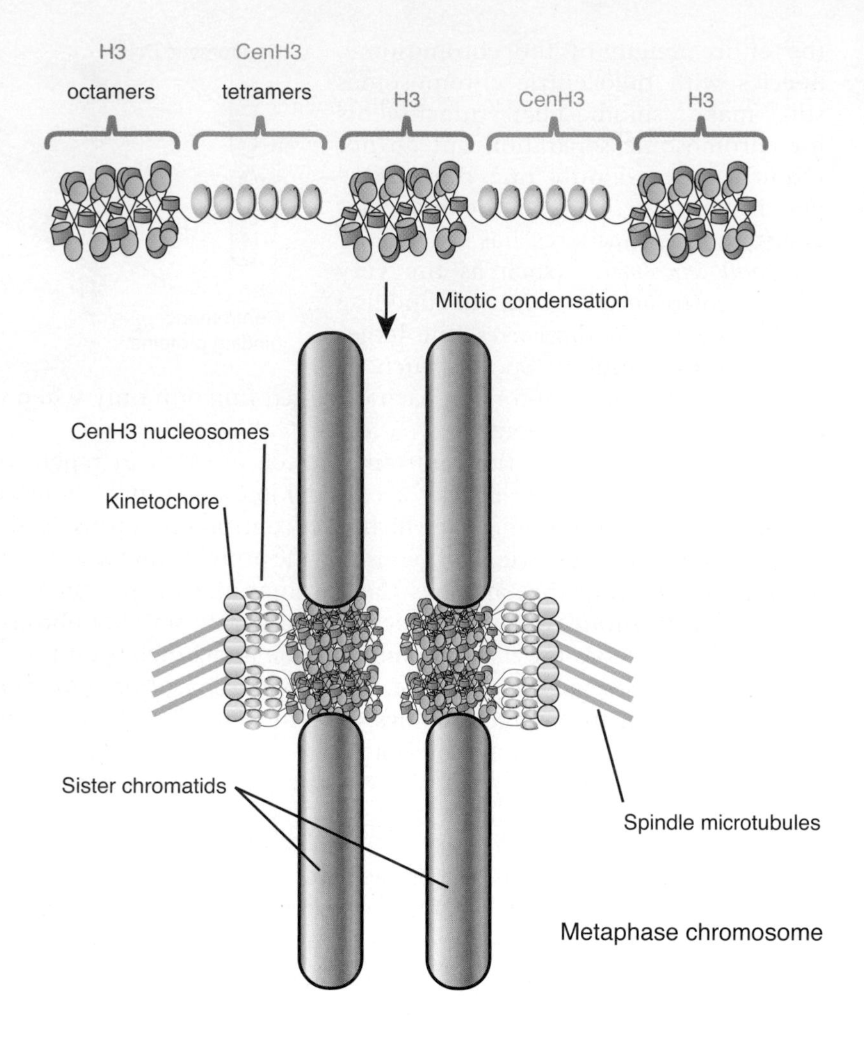

FIGURE 9.19 A model of the overall structure of a regional centromere. The CenH3-containing nucleosomes (orange) occur in clusters that protrude from the chromosome and bind to kinetochore proteins that in turn connect to spindle microtubules. Adapted from Y. Datal et al., *Proc. Natl. Acad. Sci. USA* 104 (2007): 15974–15981.

The length of DNA required for centromeric function is often quite long. The short, discrete centromeres of *Saccharomyces cerevisiae*, discussed in the next section, may be an exception to the general rule. *S. cerevisiae* is the only case so far in which centromeric DNA can be identified by its ability to confer stability on plasmids. A related approach has been used to identify centromeres in the fission yeast *Schizosaccharomyces pombe*. *S. pombe* has only three chromosomes, and the region containing each centromere has been identified by deleting most of the sequences of each chromosome to create a stable minichromosome. This approach locates the centromeres within regions of 40 to 100 kb that consist largely or entirely of repetitious DNA. Attempts to localize centromeres on *Drosophila* chromosomes suggests they are dispersed in a large region of 200 to 600 kb. Similarly, the chromosomes of *Arabidopsis* contain >500 kb centromeric regions in which recombination is largely suppressed.

The predominant sequence motif in primate centromeres is α satellite DNA, which consists of tandem arrays of a 171 bp repeating unit (see *Section 7.5, Satellite DNAs Often Lie in Heterochromatin*), similar to the composition of primate centromeres.

Current models for regional centromere organization and function invoke alternating chromatin domains, as shown in Figure 9.19, with clusters of CenH3 nucleosomes interspersed among clusters of nucleosomes containing H3 and another histone variant, H2A.Z. The CenH3 nucleosomes form the chromatin foundation for recruitment and assembly of the other proteins that eventually comprise a functional kinetochore. Neocentromeres can be formed that contain CenH3 but do not contain satellite DNA, which provides strong evidence for the epigenetic determination of centromeres.

KEY CONCEPTS

- Centromeres are characterized by a centromere-specific histone H3 variant and often contain heterochromatin that is rich in satellite DNA sequences.
- The function of the repetitive DNA is not known.

CONCEPT AND REASONING CHECK

What would happen to a chromosome that had CenH3-containing nucleosomes present at regions outside the centromere?

9.10 Point Centromeres in *S. cerevisiae* Contain Short, Essential Protein-Binding DNA Sequences

S. cerevisiae chromosomes do not display large, visible kinetochores comparable to those of higher eukaryotes but otherwise use the same basic mechanisms for mitotic and meiotic divisions. Fragments of chromosomal DNA containing centromeres have been isolated by their ability to confer mitotic stability on plasmids. A *CEN* fragment is thus defined as the minimal sequence that can support accurate segregation of a plasmid at mitosis. Every chromosome has a *CEN* region, and *CEN* fragments from different chromosomes are interchangeable. This indicates that centromeres serve only to attach the chromosome to the spindle and play no role in distinguishing one chromosome from another.

The sequences required for centromeric function fall within a stretch of ~120 bp. The centromeric region is packaged into a nuclease-resistant structure and binds a single microtubule. We may, therefore, look to the *S. cerevisiae* centromeric region to identify proteins that bind centromeric DNA and proteins that connect the chromosome to the spindle.

There are three types of sequence element in the *CEN* region, summarized in **FIGURE 9.20**:

- The cell cycle-dependent element (*CDE*)-*I* is a sequence of 9 bp that is conserved with minor variations at the left boundary of all centromeres.
- *CDE-II* is a >90% A+T-rich sequence of 80–90 bp found in all centromeres; its function could depend on its length rather than exact sequence. Its base composition may cause some characteristic distortions of the DNA double-helical structure.
- *CDE-III* is an 11 bp sequence highly conserved at the right boundary of all centromeres. Sequences on either side of the element are less well conserved but may also be needed for centromeric function. (*CDE-III* could be longer than 11 bp if it turns out that the flanking sequences are essential.)

Mutations in *CDE-I* or *CDE-II* reduce, but do not inactivate, centromere function; point mutations in the central CCG of *CDE-III* completely inactivate the centromere.

A large protein complex assembles at the *CDE* sequences and connects the chromosome to microtubules. The structure is summarized in **FIGURE 9.21**.

The *CEN* region recruits three DNA-binding factors, Cbf1, CBF3 (an essential four-protein complex) and Mif2 (CENP-C in multicellular eukaryotes). In addition, a specialized chromatin structure is built by binding the *CDE-II* region to a protein called Cse4, which is the yeast CenH3 histone variant. A protein called Scm3 is required for proper association of Cse4 with *CEN*. Inclusion of histone variants related to Cse4

TCACATGATGATATTTGATTTTATTATATTTTTAAAAAAAGTAAAAAATAAAAAGTAGTTTATTTTTAAAAAATAAAATTTAAAATATTTCACAAAATGATTTCCGAA
AGTGTACTACTATAAACTAAAATAATATAAAAATTTTTTTCATTTTTTATTTTTCATCAAATAAAAATTTTTTATTTTAAATTTTATAAAGTGTTTTACTAAAGGCTT

CDE-I *CDE-II* 80–90 bp, >90% A + T *CDE-III*

FIGURE 9.20 Three conserved regions can be identified by the sequence homologies between yeast CEN elements.

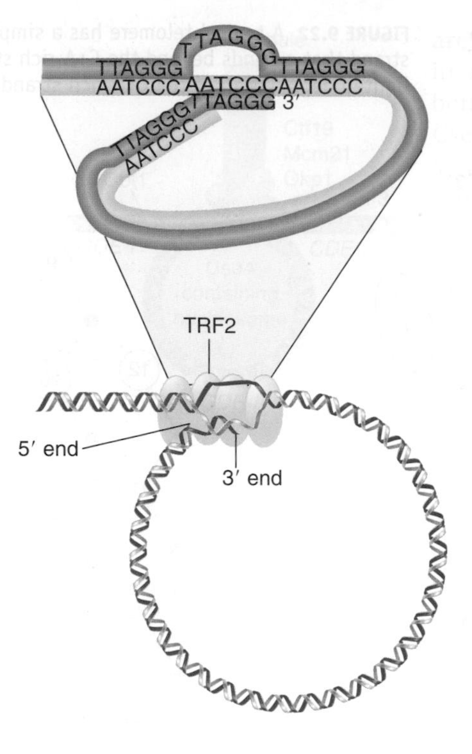

FIGURE 9.25 The 3′ single-stranded end of the telomere $(TTAGGG)_n$ displaces the homologous repeats from duplex DNA to form a t-loop. The reaction is catalyzed by Trf2.

FIGURE 9.25 shows that the loop is formed when the 3′ single-stranded end of the telomere $(TTAGGG)_n$ displaces the same sequence in an upstream region of the telomere. This converts the duplex region into a structure like a D-loop (see *Section 15.3, Double-Strand Breaks Initiate Recombination*), where a series of TTAGGG repeats are displaced to form a single-stranded region, and the tail of the telomere is paired with the homologous strand. Loop formation is catalyzed by the telomere-binding protein Trf2, which together with other proteins forms a complex that stabilizes the chromosome ends. Its importance in protecting the ends is indicated by the fact that the deletion of *TRF2* causes chromosome rearrangements to occur.

In mammals, six telomeric proteins (Trf1, Trf2, Rap1, Tin2, Tpp1 and Pot1) comprise a complex called shelterin, depicted in FIGURE 9.26. Shelterin functions to protect telomeres from DNA damage repair pathways and to regulate telomere length control. Increasing understanding of the roles of telomeres in aging, cancer, and cell differentiation reveal that telomeres are more than static caps at the ends of linear chromosomes.

A key aspect of the telomere is its ability to be extended, which can be achieved by addition of telomeric repeats to the end of the chromosome in every replication cycle. This can solve the difficulty of replicating linear DNA molecules discussed in *Section 14.2, The Ends of Linear DNA Are a Problem for Replication*. The addition of repeats by *de novo* synthesis would counteract the loss of repeats resulting from failure to replicate up to the end of the chromosome. Extension and shortening would be in dynamic equilibrium.

▸ **telomerase** The ribonucleoprotein enzyme that creates repeating units of one strand at the telomere by adding individual bases to the DNA 3′ end, as directed by an RNA sequence in the RNA component of the enzyme.

A **telomerase** enzyme extends the C+A-rich strand. Telomerase uses the 3′–OH of the G+T telomeric strand as a primer for synthesis of tandem TTGGGG repeats. Only dGTP and dTTP are needed for the activity. The telomerase is a large ribonucleoprotein that consists of a *templating RNA* and a protein with polymerase activity. The RNA component (ranging from ~150 nt to >1300 nt, depending on species) includes a sequence of 15 to 22 bases that is identical to two repeats of the C-rich repeating sequence. This RNA provides the template for synthesizing the G-rich repeating sequence. The protein component of the telomerase is a catalytic subunit that can act only upon the RNA template provided by the nucleic acid component.

FIGURE 9.27 shows the action of telomerase. The enzyme progresses discontinuously: the template RNA is positioned on the DNA primer, several nucleotides are added to the primer, and then the enzyme translocates to begin again. The telomerase

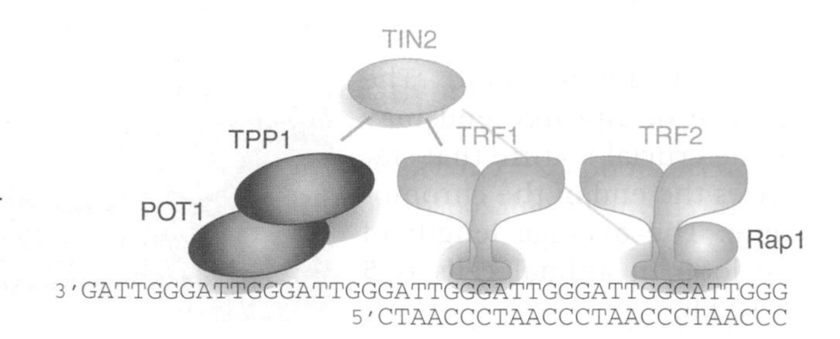

FIGURE 9.26 A schematic of how shelterin might be positioned on telomeric DNA. Note that Pot1 binds to the single-stranded TTAGGG repeats. Although one of the shelterin complexes may have the depicted structure, telomeres contain numerous copies of the complex bound along the double-stranded TTAGGG repeat array. It is not known whether all (or even most) shelterin is present as a six-protein complex. Nucleosomes are omitted for simplicity. Reprinted, with permission, from the Annual Review of Genetics, Volume 42, copyright 2008 by Annual Reviews www.annualreviews.org. Courtesy of Titia de Lange, The Rockefeller University.

is a specialized example of a reverse transcriptase, an enzyme that synthesizes a DNA sequence using an RNA template. We do not know how the complementary (C-A-rich) strand of the telomere is assembled, but we may speculate that it could be synthesized by using the 3′–OH of a terminal G-T hairpin as a primer for DNA synthesis.

Telomerase synthesizes the individual repeats that are added to the chromosome ends but does not itself control the number of repeats. Other proteins are involved in determining the length of the telomere, such as the Est1 and Est3 proteins in yeast. These proteins may bind telomerase and also influence the length of the telomere by controlling the access of telomerase to its substrate. Proteins that bind telomeres in mammalian cells have been found, but less is known about their functions. As a result of the actions of these types of proteins, each organism has a characteristic range of telomere lengths. They are long in mammals (typically 5 to 15 kb in humans) and short in yeast (typically ~300 bp in *S. cerevisiae*).

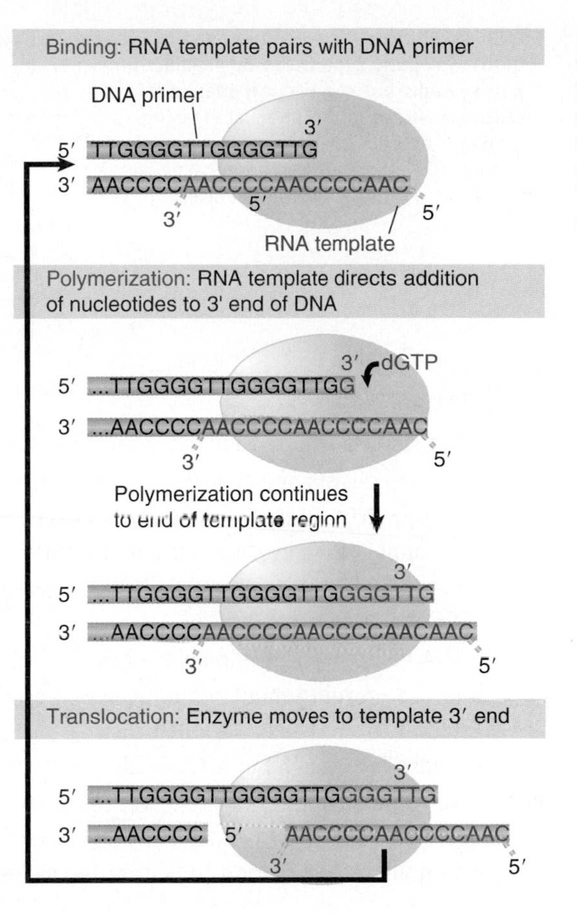

FIGURE 9.27 Telomerase positions itself by base pairing between the RNA template and the protruding single-stranded DNA primer. It adds G and T bases one at a time to the primer, as directed by the template. The cycle starts again when one repeating unit has been added.

Telomerase activity is found in all dividing cells and is generally turned off in terminally differentiated cells that do not divide. **FIGURE. 9.28** shows that if telomerase is mutated in a dividing cell, the telomeres become gradually shorter with each cell division. Loss of telomeres has dire effects. When the telomere length reaches zero, it becomes difficult for the cells to divide successfully. Attempts to divide typically result in chromosome breaks and translocations, causing an increased mutation rate. When cells from a multicellular eukaryote are placed in culture, they usually divide for a fixed number of generations and then enter senescence. The reason appears to be due to a decline in telomere length because of the absence of telomerase expression.

Some cells grow out of senescing cultures. They pass through a "crisis point," during which the telomeres become extremely short but then manage to continue dividing. These cells have undergone one of two alternative changes: they have either reactivated telomerase, or they have managed to become telomerase independent. A minority of the latter class have circularized their chromosomes and thus no longer need telomeres. More often, these cells have acquired the ability to use homologous recombination to maintain telomeres through unequal crossing over (**FIGURE. 9.29**). Recombination between two misaligned telomeres results in the increase in length of one telomere, whereas the length of the other decreases.

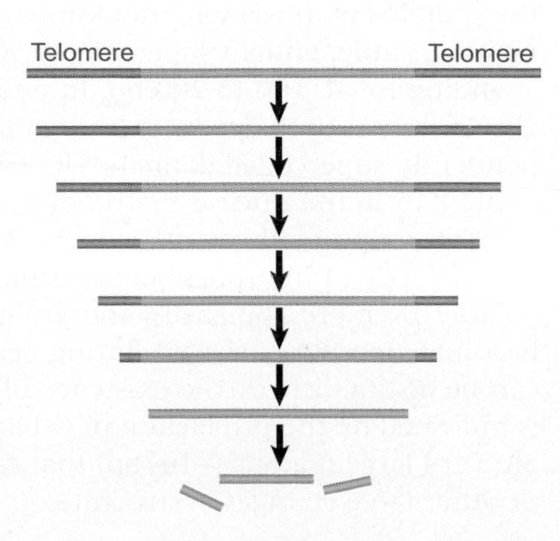

FIGURE 9.28 Mutation in telomerase causes telomeres to shorten in each cell division. Eventual loss of the telomere causes chromosome breaks and rearrangements.

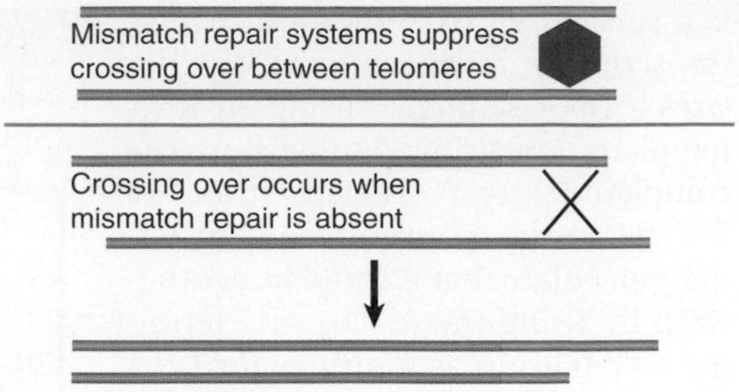

FIGURE 9.29 Crossing over in telomeric regions is usually suppressed by mismatch-repair systems but can occur if these are mutated or silenced. An unequal crossing-over event extends the telomere of one of the products, allowing the chromosome to survive in the absence of telomerase.

Mismatch repair systems suppress crossing over between telomeres

Crossing over occurs when mismatch repair is absent

KEY CONCEPTS

- The telomere is required for the stability of the chromosome end.
- A telomere consists of a simple repeat where a C+A-rich strand has the sequence $C_{>1}(A/T)_{1-4}$.
- The protein Trf2 catalyzes a reaction in which the 3′ repeating unit of the G+T-rich strand forms a loop by displacing its homolog in an upstream region of the telomere.
- Telomerase uses the 3′–OH of the G+T telomeric strand to prime synthesis of tandem TTGGGG repeats.
- The RNA component of telomerase has a sequence that pairs with the C+A-rich repeats.
- One of the protein subunits is a reverse transcriptase that uses the RNA as a template to synthesize the G+T-rich sequence.
- Telomerase is expressed in actively dividing cells and is not expressed in differentiated cells.
- Loss of telomeres results in senescence.
- Escape from senescence can occur if telomerase is reactivated, or via unequal homologous recombination to restore telomeres.

CONCEPT AND REASONING CHECK

Most cancer cells, even when derived from differentiated cells, contain active telomerase. Why might this be?

9.12 Summary

The genetic material of all organisms and viruses takes the form of tightly packaged nucleoprotein. Some virus genomes are inserted into preformed viral particles, whereas others assemble a protein coat around the nucleic acid. The bacterial genome forms a dense nucleoid, with ~20% protein by mass, but details of the interaction of the proteins with DNA are not known. The DNA is organized into ~400 domains that maintain independent supercoiling, with a density of unrestrained supercoils corresponding to ~1/100 to 200 bp. In eukaryotes, interphase chromatin and metaphase chromosomes both appear to be organized into large loops. Each loop may be an independently supercoiled domain. The bases of the loops are connected to a metaphase scaffold or to the nuclear matrix by specific DNA sites.

Transcriptionally active sequences reside within the euchromatin that comprises the majority of interphase chromatin. The regions of heterochromatin are packaged ~5 to 10× more compactly and are generally transcriptionally inert. All chromatin becomes densely packaged during cell division, when the individual chromosomes can be distinguished. The existence of a reproducible ultrastructure in chromosomes is indicated by the production of G-bands by treatment with Giemsa stain. The bands are very large regions (~10^7 bp) that can be used to map chromosomal translocations or other large changes in structure.

Polytene chromosomes of insects have unusually extended structures. Polytene chromosomes of *D. melanogaster* are divided into ~5000 bands. These bands vary in size by an order of magnitude, with an average of ~25 kb. Transcriptionally active regions can be visualized in even more unfolded ("puffed") structures, in which material is extruded from the axis of the chromosome. This may resemble the changes that occur on a smaller scale when a sequence in euchromatin is transcribed.

The centromeric region contains the assembly site for the kinetochore, which is responsible for attaching a chromosome to the mitotic spindle. The centromere often is surrounded by heterochromatin. Centromeres can be limited to short regions (point centromeres) or extend over many kilobases (regional centromeres). Centromeric sequences have been identified in the yeast *S. cerevisiae,* where they consist of short conserved elements. The elements *CDE-I* and *CDE-III* bind Cbf1 and the CBF3 complex, respectively, and a long A+T-rich region called *CDE-II* binds Cse4 to form a specialized nucleosome. Another group of proteins that binds to this assembly provides the connection to microtubules. All centromeres contain a relative of the CenH3 histone variant, and centromere function appears to be epigenetically determined.

Telomeres make the ends of chromosomes stable. Almost all known telomeres consist of multiple repeats in which one strand has the general sequence $C_n(A/T)_m$, where $n > 1$ and $m = 1$ to 4. The other strand, $G_n(T/A)_m$, has a single protruding end that provides a template for addition of individual bases in defined order. The enzyme telomerase is a ribonucleoprotein whose RNA component provides the template for synthesizing the G-rich strand. This overcomes the problem of the inability to replicate at the very end of a duplex. The telomere stabilizes the chromosome end because the overhanging single strand $G_n(T/A)_m$ displaces its homolog in earlier repeating units in the telomere to form a loop, so there are no free ends. Telomerase is active in dividing cells but not in differentiated cells. Cells dividing in culture gradually lose telomere length, until they enter senescence. Cells can escape senescence by reactivating telomerase or by maintaining telomeres via unequal crossing over.

CHAPTER QUESTIONS

What are the two ways that bacteriophage capsids containing DNA or RNA are assembled, and what shape of capsid is associated with each?

1. ______________________ **2.** shape:______________________

3. ______________________ **4.** shape:______________________

What are the four common features of constitutive heterochromatin?

5. ______________________ **6.** ______________________

7. ______________________ **8.** ______________________

9. The density of DNA in a eukaryotic nucleus is about:

A. 1 mg/ml.
B. 10 mg/ml.
C. 100 mg/ml.
D. 500 mg/ml.

10. Which of the following viruses has a filamentous shape?

A. adenovirus
B. bacteriophage T4
C. tobacco mosaic virus
D. HIV

11. For spherical capsids of DNA viruses such as lambda phage, assembly occurs:

A. by capsid coat proteins assembling around the linear DNA.
B. by capsid coat proteins assembling around a tightly wound DNA.
C. by a coordinated DNA and capsid protein monomer assembly process.
D. by capsid proteins forming a shell into which the DNA is inserted.

12. Centromeric regions in chromosomes contain:
   A. euchromatin.
   B. constitutive heterochromatin.
   C. facultative heterochromatin.
   D. acentric fragments.
13. Matrix attachment regions (MARs) are DNA sequences that attach to the nuclear matrix, and are characterized by:
   A. A-T rich regions with a specific consensus sequence.
   B. G-C rich regions with a specific consensus sequence.
   C. A-T rich regions without any specific consensus sequence.
   D. G-C rich regions without any specific consensus sequence.
14. Giemsa staining of eukaryotic chromosomes allows visualization of dense bands with less dense interbands. This is due to:
   A. higher G-C content in the interbands.
   B. higher G-C content in the bands.
   C. higher purine content in the interbands.
   D. higher purine content in the bands.
15. Bands in an expanded state observed on chromosomes are sites of:
   A. DNA replication.
   B. DNA recombination.
   C. gene expression.
   D. protein binding.
16. Telomeres consist of:
   A. simple sequence repeats of a C+A rich strand that interact with proteins.
   B. simple sequence repeats of a G+C rich strand that interact with proteins.
   C. simple sequence repeats of a C+T rich strand that interact with proteins.
   D. simple sequence repeats of a T+A rich strand that interact with proteins.

KEY TERMS

acentric fragment
bands
capsid
centromere
chromatin
chromocenter
chromomeres
chromosome
domain
euchromatin
G-bands
heterochromatin
***in situ* hybridization**
interbands
kinetochore
matrix attachment region (MAR)
metaphase (or mitotic) scaffold
microtubule organizing center (MTOC)
nucleation center
nucleoid
polytene chromosomes
puff
spindle
telomerase
telomere
terminase

FURTHER READING

Bailey, S. M., and Murname, J. P. (2006). Telomeres, chromosome instability and cancer. *Nucleic Acids Research* **34**(8), 2408–2417. An overview of telomere structure and function, including the role of telomeres in maintaining chromosome stability.

Blackburn, E. H., Greider, C. W., and Szostak, J. W. (2006). Telomeres and telomerase: the path from maize, *Tetrahymena* and yeast to human cancer and aging. *Nature Medicine* **12**, 1133–1138.

Bloom, K. (2007). Centromere dynamics. *Current Opinion in Genetics and Development* **17**(2), 151–156. A review of eukaryotic kinetochore structure and function, focusing on the conserved role of the centromeric H3 variant.

Chattopadhyay, S., and Pavithra, L. (2007). MARs and MARBPs: key modulators of gene regulation and disease manifestation. *Subcellular Biochemistry* **41**, 213–230. A review of MARs/SARs and their known interacting proteins, with an emphasis on the role MARs/SARs play in gene regulation and cancer.

Dalal, Y. (2009). Epigentic specification of centromeres. *Biochemistry and Cell Biology* **87**, 273–282.

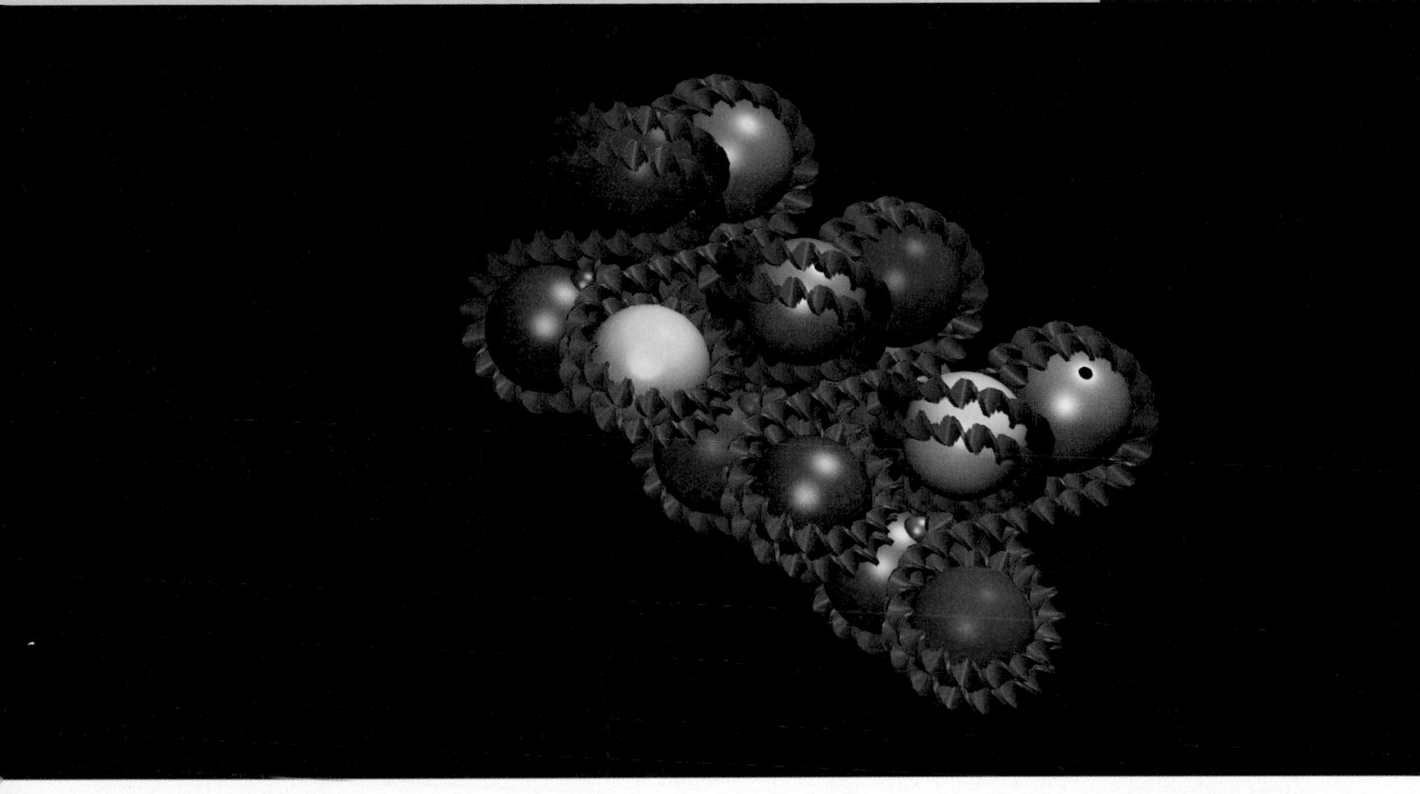

A three-dimensional model of chromatin showing one possible arrangement for the nucleosomes in the 30 nm fiber. Photo courtesy of Thomas Bishop, Tulane University, using VDNA plug-in for VMD (http://www.ks.uiuc.edu/research/vmd/plugins/vdna/).

Chromatin

CHAPTER OUTLINE

10.1 Introduction

Chromatin has a compact organization in which most DNA sequences are structurally inaccessible and functionally inactive. Within this mass is the minority of active sequences. What is the general structure of chromatin, and what is the difference between active and inactive sequences? The high overall packing ratio of the genetic material immediately suggests that DNA cannot be directly packaged into the final structure of chromatin. There must be *hierarchies* of organization.

The fundamental subunit of chromatin has the same basic structure in all eukaryotes. The **nucleosome** contains ~200 bp of DNA, organized by an octamer of small, basic proteins into a beadlike structure. The proteins are called **histones**. They form an interior core; the DNA lies on the surface of the particle. Additional regions of the histones, known as the **histone tails**, extend from the surface. Nucleosomes are the underlying component of euchromatin and heterochromatin in the interphase nucleus and of mitotic chromosomes. The nucleosome provides the first level of organization, compacting the DNA ~6-fold over the length of naked DNA, resulting in a fiber ~10 nm in diameter. Its components and structure are well characterized.

The second level of organization is the coiling of the **10 nm fiber** into a helical array to constitute the fiber of diameter ~30 nm that is found in both interphase chromatin and mitotic chromosomes. This compacts the DNA ~40-fold.

This **30 nm fiber** is then further folded and compacted into interphase chromatin or into mitotic chromosomes. This results in ~1000-fold compaction in euchromatin, cyclically interchangeable with packing into mitotic chromosomes to achieve ~10,000-fold compaction. Heterochromatin is generally as compact as mitotic chromatin.

We need to work through these levels of organization to characterize the events involved in cyclical packaging, replication, and transcription. Association with additional proteins, or modification of existing chromosomal proteins, is involved in changing the structure of chromatin. Both replication and transcription require unwinding of DNA, and thus must involve an unfolding of the structure that allows the relevant enzymes to manipulate the DNA. This is likely to involve changes in all levels of organization.

The mass of chromatin contains up to twice as much protein as DNA. Approximately half of the protein mass is accounted for by the nucleosomes. The mass of RNA is <10% of the mass of DNA. Much of the RNA consists of nascent transcripts still associated with the template DNA.

The term **nonhistones** is used to describe all the proteins of chromatin except the histones. They are more variable between tissues and species, and they comprise a smaller proportion of the mass than the histones. They also comprise a much larger number of proteins, so that any individual protein is present in amounts much smaller than any histone.

- **nucleosome** The basic structural subunit of chromatin, consisting of ~200 bp of DNA and an octamer of histone proteins.
- **histones** Conserved DNA-binding proteins that form the basic subunit of chromatin in eukaryotes. H2A, H2B, H3, and H4 form an octameric core around which DNA coils to form a nucleosome. Linker histones are external to the nucleosome.
- **histone tails** Flexible amino- or carboxy-terminal regions of the core histones that extend beyond the surface of the nucleosome; histone tails are sites of extensive post-translational modification.
- **10 nm fiber** A linear array of nucleosomes, generated by unfolding from the natural condition of chromatin.
- **30 nm fiber** A coil of nucleosomes. It is the basic level of organization of nucleosomes in chromatin.
- **nonhistone** Any structural protein found in a chromosome except one of the histones.

FIGURE 10.1 Chromatin spilling out of lysed nuclei consists of a compactly organized series of particles. The bar is 100 nm. Reproduced from P. Oudet et al., Electron microscopic and biochemical evidence that chromatin structure is a repeating unit, *Cell* 4, pp. 281–300. Copyright 1975, with permission from Elsevier (http://www.sciencedirect.com/science/journal/00928674). Photo courtesy of Pierre Chambon.

10.2 The Nucleosome Is the Subunit of All Chromatin

When interphase nuclei are suspended in a solution of low ionic strength, they swell and rupture to release fibers of chromatin. **FIGURE 10.1** shows a lysed nucleus in which fibers are streaming out. In some regions, the fibers consist of tightly packed material, but in regions that have become stretched, they can be seen to consist of discrete particles. These particles are the nucleosomes. In especially extended regions, individual nucleosomes are visibly connected by a fine thread, which is a free duplex of DNA. A continuous duplex thread of DNA runs through the series of particles.

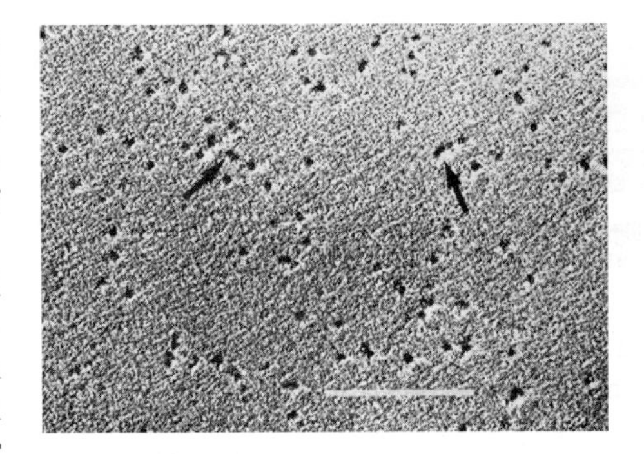

FIGURE 10.2 Individual nucleosomes are released by digestion of chromatin with micrococcal nuclease. The bar is 100 nm. Reproduced from P. Oudet et al., Electron microscopic and biochemical evidence that chromatin structure is a repeating unit, *Cell* 4, pp. 281–300. Copyright 1975, with permission from Elsevier (http://www.sciencedirect.com/science/journal/00928674). Photo courtesy of Pierre Chambon.

Individual nucleosomes can be obtained by treating chromatin with the endonuclease **micrococcal nuclease (MNase)**, which cuts the DNA duplex at the junction between nucleosomes, a region known as **linker DNA**. Ongoing digestion with MNase releases groups of particles and, eventually, single nucleosomes. Individual nucleosomes can be seen in **FIGURE 10.2** as compact particles. When chromatin is digested with MNase, the DNA is cleaved into integral multiples of a unit length. Fractionation of the DNA fragments by gel electrophoresis (after removal of the proteins) reveals the "ladder" presented in **FIGURE 10.3**. Such ladders extend for ~10 steps, and the unit length, determined by the increments between successive steps, is ~200 bp.

The ladder is generated by DNA corresponding to groups of nucleosomes. The fastest-migrating band contains DNA from single nucleosomes (monomers), whereas the higher bands correspond to DNA from nucleosome dimers, trimers, and so on, still connected by linker DNA. A monomeric nucleosome contains DNA of the ~200 bp unit length, a nucleosome dimer contains DNA of twice the unit length, and so on. (Intact nucleosomes can also be separated on nondenaturing gels, where they form a comparable ladder.) More than 95% of the DNA of chromatin can be recovered in the form of the 200 bp ladder, indicating that almost all DNA must be organized in nucleosomes.

The length of DNA present in the nucleosome can vary from the "typical" value of 200 bp. The chromatin of any particular cell type has a characteristic average value (±5 bp). The average most often is between 180 and 200, but there are extremes as low as 154 bp (in a fungus) or as high as 260 bp (in a sea urchin sperm). The average value may be different in individual tissues of the adult organism, and there can be differences between different parts of the genome in a single cell type. The differences are primarily a function of how much linker DNA is present between nucleosomes.

What is the nature of the nucleosome itself? The nucleosome contains ~200 bp of DNA associated with a **histone octamer** that consists of two copies each of H2A, H2B, H3, and H4. These are known as the **core histones**. Their association is illustrated diagrammatically in **FIGURE 10.4**.

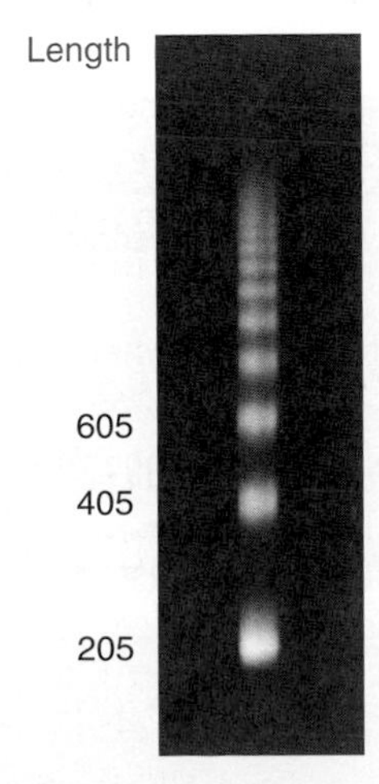

FIGURE 10.3 Micrococcal nuclease digests chromatin into a multimeric series of DNA bands that can be separated by gel electrophoresis. Photo courtesy of Markus Noll, Universität Zürich.

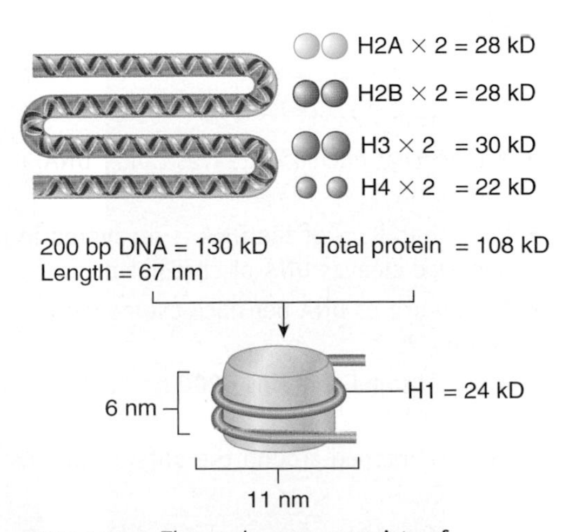

FIGURE 10.4 The nucleosome consists of approximately equal masses of DNA and histones (including linker histones such as H1). The predicted mass of the nucleosome is 262 kD.

- **micrococcal nuclease (MNase)** An endonuclease that cleaves DNA; in chromatin, DNA is cleaved preferentially between nucleosomes.
- **linker DNA** Nonnucleosomal DNA present between nucleosomes.
- **histone octamer** The complex of two copies each of the four different core histones (H2A, H2B, H3, and H4); DNA wraps around the histone octamer to form the nucleosome.
- **core histone** One of the four types of histone (H2A, H2B, H3, and H4 and their variants) found in the core particle derived from the nucleosome. (This excludes linker histones.)

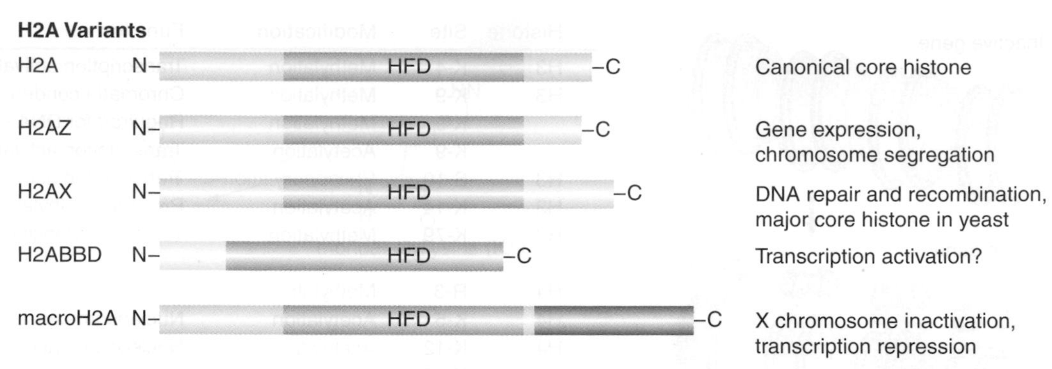

FIGURE 10.18 The major core histones contain a conserved histone-fold domain (HFD). In the histone H3.3 variant, the residues that differ from the major histone H3 (also known as H3.1) are highlighted in yellow. The centromeric histone CenH3 has a unique N-terminus, which does not resemble other core histones. Most H2A variants contain alternative C-termini, except H2ABBD, which contains a distinct N-terminus. The sperm-specific spH2B has a long N-terminus. Proposed functions of the variants are listed.

10.5 Histone Variants Produce Alternative Nucleosomes

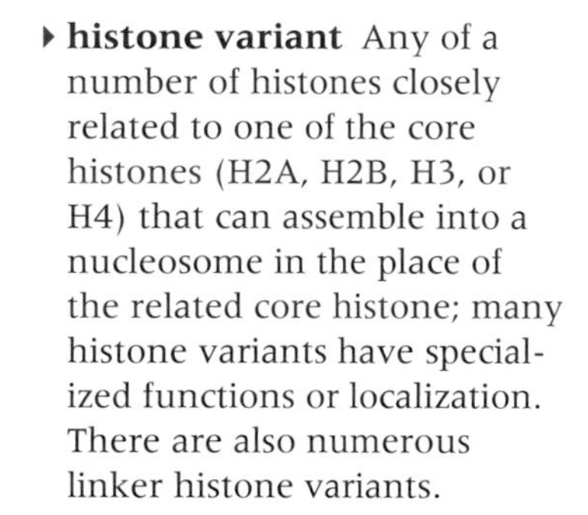

▸ **histone variant** Any of a number of histones closely related to one of the core histones (H2A, H2B, H3, or H4) that can assemble into a nucleosome in the place of the related core histone; many histone variants have specialized functions or localization. There are also numerous linker histone variants.

While all nucleosomes share a related core structure, some nucleosomes exhibit subtle or dramatic differences resulting from the incorporation of **histone variants**. Histone variants comprise a large group of histones that are related to the histones we have already discussed but have differences in sequence from the so-called canonical histones. These sequence differences can be small (as few as four amino acid differences) or extensive (such as alternative tail sequences).

Variants have been identified for all core histones except histone H4. The best-characterized histone variants are summarized in **FIGURE 10.18**. Most variants have significant differences between them, particularly in the N- and C-terminal tails. At one extreme, macroH2A is nearly three times larger than conventional H2A and contains a large C-terminal tail that is not related to any other histone. At the other end of the spectrum, canonical H3 (also known as H3.1) differs from the H3.3 variant at only four amino acid positions, three in the histone core and one in the N-terminal tail.

Histone variants have been implicated in a number of different functions, and their incorporation changes the nature of the chromatin containing the variant. We have already discussed one type of histone variant, the centromeric H3 (or CenH3) histone, known as Cse4 in yeast. CenH3 histones are incorporated into specialized nucleosomes present at centromeres in all eukaryotes (see *Sections 9.9, Regional Centromeres Contain a Centromeric Histone H3 Variant and Repetitive DNA*, and *9.10, Point Centromeres in* S. cerevisiae *Contain Short, Essential Protein-Binding DNA Sequences*). In yeast, it has been shown that these centromeric nucleosomes consist of Cse4, H4, and a nonhistone protein Scm3, which replaces H2A/H2B dimers. In *Drosophila*, the centromeric chromatin appears to consist of "hemisomes" containing one copy each of CenH3, H4, H2A, and H2B. It is not known whether any centromeric chromatin in higher eukaryotes contains an Scm3-like protein at a subset of centromeric nucleosomes.

The other major H3 variant is histone H3.3. In multicellular eukaryotes this variant is a minority component of the total H3 in the cell, but in yeast, the major H3 is actually of the H3.3 type. H3.3 is expressed throughout the cell cycle, in contrast to most histones that are expressed during S phase, when new chromatin assembly is required during DNA replication. As a result, H3.3 is available for assembly at any time in the cell cycle and is incorporated at sites of active transcription, where nucleosomes become disrupted. Because of this, H3.3 is often referred to as a "replacement" histone, in contrast to the "replicative" histone H3.1.

The H2A variants are the largest and most diverse family of core histone variants and have been implicated in a variety of distinct functions. One of the best studied is the variant H2AX. H2AX is normally present in only 5% to 15% of the nucleosomes

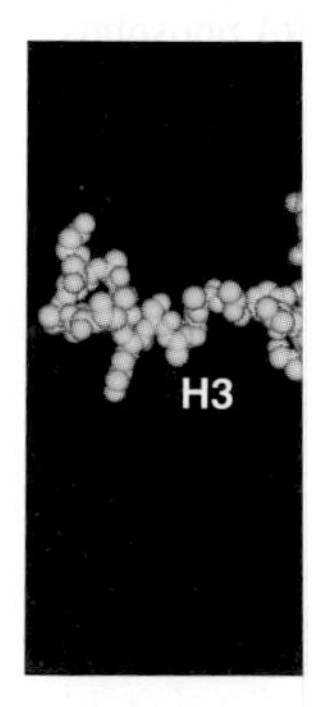

FIGURE 10.11 faces of the nu figure shows o complete tails purple: H4; tu

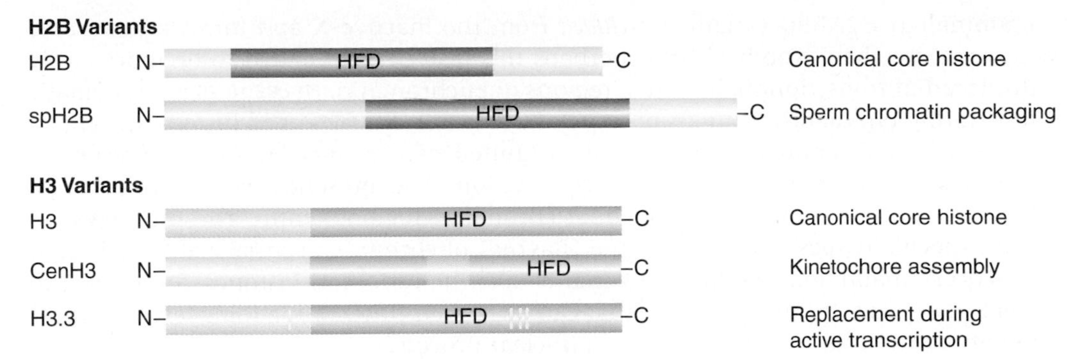

FIGURE 10.18 *(continued)*

in multicellular eukaryotes, although again (like H3.3) this subtype is the major H2A present in yeast. This variant has a C-terminal tail that is distinct from the canonical H2A, which is characterized by a SQEL/Y motif at the end. This motif is the target of phosphorylation by ATM/ATR kinases, activated by DNA damage, and this histone variant is involved in DNA repair, particularly repair of double strand breaks (see *Chapter 16, Repair Systems*). H2AX phosphorylated at the SQEL/Y motif is referred to as "γ-H2AX" and is required to stabilize binding of various repair factors at DNA breaks and to maintain checkpoint arrest. γ-H2AX appears within moments at broken DNA ends, as can be seen in **FIGURE 10.19**, which shows a cartoon depicting γ-H2AX foci forming along the path of double-strand breaks induced by a laser. The experiment depicted in this cartoon, performed in the Bonner laboratory, was one of the first indications that this modification occurs specifically at double-strand breaks.

Other H2A variants have different roles. The H2AZ variant, which has ~60% sequence identity with canonical H2A, has been shown to be important in several processes, such as gene activation, heterochromatin-euchromatin boundary formation, and cell cycle progression. The vertebrate-specific macroH2A is named for its extremely long C-terminal tail, which contains a leucine-zipper dimerization motif that may mediate chromatin compaction by facilitating internucleosome interactions. Mammalian macroH2A is enriched in the inactive X chromosome in females, which is assembled into a silent, heterochromatic state and can also be found at heterochromatic regions on autosomes (and occasionally in active euchromatin). In contrast, the

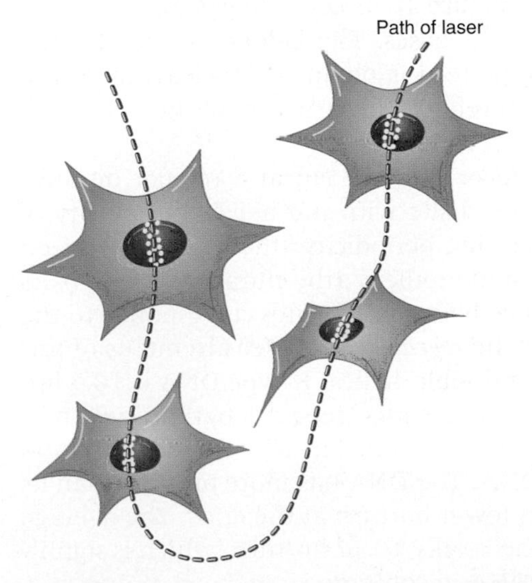

FIGURE 10.19 γ-H2AX is detected by an antibody (yellow dots) and appears as foci along the path traced by a laser that produces double-strand breaks (indicated by dotted red line).

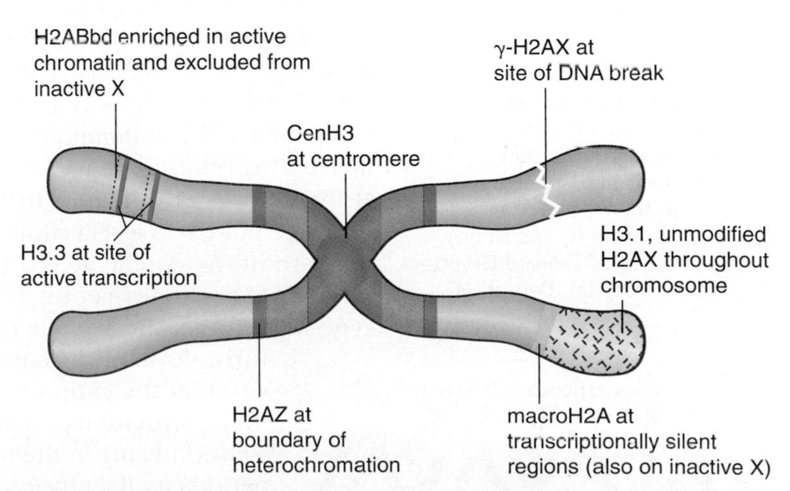

FIGURE 10.20 Some histone variants are spread throughout all or most of the chromosome, whereas others show specific distribution patterns. Characteristic patterns are shown for several histone variants on a cartoon autosome. Note that histone variant distributions are dramatically different on the inactive X in mammals and in sperm chromatin.

mammalian H2ABbd variant is *excluded* from the inactive X and forms a less stable nucleosome than canonical H2A; perhaps this histone is designed to be more easily displaced in transcriptionally active regions of euchromatin. **FIGURE 10.20** is a schematic illustrating typical distributions of some of the better-characterized histone variants.

Still other variants are expressed in limited tissues, such as spH2B, present in sperm and required for chromatin compaction. The presence and distribution of histone variants shows that individual chromatin regions, entire chromosomes, or even specific tissues can have unique "flavors" of chromatin specialized for different function. In addition, the histone variants, like the canonical histones, are subject to numerous covalent modifications that can alter their functions, adding levels of complexity to the roles chromatin plays in nuclear processes.

KEY CONCEPTS

- All core histones except H4 are members of families of related variants.
- Histone variants can be closely related or highly divergent from canonical histones.
- Different variants serve different functions in the cell.

CONCEPT AND REASONING CHECK

Why would a less stable nucleosome (such as one containing H2ABbd) be more likely to be present in euchromatin than in heterochromatin?

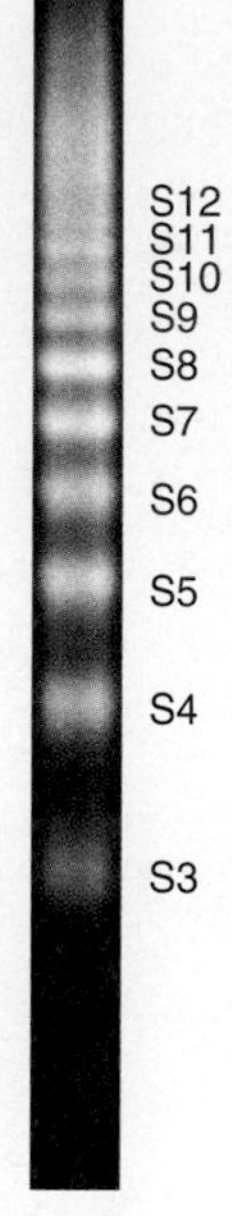

FIGURE 10.21 Sites for nicking lie at regular intervals along core DNA, as seen in a DNase I digest of nuclei. Cleavage sites are numbered as S1 through S13 (where S1 is ~10 bases from the labeled 5′ end, S2 is ~20 bases from it, and so on). Photo courtesy of Leonard C. Lutter, Henry Ford Hospital, Detroit, MI.

10.6 DNA Structure Varies on the Nucleosomal Surface

So far we have focused on the protein components of the nucleosome. The DNA wrapped around these proteins assumes an unusual conformation. The exposure of DNA on the surface of the nucleosome results in a distinctive pattern of accessibility to cleavage by certain nucleases. The reaction with nucleases that attack single strands has been especially informative. The enzymes DNase I and DNase II make single-strand nicks in DNA; they cleave a bond in one strand, but the other strand remains intact. When DNA is free in solution, it is nicked (relatively) at random. The DNA on nucleosomes also can be nicked by the enzymes, but in this case the nicking occurs only at regular intervals. When the points of cutting are determined by using radioactively end-labeled DNA and then DNA is denatured and electrophoresed, a ladder of the sort displayed in **FIGURE 10.21** is obtained.

Compared to the MNase ladder shown in Figure 10.3, DNase digestion generates products with a much shorter interval: 10 to 11 bases. The ladder extends for the full distance of core DNA. The same cutting pattern is obtained by cleaving with a hydroxyl radical, which argues that the pattern reflects the structure of the DNA itself rather than any sequence preference.

When DNA is immobilized on a flat surface, sites are cut at a regular distance apart. This reflects the recurrence of the exposed site with the helical periodicity of B-form DNA, as shown in **FIGURE 10.22**. The cutting periodicity (the spacing between cleavage points) is a reflection of the structural periodicity (the number of base pairs per turn of the double helix). Thus the distance between the sites corresponds to the number of base pairs per turn (also known as the *repeat length*). Measurements of this type suggest that the average repeat length for double-helical B-type DNA is 10.5 bp/turn. On the nucleosome, however, the average repeat length varies from ~10.0 at the ends, to ~10.7 in the middle. This means that there is variation in the structural periodicity of core DNA. The DNA has more bp/turn than its solution value in the middle, but has fewer bp/turn at the ends. The average periodicity over the entire nucleosome is only 10.17 bp/turn, which is significantly less than the 10.5 bp/turn of DNA in solution.

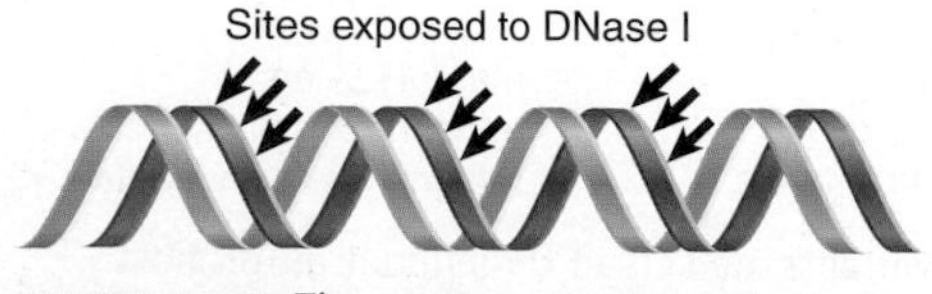

FIGURE 10.22 The most exposed positions on DNA recur with a periodicity that reflects the structure of the double helix (for clarity, sites are shown for only one strand).

The crystal structure of the core particle shows that DNA is wound into a *solenoidal* (spring-shaped) supercoil, with 1.65 turns wound around the histone octamer. The structure of DNA is distorted. The central 129 bp are in the form of B-DNA, but with a substantial curvature that is needed to form the superhelix. The major groove is smoothly bent, but the minor groove has abrupt kinks. These conformational changes may explain why the central part of nucleosomal DNA is not usually a target for binding by regulatory proteins, which typically bind to the terminal parts of the core DNA or to the linker sequences.

Some insights into the structure of nucleosomal DNA emerge when we compare predictions for supercoiling in the path that DNA follows with actual measurements of supercoiling of nucleosomal DNA. Much work on the structure of sets of nucleosomes has been carried out with the virus SV40, which infects mammalian cells. The DNA of SV40 is a circular molecule of 5200 bp. In both the viral particle and infected nucleus, it is packaged into a series of nucleosomes, which together are called a *minichromosome*.

When isolated, the minichromosome is compacted ~6-fold over naked DNA (essentially the same as the ~6 of the nucleosome itself). The degree of supercoiling constrained by the individual nucleosomes of the minichromosome can be measured as illustrated in **FIGURE 10.23**. First, the free supercoils (those not constrained in the nucleosomes) of the minichromosome itself are relaxed, so that the nucleosomes form a circular string with a superhelical density of 0. Next, the histone octamers are extracted. This releases the DNA to follow a free path. Every supercoil that was present but constrained in the nucleosomes will appear in the deproteinized DNA as −1 turn. Finally, the total number of supercoils in the SV40 DNA is measured.

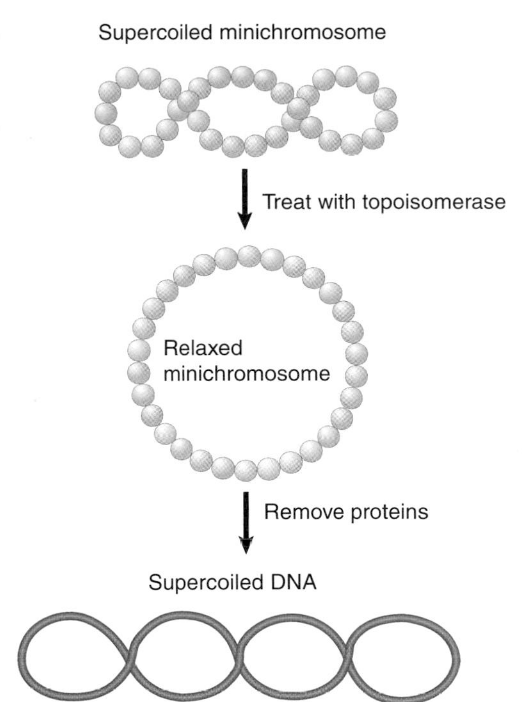

FIGURE 10.23 The supercoils of the SV40 minichromosome can be relaxed to generate a circular structure, whose loss of histones then generates supercoils in the free DNA.

The observed value is close to the number of nucleosomes. Thus the DNA follows a path on the nucleosomal surface that generates ~1 negative supercoiled turn when the restraining protein is removed. The path that DNA follows on the nucleosome, though, corresponds to −1.67 superhelical turns (see Figure 10.5). This discrepancy is sometimes called the *linking number paradox.*

The discrepancy is explained by the difference between the 10.17 average bp/turn of nucleosomal DNA and the 10.5 bp/turn of free DNA. In a nucleosome of 200 bp, there are 200/10.17 = 19.67 turns. When DNA is released from the nucleosome, it now has 200/10.5 = 19.0 turns. The path of the less tightly wound DNA on the nucleosome absorbs −0.67 turns, which explains the discrepancy between the physical path of −1.67 and the measurement of −1.0 superhelical turns. In effect, some of the torsional strain in nucleosomal DNA goes into increasing the number of bp/turn; only the remainder is measured as a supercoil.

KEY CONCEPTS

- DNA is wrapped 1.65 times around the histone octamer.
- The structure of the DNA is altered so that it has an increased number of base pairs/turn in the middle, but a decreased number at the ends.
- Approximately 0.6 negative turns of DNA are absorbed by the change in bp/turn from 10.5 in solution to an average of 10.17 on the nucleosomal surface, which explains the linking number paradox.

CONCEPT AND REASONING CHECK

Explain the different digestion patterns generated by MNase and DNase on nucleosomal DNA.

10.7 The Path of Nucleosomes in the Chromatin Fiber

When chromatin is examined with the electron microscope, two types of fibers are seen: the 10 nm fiber and 30 nm fiber. They are described by the approximate diameter of the thread (that of the 30 nm fiber actually varies from ~25–30 nm).

METHODS AND TECHNIQUES

Position Effect Variegation (PEV) and the Discovery of Insulators

Genes near the breakpoints of chromosomal rearrangement become repositioned in the genome and flanked by new neighboring genes or regulatory sequences. In many cases, the repositioning of a gene affects its level of expression or, in some cases, its ability to function; this is called *position effect*. Such effects have been studied extensively in *Drosophila* and yeast. In *Drosophila*, the most common type of position effect results in a mottled (mosaic) phenotype that is observed as interspersed patches of wild-type cells, in which the wild-type allele is expressed, and mutant cells, in which the wild-type allele is inactivated. The phenotype is said to show *variegation*, and the phenomenon is called *position-effect variegation (PEV)*.

PEV can result from a chromosome aberration that moves a wild-type gene from a position in euchromatin to a new position in or near heterochromatin (**FIGURE B10.1**). **FIGURE B10.2** illustrates some of the patterns of wild-type (red) and mutant (white) facets that are observed in male flies that carry a rearranged X chromosome in which an inversion repositions the wild-type *white* (w^+) allele into heterochromatin. The same types of patterns are found in females heterozygous for the rearranged X chromosome and an X chromosome carrying the w allele. The patterns of w^+ expression coincide with the clonal lineages in the eye; that is, all the red cells in a particular patch derive from a single ancestral cell in the embryo in which the w^+ allele was activated.

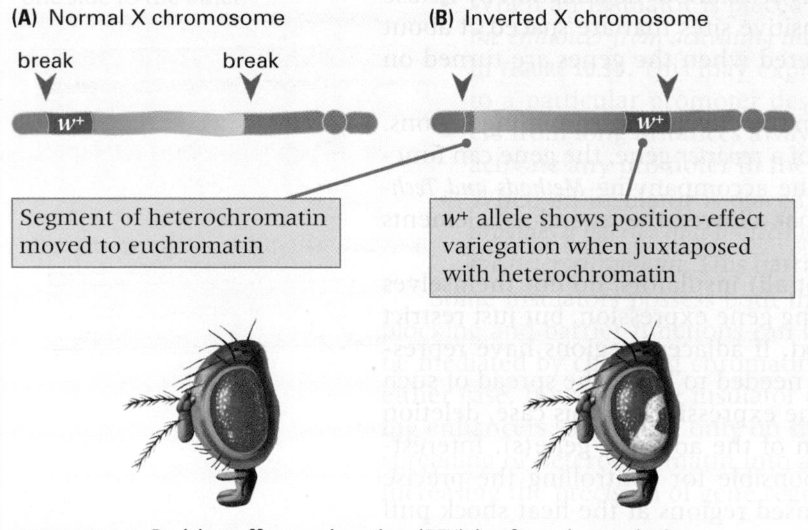

FIGURE B10.1 Position-effect variegation (PEV) is often observed when an inversion or other chromosome rearrangement repositions a gene normally in euchromatin to a new location in or near heterochromatin. In this example, an inversion in the X chromosome of *Drosophila melanogaster* repositions the wild-type allele of the *white* gene near heterochromatin. PEV of the w^+ allele is observed as mottled red and white eyes.

Although the mechanism of PEV is not understood in detail, it is thought to result from the unusual chromatin structure of heterochromatin interfering with gene activation. The determination of gene expression or nonexpression is thought to take place when the boundary between condensed heterochromatin and euchromatin is established. Where heterochromatin is juxtaposed with euchromatin, the chromatin condensation characteristic of heterochromatin may spread into the adjacent euchromatin, inactivating euchromatic genes in the cell and all of its descendants.

The phenomenon of position effects has proven to be a useful system for scientists studying various aspects of chromosome organization. For example, genes encoding products involved in heterochromatin formation have been identified in screens for factors that alter levels of PEV when deleted or overexpressed. Deletion of a gene that promotes heterochromatin spreading, such that encoding the structural heterochromatin protein 1 (HP1), would result in *more* expression of a gene located close to heterochromatin.

The first insulators to be functionally characterized, *scs* and *scs′*, were investigated using PEV as a tool. Researchers placed *scs* and *scs′* elements on both sides of a *white* reporter gene and then randomly integrated these reporters

an anti-BEAF-32 antibody stains ~50% of the interbands of polytene chromosomes. This suggests that there are many insulators in the genome (though BEAF-32 may bind noninsulators as well) and that BEAF-32 is a common part of the insulating apparatus. This result also implies that the band is a functional unit, and that interbands often have insulators that block the propagation of activating or inactivating effects. BEAF-32 is one of a group of insulator binding proteins that are unique to *Drosophila*.

Insulators have been identified in all eukaryotes. The β-globin domains in mammal and chicken are flanked by insulators. (As in mammals, the chicken β-globin cluster

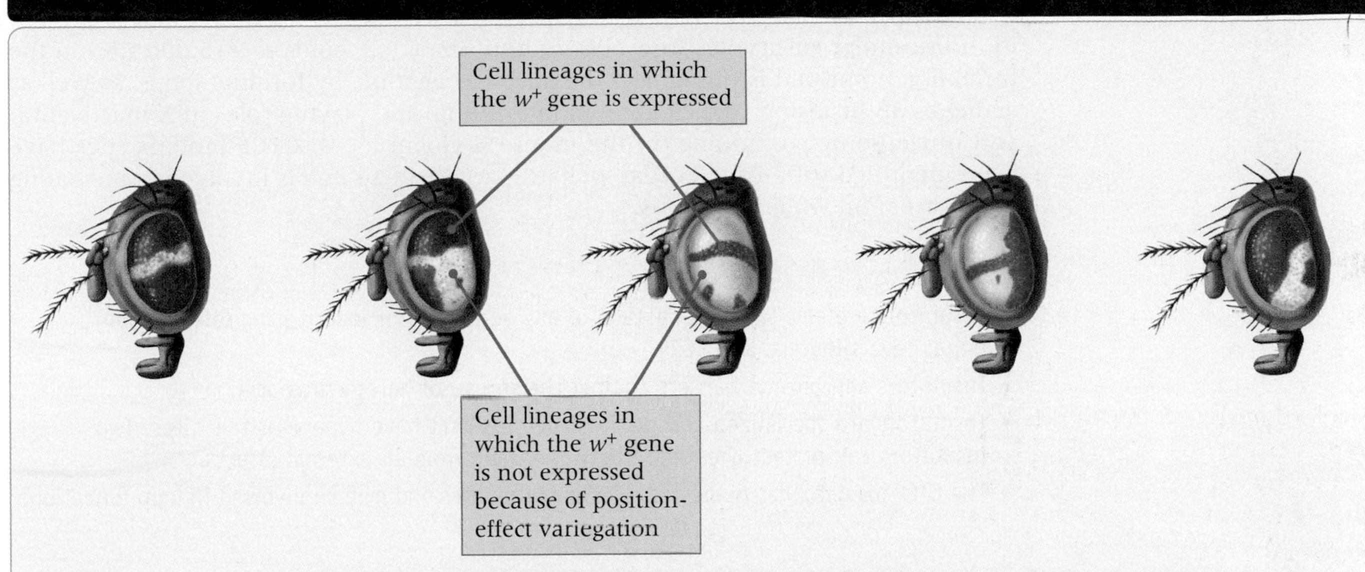

FIGURE B10.2 Patterns of red and white sectors in the eye of *Drosophila melanogaster* resulting from position-effect variegation. Each group of contiguous facets of the same color derives from a single cell in development. Such large patches of red are atypical. More often, one observes numerous very small patches of red or a mixture of many small and a few large patches.

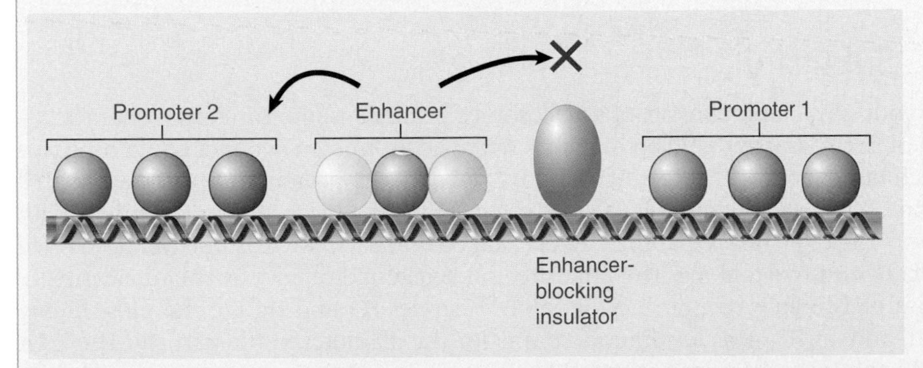

FIGURE B10.3 An enhancer-blocking insulator interferes with enhancer-promoter communication in a position-dependent manner. An enhancer-blocking insulator blocks transcriptional activation only when it lies between a promoter and an enhancer (as in the case of promoter 1); in other situations (such as for promoter 2), activation is not blocked. A transcriptional repressor, by contrast, would reduce the level of transcription from both promoters when placed in the same position. Adapted from M. Gaszner and G. Felsenfeld, *Nat. Rev. Genet.* 7 (2006): 703–713.

into the *Drosophila* genome. When the *white* gene *lacking* flanking insulators was used, the resulting transgenic flies exhibited a wide array of expression levels of the *white* gene, including silencing of the gene, reflecting its integration into or near heterochromatin, as well as high expression resulting from integration near strong enhancers. In contrast, when the *white* reporter was flanked by insulators, both these negative and positive position effects were blocked, and the transgenic flies showed remarkably uniform levels of *white* expression no matter where the gene had integrated. The discovery that *positive* position effects were also prevented led to a now-commonly used assay for insulator function: the "enhancer-blocking" assay. This assay provides a simple test for identifying insulators, as shown in **FIGURE B10-3**, when placed between an enhancer and a promoter, an insulator will prevent the enhancer from activating the promoter without actually repressing the function of the enhancer (note in the figure the enhancer can still act on promoters in the "other direction"). This assay and variations of PEV assays are still the only way to positively identify insulator elements.

is regulated by an LCR.) The furthest upstream hypersensitive site of the chicken LCR (HS4) is an insulator that ma rks the 5′ end of the functional domain. This restricts the LCR to acting only on the globin genes in the domain. It also prevents silencing of the locus by a region of condensed chromatin located upstream of this insulator. A 3′ hypersensitive site downstream of the cluster corresponds to a second insulator, which isolates the α-globin locus from a cluster of olfactory receptor genes. These two insulators combine to define a transcriptionally independent domain, in which the LCR regulates promoters within the domain, and signals from beyond the insulators are blocked.

The β-globin insulators and other vertebrate insulators are bound by the protein CTCF, which (unlike BEAF-32) is highly conserved across evolution and is present in multicellular eukaryotes from flies to humans. CTCF binds to ~15,000 sites in the human genome and is thought to organize the genome by forming loops, as well as acting as an insulator, participating in imprinting, and playing roles in X inactivation and nucleosome positioning. In the mouse β-globin locus, CTCF binding sites have been identified within the LCR as well as 3′ to the locus and is involved in mediating loop formation within the locus.

KEY CONCEPTS

- Insulators are able to block passage of any activating or inactivating effects from enhancers, silencers, and LCRs.
- Insulators can provide barriers against the spread of heterochromatin.
- Insulators are specialized chromatin structures that have hypersensitive sites. Two insulators can protect the region between them from all external effects.
- The CTCF insulator has numerous roles in vertebrates and may be involved in loop formation.

CONCEPT AND REASONING CHECK

Why are two insulators needed to define a domain?

10.13 Summary

All eukaryotic chromatin consists of nucleosomes. A nucleosome contains a characteristic length of DNA, usually ~200 bp, which is wrapped around an octamer containing two copies each of histones H2A, H2B, H3, and H4. A single H1 protein may associate with a nucleosome. Virtually all genomic DNA is organized into nucleosomes. Treatment with micrococcal nuclease shows that the DNA packaged into each nucleosome can be divided operationally into two regions. The linker region is digested rapidly by the nuclease; the core region of 146 bp is resistant to digestion. Histones H3 and H4 are the most highly conserved, and an $H3_2$–$H4_2$ tetramer accounts for the diameter of the particle. The H2A and H2B histones are organized as two H2A–H2B dimers. Octamers are assembled by the successive addition of two H2A–H2B dimers to the $H3_2$–$H4_2$ tetramer.

The path of DNA around the histone octamer creates –1.65 supercoils. The DNA "enters" and "leaves" the nucleosome in the same vicinity and could be "sealed" by histone H1. Removal of the core histones releases –1.0 supercoils. The difference can be largely explained by a change in the helical pitch of DNA, from an average of 10.2 bp/turn in nucleosomal form to 10.5 bp/turn when free in solution. There is variation in the structure of DNA from a periodicity of 10.0 bp/turn at the nucleosome ends to 10.7 bp/turn in the center. There are kinks in the path of DNA on the nucleosome.

Nucleosomes are organized into a fiber of 30 nm diameter that has six nucleosomes per turn and a compaction level of 40-fold. Removal of H1 allows this fiber to unfold into a 10 nm fiber that consists of a linear string of nucleosomes. The 30 nm fiber probably consists of the 10 nm fiber wound into a two-start solenoid. The 30 nm fiber is the basic constituent of both euchromatin and heterochromatin; nonhistone proteins are responsible for further organization of the fiber into chromatin or chromosome ultrastructure.

There are two pathways for nucleosome assembly. In the replication-coupled pathway, the PCNA processivity subunit of the replisome recruits CAF-1, which is a nucleosome assembly factor. CAF-1 assists the deposition of $H3_2$–$H4_2$ tetramers onto the daughter duplexes resulting from replication. The tetramers may be produced either by disruption of existing nucleosomes by the replic ation fork or as the result of assembly from newly synthesized histones. Similar sources provide the H2A–H2B dimers that then assemble with the $H3_2$–$H4_2$ tetramer to complete the nucleosome. The $H3_2$–$H4_2$ tetramer and the H2A–H2B dimers assemble at random, so the new nucleosomes may include both preexisting and newly synthesized histones. HIRA assembles nucleosomes outside of S phase, and ASF1 acts both during and outside replication to assemble chromatin.

Two types of changes in sensitivity to nucleases are associated with gene activity. Chromatin capable of being transcribed has a generally increased sensitivity to DNase I, reflecting a change in structure over an extensive region that can be defined as a domain containing active or potentially active genes. Hypersensitive sites in DNA occur at discrete locations and are identified by greatly increased sensitivity to DNase I.

A hypersensitive site consists of a sequence of ~200 bp from which nucleosomes are excluded by the presence of other proteins. A hypersensitive site forms a boundary that may cause adjacent nucleosomes to be restricted in position. Nucleosome positioning may be important in controlling access of regulatory proteins to DNA.

Hypersensitive sites occur at several types of regulators. Those that regulate transcription include promoters, enhancers, and LCRs. Other sites include origins for replication and centromeres. A promoter or enhancer typically acts on a single gene, but an LCR contains a group of hypersensitive sites and may regulate a domain containing several genes.

An insulator blocks the transmission of activating or inactivating effects in chromatin. An insulator that is located between an enhancer and a promoter prevents the enhancer from activating the promoter. Two insulators define the region between them as a regulatory domain; regulatory interactions within the domain are limited to it, and the domain is insulated from outside effects. Most insulators block regulatory effects from passing in either direction, but some are directional. Insulators usually can block both activating effects (enhancer–promoter interactions) and inactivating effects (mediated by the spread of heterochromatin), but some are limited to one or the other. Insulators are thought to act via changing higher-order chromatin structure, but the details are not certain and different insulators are likely to have different mechanisms.

CHAPTER QUESTIONS

Insulators have either or both of two key properties. When an insulator is placed between a(n) **1.** __________ and a promoter, it **2.** __________ the enhancer from activating the **3.** __________. When an insulator is placed between an active **4.** __________ and heterochromatin, it provides a **5.** __________ that protects the gene against the **6.** __________ effect that spreads from the **7.** __________.

8. An average nucleosome contains about:
- **A.** 100 bp of DNA.
- **B.** 200 bp of DNA.
- **C.** 300 bp of DNA.
- **D.** 500 bp of DNA.

9. Which histone is not part of the core particle?
- **A.** H1
- **B.** H2A
- **C.** H3
- **D.** H4

10. Which two histones are among the most conserved of all known proteins?
- **A.** H1 and H2B
- **B.** H2A and H3
- **C.** H3 and H4
- **D.** H2B and H4

11. One turn of the DNA double helix around a nucleosome takes about:
- **A.** 50 bp of DNA.
- **B.** 80 bp of DNA.
- **C.** 140 bp of DNA.
- **D.** 200 bp of DNA.

12. When nucleosomes are treated with DNase I and the products are separated in a polyacrylamide gel, a ladder of bands is observed, with the average distance between bands about:

A. 6 base pairs.
B. 10 base pairs.
C. 12.5 base pairs.
D. 16 base pairs.

13. A chromosomal domain containing a transcribed gene is defined by:

A. decreased sensitivity to restriction enzymes.
B. decreased sensitivity to micrococcal nuclease.
C. increased sensitivity to DNase I.
D. decreased sensitivity to DNase I.

KEY TERMS

10 nm fiber
30 nm fiber
core histone
domain
histone code
histone fold
histone octamer
histones
histone tails
histone variant
hypersensitive site
indirect end labeling
insulator
linker DNA
linker histones
locus control region (LCR)
micrococcal nuclease (MNase)
nonhistone
nucleosome
nucleosome positioning
rotational positioning
translational positioning

FURTHER READING

Boulard, M., Bouvet, P., Kundu, T. K., and Dmitrov, S. 2007. Histone variant nucleosomes: structure, function and implication in disease. *Subcell. Biochem.* **41**, 91–109. A review of core histone variants, the structural/functional consequences of histone variant incorporation into nucleosomes, and links to human disease.

Chodaparambil, J. V., Edayathumangalam, R. S., Bao, Y., Park, Y. J., and Luger, K. 2006. Nucleosome structure and function. *Ernst Schering Res. Found. Workshop* **57**, 29–46. A review of nucleosome structure and dynamics.

Eitoku, M., Sato, L., Senda, T., and Horikoshi, M. 2008. Histone chaperones: 30 years from isolation to elucidation of the mechanisms of nucleosome assembly and disassembly. *Cell Mol. Life Sci.* **65**, 414–444. An overview of the best-studied histone chaperones, including insights from structural studies.

Gaszner, M., and Felsenfeld, G. 2006. Insulators: exploiting transcriptional and epigenetic mechanisms. *Nature Revs. Gen.* **7**, 703–713. A review of insulator function for both enhancer blocking and heterochromatin barrier activities.

Izzo, A., Kamieniarz, K., and Schneider, R. 2008. The histone H1 family: specific members, specific functions? *Biological Chemistry* **389**, 333–343. A review of the linker histone H1 and the H1 variants, including a review of the differences in specificity and function of different H1 family members.

Palstra, R. J., de Laat,W., and Grosvel, F. 2008. Beta-globin regulation and long-range interactions. *Adv. Gen.* **61**, 107–142. A review of the role of the beta-globin LCR in controlling long-range, developmentally-regulated activation of globin genes in the beta-globin locus.

Talbert, P. B., and Henikoff, S. (2010). Histone variants—ancient wrap artists of the epigenome. *Nature Revs. Mol. Cell Biol.* **11**, 264–275.

II

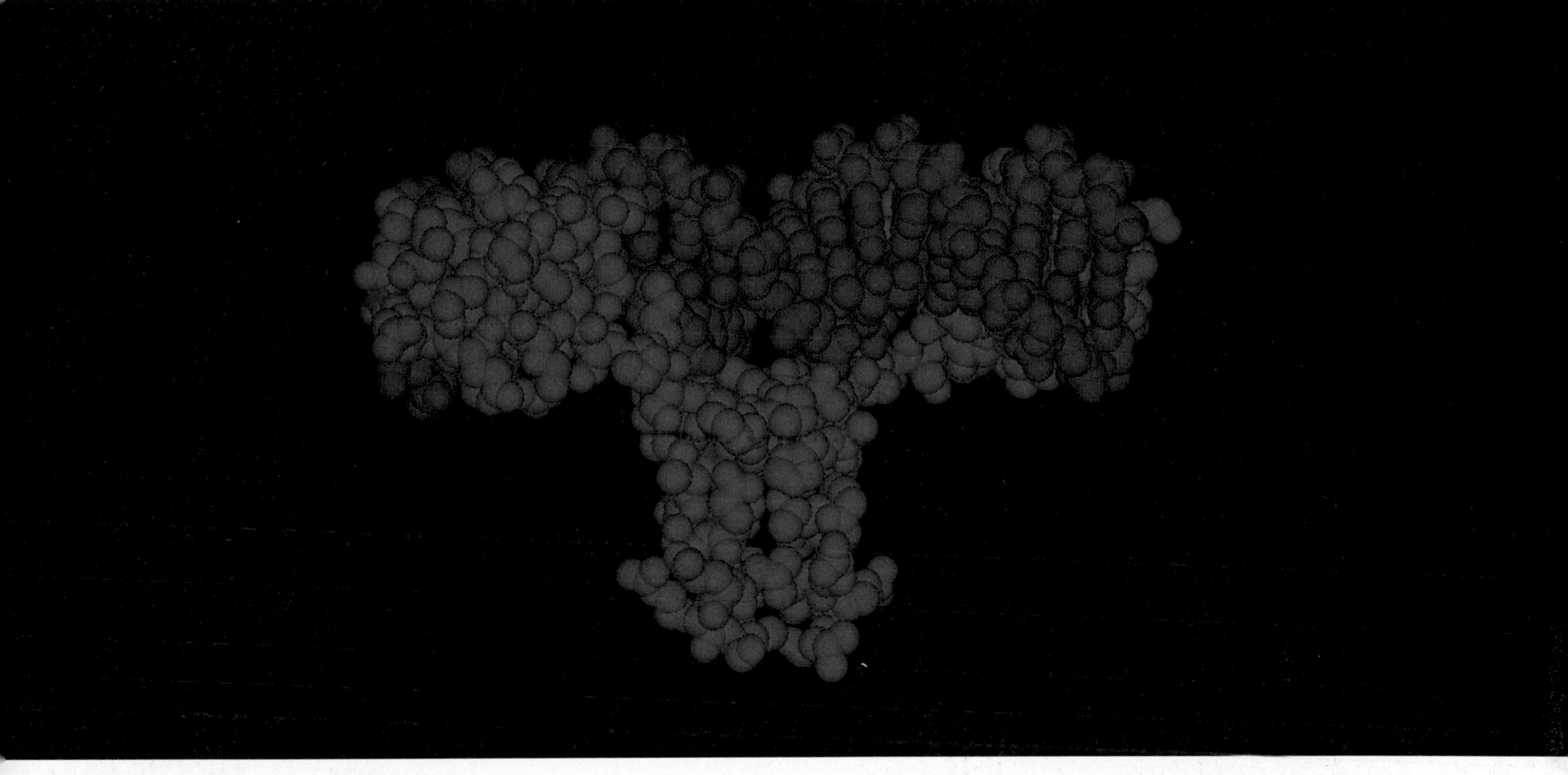

A three-dimensional structure of the GAL4 protein (blue) bound to DNA (red). The protein is composed of subunits held together by the coiled regions in the middle. The DNA-binding domains are at the extreme ends, and each physically contacts three base pairs in the major groove of the DNA. Protein Data Bank 1D66. R. Marmorstein, et al., *Nature* 356 (1992): 408–414.

DNA Replication and Recombination

11

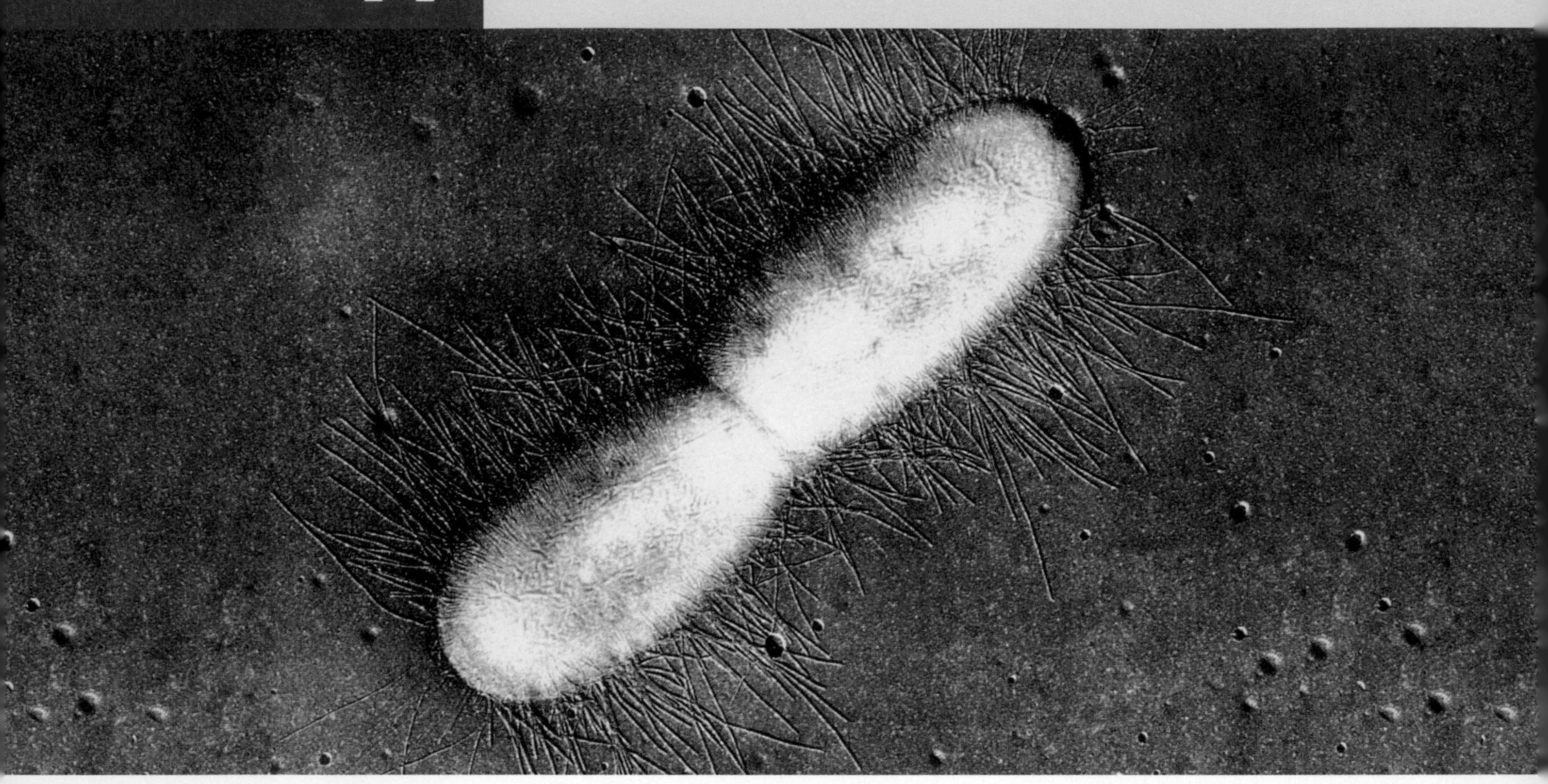

Replication Is Connected to the Cell Cycle

A transmission electron micrograph of an *E. coli* bacterium in the mid to late stages of binary fission, the process by which the bacterium divides. The hair-like appendages around the bacterium are pili, structures used for bacterial conjugation. (Magnification: 17,500×.) © CNRI/Photo Researchers, Inc.

CHAPTER OUTLINE

11.1 Introduction

A major difference between prokaryotes and eukaryotes is the way in which replication is controlled and linked to the cell cycle.

In eukaryotes, the following are true:

- chromosomes reside in the nucleus,
- each chromosome consists of many units of replication called replicons,
- replication requires coordination of these replicons to reproduce DNA during a discrete period of the cell cycle,
- the decision about whether to replicate is determined by a complex pathway that regulates the cell cycle, and
- duplicated chromosomes are segregated to daughter cells during mitosis by means of a special apparatus.

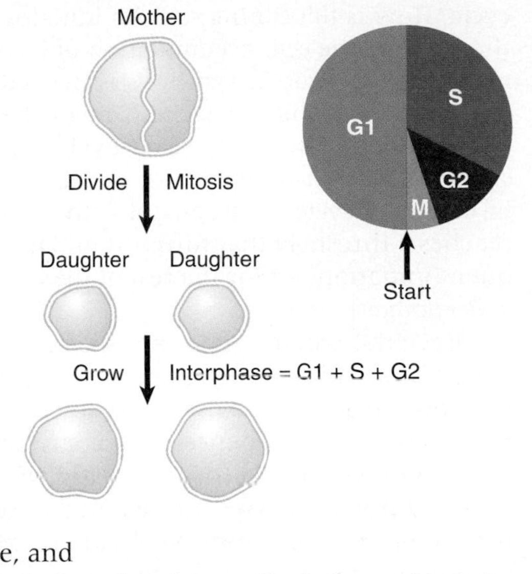

FIGURE 11.1 A growing cell alternates between cell division of a mother cell into two daughter cells and growth back to the original size.

In eukaryote cells, the replication of DNA is confined to the second part of the cell cycle called **S phase**, which follows the G1 phase (see **FIGURE 11.1**). The eukaryotic cell cycle is composed of alternating rounds of growth followed by DNA replication and cell division. After the cell divides into two daughter cells, each has the option to continue dividing or stop and enter G0. If the decision is to continue to divide, then the cell must grow back to the size of the original mother cell before cell division can occur again.

▸ **S phase** The restricted part of the eukaryotic cell cycle during which synthesis of DNA occurs.

The G1 phase of the cell cycle is concerned primarily with growth (although G1 is an abbreviation for *first gap* because the early cytologists could not see any activity). In G1, everything except DNA begins to be doubled: RNA, protein, lipids, and carbohydrates. The progression from G1 into S is very tightly regulated and is controlled by a **checkpoint**. In order for a cell to be allowed to progress into S phase, there must be a certain minimum amount of growth that is biochemically monitored. In addition, there must not be any damage to the DNA. Damaged DNA or too little growth prevents the cell from progressing into S phase. When S phase is complete, G2 phase commences. There is no control point and no sharp demarcation.

▸ **checkpoint** A biochemical control mechanism that prevents the cell from progressing from one stage to the next unless specific goals and requirements have been met.

The start of S phase is signaled by the activation of the first replicon, usually in euchromatin in areas of active genes. Over the next few hours, initiation events occur at other replicons in an ordered manner.

However, in bacteria, as seen in **FIGURE 11.2**, replication is triggered at a single origin when the cell mass increases past a threshold level, and the segregation of the daughter chromosomes is accomplished by ensuring that they find themselves on opposite sides of the septum that grows to divide the bacterium into two.

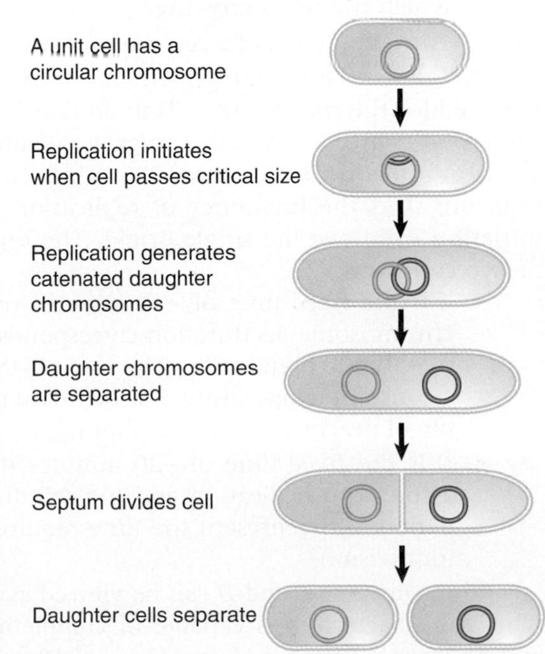

FIGURE 11.2 Replication initiates at the bacterial origin when a cell passes a critical threshold of size. Completion of replication produces daughter chromosomes, which may be linked by recombination or that may be catenated. They are separated and moved to opposite sides of the septum before the bacterium is divided into two.

How does the cell know when to initiate the replication cycle? The initiation event occurs once in each cell cycle and at the same time in every cell cycle. How is this timing set? An initiator protein could be synthesized continuously throughout the cell; accumulation of a critical amount would trigger initiation. This explains why protein synthesis is needed for the initiation event. Another possibility is that an inhibitor protein might be synthesized or activated at a fixed point and diluted below an effective level by the increase in cell volume. Current models suggest that variations of both possibilities operate to turn on and then off precisely in each cell cycle. Synthesis of active DnaA protein, the bacterial initiator protein, reaches a threshold that turns on initiation and the activity of inhibitors turns subsequent initiations off for the rest of the cell cycle (see *Chapter 12, The Replicon: Initiation of Replication*).

Bacterial chromosomes are specifically compacted and arranged inside the cell, and this organization is important for proper segregation, or partition, of daughter chromosomes at cell division. Some of the events in partitioning the daughter chromosomes are consequences of the circularity of the bacterial chromosome. Circular chromosomes are said to be *catenated* when one passes through another, connecting them. **Topoisomerases** are required to separate them. An alternative type of structure is formed when a recombination event occurs: a single recombination between two monomers converts them into a single dimer. This is resolved by a specialized recombination system that recreates the independent monomers. Essentially the partitioning process is handled by enzyme systems that act directly on discrete DNA sequences.

▸ **topoisomerase** An enzyme that changes the number of times the two strands in a closed DNA molecule cross each other. It does this by cutting the DNA, passing DNA through the break, and resealing the DNA.

The key questions in the chapters to follow are to define the DNA sequences that function in replication and to determine how they are recognized by appropriate proteins of the replication apparatus. In *Chapter 12, The Replicon: Initiation of Replication*, we examine the unit of replication and how that unit is regulated to initiate replication. In *Chapter 13, DNA Replication*, we examine the biochemistry and mechanism of DNA synthesis. In *Chapter 14, Extrachromosomal Replication*, we consider autonomously replicating units.

11.2 Bacterial Replication Is Connected to the Cell Cycle

Bacteria have two links between replication and cell growth:

- The frequency of initiation of cycles of replication is adjusted to fit the rate at which the cell is growing.
- The completion of a replication cycle is connected with division of the cell.

▸ **doubling time** The period (usually measured in minutes) that it takes for a bacterial cell to reproduce.

The rate of bacterial growth is assessed by the **doubling time**, the period required for the number of cells to double. The shorter the doubling time, the faster the growth rate. *E. coli* cells can grow at rates ranging from doubling times as fast as 18 minutes to slower than 180 minutes. The bacterial chromosome is a single replicon; thus the frequency of replication cycles is controlled by the number of initiation events at the single origin. The replication cycle can be defined in terms of two constants:

- C is the fixed time of ~40 minutes required to replicate the entire bacterial chromosome. Its duration corresponds to a rate of replication fork movement of ~50,000 bp/minute. (The rate of DNA synthesis is more or less invariant at a constant temperature; it proceeds at the same speed unless and until the supply of precursors becomes limiting.)
- D is the fixed time of ~20 minutes that elapses between the completion of a round of replication and the cell division with which it is connected. This period may represent the time required to assemble the components needed for division.

(The constants C and D can be viewed as representing the maximum speed with which the bacterium is capable of completing these processes. They apply for all growth rates between doubling times of 18 and 60 minutes, but both constant phases become longer when the cell cycle occupies >60 minutes.)

A cycle of chromosome replication must be initiated at a fixed time of $C + D =$ 60 minutes before a cell division. For bacteria dividing more frequently than every 60 minutes, a cycle of replication must be initiated before the end of the preceding division cycle. You might say that a cell is "born already pregnant" with the next generation.

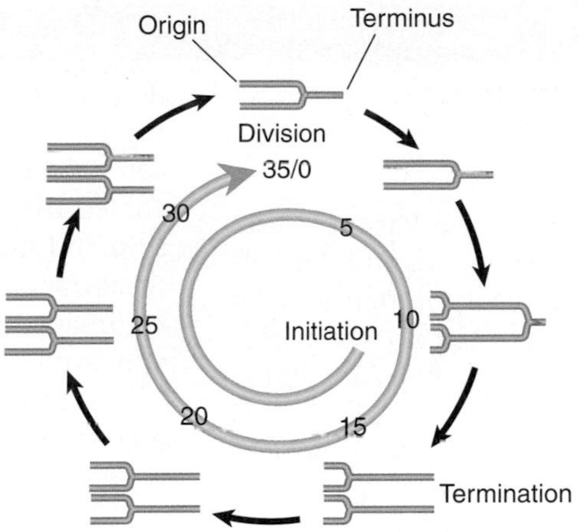

Consider the example of cells dividing every 35 minutes. The cycle of replication connected with a division must have been initiated 25 minutes before the preceding division. This situation is illustrated in **FIGURE 11.3**, which shows the chromosomal complement of a bacterial cell at 5-minute intervals throughout the cycle.

FIGURE 11.3 The fixed interval of 60 minutes between initiation of replication and cell division produces multiforked chromosomes in rapidly growing cells. Note that only the replication forks moving in one direction are shown; the chromosome actually is replicated symmetrically by two sets of forks moving in opposite directions on circular chromosomes.

At division (35/0 minutes), the cell receives a partially replicated chromosome. The replication fork continues to advance. At 10 minutes, when this "old" replication fork has not yet reached the terminus, initiation occurs at both origins on the partially replicated chromosome. The start of these "new" replication forks creates a **multiforked chromosome**.

▸ **multiforked chromosome** A bacterial chromosome that has more than one set of replication forks, because a second initiation has occurred before the first cycle of replication has been completed.

At 15 minutes—that is, at 20 minutes before the next division—the old replication fork reaches the terminus. Its arrival allows the two daughter chromosomes to separate; each of them has already been partially replicated by the new replication forks (which now are the only replication forks). These forks continue to advance.

At the point of division, the two partially replicated chromosomes segregate. This recreates the point at which we started. The single replication fork becomes "old," it terminates at 15 minutes, and 20 minutes later, there is a division. We see that the initiation event occurs 1 $^{25}/_{35}$ cell cycles before the division event with which it is associated.

The general principle of the link between initiation and the cell cycle is that as cells grow more rapidly (the cycle is shorter), the initiation event occurs an increasing number of cycles before the related division. There are correspondingly more chromosomes in the individual bacterium. This relationship can be viewed as the cell's response to its inability to reduce the periods of C and D to keep pace with the shorter cycle.

KEY CONCEPTS

- The doubling time of *E. coli* can vary over a 10x range, depending on growth conditions.
- It requires 40 minutes to replicate the bacterial chromosome (at normal temperature).
- Completion of a replication cycle triggers a bacterial division 20 minutes later.
- If the doubling time is ~60 minutes, a replication cycle is initiated before the division resulting from the previous replication cycle.
- Fast rates of growth therefore produce multiforked chromosomes.

CONCEPT AND REASONING CHECK

Since it takes 40 minutes to replicate the *E. coli* chromosome, how can the cell have a cell cycle of 35 minutes?

HISTORICAL PERSPECTIVES

John Cairns and the Bacterial Chromosome

The intact DNA molecules of most prokaryotes and many viruses are circular. The existence of circular cellular DNA molecules was not noticed for many years after the discovery of DNA because large DNA molecules usually break during isolation (eukaryotic chromosomes do not break because they are held together as chromatin). Eventually, many people began to believe that the small fragments seen in bacterial DNA preparations were broken fragments. In 1963, while he was working in Australia, the British physician and molecular biologist John Cairns used autoradiography to obtain the first image of an intact bacterial DNA chromosome. This technique takes advantage of the fact that radioactive tritium-labeled DNA emits β-particles, which, upon striking a photographic emulsion, produce an image of the DNA. Cairns cultured *E. coli* in a medium containing [^{3}H]thymidine, a specific precursor for DNA, and then gently released labeled DNA from the bacteria by treating the cells with a combination of lysozyme (an egg-white enzyme) to digest the bacterial cell wall and detergent to disrupt the cell membrane. After collecting the released DNA on a dialysis membrane (which can be used to separate molecules of different sizes), he coated the dried membrane with a photographic emulsion and stored the preparation in the dark for 2 months to allow sufficient time for the β-particles to produce an image. Analysis of the array of dark spots, which appeared after developing the emulsion, revealed that *E. coli* DNA is a double-stranded circular molecule with a contour length of approximately 1 mm, about 1000 times longer than the bacteria itself.

Cairns's work was important for a number of reasons. First, it demonstrated that the *E. coli* chromosome is a single molecule. Second, it demonstrated a circular cellular chromosome for the first time. And, third, it confirmed the prediction made by Watson and Crick that semiconservative replication proceeds via a Y fork. In fact, Cairns observed two Y forks. What could not be determined from this experiment was the nature of the two Y forks. Are they both moving around the chromosome (giving bidirectional replication) or is one moving (giving unidirectional replication) and the second acting as a fixed swivel?

11.3 The Septum Divides a Bacterium into Progeny That Each Contain a Chromosome

Chromosome segregation in bacteria is especially interesting because the DNA itself is involved in the mechanism for partition. (This contrasts with eukaryotic cells, in which segregation is achieved by the complex apparatus of mitosis.) The bacterial apparatus is quite accurate; however, **anucleate cells**, which lack a **nucleoid**, form <0.03% of a bacterial population.

- **anucleate cell** Bacteria that lack a nucleoid but are of similar shape to wild-type bacteria.
- **nucleoid** The structure in a prokaryotic cell that contains the genome. The DNA is bound to proteins and is not enclosed by a membrane.

E. coli cells are shaped as cylindrical rods that end in two curved poles. The bacterial chromosome is compacted into a dense protein-DNA structure called the *nucleoid*, which takes up most of the space inside the cell. It is not a disorganized mass of DNA; instead, specific DNA regions are localized to specific regions in the cell, and this positioning depends on the cell cycle. The arrangement is summarized in **FIGURE 11.4.** In newborn cells, the origin and terminus regions of the chromosome are at midcell. Following initiation, the new origins move toward the poles, or the 1/4 and 3/4 positions, and the terminus remains at midcell. Following cell division, the origins and terminus reorient to midcell.

The division of a bacterium into two daughter cells is accomplished by the formation of a **septum**, a structure that forms in the center of the cell as an invagination from the surrounding envelope. The septum forms an impenetrable barrier between the two parts of the cell and provides the site at which the two daughter cells eventually separate entirely. Two related questions address the role of the septum in division: What determines the location at which it forms? What ensures that the daughter chromosomes lie on opposite sides of it?

- **septum** The structure that forms in the center of a dividing bacterium, providing the site at which the daughter bacteria will separate. The same term is used to describe the cell wall that forms between plant cells at the end of mitosis.

The septum consists of the same components as the cell envelope: there is a rigid layer of peptidoglycan in the periplasm, between the inner and outer membranes. The peptidoglycan is made by polymerization of tri- or pentapeptide-disaccharide units in a reaction involving connections between both types of subunit (transpeptidation and transglycosylation). The rodlike shape of the bacterium is maintained by several activities, MreB, PBP2 and RodA.

MreB is a bacterial cytoskeletal element. The structure of MreB protein resembles that of the eukaryote protein actin. Indeed, MreB polymerizes to form filaments that traverse a helical path along the inner membrane following the long axis of the cell. This network forms a scaffold that recruits the biosynthetic machinery for peptidoglycan synthesis. RodA is a member of the SEDS family (SEDS stands for *s*hape, *e*longation, *d*ivision, and *s*porulation) present in all bacteria that have a peptidoglycan cell wall. Each SEDS protein functions together with a specific transpeptidase, which catalyzes the formation of the crosslinks in the peptidoglycan. PBP2 (*p*enicillin-*b*inding *p*rotein 2) is the transpeptidase that interacts with RodA. Mutations in the gene for either protein cause the bacterium to lose its extended shape and become round. This demonstrates the important principle that shape and rigidity can be determined by the simple extension of a polymeric structure.

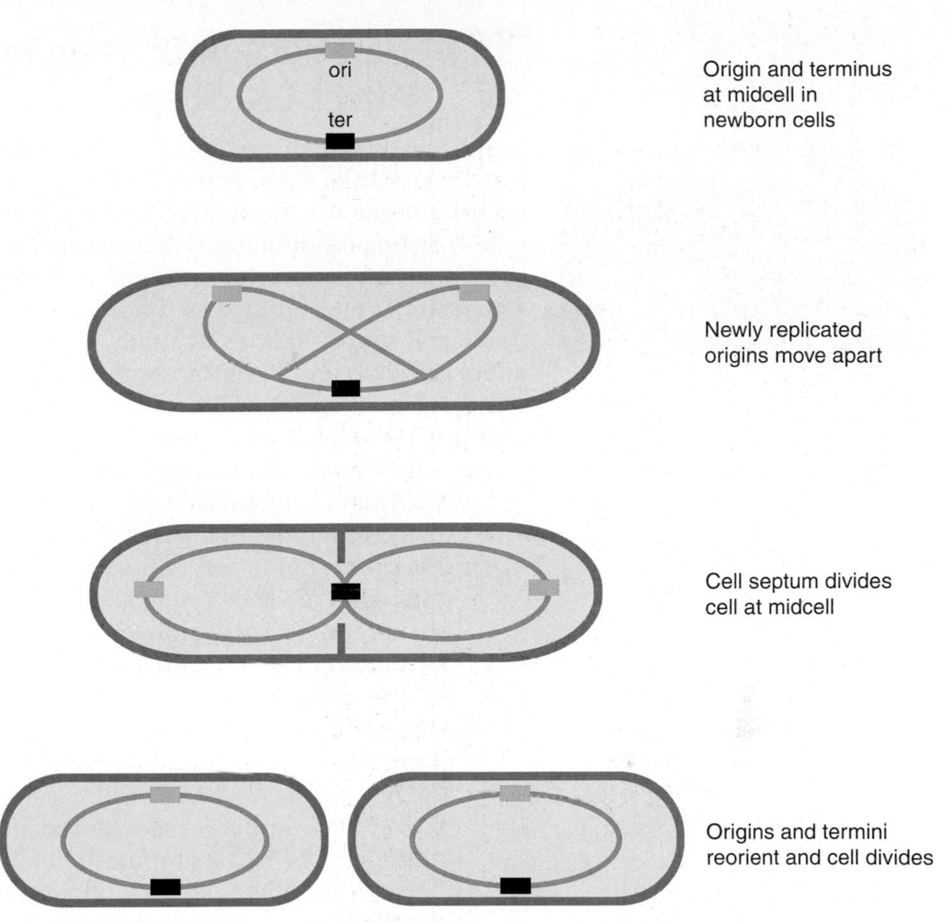

FIGURE 11.4 Attachment of bacterial DNA to the membrane could provide a mechanism for segregation.

Another enzyme, FtsZ, is responsible for generating the peptidoglycan in the septum (see *Section 11.5, FtsZ Is Necessary for Septum Formation*). The septum initially forms as a double layer of peptidoglycan, and the protein EnvA is required to split the covalent links between the layers so that the daughter cells may separate.

KEY CONCEPTS

- Bacterial chromosomes are specifically arranged and positioned inside cells.
- Septum formation is initiated midcell, 50% of the distance from the septum to each end of the bacterium.
- The septum consists of the same peptidoglycans that comprise the bacterial envelope.
- The rod shape of *E. coli* is dependent on MreB, PBP2, and RodA.
- FtsZ is necessary to recruit the enzymes needed to form the septum.

CONCEPT AND REASONING CHECKS

1. What would happen to a bacterial chromosome during cell division if it were not connected to the membrane?
2. How do eukaryotes divide their pairs of chromosomes into two separate cells?

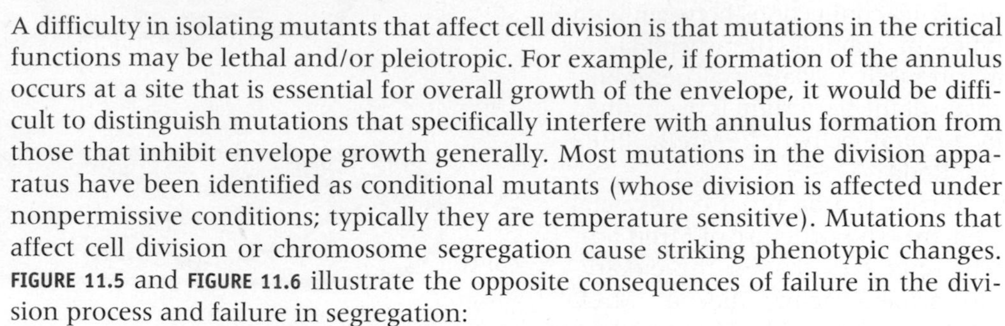

11.4 Mutations in Division or Segregation Affect Cell Shape

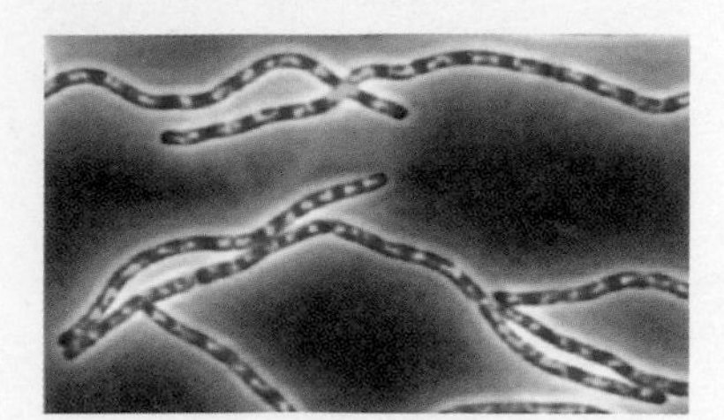

FIGURE 11.5 Failure of cell division under nonpermissive temperatures generates multinucleated filaments. Photo courtesy of Sota Hiraga, Kyoto University.

A difficulty in isolating mutants that affect cell division is that mutations in the critical functions may be lethal and/or pleiotropic. For example, if formation of the annulus occurs at a site that is essential for overall growth of the envelope, it would be difficult to distinguish mutations that specifically interfere with annulus formation from those that inhibit envelope growth generally. Most mutations in the division apparatus have been identified as conditional mutants (whose division is affected under nonpermissive conditions; typically they are temperature sensitive). Mutations that affect cell division or chromosome segregation cause striking phenotypic changes. **FIGURE 11.5** and **FIGURE 11.6** illustrate the opposite consequences of failure in the division process and failure in segregation:

- Long *filaments* form when septum formation is inhibited, but chromosome replication is unaffected. The bacteria continue to grow—and even continue to segregate their daughter chromosomes—but septa do not form. Thus the cell consists of a very long filamentous structure, with the nucleoids (bacterial chromosomes) regularly distributed along the length of the cell. This phenotype is displayed by *fts* mutants (named for temperature-sensitive filamentation), which identify a defect or multiple defects that lie in the division process itself.
- **Minicells** form when septum formation occurs too frequently or in the wrong place, with the result that one of the new daughter cells lacks a chromosome. The minicell has a rather small size and lacks DNA but otherwise appears morphologically normal. *Anucleate* cells form when segregation is aberrant; like minicells, they lack a chromosome, but because septum formation is normal, their size is unaltered. This phenotype is caused by *par* (partition) mutants (named because they are defective in chromosome segregation).

▸ **minicell** An anucleate bacterial (*E. coli*) cell produced by a division that generates a cytoplasm without a nucleus.

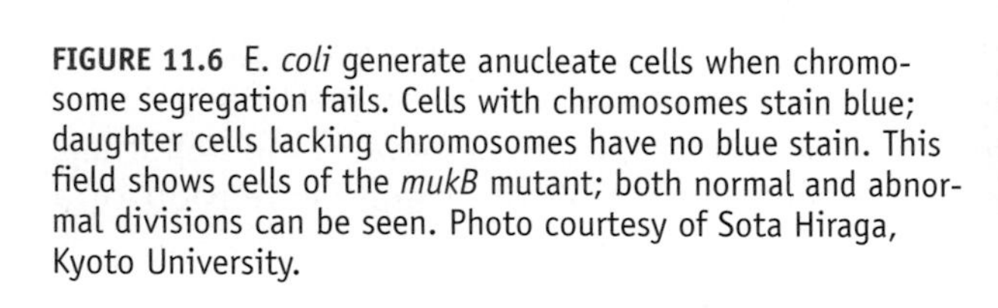

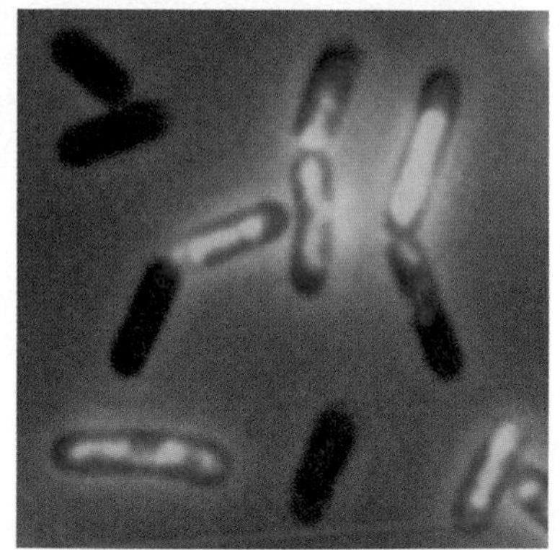

FIGURE 11.6 E. *coli* generate anucleate cells when chromosome segregation fails. Cells with chromosomes stain blue; daughter cells lacking chromosomes have no blue stain. This field shows cells of the *mukB* mutant; both normal and abnormal divisions can be seen. Photo courtesy of Sota Hiraga, Kyoto University.

KEY CONCEPTS

- *fts* mutants form long filaments because the septum that divides the daughter bacteria fails to form.
- Minicells form in mutants that produce too many septa; they are small and lack DNA.
- Anucleate cells of normal size are generated by partition mutants in which the duplicate chromosomes fail to separate.

CONCEPT AND REASONING CHECK

How could you isolate a loss of function mutation in an essential gene when mutating the gene can kill the cell?

11.5 FtsZ Is Necessary for Septum Formation

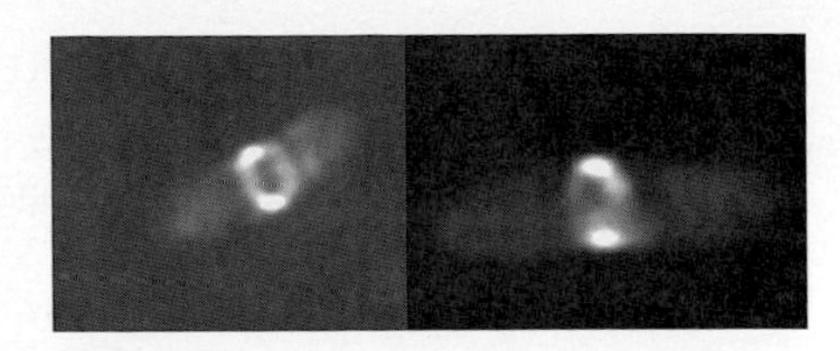

FIGURE 11.7 Immunofluorescence with an antibody against FtsZ shows that it is localized at the midcell. Photo courtesy of William Margolin, University of Texas Medical School at Houston.

The gene *ftsZ* plays a central role in division. Mutations in *ftsZ* block septum formation and generate filaments. Overexpression induces minicells by causing an increased number of septation events per unit cell mass. *ftsZ* mutants act at stages varying from the displacement of the periseptal annuli to septal morphogenesis. FtsZ recruits a battery of cell division proteins that are responsible for synthesis of the new septum.

FtsZ functions at an early stage of septum formation. Early in the division cycle, FtsZ is localized throughout the cytoplasm, but prior to cell division, FtsZ becomes localized in a ring around the circumference at the midcell position. The structure is called the **Z-ring** (or septal ring) which is shown in **FIGURE 11.7**. The formation of the Z-ring is the rate-limiting step in septum formation, and its assembly defines the position of the septum. In a typical division cycle, it forms in the center of the cell 1 to 5 minutes after division, remains for 15 minutes, and then quickly constricts to pinch the cell into two.

▸ **Z-ring** *See* septal ring.

The structure of FtsZ resembles tubulin, suggesting that assembly of the ring could resemble the formation of microtubules in eukaryotic cells. FtsZ has GTPase activity, and GTP cleavage is used to support the oligomerization of FtsZ monomers into the ring structure. The Z-ring is a dynamic structure, in which there is continuous exchange of subunits with a cytoplasmic pool.

Two other proteins needed for division, ZipA and FtsA, interact directly and independently with FtsZ. ZipA is an integral membrane protein that is located in the inner bacterial membrane. It provides the means for linking FtsZ to the membrane. FtsA is a cytosolic protein but is often found associated with the membrane. The Z-ring can form in the absence of either ZipA or FtsA, but it cannot form if both are absent. Both are needed for subsequent steps. This suggests that they have overlapping roles in stabilizing the Z-ring and perhaps in linking it to the membrane.

The products of several other *fts* genes join the Z-ring in a defined order after FtsA has been incorporated. They are all transmembrane proteins. The final structure is sometimes called the **septal ring**. It consists of a multiprotein complex that is presumed to have the ability to constrict the membrane. One of the last components to be incorporated into the septal ring is FtsW, which is a protein belonging to the SEDS family. *ftsW* is expressed as part of an operon with *ftsI*, which codes for a transpeptidase (also called PBP3 for penicillin-binding protein 3), a membrane-bound protein that has its catalytic site in the periplasm. FtsW is responsible for incorporating FtsI into the septal ring. This suggests a model for septum formation in which the transpeptidase activity then causes the peptidoglycan to grow inward, thus pushing the inner membrane and pulling the outer membrane.

▸ **septal ring** A complex of several proteins coded by *fts* genes of *E. coli* that forms at the midpoint of the cell. It gives rise to the septum at cell division. The first of the proteins to be incorporated is FtsZ, which gave rise to the original name of the Z-ring.

FtsZ is the major cytoskeletal component of septation. It is common in bacteria and is also found in chloroplasts. Mitochondria, which also share an evolutionary origin with bacteria, usually do not have FtsZ. Instead, they use a variant of the protein dynamin, which is involved in pinching off vesicles from membranes of eukaryotic cytoplasm. This functions from the outside of the organelle, squeezing the membrane to generate a constriction.

The common feature, then, in the division of bacteria, chloroplasts, and mitochondria is the use of a cytoskeletal protein that forms a ring around the organelle and either pulls or pushes the membrane to form a constriction.

KEY CONCEPTS

- The product of *ftsZ* is required for septum formation at pre-existing sites.
- FtsZ is a GTPase that forms a ring on the inside of the bacterial envelope. It is connected to other cytoskeletal components.

CONCEPT AND REASONING CHECK

How is it possible that a single gene like *ftsZ* can have mutations that have different phenotypes?

11.6 *min* and *noc/slm* Genes Regulate the Location of the Septum

Clues to the localization of the septum were first provided by minicell mutants. The original minicell mutation lies in the locus *minB;* deletion of *minB* generates minicells by allowing septation to occur at the poles as well as (or instead of) at midcell. As a result, the cell possesses the ability to initiate septum formation at midcell or at the poles, and the role of the wild-type *minB* locus is to suppress septation at the poles. The *minB* locus consists of three genes, *minC*, *-D*, and *-E*. Their roles are summarized in **FIGURE 11.8**. The products of *minC* and *minD* form a division inhibitor. (MinD is required to activate MinC, which prevents FtsZ from polymerizing into the Z-ring).

Expression of MinCD in the absence of MinE, or overexpression even in the presence of MinE, causes a generalized inhibition of division. The resulting cells grow as long filaments without septa (similar to those shown in Figure 11.5). Expression of MinE at levels comparable to MinCD confines the inhibition to the polar regions, thus restoring normal growth. The determinant of septation at the proper (midcell) site is, therefore, the ratio of MinCD to MinE.

The localization activities of the Min system are due to a remarkable dynamic behavior of MinD and MinE, which is shown in Figure 11.8. MinD, an ATPase, oscillates from one end of the cell to the other on a rapid time scale. MinD binds to and accumulates at the bacterial membrane at one pole of the cell, is released, and then rebinds to the opposite pole. The periodicity of this process takes about 30 seconds, so that multiple oscillations occur within one bacterial cell generation. MinC, which cannot move on its own, oscillates as a passenger protein bound to MinD. MinE forms a ring around the cell at the edge of the zone on MinD. The MinE ring moves towards MinD at the poles and is necessary for the release of MinD from the membrane. The MinE ring then disassembles and reforms at the edge of the MinD zone that forms at the opposite pole. MinD and MinE are each required for the dynamics of the other. The consequence of this dynamic behavior is that the concentration of the MinC inhibitor is lower at midcell and highest at the poles, which directs FtsZ assembly at midcell and inhibits its assembly at the poles.

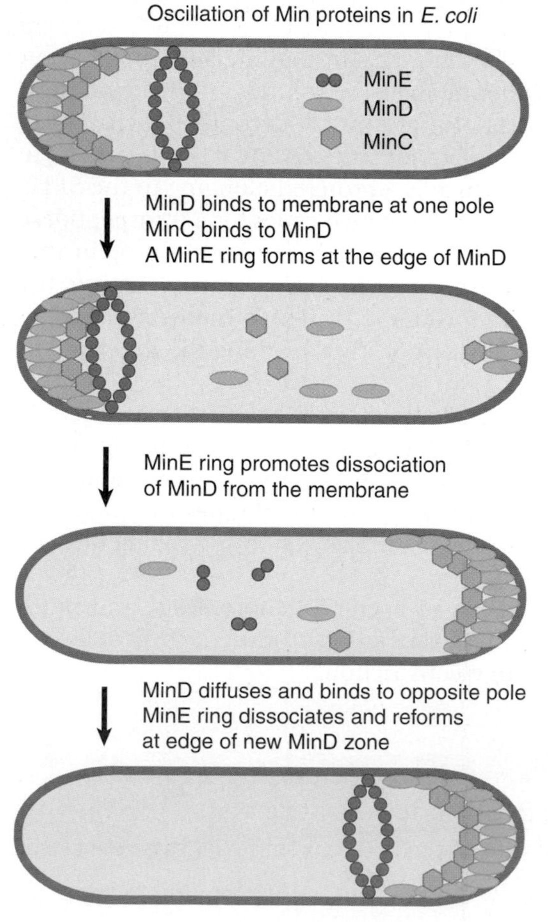

FIGURE 11.8 MinC/D is a division inhibitor whose action is confined to the poles by MinE.

Another process, called nucleoid occlusion, prevents Z-ring formation over the bacterial chromosome and thus prevents the septum from bisecting an individual chromosome at cell division. A protein called SlmA, which is an inhibitor of FtsZ, is necessary for nucleoid occlusion in *E. coli*. SlmA is a general DNA binding protein, so SlmA bound to the bacterial chromosome acts on FtsZ to prevent septum formation in this region of the cell. Because the bacterial nucleoid takes up a large volume of the cell, this process restricts Z-ring assembly to the limited nucleoid-free spaces at the poles and midcell. The combination of nucleoid occlusion and the Min system promotes the Z-rings to form, and thus cell division to occur, at midcell.

KEY CONCEPTS

- The location of the septum is controlled by *minC*, *-D*, and *-E* and by *noc/slm*.
- The number and location of septa is determined by the ratio of MinE/MinC,D.
- Dynamic movement of the Min proteins in the cell sets up a pattern in which inhibition of Z-ring assembly is highest at the poles and lowest at midcell.
- Slm/Noc proteins prevent septation from occurring in the space occupied by the bacterial chromosome.

CONCEPT AND REASONING CHECK

What would *minC* mutant bacteria look like?

11.7 Chromosomal Segregation May Require Site-Specific Recombination

After replication has created duplicate copies of a bacterial chromosome or plasmid, the copies can recombine. **FIGURE 11.9** demonstrates the consequences. A single intermolecular recombination event between two circles generates a dimeric circle. To counter this effect, cells often have **site-specific recombination** systems that act upon particular sequences to sponsor an intramolecular recombination that restores the monomeric condition.

▸ **site-specific recombination** Recombination that occurs between two specific sequences, as in phage integration/excision or resolution of cointegrated structures during transposition.

FIGURE 11.10 shows how this affects chromosomal segregation. If no recombination occurs, there is no problem, and the separate daughter chromosomes can segregate to the daughter cells. A dimer will be produced, however, if homologous recombination occurs between the daughter chromosomes produced by a replication cycle. If there has been such a recombination event, the daughter chromosomes cannot separate. In this case, a second recombination is required to achieve resolution in the same way as a plasmid dimer.

Most bacteria with circular chromosomes possess the Xer site-specific recombination system. In *E. coli*, this consists of two recombinases, XerC and XerD, which act on a 28-bp target site, called *dif*, which is located in the region of the chromosome where replication termination occurs (see *Section 13.15, Termination of Replication*). The use of the Xer system is related to cell division in an interesting way. The relevant events are summarized in **FIGURE 11.11**. XerC can bind to a pair of *dif* sequences and form a Holliday junction between them. The complex may form soon after the replication fork passes over the *dif* sequence, which explains how the two copies of the target sequence can find one another consistently. Resolution of the junction to give recombinants, however, occurs only in the presence of FtsK, a protein located in the septum that is required for chromosome segregation and cell division. In addition, the *dif*

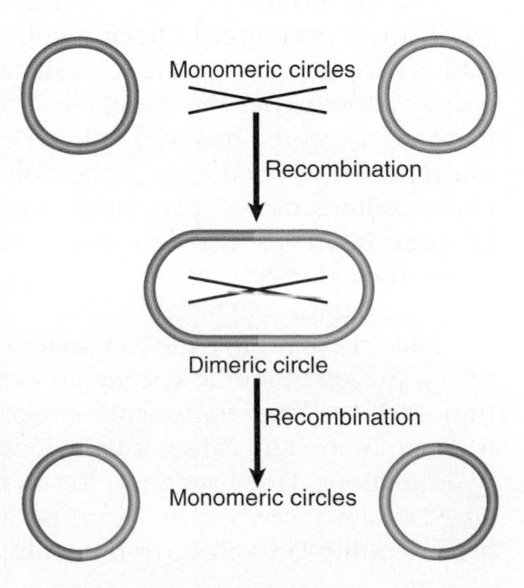

FIGURE 11.9 Intermolecular recombination merges monomers into dimers, and intramolecular recombination releases individual units from oligomers.

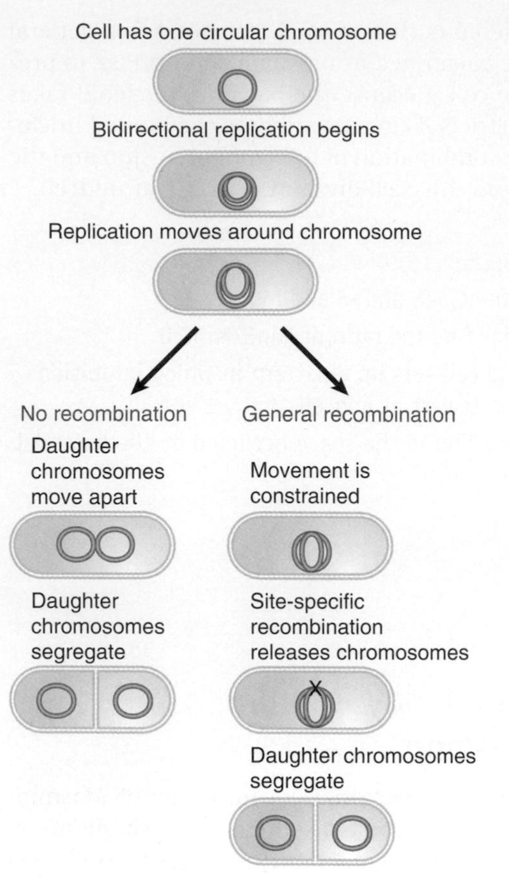

FIGURE 11.10 A circular chromosome replicates to produce two monomeric daughters that segregate to daughter cells. A generalized recombination event, however, generates a single monomeric molecule. This can be resolved into two monomers by a site-specific recombination.

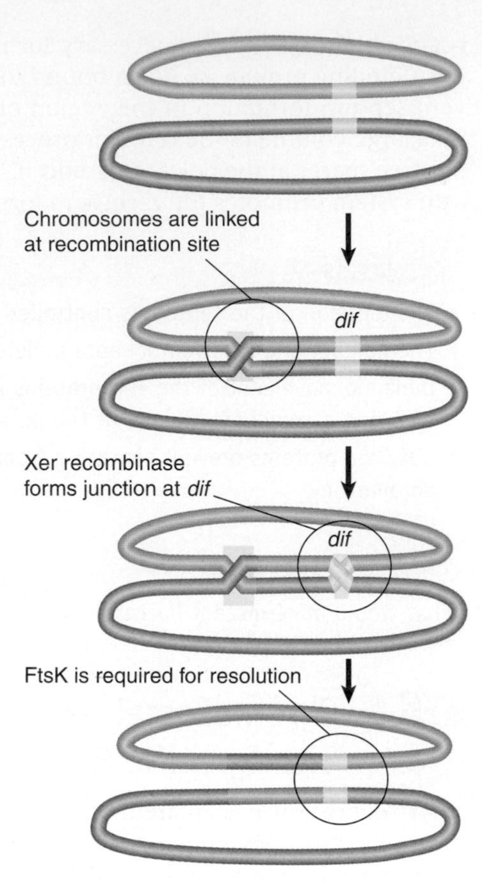

FIGURE 11.11 A recombination event creates two linked chromosomes. Xer creates a Holliday junction at the *dif* site, but can resolve it only in the presence of FtsK.

target sequence must be located in a specific region of ~30 kb; if it is moved outside of this region, it cannot support the reaction. Remember that the terminus region of the chromosome is located near the septum prior to cell division (see *Section 11.3, The Septum Divides a Bacterium into Progeny That Each Contain a Chromosome*).

The bacterium, however, wants to have site-specific recombination at *dif* only when there has already been a general recombination event to generate a dimer. (Otherwise the site-specific recombination would create a dimer!) How does the system know whether the daughter chromosomes exist as independent monomers or have been recombined into a dimer? One answer may be that segregation of chromosomes starts soon after replication. If there has been no recombination, the two chromosomes move apart from one another. The ability to move apart from one another, however, will be constrained if a dimer has been formed. This forces the terminus region to remain in the vicinity of the septum, where sites are exposed to the Xer system.

Bacteria that have the Xer system always have an FtsK homolog, and vice versa, which suggests that the system has evolved so that resolution is connected to the septum. FtsK is a large transmembrane protein. Its N-terminal domain is associated with the membrane and causes it to be localized to the septum. Its C-terminal domain has two functions. One is to cause Xer to resolve a dimer into two monomers. It also has an ATPase activity, which it can use to translocate along DNA *in vitro*. This could be used to pump DNA through the septum.

KEY CONCEPT

- The Xer site-specific recombination system acts on a target sequence near the chromosome terminus to recreate monomers if a generalized recombination event has converted the bacterial chromosome to a dimer.

CONCEPT AND REASONING CHECK

Why is recombination after replication only a problem for a bacterium that has a circular chromosome and not a linear chromosome?

11.8 Partitioning Separates the Chromosomes

Partition is the process by which the two daughter chromosomes find themselves on either side of the position at which the septum forms. Two types of event are required for proper partition:

- The two daughter chromosomes must be released from one another so that they can segregate following termination. This requires disentangling of DNA regions that are coiled around each other in the vicinity of the terminus. Most mutations affecting partition map in genes coding for topoisomerases—enzymes with the ability to pass DNA strands through one another. The mutations prevent the daughter chromosomes from segregating, with the result that the DNA is located in a single large mass at midcell. Septum formation then releases an anucleate cell and a cell containing both daughter chromosomes. This tells us that the bacterium must be able to disentangle its chromosomes topologically in order to be able to segregate them into different daughter cells.
- Mutations that affect the partition process itself are rare. We expect to find two classes: (1) *cis*-acting mutations should occur in DNA sequences that are the targets for the partition process; and (2) *trans*-acting mutations should occur in genes that code for the protein(s) that cause segregation, which could include proteins that bind to DNA or activities that control the locations in the cell. Both types of mutation have been found in the systems responsible for partitioning plasmids (see *Chapter 14, Extrachromosomal Replication*), but only *trans*-acting functions have been found in the bacterial chromosome. In addition, mutations in plasmid site-specific recombination systems increase plasmid loss (because the dividing cell has only one dimer to partition instead of two monomers), and therefore have a phenotype that is similar to partition mutants.

The original models for chromosome segregation suggested that the cell envelope grows by insertion of material between membrane-attachment sites of the two chromosomes, thus pushing them apart. In fact, the cell wall and membrane grow heterogeneously over the whole cell surface. Furthermore, the replicated chromosomes are capable of abrupt movements to their final positions at one and three quarters of the cell length. If protein synthesis is inhibited before the termination of replication, the chromosomes fail to segregate and remain close to the midcell position. When protein synthesis is allowed to resume, though, the chromosomes move to the quarter positions in the absence of any further envelope elongation. This suggests that an active process—one that requires protein synthesis—may move the chromosomes to specific locations.

Segregation is interrupted by mutations of the *muk* class, which give rise to anucleate progeny at a much increased frequency: both daughter chromosomes remain on the same side of the septum instead of segregating. Mutations in the *muk* genes are not lethal, and they may identify components of the apparatus that segregates the chromosomes. The gene *mukA* is identical to the gene for a known outer membrane protein (*tolC*) whose product could be involved with attaching the chromosome to

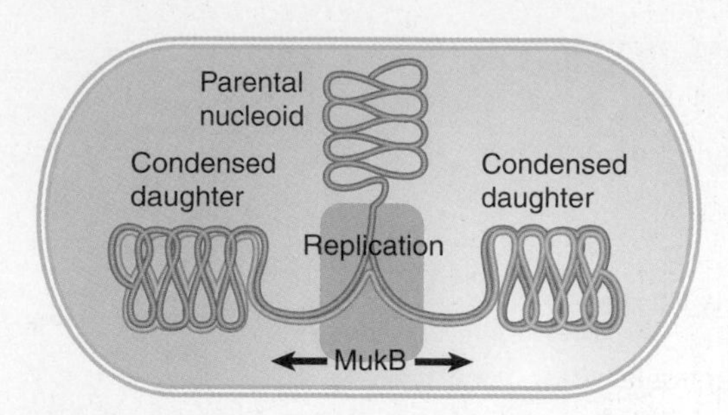

FIGURE 11.12 The DNA of a single parental nucleoid becomes decondensed during replication. MukB is an essential component of the apparatus that recondenses the daughter nucleoids.

the envelope. The gene *mukB* codes for a large (180 kD) globular protein, which has the same general type of organization as the two groups of *s*tructural *m*aintenance of *c*hromosomes (SMC) proteins that are involved in condensing and in holding together eukaryotic chromosomes. SMC-like proteins have also been found in other bacteria and mutations in their genes also increases the frequency of anuceate cells.

The insight into the role of MukB was the discovery that some mutations in *mukB* can be suppressed by mutations in *topA*, the gene that codes for topoisomerase I. MukB forms a complex with two other proteins, MukE and MukF, and the MukBEF complex is considered to be a condensin analogous to eukaryotic condensins. It uses a supercoiling mechanism to condense the chromosome. A defect in this function is the cause of failure to segregate properly. The defect can be compensated for by preventing topoisomerases from relaxing negative supercoils; the resulting increase in supercoil density helps to restore the proper state of condensation and thus allows segregation.

We still do not understand how genomes are positioned in the cell, but the process may be connected with condensation. **FIGURE 11.12** shows a current model. The parental genome is centrally positioned. It must be decondensed in order to pass through the replication apparatus. The daughter chromosomes emerge from replication, are disentangled by topoisomerases, and then are passed in an uncondensed state to MukBEF, which causes them to form condensed masses at the positions that will become the centers of the daughter cells.

A physical link, either directly or indirectly through chromosome-bound proteins, exists between bacterial DNA and the membrane. Bacterial DNA can be found in membrane fractions, which tend to be enriched in genetic markers near the origin, the replication fork, and the terminus. The proteins present in these membrane fractions may be affected by mutations that interfere with the initiation of replication. The growth site could be a structure on the membrane to which the origin must be attached for initiation.

KEY CONCEPTS

- Replicon origins are attached to the inner bacterial membrane.
- Chromosomes make abrupt movements from the midcenter to the one-quarter and three-quarter positions.

CONCEPT AND REASONING CHECK

Why do bacteria have to attach their chromosomes to the membrane for effective separation, whereas eukaryotes do not?

11.9 The Eukaryotic Growth Factor Signal Transduction Pathway

The vast majority of eukaryotic cells in a multicellular individual are not growing; that is, they are in the cell cycle stage of G0 (see Figure 11.1). Stem cells and most embryonic cells, however, are actively growing. A growing cell exiting mitosis has two choices: it can enter G1 and begin a new round of cell division, or it can stop dividing and enter G0, a quiescent stage and, if programmed, begin differentiation. This decision is controlled by growth factor hormones and their receptors.

In order for a cell to reenter the cell cycle from G0, or to continue to divide after M phase, it must be programmed to express the proper *growth factor receptor* gene. Elsewhere in the organism, typically in a master gland, the gene for the proper *growth factor* must be expressed. The **signal transduction pathway** is the biochemical process by which the growth factor signal to grow is communicated from its source outside of the cell into the nucleus to ultimately cause the cell to begin replication and growth. The general pathway that we will describe below is universal in eukaryotes ranging from yeast to humans.

▸ **signal transduction pathway** The process by which a stimulus or cellular state is sensed by and transmitted to pathways within the cell.

The genes that code for most of the elements of many signal transduction pathways are known as **oncogenes**, genes that when altered may cause cancer. As an example of this pathway, we will examine epidermal growth factor (EGF) and its receptor, EGFR. These two proteins and the genes that code for them are the first two elements in the pathway. EGF is a peptide hormone (as opposed to a steroid hormone like estrogen). The EGFR specifically binds EGF in a lock and key type of mechanism. EGFR is a one-pass membrane protein in the family known as *receptor tyrosine kinases* (RTK) as shown in **FIGURE 11.13**. The receptor has an external domain (which is outside the cell) that binds EGF, a short membrane-spanning domain, and an internal cytoplasmic domain with intrinsic tyrosine kinase activity (i.e., a domain that phosphorylates specific tyrosines).

▸ **oncogene** A gene that when mutated may cause cancer.

Hormone binding to receptor causes receptor dimerization, which leads to multiple cross-phosphorylation events of each receptor's cytoplasmic domain, as seen in Figure 11.13b; here, only one phosphorylation event is shown for simplicity. Each receptor phosphorylates the other on a set of tyrosine amino acid residues in the cytoplasmic domain. Each phosphorylated tyrosine (Tyr-P) serves as a docking site for a specific adapter protein to bind to the receptor, in this example, the Grb2/SOS

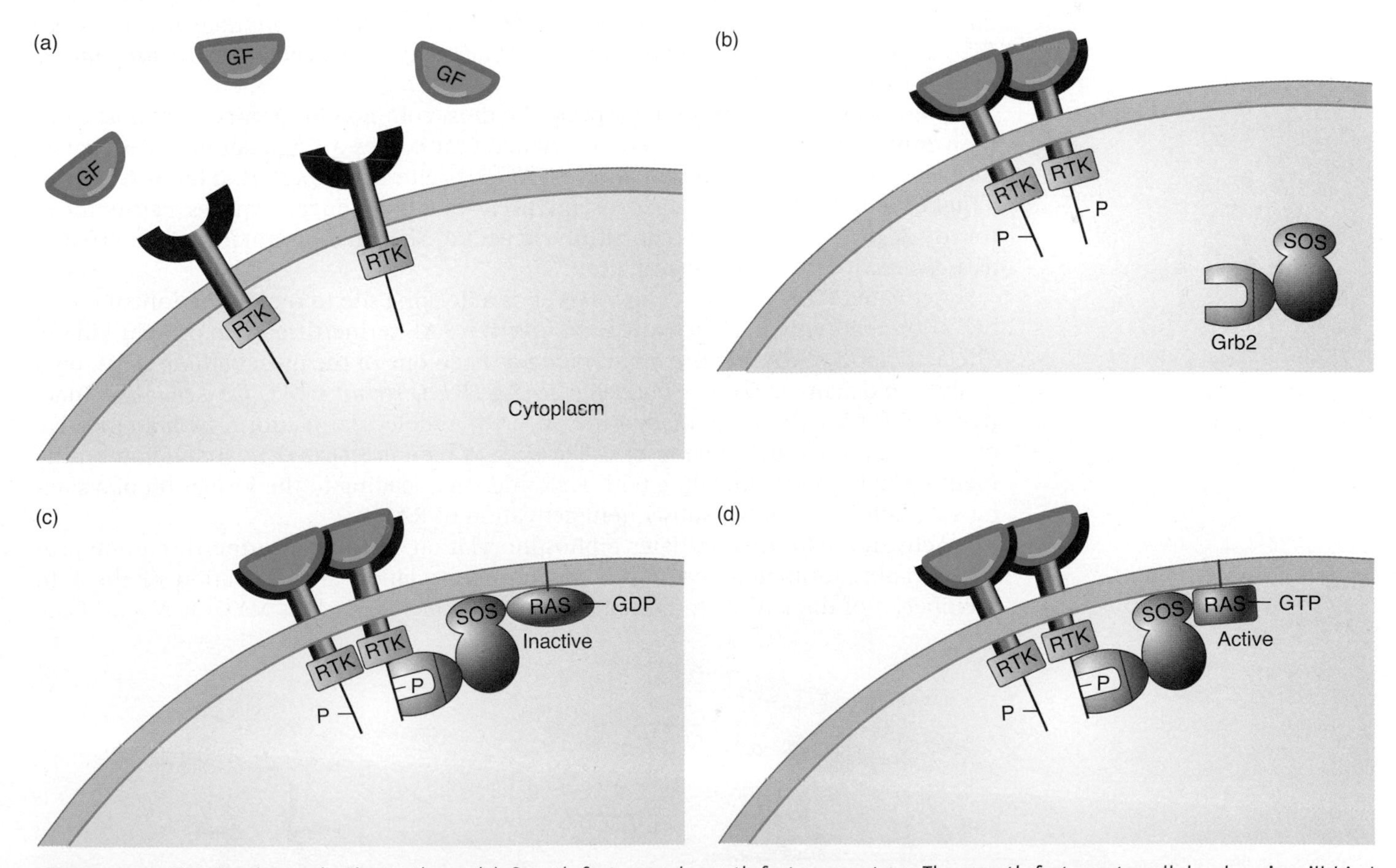

FIGURE 11.13 The signal transduction pathway (a) Growth factors and growth factor receptors: The growth factor extracellular domain will bind the growth factor (GF) in a lock-and-key fashion. The growth factor receptor intracellular domain contains an intrinsic protein kinase domain called RTK (*r*eceptor *t*yrosine *k*inase). (b) Growth factor binding to its receptor will cause receptor dimerization, leading to phosporylation of each cytoplasmic domain on tyrosine. The phosphotyrosine residues can serve as binding sites for proteins such as Grb2, shown here. (c) Grb2 binds the Tyr-P so that its binding partner SOS, a guanosine nucleotide exchange factor, is brought to the membrane and can activate the inactive RAS-GDP. (d) SOS removes the GDP, replacing it with GTP, activating RAS.

complex. We will examine a single pathway, but it is important to keep in mind that cells contain many different receptors active at the same time and each receptor has multiple docking sites for multiple proteins (and thus potentially multiple pathways). The reality is not a simple signal transduction pathway but rather an *information network*.

The third member of the signal transduction pathway is the RAS protein (encoded by the *ras* gene). RAS is a G-protein, a protein that binds a guanosine nucleotide, either GTP (for the active form of RAS) or GDP (for the inactive form). RAS is connected to the membrane by a prenylated (lipid) tail as seen in Figure 11.13c. To continue the flow of information from EGF/EGFR, inactive RAS must be converted from RAS-GDP to RAS-GTP by the cytoplasmic protein called SOS, a guanosine nucleotide exchange factor (GEF) that exchanges GTP for GDP. Its function is to remove the GDP from RAS and replace it with GTP as shown in Figure 11.13d and described in more detail shortly. RAS has a weak intrinsic phosphatase activity that slowly converts GTP to GDP. This means that growth factor must be present continually to propagate a signal.

To activate RAS, SOS must be specifically brought to the membrane in order to interact with RAS-GDP. It is the membrane phospholipids themselves which serve to unlock an autoinhibitory domain so that SOS can interact with RAS. SOS is in a complex with an adapter protein called Grb2, an interesting protein with two domains, an SH2 domain that binds Tyr-P (seen here on the activated EGFR) and an SH3 domain that binds other proteins containing another SH3 domain such as SOS. The specificity for binding to the receptor lies in the amino acids surrounding each Tyr-P. *The only function of the growth hormone is to cause dimerization of the growth factor receptor which leads to its phosphorylation, which in turn then leads to recruitment of SOS to the membrane to activate RAS.*

ras oncogene mutations are among the most common in tumors. The most common mutation is a single nucleotide change that causes a single amino acid change that alters the RAS protein structure. RAS^{ONC} has altered properties: it binds GTP with a higher affinity than GDP. The consequence is that it no longer requires a growth factor to trigger activation; it is constitutively active. This kind of mutation is referred to as a *dominant gain of function* mutation.

Activated RAS, RAS-GTP, now serves as a docking site to recruit the fourth member of the pathway to be activated: an inactive RAF serine/threonine protein kinase. The activation of RAF on the membrane has been one of the most baffling steps, only recently elucidated. *The only function of RAS-GTP is to recruit RAF to the membrane.* Inactive RAF is brought to the membrane on CNK, a molecular platform with an inactive protein kinase (or pseudokinase) called KSR as seen in **FIGURE 11.14**. Activation on the membrane involves unfolding both KSR and RAF, leading to the formation of a side-by-side heterodimer and subsequent activation of RAF.

Activated RAF then initiates a phosphorylation cascade of serine/threonine protein kinases, ultimately leading to the phosphorylation and activation of the fifth member(s) of the pathway, a set of transcription factors such as MYC, JUN, and FOS.

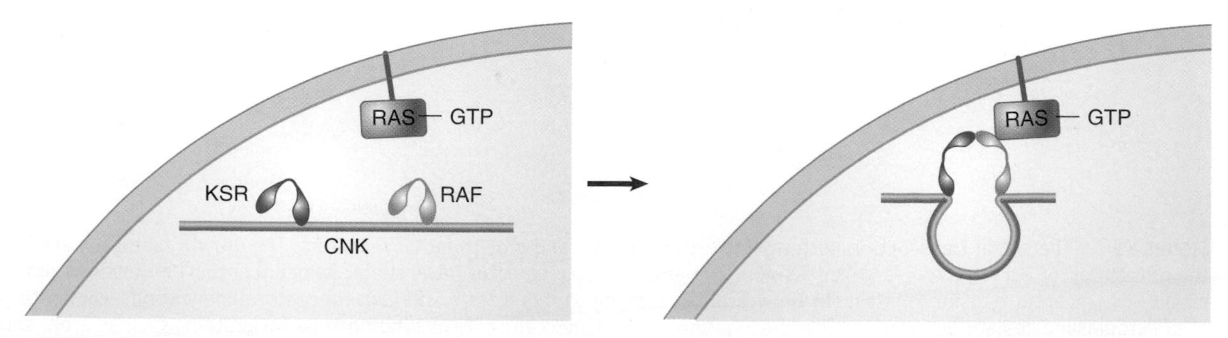

FIGURE 11.14 CNK serves as a molecular scaffold platform for KSR-mediated RAF activation on the membrane after being recruited by active RAS-GTP. Adapted from *Genes & Development* 20 (2006): 807.

This allows their entry into the nucleus to begin transcribing the genes needed to prepare for G1 and then S phase. Again note that this is a description of a single pathway within a network that has extensive crosstalk between members. In addition, this kinase cascade is negatively modulated by an extensive network of phosphatases.

KEY CONCEPTS

- The function of a growth factor is to cause dimerization of its receptor and subsequent phosphorylation of the cytoplasmic domain of the receptor.
- The function of the growth factor receptor is to recruit the exchange factor SOS to the membrane to activate RAS.
- The function of activated RAS is to recruit RAF to the membrane to become activated.
- The function of RAF is to initiate a phosphorylation cascade leading to the phosphorylation of a set of transcription factors that can enter the nucleus and begin S phase.

CONCEPT AND REASONING CHECKS

1. Why is the classical signal transduction pathway so complex; why not simply have the growth factor receptor activate the transcription factor?
2. What is the phenotype of a cell that is heterozygous for a RAS oncogenic mutation?

11.10 Checkpoint Control for Entry into S Phase: p53, A Guardian of the Checkpoint

Progression through the cell cycle, after the initial activation by growth factor, requires continuous growth factor presence and is tightly controlled by a second set of serine/threonine protein kinases called **cyclin-dependent kinases** (CDKs). The CDKs themselves are controlled in a very complex fashion as seen in **FIGURE 11.15**. They are inactive on their own and are activated by the binding of cell cycle-specific proteins called **cyclins**. This means that the CDKs can be synthesized in advance and left in the cytoplasm in an inactive state. In order for a cell to be allowed to progress from G1 to S phase, two major requirements must be met: the cell must have grown a specific amount in size, and there must be no DNA damage. The worst thing that a cell can do is to replicate damaged DNA.

▸ **cyclin-dependent kinases** A family of kinases that are inactive unless bound to a cyclin molecule. Most CDKs participate in cell cycle control.

▸ **cyclins** Proteins that bind and help activate cyclin-dependent kinases. Cyclin concentration varies throughout the cell cycle and their periodic availability plays an important role in regulating cell cycle progression.

In addition to cyclins, the CDKs are regulated by multiple phosphorylation events. One set of kinases in the general class of CKIs (*c*yclin-dependent *k*inase *i*nhibitors), the Wee1 family of serine/threonine kinases, inhibit the CDKs. (Wee1 kinases inhibit cell cycle progression and if they are mutated, premature cell cycle progression results in wee tiny cells.) There are multiple CKIs, including p21 and p27, that act on the CDK/cyclin complex itself (see *Section 11.11, Checkpoint Control For Entry into S Phase: Rb, A Guardian of the Checkpoint*). Another general class, the CAKs (*C*dk-*a*ctivating *k*inases) activates progression. This also means that *the balance of kinases and phosphatases regulates the activity of the CDKs*. (There is similar tight control at the G2-to-M transition and within various stages of mitosis and meiosis.) Cdc25 (*c*ell-*d*ivision *c*ycle) is a phosphatase required

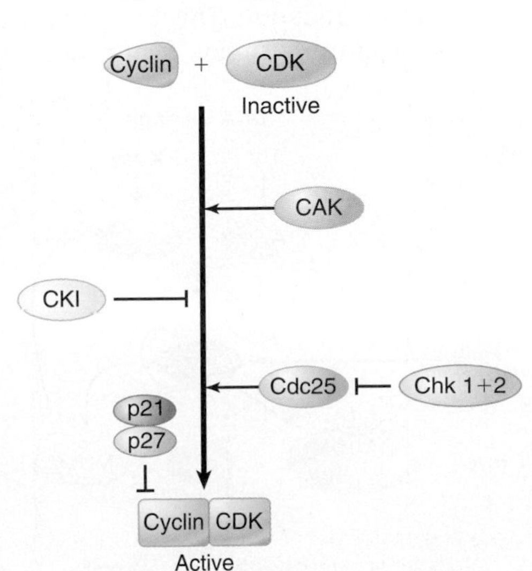

FIGURE 11.15 Formation of an active CDK requires binding to a cyclin. The process is regulated by positive and negative factors.

to remove an inhibitory phosphate group in order for the CDK/cyclin complex to become activated. Cdc25, in turn, is inhibited by Chk1 and 2 (described shortly; see Figure 11.16). The signal for entry into S phase is a positive signal controlled by negative regulators. There is no regulation at the S-to-G2 transition; it occurs when replication is complete.

In order to ensure that both the growth and no DNA damage requirements are met, the CDK/cyclin complexes work together with a pair of checkpoint proteins, the transcription factors p53 and Rb. These two proteins are in a class called **tumor-suppressor** proteins. As guardians of the cell cycle, these proteins ensure that the cell size and absence of DNA damage criteria are met. Even in the presence of an oncogenic mutant RAS protein, tumor suppressors will prevent the cell from progressing from G1 to S phase; they are the brakes on the cell cycle. Mutations in tumor-suppressor proteins allow damaged cells to replicate. These *recessive loss of function* mutations, especially in p53 and Rb genes, are the most common tumor-suppressor mutations in cancer; frequently both are seen together in the same cell.

▸ **tumor suppressor** Proteins that usually act by blocking cell proliferation or promoting cell death. Cancer may result when a tumor-suppressor gene is inactivated by a loss-of-function mutation.

The DNA damage checkpoint manned by p53 is the best understood and is shown in **FIGURE 11.16** The function of p53 is to relay information that DNA damage has occurred (especially profound damage such as radiation-induced double-strand breaks), from the protein kinases ATR and ATM in the nucleus through p21 to the next checkpoint protein, Rb (see Figures 11.15 and 11.17 and *Section 11.11, Checkpoint Control For Entry into S Phase: Rb, a Guardian of the Checkpoint*). The purpose is to prevent entry into S phase; that is, it ultimately causes cell cycle arrest until the damage can be repaired. In addition, in the event that damage is very extensive, p53 will initiate an alternate pathway, **apoptosis**, or programmed call death (PCD). p53 transcription is upregulated by growth factor stimulation, as the cell begins preparation for its trip through G1 and the important G1-to-S-phase transition (see also Figures 11.13 and 11.14).

▸ **apoptosis** The capacity of a cell to respond to a stimulus by initiating a signal transduction pathway that leads to its death through the activation of a characteristic set of reactions.

The p53 protein product is regulated by multiple complex pathways. The major negative regulator is a protein called MDM2. MDM2 transcription is stimulated by p53; it inhibits p53 in a negative feedback loop by targeting it to the ubiquitin-dependent proteosomal degradation pathway. It also binds to p53 and prevents it from activating transcription. DNA damage leads to phosphorylation of MDM2, which inhibits its ability to promote p53 degradation, allowing p53 levels to increase. Growth factor stimulation of cell cycle progression also leads to an increase in transcription of the p19^{arf} protein, which binds to and inhibits MDM2 inhibition of p53.

p53 is activated by DNA damage through the ATM/ATR relay kinases or by different kinds of stress (see *Chapter 16, Repair Systems,* Figure B20.1) through a protein kinase relay system from the nucleus which ultimately phosphorylates and stabilizes p53 from degradation. This leads to an increased level of p53 and activates its ability to serve as a transcription factor. Activated p53 turns on a number of genes, including those coding for GADD45 to stimulate DNA repair and p21/WAF-1, which inhibits the CDK/cyclin complexes, thus inhibiting Rb (see Figure 11.17). This leads to G1 arrest (or promotes apoptosis if the DNA damage is too great). DNA damage also independently activates a pair of protein kinases, Chk1 and Chk2, which phosphorylate and inhibit CDKs and phosphorylate and inhibit the phosphatase Cdc25, which is required to activate the CDKs.

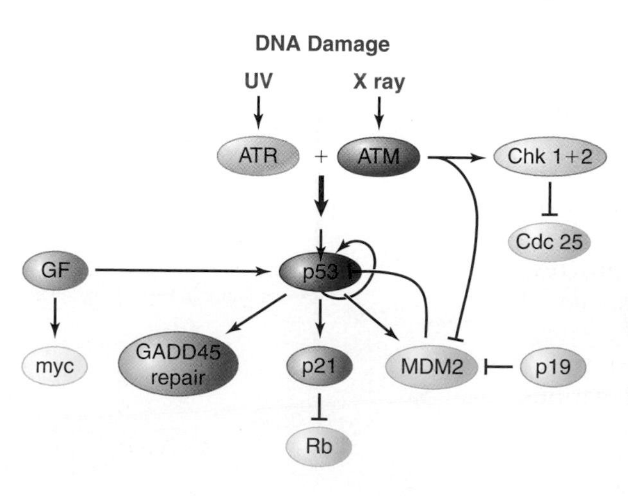

FIGURE 11.16 DNA damage pathway. p53 is activated by DNA damage. Activated p53 halts the cell cycle through Rb and stimulates DNA repair. p53 is regulated by a complex set of activators and inhibitors.

KEY CONCEPTS

- The tumor suppressor proteins p53 and Rb act as guardians of the cell.
- A set of serine/threonine protein kinases called CDKs control cell cycle progression.
- Cyclin proteins are required to activate CDK proteins.
- A set of inhibitor proteins negatively regulate the cyclin/CDKs.
- A set of activator proteins, CAKs positively regulate the cyclin/CDKs.

CONCEPT AND REASONING CHECKS

1. How does p53 communicate with Rb?
2. What is the phenotype of a cell that is heterozygous for a loss of function p53 mutant?

11.11 Checkpoint Control for Entry into S Phase: Rb, a Guardian of the Checkpoint

Let's now examine how an undamaged cell progresses through G1 (**FIGURE 11.17**). A growth factor signal, executed through a signal transduction pathway (see *Section 11.9, Eukaryotic Growth Factor Signal Transduction Pathway*), is required to turn on the gene for the first cyclin expressed, Cyclin D (humans have three different forms of this gene, while *Drosophila* has one). Its partners, already in the cytoplasm, are cdk4 and 6. Cyclins are the positive regulators of the CDK protein kinases; by themselves CDKs are inactive. Cyclin D is required for entry into S phase. Growth factor must be continuously present for at least the first half of G1.

The key for cell cycle progression is the tumor-suppressor protein Rb. Rb binds the transcription factor E2F which prevents it from turning on those genes required for progression through G1 and subsequent entry into S phase. Within G1 is a critical point called the **restriction point** (different in different species). Once through this point the cell is committed to continue onto S phase, even in the absence of continued growth factor. Ultimately, Rb integrates two sets of signals, one from p53 concerning DNA damage and a second concerning cell size (or growth of the cell) and is thus the key guardian of progression to S phase (see Figures 11.15 and 11.16).

▸ **restriction point** The point during G1 at which a cell becomes committed to division. (In yeast this point is known as START.)

In order for cell cycle progression to occur, Rb must be phosphorylated by the CDK-cyclin complex; phosphorylation of Rb releases E2F which allows it to activate transcription of its target genes. *The ultimate control of cell cycle progression is thus the regulation of CDK-cyclin activity by a set of inhibitor proteins, CKI and activator proteins.* p21, induced by DNA damage through p53 is a CKI (see Figure 11.16). Another major CKI is p27, a member of the Cip/Kip family. p27 is present in fairly high levels in G0 cells to prevent accidental

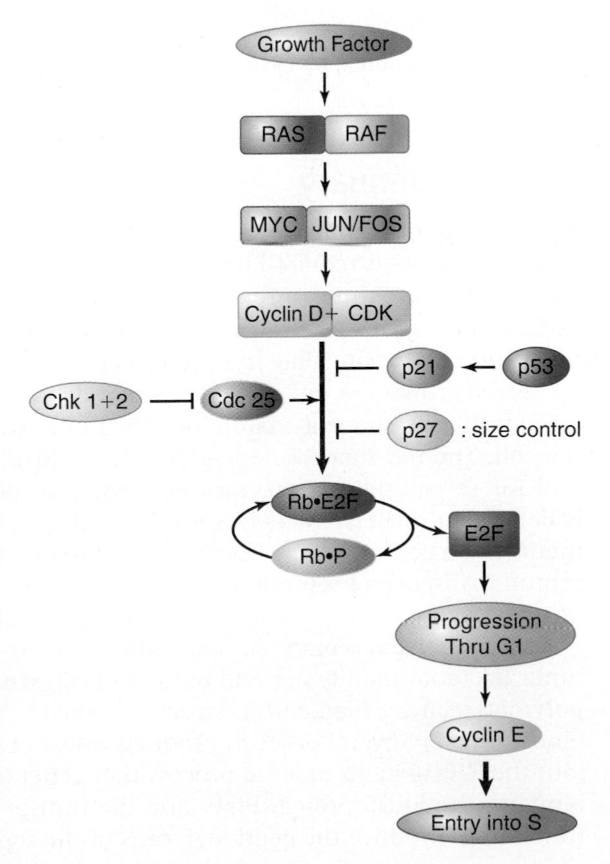

FIGURE 11.17 Growth factors are required to start the cell cycle and continue into S phase. The CDK-cyclin complex phosphorylates Rb to cause it to release the transcription factor E2F to go into the nucleus to turn on genes for progression through G1 and into S phase.

activation to G1. EGFR activation leads to its reduction. p27 is also activated in G1 by the cytokine TGF-β, a major growth inhibitor. p16/p19/INK/ARF is another major class of CKI proteins that controls Cyclin D activity (these two different proteins, INK and ARF, are made from the same gene using alternate promoters; see Figure 11.16).

Cell size or growth of the cell is monitored by a titration mechanism. A cell entering G1 has a fixed set of different classes of CKI proteins to prevent cell cycle progression. In order for the cell to progress through G1, this inhibition must be overcome by the synthesis, first, of more Cyclin D and then Cyclin E. *The length of G1 is determined by how long it takes to synthesize a sufficient level of cyclins to overcome the level of CKIs.*

During G1, three different cyclins are made. Cyclin D, as described before, is the first synthesized, activated by growth factor. As the cell continues to grow, the level of Cyclin D reaches a point of titrating out the CKIs and the CyclinD/cdk4/6 complex can begin phosphorylating Rb/E2F. This will cause Rb to begin to release E2F, which can then activate genes for progression through G1. Among the genes activated are the E2F gene itself, to increase the abundance of E2F protein, and Cyclin E. Cyclin E is activated by the middle of G1 and is also required for progression into S phase, adding to and amplifying the initial phosphorylation of Rb. Finally, just before S phase begins, Cyclin A is synthesized; it is also required for entry and continuation through S phase.

KEY CONCEPTS

- Rb is the major guardian of the cell cycle, integrating information about DNA damage and cell growth.
- Rb binds to an essential transcription factor, E2F, to prevent it from turning on the genes required for cell cycle progression.
- When Rb is phosphorylated by a Cyclin/CDK complex, it releases E2F to permit cell progression.

CONCEPT AND REASONING CHECK

Rb is only acted upon by Cyclin/CDK, how can it integrate DNA damage signals from p53?

11.12 Summary

A fixed time of 40 minutes is required to replicate the *E. coli* chromosome and a further 20 minutes is required before the cell can divide. When cells divide more rapidly than every 60 minutes, a replication cycle is initiated before the end of the preceding division cycle. This generates multiforked chromosomes. The initiation event occurs once and at a specific time in each cell cycle.

E. coli grows as a rod-shaped cell that divides by formation of a septum that forms at midcell. The shape is maintained by an envelope of peptidoglycan that surrounds the cell. The rod shape is dependent on the MreB actinlike protein that forms a scaffold for recruiting the enzymes necessary for peptidoglycan synthesis. The septum is dependent on FtsZ, which is a tubulinlike protein that can polymerize into a filamentous structure called a Z-ring. FtsZ recruits the enzymes necessary to make the septum. Absence of septum formation generates multinucleated filaments; an excess of septum formation generates anucleate minicells.

Many transmembrane proteins interact to form the septum. ZipA is located in the inner bacterial membrane and binds to FtsZ, which is a tubulinlike protein that can polymerize into a filamentous structure called a Z-ring. FtsA is a cytosolic protein that binds to FtsZ. Several other *fts* products, most of which are transmembrane proteins, join the Z-ring in an ordered process that generates a septal ring. The last proteins to bind are the SEDS protein FtsW and the transpeptidase FtsI (PBP3), which together function to produce the peptidoglycans of the septum.

Bacteria have site-specific recombination systems that regenerate pairs of monomers by resolving dimers created by general recombination. The Xer system acts on a target sequence located in the terminus region of the chromosome. The system is active only in the presence of the FtsK protein of the septum, which may ensure that it acts only when a dimer needs to be resolved.

The eukaryotic cell cycle is governed by a complex set of regulatory factors. Licensing to begin the cell cycle, as opposed to enter or remain in G0, requires a positive growth factor signal to initiate the signal transduction pathway. This biochemical relay of information from outside the cell ultimately results in the activation of a set of transcription factors in the cytoplasm. These can then enter the nucleus to begin the transcription of genes required for the progression through G1 and ultimate entry into S phase and replication of the chromosomes.

The cell cycle, that is, progression from G1 to S phase and beyond, is regulated by phosphorylation events carried out by a set of protein kinases, the CDKs, and balanced by phosphatases. The kinases are controlled by a set of cell cycle stage specific proteins called cyclins that bind to the CDKs and convert an inactive CDK into an active kinase. Progression through G1 into S phase is allowed only if there is no DNA damage and the cell has grown a sufficient amount in size. These two requirements are enforced by a pair of tumor-suppressor proteins. p53 guards the DNA damage checkpoint to prevent the replication of damaged DNA. Rb is the guardian that integrates DNA damage and cell-size information to ultimately control whether the gene regulator E2F is allowed to begin transcription.

CHAPTER QUESTIONS

1. How many replication forks would be present in the *E. coli* chromosome for cells that are growing under optimal conditions?
   A. one
   B. two
   C. four
   D. more than four
2. How long does it take to replicate the full *E. coli* chromosome at normal growth temperature?
   A. 10 minutes
   B. 20 minutes
   C. 40 minutes
   D. 60 minutes
3. What is the fixed interval of time required between initiation of DNA replication and cell division in *E. coli*?
   A. 10 minutes
   B. 20 minutes
   C. 40 minutes
   D. 60 minutes
4. The bacterial *ftsZ* gene is required for:
   A. septum formation.
   B. periseptal annulus formation and localization.
   C. DNA replication.
   D. partitioning of DNA.
5. The location of septum formation in a bacterial cell is controlled by the:
   A. *par* locus.
   B. *min* locus.
   C. *xer* locus.
   D. *fts* locus.

6. Current evidence suggests that the bacterial chromosome is:
   - **A.** freely soluble in the cytoplasm of the cell.
   - **B.** attached to a specific cytoskeleton structure in the cytoplasm.
   - **C.** attached to random sites on the inner membrane of the cell.
   - **D.** attached to one specific site on the inner membrane of the cell.
7. Occasionally two new daughter molecules can undergo homologous recombination to form a dimer of the bacterial genome. How does the bacterial genome deal with this to undergo cell division?
   - **A.** It cannot; it is lethal and the cell dies.
   - **B.** One daughter cell receives the dimer and survives; the other daughter cell lacks DNA.
   - **C.** Site-specific DNA recombination separates the two monomers, one for each new cell.
   - **D.** A reversal of the homologous recombination event occurs at random sites to separate the daughter molecules.
8. p53 communicates DNA damage to Rb through:
   - **A.** p21
   - **B.** p27
9. Growth of the cell and cell size information is communicated to Rb through:
   - **A.** p21
   - **B.** p27

KEY TERMS

anucleate cell
apoptosis
checkpoint
cyclin
cyclin-dependent kinase
doubling time
minicell
multiforked chromosome
nucleoid
oncogene
restriction point
S phase
septal ring
septum
signal transduction pathway
site-specific recombination
topoisomerase
tumor suppressor gene
Z-ring

FURTHER READING

Ghosh, S. K., Hajra, S., Paek, A., and Jayaram, M. (2006). Mechanisms for chromosomal and plasmid segregation. *Annu. Rev. Biochem.* **75**, 211–241.

Haeusser, D. P., and Levin, P. A. (2008). The great divide: coordinating cell cycle events during bacterial growth and division. *Curr. Opin. Microbiol.* **11**, 94–99.

Lutkenhaus, J. (2007). Assembly dynamics of the bacterial MinCDE system and special resolution of the Z ring. *Annu. Rev. Biochem.* **76**, 539–562.

Sancar, A., Lindsey-Boltz, L. A., Unsal-Kacmaz, K., and Linn, S. (2004). Molecular mechanisms of mammalian DNA repair and the DNA damage checkpoints. *Annu. Rev. Biochem.* **73**, 39–85.

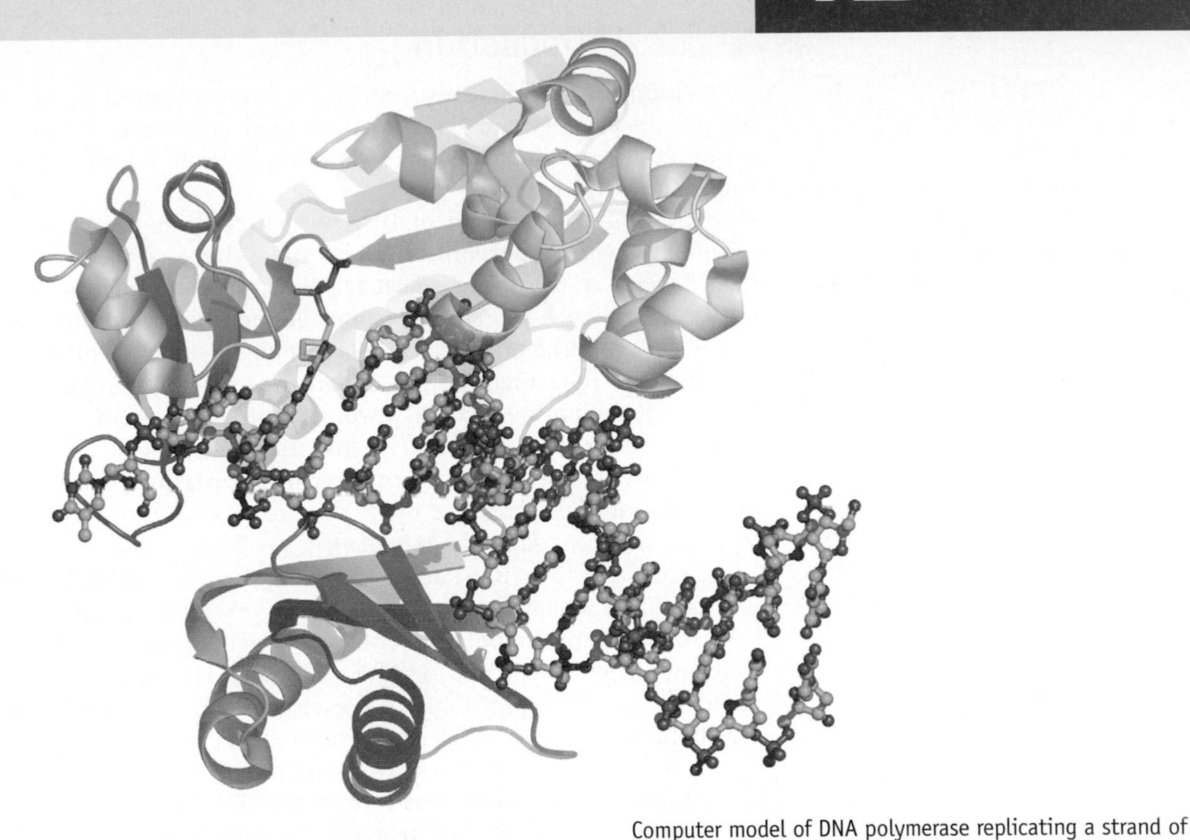

Computer model of DNA polymerase replicating a strand of DNA (across center). The secondary structure of the DNA polymerase and the primary structure of the DNA molecule are shown. DNA polymerases are enzymes that synthesize strands of DNA from a complementary template strand during DNA replication. This molecule is in the Y family of DNA polymerases, which are translesion synthesis polymerases, that is, they are able to replicate damaged areas of DNA that stall other DNA polymerases. The replication is not always accurate and can lead to mutagenesis or cancer. © Laguna Design/Photo Researchers, Inc.

The Replicon: Initiation of Replication

CHAPTER OUTLINE

12.1 Introduction

Replication of DNA is a key regulatory event in cell division. Every time that a cell divides, its entire set of DNA must be replicated once and only once. This is accomplished by controlling the initiation of replication, which occurs only at a unique site called the **origin,** or *ori*.

- **origin** A sequence of DNA at which replication is initiated.
- **replicon** A unit of the genome in which DNA is replicated. Each contains an origin for initiation of replication.

The origin is the unit of DNA which controls the initiation of replication. The origin lies within the **replicon**, the DNA that is replicated whenever an initiation event occurs at the origin. FIGURE 12.1 illustrates the general nature of the relationship between replicons and chromosomes or genomes in prokaryotes and eukaryotes.

A genome in a prokaryotic cell (usually a single circular molecule of DNA) has a single replication origin and thus constitutes a single replicon. This means that replication of the entire bacterial chromosome depends on a single initiation event that occurs at the unique origin. The initiation event occurs once for every cell division and is known as **single-copy replication control**. The frequency of initiation at the bacterial origin is controlled by its state of methylation (see *Section 12.4, Methylation of the Bacterial Origin Regulates Initiation*).

- **single-copy replication control** A control system in which there is only one copy of a replicon per unit bacterium. The bacterial chromosome and some plasmids have this type of regulation.
- **multicopy replication control** Replication occurs when the control system allows the plasmid to exist in more than one copy per individual bacterial cell.

Bacteria may contain additional genetic information in the form of plasmids. *A plasmid is an autonomous circular DNA that constitutes a separate replicon.* Plasmid replicons may show single copy control, which means that they replicate once every time the bacterial chromosome replicates, or they may be under **multicopy replication control**, when they are present in a greater number of copies than the bacterial chromosome. Each phage or virus DNA also constitutes a replicon and thus is able to initiate many times during an infection cycle. Perhaps a better way to view the prokaryotic replicon, therefore, is to reverse the definition: *any DNA molecule that contains an origin can be replicated autonomously in the cell* (see Chapter 14, *Extrachromosomal Replication*).

A major difference in the organization of bacterial and eukaryotic genomes is seen in their replication. Each eukaryotic chromosome (usually a very long linear molecule of DNA) contains a large number of replicons spaced unevenly throughout the chromosome. We will see later why the origin is placed in the center of the replicon. Like the single replicon that constitutes the bacterial chromosome, each eukaryotic origin "fires" once and only once in each cell cycle. There are exceptions to this rule for situations like regulated gene amplification and polytene chromosome formation. Eukaryotic origin usage is controlled by the ability of regulator proteins to bind to it (see *Section 12.8, Licensing Factor Controls Eukaryotic Rereplication*). The replicons of a eukaryotic genome are not active simultaneously but are activated over a fairly protracted period, which is called the *S phase* (see *Chapter 11, Replication Is Connected to the Cell Cycle*). This implies the existence of additional levels of control to ensure that each replicon does fire once and does not fire a second time. There are additional rules that control when each origin is allowed to fire. Because many replicons are activated independently, another signal must exist to indicate when the entire process has been completed. In contrast to nuclear chromosomes, which have a single-copy type of control, the DNA of mitochondria and chloroplasts may be regulated more like plasmids that exist in multiple copies per bacterium. There are multiple copies of each organelle per cell, and each organelle contains multiple copies of DNA. The control of organelle DNA replication must be coordinated with the cell cycle.

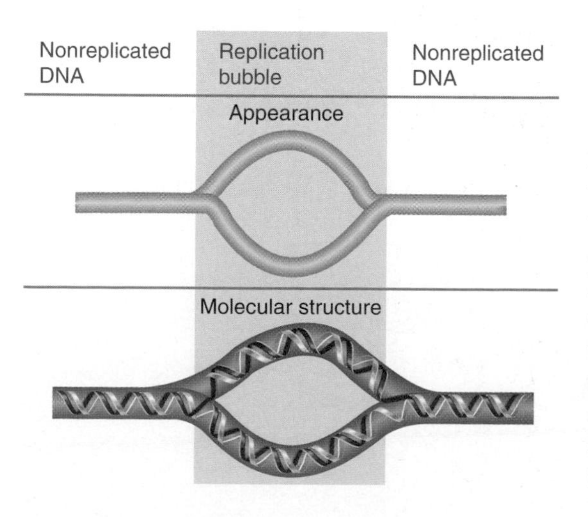

FIGURE 12.1 Replicated DNA is seen as a replication bubble flanked by nonreplicated DNA.

In all these systems, the key question is to define the DNA sequences that function as origins and to determine how they are recognized by the appropriate proteins of the replication apparatus. We start by considering the basic construction of replicons and the various forms that they take in bacteria and eukaryotic cells. We will consider how replicons are controlled during the process of initiation of replication. In the next chapter, *Chapter 13, DNA Replication*, we examine the biochemistry of DNA synthesis—the elongation and termination processes.

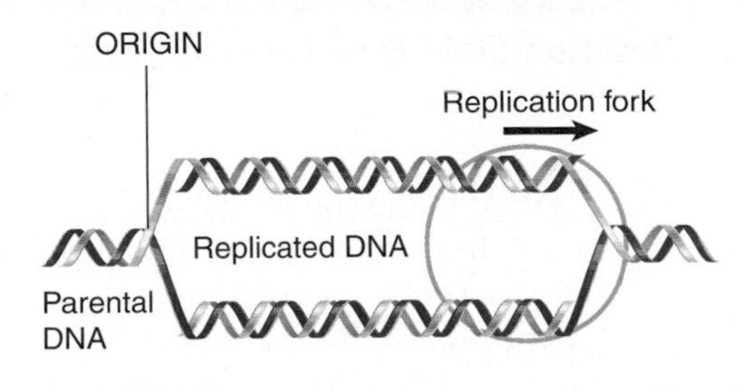

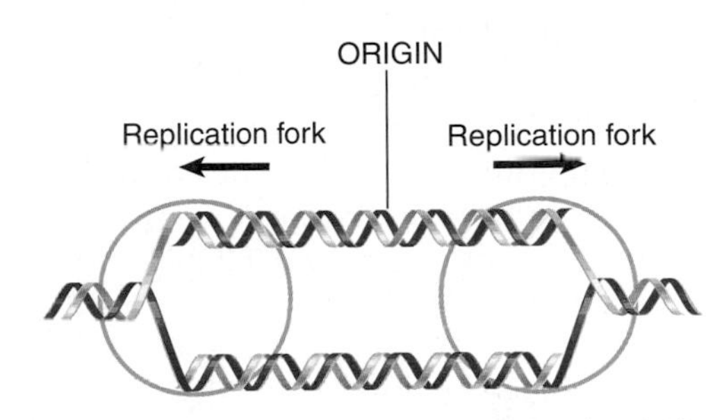

FIGURE 12.2 Replicons may be unidirectional or bidirectional, depending on whether one or two replication forks are formed at the origin.

12.2 An Origin Usually Initiates Bidirectional Replication

Replication starts at an origin by separating or melting the two strands of the DNA duplex. **FIGURE 12.2** shows that each of the parental strands then acts as a template to synthesize a complementary daughter strand. This model of replication, in which a parental duplex gives rise to two daughter duplexes, each containing one original parental strand and one new strand, is called **semiconservative replication** (see the accompanying box, *Historical Perspectives: The Meselson-Stahl Experiment*, showing semiconservative replication).

A molecule of DNA engaged in replication has two types of regions. **FIGURE 12.3** shows that when replicating DNA is viewed by electron microscopy, the replicated region appears as a **replication bubble** within the nonreplicated DNA. The nonreplicated region consists of the parental duplex; this opens into the replicated region where the two daughter duplexes have formed. When a replicon is circular, the presence of a bubble forms the θ (theta) structure.

The point at which replication occurs is called the **replication fork** (sometimes also known as the **growing point**). *A replication fork moves sequentially along the DNA from its starting point at the origin.* The origin may be used to start either **unidirectional replication** or **bidirectional replication** as shown in Figure 12.2. The type of event is determined by whether one or two replication forks set out from the origin. In unidirectional replication, only one replication fork leaves the origin and proceeds along the DNA; the other fork acts as a fixed swivel. In bidirectional replication, two replication forks are formed; they proceed away from the origin in opposite directions. The appearance of a replication bubble does not distinguish between unidirectional and bidirectional replication.

The form of replication used by both bacteria and eukaryotic nuclear chromosomes is bidirectional replication, as shown in Figure 12.2. As the replication forks continue to move apart, the replication bubble increases in size and eventually it becomes larger than the nonreplicated region. When a replicon is circular, the presence of a bubble forms the θ structure shown in Figure 12.3.

- **semiconservative replication** Replication accomplished by separation of the strands of a parental duplex, with each strand then acting as a template for synthesis of a new complementary strand.
- **replication bubble** A region in which DNA has been replicated within a longer, unreplicated region.
- **replication fork** The point at which strands of parental duplex DNA are separated so that replication can proceed. A complex of proteins including DNA polymerase is found there.
- **growing point** *See* replication fork.
- **unidirectional replication** The movement of a single replication fork from a given origin.
- **bidirectional replication** A system in which an origin generates two replication forks that proceed away from the origin in opposite directions.

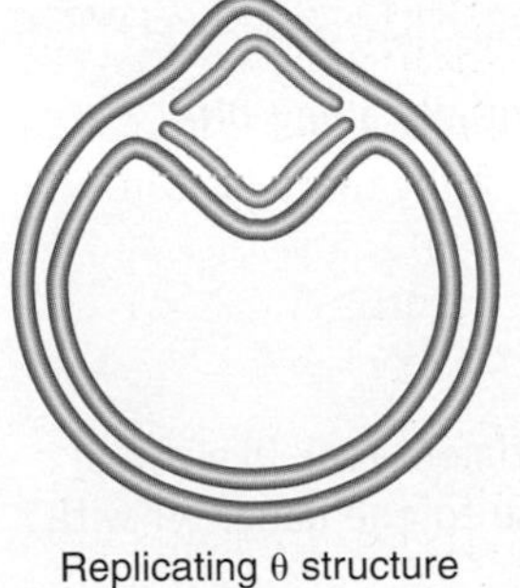

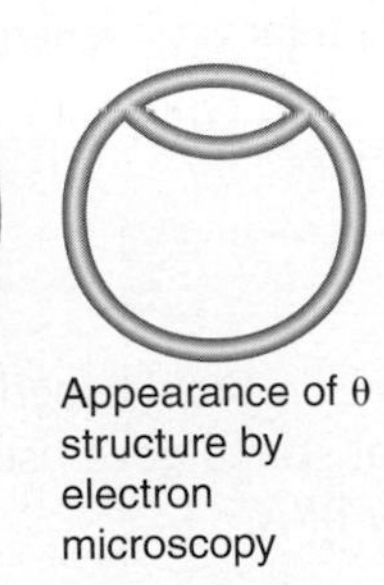

FIGURE 12.3 A replication bubble forms a θ structure in circular DNA.

HISTORICAL PERSPECTIVES

The Meselson-Stahl Experiment

In theory, DNA could be replicated by a number of mechanisms other than the semiconservative mode, but the reality of semiconservative replication was demonstrated experimentally by Matthew Meselson and Franklin Stahl in 1958. The experiment made use of a newly developed high-speed centrifuge (an *ultracentrifuge*) that could spin a solution so fast that molecules differing only slightly in density could be separated. In their experiment, the heavy ^{15}N isotope of nitrogen was used to alter the density of DNA and allow the parental and daughter DNA molecules to be separated. DNA isolated from the bacterium *E. coli* grown in a medium containing ^{15}N as the only available source of nitrogen is denser than DNA from bacteria grown in medium with the normal ^{14}N isotope. These DNA molecules can be separated in an ultracentrifuge, because they have about the same density as a very concentrated solution of cesium chloride (CsCl).

When a CsCl solution containing DNA is centrifuged at high speed, the Cs^+ ions gradually sediment toward the bottom of the centrifuge tube. This movement is counteracted by diffusion (the random movement of molecules), which prevents complete sedimentation. At equilibrium, a linear gradient of CsCl concentration (and, therefore, density) is present, increasing in cesium concentration and density from the top of the centrifuge tube to the bottom. DNA in the tube moves upward or downward to a position in the gradient at which the local density of the solution is equal to its own density. At equilibrium, a mixture of ^{14}N-containing ("light") and ^{15}N-containing ("heavy") *E. coli* DNA will separate into two distinct zones in a density gradient, even though they differ only slightly in density. DNA from *E. coli* containing ^{14}N in the purine and pyrimidine rings has a density of 1.708 g/cm^3, whereas DNA with ^{15}N in the purine and pyrimidine rings has a density of 1.722 g/cm^3. These molecules can be separated because a solution of 6.6 molar CsCl has a density of 1.700 g/cm^3. When spun in a centrifuge, the CsCl solution forms a gradient of density that brackets the densities of the light and heavy DNA molecules. It is for this reason that the separation technique is called **equilibrium density-gradient centrifugation**.

Imagine an experiment using bacteria grown for many generations in a ^{15}N-containing medium so that all parental DNA strands are "heavy." At the beginning of the experiment, the cells are transferred to a ^{14}N-containing medium so that newly synthesized DNA strands are "light." Duplex DNA is isolated from samples of cells taken from the culture at intervals, and equilibrium density-gradient centrifugation is carried out to determine the density of the molecules. With semiconservative replication, the expected result of the experiment is as follows. After one round of replication, each duplex consists of one heavy and one light strand, so all daughter molecules are of intermediate density. After two rounds of replication, the duplexes containing an original parental strand are again intermediate in density, but now there are an equal number of duplexes consisting of two light strands, so two bands differing in density result from centrifugation.

The actual result of the Meselson-Stahl experiment is shown in **FIGURE B12.1**. The bottom part of the figure is that of the CsCl solution in centrifuge tubes oriented vertically. The positions of the DNA molecules in the density gradient are, therefore, indicated by dark bands.

At the start of the experiment ("Parental DNA"), all the DNA was heavy (^{15}N). After the transfer to ^{14}N, a band of lighter density began to appear, and it gradually became more prominent as the cells replicated their DNA and divided. After one generation of growth (one round of replication of the DNA molecules and a doubling of the number of cells), all the DNA had a "hybrid" density exactly intermediate between the densities of ^{15}N-DNA and ^{14}N-DNA. The observation of molecules with a hybrid density indicates that the replicated molecules contain equal amounts of the two nitrogen isotopes.

▸ **equilibrium density-gradient centrifugation** A gradient method used to separate macromolecules on the basis of differences in their density. For DNA, it is prepared from a heavy soluble compound such as CsCl.

KEY CONCEPTS

- A replication fork is initiated at the origin and then moves sequentially along DNA.
- Replication is bidirectional when an origin creates two replication forks that move in opposite directions.

CONCEPT AND REASONING CHECK

What would the CsCl gradient pattern from the Meselson-Stahl experiment look like if replication were fully conservative, that is, if a parental strand gave rise to one daughter with all original DNA and a second daughter with all new DNA?

After two generations of replication in the ^{14}N medium, approximately half of the DNA had the density of DNA with ^{14}N in both strands ("light" DNA) and the other half had the hybrid density. This distribution of ^{15}N atoms is precisely the result predicted from semiconservative replication. Similar experiments with replicating DNA from numerous viruses and bacteria later confirmed the semiconservative mode of replication in these groups.

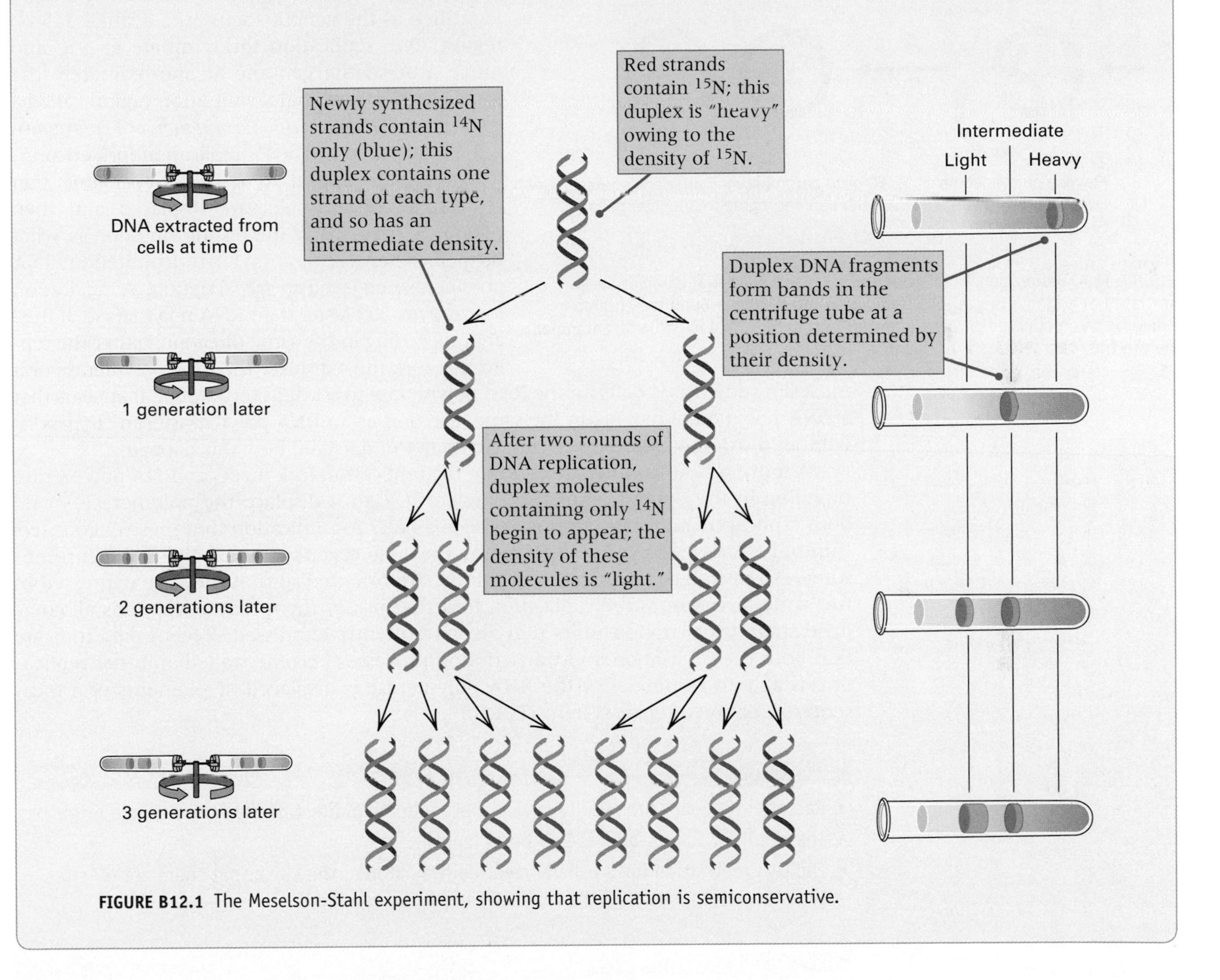

FIGURE B12.1 The Meselson-Stahl experiment, showing that replication is semiconservative.

12.3 The Bacterial Genome Is (Usually) a Single Circular Replicon

Prokaryotic chromosomes and therefore replicons are usually circular, so that the DNA forms a closed circle with no free ends. Circular structures include the bacterial chromosome itself, all plasmids, and many bacteriophages and are also common in chloroplast DNA and mitochondrial DNA. **FIGURE 12.4** summarizes the stages of replicating a circular chromosome. After replication has initiated at the origin, two replication forks proceed in opposite directions. The circular chromosome is sometimes described as a θ structure at this stage because of its appearance. An important consequence of circularity is that

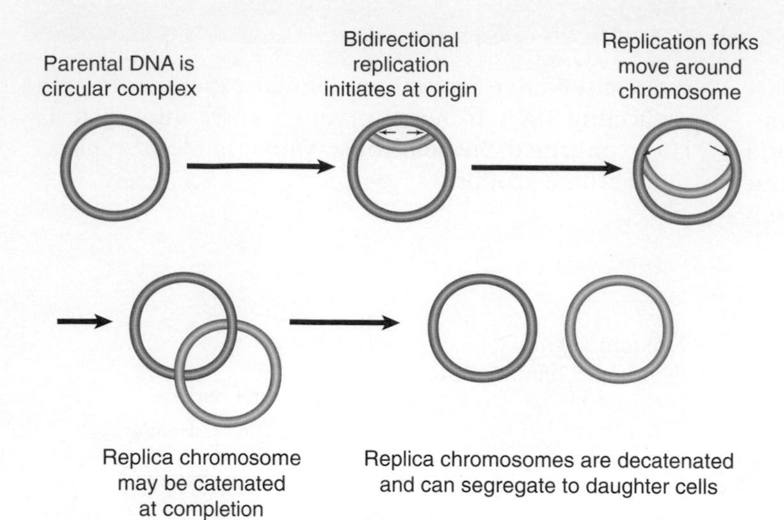

FIGURE 12.4 Bidirectional replication of a circular bacterial chromosome is initiated at a single origin. The replication forks move around the chromosome. If the replicated chromosomes are catenated, they must be disentangled before they can segregate to daughter cells.

the completion of the process can generate two chromosomes that are linked because one passes through the other (they are said to be *catenated*), and specific enzyme systems may be required to separate them (see *Section 11.7, Chromosomal Segregation May Require Site-Specific Recombination*).

The genome of *E. coli* is replicated bidirectionally from a single unique site called the origin, identified as the genetic locus *oriC*, a small 245 bp region. Two replication forks initiate at *oriC* and move around the genome at approximately the same speed to a special termination region called a *ter* site (see *Section 13.15, Termination of Replication*).

What happens when a replication fork encounters a protein bound to DNA? We assume that repressors, for example, are displaced and then rebind. A particularly interesting question is what happens when a replication fork encounters an RNA polymerase engaged in transcription. A replication fork moves 10× faster than RNA polymerase. If they are proceeding in the same direction, either the replication fork must displace the RNA polymerase or it must slow down as it waits for the RNA polymerase to reach its terminator. It appears that a DNA polymerase moving in the same direction as an RNA polymerase can "bypass" it without disrupting transcription, but we do not understand how this happens.

A more serious conflict arises when the replication fork meets an RNA polymerase traveling in the opposite direction, toward it. Can it displace the polymerase, or do both replication and transcription come to a halt? An indication that these encounters cannot be easily resolved is provided by the gene organization on the *E. coli* chromosome. Almost all active transcription units are oriented so that they are expressed in the same direction as the replication fork that passes them. The exceptions all comprise small transcription units that are infrequently expressed. Recent data indicate that both the replication and transcription processes become stalled and that replication is able to resume after the RNA polymerase is displaced by elements of a transcription-coupled repair system (TCR).

KEY CONCEPTS

- Bacterial replicons are usually circles that replicate bidirectionally from a single origin.
- The origin of *E. coli, oriC,* is 245 bp in length.
- The two replication forks usually meet halfway around the circle, but there are *ter* sites that cause termination if the replication forks go too far.

CONCEPT AND REASONING CHECK

Almost all active transcription units are oriented so that they are expressed in the same direction as the replication fork. Why do we think the other transcription units are in the opposite orientation and infrequently expressed?

12.4 Methylation of the Bacterial Origin Regulates Initiation

The bacterial origin contains sequences that are methylated and that are in different methylation states before and after replication. This difference is used as a mark to distinguish a replicated origin from a nonreplicated origin.

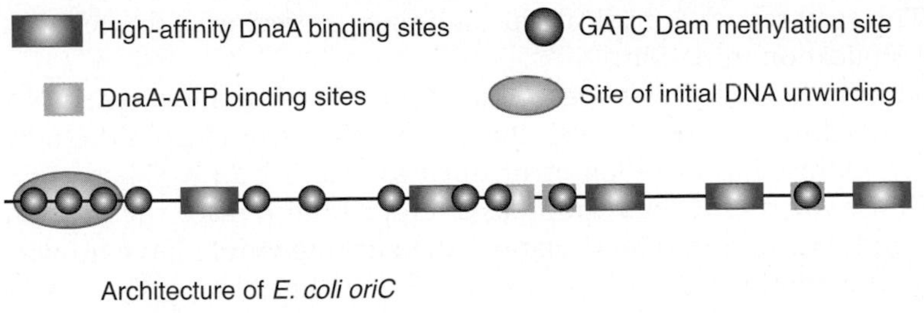

FIGURE 12.5 The *E. coli* origin of replication, *oriC* contains multiple binding sites for the DnaA initiator protein. In a number of cases these sites overlap Dam methylation sites.

The *E. coli oriC* contains 11 copies of the palindromic sequence $\frac{\text{GATC}}{\text{CTAG}}$, which is a target for methylation at the N^6 position of adenine by the Dam methylase enzyme. These sites are also found throughout the genome. This is illustrated in **FIGURE 12.5**.

Before replication, the palindromic target site is methylated on the adenines of each strand. Replication inserts the normal (unmodified) bases into the daughter strands. This generates **hemimethylated DNA**, in which one strand is methylated and one strand is unmethylated. Thus the replication event converts Dam target sites from fully methylated to hemimethylated condition.

hemimethylated DNA DNA that is methylated on one strand of a target sequence that has a cytosine on each strand.

What is the consequence for replication? The ability of a plasmid relying upon *oriC* to replicate in *dam⁻ E. coli* depends on its state of methylation. If the plasmid is methylated it undergoes a single round of replication. The hemimethylated plasmids then accumulate, as described in **FIGURE 12.6**, rather than being replaced by unmethylated plasmids, suggesting that a hemimethylated origin cannot be used to initiate a replication cycle.

This suggests two explanations. Initiation may require full methylation of the Dam target sites in the origin, or it may be inhibited by hemimethylation of these sites. The latter seems to be the case, because an origin of unmethylated DNA can function effectively.

Thus hemimethylated origins cannot initiate again until the Dam methylase has converted them into fully methylated origins. The GATC sites at the origin remain hemimethylated for ~13 minutes after replication. This long period is unusual because at typical GATC sites elsewhere in the genome, remethylation begins immediately (<1.5 minutes) following replication.

What is responsible for the delay in methylation at *oriC*? The most likely explanation is that these regions are sequestered in a form in which they are inaccessible to the Dam methylase. The key for controlling reuse of origins is the gene *seqA* which is part of a negative regulatory circuit that prevents origins from being remethylated. SeqA binds to hemimethylated DNA more strongly that to fully methylated DNA. It may initiate binding when the DNA becomes hemimethylated, at which point its continued presence prevents formation of an open complex at the origin. SeqA does not have specificity for the *oriC* sequence, and it seems likely that this is conferred by the initiation protein DnaA (see *Section 12.5, Initiation: Creating the Replication Forks at the Origin*).

The full scope of the system used to control reinitiation is not clear, but several mechanisms may be involved: physical sequestration of the origin, delay in remethylation, inhibition of DnaA binding, and repression of *dnaA* transcription. It is not immediately obvious which of these events cause the others and whether their effects on initiation are direct or indirect. The period of sequestration appears to increase with the length of the cell cycle, which suggests that it directly reflects the clock that controls reinitiation. One aspect of the control may lie in the observation that hemimethylation of *oriC* is required for its association with cell

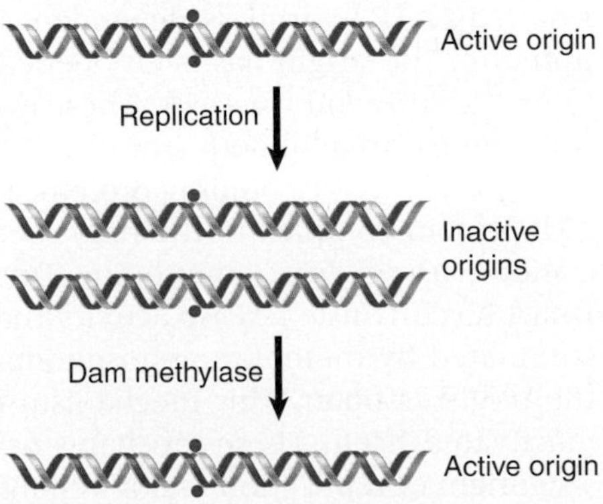

FIGURE 12.6 Only fully methylated origins can initiate replication; hemimethylated daughter origins cannot be used again until they have been restored to the fully methylated state.

membranes *in vitro*. This may reflect a physical repositioning to a region of the cell that is not permissive for replication initiation.

DNA methylation in bacteria serves a second function as well. It allows the DNA mismatch recognition machinery to distinguish the old template strand from the new strand. If the DNA polymerase has made an error, such as creating an A-C base pair, the repair system will use the methylated strand as a template to replace the base on the unmethylated strand. Without that methylation, the enzyme would have no way to determine which is the new strand.

KEY CONCEPTS

- *oriC* contains eleven $\substack{\text{GATC}\\\text{CTAG}}$ repeats that are methylated on adenine on both strands.
- Replication generates hemimethylated DNA, which cannot initiate replication.
- There is a 13 minute delay before the $\substack{\text{GATC}\\\text{CTAG}}$ repeats are remethylated.
- SeqA binds to hemimethylated DNA and is required for delaying replication.

CONCEPT AND REASONING CHECK

Why is it important for the cell to detect whether or not a replication origin has replicated?

12.5 Initiation: Creating the Replication Forks at the Origin

Initiation of replication of duplex DNA in *E. coli* at the origin of replication, *oriC*, requires several successive activities:

- Protein synthesis is required to synthesize the origin recognition protein, DnaA. This is the *E. coli* **licensing factor**, which must be made anew for each round of replication. Drugs that block protein synthesis block a new round of replication, but not continuation of replication.
- There is a requirement for transcription activation. This is not synthesis of the mRNA for DnaA but rather either one of two genes that flank *oriC* must be transcribed. This transcription near the origin aids DnaA in twisting open the origin.
- There must be membrane/cell wall synthesis. Drugs (such as penicillin) that inhibit cell wall synthesis block initiation of replication.

▸ **licensing factor** A factor necessary for replication; it is inactivated or destroyed after one round of replication. New factors must be provided for further rounds of replication to occur.

Most events that are required for initiation therefore occur uniquely at the origin; others recur with the initiation of each Okazaki fragment (see *Section 13.6, The Two New DNA Strands Have Different Modes of Synthesis*) during the elongation phase.

Initiation of replication at *oriC* starts with formation of a complex that ultimately requires six proteins: DnaA, DnaB, DnaC, HU, Gyrase, and SSB. Of the six proteins, DnaA draws our attention as the one uniquely involved in the initiation process. DnaB, an ATP hydrolysis-dependent 5′ to 3′ **helicase**, provides the "engine" of initiation after the origin has been opened (and the DNA is single stranded), by its ability to further unwind the DNA. These events will happen only if the DNA at the origin is fully methylated on both strands.

▸ **helicase** An enzyme that uses energy provided by ATP hydrolysis to separate the strands of a nucleic acid duplex.

DnaA is an ATP binding protein. The first stage in initiation is binding of the DnaA-ATP protein complex to the fully methylated *oriC* sequence. This takes place in association with the inner membrane. DnaA is in the active form only when bound to ATP. DnaA has intrinsic ATPase activity that hydrolyzes ATP to ADP. This ATPase activity is stimulated by membrane phospholipids and single-stranded DNA which forms once the origin is open. This mechanism ensures that once initiation is complete, DnaA inactivated itself. These mechanisms are used to prevent reinitiation of replication. The origin of replication region remains attached to the membrane for about one third of the cell cycle as part of the mechanism to prevent reinitiation. While sequestered in the membrane, the newly synthesized strand of *oriC* cannot be methylated and so remains hemimethylated until DnaA is degraded.

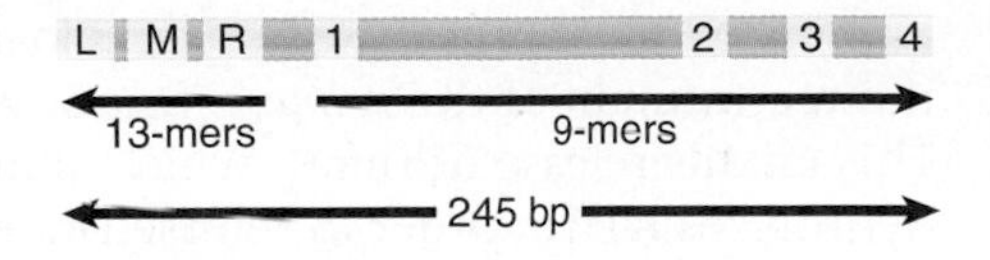

FIGURE 12.7 The minimal origin is defined by the distance between the outside members of the 13-mer and 9-mer repeats.

Opening *oriC* involves action at two types of sequence in the origin: 9 bp and 13 bp repeats. Together the 9 bp and 13 bp repeats define the limits of the 245 bp minimal origin, as indicated in **FIGURE 12.7**. An origin is activated by the sequence of events summerized in **FIGURE 12.8**, in which binding of DnaA-ATP is succeeded by association with the other proteins.

The four 9 bp consensus sequences on the right side of *oriC* provide the initial binding sites for DnaA-ATP. It binds cooperatively to form a central core around which *oriC* DNA is wrapped. DnaA then acts at three A-T-rich 13 bp tandem repeats located on the left side of *oriC*. In its active form, DnaA-ATP twists open the DNA strands at each of these sites to form an open bubble complex. All three 13 bp repeats must be opened for the reaction to proceed to the next stage. Transcription of either gene flanking *oriC* provides additional torsional stress to help snap apart the double-stranded DNA.

Altogether, two to four monomers of DnaA bind at the origin, and they recruit two "prepriming" complexes of DnaB helicase bound to DnaC, so that there is one DnaB-DnaC complex for each of the two (bidirectional) replication forks. The only function of DnaC is that of a chaperone to repress the helicase activity of DnaB until it is needed. Each DnaB-DnaC complex consists of six DnaC monomers bound to a hexamer of DnaB. Note that the DnaB helicase cannot open double stranded DNA; it can only unwind DNA that has already been opened, in this case by DnaA.

The region of strand separation in the open complex is large enough for both DnaB hexamers to bind, which initiates the two replication forks. As DnaB binds, it displaces DnaA from the 13 bp repeats and extends the length of the open region using its helicase activity.

▸ **single-strand binding protein (SSB)** The protein that attaches to single-stranded DNA, thereby preventing the DNA from forming a duplex.

Some additional proteins are required to support the unwinding reaction. *Gyrase*, a type II topoisiomerase, provides a swivel that allows one DNA strand to rotate around the other. Without this reaction, unwinding would generate torsional strain (overwinding) in the DNA that would resist unwinding by the helicase. The **single-strand binding protein (SSB)** stabilizes the single-stranded DNA as it is formed and modulates the helicase activity. The length of duplex DNA that usually is unwound to initiate replication is probably <60 bp. The protein *HU* is a general DNA-binding protein in *E. coli*. Its presence is not absolutely required to initiate replication *in vitro*, but it stimulates the reaction. HU has the capacity to bend DNA and is involved in building the structure that leads to the formation of the open complex.

Input of energy in the form of ATP is required at several stages for the prepriming reaction, and it is required for unwinding DNA. The helicase action of DnaB depends on ATP hydrolysis and the swivel action of gyrase requires ATP hydrolysis.

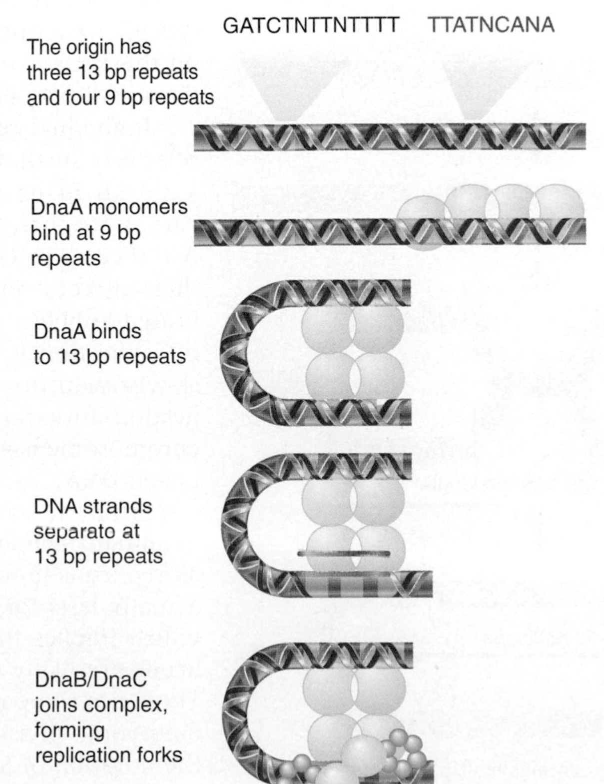

FIGURE 12.8 Prepriming involves formation of a complex by sequential association of proteins, which leads to the separation of DNA strands.

Once the prepriming complex is loaded onto the replication forks, the next step is the recruitment of the *primase*, DnaG, which is then loaded onto the DnaB hexamer. This entails release of DnaC, which allows the DnaB helicase to become active. DnaC hydrolyzes ATP in order to release DnaB. This step marks the trasition from initiation to elongation (see *Chapter 13, DNA Replication*).

KEY CONCEPTS

- Initiation at *oriC* requires the sequential assembly of a large protein complex on the membrane.
- *oriC* must be fully methylated for replication to initiate.
- DnaA-ATP binds to short repeated sequences and forms an oligomeric complex that melts DNA.
- Six DnaC monomers bind each hexamer of DnaB, and this complex binds the origin.
- A hexamer of DnaB forms the replication fork. Gyrase and SSB are also required.
- DnaG is bound to the helicase complex and creates the replication fork.

CONCEPT AND REASONING CHECK

How does DnaA initiate replication, yet prevent a second round of replication?

12.6 Each Eukaryotic Chromosome Contains Many Replicons

Replication of the large amount of DNA contained in a eukaryotic chromosome takes place in *S phase*, which usually lasts a few hours in a multicellular eukaryote (see Figure 11.1). Replication is accomplished by dividing the chromosome into many individual replicons. Only some of these replicons are engaged in replication at any point in S phase. Presumably each replicon is activated at a specific time during S phase, although the evidence on this issue is not decisive. Chromosomal replicons usually display bidirectional replication.

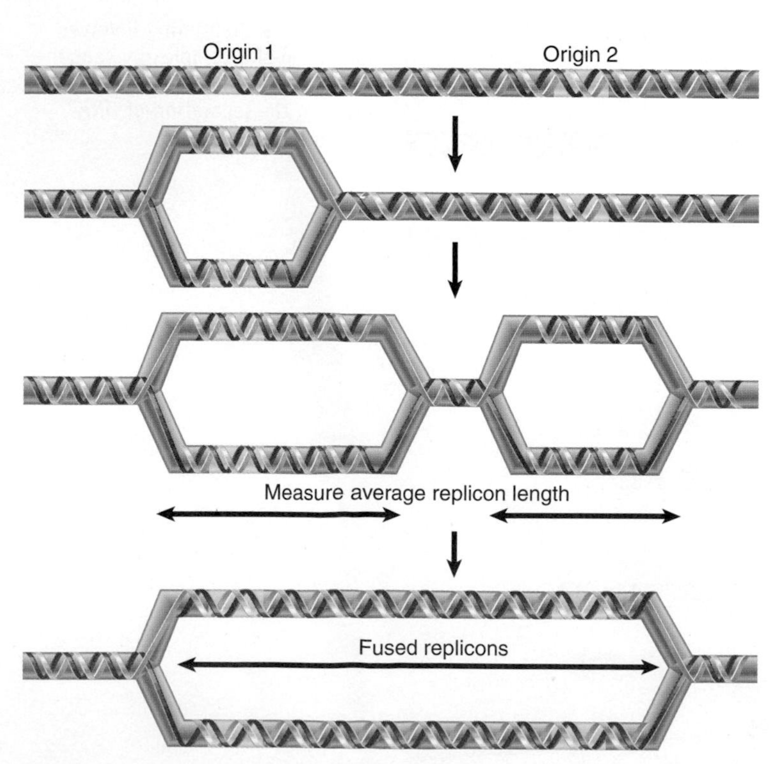

FIGURE 12.9 A eukaryotic chromosome contains multiple origins of replication that ultimately merge during replication.

Individual replicons in eukaryotic genomes are relatively small, typically ~40 kb in yeast or flies and ~100 kb in mammalian cells. However, they can vary more than tenfold in length within a genome. A difficulty in characterizing the individual unit is that adjacent replicons may fuse to give large replicated bubbles, as illustrated in **FIGURE 12.9**. The rate of replication is ~2000 bp/minute, which is much slower than the 50,000 bp/minute of bacterial replication fork movement, presumably because the chromosome is assembled into chromatin and is not naked DNA.

From the speed of replication, it is evident that a mammalian genome could be replicated in ~1 hour if all replicons functioned simultaneously. But S phase actually lasts for >6 hours in a typical somatic cell, which implies that no more than 15% of the replicons are likely to be active at any given moment. There are some exceptional cases, such as the early embryonic divisions of *Drosophila* embryos, where the duration of S phase is compressed by the simultaneous functioning of a large number of replicons.

How are origins selected for initiation at different times during S phase? There is a general hierarchy to the order of replication. Replicons near active genes are replicated earliest, and replicons in heterochromatin replicate last.

Available evidence suggests that most chromosomal replicons do not have a termination region like that of bacteria, at which the replication forks cease movement and (presumably) dissociate from the DNA. It seems more likely that a replication fork continues from its origin until it meets a fork proceeding toward it from the adjacent replicon. We have already mentioned the potential topological problem of joining the newly synthesized DNA at the junction of the replication forks.

The propensity of replicons located in the same vicinity to be active at the same time could be explained by "regional" controls, in which groups of replicons are initiated more or less coordinately, as opposed to a mechanism in which individual replicons are activated one by one in dispersed areas of the genome. Two structural features suggest the possibility of large scale organization. Quite large regions of the chromosome can be characterized as "early replicating" or "late replicating," implying that there is little interspersion of replicons that fire at early or late times. Visualization of replicating forks by labeling with DNA precursors identifies 100 to 300 "foci" instead of uniform staining; each focus shown in Figure 12.9 probably contains >300 replication forks. The foci could represent fixed structures through which replicating DNA must move.

KEY CONCEPTS

- A eukaryotic chromosome is divided into many replicons.
- The progression into S phase is tightly controlled.
- Eukaryotic replicons are 40 to 100 kb in length.
- Individual replicons are activated at characteristic times during S phase.
- Regional activation patterns suggest that replicons near one another are activated at the same time.

CONCEPT AND REASONING CHECKS

1. Why do bacteria have only one *ori* whereas eukaryotes have many per chromosome?
2. Why would organisms like *Drosophila* need such high rates of embryonic replication?

12.7 Replication Origins Bind the ORC

Because a eukaryotic chromosome contains many replicons, we cannot identify origins directly by mutations that prevent replication, as we can with a bacterial chromosome. However, any segment of DNA that has an origin should be able to replicate independently. This provided the first means to identify an origin by testing sequences for their ability to support replication of artificially created independent DNA molecules. A sequence that confers the ability to replicate efficiently in yeast is called an ***ARS*** (*a*utonomously *r*eplicating *s*equence). *ARS* elements are derived from origins of replication.

▸ ***ARS*** An origin for replication in yeast. The common feature among different examples of these sequences is a conserved 11 bp sequence called the A domain.

The yeast *ARS* element consists of an A-T-rich region that contains discrete sites in which mutations affect origin function. Base composition rather than sequence may be important in the rest of the region. **FIGURE 12.10** shows a systematic mutational analysis along the length of an origin. Origin function is abolished completely by mutations in a 14-bp "core" region, called the *A domain,* which contains an 11 bp consensus sequence consisting of A-T base pairs. This consensus sequence sometimes called the *ACS* (*A*RS *c*onsensus *s*equence) is the only homology between known *ARS* elements.

FIGURE 12.10 An *ARS* extends for ~50 bp and includes a consensus sequence (A) and additional elements (B1–B3).

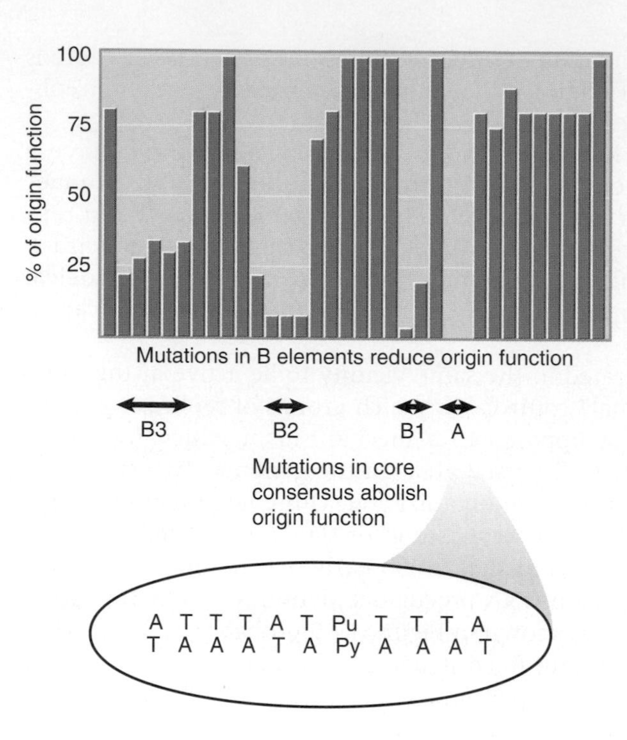

▸ **ORC** Origin recognition complex, a multiprotein complex found in eukaryotes that binds to the replication origin (the *ARS* in yeast) and remains associated with it throughout the cell cycle.

Mutations in three adjacent elements, numbered B1 to B3, reduce origin function. An origin can function effectively with any two of the B elements, so long as a functional A element is present. (Imperfect copies of the core consensus, typically conforming at 9 out of 11 positions, are found close to, or overlapping with, each B element, but they do not appear to be necessary for origin function.)

The **ORC** (*o*rigin *r*ecognition *c*omplex) is a complex of six proteins with a mass of ~400 kD. ORC binds to the A and B1 elements on the A-T-rich strand and is associated with *ARS* elements throughout the cell cycle. This means that initiation depends on changes in its condition rather than *de novo* association with an origin (see *Section 12.8, Licensing Factor Controls Eukaryotic Rereplication*). By counting the number of sites to which ORC binds, we can estimate that there are about 400 origins of replication in the yeast genome. This means that the average length of a replicon is ~35,000 bp. Counterparts to ORC are found in multicellular eukaryotic cells.

ORC was first found in *S. cerevisiae* (where it is called scORC), but similar complexes have now been characterized in *Schizosaccharomyces pombe* (spORC), *Drosophila* (DmORC), and *Xenopus* (XlORC). All the ORC complexes bind to DNA. Although none of the binding sites have been characterized in the same detail as in *S. cerevisiae,* in several cases they are at locations associated with the initiation of replication. It seems clear that ORC is an initiation complex whose binding identifies an origin of replication. Although the ORC is conserved, thus far it has not been possible to identify origins of replication in multicellular eukaryotes DNA that function like *ARS* elements. The conservation of the ORC suggests that origins are likely to take the same sort of form in other eukaryotes, but in spite of this, there is little conservation of sequence among putative origins in different organisms. Difficulties in finding consensus origin sequences suggest the possibility that origins may be more complex.

KEY CONCEPTS

- Origins in *S. cerevisiae* are short A-T-rich sequences that have an essential 11 bp sequence.
- The ORC is a complex of six proteins that binds to an *ARS*. It remains bound through the cell cycle but is activated only during S phase.
- Related ORC complexes are found in multicellular eukaryotes.

CONCEPT AND REASONING CHECK

Why do origins of replication contain a core of A-T-rich DNA?

12.8 Licensing Factor Controls Eukaryotic Rereplication

A eukaryotic genome is divided into multiple replicons, and the origin in each replicon is activated once and only once in a single division cycle. This could be achieved by the provision of some rate-limiting component that functions only once at an origin or by the presence of a repressor that prevents rereplication at origins that have been used. The critical questions about the nature of this regulatory system are how the system determines whether any particular origin has been replicated and what protein components are involved.

Insights into the nature of the protein components have been provided by using a system in which a substrate DNA undergoes only one cycle of replication. *Xenopus* eggs have all the components needed to replicate DNA—in the first few hours after fertilization they undertake 11 division cycles without new gene expression—and they can replicate the DNA in a nucleus that is injected into the egg. **FIGURE 12.11** summarizes the features of this system.

When a sperm or interphase nucleus is injected into the egg, its DNA is replicated only once (this can be followed by use of a density label, just like the original experiment that characterized semiconservative replication, shown in the accompanying box, *Historical Perspectives*, in this chapter). If protein synthesis is blocked in the egg, the membrane around the injected material remains intact and the DNA cannot replicate again. In the presence of protein synthesis, however, the nuclear membrane breaks down just as it would for a normal cell division, and in this case subsequent replication cycles can occur. The same result can be achieved by using agents that permeabilize the nuclear membrane. This suggests that the nucleus contains a protein(s) needed for replication that is used up in some way by a replication cycle, so even though more of the protein is present in the egg cytoplasm, it can enter the nucleus only if the nuclear membrane breaks down. The system can, in principle, be taken further by developing an *in vitro* extract that supports nuclear replication, thus allowing the components of the extract to be isolated and the relevant factors identified.

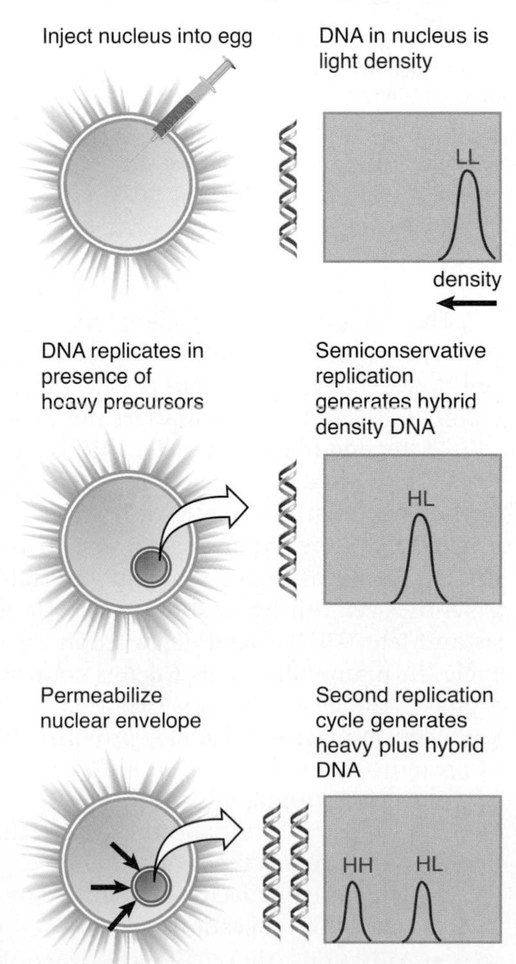

FIGURE 12.11 A nucleus injected into a *Xenopus* egg can replicate only once unless the nuclear membrane is permeabilized to allow subsequent replication cycles.

FIGURE 12.12 explains the control of reinitiation by proposing that this protein is a *licensing factor*. It is present in the nucleus prior to replication. One round of replication either inactivates or destroys the factor, and another round cannot occur until further factor is provided. Factor in the cytoplasm can gain access to the nuclear material only at the subsequent mitosis when the nuclear envelope breaks down. This regulatory system achieves two purposes. By removing a necessary component after replication, it prevents more than one cycle of replication from occurring. It also provides a feedback loop that makes the initiation of replication dependent on passing through cell division.

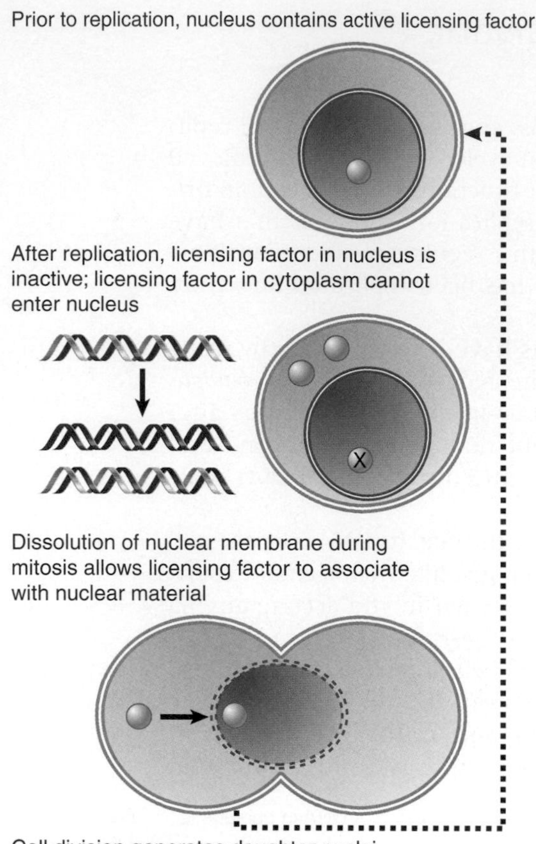

FIGURE 12.12 Licensing factor in the nucleus is inactivated after replication. A new supply of licensing factor can enter only when the nuclear membrane breaks down at mitosis.

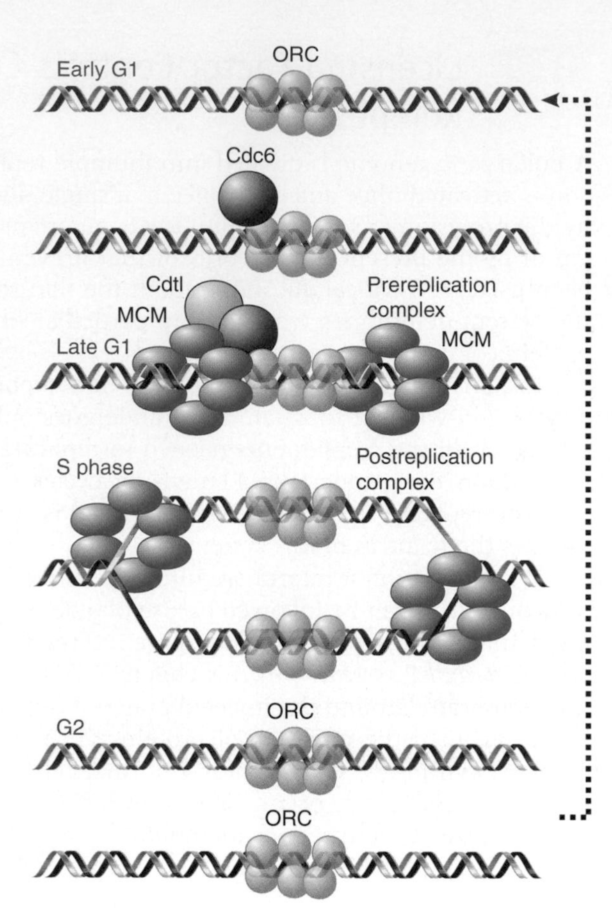

FIGURE 12.13 Proteins at the origin control susceptibility to initiation.

The key event in controlling replication is the behavior of the ORC complex at the origin. The striking feature is that ORC remains bound at the origin through the entire cell cycle. However, changes occur in the binding of other proteins to the ORC-origin complex. **FIGURE 12.13** summarizes the cycle of events at the origin.

At the end of the cell cycle, ORC is bound to A–B1 elements of the origin and protects the origin against degradation by DNase, but there is a site that is hypersensitive to the enzyme in the center of B1. There is a change during G1, seen most strikingly by the loss of the hypersensitive site. This results from the binding of Cdc6 protein to the ORC. In yeast, Cdc6 is a highly unstable protein, with a half-life of <5 minutes. It is synthesized during G1 and typically binds to the ORC between the exit from mitosis and late G1. Its rapid degradation means that no protein is available later in the cycle. In mammalian cells Cdc6 is controlled differently; it is phosphorylated during S phase, and as a result it is exported from the nucleus. This feature makes Cdc6 the key licensing factor. Cdc6 also provides the connection between ORC and a complex of proteins that is involved in initiation. Cdc6 has an ATPase activity that is required for it to support initiation.

The licensing factor and the system that controls its availability in yeast is identified by two different types of mutations:

- The licensing factor is identified by mutations in *MCM2,3,5,* which prevent initiation of replication.
- Mutations that have the opposite effect and allow the accumulation of excess quantities of DNA are found in genes that encode components of the

ubiquitination system that is responsible for programmed degradation of specific proteins. This suggests that licensing factor may be destroyed after the start of the replication cycle.

In yeast, free MCM2,3,5 enter the nucleus only during mitosis. Homologs are found in animal cells, where MCM3 is bound to chromosomal material before replication but is released after replication. The animal cell MCM2,3,5 complex remains in the nucleus throughout the cell cycle, suggesting that it may be only one component of the licensing factor. Another component, able to enter only at mitosis, may be necessary for MCM2,3,5 to associate with chromosomal material.

The presence of Cdc6 at the yeast origin allows Cdt1 and MCM proteins to bind to the complex. Their presence is necessary for initiation. The origin therefore enters S phase in the condition of a **prereplication complex**, which contains ORC, Cdc6, Cdt1 and MCM proteins. When initiation occurs, Cdc6, Cdt1 and MCM are displaced, returning the origin to the state of the **postreplication complex**, which contains only ORC. Because Cdc6 is rapidly degraded during S phase, it is not available to support reloading of MCM proteins, and so the origin cannot be used for a second cycle of initiation during S phase. Cdt1 is also targeted for degradation at the beginning of DNA synthesis.

The MCM2-7 proteins form a six-member ring-shaped complex around DNA. This complex is believed to be the eukaryotic version of the bacterial helicase (see *Section 13.7, Replication Requires a Helicase and Single-Strand Binding Protein*). Some of the ORC proteins have similarities to replication proteins that load DNA polymerase onto DNA. It is possible that ORC uses hydrolysis of ATP to load the MCM ring onto DNA. In *Xenopus* extracts, replication can be initiated if ORC is removed after it has loaded Cdc6 and MCM proteins. This shows that the major role of ORC is to identify the origin to the Cdc6 and MCM proteins that control initiation and licensing.

- **prereplication complex** A protein-DNA complex at the origin in *S. cerevisiae* that is required for DNA replication. The complex contains the ORC complex, Cdc6, and the MCM proteins.
- **postreplication complex** A protein-DNA-complex in *S. cerevisiae* that consists of the ORC complex bound to the origin.

KEY CONCEPTS

- Licensing factor is necessary for initiation of replication at each origin.
- Licensing factor is present in the nucleus prior to replication, but it is removed, inactivated, or destroyed by replication.
- Initiation of another replication cycle becomes possible only after licensing factor re-enters the nucleus after mitosis.
- The ORC is a protein complex that is associated with yeast origins throughout the cell cycle.
- Cdc6 protein is an unstable protein that is synthesized only in G1.
- Cdc6 binds to ORC and allows MCM proteins to bind.
- When replication is initiated, Cdc6, Cdt1 and MCM proteins are displaced. The degradation of Cdc6 prevents reinitiation.
- Some MCM proteins are in the nucleus throughout the cycle, but others may enter only after mitosis.

CONCEPT AND REASONING CHECK

What mechanism do you think could allow some proteins to have long half-lives and other proteins to have short half-lives?

12.9 Summary

Replicons in bacterial or eukaryotic chromosomes have a single unifying feature: replication is initiated at an origin once and only once in each cell cycle. The origin is located within the replicon, and replication typically is bidirectional, with replication forks proceeding away from the origin in both directions.

An origin consists of a discrete sequence at which replication of DNA is initiated. Origins of replication tend to be rich in A-T base pairs. A bacterial chromosome contains a single origin, which is responsible for initiating replication once every cell cycle. The *oriC* in *E. coli* is a sequence of 245 bp. Any DNA molecule with this sequence can replicate in *E. coli*. Replication of the circular bacterial chromosome produces a θ structure, in which the replicated DNA starts out as a small replicating eye. Replication proceeds until the eye occupies the whole chromosome.

The bacterial origin contains sequences that are methylated on both strands of DNA. Replication produces hemimethylated DNA, which cannot function as an origin. There is a delay of ~10 minutes before the hemimethylated origins are remethylated to convert them to a functional state, and this is responsible for preventing improper reinitiation.

The common mode of origin activation involves an initial limited melting of the double helix, followed by more general unwinding to create single strands. Several proteins act sequentially at the *E. coli* origin. Replication is initiated at *oriC* in *E. coli* when DnaA binds to a series of 9 bp repeats. This is followed by binding to a series of 13 bp repeats, where it uses hydrolysis of ATP to generate the energy to separate the DNA strands. The prepriming complex of DnaB-DnaC displaces DnaA. DnaC is released in a reaction that depends on ATP hydrolysis; DnaB is joined by the replicase enzyme, and replication is initiated at two forks that set out in opposite directions. The availability of DnaA at the origin is an important component of the system that determines when replication cycles should initiate.

A eukaryotic chromosome is divided into many individual replicons. Replication occurs during a discrete part of the cell cycle called S phase, but because not all replicons are active simultaneously, the process may take several hours. Eukaryotic replication is at least an order of magnitude slower than bacterial replication. Origins sponsor bidirectional replication and are probably used in a fixed order during S phase. Each replicon is activated only once in each cycle. Origins of replication were isolated as *ARS* sequences in yeast by virtue of their ability to support replication of any sequence attached to them. The core of an *ARS* is an 11 bp A-T-rich sequence that is bound by the ORC protein complex, which remains bound throughout the cell cycle. Utilization of the origin is controlled by the MCM licensing factors that associate with the ORC.

CHAPTER QUESTIONS

1. For cell division to occur, DNA replication in the cell must be:

A. initiated and have cleared the origin.
B. at least halfway completed.
C. at least 80% completed.
D. 100% completed.

2. Replication origins are:

A. *cis*-acting elements.
B. *trans*-acting elements.
C. both *cis*- and *trans*-acting elements.
D. not considered either *cis*- or *trans*-acting elements.

3. Bacterial plasmids may be present in:

A. a single copy.
B. a multicopy.
C. the host chromosome as an integrated copy.
D. any of the above.

4. What Greek letter is used to describe how DNA replicates?

A. θ
B. γ
C. β
D. ω

5. Differential labeling of replicating eukaryotic cell DNA showed that this genome replicates by:
   A. bidirectional replication from multiple origins.
   B. unidirectional replication from multiple origins.
   C. bidirectional replication from a single origin.
   D. unidirectional replication from a single origin.
6. In both *E. coli* and mammalian chromosomes, DNA replication is:
   A. unidirectional and semiconservative.
   B. bidirectional and semiconservative.
   C. unidirectional and conservative.
   D. bidirectional and conservative.
7. Replication origins in the yeast chromosome have a tendency to be:
   A. AT-rich.
   B. GC-rich.
   C. AG-rich.
   D. TC-rich.
8. In yeast, what protein binds to ORC to support initiation of DNA replication?
   A. SeqA
   B. Cdc2
   C. Cdc6
   D. DnaA
9. What proportion of replicons in mammalian cells are actively replicating at any given time?
   A. ~10%
   B. ~15%
   C. ~25%
   D. ~60%

KEY TERMS

ARS
bidirectional replication
equilibrium density-gradient centrifugation
growing point
helicase
hemimethylated DNA
licensing factor
multicopy replication control
ORC
origin
postreplication complex
prereplication complex
replication bubble
replication fork
replicon
semiconservative replication
single-copy replication control
single-strand binding protein (SSB)
unidirectional replication

FURTHER READING

Gilbert, D. M. (2001). Making sense of eukaryote DNA replication origins. *Science* **294**, 96–100. A review describing how origins of replication differ between eukaryotes and prokaryotes.

Kaguni, J. M. (2006). DnaA: controlling the initiation of bacterial DNA replication and more. *Annu. Rev. Micro* **61**, 351–373.

Meselson, M., and Stahl, F. W. (1958). The replication of DNA in *E. coli*. *Proc. Natl. Acad. Sci. USA* 44, pp. 671–682.

13

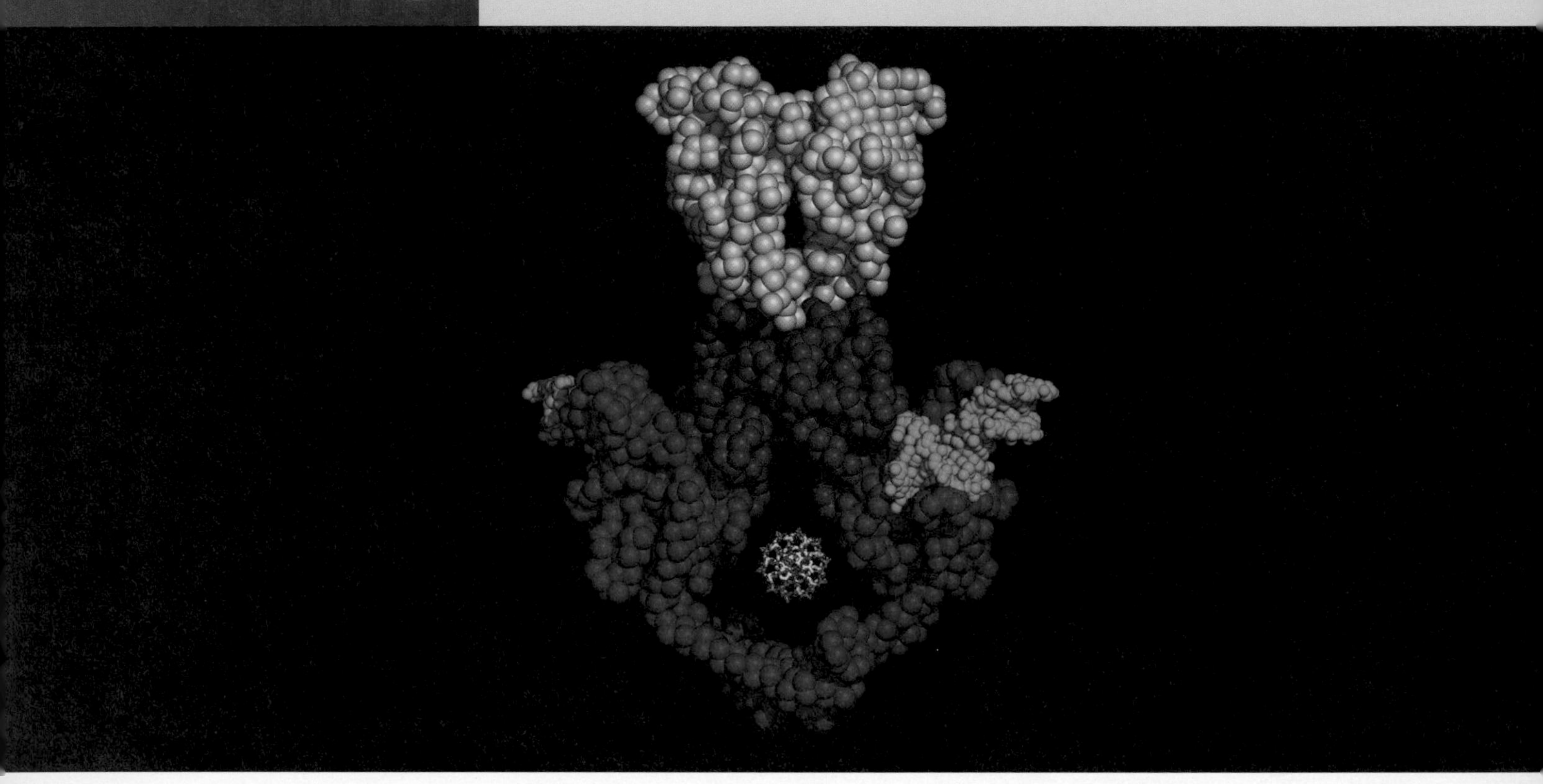

A type II topoisomerase acting on a knotted or supercoiled DNA substrate. The cleaved DNA is shown in green. The DNA that is being passed through the break is viewed end-on. Photo courtesy of James M. Berger, California Institute for Quantitative Biology, University of California, Berkeley.

DNA Replication

CHAPTER OUTLINE

13.1 Introduction

Replication of duplex DNA is a complicated endeavor involving multiple enzyme complexes. Different activities are involved in the stages of initiation, elongation, and termination. Before initiation can occur, however, the supercoiled chromosome must be in a negative supercoiled state or underwound (see *Section 1.5, Supercoiling Affects the Structure of DNA*). This occurs in segments beginning with the replication origin region. This alteration to the structure of the chromosome is accomplished by the enzyme **topoisomerase**. **FIGURE 13.1** shows an overview of the first stages of the process.

- *Initiation* involves recognition of an origin by a large protein complex. Before DNA synthesis can begin, the parental strands must be separated and transiently stabilized in the single-stranded state, creating a replication bubble. After this stage, synthesis of daughter strands can be initiated at the replication fork (see *Chapter 12, The Replicon: Initiation of Replication*).
- *Elongation* is undertaken by another complex of proteins. The **replisome** exists only as a protein complex associated with the particular structure that DNA takes at the replication fork. It does not exist as an independent unit (for example, analogous to the ribosome) but assembles *de novo* at the origin for each replication cycle. As the replisome moves along DNA, the parental strands unwind and daughter strands are synthesized.
- At the end of the replicon, *joining of replicons* and/or *termination* reactions are necessary. Following termination, the duplicate chromosomes must be separated from one another, which requires manipulation of higher-order DNA structure.

The inability to replicate DNA is fatal for a growing cell. Mutants for replication must therefore be obtained as **conditional lethals**. These are able to accomplish replication under *permissive* conditions (provided by the normal temperature of incubation), but they are defective under *nonpermissive* conditions (provided by the higher temperature of 42°C). A comprehensive series of such temperature-sensitive mutants in *E. coli* identifies a set of loci called the *dna* genes.

▸ **topoisomerase** An enzyme that changes the number of times the two strands in a closed DNA molecule cross each other. It does this by cutting the DNA, passing DNA through the break, and resealing the DNA.

▸ **replisome** The multiprotein structure that assembles at the bacterial replication fork to undertake synthesis of DNA. It contains DNA polymerase and other enzymes.

▸ **conditional lethal** A mutation that is lethal under one set of conditions but not lethal under a second set of permissive conditions, such as temperature.

13.2 DNA Polymerases Are the Enzymes That Make DNA

There are two basic types of DNA synthesis. Figure 13.1 shows the result of **semiconservative replication**. The two strands of the parental duplex are separated, and each serves as a template for synthesis of a new strand. The parental duplex is replaced with two daughter duplexes, each of which has one parental strand and one newly synthesized strand.

▸ **semiconservative replication** Replication that is accomplished by separation of the strands of a parental duplex, each strand then acting as a template for synthesis of a complementary strand.

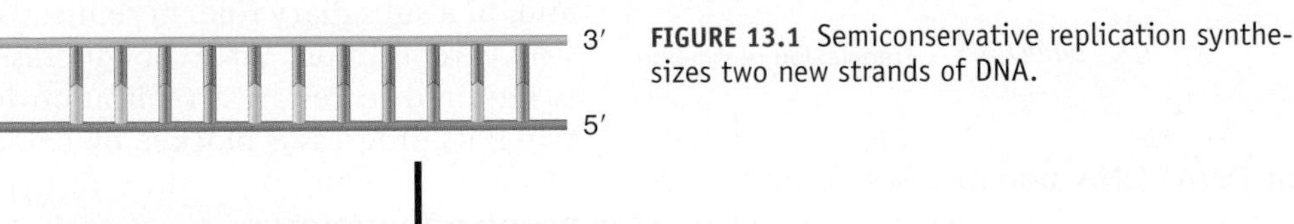

FIGURE 13.1 Semiconservative replication synthesizes two new strands of DNA.

FIGURE 13.2 shows the consequences of a **DNA repair** reaction. One strand of DNA has been damaged. A portion of it is excised, and new material is synthesized to replace the excised portion. An enzyme that can synthesize a new DNA strand on a template strand is called a **DNA polymerase** (or more properly, *DNA-dependent DNA polymerase*). Both prokaryotic and eukaryotic cells contain multiple DNA polymerase activities. Only a few of these enzymes actually undertake replication. The remaining enzymes are involved in repair synthesis or participate in subsidiary roles in replication.

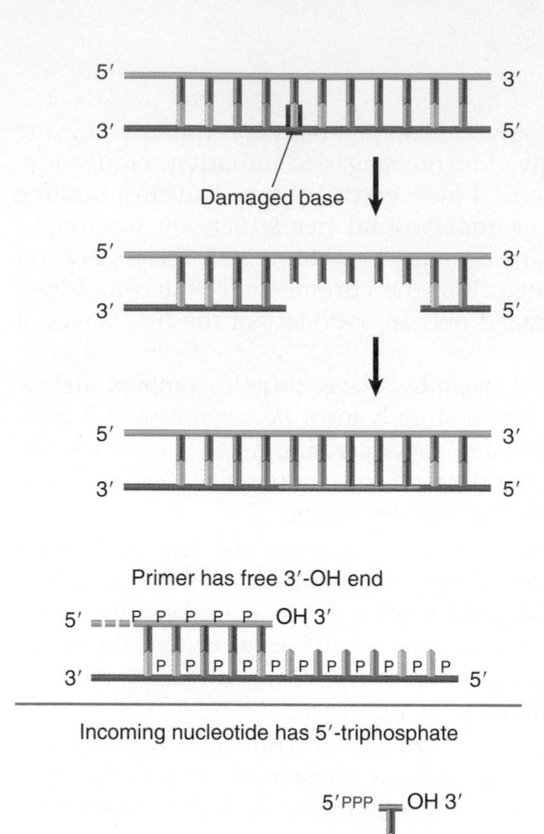

FIGURE 13.2 Repair synthesis replaces a short stretch of one strand of DNA containing a damaged base.

▸ **DNA repair** The removal and replacement of damaged DNA by the correct sequence.

▸ **DNA polymerase** An enzyme that synthesizes a daughter strand(s) of DNA (under direction from a DNA template). Any particular enzyme may be involved in repair or replication (or both).

All prokaryotic and eukaryotic DNA polymerases share the same fundamental type of synthetic activity, synthesis from 5′ to 3′ from a template that is 3′ to 5′. This means adding nucleotides one at a time to a 3′–OH end, as illustrated diagrammatically in **FIGURE 13.3**. The choice of the nucleotide to add to the chain is dictated by base pairing with the template strand.

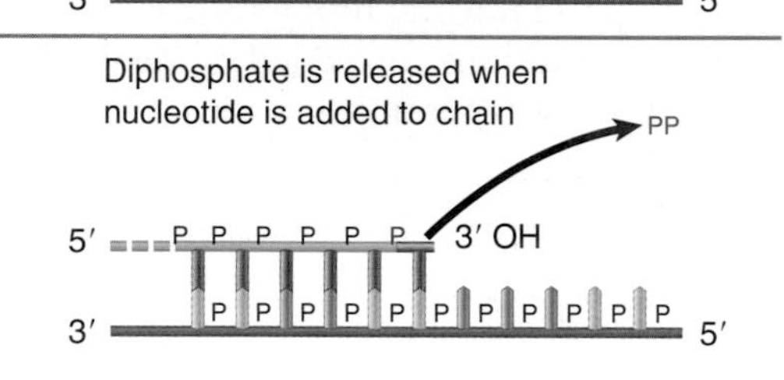

FIGURE 13.3 DNA is synthesized by adding nucleotides to the 3′–OH end of the growing chain, so that the new chain grows in the 5′→3′ direction. The precursor for DNA synthesis is a nucleoside triphosphate, which loses the terminal two phosphate groups in the reaction.

Some DNA polymerases, such as the repair polymerases, function as independent enzymes, but others, notably the replication polymerases, are incorporated into large protein assemblies called **holoenzymes**. The DNA-synthesizing subunit is only one of several functions of the holoenzyme, which typically contains other activities concerned with fidelity.

▸ **holoenzyme** The DNA polymerase complex that is competent to initiate replication.

FIGURE 13.4 summarizes the DNA polymerases that have been characterized in *E. coli*. DNA polymerase III, a multisubunit protein, is the replication polymerase responsible for *de novo* synthesis of new strands of DNA. DNA polymerase I (coded by *polA*) is involved in the repair of damaged DNA and, in a subsidiary role, in semiconservative replication. DNA polymerase II is required to restart a replication fork when its progress is blocked by damage in DNA. DNA polymerases IV and V are involved in allowing replication to bypass certain types of damage and are called **error-prone polymerases**.

Enzyme	Gene	Function
I	*polA*	major repair enzyme
II	*polB*	replication restart
III	*polC*	replicase
IV	*dinB*	translesion replication
V	*umuD*$'_2$*C*	translesion replication

FIGURE 13.4 Only one DNA polymerase is the replication enzyme. The others participate in repair of damaged DNA, restarting stalled replication forks, or bypassing damage in DNA.

▸ **error-prone polymerase** A DNA polymerase that incorporates noncomplementary bases into the daughter strand.

When extracts of *E. coli* are assayed for their ability to synthesize DNA, the predominant enzyme activity is DNA polymerase I. Its activity is so great that it makes it impossible to detect the activities of the enzymes actually responsible for DNA replication! To develop *in vitro* systems in which replication can be followed, extracts are, therefore, prepared from *polA* mutant cells.

Several classes of eukaryotic DNA polymerases have been identified. DNA polymerase δ and ε are required for nuclear replication; DNA polymerase α is concerned with "priming" (initiating) replication. Other DNA polymerases are involved in repairing damaged nuclear DNA or in translesion replication of damaged DNA when repair is impossible. Mitochondrial DNA replication is carried out by DNA polymerase γ (see *Section 13.13, Separate Eukaryote DNA Polymerases Undertake Initiation and Elongation*).

KEY CONCEPTS

- DNA is synthesized in both semiconservative replication and repair reactions.
- A bacterium or eukaryotic cell has several different DNA polymerase enzymes.
- One bacterial DNA polymerase undertakes semiconservative replication; the others are involved in repair reactions.

CONCEPT AND REASONING CHECK

How does repair DNA synthesis differ from replication DNA synthesis?

13.3 DNA Polymerases Have Various Nuclease Activities

Replicases often have nuclease activities as well as the ability to synthesis DNA. A 3′–5′ exonuclease activity is typically used to excise bases that have been added to DNA incorrectly. This provides a "proofreading" error-control system (see *Section 13.4, DNA Polymerases Control the Fidelity of Replication*).

The first DNA-synthesizing enzyme to be characterized was the *E. coli* DNA polymerase I, which is a single polypeptide of 103 kD that can be cleaved into two parts by proteolytic treatment. The larger cleavage product (68kD) is called the Klenow fragment. It is used in synthetic reactions *in vitro*. It contains the polymerase and the 3′–5′ exonuclease activities. The active sites are ~30 Å apart in the protein, which indicates that there is spacial separation between the area adding a base and the area removing one.

The small fragment (35kD) posses a 5′–3′ exonuclease activity, which excises small groups of nucleotide, up to ~10 bases at a time. This activity is coordinated with the synthetic/proofreading activity. It provides DNA polymerase I with a unique ability to start replication *in vitro* at a nick in DNA. (No other DNA polymerase has this ability.) At a point where a phosphodiester bond has been broken in a double-stranded DNA, the enzyme extends the 3′–OH end. As the new segment of DNA is synthesized, it displaces the existing homologous strand in the duplex. The displaced strand is degraded by the 5′–3′ exonuclease activity of the enzyme.

The coupled 5′–3′ synthetic/5′–3′ exonucleolytic action is used most extensively for filling in short single-stranded regions in double-stranded DNA. These regions arise during lagging-strand DNA replication (see *Section 13.6, The Two New DNA Strands Have Different Modes of Synthesis*) and during DNA repair (Figure 13.3).

KEY CONCEPT

- DNA polymerase I has a unique 5′ and 3′ exonuclease activity.

CONCEPT AND REASONING CHECK

Why would a DNA polymerase enzyme need to have nuclease activity?

13.4 DNA Polymerases Control the Fidelity of Replication

The fidelity of replication poses the same sort of problem we have encountered already in considering (for example) the accuracy of translation. It relies on the specificity of base pairing. Yet when we consider the interactions involved in base pairing, we would expect errors to occur with a frequency of ~10^{-2} per base pair replicated. The actual rate in bacteria seems to be ~10^{-8} to 10^{-10}. This corresponds to ~1 error per genome per 1000 bacterial replication cycles, or ~10^{-6} per gene per generation.

We can divide the errors that DNA polymerase makes during replication into two classes:

- *Substitutions* occur when the wrong (improperly paired) nucleotide is incorporated. The error level is determined by the efficiency of **proofreading**, in which the enzyme scrutinizes the newly formed base pair and removes the nucleotide if it is mispaired.
- *Frameshifts* occur when an extra nucleotide is inserted or omitted. Fidelity with regard to frameshifts is affected by the **processivity** of the enzyme: the tendency to remain on a single template rather than to dissociate and reassociate. This is particularly important for the replication of a homopolymeric stretch—for example, a long sequence of dT_n:dA_n, in which "replication slippage" can change the length of the homopolymeric run. As a general rule, increased processivity reduces the likelihood of such events. In multimeric DNA polymerases, processivity is usually increased by a particular subunit that is not needed for catalytic activity *per se*.

▸ **proofreading** A mechanism for correcting errors in DNA synthesis that involves scrutiny of individual units after they have been added to the chain.

▸ **processivity** The ability of an enzyme to perform multiple catalytic cycles with a single template instead of dissociating after each cycle.

Bacterial replication enzymes have multiple error-reduction systems. As discussed in *Chapter 1 (Genes are DNA)*, the geometry of an A-T base pair is very similar to that of a G-C base pair. This geometry is used by high-fidelity DNA polymerases as a fidelity mechanism. Only an incoming dNTP that base-pairs properly with the template nucleotide fits in the active site, while mispairs such as A-C or A-A have the wrong geometry to fit in the active site. On the other hand, low-fidelity DNA polymerases (such as *E. coli* DNA polymerase IV used for damage bypass replication) have a more open active site that accommodates damaged nucleotides but also mispairs. Therefore, either the expression or activity of these error-prone polymerases is tightly regulated so that they are active only after DNA damage.

All the bacterial enzymes possess a 3′ to 5′ exonucleolytic activity that proceeds in the reverse direction from DNA synthesis. This provides a proofreading function illustrated diagrammatically in **FIGURE 13.5**. In the chain-elongation step, a precursor nucleotide enters the position at the end of the growing chain. A bond is formed. The enzyme moves one base pair farther and then is ready for the next precursor nucleotide to enter. If a mistake has been made, the DNA is structurally warped by the incorporation of the incorrect base that will cause the polymerase to pause or slow down. This will allow the enzyme to back up and remove the incorrect base (see *Section 13.5, DNA Polymerases Have a Common Structure*). In some regions errors occur more frequently than in others; that is, **mutation hotspots** occur in the DNA. This is caused by the underlying sequence context; that is, some sequences cause the polymerase to move faster or slower, which affects the ability to catch an error.

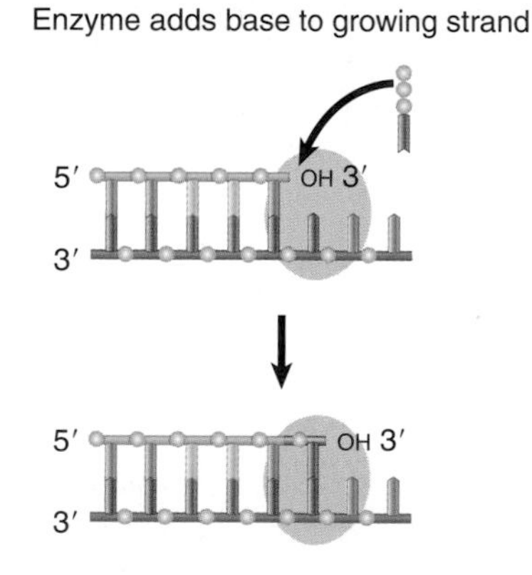

Enzyme moves on if new base is correct

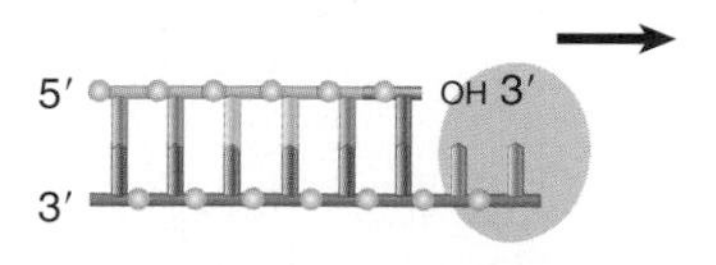

Base is hydrolyzed and expelled if incorrect

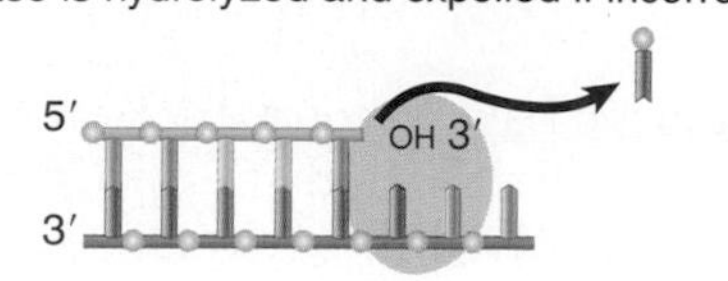

FIGURE 13.5 Bacterial DNA polymerases scrutinize the base pair at the end of the growing chain and excise the nucleotide added in the case of a misfit.

▸ **mutation hotspot** A site in the genome at which the frequency of mutation (or recombination) is very much increased, usually by at least an order of magnitude relative to neighboring sites.

As noted in *Section 13.2, DNA Polymerases Are the Enzymes That Make DNA*, replication enzymes typically are found as multisubunit holoenzyme complexes, whereas repair DNA polymerases are typically found as single subunit enzymes. An advantage to a holoenzyme system is the availability of a specialized subunit responsible for error correction. In *E. coli* DNA polymerase III, this activity, a 3′–5′ exonuclease, resides in a separate subunit, the ε subunit. This subunit gives the replication enzyme a greater fidelity than the repair enzymes.

Different DNA polymerases handle the relationship between the polymerizing and proofreading activities in different ways. In some cases, the activities are part of the same protein subunit, but in others they are contained in different subunits. Each DNA polymerase has a characteristic error rate that is reduced by its proofreading activity. Proofreading typically decreases the error rate in replication from $\sim 10^{-5}$ to $\sim 10^{-7}$ per base pair replicated. Systems that recognize errors and correct them following replication then eliminate some of the errors, bringing the overall rate to $<10^{-9}$ per base pair replicated (see *Section 16.6, Controlling the Direction of Mismatch Repair*).

KEY CONCEPTS

- DNA polymerases often have a 3′ to 5′ exonuclease activity that is used to excise incorrectly paired bases.
- The fidelity of replication is improved by proofreading by a factor of ~100.

CONCEPT AND REASONING CHECK

If the location of a mutation is generally random, why are there mutation hotspots in a genome?

13.5 DNA Polymerases Have a Common Structure

FIGURE 13.6 shows the common structural feature that all DNA polymerases share. The enzyme structure can be divided into several independent domains, which are described by analogy with a human right hand. DNA binds in a large cleft composed of three domains. The "palm" domain has important conserved sequence motifs that provide the catalytic active site. The "fingers" are involved in positioning the template correctly at the active site. The "thumb" binds the DNA as it exits the enzyme and is important in processivity. The most important conserved regions of each of these three domains converge to form a continuous surface at the catalytic site. The exonuclease activity resides in an independent domain with its own catalytic site. The N-terminal domain extends into the nuclease domain. DNA polymerases fall into five families based on sequence homologies; the palm is well conserved among them, but the thumb and fingers provide analogous secondary structure elements from different sequences.

The catalytic reaction in a DNA polymerase occurs at an active site in which a nucleotide triphosphate pairs with an (unpaired) single strand of DNA. The DNA lies across the palm in a groove that is created by the thumb and fingers. **FIGURE 13.7** shows the crystal structure of the phage T7 enzyme complexed with DNA (in the form of a primer annealed to a template strand) and an incoming nucleotide that is about to be added to the primer. The DNA is in the classic B-form duplex up to the last two base pairs at the 3′ end of the primer, which are in the more open A-form. A sharp turn in the DNA exposes the

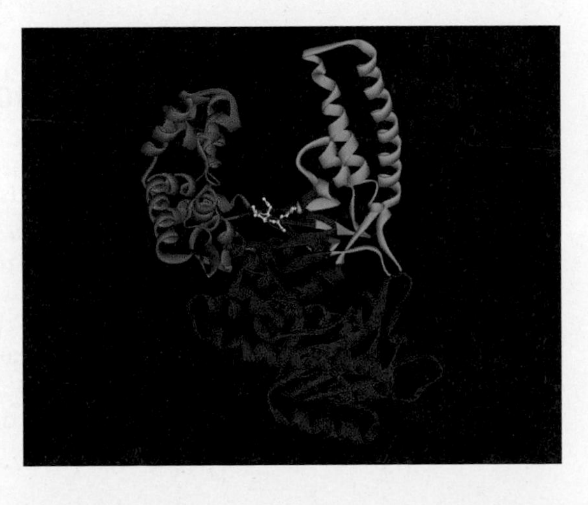

FIGURE 13.6 The structure of the Klenow fragment from *E. coli* DNA polymerase I. It has the form of a right hand with fingers (blue), a palm (red), and a thumb (green). The Klenow fragment also includes an exonuclease domain. Adapted from Protein Data Bank 1KFD. L. S. Breese, J. M. Friedman, and T. A. Steitz, *Biochemistry* 32 (1993): 14095–14101.

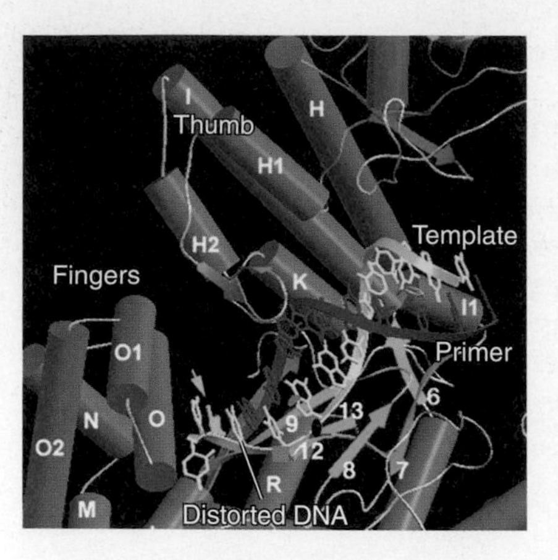

FIGURE 13.7 The crystal structure of phage T7 DNA polymerase shows that the template strand takes a sharp turn that exposes it to the incoming nucleotide. Reprinted with permission from "Crystal structures of the Klenow fragment of DNA polymerase I complexed with deoxynucleoside triphosphate and pyrophosphate" by Lorena S. Beese. Copyright 1993 American Chemical Society.

template base to the incoming nucleotide. The 3′ end of the primer (to which bases are added) is anchored by the fingers and palm. The DNA is held in position by contacts that are made principally with the phosphodiester backbone (thus enabling the polymerase to function with DNA of any sequence).

In structures of DNA polymerases of this family complexed only with DNA (that is, lacking the incoming nucleotide), the orientation of the fingers and thumb relative to the palm is more open, with the O helix (O, O1, O2; see Figure 13.8) rotated away from the palm. This suggests that an inward rotation of the O helix occurs to grasp the incoming nucleotide and create the active catalytic site. When a nucleotide binds, the fingers domain rotates 60° toward the palm, with the tops of the fingers moving by 30 Å. The thumb domain also rotates toward the palm by 8°. These changes are cyclical: they are reversed when the nucleotide is incorporated into the DNA chain, which then translocates through the enzyme to recreate an empty site.

The exonuclease activity is responsible for removing mispaired bases. The catalytic site of the exonuclease domain is distant from the active site of the catalytic domain, though. The enzyme alternates between polymerizing and editing modes, as determined by a competition between the two active sites for the 3′ primer end of the DNA. Amino acids in the active site contact the incoming base in such a way that the enzyme structure is affected by the structure of a mismatched base. When a mismatched base pair occupies the catalytic site, the fingers cannot rotate toward the palm to bind the incoming nucleotide. This leaves the 3′ end free to bind to the active site in the exonuclease domain, which is accomplished by a rotation of the DNA in the enzyme structure.

KEY CONCEPTS

- Many DNA polymerases have a large cleft composed of three domains that resemble a hand.
- DNA lies across the palm in a groove created by the fingers and thumb.

CONCEPT AND REASONING CHECK

How does the DNA polymerase "know" that it has incorporated an incorrect base?

13.6 The Two New DNA Strands Have Different Modes of Synthesis

The antiparallel structure of the two strands of duplex DNA poses a problem for replication. As the replication fork advances, daughter strands must be synthesized on both of the exposed parental single strands. The fork template strand moves in the direction from 5′ to 3′ on one strand, and in the direction from 3′ to 5′ on the other strand. Yet DNA is synthesized only from a 5′ end toward a 3′ end (by adding new nucleotide to the growing 3′ end) on a template that is 3′ to 5′. The problem is solved by synthesizing the new strand on the 5′ to 3′ template in a series of short fragments, each actually synthesized in the "backward" direction, that is, with the customary 5′ to 3′ polarity.

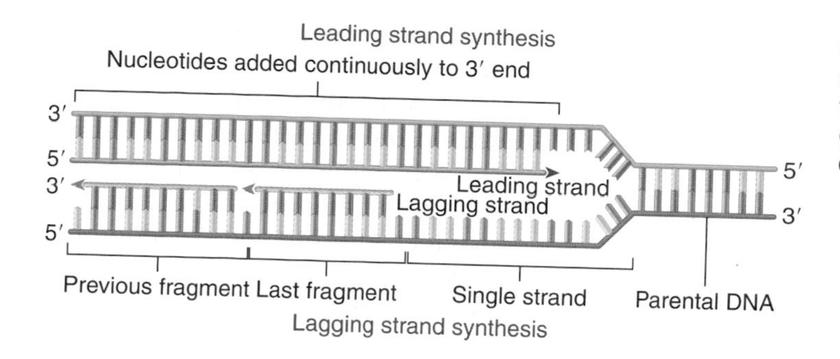

FIGURE 13.8 The leading strand is synthesized continuously, whereas the lagging strand is synthesized discontinuously.

Consider the region immediately behind the replication fork, as illustrated in **FIGURE 13.8**. We describe events in terms of the different properties of each of the newly synthesized strands:

- On the **leading strand** (sometimes called the *forward strand*) DNA synthesis can proceed continuously in the 5′ to 3′ direction as the parental duplex is unwound.
- On the **lagging strand** a stretch of single-stranded parental DNA must be exposed, and then a segment is synthesized in the reverse direction (relative to fork movement). A series of these fragments are synthesized, each 5′ to 3′; they then are joined together to create an intact lagging strand.

Discontinuous replication can be followed by the fate of a very brief label of radioactivity. The label enters newly synthesized DNA in the form of short fragments, of ~1000 to 2000 bases in length. These **Okazaki fragments** are found in replicating DNA in both prokaryotes and eukaryotes. After longer periods of incubation, the label enters larger segments of DNA. The transition results from covalent linkages between Okazaki fragments.

The lagging strand is synthesized discontinuously and the leading strand is synthesized continuously. This is called **semidiscontinuous replication**.

‣ **leading strand** The strand of DNA that is synthesized continuously in the 5′ to 3′ direction.

‣ **lagging strand** The strand of DNA that must grow overall in the 3′ to 5′ direction and is synthesized discontinuously in the form of short fragments (5′–3′) that are later connected covalently.

‣ **Okazaki fragment** Short stretches of 1000 to 2000 bases produced during discontinuous replication; they are later joined into a covalently intact strand.

‣ **semidiscontinuous replication** The mode of replication in which one new strand is synthesized continuously while the other is synthesized discontinuously.

KEY CONCEPT

- The DNA polymerase advances continuously when it synthesizes the leading strand (5′–3′) but synthesizes the lagging strand by making short fragments that are subsequently joined together.

CONCEPT AND REASONING CHECK

Why can't the DNA polymerase move continuously on both strands?

13.7 Replication Requires a Helicase and Single-Strand Binding Protein

As the replication fork advances, it unwinds the duplex DNA. One of the template strands is rapidly converted to duplex DNA as the leading daughter strand is synthesized. The other remains single stranded until a sufficient length has been exposed to initiate synthesis of an Okazaki fragment complementary to the lagging strand in the backward direction. The generation and maintenance of single-stranded DNA is, therefore, a crucial aspect of replication. Two types of function are needed to convert double-stranded DNA to the single-stranded state:

- A **helicase** is an enzyme that separates (or melts) the strands of duplex DNA, usually using the hydrolysis of ATP to provide the necessary energy. Note that the helicase cannot initiate the melting, all it can do is to extend and expand an already open segment.

‣ **helicase** An enzyme that uses energy provided by ATP hydrolysis to separate strands of a nucleic acid.

5′-phosphate of the nick and then a phosphodiester bond is formed with the 3′–OH terminus of the nick, releasing the enzyme and the AMP. Ligases are present in both prokaryotes and eukaryotes.

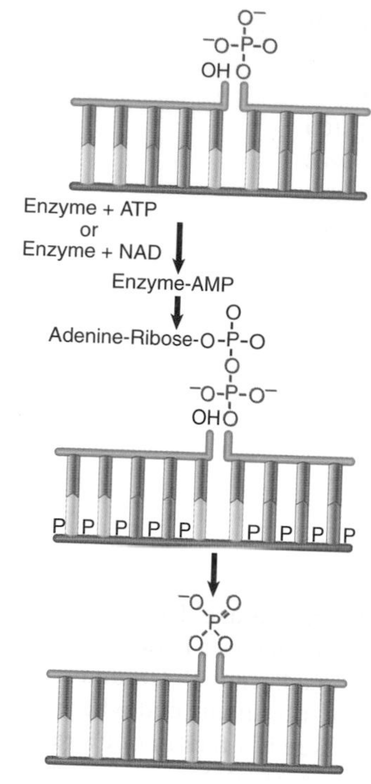

FIGURE 13.20 DNA ligase seals nicks between adjacent nucleotides by employing an enzyme-AMP intermediate.

KEY CONCEPTS

- Each Okazaki fragment starts with a primer and stops before the next fragment.
- In *E. coli,* DNA polymerase I removes the primer and replaces it with DNA.
- DNA ligase makes the bond that connects the 3′ end of one Okazaki fragment to the 5′ beginning of the next fragment.

CONCEPT AND REASONING CHECK

In what processes other than replication would DNA ligase participate?

13.13 Separate Eukaryotic DNA Polymerases Undertake Initiation and Elongation

Eukaryotic replication is similar in most aspects to bacterial replication. It is semiconservative, bidirectional, and semidiscontinuous. Because of the greater amount of DNA in a eukaryote, the genome has multiple replicons. Replication takes place during S phase of the cell cycle. Replicons in euchromatin initiate before replicons in heterochromatin; replicons near active genes initiate before replicons near inactive genes. Origins of replication in eukaryotes are not well defined, except for those in yeast (called *ARS*, *a*utonomously *r*eplicating *s*equences in *S. cerevisiae*). The number of replicons used in any one cycle is tightly controlled. During embryonic development more are activated than in slower growing adult cells.

Eukaryotes have a much larger number of DNA polymerases. They can be broadly divided into those required for replication and repair polymerases involved in repairing damaged DNA. Nuclear DNA replication requires DNA polymerases α, β, and ε. All the other nuclear DNA polymerases are concerned with synthesizing stretches of new DNA to replace damaged material or using damaged DNA as a template (see *Section 13.14, The Primosome Is Needed to Restart Replication,* for the error-prone DNA polymerases). **FIGURE 13.21** shows that most of the nuclear replicases are large heterotetrameric enzymes. In each case, one of the subunits has the responsibility for catalysis, and the others are concerned with ancillary functions, such as priming, processivity, or proofreading. These enzymes all replicate DNA with high fidelity, as does the slightly less complex mitochondrial enzyme. The repair polymerases have much simpler structures, which often consist of a single monomeric subunit (although it may function in the context of a complex of other repair enzymes). Of the enzymes involved in repair, only DNA polymerase β has a fidelity approaching the replication polymerases; all of the others have much greater error rates and are called *error-prone polymerases.* All mitochondrial DNA synthesis, including repair and recombination reactions as well as replication, is undertaken by DNA polymerase γ.

DNA polymerase	Function	Structure
	High-fidelity replicases	
α	Nuclear replication	350 kD tetramer
δ	Lagging strand	250 kD tetramer
ε	Leading strand	350 kD tetramer
γ	Mitochondrial replication	200 kD dimer
	High-fidelity repair	
β	Base excision repair	39 kD monomer
	Low-fidelity repair	
ζ	Base damage bypass	heteromer
η	Thymine dimer bypass	monomer
ι	Required in meiosis	monomer
κ	Deletion and base substitution	monomer

FIGURE 13.21 Eukaryotic cells have many DNA polymerases. The replication enzymes operate with high fidelity. Except for the β enzyme, the repair enzymes all have low fidelity. Replication enzymes have large structures, with separate subunits for different activities. Repair enzymes have much simpler structures.

Function	E. coli	Eukaryote	Phage T4
Helicase	DnaB	MCM complex	41
Loading helicase/primase	DnaC	cdc6	59
Single-strand maintenance	SSB	RPA	32
Priming	DnaG	Polα/primase	61
Sliding clamp	β	PCNA	45
Clamp loading (ATPase)	γδ complex	RFC	44/62
Catalysis	Pol III core	Polδ + Pol ϵ	43
Holoenzyme dimerization	τ	?	43
RNA removal	Pol I	FEN1	43
Ligation	Ligase	Ligase 1	T4 ligase

FIGURE 13.22 Similar functions are required at all replication forks.

Each of the three nuclear DNA replication polymerases has a different function, as summarized in **FIGURE 13.22**:

- DNA polymerase α/primase initiates the synthesis of new strands of DNA by elongating the primer.
- DNA polymerase ε then elongates the leading strand.
- DNA polymerase δ then elongates the lagging strand.

DNA polymerase α is unusual because it has the ability to initiate a new strand. It is used to initiate both the leading and lagging strands. The enzyme exists as a complex consisting of a 180 kD catalytic (DNA polymerase) subunit, which is associated with three other subunits: the B subunit that appears necessary for assembly, and two small subunits that provide the primase (RNA polymerase) activity. Reflecting its dual capacity to prime and extend chains, this complex is sometimes called pol α/primase.

As shown in **FIGURE 13.23**, the pol α/primase complex binds to the initiation complex at the origin and synthesizes a short strand consisting of ~10 bases of RNA followed by 20 to 30 bases of DNA. It is then replaced by an enzyme that will extend the chain. On the leading strand, this is DNA polymerase ε; on the lagging strand, this is DNA polymerase δ. This event is called the *polymerase switch*. It involves interactions among several components of the initiation complex.

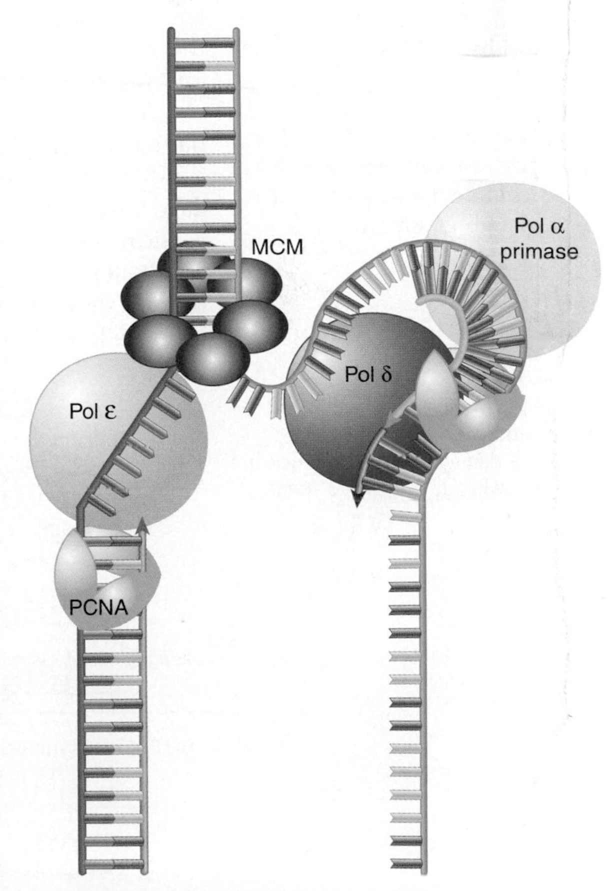

FIGURE 13.23 Three different DNA polymerases make up the eukaryote replication fork. Pol α/primase is responsible for primer synthesis on the lagging strand. The MCM helicase (the eukaryote homolog of DnaB) unwinds the dsDNA, whereas PCNA (a homolog of β) endows the complex with processivity.

DNA polymerase ε is a highly processive enzyme that continuously synthesizes the leading strand. Its processivity results from its interaction with two other proteins, RF-C and PCNA (PCNA stands for proliferating cell nuclear antigen, a historical name resulting from the fact that PCNA is enriched in dividing cells). The roles of RF-C and PCNA are analogous to the *E. coli* γ clamp loader and β processivity unit (see *Section 13.11, The Clamp Controls Association of Core Enzyme with DNA*). RF-C is a clamp loader that catalyzes the loading of PCNA on to DNA. It binds to the 3′ end of the DNA and uses ATP hydrolysis to open the ring of PCNA so that it can encircle the DNA.

The processivity of DNA polymerase δ is maintained by PCNA, which tethers DNA polymerase δ to the template. The crystal structure of PCNA closely resembles the *E. coli* β subunit: a trimer forms a ring that surrounds the DNA. The sequence and subunit organization are different from the dimeric β clamp; however, the function is likely to be similar. DNA polymerase δ elongates the lagging strand, in a manner similar to that of bacterial replication.

A general model suggests that a replication fork contains one complex of DNA polymerase α/primase and two additional DNA polymerase complexes. One is DNA polymerase δ and the other is DNA polymerase ε. The two complexes of DNA polymerase behave in the same ways as the two complexes of DNA polymerase III in the *E. coli* replisome: one synthesizes the leading strand, and the other synthesizes Okazaki fragments on the lagging strand. The exonuclease FEN1 removes the RNA primers of Okazaki fragments. The enzyme DNA ligase I is specifically required to seal the nicks between the completed Okazaki fragments.

KEY CONCEPTS

- A replication fork has one complex of DNA polymerase α/primase and two complexes of DNA polymerase δ and/or ε.
- The DNA polymerase α/primase complex initiates the synthesis of both DNA strands.
- DNA polymerase ε elongates the leading strand and a second DNA polymerase δ elongates the lagging strand.

CONCEPT AND REASONING CHECK

How is priming Okazaki fragments different in eukaryotes than in bacteria?

13.14 Lesion Bypass Requires Polymerase Replacement

Damage to chromosomes that is not repaired before replication can be catastrophic and lethal. When the replication complex encounters damaged and modified bases such that it cannot place a complementary base opposite it, the polymerase stops and the replication fork collapses. A cell has two options to avoid death, **lesion bypass** and recombination (see *Chapter 15, Homologous and Site-Specific Recombination*).

lesion bypass Replication by an error-prone DNA polymerase on a template that contains a damaged base. The polymerase can incorporate a noncomplementary base into the daughter strand.

Both bacteria and eukaryotes have multiple error-prone DNA polymerases that have the ability to synthesize past a lesion on the template (see *Chapter 16, Repair Systems*). These enzymes have this ability because they are not constrained to follow standard base pairing rules. Note that this is not to repair the lesion, but simply to continue replication. This will allow the cell to return to the lesion to repair it.

FIGURE 13.24 compares an advancing replication fork with what happens when there is damage to a base in the DNA or a nick in one strand. In either case, DNA synthesis is halted, and the replication fork either is stalled or is disrupted and collapses. Replication-fork stalling appears to be quite common; estimates for the frequency in *E. coli* suggest that 18% to 50% of bacteria encounter a problem during a replication cycle.

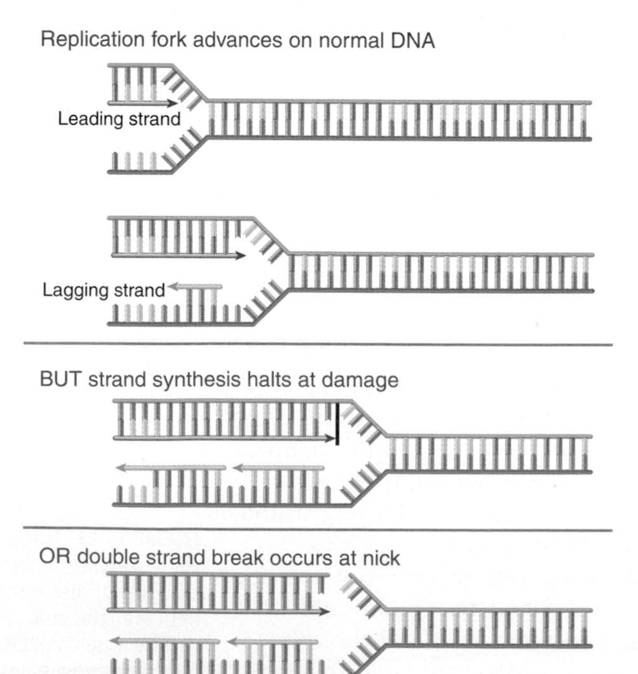

FIGURE 13.24 The replication fork stalls and may collapse when it reaches a damaged base or a nick in DNA. Arrowheads indicate 3′ ends.

E. coli has two error-prone DNA polymerases that can replicate through a lesion, DNA polymerases IV and V (see *Section 16.5, Error-Prone Repair and Translesion Syntheses*). Eukaryotes have five error-prone DNA polymerases with different specificities. When used for lesion bypass during replication, these replace the replisome and are connected to the β-ring temporarily to allow the lesion bypass polymerase to insert nucleotides opposite the lesion. DNA polymerase III then replaces the error-prone polymerase.

Alternatively, the situation can be rescued by a recombination event that excises and replaces the damage or provides a new duplex to replace the region containing the double-strand break. The principle of the repair event is to use the built-in redundancy of information between the two DNA strands. **FIGURE 13.25** shows the key events in such a repair event. Basically, information from the undamaged DNA daughter duplex is used to repair the damaged sequence. This creates a typical recombination junction that is resolved by the same systems that perform homologous recombination. In fact, one view is that the major importance of these systems for the cell is in repairing damaged DNA at stalled replication forks.

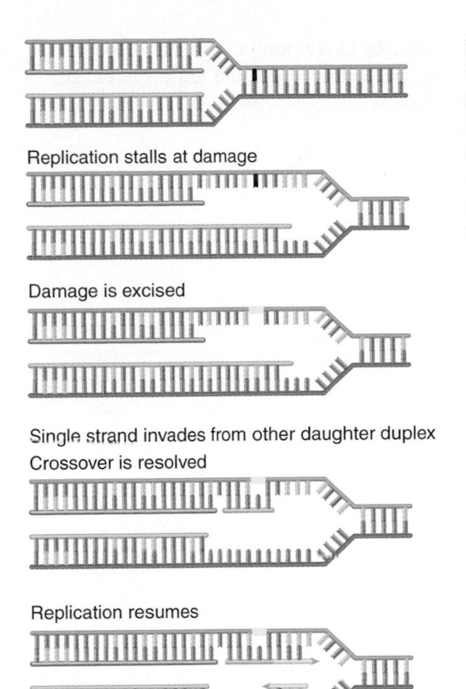

FIGURE 13.25 When replication halts at damaged DNA, the damaged sequence is excised and the complementary (newly synthesized strand) of the other daughter duplex crosses over to repair the gap. Replication can now resume, and the gaps are filled in.

After the damage has been repaired, the replication fork must be restarted. **FIGURE 13.26** shows that this may be accomplished by assembly of the primosome, which in effect reloads DnaB so that helicase action can continue.

Replication fork reactivation is a common (and therefore important) reaction. It may be required in most chromosomal replication cycles. It is impeded by mutations in either the retrieval systems that replace the damaged DNA or in the components of the primosome.

KEY CONCEPTS

- A replication fork stalls when it arrives at damaged DNA.
- After the damage has been repaired, the primosome is required to reinitiate replication.

CONCEPT AND REASONING CHECK

Why do all cells have error-prone DNA polymerase(s) since their use is guaranteed to make mistakes?

13.15 Termination of Replication

Sequences that are involved with termination of replication in *E. coli* are called *ter* sites. A *ter* site contains a short, ~23 bp sequence. The termination sequences are unidirectional; that is, they function in only one orientation. The *ter* site is recognized by a protein called Tus in *E. coli* that recognizes the consensus sequence and prevents the replication fork from proceeding. However, deletion of the *ter* sites does not prevent normal replication cycles from occurring, although it does affect segregation of the daughter chromosomes.

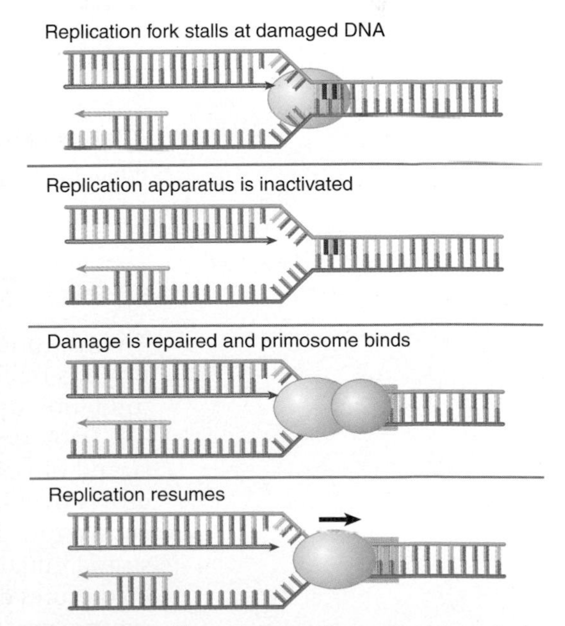

FIGURE 13.26 The primosome is required to restart a stalled replication fork after the DNA has been repaired.

Termination in *E. coli* has the interesting features shown in **FIGURE 13.27**. The two replication forks meet and halt in a region approximately halfway around the chromosome from the origin. In *E. coli*, two clusters of five *ter* sites each, including *terA, D, E, I,* and *K* on one side and *terB, C, F, G,* and *J* on the other side are located ~100 kb on either side of this termination region. Each set of *ter* sites is specific for one direction of fork movement; that is, each set of *ter* sites allows a replication fork into the termination region, but does not allow it out the other side. For example, replication fork 1 can pass through *terB, C, F, G* and *J* into the region but cannot continue pass *terA, D, E, I* and *K*. This arrangement

FIGURE 13.27 Replication termini in *E. coli* are located in a region between two sets of *ter* sites.

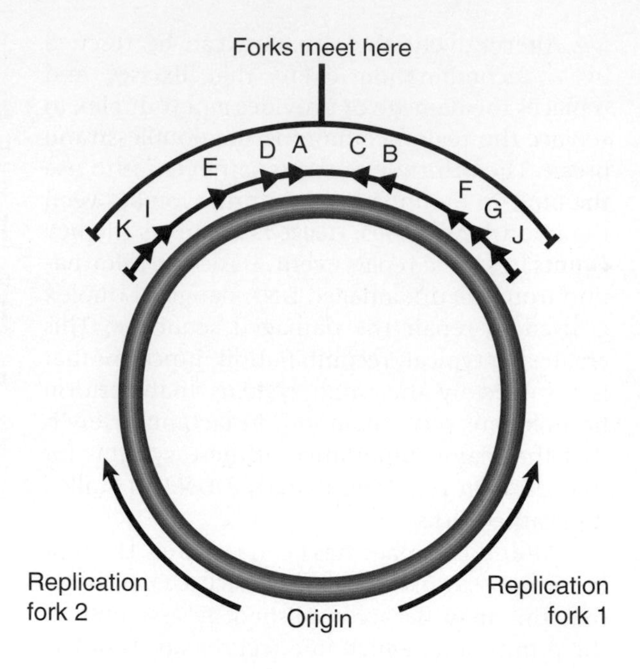

creates a "replication fork trap". If for some reason, one fork is delayed, so that the forks fail to meet in the middle, the faster fork will be trapped at the distal *ter* sites to wait for the slower fork.

The situation is different in eukaryotes that have large linear chromosomes with multiple replicons. The precise mechanism of the merger of two Y forks is not known, but is most likely a modification of the mechanism when two Okazaki fragments are ligated together. At the end of the linear chromosome is the telomere, which has a special mechanism of replication (see *Section 9.11, Telomeres Have Simple Repeating Sequences That Seal the Chromosome Ends*).

KEY CONCEPT

- The two *E. coli* replication forks usually meet halfway around the circle, but there are *ter* sites that halt the replication fork if it advances too far.

CONCEPT AND REASONING CHECK

Why don't eukaryotes use the *ter*/Tus system as *E. coli* does?

13.16 Summary

The transition from initiation of replication to the elongation phase, or semidiscontinuous replication, is complete in *E. coli* when the DnaB helicase is released from its inhibition by DnaC and the DnaG primase is brought in and complexed to DnaB.

DNA synthesis occurs by semidiscontinuous replication, in which the leading strand of DNA growing 5′ to 3′ is extended continuously, but the lagging strand that grows overall in the opposite 3′ to 5′ direction is made as short Okazaki fragments, each synthesized 5′ to 3′. The leading strand and each Okazaki fragment of the lagging strand initiate with an RNA primer that is extended by DNA polymerase. Bacteria and eukaryotes each possess more than one DNA polymerase activity. DNA polymerase III synthesizes both lagging and leading strands in *E. coli*. Many proteins are required for DNA polymerase III action and several constitute part of the replisome within which it functions.

The replisome contains an asymmetric dimer of DNA polymerase III; each new DNA strand is synthesized by a different core complex containing a catalytic (α) subunit. Processivity of the core complex is maintained by the β clamp, which forms a ring around DNA. The clamp is loaded onto DNA by the clamp loader complex. Clamp/clamp loader pairs with similar structural features are widely found in both prokaryotic and eukaryotic replication systems.

The looping model for the replication fork proposes that, as one half of the dimer advances to synthesize the leading strand, the other half of the dimer pulls DNA through as a single loop that provides the template for the lagging strand. The transition from completion of one Okazaki fragment to the start of the next requires the lagging strand catalytic subunit to dissociate from DNA and then reattach to a β clamp at the priming site for the next Okazaki fragment.

The replication apparatus in eukaryotes is very similar to the bacterial model. Different enzymes initiate and elongate the new strands of DNA. DNA polymerase α/primase primes the leading strands at origins and primes the Okazaki fragments. DNA polymerase δ synthesizes Okazaki fragments and DNA polymerase ε synthesizes the leading strand.

Replication is not terminated at specific sequences but continues until DNA polymerase meets another DNA polymerase halfway around a circular replicon, in the *E. coli* chromosome, or at the junction between two linear eukaryote replicons.

CHAPTER QUESTIONS

1. The predominant DNA synthesis activity in an *E. coli* cell extract is by:
A. DNA polymerase I.
B. DNA polymerase II.
C. DNA polymerase III.
D. DNA polymerase IV.

2. A mutation that greatly reduces the activity of *E. coli* DNA polymerase I would result in:
A. quick cell death.
B. continued growth but with a significant increase in mutation rate.
C. continued growth but with a significant decrease in mutation rate.
D. very little or no effect on cell growth and viability.

3. Replication slippage is reduced for DNA polymerases that have:
A. low processivity.
B. high processivity.
C. low proofreading.
D. high proofreading.

4. Proofreading improves the accuracy or fidelity of DNA replication by a factor of about:
A. 2-fold.
B. 10-fold.
C. 50-fold.
D. 100-fold.

5. Which of the following genes encodes the DNA primase protein for bacterial DNA replication?
A *dnaA*
B. *dnaB*
C. *dnaG*
D. *dnaH*

Match each of the bacterial DNA polymerase enzymes from the list at the left with the correct function from the list at the right. Choices may be used more than once.

6. DNA polymerase I	**A.** translesion replication
7. DNA polymerase II	**B.** main replicase for the cell
8. DNA polymerase III	**C.** replication restart
9. DNA polymerase IV	**D.** major repair enzyme
10. DNA polymerase V	**E.** specific replication of Okazaki fragments

KEY TERMS

clamp
clamp loader
conditional lethal
DNA ligase
DNA polymerase
DNA repair
error-prone polymerase
helicase
holoenzyme
lagging strand
leading strand
lesion bypass
mutation hotspot
Okazaki fragment
primase
primer
processivity
proofreading
replisome
semiconservative replication
semidiscontinuous replication
single-strand binding protein (SSB)
topoisomerase

FURTHER READING

Batista, D., Zzaman, S., Prakash, L., and Krings, G. (2008). Replication termination mechanism as revealed by Tus-mediated polar arrest of a sliding helicase. *Proc. Natl. Acad. Sci. USA* **105**, 12831–12836.

Johnson, A., and O'Donnell, M. (2005). Cellular DNA replicases: components and dynamics at the replication fork. *Annu. Rev. Biochem.* **74**, 283–315. A review of both bacterial and eukaryotic DNA replication polymerases.

Prakash, S., Johnson, R. E., and Prakash, L. (2005). Eukaryote translesion synthesis DNA polymerases: specificity of structure and function. *Annu. Rev. Biochem.* **74**, 317–353.

Rursell, Z. F., Isoz, I., Lundström, E.-B., Johansson, E., and Kunkel, T. A. (2007). Yeast DNA polymerase ε participates in leading-strand DNA replication. *Science* **317**, 127–130.

14

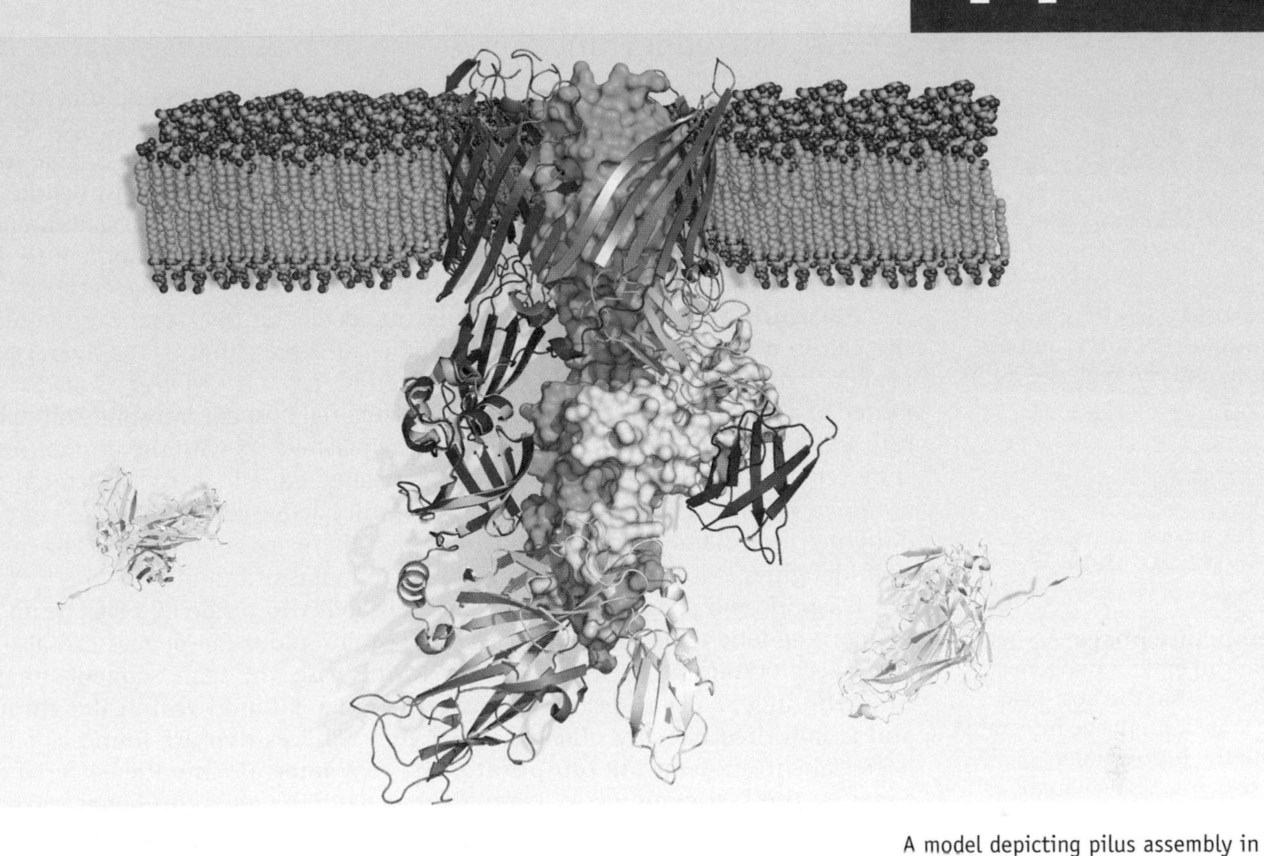

Extrachromosomal Replication

A model depicting pilus assembly in bacteria. The pilus is being built at the bacterial outer membrane on an assembly platform consisting of two "usher" complexes, which include membrane channels through which the pilus components pass. Photo courtesy of Dr. Han Remaut (Structural and Microbiology, VUB/VIB, Belgium) and Gabriel Waksman (ISMB, UK).

CHAPTER OUTLINE

14.1 Introduction

A bacterium may be a host for independently replicating genetic units in addition to its chromosome. These extrachromosomal genomes fall into two general types: plasmids and bacteriophages (phages). Some plasmids and all phages have the ability to transfer from a donor bacterium to a recipient. An important distinction between them is that plasmids exist only as free DNA genomes, whereas bacteriophages are viruses that contain genes required to package a nucleic acid genome into a protein coat and are released from the bacterium at the end of an infective cycle.

- **plasmid** Circular, extrachromosomal DNA. It is autonomous and can replicate itself.

Plasmids are self-replicating circular molecules of DNA that are maintained in the cell in a stable and characteristic number of copies; that is, the average number remains constant from generation to generation. *Low-copy number plasmids* are maintained at a constant quantity relative to the bacterial host chromosome, often between one and ten per bacterium, depending on the plasmid. As with the host chromosome, they rely on a specific apparatus to be segregated equally at each bacterial division. *Multicopy plasmids* exist in many copies per unit bacterium and may be segregated to daughter bacteria stochastically (meaning that there are enough copies to ensure that each daughter cell always gains some by a random distribution).

- **temperate phage** A phage that can enter a lysogenic cycle within the host (can become a prophage integrated into the host genome).
- **lysogenic** The ability of a phage to survive in a bacterium as a stable prophage component of the bacterial genome.
- **episome** A plasmid able to integrate into bacterial DNA.
- **immunity** In phages, the ability of a prophage to prevent another phage of the same type from infecting a cell. In plasmids, the ability of a plasmid to prevent another of the same type from becoming established in a cell. It can also refer to the ability of certain transposons to prevent others of the same type from transposing to the same DNA molecule.

Plasmids and phages are defined by their ability to reside in a bacterium as independent genetic units. However, certain plasmids and some phages can also exist as sequences within the bacterial genome. In this case, the same sequence that constitutes the independent plasmid or phage genome is found within the chromosome and is inherited like any other bacterial gene. Phages that are found as part of the bacterial chromosome are **temperate,** and they integrate into the bacterial chromosome by the **lysogenic** cycle. Plasmids that have the ability to behave like this are called **episomes**. All episomes are plasmids, but not all plasmids are episomes. Related processes are used by phages and episomes to insert into and excise from the bacterial chromosome.

A parallel between lysogenic phages and plasmids and episomes is that they maintain a selfish possession of their bacterium and often make it impossible for another element of the same type to become established. This effect is called **immunity**, although the molecular basis for plasmid immunity is different from lysogenic immunity, and is a consequence of the replication control system.

FIGURE 14.1 summarizes the types of genetic units that can be propagated in bacteria as independent genomes. Virulent phages may have genomes of any type of nucleic acid, DNA or RNA, double stranded or single stranded; they transfer between cells by release of infective particles. Temperate phages have double-stranded DNA genomes, as do plasmids and episomes. Some plasmids transfer between cells by a conjugative process (with direct contact between donor and recipient cells). A feature of the transfer process in both cases is that on occasion some bacterial host genes are transferred with the phage or plasmid DNA, so these events play a role in allowing exchange of genetic information between bacteria.

FIGURE 14.1 Several types of independent genetic units exist in bacteria.

Type of unit	Genome structure	Mode of propagation	Consequences
Lytic phage	ds- or ss-DNA or RNA; linear or circular	Infects susceptible host	Usually kills host
Lysogenic phage	dsDNA	Linear sequence in host chromosome	Immunity to infection
Plasmid	dsDNA circle	Replicates at defined copy number; may be transmittable	Immunity to plasmids in same group
Episome	dsDNA circle	Free circle or linear integration	May transfer host DNA

The key feature in determining the behavior of each type of unit is how its origin is used. An origin in a bacterial or eukaryotic chromosome is used to initiate a single replication event that extends across the replicon. Replicons, however, can also be used to sponsor other forms of replication. The most common alternative is used by the small, independently replicating units of viruses. The objective of a viral replication cycle is to produce many copies of the viral genome before the host cell is lysed to release them. Some viruses replicate in the same way as a host genome, with an initiation event leading to production of duplicate copies, each of which then replicates again, and so on. Others use a mode of replication in which many copies are produced as a tandem array following a single initiation event. A similar type of event is triggered by episomes when an integrated plasmid DNA ceases to be inert and initiates a replication cycle.

Many prokaryotic replicons are circular, and this indeed is a necessary feature for replication modes that produce multiple tandem copies. Some extrachromosomal replicons are linear, though, and in such cases we have to account for the ability to replicate the end of the replicon. (Of course, eukaryotic chromosomes are linear, so the same problem applies to the replicons at each end. These replicons, however, have a special system for resolving the problem.)

14.2 The Ends of Linear DNA Are a Problem for Replication

None of the replicons that we have considered so far have a linear end: either they are circular (as in the *E. coli* genome) or they are part of longer segregation units (as in eukaryotic chromosomes). Linear replicons do occur, though—in some cases as single extrachromosomal units and, of course, at the telomere ends of eukaryotic chromosomes (see *Section 13.15, Termination of Replication*).

The end of a linear replicon poses a problem for DNA polymerases. First, they can synthesize only from 5′ to 3′; second, there must be a primer. Consider the two parental strands depicted in **FIGURE 14.2**. The lower strand presents no problem. It can act as a template to synthesize a daughter strand that runs right up to the end, where presumably the polymerase falls off. To synthesize a complement at the end of the upper strand, however, synthesis must start right at the very last base, or else this strand would become shorter in successive cycles of replication.

No DNA replication polymerase can initiate replication. It can only extend what is there already. We usually think of a polymerase as binding at a site *surrounding* the position at which a base is to be incorporated. Thus a special mechanism must be employed for replication at the ends of linear replicons to provide a 3′–OH group. Several types of solution may be imagined to accommodate the need to copy a terminus:

- The problem may be circumvented by converting a linear replicon into a circular or multimeric molecule. Phages such as T4 or lambda use such mechanisms (see *Section 14.4, Rolling Circles Produce Multimers of a Replicon*).
- The DNA may form an unusual structure—for example, by creating a hairpin at the terminus so that there is no free end. Formation of a crosslink is involved in replication of the linear mitochondrial DNA of *Paramecium*.
- Instead of being precisely determined, the end may be variable. Eukaryotic chromosomes may adopt this solution at the telomere, in which the number of copies of a short repeating unit at the end of the DNA changes (see *Section 9.11, Telomeres Have*

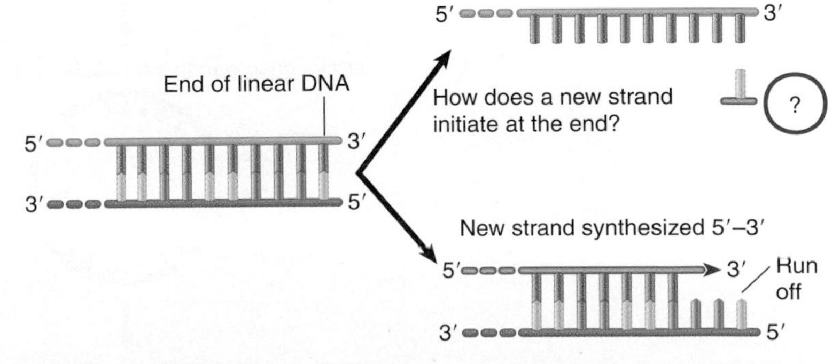

FIGURE 14.2 Replication could run off the 3′ end of a newly synthesized linear strand, but could it initiate at a 5′ end?

Simple Repeating Sequences That Seal the Chromosome Ends). A mechanism to add or remove units makes it unnecessary to replicate right up to the very end.

- A protein may intervene to make initiation possible at the actual terminus. Several linear viral nucleic acids have proteins that are *covalently linked to the 5′ terminal base*. The best characterized examples are adenovirus DNA, phage φ29 DNA, and poliovirus RNA.

KEY CONCEPT

- Special arrangements must be made to replicate the DNA strand with a 5′ end.

CONCEPT AND REASONING CHECK

What is the problem during replication at the end of a linear chromosome?

14.3 Terminal Proteins Enable Initiation at the Ends of Viral DNAs

▸ **strand displacement** A mode of replication of some viruses in which a new DNA strand grows by displacing the previous (homologous) strand of the duplex.

▸ **terminal protein** A protein that allows replication of a linear phage genome to start at the very end. It attaches to the 5′ end of the genome through a covalent bond, is associated with a DNA polymerase, and contains a cytosine residue that serves as a primer.

An example of initiation at a linear end is provided by adenovirus and φ29 DNAs, which actually replicate from both ends using the mechanism of **strand displacement** illustrated in **FIGURE 14.3**. The same events can occur independently at either end. Synthesis of a new strand starts at one end, displacing the homologous strand that was previously paired in the duplex. When the replication fork reaches the other end of the molecule, the displaced strand is released as a free single strand. It is then replicated independently; this requires the formation of a duplex origin by base pairing between some short complementary sequences at the ends of the molecule.

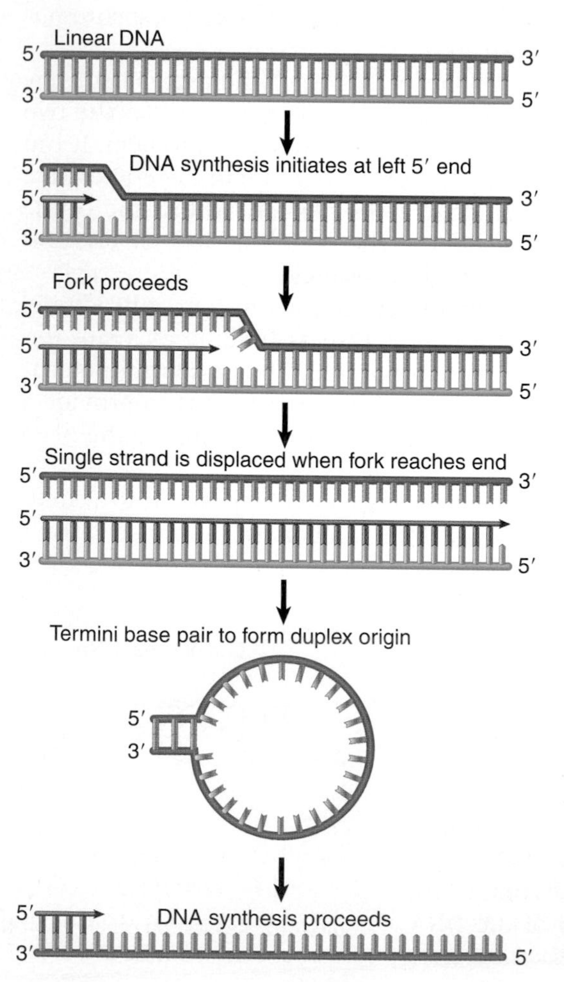

FIGURE 14.3 Adenovirus DNA replication is initiated separately at the two ends of the molecule and proceeds by strand displacement.

In several viruses that use such mechanisms, a protein is found covalently attached to each 5′ end. In the case of adenovirus, a **terminal protein** is linked to the mature viral DNA via a phosphodiester bond to serine, as indicated in **FIGURE 14.4.**

How does the attachment of the protein overcome the initiation problem? The terminal protein has a dual role: it carries a cytidine nucleotide that provides the primer, and it is associated with DNA polymerase. In fact, linkage of terminal protein to a nucleotide is undertaken by DNA polymerase in the presence of adenovirus DNA. This suggests the model illustrated in **FIGURE 14.5**. The complex of polymerase and terminal protein, bearing the priming C nucleotide, binds to the end of the adenovirus DNA. The free 3′–OH end of the C nucleotide is used to prime the elongation reaction by the DNA polymerase. This generates a new strand whose 5′ end is covalently linked to the initiating C nucleotide. (The reaction actually involves displacement of

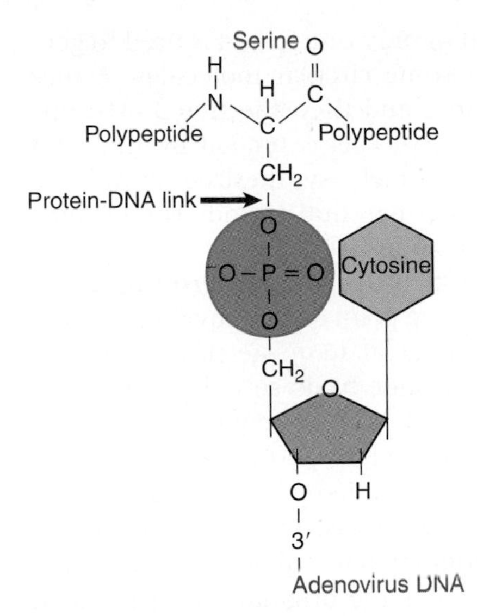

FIGURE 14.4 The 5′ terminal phosphate at each end of adenovirus DNA is covalently linked to serine in the 55 kd Ad-binding protein.

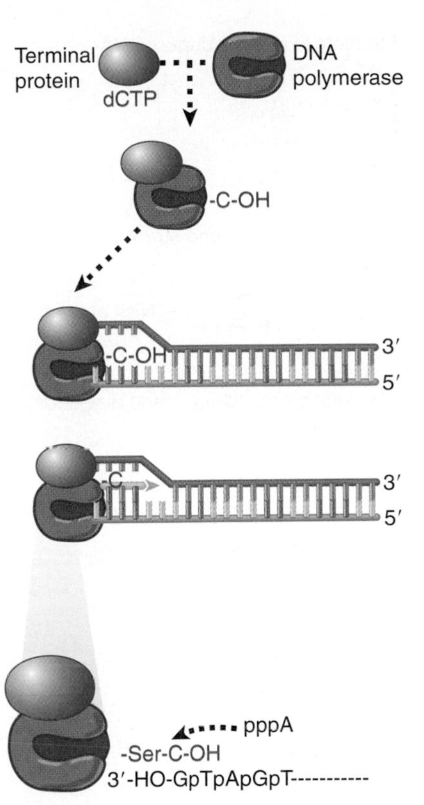

FIGURE 14.5 Adenovirus terminal protein binds to the 5′ end of the DNA and provides a C-OH end to prime synthesis of a new strand.

protein from DNA rather than binding *de novo*. The 5′ end of adenovirus DNA is bound to the terminal protein that was used in the previous replication cycle. The old terminal protein is displaced by the new terminal protein for each new replication cycle.)

Terminal protein binds to the region located between 9 and 18 bp from the end of the DNA. The adjacent region, between positions 17 and 48, is essential for the binding of a host protein, nuclear factor I, which is also required for the initiation reaction. The initiation complex may, therefore, form between positions 9 and 48, a fixed distance from the actual end of the DNA.

KEY CONCEPTS

- The dsDNA viruses adenovirus and φ29 have terminal proteins that initiate replication by generating a new 5′ end.
- The newly synthesized strand displaces the corresponding strand of the original duplex.
- The released strand base pairs at the ends to form a duplex origin that initiates synthesis of the complementary strand.

CONCEPT AND REASONING CHECK

How can a DNA polymerase use a protein as a primer for replication?

14.4 Rolling Circles Produce Multimers of a Replicon

The structures generated by replication depend on the relationship between the template and the replication fork. The critical features are whether the template is circular or linear and whether the replication fork is engaged in synthesizing both strands of DNA or only one.

FIGURE 14.6 The rolling circle generates a multimeric single-stranded tail.

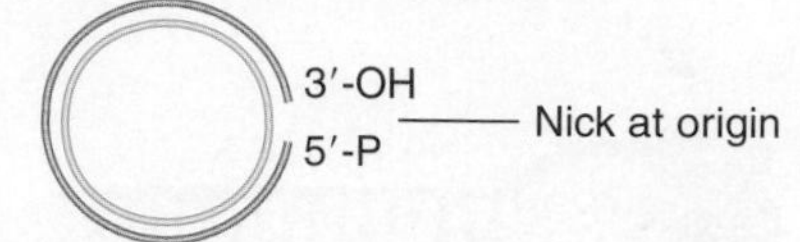

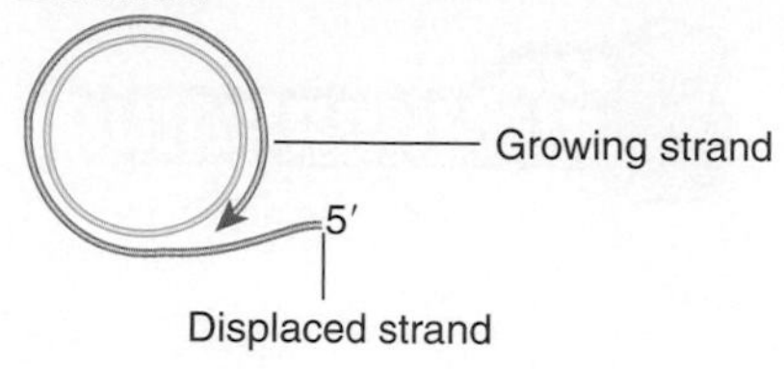

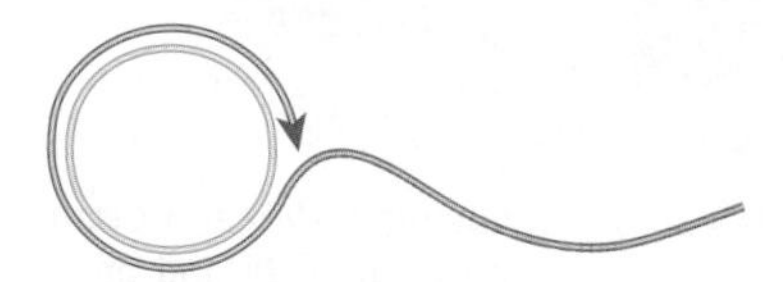

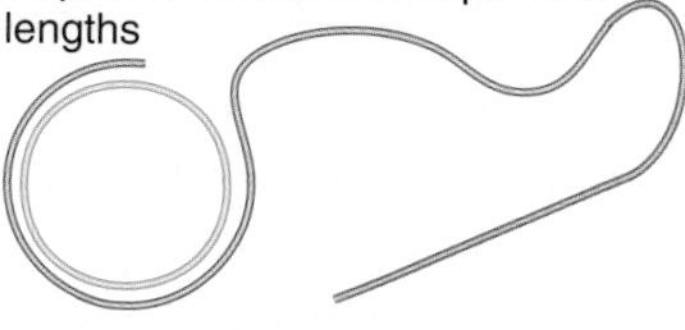

▸ **rolling circle** A mode of replication in which a replication fork proceeds around a circular template for an indefinite number of revolutions; the DNA strand newly synthesized in each revolution displaces the strand synthesized in the previous revolution, giving a tail containing a linear series of sequences complementary to the circular template strand.

Replication of only one strand is used to generate copies of some circular molecules. A nick opens one strand, and then the free 3′–OH end generated by the nick is extended by the DNA polymerase. The newly synthesized strand displaces the original parental strand. The ensuing events are depicted in **FIGURE 14.6.**

This type of structure is called a **rolling circle** because the growing point can be envisaged as rolling around the circular template strand. In principle, it could continue to do so indefinitely. As it moves, the replication fork extends the outer strand and displaces the previous partner. An example is shown in the electron micrograph of **FIGURE 14.7**.

The newly synthesized material is covalently linked to the original material, and as a result the displaced strand has the original unit genome at its 5′ end. The original unit is followed by any number of unit genomes, synthesized by continuing revolutions of the template. Each revolution displaces the material synthesized in the previous cycle.

The rolling circle is used in several ways *in vivo*. Some pathways that are used to replicate DNA are depicted in **FIGURE 14.8.**

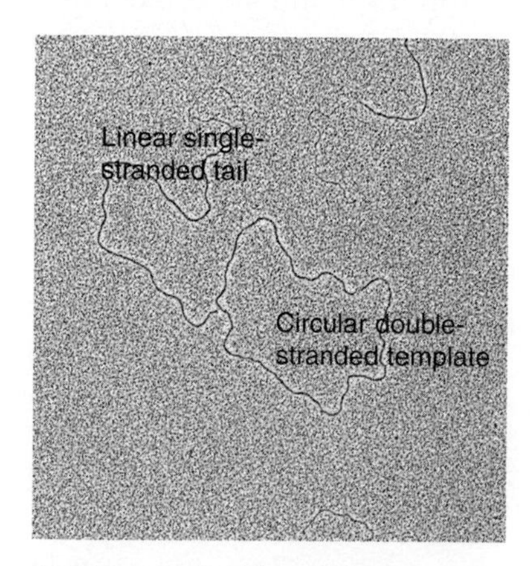

FIGURE 14.7 A rolling circle appears as a circular molecule with a linear tail by electron microscopy. Photo courtesy of Ross B. Inman, Institute of Molecular Virology, Bock Laboratory and Department of Biochemistry, University of Wisconsin, Madison, Wisconsin, USA.

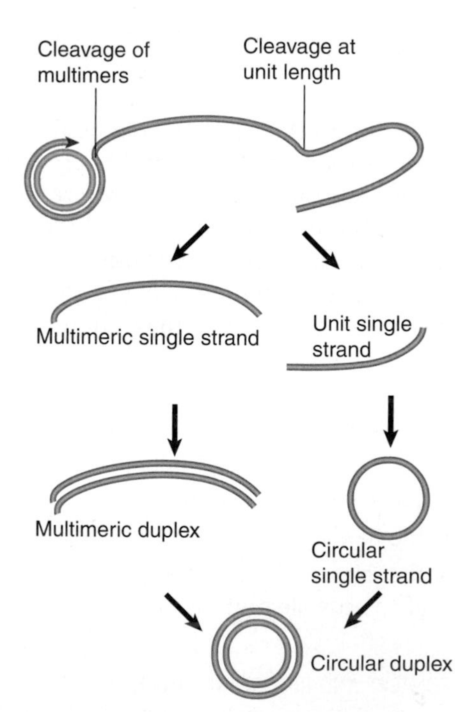

FIGURE 14.8 The fate of the displaced tail determines the types of products generated by rolling circles. Cleavage at unit length generates monomers, which can be converted to duplex and circular forms. Cleavage of multimers generates a series of tandemly repeated copies of the original unit. Note that the conversion to double-stranded form could occur earlier, before the tail is cleaved from the rolling circle.

Cleavage of a unit-length tail generates a copy of the original circular replicon in linear form. The linear form may be maintained as a single strand or may be converted into a duplex by synthesis of the complementary strand (which is identical in sequence to the template strand of the original rolling circle).

The rolling circle provides a means for amplifying the original (unit) replicon. This mechanism is used to generate amplified ribosomal DNA (rDNA) in the *Xenopus* oocyte. The genes for ribosomal RNA (rRNA) are organized as a large number of contiguous repeats in the genome. A single repeating unit from the genome is converted into a rolling circle. The displaced tail, which contains many units, is converted into duplex DNA; later it is cleaved from the circle so that the two ends can be joined together to generate a large circle of amplified rDNA. The amplified material, therefore, consists of a large number of identical repeating units.

KEY CONCEPT

- A rolling circle generates single-stranded multimers of the original sequence.

CONCEPT AND REASONING CHECK

Why is a rolling circle model of replication advantageous for viral replication but not bacterial replication?

14.5 Rolling Circles Are Used to Replicate Phage Genomes

Replication by rolling circles is common among bacteriophages. Unit genomes can be cleaved from the displaced tail, generating monomers that can be packaged into phage particles or used for further replication cycles. Phage φX174 consists of a single-stranded circular DNA known as the plus (+) strand. A complementary strand, called the minus (–) strand, is synthesized as the first step in replication. This action generates the duplex circle shown at the top of **FIGURE 14.9**. Replication then proceeds by a rolling circle mechanism.

The duplex circle is converted to a covalently closed form, which becomes supercoiled. A protein coded by the phage genome, the *A protein*, nicks the (+) strand of the duplex DNA at a specific site that defines the origin for replication. After nicking the origin, the A protein remains connected to the 5′ end that it generates, while the 3′ end is extended by DNA polymerase.

The structure of the DNA plays an important role in this reaction because the DNA can be nicked *only when it is negatively supercoiled* (i.e., wound about its axis in space in the opposite sense from the handedness of the double helix; see *Section 1.5, Supercoiling Affects the Structure of DNA*). The A protein is able to bind to a single-stranded decamer

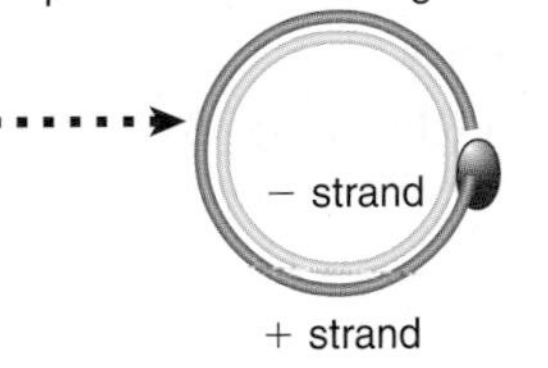

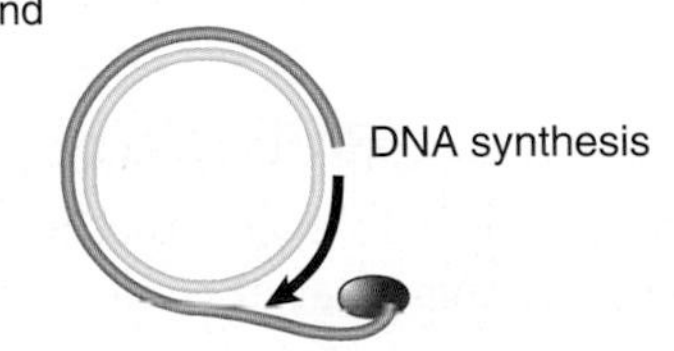

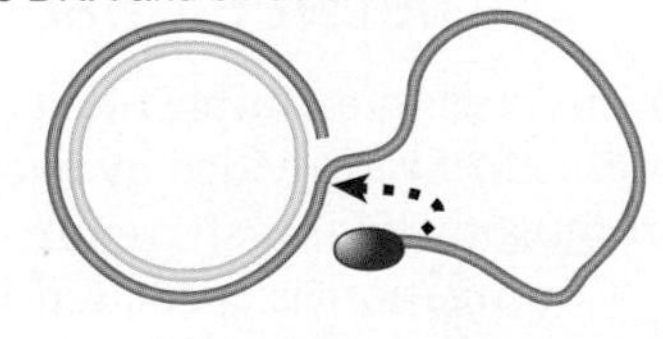

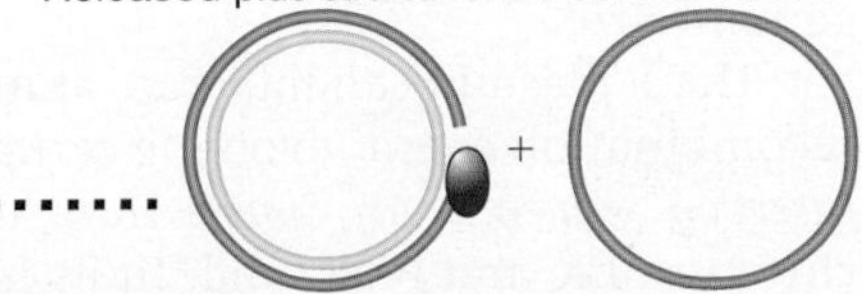

FIGURE 14.9 φX174 RF DNA is a template for synthesizing single-stranded viral circles. The A protein remains attached to the same genome through indefinite revolutions, each time nicking the origin on the viral (+) strand and transferring to the new 5′ end. At the same time, the released viral strand is circularized.

fragment of DNA that surrounds the site of the nick. This suggests that the supercoiling is needed to assist the formation of a single-stranded region that provides the A protein with its binding site. (An enzymatic activity in which a protein cleaves duplex DNA and binds to a released 5′ end is sometimes called a **relaxase**.) The nick generates a 3′–OH end and a 5′–phosphate end (covalently attached to the A protein), both of which have roles to play in φX174 replication.

▸ **relaxase** An enzyme that cuts one strand of DNA and binds to the free 5′ end.

Using the rolling circle, the 3′–OH end of the nick is extended into a new chain. The chain is elongated around the circular (–) strand template until it reaches the starting point and displaces the origin. Now the A protein functions again. It remains connected with the rolling circle as well as to the 5′ end of the displaced tail, and is, therefore, in the vicinity as the growing point returns past the origin. Thus the same A protein is available again to recognize the origin and nick it, now attaching to the end generated by the new nick. The cycle can be repeated indefinitely.

Following this nicking event, the displaced single (+) strand is freed as a circle. The A protein is involved in the circularization. In fact, the joining of the 3′ and 5′ ends of the (+) strand product is accomplished by the A protein as part of the reaction by which it is released at the end of one cycle of replication, and starts another cycle.

The A protein has an unusual property that may be connected with these activities. It is *cis*-acting *in vivo*. (This behavior is not reproduced *in vitro*, as can be seen from its activity on any DNA template in a cell-free system.) *The implication is that in vivo the A protein synthesized by a particular genome can attach only to the DNA of that genome.* We do not know how this is accomplished. Its activity *in vitro*, however, shows how it remains associated with the same parental (–) strand template. The A protein has two active sites; this may allow it to cleave the "new" origin while still retaining the "old" origin; it then ligates the displaced strand into a circle.

The displaced (+) strand may follow either of two fates after circularization. During the replication phase of viral infection, it may be used as a template to synthesize the complementary (–) strand. The duplex circle may then be used as a rolling circle to generate more progeny. During phage morphogenesis, the displaced (+) strand is packaged into the phage virion.

KEY CONCEPT

- The φX174 A protein is a *cis*-acting relaxase that generates single-stranded circles from the tail produced by rolling circle replication.

CONCEPT AND REASONING CHECK

What happens to a virus that uses the rolling circle model of replication but has no means to cut monomers at discrete sites and cuts only unit lengths?

14.6 The F Plasmid Is Transferred by Conjugation between Bacteria

An interesting example of a connection between replication and the propagation of a genetic unit is provided by bacterial **conjugation**, in which a plasmid genome or host chromosome is transferred from one bacterium to another.

▸ **conjugation** A process in which two cells come in contact and transfer genetic material. In bacteria, DNA is transferred from a donor to a recipient cell. In protozoa, DNA passes from each cell to the other.

▸ **F plasmid** An episome that can be free or integrated in *E. coli* and that can sponsor conjugation in either form.

Conjugation is mediated by the **F plasmid**, which is the classic example of an episome—an element that may exist as a free circular plasmid or that may become integrated into the bacterial chromosome as a linear sequence like a temperate bacteriophage. The F plasmid is a large circular DNA ~100 kb in length, 5% of the *E. coli* genome size.

The F plasmid can integrate at multiple sites in the *E. coli* chromosome, often by a recombination event involving certain sequences (called IS sequences; see *Section 17.2, Insertion Sequences Are Simple Transposition Modules*) that are present on both the host chromosome and F plasmid. In its free (plasmid) form, the F plasmid utilizes its own replication origin (*oriV*) and control system and is maintained at a level of one copy per bacterial chromosome. When it is integrated into the bacterial chromosome, this system is suppressed and F DNA is replicated as a part of the chromosome.

The presence of the F plasmid, whether free or integrated, has important consequences for the host bacterium. Bacteria that are F^+, meaning that they contain the episome, are able to conjugate (or mate) with bacteria that are F^-, and do not contain the episome. Conjugation involves physical contact between donor F^+ and recipient F^- bacteria; contact is followed by transfer of the F plasmid. If the F plasmid exists as a free plasmid in the donor bacterium, it is transferred as a plasmid, and the infective process converts the F^- recipient into an F^+ state. If the F plasmid is present in an integrated form in the donor, the transfer process may also cause some or all of the bacterial chromosome to be transferred. Many plasmids have conjugation systems that operate in a generally similar manner, but the F factor was the first to be discovered and remains the paradigm for this type of genetic transfer.

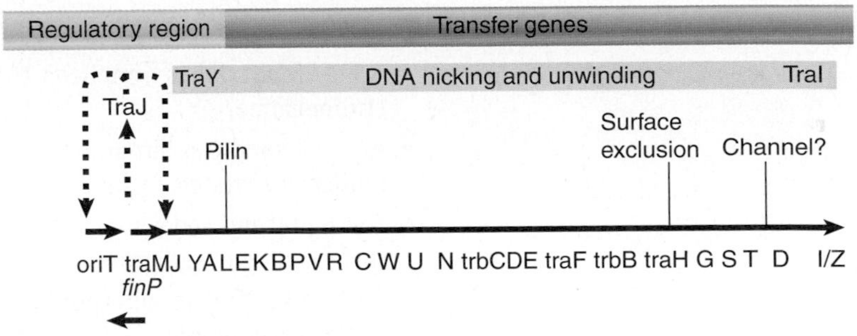

Regulatory region
Transfer genes
TraY
DNA nicking and unwinding
TraI
TraJ
Surface exclusion
Pilin
Channel?
oriT traMJ YALEKBPVR C W U N trbCDE traF trbB traH G S T D I/Z
finP

FIGURE 14.10 The *tra* region of the F plasmid contains the genes needed for bacterial conjugation.

A large (~33 kb) region of the F plasmid called the **transfer region** is required for conjugation. It contains ~40 genes that are required for the transmission of DNA; their organization is summarized in **FIGURE 14.10**. The genes are arranged in loci named *tra* and *trb*. Only four of the *tra* genes, *traD, traI, traM* and *traY*, in the major transcription unit are concerned directly with the transfer of DNA; most of these genes encode proteins that form a large membrane-spanning complex called a type 4 secretion system (T4SS) and are also concerned with the properties of the bacterial cell surface and with maintaining contacts between mating bacteria.

▸ **transfer region** A segment on the F plasmid that is required for bacterial conjugation.

F^+ bacteria possess surface appendages called **pili** (singular *pilus*) that are encoded by the F plasmid. The gene *traA* codes for the single subunit protein, **pilin**, that is polymerized into the pilus. At least 12 *tra* genes are required for the modification and assembly of pilin into the pilus and the stabilization of the T4SS. The F-pili are hairlike structures, 2 to 3 μm long, that protrude from the bacterial surface. A typical F-positive cell has two to three pili. The pilin subunits are polymerized into a hollow cylinder, ~8 nm in diameter, with a 2 nm axial hole.

▸ **pili** A surface appendage on a bacterium that allows the bacterium to attach to other bacterial cells. It appears as a short, thin, flexible rod. During conjugation, it is used to transfer DNA from one bacterium to another.

▸ **pilin** The subunit that is polymerized into the pilus in bacteria.

Mating is initiated when the tip of the F-pilus contacts the surface of the recipient F^- cell. **FIGURE 14.11** shows an example of *E. coli* cells beginning to mate. An F^+ donor cell does not contact other F^+ cells because the genes *traS* and *traT* code for "surface exclusion" proteins that make the cell a poor recipient in such contacts. This effectively restricts donor cells to mating with F^- cells. (The presence of F-pili has secondary consequences: they provide the sites to which RNA phages and some single-stranded DNA phages attach, so F^+ bacteria are susceptible to infection by these phages, whereas F^- bacteria are resistant.)

The initial contact between donor and recipient cells is easily broken, but other *tra* genes act to stabilize the association; this brings the mating cells closer together. The F-pili are essential for initiating pairing but retract or disassemble as part of the process by which the mating cells are brought into close contact. There must be a channel through which DNA is transferred, but the pilus itself does not appear to provide it. It has been proposed that the T4SS provides the channel through which DNA is transferred. TraD is a so-called coupling protein that is necessary for recruitment of plasmid DNA to the T4SS.

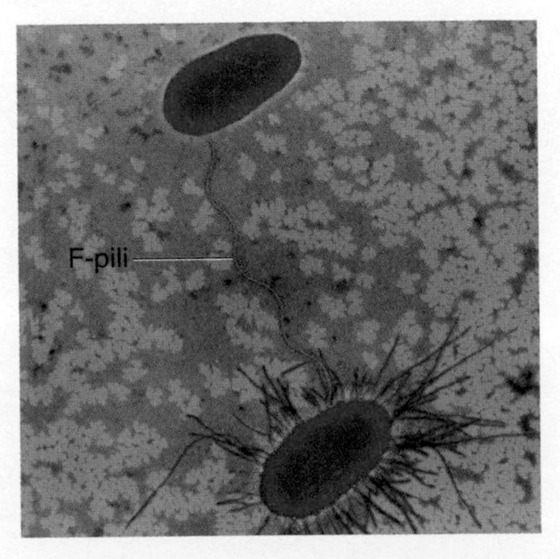

FIGURE 14.11 Mating bacteria are initially connected when donor F-pili contact the recipient bacterium. Photo courtesy of Ron Skurray, School of Biological Sciences, University of Sydney. © Dennis Kunkel/Phototake, Inc./Alamy Images.

KEY CONCEPTS

- A free F plasmid is a replicon that is maintained at the level of one plasmid per bacterial chromosome.
- An F plasmid can integrate into the bacterial chromosome, in which case its own replication system is suppressed.
- The F plasmid codes for a DNA translocation complex and specific pili that form on the surface of the bacterium.
- An F-pilus enables an F-positive bacterium to contact an F-negative bacterium and to initiate conjugation.

CONCEPT AND REASONING CHECK

How can an integrated F factor be used to map the bacterial chromosome?

14.7 Conjugation Transfers Single-Stranded DNA

Transfer of the F plasmid is initiated at a site called *oriT*, the origin of transfer, which is located at one end of the transfer region. The transfer process may be initiated when TraM recognizes that a mating pair has formed. TraY then binds near *oriT* and causes TraI to bind to form the relaxosome in conjunction with host encoded proteins called integration-host factor (IHF). TraI is a relaxase, like φX174 A protein. TraI nicks *oriT* at a unique site (called *nic*) and then forms a covalent link to the 5′ end that has been generated. TraI also catalyzes the unwinding of ~200 bp of DNA and remains attached to the DNA 5′ end throughout the conjugation process (this is a helicase activity). The TraI-bound DNA is then transferred to the T4SS by the coupling protein TraD, where it is exported to the recipient cell. **FIGURE 14.12** shows that the relaxase-bound 5′ end leads the way into the recipient bacterium. The transferred strand is circularized and a complement for the transferred single strand is synthesized in the recipient bacterium, which as a result is converted to the F-positive state.

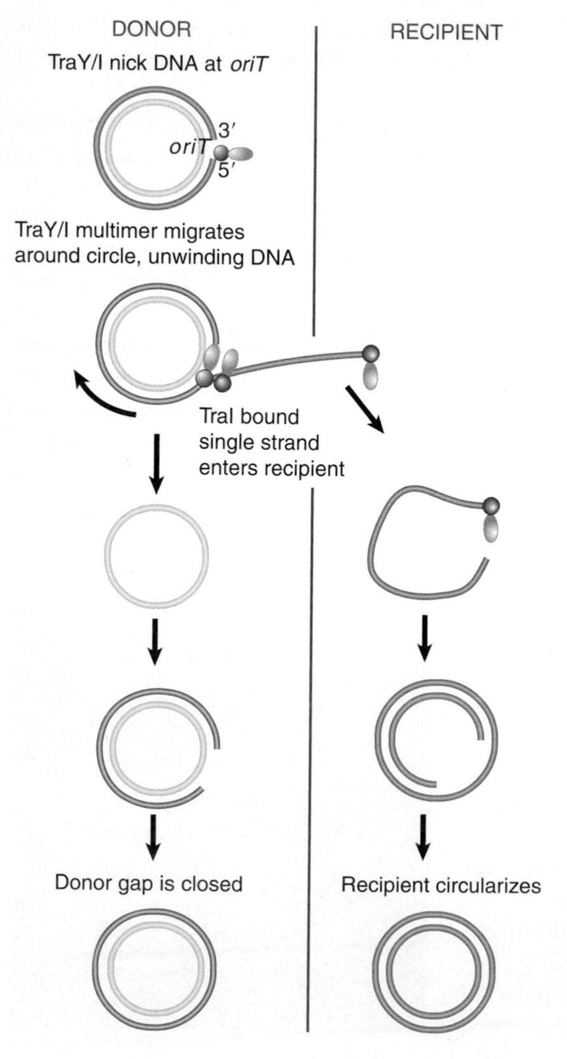

FIGURE 14.12 Transfer of DNA occurs when the F plasmid is nicked at *oriT* and a single strand is led by the 5′ end bound to TraI into the recipient. Only one unit length is transferred. Complementary strands are synthesized to the single strand remaining in the donor and to the strand transferred into the recipient.

A complementary strand must be synthesized in the donor bacterium to replace the strand that has been transferred. If this happens concomitantly with the transfer process, the state of the F plasmid will resemble the rolling circle of **FIGURE 14.13**. DNA synthesis could occur instantly, using the freed 3′ end as a starting point. Conjugating DNA usually appears like a rolling circle, but replication as such is not necessary to provide the driving energy, and single-strand transfer is independent of DNA synthesis.

Only a single unit length of the F factor is transferred to the recipient bacterium. This implies that some feature terminates the process after one revolution, after which the covalent integrity of the F plasmid is restored.

When an integrated F plasmid initiates conjugation, the orientation of transfer is directed away from the transfer region and into the bacterial chromosome. Figure 14.13 shows that, following a short leading sequence of F DNA, bacterial DNA is transferred. The process continues until it is interrupted by the breaking of contacts between the mating bacteria. It takes ~100 minutes to transfer the entire bacterial chromosome, and under standard conditions contact is often broken before the completion of transfer.

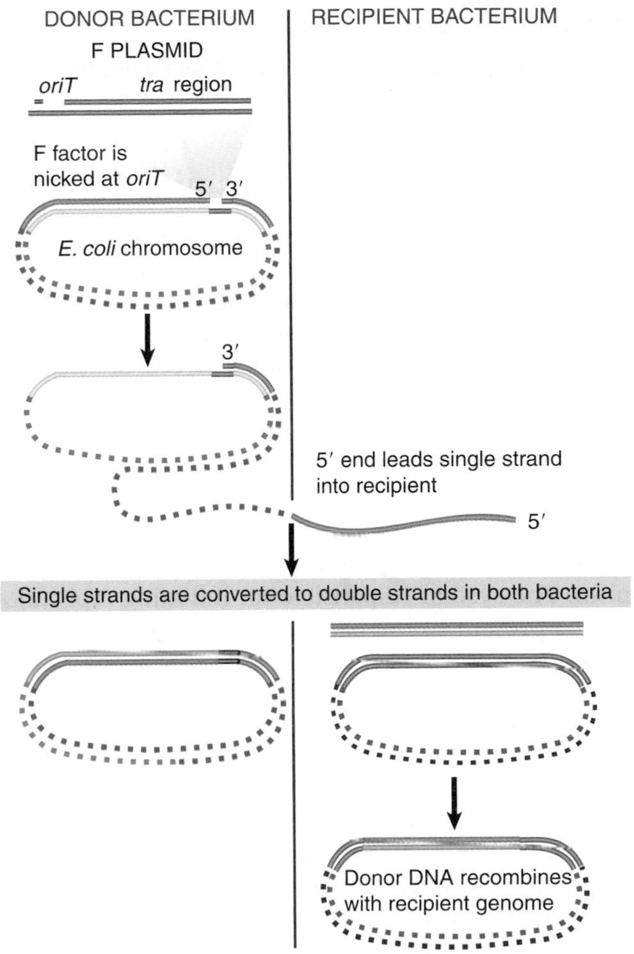

FIGURE 14.13 Transfer of chromosomal DNA occurs when an integrated F plasmid is nicked at *oriT*. Transfer of DNA starts with a short sequence of F DNA and continues until prevented by loss of contact between the bacteria.

Donor DNA that enters a recipient bacterium is converted to double-stranded form and may recombine with the recipient chromosome. (Note that two recombination events are required to insert the donor DNA.) Thus conjugation affords a means to exchange genetic material between bacteria, a contrast with their usual asexual growth (hence the original name fertility factor, or F factor). A strain of *E. coli* with an integrated F plasmid supports such recombination at relatively high frequencies (compared to strains that lack integrated F plasmid); such strains are described as **Hfr** (for *h*igh-*f*requency *r*ecombination). Each position of integration for the F factor gives rise to a different Hfr strain, with a characteristic pattern of transferring bacterial markers to a recipient chromosome.

▸ **Hfr** A bacterium that has an integrated F plasmid within its chromosome. Hfr stands for *high frequency recombination*, referring to the fact that chromosomal genes are transferred from an Hfr cell to an F^- cell much more frequently than from an F^+ cell.

Contact between conjugating bacteria is usually broken before transfer of DNA is complete. As a result, the probability that a region of the bacterial chromosome will be transferred depends upon its distance from *oriT*. Bacterial genes located close to the site of F integration (in the direction of transfer) enter recipient bacteria first, and are therefore found at greater frequency than those that are located farther away and enter later.

KEY CONCEPTS

- Transfer of an F plasmid is initiated when rolling circle replication begins at *oriT*.
- The formation of a relaxosome initiates transfer into the recipient bacterium.
- The transferred DNA is converted into double-stranded form in the recipient bacterium.
- When an F plasmid is free, conjugation "infects" the recipient bacterium with a copy of the F plasmid.
- When an F plasmid is integrated, conjugation causes transfer of the bacterial chromosome until the process is interrupted by (random) breakage of the contact between donor and recipient bacteria.

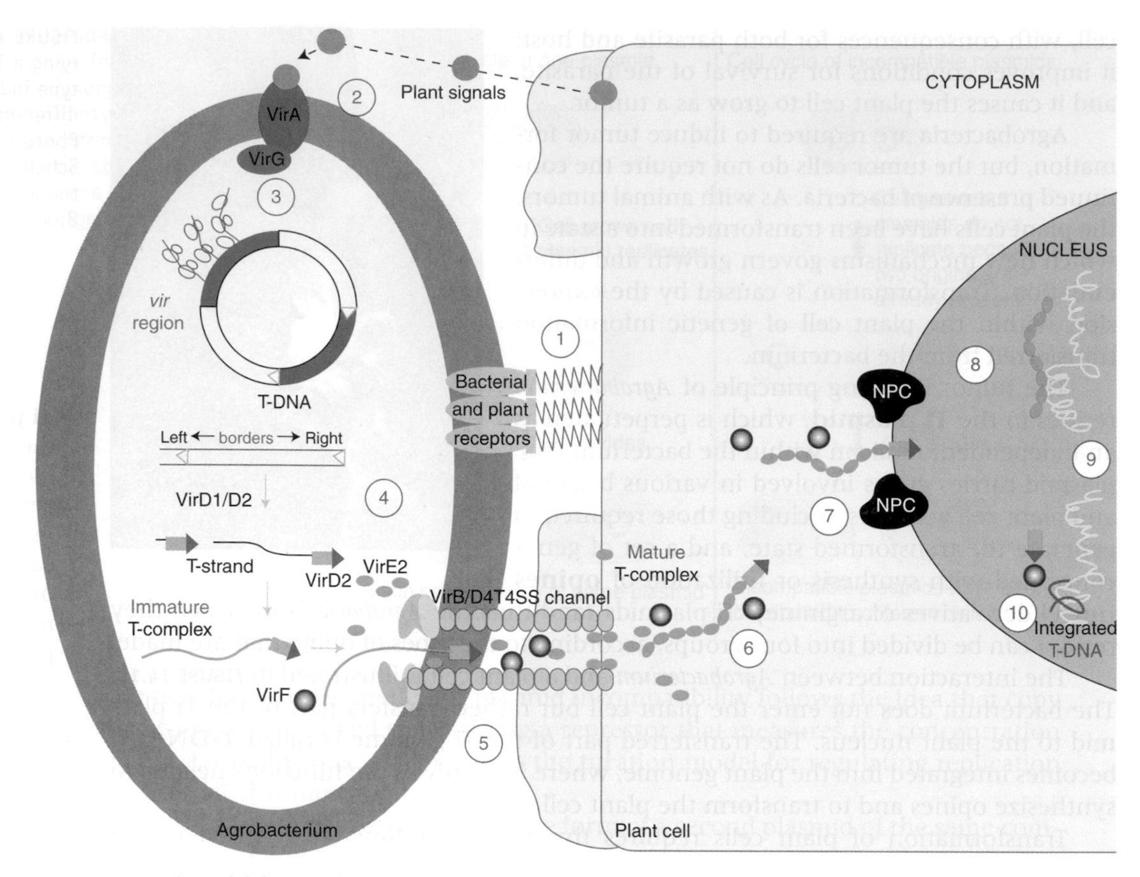

FIGURE 14.19 A model for *Agrobacterium*-mediated genetic transformation. The transformation process comprises ten major steps and begins with recognition and attachment of the *Agrobacterium* to the host cell (1) and the sensing of specific plant signals by the *Agrobacterium* VirA/VirG two component signal transduction system (2). Following activation of the *vir* gene region (3), a mobile copy of the T-DNA is generated by the VirD1/VirD2 protein complex (4) and delivered as a VirD2-DNA complex (immature T-complex), together with several other Vir proteins, into the host cell cytoplasm (5). Following the association of VirE2 with the T-strand, the mature T-complex forms, travels through the host cell cytoplasm (6) and is actively imported into the host cell nucleus (7). Once inside the nucleus, the T-DNA is recruited to the site of integration (8), stripped of its escorting proteins (9), and integrated into the host genome (10). Reprinted from T. Tzfira and V. Citovsky, Agrobacterium-mediated genetic transformation of plants, *Curr. Opin. Biotechnol.* 17, pp. 147–154. Copyright 2006, and with permission from Elsevier (http://www.sciencedirect.com/science/journal/09581669).

bacteria are encoded by the *tra* region). Six loci (*virA, -B, -C, -D, -E,* and *-G*) reside in a 40 kb region outside the T-DNA. Each locus is transcribed as an individual unit; some contain more than one open reading frame. Some of the most important components and their roles are illustrated in **FIGURE 14.19**.

We may divide the transforming process into (at least) two stages:

- *Agrobacterium* contacts a plant cell and the *vir* genes are induced.
- *vir* gene products cause T-DNA to be transferred to the plant cell nucleus, where it integrates into the genome.

The *vir* genes fall into two groups, which correspond to these stages. Genes *virA* and *virG* are regulators that respond to a change in a plant by inducing the other genes. Genes *virB, -C, -D,* and *-E* code for proteins involved in the transfer of DNA.

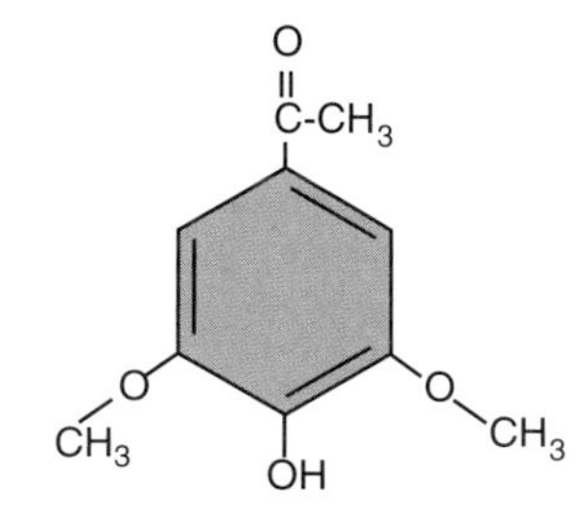

FIGURE 14.20 Acetosyringone (4-acetyl-2,6-dimethoxyphenol) is produced by *N. tabacum* upon wounding and induces transfer of T-DNA from *Agrobacterium*.

The genes virA and virG are expressed constitutively at a low level. The signal to which they respond is provided by phenolic compounds generated by plants as a response to wounding. **FIGURE 14.20** presents an example. *Nicotiana tabacum* (tobacco) generates the molecules acetosyringone and α-hydroxyacetosyringone. Exposure to these compounds activates virA, which acts on virG, which in turn induces the expression of virB, C, D, E. This reaction explains why Agrobacterium infection succeeds only on wounded plants.

KEY CONCEPTS

- Infection with the bacterium *A. tumefaciens* can transform plant cells into tumors.
- The infectious agent is a plasmid carried by the bacterium.
- The plasmid also carries genes for synthesizing and metabolizing opines (arginine derivatives) that are used by the bacterium.
- Part of the DNA of the Ti plasmid is transferred to the plant cell nucleus, but the *vir* genes outside this region are required for the transfer process.

CONCEPT AND REASONING CHECK

How is crown gall disease similar to cancer in humans?

14.11 Transfer of T-DNA Resembles Bacterial Conjugation

The transfer process selects the T-region for entry into the plant. **FIGURE 14.21** shows that the T-DNA of a nopaline plasmid is demarcated from the flanking regions in the Ti plasmid by repeats of 25 bp, which differ at only two positions between the left and right ends. When T-DNA is integrated into a plant genome, it has a well-defined right junction, which retains 1 to 2 bp of the right repeat. The left junction is variable; the boundary of T-DNA in the plant genome may be located at the 25 bp repeat or at one of a series of sites extending over ~100 bp within the T-DNA. Sometimes multiple tandem copies of T-DNA are integrated at a single site.

The *virD* locus has four open reading frames. Two of the proteins coded at *virD*, VirD1 and VirD2, provide an endonuclease that initiates the transfer process nicking T-DNA at a specific site. A model for transfer is illustrated in **FIGURE 14.22**. A nick is made at the right 25 bp repeat. It provides a priming end for synthesis of a DNA single strand. Synthesis of the new strand displaces the old strand, which is used in the transfer process. The transfer is terminated when the DNA synthesis reaches a nick at the left repeat. This model explains why the right repeat is essential, and it accounts for the polarity of the process. If the left repeat fails to be nicked, production of DNA for transfer could continue farther along the Ti plasmid.

The single molecule of single-stranded DNA produced for transfer in the infecting bacterium is transferred as a DNA-protein complex, sometimes called the T-complex. The DNA is covered by the VirE2 single-strand binding protein, which has a nuclear localization signal and is responsible for transporting T-DNA into the plant cell nucleus. A single molecule of the D2 subunit of the endonuclease remains bound at the 5′ end. The *virB* operon codes for 11 products that are involved in the transfer process.

This model for transfer of T-DNA closely resembles the events involved in bacterial conjugation, when the *E. coli* chromosome is transferred from one cell to another in single-stranded form. The genes of the *virB* operon are homologous to the *tra* genes

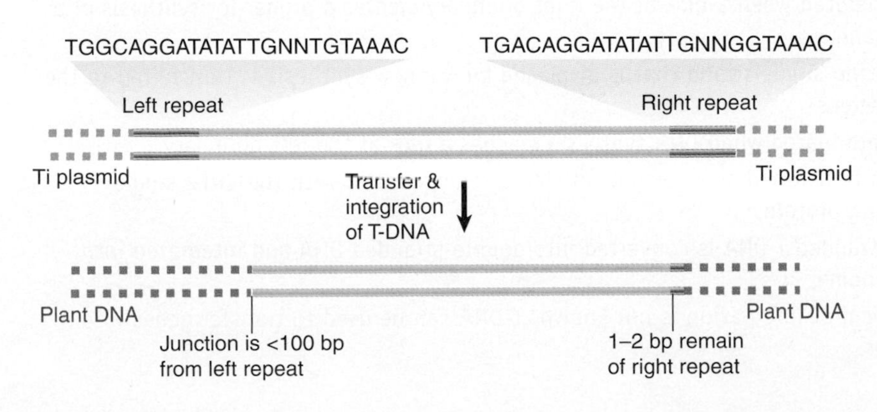

FIGURE 14.21 T-DNA has almost identical repeats of 25 bp at each end in the Ti plasmid. The right repeat is necessary for transfer and integration to a plant genome. T-DNA that is integrated in a plant genome has a precise junction that retains 1 to 2 bp of the right repeat, but the left junction varies and may be up to 100 bp short of the left repeat.

FIGURE 14.22 T-DNA is generated by displacement when DNA synthesis starts at a nick made at the right repeat. The reaction is terminated by a nick at the left repeat.

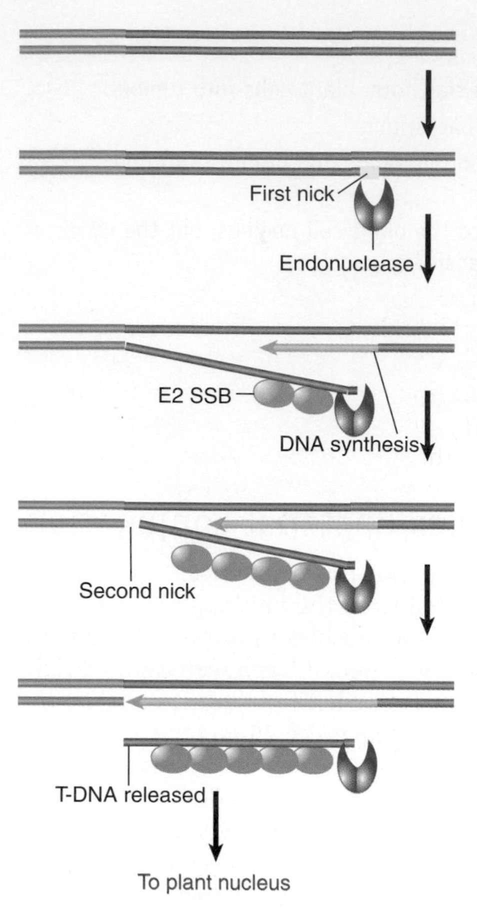

of certain bacterial plasmids that are involved in conjugation (see *Section 14.7, Conjugation Transfers Single-Stranded DNA*). Together with VirD4 (a coupling protein), the gene products of the *virB* genes form a T4SS.

We do not know how the transferred DNA is integrated into the plant genome. At some stage, the newly generated single strand must be converted into duplex DNA. Circles of T-DNA that are found in infected plant cells appear to be generated by recombination between the left and right 25 bp repeats, but we do not know if they are intermediates. The actual event is likely to involve a nonhomologous recombination, because there is no homology between the T-DNA and the sites of integration.

The Ti plasmid presents an interesting organization of functions. Outside the T-region, it carries genes needed to initiate oncogenesis; at least some are concerned with the transfer of T-DNA, and we would like to know whether others function in the plant cell to affect its behavior at this stage. Also outside the T-region are the genes that enable the *Agrobacterium* to catabolize the opine that the transformed plant cell will produce. Within the T-region are the genes that control the transformed state of the plant, as well as the genes that cause it to synthesize the opines that will benefit the *Agrobacterium* that originally provided the T-DNA.

As a practical matter, the ability of *Agrobacterium* to transfer T-DNA to the plant genome makes it possible to introduce new genes into plants. The transfer/integration and oncogenic functions are separate; thus it is possible to engineer new Ti plasmids in which the oncogenic functions have been replaced by other genes whose effect on the plant we wish to test. The existence of a natural system for delivering genes to the plant genome has greatly facilitated genetic engineering of plants.

KEY CONCEPTS

- The *vir* genes are induced by phenolic compounds released by plants in response to wounding.
- The membrane protein VirA is autophosphorylated on histidine when it binds an inducer and activates VirG by transferring the phosphate to it.
- T-DNA is generated when a nick at the right boundary creates a primer for synthesis of a new DNA strand.
- The pre-existing single strand that is displaced by the new synthesis is transferred to the plant cell nucleus.
- Transfer is terminated when DNA synthesis reaches a nick at the left boundary.
- The T-DNA is transferred as a complex of single-stranded DNA with the VirE2 single strand-binding protein.
- The single-stranded T-DNA is converted into double-stranded DNA and integrated into the plant genome.
- The mechanism of integration is not known. T-DNA can be used to transfer genes into a plant nucleus.

CONCEPT AND REASONING CHECK

What features of the Ti plasmid make it attractive to use as a system to transfer genes into plants?

14.12 How Do Mitochondria Replicate and Segregate?

Mitochondria must be duplicated during the cell cycle and segregated to the daughter cells. We understand some of the mechanics of this process but not its regulation.

At each stage in the duplication of mitochondria—DNA replication, DNA segregation to duplicate mitochondria, and organelle segregation to daughter cells—the process appears to be stochastic, governed by a random distribution of each copy. The theory of distribution in this case is analogous to that of multicopy bacterial plasmids, with the same conclusion that >10 copies are required to ensure that each daughter gains at least one copy (see *Section 14.8, Single-Copy Plasmids Have a Partition System*). When there are mtDNAs with allelic variations in the same cell, called **heteroplasmy** (either because of inheritance from different parents or because of mutation), the stochastic distribution may generate cells that have only one of the alleles.

▸ **heteroplasmy** Having more than one mitochondrial allelic variant in a cell.

Replication of mitochondrial DNA may be stochastic because there is no control over which particular copies are replicated, so that in any cycle some mtDNA molecules may replicate more times than others. The total number of copies of the genome may be controlled by titrating mass in a way similar to bacteria (see *Section 11.2, Bacterial Replication Is Connected to the Cell Cycle*).

A mitochondrion divides by developing a ring around the organelle that constricts to pinch it into two halves. The mechanism is similar in principle to that involved in bacterial division. The apparatus that is used in plant cell mitochondria is similar to that used in bacteria and uses a homolog of the bacterial protein FtsZ (see *Section 11.5, FtsZ Is Necessary for Septum Formation*). The molecular apparatus is different in animal cell mitochondria and uses the protein dynamin, which is involved in formation of membrane vesicles. An individual organelle may have more than one copy of its genome.

We do not know whether there is a partition mechanism for segregating mtDNA molecules within the mitochondrion or whether they are simply inherited by daughter mitochondria according to which half of the mitochondrion they happen to lie in. **FIGURE 14.23** shows that the combination of replication and segregation mechanisms can result in a stochastic assignment of DNA to each of the copies, that is, so that the distribution of mitochondrial genomes to daughter mitochondria does not depend on their parental origins.

The assignment of mitochondria to daughter cells at mitosis also appears to be random. Indeed, it was the observation of somatic variation in plants that first suggested the existence of genes

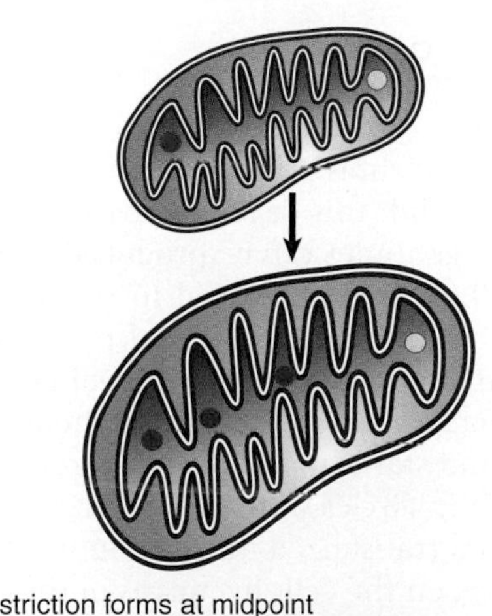

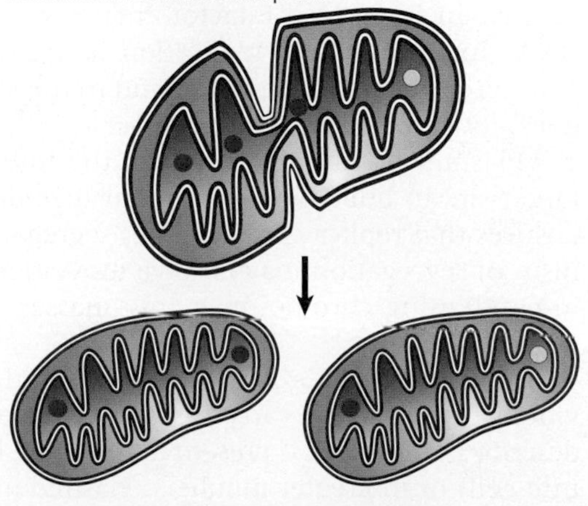

FIGURE 14.23 Mitochondrial DNA replicates by increasing the number of genomes in proportion to mitochondrial mass but without ensuring that each genome replicates the same number of times. This can lead to changes in the representation of alleles in the daughter mitochondria.

that could be lost from one of the daughter cells because they were not inherited according to Mendel's laws.

In some situations a mitochondrion has both paternal and maternal alleles. This has two requirements: that both parents provide alleles to the zygote (which, of course, is not the case when there is maternal inheritance; see *Section 5.8, Some Organelles Have DNA*) and that the parental alleles are found in the same mitochondrion. For this to happen, parental mitochondria must have fused.

The size of the individual mitochondrion may not be precisely defined. Indeed, there is a continuing question as to whether an individual mitochondrion represents a unique and discrete copy of the organelle or whether it is in dynamic flux in which it can fuse with other mitochondria. We know that mitochondria can fuse in yeast, because recombination between mtDNAs can occur after two haploid yeast strains have mated to produce a diploid strain. This implies that the two mtDNAs must have been exposed to one another in the same mitochondrial compartment. Attempts have been made to test for the occurrence of similar events in animal cells by looking for complementation between alleles after two cells have fused, but the results are not clear.

KEY CONCEPTS

- mtDNA replication and segregation to daughter mitochondria is stochastic.
- Mitochondrial segregation to daughter cells is also stochastic.

CONCEPT AND REASONING CHECK

Why is the mitochondrial replication system not identical to eukaryotic nuclear replication?

14.13 Summary

The rolling circle is an alternative form of replication for circular DNA molecules in which an origin is nicked to provide a priming end. One strand of DNA is synthesized from this end; this displaces the original partner strand, which is extruded as a tail. Multiple genomes can be produced by continuing revolutions of the circle.

Rolling circles are used to replicate some phages. The A protein that nicks the φX174 origin has the unusual property of *cis*-action. It acts only on the DNA from which it was synthesized. It remains attached to the displaced strand until an entire strand has been synthesized and then nicks the origin again; this releases the displaced strand and starts another cycle of replication.

Rolling circles also characterize bacterial conjugation, which occurs when an F plasmid is transferred from a donor to a recipient cell following the initiation of contact between the cells by means of the F-pili. A free F plasmid infects new cells by this means; an integrated F factor creates an Hfr strain that may transfer chromosomal DNA. In conjugation, replication is used to synthesize complements to the single strand remaining in the donor and to the single strand transferred to the recipient but does not provide the motive power.

Plasmid partitioning involves the interaction of the ParB protein with the *parS* target site to build a structure that includes the IHF protein. This partition complex ensures that replica chromosomes segregate into different daughter cells. The mechanism of segregation may involve movement of DNA, possibly by the action of MukB in condensing chromosomes into masses at different locations as they emerge from replication.

Plasmids have a variety of systems that ensure or assist partition, and an individual plasmid may carry systems of several types. The copy number of a plasmid describes whether it is present at the same level as the bacterial chromosome (one per unit cell) or in greater numbers. Plasmid incompatibility can be a consequence of the mechanism involved in either replication or partition (for single-copy plasmids). Two plasmids that share the same control system for replication are incompatible because the number of replication events ensures that there is only one plasmid for each bacterial genome.

Agrobacteria induce tumor formation in wounded plant cells. The wounded cells secrete phenolic compounds that activate *vir* genes carried by the Ti plasmid of the bacterium. The *vir* gene products cause a single strand of DNA from the T-DNA region of the plasmid to be transferred to the plant cell nucleus. Transfer is initiated at one boundary of T-DNA, but ends at variable sites. The single strand is converted into a double strand and integrated into the plant genome. Genes within the T-DNA transform the plant cell and cause it to produce particular opines (derivatives of arginine). Genes in the Ti plasmid allow Agrobacteria to metabolize the opines produced by the transformed plant cell. T-DNA has been used to develop vectors for transferring genes into plant cells.

Mitochondrial replication and segregation appears to be stochastic rather than tightly controlled.

CHAPTER QUESTIONS

How does each of the linear genomes in the list at the left replicate its full genome? Choose the correct mechanism for each from the list at the right. Choices may be used more than once.

1. Lambda phage

2. Adenovirus

3. T4 phage

A. A hairpin is created at the terminus so there is no free end.

B. A protein covalently binds to the 5′ terminal base to provide an OH group for DNA synthesis.

C. The linear genome may convert to a circular or multimeric form for replication.

4. Lytic phages may have genomes composed of:
- **A.** single-stranded RNA.
- **B.** single-stranded DNA.
- **C.** double-stranded DNA.
- **D.** any of the above.

5. Bacterial plasmids may have genomes composed of:
- **A.** single-stranded RNA.
- **B.** single-stranded DNA.
- **C.** double-stranded DNA.
- **D.** any of the above.

6. Bacterial conjugation can result in transfer of:
- **A.** only mRNA transcripts.
- **B.** plasmid DNA sequences only.
- **C.** plasmid and host chromosome DNA sequences.
- **D.** host chromosome DNA sequences only.

7. What DNA replication mechanism is commonly used by lytic bacteriophages to enable rapid generation of new phage genomes?
- **A.** terminal-protein mediated replication of a linear genome
- **B.** bidirectional DNA replication from a specific origin
- **C.** unidirectional DNA replication from a specific origin
- **D.** rolling circle DNA replication

8. F-positive bacteria possess surface appendages called:
- **A.** flagella, hairlike structures used for mobility.
- **B.** flagella, hairlike structures used to conjugate with other cells.
- **C.** pili, hairlike structures used for mobility.
- **D.** pili, hairlike structures used to conjugate with other cells.

9. The F plasmid is present in how many copies per cell?
   A. one
   B. two
   C. variable, from one to several
   D. high-copy number
10. T-DNA is transferred from the bacterial cell to the plant cell as a:
   A. single-stranded DNA beginning from a nick at the leftward T-DNA repeat.
   B. single-stranded DNA beginning from a nick at the rightward T-DNA repeat.
   C. double-stranded DNA beginning from a double-strand cut at the leftward T-DNA repeat.
   D. double-stranded DNA beginning from a double-strand cut at the rightward T-DNA repeat.

KEY TERMS

compatibility group
conjugation
copy number
crown gall disease
episome
F plasmid
Hfr
heteroplasmy
immunity
lysogenic
opine
pili
pilin
plasmid
relaxase
rolling circle
strand displacement
temperate phage
terminal protein
T-DNA
Ti plasmid
transfer region

FURTHER READING

Falkenberg, M., Larsson, N.-G., and Gutafsson, C. M. (2007). DNA replication and transcription in mammalian mitochodria. *Annu. Rev. Biochem.* **76**, 679–699.

Ghosh, S. K., Hajra, S., Paek, A., and Jayaram, M. (2006). Mechanisms for chromosomal and plasmid segregation. *Annu. Rev. Biochem.* **75**, 211–241.

Lutkenhaus, J. (2007). Assembly dynamics of the bacterial MinCDE system and special resolution of the Z ring. *Annu. Rev. Biochem.* **76**, 539–562.

Ulher, B., Li, Y., Logemann, E., Somssich, I. E., and Weisshaas, B. (2008). T-DNA mediated transfer of *Agrobacterium tumefaciens* chromosomal DNA into plants. *Nat. Biotechnol.* **26**, 1015–1017.

Zzaman, S., Abhyanker, M. M., and Batistia, D. (2004). Reconstruction of F factor DNA replication *in vitro* with purified proteins. *J. Biolog. Chem.* **279**, 17404–17410.

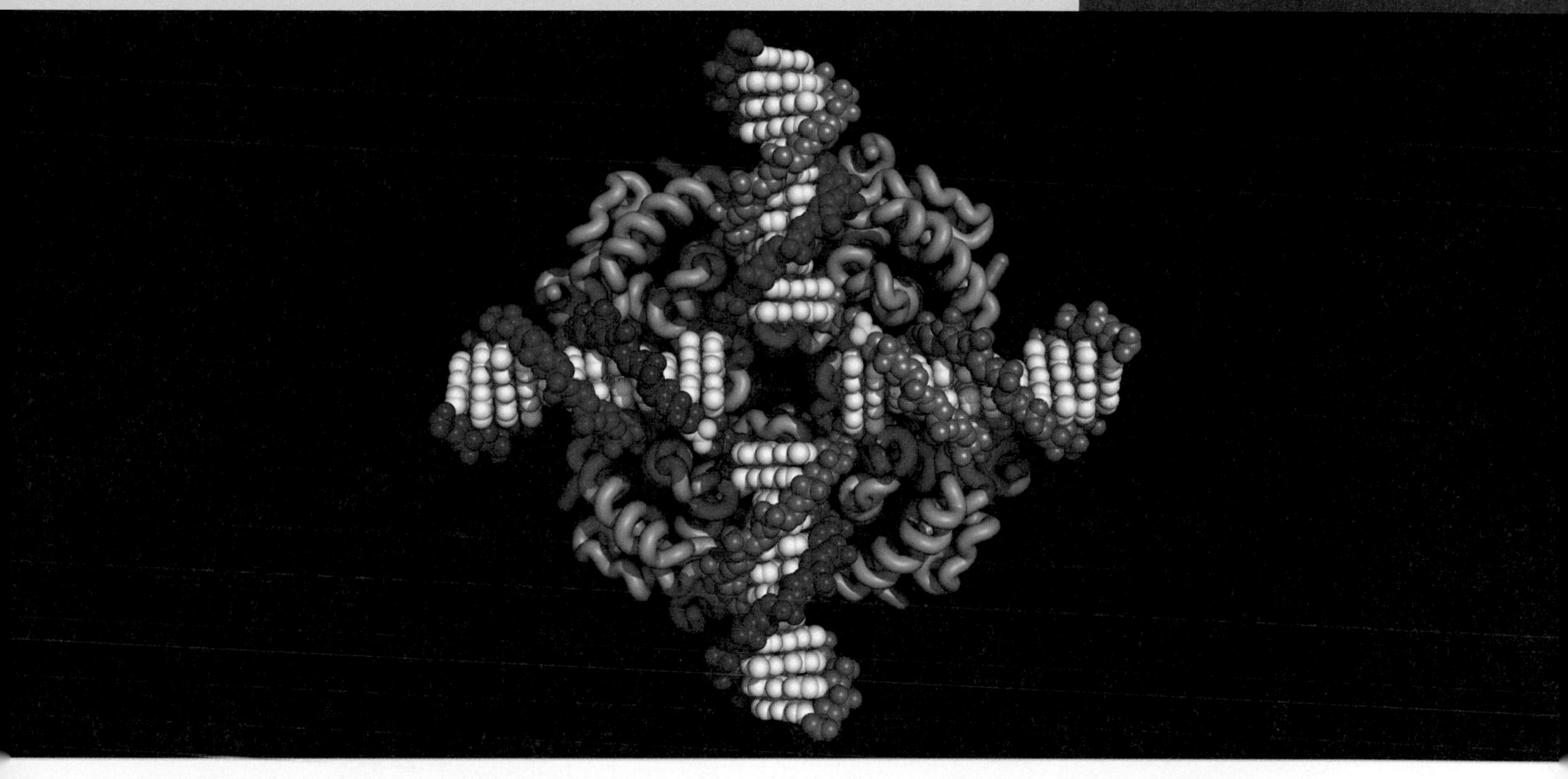

A model demonstrating the interaction of a RuvA tetramer with a DNA Holliday junction, an intermediate of recombination. Reproduced with permission from J. B. Rafferty et al, *Science* 274 (1996): 415–421. © 1996 AAAS. Photo courtesy of David W. Rice and John B. Rafferty, University of Sheffield.

Homologous and Site-Specific Recombination

CHAPTER OUTLINE

15.1 Introduction

Homologous recombination is an essential cellular process required for creating genetic diversity, ensuring proper chromosome segregation, and repairing certain types of DNA damage. Genetic recombination is essential for the process of evolution. If it were not possible to exchange material between (homologous) chromosomes, the content of each individual chromosome would be irretrievably fixed (except for mutation) in its particular alleles. When mutations occurred, it would not be possible to separate favorable and unfavorable changes. The length of the target for mutation damage would effectively be increased from the gene to the chromosome. Ultimately a chromosome would accumulate so many deleterious mutations that it would fail to function.

By shuffling the genes between chromosomes, recombination allows favorable and unfavorable mutations to be separated and tested as individual units in new assortments. It provides a means of escape and spreading for favorable alleles and a means to eliminate an unfavorable allele without changing allele frequencies for all the other genes with which this allele is linked. This is the basis for natural selection.

In addition to its role in genetic diversity, homologous recombination is also required for repair of lesions at replication forks and for restarting replication that has stalled at these lesions. The importance of these recombination events is highlighted by examples of human diseases that result from defects in recombination repair of DNA damage. Homologous recombination is also essential for a process known as antigenic switching, which allows disease-causing parasites known as trypanosomes to evade the human immune system. This is discussed in the accompanying box, *Medical Applications*, in this chapter.

Recombination occurs between precisely corresponding sequences, so that not a single base pair is added or lost from the recombinant chromosomes (although the presence of repeated sequences can lead to unequal crossovers, which result in deletions and insertions; see Figure 17.1 for an example). Three types of recombination share the feature that the process involves physical exchange of material between duplex DNAs:

- Recombination involving a reaction between homologous sequences of DNA is called generalized or **homologous recombination**. In eukaryotes, it occurs during meiosis, usually both in males (during spermatogenesis) and females (during oogenesis) in higher eukaryotes. This occurs at the "four-strand" stage of meiosis and involves only two of the four strands (see *Section 2.6, Recombination Occurs by Physical Exchange of DNA*). The same basic mechanisms are used for certain types of DNA repair (discussed in *Chapter 16, Repair Systems*).
- Another type of event sponsors recombination between specific pairs of sequences. This was first characterized in prokaryotes, where specialized recombination, also known as **site-specific recombination**, is responsible for the integration of phage genomes into the bacterial chromosome. The recombination event involves specific sequences of the phage DNA and bacterial DNA, which include a short stretch of homology. The enzymes involved in this event act only on the particular pair of target sequences in an intermolecular reaction.
- In special circumstances, gene rearrangement is used to control gene expression. Rearrangement may create new genes, which are needed for expression in particular circumstances, as in the case of the immunoglobulins. This example of **somatic recombination** will be discussed in *Chapter 18, Somatic Recombination and Hypermutation in the Immune System*. Recombination events can also be responsible for switching expression from one preexisting gene to another, as in the example of yeast mating type, where the sequence at an active locus can be replaced by a sequence from a silent locus. Rearrangements are also involved in controlling expression of surface antigens in trypanosomes, in which silent alleles of surface antigen genes are duplicated into active expression sites. Some of these types of rearrangement share mechanistic similarities with transposition (see *Chapter 17, Transposable Elements and Retroviruses*).

▸ **homologous recombination** Recombination involving a reciprocal exchange of sequences of DNA, e.g., between two chromosomes that carry the same genetic loci.

▸ **site-specific recombination** Recombination that occurs between two specific sequences, as in phage integration/excision or resolution of cointegrated structures during transposition.

▸ **somatic recombination** Recombination that occurs in nongerm cells (i.e., it does not occur during meiosis); most commonly used to refer to recombination in the immune system.

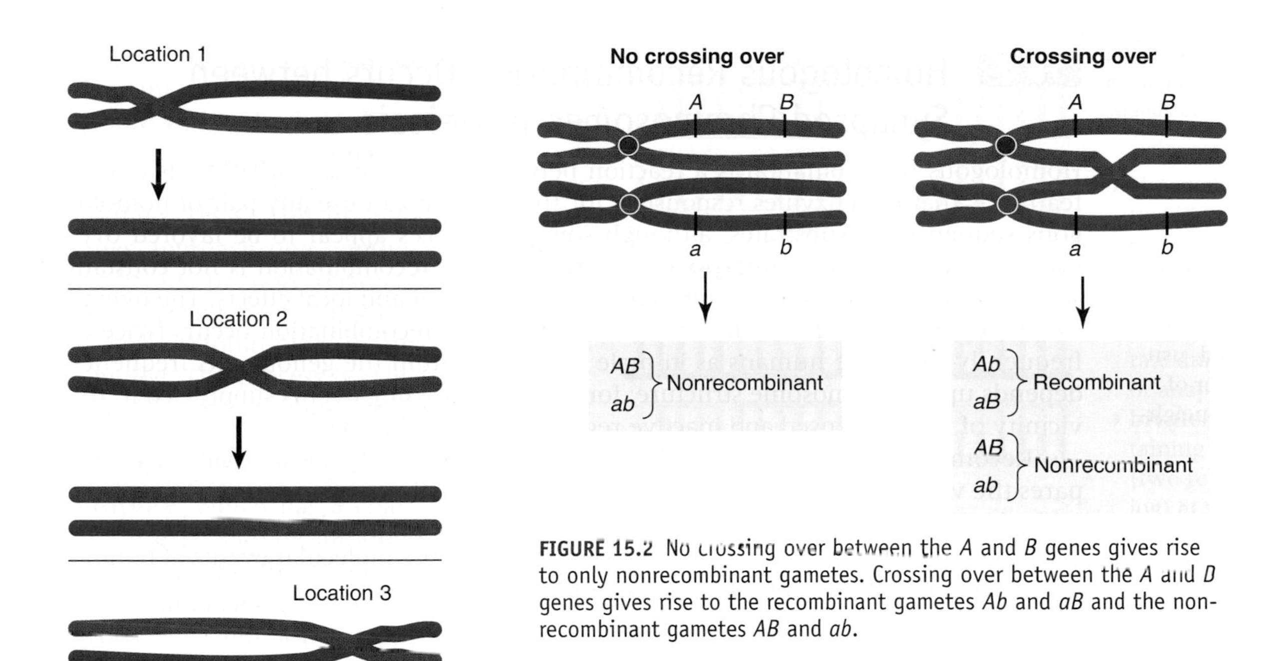

FIGURE 15.1 Homologous recombination can occur at any point along the lengths of two homologous DNAs.

FIGURE 15.2 No crossing over between the *A* and *B* genes gives rise to only nonrecombinant gametes. Crossing over between the *A* and *B* genes gives rise to the recombinant gametes *Ab* and *aB* and the nonrecombinant gametes *AB* and *ab*.

Let's consider the nature and consequences of homologous and site-specific recombination reactions.

FIGURE 15.1 shows that homologous recombination occurs between two homologous DNA duplexes and can take place at any point along their length. **FIGURE 15.2** shows the results of a crossover with respect to two loci, *A* and *B*. The two chromosomes are cut at equivalent points, and then each is joined to the other to generate reciprocal recombinants. The crossover is the point at which each becomes joined to the other. There is no change in the overall organization of DNA; the products have the same structure as the parents, and both parents and products are homologous.

Site-specific, or specialized, recombination occurs *only* between specific sites. The results depend on the locations of the two recombining sites. **FIGURE 15.3** shows that an intermolecular recombination between a circular DNA and a linear DNA inserts the circular DNA into the linear DNA. Similar reactions can result excision of a region of DNA, or conversion of a single circular DNA into two circles. Site-specific recombination is often used to make changes such as these in the organization of DNA. We have a large amount of information about the enzymes that undertake site-specific recombination, which are related to the topoisomerases that act to change the supercoiling of DNA (see *Section 1.5, Supercoiling Affects the Structure of DNA*).

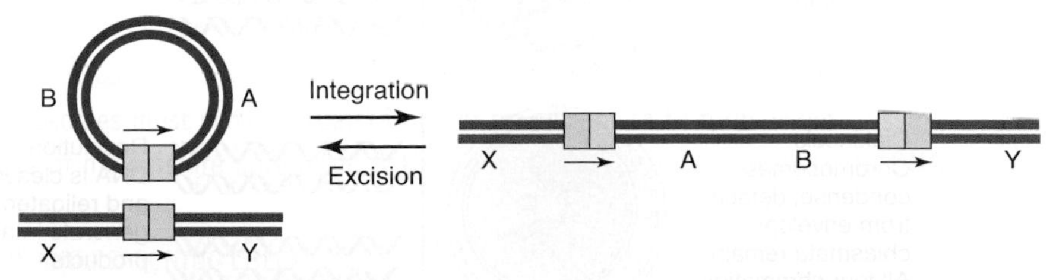

FIGURE 15.3 Site-specific recombination occurs between two specific sequences (identified by boxes). The other sequences in the two recombining DNAs are not homologous.

KEY CONCEPTS

- Recombination is initiated by making a double-strand break in one (recipient) DNA duplex.
- Exonuclease action generates 3′ single-stranded ends that invade the other (donor) duplex.
- When a single strand from one duplex displaces its counterpart in the other duplex, it creates a D-loop.
- The exchange generates a stretch of heteroduplex DNA consisting of one strand from each parent.
- New DNA synthesis replaces any material that has been degraded.
- This generates a recombinant joint molecule in which the two DNA duplexes are connected by heteroduplex DNA.
- Whether recombinants are formed depends on whether the strands involved in the original exchange or the other pair of strands are nicked during resolution.

CONCEPT AND REASONING CHECK

Is it possible to tell whether a DNA gap was generated during a homologous recombination event that occurs at a heterozygous locus? Why or why not?

15.4 Recombining Chromosomes Are Connected by the Synaptonemal Complex

A basic paradox in recombination is that the parental chromosomes never seem to be in close enough contact for recombination of DNA to occur. The chromosomes enter meiosis in the form of replicated (sister chromatid) pairs, which are visible as a mass of chromatin. They pair to form the synaptonemal complex, and it has been assumed for many years that this represents some stage involved with recombination—possibly a necessary preliminary to exchange of DNA. A more recent view is that the synaptonemal complex is a consequence rather than a cause of recombination, but we have yet to define how the structure of the synaptonemal complex relates to molecular contacts between DNA molecules.

Synapsis begins when each chromosome (sister chromatid pair) condenses around a proteinaceous structure called the **axial element**. The axial elements of corresponding chromosomes then become aligned, and the synaptonemal complex forms as a tripartite structure, in which the axial elements, now called **lateral elements**, are separated from each other by a **central element**. **FIGURE 15.9** shows an example.

▸ **axial element** A proteinaceous structure around which the chromosomes condense at the start of synapsis.

▸ **lateral element** A structure in the synaptonemal complex that forms when a pair of sister chromatids condenses on to an axial element.

▸ **central element** A structure that lies in the middle of the synaptonemal complex, along which the lateral elements of homologous chromosomes align. It is formed from Zip proteins.

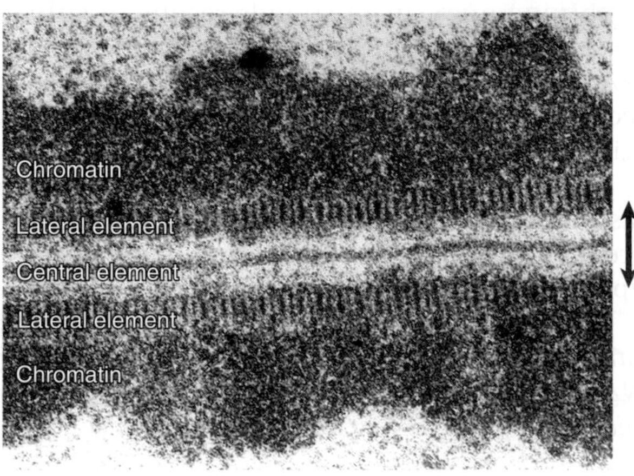

FIGURE 15.9 The synaptonemal complex brings chromosomes into juxtaposition. Reproduced from D. Von Wettstein, *Proc. Natl. Acad. Sci. USA* 68 (1971): 851–855. Photo courtesy of D. Von Wettstein, Washington State University.

Each chromosome at this stage appears as a mass of chromatin bounded by a lateral element. The two lateral elements are separated from each other by a fine, but dense, central element. The triplet of parallel dense strands lies in a single plane that curves and twists along its axis. The distance between the homologous chromosomes is greater than 200 nm, which is considerable in molecular terms (the diameter of DNA is 2 nm). Thus a major problem in understanding the role of the complex is that although it aligns homologous chromosomes, it is far from bringing homologous DNA molecules into contact.

The only visible link between the two sides of the synaptonemal complex is provided by spherical or cylindrical structures observed in fungi and insects. They lie across the complex and are called **recombination nodules (nodes)**; they occur with the same frequency and distribution as the chiasmata. Their name reflects the possibility that they may prove to be the sites of recombination.

▸ **recombination nodules (nodes)** Dense objects present on the synaptonemal complex; they may represent protein complexes involved in crossing over.

From mutations that affect synaptonemal complex formation, we can identify the types of proteins that are involved in its structure. **FIGURE 15.10** presents a molecular view of the synaptonemal complex. Its distinctive structural features are due to two groups of proteins:

- The *cohesins* form a single linear axis for each pair of sister chromatids from which loops of chromatin extend. This is equivalent to the lateral element of Figure 15.9. (The cohesins belong to a general group of proteins involved in connecting sister chromatids so that they segregate properly at mitosis or meiosis.)
- The lateral elements are connected by transverse filaments that are equivalent to the central element of Figure 15.9. These are formed from *Zip proteins*.

Mutations in genes for proteins that are needed for lateral elements to form are found in the genes coding for cohesins. The cohesins that are used in meiosis include Smc3 (which is also used in mitosis) and Rec8 (which is specific to meiosis and is related to the mitotic cohesin Scc1). The cohesins appear to bind to specific sites along the chromosomes in both mitosis and meiosis. They are likely to play a structural role in chromosome segregation. At meiosis, the formation of the lateral elements may be necessary for the later stages of recombination, because although these mutations do not prevent the formation of double-strand breaks, they do block formation of recombinants.

Mutation of the *ZIP1* gene allows lateral elements to form and to become aligned, but they do not become closely synapsed. The N-terminal domain of the Zip1 protein is localized in the central element, but the C-terminal domain is localized in the lateral elements. Zip1 also interacts with a number of other proteins, including Zip2–4, Mer3, and Msh4/5, collectively known as the ZMM proteins. The Zip proteins form transverse filaments that connect the lateral elements of the sister chromatid pairs, while the Mer3, Msh4, and Msh5 proteins promote recombination.

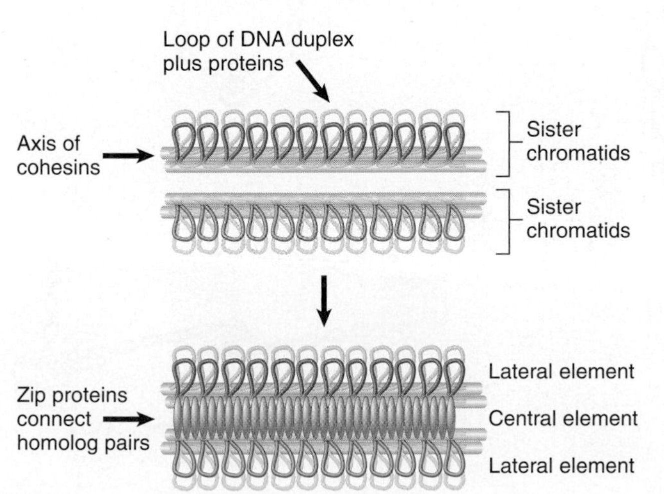

FIGURE 15.10 Each pair of sister chromatids has an axis made of cohesins. Loops of chromatin project from the axis. The synaptonemal complex is formed by linking together the axes via Zip proteins.

KEY CONCEPTS

- During the early part of meiosis, homologous chromosomes are paired in the synaptonemal complex.
- The mass of chromatin of each homolog is separated from the other by a proteinaceous complex.
- Cohesins and Zip proteins form the lateral elements and transverse filaments/central elements.

CONCEPT AND REASONING CHECK

Based on its structure, explain how the synaptonemal complex could actually represent an obstacle to recombination.

15.5 Specialized Enzymes Catalyze 5′ End Resection and Single-Strand Invasion

The central unique feature of homologous recombination is the generation of single-stranded regions that are used to align with and invade homologous donor sequences. The conserved Mre11 complex is required for generation of single strands at meiotic breaks (and at sites of DSBs in mitotic cells) in eukaryotes, and this complex also dissociates Spo11 from the DNA ends by cleaving off a short oligonucleotide containing the covalently linked Spo11. The *mre11* mutation was first discovered in cells that were *m*eiotic *re*combination deficient. The conserved complex contains three components: Mre11, Rad50, and Nbs1 (Xrs2 in *S. cerevisiae*) and is usually referred to as the MRN (or MRX in yeast) complex. This complex, which also interacts with the CtIP endonuclease (Sae2 in yeast), is required for 5′ end resection. The exonuclease Exo1 has been implicated in resection, and Rad50 and Mre11 are related to bacterial proteins known to have both exo- and endonuclease activities. At least half a dozen other proteins interact with MRN and are required for meiotic DSB generation and end resection. A critical role of MRN appears to be as a "chromosomal glue" to prevent separation of DNA ends. This is mediated by Rad50, which forms a zinc-dependent dimer and connects to the DNA ends, as shown in **FIGURE 15.11.**

Once the single-stranded regions have been generated, the next step is to use these regions to search for and invade a homologous double-stranded donor sequence.

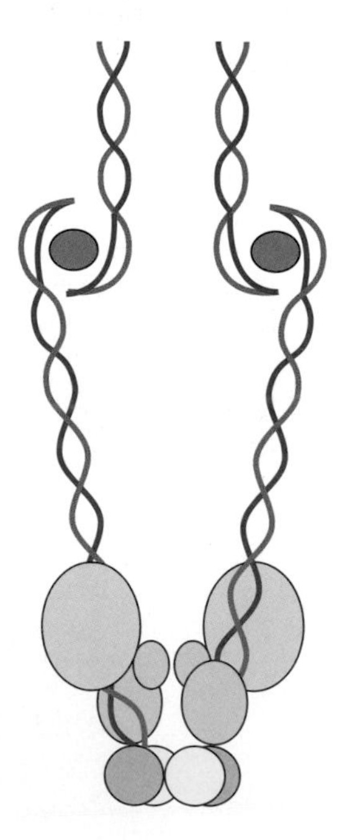

FIGURE 15.11 Structure of Rad50 and model for the MRN/X complex, which serves as a DNA bridge to prevent broken ends from separating. Rad50 has a coiled coil domain (pink and purple ribbons) and the globular "head" region of Rad50 contains ATP binding and hydrolysis domains (orange and yellow circles) and forms a complex with Mre11 (green) and Nbs1 or Xrs2 (pink). The other end of the coil binds zinc and forms a dimer through the "zinc hook" with another MRN/X molecule. The globular end binds to chromatin. The complex binds to double-strand breaks and can bring them together in a reaction involving two DNA ends and one MRN/X complex (top right), or through an interaction between two MRN/X dimers as depicted in the bottom right. (Only one possible model is shown for simplicity.)

The *E. coli* protein RecA was the first example of a DNA strand-transfer protein to be discovered. It is the paradigm for a group that includes several other bacterial and archaeal proteins, and the Rad51 and Dmc1 proteins in eukaryotes. Analysis of yeast *rad51* mutants shows that this class of protein plays a central role in recombination. They accumulate double-strand breaks and fail to form normal synaptonemal complexes. Dmc1 is meiosis-specific.

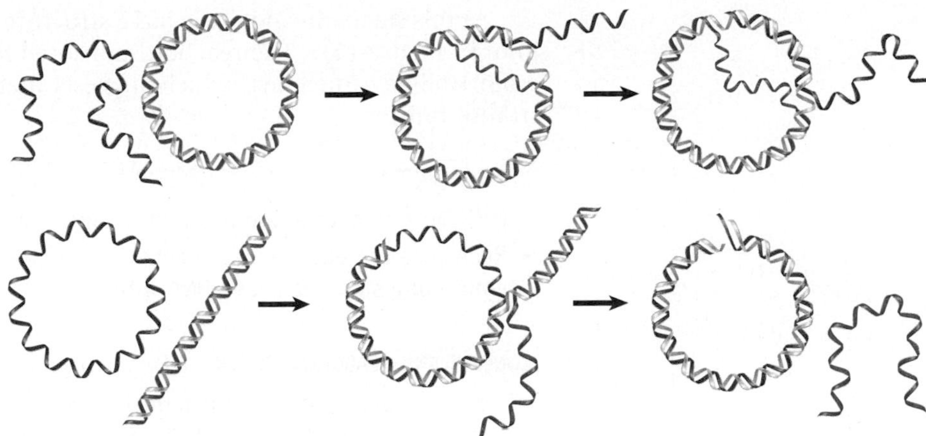

FIGURE 15.12 RecA promotes the assimilation of invading single strands into duplex DNA as long as one of the reacting strands has a free end.

Both Rad51/Dmc1 and their prokaryotic counterpart RecA assemble onto single-stranded DNA to form helical nucleoprotein filaments in the presence of ATP. These filaments hold the single strand in an extended conformation, stretched about 50% longer than DNA in a normal duplex. They are sometimes referred to as **presynaptic filaments**, as filament formation can occur prior to any interaction with homologous donor sequences. When duplex DNA is bound, it contacts the RecA-class protein via its minor groove, leaving the major groove accessible for possible reaction with a second DNA molecule. These filaments can then promote base pairing between a single strand of DNA and its complement in a duplex molecule, creating a three-stranded intermediate.

▸ **presynaptic filaments** Single-stranded DNA bound in a helical nucleoprotein filament with a strand-transfer protein such as Rad51 or RecA.

Once the homologous duplex has been identified, the nucleoprotein filaments promote the displacement of the corresponding strand of the duplex, and lead to base pairing between the invading strand and its complementary strand in the donor. The displacement reaction can occur between DNA molecules in several configurations and has three general conditions:

- One of the DNA molecules must have a single-stranded region.
- One of the molecules must have a free 3′ end.
- The single-stranded region and the 3′ end must be located within a region that is complementary between the molecules.

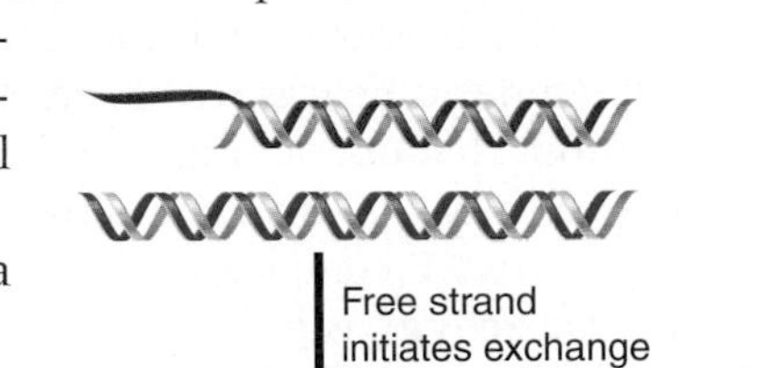

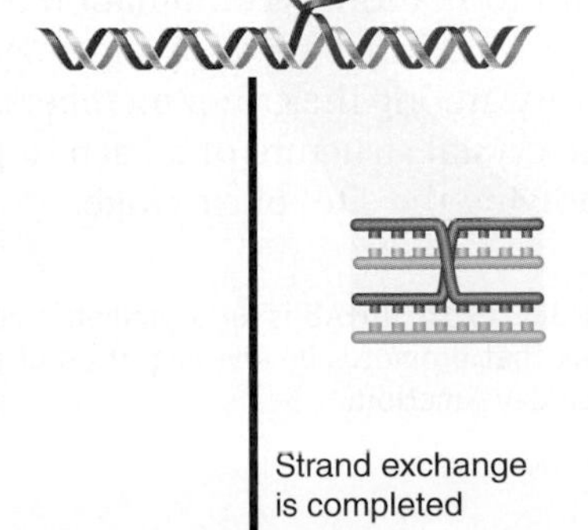

FIGURE 15.13 RecA-mediated strand exchange between partially duplex and entirely duplex DNA generates a joint molecule with the same structure as a recombination intermediate.

The reaction is illustrated in **FIGURE 15.12**. When a linear single strand invades a duplex, it displaces the original partner to its complementary strand. The reaction can be followed most easily by making either the donor or the recipient a circular molecule. The reaction proceeds 5′ to 3′ along the strand whose partner is being displaced and replaced; that is, the reaction involves an exchange in which (at least) one of the exchanging strands has a free 3′ end.

The reaction between a partially duplex molecule and an entirely duplex molecule leads to the exchange of strands. An example is illustrated in **FIGURE 15.13**. Strand invasion starts with one end of the linear molecule, where the invading single strand displaces its homolog in the duplex in the customary way. When the reaction reaches the region that is duplex in both molecules, the invading strand unpairs from its partner, which then pairs with the other displaced strand.

At this stage, the molecule has a structure indistinguishable from the recombinant joint in Figure 15.5. The reaction sponsored *in vitro* by RecA-family proteins can generate Holliday junctions, which suggests that these enzymes can mediate reciprocal strand transfer.

KEY CONCEPTS

- MRN/MRX complexes are required for Spo11 displacement and 5′ end resection.
- RecA-type proteins form filaments with single-stranded or duplex DNA and catalyze the ability of a single-stranded DNA with a free 3′ to displace its counterpart in a DNA duplex.

CONCEPT AND REASONING CHECK

ATP binding promotes filament formation by RecA-family proteins, but ATP hydrolysis appears to promote turnover of the filament. What would be the effect of the nonhydrolyzable ATP analog ATP-γ-S on presynaptic filament formation?

15.6 Holliday Junctions Must Be Resolved

One of the most critical steps in recombination is the resolution of the Holliday junction, which determines whether there is a reciprocal recombination or a reversal of the structure that leaves only a short stretch of hybrid DNA (see Figure 15.6). Branch migration from the exchange site (see Figure 15.8) determines the length of the region of hybrid DNA (with or without recombination). The proteins involved in stabilizing and resolving Holliday junctions have been identified as the products of the *ruv* genes in *E. coli.* RuvA and RuvB increase the formation of heteroduplex structures. RuvA recognizes the structure of the Holliday junction. RuvA binds to all four strands of DNA at the crossover point and forms two tetramers that sandwich the DNA. RuvB is a hexameric helicase with an ATPase activity that provides the motor for branch migration. Hexameric rings of RuvB bind around each duplex of DNA upstream of the crossover point. A diagram of the complex is shown in **FIGURE 15.14**.

The RuvAB complex can cause the branch to migrate as fast as 10 to 20 bp/second. A similar activity is provided by another helicase, RecG. RuvAB displaces RecA from DNA during its action. The RuvAB and RecG activities both can act on Holliday junctions, but if both are mutant, *E. coli* is completely defective in recombination activity.

The third gene, *ruvC,* codes for an endonuclease that specifically recognizes Holliday junctions. It can cleave the junctions *in vitro* to resolve recombination intermediates. A common tetranucleotide sequence provides a hotspot for RuvC to resolve the Holliday junction. The tetranucleotide (ATTG) is asymmetric and thus may direct resolution with regard to which pair of strands is nicked. This determines whether the outcome is patch recombinant formation (no overall recombination) or splice recombinant formation (recombination between flanking markers). Crystal structures of RuvC and other junction-resolving enzymes show that there is only moderate structural similarity among the group members, in spite of their common function. **FIGURE 15.15** shows the crystal structure of a bacteriophage resolvase complexed with a Holliday junction, showing the sites of cleavage.

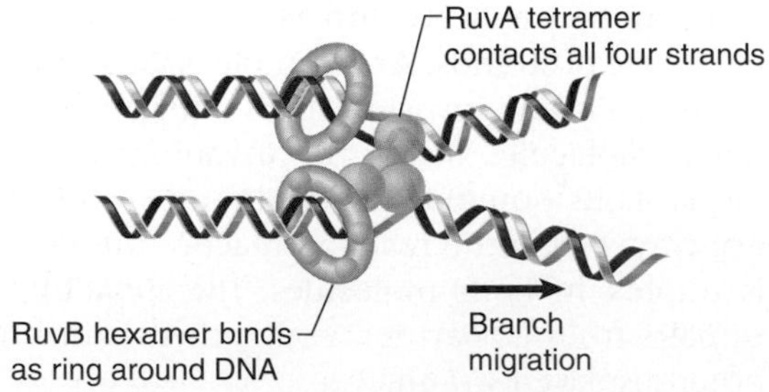

FIGURE 15.14 RuvAB is an asymmetric complex that promotes branch migration of a Holliday junction.

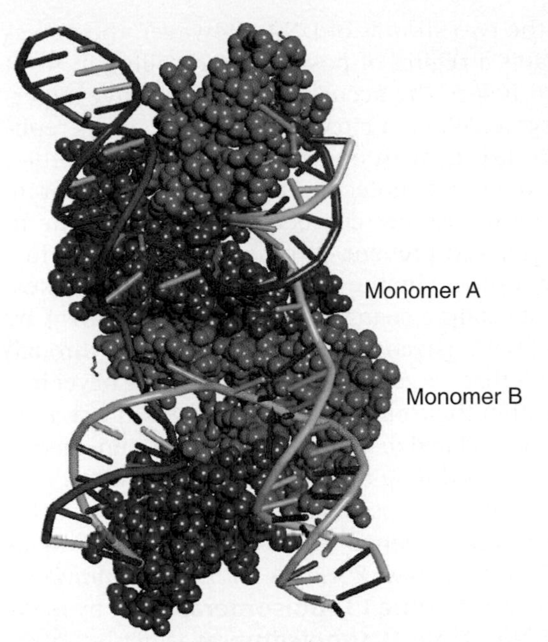

FIGURE 15.15 Complex of T4 endo VII, with the two subunits colored differently (blue and green) bound to a Holliday junction (each DNA strand is color-coded). The branch point of the junction is open. Arrows indicate the sites of cleavage.

All this suggests that recombination uses a "resolvasome" complex that includes enzymes catalyzing branch migration as well as junction-resolving activity. It is likely that mammalian cells contain a similar complex.

Although resolution in eukaryotic cells is less well understood, a number of proteins have been implicated in mitotic and meiotic resolution. *S. cerevisiae* strains that contain *mus81* mutations are defective in recombination. Mus81 is a component of an endonuclease that resolves Holliday junctions into duplex structures. This resolvase is important both in meiosis and for restarting stalled replication forks (see *Section 16.7, Recombination-Repair Systems*).

KEY CONCEPTS

- The Ruv complex acts on recombinant junctions.
- RuvA recognizes the structure of the junction and RuvB is a helicase that catalyzes branch migration.
- RuvC cleaves junctions to resolve Holliday junctions.

CONCEPT AND REASONING CHECK

Note the symmetry of the resolvase depicted in Figure 15.15. Why is it important that resolvases function as symmetrical molecules with two active sites?

15.7 Topoisomerases Relax or Introduce Supercoils in DNA

Topological manipulation of DNA is a central aspect of all its functional activities—recombination, replication, and transcription—as well as of the organization of higher-order structure. All synthetic activities involving double-stranded DNA require the strands to separate, and as we have seen, recombination generates topologically complex structures that must be resolved. Because the DNA strands do not simply lie side by side but are intertwined, their separation requires the strands to rotate about each other in space. The principle of **supercoiling** and the implications of over- and underwinding DNA were introduced in *Chapter 1* (see *Section 1.5, Supercoiling Affects the Structure of DNA*).

Changes in the topology of DNA can be caused in several ways, some of which can be detrimental to cellular processes if not properly resolved. The initiation of both replication and transcription can actually be facilitated by negative supercoiling,

▸ **supercoiling** The coiling of a closed duplex DNA in space so that it crosses over its own axis.

as this tends to promote unwinding of the two strands of DNA. However, the movement of RNA or DNA polymerases creates a region of positive supercoiling in front of the enzyme, which must be resolved before the accumulation of positive supercoils impedes the movement of the enzyme. When a circular DNA molecule is replicated, the circular products may be **catenated**, with one passed through the other, and must be separated in order for the daughter molecules to segregate to separate daughter cells. Even linear eukaryotic chromosomes can become entangled due to their lengths and must be similarly resolved to prevent chromosome breakage during cell division. All these situations are resolved by the actions of **topoisomerases**.

- **catenate** To link together two circular molecules, as in a chain.
- **topoisomerase** An enzyme that changes the number of times the two strands in a closed DNA molecule cross each other. It does this by cutting the DNA, passing DNA through the break, and resealing the DNA.

DNA topoisomerases are enzymes that catalyze changes in the topology of DNA by transiently breaking one or both strands of DNA, passing the unbroken strand(s) through the gap, and then resealing the gap. The ends that are generated by the break are never free but instead are manipulated exclusively within the confines of the enzyme—in fact, they are covalently linked to the enzyme. Spo11 is related to topoisomerases and undergoes a similar covalent attachment when it forms DSBs during meiosis (discussed in *Section 15.3, Double-Strand Breaks Initiate Recombination*). Topoisomerases act on any sequence of DNA, but some enzymes involved in site-specific recombination function in the same way as topoisomerases (see *Section 15.8, Site-Specific Recombination Resembles Topoisomerase Activity*).

- **type I topoisomerase** An enzyme that changes the topology of DNA by nicking and resealing one strand of DNA.
- **type II topoisomerase** An enzyme that changes the topology of DNA by nicking and resealing both strands of DNA.
- **gyrase** Enzymes that introduce negative supercoils into DNA.
- **reverse gyrase** Enzyme that introduces positive supercoils into DNA.

Topoisomerases are divided into two classes: **Type I topoisomerases** act by making a transient break in one strand of DNA. **Type II topoisomerases** act by introducing a transient double-strand break. Topoisomerases, in general, vary with regard to the types of topological change they introduce. Some topoisomerases can relax (remove) only negative supercoils from DNA; others can relax both negative and positive supercoils. Enzymes that can introduce negative supercoils are called **gyrases**; those that can introduce positive supercoils are called **reverse gyrases**.

There are four topoisomerase enzymes in *E. coli*: topoisomerases I, III, and IV and DNA gyrase. DNA topoisomerases I and III are type I enzymes. Gyrase and DNA topoisomerase IV are type II enzymes. Each of the four enzymes is important in one or more of the following functions:

- The overall level of negative supercoiling in the bacterial nucleoid is the result of a balance between the introduction of supercoils by gyrase and their relaxation by topoisomerases I and IV. This is a crucial aspect of nucleoid structure (see *Section 9.3, The Bacterial Genome Is a Supercoiled Nucleoid*), and it affects initiation of transcription at certain promoters (see *Section 19.14, Supercoiling Is an Important Feature of Transcription*).
- The same enzymes are involved in resolving the problems created by transcription; gyrase converts the positive supercoils that are generated ahead of RNA polymerase into negative supercoils, and topoisomerases I and IV remove the negative supercoils that are left behind the enzyme. Gyrase is also critical for removing positive supercoils produced during replication.
- As replication proceeds, the daughter duplexes can become twisted around one another in a stage known as *precatenation*. The precatenanes are removed by topoisomerase IV, which also decatenates any catenated genomes that are left at the end of replication.

The enzymes in eukaryotes follow the same principles, although the detailed division of responsibilities may be different. They do not show sequence or structural similarity with the prokaryotic enzymes. Most eukaryotes contain a single topoisomerase I enzyme that is required both for replication fork movement and for relaxing supercoils generated by transcription. A topoisomerase II enzyme(s) is required to unlink entangled chromosomes following replication. Other individual topoisomerases have been implicated in recombination and repair activities.

The common mechanism for all topoisomerases is to link one end of each broken strand to a tyrosine residue in the enzyme. A type I enzyme links to the single broken strand; a type II enzyme links to one end of each broken strand. The topoisomerases are further divided into the A and B groups according to whether the linkage is to a 5′ phosphate or 3′ phosphate. The use of the transient phosphodiester–tyrosine bond suggests a mechanism for the action of the enzyme; it transfers a phosphodiester bond(s) in DNA to the protein, manipulates the structure of one or both DNA strands, and then rejoins the bond(s) in the original strand.

The *E. coli* enzymes are all of type A and use links to 5′ phosphate. This is the general pattern for bacteria, where there are almost no type B topoisomerases. All four possible types of topoisomerase (IA, IB, IIA, and IIB) are found in eukaryotes.

A model for the action of topoisomerase IA is illustrated in **FIGURE 15.16**. The enzyme binds to a region in which duplex DNA becomes separated into its single strands. The enzyme then breaks one strand, pulls the other strand through the gap, and, finally, seals the gap. The transfer of bonds from nucleic acid to protein explains how the enzyme can function without requiring any input of energy. There has been no irreversible hydrolysis of bonds; their energy has been conserved through the transfer reactions. The model is supported by the crystal structure of the enzyme.

Type II topoisomerases generally relax both negative and positive supercoils. The reaction requires ATP, with one ATP hydrolyzed for each catalytic event. As illustrated in **FIGURE 15.17**, the reaction is mediated by making a double-strand break in one DNA

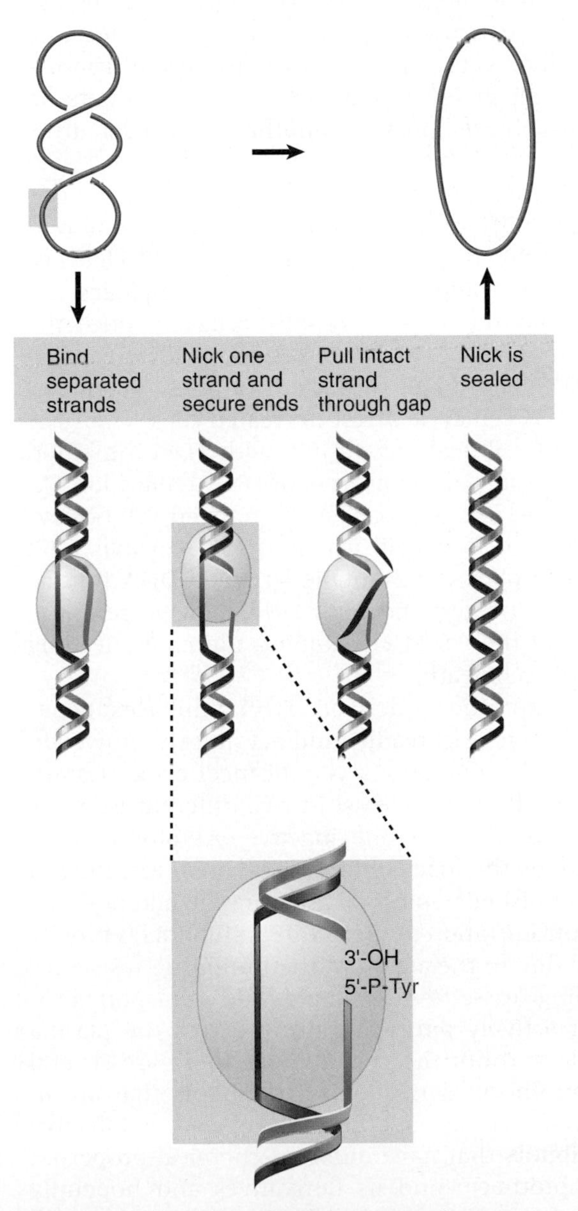

FIGURE 15.16 Type I topoisomerases recognize partially unwound segments of DNA and pass one strand through a break made in the other.

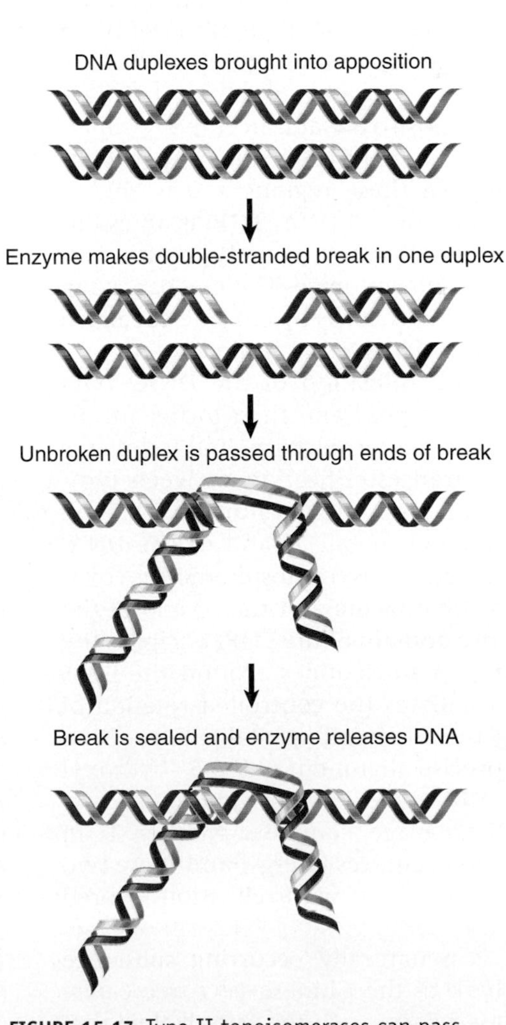

FIGURE 15.17 Type II topoisomerases can pass a duplex DNA through a double-strand break in another duplex.

CLINICAL APPLICATIONS

Camptothecin and Topoisomerase I Inhibition

Topoisomerases are enzymes that catalyze strand breakage of DNA to relieve or introduce supercoiling into double-stranded DNA. The relaxation of supercoiled DNA is a prerequisite for essential nuclear processes including DNA replication, recombination, transcription, and chromatin condensation. There are two general categories of topoisomerases: type I enzymes that act by introducing nicks in one strand of DNA and type II enzymes that act by ATP-dependent double-stranded cleavage. In mammals, there are four genes encoding distinct type I topoisomerases: type IA nuclear topoisomerase and mitochondrial topoisomerase and type IB topoisomerases IIIα and IIIβ. Only the topoisomerase type IB can relax both positive and negative supercoils in double-stranded DNA.

Camptothecin and its derivatives are chemotherapy drugs that target dividing cells by specific inhibition of nuclear topoisomerase type IB (TOPI). In regions of the genome that are undergoing replication and transcription, the two strands of duplex DNA are separated, generating positively and negatively supercoiled DNA upstream and downstream, respectively. TOPI is concentrated in these regions and is able to relax both topological forms of DNA. TOPI is an essential protein, as demonstrated by the embryonic lethality observed in TOPI-deficient mice. The mechanism of DNA relaxation by TOPI proceeds by enzyme-mediated nicking of one strand of DNA, controlled rotation of the nicked strand, and religation of the DNA. TOPI nicks the DNA on one strand and then forms a transient covalent complex, referred to as TOPI cleavage complex. There is a transesterification between tyrosine 723 in the C-terminal portion of human TOPI and the phosphodiester bond of one strand of the DNA, creating an intermediate tyrosyl-phosphodiester covalent bond between the enzyme and the 3′ end of the nicked DNA. The core domain of the TOPI encircles the DNA, encompassing 14 nucleotides around the DNA cleavage site, and mediates the controlled rotation of the DNA, relaxing the supercoiling. Religation of the DNA requires the precise alignment of the 5′-hydroxyl end of the nicked DNA and the tyrosyl-phosphodiester bond of the TOPI cleavage complex. Neither TOPI cleavage nor religation requires energy, and these two processes exist in equilibrium, with religation favored over cleavage.

Camptothecin is a naturally occurring substance isolated from the bark of the Chinese yew tree *Camptotheca acuminata.* It is a 5-ring heterocyclic alkaloid that was first isolated and demonstrated to have anticancer activity by the National Cancer Institute in 1966. After clinical testing, it was abandoned as a treatment for cancer because of the severity of the side effects, including hemorrhagic cystitis (inflammation and bleeding of the bladder). It was later shown that the mechanism of camptothecin's cytotoxicity is through the inhibition of TOPI and that this enzyme is the sole cellular target for the drug. Currently, there are two camptothecin derivatives, topotecan and irinotecan, that are also TOPI inhibitors being used to treat cancer. Topotecan is FDA approved for treating recurrent ovarian, cervical, and small-cell lung cancer, and irinotecan is approved for treating colorectal cancer. The side effects of these camptothecin analogs include neutropenia, a disruption of bone marrow function resulting in a decrease in white blood cells. In some cases, the use of irinotecan can result in severe diarrhea. Typically, topotecan is used in conjunction with another anticancer drug, cisplatin.

Camptothecin and its two derivatives bind reversibly to the TOPI cleavage complex and prevent religation, thereby trapping the transient TOPI cleavage complex. The trapped TOPI cleavage complexes are reversed, and the DNA is rapidly religated once the drugs are removed. TOPI inhibitors do not directly damage the DNA, although they cause S phase cytotoxicity and G2-M cell cycle arrest in treated cells. The DNA damage and cell death associated with TOPI inhibition occurs as a result of the process of DNA replication. As the replication fork approaches the trapped TOPI cleavage complex, it is stalled, leading to irreversible TOPI cleavage complexes and double-stranded DNA breaks. This in turn, triggers the cell's DNA damage response and arrest at the GS-M checkpoint, ultimately leading to apoptotic cell death.

The camptothecin-derived TOPI inhibitors penetrate eukaryotic cells readily and act quickly. However, they have some limitations as anticancer drugs. Camptothecin and its analogs exist in two different isomeric forms, one active and one inactive. At physiological pH conditions the inactive form is favored, resulting in greatly diminished potency of the drugs. In addition, the cellular concentration of these TOPI inhibitors is rapidly decreased due to the action of the multidrug resistance ATP-binding cassette transmembrane transport proteins that actively pump the drugs across the plasma membrane and out the cell. To date there are several non-camptothecin-derived TOPI inhibitors that are in clinical trials. These have been developed as effective TOPI inhibitors that have different chemical properties than camptothecin and its derivatives and hopefully will prove to be more stable and more effective anticancer drugs.

duplex. The double strand is cleaved with a four-base stagger between the ends, and each subunit of the dimeric enzyme attaches to a protruding broken end. Another duplex region is then passed through the break. The ATP is used in the following religation/release step, when the ends are rejoined and the DNA duplexes are released. This is why inhibiting the ATPase activity of the enzyme results in a "cleavable complex" that contains broken DNA.

Type II topoisomerases recognize a variety of DNA structures, most commonly sites where two double-stranded segments cross each other. The hydrolysis of ATP may be used to drive the enzyme through conformational changes that provide the force needed to push one DNA duplex through the break made in the other. As a result of the topology of supercoiled DNA, the relationship of the crossing segments allows supercoils to be removed from either positively or negatively supercoiled circles.

KEY CONCEPTS

- Topoisomerases alter supercoiling by breaking bonds in DNA, changing the conformation of the double helix in space and remaking the bonds.
- Type I enzymes act by breaking a single strand of DNA; type II enzymes act by making double-strand breaks.
- Type I topoisomerases function by forming a covalent bond to one of the broken ends, moving one strand around the other and then transferring the bound end to the other broken end. Bonds are conserved; as a result, no input of energy is required.
- Type II topoisomerases also form covalent bonds to the broken ends, and pass a duplex DNA region through the double-strand break. ATP is required to complete the reaction and reseal the break.

CONCEPT AND REASONING CHECK

Explain why either a single- or double-strand break can result in a change in supercoiling.

15.8 Site-Specific Recombination Resembles Topoisomerase Activity

Site-specific, or specialized, recombination involves a reaction between two specific sites. The lengths of target sites are short, and are typically in a range of 14 to 50 bp. In some cases the two sites have the same sequence, but in other cases they are nonhomologous. The reaction can be used to insert a free phage DNA into the bacterial chromosome or to excise an integrated phage DNA from the chromosome, and in this case the two recombining sequences are different from one another.

The enzymes that catalyze site-specific recombination are generally called **recombinases**, and more than 100 of them are now known. Those involved in phage integration or that are related to these enzymes are also known as the **integrase** family. Prominent members of the integrase family are the prototype Int from phage lambda, Cre from phage P1, and the yeast FLP enzyme (which catalyzes a chromosomal inversion).

The classic model for site-specific recombination is illustrated by phage lambda. The conversion of lambda DNA between its different forms involves two types of event. (The regulation of these events is described in *Chapter 27, Phage Strategies*.) The physical condition of the DNA is different in the lysogenic and lytic states:

- In the lytic state lambda DNA exists as an independent, circular molecule in the infected bacterium.
- In the lysogenic state, the phage DNA is an integral part of the bacterial chromosome (called **prophage**).

▸ **recombinase** Enzyme that catalyzes site-specific recombination.

▸ **integrase** Enzyme that is responsible for a site-specific recombination event that inserts one molecule of DNA (e.g., a phage genome) into another.

▸ **prophage** A phage genome covalently integrated as a linear part of the bacterial chromosome.

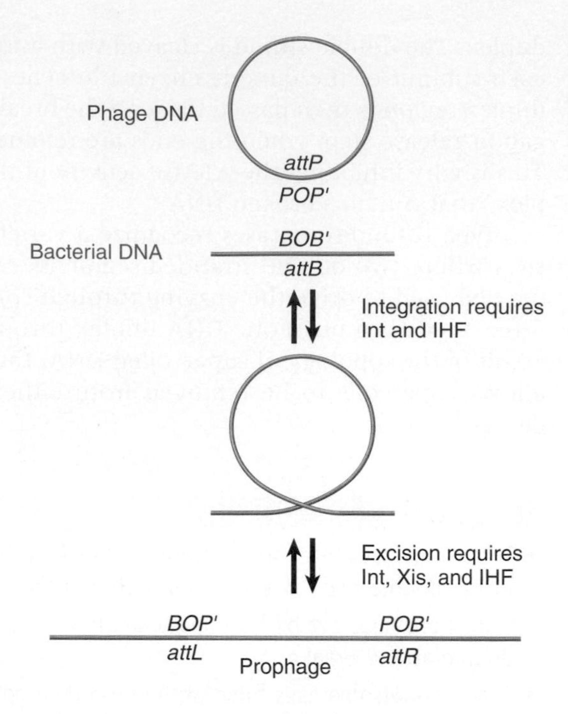

FIGURE 15.18 Circular phage DNA is converted to an integrated prophage by a reciprocal recombination between *attP* and *attB*; the prophage is excised by reciprocal recombination between *attL* and *attR*.

Transition between these states involves site-specific recombination:

- To enter the lysogenic condition, free lambda DNA must be inserted into the host DNA. This is called **integration**.
- To be released from lysogeny into the lytic cycle, prophage DNA must be released from the chromosome. This is called **excision**.

Integration and excision occur by recombination at specific loci on the bacterial and phage DNAs called **attachment (*att*) sites**. When describing the integration/excision reactions, the bacterial attachment site is called *attB,* consisting of the sequence components *BOB′*. The attachment site on the phage, *attP,* consists of the components *POP′*. **FIGURE 15.18** outlines the recombination reaction between these sites. The sequence *O* is common to *attB* and *attP.* It is called the **core sequence**, and the recombination event occurs within it. The flanking regions *B, B′* and *P, P′* are referred to as the *arms*; each is distinct in sequence. The phage DNA is circular, so the recombination event inserts it into the bacterial chromosome as a linear sequence. The prophage is bounded by two new *att* sites—the products of the recombination—called *attL* and *attR.*

▸ **integration** Insertion of a viral or another DNA sequence into a host genome as a region covalently linked on either side to the host sequences.

▸ **excision** Release of phage or episome or other sequences from the host chromosome as an autonomous DNA molecule.

▸ **attachment (*att*) sites** The loci on a lambda phage and the bacterial chromosome at which recombination integrates the phage into, or excises it from, the bacterial chromosome.

▸ **core sequence** The segment of DNA that is common to the attachment sites on both the phage lambda and bacterial genomes. It is the location of the recombination event that allows phage lambda to integrate.

An important consequence of the constitution of the *att* sites is that the integration and excision reactions do not involve the same pair of reacting sequences. Integration requires recognition between *attP* and *attB,* whereas excision requires recognition between *attL* and *attR.* The directional character of site-specific recombination is controlled by the identity of the recombining sites. The recombination event is reversible, but different conditions prevail for each direction of the reaction. This is an important feature in the life of the phage because it offers a means to ensure that an integration event is not immediately reversed by an excision, and vice versa.

The difference in the pairs of sites reacting at integration and excision is reflected by a difference in the proteins that mediate the two reactions:

- Integration (*attB* × *attP*) requires the product of the phage gene *int,* which codes for the integrase Int, and a bacterial protein called integration host factor (IHF).
- Excision (*attL* × *attR*) requires the product of phage gene *xis* in addition to Int and IHF.

Thus Int and IHF are required for both reactions. Xis plays an important role in controlling the direction; it is required for excision but inhibits integration.

A similar system, but with somewhat simpler requirements for both sequence and protein components, is found in the bacteriophage P1. The Cre recombinase encoded by the phage catalyzes a recombination between two target sequences. Unlike phage lambda, for which the recombining sequences are different, in phage P1 they are identical. Each consists of a 34 bp–long sequence called *loxP.* The Cre recombinase is sufficient for the reaction; no accessory proteins are required. As a result of its simplicity and its efficiency, what is now known as the Cre/*lox* system has been adapted for use in eukaryotic cells, where it has become one of the standard techniques for undertaking site-specific recombination (see *Section 3.12, Gene Knockouts and Transgenics*).

Enzymes such as Int and Cre use a mechanism similar to that of type I topoisomerases, in which a break is made in one DNA strand at a time. The difference is that a recombinase reconnects the ends *crosswise,* whereas a topoisomerase makes a break, manipulates the ends, and then rejoins the *original* ends. The basic principle of the system is that four molecules of the recombinase are required, one to cut each of the four strands of the two duplexes that are recombining.

FIGURE 15.19 shows the nature of the reaction catalyzed by an integrase. The enzyme is a monomeric protein that has an active site capable of cutting and ligating DNA. The reaction involves an attack by a tyrosine on a phosphodiester bond. The 3′ end of the DNA chain is linked through a phosphodiester bond to a tyrosine in the enzyme. This releases a free 5′ hydroxyl end.

1. Two enzyme subunits bind to each duplex DNA

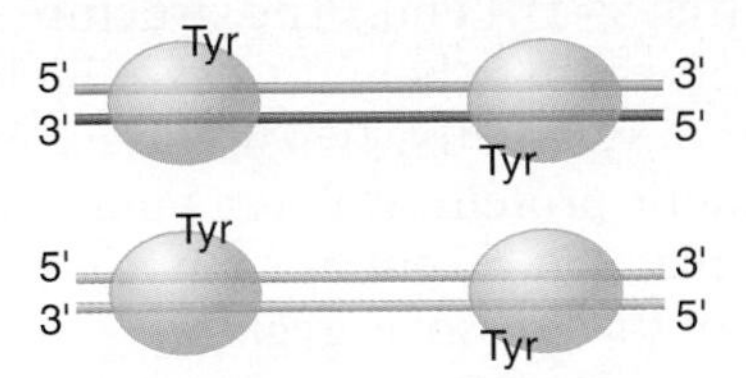

2. Each duplex is cleaved on one strand to generate a P-Tyr bond and an -OH end

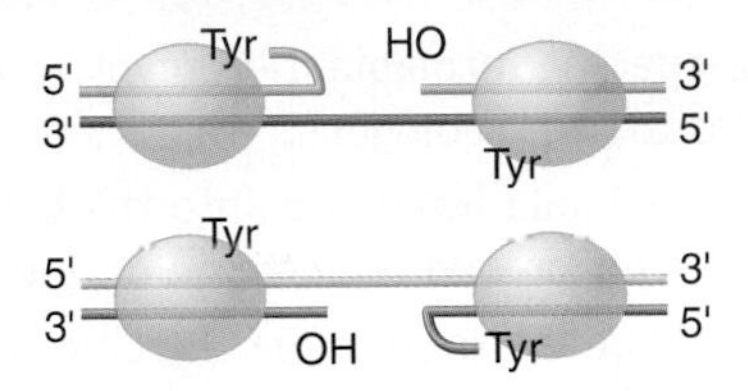

3. Each hydroxyl attacks the Tyr-phosphate link in the other duplex

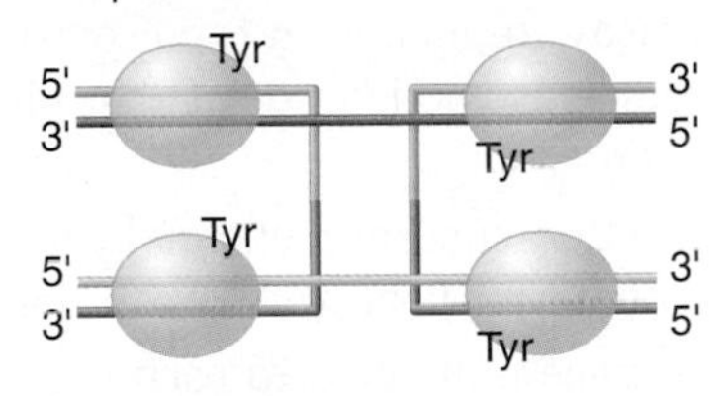

4. The reactions are repeated by the other subunits to join the other strands

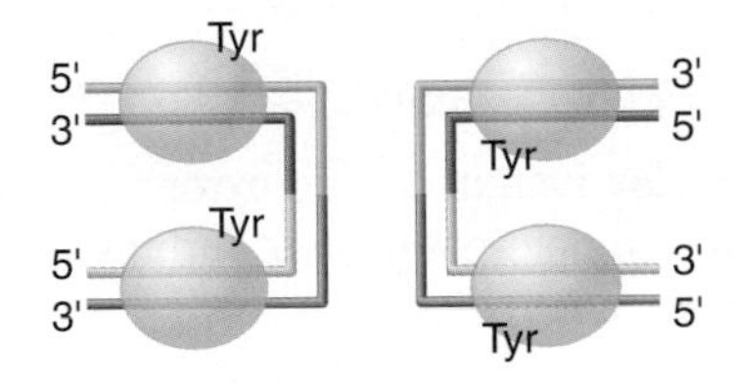

FIGURE 15.19 Integrases catalyze recombination by a mechanism similar to topoisomerases. Staggered cuts are made in DNA, and the 3′-phosphate end is covalently linked to a tyrosine in the enzyme. The free hydroxyl group of each strand then attacks the P-Tyr link of the other strand. The first exchange shown in the figure generates a Holliday junction. The junction is resolved by repeating the process with the other pair of strands.

Two enzyme units are bound to each of the recombination sites. At each site, only one of the units attacks the DNA. The symmetry of the system ensures that complementary strands are broken in each recombination site. The free 5′–OH end in each site attacks the 3′–phosphotyrosine link in the other site. This generates a Holliday junction.

The structure is resolved when the other two enzyme units (which had not been involved in the first cycle of breakage and reunion) act on the other pair of complementary strands.

The successive interactions accomplish a conservative strand exchange, in which there are no deletions or additions of nucleotides at the exchange site and there is no need for input of energy. The transient 3′–phosphotyrosine link between protein and DNA conserves the energy of the cleaved phosphodiester bond.

FIGURE 15.20 shows the reaction intermediate, based on the crystal structure. (Trapping the intermediate was made possible by using a "suicide substrate," which consists of a synthetic DNA duplex

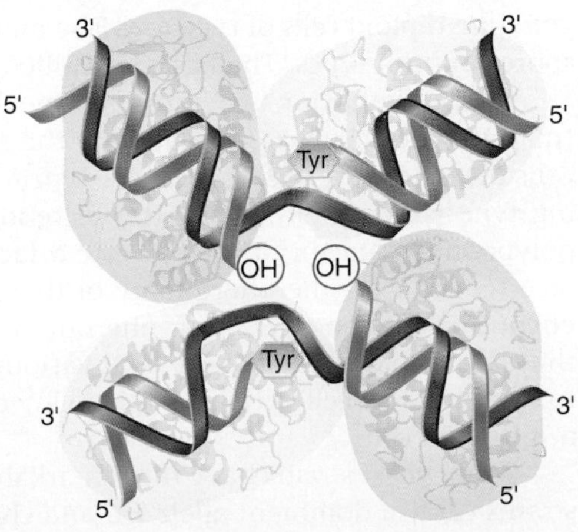

FIGURE 15.20 A synapsed *loxA* recombination complex has a tetramer of Cre recombinases, with one enzyme monomer bound to each half site. Two of the four active sites are in use, acting on complementary strands of the two DNA sites.

with a missing phosphodiester bond, so that the attack by the enzyme does not generate a free 5′–OH end.) The structure of the Cre-*lox* complex shows two Cre molecules, each of which is bound to a 15 bp length of DNA. The DNA is bent by ~100° at the center of symmetry. Two of these complexes assemble in an antiparallel way to form a tetrameric protein structure bound to two synapsed DNA molecules. Strand exchange takes place in a central cavity of the protein structure that contains the central six bases of the crossover region.

KEY CONCEPTS

- Site-specific recombination involves reaction between specific sites that are not necessarily homologous.
- Phage lambda integrates into the bacterial chromosome by recombination between a site on the phage and the *att* site on the *E. coli* chromosome.
- The phage is excised from the chromosome by recombination between the sites at the end of the linear prophage.
- Phage lambda *int* codes for an integrase that catalyzes the integration reaction.
- Integrases are related to topoisomerases, and the recombination reaction resembles topoisomerase action except that nicked strands from different duplexes are sealed together.
- The reaction conserves energy by using a catalytic tyrosine in the enzyme to break a phosphodiester bond and link to the broken 3′ end.
- Two enzyme units bind to each recombination site and the two dimers synapse to form a complex in which the transfer reactions occur.

CONCEPT AND REASONING CHECK

The Cre/*lox* system is used extensively to create "conditional knockouts" in mice, for example to delete a gene of interest in a particular tissue type. Describe how such a system might work.

15.9 Yeast Use a Specialized Recombination Mechanism to Switch Mating Type

The yeast *S. cerevisiae* can propagate in either the haploid or diploid condition. Conversion between these states takes place by mating (fusion of haploid cells to give a diploid) and by sporulation (meiosis of diploids to give haploid spores). The ability to engage in these activities is determined by the mating type of the strain, which can be either **a** or α. Haploid cells of type **a** can mate only with haploid cells of type α to generate diploid cells of type **a**/α. The diploid cells can sporulate to regenerate haploid spores of both types. The basic yeast life cycle is shown in **FIGURE 15.21**.

Mating behavior is determined by the genetic information present at the *MAT* (mating type) locus. Cells that carry the *MAT***a** allele at this locus are type **a**; likewise, cells that carry the *MAT*α allele are type α. Recognition between cells of opposite mating type is accomplished by the secretion of pheromones: α cells secrete the small polypeptide α-factor; **a** cells secrete **a**-factor. A cell of one mating type carries a surface receptor for the pheromone of the opposite type. When an **a** cell and an α cell encounter one another, their pheromones act on their receptors to arrest the cells in the G1 phase of the cell cycle, and various morphological changes occur. In a successful mating, the cell cycle arrest is followed by cell and nuclear fusion to produce an **a**/α diploid cell.

Some yeast strains have the remarkable ability to switch their mating types. These strains carry a dominant allele *HO* and change their mating type frequently—as often

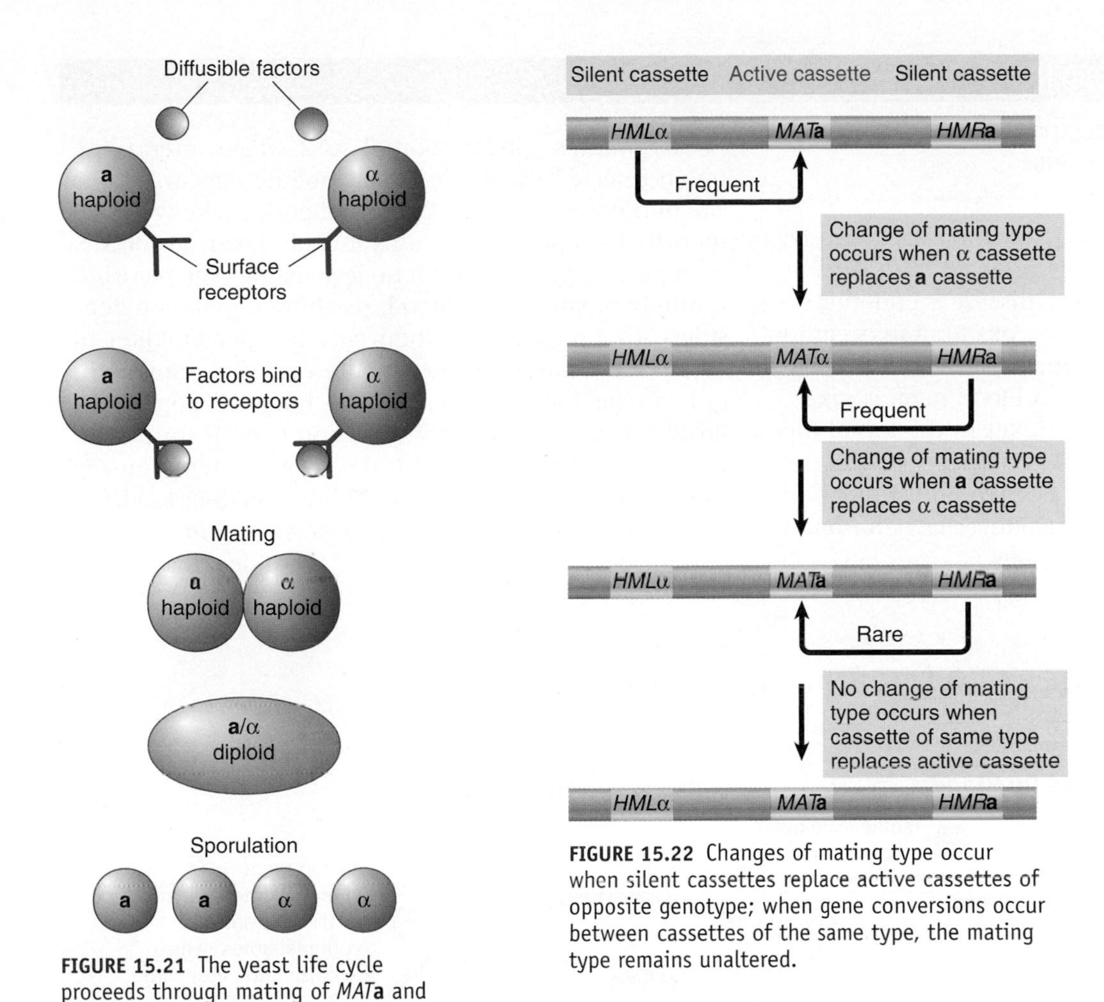

FIGURE 15.21 The yeast life cycle proceeds through mating of *MAT***a** and *MAT*α haploids to give heterozygous diploids that sporulate to give haploid spores.

FIGURE 15.22 Changes of mating type occur when silent cassettes replace active cassettes of opposite genotype; when gene conversions occur between cassettes of the same type, the mating type remains unaltered.

as once every generation. Strains with the recessive loss-of-function allele *ho* have a stable mating type, which is subject to change with a frequency of ~10^{-6}.

The presence of *HO* causes the genotype of a yeast population to change. Irrespective of the initial mating type, within a very few generations there are large numbers of cells of both mating types, leading to the formation of *MAT***a**/*MAT*α diploids. Mating type switching allows sexual reproduction and the benefits of meiotic recombination in populations that may have begun as only a single mating type.

The existence of switching suggests that all cells contain the potential information needed to be either *MAT***a** or *MAT*α but that haploids express only one type. Where does the information to change mating types come from? Two additional loci are needed for switching. *HML*α is needed for switching to give a *MAT*α type; *HMR***a** is needed for switching to give a *MAT***a** type. These loci lie on the same chromosome that carries *MAT*. *HML* is far to the left, *HMR* is far to the right.

In addition to the active **a** or α **mating-type cassette** present at the *MAT* locus, yeast also carry silent mating-type cassettes at *HML* and *HMR*, as illustrated in **FIGURE 15.22**. All cassettes carry information that codes for mating type, but only the active cassette at *MAT* is expressed. The cassettes at *HML* and *HMR* are silent because they are assembled into heterochromatin (see *Chapter 29, Epigenetic Effects Are Inherited*). Mating-type switching occurs when the active cassette is replaced by information from a silent cassette. The newly installed cassette is then expressed.

Switching is nonreciprocal; the copy at *HML* or *HMR* replaces the allele at *MAT*. We know this because a mutation at *MAT* is lost permanently when it is replaced

▸ **mating-type cassette** A locus containing the genes required for mating type in yeast (either **a** or α), which can either be active or inactive (silent) copies of the locus. Mating type is changed when an active cassette of one type is replaced by a silent cassette of the other type.

MEDICAL APPLICATIONS

Trypanosomes Use Gene Switching to Evade the Host Immune System

African sleeping sickness, or trypanosomiasis, is caused by the bite of a tsetse fly that is infected with a parasite of the genus *Trypanosoma*. The disease is endemic on the African continent. Major outbreaks tend to occur in Sudan, Uganda, Angola, and Congo. Tsetse flies become infected with *Trypanosoma* when a blood meal is taken. The parasite is ingested and multiplies in the gut of the fly. After two weeks, the parasites migrate back to the salivary glands. The tsetse fly is then infectious and remains infectious for life as it introduces trypanosomes into a host at subsequent blood meals.

Symptoms appear about 1 to 2 weeks after a bite by an infected fly. An inflamed nodule appears at the site of the tsetse fly bite. Other symptoms take weeks to months to appear. Symptoms include fever, headache, joint pain, rash, and lymph node swelling. The parasites multiply rapidly in the blood, reaching population densities of more than 1 million parasites per milliliter of blood. This number drops rapidly (called a remission), only to be followed by another burst of population growth (relapse). This pattern continues. If untreated, the parasite invades the central nervous system, causing sleeping sickness (**FIGURE B15.1**). These infections last from weeks to years. Sleeping sickness symptoms include chronic fever, severe headaches, and sleeping during

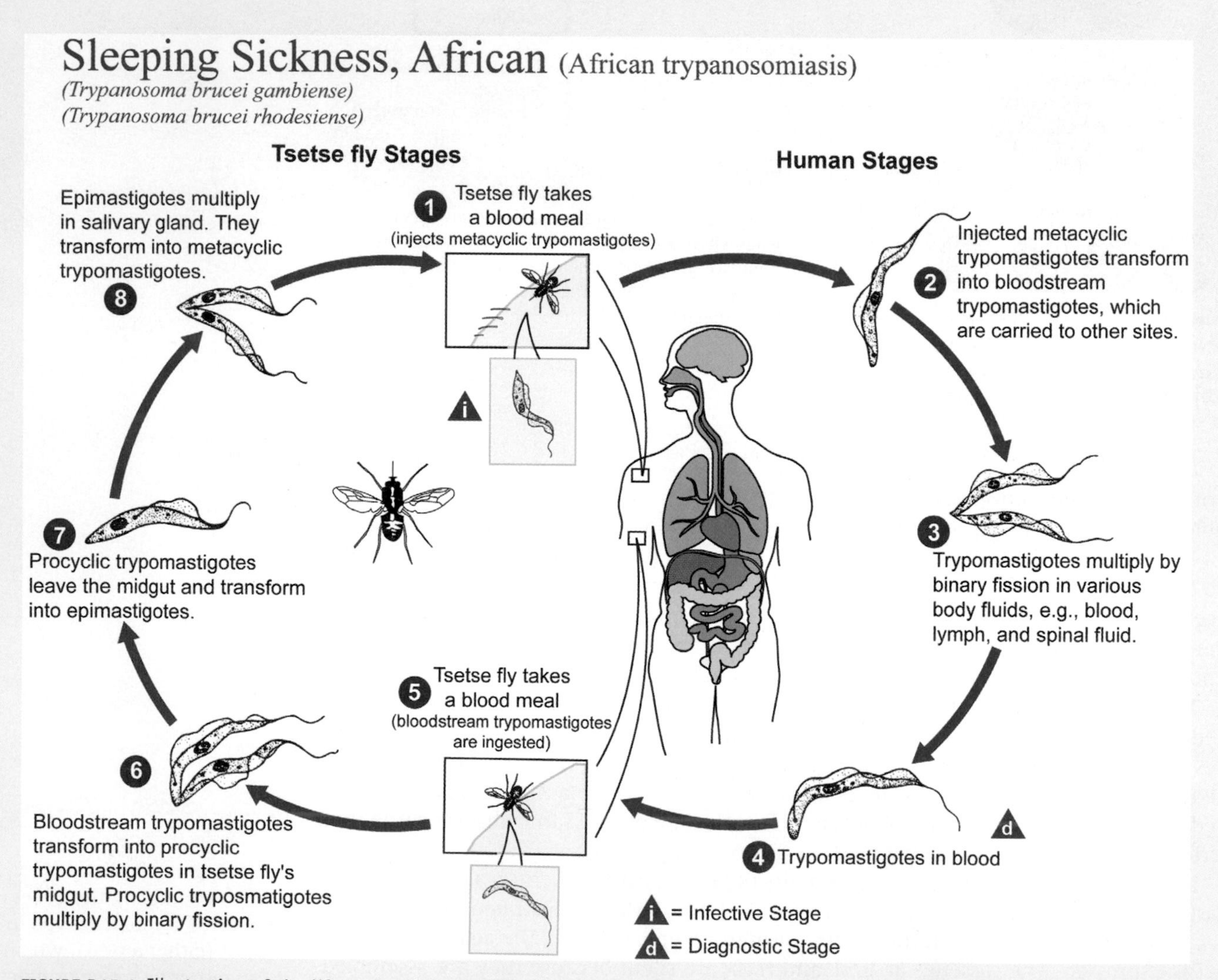

FIGURE B15.1 Illustration of the life cycle of the parasite that causes African sleeping sickness. Courtesy of Alexander J. da Silva, PhD, and Melanie Moser/CDC.

long periods of the day, as well as nighttime insomnia. Trypanosomes can invade the brain several months to years after infection, leading to coma and death.

The African trypanosomes are able to change the surface coat proteins of their outer membranes. This phenomenon is referred to as *antigenic variation*. In doing so, the parasite is always one step ahead of the immune system. *Antibodies* are highly specific proteins produced by the body in response to a foreign substance such as a parasite and are capable of binding to it. Antibodies are produced against the parasites *v*ariant *s*urface *g*lycoproteins coat (*VSG*). The host's antibodies can neutralize and kill approximately 99% of the original parasite population within the host (remission), but a few trypanosomes switch VSGs, gaining a new antigenically distinct VSG coat (causing a relapse). The trypanosomes undergo a type of gene switching that is similar to the yeast mating-type switching mechanism. Approximately 1000 different genes encode VSGs. VSG genes represent about 10% of the total DNA of a trypanosome. The VSG genes are located either in the chromosome's interior or near the telomeres. Expressed VSG genes are always near telomeres. Only one of the VSG genes is expressed at a time; all other VSG genes are transcriptionally silent. Therefore, only one type of VSG molecule is present within the trypanosome surface coat, resulting in a homogenous display of identical surface antigens that is recognized by the immune system.

The mechanism of this type of antigen variation in trypanosomes involves a silent, or basic, copy of the VSG gene displacing a previously active VSG gene, located in a VSG expression site. The trigger for the switching is unknown. The gene that is to be expressed is copied onto a cassette and the copy is unidirectionally moved or recombined into one of potentially 20 different active *b*loodstream-form *e*xpression *s*ites (BESs) that contain *e*xpression *s*ite-*a*ssociated *g*enes (ESAGs) on the chromosome near a telomere where the gene is transcribed and subsequently translated into the VSG protein. The VSG proteins are densely packed as monolayer, or protective, coat around the parasite. As the antibody response against the VSG mounts, some trypanosomes switch to expression of another VSG and proliferate, causing the next wave of disease. The trypanosomes that are ingested by the tsetse fly lose their coat and revert to a basic coat type.

Drugs used to treat late-stage disease are highly toxic; there is no prophylactic chemotherapy and little or no prospect of a vaccine. A collaboration to sequence and analyze the genomes of *Trypanosoma* species that cause sleeping sickness has been under way at institutions in Africa and South America. Findings of the initial sequencing shows that the most sequenced silent VSGs are defective or pseudogenes and almost all VSGs form arrays. More than 90% of the VSGs contain one or more 70 bp repeats upstream of the VSG (**FIGURE B15.2**). It is to be hoped that this new genomic information will help scientists identify novel potential drug targets that will accelerate the development of new therapeutics.

References

Berriman et al. (2005). The Genome of the African Trypanosome Trypanosoma brucei. *Science* **309**, 416–422.

Dubois, Demick, and Mansfield. (2005). Trypanosomes Expressing a Mosaic Variant Surface Glycoprotein Coat Escape Early Detection by the Immune System. *Infection and Immunity* **73**, 5; 2690–2697.

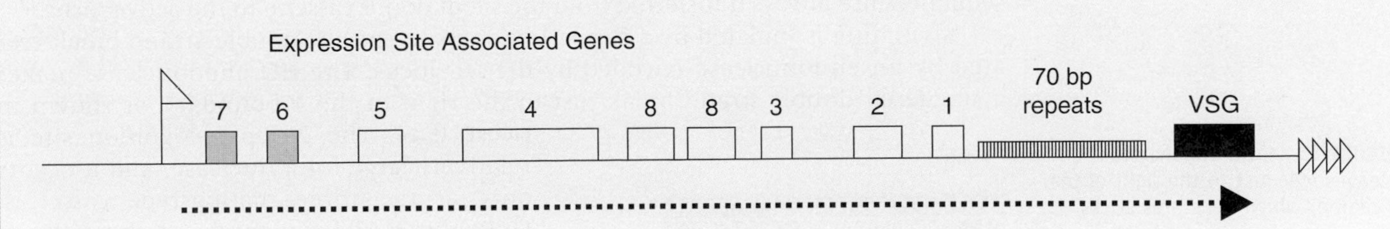

FIGURE B15.2 Organization of a bloodstream-form VSG expression site (BES) from *T. brucei*. The black box indicates the expressed VSG, located near the telomere. These sites also contain expression-site-associated genes (ESAGs) and characteristic 70-bp repeats. Adapted from G. Rudenko, *Clin. Sci.* 28 (2000): 536–540.

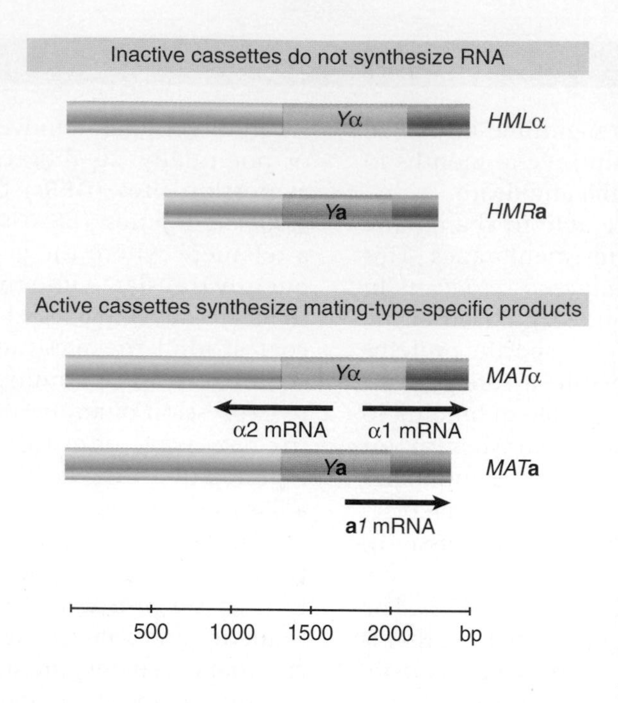

FIGURE 15.23 Silent cassettes have the same sequences as the corresponding active cassettes, except for the absence of the extreme flanking sequences in *HMR***a**. Only the *Y* region changes between **a** and α types.

by switching: it does not exchange with the copy that replaces it. Likewise, if the silent copy present at *HML* or *HMR* is mutated, switching introduces a mutant allele into the *MAT* locus. The mutant copy at *HML* or *HMR* remains there through an indefinite number of switches. The donor element generates a new copy at the recipient site while itself remaining inviolate.

Mating-type switching is a directional event, in which there is only one recipient (*MAT*) but two potential donors (*HML* and *HMR*). Switching usually involves replacement of *MAT***a** by the copy at *HML*α or replacement of *MAT*α by the copy at *HMR***a**. In 80% to 90% of switches, the *MAT* allele is replaced by one of the opposite type. This is determined by the mating type of the cell: **a** cells preferentially choose *HML* as the donor; α cells preferentially choose *HMR*.

Several groups of genes are involved in establishing and switching mating type. In addition to the genes that directly determine mating type, they include genes needed to repress the silent cassettes, switch mating type, or execute the functions involved in mating and, most importantly, homologous recombination factors.

By comparing the sequences of the two silent cassettes (*HML*α and *HMR***a**) with the sequences of the two types of active cassette (*MAT***a** and *MAT*α), we can delineate the sequences that determine mating type. The organization of the mating type loci is summarized in **FIGURE 15.23**. Each cassette contains common sequences that flank a central region that differs in the **a** and α types of cassette (called *Y***a** or *Y*α). On either side of this region, the flanking sequences are virtually identical, although they are shorter at *HMR*. The active cassette at *MAT* is transcribed from a promoter within the *Y* region. Note that the same *Y* region is present in the cassettes at *HML* or *HMR*, but these cassettes are maintained in a transcriptionally silent state so that no mating type information can be expressed from these copies.

A switch in mating type is accomplished by a gene conversion event in which the recipient site (*MAT*) acquires the sequence of the donor type (*HML* or *HMR*). This is a modified homologous recombination mechanism, which uses many of the same enzymes required for meiotic recombination or repair of DNA double-strand breaks; what is unique about mating type switching is the directionality by which homologous sequences are always transferred from the silent donor cassette to the active locus.

Switching is initiated by a formation of a site-specific double-strand break created by an endonuclease encoded by the *HO* locus. The HO endonuclease makes a staggered double-strand break just to the right of the *Y* boundary, as shown in **FIGURE 15.24**. The 24-bp recognition site is relatively large for a nuclease, and it occurs only at the three mating-type cassettes. Only the *MAT* locus, and not the *HML* or *HMR* loci, is a target for the endonuclease. The same mechanisms that keep the silent cassettes from being transcribed also keep them inaccessible to the HO endonuclease. This inaccessibility ensures that switching is unidirectional.

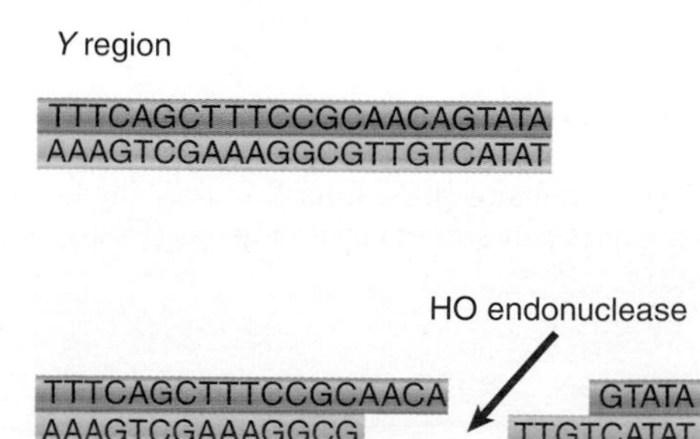

FIGURE 15.24 HO endonuclease cleaves *MAT* just to the right of the *Y* region, which generates sticky ends with a 4-base overhang.

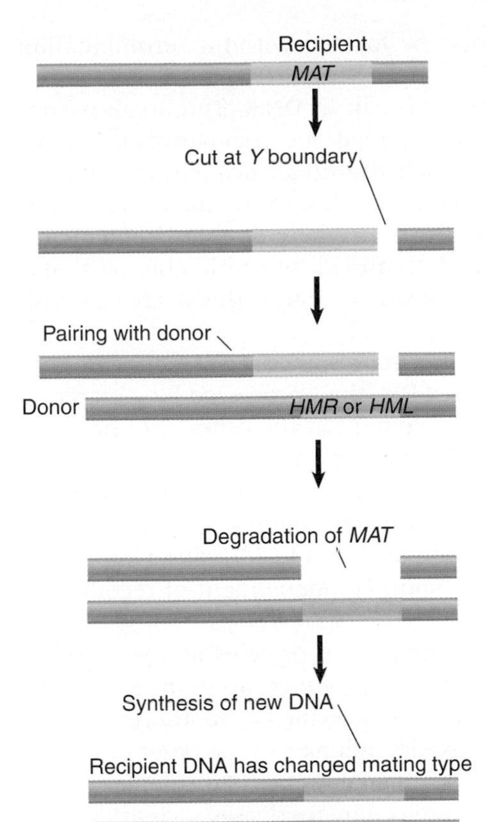

FIGURE 15.25 Cassette substitution is initiated by a double-strand break in the recipient (*MAT*) locus and may involve pairing on either side of the *Y* region with the donor (*HMR* or *HML*) locus.

The reaction triggered by the cleavage is illustrated schematically in **FIGURE 15.25** in terms of the general reaction between donor and recipient regions. As expected, the stages following the initial cut require the enzymes involved in homologous recombination.

KEY CONCEPTS

- The yeast mating type locus *MAT* has either the *MAT***a** or *MAT*α genotype.
- Yeast with the dominant allele *HO* switch their mating type at a frequency of ~10^{-6}.
- The allele at *MAT* is called the active cassette.
- There are also two silent cassettes, *HML*α and *HMR***a**.
- Switching occurs if *MAT***a** is replaced by *HML*α or *MAT*α is replaced by *HMR***a**.
- Mating type switching is initiated by a double-strand break made at the *MAT* locus by the HO endonuclease.
- Mating type switching is achieved by a special homologous recombination event that copies information from *HML*α or *HMR***a** to the active *MAT* locus.

CONCEPT AND REASONING CHECK

An α yeast strain is mutated so that the mating cassettes at *HML* and *HMR* are both **a** type cassettes. Can this strain still switch mating type? How many times? What sequences will be present at *HML* and *HMR* after a switching event?

15.10 Summary

Recombination involves the physical exchange of parts between corresponding DNA molecules. This results in a duplex DNA in which two regions of opposite parental origins are connected by a stretch of hybrid (heteroduplex) DNA, with one strand derived from each parent. Hybrid DNA can also be formed without recombination occurring between markers on either side.

Recombination is initiated by a double-strand break in DNA. The break is converted to a single-stranded region; a free single-stranded end then forms a heteroduplex with a donor sequence. The DNA in which the break occurs may actually incorporate the sequence of the chromosome that it invades, so the initiating DNA is called the recipient. Hotspots for recombination are sites where double-strand breaks are initiated. A gradient of gene conversion is determined by the likelihood that a sequence near the free end will be converted to a single strand; this decreases with distance from the break.

Recombination is initiated in yeast by Spo11, a topoisomerase-like enzyme that becomes linked to the free 5′ ends of DNA. The DSB is then processed by generating single-stranded DNA that can anneal with its complement in the other chromosome. Yeast mutations that block synaptonemal complex formation show that recombination is required for its formation. Formation of the synaptonemal complex may be initiated by double-strand breaks, and it may persist until recombination is completed. Mutations in components of the synaptonemal complex block its formation but do not prevent chromosome pairing, so homolog recognition is independent of recombination and synaptonemal complex formation.

The single-strand regions are generated by a complex of proteins and require the MRN/MRX complex. The single strands provide a substrate for proteins in the RecA family, which have the ability to synapse homologous DNA molecules by sponsoring a reaction in which a single strand from one molecule invades a duplex of the other molecule. Heteroduplex DNA is formed by displacing one of the original strands of the duplex. These actions create a recombination junction, which is resolved by resolvases such as the Ruv proteins. RuvA and RuvB act at a heteroduplex, and RuvC cleaves Holliday junctions.

Recombination, like replication and transcription, requires topological manipulation of DNA. Topoisomerases may relax (or introduce) supercoils in DNA, and are required to disentangle DNA molecules that have become catenated by recombination or by replication. Type I topoisomerases introduce a break in one strand of a DNA duplex; type II topoisomerases make double-strand breaks. The enzyme becomes linked to the DNA by a bond from tyrosine to either 5′ phosphate (type A enzymes) or 3′ phosphate (type B enzymes).

The enzymes involved in site-specific recombination have actions related to those of topoisomerases. Among this general class of recombinases, those concerned with phage integration form the subclass of integrases. The Cre/*lox* system uses two molecules of Cre to bind to each *lox* site, so that the recombining complex is a tetramer. This is one of the standard systems for inserting DNA into a foreign genome. Phage lambda integration requires the phage Int protein and host IHF protein and involves a precise breakage and reunion in the absence of any synthesis of DNA. Reaction in the reverse direction requires the phage protein Xis. Some integrases function by *cis*-cleavage, where the tyrosine that reacts with DNA in a half site is provided by the enzyme subunit bound to that half site; others function by *trans*-cleavage, for which a different protein subunit provides the tyrosine.

The yeast *S. cerevisiae* can propagate in either the haploid or diploid condition. Conversion between these states takes place by mating (fusion of haploid cells to give a diploid) and by sporulation (meiosis of diploids to give haploid spores). The ability to engage in these activities is determined by the mating type of the strain. The mating type is determined by the sequence of the *MAT* locus and can be changed by a recombination event that substitutes a different sequence at this locus. The recombination event is initiated by a double-strand break—such as a homologous recombination event—but then the subsequent events ensure a unidirectional replacement of the sequence at the *MAT* locus.

Replacement is regulated so that *MAT***a** is usually replaced by the sequence from *HML*α, whereas *MAT*α is usually replaced by the sequence from *HMR***a**. The endonuclease *HO* triggers the reaction by recognizing a unique target site at *MAT*.

CHAPTER QUESTIONS

1. What are the three requirements for RecA-mediated strand invasion?
   A. ______________________________
   B. ______________________________
   C. ______________________________
2. What four components are required for bacteriophage lambda integration into the bacterial chromosome, and where do these components come from?
   A. ______________________________
   B. ______________________________
   C. ______________________________
   D. ______________________________
3. Excision of an integrated lambda phage genome from the bacterial chromosome requires what three proteins?
   A. ______________________________
   B. ______________________________
   C. ______________________________
4. Recombination between homologous DNA sequences is called:
   A. homologous recombination.
   B. excision.
   C. site-specific recombination.
   D. transposition.
5. A classic example of site-specific recombination is:
   A. HIV virus integration.
   B. bacteriophage lambda integration.
   C. meiotic recombination.
   D. recombination repair.
6. When chromosomes begin to separate during meiosis, they can be observed to be held together at discrete sites called:
   A. bivalents.
   B. synaptonemal complexes.
   C. chiasmata.
   D. Holliday junctions.
7. Recombinant DNA molecules where the duplex of one DNA parent is covalently linked to the duplex of the other parent by a stretch of heteroduplex DNA are called:
   A. patch recombinants.
   B. splice recombinants.
   C. Holliday junctions.
   D. joint molecules.
8. What typically initiates homologous DNA recombination?
   A. a site-specific nick in one strand of duplex DNA
   B. a site-specific double-strand break in duplex DNA
   C. a random nick in one strand of duplex DNA
   D. a random double-strand break in duplex DNA
9. Which of the following enzymes remove positive supercoils that occur ahead of a replicating DNA polymerase?
   A. Topoisomerase I
   B. Topoisomerase II
   C. DNA gyrase
   D. Reverse gyrase

10. Formation of diploid *S. cerevisiae* involves mating between which pair of haploid cells?
    A. two type **a** cells
    B. two type α cells
    C. a type **a** cell and a type α cell
    D. any of the above combinations
11. What is responsible for initiation of yeast mating type switching?
    A. a double-strand break made at the *MAT* locus by the Spo11 endonuclease
    B. a single-strand nick at a silent mating cassette by the Spo11 endonuclease
    C. a single-strand nick made at the *MAT* locus by the HO endonuclease
    D. a double-strand break made at the *MAT* locus by the HO endonuclease

KEY TERMS

5′ end resection
attachment (*att*) sites
axial element
bivalent
branch migration
catenate
central element
chiasma (pl. chiasmata)
chromosome pairing
core sequence
D loop (displacement loop)
double-strand breaks (DSB)
excision
gene conversion
gyrase
heteroduplex DNA
Holliday junction
homologous recombination
hotspot
integrase
integration
joint molecule
lateral element
mating type cassette
patch recombinant
presynaptic filaments
prophage
recombinant joint
recombinase
recombination nodules (nodes)
resolution
reverse gyrase
single-strand invasion
sister chromatid
site-specific recombination
somatic recombination
splice recombinant
supercoiling
synapsis
synaptonemal complex
topoisomerase
type I topoisomerase
type II topoisomerase

FURTHER READING

Cromie, G. A., and Smith, G. R. (2007). Branching out: meiotic recombination and its regulation. *Trends Cell Biol.* **17**(9), 448–455. A review of the multiple pathways controlling meiotic recombination.

Haber, J. E. (1998). Mating-type gene switching in *Saccharomyces cerevisiae*. *Ann. Rev. Gen.* **32**, 561–599. An introduction to mating type switching in yeast.

San Filippo, J., Sung, P., and Klein, H. (2008). Mechanism of eukaryotic homologous recombination. *Ann. Rev. Biochem.* **77**, 229–257. A review of the mechanisms of homologous recombination, with a focus on the role of the Rad51 and Dmc1 recombinases.

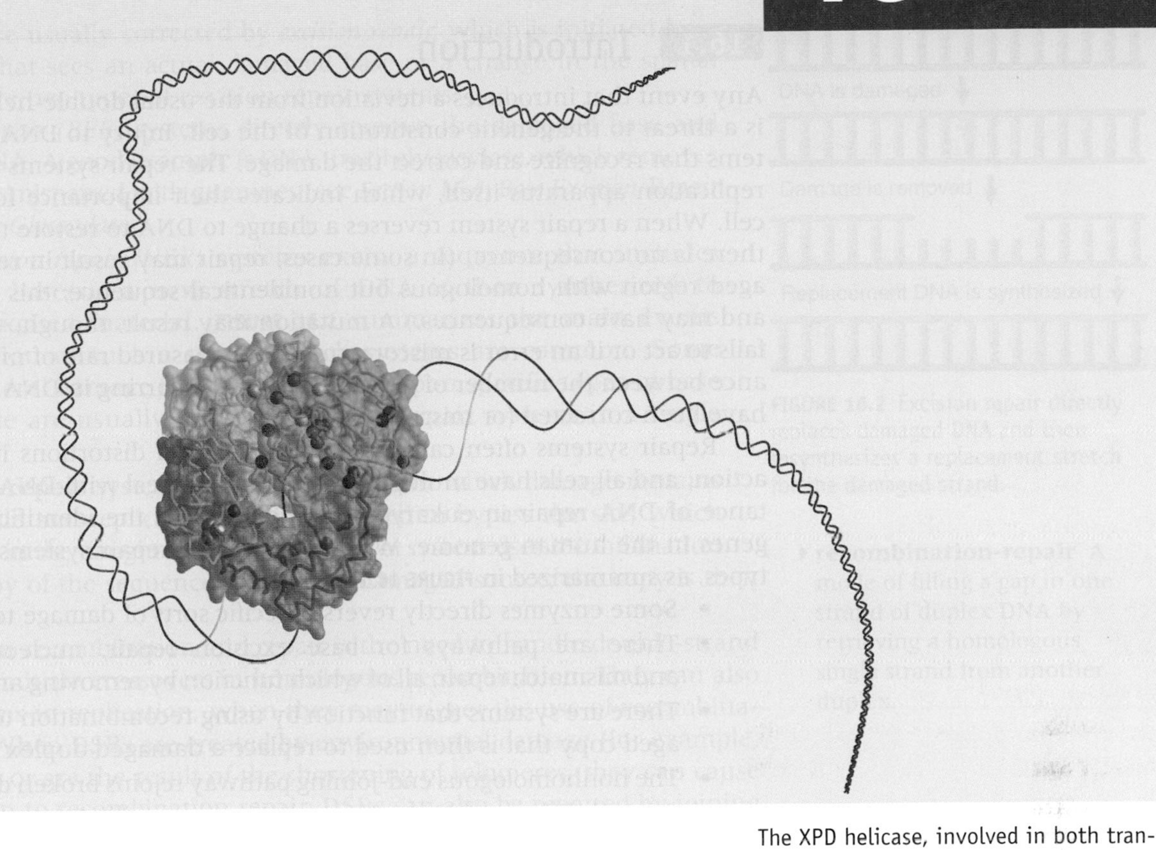

The XPD helicase, involved in both transcription and nucleotide excision repair, is shown unwinding DNA. Colors indicate different domains of XPD; the helicase domain is green. Photo courtesy of M. Pique, Li Fan, Jill Fuss, and John Tainer, The Scripps Research Institute and Lawrence Berkeley National Laboratory. More information available at L. Fan et al., *Cell* 133 (2008): 789–800.

Repair Systems

CHAPTER OUTLINE

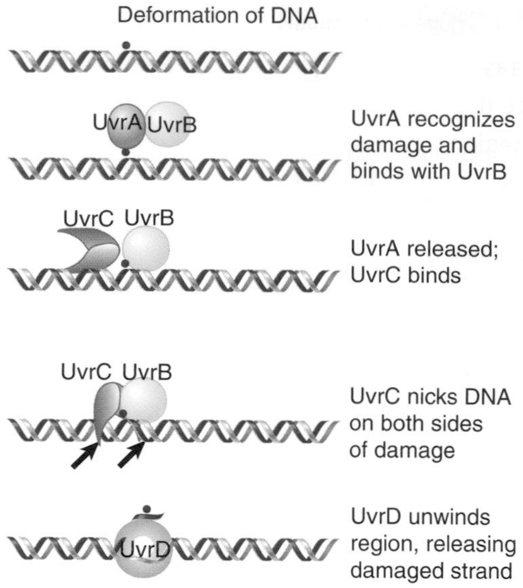

enzyme involved in the repair synthesis also is likely to be DNA polymerase I (although DNA polymerases II and III can substitute for it).

UvrABC repair accounts for virtually all the excision repair events in *E. coli*. In almost all (99%) of cases, the average length of replaced DNA is ~12 nucleotides. (For this reason, the process is sometimes described as *short-patch repair*.) The remaining 1% of cases involve the replacement of stretches of DNA mostly ~1500 nucleotides long, but extending as much as >9000 nucleotides (sometimes called *long-patch repair*). We do not know why some events trigger the long-patch rather than the short-patch mode.

The Uvr complex can be directed to sites of damage by other proteins. Damage to DNA can result in stalled transcription, in which case a protein called Mfd displaces the RNA polymerase and recruits the Uvr complex. **FIGURE 16.10** shows a model for the link between transcription and repair. When RNA polymerase encounters DNA damage on the template strand, it stalls because it cannot use the damaged sequences as a template to direct complementary base pairing. This effect is specific for damage in the template strand; damage in the nontemplate strand does not impede progress of the RNA polymerase.

Deformation of DNA

UvrA UvrB — UvrA recognizes damage and binds with UvrB

UvrC UvrB — UvrA released; UvrC binds

UvrC UvrB — UvrC nicks DNA on both sides of damage

UvrD — UvrD unwinds region, releasing damaged strand

FIGURE 16.9 The Uvr system operates in stages in which UvrAB recognizes damage, UvrBC nicks the DNA, and UvrD unwinds the marked region.

The Mfd protein has two roles. First, it displaces the ternary complex of RNA polymerase from DNA. Second, it causes the UvrABC complex to bind to the damaged DNA, directing excision repair to the damaged strand. After the DNA has been repaired, the next RNA polymerase to traverse the gene is able to produce a normal transcript.

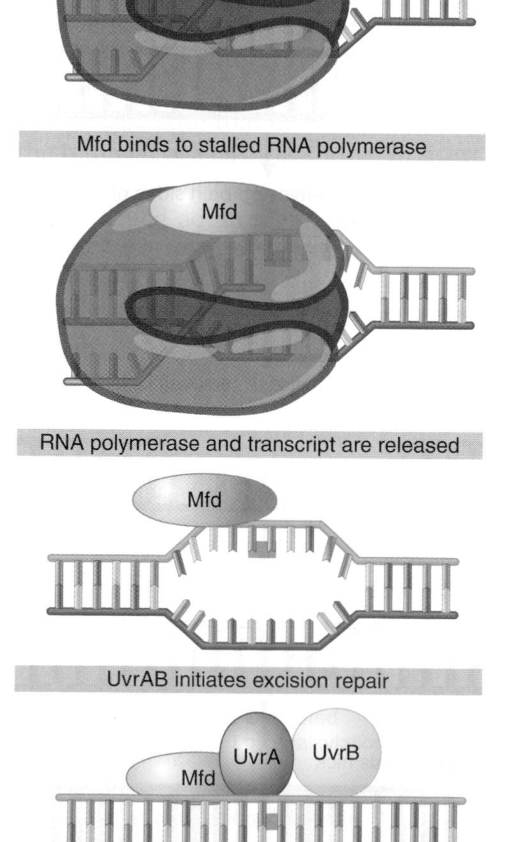

FIGURE 16.10 Mfd recognizes a stalled RNA polymerase and directs the Uvr complex to the damaged template strand.

▸ **xeroderma pigmentosum (XP)** A disease caused by mutation in one of the XP genes that results in hypersensitivity to sunlight (particularly ultraviolet light), skin disorders, and cancer predisposition.

The general principle of excision repair in eukaryotic cells is similar to that of bacteria. Bulky lesions, such as those created by UV damage, are also recognized and repaired by a nucleotide excision repair system. The critical role of mammalian nucleotide excision repair is seen in certain human hereditary disorders. The best investigated of these is **xeroderma pigmentosum (XP)**, a recessive disease resulting in hypersensitivity to sunlight, and in particular, ultraviolet light. The deficiency results in skin disorders and cancer predisposition (for more discussion, see the accompanying box, *Medical Applications*).

The disease is caused by a deficiency in nucleotide excision repair. XP patients cannot excise pyrimidine dimers and other bulky adducts. Mutations occur in one of eight genes called *XP-A* to *XP-G*. A protein complex that includes products of several of the *XP* genes is responsible for excision of thymine dimers. There are actually two major pathways of nucleotide excision repair in eukaryotes, illustrated in **FIGURE 16.11**. The major difference between the two pathways is how the damage is initially recognized. In *global genome repair*, a protein called XPC detects the damage and initiates the repair pathway. XPC can recognize damage anywhere in the genome. XPC is typically a component of a lesion-sensing complex with other proteins that

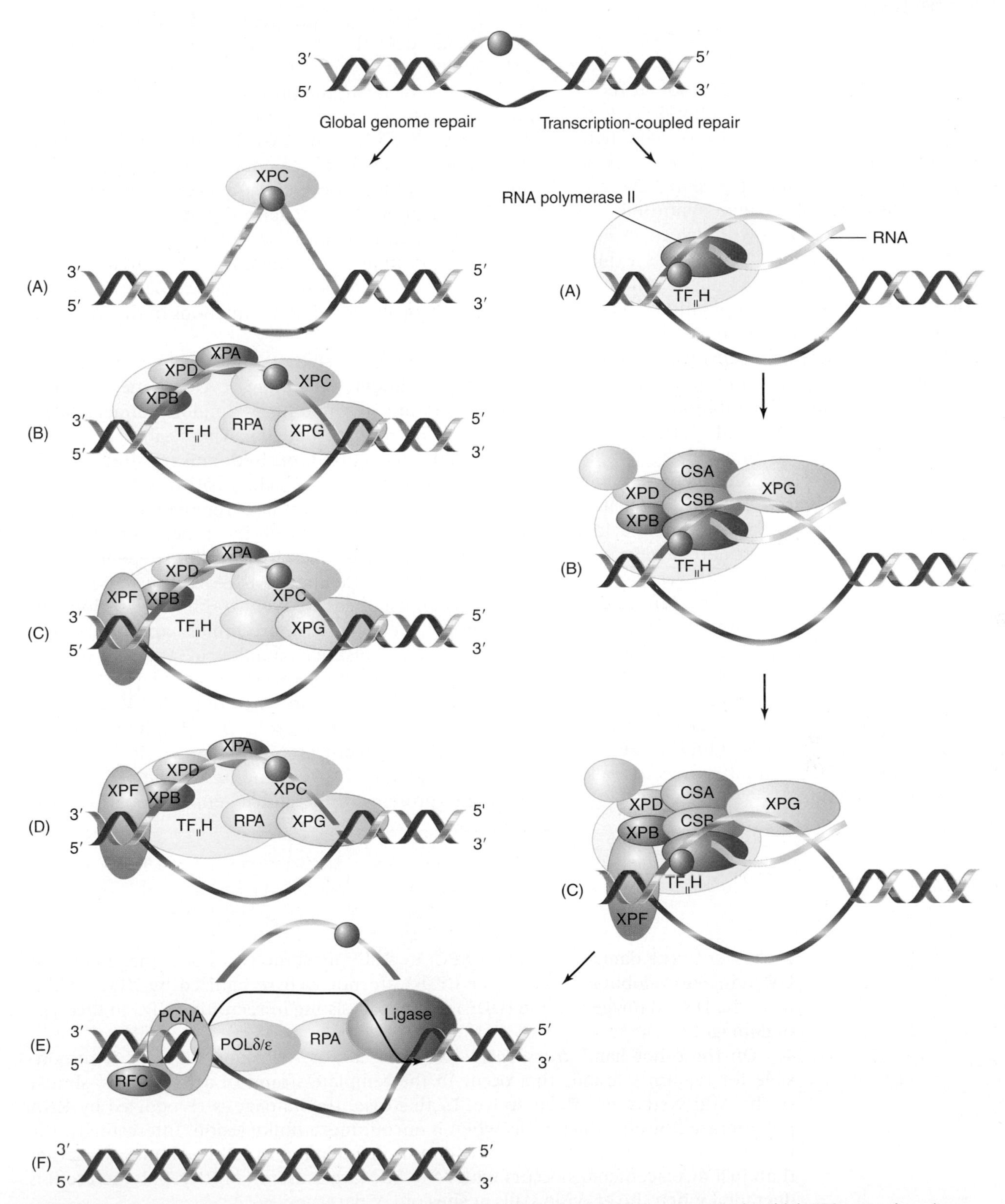

FIGURE 16.11 Nucleotide excision repair occurs via two major pathways: global genome repair, in which XPC recognizes damage anywhere in the genome, and transcription-coupled repair, in which the transcribed strand of active genes is preferentially repaired, and the damage is recognized by an elongating RNA polymerase. Adapted from E. C. Friedberg et al., *Nature Rev. Cancer* 1 (2001): 22–23.

MEDICAL APPLICATIONS

Xeroderma Pigmentosum

Xeroderma pigmentosum (XP) is an autosomal recessive disorder associated with extreme ultraviolet (UV) light sensitivity, increased risk of skin cancer, and neurological degeneration. The name *xeroderma pigmentosum* literally means "dry pigmented skin." XP and two other related disorders, Cockayne syndrome and trichothiodystrophy, are associated with defects in DNA nucleotide excision repair (NER). XP patients exhibit sun-induced freckles and other skin pigmentation as infants, which is rare in the general population, and this is often the basis of the clinical diagnosis. About half the patients with XP are extremely light sensitive and many exhibit sunburn and blistering with minimal sun exposure; others tan normally but still exhibit pigmentation on sun-exposed skin. With continued sun exposure, the skin becomes dry and looks prematurely aged. Children with XP frequently develop precancerous lesions and have a 1000-fold increased risk of developing either melanoma or non-melanoma skin cancer before age 20. The average age of XP patients that have skin cancer is under 10 years old, which is 50 years earlier than the average age of the appearance of skin cancers in the general population. With continued exposure to sunlight, XP patients may develop chronic UV-induced eye inflammation, corneal damage, and the loss of eyelashes. About 30% of XP patients also exhibit neurological defects, including progressive mental retardation, loss of deep tendon reflexes, and loss of high-frequency hearing. Patients may also develop difficulties in walking and swallowing over time.

XP is a relatively rare genetic disorder; the overall prevalence in the United States is estimated to be 1/1,000,000 but is somewhat more common in some populations. In Japan the prevalence is 1/22,000. There are seven distinct complementation groups (*XPA, XPB, XPC, XPD, XPE, XPF,* and *XPG*) that are associated with this disease. Each complementation group represents mutations in different proteins involved in the recognition and repair of DNA damage induced by UV radiation, and the proteins have been named for the complementation groups. In addition, there is an XP variant (XPV) that is defective in post-replication repair. Approximately 50% of all XP is due to mutations in the *XPA* and *XPC* genes.

UV radiation damages DNA by the formation of pyrimidine dimers, including cyclobutane pyrimidine dimers and pyrimidine-pyrimidone dimers (also known as 6,4-photoproducts). These photoproducts are normally recognized, excised, and repaired by the nucleotide excision repair (NER) pathway (see Figure 16.11). The NER pathway also removes other bulky DNA modifications induced by other carcinogens. DNA repair by the NER pathway differs depending on whether or not the damaged DNA is in a region that is actively transcribed. Actively transcribed regions of the genome are repaired more rapidly and are recognized through a distinct complex of proteins of the transcription-coupled repair (TCR) pathway. Damaged DNA is detected by the presence of stalled RNA polymerase, which, in turn, recruits DNA binding proteins CSA and CSB. In contrast, DNA damage in the nontranscribed regions of the genome is detected by the binding of proteins of the global genome repair (GGR)

act to verify the damage bound by XPC. In addition, some types of damage (such as UV-induced cyclobutane dimers, or CPDs) are not well recognized by XPC. In this case, the DNA damage binding (DDB) complex assisting in recruiting XPC to this type of damage.

On the other hand, *transcription-coupled repair*, as the name suggests, is responsible for repairing lesions that occur in the template strand of active genes, similar to the Mfd system described above. In this case, the damage is recognized by RNA polymerase II itself, which stalls when it encounters a bulky lesion. Interestingly, the repair function may require modification or degradation of RNA polymerase, rather than just displacement as occurs with Mfd. The large subunit of RNA polymerase II is degraded when the enzyme stalls at sites of UV damage.

The two pathways eventually merge and use a common set of proteins to effect the repair. The strands of DNA are unwound for ~20 bp around the damaged site. This action is performed by the helicase activity of the transcription factor $TF_{II}H$, itself a large complex, which includes the products of two *XP* genes (*XPB* and *XPD*). XPB and

pathway, XPE and XPC proteins. After the recognition of the DNA lesions by these two distinct mechanisms, the pathways converge and the damaged DNA is unwound, excised, and repaired by common proteins. XPA protein binds the flagged, damaged DNA and recruits two additional proteins, XPB and XPD. Both of these proteins are helicases that are components of the transcription factor $TF_{II}H$ (and thus already present at sites of damage recognized by stalled RNA polymerase). Once the DNA is unwound, exonucleases are recruited to regions surrounding the damaged DNA. XPF and XPG are 5′ and 3′ exonucleases, respectively, and are cut on either side of the damaged DNA to generate a fragment of approximately 30 nucleotides of single-stranded DNA. The resulting gap in the genomic DNA is repaired by *de novo* DNA synthesis. XPV is not a component of the NER pathway but is a member of the Y family of error-prone DNA polymerases that can replicate DNA containing photoproducts such as pyrimidine dimers, which cannot be replicated by the polymerases of the replication fork.

There is a very interesting genotype/phenotype correlation between mutations in the NER pathway and the sun sensitivity, cancer risk, and neurological manifestations of the XP disease. Patients with mutations in the proteins that are unique to the GGR pathway (XPC and XPE) and mutations in XPV, exhibit XP without neurological defects. Patients with mutations in the other XP proteins that are common to GGR and TCR exhibit XP with the associated neurological defects. Interestingly, mutations in CSA and CSB, the DNA binding proteins in the TCR pathway, result in the disorder Cockayne syndrome. Cockayne syndrome is characterized by sun sensitivity, mental retardation, and developmental retardation; however, there is no increased risk of skin cancer. Cockayne syndrome can also result from specific mutations in the XPB and XPD helicases. These mutations are at different sites from those that result in XP.

Trichothiodystrophy is another disease resulting from defects in proteins involved in NER; specifically, another subset of mutations in XPB and XPD which may affect their roles as components of the transcription factor $TF_{II}H$. This disorder is associated with photosensitivity, brittle hair and nails, and growth retardation but not with sun-induced pigmentation or cancer, and these patients appear to have specific transcriptional defects rather than repair deficiency. These observations of XP and other NER defective disorders has led to the suggestion that the neurological defects associated with these diseases may arise due to disruption of the TCR pathway, resulting in a blockage of transcription and the apoptosis of neural cells. Alternatively, the cancer risk associated with XP, but not the other two disorders, may arise due to genomic instability resulting from the disruption of the GGR pathway in cells.

There is no cure or any treatment for XP, other than aggressive avoidance of UV exposure. This includes limiting time spent outdoors or in the car, wearing protective clothing, using high SPF sunscreen at all times and wearing UV absorbing glasses. Patients need to be continuously monitored and treated for precancerous and cancerous skin lesions. Recently, several promising clinical trials have tested the effects of topical lotions to treat photosensitivity in XP patients. These lotions contain liposomes designed to deliver DNA repair enzymes to

XPD are both helicases; the XPB helicase is required for promoter melting during transcription, whereas the XPD helicase performs the primary unwinding function in NER (although the ATPase activity of XPB is also required during NER). $TF_{II}H$ is already present in a stalled transcription complex, as a result, repair of transcribed strands is extremely efficient compared to repair of nontranscribed regions.

In the next step, cleavages are made on either side of the lesion by endonucleases encoded by the *XPG* and *XPH* genes. Typically, about 25 to 30 nucleotides are excised during NER. Finally, the single-stranded stretch including the damaged bases can then be replaced by new synthesis, and the final remaining nick is ligated by a complex of ligase III and XRCC1.

In cases where replication encounters a thymine dimer that has not been removed, replication requires DNA polymerase η (eta) activity in order to proceed past the dimer. This polymerase is encoded by *XPV*. This bypass mechanism allows cell division to proceed even in the presence of unrepaired damage, but this is generally a last resort as cells prefer to put a hold on cell division until all damage is repaired.

replication, DSBs can be generated in a number of other ways, including ionizing radiation, oxygen radicals generated by cellular metabolism, or action of endonucleases. The preferred mechanism for repairing DSBs is to use recombination repair, as this ensures that no critical genetic information is lost due to sequence loss at the breakpoint.

Several of the genes required for recombination repair in eukaryotes have already been discussed in the context of meiotic homologous recombination (see *Section 15.5, Specialized Enzymes Catalyze 5′ End Resection and Single-Strand Invasion*). Many eukaryotic repair genes are named *RAD* genes; they were initially characterized genetically in yeast by virtue of their sensitivity to *rad*iation. There are three general groups of repair genes in the yeast *S. cerevisiae*, identified by the *RAD3* group (involved in excision repair), the *RAD6* group (required for postreplication repair), and the *RAD52* group (concerned with recombination-like mechanisms). Homologs of these genes are present in higher eukaryotes as well.

The *RAD52* group plays essential roles in homologous recombination and includes a large number of genes such as *RAD50, RAD51, RAD54, RAD55, RAD57*, and *RAD59*. These Rad proteins all assemble at a double-strand break. As occurs during meiotic recombination, the Mre11/Rad50/Xbs1 (MRX)-dependent exonuclease acts on the free ends to generate single-stranded tails, and serves to tether the free ends together. The single-stranded DNA serves to activate a DNA damage checkpoint, stopping cell division until the damage can be repaired. The RecA homolog Rad51 binds to the single-stranded DNA to form a nucleoprotein filament, which is used for strand invasion of a homologous sequence. Rad52, Rad55, and Rad54 are required to form a stable Rad51 filament and also assist in the homology search and strand invasion. Following repair synthesis, the resulting structure (which resembles a Holliday junction) is resolved.

KEY CONCEPTS

- The *rec* genes of *E. coli* code for the principal recombination-repair system.
- Recombination-repair functions when replication leaves a gap in a newly synthesized strand that is opposite a damaged sequence.
- The single strand of another duplex is used to replace the gap.
- The damaged sequence is then removed and resynthesized.
- A replication fork may stall when it encounters a damaged site or a nick in DNA.
- A stalled fork may reverse by pairing between the two newly synthesized strands.
- A stalled fork may restart after repairing the damage and use a helicase to move the fork forward.
- The yeast *RAD* mutations, identified by radiation sensitive phenotypes, are in genes that code for repair systems.
- The *RAD52* group of genes are required for recombination repair.

CONCEPT AND REASONING CHECK

Explain why large gaps and double-strand breaks are preferentially repaired by recombination, but a stalled replication fork can utilize either excision repair or recombination-repair pathways.

16.8 Nonhomologous End Joining Also Repairs Double-Strand Breaks

Repair of DSBs by homologous recombination ensures no genetic information is lost from a broken DNA end. However, in many cases a sister chromatid or homologous chromosome is not easily available to use as a template for repair. In addition, some

DSBs are specifically repaired using error-prone mechanisms as intermediate in the recombination of immunoglobulin genes (see *Section 18.6, The RAG1/RAG2 Catalyze Breakage and Religation of V(D)J Gene Segments*). In these cases, the mechanism used to repair these breaks is called **nonhomologous end joining (NHEJ)**, and consists of ligating the blunt ends together.

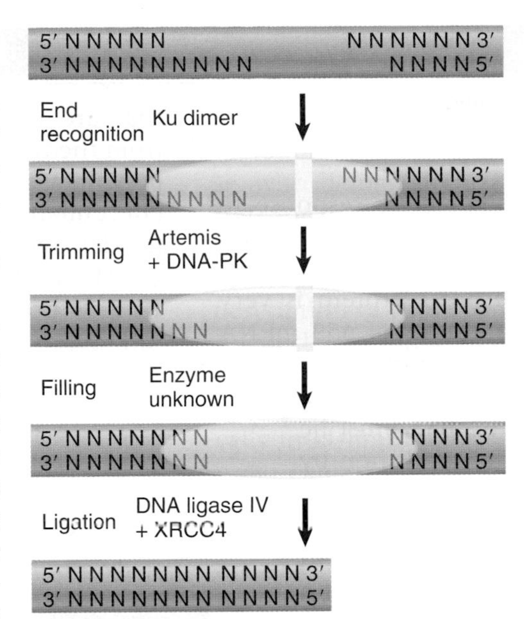

FIGURE 16.22 Nonhomologous end joining requires recognition of the broken ends, trimming of overhanging ends, and/or filling, followed by ligation.

▸ **nonhomologous end joining (NHEJ)** The pathway that ligates blunt ends. It is used to repair DNA double-strand breaks and in certain recombination pathways (such as immunoglobulin recombination).

The steps involved in NHEJ are summarized in **FIGURE 16.22**. The same enzyme complex undertakes the process in both NHEJ and immune recombination. The first stage is recognition of the broken ends by a heterodimer consisting of the proteins Ku70 and Ku80. They form a scaffold that holds the ends together and allows other enzymes to act on them. A key component is the DNA-dependent protein kinase (DNA-PKcs), which is activated by DNA to phosphorylate protein targets. One of these targets is the protein Artemis, which in its activated form has both exonuclease and endonuclease activities and can both trim overhanging ends and cleave the hairpins generated by recombination of immunoglobulin genes. The DNA polymerase activity that fills in any remaining single-stranded protrusions is not known. The actual joining of the double-stranded ends is performed by DNA ligase IV, which functions in conjunction with the protein XRCC4. Mutations in any of these components may render eukaryotic cells more sensitive to radiation. Some of the genes for these proteins are mutated in patients who have diseases due to deficiencies in DNA repair.

The Ku heterodimer is the sensor that detects DNA damage by binding to the broken ends. The crystal structure in **FIGURE 16.23** shows why it binds only to ends. The bulk of the protein extends for about two turns along one face of DNA (lower), but a narrow bridge between the subunits, located in the center of the structure, completely encircles DNA. This means that the heterodimer needs to slip onto a free end.

Ku can bring broken ends together by binding two DNA molecules. The ability of Ku heterodimers to associate with one another suggests that the reaction might

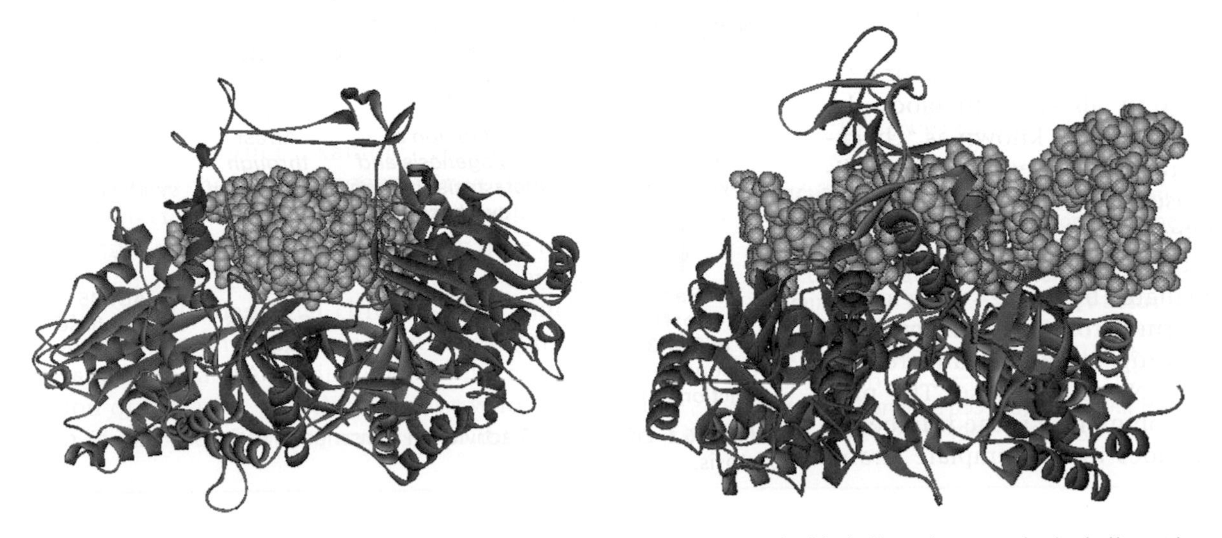

FIGURE 16.23 The Ku70-Ku80 heterodimer binds along two turns of the DNA double helix and surrounds the helix at the center of the binding site.

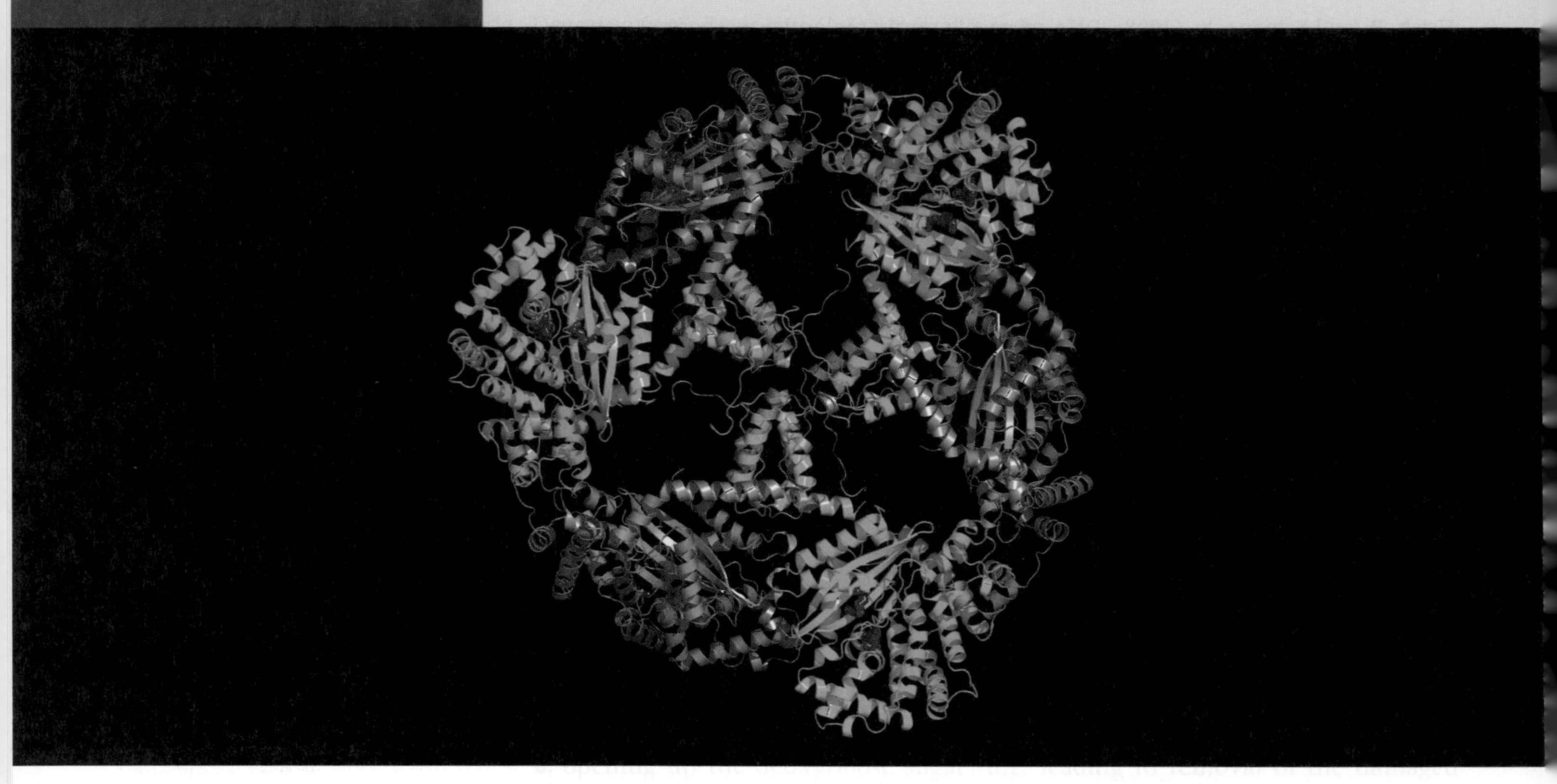

The Hermes DNA transposase, with adjacent monomers of the hexamer colored green and orange and active sites colored red. Photo courtesy of Fred Dyda, NIDDK, National Institutes of Health.

Transposable Elements and Retroviruses

CHAPTER OUTLINE

17.1 Introduction

A major cause of variation in nearly all genomes is provided by **transposable elements**, or **transposons**: these are discrete sequences in the genome that are mobile—they are able to transport themselves to other locations within the genome. The mark of a transposon is that it does not utilize an independent form of the element (such as phage or plasmid DNA) but moves directly from one site in the genome to another. Unlike most other processes involved in genome restructuring, transposition does not rely on any relationship between the sequences at the donor and recipient sites. Transposons are restricted to moving themselves, and sometimes additional sequences, to new sites elsewhere within the same genome; they are, therefore, an internal counterpart to the vectors that can transport sequences from one genome to another. They can be a major source of mutations in the genome, as shown in **FIGURE 17.1**, and have had a significant impact on the overall size of many genomes, including our own.

▸ **transposable element (transposon)** A DNA sequence able to insert itself (or a copy of itself) at a new location in the genome without having any sequence relationship with the target locus.

Unequal recombination results from mispairing by the cellular systems for homologous recombination. Nonreciprocal recombination results in duplication or rearrangement of loci (see *Section 7.2, Unequal Crossing Over Rearranges Gene Clusters*). Duplication of sequences within a genome provides a major source of new sequences. One copy of the sequence can retain its original function, whereas the other may evolve into a new function. Furthermore, significant differences between individual genomes are found at the molecular level because of polymorphic variations caused by recombination. We saw in *Section 7.8, Minisatellites Are Useful for Genetic Mapping*, that recombination between minisatellites adjusts their lengths so that every individual genome is distinct.

Transposons fall into two general classes: those that are able to directly manipulate DNA so as to propagate themselves within the genome (Class II elements, or *DNA-type elements*) and those whose source of mobility is the ability to make DNA copies of their RNA transcripts, which are then integrated into new sites in the genome (Class I elements, or *retroelements*).

Transposons that mobilize via DNA are found in both prokaryotes and eukaryotes. Each bacterial transposon carries gene(s) that code for the enzyme activities required for its own transposition, but it may also require ancillary functions of the genome in which it resides (such as DNA polymerase or DNA gyrase). Comparable systems exist in eukaryotes, although their enzymatic functions are not so well characterized.

Transposition that involves an obligatory intermediate of RNA is primarily confined to eukaryotes. Transposons that employ an RNA intermediate all use some form of reverse transcriptase to convert RNA into DNA. Some of these elements are closely related to retroviral proviruses in their general organization and mechanism of transposition (discussed shortly and further in *Section 17.7, The Retrovirus Life Cycle Involves Transposition-like Events*). As a class, these elements are called **LTR (*l*ong-*t*erminal *r*epeat) retrotransposons**, or simply **retrotransposons**. Members of a second class of elements that also use reverse transcriptase but lack LTRs, and that employ a distinct mode of transposition, are referred to as **non-LTR retrotransposons**, or simply **retroposons**. [The nomenclature of transposable elements is somewhat confusing in the literature, but this system of distinguishing elements by the presence or lack of LTRs reflects the modern understanding of both the evolution and transposition mechanism of these elements.]

▸ **long-terminal repeat (LTR)** The sequence that is repeated at each end of the provirus (integrated retroviral sequence).

▸ **retrotransposon (LTR retrotransposon)** A transposon that mobilizes via an RNA form; the DNA element is transcribed into RNA and then reverse transcribed into DNA, which is inserted at a new site in the genome. It does not have an infective (viral) form. This name typically refers to retroelements that contain retrovirus-like LTRs and resemble retroviruses.

▸ **retroposon (non-LTR retrotransposon)** A transposon that mobilizes via an RNA intermediate, similar to an LTR retrotransposon, but that lacks LTRs and uses a distinct transposition mechanism.

The very simplest retrotransposons do not themselves have transposition activity but have sequences that are recognized as substrates for transposition by active elements. Thus elements that use RNA-dependent transposition range from the

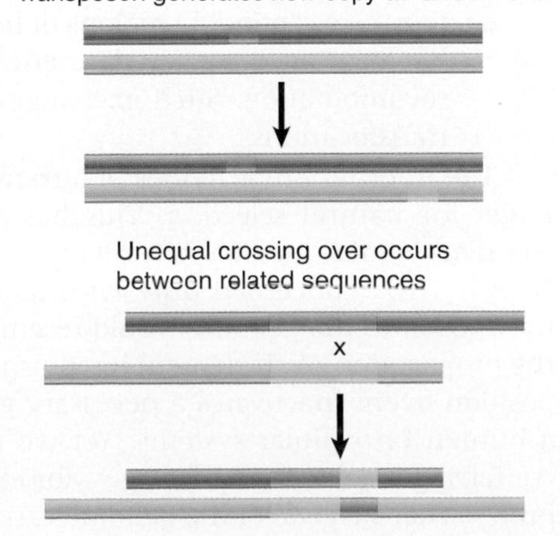

FIGURE 17.1 A major cause of sequence change within a genome is the movement of a transposon to a new site. This may have direct consequences on gene expression. Unequal crossing over between related sequences causes rearrangements. Copies of transposons can provide targets for such events.

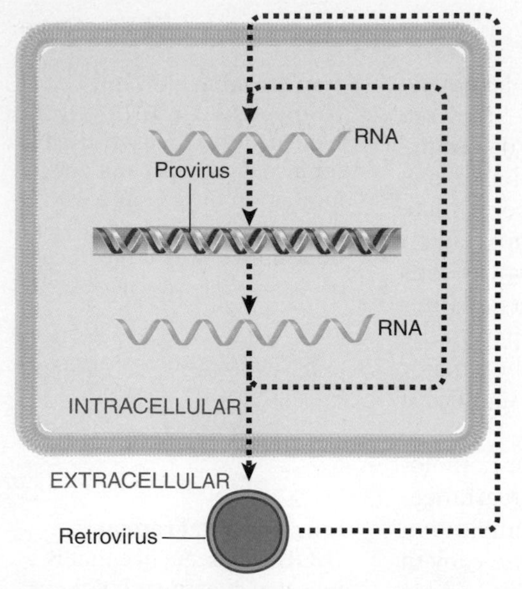

FIGURE 17.2 The reproductive cycles of retroviruses and retrotransposons alternate reverse transcription from RNA to DNA with transcription from DNA to RNA. Only retroviruses can generate infectious particles. Retrotransposons are confined to an intracellular cycle.

retroviruses themselves (which are able to infect host cells freely), to sequences that transpose via RNA, to those that do not themselves possess the ability to transpose. They share with all transposons the diagnostic feature of generating short direct repeats of target DNA at the site of an insertion. Even in genomes where active transposons have not been detected, footprints of ancient transposition events are found in the form of direct target repeats flanking dispersed repetitive sequences. The features of these sequences sometimes implicate an RNA sequence as the progenitor of the genomic (DNA) sequence. This suggests that the RNA must have been converted into a duplex DNA copy that was inserted into the genome by a transposition-like event.

Like any other reproductive cycle, the cycle of a **retrovirus** or retrotransposon is continuous; it is arbitrary to consider the point at which we interrupt it a "beginning." Our perspectives of these elements are biased, though, by the forms in which we usually observe them. The interlinked cycles of retroviruses and retrotransposons are depicted in **FIGURE 17.2**. Retroviruses were first observed as infectious virus particles that were capable of transmission between cells, and so the intracellular cycle (involving duplex DNA) is thought of as the means of reproducing the RNA virus. Retrotransposons were discovered as components of the genome, and the RNA forms have been mostly characterized for their functions as mRNAs and transposition intermediates. Thus we think of retroposons as genomic (duplex DNA) sequences and retroviruses as RNA/protein complexes, but this obscures the close relationship between these elements. Indeed, recent phylogenetic evidence suggests that retroviruses as a class are simply retroviruses that have acquired envelope proteins. This is the inverse of the previously assumed relationship, which was that retrotransposons were retroviruses that had lost the ability to exit the cell.

▸ **retrovirus** An RNA virus with the ability to convert its sequence into DNA by reverse transcription.

A genome may contain both functional and nonfunctional (defective) elements of either class of element. In most cases the majority of elements in a eukaryotic genome are defective and have lost the ability to transpose independently, although they may still be recognized as substrates for transposition by the enzymes produced by functional transposons. A eukaryotic genome contains a large number and variety of transposons. The relatively small fly genome has 1572 identified transposons belonging to 96 distinct families. Larger genomes, such as those of maize and humans, can harbor hundreds of thousands of individual transposons. Roughly half of the genetic material of each of these species is composed of transposons.

Transposable elements can promote rearrangements of the genome either directly or indirectly:

- The transposition event itself may cause deletions or inversions or lead to the movement of a host sequence to a new location.
- Transposons serve as substrates for cellular recombination systems by functioning as "portable regions of homology"; two copies of a transposon at different locations (even on different chromosomes) may provide sites for reciprocal recombination. Such exchanges result in deletions, insertions, inversions, or translocations.

The intermittent activities of a transposon seem to provide a somewhat nebulous target for natural selection. This has prompted suggestions that most transposable elements confer neither advantage nor disadvantage but could constitute "selfish DNA"—DNA concerned only with its own propagation. Such a relationship of the transposon to the genome would resemble that of a parasite with its host. Presumably the propagation of an element by transposition is balanced by the harm done if a transposition event inactivates a necessary gene or if the number of transposons becomes a burden on cellular systems. Yet we must remember that any transposition event conferring a selective advantage—for example, a genetic rearrangement—will lead to preferential survival of the genome carrying the active transposon.

17.2 Insertion Sequences Are Simple Transposition Modules

Transposable elements were first identified at the molecular level in the form of spontaneous insertions in bacterial operons. Such an insertion prevents transcription and/or translation of the gene in which it is inserted. Many different types of transposable elements have now been characterized in both prokaryotes and eukaryotes (in which they are far more abundant), but the basic principles and biochemistry of elements first described in bacteria apply to DNA-type elements in many species.

The simplest bacterial transposons are called **insertion sequences** (reflecting the way in which they were detected). Each type is given the prefix **IS**, followed by a number that identifies the type. (The original classes were numbered IS1 to IS4; later classes have numbers reflecting the history of their isolation but not corresponding to the more than 700 elements so far identified!)

▸ **insertion sequence (IS)** A small bacterial transposon that carries only the genes needed for its own transposition.

The IS elements are normal constituents of bacterial chromosomes and plasmids. A standard strain of *E. coli* is likely to contain several (<10) copies of any one of the more common IS elements. To describe an insertion into a particular site, a double colon is used; so λ::IS1 describes an IS1 element inserted into phage lambda. Most IS elements insert at a variety of sites within host DNA. Some, though, show varying degrees of preference for particular insertion hotspots.

The IS elements are autonomous units, each of which codes only for the proteins needed to sponsor its own transposition. Each IS element is different in sequence, but there are some common features in organization. The structure of a generic transposon before and after insertion at a target site is illustrated in **FIGURE 17.3**, which also summarizes the details of some common IS elements.

An IS element ends in short **inverted-terminal repeats**; usually the two copies of the repeat are closely related rather than identical. As illustrated in the figure, the presence of the inverted terminal repeats means that the same sequence is encountered proceeding toward the element from the flanking DNA on either side of it.

▸ **inverted-terminal repeats** The short related or identical sequences present in reverse orientation at the ends of some transposons.

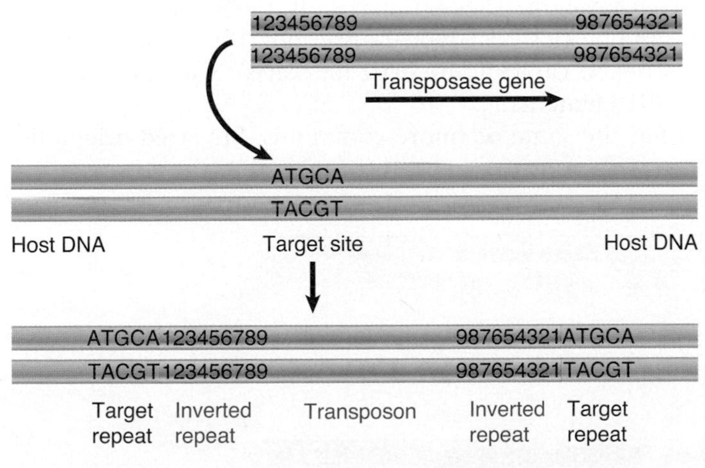

Transposon	Target repeat (bp)	Inverted repeat (bp)	Overall length (bp)	Target selection
IS1	9	23	768	random
IS2	5	41	1327	hotspots
IS4	11–13	18	1428	$AAAN_{20}TTT$
IS5	4	16	1195	hotspots
IS10R	9	22	1329	NGCTNAGCN
IS50R	9	9	1531	hotspots
IS903	9	18	1057	random

FIGURE 17.3 Transposons have inverted terminal repeats and generate direct repeats of flanking DNA at the target site. In this example, the target is a 5 bp sequence. The ends of the transposon consist of inverted repeats of 9 bp, where the numbers 1 through 9 indicate a sequence of base pairs.

When an IS element transposes, a sequence of host DNA at the site of insertion is duplicated. The nature of the duplication is revealed by comparing the sequence of the target site before and after an insertion has occurred. Figure 17.3 shows that at the site of insertion, the IS DNA is always flanked by very short **direct repeats**. (In this context, "direct" indicates that two copies of a sequence are repeated in the same orientation, not that the repeats are adjacent.) In the original gene (prior to insertion), however, the target site has the sequence of only one of these repeats. In the figure, the target site consists of the sequence ATGCA/TACGT. After transposition, one copy of this sequence is present on either side of the transposon. The sequence of the direct repeat varies among individual transposition events undertaken by a transposon, but the length is constant for any particular IS element (a reflection of the mechanism of transposition).

▸ **direct repeats** Identical (or closely related) sequences present in two or more copies in the same orientation in the same molecule of DNA.

An IS element therefore displays a characteristic structure in which its ends are identified by the inverted terminal repeats, whereas the adjacent ends of the flanking host DNA are identified by the short direct repeats. The inverted repeats define the ends of the transposon. Recognition of the ends is common to transposition events sponsored by all DNA-type transposons. The protein(s) that recognize the ends and are responsible for transposition are called **transposases**.

▸ **transposase** The enzyme activity involved in insertion of transposon at a new site.

Many of the IS elements contain a single long coding region, which starts just inside the inverted repeat at one end and terminates just before or within the inverted repeat at the other end. This codes for the transposase. Some elements have a more complex organization. IS1, for instance, has two separate reading frames; the transposase is produced by making a frameshift during translation to allow both reading frames to be used.

The frequency of transposition varies among different elements. The overall rate of transposition is ~10^{-3} to 10^{-4} per element per generation. Insertions in individual targets occur at a level comparable with the spontaneous mutation rate, usually ~10^{-5} to 10^{-7} per generation. Reversion (by precise excision of the IS element) is usually infrequent, with a range of rates of 10^{-6} to 10^{-10} per generation, which is ~10^{3} times less frequent than insertion.

Some transposons carry drug resistance (or other) markers in addition to the functions concerned with transposition. These transposons are named **Tn** followed by a number. The members of one class of larger transposons are called **composite elements** because a central region carrying the drug marker(s) is flanked on either side by "arms" that consist of IS elements.

▸ **Tn** Followed by a number, a means of denoting bacterial transposons carrying markers that are not related to their function, e.g., drug resistance.

▸ **composite elements** Transposable elements consisting of two IS elements (can be the same or different) and the DNA sequences between the IS elements; the non-IS sequences often include gene(s) conferring antibiotic resistance.

The arms may be in either the same or (more commonly) inverted orientation. Thus a composite transposon with arms that are direct repeats has the structure

If the arms are inverted repeats, the structure is

The arrows indicate the orientation of the arms, which are identified as L and R according to an (arbitrary) orientation of the genetic map of the transposon from left to right. The structure of a composite transposon is illustrated in more detail in **FIGURE 17.4**, which also summarizes the properties of some common composite transposons.

Arms consist of IS modules, and each module has the usual structure ending in inverted repeats; as a result the composite transposon also ends in the same short, inverted repeats. A composite transposon can have two identical IS modules, such as

Tn9 (direct repeats of IS1), or two closely related but not identical modules. Thus we can distinguish the L and R modules in Tn10 or in Tn5. Functional IS elements code for transposase activities that are responsible both for creating a target site and for recognizing the ends of the transposon, so only the ends are needed for a transposon to serve as a substrate for transposition.

A functional IS module can transpose either itself or the entire transposon. When the modules of a composite transposon are identical, presumably either module can sponsor movement of the transposon. When the modules are different, they may differ in functional ability, so transposition can depend entirely or principally on one of the modules.

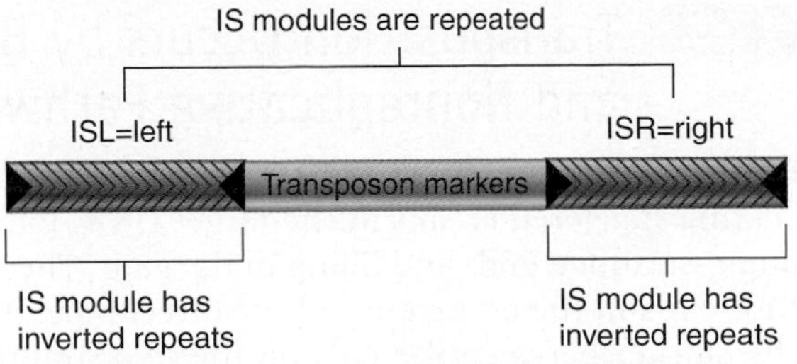

What is responsible for transposing a composite transposon instead of just the individual module? This question is especially pressing in cases where both modules are functional. In the example of Tn9, where the modules are IS1 elements, presumably each is active in its own right as well as on behalf of the composite transposon. Why is the transposon preserved as a whole, instead of each insertion sequence looking out for itself?

IS modules are repeated

ISL=left — Transposon markers — ISR=right

IS module has inverted repeats | IS module has inverted repeats

Example	Left end	Markers	Right end
Tn9	IS1	*cam*R	IS modules identical both functional

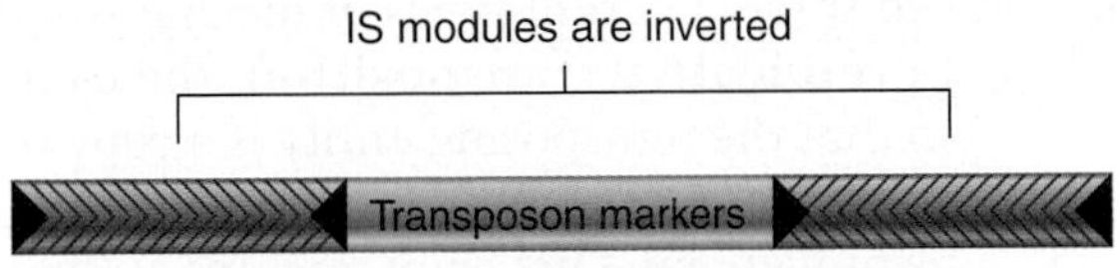

IS modules are inverted

Transposon markers

Example	Left end	Markers	Right end
Tn903	IS903	*kan*R	Both IS ends functional
Tn10	IS10L nonfunctional	*tet*R	IS10R functional
Tn5	IS50L nonfunctional	*kan*R	IS50R functional

FIGURE 17.4 A composite transposon has a central region carrying markers (such as drug resistance) flanked by IS modules. The modules have short inverted terminal repeats. If the modules themselves are in inverted orientation, the short inverted terminal repeats at the ends of the transposon are identical.

A major force supporting the transposition of composite transposons is selection for the marker(s) carried in the central region. For example, Tn10 consists of two IS10 modules flanking a *tet*R gene conferring tetracycline resistance. An IS10 module is free to move around on its own and mobilizes an order of magnitude more frequently than Tn10. Tn10 is held together by selection for *tet*R, though, so that under selective conditions, the relative frequency of intact Tn10 transposition is much increased.

KEY CONCEPTS

- An insertion sequence is a transposon that codes for the enzyme(s) needed for transposition flanked by short inverted terminal repeats.
- The target site at which a transposon is inserted is duplicated during the insertion process to form two repeats in direct orientation at the ends of the transposon.
- The length of the direct repeat is 5 to 9 bp and is characteristic for any particular transposon.
- Transposons can carry other genes in addition to those coding for transposition.
- Composite transposons have a central region flanked by an IS element at each end.
- Either one or both of the IS elements of a composite transposon may be able to undertake transposition.
- A composite transposon may transpose as a unit, but an active IS element at either end may also transpose independently.

CONCEPT AND REASONING CHECK

Can a transposon with intact terminal repeats but a mutant transposase gene ever transpose? Why or why not?

17.3 Transposition Occurs by Both Replicative and Nonreplicative Pathways

The insertion of a transposon into a new site is illustrated in **FIGURE 17.5**. It consists of making staggered breaks in the target DNA, joining the transposon to the protruding single-stranded ends and filling in the gaps. The generation and filling of the staggered ends explain the occurrence of the direct repeats of target DNA at the site of insertion. The stagger between the cuts on the two strands determines the length of the direct repeats; thus the target repeat characteristic of each transposon reflects the geometry of the enzyme involved in cutting target DNA.

The use of staggered ends is common to all means of transposition, but we can distinguish three different types of mechanism by which a transposon moves:

- In **replicative transposition**, the element is duplicated during the reaction, so that the transposing entity is a copy of the original element. **FIGURE 17.6** summarizes the results of such a transposition. The transposon is copied as part of its movement. One copy remains at the original site, whereas the other inserts at the new site. Thus transposition is accompanied by an increase in the number of copies of the transposon. Replicative transposition involves two types of enzymatic activity: a transposase that acts on the ends of the original transposon and a **resolvase** that acts on the duplicated copies.
- In **nonreplicative transposition**, the transposing element moves as a physical entity directly from one site to another and is conserved. The insertion sequences and composite transposons Tn10 and Tn5 use the mechanism shown in **FIGURE 17.7**, which involves the release of the transposon from the flanking donor DNA during transfer. This type of mechanism, often referred to as "cut and paste," requires only a transposase. Another mechanism utilizes the connection of donor and target DNA sequences and shares some steps with replicative transposition. Both mechanisms of nonreplicative transposition cause the element to be inserted at the target site and lost from the donor site. What happens to the donor molecule after a nonreplicative transposition? Its survival requires that host repair systems recognize the double-strand break and repair it.

- **replicative transposition** The movement of a transposon by a mechanism in which first it is replicated, and then one copy is transferred to a new site.
- **resolvase** The enzyme activity involved in site-specific recombination between two copies of a transposon that has been duplicated.
- **nonreplicative transposition** The movement of a transposon that leaves a donor site (usually generating a double-strand break) and moves to a new site.

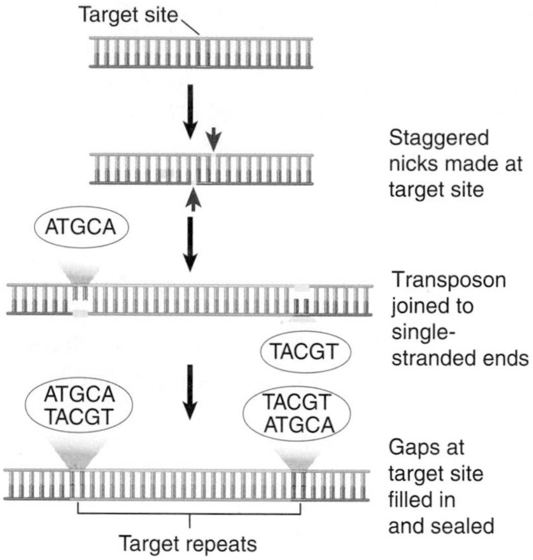

Target site
Staggered nicks made at target site
ATGCA
Transposon joined to single-stranded ends
TACGT
ATGCA TACGT
TACGT ATGCA
Gaps at target site filled in and sealed
Target repeats

FIGURE 17.5 The direct repeats of target DNA flanking a transposon are generated by the introduction of staggered cuts whose protruding ends are linked to the transposon.

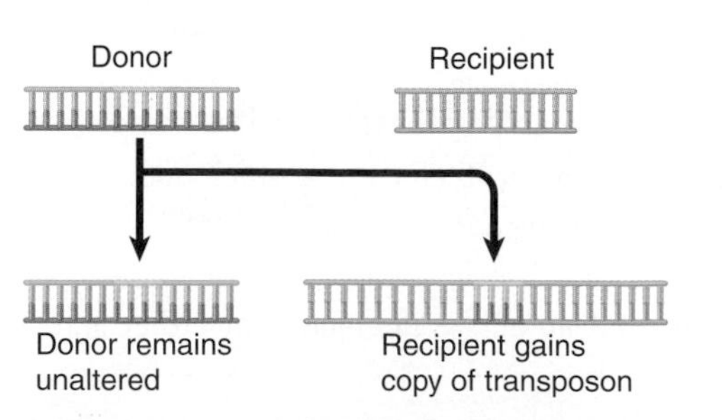

FIGURE 17.6 Replicative transposition creates a copy of the transposon, which inserts at a recipient site. The donor site remains unchanged, so both donor and recipient have a copy of the transposon.

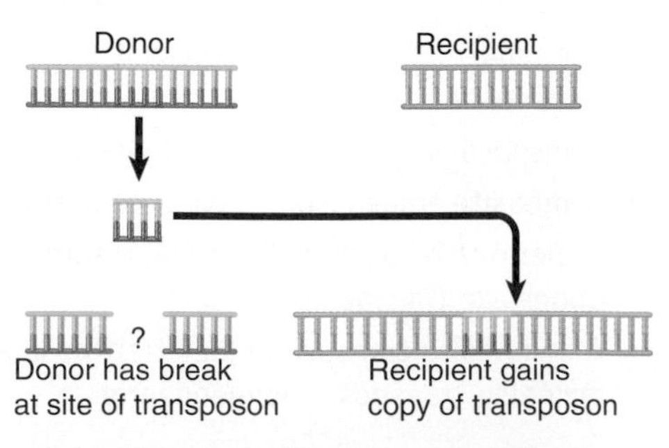

FIGURE 17.7 Nonreplicative transposition allows a transposon to move as a physical entity from a donor to a recipient site. This leaves a break at the donor site, which is lethal unless it can be repaired.

Some transposons use only one type of pathway for transposition, whereas others may be able to use multiple pathways. The elements IS1 and IS903 use both nonreplicative and replicative pathways, and the phage Mu can enter either type of pathway from a common intermediate.

The same basic types of reaction are involved in all classes of transposition event. The ends of the transposon are disconnected from the donor DNA by cleavage reactions that generate 3′–OH ends. The exposed ends are then joined to the target DNA by transfer reactions, involving transesterification in which the 3′–OH end directly attacks the target DNA. These reactions take place within a nucleoprotein complex that contains the necessary enzymes and both ends of the transposon. Transposons differ as to whether the target DNA is recognized before or after the cleavage of the transposon itself, and whether one or both strands at the end of the transposon are cleaved prior to integration.

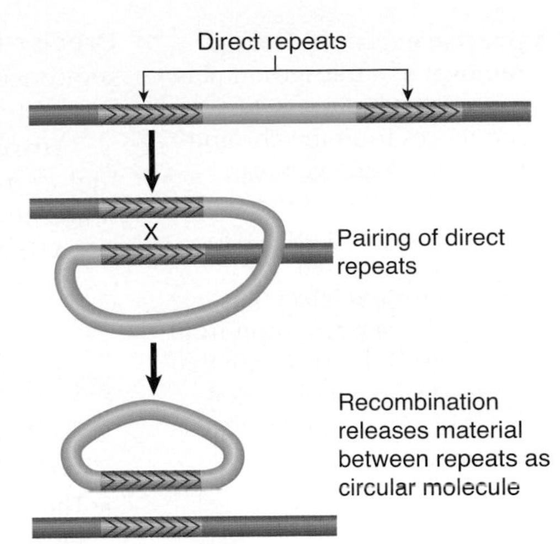

FIGURE 17.8 Reciprocal recombination between direct repeats excises the material between them; each product of recombination has one copy of the direct repeat.

The choice of target site is in effect made by the transposase. In some cases, the target is chosen virtually at random. In others, there is specificity for a consensus sequence or for some other feature in DNA. The feature can take the form of a structure in DNA, such as DNA containing an intrinsic bend or a protein-DNA complex. In the latter case, the nature of the target complex can cause the transposon to insert at specific promoters (such as Ty1 or Ty3, which select RNA polymerase III promoters in yeast), inactive regions of the chromosome, or replicating DNA.

In addition to the "simple" intermolecular transposition that results in insertion at a new site, transposons promote other types of DNA rearrangements. Some of these events are consequences of the relationship between the multiple copies of the transposon. Others represent alternative outcomes of the transposition mechanism, and they leave clues about the nature of the underlying events.

Rearrangements of host DNA may result when a transposon inserts a copy at a second site near its original location. Host systems may undertake reciprocal recombination between the two copies of the transposon; the consequences are determined by whether the repeats are oriented in the same or inverted direction.

FIGURE 17.8 illustrates the general rule that recombination between any pair of direct repeats in *cis* will delete the material between them. The intervening region is excised as a circle of DNA (which is lost from the cell); the chromosome retains a single copy of the direct repeat. A recombination between the directly repeated IS1 modules of the composite transposon Tn9 would replace the transposon with a single IS1 module.

Deletion of sequences adjacent to a transposon could, therefore, result from a two-stage process; transposition generates a direct repeat of a transposon, and recombination occurs between the repeats. The majority of deletions that arise in the vicinity of transposons, however, probably result from a variation in the pathway followed in the transposition event itself.

FIGURE 17.9 depicts the consequences of a reciprocal recombination between a pair of inverted repeats. The region between the repeats becomes inverted; the repeats themselves remain available to sponsor further inversions. A composite transposon whose modules are inverted is a stable component of the genome, although the direction of the central region with regard to the modules could be inverted by recombination.

Excision as in the case of Figure 17.8 is not supported by transposons themselves but occurs when bacterial enzymes recognize homologous regions in the transposons. This is important because the loss of a transposon may restore function at the site of insertion.

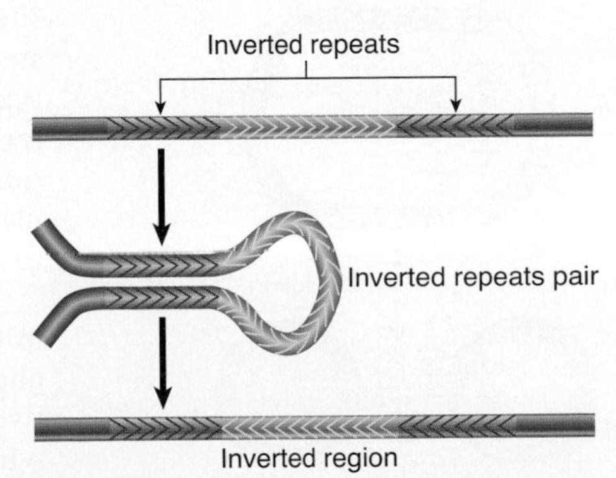

FIGURE 17.9 Reciprocal recombination between inverted repeats inverts the region between them.

- **precise excision** The removal of a transposon plus one of the duplicated target sequences from the chromosome. Such an event can restore function at the site where the transposon was previously inserted.
- **imprecise excision** The removal of a transposon from the original insertion site that results in some transposon sequence remaining at the insertion site.

Precise excision requires removal of the transposon plus one copy of the duplicated sequence. This is rare; it occurs at a frequency of $\sim 10^{-6}$ for Tn5 and $\sim 10^{-9}$ for Tn10. It probably involves a recombination between the duplicated target sites.

Imprecise excision leaves a remnant of the transposon and occurs at a much higher frequency than precise excision (e.g., $\sim 10^{-6}$ for Tn10). Neither type of excision relies on transposon-encoded functions, but the mechanism is not known. Excision is RecA-independent and could occur by some cellular mechanism that generates spontaneous deletions between closely spaced repeated sequences.

KEY CONCEPTS

- All transposons use a common mechanism in which staggered nicks are made in target DNA, the transposon is joined to the protruding ends, and the gaps are filled.
- The order of events and exact nature of the connections between transposon and target DNA determine whether transposition is replicative or nonreplicative.
- Homologous recombination between multiple copies of a transposon causes rearrangement of host DNA.
- Homologous recombination between the repeats of a transposon may lead to precise or imprecise excision.

CONCEPT AND REASONING CHECK

Which is more likely to cause detrimental mutations in a host, a transposon that uses random insertion or one that shows site selectivity? Why?

17.4 Mechanisms of Transposition

Many mobile DNA elements transpose from one chromosomal location to another by a fundamentally similar mechanism. They include IS elements, prokaryotic and eukaryotic transposons, and certain bacteriophages. Insertion of the DNA copy of retroviral RNA uses a similar mechanism (see *Section 17.7, The Retrovirus Life Cycle Involves Transposition-like Events*). The first stages of immunoglobulin recombination also are similar (see *Section 18.6, RAG1/RAG2 Catalyze Breakage and Religation of V(D)J Gene Segments*).

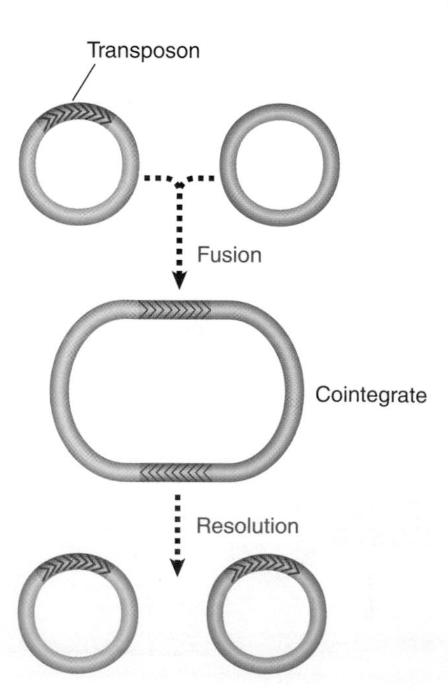

FIGURE 17.10 Transposition may fuse a donor and recipient replicon into a cointegrate. Resolution releases two replicons, each containing a copy of the transposon.

Transposition starts with a common mechanism for joining the transposon to its target. In the course of the reaction, the transposon is nicked at both ends and the target site is nicked on both strands. The nicked ends are joined between transposon and target to generate a covalent connection between the two DNAs. The product of these reactions is a strand transfer complex in which the transposon is connected to the target site through one strand at each end. The next step of the reaction differs and determines the type of transposition. The common strand-transfer structure can be a substrate for replication (leading to replicative transposition) or it can be used directly for breakage and reunion (leading to nonreplicative transposition).

The basic structures involved in replicative transposition are illustrated in **FIGURE 17.10**. For simplicity, this example shows transposition between two circular replicons, resulting in a **cointegrate**, which represents a fusion of the two original

- **cointegrate** A structure that is produced by fusion of two replicons, one originally possessing a transposon and the other lacking it; the cointegrate has copies of the transposon present at both junctions of the replicons, oriented as direct repeats.

molecules. Transposition by the same mechanism *within* a replicon can result in inversions or deletions. The crossover is formed by the transposase, whereas its conversion to a cointegrate requires host replication function. Next, homologous recombination between the two copies of the transposon releases two individual replicons, each of which has a copy of the transposon. One of the replicons is the original donor replicon. The other is a target replicon that has gained a transposon flanked by short direct repeats of the host target sequence. The recombination reaction is called **resolution**.

The principle of replicative transposition is that replication through the transposon duplicates it, which creates copies at both the target and donor sites, illustrated in **FIGURE 17.11**. The strand-transfer complex (sometimes also called a *crossover complex*) contains *pseudoreplication forks* that provide a template for DNA synthesis. Replication from both the pseudoreplication forks will proceed through the transposon, separating its strands and terminating at its ends. Replication is accomplished by host-encoded functions. At this juncture, the structure has become a cointegrate, possessing direct repeats of the transposon at the junctions between the replicons (as can be seen by tracing the path around the cointegrate).

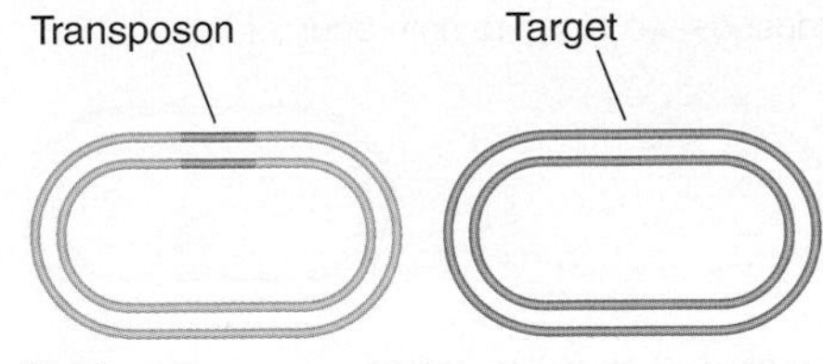

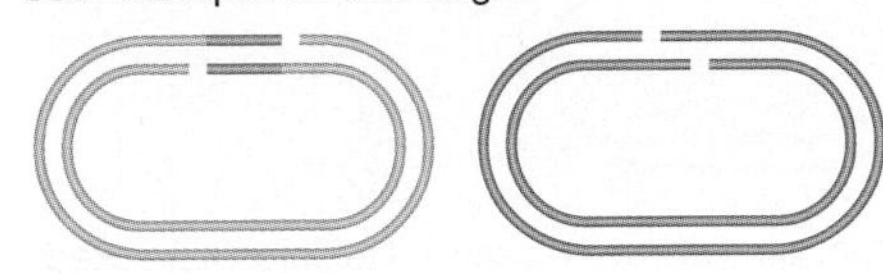

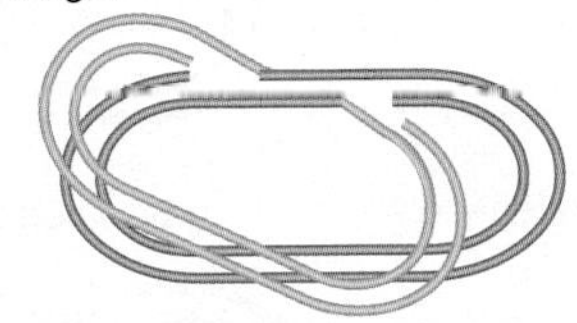

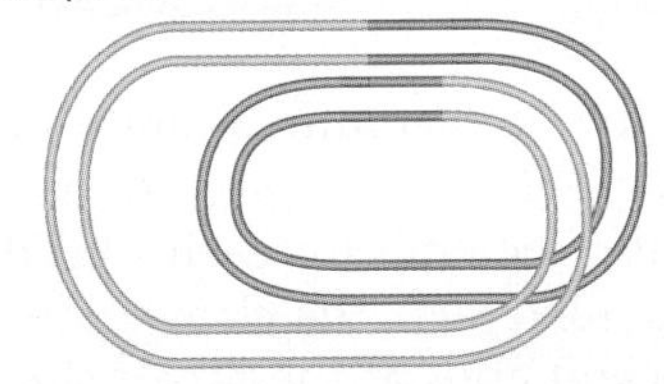

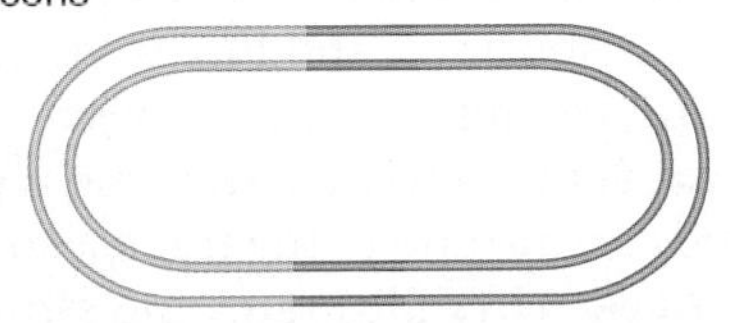

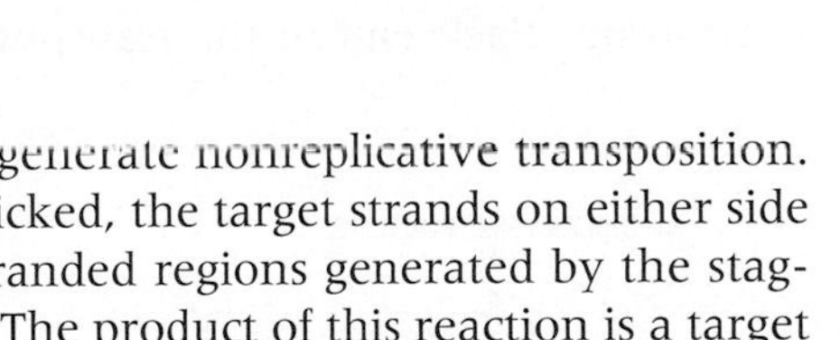

FIGURE 17.11 Transposition generates a crossover structure, which is converted by replication into a cointegrate.

▸ **resolution** A process occurring by a homologous recombination reaction between the two copies of the transposon in a cointegrate. The reaction generates the donor and target replicons, each with a copy of the transposon.

The crossover structure can also be used in nonreplicative transposition. The principle of nonreplicative transposition by this mechanism is that a breakage and reunion reaction allows the target to be reconstructed with the insertion of the transposon; the donor remains broken. No cointegrate is formed.

FIGURE 17.12 shows the cleavage events that generate nonreplicative transposition. Once the unbroken donor strands have been nicked, the target strands on either side of the transposon can be ligated. The single-stranded regions generated by the staggered cuts must be filled in by repair synthesis. The product of this reaction is a target replicon in which the transposon has been inserted between repeats of the sequence created by the original single-strand nicks. The donor replicon has a double-strand break across the site where the transposon was originally located.

Nonreplicative transposition can also occur by an alternative pathway in which nicks are made in target DNA, but a double-strand break is made on either side of the transposon, releasing it entirely from flanking donor sequences (as depicted in Figure 17.7). This cut-and-paste pathway is used by Tn10, as illustrated in **FIGURE 17.13**.

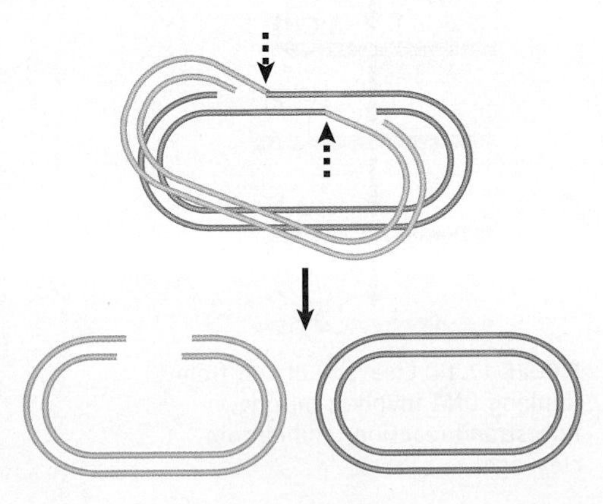

FIGURE 17.12 Nonreplicative transposition results when a crossover structure is released by nicking. This inserts the transposon into the target DNA, flanked by the direct repeats of the target, and the donor is left with a double-strand break.

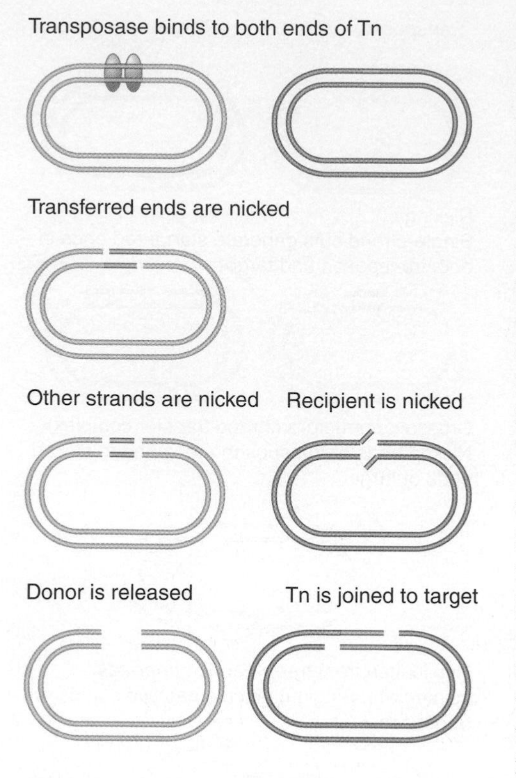

FIGURE 17.13 Both strands of Tn10 are cleaved sequentially, and then the transposon is joined to the nicked target site.

The basic difference between the mechanism shown for Tn10 in Figure 17.13 and the crossover nicking mechanism of Figure 17.12 is that both strands of Tn10 are cleaved before any connection is made to the target site. The first step in the reaction is recognition of the transposon ends by the transposase, forming a proteinaceous structure within which the reaction occurs. At each end of the transposon, the strands are cleaved in a specific order: the transferred strand (the one to be connected to the target site) is cleaved first, followed by the other strand.

Tn5 also transposes by nonreplicative transposition, and **FIGURE 17.14** shows the interesting cleavage reaction that separates the transposon from the flanking sequences. First one DNA strand is nicked. The 3′–OH end that is released then attacks the other strand of DNA. This releases the flanking sequence and joins the two strands of the transposon in a *hairpin*. A similar hairpin-forming step occurs during somatic recombination in the immune system (see *Section 18.6, RAG1/RAG2 Catalyze Breakage and Religation of V(D)J Gene Segments*). An activated water molecule then attacks the hairpin to generate free ends for each strand of the transposon.

In the next step, the cleaved donor DNA is released, and the transposon is joined to the nicked ends at the target site. The transposon and the target site remain constrained in the proteinaceous structure created by the transposase (and other proteins). The double-strand cleavage at each end of the transposon precludes any replicative-type transposition and forces the reaction to proceed by nonreplicative transposition.

The Tn5 and Tn10 transposases both function as dimers. Each subunit in the dimer has an active site that successively catalyzes the double-strand breakage of the two strands at one end of the transposon and then catalyzes staggered cleavage of the target site. **FIGURE 17.15** illustrates the structure of the Tn5 transposase bound to the cleaved transposon. Each end of the transposon is located in the active site of one subunit.

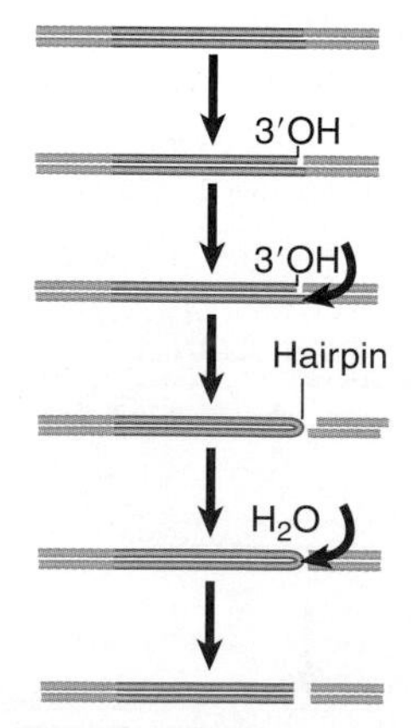

FIGURE 17.14 Cleavage of Tn5 from flanking DNA involves nicking, interstrand reaction, and hairpin cleavage.

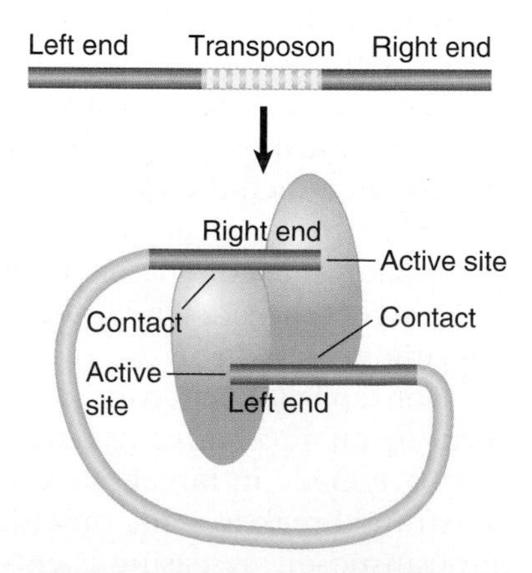

FIGURE 17.15 Each subunit of the Tn5 transposase has one end of the transposon located in its active site and also makes contact at a different site with the other end of the transposon.

One end of the subunit also contacts the other end of the transposon. This controls the geometry of the transposition reaction. Each of the active sites will cleave one strand of the target DNA. It is the geometry of the complex that determines the distance between these sites on the two target strands (nine base pairs in the case of Tn5).

KEY CONCEPTS

- Transposition starts by forming a strand-transfer complex in which the transposon is connected to the target site through one strand at each end.
- Replicative transposition follows if the complex is replicated and nonreplicative transposition follows if it is repaired.
- Replication of a strand transfer complex generates a cointegrate, which is a fusion of the donor and target replicons.
- The cointegrate has two copies of the transposon, which lie between the original replicons.
- Recombination between the transposon copies regenerates the original replicons, but the recipient has gained a copy of the transposon.
- The recombination reaction is catalyzed by a resolvase encoded by the transposon.
- Nonreplicative transposition results if a crossover structure is nicked on the unbroken pair of donor strands and the target strands on either side of the transposon are ligated.
- Two pathways for nonreplicative transposition differ according to whether the first pair of transposon strands are joined to the target before the second pair are cut (Tn5) or whether all four strands are cut before joining to the target (Tn10).

CONCEPT AND REASONING CHECK

Why do transposases typically function as dimers or tetramers? (Why might a monomeric transposase cause problems?)

17.5 Families of Transposons in Maize Can Cause Breakage and Rearrangement

One of the most visible consequences of the existence and mobility of transposons occurs during plant development, when somatic variation occurs. This is due to changes in the location or behavior of **controlling elements** (the name that transposons were given in maize before their molecular nature was discovered).

▸ **controlling elements** Transposable units in maize originally identified solely by their genetic properties. They may be autonomous (able to transpose independently) or nonautonomous (able to transpose only in the presence of an autonomous element).

Two features of maize have helped to follow transposition events. Transposons in eukaryotes often insert near genes that have visible but nonlethal effects on the phenotype. Maize displays clonal development, which means that the occurrence and timing of a transposition event can be visualized in individual kernels. This is seen most vividly in the variation in kernel color, when patches of one color appear within another color. (This is described in more detail in the accompanying box, *Historical Perspectives*.) The nature of the event does not matter: It may be a point mutation, insertion, excision, or chromosome break. What is important is that it occurs in a heterozygote to alter the expression of one allele. The descendants of a cell that has suffered the event then display a new phenotype, whereas the descendants of cells not affected by the event continue to display the original phenotype.

Insertion of a transposon may affect the activity of adjacent genes. Deletions, duplications, inversions and translocations all occur at the sites where transposons are present. Chromosome breakage is a common consequence of the presence of some elements. In maize and other plants, the activities of the controlling elements are often regulated during development. The elements transpose and promote genetic rearrangements at characteristic times and frequencies during plant development.

transposons are known to be active in maize. The members of each family are divided into two classes:

- **autonomous controlling element** An active transposon in maize with the ability to transpose.
- ***Ac* element** Activator element; an autonomous transposable element in maize.
- **nonautonomous controlling element** A transposon in maize that encodes a nonfunctional transposase; it can transpose only in the presence of a *trans*-acting autonomous member of the same family.

- **Autonomous controlling elements** have the intrinsic ability to excise and transpose. As a result of the continuing activity of an autonomous element, its insertion at any locus creates an unstable, or "mutable," allele. Loss of the autonomous element itself or of its ability to transpose, converts a mutable allele to a stable allele. The ***Ac* element** is a major autonomous element in maize.
- **Nonautonomous controlling elements** are stable; they do not spontaneously transpose. They become unstable *only* when an autonomous member of the same family is present elsewhere in the genome. When complemented in *trans* by an autonomous element, a nonautonomous element displays the usual range of activities associated with autonomous elements, including the ability to transpose to new sites. Nonautonomous elements are derived from autonomous elements by loss of *trans*-acting functions needed for transposition. *Ds* elements are nonautonomous elements.

Families of controlling elements are defined by the interactions between autonomous and nonautonomous elements. A family consists of a single type of autonomous element accompanied by many varieties of nonautonomous elements. A nonautonomous element is placed in a family by its ability to be activated in *trans* by the autonomous elements. The major families of controlling elements in maize are summarized in **FIGURE 17.17**.

Characterized at the molecular level, the maize transposons exhibit the usual form of organization—inverted repeats at the ends and short direct repeats in the adjacent target DNA—but otherwise they vary in size and coding capacity. All families of transposons share the same type of relationship between the autonomous and nonautonomous elements. The autonomous elements have open reading frames between the terminal repeats, whereas the nonautonomous elements do not code for functional proteins. Nonautonomous elements are derived from autonomous elements by deletions (or other changes) that inactivate the *trans*-acting transposase, but leave intact the sites on which the transposase acts. Sometimes the internal sequences are related to those of autonomous elements; at other times they have diverged completely.

There are typically several members of each transposon family in a plant genome. The Mutator transposon is the most active and mutagenic of all maize transposons. The autonomous element *MuDR* encodes the genes *mudrA* (which codes for the transposase) and *mudrB* (which codes for an accessory protein required for integration). The ends of the elements are marked by 200 bp inverted repeats. Nonautonomous Mutator elements—basically any element that has the inverted repeats—are also mobilized by the products of the autonomous element. Mutator elements are the founding members of

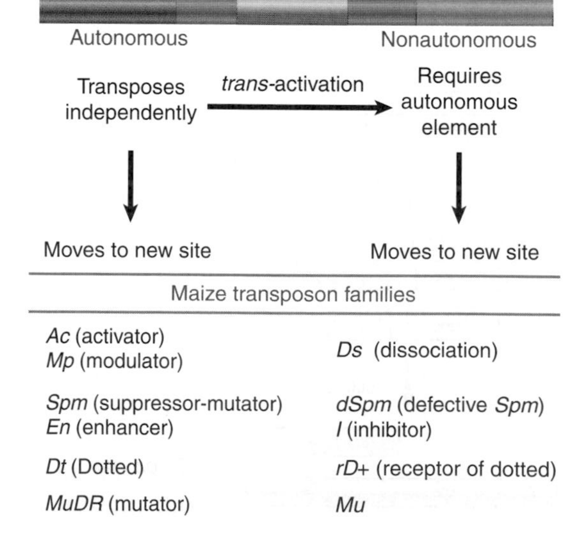

FIGURE 17.17 Each controlling element family has both autonomous and nonautonomous members. Autonomous elements are capable of transposition. Nonautonomous elements are deficient in transposition. Pairs of autonomous and nonautonomous elements can be classified in >4 families.

the MULE (*mu-like element*) superfamily of transposons, which are present in bacteria, fungi, plants and animals.

The autonomous and nonautonomous elements of the *Ac/Ds* family are among the best-characterized transposons. **FIGURE 17.18** summarizes the structures of *Ac* and a subset of its *Ds* derivatives.

Most of the length of the autonomous *Ac* element is occupied by a single gene consisting of five exons. The product is the transposase. The element itself ends in inverted repeats of 11 bp, and a target sequence of 8 bp is duplicated at the site of insertion.

Ds elements vary in both length and sequence but are related to *Ac*. They end in the same 11 bp inverted repeats. They are shorter than *Ac*, and the length of deletion varies. At one extreme, the element *Ds9* has a deletion of only 194 bp. In a more extensive deletion, the *Ds6* element retains only 2 kb of the original *Ac* sequence, representing 1 kb from each end of *Ac*.

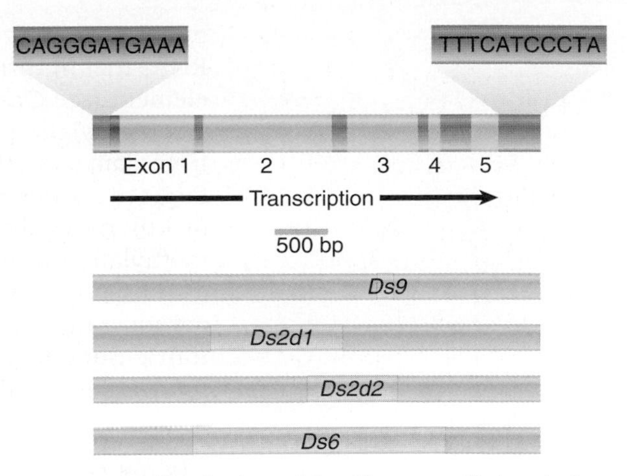

FIGURE 17.18 The *Ac* element has five exons that code for a transposase; *Ds* elements have internal deletions.

At another extreme, the *Ds1* family members comprise short sequences whose only relationship to *Ac* lies in the possession of terminal inverted repeats. Elements of this class need not be directly derived from *Ac* but could be derived by any event that generates the inverted repeats. Their existence suggests that the transposase recognizes only the terminal inverted repeats, or possibly the terminal repeats in conjunction with some short internal sequence.

Ds1 elements are just one example of a widespread form of DNA-type elements called MITEs (*miniature inverted repeat transposable elements*). These are very short derivatives of autonomous elements found in many eukaryotes that can be present in tens or hundreds of thousands of copies in a given genome. They range from 300 to 500 bp, and generate 2 to 3 bp target-site duplications. Unlike many other classes of transposons in plants, MITEs are often found in or near genes.

Transposition of *Ac/Ds* occurs by a nonreplicative cut-and-paste mechanism that involves double-strand breaks followed by integration of the released element, and is accompanied by its disappearance from the donor location. Clonal analysis suggests that transposition of*Ac/Ds* almost always occurs soon after the donor element has been replicated. This is because transposition does not occur when the DNA of the transposon in methylated on both strands, and is activated when the DNA is hemimethylated (the typical state immediately following replication). The recipient site is frequently on the same chromosome as the donor site, and often is quite close to it. Note that if transposition is from a replicated region of a chromosome into an unreplicated region, the transposition event will result in a net increase in the copy number of the element: one chromatid will carry a single copy of the transposon, and the second chromatid will carry two copies. This ensures that elements such as *Ac* can increase their copy number even though transposition is not duplicative.

Autonomous and nonautonomous elements are subject to a variety of changes in their condition. Some of these changes are genetic; others are epigenetic. The major change is (of course) the conversion of an autonomous element into a nonautonomous element, but further changes may occur in the nonautonomous element. *cis*-acting defects may render a nonautonomous element impervious to autonomous elements. Thus a nonautonomous element may become permanently stable because it can no longer be activated to transpose.

Autonomous elements are subject to "changes of phase," which are heritable but relatively unstable alterations in their properties. These can take the form of a reversible inactivation in which the element cycles between an active and inactive condition during plant development. Such phase changes result from changes in the methylation of DNA, in which the inactive form of the element is methylated in the target sequence $\substack{\text{CAG}\\\text{GTC}}$. In most cases, it is not known what triggers this inactivation, but in the case of *MuDR*, epigenetic silencing can be triggered by a derivative of *MuDR* that is dupicated and inverted relative to itself. This rearrangement results in the production of a hairpin RNA, in which two parts of the transcript are perfect complements to each other. The resulting double-stranded RNA is processed by cellular factors into small RNAs that in turn trigger methylation and transcriptional gene silencing of the *MuDR* element (see *Chapter 30, Regulatory RNA*).

Methylation is probably the major mechanism that is used to prevent transposons from damaging the genome by transposing too frequently. Transposons appear to be targeted for methylation because they are far more likely to produce double-stranded or otherwise aberrant transcripts that can be used to guide sequence-specific DNA methylation using small RNA produced from those transcripts. Once methylation of a transposon has been established, it can be heritably maintained over many generations. In both plants and animals that methylate their DNA, the vast majority of transposons are epigenetically silenced in this way (see *Chapter 29, Epigenetic Effects Are Inherited*, for further discussion of this topic).

KEY CONCEPTS

- Transposition in maize was discovered because of the effects of the chromosome breaks and other genetic changes generated by transposition of "controlling elements."
- Chromosome breaks generate one chromosome that has a centromere and a broken end and one acentric fragment.
- The acentric fragment is lost during mitosis; this can be detected by the disappearance of dominant alleles in a heterozygote.
- Each family of transposons in maize has both autonomous and nonautonomous controlling elements.
- Autonomous controlling elements code for proteins that enable them to transpose.
- Nonautonomous controlling elements have mutations that eliminate their capacity to catalyze transposition, but they can transpose when an autonomous element provides the necessary proteins.
- Autonomous controlling elements have changes of phase, when their properties alter as a result of changes in the state of methylation.
- Transposons can be epigenetically silenced by DNA methylation triggered by the production of double-stranded RNAs.

CONCEPT AND REASONING CHECK

Is an *Ac* element that has lost one of its terminal repeats a *Ds* element? Why or why not?

17.6 Transposition of P Elements Causes Hybrid Dysgenesis

Certain strains of *D. melanogaster* encounter difficulties in interbreeding. When flies from two of these strains are crossed, the progeny display "dysgenic traits"—a series of defects including mutations, chromosomal aberrations, distorted segregation at meiosis, and sterility. The appearance of these correlated defects is called **hybrid dysgenesis**.

▸ **hybrid dysgenesis** The inability of certain strains of *D. melanogaster* to interbreed, because the hybrids are sterile (although otherwise they may be phenotypically normal).

In one of the systems responsible for hybrid dysgenesis, flies are divided into the two types P (paternal contributing) and M (maternal contributing). **FIGURE 17.19** illustrates the asymmetry of the system: a cross between a P male and an M female causes dysgenesis, but the reverse cross does not.

Dysgenesis is principally a phenomenon of the germ cells. In crosses involving the P–M system, the F_1 hybrid flies have normal somatic tissues. Their gonads, however, do not develop. The morphological defect in gamete development dates from the stage at which rapid cell divisions commence in the germline.

Any one of the chromosomes of a P male can induce dysgenesis in a cross with an M female. The construction of recombinant chromosomes shows that several regions

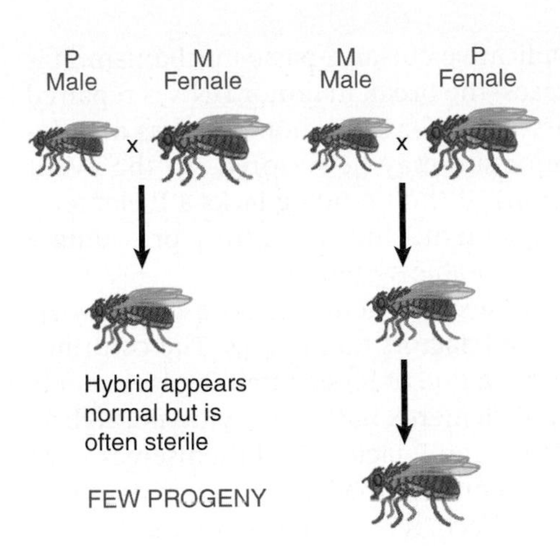

FIGURE 17.19 Hybrid dysgenesis is asymmetrical; it is induced by P male × M female crosses but not by M male × P female crosses.

within each P chromosome are able to cause dysgenesis. This suggests that a P male has sequences at many different chromosomal locations that can induce dysgenesis. The locations differ between individual P strains. The P-specific sequences are absent from chromosomes of M flies. The P-specific sequence is called the **P element**.

▸ **P element** A type of transposon in *D. melanogaster*.

The P element insertions form a classic transposable system. Individual elements vary in length but are homologous in sequence. All P elements possess inverted terminal repeats of 31 bp and generate direct repeats of target DNA of 8 bp upon transposition. The longest P elements are ~2.9 kb long and have four open reading frames. The shorter elements arise, apparently rather frequently, by internal deletions of a full-length P factor. At least some of the shorter P elements have lost the capacity to produce the transposase, resulting in nonautonomous elements, but they may be activated in *trans* by the enzyme coded by a complete P element.

A P strain carries 30 to 50 copies of the P element, about a third of them full length. The elements are absent from M strains. In a P strain the elements are carried as inert components of the genome, but they become activated to transpose when a P male is crossed with an M female.

Chromosomes from P–M hybrid dysgenic flies have P elements inserted at many new sites. The insertions inactivate the genes in which they are located and often cause chromosomal breaks. The result of the transpositions is therefore to inactivate the genome.

Activation of P elements is tissue specific: it occurs only in the germline. P elements are transcribed, though, in both germline and somatic tissues. Tissue-specificity is conferred by a change in the splicing pattern of the P element transcript.

FIGURE 17.20 depicts the organization of the element and its transcripts. The primary transcript extends for 2.5 kb or 3.0 kb, the difference probably reflecting merely the leakiness of the termination site. Two protein products can be produced:

- In somatic tissues, only the first two introns are excised, leaving intron 3, which contains a stop codon. This creates a coding region of ORF0-ORF1-ORF2. Translation of this RNA yields a protein of 66 kD. This protein is a repressor of transposon activity.
- In germline tissues, an additional splicing event occurs to remove intron 3. This connects all four open reading frames into an mRNA that is translated to generate a protein of 87 kD. This protein is the transposase.

What is responsible for the tissue-specific splicing? Somatic cells contain a protein that binds to sequences in exon 3 to prevent splicing of the last intron (see *Section 21.11, Alternative Splicing Is a Rule, Rather Than an Exception, in Multicellular Eukaryotes*). The absence of this protein in germline cells allows splicing to generate the mRNA that codes for the transposase.

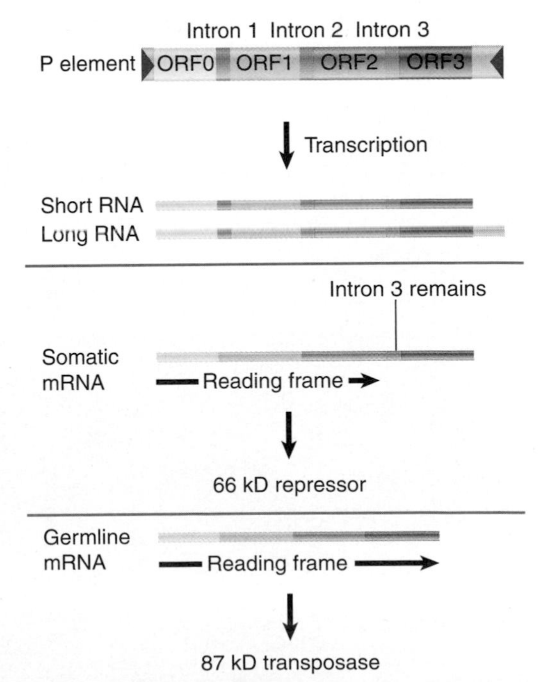

FIGURE 17.20 The P element has four exons. The first three are spliced together in somatic expression; all four are spliced together in germline expression.

P element transposition occurs by a nonreplicative cut-and-paste mechanism. It is interesting that in a significant proportion of cases, the break in donor DNA is repaired by using the sequence of the homologous chromosome. If the homolog has a P element, the presence of a P element at the donor site may be restored (so the event resembles the result of a replicative transposition). If the homolog lacks a P element, repair may generate a sequence lacking the P element, thus apparently providing a precise excision (an unusual event in other transposable systems).

▸ **cytotype** A cytoplasmic condition that affects P element activity. The effect of cytotype is due to the presence or absence of a repressor of transposition, which is provided by the mother to the egg.

The dependence of hybrid dysgenesis on the sexual orientation of a cross shows that the cytoplasm is important, in addition to the P factors themselves. The contribution of the cytoplasm is described as the **cytotype**; a line of flies containing P elements has a P cytotype, whereas a line of flies lacking P elements has an M cytotype. Hybrid dysgenesis occurs only when chromosomes containing P factors find themselves in M cytotype, that is, when the male parent has P elements and the female parent does not.

Cytotype shows an inheritable cytoplasmic effect; when a cross occurs through P cytotype (the female parent has P elements), hybrid dysgenesis is suppressed for several generations of crosses with M female parents. Thus something in P cytotype, which can be diluted out over some generations, suppresses hybrid dysgenesis.

The effect of cytotype is explained in molecular terms by the model of **FIGURE 17.21**. It depends on the assumption that a repressor molecule is deposited into the egg cell cytoplasm. The repressor is provided as a maternal factor in the egg. In a P line, there must be sufficient repressor to prevent transposition from occurring, even though the P elements are present. In any cross involving a P female, its presence prevents either synthesis or activity of the transposase. When the female parent is M type, though, there is no repressor in the egg, and the introduction of a P element from the male parent results in activity of transposase in the germline. The ability of P cytotype to exert an effect through more than one generation suggests that there must be enough repressor protein in the egg, and it must be stable enough, to be passed on through the adult to be present in the eggs of the next generation.

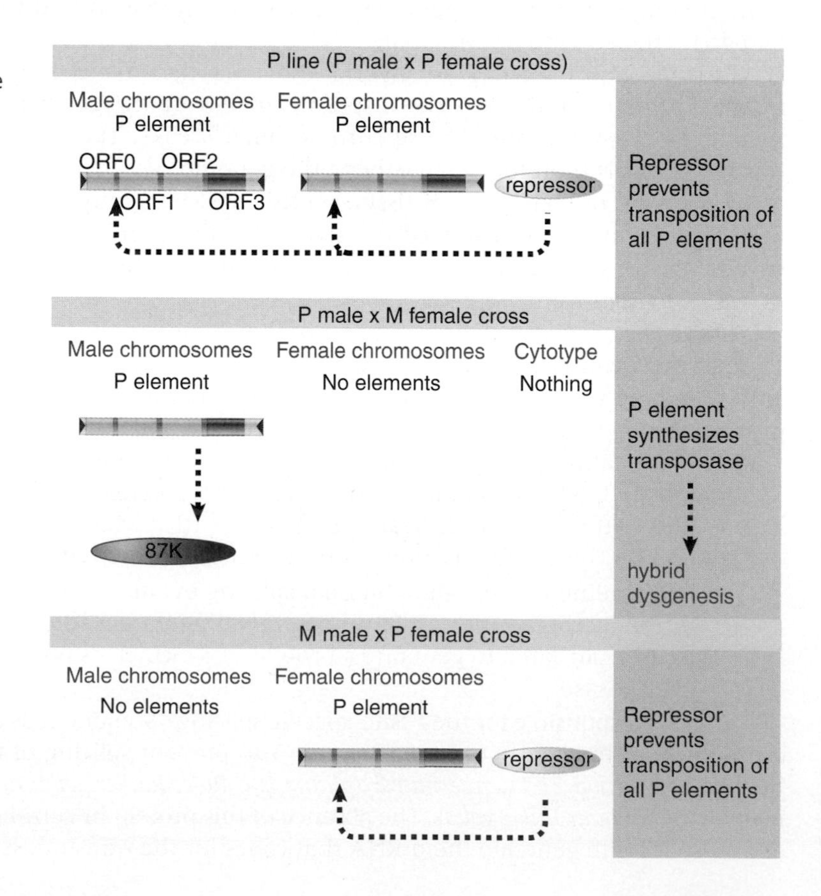

FIGURE 17.21 Hybrid dysgenesis is determined by the interactions between P elements in the genome and repressors in the cytotype.

For many years, the best candidate for the repressor was the 66 kD protein. There are, however, strains of flies that lack P elements capable of producing the 66 kD protein but that do exhibit a P cytotype. More recent evidence has implicated small RNAs in P element repression; genes important in processing small RNAs derived from P element transcripts (and those of several other transposons as well) are also required for efficient transposon silencing. This has led to a model in which P cytotype is conditioned by P elements at particular positions that produce transcripts that are processed into a specific class of small RNAs, called piRNAs (see *Section 30.5, MicroRNAs are Widespread Regulators in Eukaryotes*). In this case, it is the presence of these small RNAs in the cytoplasm that are responsible for P element cytotype repression. Like the small RNAs involved in RNA interference (see *Chapter 30, Regulatory RNA*), piRNAs are hypothesized to direct the degradation of P element transcripts. An appealing feature of this model is that it suggests that P element cytotype repression is a particular example of a widespread mechanism by which tranposon activity is repressed in plants, fungi and animals.

KEY CONCEPTS

- P elements are transposons that are carried in P strains of *Drosophila melanogaster* but not in M strains.
- When a P male is crossed with an M female, transposition is activated.
- The insertion of P elements at new sites in these crosses inactivates many genes and makes the cross infertile.
- P elements are activated in the germline of P male × M female crosses because a tissue-specific splicing event removes one intron, which generates the coding sequence for the transposase.
- The P element also produces a repressor of transposition, which is inherited maternally in the cytoplasm.
- The presence of the repressor explains why M male × P female crosses remain fertile.
- The repressor may consist of piRNAs derived from P element transcripts that target the destruction of other P element transcripts.

CONCEPT AND REASONING CHECK

Explain how hybrid dysgenesis could lead to speciation of two dysgenic populations of flies.

17.7 The Retrovirus Life Cycle Involves Transposition-like Events

Retroviruses have genomes of single-stranded RNA that are replicated through a double-stranded DNA intermediate. The life cycle of the virus involves an obligatory stage in which the double-stranded DNA is inserted into the host genome by a transposition-like event that generates short direct repeats of target DNA. This similarity is not surprising, given evidence that new retroviruses have arisen repeatedly over evolutionary time as a consequence of the capture by retrotransposons of genes encoding envelope proteins, which makes infection possible.

The significance of this reaction extends beyond the perpetuation of the virus. Some of its consequences are as follows:

- A retroviral sequence that is integrated in the germline remains in the cellular genome as an endogenous **provirus**. Like a lysogenic bacteriophage, a provirus behaves as part of the genetic material of the organism.
- Cellular sequences occasionally recombine with the retroviral sequence and then are transposed with it; these sequences may be inserted into the genome as duplex sequences in new locations.

provirus A duplex sequence of DNA integrated into a eukaryotic genome that represents the sequence of the RNA genome of a retrovirus.

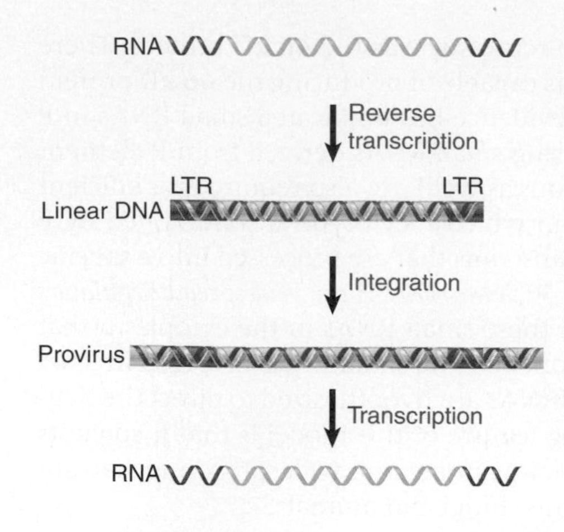

FIGURE 17.22 The retroviral life cycle proceeds by reverse transcribing the RNA genome into duplex DNA, which is inserted into the host genome in order to be transcribed into RNA.

- **reverse transcriptase** An enzyme that uses single-stranded RNA as a template to synthesize a complementary DNA strand.
- **integrase** An enzyme that is responsible for a site-specific recombination that inserts one molecule of DNA into another.

- Cellular sequences that are transposed by a retrovirus may change the properties of a cell that becomes infected with the virus.

The particulars of the retroviral life cycle are expanded in FIGURE 17.22. The crucial steps are that the viral RNA is converted into DNA, the DNA becomes integrated into the host genome, and then the DNA provirus is transcribed into RNA.

The enzyme responsible for generating the initial DNA copy of the RNA is **reverse transcriptase**. The enzyme converts the RNA into a linear duplex of DNA in the cytoplasm of the infected cell. The linear DNA then makes its way to the nucleus, where one or more DNA copies become integrated into the host genome. A single enzyme called **integrase** is responsible for integration. Retroviral integrases are related by sequence, structure and function to the transposases encoded by transposons. The provirus is transcribed by the host machinery to produce viral RNAs, which serve both as mRNAs and as genomes for packaging into virions. Integration is a normal part of the life cycle and is necessary for transcription.

Two copies of the RNA genome are packaged into each virion, making the individual virus particle effectively diploid. When a cell is simultaneously infected by two different but related viruses, it is possible to generate heterozygous virus particles carrying one genome of each type. The diploidy may be important in allowing the virus to acquire cellular sequences. The enzymes reverse transcriptase and integrase are carried with the genome in the viral particle.

A typical retroviral genome contains three or four "genes." (In this context, the term *genes* is used to identify coding regions, each of which actually gives rise to multiple proteins by processing reactions.) A typical retrovirus genome with three genes is organized in the sequence *gag-pol-env*, as indicated in FIGURE 17.23. The *gag* gene gives rise to the protein components of the nucleoprotein core of the virion. The *pol* gene

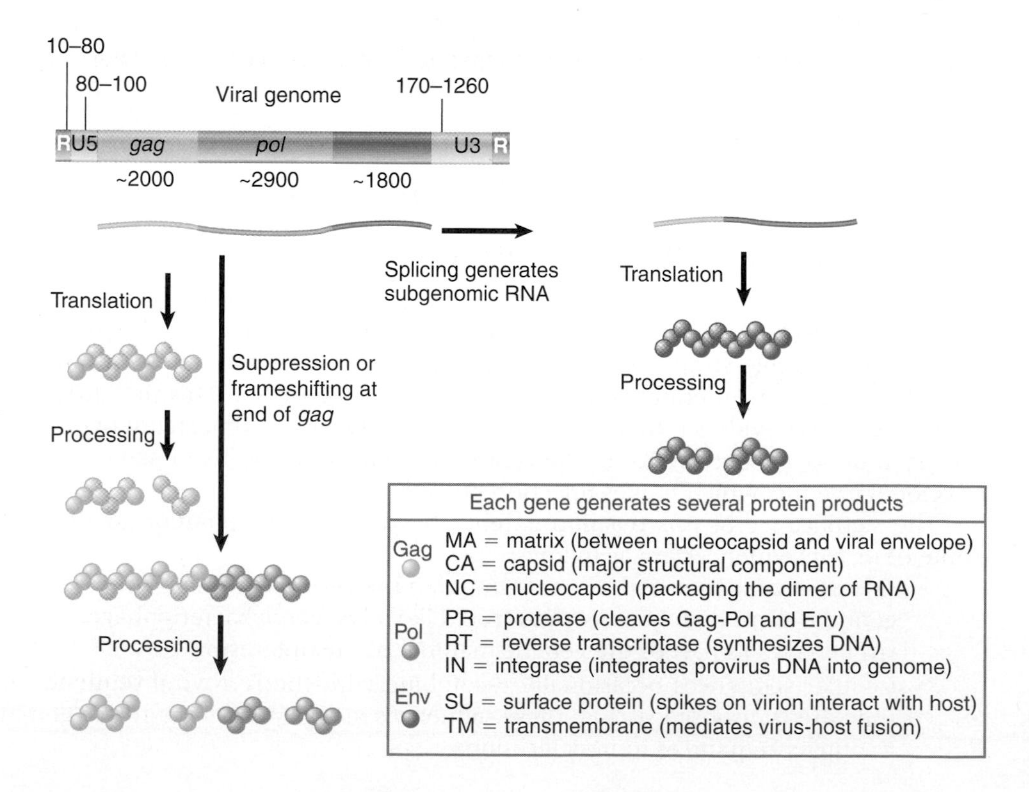

FIGURE 17.23 The genes of the retrovirus are expressed as polyproteins that are processed into individual products.

codes for functions concerned with nucleic acid synthesis and recombination. The *env* gene codes for components of the envelope of the particle, which also sequesters components from the cellular cytoplasmic membrane.

Retroviral mRNA, transcribed by the host polymerase, has a conventional structure: it is capped at the 5′ end and polyadenylated at the 3′ end. The retroviral genome is expressed as two mRNAs. The full-length mRNA is translated to give the Gag and Pol polyproteins. The Gag product is translated by reading from the initiation codon to the first termination codon. This termination codon must be bypassed to express Pol.

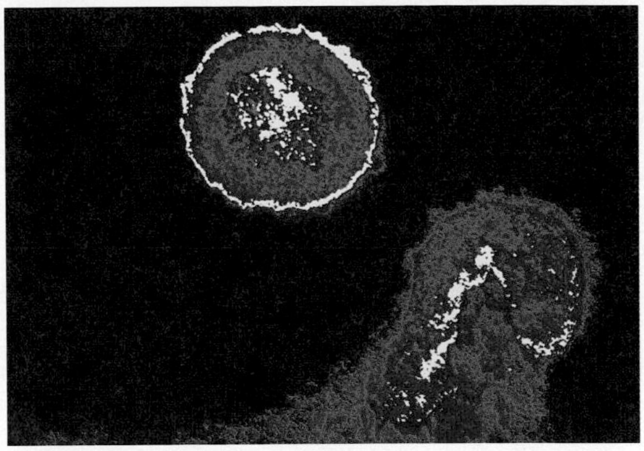

Different mechanisms are used in different viruses to proceed beyond the *gag* termination codon, depending on the relationship between the *gag* and *pol* reading frames. When *gag* and *pol* follow continuously, suppression by a glutamyl-tRNA that recognizes the termination codon allows a single protein to be generated. When *gag* and *pol* are in different reading frames, a ribosomal frameshift occurs to generate a single protein. Usually the readthrough is ~5% efficient, so Gag protein outnumbers Gag-Pol protein about 20-fold.

FIGURE 17.24 Retroviruses (HIV) bud from the plasma membrane of an infected cell. © Photodisc.

The Env polyprotein is expressed by another means: splicing generates a shorter *subgenomic* messenger that is translated into the Env product.

Both the Gag or Gag-Pol and the Env products are polyproteins that are cleaved by a protease to release the individual proteins that are found in mature virions. The protease activity is coded by the virus in different ways: in some viruses it is part of Gag or Pol, and in other viruses it exists as an additional independent reading frame.

The production of a retroviral particle involves packaging the RNA into a core, surrounding it with capsid proteins, and pinching off a segment of membrane from the host cell. The release of infective particles by such means is shown in **FIGURE 17.24**. The process is reversed during infection: a virus infects a new host cell by fusing with the plasma membrane and then releasing the contents of the virion.

KEY CONCEPTS

- A retrovirus has two copies of its genome of single-stranded RNA.
- An integrated provirus is a double-stranded DNA sequence.
- A retrovirus generates a provirus by reverse transcription of the retroviral genome.
- A typical retrovirus has three genes: *gag, pol,* and *env.*
- Gag and Pol proteins are translated from a full-length transcript of the genome.
- Translation of Pol requires a readthrough or a frameshift by the ribosome.
- Env is translated from a separate mRNA that is generated by splicing.
- Each of the three protein products is processed by proteases to give multiple proteins.

CONCEPT AND REASONING CHECK

Predict the outcomes on the viral life cycle of mutation of *gag, pol,* or *env.*

17.8 Retroviral RNA Is Converted to DNA and Integrates into the Host Genome

Retroviruses are called **plus-strand viruses** because the viral RNA itself codes for the protein products. As its name implies, reverse transcriptase is responsible for converting the genome (plus strand RNA) into a complementary DNA strand, which is called the **minus-strand DNA**. Reverse transcriptase also catalyzes subsequent stages in the production of duplex DNA. It has a DNA polymerase activity, which enables it to synthesize a duplex DNA from the single-stranded reverse

▸ **plus-strand virus** A virus with a single-stranded nucleic acid genome whose sequence directly codes for the protein products.

▸ **minus-strand DNA** The single-stranded DNA sequence that is complementary to the viral RNA genome of a plus-strand virus.

FIGURE 17.25 Retroviral RNA ends in direct repeats (R), the free linear DNA ends in LTRs, and the provirus ends in LTRs that are shortened by two bases each.

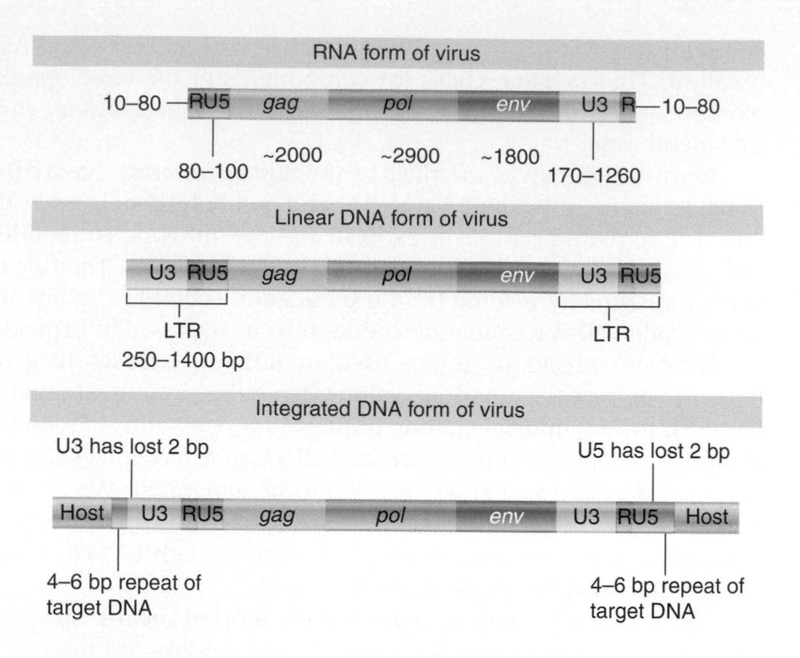

transcript of the RNA. The second DNA strand in this duplex is called the **plus-strand DNA**. As a necessary adjunct to this activity, the enzyme has an RNase H activity, which can degrade the RNA part of the RNA–DNA hybrid. All retroviral reverse transcriptases share considerable similarities of amino acid sequence, and homologous sequences can be recognized in some other retroposons.

▸ **plus-strand DNA** The strand of the duplex sequence representing a retrovirus that has the same sequence as that of the RNA.

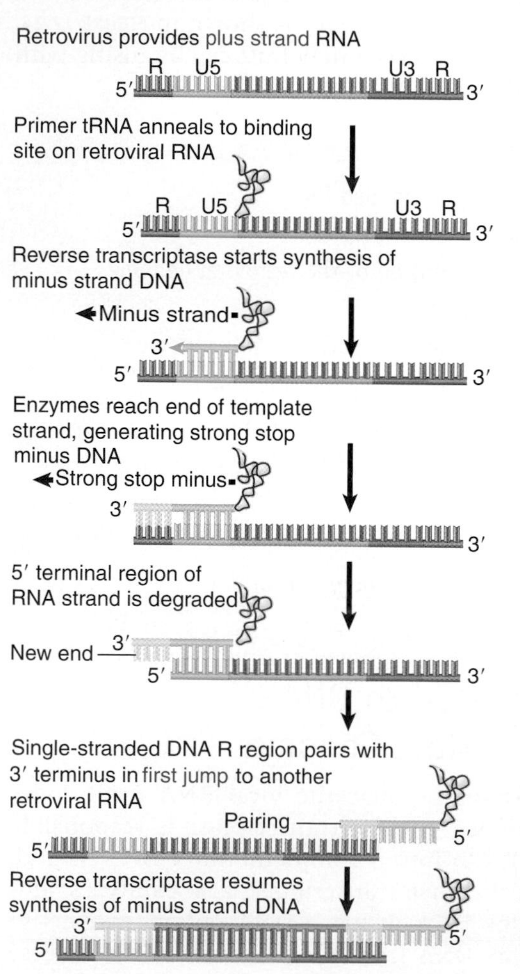

FIGURE 17.26 Minus strand DNA is generated by switching templates during reverse transcription.

▸ **R segments** The sequences that are repeated at the ends of a retroviral RNA. They are called R-U5 and U3-R.

▸ **U5** The repeated sequence at the 5′ end of a retroviral RNA.

▸ **U3** The repeated sequence at the 3′ end of a retroviral RNA.

The structures of the DNA forms of the virus are compared with the RNA in FIGURE 17.25. The viral RNA has direct repeats at its ends. These **R segments** vary in different strains between 10 and 80 nucleotides. The sequence at the 5′ end of the virus is **U5**, and the sequence at the 3′ end is **U3**. The R segments are used during the conversion from the RNA to the DNA form to generate the more extensive direct repeats that are found in linear DNA (FIGURE 17.26 and FIGURE 17.27). The shortening of 2 bp at each end in the integrated form is a consequence of the mechanism of integration (FIGURE 17.28).

Like all DNA polymerases, reverse transcriptase requires a primer to provide a 3′-OH. Retroviruses use an uncharged host tRNA as a primer; this tRNA is present in the virion. A sequence of 18 bases at the 3′ end of the tRNA is base paired to a site 100 to 200 bases from the 5′ end of one of the viral RNA molecules.

Here is a dilemma: if reverse transcriptase starts to synthesize DNA at a site only 100 to 200 bases downstream from the 5′ end, how can DNA be generated to represent the intact RNA genome? (This is an extreme variant of the general

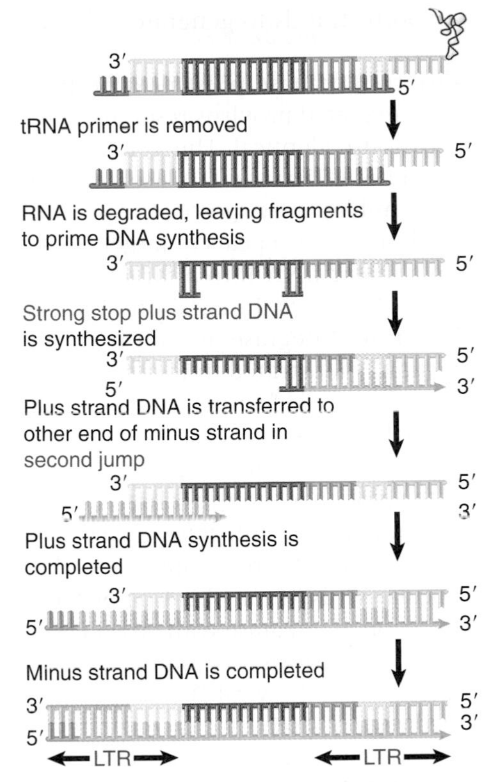

FIGURE 17.27 Synthesis of plus-strand DNA requires a second jump.

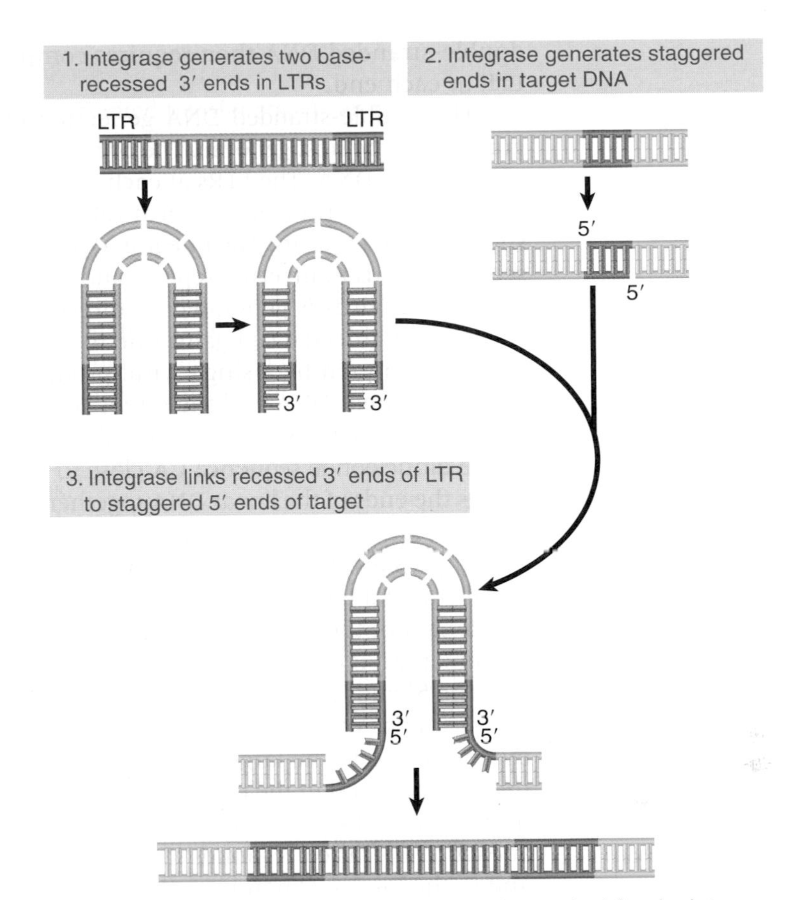

FIGURE 17.28 Integrase is the only viral protein required for the integration reaction, in which each LTR loses 2 bp and is inserted between 4 bp repeats of target DNA.

problem in replicating the ends of any linear nucleic acid; see *Section 14.2, The Ends of Linear DNA Are a Problem for Replication.*)

Reverse transcriptase solves this problem by switching templates (known as "strand switching"), carrying the nascent DNA with it to the new template (see Figure 17.26). This is the first of two jumps between templates. In this reaction, synthesis proceeds from the tRNA primer to the end of the template, generating a short DNA sequence called *strong stop minus DNA*. Then, the R region at the 5′ terminus of the RNA template is degraded by the RNase H activity of reverse transcriptase. Its removal allows the R region at a 3′ end to base pair with the newly synthesized DNA. Reverse transcription then continues through the U3 region into the body of the RNA. The source of the R region that pairs with the strong stop minus DNA can be either the 3′ end of the same RNA molecule (intramolecular pairing) or the 3′ end of a different RNA molecule (intermolecular pairing).

The result of the switch and extension is to add a U3 segment to the 5′ end. The stretch of sequence U3-R-U5 is called the long-terminal repeat (LTR)because a similar series of events adds a U5 segment to the 3′ end, giving it the same structure of U5-R-U3. Its length varies from 250 to 1400 bp (see Figure 17.25).

We now need to generate the plus strand of DNA and to generate the LTR at the other end. The reaction is shown in Figure 17.27. Reverse transcriptase primes synthesis of plus strand DNA from a fragment of RNA that is left after degrading the original RNA molecule. A *strong-stop plus-strand* DNA is generated when the enzyme reaches the end of the template. This DNA is then transferred to the other end of a minus strand, where it is probably released by a displacement reaction when a second round of DNA synthesis occurs from a primer fragment farther upstream (to its left in the figure). It uses the R region to pair with the 3′ end of a minus strand DNA. This

elements that have suffered deletions eliminating parts of the reading frames coding for the protein(s) needed for transposition.

The most common LINE in mammalian genomes is called L1. The typical member is ~6500 bp long and terminates in an A-rich tract. The two open reading frames of a full-length element are called ORF1 and ORF2. The number of full-length elements is usually small (~50), and the remainder of the copies are truncated. Transcripts corresponding to these elements can be found. The LINE family shows sequence variation among individual members. The members of the family within a species, however, are relatively homogeneous compared to the variation shown between species. L1 is the only member of the LINE family that has been active in either the mouse or human lineages. It seems to have remained highly active in the mouse, but has declined in the human lineage.

In the human genome, a large part of the moderately repetitive DNA exists as sequences of ~300 bp that are interspersed with nonrepetitive DNA. Many of these repeats contain a recognition site for the restriction enzyme AluI at a position located 170 bp along the sequence, giving this SINE family its name: the common **Alu element**. Alu is the only SINE that has been active in the human lineage. The mouse genome has a counterpart to this element (B1), and also other SINEs (B2, ID, B4) that have been active. The transposition of the SINEs probably results from their recognition as substrates by an active L1 element.

‣ **Alu element** One of a set of dispersed, related sequences, each ~300 bp long, in the human genome (members of the SINE family). The individual members have Alu cleavage sites at each end.

The individual members of the Alu family are related rather than identical. The human family seems to have originated by means of a 130 bp tandem duplication, with an unrelated sequence of 31 bp inserted in the right half of the duplicated region. The two repeats are sometimes called the "left half" and the "right half" of the Alu sequence. The individual members of the Alu family have an average identity with the consensus sequence of 87%. The mouse B1 repeating unit is 130 bp long and corresponds to a monomer of the human unit. It has 70% to 80% homology with the human sequence.

The Alu sequence is related to 7SL RNA, a component of the signal recognition particle involved in protein targeting to the endoplasmic reticulum. The 7SL RNA corresponds to the left half of an Alu sequence with an insertion in the middle. Thus, the ninety 5′ terminal bases of 7SL RNA are homologous to the left end of Alu, the central 160 bases of 7SL RNA have no homology to Alu, and the forty 3′ terminal bases of 7SL RNA are homologous to the right end of Alu. The 7SL RNA genes are transcribed by RNA polymerase III. It is possible that these genes (or genes related to them) gave rise to the inactive Alu sequences.

The members of the Alu family resemble transposons in being flanked by short direct repeats. However, they also display the curious feature that the lengths of the repeats are different for individual members of the family. They derive from RNA polymerase III transcripts, and as a result it is possible that individual members carry internal active promoters. At least some members of the family can be transcribed into independent RNAs, and other members of the Alu family can be included within structural gene transcription units, as seen by their presence in long nuclear RNA. The presence of multiple copies of the Alu sequence in a single nuclear molecule can generate secondary structure. In fact, the presence of Alu family members in the form of inverted repeats is responsible for most of the secondary structure found in mammalian nuclear RNA.

KEY CONCEPTS

- LTR retroposons mobilize via an RNA that is similar to retroviral RNA, but does not form an infectious particle.
- Some retroposons directly resemble retroviruses in their use of LTRs, whereas others do not have LTRs.
- Other elements can be found that were generated by an RNA-mediated transposition event, but they do not themselves code for enzymes that can catalyze transposition.
- Transposons and retroposons constitute almost half of the human genome.
- A major part of repetitive DNA in mammalian genomes consists of repeats of a single family organized like transposons and derived from RNA polymerase III transcripts.

CONCEPT AND REASONING CHECK

Inactive SINEs have been known to cause mutations in humans not by transposition but by unequal homologous recombination. Explain how this might occur.

17.11 Summary

Prokaryotic and eukaryotic cells contain a variety of transposons that mobilize by moving or copying DNA sequences. The transposon can be identified only as an entity within the genome; its mobility does not involve an independent form.

The archetypal transposon has inverted repeats at its termini and generates direct repeats of a short sequence at the site of insertion. The simplest types are the bacterial insertion sequences (IS), which consist essentially of the inverted terminal repeats flanking a coding frame(s) whose product(s) provide transposition activity. Composite transposons have terminal modules that consist of IS elements; one or both of the IS modules provides transposase activity, and the sequences between them (often carrying antibiotic resistance) are treated as passengers.

The generation of target repeats flanking a transposon reflects a common feature of transposition. The target site is cleaved at points that are staggered on each DNA strand by a fixed distance (often five or nine base pairs). The transposon is in effect inserted between protruding single-stranded ends generated by the staggered cuts. Target repeats are generated by filling in the single-stranded regions.

IS elements, composite transposons, and P elements mobilize by nonreplicative transposition, in which the element moves directly from a donor site to a recipient site. A single transposase enzyme undertakes the reaction. It occurs by a cut-and-paste mechanism in which the transposon is separated from flanking DNA. Cleavage of the transposon ends, nicking of the target site, and connection of the transposon ends to the staggered nicks all occur in a nucleoprotein complex containing the transposase. Loss of the transposon from the donor creates a double-strand break that must be repaired by the cell.

The best-characterized transposons in plants are the controlling elements of maize, which fall into several families. Each family contains a single type of autonomous element that is analogous to bacterial transposons in its ability to mobilize. A family also contains many different nonautonomous elements that are derived by mutations (usually deletions) of the autonomous element. The nonautonomous elements lack the ability to transpose, but display transposition activity and other abilities of the autonomous element when an autonomous element is present to provide the necessary *trans*-acting functions.

In addition to the direct consequences of insertion and excision, the maize elements may also control the activities of genes at or near the sites where they are inserted; this control may be subject to developmental regulation. Maize elements inserted into genes may be excised from the transcripts, which explains why they do not simply impede gene activity. Control of target gene expression involves a variety of molecular effects, including activation by provision of an enhancer and suppression by interference with posttranscriptional events.

Transposition of maize elements (in particular *Ac*) is nonreplicative and probably requires only a single transposase enzyme coded by the element. Transposition occurs preferentially after replication of the element. It is likely that there are mechanisms to limit the frequency of transposition. Advantageous rearrangements of the maize genome may have been connected with the presence of the elements.

P elements in *D. melanogaster* are responsible for hybrid dysgenesis, which could be a forerunner of speciation. A cross between a male carrying P elements and a female lacking them generates hybrids that are sterile. A P element has four open reading frames, which are separated by introns. Splicing of the first three ORFs generates a 66 kD repressor and occurs in all cells. Splicing of all four ORFs to generate the 87 kD transposase occurs only in the germline by a tissue-specific splicing event.

P elements mobilize when exposed to cytoplasm lacking the repressor. The burst of transposition events inactivates the genome by random insertions. Only a complete P element can generate transposase, but defective elements can be mobilized in *trans* by the enzyme.

Reverse transcription is the unifying mechanism for reproduction of retroviruses and perpetuation of retrotransposons. Retroviruses have genomes of single-stranded RNA that are replicated through a double-stranded DNA intermediate. An individual retrovirus contains two copies of its genome. The genome contains the *gag, pol,* and *env* genes that are translated into polyproteins, each of which is cleaved into smaller functional proteins. The Gag and Env components are concerned with packing RNA and generating the virion; the Pol components are concerned with nucleic acid synthesis.

Reverse transcriptase is the major component of Pol and is responsible for synthesizing a DNA (minus-strand) copy of the viral (plus-strand) RNA. The DNA product is longer than the RNA template; by switching template strands, reverse transcriptase copies the 3′ sequence of the RNA to the 5′ end of the DNA, and copies the 5′ sequence of the RNA to the 3′ end of the DNA. This generates the characteristic LTRs (long terminal repeats) of the DNA. A similar switch of templates occurs when the plus strand of DNA is synthesized using the minus strand as a template. Linear duplex DNA is inserted into a host genome by the integrase enzyme. Transcription of the integrated DNA from a promoter in the left LTR generates further copies of the RNA sequence.

During an infective cycle, a retrovirus may exchange part of its usual sequence for a cellular sequence; the resulting virus is usually replication defective but can be perpetuated in the course of a joint infection with a helper virus. Many of the defective viruses have gained an RNA version (*v-onc*) of a cellular gene (*c-onc*). The *onc* sequence may be any one of a number of genes whose expression in *v-onc* form causes the cell to be transformed into a tumorigenic phenotype.

The integration event generates direct target repeats (like transposons that mobilize via DNA). An inserted provirus, therefore, has direct terminal repeats of the LTRs, flanked by short repeats of target DNA. Mammalian and avian genomes have endogenous (inactive) proviruses with such structures. Other elements with this organization have been found in a variety of genomes, most notably in *S. cerevisiae* (*Ty* elements) and *D. melanogaster* (*copia* elements). They may generate particles resembling viruses but do not have infectious capability. The LINE sequences of mammalian genomes are further removed from the retroviruses but retain enough similarities to suggest a common origin. LINES lack LTRs and use a different type of priming event to initiate reverse transcription, and both autonomous and nonautonomous LINEs exist. Autonomous LINEs encode functional reverse transcriptase and endonuclease activities.

The members of another class of retroposons have the hallmarks of transposition via RNA but have no coding sequences (or at least none resembling retroviral functions). A particularly prominent family that appears to have originated from a processing event is the mammalian SINE; it includes the human Alu family. Some snRNAs, including 7SL snRNA (a component of the SRP), are related to this family. Both SINEs and LINEs comprise a large fraction of mammalian genomes.

CHAPTER QUESTIONS

1. Recombination between two transposons that are present in the same replicon in direct repeat orientation results in:

A. degradation of the sequence between the two transposons.
B. inversion of the sequence between the two transposons.
C. excision and circularization of the sequence between the two transposons.
D. any of the above are possible.

2. Which occurs least frequently: transposon insertion, imprecise transposon excision, or precise transposon excision?
   A. transposon insertion
   B. imprecise transposon excision
   C. precise transposon excision
   D. These all occur at about the same rate.
3. Which of the following maize transposons is a nonautonomous element of the same family as the activator (Ac) element?
   A. Spm
   B. Dt
   C. Mu
   D. Ds
4. The P element produces a:
   A. repressor of transposition, which is inherited paternally in the cytoplasm.
   B. activator of transposition, which is inherited paternally in the cytoplasm.
   C. repressor of transposition, which is inherited maternally in the cytoplasm.
   D. activator of transposition, which is inherited maternally in the cytoplasm.

List two activities for each of the following retroviral proteins

Gag 5. ______________________ 6. ______________________
Pol 7. ______________________ 8. ______________________
Env 9. ______________________ 10. ______________________

11. The enzyme responsible for generating the initial DNA copy from RNA in a retrovirus is:
   A. DNA polymerase.
   B. RNA polymerase.
   C. reverse transcriptase.
   D. integrase.
12. Translation of which retrovirus protein(s) requires ribosome frameshifting?
   A. Gag
   B. Pol
   C. Env
   D. all of the above
13. Retroviral RNA ends in:
   A. direct repeats.
   B. indirect repeats.
   C. long terminal repeats.
   D. unique sequences—no repeats.
14. Replication-defective retroviruses are most commonly generated by:
   A. recombination and rearrangement of sequences.
   B. mutation at critical sites in viral genes.
   C. deletion of a segment of the viral genome.
   D. insertion of sequences into viral genes.
15. Which of the following types of element makes up the greatest proportion of the human genome?
   A. retroviruses and retroposons
   B. LINEs
   C. SINEs
   D. DNA transposons
16. Alu elements are what type of transposable element?
   A. SINEs
   B. LINEs

C. retroposon
D. DNA transposon

17. About how many copies of Alu family elements are present in the haploid human genome?
    A. 20,000
    B. 50,000
    C. 160,000
    D. 300,000

18. Alu family elements are characterized by which of the following?
    A. identical elements of 170 bp with terminal inverted repeats and containing a single AluI restriction site
    B. related elements of 170 bp with terminal tandem repeats and containing a single AluI restriction site
    C. identical elements of 300 bp with terminal inverted repeats and containing a single AluI restriction site
    D. related elements of 300 bp with terminal tandem repeats and containing a single AluI restriction site

KEY TERMS

***Ac* element**
acentric fragment
Alu element
autonomous controlling element
cointegrate
composite elements
controlling elements
cytotype
direct repeats
***Ds* element**
helper virus
hybrid dysgenesis
imprecise excision
insertion sequences (IS)
integrase
inverted-terminal repeats
long interspersed elements (LINEs)
long-terminal repeat (LTR)
minus strand DNA
nonautonomous controlling element
nonreplicative transposition
P element
plus-strand DNA
plus-strand virus
precise excision
provirus
R segments
replication-defective virus
replicative transposition
resolution
resolvase
retroposon (non-LTR retrotransposon)
retrotransposon (LTR retrotransposon)
retrovirus
reverse transcriptase
short interspersed elements (SINEs)
Tn
transducing virus
transposase
transposable element (transposon)
U3
U5

FURTHER READING

Belancio, V. P., Hedges, D. J., and Deininger, P. (2008). Mammalian non-LTR retrotransposons: for better or worse, in sickness and in health. *Genome Res.* **18**(3), 343–358. A review of retrotransposons in mammals, including LINEs and SINES, with a discussion of the effects of transposition in human biology.

Castro, J. P., and Carareto, C. M. (2004). *Drosophila melanogaster* P transposable elements: mechanisms of transposition and regulation. *Genetica* **121**(2), 107–118. A review of the mechanisms of P element transposition and hybrid dysgenesis.

Haniford, D. B. (2006). Transpososome dynamics and regulation in Tn10 transposition. *Crit. Rev. Biochem. Mol. Biol.* **41**(6), 407–424. A review of the nonreplicative transposition mechanism typified by the bacterial Tn10 transposon.

18

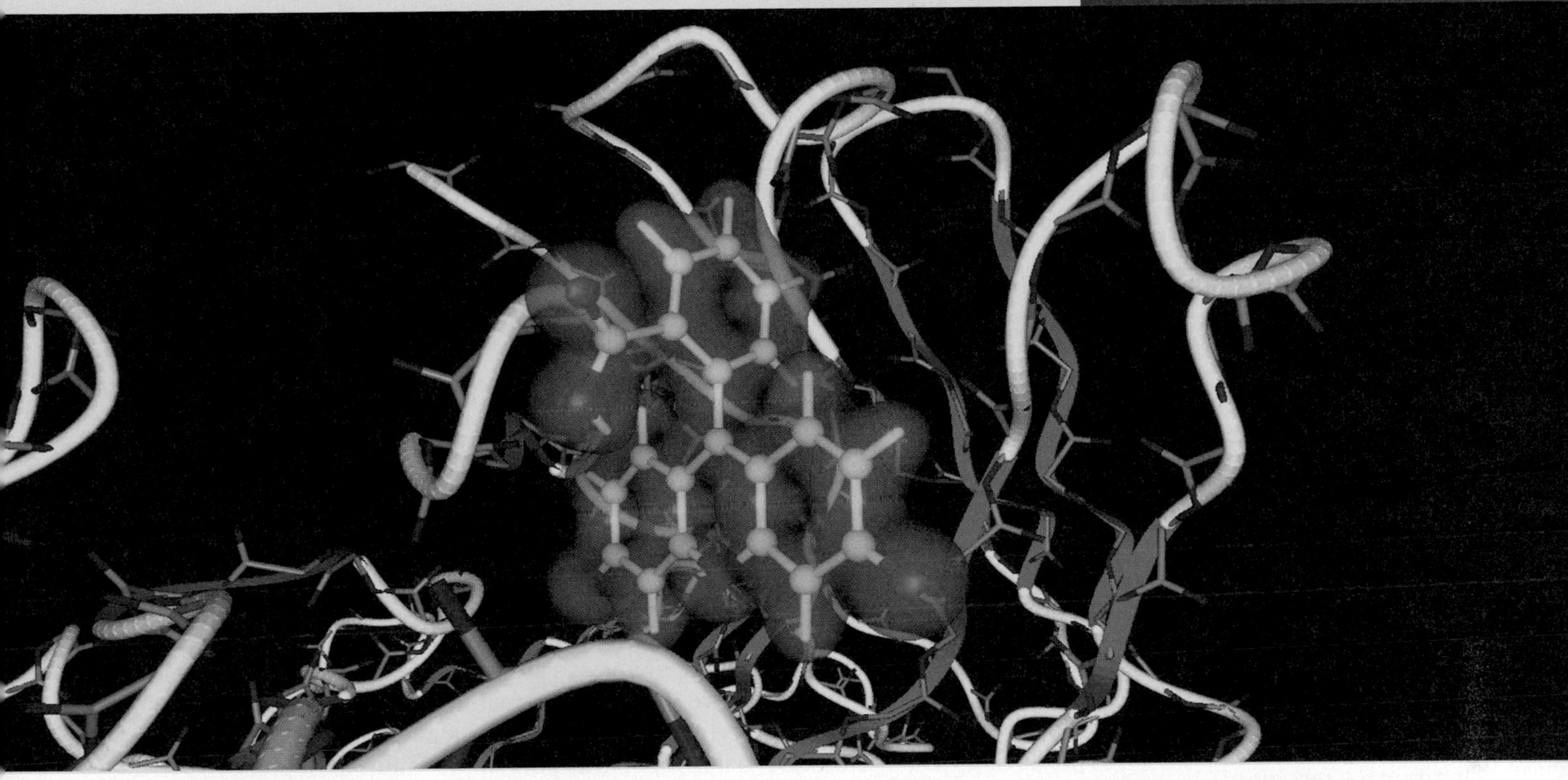

The structure of the antigen-binding site of an antibody, bound to the fluorescent molecule fluorescein (in green and orange). Photo courtesy of Kim Baldridge, University of Zürich.

Somatic Recombination and Hypermutation in the Immune System

CHAPTER OUTLINE

18.1 Introduction

In general, differential control of gene expression, rather than changes in DNA sequence, explains the different phenotypes of particular somatic cells. The immune system is a most important exception to the axiom of genetics that the genetic constitution created in the zygote by the combination of sperm and egg is inherited by all somatic cells of the organism. In developing immune cells, the genome changes through extensive somatic DNA recombination to create functional genes. Other cases of somatic recombination are represented by the substitution of one sequence for another to change the mating type of yeast or to generate new surface antigens by trypanosomes. In mature **B cells**, additional DNA recombination and mutation of recombined DNA segments occur to further diversify the function of these cells.

The **immune response** of vertebrates provides a protective system that distinguishes in general foreign ("nonself") soluble molecules or molecules on microorganisms from components (molecules or cells) of the organism itself ("self"). Nonself- or self-components capable of inducing a specific immune response are referred to as **antigens**. Typically, the antigen is a protein that has entered the bloodstream of the animal—for example, the coat protein of an infecting virus or bacterium. Exposure to an antigen initiates production of an immune response aimed at *specifically recognizing the antigen,* thereby destroying the infecting virus or bacterium expressing it.

Immune reactions are performed by white blood cells: B and T lymphocytes, macrophages, and dendritic cells. The lymphocytes are named after the tissues that produce them. In mammals, B cells mature in the bone marrow, whereas **T cells** mature in the thymus. Each class of lymphocyte uses the rearrangement of DNA as a mechanism for producing the proteins that enable it to participate in the immune response.

Responses to antigens on viruses and bacteria, such as a B cell response to *Streptococcus (Pneumococcus) pneumoniae* or a killer T lymphocyte–mediated response to influenza virus-infected cells, are highly specific and are the expression of **adaptive (acquired) immunity**. The adaptive immune response is characterized by a latency period—in general a few days—which is required for the clonal selection and expansions of the B and/or T cells specific for the antigen. Clonal selection of B cells or T cells relies on binding of the antigen to **B cell receptors (BCR)** and **T cell receptors (TCR)**, both of which possess a high affinity for that antigen. The structural basis for this selection process is provided by the generation of a very large number of BCRs/TCRs in order to create a high probability of recognizing any foreign molecule. BCRs/TCRs that recognize the body's own proteins are screened out early in the process.

Activation of the BCR on B cells triggers the pathways of the **humoral response**. It is mediated by the secretion of **immunoglobulin (antibody)** proteins. Production of an antibody specific for a foreign molecule is the primary event responsible for recognition of an antigen. Recognition requires the antibody to bind to a small region or structure on the antigen.

The function of antibodies is shown in **FIGURE 18.1**. Foreign material circulating in the

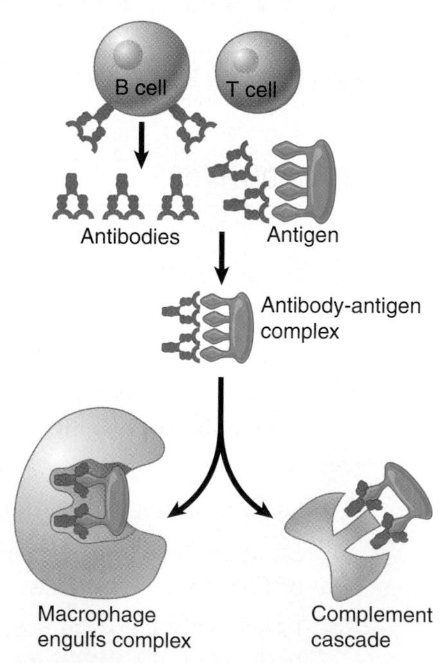

FIGURE 18.1 Humoral immunity is conferred by the binding of free antibodies to antigens to form antigen-antibody complexes that are removed from the bloodstream by macrophages or that are attacked directly by the activated complement cascade.

- **B cell** A lymphocyte that produces antibodies. Development occurs primarily in bone marrow.
- **immune response** An organism's reaction, mediated by components of the immune system, to an antigen.
- **antigen** A molecule that can bind specifically to an antigen receptor, such as an antibody.
- **T cells** Lymphocytes of the T (thymic) lineage; they may be subdivided into several functional types. They carry T cell receptors and are involved in the cell-mediated immune response.
- **adaptive (acquired) immunity** The response mediated by lymphocytes that are activated by their specific interaction with antigen.
- **B cell receptor (BCR)** The antigen receptor on B lymphocytes.
- **T cell receptor (TCR)** The antigen receptor on T lymphocytes. It is clonally expressed and binds to a complex of MHC class I or class II protein and antigen-derived peptide.
- **humoral response** An immune response that is mediated primarily by antibodies. It is defined as immunity that can be transferred from one organism to another by serum antibody.
- **immunoglobulin (antibody)** A protein that is produced by B cells and that binds to a particular antigen. They are synthesized in membrane-bound and secreted forms. Those produced during an immune response recruit effector functions to help neutralize and eliminate the pathogen.

18.2 Immunoglobulin Genes Are Assembled from Discrete DNA Segments in B Lymphocytes

Sophisticated evolutionary mechanisms have evolved to guarantee that an organism is prepared to produce specific antibodies for a broad variety of naturally occurring and anthropogenic components that it has never encountered before. Each antibody is a tetramer consisting of two identical immunoglobulin **light chains** (L) and two identical immunoglobulin **heavy chains** (H). In humans, there are two types of L chain (λ and κ) and nine types of H chain. The structure of the immunoglobulin tetramer is illustrated in **FIGURE 18.4**. Light chains and heavy chains share the same general type of organization in which each protein chain consists of two principal regions: the N-terminal **variable region (V region)** and the C-terminal **constant region (C region)**. As the names suggest, the variable regions show considerable changes in sequence from one protein to the next, whereas the constant regions show substantial homology. Different *classes* of immunoglobulins, which serve different functions, are determined by the heavy-chain constant region (see *Section 18.7, Class Switching Is Effected by DNA Recombination*).

Corresponding regions of the light chains and heavy chains associate to generate distinct domains in the immunoglobulin protein. The variable (V) domain is generated by association between the variable regions of the light chain and heavy chain. *The V domain is responsible for recognizing the antigen.* An immunoglobulin has a Y-shaped structure in which the arms of the Y are identical, and each arm has a copy of the V domain. Production of V domains of different specificities creates the ability to respond to diverse antigens. The total number of variable regions for either light- or heavy-chain proteins is measured in hundreds. Thus the protein displays the maximum versatility in the region responsible for binding the antigen.

The number of constant regions is vastly smaller than the number of variable regions; typically there are only one to ten C regions for any particular type of chain. The constant regions in the subunits of the immunoglobulin tetramer associate to generate several individual *C domains*. The first domain results from association of the single constant region of the light-chain (C_L) with the C_{H1} part of the heavy-chain constant region. The two copies of this domain complete the arms of the Y-shaped molecule. Association between the C regions of the heavy chains generates the remaining C domains, which vary in number depending on the type of heavy chain.

Comparing the characteristics of the variable and constant regions, we see the central dilemma in immunoglobulin gene structure. How does the genome code for a set of proteins in which any individual polypeptide chain must have one of <10 possible C regions, but can have any one of several hundred possible V regions? It turns

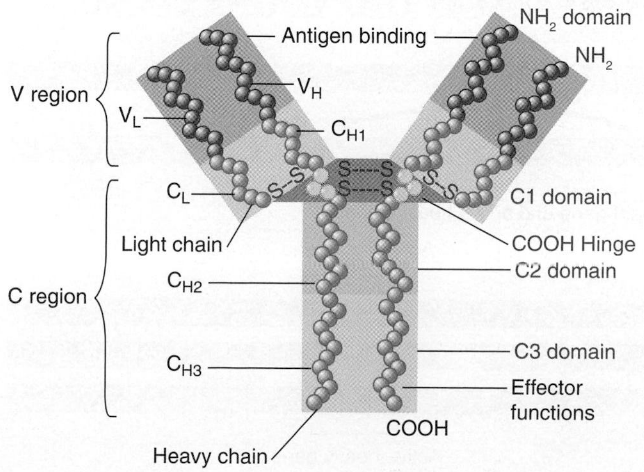

FIGURE 18.4 An antibody (immunoglobulin, or Ig) molecule consists of two identical heavy chains and two identical light chains. The IgG shown here contains an N-terminal variable (V) region and a C-terminal constant (C) region.

- **light chain** The smaller of the two types of subunits that make up an antibody tetramer. Each antibody contains two light chains. The N-terminus of the light chain forms part of the antigen recognition site.
- **heavy chain** The larger of the two types of subunits that make up an antibody tetramer. Each antibody contains two heavy chains. The N-terminus of the heavy chain forms part of the antigen recognition site, whereas the C-terminus determines the antibody subclass.
- **variable region (V region)** An antigen-binding site of an immunoglobulin or T cell receptor molecule. They are composed of the variable domains of the component chains. They are coded by V gene segments and vary extensively among antigen receptors as the result of multiple, different genomic copies and of changes introduced during synthesis.
- **constant region (C region)** The part of an immunoglobulin, or T cell receptor, that varies least in amino acid sequence between different molecules. They are coded by C gene segments. The heavy chain regions identify the type of immunoglobulin and recruits effector functions.

bloodstream—for example, a pathogenic bacterium—has a surface that presents antigens. The antigen(s) are recognized by the BCR expressed on the surface of the B cell. This leads to B cell activation and proliferation, resulting in the production and secretion of large amounts of antibodies specific for the same antigen. (The structure and specificity of the antibody produced by a given B cell are identical to the BCR of that same B cell.) Binding of secreted antibodies to antigen forms an antigen-antibody complex. This complex then recruits other components of the immune system.

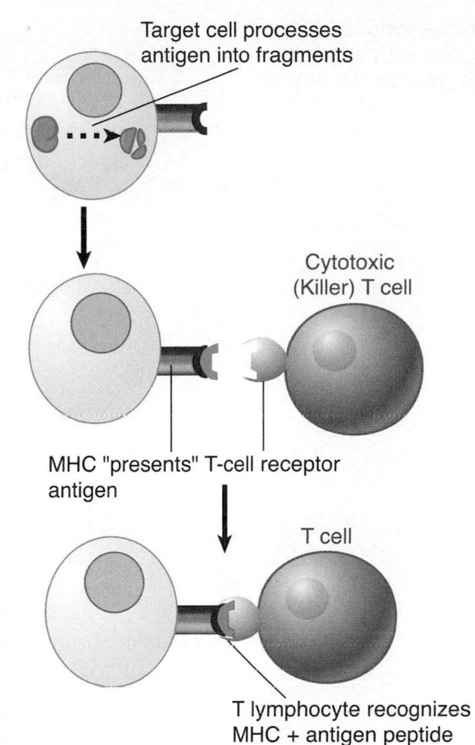

FIGURE 18.2 In cell-mediated immunity, cytotoxic T cells use the T cell receptor (TCR) to recognize a peptide fragment of the antigen that is presented on the surface of the target cell by the MHC protein.

The humoral response depends on these other components in two ways. First, B cells need signals provided by T cells to enable them to secrete antibodies. These T cells are called **helper T (T_h) cells**, because they assist the B cells. Second, antigen-antibody formation is a trigger for the antigen to be destroyed. The major pathway is provided by the action of **complement**, a multiprotein/enzymatic cascade whose name reflects its ability to "complement," or complete, the action of the antibody itself. Complement consists of a set of ~20 proteins that function through a cascade of proteolytic actions. If the target antigen is part of a cell—for example, an infecting bacterium—the action of complement culminates in lysing the target cell. The action of complement also provides a means of attracting macrophages, which scavenge the target cells or their products. Alternatively, the antigen-antibody complex may be taken up directly by macrophages (scavenger cells) and destroyed.

The **cell-mediated response** is executed by a class of T lymphocytes called **cytotoxic T cells** (CTLs; also called *killer T cells*). The basic function of the T cell in recognizing a target antigen is indicated in **FIGURE 18.2**. A cell-mediated response typically is elicited by an intracellular parasite, such as a virus that infects the body's own cells. As a result of the viral infection, fragments of foreign (viral) antigens are displayed on the surface of the cell. These fragments are recognized by the TCR expressed on the surface of T cells.

A crucial feature of this recognition reaction is that the antigen must be presented by a cellular protein that is a member of the **major histocompatibility complex (MHC)**. The MHC protein has a groove on its surface that binds a peptide fragment derived from the foreign antigen. The TCR recognizes the combination of peptide fragment and MHC protein. The requirement that T lymphocytes recognize (foreign) antigen in the context of (self) MHC protein ensures that the cell-mediated response acts only on host cells that have been infected with a foreign antigen.

The purpose of each type of immune response is to attack a foreign target. Target recognition is the prerogative of B cell immunoglobulins and T cell receptors. A crucial aspect of their function lies in the ability to distinguish "self" from "nonself." Proteins and cells of the body itself must never be attacked. Foreign targets must be destroyed entirely. The property of failing to attack "self" is called *tolerance*. Loss of this ability results in an **autoimmune disease**, in which a body's immune system attacks itself, often with disastrous consequences.

When an antigen is recognized by an antibody or TCR, the recognition triggers a signal in the B or T lymphocyte that causes it to divide. Numerous cell divisions lead

- **helper T (T_h) cell** A T lymphocyte that activates macrophages and stimulates B cell proliferation and antibody production.
- **complement** A set of ~20 proteins that function through a cascade of proteolytic actions to lyse infected target cells or to attract macrophages.
- **cell-mediated response** The immune response that is mediated primarily by T lymphocytes. It is defined based on immunity that cannot be transferred from one organism to another by serum antibody.
- **cytotoxic T cell** A T lymphocyte that can be stimulated to kill cells containing intracellular pathogens, such as viruses.
- **major histocompatibility complex (MHC)** A chromosomal region containing genes that are involved in the immune response. The genes encode proteins for antigen presentation, cytokines, and complement, as well as other functions. It is highly polymorphic. Its genes and proteins are divided into three classes.
- **autoimmune disease** A pathological condition in which the immune response is directed against self-antigens.

FIGURE 18.3 The B cell and T cell repertoires include BCRs and TCRs with a variety of specificities. Reaction with an antigen leads to clonal expansion of the lymphocyte with the BCR or TCR that can recognize the antigen.

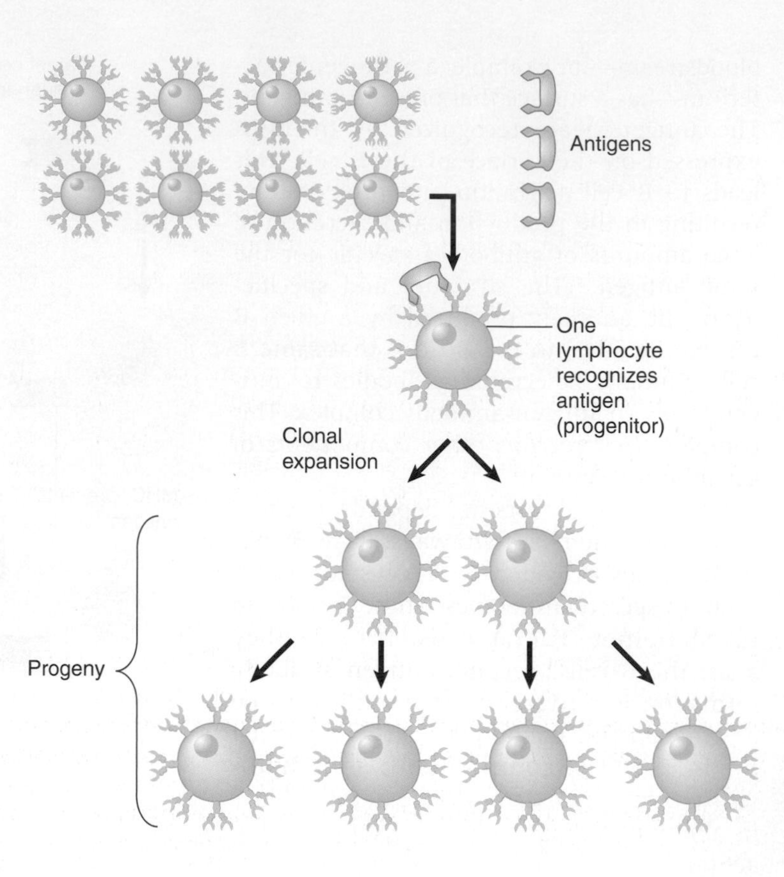

▸ **clonal expansion** The production of numerous daughter cells all arising from a single cell.

to a large population of B or T cells, all producing the same antibody or TCR. This **clonal expansion** provides the organism with an army of cells equipped to fight the infection, and confers immunity to future exposure to the same antigen, as shown in FIGURE 18.3.

Each of the three groups of proteins required for the immune response—immunoglobulins, T cell receptors, and MHC proteins—is diverse. Examining a large number of individuals, we find many variants of each protein. Each protein is coded by a large family of genes; in the case of antibodies and the T cell receptors, the diversity of the population is increased by DNA rearrangements that occur in the relevant lymphocytes.

Immunoglobulins and T cell receptors are direct counterparts, each produced by its own type of lymphocyte. The proteins are related in structure, and their genes are related in organization. The sources of variability are similar. The MHC proteins also share some common features with the antibodies, as do other lymphocyte-specific proteins. This indicates that the genetic organization of the immune system entails a series of related gene families, all members of a **superfamily** that may have evolved from some common ancestor representing a primitive immune response.

▸ **superfamily** A set of genes all related by presumed descent from a common ancestor, but now showing considerable variation.

▸ **innate immunity** An immune response that depends on recognition of predefined patterns in pathogens.

In contrast to the adaptive immunity we have been discussing, there is another important arm of the immune system that provides **innate immunity**. The innate immune response provides a first line of defense against microbial pathogens, producing an immediate defense without the latency period of the adaptive immune response. Innate immunity is based on recognition of a set of common patterns or motifs that are conserved in microorganisms, but are not found in multicellular eukaryotes. This allows the immune system to quickly and with high probability distinguish dangerous nonself patterns from self-patterns. In this chapter, we focus specifically on the mechanisms of generating antibody and TCR diversity in the cells of the adaptive immune response, and will not further discuss innate immunity.

out that the number of coding sequences for each type of region reflects its variability. There are many genes coding for V regions, but only a few genes coding for C regions.

▸ **V gene** A sequence coding for the major part of the variable (N-terminal) region of an immunoglobulin chain.

▸ **C genes** Genes that code for the constant regions of immunoglobulin protein chains.

In this context, "gene" means a sequence of DNA coding for a discrete part of the final immunoglobulin polypeptide (heavy or light chain). Thus **V genes** code for variable regions and **C genes** code for constant regions, but *neither type of gene is expressed as an independent unit.* To construct a unit that can be expressed in the form of an authentic light or heavy chain, a V gene must be joined physically to a C gene. In this system, two "genes" code for one polypeptide. To avoid confusion, we will refer to these units as "gene segments" rather than "genes."

▸ **somatic recombination** The process of joining a V gene to a C gene in a lymphocyte to generate an immunoglobulin or T cell receptor.

▸ **somatic hypermutation (SHM)** The introduction of somatic mutations in a rearranged immunoglobulin gene. The mutations can change the sequence of the corresponding antibody, especially in its antigen-binding site.

The sequences coding for light chains and heavy chains are assembled in the same way: any one of many V gene segments may be joined to any one of a few C gene segments. This **somatic recombination** occurs *in the B lymphocyte in which the antibody is expressed.* The large number of available V gene segments is responsible for a major part of the diversity of immunoglobulins. Not all diversity is encoded in the genome, though; some is generated by changes that occur during the process of constructing a functional gene, and more is generated by a process of **somatic hypermutation (SHM)**.

Essentially the same description applies to the formation of functional genes coding for the protein chains of the T cell receptor. Two types of receptor are found on T cells—one consisting of two types of chain called α and β and the other consisting of γ and δ chains. Like the genes coding for immunoglobulins, the genes coding for the individual chains in T cell receptors consist of separate parts, including V and C regions, that are brought together in an active T cell (see *Section 18.10, T Cell Receptors Are Related to Immunoglobulins*).

The crucial fact about the synthesis of immunoglobulins, therefore, is that the arrangement of V gene segments and C gene segments is different in the cells producing the immunoglobulins (or T cell receptors) from all other somatic cells or germ cells. The entire process occurs in somatic cells and does not affect the germline; thus the response to an antigen is not inherited by progeny of the organism.

FIGURE 18.5 summarizes the overall process of creating a functional immunoglobulin. There are two families of immunoglobulin light chains, κ and λ, and one family

FIGURE 18.5 The germline genome has three separate clusters of V gene segments separated from C gene segments. Recombination can occur in each of the clusters to create an active gene by linking a V segment to a C segment. To produce a functional immunoglobulin, a lymphocyte must successfully recombine the heavy cluster and one of the light clusters.

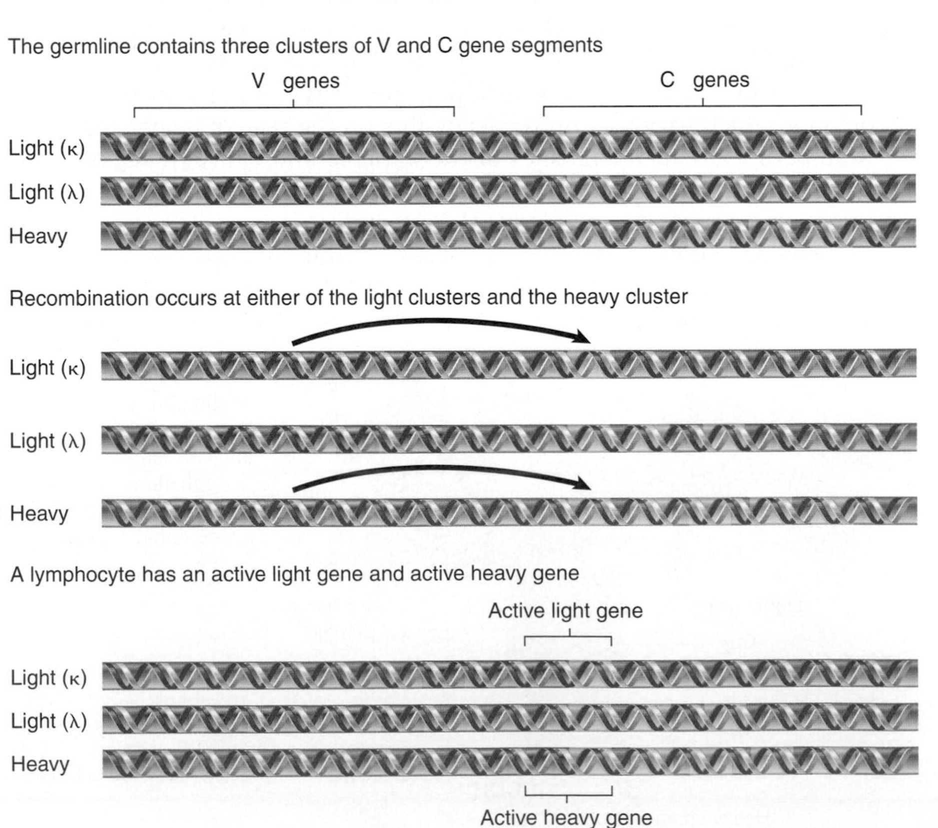

containing all the types of heavy chain (H). Each family resides on a different chromosome and consists of its own set of both V gene segments and C gene segments. This is called the *germline pattern* and is found in the germline and in somatic cells of all lineages other than the immune system.

In a cell expressing an antibody, though, each of its chains—one light type (either κ or λ) and one heavy type—is encoded by a single intact gene. The recombination event that brings a V gene segment to partner with a C gene segment creates an active gene consisting of exons that correspond precisely with the functional domains of the protein. The introns are removed in the usual way by RNA splicing.

The principles by which functional genes are assembled are the same in each family, but there are differences in the details of the organization of the V gene segments and C gene segments and, correspondingly, of the recombination reaction between them. In addition to the V gene segments and C gene segments, other short DNA sequences (including J segments and D segments) are included in the functional somatic loci.

KEY CONCEPTS

- An immunoglobulin is a tetramer of two light chains and two heavy chains.
- Light chains fall into the λ and κ families; heavy chains form a single family.
- Each chain has an N-terminal variable region (V) and a C-terminal constant region (C).
- The V domain recognizes the antigen and the C domain provides the effector response.
- V domains and C domains are separately coded by V gene segments and C gene segments, respectively.
- A gene coding for an intact immunoglobulin chain is generated by somatic recombination to join a V gene segment with a C gene segment.

CONCEPT AND REASONING CHECK

Why are there many more V gene segments than C gene segments?

18.3 Light Chains Are Assembled by a Single Recombination Event

A λ light chain is assembled from two segments, as illustrated in **FIGURE 18.6**. The V_λ gene segment consists of the leader exon (L) separated by a single intron from the variable (V) segment. The $J_\lambda C_\lambda$ gene segment consists of the J_λ segment separated by a single intron from the C_λ exon.

J is an abbreviation for joining, because the **J segment** identifies the region to which the V segment becomes connected. Thus the joining reaction does not directly involve V_λ and C_λ gene segments but occurs via the J_λ segment; by the joining of "V and C gene segments" for light chains, we really mean V_λ-$J_\lambda C_\lambda$ joining.

▸ **J segments** Coding sequences in the immunoglobulin and T cell receptor loci. They are between the variable (V) and constant (C) gene segments.

The J_λ segment is short and codes for the last few amino acids of the variable region. In the whole gene generated by recombination, the V_λ-J_λ segment constitutes a single exon coding for the entire variable region.

A κ chain is also assembled from two DNA segments, as illustrated in **FIGURE 18.7**. There are, however, differences in the organization of the C_κ locus compared to the C_λ locus. A group of five J_κ segments is spread over a region of 500 to 700 bp, separated from the C_κ exon by an intron of 2–3 kb. A V_κ segment (which contains a leader exon like V_λ) may be joined to any one of the J_κ segments.

Whichever J_κ segment is used becomes the terminal part of the intact variable exon. Any J_κ segments upstream of the recombining J_κ segment are lost (J1 has been lost in the figure). Any J segment downstream of the recombining J_κ segment is treated as part of the intron between the variable and constant exons (J3, J4, and J5 are included in the intron that is spliced out in the figure).

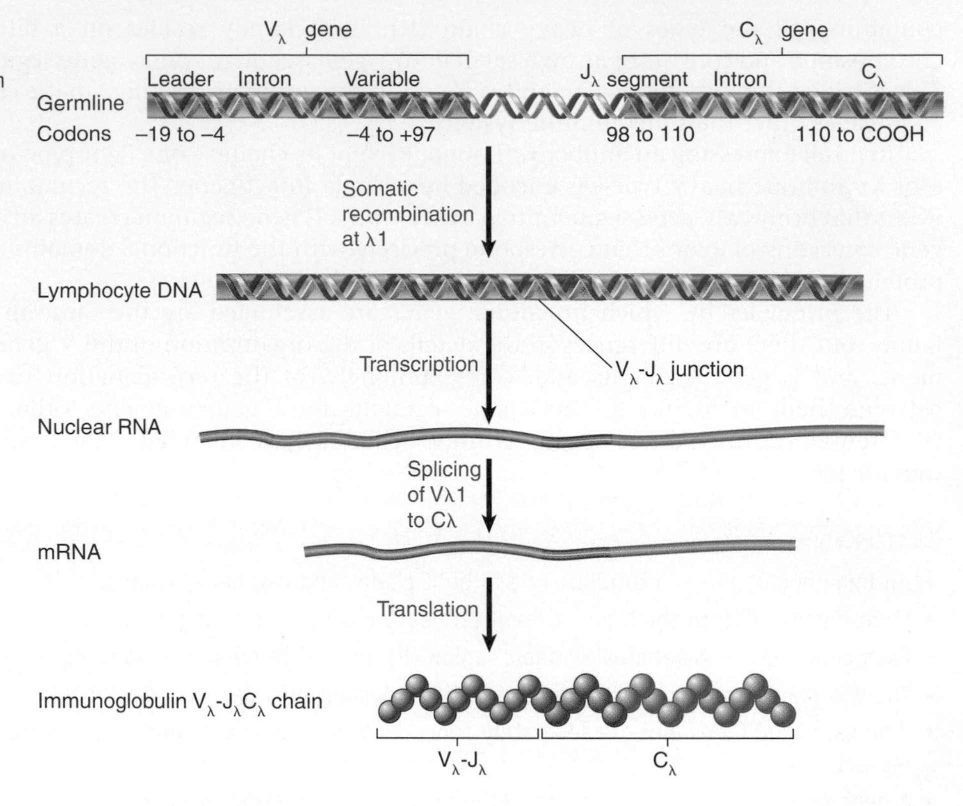

FIGURE 18.6 The lambda C gene segment is preceded by a J segment, so V-J recombination generates a functional lambda light-chain gene. Only one V gene segment is shown for simplicity.

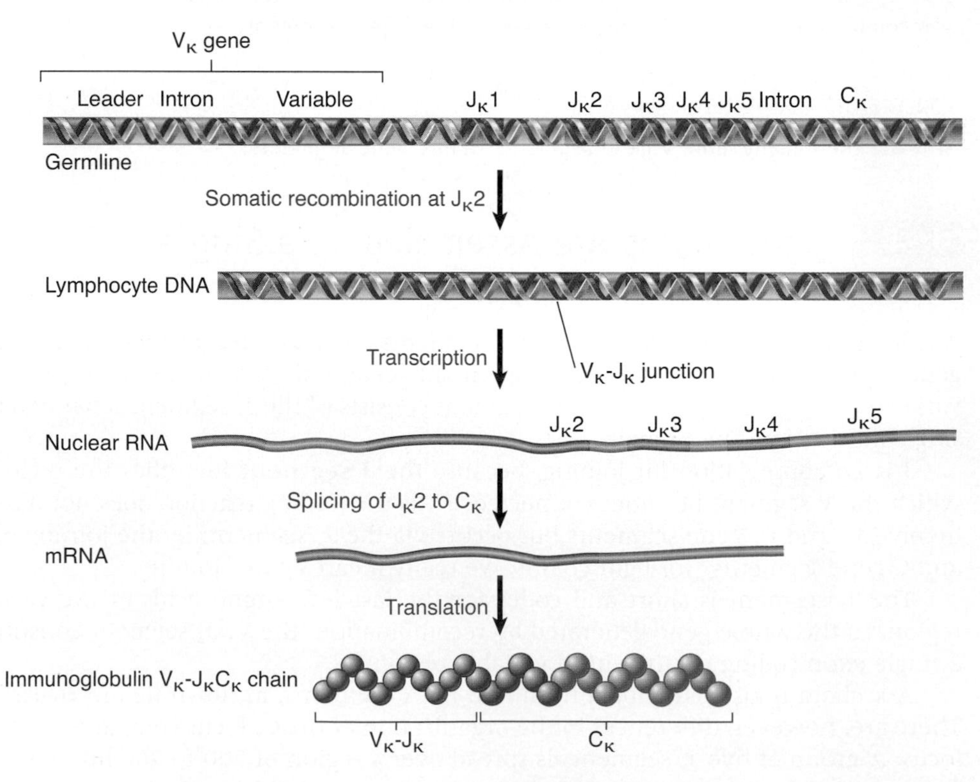

FIGURE 18.7 The kappa C gene segment is preceded by multiple J segments in the germline. V-J joining may recognize any one of the J segments, which is then spliced to the C gene segment during RNA processing. Only one V gene segment is shown for simplicity.

All functional J segments possess a signal at the left boundary that makes it possible to recombine with the V segment; they also possess a signal at the right boundary that can be used for splicing to the C exon. Whichever J segment is recognized in DNA V-J joining uses its splicing signal in RNA processing.

The human λ locus has ~300 V_λ gene segments, along with ~6 C_λ gene segments, each preceded by its own J_λ segment, as shown in **FIGURE 18.8**. The large number of

FIGURE 18.8 The lambda family consists of V gene segments linked to a small number of J-C gene segments.

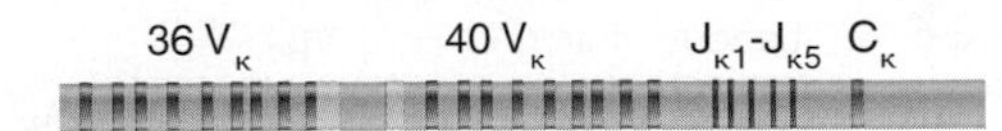

FIGURE 18.9 The human and mouse kappa families consist of V_κ gene segments linked to five functional J_κ segments connected to a single C_κ gene segment.

human V_λ gene segments are likely to include some pseudogenes, and it is not known precisely how many of the sequences are functional. The λ locus in mice is much less diverse than the λ locus in humans. The main difference is that in a mouse there are only two V_λ gene segments; each is linked to two J-C regions. Of the four C_λ gene segments, one is inactive. At some time in the past, the mouse suffered a dramatic deletion of most of its germline V_λ gene segments.

FIGURE 18.9 shows that both the human and mouse κ loci have only one C_κ gene segment, preceded by six J_κ segments (one of them inactive). The V_κ gene segments occupy a large cluster on the chromosome, upstream of the constant region. The human cluster has two regions. Just preceding the C_κ gene segment, a region of 600 kb contains the five J_κ segments and 40 V_κ gene segments. A gap of 800 kb separates this region from another group of 36 V_κ gene segments.

The V_κ gene segments can be subdivided into families, which are defined by the criterion that members of a family have >80% amino acid identity. The mouse family is unusually large (~1000 genes), and there are ~18 V_κ families that vary in size from 2 to 100 members. Like other families of related genes, related V gene segments form subclusters, which are generated by duplication and divergence of individual ancestral members. Many of the V segments are inactive pseudogenes, though, and <50 are likely to be used to generate immunoglobulins.

A given lymphocyte generates *either* a κ *or* a λ light chain to associate with the heavy chain. In humans, ~60% of the light chains are κ and ~40% are λ. In mouse, 95% of B cells express the κ type of light chain, presumably because of the reduced number of λ gene segments.

KEY CONCEPTS

- A λ light chain is assembled by a single recombination between a V_λ gene and a J_λ-C_λ gene segment.
- The V_λ gene segment has a leader exon, intron, and variable-coding region.
- The J_λ-C_λ gene segment has a short J_λ-coding exon, intron, and C_λ-coding region.
- A κ light chain is assembled by a single recombination between a V_κ gene segment and one of five J_κ segments preceding the C_κ gene.
- A light-chain locus can produce >1000 chains by combining 300 V gene segments with 4 to 5 C gene segments.

CONCEPT AND REASONING CHECK

The J_λ segment is part of the C_λ gene segments but also is part of the variable region in the immunoglobulin. Explain.

18.4 Heavy Chains Are Assembled by Two Sequential Recombination Events

The assembly of a complete H chain involves an additional segment. The **D segment** (for diversity) provides an extra two to thirteen amino acids between the sequences encoded by the V_H segment and the J_H segment. A large array of D segments lies on the chromosome between the V_H segments and the four J_H segments.

▸ **D segment** An additional sequence that is found between the V and J regions of an immunoglobulin heavy chain.

FIGURE 18.10 Heavy genes are assembled by sequential joining reactions. First a D segment is joined to a J_H segment, and then a V_H gene segment is joined to the D segment.

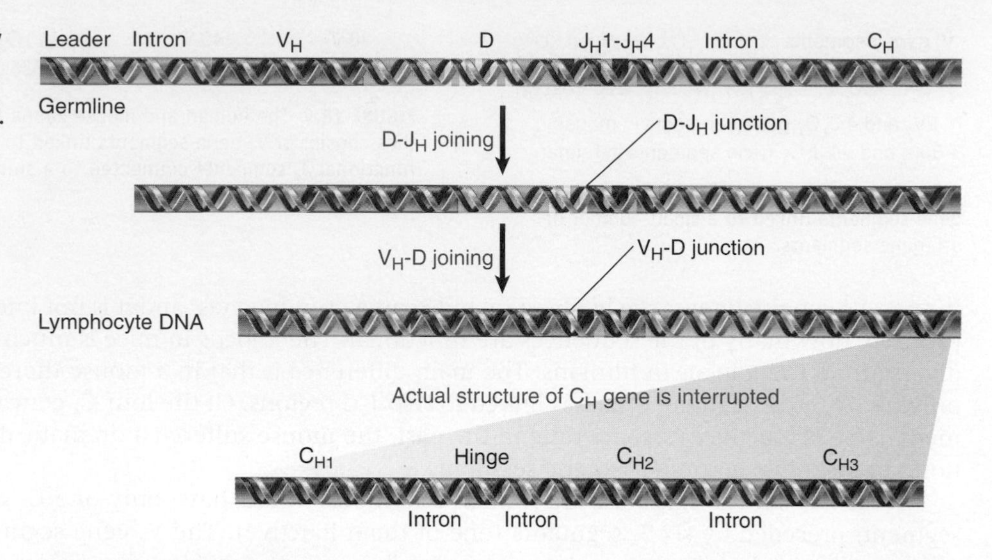

V_H-D-J_H joining takes place in two stages, as illustrated in **FIGURE 18.10**. First one of the D segments recombines with a J_H segment; a V_H segment then recombines with the already combined D-J_H segment. The resulting V_H-D-J_H DNA sequence is then expressed with the nearest downstream C_H gene (initially this is always C_μ), which consists of a cluster of four exons. (The use of different C_H genes is discussed in *Section 18.7, Class Switching Is Effected by DNA Recombination*.)

The D segments are organized in a tandem array. The human heavy-chain locus contains ~30 D segments, followed by a cluster of six J_H gene segments. Unknown mechanisms ensure that the *same* D segment is involved in the D-J_H recombination and V_H-D recombination.

The structure of recombined V(D)J segments is similar in organization in the H chain and λ and κ chain loci. The leader exon codes for a signal sequence involved in membrane attachment, and the second exon codes for the major part of the variable region itself (~100 codons long). The remainder of the variable region is provided by the D segment (in the H chain locus only) and by a J segment (in all three loci).

The structure of the C region is different in different H and L chains. For both κ and λ light chains, the C region is encoded by a single exon (which becomes the third exon of the reconstructed, active gene). For H chains, the constant region is encoded by several exons corresponding to four regions as shown in Figure 18.10: C_{H1}, hinge, C_{H2}, and C_{H3}. Each C_H exon is ~100 codons long; the hinge is shorter. The introns usually are relatively small (~300 bp).

The single locus for heavy-chain production in humans consists of several discrete sections, as summarized in **FIGURE 18.11**. The mouse shows similar organization but different numbers of segments: there are more V_H gene segments, fewer D and J segments, and a slight difference in the number and organization of C gene segments. The 3′-most member of the V_H cluster is separated by only 20 kb from the first D segment. The D segments are spread over ~50 kb, followed by the cluster of J_H segments. Over the next 220 kb lie all the C_H gene segments. There are nine functional C_H gene segments and two pseudogenes. The organization suggests that a γ gene segment must

FIGURE 18.11 A single gene cluster in humans contains all the information for heavy-chain gene assembly.

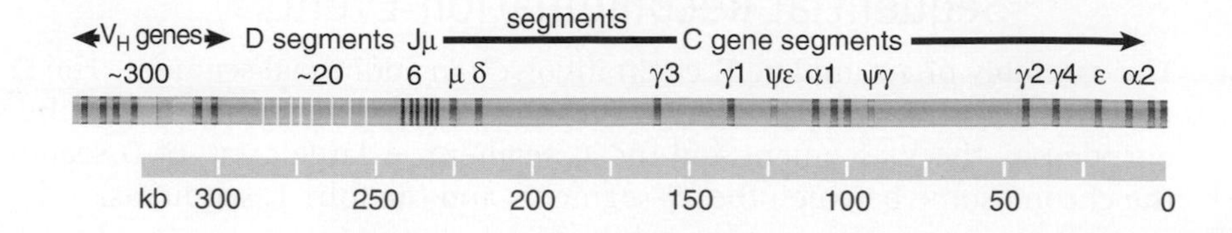

have been duplicated to give the subcluster of γ-γ-ε-α, after which the entire group was then tandemly duplicated. By combining any one of ~51 functional V_H gene segments, ~30 D segments, and 6 J_H segments, the IgH locus alone can potentially produce more than 10^4 variable regions to accompany any C_H gene segment.

KEY CONCEPTS

- The units for H chain recombination are a V_H gene, a D segment, and a J_H-C_H gene segment.
- The first recombination joins D to J_H-C_H.
- The second recombination joins V to D-J_H-C_H.
- The C_H segment consists of four exons.

CONCEPT AND REASONING CHECK

Based on the lymphocyte DNA structure shown at the bottom of Figure 18.10, what would be the composition of the resulting mRNA?

18.5 Immune Recombination Uses Two Types of Consensus Sequence

The recombination of Igλ, Igκ, and IgH chain genes involves the same mechanism, although the number and nature of recombining elements are different. The same consensus sequences are found at the boundaries of all germline segments that participate in joining reactions. Each consensus sequence (often called a *recombination signal sequence*) consists of a *heptamer* (7 bp element) separated by either 12 or 23 bp from a *nonamer* (9 bp element). These sequences are referred to as **recombination signal sequences (RSS)**.

recombination signal sequences (RSS) The combinations of heptamer-spacer-nonamer consensus sequence that are required for accurate Ig locus recombination.

FIGURE 18.12 illustrates the relationship between the consensus sequences at Ig loci. At the κ locus, each V_κ gene segment is followed by a consensus sequence with a 12 bp spacer. Each J_κ segment is preceded by a consensus sequence with a 23 bp spacer. The V_κ and J_κ consensus sequences are inverted in orientation. At the λ locus, each V_λ gene segment is followed by a consensus sequence with a 23 bp spacer; each J_λ gene segment is preceded by a consensus of the 12 bp spacer type.

The orientation of the heptamer and nonamer sequences indicates the positions of the coding regions of the gene segments. The heptamer end of the signal sequence is adjacent to the coding region of a given gene segment, whereas the nonamer end is oriented toward the intervening sequences between gene segments.

The rule that governs the joining reaction is that *a consensus sequence with one type of spacer can be joined only to a consensus sequence with the other type of spacer.* The

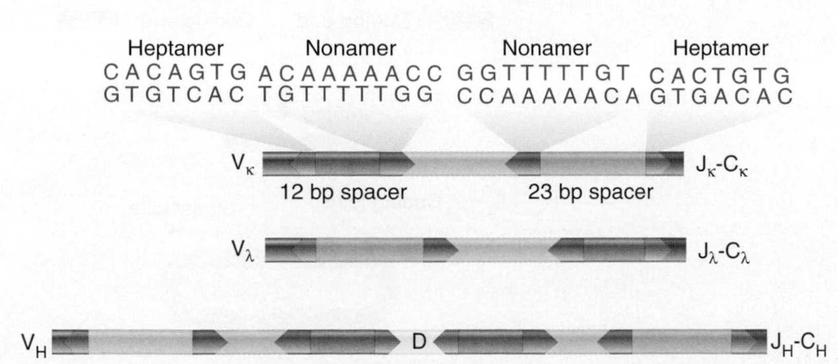

FIGURE 18.12 RSS sequences are present in inverted orientation at each pair of recombining sites. One member of each pair has a 12 bp spacer between its components; the other has a 23 bp spacer.

consensus sequences at V and J segments can lie in either order; thus (unlike the heptamer and nonamer sequences) the different spacers do not impart any directional information. Instead, the spacers serve to prevent one V or J gene segment from recombining with another of the same type of gene segment.

This concept is borne out by the structure of the components of the heavy gene segments, shown at the bottom of Figure 18.12. Each V_H gene segment is followed by a consensus sequence of the 23 bp spacer type. The D segments are flanked on either side by consensus sequences of the 12 bp spacer type. The J_H segments are preceded by consensus sequences of the 23 bp spacer type. Thus the V_H gene segment must be joined to a D segment, and the D segment must be joined to a J_H segment. A V_H gene segment cannot be joined directly to a J_H segment, because both possess the same type of consensus sequence.

- **signal end** The free DNA end of excised sequences produced during recombination of immunoglobulin and T cell receptor genes. Signal ends lie at the termini of the cleaved fragment containing the recombination signal sequences. Their subsequent joining yields a signal joint.
- **coding end** The free DNA end of coding sequences produced during recombination of immunoglobulin and T cell receptor genes. Coding ends are at the termini of the cleaved V and (D)J coding regions. Their subsequent joining yields a coding joint.
- **coding joint** The DNA junction created by the joining of two coding ends during V(D)J recombination.

The spacing between the components of the consensus sequences corresponds to close to one (12 bp) or two (23 bp) turns of the double helix. This may reflect geometric constraints in the recombination reaction. The recombination protein(s) may approach the DNA from one side, in the same way that RNA polymerase and repressors approach recognition elements such as promoters and operators.

Recombination of the components of Ig genes is accomplished by a physical rearrangement of sequences, involving DNA breakage and reunion. The general nature of the reaction for a λ light chain is illustrated in **FIGURE 18.13**. (The reaction is similar at a heavy-chain locus, with the exception that there are two recombination events: first D-J, then V-DJ.)

Breakage and reunion occur as separate reactions. A double-strand break is made at the heptamers that lie at the ends of the coding units. This releases the entire fragment between the V gene segment and J-C gene segment; the cleaved termini of this fragment are called **signal ends**. The cleaved termini of the V and J-C loci are called **coding ends**. The two coding ends are covalently linked to form a **coding joint**; this is the connection that links the V and J segments. The two signal ends are also religated such that the excised fragment forms a circular molecule.

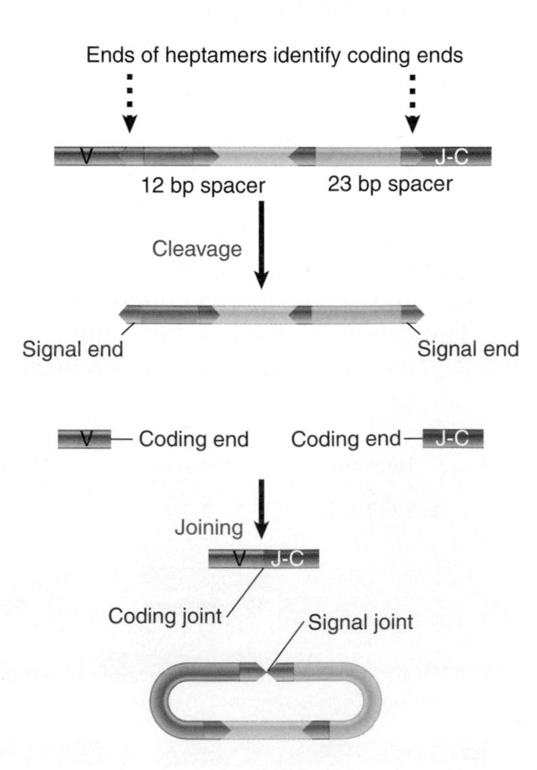

FIGURE 18.13 Breakage and reunion at RSS generates VJC sequences and releases the intervening sequence.

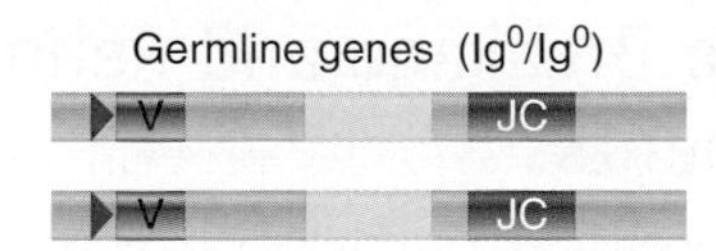

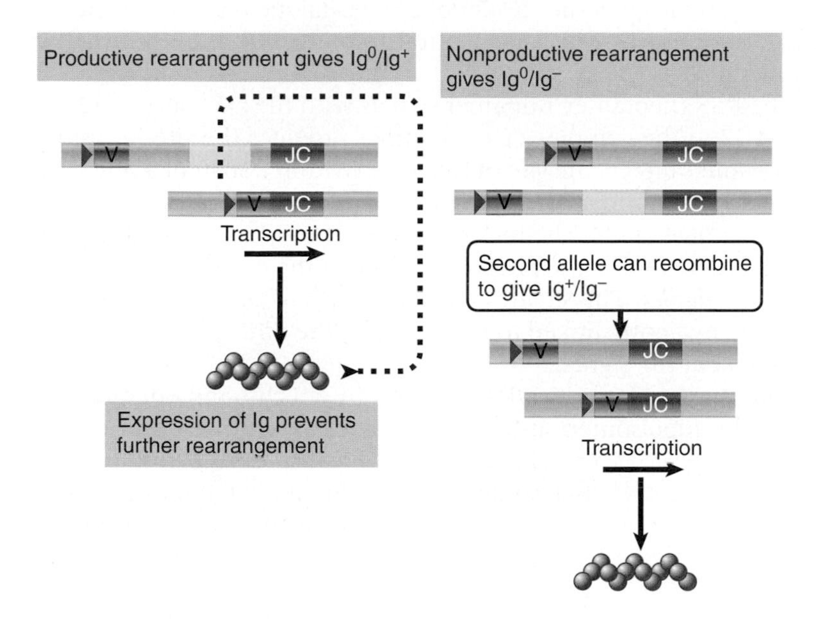

FIGURE 18.14 A successful rearrangement to produce an active light (depicted) or heavy chain suppresses further rearrangements of the same type, resulting in allelic exclusion. The pink box indicates intervening sequences (e.g., V gene segments) that are present in the absence of recombination.

Each B cell expresses a single type of light chain and a single type of heavy chain because only a single **productive rearrangement** of each type occurs in a given lymphocyte in order to produce one light- and one heavy-chain gene. Each event involves the genes of only *one* of the homologous chromosomes, and as a result *the alleles on the other chromosome are not expressed in the same cell* (although they may be nonproductively rearranged). This phenomenon is called **allelic exclusion**, illustrated in **FIGURE 18.14**. The cell will continue to recombine V gene segments and C gene segments until a productive rearrangement is achieved. Allelic exclusion is caused by the suppression of further rearrangement as soon as an active chain is produced.

- **productive rearrangement** Occurs as a result of the recombination of V, (D), and J gene segments if all the rearranged gene segments are in the correct reading frame.
- **allelic exclusion** The expression in any particular lymphocyte of only one allele coding for the expressed immunoglobulin. This is caused by feedback from the first immunoglobulin allele to be expressed that prevents activation of a copy on the other chromosome.

KEY CONCEPTS

- The consensus sequence used for recombination is a heptamer separated by either 12 or 23 base pairs from a nonamer.
- Recombination occurs between two consensus sequences that have different spacers.
- Recombination initiates with double-strand breaks at the heptamers of two consensus sequences.
- The signal ends of the fragment between the breaks usually join to generate an excised circular fragment.
- The coding ends are covalently linked to join V to J-C (L chain), or D to J-C, and V to D-J-C (H chain).
- A productive rearrangement (resulting in an active immunoglobulin protein) prevents any further rearrangement from occurring.

CONCEPT AND REASONING CHECK

What are the different functions of the heptamer/nonamer versus the 12 and 23 bp spacers in the recombination signal sequence?

18.6 RAG1/RAG2 Catalyze Breakage and Religation of V(D)J Gene Segments

The proteins RAG1 and RAG2 are necessary and sufficient to cleave DNA for V(D)J recombination. The RAG proteins together undertake the catalytic reactions of cleaving and rejoining DNA, and also provide a structural framework within which the reactions occur, as shown in **FIGURE 18.15**.

RAG1 recognizes the RSS (heptamer/nonamer signals with the appropriate 12/23 spacers) and recruits RAG2 to the complex. The nonamer provides the site for initial recognition, and the heptamer directs the site of cleavage (within a span of a few base pairs). The complex nicks one strand at each junction. The reactions of Figure 18.15 are shown in more detail in **FIGURE 18.16**. The nick created by RAG1 has 3′–OH and 5′–P ends. The free 3′–OH end then attacks the phosphate bond at the corresponding position *in the other strand of the duplex*. This creates a *hairpin* at the coding end, in which the 3′ end of one strand is covalently linked to the 5′ end of the other strand; it leaves a blunt double-strand break at the signal end.

This second cleavage is a transesterification reaction in which bond energies are conserved. It resembles the topoisomerase-like reactions catalyzed by the resolvase proteins of bacterial transposons. This suggests that somatic recombination of immune genes may have evolved from an ancestral transposon.

The hairpins at the coding ends provide the substrate for the next stage of reaction. If a single-strand break is introduced into one strand close to the hairpin, an unpairing reaction at the end generates a single-stranded overhang. Synthesis of a complement to the exposed single strand then converts the coding end to an extended duplex. This reaction explains the introduction of **P nucleotides** at coding ends; they consist of a few extra base pairs related to, but reversed in orientation from, the original coding end.

Some extra bases also may be inserted, apparently with random sequences, between the coding ends. They are called **N nucleotides**. Their insertion occurs via the activity of the enzyme deoxynucleoside transferase, active in lymphocytes, at a free 3′ coding end generated during the joining process.

Changes in sequence during recombination are, therefore, a consequence of the enzymatic mechanisms involved in breaking and rejoining the DNA. In heavy-chain recombination, base pairs are lost or inserted at the V_H-D or D-J_H junctions or both. Deletion also occurs in V_λ-J_λ joining, but insertion at these joints is unusual. The changes in sequence affect the amino acid encoded at V_H-D junctions and D-J_H junctions in heavy chains or at the V_L-J_L junction in light chains.

These various mechanisms together ensure that a coding joint may have a sequence that is different from what would be predicted by a direct joining of the coding ends of the V, D, and J regions.

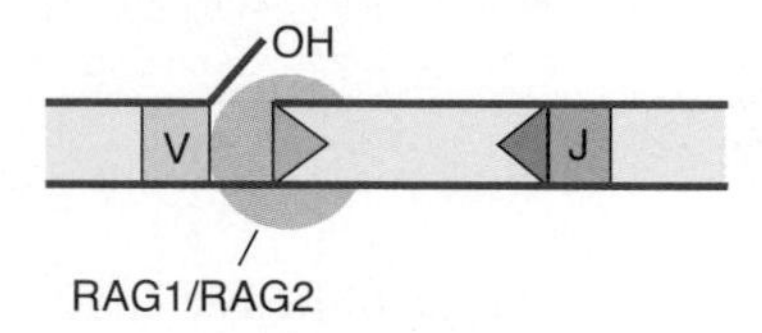

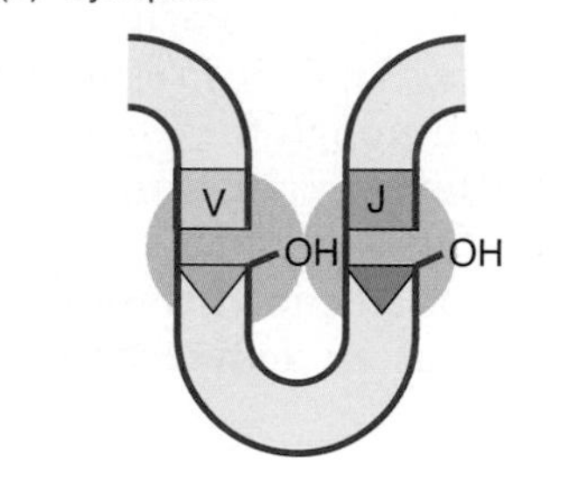

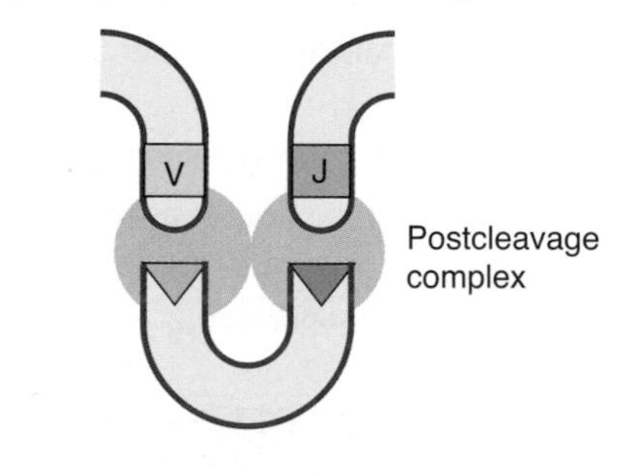

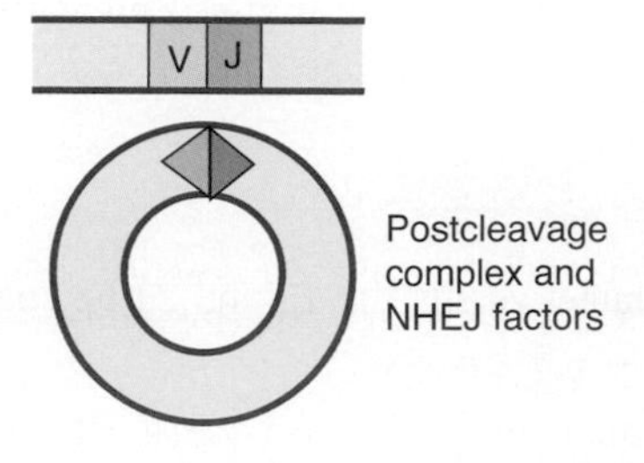

FIGURE 18.15 RAG1/RAG2 proteins are responsible for breakage and recombination at RSSs. A generic V-J rearrangement is shown for simplicity.

- **P nucleotide** A short palindromic (inverted repeat) sequence that is generated during rearrangement of immunoglobulin and T cell receptor V, (D), and J gene segments. They are generated at coding joints when proteins cleave the hairpin ends generated during rearrangement.
- **N nucleotide** A short nontemplated sequence that is added randomly by the enzyme at coding joints during rearrangement of immunoglobulin and T cell receptor genes. They augment the diversity of antigen receptors.

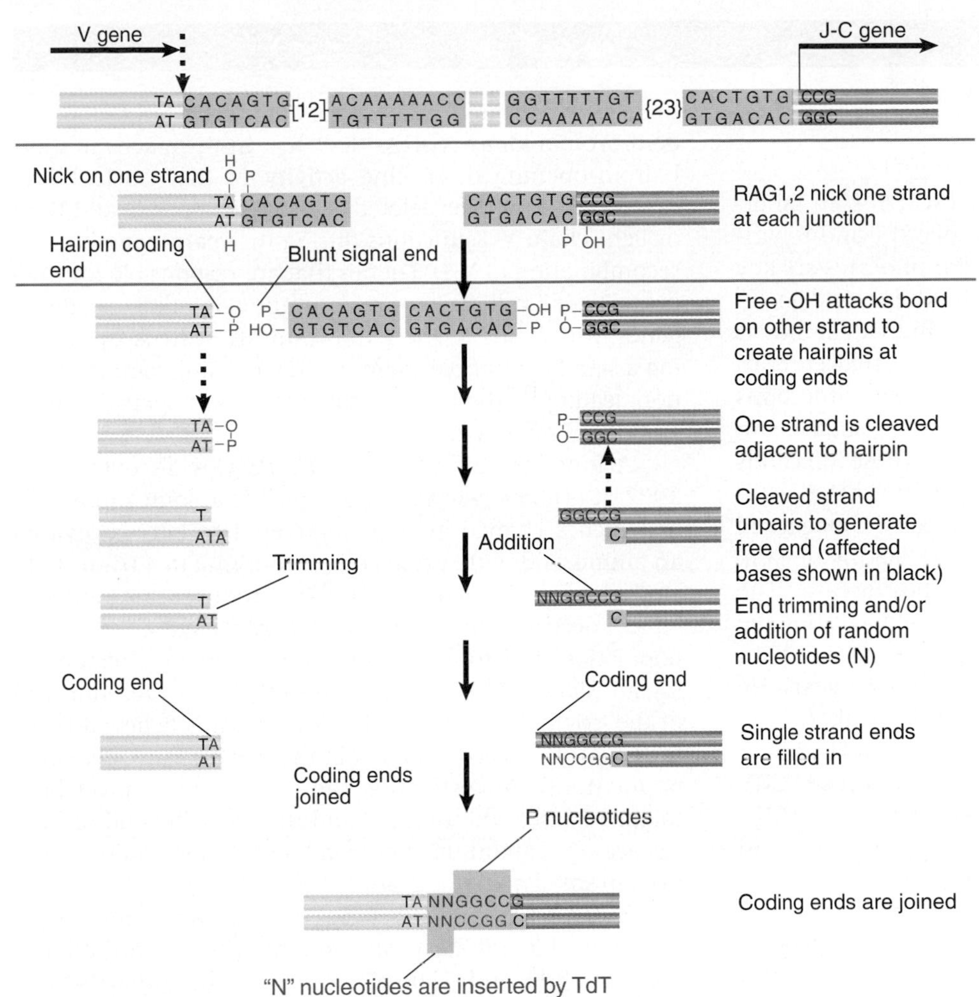

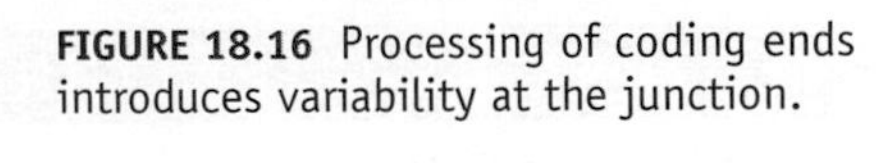

FIGURE 18.16 Processing of coding ends introduces variability at the junction.

Changes in the sequence at the junction make it possible for a great variety of amino acids to be encoded at this site. It is interesting that the amino acid at position 96 is created by the V-J joining reaction. It forms part of the antigen-binding site and also is involved in making contacts between the light chains and the heavy chains. Thus the maximum diversity is generated at the site that contacts the target antigen.

Changes in the number of base pairs at the coding joint affect the reading frame. The joining process appears to be random with regard to reading frame, so that probably only one-third of the joined sequences retain the proper frame of reading through the junctions. If the V-J region is joined so that the J segment is out of frame, translation is usually terminated prematurely by a nonsense codon in the incorrect frame. We may think of the formation of aberrant genes as the price the cell must pay for the increased diversity that it gains by being able to adjust the sequence at the joining site.

Similar—although even greater—diversity is generated in the joining reactions that involve the D segment of the heavy chain. The same result is seen with regard to reading frame; nonproductive genes are generated by joining events that place J and C out of frame with the preceding V gene segment.

The joining reaction that works on the coding end uses the same pathway of nonhomologous end joining (NHEJ) that repairs double-strand breaks in cells (see *Section 16.8, Nonhomologous End Joining Also Repairs Double-Strand Breaks*). Just as in repair, DNA-dependent protein kinase (DNA-PK) is recruited to the DNA by the Ku70 and Ku80 proteins, which bind to the DNA ends. DNA-PK phosphorylates and thereby activates the protein Artemis, which nicks the hairpin ends (it also has

MEDICAL APPLICATIONS

ARTEMIS and SCID-A

Severe combined immunodeficiency (SCID) is an inherited immune disorder that results in severe T and B lymphocyte immunodeficiency. T and B lymphocytes are key players of the immune system. They are the second line of defense, tailoring their activities toward individual threats created by invading microbial pathogens. Children born with SCID develop bacterial, viral, and fungal infections that may cause pneumonia, meningitis, or bloodstream infections during the first 6 months of life. These infections are usually severe and life threatening. If untreated, children die before they reach their first birthday. If diagnosed in time, there are effective treatments to treat the disease. SCID has been referred to as the "bubble boy disease." The disease received wide recognition in the media during the 1970s and 1980s when David Vetter, a young boy with SCID, lived in a plastic germ-free bubble for 12 years. He spent most of his life at Texas Children's Hospital.

Mutations in any of eight known genes (*IL2RG, RAG1, RAG2, ADA, CD45, IL7R, JAK3,* and *ARTEMIS*) cause SCID. Mutations in unidentified genes may also cause SCID. SCID is very rare in the U.S. population, affecting 1 in 100,000 children. Newborn screening programs are now being offered in some states. The focus of this Medical Applications box is on the role of the *ARTEMIS* gene in the development of SCID. SCID-A has a high incidence in Athabascan-speaking children. About 1 in 2000 Navajo and Apache Native Americans are born with SCID-A. The pedigrees and genetic analyses of Athabascan-speaking Navajo and Apache Native Americans allowed researchers to map the *ARTEMIS* gene for SCID-A (SCID-Athabascan) to the short arm of chromosome 10 in 1998.

ARTEMIS encodes a 77.6 kDa protein that is postulated to be a novel DNA double-stranded break repair/V(D)J recombination protein. *In vitro* studies demonstrate that the Artemis protein is phosphorylated by a DNA-dependent protein kinase (DNA-PKs) that, in turn, activates its hairpin-opening or nicking activity so that nucleotides can be added and/or deleted at heavy chain V-D and D-J or light chain V-J junctions during the rearrangement or recombination of V(D)J genes that are responsible for the diversity of T cell receptor and immunoglobulin-encoding genes. Mutations in *ARTEMIS* result in SCID-A by causing a significant impairment in V(D)J coding joint formation, leading to the inability to develop pathogen-fighting T and B lymphocytes.

A founder mutation in *ARTEMIS* was discovered in 2002; it is a *nonsense mutation,* which is a point mutation in which a change in DNA replaced a codon specifying an amino acid with a stop codon, resulting in a truncated and often nonfunctional protein product. A "founder effect" occurs when a few individuals leave an original population and found a new population (e.g., colonizing an island). The founders are likely not representative of the original population. In this case, it is believed that the SCID-A defect originated in the Athabascan-speaking populations. Ancestors of the Athabascans probably migrated across the Bering Land Bridge at the end of the last ice age, about 12,000 years ago. They traditionally lived in small nomadic groups in Alaska and the Canadian Northwest territories. Dwellings were usually temporary, and they did not have any formal tribal organization. Between 700 and 1300 CE some of the Athabascans migrated to the Southwestern United States. It is believed that the Navajo and Apache are descendants from these small migratory groups of Athabascan people because their languages share Athabascan roots.

Because SCID-A occurs at a higher rate in families of Navajo Native Americans, geneticists have been studying these families for generations. The SCID-A condition is autosomal recessive. An autosomal recessive allele can be passed from one generation to the next without

exonuclease and endonuclease activities that function in the NHEJ pathway). The actual ligation is performed by DNA ligase IV and also requires the protein XRCC4. Mutations in all these proteins have been found among human patients who have diseases caused by deficiencies in DNA repair that result in increased sensitivity to radiation, and most of these mutations also result in immunodeficiency. (See the accompanying box, *Medical Applications,* for a description of SCID-A, caused by a mutation in the *ARTEMIS* gene.)

What is the connection between joining of V and C gene segments and their transcriptional activation? Unrearranged V gene segments are not actively transcribed. When a V gene segment is joined productively to a C_κ gene segment, however, the resulting unit is transcribed. The sequence upstream of a V gene segment is not altered by the joining reaction, though, and as a result *the promoter must be the same in unrearranged, nonproductively rearranged, and productively rearranged genes.*

causing harm. But if both parents are carriers of the mutant recessive *ARTEMIS* allele, each child has a 1 in 4 chance of being born with SCID-A (**FIGURE B18.1**). Counselors are working with affected families of the Navajo Nation to help them understand and cope with this information. Support sessions unraveled family histories that were told by grandparents about newborn children who only lived a short time. Written on death certificates were the words "severe infection." Today we know that this may be because the newborns likely suffered from SCID-A. Genetic testing results can have a powerful emotional impact on individuals, their families, and community.

Treatment of SCID-A children involves a bone marrow transplant that requires suitable matched donors. An alternative approach to the treatment of SCID-A is the application of gene therapy. Gene therapy involves the use of a viral vector that carries a functional copy of the defective gene that will complement the defective *ARTEMIS* gene. Researchers have shown that the mutant allele can be complemented by a wild-type *ARTEMIS* allele in a mouse model system.

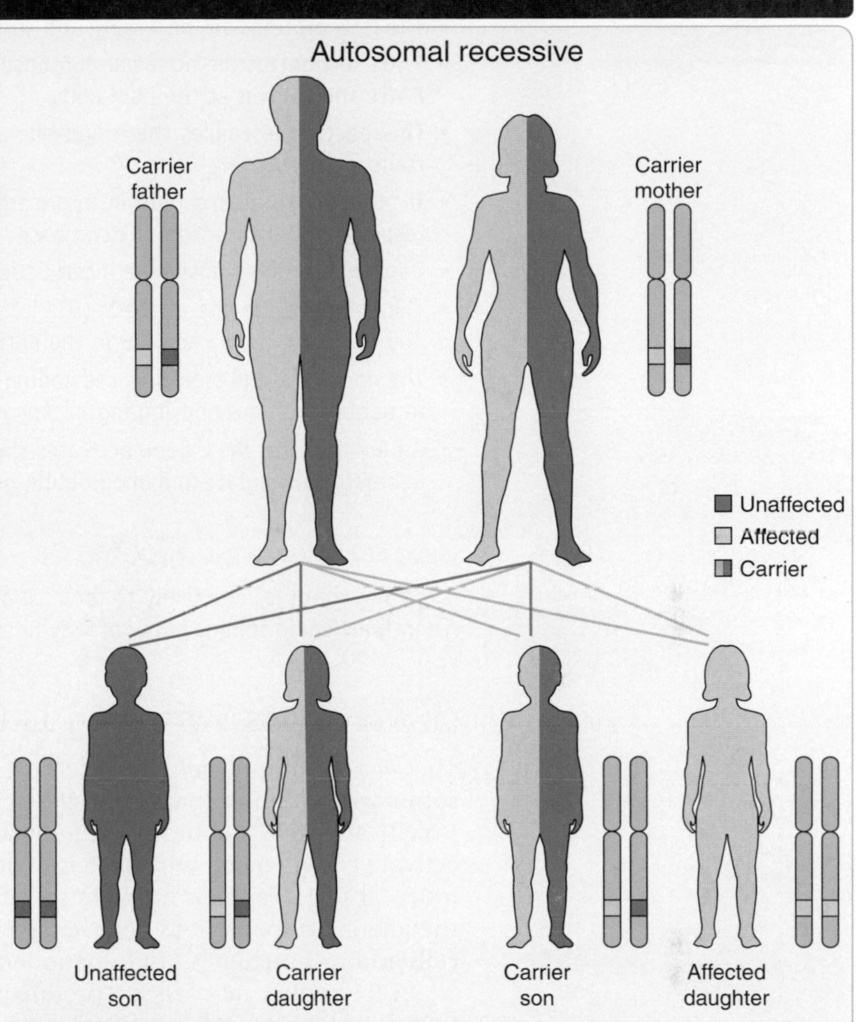

FIGURE B18.1 Two unaffected parents each carry one copy of the defective autosomal recessive *ARTEMIS* allele. They have one child with SCID-A and three unaffected children, two of which carry one copy of the defective *ARTEMIS* allele. Reproduced from U.S. National Library of Medicine. *Genetics Home Reference Handbook*, 2008. (http://ghr.nlm.nih.gov/handbook/illustrations/autorecessive)

A promoter lies upstream of every V gene segment but is inactive. It is activated by its relocation to the C region. The effect must depend on sequences downstream. What role might they play? An enhancer located within or downstream of the C gene segment activates the promoter at the V gene segment. The enhancer is tissue specific; it is active only in B cells. Its existence suggests the model illustrated in **FIGURE 18.17**, in which the V gene segment promoter is activated when it is brought within the range of the enhancer.

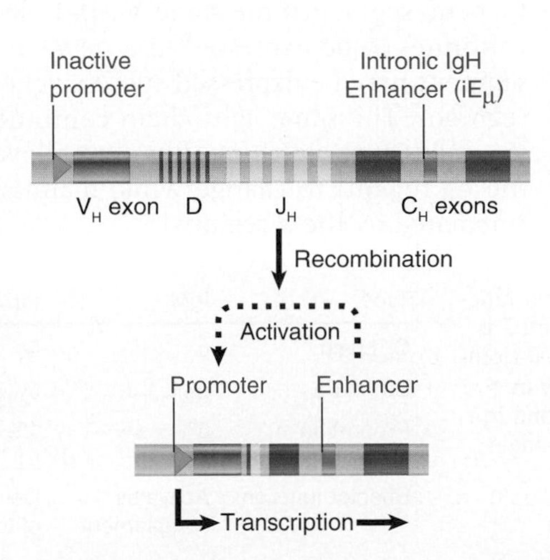

FIGURE 18.17 A V gene promoter is inactive until recombination brings it into the proximity of an enhancer in the C gene segment. The enhancer is active only in B lymphocytes.

KEY CONCEPTS

- The RAG proteins are necessary and sufficient for the cleavage reaction.
- RAG1 recognizes the nonamer consensus sequences for recombination. RAG2 binds to RAG1 and cleaves at the heptamer.
- The reaction resembles the topoisomerase-like resolution reaction that occurs in transposition.
- It proceeds through a hairpin intermediate at the coding end; opening of the hairpin is responsible for insertion of extra bases (P nucleotides) in the recombined gene.
- Deoxynucleoside transferase inserts additional N nucleotides at the coding end.
- The codon at the site of the V-(D)J joining reaction has an extremely variable sequence and codes for amino acid 96 in the antigen-binding site.
- The double-strand breaks at the coding joints are repaired by the same system involved in nonhomologous end joining of damaged DNA.
- An enhancer in the C gene activates the promoter of the V gene after recombination has generated the intact immunoglobulin gene.

CONCEPT AND REASONING CHECK

Is heavy-chain joining likely to produce a higher percentage of nonproductive products than light chain joining? Why or why not?

18.7 Class Switching Is Caused by DNA Recombination

The *class* of immunoglobulin is defined by the type of C_H region it contains. **FIGURE 18.18** summarizes the five Ig classes. IgM (the first immunoglobulin to be produced by any B cell) and IgG (the most common immunoglobulin) possess the central ability to activate complement, which leads to destruction of invading cells. IgA is abundant on mucosal surfaces and is found in secretions (such as saliva), and IgE is associated with the allergic response and defense against parasites. IgD is found on the surface of B cells and its function is not fully understood.

A B lymphocyte starts its "productive" life as an immature cell expressing IgM and IgD on its surface. A B lymphocyte expresses only a single class of Ig at a time (except for the early stage in which IgD is also present), but after encountering antigen, a B cell undergoes activation, proliferation, and differentiation from an IgM- to an IgG-, IgA- or IgE-producing cell. This change in expression is called **class switching**. It is accomplished by a substitution in the type of C_H region that is expressed. Switching can be stimulated by environmental effects; for example, interaction with a helper T cell (T_H) and T_H release of the growth factor TGFβ causes switching from C_μ to C_α.

▸ **class switching** A change in Ig gene organization in which the C region of the heavy chain is changed, but the V region remains the same.

Class switching is effected by *class switch recombination (CSR)* and involves only the C_H gene segments; the same V_H-D-J_H segment originally expressed as part of an IgM continues to be expressed in a new context (IgG, IgA or IgE). Thus a given V_H gene segment may be expressed successively in combination with more than one C_H gene segment. The same light chain continues to be expressed throughout the lineage of the cell. Class switching therefore allows the type of effector response (mediated by the C_H region) to change while maintaining the same capacity to recognize antigen (mediated by the V regions).

Type	IgM	IgD	IgG	IgA	IgE
C_H chain	μ	δ	γ	α	ε
Structure	$(\mu_2L_2)_5J$	δ_2L_2	γ_2L_2	$(\alpha_2L_2)_2J$	ε_2L_2
Proportion in circulating blood	5%	1%	80%	14%	<1%
Effector function	Activates complement	Development of tolerance (?)	Activates complement	Found in secretions	Allergic response

FIGURE 18.18 Immunoglobulin type and function are determined by the heavy chain. J is a joining protein in IgM, unrelated to J (joining) gene segments. IgM exists mainly as a pentamer (i.e. 5 IgM μ_2L_2 tetramers) and IgA as a dimer. IgD, IgG and IgE exist as single H_2L_2 tetramers..

Changes in the expression of C_H gene segments primarily occur via further DNA recombination events, which involve a system different from that concerned with V-D-J joining. Class switching is accomplished by a recombination to bring a new C_H gene segment into juxtaposition with the expressed V_H-D-J_H unit, resulting in deletion of the previously expressed C_H gene segment as well as any C gene segments located between the old and new C gene segments. The sequences of switched V_H-D-J_H-C_H units show that the sites of switching lie upstream of the C_H gene segments themselves. The switching sites are called **S regions.** **FIGURE 18.19** depicts two successive switches at the mouse IgH locus.

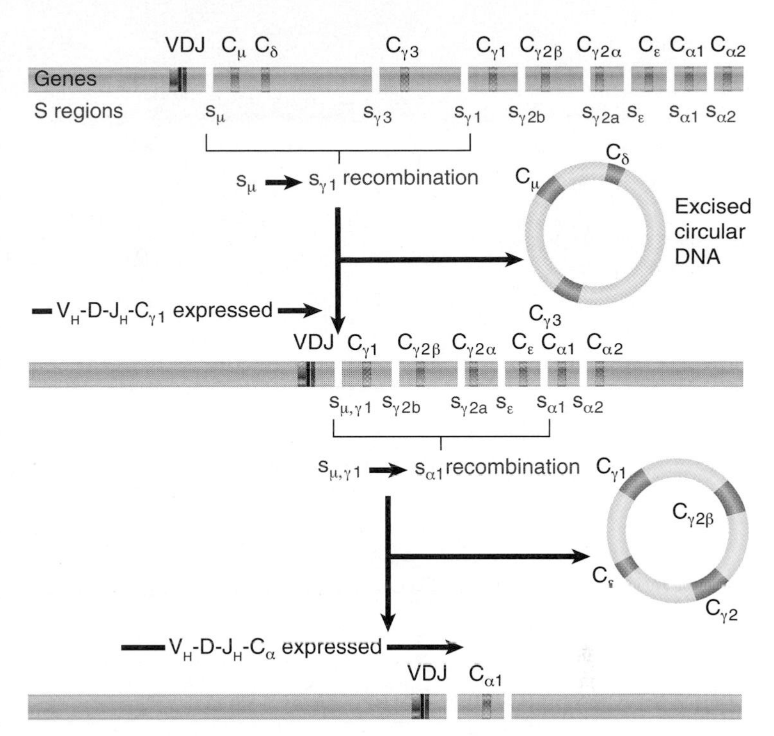

FIGURE 18.19 Class switching of C_H genes occurs by recombination between switch regions (S), deleting the material between the recombining S sites as S circles. Circles are transiently transcribed in the switching cell. Sequential recombinations can occur. The mouse IgH locus is depicted.

In the first switch, expression of C_μ is succeeded by expression of $C_{\gamma1}$. The $C_{\gamma1}$ gene segment is brought into the expressed position by recombination between the sites S_μ and $S_{\gamma1}$. The S_μ site lies between V-D-J and the C_μ gene segment. The $S_{\gamma1}$ site lies upstream of the $C_{\gamma1}$ gene segment. The DNA sequence between the two switch sites is excised as a circular molecule.

The linear deletion model imposes a restriction on the heavy-gene locus: *once a class switch has been made, it becomes impossible to express any C_H gene segment that used to reside between C_μ and the new C_H gene segment.* In the example of Figure 18.19, after the S_μ to $S_{\gamma1}$ recombination event, cells expressing $C_{\gamma1}$ will then be unable to give rise to cells expressing $C_{\gamma3}$, which have been deleted.

It is possible, however, to undertake *another* switch to any C_H gene segment *downstream* of the expressed gene. Figure 18.19 shows a second switch to C_α expression, which is accomplished by recombination between $S_{\alpha1}$ and the switch region $S_{\mu,\gamma1}$ that was generated by the original switch.

▸ **S region** A sequence involved in immunoglobulin class switching. They consist of repetitive sequences at the 5′ ends of gene segments encoding the heavy-chain constant regions.

We know that switch sites are not uniquely defined because different cells expressing the same C_H gene segment prove to have recombined at different points. Switch regions vary in length (as defined by the limits of the sites involved in recombination) from 1 to 10 kb. They contain groups of short inverted repeats, with repeating units that vary from 20 to 80 nucleotides in length. The primary sequence of the switch region does not seem to be important; what matters is the presence of the inverted repeats.

An S region typically is located ~2 kb upstream of a C_H gene segment. The switching reaction releases the excised material between the switch sites as a circular DNA molecule. Two of the factors required for the joining phase of VDJ recombination, Ku70/80 and DNA-PKcs, are also required for NHEJ, indicating that the joining reaction uses the NHEJ repair pathway. An alternative (but less efficient) pathway can be used in the absence of NHEJ components such as XRCC4 or DNA ligase IV, suggesting that at least two pathways can be used to ligate S region DNA ends.

We can put together the features of the reaction to propose a model for the generation of the double-strand break. The critical points are:

- transcription through the S region is required,
- the inverted repeats are crucial, and
- the break can occur at many different places within the S region.

FIGURE 18.20 shows the stages of the class-switching reaction. A promoter ("I") lies immediately upstream of each switch region. Switching requires transcription from this promoter. The promoter may respond to activators that respond to environmental conditions, such as stimulation by cytokines, thus creating a mechanism to regulate switching. The first stage in switching is therefore to activate the I promoters that are upstream

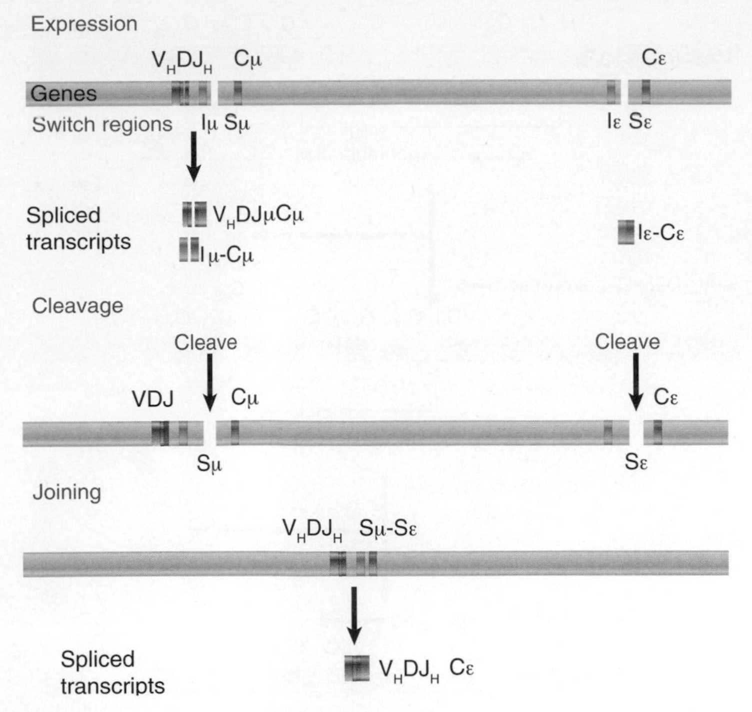

FIGURE 18.20 Class switching occurs through sequential and discrete stages. The I_H promoters initiate transcription of noncoding transcripts. The S regions are cleaved and recombination occurs at the cleaved regions. Depicted is class-switch recombination from S_μ to S_ε.

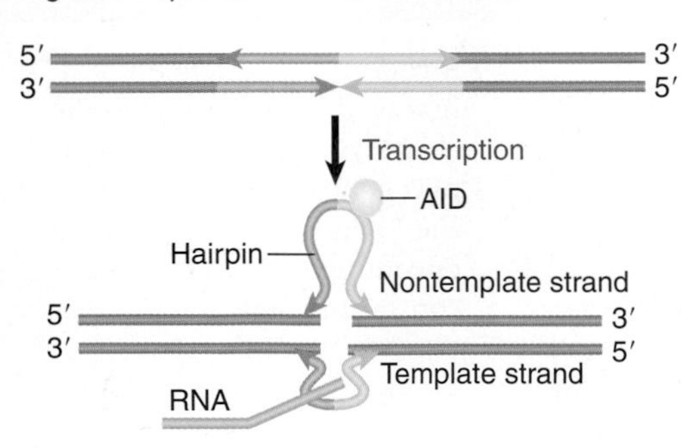

FIGURE 18.21 When transcription separates the strands of DNA, one strand may form a single stranded loop if 5′-AGCT-3′ motifs in the same strand are juxtaposed.

of each of the switch regions that will be involved. When these promoters are activated, they generate noncoding transcripts that are spliced to join the I region with the corresponding heavy constant region.

The key insight into the mechanism of switching was the discovery of the requirement for the enzyme AID (activation-induced cytidine deaminase). In the absence of AID, class switching is blocked before the nicking stage. AID is a member of a class of enzymes that act on RNA to change a cytidine to a uridine (see *Section 23.9, RNA Editing Occurs at Individual Bases*). AID has a different specificity, however, and acts on single-stranded DNA. After AID acts, UNG, a uracil DNA glycosylase, removes the uracil that AID generates by deaminating cytidine. Mice that are deficient in UNG have a tenfold reduction in class switching. This suggests a model in which the successive actions of AID and UNG create sites from which a base has been removed in DNA. The source of the single-stranded DNA target for AID is generated by the process of noncoding transcription, most likely by exposing the nontemplate strand of DNA that is displaced when the other strand is used as a template for RNA synthesis. This is supported by the observation that AID preferentially targets cytidines in the nontemplate strand.

In order to cause class switching, these sites are converted by an apyridinic/apurinic endonuclease (APE) into breaks in the nucleotide chain that provide the cleavage events shown in Figure 18.20. Nearby breaks in the opposite DNA strand would give rise to DSBs in the S regions, but the mechanism of generation of the second nick is not understood. The broken ends are joined by the NHEJ pathway.

One unexplained feature is the involvement of inverted repeats. One possibility is that hairpins are formed by an interaction between the inverted repeats on the displaced nontemplate strand, as shown in **FIGURE 18.21**. In conjunction with the generation of abasic sites on this strand, this might lead to breakage.

Two critical questions that remain unanswered are how the system is targeted to the appropriate regions in the heavy-chain locus and what controls the use of switching sites.

KEY CONCEPTS

- Immunoglobulins are divided into five classes according to the type of constant region in the heavy chain.
- Class switching to change the C_H region occurs by a recombination between S regions that deletes the region between the old C_H region and the new C_H region.
- Multiple successive switch recombinations can occur.
- Switching occurs by a double-strand break followed by the nonhomologous end-joining reaction.
- The important feature of a switch region is the presence of inverted repeats.
- Switching requires activation of promoters that are upstream of the switch sites.

CONCEPT AND REASONING CHECK

Why can't a lymphocyte resume expression of an earlier immunoglobulin class, after it has switched from that class to another?

18.8 Somatic Hypermutation Generates Additional Diversity

Comparisons between the sequences of expressed immunoglobulin genes and the corresponding V gene segments of the germline show that new sequences appear in the expressed population. Some of this additional diversity results from sequence changes at the V-J or V-D-J junctions that occur during the recombination process. Other changes occur upstream, however, at locations within the variable domain.

Two types of mechanisms can generate changes in V gene sequences after rearrangement has generated a functional immunoglobulin gene. In the mouse and human, the mechanism is somatic hypermutation (SHM). In chicken, rabbit, and pig, a different mechanism—gene conversion—is used to change a segment of the expressed V gene into the corresponding sequence from a different V gene (see *Section 18.9, Avian Immunoglobulins Are Assembled from Pseudogenes*). SHM inserts mostly point mutations in the expressed V(D)J sequence. The process is referred to as hypermutation because it introduces mutations at a rate that is at least 10^6-fold higher (10^{-3} change/base/cell division) than that of the spontaneous mutations rate of the genome at large (10^{-9} change/base/cell division).

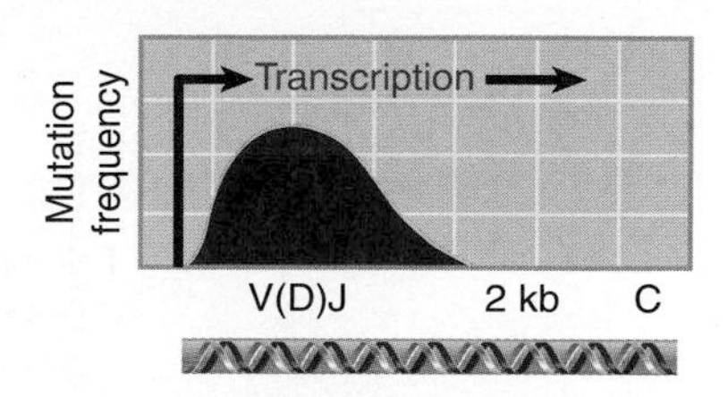

FIGURE 18.22 Somatic mutation occurs in the region surrounding the V segment and extends over the recombined V(D)J segment.

FIGURE 18.22 shows that sequence changes are localized around the V(D)J gene segment, extending in a region from ~150 bp downstream of the V gene promoter for ~1.5 kb. They take the form of substitutions of individual nucleotide pairs. Usually there are ~3 to ~15 substitutions, corresponding to <10 amino acid changes in the protein. They are concentrated in the antigen-binding site (thus generating the maximum diversity for recognizing new antigens). Only some of the mutations affect the amino acid sequence, since others lie in third-base coding positions as well as in nontranslated regions.

The large proportion of silent mutations suggests that SHM occurs more or less at random in a region including the V gene segment and extending beyond it. There is a tendency for some mutations to recur on multiple occasions. These may represent hotspots as a result of some intrinsic preference in the SHM machinery.

Upon exposure to an antigen, B cells expressing a BCR with the highest intrinsic affinity to that antigen are selected, activated, and proliferate. SHM occurs during B clonal proliferation. It randomly inserts one point mutation in the V(D)J sequence of approximately half of the progeny cells; as a result, B cells expressing mutated antibodies become a high fraction of the clonal population within a few divisions. Random mutations have unpredictable effects on protein function; some decrease the affinity of the BCR for the antigen driving the response, whereas others increase the intrinsic affinity for that antigen. The proportion and effectiveness of the lymphocytes that respond is increased by selection among the lymphocyte population for those cells bearing antibodies in which mutation has increased the affinity for the antigen.

Somatic mutation has many of the same requirements as class switching (see *Section 18.7, Class Switching Is Caused by DNA Recombination*), including that transcription must occur in the target region and that it requires the enzymes AID and UNG. The mismatch-repair (MMR) system is also involved.

When AID deaminates cytosine, it generates uracil. Uracil is not normally found in DNA and can be handled by the cell in different ways, summarized in **FIGURE 18.23**. The uracil can be "replicated over," in which the replication machinery will insert an A in the complementary daughter strand. The net result is the replacement of the original G-C pair with T-A in half of the progeny cells. Alternatively, the uracil is removed from the DNA by UNG, giving rise to an abasic site. This can be replicated over by one of several error-prone translesion polymerases, which can insert any of the three possible mismatches (mutations) across from the abasic site. Finally, the MMR system can be recruited to excise and replace the stretch of DNA containing the U:G mismatch; again it appears that an error-prone polymerase is used to fill the gap resulting from MMR-dependent excision of the region containing the mispair. We don't know yet what restricts the action of the SHM machinery to the target region for hypermutation.

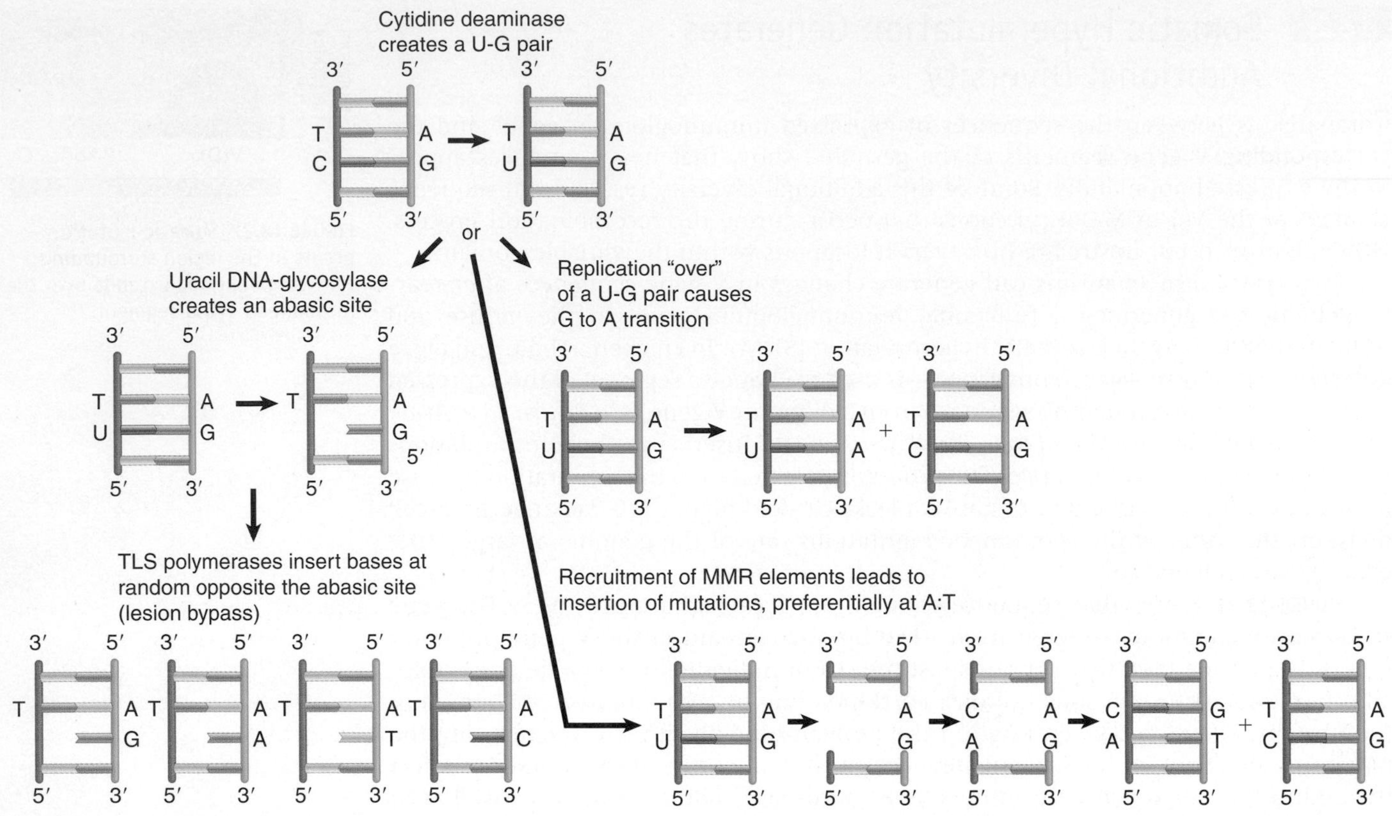

FIGURE 18.23 Deamination of C by AID gives rise to a U:G mispair. U can be replicated over, resulting in C:G→A:T transitions in 50% of progeny B cells. When the action of cytidine deaminase (top) is followed by that of uracil DNA-glycosylase, an abasic site is created (center). Replication past this site should insert all four bases at random into the daughter strand (bottom left). If the uracil is not removed from the DNA, the U:G mispair can be recognized by the MMR machinery, which excises a region of DNA containing the mismatch and then fills in the resulting gap using an error-prone DNA polymerase (bottom right). This will lead to the insert of further mismatches.

The main difference between CSR and SHM is at the end of the process, when double-strand breaks are introduced in CSR, but individual point mutations are created during SHM. We do not yet know where the systems diverge. One possibility is that breaks are introduced at abasic sites in CSR, but the sites are erratically repaired in SHM. Another possibility is that breaks are introduced in both cases, but are repaired in an error-prone manner in SHM.

KEY CONCEPTS

- Active immunoglobulin genes have V regions with sequences that are changed from the germline because of somatic hypermutation (SHM).
- The mutations occur as substitutions of individual bases.
- The sites of mutation are concentrated in the antigen-binding site.
- A cytidine deaminase is required for somatic mutation as well as for class switching.
- Uracil-DNA glycosylase activity influences the pattern of SHM.
- Hypermutation may be initiated by the sequential action of these enzymes.
- Mismatch repair also contributes to patterns of SHM.

CONCEPT AND REASONING CHECK

Why is somatic hypermutation useful even though most of the mutants generated may be nonfunctional or produce weaker antibodies?

18.9 Avian Immunoglobulins Are Assembled from Pseudogenes

The chicken Ig locus is the paradigm for the Ig diversification mechanism utilized by rabbits, cows, and pigs, which rely upon using the diversity that is encoded in the genome. A similar mechanism is used for both the single light-chain locus (of the λ type) and the H-chain loci. The organization of the chicken λ locus is shown in **FIGURE 18.24**. It has only one functional V gene segment, one J segment, and one C gene segment. Upstream of the functional $V_{\lambda 1}$ gene segment lie 25 V_{λ} pseudogenes, organized in either orientation. They are classified as pseudogenes because either the coding segment is deleted at one or both ends or proper signals for recombination are missing or both. This assignment is confirmed by the fact that only the $V_{\lambda 1}$ gene segment recombines with the J_{λ}-C_{λ} gene segment.

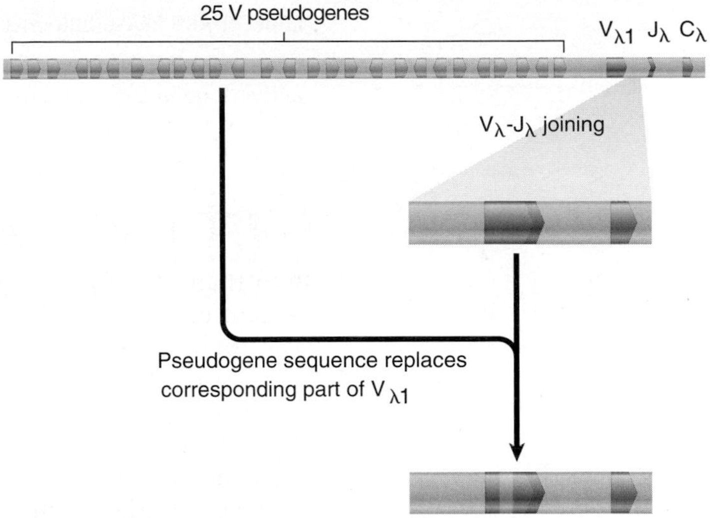

FIGURE 18.24 The chicken lambda light locus has 25 V pseudogenes upstream of the single functional V-J-C region. Sequences derived from the pseudogenes, however, are found in active rearranged V-J-C genes.

Nevertheless, sequences of active rearranged V_{λ}-J-C_{λ} gene segments show considerable diversity. A rearranged gene has one or more positions at which a cluster of changes has occurred in the sequence. A sequence identical to the new sequence can almost always be found in one of the pseudogenes (which themselves remain unchanged). The exceptional sequences that are not found in a pseudogene always represent changes at the junction between the original sequence and the altered sequence.

Thus a novel mechanism is employed to generate diversity. Sequences from the pseudogenes, between 10 and 120 bp in length, are *substituted* into the active $V_{\lambda 1}$ region by gene conversion. The unmodified $V_{\lambda 1}$ sequence is not expressed, even at early times during the immune response. A successful conversion event probably occurs every 10 to 20 cell divisions to every rearranged $V_{\lambda 1}$ sequence. At the end of the immune maturation period, a rearranged $V_{\lambda 1}$ sequence has 4 to 6 converted segments spanning its entire length, which are derived from different donor pseudogenes. If all pseudogenes participate, this allows 2.5×10^8 possible combinations!

The enzymatic basis for copying pseudogene sequences into the expressed locus depends on enzymes involved in recombination and is related to the mechanism for somatic hypermutation that introduces diversity in mouse and human. Some of the genes involved in homologous recombination are required for the gene conversion process; for example, it is prevented by deletion of *RAD54*. Deletion of other recombination genes (*XRCC2, XRCC3,* and *RAD51B*) has another, very interesting effect: somatic mutation occurs at the V gene in the expressed locus. The frequency of the somatic mutation is ~10× greater than the usual rate of gene conversion.

Thus, the absence of somatic mutation in chickens is not due to a deficiency in the enzymatic systems that are responsible for SHM in mice and humans. The most likely explanation for a connection between (lack of) recombination and somatic mutation is that unrepaired breaks at the locus trigger the induction of mutations. The reason SHM occurs in mice and humans but not in chickens may therefore lie with the nature of the repair system that operates on breaks in the Ig locus. It is more efficient in chickens, so that the gene is repaired by gene conversion before mutations can be induced.

KEY CONCEPT

- An immunoglobulin gene in chickens is generated by copying sequences from one of 25 pseudogenes into the V gene at a single active locus.

CONCEPT AND REASONING CHECK

Given that V-J-C joining occurs in chickens, do you think the V pseudogenes were once active V gene segments? Why or why not?

18.10 T Cell Receptors Are Related to Immunoglobulins

Both B and T cells use similar evolutionarily conserved mechanisms to generate significant diversity in BCR/Ig and TCR variable regions. T cells produce either of two types of T cell receptor. The $\gamma\delta$ receptor is found on <5% of T lymphocytes. It is synthesized only at an early stage of T cell development. TCR$\alpha\beta$ is found on >95% of lymphocytes. It is synthesized later in T cell development than in $\gamma\delta$. It is synthesized by a separate lineage of cells from those involved in TCR$\gamma\delta$ synthesis and involves independent rearrangement events.

Like the BCR and immunoglobulins, a TCR must recognize a foreign antigen of virtually any possible structure. The problem of antigen recognition by B cells and T cells is resolved in the same way, and the organization of the T cell receptor genes resembles the immunoglobulin genes in the use of variable and constant regions. Each locus is organized in the same way as the immunoglobulin genes, with separate segments that are brought together by a recombination reaction specific to the lymphocyte. The components are the same as those found in the Ig heavy- and light-chain families. TCRα resembles the Ig light chain, whereas TCRβ resembles a heavy chain. The organization of the TCR proteins resembles that of the immunoglobulins. The V regions have the same general internal organization in both Ig and TCR proteins. The TCR C region is related to the constant Ig regions and has a single constant domain followed by transmembrane and cytoplasmic portions. Exon-intron structure is related to protein function.

As summarized in **FIGURE 18.25**, the genomic organization of TCRα resembles that of Igκ, with V_α gene segments separated from a cluster of J_α segments that precedes a single C_α gene segment. The organization of the locus is similar in both human and mouse, with some differences only in the number of V_α gene segments and J_α segments. Note the presence of δ genes in this locus as well (see following text), which are not incorporated into TCRα.

The components of TCRβ resemble those of IgH. **FIGURE 18.26** shows that the organization is different, with V_β gene segments separated from two clusters each containing a D segment, several J_β segments, and a C_β gene segment. Again, the only differences between human and mouse are in the numbers of the V_β and J_β units.

Diversity is generated by the same mechanisms as in the BCR/Ig. Intrinsic (germline encoded) diversity results from the combination of a variety of V, D, J, and C segments; some additional diversity results from the introduction of new sequences at the junctions between these components (in the form of P and N nucleotides). The same

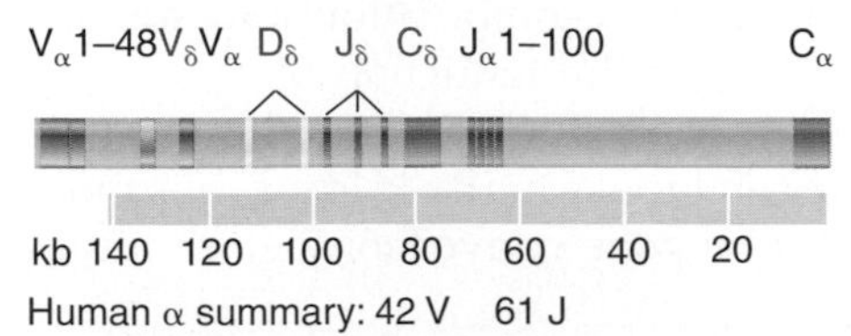

FIGURE 18.25 The human TCRα locus has interspersed α and δ segments. A V_δ segment is located within the V_α cluster. The D-J-C_δ segments lie between the V gene segments and the J-C_α segments. The mouse locus is similar, but has more V_δ segments.

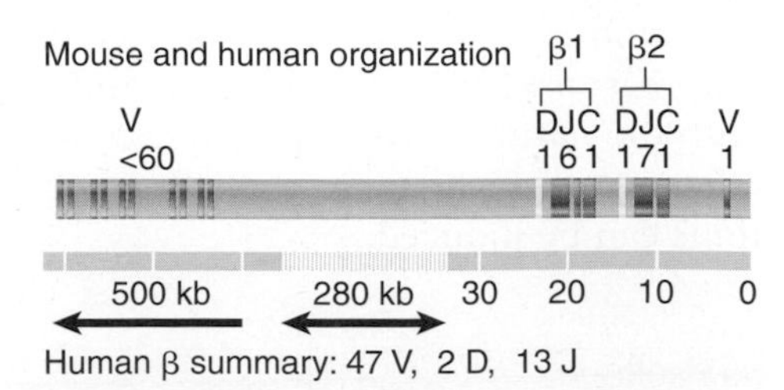

FIGURE 18.26 The TCRβ locus contains many V gene segments spread over ~500 kb that lie ~280 kb upstream of the two D-J-C clusters.

mechanisms are involved in the reactions that recombine Ig genes in B cells and TCR genes in T cells. The recombining TCR segments are surrounded by nonamer and heptamer RSS consensus sequences identical to those used by the Ig genes. Some TCRβ chains incorporate two D segments, which are generated by D-D joins (directed by an appropriate organization of the RSS sequences). A difference between TCR and Ig is that SHM does not occur at the TCR loci.

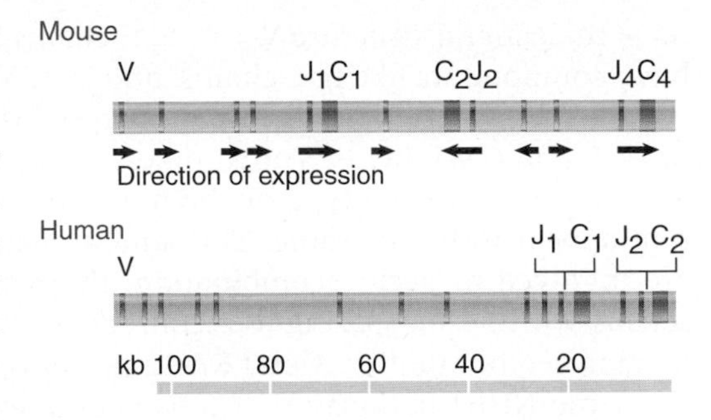

FIGURE 18.27 The TCRγ locus contains a small number of functional V gene segments (and also some pseudogenes not shown) that lie upstream of the J-C loci.

The organization of the γ locus resembles that of Igλ, with V_γ gene segments separated from a series of J_γ-C_γ segments. **FIGURE 18.27** shows that this locus has relatively little diversity, with ~8 functional V segments. The organization is different in human and mouse. The mouse locus has three functional J_γ-C_γ loci, but some segments are inverted in orientation. The human has multiple J_γ segments for each C_γ gene segment.

The δ subunit is encoded by segments that lie at the TCRα locus, as illustrated previously in Figure 18.25. The segments D_δ-D_δ-J_δ-C_δ lie between the V gene segments and the J_α-C_α segments. Both of the D segments may be incorporated into the δ chain to give the structure VDDJ. The basis for specificity in choosing V segments in α and δ rearrangement is not known. One possibility is that many of the V_α gene segments can be joined to the DDJδ segment, but that only some (therefore defined as V_δ) can give active proteins.

Rearrangements at the TCR loci, like those of immunoglobulin genes, may be productive or nonproductive. The β locus shows allelic exclusion in much the same way as immunoglobulin loci; rearrangement is suppressed once a productive allele has been generated. The α locus may be different; several cases of continued rearrangement suggest the possibility that substitution of V_α sequences may continue after a productive allele has been generated.

KEY CONCEPTS

- T cells use a similar mechanism of V(D)J-C joining to B cells to produce either of two types of T cell receptor.
- TCRαβ is found on >95% and TCRγδ is found on <5% of T lymphocytes in the adult.

CONCEPT AND REASONING CHECK

Why do T cell receptors need to be as diverse as immunoglobulins?

18.11 Summary

Immunoglobulins and T cell receptors are proteins that play analogous functions in the roles of B cells and T cells in the immune system. An Ig or TCR protein is generated by rearrangement of DNA in a single lymphocyte; exposure to an antigen recognized by the Ig or TCR leads to clonal expansion to generate many cells that have the same specificity as the original cell. Many different rearrangements occur early in the development of the immune system, thereby creating a large repertoire of cells of different specificities.

Each immunoglobulin protein is a tetramer containing two identical light chains and two identical heavy chains. A TCR is a dimer containing two different chains. Each polypeptide chain is expressed from a gene created by linking one of many V segments via D segments and J segments to one of a few C segments. Ig L chains (either κ or λ) have the general structure V-J-C, Ig H chains have the structure V-D-J-C, TCR α and γ have components like Ig L chains, and TCR δ and β are like Ig H chains.

Each type of chain is coded by a large cluster of V genes separated from the cluster of D, J, and C segments. The numbers of each type of segment and their organization are different for each type of chain, but the principle and mechanism of recombination appear to be the same. The same nonamer and heptamer consensus sequences are involved in each recombination; the reaction always involves joining of a consensus with 23 bp spacing to a consensus with 12 bp spacing. The cleavage reaction is catalyzed by the RAG1 and RAG2 proteins, and the joining reaction is catalyzed by the same NHEJ pathway that repairs double-strand breaks in cells. The mechanism of action of the RAG proteins is related to the action of site-specific recombination catalyzed by resolvases.

Considerable diversity is generated by joining different V, D, and J segments to a C segment; however, additional variations are introduced in the form of changes at the junctions between segments during the recombination process. Changes are also induced in immunoglobulin genes by somatic hypermutation, which requires the actions of cytidine deaminase and uracil glycosylase. Mutations induced by cytidine deaminase probably lead to removal of uracil by uracil glycosylase, followed by the induction of mutations at the sites where bases are missing.

Allelic exclusion ensures that a given lymphocyte synthesizes only a single Ig or TCR. A productive rearrangement inhibits the occurrence of further rearrangements. The use of the V region is fixed by the first productive rearrangement, but B cells switch use of C_H genes from the initial μ chain to one of the H chains coded farther downstream. This process involves a different type of recombination in which the sequences between the VDJ region and the new C_H gene are deleted. More than one switch occurs in C_H gene usage. Class switching requires the same cytidine deaminase that is required for somatic hypermutation, but its role is not known.

CHAPTER QUESTIONS

1. A set of about 20 cellular proteins that function through a cascade of proteolytic actions is:
A. major histocompatibility complex.
B. complement.
C. immunoglobulins.
D. haptens.

2. Somatic recombination to join one of many immunoglobulin V gene segments with one of a few C gene segments occurs in:
A. B cells.
B. T cells.
C. germline cells.
D. any of the above.

3. The J segment of an immunoglobulin gene is:
A. relatively short and codes for the first few amino acids of the constant region.
B. relatively short and codes for the last few amino acids of the constant region.
C. relatively short and codes for the first few amino acids of the variable region.
D. relatively short and codes for the last few amino acids of the variable region.

4. The target sequence for somatic recombination in generation of immune diversity is what type of arrangement?
A. a heptamer-intron-nonamer
B. a heptamer-spacer-nonamer
C. a nonamer-intron-heptamer
D. a nonamer-spacer-heptamer

5. Somatic recombination of immunoglobulin genes occurs by:
   - A. single-strand nick and exonuclease action at the heptamers of two consensus sequences.
   - B. single-strand nick and exonuclease action at the nonomers of two consensus sequences.
   - C. double-strand breaks at the heptamers of two consensus sequences.
   - D. double-strand breaks at the nonamers of two consensus sequences.
6. B cells that are not blocked from further rearrangement due to recombination are said to have undergone:
   - A. class switching.
   - B. allelic exclusion.
   - C. productive rearrangement.
   - D. nonproductive rearrangement.
7. _______ proteins are necessary and sufficient for the cleavage reaction during V(D)J recombination.
8. The _______ portion of the RSS is adjacent to the coding region, whereas the
9. _______ sequence is adjacent to the **10.** _______.

KEY TERMS

adaptive (acquired) immunity
allelic exclusion
antigen
autoimmune disease
B cell
B cell receptor (BCR)
C genes
cell-mediated response
class switching
clonal expansion
coding end
coding joint
complement
constant region (C region)
cytotoxic T cell
D segment
heavy chain
helper T (T_h) cell
humoral response
immune response
immunoglobulin (antibody)
innate immunity
J segments
light chain
major histocompatibility complex (MHC)
N nucleotide
P nucleotide
productive rearrangement
recombination signal sequence (RSS)
S region
signal end
somatic hypermutation (SHM)
somatic recombination
superfamily
T cell receptor (TCR)
T cells
V gene
variable region (V region)

FURTHER READING

Dudley, D. D., Chaudhuri, J., Bassing, C. H., and Alt, F. W. (2005). Mechanism and control of V(D)J recombination versus class switch recombination: similarities and differences. *Adv. Immunol.* **86**, 43–112. A review of both V(D)J recombinations and class switch recombination, comparing and contrasting the enzymes involved in each mechanism.

Li, Moshous, Zhou, Wang, Xie, Salido, Hu, de Villartay, and Cowan. (2002). A Founder Mutation in Artemis, an SNM1-Like Protein, Causes SCID in Athabascan-Speaking Native Americans. *J. Immunology* **168**, 6323–6329.

Li, Salido, Zhou, Bhattacharyya, Yannone, Dunn, Meneses, Feeney, and Cowan. (2005). Targeted Disruption of the Artemis Murine Counterpart Results in SCID and Defective V(D)J Recombination That Is Partially Corrected with Bone Marrow Transplantation. *J. Immunology* **174**, 2420–2428.

Maul, R. W., and Gearhart, P. J. (2010). AID and somatic hypermutation. *Adv. Immunol.* **105**, 159–191.

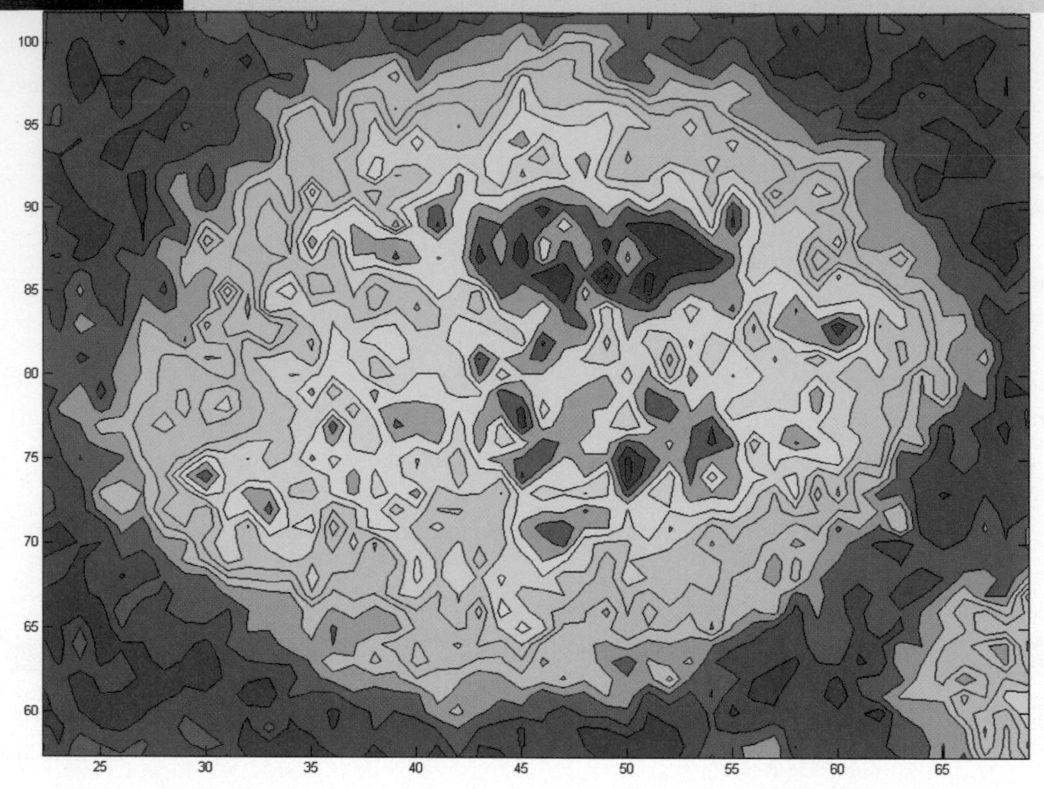

A computer-generated intensity contour map of the *Drosophila* activator bicoid detected by fluorescence confocal imaging of a single slice of a nucleus in an early embryo.

Gene Expression

19

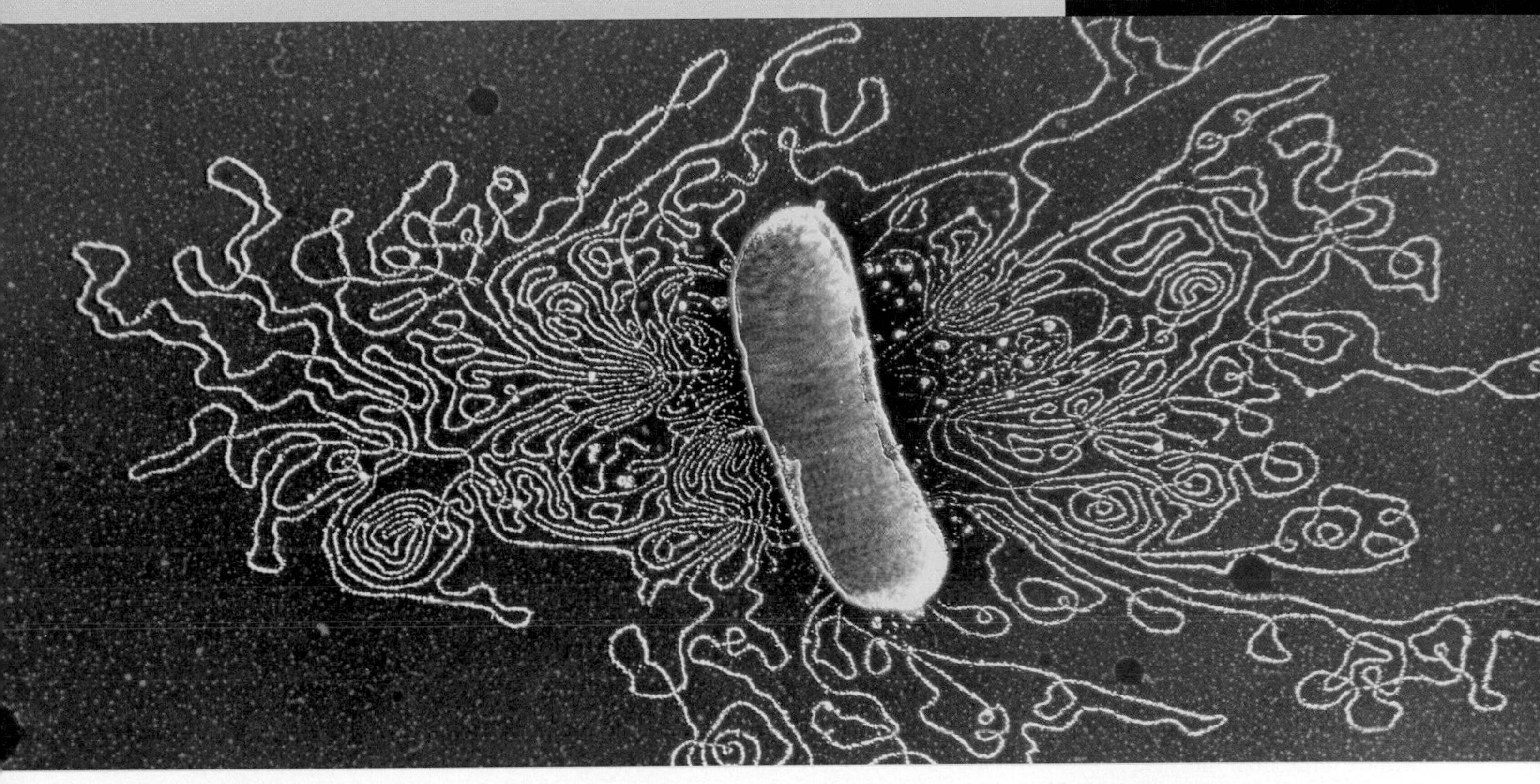

An electron micrograph of the bacterium *E. coli* (after treatment to break its cell wall) surrounded by its DNA (colored gold). © Dr. Gopal Murti/Photo Researchers, Inc.

Prokaryotic Transcription

CHAPTER OUTLINE

19.1 Introduction

- **template strand** The DNA strand that is copied by the polymerase.
- **coding strand** The DNA strand that has the same sequence as the mRNA and is related by the genetic code to the protein sequence that it represents.
- **RNA polymerase** An enzyme that synthesizes RNA using a DNA template (formally described as a DNA-dependent RNA polymerase).
- **promoter** A region of DNA where RNA polymerase binds to initiate transcription.
- **terminator** A sequence of DNA that causes RNA polymerase to terminate transcription.
- **transcription unit** The sequence between sites of initiation and termination by RNA polymerase; it may include more than one gene.
- **startpoint** The position on DNA corresponding to the first base incorporated into RNA.
- **upstream** Sequences in the opposite direction from expression.
- **downstream** Sequences proceeding farther in the direction of expression within the transcription unit.
- **primary transcript** The original unmodified RNA product corresponding to a transcription unit.

Transcription produces an RNA chain representing a copy of one strand of a DNA duplex. The newly synthesized RNA transcript is made 5′ to 3′ from a **template strand** that is 3′ to 5′ (**FIGURE 19.1**). Note that the RNA transcript is *complementary* to (i.e., it base pairs with) the template and has the same sequence as the **coding strand**.

RNA synthesis is catalyzed by the enzyme **RNA polymerase**. Transcription starts when RNA polymerase binds to a special region, the **promoter**, at the start of the gene. From the promoter, RNA polymerase moves along the template, synthesizing RNA, until it reaches a **terminator** (t) sequence. This action defines a **transcription unit** that extends from the promoter to the terminator. The critical feature of the transcription unit, depicted in **FIGURE 19.2**, is that it constitutes a stretch of DNA used as a template for the production of a *single* RNA molecule. A transcription unit may include more than one cistron.

Sequences prior to the **startpoint**, the first base that is transcribed into RNA, are described as **upstream** of it; those after the startpoint (within the transcribed sequence) are **downstream** of it. Sequences are conventionally written so that transcription proceeds from left (upstream) to right (downstream). This corresponds to writing the mRNA in the usual 5′ to 3′ direction.

The DNA sequence often is written to show only the coding strand, which has the same sequence as the RNA. Base positions are numbered in both directions away from the startpoint, which is assigned the value +1; numbers increase as they go downstream. The base before the startpoint is numbered −1, and the negative numbers increase going upstream. (There is no base assigned the number 0.)

The initial product of transcription is called the **primary transcript**. It consists of an RNA extending from the promoter to the terminator and possesses the original 5′ and 3′ ends. The primary transcript is, however, almost always immediately modified. In prokaryotes, it is rapidly degraded (mRNA) or cleaved to give mature products (rRNA and tRNA). Transcription is the first stage in gene expression and the principal step at which it is controlled. Regulatory proteins determine whether a particular gene is available to be transcribed by RNA polymerase. The initial (and often the only) step in regulation is the decision of whether or not to transcribe a gene. Most regulatory events occur at the initiation of transcription, although subsequent stages in transcription (or other stages of gene expression) are sometimes regulated.

Within this context, there are two basic questions in gene expression:

- How does RNA polymerase find promoters on DNA? This is a particular example of a more general question: how do proteins read the DNA sequence to find their specific binding sites in DNA?
- How do regulatory proteins interact with RNA polymerase (and with one another) to activate or to repress specific steps in the initiation, elongation, or termination of transcription?

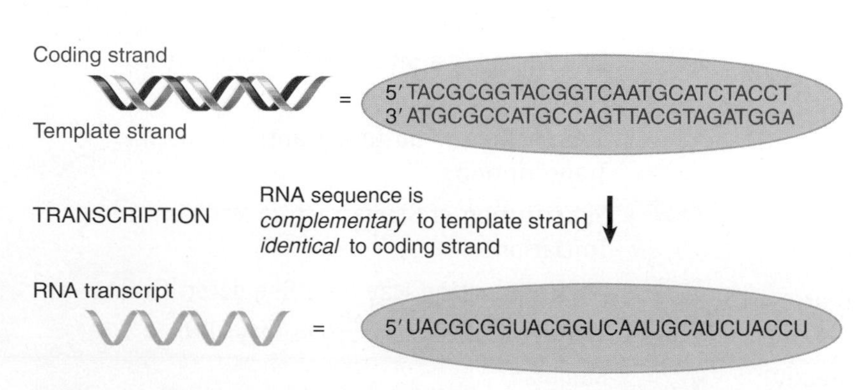

FIGURE 19.1 The function of RNA polymerase is to copy one strand of the duplex DNA into RNA.

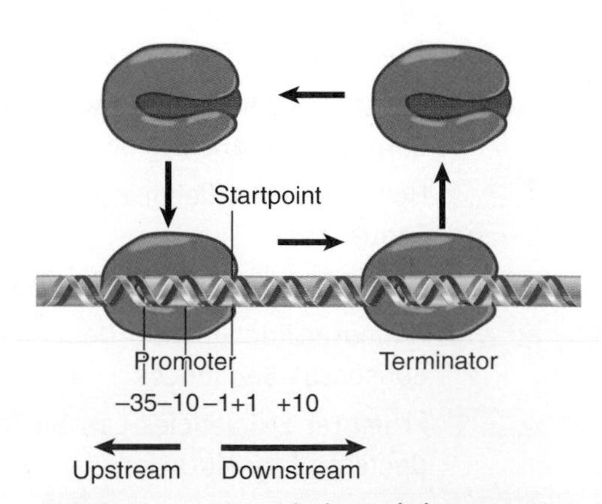

FIGURE 19.2 A transcription unit is a sequence of DNA transcribed into a single RNA, starting at the promoter and ending at the terminator.

In this chapter, we analyze the interactions of bacterial RNA polymerase with DNA from its initial contact with a gene, through the act of transcription, and then finally its release when the transcript has been completed. *Chapter 26, The Operon,* describes the various means by which regulatory proteins can assist or prevent bacterial RNA polymerase from recognizing a particular gene for transcription. In *Chapter 27, Phage Strategies,* we consider how individual regulatory interactions can be connected into more complex networks. In *Chapter 20, Eukaryotic Transcription* and *Chapter 28, Eukaryotic Transcription Regulation,* we consider the analogous reactions between eukaryotic RNA polymerases and their templates. *Chapter 30, Regulatory RNA,* discusses other means of regulation, including the use of small RNAs, and considers how these interactions can be connected into larger regulatory networks.

CONCEPT AND REASONING CHECK

RNA is synthesized from 5′ to 3′. Why is the template 3′ to 5′?

19.2 Transcription Occurs by Base Pairing in a "Bubble" of Unpaired DNA

Transcription takes place by the usual process of complementary base pairing, in common with the other polymerization reactions: replication and translation. **FIGURE 19.3** illustrates the general principle of transcription. RNA synthesis takes place within a "transcription bubble," in which DNA is transiently separated into its single strands and the template strand is used to direct synthesis of the RNA strand.

The RNA chain is synthesized from the 5′ end toward the 3′ end from a template that runs from 3′ to 5′ by adding new nucleotides to the 3′ end of the growing chain. The 3′-OH group of the last nucleotide added to the chain reacts with an incoming nucleoside 5′ triphosphate. The incoming nucleotide loses its terminal two phosphate groups (γ and β); its α phosphate is used in the phosphodiester bond linking it to the growing chain. The overall reaction rate can be as fast as ~40 to 50 nucleotides/second

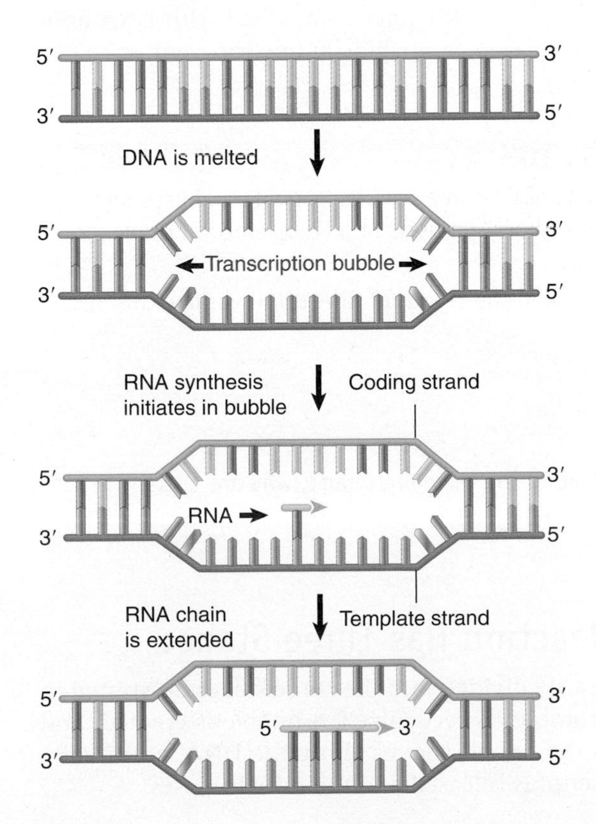

FIGURE 19.3 DNA strands separate to form a transcription bubble. RNA is synthesized by complementary base pairing with one of the DNA strands.

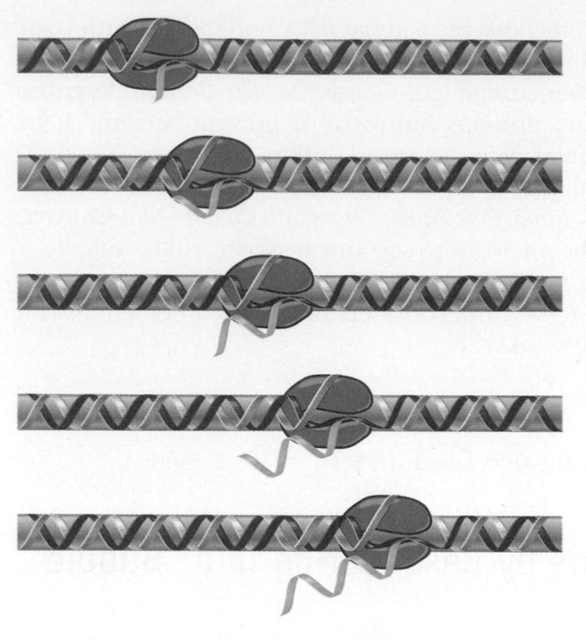

FIGURE 19.4 Transcription takes place in a bubble, in which RNA is synthesized by base pairing with one strand of DNA in the transiently unwound region. As the bubble progresses, the DNA duplex reforms behind it, displacing RNA in the form of a single polynucleotide chain.

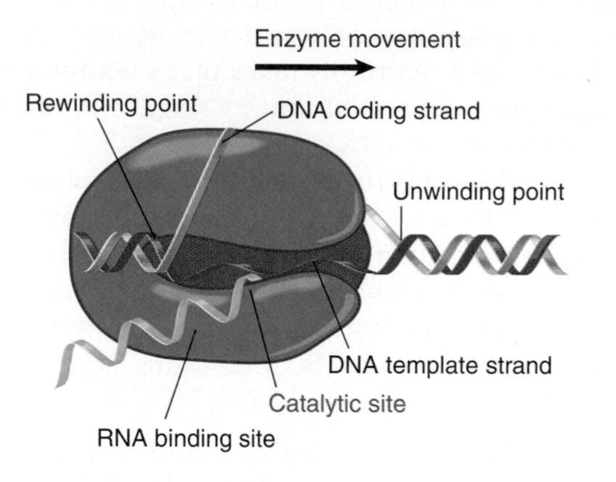

FIGURE 19.5 During transcription, the bubble is maintained within bacterial RNA polymerase, which unwinds and rewinds DNA and synthesizes RNA.

at 37°C for most transcripts; this is about the same as the rate of translation (15 amino acids/second) but much slower than the rate of DNA replication (~800 bp/second).

RNA polymerase creates the transcription bubble when it binds to a promoter and melts or separates the two strands. FIGURE 19.4 shows that as RNA polymerase moves along the DNA, the bubble moves with it and the RNA chain grows longer. The process of base pairing and base addition within the bubble is catalyzed and scrutinized by the RNA polymerase itself.

The structure of the bubble within RNA polymerase is shown in the expanded view of FIGURE 19.5. As RNA polymerase moves along the DNA template, it unwinds the duplex at the front of the bubble and rewinds the DNA at the back. The length of the transcription bubble is ~12 to 14 bp, but the length of the RNA-DNA hybrid region within it is ~8 to 9 bp. As the enzyme moves along the template, the DNA duplex reforms, and the RNA is displaced as a free polynucleotide chain. The last 14 ribonucleotides in the growing RNA are complexed with DNA and/or enzyme at any moment.

KEY CONCEPTS

- RNA polymerase separates the two strands of DNA in a transient bubble and uses one strand that runs 3′ to 5′ as a template to direct synthesis of a complementary sequence of RNA running 5′ to 3′.
- The length of the bubble is ~12 to 14 bp, and the length of RNA-DNA hybrid within it is ~8 to 9 bp.

CONCEPT AND REASONING CHECK

If the RNA polymerase is moving 3′ to 5′ along the template strand, why are nucleotides being added to the 3′ end of the growing chain?

19.3 The Transcription Reaction Has Three Stages

The transcription reaction can be arbitrarily divided into the three stages illustrated in FIGURE 19.6: **initiation**, in which the promoter is recognized, a bubble is created, and RNA synthesis begins; **elongation**, as the bubble moves along the DNA; and finally **termination**, in which the RNA transcript is released and the bubble closes.

- **initiation** The stages of transcription up to synthesis of the first bond in RNA. These include binding of RNA polymerase to the promoter (this is sometimes referred to as preinitiation) and melting a short region of DNA into single strands.
- **elongation** The stage in a macromolecular synthesis reaction (replication, transcription, or translation) when the nucleotide or polypeptide chain is extended by the addition of individual subunits.
- **termination** A separate reaction that ends a macromolecular synthesis reaction (replication, transcription, or translation) by stopping the addition of subunits and (typically) causing disassembly of the synthetic apparatus.

Initiation is the most complex part of transcription and consists of multiple steps. *Template recognition* begins with the binding of RNA polymerase to the double-stranded DNA at a DNA sequence called the *promoter.* The enzyme first forms a **closed complex,** in which the DNA remains double stranded. The polymerase then locally unwinds the section of promoter DNA that includes the transcription start site to form the **open complex**. Separation of the DNA double strands makes the template strand available for base pairing with incoming ribonucleotides and synthesis of the first nucleotide bonds in RNA. The initiation phase can be protracted by the occurrence of abortive events, in which the enzyme makes short transcripts, typically shorter than ~10 nucleotides (nt) while still bound at the promoter. The enzyme often makes successive rounds of abortive transcripts by releasing them and starting synthesis of RNA again. The initiation phase ends when the enzyme succeeds in extending the chain and clears the promoter. The sequence of DNA needed for RNA polymerase to bind to the template and accomplish the initiation reaction defines the promoter.

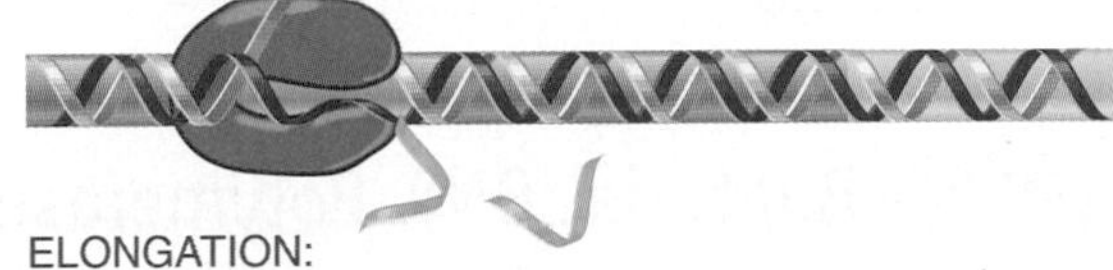

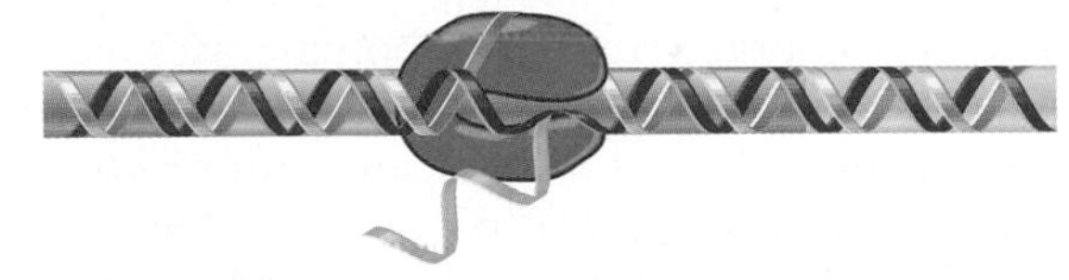

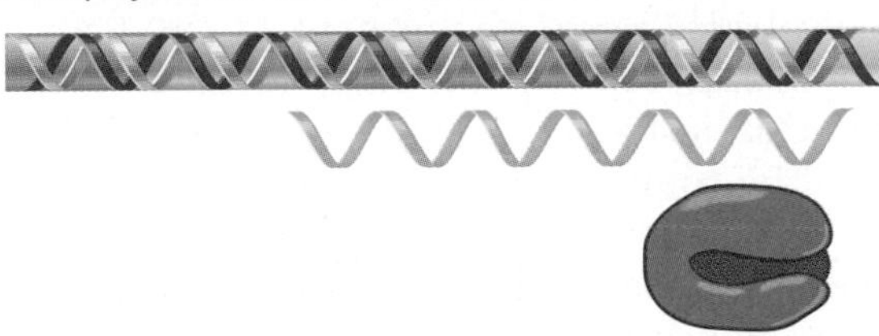

FIGURE 19.6 Transcription has three stages: The enzyme binds to the promoter and melts DNA and remains stationary during initiation; moves along the template during elongation; and dissociates at termination.

- **closed complex** The stage of initiation of transcription before RNA polymerase causes the two strands of DNA to separate to form the transcription bubble. The DNA is double stranded.
- **open complex** The stage of initiation of transcription when RNA polymerase causes the two strands of DNA to separate to form the transcription bubble.

Elongation involves processive movement of the enzyme by disruption of base pairing in double-stranded DNA, exposing the template strand for nucleotide addition, and translocation of the transcription bubble downstream. As the enzyme moves, the template strand of the transiently unwound region is paired with the nascent RNA at the point of growth. Nucleotides are covalently added to the 3′ end of the growing RNA chain, which forms an RNA-DNA hybrid within the unwound region. Behind the unwound region, the DNA template strand pairs with its original partner to reform the double helix as the bubble moves. The RNA emerges from the enzyme as a free single strand.

The traditional view of elongation was that it is a monotonic process, in which the enzyme moves forward at a constant rate. We now know that this is not correct. Certain DNA sequences can cause the polymerase to slow down or even pause indefinitely. As we will see in *Section 19.11, Bacterial Transcription Termination,* prolonged pausing can lead to transcription termination. In addition, the insertion of an incorrect base will alter the DNA structure and cause the polymerase to "backtrack." This is part of an error-editing mechanism.

Termination involves recognition of the point at which no further bases should be added to the chain; this point is, like the promoter, a specific DNA sequence. To terminate transcription, the formation of phosphodiester bonds must cease, and the transcription complex must come apart. When the last base is added to the RNA chain, the transcription bubble collapses as the RNA-DNA hybrid is disrupted, the DNA reforms in duplex state, and the enzyme and RNA are both released. *The sequence of DNA required for these reactions defines the terminator.*

KEY CONCEPTS

- RNA polymerase binds to the promoter on the DNA to form a closed complex.
- RNA polymerase initiates transcription after opening the DNA duplex to form a transcription bubble.
- During elongation the transcription bubble moves along DNA and the RNA chain is extended in the 5′ to 3′ direction by adding nucleotides to the 3′ end of the growing chain.
- When transcription stops, the DNA duplex reforms and RNA polymerase dissociates at a terminator site.

CONCEPT AND REASONING CHECK

DNA polymerase requires a helicase to move the replication fork. How does RNA polymerase move the bubble?

19.4 Bacterial RNA Polymerase Consists of the Core Enzyme and Sigma Factor

The best-characterized RNA polymerases are those of bacteria, especially *Escherichia coli*. The only bacterial RNA polymerases for which high-resolution crystal structures have been solved, however, are from two thermophilic bacterial species, *Thermus aquaticus* and *Thermus thermophilus*. A single type of RNA polymerase is responsible for the synthesis of mRNA, rRNA and tRNA, unlike the situation in eukaryotes, where mRNA, tRNA, and rRNA are transcribed by different polymerases (see *Chapter 20, Eukaryotic Transcription*). About 13,000 RNA polymerase molecules are usually present in an *E. coli* cell. Although not all of them are engaged in transcription at any one time, almost all are bound either specifically or nonspecifically to DNA.

The complete enzyme, or **holoenzyme,** in *E. coli* consists of six subunits and has a molecular weight of ~460 kD. The holoenzyme ($\alpha_2\beta\beta'\omega\sigma$) can be separated into two components, the **core enzyme** ($\alpha_2\beta\beta'\omega$) and the **sigma factor** (the σ polypeptide), which is concerned specifically with promoter recognition. The β and β' subunits together account for catalysis and make up most of the enzyme by mass. Their sequences are related to those of the largest subunits of eukaryotic RNA polymerases (see *Section 25.2, Eukaryotic RNA Polymerases Consist of Many Subunits*). The α subunits serve as a scaffold for assembly of the core enzyme, and also play a role in the interaction of RNA polymerase with the promoter and some regulatory factors through its carboxy terminal domain (**C-terminal domain**). The α and σ subunits are the major surfaces on RNA polymerase for interaction of the enzyme with factors that regulate transcription initiation. The ω subunit also plays a role in assembly and may also play a role in certain regulatory functions.

The σ subunit is primarily responsible for promoter recognition. Only the holoenzyme can recognize a gene at its promoter and initiate transcription. Sigma factor ensures that bacterial RNA polymerase binds in a stable manner to DNA only at the promoter. The sigma factor is sometimes released when the RNA chain reaches eight to nine bases, leaving the core enzyme to undertake elongation. The core enzyme has the ability to synthesize RNA on a DNA template, but it cannot initiate transcription at the proper sites.

The core enzyme has a general affinity for DNA, in which electrostatic attraction between the basic protein and the acidic nucleic acid plays a major role. Any (random) sequence of DNA that is bound by core polymerase in this general binding reaction is described as a loose binding site. No change occurs in the DNA, which remains duplex. The complex at such a site is stable, with a half-life for dissociation of the enzyme from DNA of ~60 minutes. Core enzyme does not distinguish between promoters and other sequences of DNA.

The core enzyme has a general affinity for binding to DNA. **FIGURE 19.7** shows that the sigma factor introduces a major change in the affinity of RNA polymerase for DNA.

- **holoenzyme** The RNA polymerase form that is competent to initiate transcription. It consists of the five subunits of the core enzyme ($\alpha_2\beta\beta'\omega$) and σ factor.
- **core enzyme** The complex of RNA polymerase subunits needed for elongation. It does not include additional subunits or factors that may be needed for initiation or termination.
- **sigma factor** The subunit of bacterial RNA polymerase needed for initiation; it is the major influence on selection of promoters.
- **C-terminal domain** The domain of RNA polymerase α subunit that is involved in stimulating transcription by contact with regulatory proteins.

The holoenzyme has a drastically reduced ability to recognize loose binding sites—that is, to bind to any general sequence of DNA. The half-life of the holoenzyme-DNA complex is <1 second, so sigma factor destabilizes the general binding ability dramatically.

Note that sigma factor also confers the ability to recognize specific DNA binding sites. The holoenzyme binds to promoters very tightly, with an association constant increased from that of core enzyme by (on average) 1000 times, and with a half-life of several hours.

The specificity of holoenzyme for promoters compared to other sequences is $\sim 10^7$, but this is only an average because there is wide variation in the affinity with which the holoenzyme binds to different promoter sequences. This is an important parameter in determining the efficiency of an individual promoter in initiating transcription.

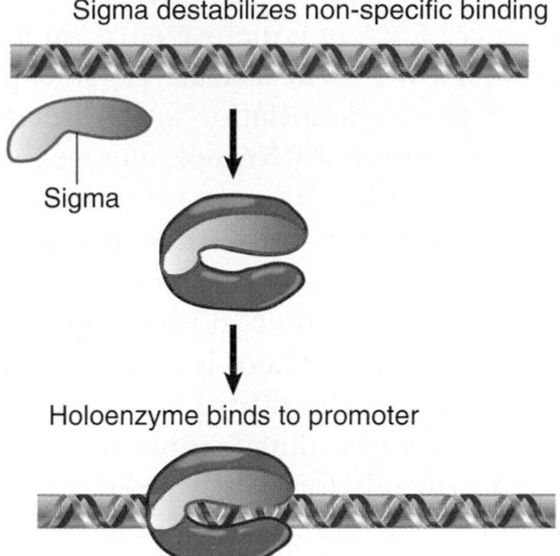

FIGURE 19.7 Core enzyme binds indiscriminately to any DNA. Sigma factor reduces the affinity for sequence-independent binding and confers specificity for promoters.

KEY CONCEPTS

- Bacterial RNA polymerase holoenzyme can be divided into an $\alpha_2\beta\beta'\omega$ core enzyme that catalyzes transcription and a sigma (σ) subunit that is required only for initiation.
- Sigma factor changes the DNA-binding properties of RNA polymerase so that its affinity for general DNA is reduced and its affinity for promoters is increased.

CONCEPT AND REASONING CHECK

How does sigma factor provide promoter specificity?

19.5 How Does RNA Polymerase Find Promoter Sequences?

RNA polymerase must find promoters within the context of the genome. How are promoter sequences distinguished from the several million base pairs in the cell? Virtually all of it is bound to DNA, with no free core enzyme or holoenzyme. **FIGURE 19.8**

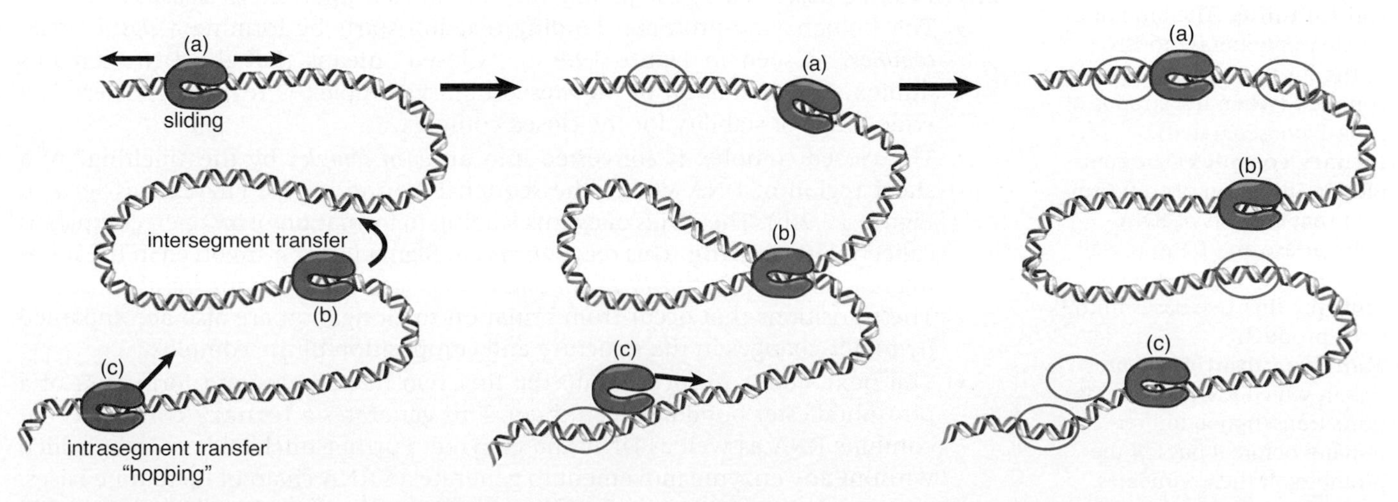

FIGURE 19.8 Proposed mechanisms for how RNA polymerase finds a promoter. (a) Sliding; (b) intersegment transport; (c) intradomain association and dissociation or hopping. Adapted from C. Bustamante et al., *J. Biol. Chem.* 274 (1999): 16665–16668.

illustrates simple models for how RNA polymerase might find promoter sequences from among all the sequences it can access. RNA polymerase holoenzyme locates the chromosome by random diffusion and binds sequence nonspecifically to the negatively charged DNA. In this mode the holoenzyme dissociates very rapidly and then rebinds at another site. However, making and breaking a series of complexes until (by chance) RNA polymerase encounters a promoter sequence and progresses to an open complex would be a relatively slow process. Thus the time required for random cycles of successive association and dissociation at loose binding sites is too great to account for the way that RNA polymerase finds a promoter. RNA polymerase must, therefore, use some other means to seek its binding sites.

Figure 19.8 shows that the process is likely to be speeded up because the initial target for RNA polymerase is the whole genome, not just the specific promoter sequence. How then, does the polymerase move from a random binding site on DNA to a promoter? There is considerable evidence that at least three different processes contribute to the rate of promoter search by RNA polymerase. First, the enzyme may move in a one-dimensional random walk along the DNA ("sliding"). Second, given the intricately folded nature of the chromosome in the nucleoid, having bound to one sequence on the chromosome, the enzyme is now closer to the other sites, reducing the time needed for dissociation and rebinding to another site ("intersegment transfer," or "hopping"). Third, while bound nonspecifically to one site, the enzyme may exchange DNA sites until a promoter is found ("direct transfer").

KEY CONCEPTS

- The rate at which RNA polymerase binds to promoters is too fast to be accounted for by random diffusion.
- RNA polymerase probably binds to random sites on DNA and exchanges them with other sequences very rapidly until a promoter is found.

CONCEPT AND REASONING CHECK

How does sigma factor "read" the DNA sequence of the promoter?

19.6 Sigma Factor Controls Binding to Promoters

- **tight binding** The binding of RNA polymerase to DNA in the formation of an open complex (when the strands of DNA have separated).
- **ternary complex** The complex in initiation of transcription that consists of RNA polymerase and DNA as well as a dinucleotide that represents the first two bases in the RNA product.
- **abortive initiation** A process in which RNA polymerase starts transcription but terminates before it has left the promoter. It then reinitiates. Several cycles may occur before the elongation stage begins.

We can now describe the beginning stages of transcription in terms of the interactions between different forms of RNA polymerase and the DNA template. The initiation reaction can be described by the parameters that are summarized in **FIGURE 19.9**:

- The holoenzyme-promoter binding reaction starts by forming a *closed binary complex*, as seen in Figure 19.9(a). "Closed" means that the DNA remains duplex. The formation of the closed binary complex is reversible. There is a wide range of stability for the closed complex.
- The closed complex is converted into an *open complex* by the "melting" of a short region of DNA within the sequence bound by the enzyme, as seen in Figure 19.9(b). The series of events leading to formation of an open complex is called **tight binding**. This reaction is fast. Sigma factor is involved in the melting reaction (see *Section 19.15, Substitution of Sigma Factors May Control Initiation*). The transitions that occur from initiation to elongation are also accompanied by major changes in the structure and composition of the complex.
- The next step is to incorporate the first two nucleotides and formation of a phosphodiester bond between them. This generates a **ternary complex** that contains RNA as well as DNA and enzyme. Further nucleotides can be added without any enzyme movement to generate an RNA chain of up to nine bases. After each base is added, there is a certain probability that the enzyme will release the chain. This comprises an **abortive initiation**, after which the

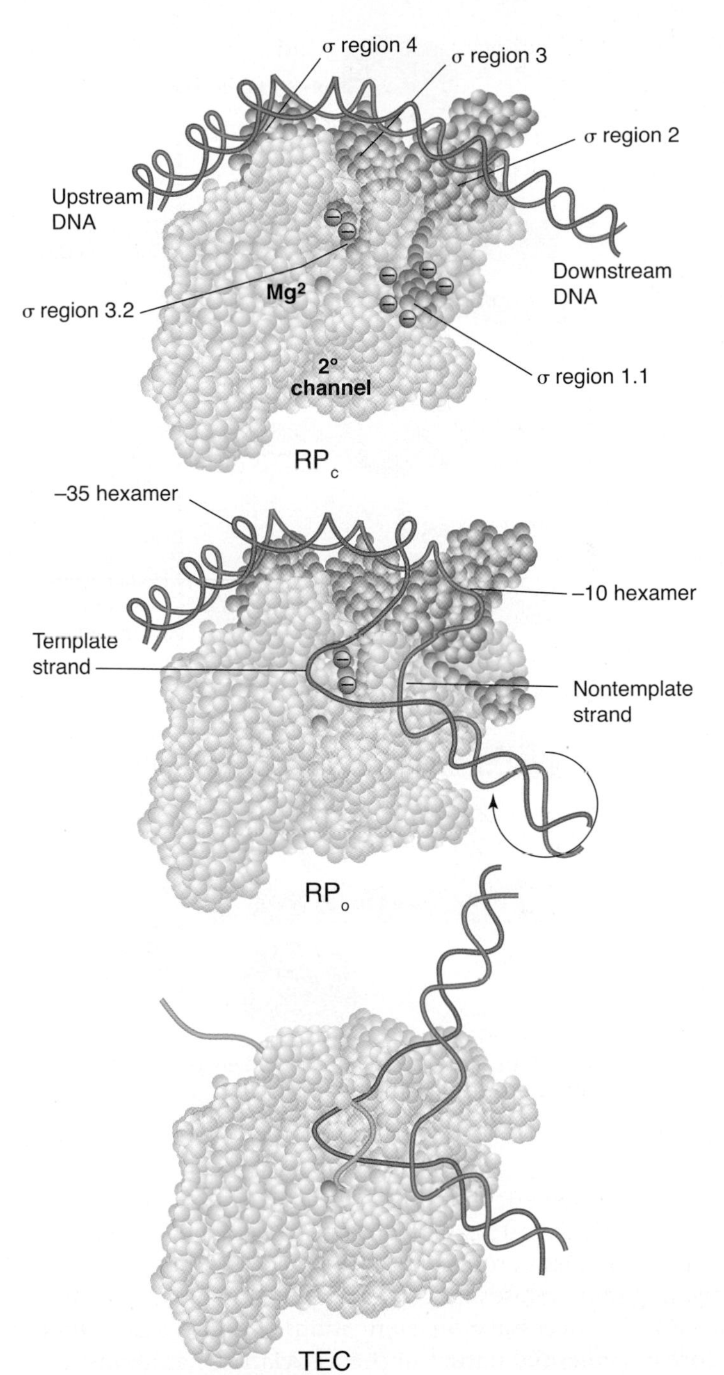

FIGURE 19.9 RNA polymerase passes through several steps prior to elongation. A closed binary complex is converted to an open form and then into a ternary complex. Adapted from S. P. Haugen, W. Ross, and R. L. Gourse, *Nat. Rev. Microbio.* 6 (2008): 507–519.

enzyme begins again with the first base. A cycle of abortive initiations usually occurs to generate a series of very short oligonucleotides.

- When initiation succeeds, sigma factor is no longer necessary, and the enzyme makes the transition to the elongation ternary complex of core polymerase-DNA-nascent RNA (see Figure 19.9 and **FIGURE 19.10**). The critical parameter here is how long it takes for the polymerase to leave the promoter so another polymerase can initiate. This parameter is the promoter clearance time; its minimum value of 1 to 2 seconds establishes the maximum frequency of initiation as <1 event per second. The enzyme then moves along the template, and the RNA chain extends beyond ten bases.
- When the RNA chain extends to 15 to 20 bases, the enzyme makes a further structural transition, to form the complex that undertakes elongation.

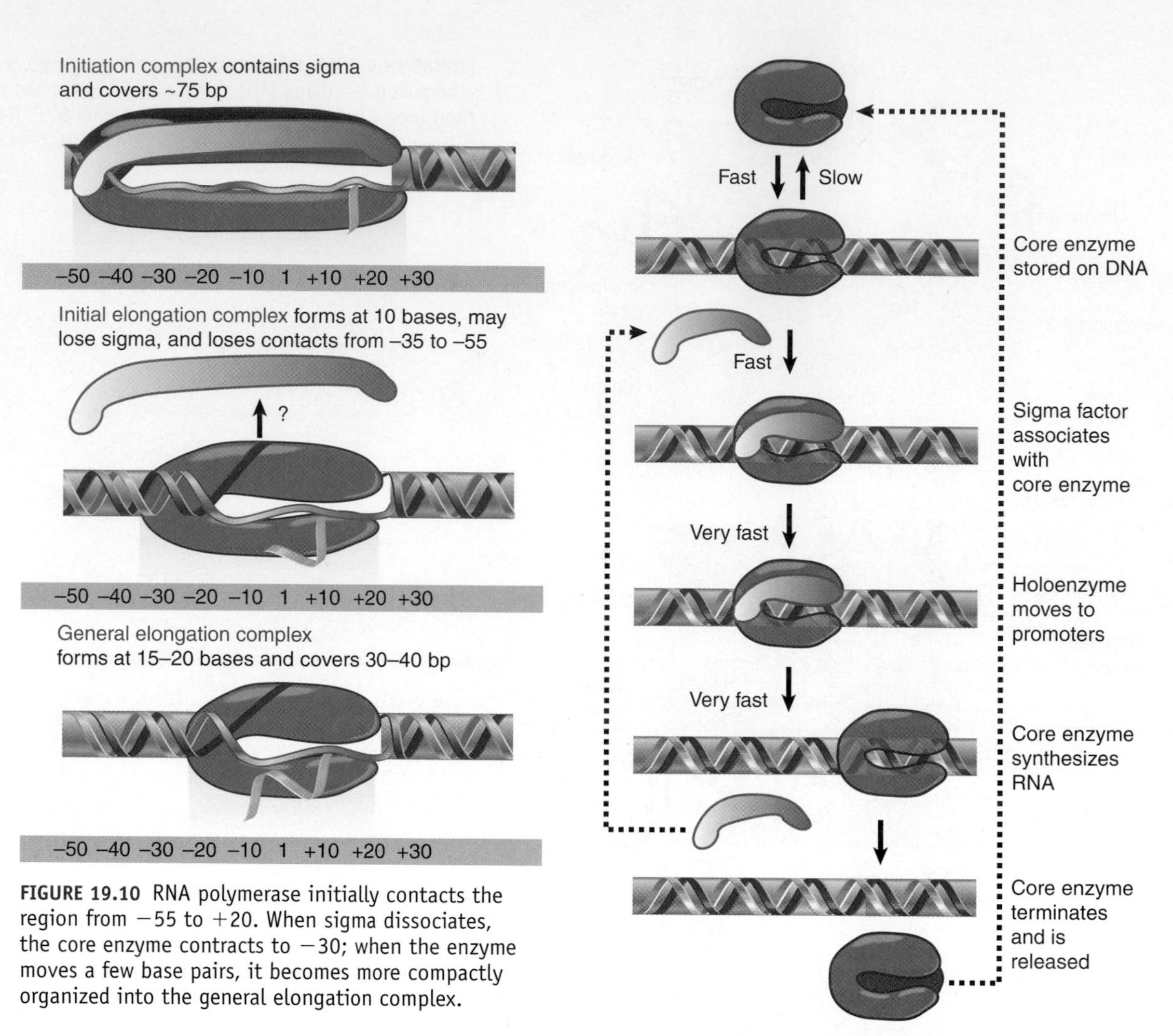

FIGURE 19.10 RNA polymerase initially contacts the region from −55 to +20. When sigma dissociates, the core enzyme contracts to −30; when the enzyme moves a few base pairs, it becomes more compactly organized into the general elongation complex.

FIGURE 19.11 Sigma factor and core enzyme recycle at different points in transcription.

Since soon after the discovery of sigma factor, it has been a tenet of transcription that sigma factor is released after initiation. This may not be strictly true. Direct measurements of elongating RNA polymerase complexes show that ~70% of them retain sigma factor. A third of elongating polymerases lack sigma; hence the original conclusion is certainly correct that it is not necessary for elongation. In those cases where it remains associated with core enzyme, the nature of the association has almost certainly changed (see *Section 19.15, Substitution of Sigma Factors May Control Initiation*).

RNA polymerase encounters a dilemma in reconciling its needs for initiation with those for elongation. Initiation requires tight binding only to particular sequences (promoters), whereas elongation requires close association with all sequences that the enzyme encounters during transcription. **FIGURE 19.11** illustrates how the dilemma is solved by the reversible association between sigma factor and core enzyme.

Sigma factor is involved only in initiation. It becomes unnecessary when abortive initiation is concluded and RNA synthesis has been successfully initiated. At this point, the core enzyme in the ternary complex is bound very tightly to DNA. It is essentially "locked in" until elongation has been completed. When transcription terminates, the core enzyme is released. It is then "stored" by binding to a loose site on DNA. Core enzyme has a high intrinsic affinity for DNA, which is increased by the presence of nascent RNA. Its affinity for loose binding sites is, however, too high to allow the enzyme to efficiently distinguish promoters from other sequences. By reducing the stability of the loose complexes, sigma factor allows the process to occur much more rapidly; by stabilizing the association at tight binding sites, the factor drives the reaction irreversibly into the formation of open complexes. When the enzyme releases sigma factor (or changes its association with it), it reverts to a general affinity for all DNA, irrespective of sequence, that allows it to continue transcription.

What is responsible for the ability of holoenzyme to bind specifically to promoters? Sigma factor has domains that recognize (read) the promoter DNA sequence. As an independent polypeptide, sigma factor does not bind to DNA, but when holoenzyme forms a tight binding complex, sigma factor contacts the DNA in the region upstream of the startpoint. This difference is due to a change in the conformation of sigma factor when it binds to core enzyme. The N-terminal region of free sigma factor suppresses the activity of the DNA-binding region; when sigma binds to core, this inhibition is released, and it becomes able to bind specifically to promoter sequences (see also *Section 19.9, Multiple Regions in RNA Polymerase Directly Contact Promoter DNA*). The inability of free sigma factor to recognize promoter sequences may be important: if sigma factor could freely bind to promoters, it might block holoenzyme from initiating transcription.

KEY CONCEPTS

- RNA polymerase binds to the promoter as a closed complex in which the DNA remains double stranded.
- RNA polymerase then separates the DNA strands to form an open complex and a transcription bubble that incorporates up to nine nucleotides into RNA.
- There may be a cycle of abortive initiations before the enzyme moves to the next phase.
- Sigma factor may be released from RNA polymerase when the nascent RNA chain reaches eight to nine bases in length.
- A change in association between sigma factor and holoenzyme changes binding affinity for DNA, so that core enzyme can move along DNA.

CONCEPT AND REASONING CHECK

What does it mean to convert a promoter from the closed complex to the open complex?

19.7 Promoter Recognition Depends on Consensus Sequences

As a sequence of DNA whose function is to be recognized by proteins, a promoter differs from sequences whose role is to be transcribed. The information for promoter function is provided directly by the DNA sequence: its structure is the signal. This is a classic example of a *cis*-acting site, as defined previously in Figure 2.16 and Figure 2.17. By contrast, expressed regions gain their meaning only after the information is transferred into the form of some other nucleic acid or protein.

A key question in examining the interaction between a protein such as RNA polymerase and the DNA sequence we call a promoter is how the protein reads the specific DNA promoter sequence. In the bacterial genome, the minimum length that could provide an adequate signal is 12 bp. (Any shorter sequence is likely to occur—just by chance—a sufficient number of additional times to provide false signals. The minimum length required for unique recognition increases with the size of genome.) The 12 bp sequence need not be contiguous. If a specific number of base pairs separate two constant shorter sequences, their combined length could be less than 12 bp because the distance of separation itself provides a part of the signal (even if the intermediate sequence is itself irrelevant).

Attempts to identify the features in DNA that are necessary for RNA polymerase binding involved comparing the sequences of different promoters. Any nucleotide sequence essential for polymerase binding should be present in all promoters. Such a

▸ **conserved sequence** Sequences in which many examples of a particular nucleic acid or protein are compared and the same individual bases or amino acids are always found at particular locations.

▸ **consensus sequence** An idealized sequence in which each position represents the base most often found when many actual sequences are compared.

▸ **−35 box** The consensus sequence centered about 35 bp before the startpoint of a bacterial gene. It is involved in initial recognition by RNA polymerase.

sequence is said to be a **conserved sequence**. However, a conserved sequence need not necessarily be identical at every single position; some variation is permitted. How do we analyze a sequence of DNA to determine whether it is sufficiently conserved to constitute a recognizable signal?

Putative DNA recognition sites can be defined in terms of an idealized sequence that represents the base most often present at each position. A **consensus sequence** is defined by aligning all known examples so as to maximize their homology. For a sequence to be accepted as a consensus, each particular base must be reasonably predominant at its position, and most of the actual examples must be related to the consensus by only one or two substitutions.

A striking feature in the sequence of promoters in *E. coli* is the lack of extensive conservation of sequences over the entire 75 bp associated with RNA polymerase. Some short stretches within the promoter are conserved, however, and they are critical for its function. *Conservation of only very short consensus sequences is a typical feature of regulatory sites (such as promoters) in both prokaryotic and eukaryotic genomes.*

There are several elements in bacterial promoters: two 6 bp elements referred to as the −10 element and the −35 element (as well as the length of the "spacer" sequences between them) are usually the most important of these recognition sequences. The promoter sequence at and directly adjacent to the transcription start site, the sequences on either side of the −10 element (referred to as the extended −10 element on the upstream side and the discriminator on the downstream side), and the 10 to 20 bp directly upstream of the −35 element (referred to as the UP element) however, also interact sequence-specifically with RNA polymerase and contribute to promoter efficiency:

- The startpoint is usually (>90% of the time) a purine. It is common for the startpoint to be the central base in the sequence CAT, but the conservation of this triplet is not great enough to regard it as an obligatory signal.
- Just upstream of the startpoint, a 6 bp region is recognizable in almost all promoters. The center of this hexamer is generally about 10 bp upstream of the startpoint; the distance varies in known promoters from position −18 to −9. Named for its location, this hexamer is called the *−10 sequence* or the *Pribnow box*. Its consensus is TATAAT and can be summarized in the form:
  T_{80} A_{95} T_{45} A_{60} A_{50} T_{96}
  where the subscript denotes the percent occurrence of the most frequently found base, which varies from 45% to 96%. (A position at which there is no discernible preference for any base is indicated by "N.") If the frequency of occurrence indicates the likely importance in binding RNA polymerase, we would expect the initial highly conserved TA and the final almost completely conserved T in the −10 sequence to be the most important bases. We now know that the −10 element makes sequence-specific contacts to the sigma factor. This region of the promoter is double stranded in the closed complex when the polymerase binds to it, though, and single stranded in the open complex, so interactions between the −10 element and the polymerase are complex and change at different stages in the process of transcription initiation.
- The conserved hexamer centered ~35 bp upstream of the startpoint is called the **−35 box**. The consensus is TTGACA; in more detailed form, the conservation is: T_{82} T_{84} G_{78} A_{65} C_{54} A_{45}. Bases in this element interact directly with the sigma factor.
- The distance separating the –35 and –10 sites is between 16 and 18 bp in 90% of promoters; in the exceptions, it is as little as 15 bp or as great as 20 bp. Although the actual sequence in the intervening region is unimportant, the distance is critical in holding the two sites at the appropriate separation for the geometry of RNA polymerase.
- Some promoters have an A-T-rich sequence located farther upstream. This is called the *UP element*. Proteins that bind the element interact with the CTDs of the two α subunits of the RNA polymerase. It is typically found in promoters for genes that are highly expressed, such as the promoters for rRNA genes.

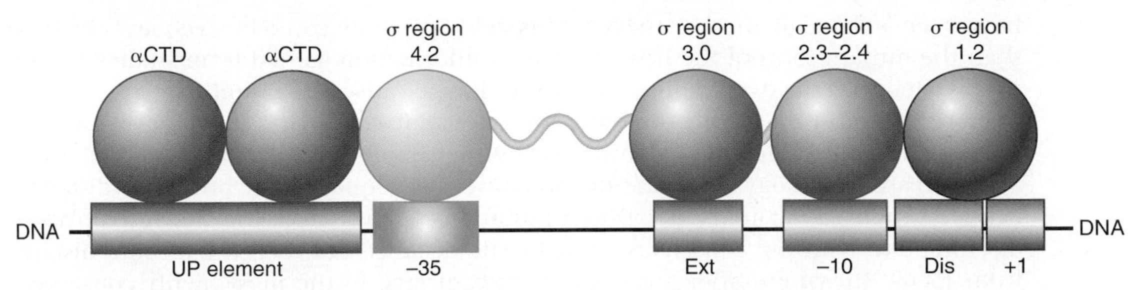

FIGURE 19.12 DNA elements (rectangles) and RNA polymerase regions (circles) that contribute to promoter recognition by sigma factor. Adapted from S. P. Haugen, W. Ross, and R. L. Gourse, *Nat. Rev. Microbiol.* 6 (2008): 507–519.

The optimal promoter is a sequence consisting of the −35 hexamer, separated by 17 bp from the −10 hexamer, lying 7 bp upstream of the startpoint. The structure of a promoter, showing the permitted range of variation from this optimum, is illustrated in **FIGURE 19.12**. The implication that the consensus promoter sequence is the optimal promoter means, first, that there are many different promoter sequences, and second, that there are "good" promoters (with sequence similar to the consensus) and poor promoters (whose sequences vary from the consensus). There is a range of promoter strengths.

RNA polymerase initially binds the promoter region from −50 to +20. Figure 19.12 shows the contacts that RNA polymerase makes with the DNA at a typical promoter. The regions at −35 and −10 contain most of the contact points for the enzyme. Viewed in three dimensions, the points of contact upstream of the −10 sequence all lie on one face of the DNA double helix.

KEY CONCEPTS

- A promoter is the site where RNA polymerase binds the DNA, defined by the presence of short consensus sequences at specific locations.
- The promoter consensus sequences consist of a purine at the startpoint, the hexamer TATAAT centered at −10, and another hexamer centered at −35.
- Individual promoters usually differ from the consensus at one or more positions.
- Promoter efficiency can be affected by additional elements as well.

CONCEPT AND REASONING CHECKS

1. When RNA polymerase binds to a promoter, how does it choose which strand is the template strand?
2. Why is the −10 box AT rich?

19.8 Promoter Efficiencies Can Be Increased or Decreased by Mutation

Mutations are a major source of information about promoter function. Mutations in promoters affect the level of expression of the gene(s) they control without altering the gene products themselves. Most are identified as bacterial mutants that have lost, or have very much reduced, transcription of the adjacent genes. They are known as **down mutations**. Less often, mutants are found in which there is increased transcription from the promoter. They are **up mutations**.

It is important to remember that "up" and "down" mutations are defined relative to the usual efficiency with which a specific promoter functions. This varies widely. Thus a change that is recognized as a down mutation in one promoter might never

- **down mutation** A mutation in a promoter that decreases the rate of transcription.
- **up mutation** A mutation in a promoter that increases the rate of transcription.

have been isolated in another (which in its wild-type state could be even less efficient than the mutant form of the first promoter). Information gained from studies *in vivo* simply identifies the overall direction of the change caused by mutation.

Are the most effective promoters the ones that come closest to the consensus sequences? This expectation is borne out by the simple rule that up mutations usually increase homology with one of the consensus sequences or bring the distance between them closer to 17 bp. Down mutations usually decrease the resemblance of either site with the consensus or make the distance between them more distant from 17 bp. Down mutations tend to be concentrated in the most highly conserved positions, which confirms their particular importance as the main determinant of promoter efficiency. There are, however, occasional exceptions to these rules. In fact, there is no promoter with a perfect match to the consensus sequence. An artificial promoter with a perfect match is actually weaker than promoters with at least one mismatch in the −10 element. This is because they bind RNA polymerase so tightly that it impedes promoter escape.

There is ~100-fold variation in the rate at which RNA polymerase binds to different promoters *in vitro*, which correlates well with the frequencies of transcription when their genes are expressed *in vivo*. By measuring the kinetic constants for formation of a closed complex and its conversion to an open complex, we can dissect two stages of the initiation reaction. Mutations in the two regions have different effects:

- Down mutations in the −35 sequence usually reduce the rate of closed complex formation, but they do not inhibit the conversion to an open complex.
- Down mutations in the −10 sequence can reduce either the initial formation of a closed complex or its conversion to the open form or affect both.

The consensus sequence of the −10 site consists exclusively of A-T base pairs, a configuration that assists the initial melting of DNA into single strands. The lower energy needed to disrupt A-T pairs compared with G-C pairs means that a stretch of A-T pairs demands the minimum amount of energy for strand separation.

The sequences immediately around and downstream from the startpoint also influence the initiation event. Furthermore, the initial transcribed region (from ~+1 to ~+20) influences the rate at which RNA polymerase clears the promoter and, therefore, has an effect upon promoter strength.

Notice that the promoter is asymmetrical; that is, the −10 box is closest to the startpoint, the −35 box furthest away. This allows the polymerase to know whether it should be facing to the right or to the left or, in other words, which strand is the template strand.

A "typical" promoter relies upon its −35 and −10 sequences to be recognized by RNA polymerase, but one or the other of these sequences can be absent from some (exceptional) promoters. In at least some of these cases, the promoter cannot be recognized by RNA polymerase alone; the reaction requires ancillary proteins, which overcome the deficiency in intrinsic interaction between RNA polymerase and the promoter (see *Chapter 26, The Operon*).

KEY CONCEPTS

- Down mutations that decrease promoter efficiency usually decrease conformance to the consensus sequences, whereas up mutations have the opposite effect.
- Mutations in the −35 sequence usually affect initial binding of RNA polymerase.
- Mutations in the −10 sequence usually affect the melting reaction that converts a closed to an open complex.

CONCEPT AND REASONING CHECK

What is the DNA sequence difference between a "good" promoter and a "bad" promoter?

19.9 Multiple Regions in RNA Polymerase Directly Contact Promoter DNA

As mentioned briefly in *Section 19.7*, several domains in the sigma factor subunit and the CTD in the alpha subunits contact promoter DNA. The fact that different holoenzymes with different sigma factors (as seen in **FIGURE 19.13**) and a common core enzyme recognize different promoter sequences implies that the sigma factor subunit must itself contact DNA in these regions. This suggests the general principle that there is a common type of relationship between sigma factor and core enzyme, in which the sigma factor is positioned in such a way as to make critical contacts with the promoter sequences in the vicinity of −35 and −10.

Subunit/gene	Size (# aa)	Approx. # of promoters	Promoter sequence recognized
Sigma 70 (*rpoD*)	613	1000	TTGACA-16 to 18 bp-TATAAT
Sigma 54 (*rpoN*)	477	5	CTGGNA-6 bp-TTGCA
Sigma S (*rpoS*)	330	100	TTGACA-16 to 18 bp-TATAAT
Sigma 32 (*rpoH*)	284	30	CCCTTGAA-13 to 15 bp-CCCGATNT
Sigma F(*rpoF*)	239	40	CTAAA-15 bp-GCCGATAA
Sigma E (*rpoE*)	202	20	GAA-16 bp-YCTGA
Sigma FecI (*fecI*)	173	1–2	?

FIGURE 19.13 *E. coli* sigma factors recognize promoters with different consensus sequences.

Further evidence that sigma factor contacts the promoter directly at both the −35 and −10 consensus sequences is provided by mutations in sigma factor that suppress mutations in the consensus sequences. When a mutation at a particular position in the promoter prevents recognition by RNA polymerase and a compensating mutation in sigma factor allows the polymerase to use the mutant promoter, the most likely explanation is that the relevant base pair in DNA is contacted by the amino acid that has been substituted.

Comparisons of the sequences of several bacterial sigma factors identify regions that have been conserved. Two short regions named 2.4 and 4.2 are involved in contacting bases in the −10 and −35 elements, respectively. Both of these regions form short stretches of α-helix in the protein. They contact bases principally on the coding strand, and continue to hold these contacts after the DNA has been unwound in this region. This suggests that sigma factor could be important in the melting reaction.

The use of α-helical motifs in proteins to recognize duplex DNA sequences is common (see *Section 27.10, Lambda Repressor Uses a Helix-Turn-Helix Motif to Bind DNA*). Amino acids separated by three to four positions lie on the same face of an α-helix and are, therefore, in a position to contact adjacent base pairs. **FIGURE 19.14** shows that amino acids lying along one face of the 2.4 region α-helix contact the bases at positions −12 to −10 of the −10 promoter sequence.

The N-terminal region of σ^{70} has important regulatory functions. If it is removed, the shortened protein becomes able to bind specifically to promoter sequences in the absence of RNA polymerase core enzyme. This suggests that the N-terminal region behaves as an autoinhibition domain. It occludes the DNA-binding domains when σ^{70} is free. Association with core enzyme changes the conformation of sigma so that

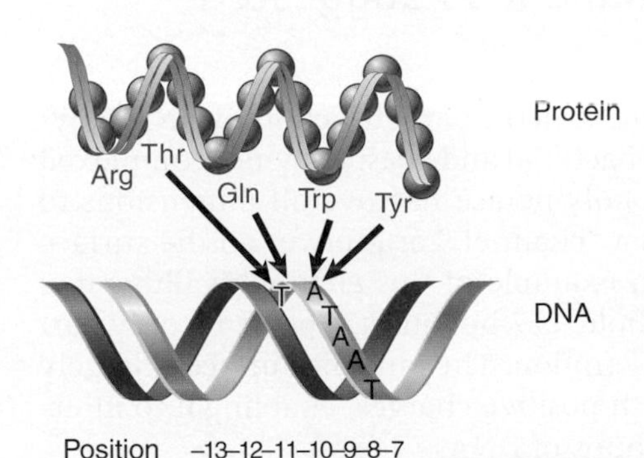

FIGURE 19.14 Amino acids in the 2.4 helix of σ^{70} contact specific bases in the coding strand of the -10 promoter element.

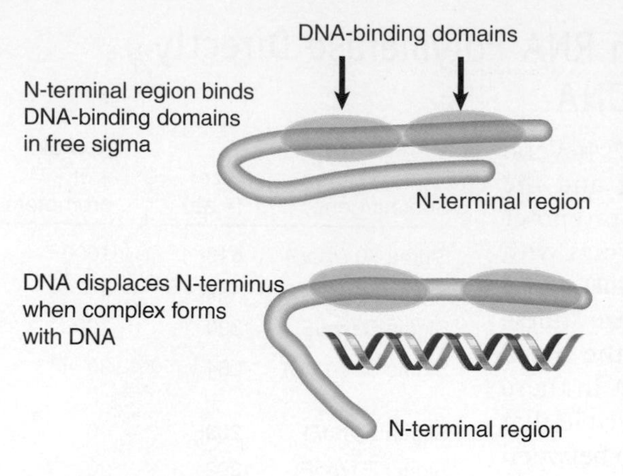

FIGURE 19.15 The N-terminus of sigma blocks the DNA-binding regions from binding to DNA. When an open complex forms, the N-terminus swings 20 Å away, and the two DNA-binding regions separate by 15 Å.

the inhibition is released, and the DNA-binding domains can contact DNA. **FIGURE 19.15** schematizes the conformational change in sigma factor at open complex formation. When sigma factor binds to the core polymerase, the N-terminal domain swings ~20 Å away from the DNA-binding domains, and the DNA-binding domains separate from one another by ~15 Å, presumably to acquire a more elongated conformation appropriate for contacting DNA. Mutations in either the −10 or −35 sequences prevent an (N-terminal-deleted) σ^{70} from binding to DNA, which suggests that σ^{70} contacts both sequences simultaneously. In the free holoenzyme, the N-terminal domain is located in the active site of the core enzyme components, essentially mimicking the location that DNA will occupy when a transcription complex is formed. When the holoenzyme forms an open complex on DNA, the N-terminal sigma domain is displaced from the active site. Its relationship with the rest of the protein is therefore very flexible; the relationship changes when sigma factor binds to core enzyme and again when the holoenzyme binds to DNA.

Although it was first thought that sigma factor is the only subunit of RNA polymerase that contributes to the promoter region binding, the C-terminal domains of the two alpha subunits can also play a major role in contacting promoter DNA by binding to UP elements (see *Section 19.7*). Because the αCTDs are tethered flexibly to the rest of the RNA polymerase, the enzyme can reach regions quite far upstream while still bound to the –10 and –35 elements. The αCTDs thereby provide mobile domains for contacting transcription factors bound at different distances upstream from the transcription start site in different promoters (see *Chapter 26, The Operon*).

KEY CONCEPTS

- σ^{70} changes its structure to expose its DNA-binding regions when it associates with core enzyme.
- σ^{70} binds both the −35 and −10 sequences.

CONCEPT AND REASONING CHECK

How can changing sigma factors change which gene is expressed?

19.10 A Model for Enzyme Movement Is Suggested by the Crystal Structure

We now have abundant information about the structure and function of RNA polymerase as the result of the crystal structures of bacterial and yeast enzymes complexed with NTPs and/or with DNA. Bacterial RNA polymerase has overall dimensions of ~90 × 95 × 160 Å. Structural analysis shows a "channel," or groove, on the surface ~25 Å wide that accommodates the DNA. An example of this channel is illustrated in **FIGURE 19.16**. This groove is long enough to hold ~17 bp, but it represents only part of the total length of DNA bound during transcription. The enzyme surface is largely negatively charged, but the groove is lined with positive charges, enabling it to interact with the negatively charged phosphate groups of DNA.

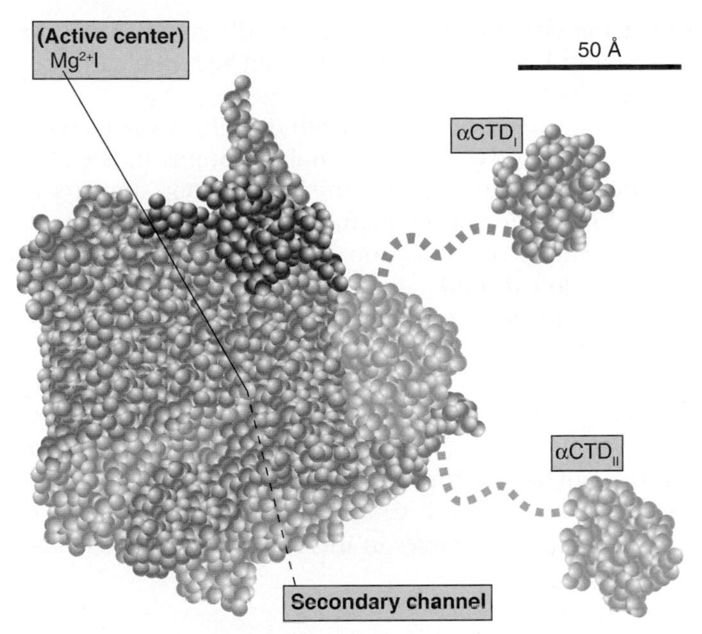

FIGURE 19.16 A model showing the structure of RNA polymerase through the main channel. Subunits are color coded as follows: β′, violet, β, cyan, αI, green, αII, yellow, ω, red. Adapted from K. M. Geszvain and R. Landick (ed. N. P. Higgins). *The Bacterial Chromosome*. American Society for Microbiology, 2004.

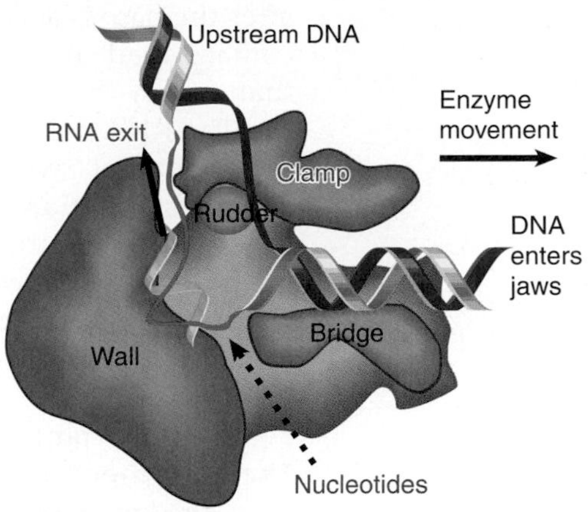

FIGURE 19.17 DNA is forced to make a turn at the active site by a wall of protein. Nucleotides may enter the active site through a pore in the protein.

The catalytic site of RNA polymerase is formed by a cleft between the two large subunits which grasp DNA downstream in its "jaws" as it enters the enzyme. RNA polymerase surrounds the DNA, and a catalytic Mg^{2+} ion is found in the active site. The DNA is held in position by the downstream clamp, another name for one of the jaws. **FIGURE 19.17** illustrates the 90° turn that the DNA takes at the entrance to the active site because of an adjacent wall of protein. The length of the RNA hybrid is limited by another protein obstruction, called the lid. Nucleotides are thought to enter the active site from below, via the secondary channel. The transcription bubble includes 8 to 9 bp of DNA-RNA hybrid.

Once DNA has been melted, the individual strands have a flexible structure in the transcription bubble. This enables DNA to make its 90° turn in the active site. Furthermore, there is a large conformational shift in the enzyme involving the clamp that makes up one of the jaws that holds the downstream DNA in place.

One of the dilemmas for any nucleic acid polymerase is that the enzyme must make tight contacts with the nucleic acid substrate and product, but must break these contacts and remake them with each cycle of nucleotide addition. Consider the situation illustrated in **FIGURE 19.18**. A polymerase makes a series of specific contacts with the bases at particular positions. For example, contact 1 is made with the base at the end of the growing chain and contact 2 is made with the base in the template strand that is complementary to the next base to be added. Note, however, that the bases that occupy these locations in the nucleic acid chains change every time a nucleotide is added!

The top and bottom panels of the figure show the same situation: a base is about to be added to the growing chain. The difference is that the growing chain has been extended by one base in the bottom panel. The geometry of both complexes is exactly the same, but contacts 1 and 2 in the bottom panel are made to bases in the nucleic acid chains that are located one position farther along the chain. The middle panel shows that this must mean that

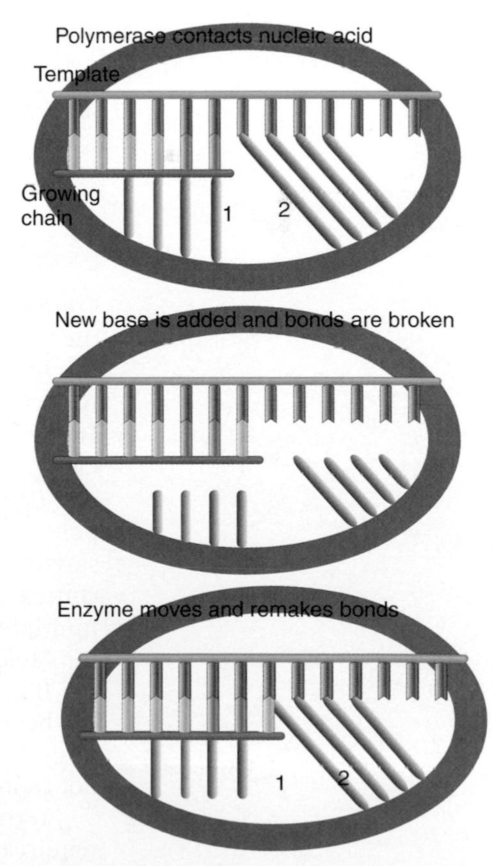

FIGURE 19.18 Movement of a nucleic acid polymerase requires breaking and remaking bonds to the nucleotides at fixed positions relative to the enzyme structure. The nucleotides in these positions change each time the enzyme moves a base along the template.

after the base is added, and before the enzyme moves relative to the nucleic acid, the contacts made to specific positions must be broken so that they can be remade to bases that occupy those positions after the movement.

The RNA polymerase crystal structure suggests an insight into how the enzyme retains contact with its substrate while breaking and remaking bonds in the process of nucleotide addition. A flexible module called the trigger loop appears to be unfolded before nucleotide addition but becomes folded once the NTP enters the active site. Once bond formation and translocation of the enzyme to the next position are complete, the trigger loop unfolds again, ready for the next cycle. Thus, a structural change in the trigger loop coordinates the sequence of events in catalysis.

An important question is: How does RNA polymerase move down the DNA? The movement of this molecular machine or motor can best be described in terms of a Brownian ratchet. Random oscillations or fluctuations in structure between a preposition conformation and a postposition conformation occur stochastically. Pairing of the proper nucleotide to the template strand locks in the postposition structure of the polymerase. The polymerase moves in increments of one base pair steps.

KEY CONCEPTS

- DNA moves through a channel in RNA polymerase and makes a sharp turn at the active site.
- Changes in the conformation of certain flexible modules within the enzyme control the entry of nucleotides to the active site.

CONCEPT AND REASONING CHECK

Why is understanding structure necessary for understanding function?

19.11 Bacterial Transcription Termination

Transcription termination at a site called the *terminator* occurs when the RNA polymerase stops adding nucleotides to the growing RNA chain, releases the completed product, and dissociates from the DNA template. Termination requires that all hydrogen bonds holding the RNA-DNA hybrid together be broken, after which the DNA duplex reforms. It is misleading, however, to consider that termination and elongation are separate processes. In fact, they are a continuum. As discussed in *Section 19.3, The Transcription Reaction Has Three Stages*, the rate of transcription by RNA polymerase is not constant. Transcription can proceed rapidly in some regions or slowly in other regions or even pause, depending on the underlying DNA sequence context. Pausing is a prerequisite for transcription termination. All organisms have multiple elongation control regulators to prevent prolonged pausing at nonterminator sites.

It is sometimes difficult to define the termination point of an RNA molecule that has been synthesized in the living cell. It is always possible that the 3′ end of the molecule has been generated by cleavage of the primary transcript and, therefore, does not represent the actual site at which RNA polymerase terminated.

Terminator sites are distinguished in *E. coli* according to whether RNA polymerase requires any additional factors to terminate transcription:

- Core enzyme can terminate *in vitro* at certain sites in the absence of any other factor. These sites are called *intrinsic terminators*.
- Rho-dependent terminators are defined by the need for addition of rho factor (ρ) *in vitro*, and mutations show that the factor is involved in termination *in vivo*.

Both types of terminators share the characteristics summarized in **FIGURE 19.19**.

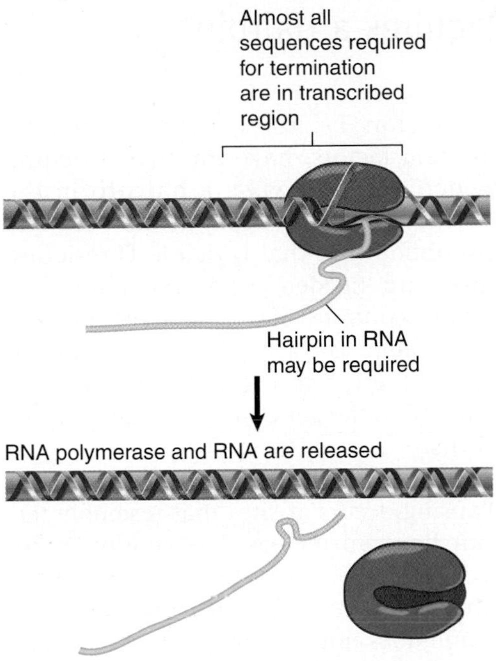

Almost all sequences required for termination are in transcribed region

Hairpin in RNA may be required

RNA polymerase and RNA are released

FIGURE 19.19 The DNA sequences required for termination are located upstream of the terminator sequence. Formation of a hairpin in the RNA may be necessary.

- Terminators are located before the point at which the last base is added to the RNA. The responsibility for termination lies with the sequences already transcribed by RNA polymerase. Thus termination relies on scrutiny of the template or product that the polymerase is currently transcribing.
- Many terminators require a hairpin to form in the secondary structure of the RNA being transcribed. This indicates that termination depends on the RNA product and is not determined simply by scrutiny of the DNA sequence during transcription.

Terminators vary widely in their efficiency. There are "strong" terminators and "weak" terminators. At some terminators, the termination event can be prevented by specific ancillary factors that interact with RNA polymerase. **Antitermination** causes the enzyme to continue transcription past the terminator sequence, an event called *readthrough* (the same term used in *Section 25.11, Suppressor tRNAs Have Mutated Anticodons That Read New Codons,* to describe a ribosome's suppression of termination codons).

In approaching the termination event, we must regard it not simply as a mechanism for generating the 3′ end of the RNA molecule but as an opportunity to control gene expression as described in *Chapter 27, Phage Strategies*. Thus the stages when RNA polymerase associates with DNA (initiation) or dissociates from it (termination) both are subject to specific control. There are interesting parallels between the systems employed in initiation and termination. Both require breaking of hydrogen bonds (initial melting of DNA at initiation and RNA-DNA dissociation at termination), and both require additional proteins to interact with the core enzyme.

▸ **antitermination** A mechanism of transcriptional control in which termination is prevented at a specific terminator site, allowing RNA polymerase to read into the genes beyond it.

KEY CONCEPT

- Termination may require both recognition of the terminator sequence in DNA and the formation of a hairpin structure in the RNA product.

CONCEPT AND REASONING CHECK

Why is pausing required for transcription termination?

19.12 Intrinsic Termination Requires a Hairpin and U-Rich Region

▸ **hairpin** An RNA sequence that can fold back on itself forming double stranded RNA.

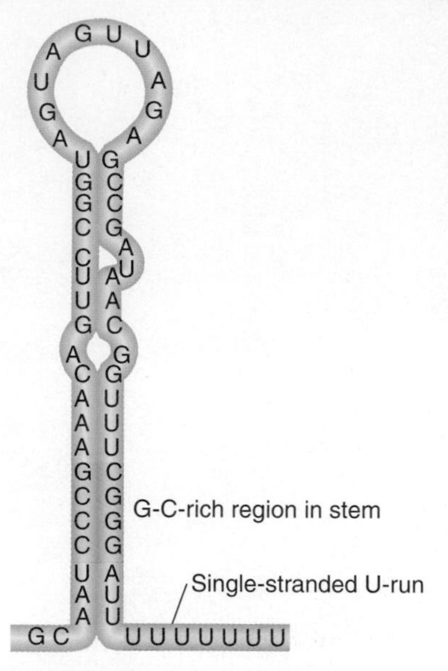

FIGURE 19.20 Intrinsic terminators include palindromic regions that can form hairpins varying in length from 7 to 20 bp. The stem-loop structure includes a G-C-rich region and is followed by a stretch of U residues.

Intrinsic terminators, i.e., those that do not require auxiliary protein factors, have the two structural features evident in **FIGURE 19.20**: a **hairpin** in the secondary structure and a region at the very end of the transcription unit that is rich in U residues. Both features are needed for termination. The hairpin usually contains a G-C-rich region near the base of the stem. The typical distance between the hairpin and the U-rich region is seven to nine bases. About half of the genes in *E. coli* have intrinsic terminators.

Pausing creates an opportunity for termination to occur. Pausing occurs at sites that resemble terminators but have an increased separation (typically 10 to 11 bases) between the hairpin and the U-run. If the pause site does not correspond to a terminator and does not last long enough, though, the enzyme usually moves on again to continue transcription. The length of the pause varies, but at a typical terminator it lasts ~60 seconds. These pause sites can serve regulatory purposes on their own (see *Section 26.13, The trp Operon Is Also Controlled by Attenuation*).

A downstream U-rich region destabilizes the RNA-DNA hybrid when RNA polymerase pauses at the hairpin. The rU-dA RNA-DNA hybrid has an unusually weak base-paired structure; it requires the least energy of any RNA-DNA hybrid to break the association between the two strands. When the polymerase pauses, the RNA-DNA hybrid unravels from the weakly bonded rU-dA terminal region. Often, the actual termination event takes place at any one of several positions toward or at the end of the U-rich region, as though the enzyme "stutters" during termination. The U-rich region in RNA corresponds to an A-T-rich region in DNA, so we see that A-T-rich regions are important in intrinsic termination as well as initiation.

Both the sequence of the hairpin and the length of the U-run influence the efficiency of termination. Termination efficiency *in vitro*, can vary widely, though, for example, from 2% to 90%. The efficiency of termination depends not only on the sequences in the hairpin and the number and position of U residues downstream of the hairpin but also on the surrounding DNA sequence context. The hairpin and U-region are, therefore necessary but not sufficient, and additional parameters influence the interaction with RNA polymerase. In particular, the sequences both upstream and downstream of the intrinsic terminator influence its efficiency.

KEY CONCEPT

- Intrinsic terminators consist of a G-C-rich hairpin in the RNA product followed by a U-rich region in which termination occurs.

CONCEPT AND REASONING CHECK

What are the two kinds of transcription terminator sites, and how do they differ?

19.13 Rho Factor Is a Site-Specific Terminator Protein

Rho factor is an essential protein in *E. coli* that functions solely at the stage of termination. It acts at **rho-dependent terminators**, which account for about half of *E. coli* terminators.

FIGURE 19.21 illustrates the model for Rho functioning. First it binds to a sequence within the transcript upstream of the site of termination. This sequence is called a ***rut*** site (an acronym for rho utilization). Rho then tracks along the RNA until it catches up to RNA polymerase. When the RNA polymerase reaches the termination site, Rho acts on the RNA-DNA hybrid in the enzyme to cause release of the RNA. Pausing by the polymerase at the site of termination allows time for Rho factor to translocate to the hybrid stretch, and is an important feature of termination.

We see an important general principle here. When we know the site on DNA at which some protein exercises its effect, we cannot assume that this coincides with the DNA sequence that it initially recognizes. They can be separate, and there need not be a fixed relationship between them. In fact, *rut* sites in different transcription units are found at varying distances preceding the sites of termination. A similar distinction is made by antitermination factors (see *Section 19.16*).

The common feature of *rut* sites is that the sequence is rich in C residues and poor in G residues and has no secondary structure. An example is given in **FIGURE 19.22.** C is by far the most common base (41%) and G is the least common base (14%).

Rho is a member of the family of hexameric ATP-dependent helicases. Each subunit has an RNA-binding domain and an ATP hydrolysis domain. The hexamer functions by passing nucleic acid through a hole in the middle of the assembly formed from the RNA-binding domains of the subunits

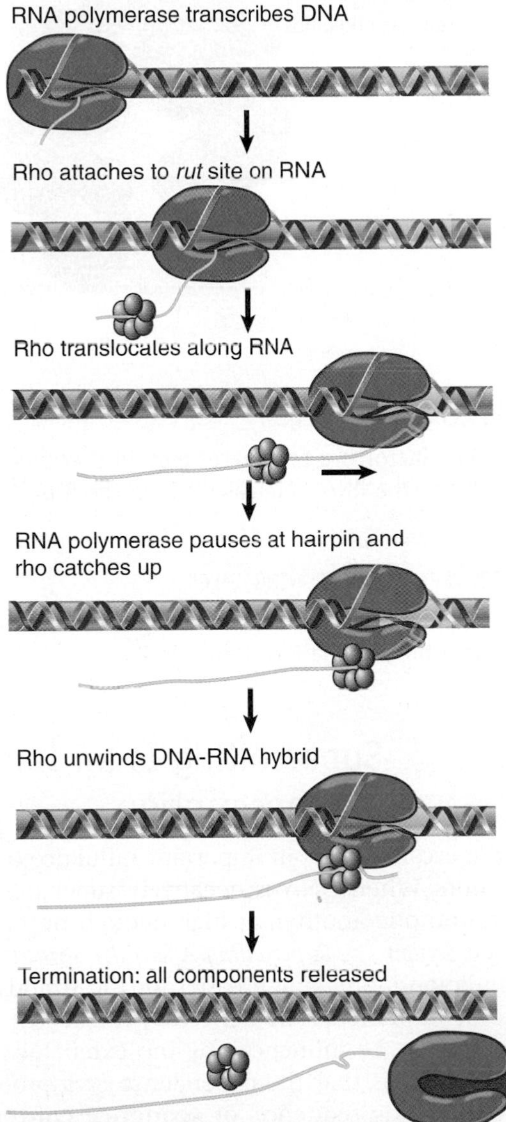

FIGURE 19.21 Rho factor binds to RNA at a *rut* site and translocates along RNA until it reaches the RNA-DNA hybrid in RNA polymerase, where it releases the RNA from the DNA.

- **Rho factor** A protein involved in assisting *E. coli* RNA polymerase to terminate transcription at certain terminators (called rho-dependent terminators).
- **rho-dependent termination** Transcriptional termination by bacterial RNA polymerase in the presence of the Rho factor.
- ***rut*** An acronym for rho utilization site, the sequence of RNA that is recognized by the rho termination factor.

AUCGCUACCUCAUAU CCGCACCUCCUCAAACGCUACCUCGACCAGAAAGGCGUCUCUU

← Deletion prevents termination →

Bases	
C	41%
A	25%
U	20%
G	14%

FIGURE 19.22 A *rut* site has a sequence rich in C and poor in G preceding the actual site(s) of termination. The sequence corresponds to the 3′ end of the RNA.

(see **FIGURE 19.23**). The structure of Rho gives some hints about how it might function. It winds RNA from the 3′ end around the exterior of the N-terminal domainsand pushes the 5′ end of the bound RNA into the interior of the hexamer, where it binds to the C-terminal domains. When it reaches the RNA-DNA hybrid region at the point of transcription, Rho may then use its helicase activity to unwind the duplex structure and to help release the RNA. Some Rho mutations can be suppressed by mutations in other genes. Studying the mutants is an excellent way to identify proteins that interact with Rho, and these studies have implicated the β subunit of RNA polymerase as interacting with the factor.

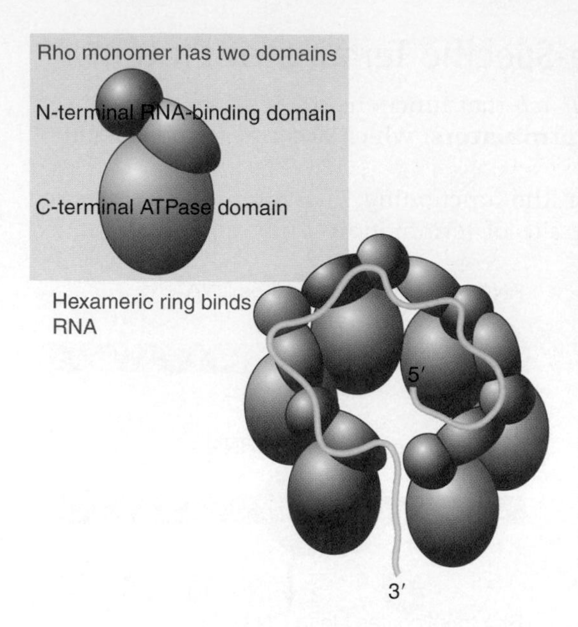

FIGURE 19.23 Rho has an N-terminal RNA binding domain and a C-terminal ATPase domain. A hexamer in the form of a gapped ring binds RNA along the exterior of the N-terminal domains. The 5′ end of the RNA is bound by a secondary binding site in the interior of the hexamer.

KEY CONCEPT

- Rho factor is a terminator protein that binds to a *rut* site on a nascent RNA and tracks along the RNA to release it from the RNA-DNA hybrid structure at the RNA polymerase.

CONCEPT AND REASONING CHECK

If transcription termination is a process that occurs on the DNA template, why does Rho factor bind to the RNA transcript?

19.14 Supercoiling Is an Important Feature of Transcription

Supercoiling has an important influence on transcription at both initiation and elongation. When DNA is negatively supercoiled, the two strands are less tightly wound around one another; at high-enough negative supercoiling density they may separate (see *Section 1.5, Supercoiling Affects the Structure of DNA*). Negative supercoiling may assist initiation by making it easier for the strands of DNA to be separated. The efficiency of some promoters is influenced by the degree of supercoiling. Why, though, should some promoters be influenced by the extent of supercoiling, whereas others are not? One possibility is that the dependence of a promoter on supercoiling is determined by its surrounding sequence, or **sequence context**. This would predict that some promoters have sequences that are easier to melt (and are therefore less dependent on supercoiling), whereas others have more difficult sequences (and have a greater need to be supercoiled). An alternative is that the location of the promoter might be important if different regions of the bacterial chromosome have different degrees of supercoiling.

sequence context The sequence surrounding a consensus sequence. It may modulate the activity of the consensus sequence.

Supercoiling also has a continuing involvement with transcription. As RNA polymerase transcribes DNA, unwinding and rewinding occurs. The consequences of the rotation of DNA are illustrated in **FIGURE 19.24**, in what is often referred to as

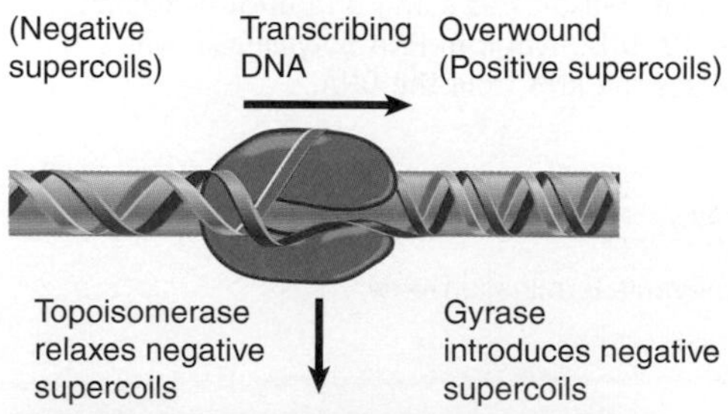

(Negative supercoils)
Transcribing DNA
Overwound (Positive supercoils)
Topoisomerase relaxes negative supercoils
Gyrase introduces negative supercoils
Duplex DNA (10.4 bp/turn)

FIGURE 19.24 Transcription generates more tightly wound (positively supercoiled) DNA ahead of RNA polymerase, while the DNA behind becomes less tightly wound (negatively supercoiled).

the twin domain model for transcription. As RNA polymerase moves with respect to the double helix, it generates positive supercoils (more tightly wound DNA) ahead and leaves negative supercoils (partially unwound DNA) behind. For each helical turn traversed by RNA polymerase, +1 turn is generated ahead and –1 turn behind. Too much positive supercoiling can actually prevent the continued forward motion of the polymerase.

Therefore, transcription is not only affected by the local structure of the DNA, but it also affects the local structure of DNA. As a result, the enzymes gyrase (introduces negative supercoils) and topoisomerase I (removes negative supercoils) are required to rectify the situation in front of and behind the polymerase, respectively. Blocking the activities of gyrase and topoisomerase causes major changes in the supercoiling of DNA.

KEY CONCEPTS

- Negative supercoiling increases the efficiency of some promoters by assisting the melting reaction.
- Transcription generates positive supercoils ahead of the enzyme and negative supercoils behind it, and these must be removed by gyrase and topoisomerase.

CONCEPT AND REASONING CHECK

Why is it more important to relieve positive supercoils rather than negative supercoils that are generated during transcription?

19.15 Substitution of Sigma Factors May Control Initiation

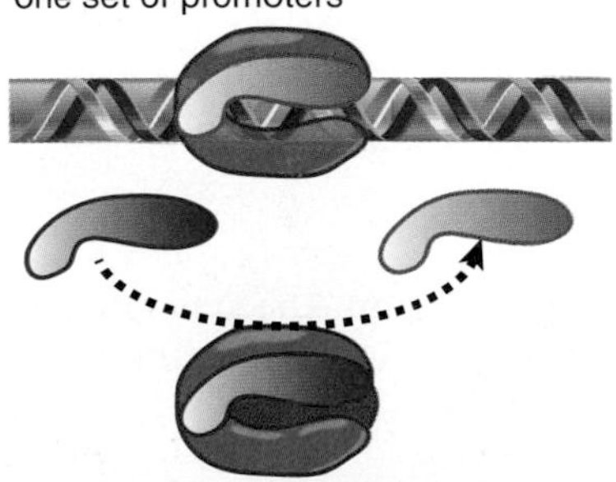

FIGURE 19.25 The sigma factor associated with core polymerase determines the set of promoters at which transcription is initiated.

Because sigma factor determines the choice of the promoter for initiation of transcription, it is possible to control the choice of which gene or group of genes is to be transcribed by changing the sigma factor. **FIGURE 19.25** illustrates the basic idea. When the sigma factor is changed, RNA polymerase behaves in exactly the same way as before, except that it binds to a different promoter sequence. *E. coli* uses alternative sigma factors to respond to changes in environmental or nutritional conditions; they are listed in **FIGURE 19.26** and are named either by molecular weight of the product or for the gene. The general factor, which is responsible for transcription of most genes under normal conditions, is σ^{70}. The alternative sigma factors σ^{S}, σ^{32}, σ^{E}, and σ^{54} are activated in response to environmental changes; σ^{28} is used for expression of flagellar genes during normal growth, but its level of expression responds to changes in the environment. All the sigma factors except σ^{54} belong to the same protein family and function in the same general manner.

Temperature fluctuation is a common type of environmental challenge. Many organisms, both prokaryotic and eukaryotic, respond in a similar way. Upon an increase in temperature, synthesis of the proteins currently being made is turned off

Gene	Factor	Use
rpoD	σ^{70}	most required functions
rpoS	σ^{S}	stationary phase/some stress responses
rpoH	σ^{32}	heat shock
rpoE	σ^{E}	periplasmic/extracellular proteins
rpoN	σ^{54}	nitrogen assimilation
rpoF	σ^{F}	flagellar synthesis/chemotaxis
fecI	σ^{fecI}	iron metabolism/transport

FIGURE 19.26 In addition to σ^{70}, *E. coli* has several sigma factors that are induced by particular environmental conditions. (A number in the name of a factor indicates its mass.)

or down, and a new set of proteins is synthesized. The new proteins are the products of the heat shock genes, which play a role in protecting the cell against environmental stress. Heat shock genes are synthesized in response to conditions other than heat shock as well. Several of the heat shock proteins are chaperones, which reduce the levels of unfolded proteins by refolding them or degrading them. In *E. coli*, the expression of 17 heat-shock proteins is triggered by changes at transcription. The gene *rpoH* is a regulator needed to switch on the heat shock response. Its product is σ^{32}, which functions as an alternative sigma factor that causes transcription of the heat shock genes.

The heat-shock response is feedback regulated and accomplished by increasing the amount of σ^{32} when the temperature increases and decreasing its activity when the temperature change is reversed. The basic signal that induces production of σ^{32} is the accumulation of unfolded (partially denatured) proteins that results from increase in temperature. When unfolded protein levels go down, either by refolding or degradation, they no longer titrate away the proteases that degrade σ^{32} and its levels return to normal. The σ^{32} protein is unstable, which is important in allowing its quantity to be increased or decreased rapidly. The proteins σ^{70} and σ^{32} can compete for the available core enzyme, so that the set of genes transcribed during heat shock depends on the balance between them. More complex regulatory circuits can be constructed by forming cascades of sigma factors; in such a cascade, a set of genes activated by one sigma factor includes a gene coding for another sigma factor that in turn activates another set of genes.

Sigma factors are used extensively to control initiation of transcription in the bacterium *B. subtilis*, for which at least 18 different sigma factors are known. Some are present in vegetative cells; others are produced only in the special circumstances of phage infection or the change from vegetative growth to sporulation.

The major RNA polymerase found in *B. subtilis* cells engaged in normal vegetative growth has the same structure as that of *E. coli*, $\alpha_2\beta\beta'\omega\sigma$. Its sigma factor (described as σ^{43} or σ^{A}) recognizes promoters with the same consensus sequences used by the *E. coli* enzyme under direction from σ^{70}. Several variants of the RNA polymerase that contain other sigma factors are found in much smaller amounts. The variant enzymes recognize different promoter DNA sequences on the basis of consensus sequences at −35 and −10.

▸ **sporulation** The generation of a spore by a bacterium (by morphological conversion) or by a yeast (as the product of meiosis).

▸ **vegetative phase** The period of normal growth and division of a bacterium. For a bacterium that can sporulate, this contrasts with the sporulation phase, when spores are being formed.

Perhaps the most dramatic example of switches in sigma factors is provided by **sporulation**, an alternative lifestyle available to some bacteria. At the end of the **vegetative phase** in a bacterial culture, logarithmic growth ceases because nutrients in the medium become depleted. This triggers sporulation. DNA is replicated, a genome is segregated to one end of the cell, and eventually the genome is surrounded by the tough spore coat. **FIGURE 19.27** shows how successive sigma factors play key regulatory roles in such a cascade.

Sporulation takes approximately 8 hours. It can be viewed as a primitive sort of differentiation, in which a parent cell (the vegetative bacterium) gives rise to two different daughter cells with distinct fates: the mother cell is eventually lysed, and the spore that is released has an entirely different structure from the original bacterium.

Sporulation involves a drastic change in the biosynthetic activities of the bacterium, in which many genes are involved. The basic level of control lies at transcription. Some of the genes that functioned in the vegetative phase are turned off during sporulation, but most continue to be expressed. In addition, the genes specific for sporulation are expressed only during this period. At the end of sporulation, ~40% of the bacterial mRNA is sporulation specific.

New forms of the RNA polymerase become active in sporulating cells; they contain the same core enzyme as vegetative cells but have different sigma factors in place of the vegetative σ^{43}. The changes in transcriptional specificity occur in both the mother cell and the spore. The principle is that in each compartment the existing sigma factor is successively displaced by a new factor that causes transcription of a different set of genes.

Each sigma factor causes RNA polymerase to initiate at a different promoter with a different DNA sequence. The promoters have the same size and location relative to the startpoint; they differ in the –10 and –35 consensus sequences.

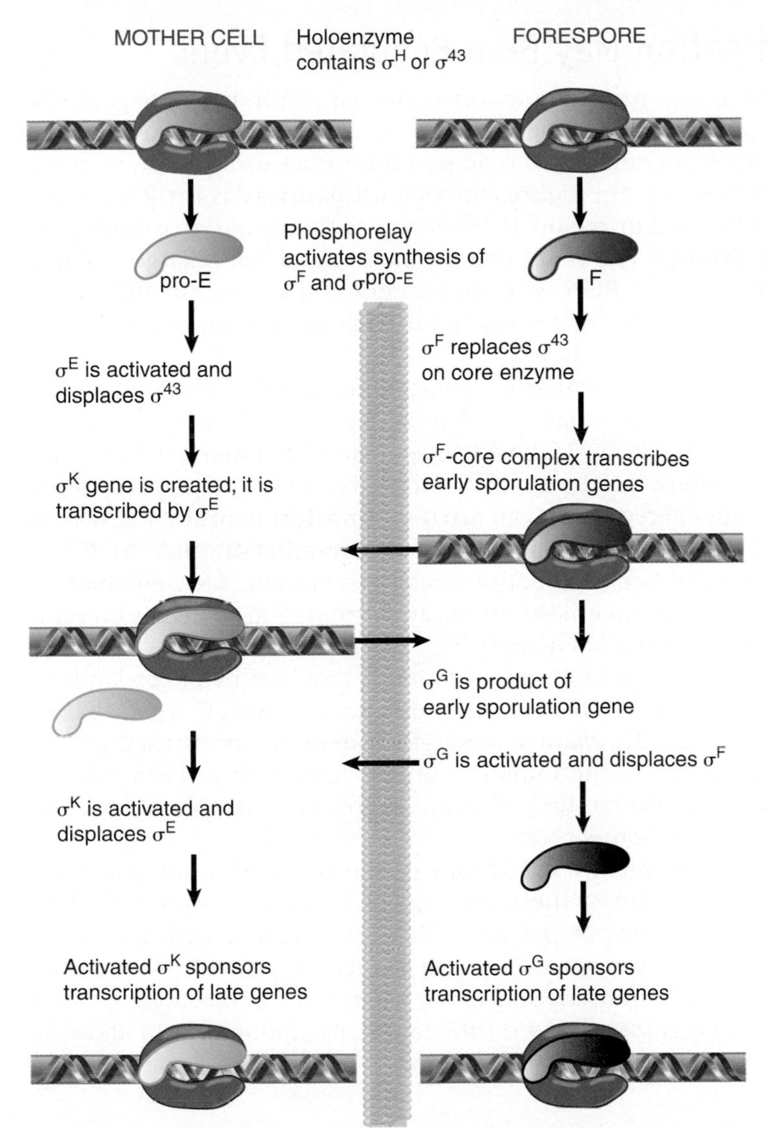

FIGURE 19.27 Sporulation involves successive changes in the sigma factors that control the initiation specificity of RNA polymerase. The cascades in the mother cell (left) and the forespore (right) are regulated by signals passed across the septum (indicated by horizontal arrows).

KEY CONCEPTS

- *E. coli* has seven sigma factors, each of which causes RNA polymerase to initiate at a set of promoters defined by specific –35 and –10 sequences.
- σ^{70} is used for general transcription, and the other sigma factors are activated by special conditions.
- A cascade of sigma factors is created when one sigma factor is required to transcribe the gene coding for the next sigma factor.
- Sporulation divides a bacterium into two compartments: a mother cell that is lysed and a spore that is released.
- Each compartment advances to the next stage of development by synthesizing a new sigma factor that displaces the previous sigma factor.

CONCEPT AND REASONING CHECK

How do different sigma factors recognize different promoters?

19.16 Antitermination May Be a Regulated Event

Antitermination is used as a transcription control mechanism in both phage regulatory circuits and bacterial operons. **FIGURE 19.28** shows that antitermination controls the ability of the RNA polymerase to read past a terminator into downstream genes. In the example shown in the figure, the default pathway is for RNA polymerase to terminate at the end of region 1; however, antitermination allows it to continue transcription through region 2. The promoter does not change, so as a result both situations produce an RNA with the same 5′ sequences; the difference is that after antitermination the RNA is extended to include new sequences at the 3′ end.

Antitermination was discovered in bacteriophage infections. A common feature in the control of phage infection is that very few of the phage genes (the "early" genes) can be transcribed by the bacterial host RNA polymerase. Among these genes, however, are regulator(s) whose product(s) allow the next set of phage genes to be expressed. One of these types of regulator is an **antitermination protein**. **FIGURE 19.29** shows that it enables RNA polymerase to read through a terminator, thus extending the RNA transcript. In the absence of the antitermination protein, RNA polymerase terminates at the terminator (top panel). When the antitermination protein is present, it continues past the terminator (middle panel).

▸ **antitermination protein** Proteins that allow RNA polymerase to transcribe through certain terminator sites.

The best-characterized example of antitermination is provided by phage lambda, in which the phenomenon was discovered. (We discuss the overall regulation of lambda development in *Chapter 27, Phage Strategies*.) Antitermination is used at two stages of phage expression. The antitermination protein produced at each stage is specific for the particular transcription units that are expressed at that stage, as summarized in the bottom panel of Figure 19.29.

The host RNA polymerase initially transcribes two genes, which are called the *immediate early genes*. The transition to the next stage of expression is controlled by preventing termination at the ends of the immediate early genes, with the result that the *delayed early genes* are expressed. The antitermination protein pN acts specifically on the immediate early transcription units. Later, during infection, another antitermination protein pQ acts specifically on the late transcription unit to allow its transcription to continue past a termination sequence.

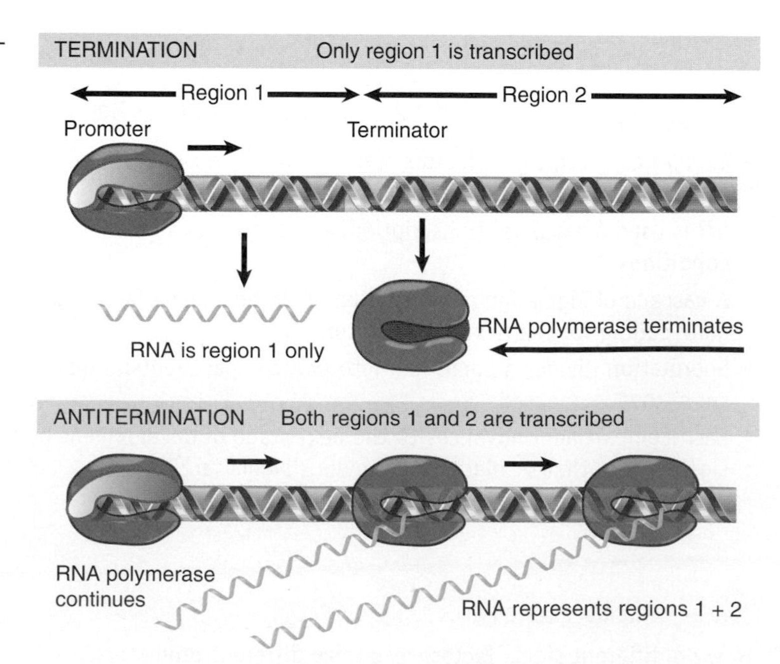

FIGURE 19.28 Antitermination can control transcription by determining whether RNA polymerase terminates or reads through a particular terminator into the following region.

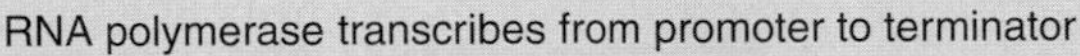

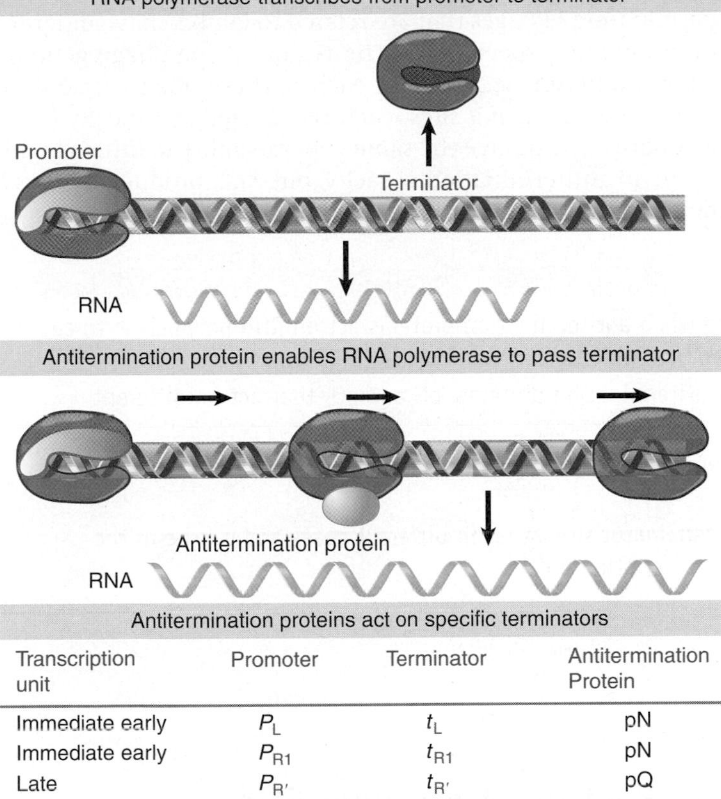

Transcription unit	Promoter	Terminator	Antitermination Protein
Immediate early	P_L	t_L	pN
Immediate early	P_{R1}	t_{R1}	pN
Late	$P_{R'}$	$t_{R'}$	pQ

FIGURE 19.29 An antitermination protein can act on RNA polymerase to enable it to read through a specific terminator.

The different specificities of pN and pQ establish an important general principle: RNA polymerase interacts with transcription units in such a way that an ancillary factor can sponsor antitermination specifically for some transcripts. Termination can be controlled with the same sort of precision as initiation.

What sites are involved in controlling the specificity of antitermination? The antitermination activity of pN is highly specific, but the antitermination event is not determined by the terminators t_{L1} and t_{R1}; the recognition site needed for antitermination lies upstream in the transcription unit, that is, at a different place from the terminator site at which the action eventually is accomplished. N forms a complex with host proteins called Nus factors (*N utilization substances*) to modify RNA polymerase in such a way that it no longer responds to terminators.

The antitermination complex actually forms on the nascent RNA transcribed from a DNA sequence called a ***nut*** (for *N utilization*) site. The complex contains N itself and four host encoded factors. The complex remains bound to the RNA site as a *persistant antitermination complex* as RNA polymerase synthesizes two transcripts to the right and the left.

▸ ***nut*** An acronym for N utilization site, the sequence of DNA that is recognized by the N antitermination factor.

How does antitermination occur? When pN recognizes the *nut* site, it nucleates the formation of the persistent antitermination complex in cooperation with a number of *E. coli* host proteins. These include four host Nus proteins, NusA, B, C and G. NusA is an interesting protein. By itself in *E. coli*, it is part of the transcription termination system. However, when co-opted by N, it participates in antitermination. The complex must act on RNA polymerase to ensure that the enzyme can no longer respond to the terminator. The variable locations of the *nut* sites indicate that this event is linked neither to initiation nor to termination but can occur to RNA polymerase as it elongates the RNA chain past the *nut* site. The polymerase then becomes a juggernaut that continues past the terminator, heedless of its signal. Is the ability of pN to recognize a short sequence within the transcription unit an example of a more widely used mechanism for antitermination? Phages that are related to lambda have different N genes and different antitermination specificities. The region of the phage genome in which the *nut* sites lie has a different sequence in each of these phages, and each phage must therefore have characteristic *nut* sites, each recognized specifically by its own pN. Each of these pN products must have the same general ability to interact with the transcription apparatus in an antitermination capacity, but each product also has a different specificity for the sequence of DNA that activates the mechanism.

KEY CONCEPTS

- Termination is prevented when antitermination proteins act on RNA polymerase to cause it to read through a specific terminator or terminators.
- Phage lambda has two antitermination proteins, pN and pQ, that act on different transcription units.
- The site where an antiterminator protein acts is upstream of the terminator site in the transcription unit.
- The location of the antiterminator site varies in different cases and can be in the promoter or within the transcription unit.

CONCEPT AND REASONING CHECK

How can termination regulation control the expression of other genes?

19.17 The Cycle of Bacterial Messenger RNA

Messenger RNA has the same function in all cells, but there are important differences in the details of the synthesis and structure of prokaryotic and eukaryotic mRNA.

A major difference in the production of mRNA depends on the locations where transcription and translation occur:

- In bacteria, mRNA is transcribed and translated in the single cellular compartment; the two processes are so closely linked that they occur simultaneously. Ribosomes attach to bacterial nascent mRNA even before its transcription has been completed so the *polysome* is likely still to be attached to DNA. Bacterial mRNA usually is unstable, and is therefore translated into polypeptides for only a few minutes. This process is called **coupled transcription/translation**.
- In a eukaryotic cell, synthesis and maturation of mRNA occur exclusively in the nucleus. Only after these events are completed is the mRNA exported to the cytoplasm, where it is translated by ribosomes. A typical eukaryotic mRNA is relatively stable and continues to be translated for several hours, although there is a great deal of variation in the stability of specific mRNAs.

coupled transcription/translation The process in bacteria where a message is simultaneously being translated while it is being transcribed.

FIGURE 19.30 shows that transcription and translation are intimately related in bacteria. Transcription begins when the enzyme RNA polymerase binds to DNA and then moves along, making a copy of one strand. Soon after transcription begins, ribosomes attach to the 5′ end of the mRNA and start translation, even before the rest of the message has been synthesized. Multiple ribosomes move along the mRNA while it is being synthesized. The 3′ end of the mRNA is generated when transcription terminates. Ribosomes continue to translate the mRNA while it survives, but it is degraded in the overall 5′ to 3′ direction quite rapidly. The mRNA is synthesized, translated by the ribosomes, and degraded, all in rapid succession. An individual molecule survives only a matter of minutes at most.

Bacterial transcription and translation take place at similar rates. At 37°C, transcription of mRNA occurs at ~40 to 50 nucleotides/second. This is very close to the rate of protein synthesis, which is roughly 15 amino acids/second. It therefore takes ~1 minute to transcribe and translate an mRNA of 2500 nucleotides, corresponding to a 90 kD polypeptide. When expression of a new gene is initiated, its mRNA typically will appear in the cell within ~1.5 minutes. The corresponding polypeptide will appear within another 0.5 minute.

Bacterial translation is very efficient, and most mRNAs are translated by a large number of tightly packed ribosomes. In one example, (*trp* mRNA), about 15 initiations of transcription occur every minute, and each of the 15 mRNAs probably is translated by ~30 ribosomes in the interval between its transcription and degradation.

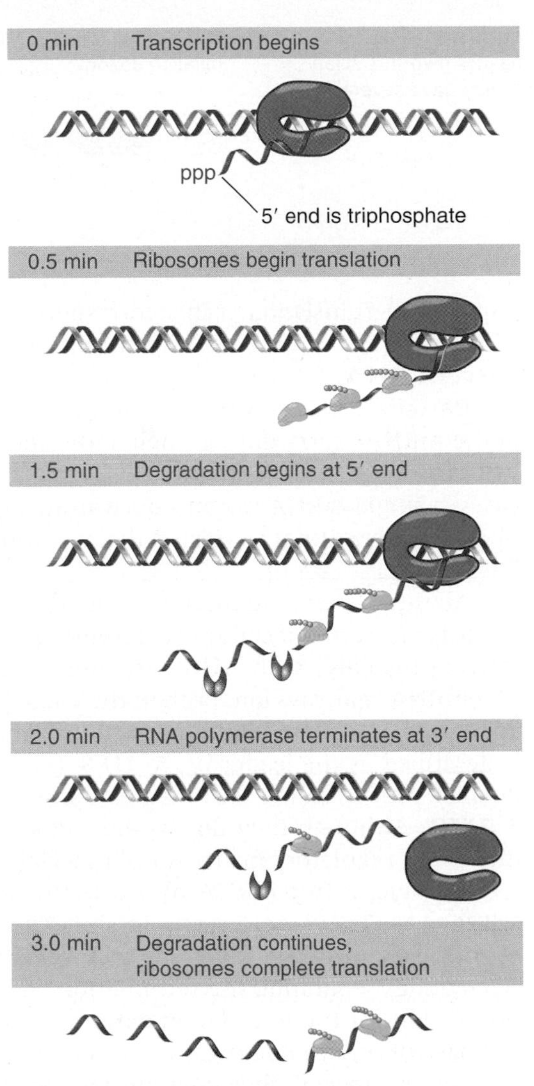

FIGURE 19.30 mRNA is transcribed, translated, and degraded simultaneously in bacteria.

The instability of most bacterial mRNAs is striking. Degradation of mRNA closely follows its translation and likely begins within one minute of the start of transcription. The 5′ end of the mRNA starts to decay before the 3′ has been synthesized or translated. Degradation seems to follow the last ribosome of the convoy along the mRNA. Degradation proceeds more slowly, though—probably at about half the speed of transcription or translation.

The stability of mRNA has a major influence on the amount of polypeptide that is produced. It is usually expressed in terms of the half-life. The mRNA representing any particular gene has a characteristic half-life, but the average is ~2 minutes in bacteria.

The series of events is only possible, of course, because transcription, translation and degradation all occur in the same direction. The dynamics of gene expression have been "caught in the act" in the electron micrograph of **FIGURE 19.31**. In these (unknown) transcription units, several mRNAs are undergoing synthesis simultaneously, and each carries many ribosomes

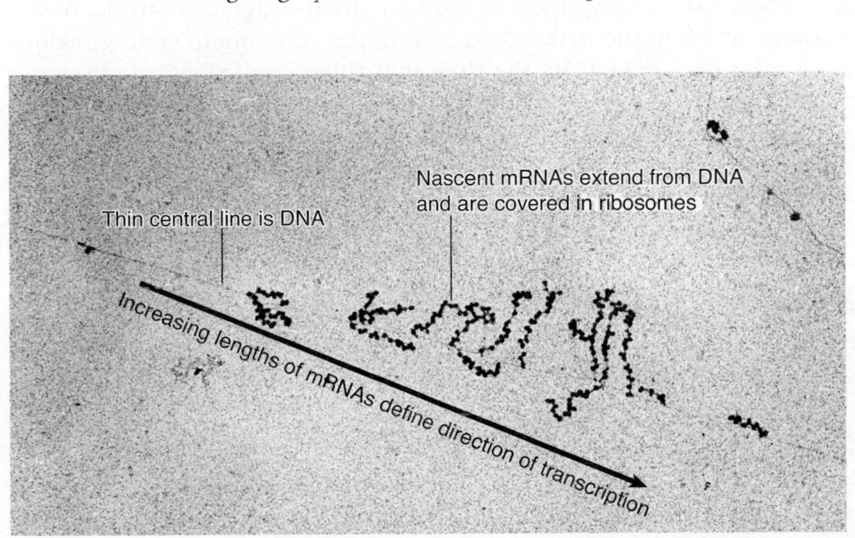

FIGURE 19.31 Transcription units can be visualized in bacteria. Photo courtesy of Oscar Miller.

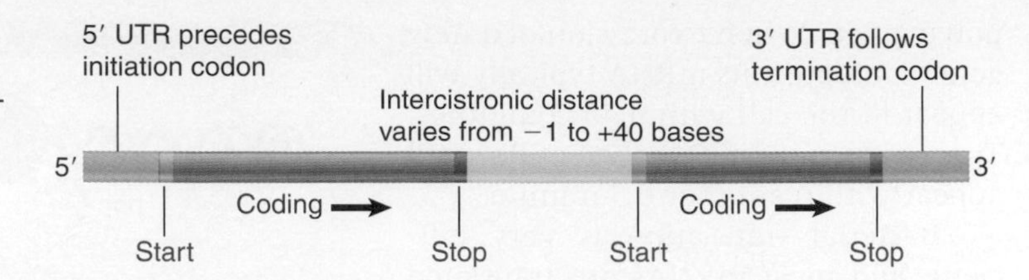

FIGURE 19.32 Bacterial mRNA includes untranslated as well as translated regions. Each coding region has its own initiation and termination codons. A typical mRNA may have several coding regions (ORFs).

engaged in translation. (This corresponds to the stage shown in the second panel of Figure 19.30.) An RNA whose synthesis has not yet been completed is called a **nascent RNA**.

Bacterial mRNAs vary greatly in the number of proteins for which they code. Some mRNAs carry only a single ORF; they are **monocistronic.** Others (the majority) carry sequences coding for several polypeptides; they are **polycistronic**. In these cases, a single mRNA is transcribed from a group of adjacent cistrons. (Such a cluster of cistrons constitutes an operon that is controlled as a single genetic unit; see *Chapter 26, The Operon.*)

All mRNAs contain three regions. The coding region or open reading frame (ORF) consists of a series of codons representing the amino acid sequence of the polypeptide, starting (usually) with AUG and ending with one of the three termination codons. The mRNA is always longer than the coding region, though, as extra regions are present at both ends. An additional sequence at the 5′ end, upstream of the coding region, is described as the leader or **5′ UTR** (untranslated region). An additional sequence downstream from the termination codon, forming the 3′ end, is called the trailer or **3′ UTR**. Although they do not encode a polypeptide, these sequences may contain important regulatory instructions, especially in eukaryotic mRNAs.

A polycistronic mRNA also contains **intercistronic regions**, as illustrated in **FIGURE 19.32**. They vary greatly in size. They may be as long as 30 nucleotides (and even longer in phage RNA), or they may be very short, with as few as one or two nucleotides separating the termination codon from one polypeptide from the initiation codon for the next. In an extreme case, two genes actually overlap, so that the last base of one coding region is also the first base of the next coding region.

The number of ribosomes engaged in translating a particular cistron depends on the efficiency of its initiation site in the 5′ UTR. The initiation site for the first cistron becomes available as soon as the 5′ end is synthesized. How are subsequent cistrons translated? Are the several coding regions in a polycistronic mRNA translated independently or is their expression connected? Is the mechanism of initiation the same for all cistrons, or is it different for the first cistron and the internal cistrons?

Translation of a bacterial mRNA proceeds sequentially through its cistrons. At the time when ribosomes attach to the first coding region, the subsequent coding regions have not yet even been transcribed. By the time that the second ribosome prepares to enter the mRNA, translation is well under way through the first cistron. Typically, ribosomes terminate translation at the end of each cistron and then a new ribosome will assemble at the next coding region. The intercistronic region and the density of ribosomes on the mRNA influence this (see *Chapter 24, Translation*).

- **nascent RNA** An RNA chain that is still being synthesized, so that its 3′ end is still paired with DNA where RNA polymerase is elongating.
- **monocistronic** mRNA that codes for one polypeptide.
- **polycistronic** mRNA that includes coding regions for more that one polypeptide.
- **5′ UTR** The untranslated sequence upstream from the coding region of an mRNA.
- **3′ UTR** The untranslated sequence downstream from the coding region of an mRNA.
- **intercistronic region** In a polycistronic mRNA, the distance between the termination codon of one cistron and the initiation codon of the next cistron.

KEY CONCEPTS

- Transcription and translation occur simultaneously in bacteria, coupled transcription/translation, as ribosomes begin translating an mRNA before its synthesis has been completed.
- Bacterial mRNA is unstable and has a half-life of only a few minutes.
- A bacterial mRNA may be polycistronic in having several coding regions that represent different cistrons.

CONCEPT AND REASONING CHECK

In bacteria, a polysome can form before transcription is complete. Why can't this happen in eukaryotes?

19.18 Summary

A transcription unit comprises the DNA between a promoter, where transcription initiates, and a terminator, where it ends. One strand of the DNA in this region serves as a template for synthesis of a complementary strand of RNA. The RNA-DNA hybrid region is short and transient, as the transcription "bubble" moves along DNA. The RNA polymerase holoenzyme that synthesizes bacterial RNA can be separated into two components. Core enzyme is a multimer of structure $\alpha_2\beta\beta'\omega$ that is responsible for elongating the RNA chain. Sigma factor (σ) is a single subunit that is required at the stage of initiation for recognizing the promoter.

Core enzyme has a general affinity for DNA. The addition of sigma factor reduces the affinity of the enzyme for nonspecific binding to DNA, but increases its affinity for promoters. The rate at which RNA polymerase finds its promoters is too great to be accounted for by diffusion and random contacts with DNA; direct exchange of DNA sequences held by the enzyme may be involved.

Bacterial promoters are identified by two short conserved sequences centered at −35 and −10 relative to the startpoint. Most promoters have sequences that are well related to the consensus sequences at these sites. The distance separating the consensus sequences is 16 to 18 bp. RNA polymerase initially "touches down" at the −35 sequence and then extends its contacts over the −10 region. The initial "closed" binary complex is converted to an "open" binary complex by melting of a sequence of ~12 bp that extends from the −10 region to the startpoint. The A-T-rich base pair composition of the −10 sequence contributes to the melting reaction.

The binary complex between RNA polymerase and DNA is converted to a ternary complex by the incorporation of ribonucleotide precursors. There are multiple cycles of abortive initiation, during which RNA polymerase synthesizes and releases very short RNA chains without moving from the promoter. At the end of this stage, there is a change in structure, and the core enzyme contracts. Sigma factor is either released (30% of cases) or it changes its form of association with the core enzyme. The core enzyme then moves along DNA, synthesizing RNA. A locally unwound region of DNA moves with the enzyme. The "strength" of a promoter describes the frequency at which RNA polymerase initiates transcription; it is related to the closeness with which its –35 and –10 sequences conform to the ideal consensus sequences but is influenced also by the sequences immediately downstream of the startpoint. Negative supercoiling increases the strength of certain promoters. Transcription generates positive supercoils ahead of RNA polymerase and leaves negative supercoils behind the enzyme. The supercoiling must be resolved by topoisomerases.

The core enzyme can be directed to recognize promoters with different consensus sequences by alternative sigma factors. In *E. coli*, these sigma factors are activated by adverse conditions, such as heat shock or nitrogen starvation. *B. subtilis* contains a single major sigma factor with the same specificity as the *E. coli* sigma factor and also contains a variety of minor sigma factors. A specific sequence of sigma factors is activated when sporulation is initiated; sporulation is regulated by two cascades in which sigma factor replacements occur in the forespore (a precursor to the final spore) and mother cell. The geometry of RNA polymerase-promoter recognition is similar for holoenzymes containing all sigma factors. Each sigma factor causes RNA polymerase to initiate transcription at a promoter that conforms to a particular consensus at −35 and −10. Direct contacts between sigma and DNA at these sites have been demonstrated for *E. coli* σ^{70}. The σ^{70} factor of *E. coli* has an N-terminal autoinhibitory domain that prevents the DNA-binding regions from recognizing DNA. The autoinhibitory domain is displaced by DNA when the holoenzyme forms an open complex.

Bacterial RNA polymerase terminates transcription at two types of sites. Intrinsic terminators contain a G-C-rich hairpin followed by a U-rich region. They are recognized *in vitro* by core enzyme alone. Rho-dependent terminators require Rho factor both *in vitro* and *in vivo*; Rho binds to *rut* sites that are rich in C and poor in G residues and that precede the actual site of termination. Rho is a hexameric ATP-dependent helicase that translocates along the RNA until it reaches the RNA-DNA hybrid region in the transcription bubble of RNA polymerase, where it dissociates the RNA from DNA. In both types of termination, pausing by RNA polymerase is important in order to allow time for the actual termination event to occur.

Antitermination is used by some phages to regulate progression from one stage of gene expression to the next. The lambda gene N codes for an antitermination protein (pN) that is necessary to allow RNA polymerase to read through the terminators located at the ends of the immediate early genes. Another antitermination protein, pQ, is required later in phage infection. pN and pQ act on RNA polymerase as it passes specific sites (*nut* and *qut*, respectively). These sites are located at different relative positions in their respective transcription units.

A typical prokaryotic mRNA contains both a nontranslated 5′ UTR (leader) and a 3′ UTR (trailer) as well as a coding region or regions. Bacterial mRNA is usually polycistronic, with untranslated regions between the cistrons. Each cistron is represented by a coding region that starts with a specific initiation codon and ends with a termination codon. Bacterial mRNA has an extremely short half-life of only a few minutes. The 5′ end starts translation even while the downstream sequences are being transcribed.

CHAPTER QUESTIONS

1. Abortive initiation usually occurs when the new transcript is less than:
A. ~5 bases.
B. ~9 bases.
C. ~12 bases.
D. ~16 bases.

2. When RNA polymerase initially binds to DNA, it covers a region of about:
A. 40 to 45 bp.
B. 55 to 60 bp.
C. 75 to 80 bp.
D. 85 to 90 bp.

3. In a bacterial genome, what is the minimal length of DNA that is required to provide a specific signal for protein binding (such as RNA polymerase)?
A. 6 bp
B. 12 bp
C. 21 bp
D. 35 bp

4. In bacteria, the first base of a transcription unit is almost always (>90%):
A. a purine.
B. a pyrimidine.
C. G.
D. C.

5. What is the most common, conserved distance between the −35 and −10 regions of a bacterial gene promoter?
A. 8–10 bp
B. 12–15 bp
C. 16–18 bp
D. 20–23 bp

6. A mutation at the −10 region of a bacterial gene from TATAGT to TCTAGT would be expected to result in:
   A. a significant increase in expression.
   B. elimination of expression.
   C. no change in expression.
   D. a decrease in expression.
7. Down mutations in the −35 sequence of a bacterial gene promoter usually affect:
   A. initial binding of RNA polymerase to form a closed complex.
   B. transition of a closed promoter complex to an open complex.
   C. initiation of transcription.
   D. clearance of the promoter and transition to an elongation complex.
8. Down mutations in the −10 sequence of a bacterial gene promoter usually affect:
   A. initial binding of RNA polymerase.
   B. transition of a closed promoter complex to an open complex.
   C. initiation of transcription.
   D. clearance of the promoter and transition to an elongation complex.
9. All transcription termination in bacteria is most often associated with:
   A. a hairpin structure caused by internal RNA pairing.
   B. a specific termination sequence.
   C. an AU-rich run of sequences.
   D. a GC-rich run of sequences.
10. The bacteriophage lambda N protein functions in transcription by binding:
   A. directly to the DNA template.
   B. to the mRNA transcript.
   C. directly to the RNA polymerase to alter its structure.
   D. to other transcription factors to alter specificity.

KEY TERMS

3′ UTR
5′ UTR
-35 box
abortive initiation
antitermination
antitermination protein
closed complex
coding strand
consensus sequence
conserved sequence
core enzyme
coupled transcription/translation
C-terminal domain
down mutation
downstream
elongation
hairpin
holoenzyme
initiation
intercistronic regions
monocistronic
nascent RNA
nut
open complex
polycistronic
primary transcript
promoter
rho-dependent termination
Rho factor
RNA polymerase
rut
sequence context
sigma factor
sporulation
startpoint
template strand
termination
terminator
ternary complex
tight binding
transcription unit
up mutation
upstream
vegetative phase

FURTHER READING

Greenblat, J. F. (2008). Transcription termination: pulling out all the stops. *Cell* **132**, 917–919. A short review of transcription termination, and see the article on page 971 of the same issue.

Herbert, K. M., Greenleaf, W. J., and Block, S. M. (2008). Single-molecule studies of RNA polymerase: motoring along. *Annu. Rev. Biochem.* **77**, 149–176. An elegant review demonstrating how studies of individual molecules can provide insight into mechanism of action.

Nudler, E. (2009). RNA polymerase Active Center: The Molecular Engine of Transcription. *Annu. Rev. Biochem.* **78**, 335–361. A review of the RNA polymerase structure and function.

Vassylyev, D. G., Vassylyev, M. N., Perederina, A., Tahirov, T. H., and Artsimovitch, I. (2007). Structural basis for transcription elongation by bacterial RNA polymerase. *Nature* **448**, 157–163. The correlation between structure and function during transcription elongation.

Wang, Q., Tullius, T. D., and Levin, J. R. (2007). Effects of discontinuities in the DNA template on abortive initiation and promoter escape by *E. coli* RNA polymerase. *J. Biol. Chem.* **282**, 26917–26927. A report on the steps required to complete the initiation process by RNA polymerase.

20

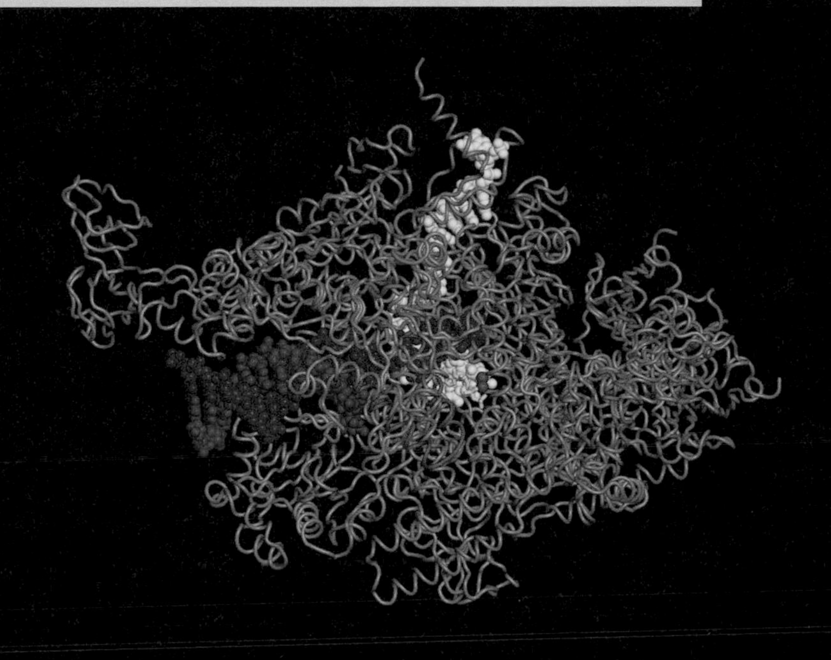

A structure of RNA polymerase in the act of transcription, showing the polymerase itself (green), the DNA template (red and blue), and the RNA transcript (yellow). The incoming nucleotide is shown in cyan. Photo courtesy of Irina Artsimovitch, Ohio State University.

Eukaryotic Transcription

CHAPTER OUTLINE

20.1 Introduction

Initiation of transcription on a chromatin template that is already opened requires the enzyme RNA polymerase to bind at the promoter and transcription factors to bind to enhancers. *In vitro* transcription on a DNA template requires a different subset of transcription factors than are needed to transcribe a chromatin template (we examine how chromatin is opened in *Chapter 28, Eukaryotic Transcription Regulation*). Any protein that is needed for the initiation of transcription, but that is not itself part of RNA polymerase, is defined as a *transcription factor*. Many transcription factors act by recognizing *cis*-acting sites on DNA. Binding to DNA, however, is not the only means of action for a transcription factor. A factor may recognize another factor, may recognize RNA polymerase, or may be incorporated into an initiation complex only in the presence of several other proteins. The ultimate test for membership in the transcription apparatus is functional: a protein must be needed for transcription to occur at a specific promoter or set of promoters.

A significant difference between the transcription of eukaryotic and prokaryotic RNAs is that in bacteria, transcription takes place on a DNA template, whereas in eukaryotes, transcription takes place on a chromatin template. Chromatin changes everything and must be taken into account at every step. The chromatin must be in an open structure and even in an open structure, there must not be a nucleosome occupying the promoter sequence.

A second major difference is that the bacterial RNA polymerase, with its sigma factor subunit, can read the DNA sequence to find and bind to its promoter. A eukaryotic RNA polymerase cannot read the DNA. Initiation at eukaryotic promoters, therefore, involves a large number of factors that must prebind to a variety of *cis*-acting elements before the RNA polymerase can bind. These factors are called **basal transcription factors**. The RNA polymerase then binds to this basal transcription factor/DNA complex. This binding region is defined as the **core promoter**, the region containing all the binding sites necessary for RNA polymerase to bind and function. RNA polymerase itself binds around the **start point** of transcription but does not directly contact the extended upstream region of the promoter. By contrast, the bacterial promoters discussed in *Chapter 19, Prokaryotic Transcription*, are largely defined in terms of the binding site for RNA polymerase in the immediate vicinity of the start point.

- **basal transcription factors** Transcription factors required by the RNA polymerase to form the initiation complex at all promoters. Factors are identified as TF_NX, where N is I, II, or III (signifying the polymerase) and X is a letter.
- **core promoter** The shortest sequence at which an RNA polymerase can initiate transcription (typically at a much lower level than that displayed by a promoter containing additional elements). For RNA polymerase II it is the minimal sequence at which the basal transcription apparatus can assemble, and it often includes one or more of three common promoter elements: the Inr, the TATA box and the DPE. It is typically ~40 bp long.
- **start point** The position on DNA corresponding to the first base incorporated into RNA.

While bacteria have a single RNA polymerase that transcribes all three major classes of genes, transcription in eukaryotic cells is divided into three classes. Each class is transcribed by a different RNA polymerase:

- RNA polymerase I transcribes 18S/28S rRNA.
- RNA polymerase II transcribes mRNA and a few small RNAs.
- RNA polymerase III transcribes tRNA, 5S ribosomal RNA, and other small RNAs.

This is the picture that we have of the major classes of genes. As we will see in *Chapter 30, Regulatory RNA*, recent discoveries using whole genome tiling arrays have uncovered a new world of antisense transcripts, intergenic transcripts, and heterochromatin transcripts. We have very little information about their promoters or their regulation, but we do know that many (possibly most) are produced by RNA polymerase II.

Basal transcription factors are needed for initiation, but most are not required subsequently. For the three eukaryotic RNA polymerases, the transcription factors, rather than the RNA polymerases themselves, are responsible for recognizing the promoter DNA sequence. For all eukaryotic RNA polymerases, the basal transcription factors create a structure at the promoter to provide the target that is recognized by the RNA polymerase. For RNA polymerases I and III, these factors are relatively simple, but for RNA polymerase II they form a sizeable group. The basal factors join with RNA polymerase II to form a complex surrounding the start point, and they determine the site of initiation. The basal factors together with RNA polymerase constitute the basal transcription apparatus.

The promoters for RNA polymerases I and II are (mostly) upstream of the start point, but a large number of promoters for RNA polymerase III lie downstream (within the transcription unit) of the start point. Each promoter contains characteristic sets of short conserved sequences that are recognized by the appropriate class of basal transcription factors. RNA polymerases I and III each recognize a relatively restricted set of promoters, and rely upon a small number of accessory factors.

Promoters utilized by RNA polymerase II show much more variation in sequence and have a modular organization. All RNA polymerase II promoters have sequence elements close to the start point that are bound by the basal apparatus and the polymerase to establish the site of initiation. Other sequences farther upstream (or downstream), called **enhancer** sequences, determine whether the promoter is expressed and, if expressed, whether this occurs in all cell types or is cell type specific. An enhancer is another type of site involved in transcription and is identified by sequences that stimulate initiation but that are located a variable distance from the core promoter. Enhancer elements are often targets for tissue-specific or temporal regulation. Some enhancers bind transcription factors that function by short-range interactions and are located near the promoter, whereas others can be located thousands of base pairs away. **FIGURE 20.1** illustrates the general properties of promoters and enhancers. A regulatory site that binds more negative regulators than positive regulators to control transcription is called a **silencer**.

- **enhancer** A *cis*-acting sequence that increases the utilization of (most) eukaryotic promoters and can function in either orientation and in any location (upstream or downstream) relative to the promoter.

- **silencer** A short sequence of DNA that can inactivate expression of a gene in its vicinity.
- **housekeeping genes** Genes that are (theoretically) expressed in all cells because they provide basic functions needed for sustenance of all cell types.

Promoters that are constitutively expressed and needed in all cells (their genes are sometimes called **housekeeping genes**) have upstream sequence elements that are recognized by ubiquitous activators. No one element/factor combination is an essential component of the promoter, which suggests that initiation by RNA polymerase II may be regulated in many different ways. Promoters that are expressed only in certain times or places have sequence elements that require activators that are available only at those times or places.

The components of an enhancer or silencer resemble those of the promoter, in that they consist of a variety of modular elements that can bind positive regulators or negative regulators in a closely packed array. Enhancers do not need to be near the promoter. They can be upstream, inside a gene, or beyond the end of a gene. Proteins bound at enhancer elements interact with proteins bound at promoter elements, very often through intermediates called **coactivators**.

- **coactivator** Factors required for transcription that do not bind DNA but are required for (DNA-binding) activators to interact with the basal transcription factors.

Eukaryotic transcription is most often under positive regulation: A transcription factor is provided under tissue-specific control to activate a promoter or set of promoters that contain a common target sequence. This is a multistep process that first involves opening the chromatin, then binding the basal transcription factors, and then binding the polymerase. Regulation by specific repression of a target promoter is less common.

A eukaryotic transcription unit generally contains a single gene, and termination occurs beyond the end of the coding region. Termination lacks the regulatory importance that applies in prokaryotic systems. RNA polymerases I and III terminate at discrete sequences in defined reactions, but the mode of termination by RNA polymerase II is not clear. The significant event in generating the 3′ end of an mRNA, however, is not the termination event itself but instead results from a cleavage reaction in the primary transcript (see *Chapter 21, RNA Splicing and Processing*).

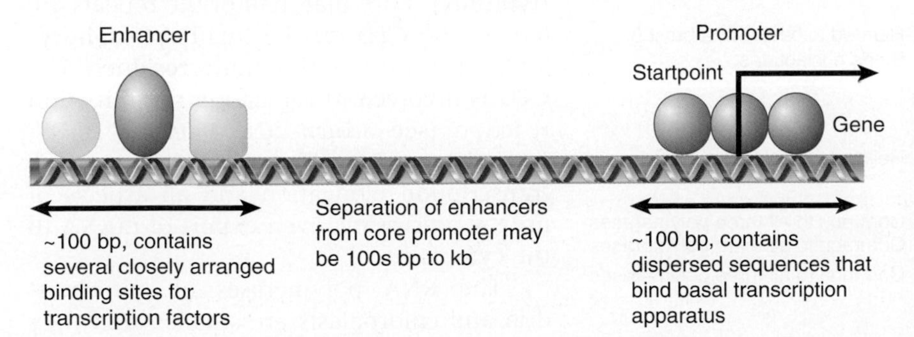

FIGURE 20.1 A typical gene transcribed by RNA polymerase II has a promoter that extends upstream from the site where transcription is initiated. The promoter contains several short (~10 bp) sequence elements that bind transcription factors, dispersed over ~100 bp. An enhancer containing a more closely packed array of elements that also bind transcription factors may be located several hundred bp to several kb distant. (DNA may be coiled or otherwise rearranged so that transcription factors at the promoter interact to form a large protein complex.)

KEY CONCEPT

- Chromatin must be opened before RNA polymerase can bind the promoter.

CONCEPT AND REASONING CHECK

What are the two major differences between bacterial and eukaryotic transcription?

20.2 Eukaryotic RNA Polymerases Consist of Many Subunits

The three eukaryotic RNA polymerases have different locations in the nucleus that correspond to the different genes that they transcribe.

The most prominent activity is the enzyme RNA polymerase I, which resides in the nucleolus and is responsible for transcribing the genes coding for the 18S and 28S rRNA. It accounts for most cellular RNA synthesis (in terms of quantity).

The other major enzyme is RNA polymerase II, which is located in the nucleoplasm (the part of the nucleus excluding the nucleolus). It represents most of the remaining cellular activity and is responsible for synthesizing most of the **heterogeneous nuclear RNA (hnRNA)**, the precursor of most mRNA and a lot more. The classical definition was that hnRNA includes everything but rRNA and tRNA in the nucleus (again, classically, mRNA is only found in the cytoplasm). With modern molecular tools, we can now look a little closer at hnRNA and find many low abundance RNAs that are very important, plus a lot that we are just now starting to understand. The mRNA is the least abundant of the three major RNAs, accounting for just 2% to 5% of the cytoplasmic RNA.

▸ **heterogeneous nuclear RNA (hnRNA)** RNA that comprises transcripts of nuclear genes made primarily by RNA polymerase II; it has a wide size distribution and variable stability.

RNA polymerase III is a minor enzyme in terms of activity, but it produces a collection of stable, essential RNAs. This nucleoplasmic enzyme synthesizes the 5S rRNA, tRNAs and other small RNAs that constitute over a quarter of the cytoplasmic RNAs.

All eukaryotic RNA polymerases are large proteins, appearing as aggregates of ~500 kD. They typically have ~12 subunits. The purified enzyme can undertake template-dependent transcription of RNA but is not able to initiate selectively at promoters. The general constitution of a eukaryotic RNA polymerase II enzyme, as typified in *Saccharomyces cerevisiae,* is illustrated in **FIGURE 20.2**. The two largest subunits are homologous to the β and β′ subunits of bacterial RNA polymerase. Three of the remaining subunits are common to all the RNA polymerases; that is, they are also components of RNA polymerases I and III. Note that there is no subunit related to the bacterial sigma factor. Its function is contained in the basal transcription factors.

The largest subunit in RNA polymerase II has a **carboxy-terminal domain (CTD)**, which consists of multiple repeats of a consensus sequence of seven amino acids. The sequence is unique to RNA polymerase II. There are ~26 repeats in yeast and ~50 in mammals. The number of repeats is important because deletions that remove (typically) more than half of the repeats are lethal. The CTD can be highly phosphorylated on serine or threonine residues. The CTD is involved in regulating the initiation reaction (see *Section 20.8, Initiation Is Followed by Promoter Clearance and Elongation*), transcription elongation and all aspects of mRNA processing, even export of mRNA to the cytoplasm.

▸ **carboxy-terminal domain (CTD)** The domain of the largest subunit of eukaryotic RNA polymerase II that is phosphorylated at initiation and is involved in coordinating several activities with transcription.

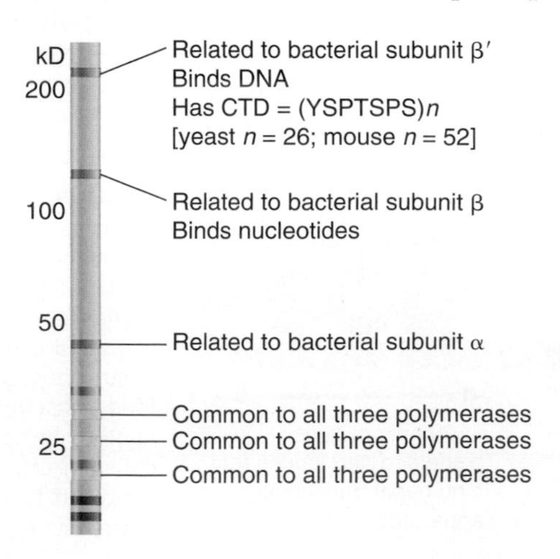

FIGURE 20.2 Some subunits are common to all classes of eukaryotic RNA polymerases and some are related to bacterial RNA polymerase. This drawing is a simulation of purified yeast RNA polymerase II run on an SDS gel to separate the subunits by size.

The RNA polymerases of mitochondria and chloroplasts are smaller, and they resemble bacterial RNA polymerase rather than any of the nuclear enzymes (since they evolved from eubacteria). Of course, the organelle genomes are much smaller, the resident polymerase needs to transcribe relatively few genes, and the control of transcription is likely to be very much simpler.

KEY CONCEPTS

- RNA polymerase I synthesizes rRNA in the nucleolus.
- RNA polymerase II synthesizes mRNA in the nucleoplasm.
- RNA polymerase III synthesizes small RNAs in the nucleoplasm.
- All eukaryotic RNA polymerases have ~12 subunits and are aggregates of ~500 kD.
- Some subunits are common to all three RNA polymerases.
- The largest subunit in RNA polymerase II has a CTD (carboxy-terminal domain) consisting of multiple repeats of a heptamer.

CONCEPT AND REASONING CHECK

Explain the apparent paradox that ribosomal genes number only a few hundreds, yet they account for about 75% of the total cytoplasmic RNA.

20.3 RNA Polymerase I Has a Bipartite Promoter

RNA polymerase I transcribes from a single type of promoter only the genes for ribosomal RNA. The precursor transcript includes the sequences of both large 28S and small 18S rRNAs (but not the 5S rRNA), which are later processed by cleavages and modifications. There are many copies of the transcription unit. They alternate with **nontranscribed spacers** and are organized in a cluster, as discussed in *Section 7.3, Genes for rRNA Form Tandem Repeats, Including an Invariant Transcription Unit*. The organization of the promoter and the events involved in initiation are illustrated in **FIGURE 20.3**. RNA polymerase I exists as a holoenzyme that contains additional factors required for initiation and is recruited by its transcription factors directly as a giant complex to the promoter.

nontranscribed spacer The region between transcription units in a tandem gene cluster.

The promoter consists of two separate regions. The core promoter surrounds the start point, extending from –45 to 20, and is sufficient for transcription to initiate. It is generally G-C-rich (unusual for a promoter), except for the only conserved sequence

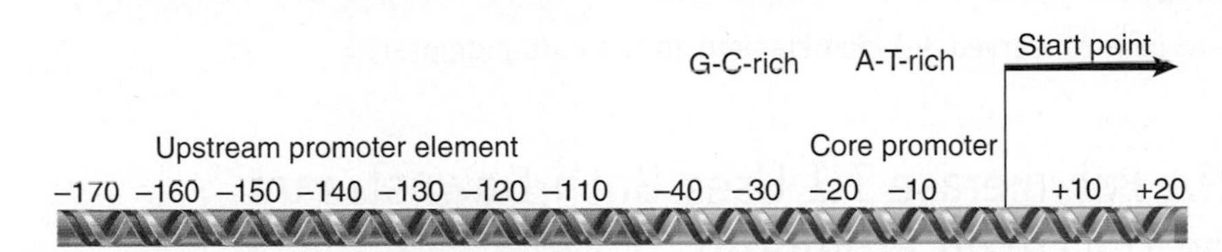

UBF binds to upstream promoter element

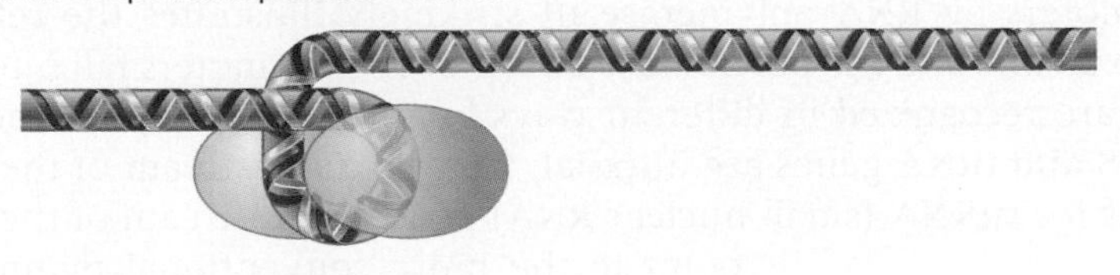

RNA polymerase I holoenzyme includes core binding factor (SL1) that binds to core promoter

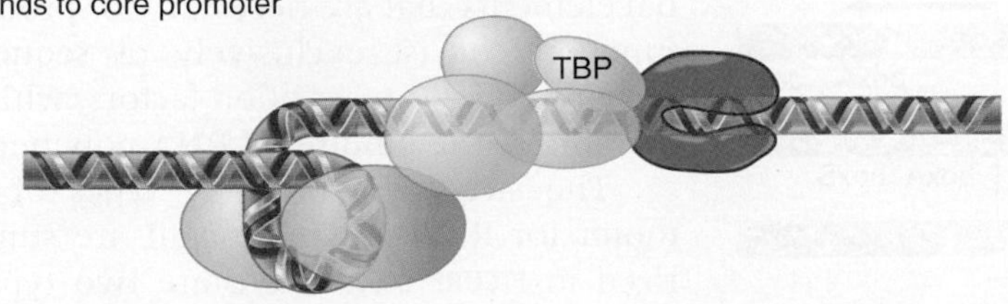

FIGURE 20.3 Transcription units for RNA polymerase I have a core promoter separated by ~70 bp from the upstream promoter element. UBF binding to the UPE increases the ability of core binding to bind to the core promoter. Core-binding factor (SL1) positions RNA polymerase I at the start point.

element, a short A-T-rich sequence around the start point. The core promoter's efficiency, however, is very much increased by the upstream promoter element (UPE, sometimes also called the upstream control element, or UCE). The UPE is another G-C-rich sequence related to the core promoter sequence and extends from –180 to –107. This type of organization is common to pol I promoters in many species, although the actual sequences vary widely.

RNA polymerase I requires two ancillary transcription factors. For high frequency initiation, the factor UBF is required. This is a single polypeptide that binds to a G-C-rich element in the UPE. UBF binds to the minor groove of DNA and wraps the DNA in a loop of almost 360° turn on the protein surface, with the result that the core promoter and UPE come into close proximity. This enables UBF to stimulate binding of the second factor, SL1 (also known as TIF-IB, Rib1 in different species), to the core promoter.

▸ **TATA-binding protein (TBP)** The subunit of transcription factor $TF_{II}D$ that binds to the TATA box in the promoter and is positioned at the promoters that do not contain a TATA box by other factors. Also present in SL1 and $TF_{III}B$.

One of the components of SL1 is the **TATA-binding protein (TBP)**, a factor that also is required for initiation by RNA polymerases II and III (see *Section 20.6, TBP Is a Universal Factor*). TBP does not bind directly to G-C-rich DNA, and DNA binding is the responsibility of the other components of SL1. It is likely that TBP interacts with RNA polymerase, probably with a common subunit or a feature that has been conserved among polymerases. SL1 enables RNA polymerase I to initiate from the promoter at a low basal frequency. SL1 has primary responsibility for ensuring that the RNA polymerase is properly localized at the start point. It consists of four proteins, one of which (TBP) is a component of "positioning factors" that are also required by RNA polymerases II and III. Its exact mode of action is different in each of the positioning factors; at the promoter for RNA polymerase I, it does not bind DNA, whereas at the promoter for RNA polymerase II it can be the principal means for locating the factor on DNA.

KEY CONCEPTS

- The RNA polymerase I promoter consists of a core promoter and an upstream promoter element (UPE).
- The factor UBF1 wraps DNA around a protein structure to bring the core and UPE into proximity.
- SL1 includes the factor TBP that is involved in initiation by all three RNA polymerases.
- RNA polymerase I binds to the UBF1-SL1 complex at the core promoter.

CONCEPT AND REASONING CHECK

Why would there be a conserved A-T rich element in the core promoter?

20.4 RNA Polymerase III Uses Both Downstream and Upstream Promoters

Recognition of promoters by RNA polymerase III strikingly illustrates the relative roles of transcription factors and the polymerase enzyme. The promoters fall into two general classes that are recognized in different ways by different groups of factors. The promoters for 5S and tRNA genes are internal; they lie downstream of the start point. The promoters for snRNA (small nuclear RNA) genes lie upstream of the start point in the more conventional manner of other promoters. In both cases, the individual elements that are necessary for promoter function consist exclusively of sequences recognized by transcription factors, which in turn direct the binding of RNA polymerase.

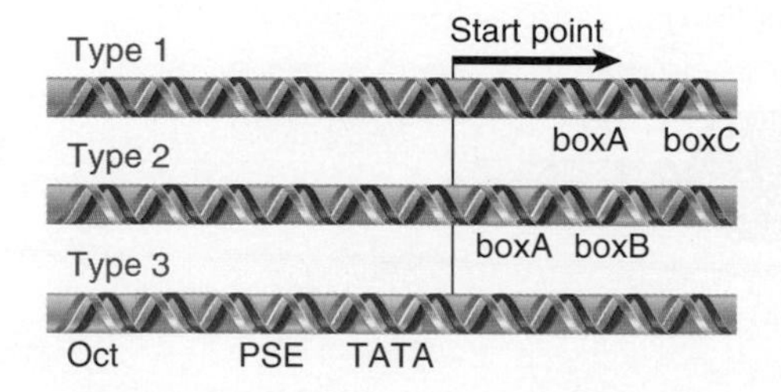

FIGURE 20.4 Promoters for RNA polymerase III may consist of bipartite sequences downstream of the start point, with *boxA* separated from either *boxC* or *boxB*, or they may consist of separated sequences upstream of the start point (Oct, PSE, TATA).

The structures of three types of promoter for RNA polymerase III are summarized in **FIGURE 20.4**. There are two types of internal promoter. Each contains a bipartite structure, in which two short sequence elements are separated by a variable sequence. The 5S ribosomal gene type 1 promoter consists of a *boxA* sequence separated from a *boxC* sequence, and the tRNA type 2 promoter consists of a *boxA* sequence separated from a *boxB* sequence. A common group of type 3 promoters coding for other small RNAs have three sequence elements that are all located upstream of the start point.

The detailed interactions are different at the two types of internal promoter, but the principle is the same. $TF_{III}C$ binds downstream of the start point, either independently (tRNA type 2 promoters) or in conjunction with $TF_{III}A$ (5S type 1 promoters). The presence of $TF_{III}C$ enables the positioning factor $TF_{III}B$ to bind at the start point. RNA polymerase is then recruited.

FIGURE 20.5 summarizes the stages of reaction at type 2 internal promoters used for tRNA genes. The distance between *boxA* and *boxB* can vary since many tRNA genes contain a small intron. $TF_{III}C$ binds to both *boxA* and *boxB*. This enables $TF_{III}B$ to bind at the start point. At this point RNA polymerase III can bind.

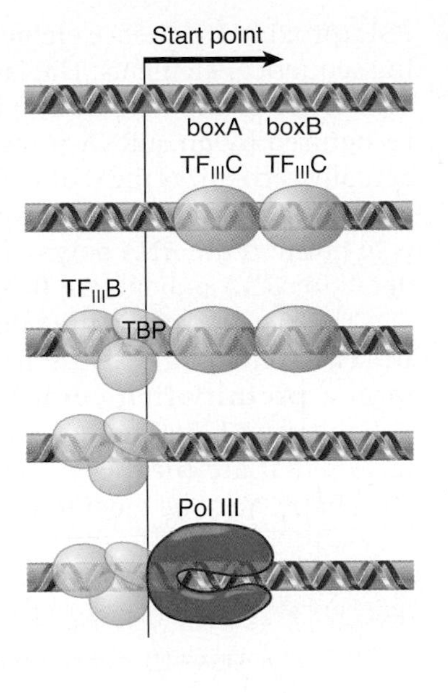

FIGURE 20.5 Internal type 2 pol III promoters use binding of $TF_{III}C$ to *boxA* and *boxB* sequences to recruit the positioning factor $TF_{III}B$, which recruits RNA polymerase III.

The difference at type 1 internal promoters (for 5S genes) is that $TF_{III}A$ must bind at *boxA* to enable $TF_{III}C$ to bind at *boxC*. $TF_{III}A$ is a 5S sequence-specific binding factor that binds to the DNA promoter and to the 5S RNA as a chaperone and gene regulator. **FIGURE 20.6** shows that once $TF_{III}C$ has bound, events follow the same course as at type 2 promoters, with $TF_{III}B$ (which contains the ubiquitous TBP) binding at the start point and RNA polymerase III joining the complex. Type 1 promoters are found only in the genes for 5S rRNA.

$TF_{III}A$ and $TF_{III}C$ are **assembly factors**, whose sole role is to assist the binding of the positioning factor $TF_{III}B$ at the correct location. Once $TF_{III}B$ has bound, $TF_{III}A$ and $TF_{III}C$ can be removed from the promoter without affecting the initiation reaction. $TF_{III}B$ remains bound in the vicinity of the start point, and its presence is sufficient to allow RNA polymerase III to identify and bind at the start point. Thus $TF_{III}B$ is the only true initiation factor required by RNA polymerase III. This sequence of events explains how the promoter boxes downstream can cause RNA polymerase to bind at the start point, farther upstream. Although the ability to transcribe these genes is conferred by the internal promoter, changes in the region immediately upstream of the start point can alter the efficiency of transcription.

▸ **assembly factors** Proteins that are required for formation of a macromolecular structure but are not themselves part of that structure.

The upstream region has a conventional role in the third class of polymerase III promoters. In the example shown in Figure 20.4, there are three upstream elements. These elements are also found in promoters for snRNA genes that are transcribed by RNA polymerase II. (Genes for some snRNAs are transcribed by RNA polymerase II, whereas others are transcribed by RNA polymerase III.) The upstream elements function in a similar manner in promoters for both RNA polymerases II and III.

Initiation at an upstream promoter for RNA polymerase III can occur on a short region that immediately precedes the start point and contains only the TATA element. Efficiency of transcription, however, is much increased by the presence of the enhancer

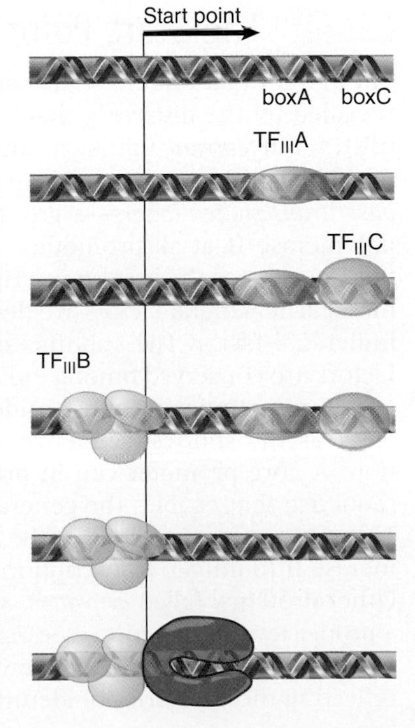

FIGURE 20.6 Internal type 1 pol III promoters use the assembly factors $TF_{III}A$ and $TF_{III}C$, at *boxA* and *boxC*, to recruit the positioning factor $TF_{III}B$, which recruits RNA polymerase III.

PSE (proximal sequence element) and Oct (named because it has an 8 base pair binding sequence) elements. The factors that bind at these elements interact cooperatively. The TATA element confers specificity for the type of polymerase (II or III) that is recognized by an snRNA promoter. It is bound by a factor that includes TBP, which actually recognizes the sequence in DNA. TBP is associated with other proteins, which are specific for the type of promoter. The function of TBP and its associated proteins is to position the RNA polymerase correctly at the start point. We discuss this in more detail for RNA polymerase II (see *Section 20.6, TBP Is a Universal Factor*).

The factors work in the same way for both types of promoters for RNA polymerase III. The factors bind at the promoter before RNA polymerase itself can bind. They form a **preinitiation complex** that directs binding of the RNA polymerase. RNA polymerase III does not itself recognize the promoter sequence but binds adjacent to factors that are themselves bound just upstream of the start point. In all cases, the chromatin must be modified and in an open configuration.

▸ **preinitiation complex** The assembly of transcription factors at the promoter before RNA polymerase binds in eukaryotic transcription.

KEY CONCEPTS

- RNA polymerase III has two types of promoters.
- Internal promoters have short consensus sequences located within the transcription unit and cause initiation to occur at a fixed distance upstream.
- Upstream promoters contain three short consensus sequences upstream of the start point that are bound by transcription factors.
- $TF_{III}A$ and $TF_{III}C$ bind to the consensus sequences and enable $TF_{III}B$ to bind at the start point.
- $TF_{III}B$ has TBP as one subunit and enables RNA polymerase to bind.

CONCEPT AND REASONING CHECK

The tRNA and 5S gene promoters are inside the gene. How can the polymerase find the start point at the beginning of the gene?

20.5 The Start Point for RNA Polymerase II

The basic organization of the apparatus for transcribing protein-coding genes was revealed by the discovery that purified RNA polymerase II can catalyze synthesis of mRNA but cannot initiate transcription unless an additional extract is added. The purification of this extract led to the definition of the general transcription factors, or *basal transcription factors*—a group of proteins that are needed for initiation by RNA polymerase II at all promoters. RNA polymerase II in conjunction with these factors constitutes the basal transcription apparatus that is needed to transcribe any promoter. The general factors are described as $TF_{II}X$, where X is a letter that identifies the individual factor. The subunits of RNA polymerase II and the general transcription factors are conserved among eukaryotes.

Our starting point for considering promoter organization is to define the *core promoter* as the shortest sequence at which RNA polymerase II can initiate transcription. A core promoter can in principle be expressed in any cell. It is the minimum sequence that enables the general transcription factors to assemble at the start point. These factors are involved in the mechanics of binding to DNA and enable RNA polymerase II to initiate transcription. A core promoter functions at only a low efficiency. Other proteins, called *activators*, another class of transcription factors, are required for a proper level of function (see *Section 20.9, Enhancers Contain Bidirectional Elements That Assist Initiation*). The activators are not described systematically, but have casual names reflecting their histories of identification.

We may expect any sequence components involved in the binding of RNA polymerase and general transcription factors to be conserved at most or all promoters. As with bacterial promoters, when promoters for RNA polymerase II are compared, homologies in the regions near the start point are restricted to rather short sequences. These elements correspond with the sequences implicated in promoter function by mutation. **FIGURE 20.7** shows the construction of a typical pol II core promoter with three of the most common pol II promoter elements. The eukaryotic pol II promoter is far more structurally diverse than the bacterial promoter, though. In addition to the three major elements, there are a number of minor elements that can also serve to define the promoter.

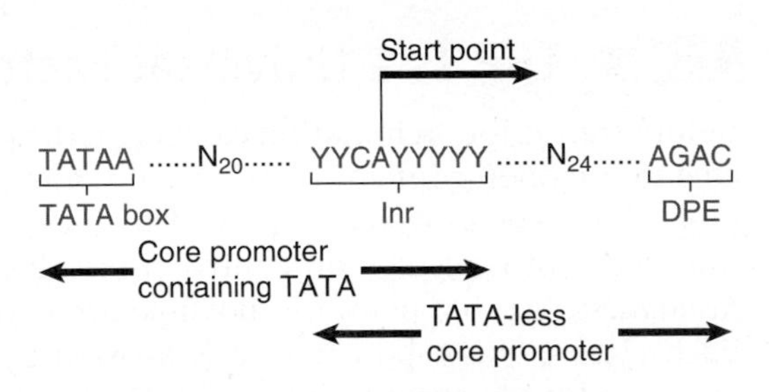

FIGURE 20.7 A minimal pol II promoter may have a TATA box ~25 bp upstream of the Inr. The TATA box has the consensus sequence of TATAA. The Inr has pyrimidines (Y) surrounding the CA at the start point. The DPE is downstream of the start point. The sequence shows the coding strand.

At the start point, there is no extensive homology of sequence, but there is a tendency for the first base of mRNA to be A, flanked on either side by pyrimidines. (This description is also valid for the CAT start sequence of bacterial promoters.) This region is called the **initiator (Inr)** and may be described in the general form Py_2CAPy_5, where Py stands for any pyrimidine. The Inr is contained between positions −3 and 5.

Many promoters have a sequence called the **TATA box**, usually located ~25 bp upstream of the start point in higher eukaryotes (in yeast it is typically ~90 bp upstream of the start point). It constitutes the only upstream promoter element that has a relatively fixed location with respect to the start point. The core sequence is TATAA, usually followed by three more A-T base pairs. The TATA box tends to be surrounded by G-C-rich sequences, which could be a factor in its function. It is almost identical with the −10 TATA box sequence found in bacterial promoters; in fact, it could pass for one except for the difference in its location at −25 instead of −10.

Promoters that do not contain a TATA element are called **TATA-less promoters**. Surveys of promoter sequences suggest that 50% or more of promoters may be TATA-less. When a promoter does not contain a TATA box, it often contains another element, the **downstream promoter element (DPE)**, which is located at +28 to 32.

Most core promoters consist either of a TATA box plus Inr, or of an Inr plus DPE, although other combinations with minor elements exist as well.

- **initiator (Inr)** The sequence of a pol II promoter between −3 and +5 and has the general sequence Py_2CAPy_5. It is the simplest possible pol II promoter.
- **TATA box** A conserved A-T-rich octamer found about 25 bp before the start point of each eukaryotic RNA polymerase II transcription unit; it is involved in positioning the enzyme for correct initiation.
- **TATA-less promoter** A promoter that does not have a TATA box in the sequence upstream of its start point.
- **downstream promoter element (DPE)** A common component of RNA polymerase II promoters that do not contain a TATA box.

KEY CONCEPTS

- RNA polymerase II requires general transcription factors (called $TF_{II}X$) to initiate transcription.
- RNA polymerase II promoters have a short conserved sequence Py_2CAPy_5 (the initiator, Inr) at the start point.
- The TATA box is a common component of RNA polymerase II promoters and consists of an A-T-rich octamer located ~25 bp upstream of the start point.
- The DPE is a common component of RNA polymerase II promoters that do not contain a TATA box.
- A core promoter for RNA polymerase II includes the Inr and commonly either a TATA box or a DPE.

CONCEPT AND REASONING CHECK

Why might RNA polymerase II promoters be more variable than RNA polymerase I promoters?

20.6 TBP Is a Universal Factor

Before transcription initiation can begin, the chromatin has to be modified and remodeled to the open configuration, and any nucleosome octamer positioned over the promoter has to be moved or removed at all classes of eukaryotic promoters (we examine this aspect of transcription control more closely in *Chapter 28, Eukaryotic Transcription Regulation*). At that point it is possible for a positioning factor to bind to the promoter. Each class of RNA polymerase is assisted by a positioning factor that contains TBP associated with other components. The name TBP has an interesting history. It was initially so named because it was identified as a protein that bound to the TATA box in RNA polymerase II genes. It was subsequently discovered to also be part of the positioning factors for both RNA polymerase I (see *Section 20.3, RNA Polymerase I Has a Bipartite Promoter*) and RNA polymerase III (see *Section 20.4, RNA Polymerase III Uses Both Downstream and Upstream Promoters*). For these latter two RNA polymerases, TBP does not recognize the TATA box sequence; thus the name is misleading (except in the case of upstream promoters for RNA polymerase III). In addition, many RNA polymerase II promoters lack TATA boxes but still require the presence of TBP.

- **$TF_{II}D$** The transcription factor that binds to the TATA sequence or Inr of promoters for RNA polymerase II. It consists of TBP (TATA binding protein) and the TAF subunits that bind to TBP.
- **TAFs** The subunits of $TF_{II}D$ that assist TBP in binding to DNA. They also provide points of contact for other components of the transcription apparatus.

The positioning factor for RNA polymerase I is SL1 and for RNA polymerase III is $TF_{III}B$. For RNA polymerase II, the positioning factor is **$TF_{II}D$**, which consists of TBP associated with ~11 other subunits called **TAFs** (for *T*BP-*a*ssociated *f*actors). Some TAFs are stoichiometric with TBP; others are present in lesser amounts. $TF_{II}Ds$ containing different TAFs could recognize different promoters. Some TAFs are tissue specific. The total mass of $TF_{II}D$ typically is ~800 kD. The TAFs in $TF_{II}D$ were originally named in the form $TAF_{II}00$, for example, where the number 00 gives the molecular mass of the subunit. Recently, the RNA polymerase II TAFs have been renamed TAF1, TAF2, etc.; in this nomenclature TAF1 is the largest TAF, TAF2 is the next largest, and homologous TAFs in different species thus have the same names.

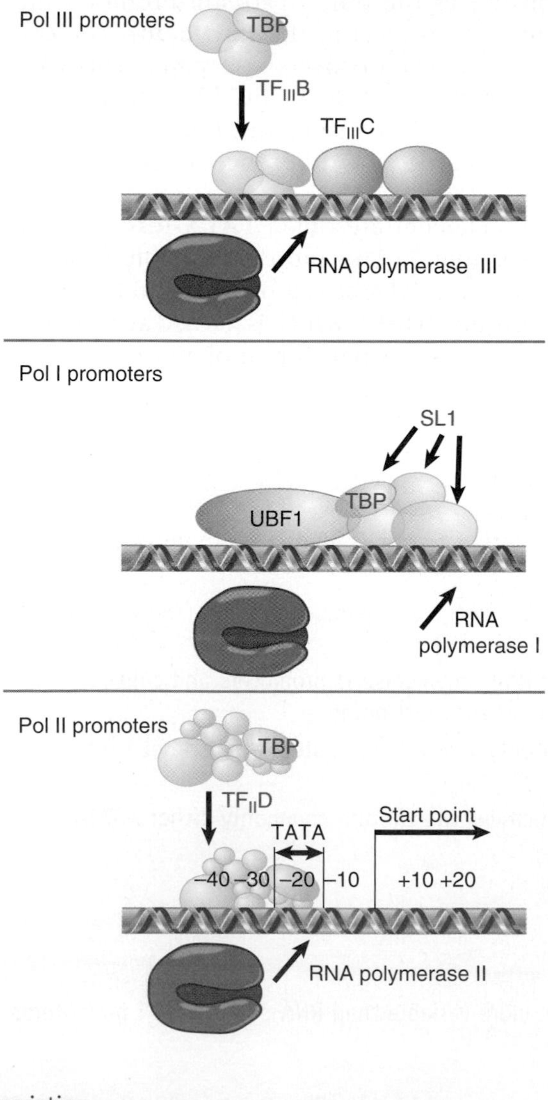

FIGURE 20.8 RNA polymerases are positioned at all promoters by a factor that contains TBP.

FIGURE 20.8 shows that the positioning factor recognizes the promoter in a different way in each case. At promoters for RNA polymerase III, $TF_{III}B$ binds adjacent to $TF_{III}C$. At promoters for RNA polymerase I, SL1 binds in conjunction with UBF. $TF_{II}D$ is solely responsible for recognizing promoters for RNA polymerase II. At a promoter that has a TATA element, TBP binds specifically to the TATA box; but, at TATA-less promoters, it may be incorporated by association with other factors that bind to DNA first. Whatever its means of entry into the initiation complex, it has the common purpose of interaction with the RNA polymerase.

TBP has the unusual property of binding to DNA in the minor groove. (The vast majority of DNA-binding proteins bind in the major groove.) The crystal structure of TBP suggests a detailed model for its binding to DNA. **FIGURE 20.9** shows that it surrounds one face of DNA, forming a "saddle" around a stretch of the minor groove, which is bent to fit into this saddle. In effect, the inner surface of TBP binds to DNA, and the larger outer surface is available to extend contacts to other proteins. The DNA-binding site consists of a C-terminal domain that is conserved between species, and the variable N-terminal tail is exposed to interact with other proteins. It is a measure of the conservation of mechanism in transcriptional initiation that the DNA-binding sequence of TBP is 80% conserved between yeast and humans.

FIGURE 20.9 A view in cross section shows that TBP surrounds DNA from the side of the narrow groove. TBP consists of two related (40% identical) conserved domains, which are shown in light and dark blue. The N-terminal region varies extensively and is shown in green. The two strands of the double helix are in white and grey. Photo courtesy of Stephen K. Burley.

Binding of TBP may be inconsistent with the presence of nucleosome octamers. Nucleosomes form preferentially by placing AT-rich sequences with the minor grooves facing inward; as a result, they could prevent binding of TBP. This may explain why the presence of a nucleosome at the promoter prevents initiation of transcription.

TBP binds to the minor groove and bends the DNA by ~80°, as illustrated in **FIGURE 20.10**. The TATA box bends toward the major groove, widening the minor groove. The distortion is restricted to the 8 bp of the TATA box; at each end of the sequence, the minor groove has its usual width of ~5 Å, but at the center of the sequence the minor groove is >9 Å. This is a deformation of the structure but does not actually separate the strands of DNA because base pairing is maintained. The extent of the bend can vary with the exact sequence of the TATA box and is correlated with the efficiency of the promoter.

Within $TF_{II}D$ as a free protein complex, the factor TAF1 binds to TBP, where it occupies the concave DNA-binding surface. In fact, the structure of the binding site, which lies in the N-terminal domain of TAF1, mimics the surface of the minor groove in DNA. This molecular mimicry allows TAF1 to control the ability of TBP to bind to DNA; the N-terminal domain of TAF1 must be displaced from the DNA-binding surface of TBP in order for $TF_{II}D$ to bind to DNA.

What happens at TATA-less promoters? The same general transcription factors, including $TF_{II}D$, are needed. The Inr, when present, can provide the positioning element; $TF_{II}D$ binds to it via the ability of one or more of the TAFs to recognize the Inr directly. Other TAFs in $TF_{II}D$ also recognize the DPE element downstream from the start point. The positioning of TBP at TATA-less promoters is more like that at internal promoters for RNA polymerase III.

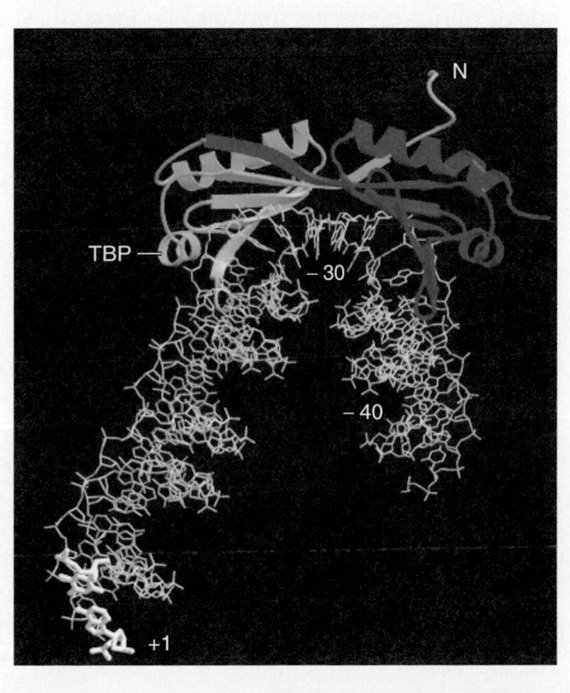

FIGURE 20.10 The cocrystal structure of TBP with DNA from –40 to the start point shows a bend at the TATA box that widens the narrow groove where TBP binds. Photo courtesy of Stephen K. Burley.

When a TATA box is present, it determines the location of the start point. Its deletion causes the site of initiation to become erratic, although any overall reduction in transcription is relatively small. Indeed, some TATA-less promoters lack unique start points, so initiation occurs within a cluster of start points. The TATA box aligns the RNA polymerase via the interaction with $TF_{II}D$ and other factors so that it initiates at the proper site. Binding of TBP to TATA is the predominant feature in the recognition of that promoter.

KEY CONCEPTS

- TBP is a component of the positioning factor that is required for each type of RNA polymerase to bind its promoter.
- The factor for RNA polymerase II is $TF_{II}D$, which consists of TBP and ~11 TAFs, with a total mass ~800 kD.
- TBP binds to the TATA box in the minor groove of DNA.
- It forms a saddle around the DNA and bends it by ~80°.

CONCEPT AND REASONING CHECK

Why does initiation at a TATA-less promoter still require the TATA binding factor?

20.7 The Basal Apparatus Assembles at the Promoter

In a cell, gene promoters can be found in three types of chromatin with respect to activity. The first is an inactive gene in closed chromatin. The second is a potentially active gene in open chromatin, called a poised gene. This class may assemble the basal apparatus but cannot proceed to transcribe the gene without a second signal to start transcription. Heat-shock genes are an extreme example, in which transcription is initiated and then paused, so that they can be activated immediately upon a rise in temperature. The third class (which we will examine shortly) is an active gene in open chromatin. Initiation requires the basal transcription factors to act in a defined order to build a complex that is joined by RNA polymerase. The series of events is summarized in **FIGURE 20.11**. Once a polymerase is bound, its activity then is controlled by enhancer-binding transcription factors.

A promoter for RNA polymerase II often consists of two types of region. The major elements of the core promoter contains the start point itself, typically identified by the Inr, and often includes either the TATA box or DPE close by; a selection from a number of minor elements may also be present. The efficiency and specificity with which a promoter is recognized, however, depend upon short sequences farther upstream, which are recognized by a different group of transcription factors, sometimes called *activators*. In general, the target sequences are 100 bp upstream of the start point, but sometimes they are more distant. Binding of activators at these sites may influence the formation of the initiation complex at (probably) any one of several stages. Promoters are organized on a principle of "mix and match." A variety of elements can contribute to promoter function, but none is essential for all promoters.

What has been largely unexplored until recently is the involvement of noncoding RNA (ncRNA) transcripts in gene activation. Numerous recent examples have been described in which transcription of ncRNAs regulate transcription of nearby or overlapping protein-coding genes. The production of these functional ncRNAs (also referred to as cryptic unstable transcripts or CUTS) may be much more common than originally believed. A significant number of active promoters have transcripts generated upstream of the promoter (known as promoter upstream transcripts or PROMPTs). PROMPTs are transcribed in both sense and antisense orientations relative to the downstream promoter and may play a role in regulation transcription. The many roles of ncRNAs in transcriptional regulation will be discussed further in *Section 30.3, Noncoding RNAs Can Be Used to Regulate Gene Expression*.

The first step in activating a promoter in open chromatin is initiated when $TF_{II}D$ binds the TATA box. This may be enhanced by upstream elements acting through a

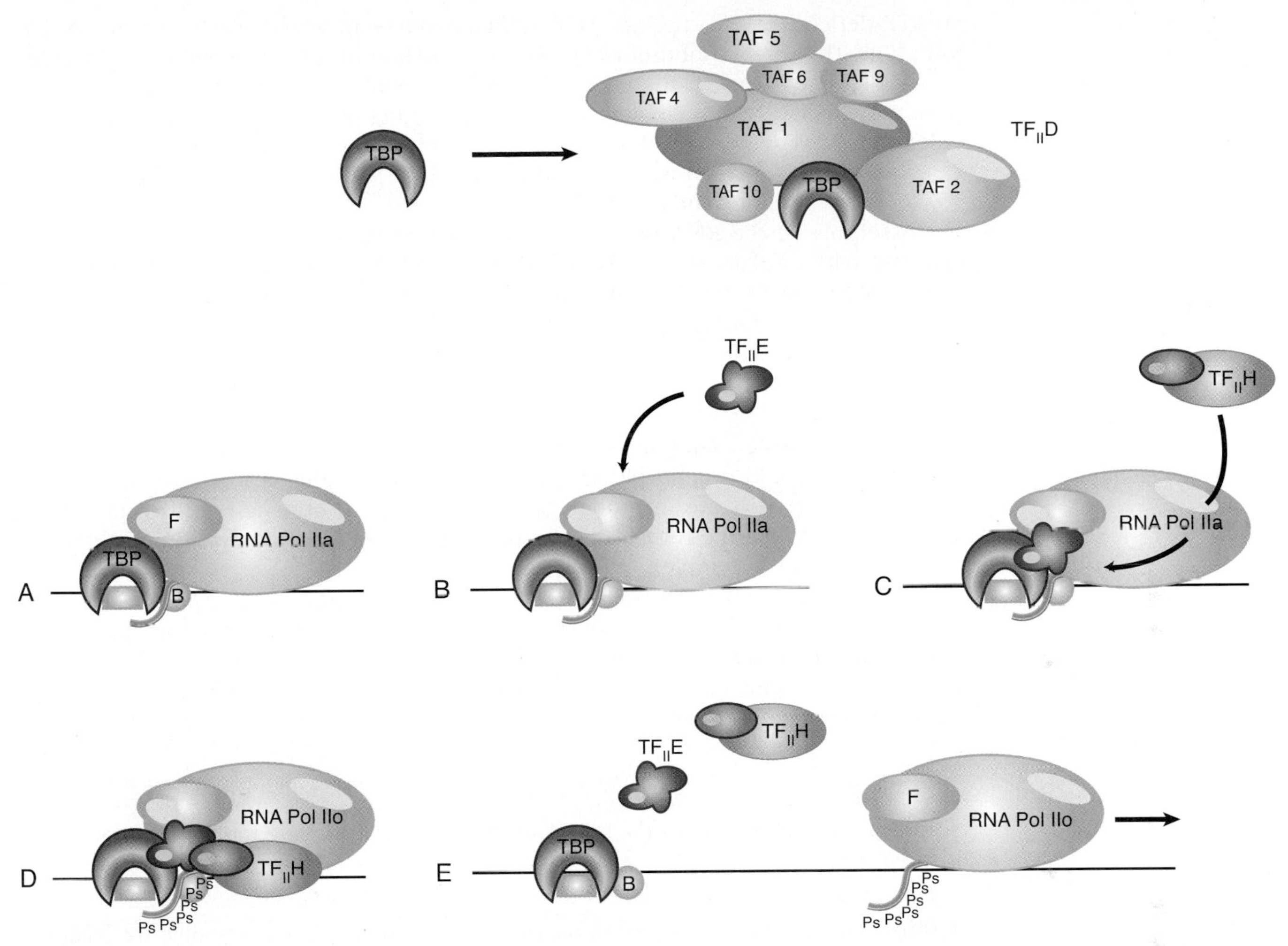

FIGURE 20.11 An initiation complex assembles at promoters for RNA polymerase II by an ordered sequence of association of transcription factors. $TF_{II}D$ consists of TBP plus its associated TAFs, as shown in the top panel; TBP alone, rather than $TF_{II}D$, is shown in the remaining panels for simplicity. Adapted from M. E. Maxon, J. A. Goodrich, and R. Tijian, *Genes Dev.* 8 (1994): 515–524.

coactivator ($TF_{II}D$ also recognizes the Inr sequence at the start point.) When $TF_{II}A$ joins the complex, $TF_{II}D$ becomes able to protect a region extending farther upstream. $TF_{II}A$ may activate TBP by relieving the repression that is caused by TAF1.

$TF_{II}B$ binds downstream of the TATA box, adjacent to TBP, extending contacts along one face of the DNA. It makes contacts in the minor groove downstream of the TATA box and contacts the major groove upstream of the TATA box in a region called the BRE, where it is involved in selecting the start site of transcription. The schematic of **FIGURE 20.12** shows the relationship between $TF_{II}B$, $TF_{II}D$, and RNA polymerase II. $TF_{II}B$ binds adjacent to $TF_{II}D$, and its N-terminal region contacts RNA polymerase near the RNA exit site. Its C-terminal region extends across the enzyme, with a protrusion into the active site. $TF_{II}B$ determines the path of the DNA where it contacts the factors $TF_{II}E$, $TF_{II}F$, and $TF_{II}H$, which may align them in the basal factor complex and determine the start point. The factor $TF_{II}F$ is a heterotetramer consisting of two types of subunit. The larger subunit (RAP74) has

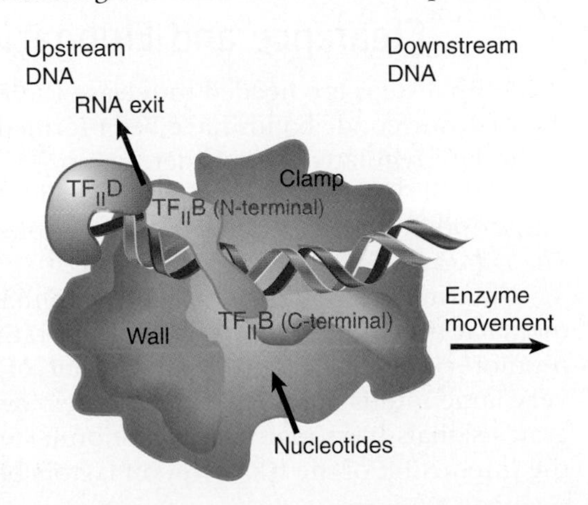

FIGURE 20.12 $TF_{II}B$ binds to DNA and contacts RNA polymerase near the RNA exit site and at the active center and orients it on DNA.

an ATP-dependent DNA helicase activity that could be involved in melting the DNA at initiation. The smaller subunit (RAP38) has some homology to the regions of bacterial sigma factor that contact the core polymerase; it binds tightly to RNA polymerase II. $TF_{II}F$ may bring RNA polymerase II to the assembling transcription complex and provide the means by which it binds. The complex of TBP and TAFs may interact with the CTD tail of RNA polymerase, and interaction with $TF_{II}B$ may also be important when $TF_{II}F$/polymerase joins the complex.

Assembly of the RNA polymerase II initiation complex provides an interesting contrast with prokaryotic transcription. Bacterial RNA polymerase is essentially a coherent aggregate with intrinsic ability to bind DNA; the sigma factor, needed for initiation but not for elongation, becomes part of the enzyme before DNA is bound, although it may be later released. RNA polymerase II can bind to the promoter, but only after separate transcription factors have bound. The factors play a role analogous to that of bacterial sigma factor—to allow the basic polymerase to recognize DNA specifically at promoter sequences—but have evolved more independence. Indeed, the factors are primarily responsible for the specificity of promoter recognition. Only some of the factors participate in protein-DNA contacts (and only TBP and certain TAFs make sequence-specific contacts); thus protein-protein interactions are important in the assembly of the complex.

Although assembly can take place just at the core promoter *in vitro*, this reaction is not sufficient for transcription *in vivo*, where interactions with activators that recognize the more upstream elements are required. The activators interact with the basal apparatus at various stages during its assembly (see *Section 28.5, Activators Interact with the Basal Apparatus*).

KEY CONCEPTS

- The upstream elements and the factors that bind to them increase the frequency of initiation.
- Binding of $TF_{II}D$ to the TATA box or Inr is the first step in initiation.
- Other transcription factors bind to the complex in a defined order, extending the length of the protected region on DNA.
- When RNA polymerase II binds to the complex, it initiates transcription.

CONCEPT AND REASONING CHECK

Why are there so many basal transcription factors for RNA polymerase II promoters compared to RNA polymerase I or III promoters?

20.8 Initiation Is Followed by Promoter Clearance and Elongation

Some final steps are needed to release the RNA polymerase from the promoter once the first nucleotide bonds have been formed. This step is called *promoter clearance* and is the key regulated step in determining if a poised gene or an active gene will be transcribed. This step is controlled by enhancers. (Remember, the key step in bacterial transcription is conversion of the closed complex to the open complex; see *Section 19.3, The Transcription Reaction Has Three Stages.*)

The transcription factors that bind enhancers usually do not directly contact elements at the promoter to control it, but rather bind to a coactivator that binds to the promoter elements. The Mediator is one of the most common coactivators. This is a very large multisubunit protein complex, conserved from yeast to humans, that integrates signals from many transcription factors. Both poised and active genes require the interaction of the transcription factors bound to enhancers with the promoter.

The last factors to join the initiation complex are TF$_{II}$E and TF$_{II}$H. They act at the later stages of initiation. TF$_{II}$H is the only general transcription factor that has multiple independent enzymatic activities. Its several activities include an ATPase, helicases of both polarities, and a kinase activity that can phosphorylate the CTD tail of RNA polymerase II. TF$_{II}$H is an exceptional factor that may also play a role in elongation. Its interaction with DNA downstream of the start point is required for RNA polymerase to escape from the promoter. TF$_{II}$H is also involved in repair of damage to DNA (see *Section 16.3, Nucleotide Excision Repair Systems Repair Several Classes of Damage*).

Phosphorylation of the CTD tail is needed to release RNA polymerase II from the promoter and transcription factors so that it can make the transition to the elongating form, as shown in **FIGURE 20.13**. The phosphorylation pattern on the CTD is dynamic during the elongation process, controlled and catalyzed by multiple protein kinases

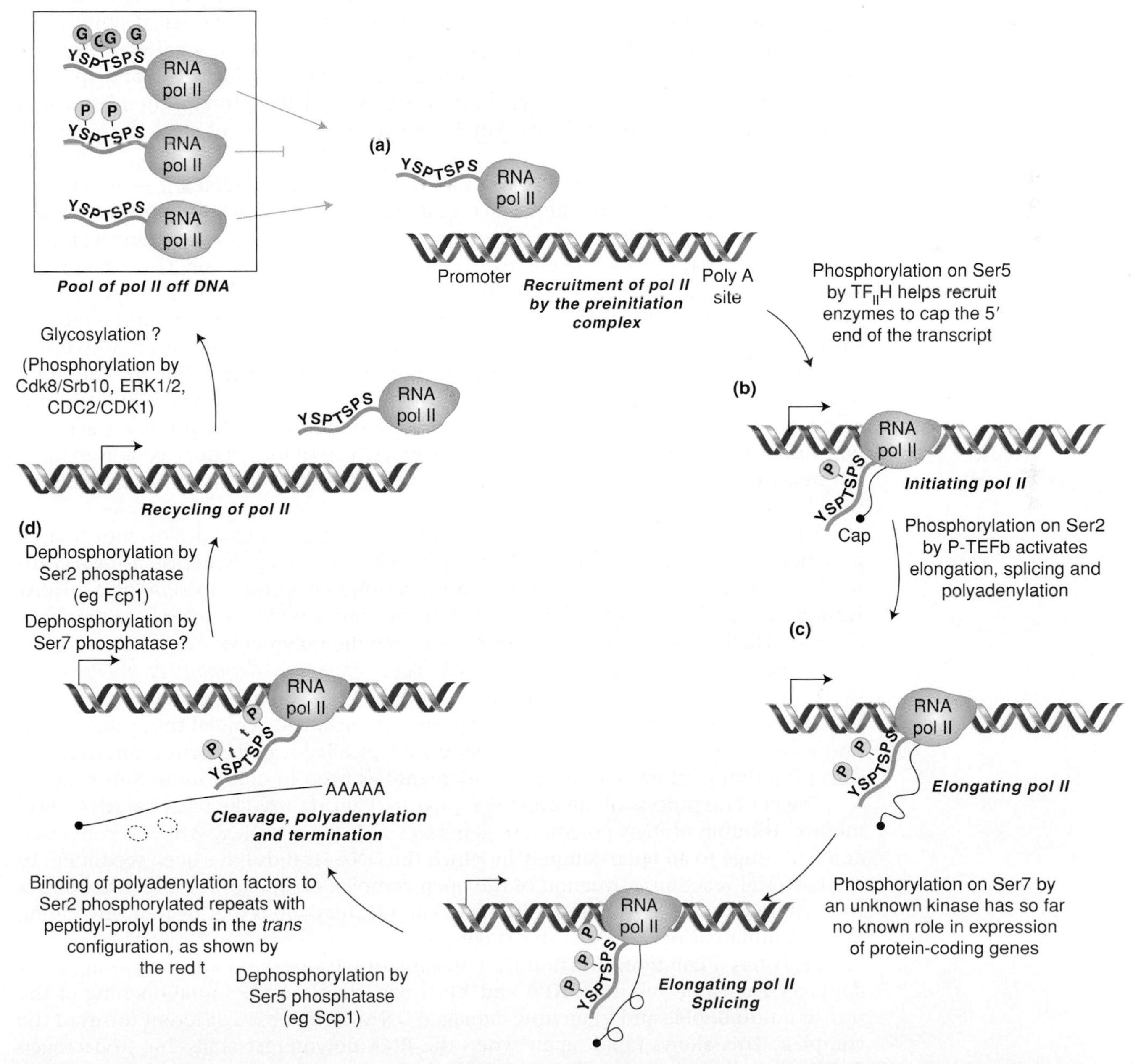

FIGURE 20.13 Modification of the RNA polymerase II CTD heptapeptide during transcription. The CTD of RNA polymerase II when it enters the preinitiation complex is unphosphorylated. Phosphorylation of Ser residues serves as binding sites for both mRNA processing enzymes and kinases that catalyze further phosphorylation as described in the figure. Reprinted from Trends Genet., Vol. 24, P.P. Gardner and J. Vinther, Mutation of miRNA target sequences . . ., pp. 262–265. Copyright 2008, with permission from Elsevier [http://www.sciencedirect.com/science/journal/01689525].

and phosphatases. Most of the basal transcription factors are released from the promoter at this stage.

The CTD is involved, directly or indirectly, in processing mRNA while it is being synthesized and after it has been released by RNA polymerase II. Each site of phosphorylation on the CTD serves as a recognition or anchor point for other proteins to dock with the polymerase. The capping enzyme (guanylyl transferase), which adds the G residue to the 5′ end of newly synthesized mRNA, binds to the phosphorylated CTD. This may be important in enabling it to modify (and thus protect) the 5′ end as soon as it is synthesized. A set of proteins called SCAFs bind to the CTD, and they may, in turn, bind to splicing factors. This may be a means of coordinating transcription and splicing. Some components of the cleavage/polyadenylation apparatus used after transcription termination also bind to the CTD. Oddly enough, they do so at the time of initiation, so that RNA polymerase is ready for the 3′ end processing reactions as soon as it sets out. Export from the nucleus through the nuclear pore is also coordinated by the CTD. All this suggests that the CTD may be a general focus for connecting other processes with transcription. In the cases of capping and splicing, the CTD functions indirectly to promote formation of the protein complexes that undertake the reactions. In the case of 3′ end generation, it may participate directly in the reaction.

The key event in determining whether a gene will be expressed is *promoter clearance*, release from the promoter. Once that has occurred and initiation factors are released, there is a transition to the elongation phase. The transcription complex now consists of the RNA polymerase II, the basal factors $TF_{II}E$ and $TF_{II}H$, and elongation factors like $TF_{II}S$ to prevent inappropriate pausing and all the enzymes and factors bound to the CTD. This complex now has to transcribe a chromatin template, through nucleosomes. The whole gene may be in open chromatin, especially if it is not too large, or only the area around the promoter. Some genes, like the muscular dystrophy gene (*DMD*), can be megabases in size and require many hours to transcribe. There is a model in which the first polymerase to leave the promoter acts as a pathfinder polymerase. Its major function is to ensure that the entire gene is in open chromatin. It carries with it enzyme complexes to modify the histones and remodel the chromatin.

The second problem is that even open chromatin contains nucleosomes that all polymerases must traverse. The most recent model has each polymerase using a chromatin-remodeling complex together with a histone chaperone to remove an H2A/H2B dimer, leaving a hexamer (in place of the octamer), which is easier to temporarily displace. Nucleosomes are then reassembled once the polymerase has passed.

As discussed above in *Section 20.7, The Basal Apparatus Assembles at the Promoter*, there can be considerable heterogeneity in the DNA sequence elements that comprise the core promoter that can lead to promoter specificity. One of these elements is known as the *pause button*, a GC-rich sequence typically located downstream from the start of initiation of transcription. This element has been found in numerous genes.

The general process of initiation is similar to that catalyzed by bacterial RNA polymerase. Binding of RNA polymerase generates a closed complex, which is converted at a later stage to an open complex in which the DNA strands have been separated. In the bacterial reaction, formation of the open complex completes the necessary structural change to DNA; a difference in the eukaryotic reaction is that further unwinding of the template is needed after this stage.

$TF_{II}H$ has a common function in both initiating transcription and repairing DNA damage. The same subunits (XPB and XPD) participate in the initial opening of the transcription bubble and in melting damaged DNA for repair, in different forms of the complex. This allows rapid repair when the RNA polymerase stalls due to damaged DNA on the template strand. Subunits with the name XP are coded for by genes in which mutations cause the disease *xeroderma pigmentosum*, which causes a predisposition to cancer (see the box *Medical Applications* in *Section 16.3, Eukaryotic Nucleotide Excision Repair Pathways*).

KEY CONCEPTS

- $TF_{II}E$ and $TF_{II}H$ are required to melt DNA to allow polymerase movement.
- Phosphorylation of the CTD is required for elongation to begin.
- Further phosphorylation of the CTD is required at some promoters to end abortive initiation.
- The CTD coordinates processing and export of RNA from the nucleus with transcription.
- Transcribed genes are preferentially repaired when DNA damage occurs.
- $TF_{II}H$ provides the link to a complex of repair enzymes.

CONCEPT AND REASONING CHECK

Why is the CTD important for transcription?

20.9 Enhancers Contain Bidirectional Elements That Assist Initiation

We have largely considered the promoter as an isolated region responsible for binding RNA polymerase. Eukaryotic promoters do not necessarily function alone, though. In most cases, the activity of a promoter is greatly increased by the presence of an enhancer located at a variable distance from the core promoter. Some enhancers function through long-range interactions of tens of kilobases; other enhancers function through short-range interactions and may lie quite close to the core promoter.

The concept that the enhancer is distinct from the promoter reflects two characteristics. The position of the enhancer relative to the promoter need not be fixed, but can vary substantially. **FIGURE 20.14** shows that it can be upstream, downstream, or within a gene (typically in introns). In addition, it can function in either orientation (that is, it can be inverted) relative to the promoter. Manipulations of DNA show that an enhancer can stimulate any promoter placed in its vicinity, even tens of kilobases away in either direction.

Like the promoter, an enhancer (or its alter ego, a silencer) is a modular element constructed of short DNA sequence elements that bind various types of transcription factors. Enhancers can be simple or complex, depending on the number of binding elements and the type of transcription factors they bind.

One way to divide up the world of enhancer-binding transcription factors is to consider positive and negative factors. Transcription factors can be positive and stimulate transcription: **activators**, or negative and repress transcription: **repressors**. At any given time in a cell, determined by its developmental history, that cell will contain

▸ **activator** A protein that stimulates the expression of a gene, typically by interacting with a promoter to stimulate RNA polymerase. In eukaryotes, the sequence to which it binds in the promoter is called an enhancer.

▸ **repressor** A protein that inhibits expression of a gene. It may act to prevent transcription by binding to an enhancer or silencer.

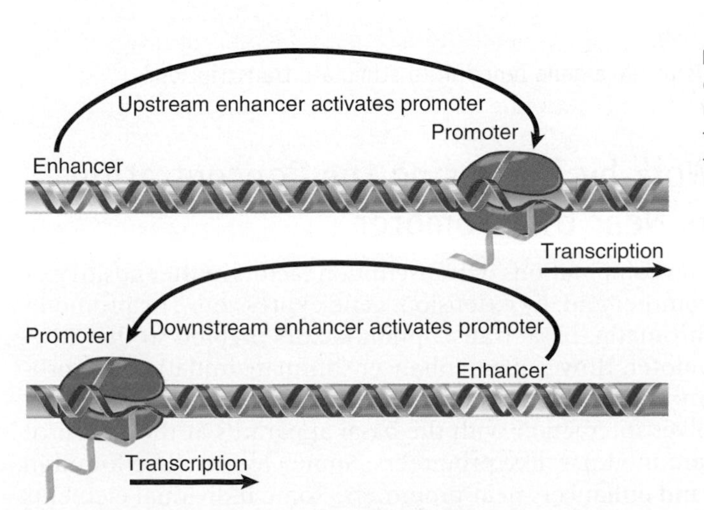

FIGURE 20.14 An enhancer can activate a promoter from upstream or downstream locations, and its sequence can be inverted relative to the promoter.

a mixture of transcription factors that can bind to an enhancer. If more activators bind than repressors, the element will be an enhancer. If more repressors bind than activators, the element will be a silencer.

A second way to examine the transcription factors that bind enhancers is by function. The first class we will consider is called *true activators*; that is, they function by both binding specific DNA sites and making contact with the basal machinery at the promoter, either directly by themselves, or, more commonly, through coactivators like Mediator. This class functions equally well on a DNA template or a chromatin template. There are two additional classes of activators that have a completely different mechanism of activation. The second class includes those that function by recruiting chromatin modification enzymes and chromatin remodeling complexes. The third class includes architectural modifiers. Their sole function is to change the structure of the DNA, typically to bend it. This can then result in bringing together two transcription factors separated by a short distance to synergize. In the next section, we will examine more closely how the different classes of activators and repressors work together in an enhancer, and in *Chapter 28, Eukaryotic Transcription Regulation*, we will examine transcription regulation in more detail.

▸ **upstream activating sequence (UAS)** The equivalent in yeast of the enhancer in higher eukaryotes; a UAS cannot function downstream of the promoter.

Elements analogous to enhancers, called **upstream activating sequences (UAS)**, are found in yeast. They can function in either orientation at variable distances upstream of the promoter (up to a limit), but cannot function when located downstream. They have a regulatory role analogous to enhancers in higher eukaryotes: the UAS is bound by the regulatory protein(s) that activates the genes downstream.

Reconstruction experiments in which the enhancer sequence is removed from the DNA and then is inserted elsewhere show that normal transcription can be sustained as long as it is present anywhere on the DNA molecule (as long as no insulators are present in the intervening DNA; see *Section 10.12, Insulators Define Independent Domains*). If a β-globin gene is placed on a DNA molecule that contains an enhancer, its transcription is increased *in vivo* more than 200-fold, even when the enhancer is several kilobases upstream or downstream of the start point, in either orientation. We have yet to discover at what distance the enhancer fails to work.

KEY CONCEPTS

- An enhancer activates the promoter nearest to itself, and can be any distance either upstream or downstream of the promoter.
- A UAS (upstream activating sequence) in yeast behaves like an enhancer, but works only upstream of the promoter.
- Enhancers form complexes of activators that interact directly or indirectly with the promoter.

CONCEPT AND REASONING CHECK

How might an enhancer that is inside a gene function to stimulate transcription?

20.10 Enhancers Work by Increasing the Concentration of Activators Near the Promoter

Enhancers function by binding combinations of transcription factors, either positive or negative, that control the promoter and, by extension, gene expression. The promoter is the site where, in open chromatin, basal transcription factors prebind so that RNA polymerase can find the promoter. How can an enhancer stimulate initiation at a promoter that can be located any distance away on either side of it?

Enhancer function involves interaction with the basal apparatus at the core promoter element. Enhancers are modular, like promoters. Some elements are found in both long-range enhancers and enhancers near promoters. Some individual elements

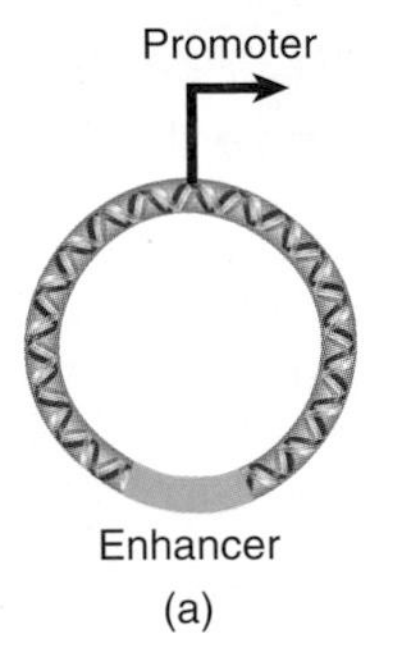

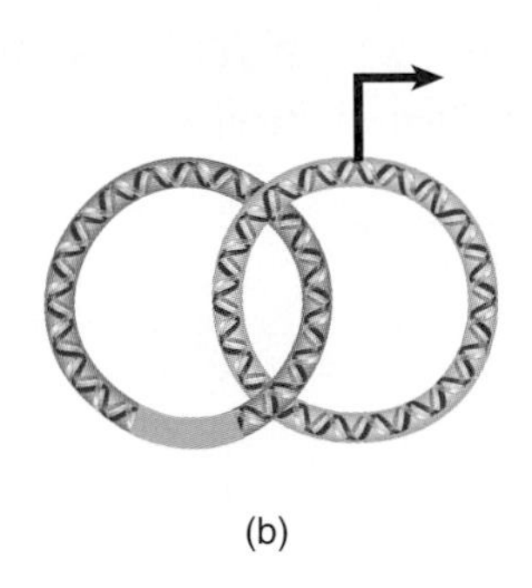

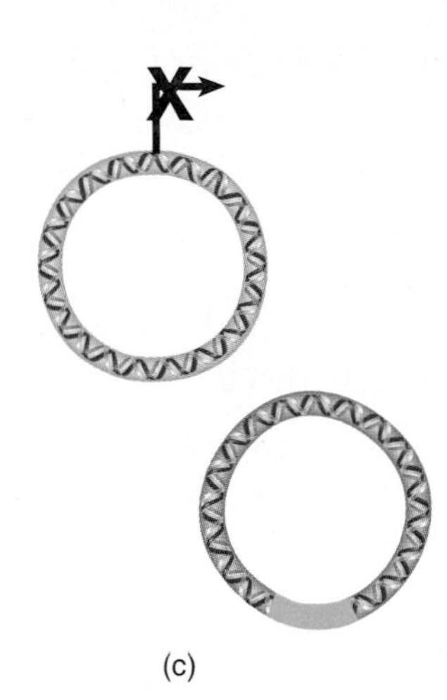

FIGURE 20.15 An enhancer may function by bringing proteins into the vicinity of the promoter. An enhancer and promoter on separate circular DNAs do not interact, as in (c), but can interact when the two molecules are catenated, as in (b).

found near promoters share with distal enhancers the ability to function at variable distance and in either orientation. Thus the distinction between long-range enhancers and short-range enhancers is blurred.

The essential role of the enhancer may be to increase the concentration of activator in the vicinity of the promoter (vicinity in this sense being a relative term) in *cis*. Numerous experiments have demonstrated that the level of gene expression (that is, the rate of transcription) is proportional to the net number of activator binding sites. The more activators bound at an enhancer site, the higher the level of expression.

The *Xenopus laevis* ribosomal RNA enhancer is able to stimulate transcription from its RNA polymerase I promoter. This stimulation is relatively independent of location and is able to function from a circle (plasmid). There is, however, no stimulation when the enhancer and promoter are on separated circles. Yet, when the enhancer is placed on a circle of DNA that is catenated (interlocked) with a circle that contains the promoter, initiation is almost as effective as when the enhancer and promoter are on the same circular molecule, as shown in **FIGURE 20.15** (even though, in this case, the enhancer is acting on its promoter in *trans*). Again, this suggests that the critical feature is localization of the protein bound at the enhancer, which increases the enhancer's chance of contacting a protein bound at the promoter.

If proteins bound at an enhancer several kilobases distant from a promoter interact directly with proteins bound in the vicinity of the start point, the organization of DNA must be flexible enough to allow the enhancer and promoter to be closely located. This requires the intervening DNA to be extruded as a large "loop." Such loops have been directly observed in the case of bacterial enhancers.

What limits the activity of an enhancer? Typically it works upon the nearest promoter. There are situations in which an enhancer is located between two promoters but activates only one of them on the basis of specific protein-protein contacts between the complexes bound at the two elements. The action of an enhancer may be limited by an insulator—an element in DNA that prevents the enhancer from acting on promoters beyond the insulator (see *Section 10.12, Insulators Define Independent Domains*).

KEY CONCEPTS

- Enhancers usually work only in *cis* configuration with a target promoter.
- Enhancers can be made to work in *trans* configuration by linking the DNA that contains the target promoter to the DNA that contains the enhancer via a protein bridge or by catenating the two molecules.
- The principle is that an enhancer works in any situation in which it is constrained to be in proximity to the promoter.

CONCEPT AND REASONING CHECK

Why would an enhancer that has bound two activators stimulate transcription more than an enhancer that has only bound one?

20.11 Summary

Of the three eukaryotic RNA polymerases, RNA polymerase I transcribes rDNA and accounts for the majority of activity, RNA polymerase II transcribes structural genes for mRNA and has the greatest diversity of products, and RNA polymerase III transcribes small RNAs. The enzymes have similar structures, with two large subunits and many smaller subunits; there are some common subunits among the enzymes.

None of the three RNA polymerases recognize their promoters directly. A unifying principle is that transcription factors have primary responsibility for recognizing the characteristic sequence elements of any particular promoter, and they serve in turn to bind the RNA polymerase and to position it correctly at the start point. At each type of promoter, the initiation complex is assembled by a series of reactions in which individual factors join (or leave) the complex. The factor TBP is required for initiation by all three RNA polymerases. In each case it provides one subunit of a transcription factor that binds in the vicinity of the start point.

An RNA polymerase II promoter consists of a number of short-sequence elements in the region upstream of the start point. Each element is bound by one or more transcription factors. The basal apparatus, which consists of the TF_{II} factors, assembles at the start point and enables RNA polymerase to bind. The TATA box (if there is one) near the start point and the initiator region immediately at the start point are responsible for selection of the exact start point at promoters for RNA polymerase II. TBP binds directly to the TATA box when there is one; in TATA-less promoters it is located near the start point by binding to the Inr or to the DPE downstream. After binding of $TF_{II}D$, the other general transcription factors for RNA polymerase II assemble the basal transcription apparatus at the promoter. Other elements in the promoter, located upstream of the TATA box, bind activators that interact with the basal apparatus. The activators and basal factors are released when RNA polymerase begins elongation.

The CTD of RNA polymerase II is phosphorylated during the initiation reaction. It provides a point of contact for proteins that modify the RNA transcript, including the 5′ capping enzyme, splicing factors, the 3′ processing complex, and mRNA export from the nucleus.

Promoters may be stimulated by enhancers, sequences that can act at great distances and in either orientation on either side of a gene. Enhancers also consist of sets of elements, although they are more compactly organized. Some elements are found both close to promoters and in distant enhancers. Enhancers probably function by assembling a protein complex that interacts with the proteins bound at the promoter, requiring that DNA between is "looped out."

CHAPTER QUESTIONS

1. List three ways the C-terminal domain of the largest RNA polymerase II subunit may be involved in posttranscriptional RNA modifications:

   A. ____________________

   B. ____________________

   C. ____________________

2. Where in the cell is each type of RNA transcribed? Choices may be used more than once or not at all.

Ribosomal RNA	**A.** cytoplasm
Small RNA	**B.** nucleolus
Messenger RNA	**C.** nucleoplasm

3. Genes transcribed from which of the following RNA polymerases often have promoters located downstream of the transcription start points?
   A. RNA polymerase I
   B. RNA polymerase II
   C. RNA polymerase III
   D. all of the above
4. Which enzyme is responsible for synthesis of heterogeneous nuclear RNA (hnRNA)?
   A. RNA polymerase I
   B. RNA polymerase II
   C. RNA polymerase III
   D. RNA polymerase IV
5. Which RNA polymerase utilizes bipartite promoters?
   A. RNA polymerase I
   B. RNA polymerase II
   C. RNA polymerase III
   D. RNA polymerase IV
6. The first base of a transcript synthesized by RNA polymerase II tends to be:
   A. an A surrounded by pyrimidines.
   B. an A surrounded by purines.
   C. a C surrounded by pyrimidines.
   D. a C surrounded by purines.
7. Which of the following promoter elements are required for RNA polymerase II transcription?
   A. a TATA box, Inr element, and DPE element
   B. a TATA box and DPE element
   C. a TATA box and Inr element or TATA box and DPE element
   D. a TATA box and Inr element or Inr element and DPE element
8. What is the first factor to bind to an RNA polymerase II–transcribed promoter containing a TATA box?
   A. $TF_{II}A$
   B. $TF_{II}D$
   C. $TF_{II}H$
   D. RNA polymerase II
9. What is thought to be responsible for releasing RNA polymerase II from its associated transcription factors to allow transition to elongation?
   A. methylation of the C-terminal domain of a large RNA polymerase II subunit
   B. phosphorylation of the C-terminal domain of a large RNA polymerase II subunit
   C. methylation of the N-terminal domain of a large RNA polymerase II subunit
   D. phosphorylation of the N-terminal domain of a large RNA polymerase II subunit
10. Enhancers function to activate or stimulate:
   A. the nearest promoter in *cis* to it.
   B. the nearest promoter in *trans* to it.
   C. the nearest promoter in *cis* or *trans* to it, either upstream or downstream.
   D. the nearest promoter in *cis* or *trans* to it, upstream of the promoter.

KEY TERMS

activator
assembly factors
basal transcription factors
carboxy-terminal domain (CTD)
coactivator
core promoter
downstream promoter element (DPE)
enhancer
heterogeneous nuclear RNA (hnRNA)
housekeeping genes
initiator (Inr)
nontranscribed spacer
preinitiation complex
repressor
silencer
start point
TAFs
TATA-binding protein (TBP)
TATA box
TATA-less promoter
$TF_{II}D$
upstream activating sequence (UAS)

FURTHER READING

Eccleston, A., and Skipper, M. (2009). Transcribing the Genome. *Nature* **461**, 185–218. A set of reviews describing RNA polymerase II transcription.

Grummt, I. (2003). Life on a small planet of its own: regulation of RNA polymerase I transcription in the nucleolus. *Genes Dev*. **17**, 1691–1702. A good review of RNA polymerase I transcription.

Schramm, L., and Hernandez, N. (2003). Recruitment of RNA polymerase III to its target promoters. *Genes Dev*. **16**, 2593–2620. A good review of RNA polymerase III transcription.

Selth, L. A., Sigurdsson, S., and Svejstrup, J. Q. (2010). Transcript Elongation by RNA polymerase II. *Annu. Rev. Biochem*. **79**, 271–293.

Smale, S. T., and Kadonaga, J. T. (2003). The RNA polymerase II core promoter. *Annu. Rev. Biochem*. **72**, 449–479.

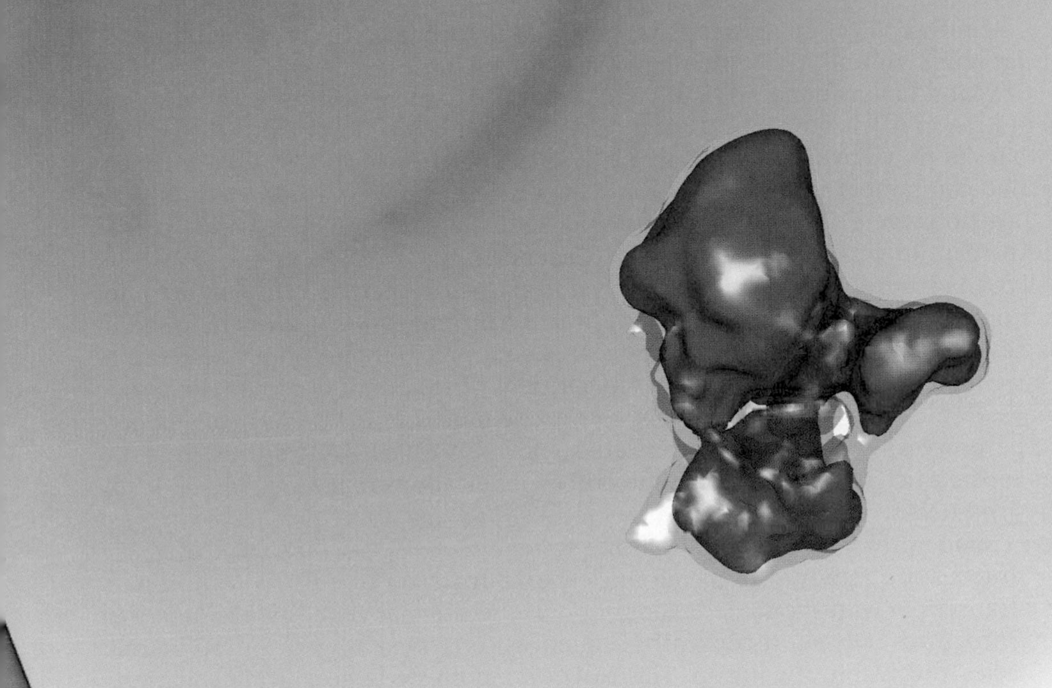

21

RNA Splicing and Processing

A three-dimensional structure of the mammalian spliceosomal C complex. Spliceosome structure courtesy of Nikolaus Grigorieff, Brandeis University. Background photo © Bella D/ ShutterStock, Inc.

CHAPTER OUTLINE

21.1 Introduction

RNA is a central player in gene expression. It was first characterized as an intermediate in protein synthesis, but since then many other RNAs have been discovered that play structural or functional roles at various stages of gene expression. The involvement of RNA in many functions concerned with gene expression supports the general view that life may have evolved from an "RNA world" in which RNA was originally the active component in maintaining and expressing genetic information. Many of these functions were subsequently assisted or taken over by proteins, with a consequent increase in versatility and probably efficiency.

All RNAs studied thus far are transcribed from their respective genes and require further processing to become mature and functional. Interrupted genes are found in all classes of eukaryotic organisms. They represent a small proportion of the genes of the unicellular eukaryotes, but the vast majority of genes in multicellular eukaryotic genomes. Removal of introns is a major part of the processing of RNAs in all eukaryotes. The process by which the introns are removed is called **RNA splicing**, and it occurs in the nucleus, together with the other modifications that are made to newly synthesized RNAs.

- **RNA splicing** The process of excising introns from RNA and connecting the exons into a continuous mRNA.
- **pre-mRNA** The nuclear primary transcript that is processed by modification and splicing to give an mRNA.
- **spliceosome** A complex formed by snRNPs and additional protein factors that is required for RNA splicing.

We can identify several types of splicing systems:

- Introns are removed from the nuclear **pre-mRNAs** of eukaryotes (primary transcripts with the same organization as the original gene) by a system that recognizes only short consensus sequences conserved at exon-intron boundaries and within the intron. This reaction requires a large splicing apparatus, which takes the form of an array of proteins and ribonucleoproteins that functions as a large particulate complex (the **spliceosome**). The mechanism of splicing involves transesterifications, and the catalytic center includes both RNA and proteins.
- Certain RNAs have the ability to excise their introns autonomously. Introns of this type fall into two groups, as distinguished by secondary/tertiary structure. Both groups use transesterification reactions in which the RNA is the catalytic agent (see *Chapter 23, Catalytic RNA*).
- The removal of introns from yeast nuclear tRNA precursors involves enzymatic activities that handle the substrate in a way resembling the tRNA processing enzymes, in which a critical feature is the conformation of the tRNA precursor. These splicing reactions are accomplished by enzymes that use cleavage and ligation.

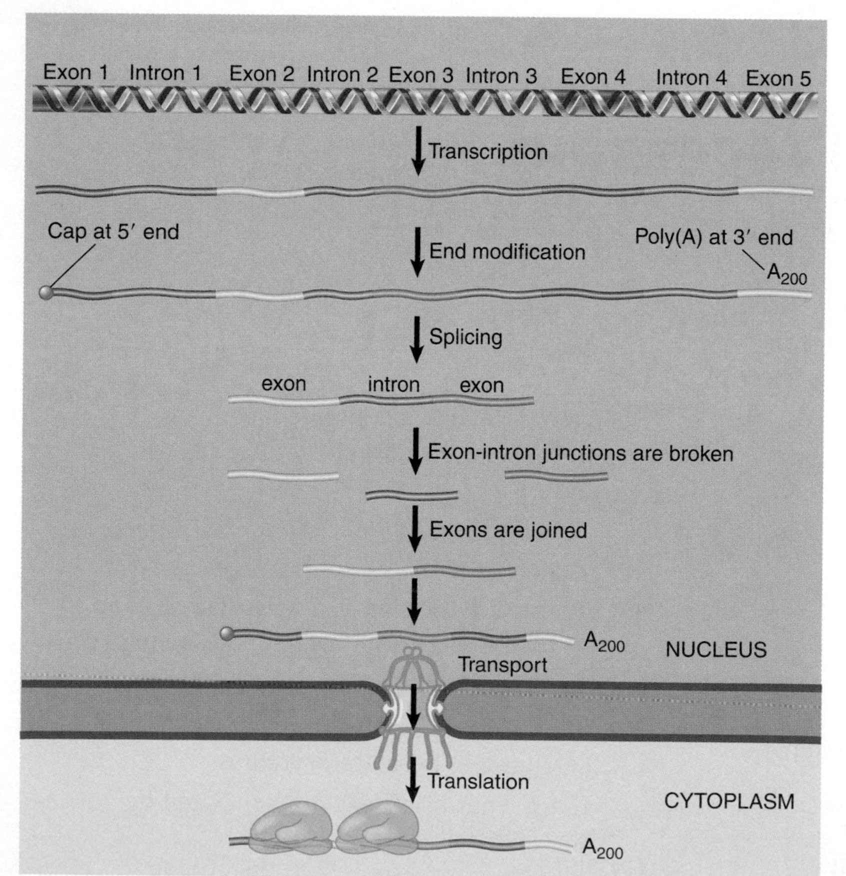

FIGURE 21.1 RNA is modified in the nucleus by additions to the 5′ and 3′ ends and by splicing to remove the introns. The splicing event requires breakage of the exon-intron junctions and joining of the ends of the exons. Mature mRNA is transported through nuclear pores to the cytoplasm, where it is translated.

The process of expressing an interrupted gene is reviewed in **FIGURE 21.1**. The transcript is capped at the 5′ end (see *Section 21.2, The 5′ End of Eukaryotic mRNA Is Capped*), has the introns removed, and is polyadenylated at the 3′ end (see *Section 21.14, The 3′ Ends of mRNAs Are Generated by Cleavage and Polyadenylation*). The processed mRNA is then transported through nuclear pores to the cytoplasm, where it is available to be translated.

When the pre-RNA is synthesized, it becomes bound by proteins to form a ribonucleoprotein particle (called **hnRNP**). (Taking its name from its broad size distribution, the RNA was originally called **heterogeneous nuclear RNA,** or **hnRNA**). Some of the proteins may have a structural role in packaging the hnRNA; several are known to shuttle between the nucleus and cytoplasm and to play roles in exporting the RNA or otherwise controlling its activity.

▸ **hnRNP** The ribonucleoprotein form of hnRNA (heterogeneous nuclear RNA), in which the hnRNA is complexed with proteins. Pre-mRNAs are not exported until processing is complete; thus they are found only in the nucleus.

▸ **heterogeneous nuclear RNA (hnRNA)** RNA that comprises transcripts of nuclear genes made by RNA polymerase II; it has a wide size distribution and low stability.

21.2 The 5′ End of Eukaryotic mRNA Is Capped

Transcription starts with a nucleoside triphosphate (usually a purine, A or G). The first nucleotide retains its 5′ triphosphate group and makes the usual phosphodiester bond from its 3′ position to the 5′ position of the next nucleotide. The initial sequence of the transcript can be represented as:

5′ppp A/GpNpNpNp . . .

When the mature mRNA is treated *in vitro* with enzymes that should degrade it into individual nucleotides, however, the 5′ end does not give rise to the expected nucleoside triphosphate. Instead it contains two nucleotides, which are connected by a 5′–5′ triphosphate linkage and also bear a methyl group. The terminal base is always a guanine that is added to the original RNA molecule after transcription.

Addition of the 5′ terminal G is catalyzed by a nuclear enzyme, guanylyl-transferase (GT). In mammals, GT has two enzymatic activities, one functioning as the triphosphatase to remove the two phosphates in GTP and the other as the quanylyl-transferase to fuse the guanine to the original 5′ triphosphate terminus of the RNA. In yeast, these two activities are carried out by two separate enzymes. The new G residue added to the end of the RNA is in the reverse orientation from all the other nucleotides:

5′Gppp + 5′ pppApNpNp . . . → Gppp5′–5′ApNpNp . . . + pp + p

This structure is called a **cap**. It is a substrate for several methylation events. **FIGURE 21.2** shows the full structure of a cap after all possible methyl groups have been added. The most important event is the addition of a single methyl group at the 7 position of the terminal guanine, which is carried out by guanine-7-methyltransferase (MT).

▸ **cap** The structure at the 5′ end of eukaryotic mRNA, which is introduced after transcription by linking the terminal phosphate of 5′ GTP to the terminal base of the mRNA.

Although the capping process can be accomplished *in vitro* using purified enzymes, the reaction normally takes place during transcription. Shortly after transcription initiation, Pol II is paused ~30 nucleotides downstream from the initiation site, waiting for the recruitment of the capping enzymes to add the cap to the 5′ end of nascent RNA. Without this protection, nascent RNA may be vulnerable to attack by 5′–3′ exonucleases, and such trimming may induce the Pol II complex to fall off from the DNA

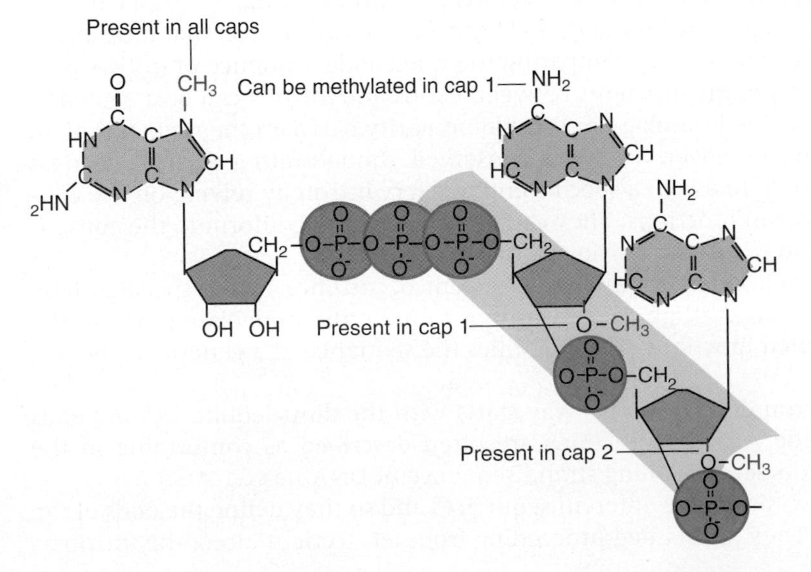

FIGURE 21.2 The cap blocks the 5′ end of mRNA and can be methylated at several positions.

template. Thus, the process of capping is important for Pol II to enter the productive mode of elongation to transcribe the rest of the gene. In this regard, the evolvement of the pausing mechanism for 5′ capping represents a checkpoint for transcription reinitiation from the initial pausing site.

In a population of eukaryotic mRNAs, every molecule contains only one methyl group in the terminal guanine, generally referred to as monomethylated cap. In contrast, some other small noncoding RNAs, such as those involved in RNA splicing in the spliceosome (see *Section 21.6, snRNAs Are Required for Splicing*) are further methylated to contain three methyl groups in the terminal guanine. This structure is called a trimethylated cap. The enzymes for these additional methyl transfers are present in the cytoplasm. This may ensure that only some specialized RNAs are further modified at their caps.

One of the major functions for the formation of a cap is to protect the mRNA from degradation. In fact, enzymatic decapping represents one of the major mechanisms in eukaryotic cells to regulate mRNA turnover (see *Section 21.9, Splicing Is Temporally and Functionally Coupled with Multiple Steps in Gene Expression*). In the nucleus, the cap is recognized and bound by the cap binding CBP20/80 heterodimer. This binding event stimulates splicing of the first intron and, via a direct interaction with the mRNA export machinery (TREX complex), facilitates mRNA export out of the nucleus. Once reaching the cytoplasm, a different set of proteins (eIF4F) binds the cap to initiate translation of the mRNA in the cytoplasm.

KEY CONCEPTS

- A 5′ cap is formed by adding a G to the terminal base of the transcript via a 5′–5′ link.
- The capping process takes place during transcription and may be important for release from pausing of transcription.
- The 5′ cap of most mRNA is monomethylated, but some small noncoding RNAs are trimethylated.
- The cap structure is recognized by protein factors to influence mRNA stability, splicing, export, and translation.

CONCEPT AND REASONING CHECK

What are the functions of the 5′ cap of eukaryotic mRNAs?

21.3 Nuclear Splice Junctions Are Short Sequences

To focus on the molecular events involved in nuclear intron splicing, we must consider the nature of the *splice sites*, the boundaries at both ends of each intron that include the sites of breakage and reunion. By comparing the nucleotide sequence of mRNA with that of the original gene, the junctions between exons and introns can be assigned.

There is no extensive homology or complementarity between the two ends of an intron. The junctions, however, have well conserved, though rather short, consensus sequences. It is possible to assign a specific end to every intron by relying on the conservation of exon-intron junctions. They can all be aligned to conform to the consensus sequence given in the upper portion of **FIGURE 21.3**.

The height of each letter indicates the percent occurrence of the specified base at each consensus position. High conservation is found only immediately within the intron at the presumed junctions. This identifies the sequence of a generic intron as:

GU . . . . . AG

Because the intron defined in this way starts with the dinucleotide GU and ends with the dinucleotide AG, the junctions are often described as conforming to the **GU-AG rule**. (Of course, the coding strand sequence of DNA has GT-AG.)

Note that the two sites have different sequences and so they define the ends of the intron *directionally*. They are named proceeding from left to right along the intron as

▸ **GU-AG rule** The rule that describes the presence of these constant dinucleotides at the first two and last two positions of introns of nuclear genes.

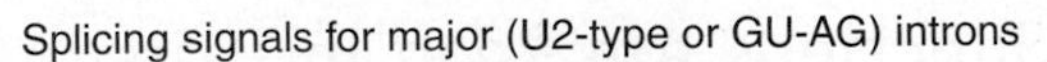

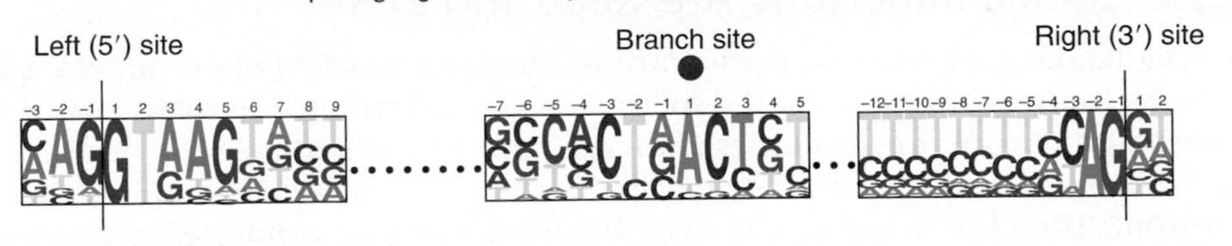

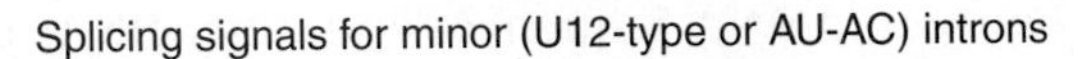

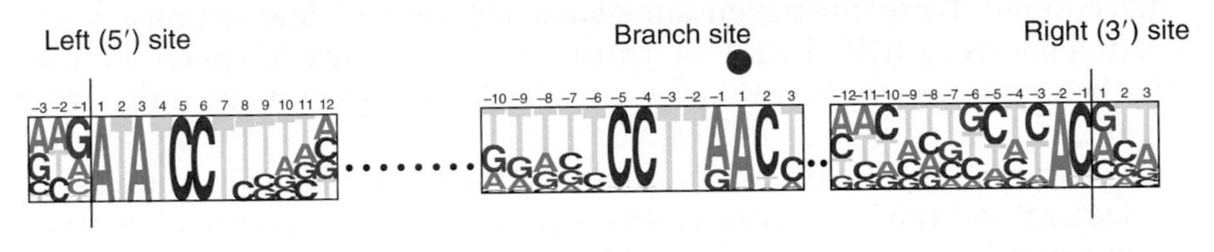

Splicing signals for minor (U12-type) introns that are flanked by GU and AG at ends

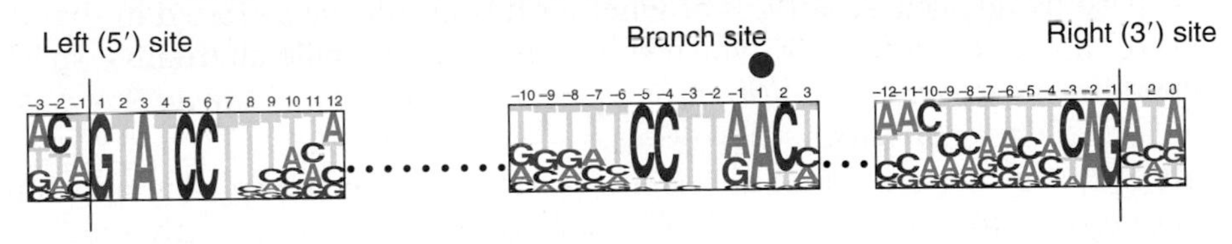

FIGURE 21.3 The ends of nuclear introns are defined by the GU-AG rule (shown here as GT-AG in the DNA sequence of the gene). Minor introns are defined by different consensus sequences at the 5′ splice site, branch site, and 3′ splice site.

the *5′ splice site* (sometimes called the *left,* or *donor, site*) and the *3′ splice site* (also called the *right,* or *acceptor, site*). The consensus sequences are implicated as the sites recognized in splicing by point mutations that prevent splicing *in vivo* and *in vitro*.

In addition to the majority of introns that follow the GU-AG rule, a small fraction of introns are exceptions with a different set of consensus sequences at the exon-intron boundaries as shown in the lower portion of Figure 21.3. These introns were initially described as minor introns that follow the AU-AC role because of the conserved AU-AC dinucleotides at both ends of each intron as shown in the middle panel of Figure 21.3. The major and minor introns, however, are better described as U2-type and U12-type introns based on the distinct splicing machineries that process them. As a result, some introns that appear to follow the GU-AG rule are actually processed as the U12-type of introns, as indicated in the lower panel of Figure 21.3.

KEY CONCEPTS

- Splice sites are the sequences immediately surrounding the exon-intron boundaries. They are named for their positions relative to the intron.
- The 5′ splice site at the 5′ (left) end of the intron includes the consensus sequence GU.
- The 3′ splice site at the 3′ (right) end of the intron includes the consensus sequence AG.
- The GU-AG rule (originally called the GT-AG rule in terms of DNA sequence) describes the requirement for these constant dinucleotides at the first two and last two positions of introns in pre-mRNAs.
- Minor introns follow a general AU-AC rule with a different set of consensus sequences at the exon-intron boundaries.

CONCEPT AND REASONING CHECK

What would be the effect of a mutation that altered the GU or AG sequence of an intron splice site?

21.4 Splice Junctions Are Read in Pairs

A typical mammalian gene has many introns. The basic problem of pre-mRNA splicing results from the simplicity of the splice sites and is illustrated in **FIGURE 21.4**. What ensures that the correct pairs of sites are recognized and spliced together in the presence of numerous sequences that match the consensus of *bona fide* splice sites in the intron? The corresponding GU-AG pairs must often be connected across great distances (some introns are >100 kb long). The recognition and splicing processes involve interactions between mRNA sequences and other RNAs and proteins as part of the spliceosome. These interactions are detailed in the next few sections.

Experiments using hybrid RNA precursors show that any 5′ splice site can, in principle, be connected to any 3′ splice site. Such experiments have two general conclusions:

- *Splice sites are generic.* They do not have specificity for individual RNA precursors, and individual precursors do not convey specific information (such as secondary structure) that is needed for splicing.
- *The apparatus for splicing is not tissue-specific.* An RNA can usually be properly spliced by any cell, regardless of whether it is usually synthesized in that cell. (We discuss exceptions in which there are tissue-specific alternative splicing patterns in *Section 21.11, Alternative Splicing Is a Rule, Rather Than an Exception, in Multicellular Eukaryotes.*)

If all 5′ splice sites and all 3′ splice sites look similar to the splicing apparatus, what rules ensure that recognition of splice sites is restricted so that only the 5′ and 3′ sites of the same intron are spliced? Are introns removed in a specific order from a particular RNA?

Splicing is temporally coupled with transcription (e.g., many splicing events are already completed before the RNA polymerase reaches the end of the gene); as a result it is reasonable to assume that transcription provides a rough order of splicing in the 5′ to 3′ direction (something like a first-come, first-served mechanism). Secondly, a functional splice site is often surrounded by a series of sequence elements that can enhance or suppress the site. Thus, sequences in both exons and introns can also function as regulatory elements for splice site selection.

We can imagine that, in order to be efficiently recognized by the splicing machinery, a functional splice site has to have the right sequence context, including specific consensus sequences and surrounding splicing enhancing elements that take precedence over splicing suppressing elements. These mechanisms together may ensure that splice signals are read in pairs in a relatively linear order.

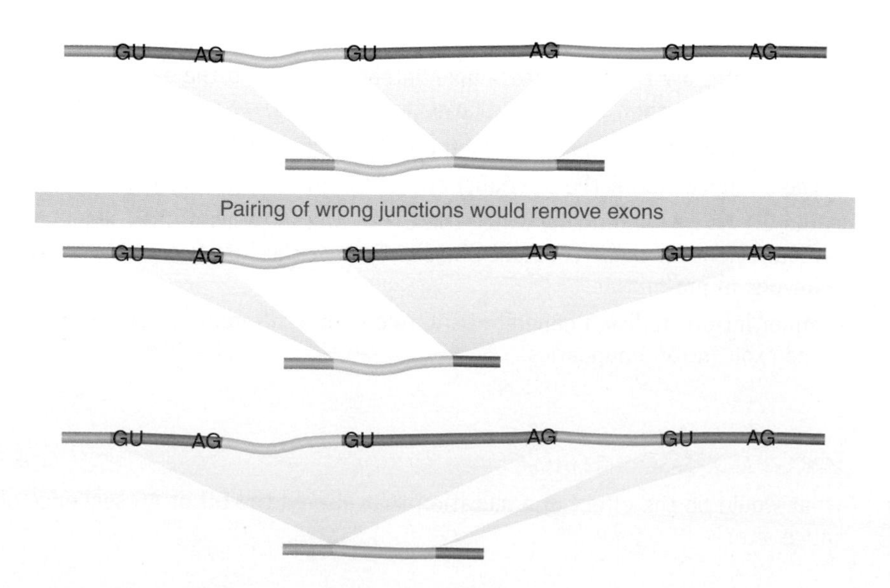

FIGURE 21.4 Splicing junctions are recognized only in the correct pairwise combinations.

KEY CONCEPTS

- Splicing depends only on recognition of pairs of splice junctions.
- All 5′ splice sites are functionally equivalent, and all 3′ splice sites are functionally equivalent.
- Additional conserved sequences at both 5′ and 3′ splice sites define functional splice sites among numerous other potential sites in the pre-mRNA.

CONCEPT AND REASONING CHECK

Explain how the correct 5′ and 3′ splice sites are recognized by the cell's splicing apparatus.

21.5 Pre-mRNA Splicing Proceeds through a Lariat

In addition to the 5′ and 3′ splice sites, the splicing apparatus recognizes a sequence downstream from the 5′ site called the **branch site**, so called because of the branched intermediate formed during the splicing reaction.

The stages of splicing are illustrated in the pathway of **FIGURE 21.5**. We discuss the reaction in terms of the individual RNA species that can be identified, but remember that *in vivo* the species containing exons are not released as free molecules but remain held together by the splicing apparatus.

FIGURE 21.6 shows that the first step of the splicing reaction is to make a cut at the 5′ splice site, separating the left exon and the right intron-exon molecule. The left exon takes the form of a linear molecule. The right intron-exon molecule forms a branched structure called the **lariat**, in which the 5′ terminus generated at the end of the intron becomes linked by a 5′–2′ bond to a base within the intron. The target base is an A in the branch site.

▸ **branch site** A short sequence just before the end of an intron at which the lariat intermediate is formed in splicing by joining the 5′ nucleotide of the intron to the 2′ position of an adenosine.

▸ **lariat** An intermediate in RNA splicing in which a circular structure with a tail is created by a 5′ to 2′ bond.

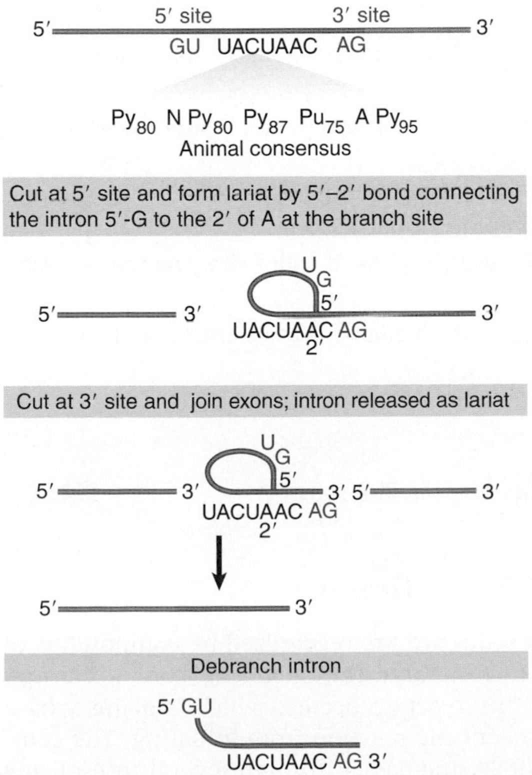

5′ site 3′ site
5′ GU UACUAAC AG 3′
Py_{80} N Py_{80} Py_{87} Pu_{75} A Py_{95}
Animal consensus

Cut at 5′ site and form lariat by 5′–2′ bond connecting the intron 5′-G to the 2′ of A at the branch site

5′ 3′ U G 5′ UACUAAC AG 2′ 3′

Cut at 3′ site and join exons; intron released as lariat

5′ 3′ U G 5′ UACUAAC AG 2′ 3′ 5′ 3′

5′ 3′

Debranch intron

5′ GU UACUAAC AG 3′

FIGURE 21.5 Splicing occurs in two stages. First the 5′ exon is cleaved off, and then it is joined to the 3′ exon.

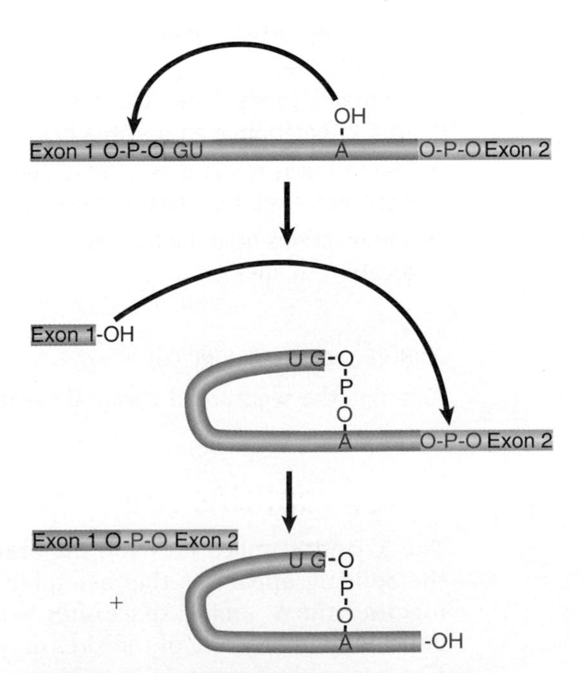

FIGURE 21.6 Nuclear splicing occurs by two transesterification reactions, in which an OH group attacks a phosphodiester bond.

▸ **transesterification** A reaction that breaks and makes chemical bonds in a coordinated transfer so that no energy is required.

The chemical reactions proceed by **transesterification**: in effect, a bond is transferred from one location to another. In the second step, the free 3′–OH of the exon that was released by the first reaction now attacks the bond at the 3′ splice site. Note that the number of phosphodiester bonds is conserved. There were originally two 5′–3′ bonds at the exon-intron splice sites; one has been replaced by the 5′–3′ bond between the exons, and the other has been replaced by the 2′–5′ bond that forms the lariat. The lariat is then "debranched" to give a linear excised intron, which is rapidly degraded.

The sequences needed for splicing are the short consensus sequences at the 5′ and 3′ splice sites and at the branch site. Together with the knowledge that most of the sequence of an intron can be deleted without impeding splicing, this indicates that there is no demand for specific conformation in the intron (or exon).

The branch site plays an important role in identifying the 3′ splice site. The branch site in yeast is highly conserved and has the consensus sequence UACUAAC. The branch site in multicellular eukaryotes is not well conserved but has a preference for purines or pyrimidines at each position and retains the target A nucleotide (see Figure 21.5).

The branch site lies 18 to 40 nucleotides upstream of the 3′ splice site. Mutations or deletions of the branch site in yeast prevent splicing. In multicellular eukaryotes, the relaxed constraints in its sequence result in the ability to use related sequences (called cryptic sites) when the authentic branch is deleted or mutated. Proximity to the 3′ splice site appears to be important because the cryptic site is always close to the authentic site. A cryptic site is used only when the branch site has been inactivated. When a cryptic branch sequence is used in this manner, splicing otherwise appears to be normal, and the exons give the same products as the use of the authentic site does. The role of the branch site, therefore, is to identify the nearest 3′ splice site as the target for connection to the 5′ splice site. This can be explained by the fact that an interaction occurs between protein complexes that bind to these two sites.

KEY CONCEPTS

- Splicing requires the 5′ and 3′ splice sites and a branch site just upstream of the 3′ splice site.
- The branch sequence is conserved in yeast but less well conserved in multicellular eukaryotes.
- A lariat is formed when the intron is cleaved at the 5′ splice site and the 5′ end is joined to a 2′ position at an A at the branch site in the intron.
- The intron is released as a lariat when it is cleaved at the 3′ splice site, and the left and right exons are then ligated together.
- The reactions occur by transesterifications, in which a bond is transferred from one location to another.

CONCEPT AND REASONING CHECK

Describe the sequence of chemical events that occur in RNA splicing.

21.6 snRNAs Are Required for Splicing

The 5′ and 3′ splice sites and the branch sequence are recognized by components of the splicing apparatus that assemble to form a large complex. This complex brings together the 5′ and 3′ splice sites before any reaction occurs, which explains why a deficiency in any one of the sites may prevent the reaction from initiating. The complex assembles sequentially on the pre-mRNA and passes through several "presplicing complexes" before forming the final, active complex, which is called the spliceosome. Splicing occurs only after all the components have assembled.

The splicing apparatus contains both proteins and RNAs (in addition to the pre-mRNA). The RNAs take the form of small molecules that exist as ribonucleoprotein particles. Both the nucleus and cytoplasm of eukaryotic cells contain many discrete small RNA species. They range in size from 100 to 300 bases in multicellular eukaryotes and extend in length to ~1000 bases in yeast. They vary considerably in abundance, from 10^5 to 10^6 molecules per cell to concentrations too low to be detected directly.

Those restricted to the nucleus are called **small nuclear RNAs (snRNAs)**; those found in the cytoplasm are called **small cytoplasmic RNAs (scRNAs)**. In their natural state, they exist as ribonucleoprotein particles (*snRNPs* and *scRNPs*). Colloquially, they are sometimes known as **snurps** and **scyrps**, respectively. There is also a class of small RNAs found in the nucleolus, called *snoRNAs*, which are involved in processing ribosomal RNA (see *Section 21.15, Production of rRNA Requires Cleavage Events and Involves Small RNAs*).

The spliceosome is a large particle, greater in mass than the ribosome, and contains five snRNPs as well as many additional proteins. The snRNPs involved in splicing are U1, U2, U5, U4, and U6. They are named according to the snRNAs that are present. Each snRNP contains a single snRNA and several (<20) proteins. The U4 and U6 snRNPs are usually found as a di-snRNP (U4/U6) particle. A common structural core for each snRNP consists of a group of eight proteins, all of which are recognized by an autoimmune antiserum called **anti-Sm**; conserved sequences in the proteins form the target for the antibodies. The other proteins in each snRNP are unique to it. The Sm proteins bind to the conserved sequence PuAU$_{3-6}$GPu, which is present in all snRNAs except U6. The U6 snRNP instead contains a set of Sm-like (Lsm) proteins.

FIGURE 21.7 summarizes the components of the spliceosome. The five snRNAs account for more than a quarter of the mass; together with their 41 associated proteins, they account for almost half of the mass. Some 70 other proteins found in the spliceosome are described as **splicing factors**. They include proteins required for assembly of the spliceosome, proteins required for it to bind to the RNA substrate, and proteins involved in constructing an RNA-based center for the transesterification reactions. In addition to these proteins, another ~30 proteins associated with the spliceosome are believed to be acting at other stages of gene expression, which suggests splicing may be connected to other steps in gene expression (see *Section 21.9, Splicing Is Temporally and Functionally Coupled with Multiple Steps in Gene Expression*).

Some of the proteins in the snRNPs may be directly involved in splicing; others may be required in structural roles or just for assembly or interactions between the snRNP particles. About one-third of the proteins involved in splicing are components of the snRNPs. Increasing evidence for a direct role of RNA in the splicing reaction suggests that relatively few of the splicing factors play a direct role in catalysis; most splicing factors may therefore provide structural or assembly roles in the spliceosome.

- **small nuclear RNAs (snRNAs; snurps)** Small RNA species confined to the nucleus; several of them are involved in splicing or other RNA-processing reactions. Snurps are the ribonucleoprotein particles that include a specific snRNA and its protein partners.
- **small cytoplasmic RNAs (scRNAs; scyrps)** RNAs that are present in the cytoplasm (and sometimes are also found in the nucleus). Scyrps are the ribonucleoprotein particles that include an scRNA and its associated proteins.
- **anti-Sm** An autoimmune antiserum that defines the Sm domain that is common to a group of proteins found in snRNPs that are involved in RNA splicing.
- **splicing factor** A protein component of the spliceosome that is not part of one of the snRNPs.

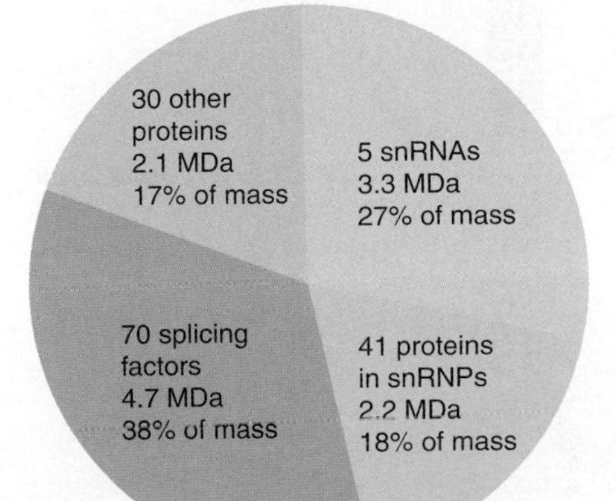

FIGURE 21.7 The spliceosome is ~12 MDa. Five snRNPs account for almost half of the mass. The remaining proteins include known splicing factors as well as proteins that are involved in other stages of gene expression.

KEY CONCEPTS

- The five snRNPs involved in splicing are U1, U2, U5, U4, and U6.
- Together with some additional proteins, the snRNPs form the spliceosome.
- All the snRNPs except U6 contain a conserved sequence that binds the Sm proteins that are recognized by antibodies generated in autoimmune disease.

CONCEPT AND REASONING CHECK

What are the probable respective roles of RNAs and proteins in RNA splicing?

21.7 Commitment of Pre-mRNA to the Splicing Pathway

Recognition of the consensus splicing signals involves both RNAs and proteins. Certain snRNAs have sequences that are complementary to the mRNA consensus sequences or to one another, and base pairing between snRNA and pre-mRNA, or between snRNAs, plays an important role in splicing.

Binding of U1 snRNP to the 5′ splice site is the first step in splicing. The human U1 snRNP contains eight proteins as well as the RNA. The secondary structure of the U1 snRNA is shown in **FIGURE 21.8**. It contains several domains. The Sm-binding site is required for interaction with the common snRNP proteins. Domains identified by the individual stem-loop structures provide binding sites for proteins that are unique to U1 snRNP.

U1 snRNA base pairs with the 5′ splice site by means of a single-stranded region at its 5′ terminus, which usually includes a stretch of four to six bases that is complementary with the splice site. **FIGURE 21.9** describes an experiment that

FIGURE 21.8 U1 snRNA has a base-paired structure that creates several domains. The 5′ end remains single stranded and can base pair with the 5′ splicing site.

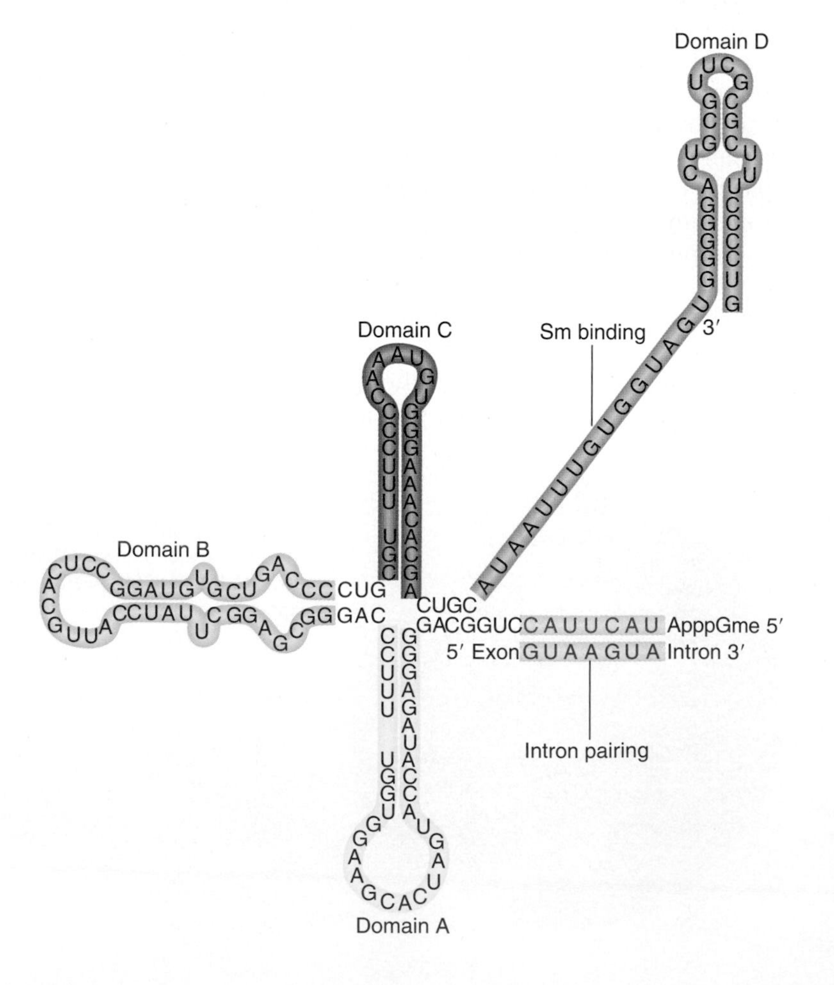

directly demonstrated the need for this base pairing. The wild-type sequence of the splice site of the 12S adenovirus pre-mRNA pairs at five out of six positions with U1 snRNA. A mutant in the 12S RNA that cannot be spliced has two sequence changes; the GG residues at positions 5 and 6 in the intron are changed to AU. When a mutation is introduced into U1 snRNA that restores pairing at position 5, normal splicing is regained.

FIGURE 21.10 shows the early stages of splicing. The first complex formed during splicing is the **E complex** ("E" for "early"), which contains U1 snRNP, the splicing factor U2AF, and members of a family called **SR proteins**, which comprise an important group of splicing factors and regulators. They take their name from the presence of a Ser-Arg-rich region that is variable in length. SR proteins interact with one another via this region. They also bind to RNA. They are an essential component of the spliceosome, forming a framework on the RNA substrate. The E complex is sometimes called the *commitment complex* because its formation identifies a pre-mRNA as a substrate for formation of the splicing complex.

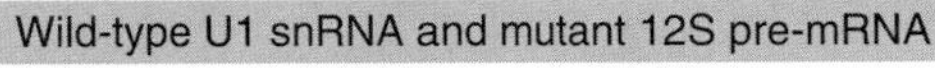

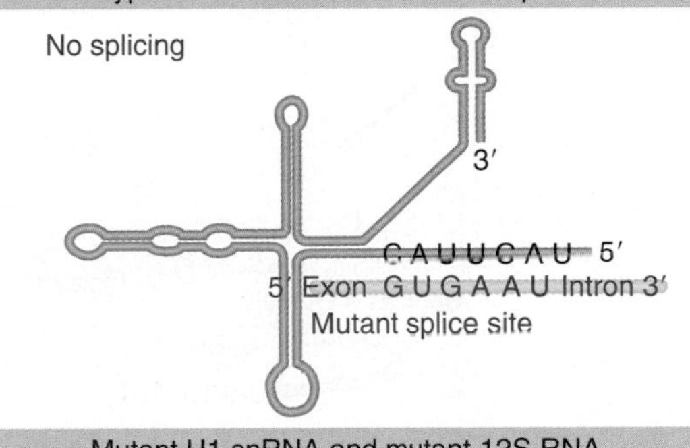

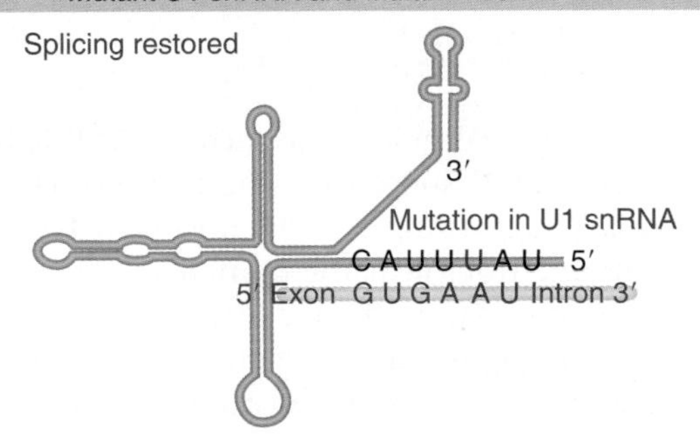

FIGURE 21.9 Mutations that abolish function of the 5′ splice site can be suppressed by compensating mutations in U1 snRNA that restore base pairing.

- **E complex** The first (early) complex to form at a splice site, consisting of U1 snRNP bound at the splice site together with factor ASF/SF2, U2AF bound at the branch site, and the bridging protein SF1/BBP.
- **SR protein** A protein that has a variable length of a Ser-Arg-rich region and is involved in splicing.

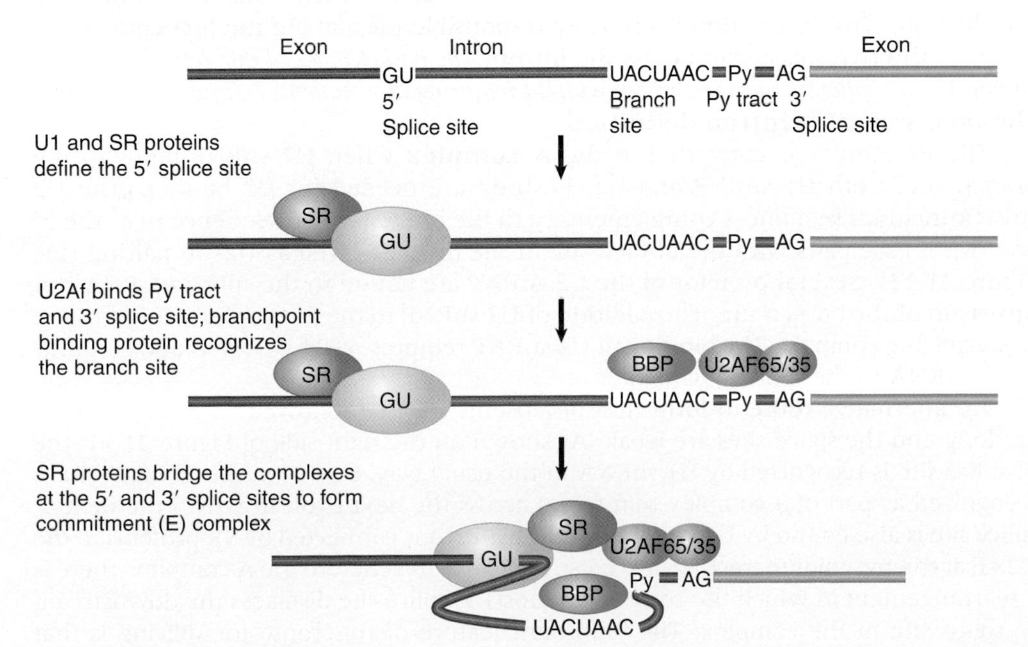

FIGURE 21.10 The commitment (E) complex forms by the successive addition of U1 snRNP to the 5′ splice site, U2AF to the pyrimidine tract/3′ splice site, and the bridging protein SF1/BBP.

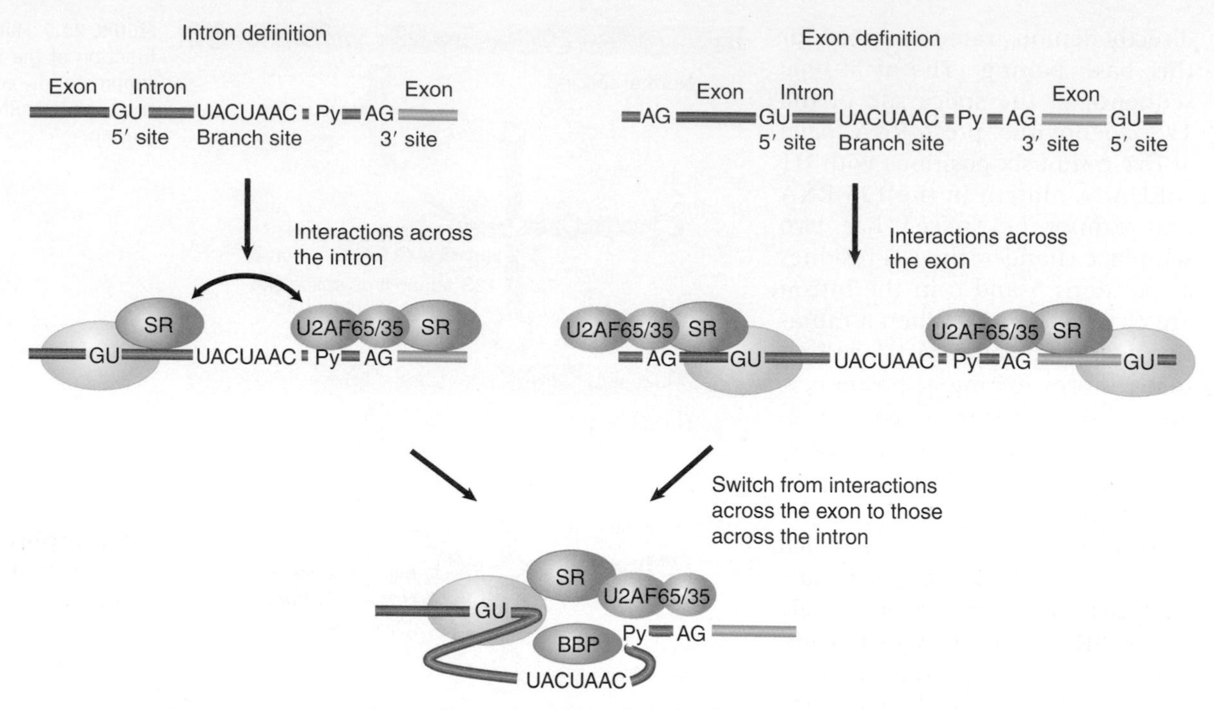

FIGURE 21.11 There are two routes for initial recognition of 5′ and 3′ splice sites by either intron definition or exon definition.

In the E complex, the factor U2AF is bound to the region between the branch site and the 3′ splice site. The name of U2AF reflects its original isolation as the U2 auxiliary factor. In most organisms, it has a large subunit (U2AF65) that contacts a pyrimidine tract near the branch site; a small subunit (U2AF35) directly contacts the dinucleotide AG at the 3′ splice site.

The recognition of functional splice sites during the formation of the E complex can take two routes, as illustrated in **FIGURE 21.11**. The most direct reaction is for both splice sites to be recognized across the intron. The presence of U1 snRNP at the 5′ splice site enables U2AF to bind at the pyrimidine tract near the branch site. A splicing factor, an SR protein called SF1 in mammals (the equivalent protein is called BBP in yeast, for branch point binding protein), connects U2AF to the U1 snRNP bound at the 5′ splice site. This interaction is probably responsible for making the first connection between the two splice sites across the intron. *The basic feature of this route for splicing is that the two splice sites are recognized without requiring any sequences outside of the intron.* This process is called **intron definition**.

- **intron definition** The process in which a pair of splicing sites are recognized by interactions involving only the 5′ site and the branchpoint/3′ site.
- **A complex** The second splicing complex, formed by the binding of U2 snRNP to the E complex.

The E complex is converted to the **A complex** when U2 snRNP binds to the branch site. Both U1 snRNP and U2AF/Mud2 are needed for U2 binding. The U2 snRNA includes sequences complementary to the branch site; a sequence near the 5′ end of U2 base pairs with the branch site in the intron as well as U2-U6 pairing (see Figure 21.13). Several proteins of the U2 snRNP are bound to the substrate RNA just upstream of the branch site. The addition of U2 snRNP to the E complex generates the A presplicing complex. The binding of U2 snRNP requires ATP hydrolysis and commits a pre-mRNA to the splicing pathway.

An alternative route to form the spliceosome may be followed when the introns are long and the splice sites are weak. As shown on the right side of Figure 21.11, the 5′ splice site is recognized by U1 snRNA in the usual way. However, the 3′ splice site is recognized as part of a complex that forms across the next exon in which the next 5′ splice site is also bound by U1 snRNA. This U1 snRNA is connected by SR proteins to the U2AF at the pyrimidine tract. When U2 snRNP joins to generate the A complex, there is a rearrangement in which the correct (leftmost) 5′ splice site displaces the downstream 5′ splice site in the complex. The important feature of this route for splicing is that sequences downstream of the intron itself are required. Usually these sequences include the next 5′-splice site. This process is called **exon definition**. This mechanism is not universal; neither SR proteins nor exon definition are found in *S. cerevisiae*.

▸ **exon definition** The process in which a pair of splicing sites are recognized by interactions involving the 5′ site of the intron and also the 5′ site of the next intron downstream.

KEY CONCEPTS

- U1 snRNP initiates splicing by binding to the 5′ splice site by means of an RNA–RNA pairing reaction.
- The direct way of forming an E complex is for U1 snRNP to bind at the 5′ splice site and U2AF to bind at a pyrimidine tract between the branch site and the 3′ splice site. This is intron definition.
- Another possibility is for the complex to form between U2AF at the pyrimidine tract and U1 snRNP at a downstream 5′ splice site. This is exon definition.

CONCEPT AND REASONING CHECKS

1. How does U1 snRNP recognize both the pre-mRNA and other snRNAs?
2. Describe the formation of the E and A presplicing complexes.

21.8 The Spliceosome Assembly Pathway

snRNPs and other factors involved in splicing associate with the presplicing complexes in a defined order. **FIGURE 21.12** shows the components of the complexes that can be identified as the reaction proceeds.

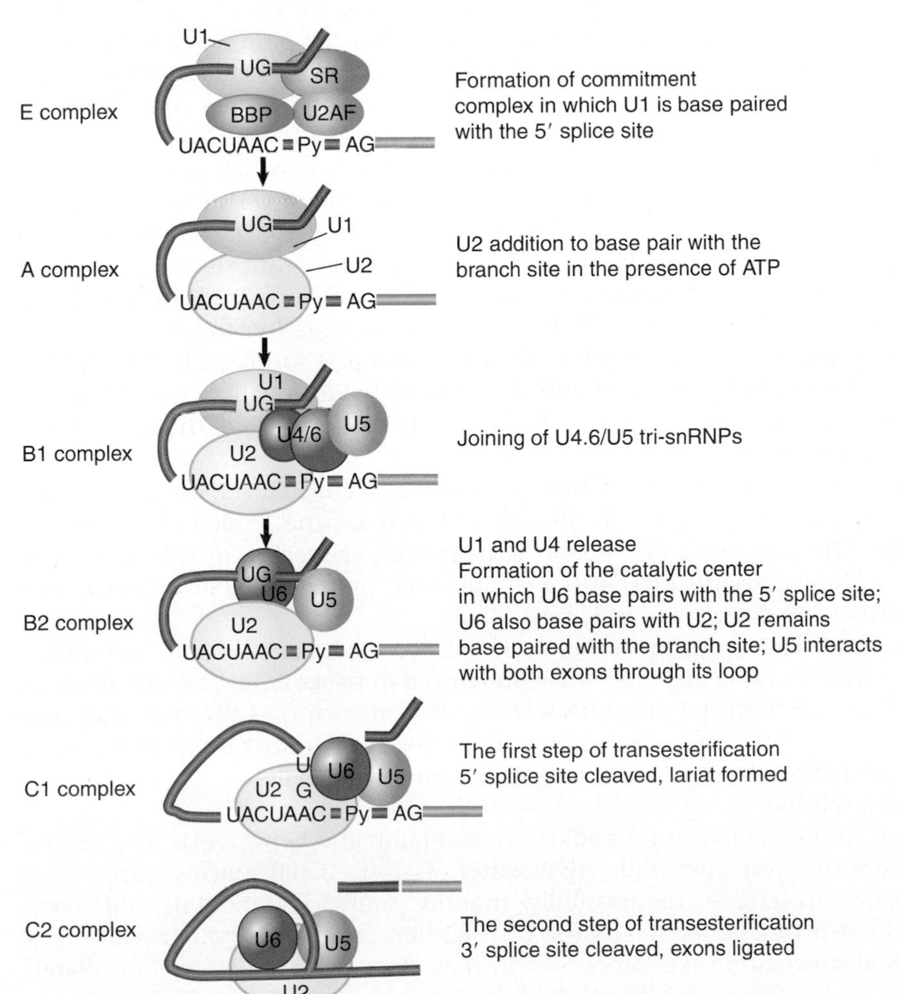

FIGURE 21.12 The splicing reaction proceeds through discrete stages in which spliceosome formation involves the interaction of components that recognize the consensus sequences.

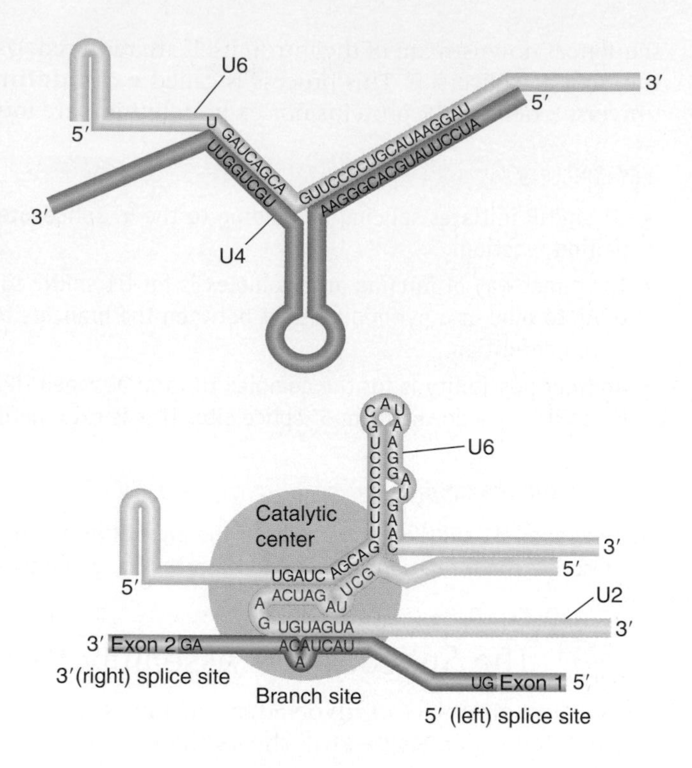

FIGURE 21.13 U6-U4 pairing is incompatible with U6-U2 pairing. When U6 joins the spliceosome it is paired with U4. Release of U4 allows a conformational change in U6; one part of the released sequence forms a hairpin, and the other part pairs with U2. An adjacent region of U2 is already paired with the branch site, which brings U6 into juxtaposition with the branch. Note that the substrate RNA is reversed from the usual orientation and is shown 3′ to 5′.

The B1 complex is formed when a trimer containing the U5 and U4/U6 snRNPs binds to the A complex. This complex is regarded as a spliceosome because it contains the components needed for the splicing reaction. It is converted to the B2 complex after U1 is released. The dissociation of U1 is necessary to allow other components to come into juxtaposition with the 5′-splice site, most notably U6 snRNA.

The catalytic reaction is triggered by the release of U4, which also takes place during the transition from the B1 to B2 complex. The role of U4 snRNA may be to sequester U6 snRNA until it is needed. **FIGURE 21.13** shows the changes that occur in the base pairing interactions between snRNAs during splicing. In the U6/U4 snRNP, a continuous length of 26 bases of U6 is paired with two separated regions of U4. When U4 dissociates, the region in U6 that is released becomes free to take up another structure. The first part of it pairs with U2; the second part forms an intramolecular hairpin. The interaction between U4 and U6 is mutually incompatible with the interaction between U2 and U6, so the release of U4 controls the ability of the spliceosome to proceed to the activated state.

For clarity, the figure shows the RNA substrate in extended form, but the 5′ splice site is actually close to the U6 sequence immediately on the 5′ side of the stretch bound to U2. This sequence in U6 snRNA pairs with sequences in the intron just downstream of the conserved GU at the 5′ splice site (mutations that enhance such pairing improve the efficiency of splicing).

Thus several pairing reactions between snRNAs and the substrate pre-mRNA occur in the course of splicing. They are summarized in **FIGURE 21.14**. The snRNPs have sequences that pair with the pre-mRNA substrate and with one another. They also have single-stranded regions in loops that are in close proximity to sequences in the substrate and that play an important role, as judged by the ability of mutations in the loops to block splicing.

The base pairings between U2 and the branch point and between U2 and U6 create a structure that resembles the active center of group II self-splicing introns (see Figure 21.18). This suggests the possibility that the catalytic component could comprise an RNA structure generated by the U2-U6 interaction. U6 is paired with the 5′ splice site, and crosslinking experiments show that a loop in U5 snRNA is immediately adjacent to the first base positions in both exons. Although the available evidence points to a RNA-based catalysis mechanism within the spliceosome, contribution(s) by proteins cannot be ruled out. One candidate protein is Prp8, a large scaffold protein that directly contacts both the 5′ and 3′ splice sites within the spliceosome.

The important conclusion suggested by these results is that *the snRNA components of the splicing apparatus interact both among themselves and with the substrate pre-mRNA by means of base pairing interactions, and these interactions allow for changes in structure that may bring reacting groups into apposition and may even create catalytic centers*. Furthermore, the conformational changes in the snRNAs are reversible; for example, U6 snRNA is not used up in a splicing reaction and at completion must be released from U2 so that it can reform the duplex structure with U4 to undertake another cycle of splicing.

A small proportion of introns is spliced by an alternative apparatus, called the U12 spliceosome, consisting of U11 and U12 (related to U1 and U2, respectively), a U5 variant, and the $U4_{atac}$ and $U6_{atac}$ snRNAs. The splicing reaction is essentially similar to that at U2-dependent introns, and the snRNAs play analogous roles. Whether there are differences in the protein components of this apparatus is not known.

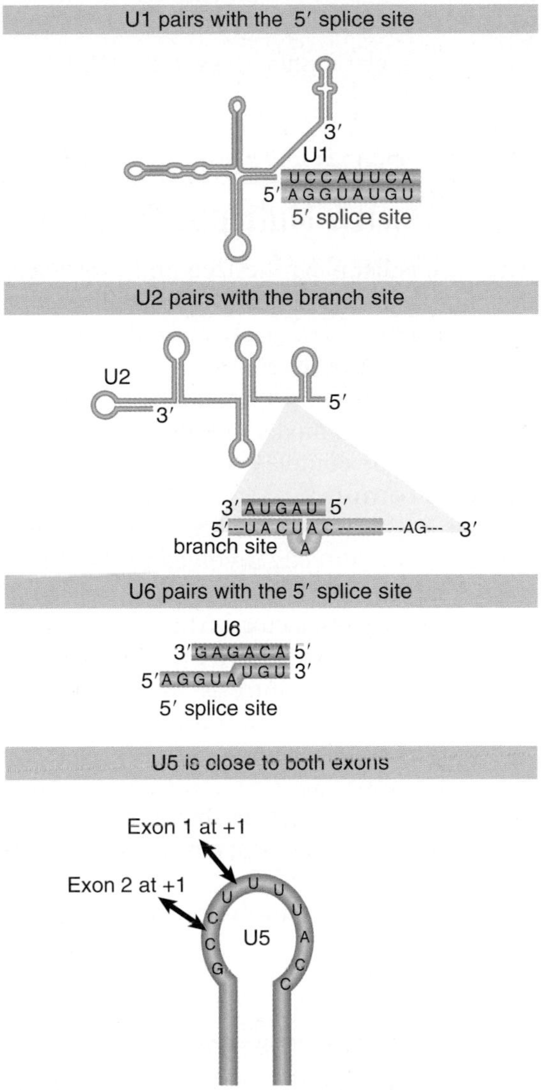

FIGURE 21.14 Splicing utilizes a series of base pairing reactions between snRNAs and splice sites.

The specific type of spliceosome used for splicing is influenced by sequences in the intron. A strong consensus sequence at the 5′ end defines the U12-dependent type of intron: $5'^{G}_{A}$UAUCCUUU . . . PyA$^{G}_{C}$3′. In addition, U12-dependent introns have a highly conserved branch point, UCCUUPuAPy, which pairs with U12. Both U12-dependent and U2-dependent introns may have either GU-AG or AU-AC termini.

KEY CONCEPTS

- Binding of U5 and U4/U6 snRNPs converts the A complex to the B1 spliceosome, which contains all the components necessary for splicing.
- The spliceosome passes through a series of further complexes as splicing proceeds.
- Release of U1 snRNP allows U6 snRNA to interact with the 5′ splice site and converts the B1 spliceosome to the B2 spliceosome.
- When U4 dissociates from U6 snRNP, U6 snRNA can pair with U2 snRNA to form the catalytic active site.
- An alternative splicing pathway uses another set of snRNPs that comprise the U12 spliceosome.
- The target introns are defined by longer consensus sequences at the splice junctions rather than strictly following the GU-AG or AU-AC rules.

CONCEPT AND REASONING CHECK

What is the relationship between the U2, U4, and U6 snRNPs?

21.9 Splicing Is Temporally and Functionally Coupled with Multiple Steps in Gene Expression

After it has been synthesized and processed, mRNA is exported from the nucleus to the cytoplasm in the form of a ribonucleoprotein complex. One means for ensuring that transport occurs only after the completion of splicing may be that introns can prevent export of mRNA because they are associated with the splicing apparatus. The spliceosome also may provide the initial point of contact for the export apparatus. **FIGURE 21.15** shows a model in which a protein complex binds to the RNA via the splicing apparatus. The complex consists of more than 9 proteins and is called the **EJC (*e*xon *j*unction *c*omplex)**.

▸ **EJC (exon junction complex)** A protein complex that assembles at exon-exon junctions during splicing and assists in RNA transport, localization, and degradation.

The EJC is involved in several functions of spliced mRNAs. Some of the proteins of the EJC are directly involved in these functions, and others recruit additional proteins for particular functions. The first contact in assembling the EJC is made with one of the splicing factors. After splicing, the EJC remains attached to the mRNA just upstream of the exon-exon junction. The EJC is not associated with RNAs transcribed from genes that lack introns, so it is uniquely involved with spliced products.

If introns are deleted from a gene, its RNA product is exported much more slowly to the cytoplasm. This suggests that the intron may provide a signal for attachment of the export apparatus. We can now account for this phenomenon in terms of a series of protein interactions, as shown in **FIGURE 21.16**. The EJC includes a group of proteins called the REF family (the best characterized member is called Aly). The REF proteins, in turn, interact with a transport protein (variously called TAP and Mex), which has direct responsibility for interaction with the nuclear pore.

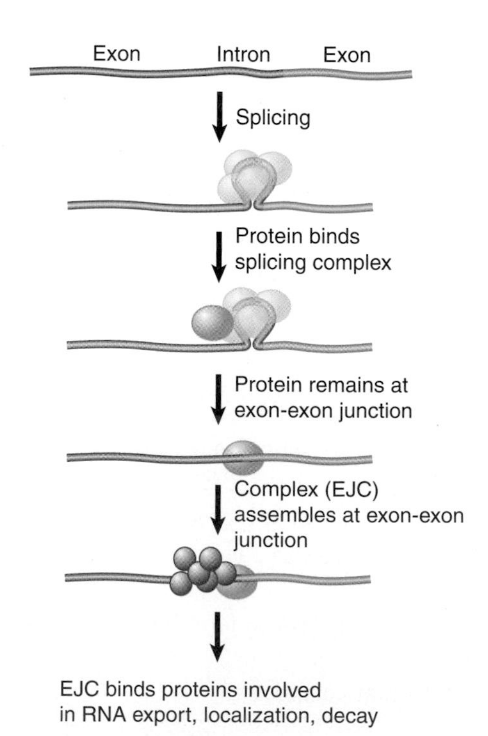

FIGURE 21.15 The EJC (exon junction complex) is deposited near the splice junction as a consequence of the splicing reaction.

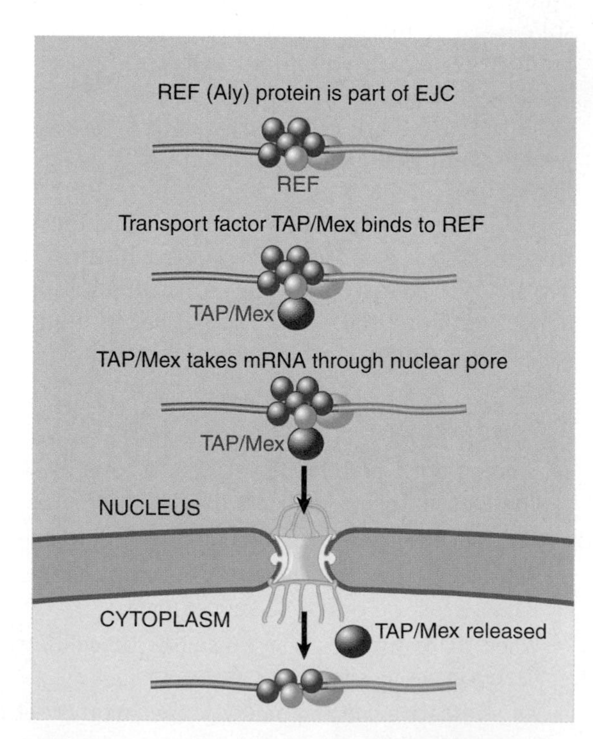

FIGURE 21.16 A REF protein (shown in green) binds to a splicing factor and remains with the spliced RNA product. REF binds to a transport protein (shown in purple) that binds to the nuclear pore.

A similar system may be used to identify a spliced RNA so that nonsense mutations prior to the last exon trigger its degradation in the cytoplasm (see *Section 22.9, Quality Control of mRNA Translation Is Performed by Cytoplasmic Surveillance Systems*).

KEY CONCEPTS

- The REF proteins bind to splicing junctions by associating with the spliceosome.
- After splicing, they remain attached to the RNA at the exon-exon junction.
- They interact with the transport protein TAP/Mex that exports the RNA through the nuclear pore.

CONCEPT AND REASONING CHECK

Explain how the presence of introns in a gene may aid in export of the mRNA product to the cytoplasm.

21.10 Pre-mRNA Splicing Likely Shares the Mechanism with Group II Autocatalytic Introns

Introns in all genes except (nuclear tRNA-encoding genes) can be divided into three general classes. Nuclear pre-mRNA introns are identified only by the possession of the GU . . . AG dinucleotides at the 5′ and 3′ ends and the branch site/pyrimidine tract near the 3′ end. They do not show any common features of secondary structure. In contrast, group I and group II introns found in organelles and in bacteria (group I introns are found also in the nucleus in unicellular/oligocellular eukaryotes) are classified according to their internal organization. Each can be folded into a typical type of secondary structure.

The group I and group II introns have the remarkable ability to excise themselves from an RNA. This is called **autosplicing**, or **self-splicing**. Group I introns are more common than group II introns. There is little relationship between the two classes, but in each case the RNA can perform the splicing reaction *in vitro* by itself, without requiring enzymatic activities provided by proteins; however, proteins are almost certainly required *in vivo* to assist with folding (see *Chapter 23, Catalytic RNA*).

FIGURE 21.17 shows that three classes of introns are excised by two successive transesterifications (shown previously for nuclear introns in Figure 21.6). In the first reaction, the 5′ exon-intron junction is attacked by a free hydroxyl

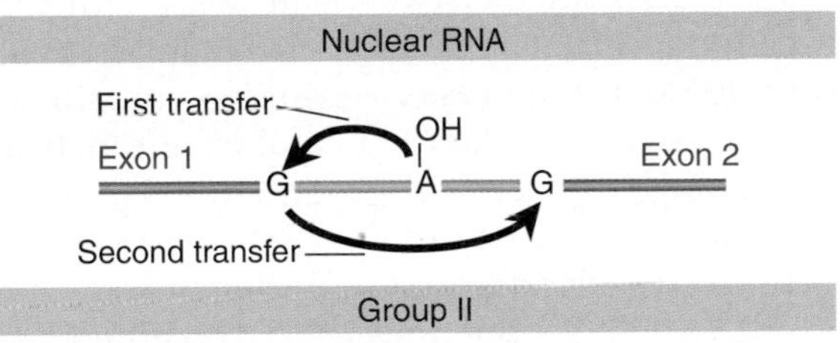

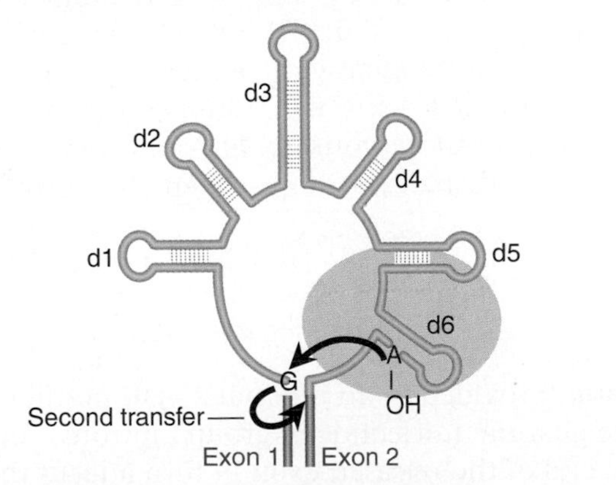

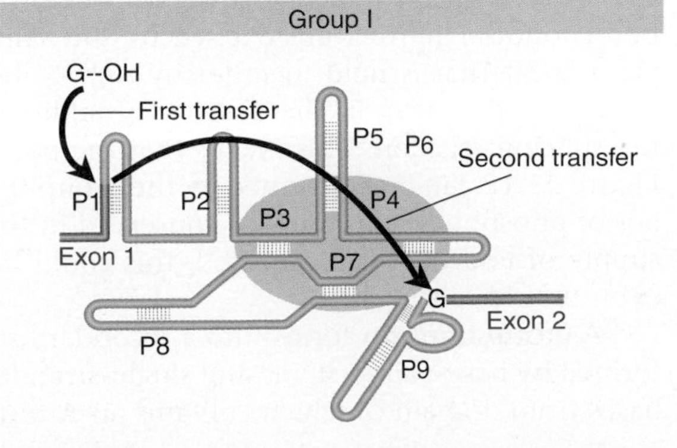

FIGURE 21.17 Three classes of splicing reactions proceed by two transesterifications. First, a free -OH group attacks the exon 1–intron junction. Second, the -OH created at the end of exon 1 attacks the intron–exon 2 junction.

▸ **autosplicing (self-splicing)** The ability of an intron to excise itself from an RNA by a catalytic action that depends only on the sequence of RNA in the intron.

MEDICAL APPLICATIONS

Alternative Splicing and Cancer

One in every two men and one in every three women will develop cancer at some time in their lives. There are more than 200 different types of cancer, and any of the different body tissues may be affected. *Cancer* is the result of the uncontrolled growth of a single cell that eventually forms a mass of tumor cells that have the potential to *metastasize* to other sites of the body. Each of us is composed of approximately 10^{14} cells, all of which have the potential to become cancerous, but the chance of an individual cell becoming cancerous is very small. We now know that during the change of a healthy cell into a cancerous cell, a series of changes slowly releases the cell from the multiple checks and balances that control its normal growth. Over the years, researchers have identified many factors that may increase the risk of developing certain types of cancers. Molecular changes in cells at the DNA and RNA levels make them more likely to become cancerous.

Breast cancer is the most commonly diagnosed type of cancer in women. It is estimated that in the United States in 2010, there were 207,090 new cases of breast cancer in women, and approximately 39,840 women died from breast cancer. Changes in certain genes (*BRCA1*, *BRCA2*, and others) make women more susceptible to breast cancer. Recent research suggests that the processing of *CD44* pre-mRNA plays a role in breast cancer. The *CD44* gene is located on the short arm of chromosome 11. It encodes a transmembrane cell-surface glycoprotein that is present in a variety of cell types, including the epidermis, central nervous system, lung, liver, and pancreas. It participates in a variety of cellular functions, including cell-cell interactions, release of cytokines, cell adhesion, and migration. It is a receptor for hyaluronic acid (a natural skin moisturizer) and can also interact with other ligands such as osteopontin (a glycoprotein that is abundant in bone mineral matrix), collagens (major component of connective tissue) and metalloproteinases. There are many variant isoforms of CD44 found in normal cells. The *CD44* gene contains in-frame constant exons and alternative variable exons. These isoforms are encoded by alternative splicing in two different regions containing exons of the gene. The *CD44* gene contains ten constant exons located at the 5′ and 3′ ends of the gene and ten variable exons located between the constant exons (**FIGURE B21.1**). Some of these variations diversify the biological function(s) in the final protein molecule.

Tremendous interest in CD44 was generated when researchers discovered that CD44 variants have metastatic properties. Research findings showed that the inclusion of *CD44* variable exons v4 through v6 correlate with both tumorigenesis and metastasis of several malignancies, including breast cancer. *CD44* exons v4 and v5 contain several copies of cytosine/adenine-rich and purine-rich sequences that are known to be strong enhancers of transcription. Overexpression of abnormal CD44 variants has been associated with breast cancer and poor prognosis.

The number of diseases identified as associated with missplicing is increasing. Missplicing can be caused by either mutations in regulatory sequences such as splice sites, mutations in enhancer/silencer sequences or alterations in *trans*-acting factors such as *splicing factors*. Given the correlation between the overexpression of

group (provided by an internal 2′–OH position in nuclear and group II introns or by a free guanine nucleotide in group I introns). In the second reaction, the free 3′–OH at the end of the released exon in turn attacks the 3′ intron-exon junction.

There are parallels between group II introns and pre-mRNA splicing. Group II mitochondrial introns are excised by the same mechanism as nuclear pre-mRNAs via a lariat that is held together by a 2′–5′ bond. When an isolated group II RNA is incubated *in vitro* in the absence of additional components, it is able to perform the splicing reaction. This means that the two transesterification reactions shown in Figure 21.18 can be performed by the group II intron RNA sequence itself. The number of phosphodiester bonds is conserved in the reaction, and as a result an external supply of energy is not required; this could have been an important feature in the evolution of splicing.

A group II intron forms into a secondary structure that contains several domains formed by base-paired stems and single-stranded loops. Domain 5 is separated by two bases from domain 6, which contains an A residue that donates the 2′–OH group for

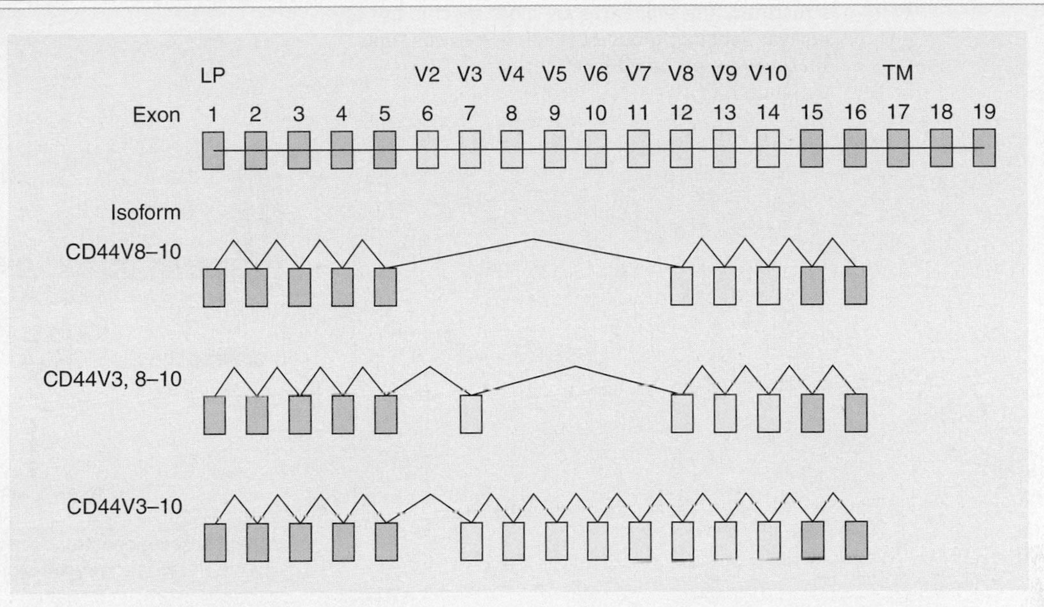

FIGURE B21.1 Top of the figure shows a schematic of the constant and variable exons of the normal human CD44 gene. Isoforms showing alternative splicing are shown below. Reproduced with permission, from Michael Piepkorn, Peter Hovingh, Kelly L. Bennett, Alejandro Aruffo and Alfred Linker, 1997, *Biochem. J.*, **327** 499–506. © the Biochemical Society. http://www.biochemj.org/bj/327/bj3270499.htm

the CD44 variants and breast tumors, a group of German researchers working on a model of breast cancer development performed experiments to determine the cause of *CD44* missplicing in breast cancers. To do this, they tested whether the expression of splicing activators or factors Tra2-α and Tra2-β expression change in mastectomy tissue removed from primary invasive adenocarcinomas of the breast compared to normal tissue. All tissue specimens showed a significant induction of Tra2-β in the invasive breast cancer at both the RNA and protein levels. This was not observed in normal breast tissue. It was also shown that Tra2-β was bound to the enhancers found in the *CD44* exon v4 and exon v5 sequences, inducing missplicing. Therefore, the induction of the Tra2-β splicing factor might explain the observed alternative splicing of CD44 isoforms associated with tumor progression and metastasis during breast cancer development.

Reference

Watermann, Tang, zur Hausen, Jäger, Stamm, and Stickeler. (2006). Splicing Factor Tra2-β1 Is Specifically Induced in Breast Cancer and Regulates Alternative Splicing of the CD44 Gene. *Cancer Research* **66**(9): 4774–4780.

the first transesterification. This constitutes a catalytic domain in the RNA. **FIGURE 21.18** compares this secondary structure with the structure formed by the combination of U6 with U2 and of U2 with the branch site. The similarity suggests that U6 may have a catalytic role in pre-mRNA splicing.

The features of group II splicing suggest that splicing evolved from an autocatalytic reaction undertaken by an individual RNA molecule, in which it accomplished a controlled deletion of an internal sequence. It is likely that such a reaction would require the RNA to fold into a specific conformation, or series of conformations, and would occur exclusively in *cis* conformation.

The ability of group II introns to remove themselves by an autocatalytic splicing event stands in great contrast to the requirement of nuclear introns for a complex apparatus. We may regard the snRNAs of the spliceosome as compensating for the lack of sequence information in the intron, and as providing the information required to form particular structures in RNA. The functions of the snRNAs may have evolved from the original autocatalytic system. These snRNAs act in *trans* upon the substrate

FIGURE 21.18 Nuclear splicing and group II splicing involve the formation of similar secondary structures. The sequences are more specific in nuclear splicing; group II splicing uses positions that may be occupied by either purine (R) or pyrimidine (Y).

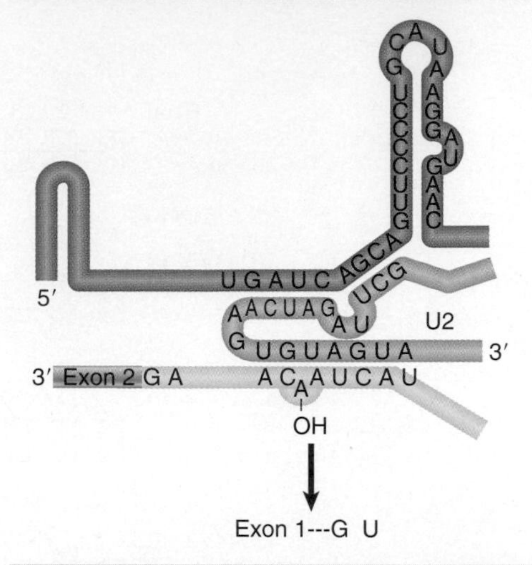

Group II splicing constructs an active center from the base paired regions of domains 5 and 6

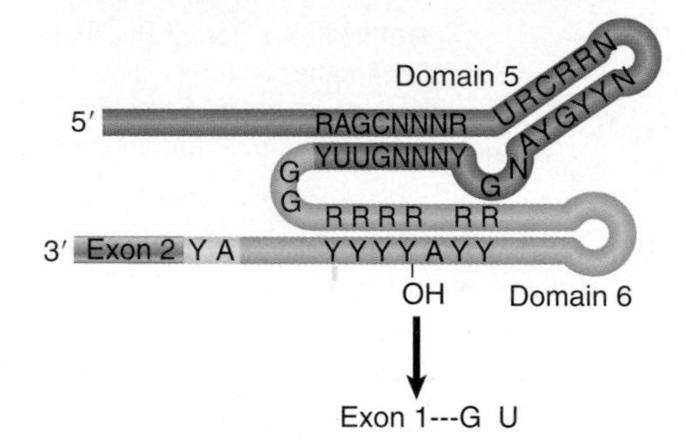

pre-mRNA; we might imagine that the ability of U1 to pair with the 5′ splice site, or of U2 to pair with the branch sequence, replaced a similar reaction that required the relevant sequence to be carried by the intron. Thus the snRNAs may undergo reactions with the pre-mRNA substrate—and with one another—that have substituted for the series of conformational changes that occur in RNAs that splice by group II mechanisms. In effect, these changes have relieved the substrate pre-mRNA of the obligation to carry the sequences needed to sponsor the reaction. As the splicing apparatus has become more complex (and as the number of potential substrates has increased), proteins have played a more important role.

KEY CONCEPTS

- Group II introns excise themselves from RNA by an autocatalytic splicing event.
- The splice junctions and mechanism of splicing of group II introns are similar to splicing of nuclear introns.
- A group II intron folds into a secondary structure that generates a catalytic site resembling the structure of U6-U2-nuclear intron.

CONCEPT AND REASONING CHECK

Compare the roles of RNA sequences in both autosplicing and the splicing of nuclear RNAs.

21.11 Alternative Splicing Is a Rule, Rather Than an Exception, in Multicellular Eukaryotes

When an interrupted gene is transcribed into an RNA that gives rise to a single type of spliced mRNA, there is no ambiguity in assignment of exons and introns. The RNAs of most mammalian genes, however, follow patterns of **alternative splicing**, which occurs when a single gene gives rise to more than one mRNA sequence. In some cases, the ultimate pattern of expression is dictated by the primary transcript because the use of different startpoints or the generation of alternative 3′ ends alters the pattern of splicing. In other cases, a single primary transcript is spliced in more than one way, and internal exons are substituted, added, or deleted. In some cases, the multiple products all are made in the same cell, but in others the process is regulated so that particular splicing patterns occur only under particular conditions. For example, in humans, different versions of a protein may be expressed in different tissues or at different developmental stages of the same tissue.

▸ **alternative splicing** The production of different RNA products from a single product by changes in the usage of splicing junctions.

There are various modes of alternative splicing, including intron retention, alternative 5′ splice-site selection, alternative 3′ splice-site selection, exon inclusion or skipping, and mutually exclusive selection of the alternative exons, as summarized in **FIGURE 21.19**. A single primary transcript may undergo more than one mode of alternative splicing. The mutually exclusive exons are normally regulated in a tissue-specific manner. Adding to this complexity in some cases, the ultimate pattern of expression is also dictated by the use of different transcription start points or the generation of alternative 3′ ends.

Although many alternative splicing events have been characterized and the biological roles of the alternatively spliced products determined, the best-understood

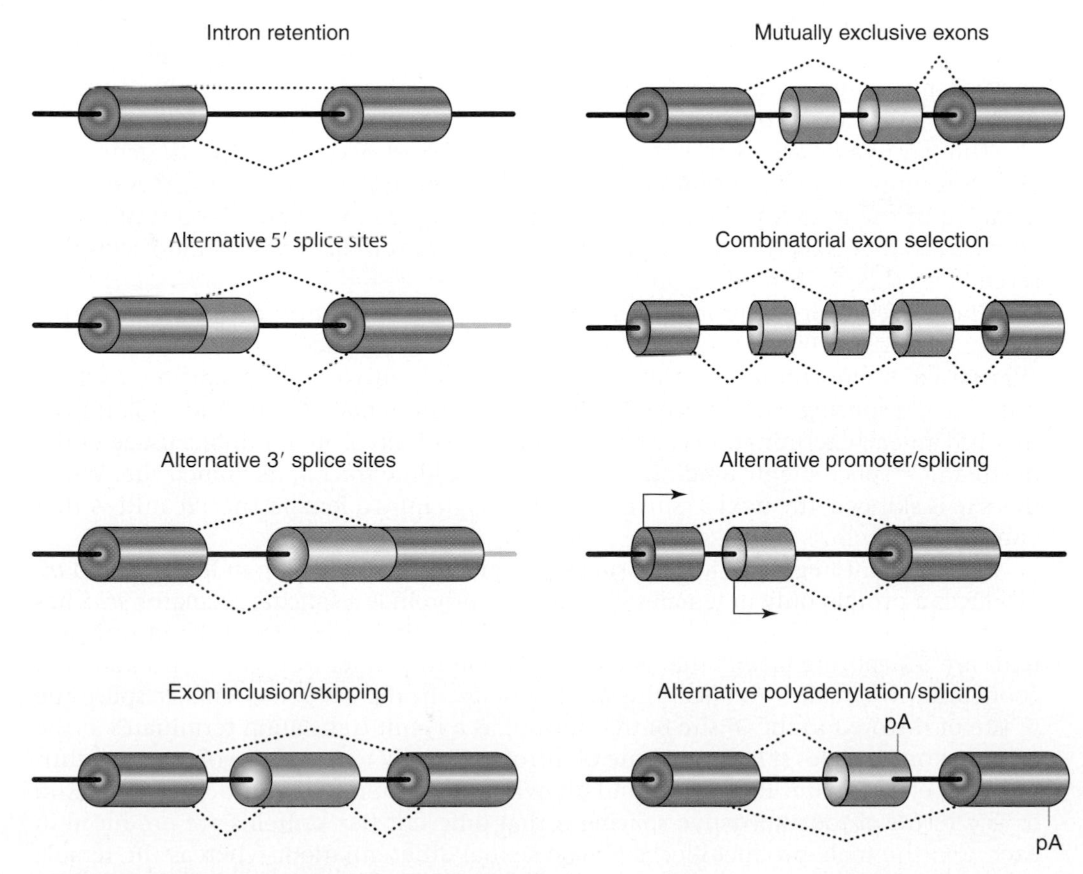

FIGURE 21.19 Different modes of alternative splicing.

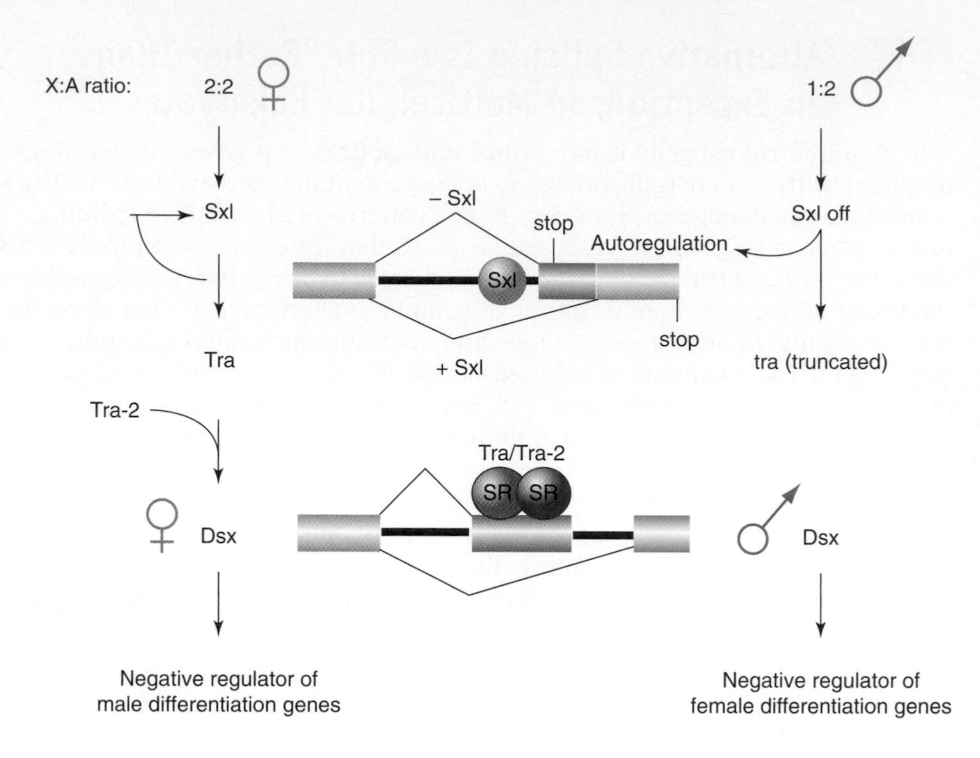

FIGURE 21.20 Sex determination in *D. melanogaster* involves a pathway in which different splicing events occur in females. Blockages at any stage of the pathway result in male development. Illustrated are *tra* pre-mRNA splicing controlled by the Sxl protein, which blocks the use of the alternative 3′ splice site, and *dsx* pre-mRNA splicing regulated by both Tra and Tra2 proteins in conjunction with other SR proteins, which positively influence the inclusion of the alternative exon.

example is still the pathway of sex determination in *D. melanogaster*, which involves interactions between a series of genes in which alternative splicing events distinguish males and females. The pathway takes the form illustrated in **FIGURE 21.20**, in which the ratio of X chromosomes to autosomes determines the expression of *sex lethal* (*sxl*), and changes in expression are passed sequentially through the other genes to *doublesex* (*dsx*), the last in the pathway.

The pathway starts with sex-specific splicing of *sxl*. Exon 3 of the *sxl* gene contains a termination codon that prevents synthesis of functional protein. This exon is included in the mRNA produced in males but is skipped in females. As a result, only females produce Sxl protein. The protein has a concentration of basic amino acids that resembles other RNA-binding proteins.

The presence of Sxl protein changes the splicing of the pre-mRNA of the transformer (*tra*) gene. Figure 21.20 shows that this involves splicing a constant 5′ site to alternative 3′ sites (note that this mode applies to both *sxl* and *tra* splicing, as illustrated). One splicing pattern occurs in both males and females, and results in an RNA that has an early termination codon. The presence of Sxl protein inhibits usage of the upstream 3′ splice site by binding to the polypyrimidine tract at its branch site. When this site is skipped, the next 3′ site is used. This generates a female-specific mRNA that encodes a protein.

Thus Sxl autoregulates its own splicing to ensure its expression in females, and *tra* produces a protein only in females; like Sxl, Tra protein is a splicing regulator. *tra2* has a similar function in females (but is also expressed in the males). The Tra and Tra2 proteins are SR splicing factors that act directly upon the target transcripts. Tra and Tra2 cooperate (in females) to affect the splicing of *dsx*. In the *dsx* gene, females splice the 5′ site of intron 3 to the 3′ site of that intron; as a result translation terminates at the end of exon 4. Males splice the 5′ site of intron 3 directly to the 3′ site of intron 4, thus omitting exon 4 from the mRNA and allowing translation to continue through exon 6. The result of the alternative splicing is that different Dsx proteins are produced in each sex: the male product blocks female sexual differentiation, whereas the female product represses expression of male-specific genes.

KEY CONCEPTS

- Specific exons or exonic sequences may be excluded or included in the mRNA products by using alternative splicing sites.
- Alternative splicing contributes to structural and functional diversity of gene products.
- Sex determination in *Drosophila* involves a series of alternative splicing events in genes encoding successive products of a pathway.

CONCEPT AND REASONING CHECK

Initial estimates of the number of human genes, based on the number of mRNAs produced in human cells, were much higher than the current count of human genes. How is this possible?

21.12 *trans*-Splicing Reactions Use Small RNAs

In both mechanistic and evolutionary terms, splicing has been viewed as an *intramolecular* reaction, essentially amounting to a controlled deletion of the intron sequences at the level of RNA. In genetic terms, splicing occurs only in *cis*. This means that *only sequences on the same molecule of RNA can be spliced together.*

The upper part of **FIGURE 21.21** shows the normal situation. The introns can be removed from each RNA molecule, allowing the exons of that RNA molecule to be spliced together, but there is no *intermolecular* splicing of exons between different RNA molecules. Although we know that *trans*-splicing between pre-mRNA transcripts of the same gene does occur, it must be exceedingly rare, because if it were prevalent the exons of a gene would be able to complement one another genetically instead of belonging to a single complementation group.

Although *trans*-splicing is rare in multicellular eukaryotes, it occurs as the primary mechanism to process precursor RNA into mature, translatable mRNAs in some organisms, such as trypanosomes and nematodes. In trypanosomes, all genes are expressed as polycistronic transcripts like those in bacteria. The transcribed RNA, however, cannot be translated without a 37-nucleotide leader brought in by *trans*-splicing to convert a polycistronic RNA into individual monocistronic mRNAs for translation. The leader sequence is not encoded upstream of the individual transcription units, though. Instead it is transcribed into an independent RNA, carrying additional sequences at its 3′ end, from a repetitive unit located elsewhere in the genome. **FIGURE 21.22** shows that this RNA carries the leader sequence followed by a 5′ splice site sequence. The sequences encoding the mRNAs carry a 3′ splice site just preceding the sequence found in the mature mRNA.

When the leader and the mRNA are connected by a *trans*-splicing reaction, the 3′ region of the leader RNA and the 5′ region of the mRNA in effect comprise the 5′ and 3′ halves of an intron. When splicing occurs, a 2′–5′ link forms by the usual reaction between the GU of the 5′ intron and the branch sequence near the AG of the 3′ intron. The two parts of the intron are covalently linked, but generate a Y-shaped molecule instead of a lariat.

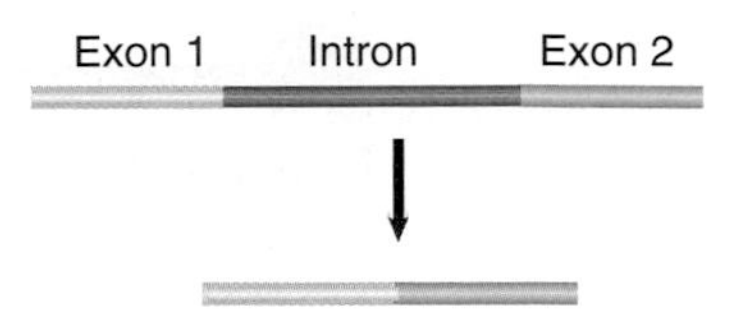

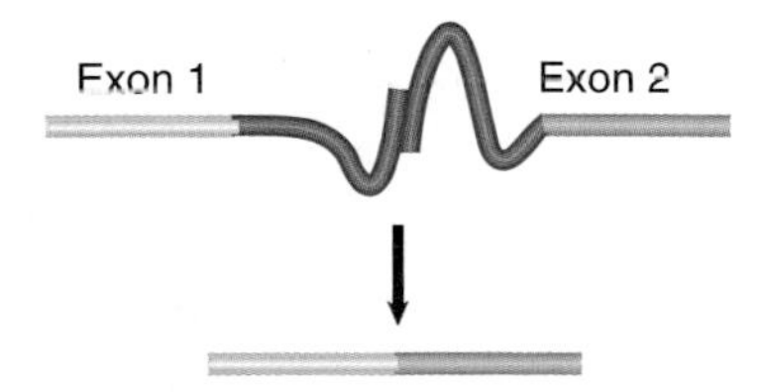

FIGURE 21.21 Splicing usually occurs only in *cis* between exons carried on the same physical RNA molecule, but *trans*-splicing can occur when special constructs are made that support base pairing between introns.

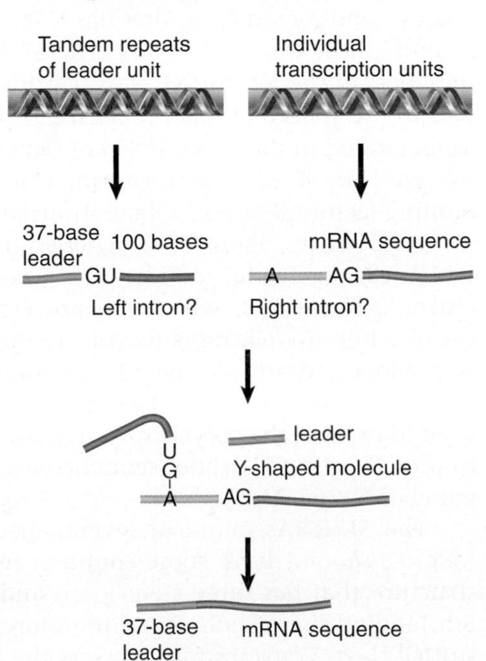

FIGURE 21.22 The SL RNA provides an exon that is connected to the first exon of an mRNA by *trans*-splicing. The reaction involves the same interactions as nuclear *cis*-splicing but generates a Y-shaped RNA instead of a lariat.

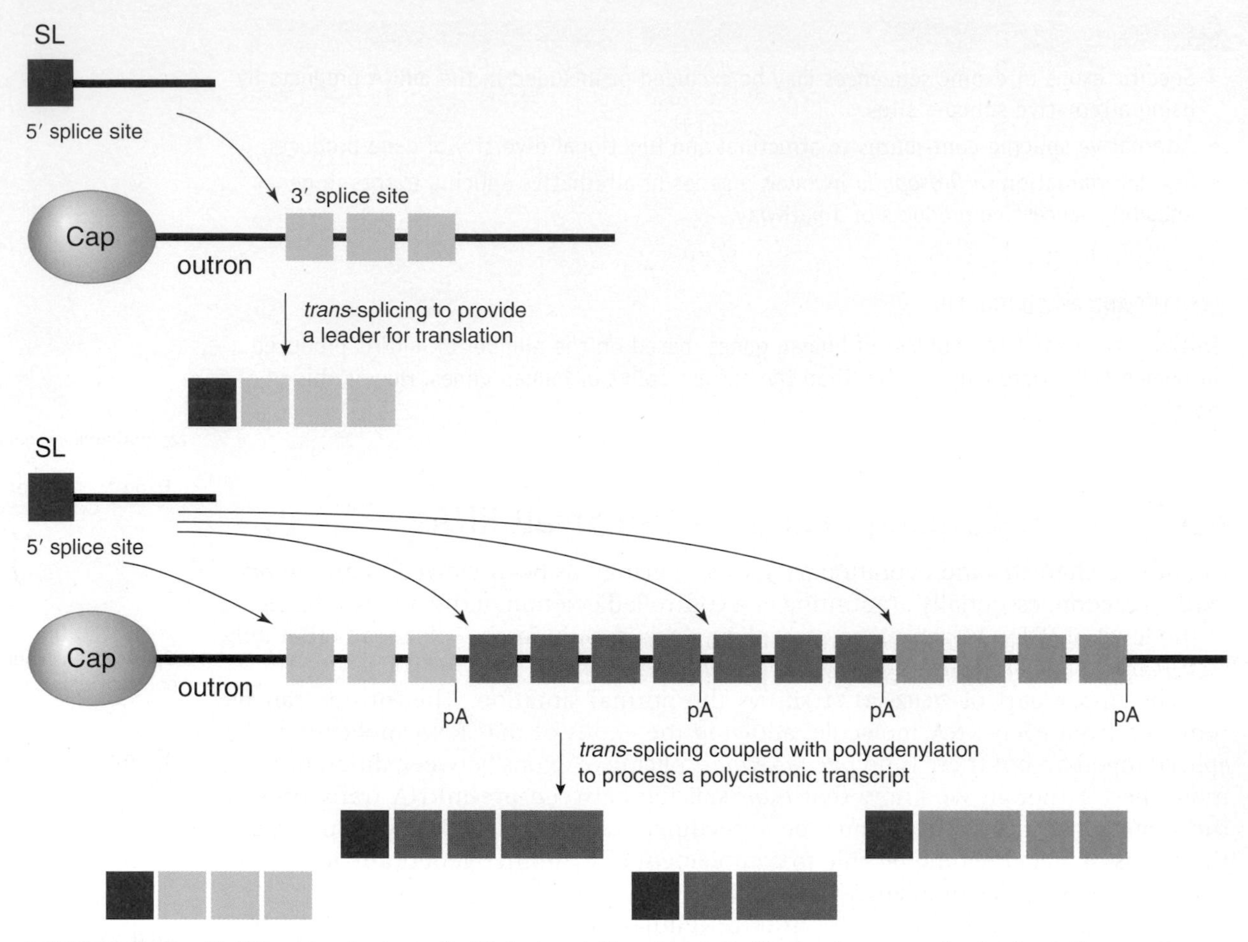

FIGURE 21.23 The SL RNA adds a leader to facilitate translation. Coupled with the cleavage and polyadenylation reactions, the addition of the SL RNA is also used to convert polycistronic transcripts to monocistronic units.

SL RNA (spliced leader RNA) A small RNA that donates an exon in the *trans*-splicing reaction of trypanosomes and nematodes.

The RNA that donates the 5′ exon for *trans*-splicing is called the **SL RNA (*s*pliced *l*eader RNA)**. The SL RNAs, which are 100 nucleotides in length, can fold into a common secondary structure that has three stem-loops and a single-stranded region that resembles the Sm-binding site. The SL RNAs therefore exist as snRNPs that count as members of the Sm snRNP class. During the *trans*-splicing reaction, SL RNA becomes part of the spliced product replacing the original cap and leader (called an "outron"), as illustrated in the upper panel of **FIGURE 21.23**. Like other snRNPs involved in splicing (except U6), SL RNA carries a trimethylated cap, which is recognized by a variant cap binding factor eIF4E to facilitate translation.

In *C. elegans*, about 70% of genes are processed by the *trans*-splicing mechanism, which can be further divided into two classes. One class of gene produces monocistronic transcripts, which are processed by both *cis*- and *trans*-splicing. In these cases, while *cis*-splicing is used to remove internal intronic sequences, *trans*-splicing is employed to provide the 22-nucleotide leader sequence derived from the SL RNA for translation. The other class of gene is polycistronic. In these cases, *trans*-splicing is used to convert the polycistronic transcripts into monocistronic transcripts in addition to providing the SL leader sequence for their translation, as illustrated in the bottom panel of Figure 21.23.

The SL RNAs found in several species of trypanosomes and also in the nematode (*C. elegans*) have some common features. They fold into a common secondary structure that has three stem-loops and a single-stranded region that resembles the Sm-binding site. The SL RNAs therefore exist as snRNPs that are members of the Sm snRNP class. Trypanosomes possess the U2, U4, and U6 snRNAs but do not have U1 or U5 snRNAs. The absence of U1 snRNA can be explained by the properties of the SL RNA, which can carry out the functions that U1 snRNA usually performs at the 5′-splice site. In effect, SL RNA consists of an snRNA sequence that possesses U1 function and is linked to the exon-intron site that it recognizes.

The *trans*-splicing reaction of the SL RNA may represent an evolutionary transition toward the pre-mRNA splicing apparatus. In *cis*, the SL RNA provides the ability to recognize the 5′ splice site, and this probably depends upon the specific conformation of the RNA. The remaining functions required for splicing are provided by independent snRNPs.

KEY CONCEPTS

- Splicing reactions usually occur only in *cis* between splice junctions on the same molecule of RNA.
- *trans*-splicing occurs in trypanosomes and worms where a short sequence (SL RNA) is spliced to the 5′ ends of many precursor mRNAs.
- SL RNA has a structure resembling the Sm-binding site of U snRNAs.

CONCEPT AND REASONING CHECK

Compare the processes of self-splicing, *trans*-splicing, and pre-mRNA splicing.

21.13 tRNA Splicing Involves Cutting and Rejoining in Separate Reactions

Most splicing reactions depend on short consensus sequences and occur by transesterification reactions in which breaking and forming bonds are coordinated. The splicing of tRNA genes is achieved by a different mechanism that relies upon separate cleavage and ligation reactions.

Some 59 of the 272 nuclear tRNA genes in the yeast *S. cerevisiae* are interrupted. Each has a single intron that is located just one nucleotide beyond the 3′ side of the anticodon. The introns vary in length from 14 to 60 bases. Those in related tRNA genes are related in sequence, but the introns in tRNA genes representing different amino acids are unrelated. *There is no consensus sequence that could be recognized by the splicing enzymes.* This is also true of interrupted nuclear tRNA genes of plants, amphibians, and mammals.

All the introns include a sequence that is complementary to the anticodon of the tRNA. This creates an alternative conformation for the anticodon arm in which the anticodon is base paired to form an extension of the usual arm. An example is shown in **FIGURE 21.24**. Only the anticodon arm is affected—the rest of the molecule retains its usual structure.

The exact sequence and size of the intron is not important. Most mutations in the intron do not prevent splicing. *Splicing of tRNA depends principally on recognition of a common secondary structure in tRNA rather than a common sequence of the intron.* Regions in various parts of the molecule are important, including the stretch between the acceptor arm and D arm, in the TψC arm, and especially the anticodon arm. This is reminiscent of the structural demands placed on tRNA for translation (see *Chapter 24, Translation*).

The intron is not entirely irrelevant, however. Pairing between a base in the intron loop and an unpaired base in the

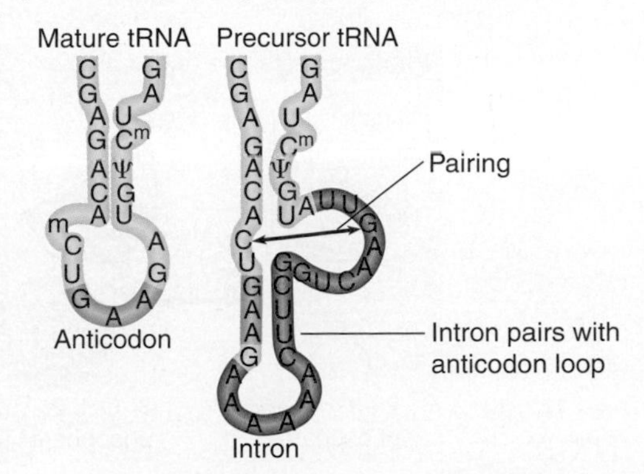

FIGURE 21.24 The intron in yeast tRNA^{Phe} base pairs with the anticodon to change the structure of the anticodon arm. Pairing between an excluded base in the stem and the intron loop in the precursor may be required for splicing.

stem is required for splicing. Mutations at other positions that influence this pairing (for example, to generate alternative patterns for pairing) influence splicing. The rules that govern availability of tRNA precursors for splicing resemble the rules that govern recognition by aminoacyl-tRNA synthetases (see *Section 25.8, tRNAs Are Charged with Amino Acids by Aminoacyl-tRNA Synthetases*).

The reaction occurs in two stages that are catalyzed by different enzymes:

- The first step does not require ATP. It involves phosphodiester bond cleavage by an atypical nuclease reaction. It is catalyzed by an endonuclease.
- The second step requires ATP and involves bond formation; it is a ligation reaction, and the responsible enzyme activity is described as an **RNA ligase**.

▸ **RNA ligase** An enzyme that functions in tRNA splicing to make a phosphodiester bond between the two exon sequences that are generated by cleavage of the intron.

The overall tRNA splicing reaction is summarized in **FIGURE 21.25**. The products of cleavage are a linear intron and two half-tRNA molecules. These intermediates have unique ends. Each 5′ terminus ends in a hydroxyl group; each 3′ terminus ends in a 2′, 3′–cyclic phosphate group. (All other known RNA splicing enzymes cleave on the other side of the phosphate bond.)

The two half-tRNAs base pair to form a tRNA-like structure. When ATP is added, the second reaction occurs, which is catalyzed by a single enzyme with multiple enzymatic activities.

- *Cyclic phosphodiesterase activity*. Both of the unusual ends generated by the endonuclease must be altered prior to the ligation reaction. The cyclic phosphate group is first opened to generate a 2′–phosphate terminus.
- *Kinase activity*. The product has a 2′–phosphate group and a 3′–OH group. The 5′–OH group generated by the endonuclease must be phosphorylated to give a 5′–phosphate. This generates a site in which the 3′–OH is next to the 5′–phosphate.
- *Ligase activity*. Covalent integrity of the polynucleotide chain is then restored by ligase activity. The spliced molecule is now uninterrupted, with a 5′–3′ phosphate linkage at the site of splicing, but it also has a 2′–phosphate group marking the event on the spliced tRNA. In the last step, this surplus group is removed by a phosphatase, which transfers the 2′–phosphate to NDP to form ADP ribose 1′,2′–cyclic phosphate.

The yeast endonuclease is a heterotetrameric protein consisting of two catalytic subunits, Sen34 and Sen2, and two structural subunits, Sen54 and Sen15. Its activities

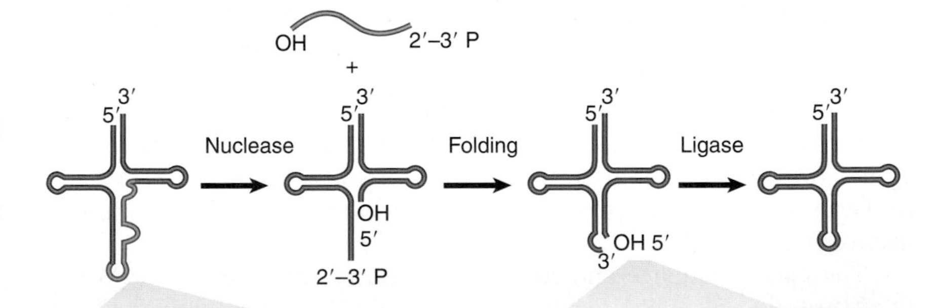

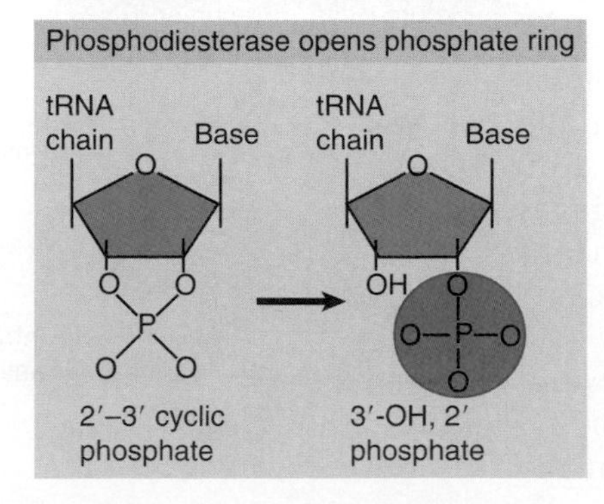

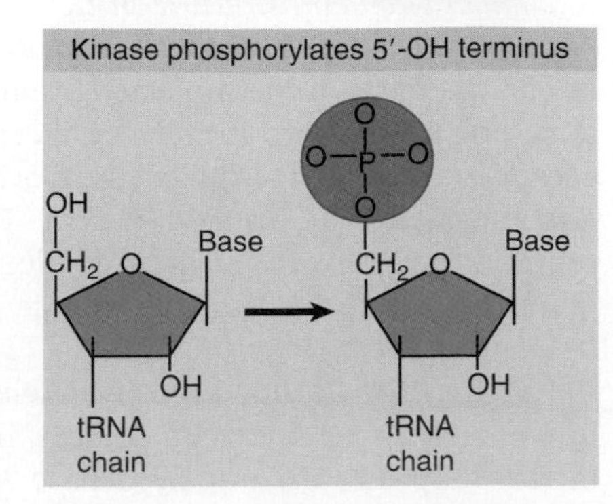

FIGURE 21.25 Splicing of tRNA requires separate nuclease and ligase activities. The exon-intron boundaries are cleaved by the nuclease to generate 2′ to 3′ cyclic phosphate and 5′ OH termini. The cyclic phosphate is opened to generate 3′-OH and 2′ phosphate groups. The 5′-OH is phosphorylated. After releasing the intron, the tRNA half molecules fold into a tRNA-like structure that now has a 3′-OH, 5′-P break. This is sealed by a ligase.

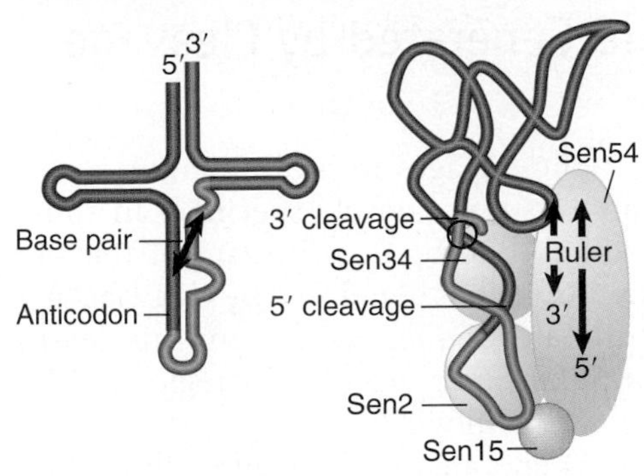

FIGURE 21.26 The 3′ and 5′ cleavages in *S. cerevisiae* pre-tRNA are catalyzed by different subunits of the endonuclease. Another subunit may determine location of the cleavage sites by measuring distance from the mature structure. The AI base pair is also important.

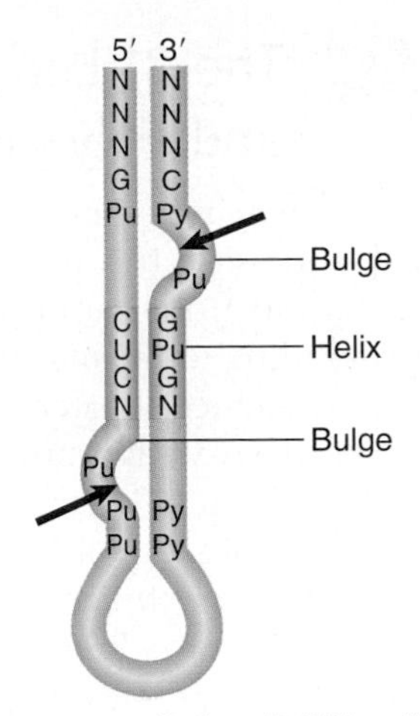

FIGURE 21.27 Archaeal tRNA splicing endonuclease cleaves each strand at a bulge in a bulge-helix-bulge motif.

are illustrated in **FIGURE 21.26**. The related subunits Sen34 and Sen2 cleave the 3′ and 5′ splice sites, respectively. Subunit Sen54 may determine the sites of cleavage by "measuring" distance from a point in the tRNA structure. This point is in the elbow of the (mature) L-shaped structure. The role of subunit Sen15 is not known, but its gene is essential in yeast. The base pair that forms between the first base in the anticodon loop and the base preceding the 3′ splice site is required for 3′ splice site cleavage.

An interesting insight into the evolution of tRNA splicing is provided by the endonucleases of Archaea. These are homodimers or homotetramers, in which each subunit has an active site (although only two of the sites function in the tetramer) that cleaves one of the splice sites. The subunit has sequences related to the sequences of the active sites in the Sen34 and Sen2 subunits of the yeast enzyme. The archaeal enzymes recognize their substrates in a different way, though. Instead of measuring distance from particular sequences, they recognize a structural feature called the bulge-helix-bulge. **FIGURE 21.27** shows that cleavage occurs in the two bulges. Thus the origin of splicing of tRNA precedes the separation of the Archaea and the eukaryotes. If it originated by insertion of the intron into tRNAs, this must have been a very ancient event.

KEY CONCEPTS

- tRNA splicing occurs by successive cleavage and ligation reactions.
- An endonuclease cleaves the tRNA precursors at both ends of the intron.
- Release of the intron generates two half-tRNAs that pair to form the mature structure.
- The halves have the unusual ends 5′ hydroxyl and 2′-3′ cyclic phosphate.
- The 5′–OH end is phosphorylated by a polynucleotide kinase, the cyclic phosphate group is opened by phosphodiesterase to generate a 2′–phosphate terminus and 3′–OH group, exon ends are joined by an RNA ligase, and the 2′-phosphate is removed by a phosphatase.
- The yeast endonuclease is a heterotetramer with two (related) catalytic subunits.
- It uses a measuring mechanism to determine the sites of cleavage by their positions relative to a point in the tRNA structure.
- The archaeal nuclease has a simpler structure and recognizes a bulge-helix-bulge structural motif in the substrate.

CONCEPT AND REASONING CHECK

Compare tRNA splicing and pre-mRNA splicing.

21.14 The 3′ Ends of mRNAs Are Generated by Cleavage and Polyadenylation

It is not clear whether RNA polymerase II actually engages in a termination event at a specific site. It is possible that its termination is only loosely specified. In some transcription units, termination occurs more than 1000 bp downstream of the site corresponding to the mature 3′ end of the mRNA (which is generated by cleavage at a specific sequence). Instead of using specific terminator sequences, the enzyme ceases RNA synthesis within multiple sites located in rather long "terminator regions." The nature of the individual termination sites is largely unknown.

The 3′ ends of Pol II transcribed mRNAs are generated by cleavage followed by polyadenylation, which is necessary for the maturation of mRNA from nuclear RNA. The poly(A) tail is known to protect the mRNA from degradation by 3′–5′ exonucleases. In eukaryotes, the poly(A) tail is also suggested to play roles in facilitating nuclear export of matured mRNA and in cap stability.

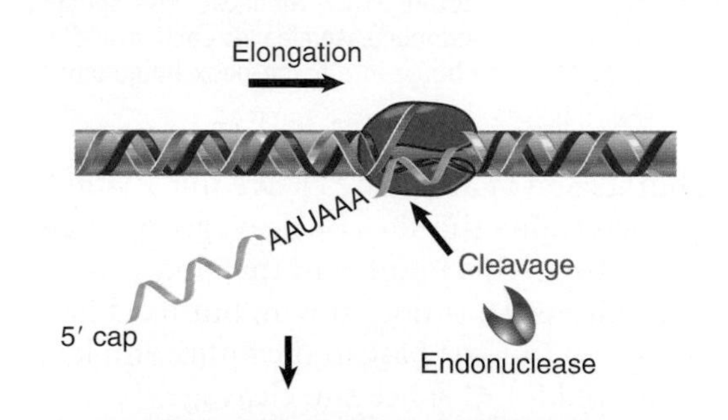

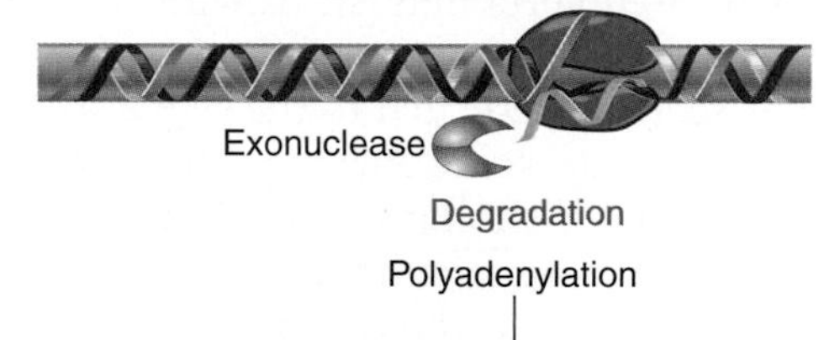

FIGURE 21.28 The sequence AAUAAA is necessary for cleavage to generate a 3′ end for polyadenylation.

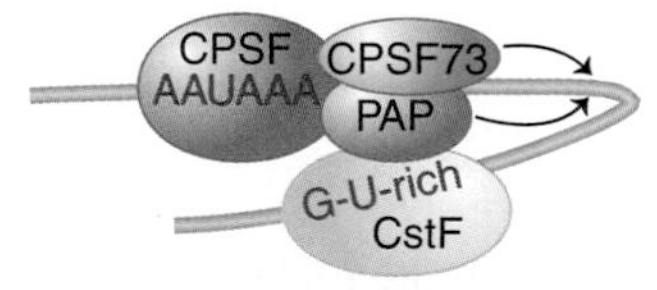

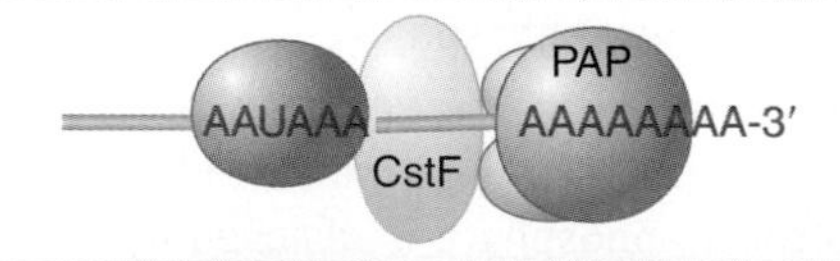

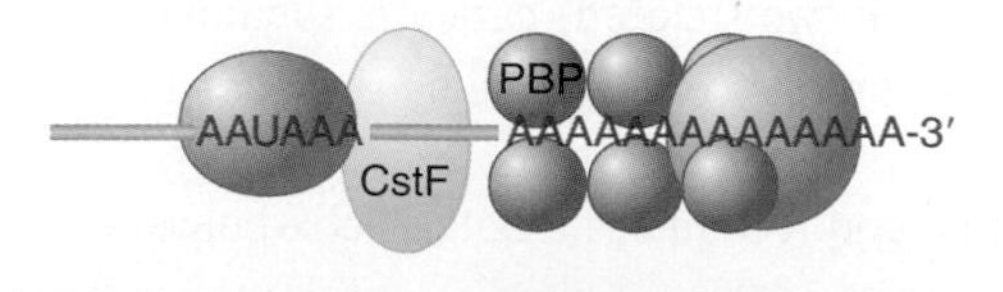

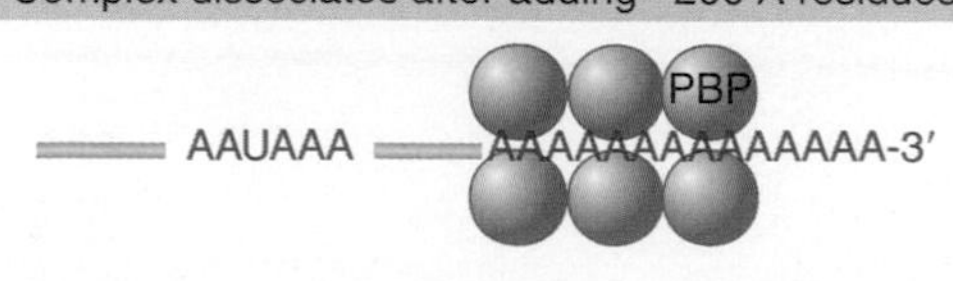

FIGURE 21.29 The 3′ processing complex consists of several activities. CPSF and CstF each consist of several subunits; the other components are monomeric. The total mass is >900 kD.

Generation of the 3′ end is illustrated in **FIGURE 21.28**. The RNA polymerase transcribes past the site corresponding to the 3′ end, and sequences in the RNA are recognized as targets for an endonucleolytic cut followed by polyadenylation. RNA polymerase continues transcription after the cleavage, but the 5′ end that is generated by the cleavage is unprotected, which signals transcriptional termination.

The site of cleavage/polyadenylation in most pre-mRNAs is flanked by two *cis*-acting signals: an upstream AAUAAA motif, which is usually located 11 to 30 nucleotides from the site, and a downstream U-rich or GU-rich element. The AAUAAA is needed for both cleavage and polyadenylation because deletion of the AAUAAA hexamer prevents generation of the polyadenylated 3′ end (though in plants and fungi there can be considerable variation from the AAUAAA motif).

The development of a system in which polyadenylation occurs *in vitro* opened the route to analyzing the reactions. The formation and functions of the complex that undertakes 3′ processing are illustrated in **FIGURE 21.29**. Generation of the proper 3′ terminal structure depends on the *cleavage and polyadenylation specific factor* (CPSF), which contains multiple subunits. One of the subunits binds directly to the AAUAAA motif and to the *cleavage stimulatory factor* (CstF), which is also a multicomponent complex. One of these components binds directly to a downstream GU-rich sequence. CPSF and CstF can enhance each other in recognizing the polyadenylation signals. The specific enzymes involved are an *endonuclease* (the 73kD subunit of CPSF) to cleave the RNA and a **poly(A) polymerase (PAP)** to synthesize the poly(A) tail.

▸ **poly(A) polymerase (PAP)** The enzyme that adds the stretch of polyadenylic acid to the 3′ end of eukaryotic mRNA. It does not use a template.

The poly(A) polymerase has a nonspecific catalytic activity. When it is combined with the other components, the synthetic reaction becomes specific for RNA containing the sequence AAUAAA. The polyadenylation reaction passes through two stages. First, a rather short oligo(A) sequence (~10 residues) is added to the 3′ end. This reaction is absolutely dependent on the AAUAAA sequence, and poly(A) polymerase performs it under the direction of the specificity factor. In the second phase, the nuclear poly(A) binding protein (PABP II) binds the oligo(A) tail to allow extension of the poly(A) tail to the full ~200 residue length. The poly(A) polymerase by itself adds A residues individually to the 3′ position. Its intrinsic mode of action is distributive; it dissociates after each nucleotide has been added. In the presence of CPSF and PABP II, however, it functions processively to extend an individual poly(A) chain. After the polyadenylation reaction, PABP II binds stoichiometrically to the poly(A) stretch, which by some unknown mechanism limits the action of poly(A) polymerase to ~200 additions of A residues.

Upon export of mature mRNAs out of the nucleus, the poly(A) tail is bound by the cytoplasmic poly(A) binding protein (PABP I). PABP I not only protects the mRNA from degradation by the 3′ to 5′ exonucleases but also binds to the translation initiation factor eIF4G to facilitate translation of the mRNA. Thus the mRNA in the cytoplasm forms a closed loop in which a protein complex contains both the 5′ and 3′ ends of the mRNA (see Figure 24.18 in *Section 24.7, Small Subunits Scan for Initiation Sites on Eukaryotic mRNA*). Polyadenylation therefore affects both stability and initiation of translation in the cytoplasm.

KEY CONCEPTS

- The sequence AAUAAA is a signal for cleavage to generate a 3′ end of mRNA that is polyadenylated.
- The reaction requires a protein complex that contains a specificity factor, an endonuclease, and poly(A) polymerase.
- The specificity factor and endonuclease cleave RNA downstream of AAUAAA.
- The specificity factor and poly(A) polymerase add ~200 A residues processively to the 3′ end.
- The poly(A) tail controls mRNA stability and influences translation.

CONCEPT AND REASONING CHECK

What would be the effect of deletion of the AATAAA consensus sequence at the 3′ end of the coding strand of a gene?

21.15 Production of rRNA Requires Cleavage Events and Involves Small RNAs

The major eukaryotic rRNAs are synthesized as part of a single primary transcript that is processed by cleavage and trimming events to generate the mature products. The precursor contains the sequences of the 18S, 5.8S, and 28S rRNAs. (The nomenclature of different ribosomal RNAs is based on early sedimentation studies conducted on sucrose gradients in the 1970s.) In multicellular eukaryotes, the precursor is named for its sedimentation rate as *45S RNA*. In unicellular/oligocellular eukaryotes it is smaller (35S in yeast).

The mature rRNAs are released from the precursor by a combination of cleavage events and trimming reactions to remove both external transcribed spacers (ETS) and

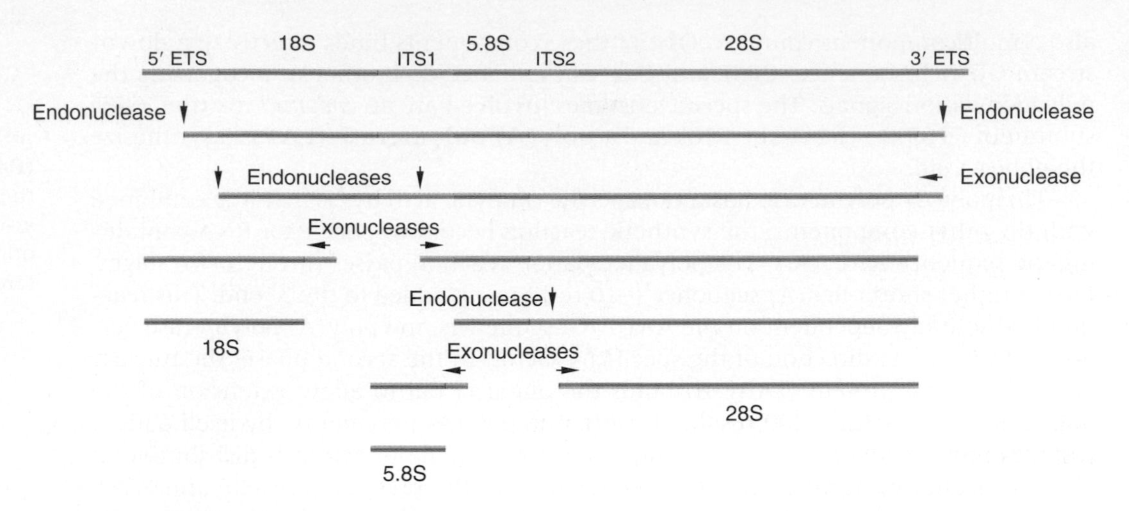

FIGURE 21.30 Mature eukaryotic rRNAs are generated by cleavage and trimming events from a primary transcript.

internal transcribed spacers (ITS). **FIGURE 21.30** shows the general pathway in yeast. There can be variations in the order of events, but basically similar reactions are involved in all eukaryotes. Most of the 5′ ends are generated directly by a cleavage event. Most of the 3′ ends are generated by cleavage followed by a 3′–5′ trimming reaction. These processes are specified by many *cis*-acting RNA motifs in both ETSs and ITSs and are acted upon by more than 150 processing factors.

There are always multiple copies of the transcription unit for the rRNAs. The copies are organized as tandem repeats (see *Section 7.3, Genes for rRNA Form Tandem Repeats, Including an Invariant Transcription Unit*). The genes encoding rRNAs are transcribed by RNA polymerase I in the nucleolus. In contrast, 5S RNA is transcribed from separate genes by RNA polymerase III. In general, the 5S genes are clustered, but are separated from the genes for the major rRNAs.

There is a difference in the organization of the precursor in bacteria. The sequence corresponding to 5.8S rRNA forms the 5′ end of the large (23S) rRNA; that is, there is no processing between these sequences. **FIGURE 21.31** shows that the precursor also contains the 5S rRNA and one or two tRNAs. In *E. coli*, the seven *rrn* operons are dispersed around the genome; four *rrn* loci contain one tRNA gene between the 16S and 23S rRNA sequences, and the other *rrn* loci contain two tRNA genes in this region. Additional tRNA genes may or may not be present between the 5S sequence and the 3′ end. Thus the processing reactions required to release the products depend on the content of the particular *rrn* locus.

In both prokaryotic and eukaryotic rRNA processing, both processing factors and ribosomal proteins (and possibly other proteins) bind to the precursor, so that the substrate for processing is not the free RNA but rather a ribonucleoprotein complex. rRNA processing takes place just after transcription. As a result, the processing factors are intertwined with ribosomal proteins in building the ribosomes, instead of first processing and then stepwise assembly on processed rRNAs.

▸ **small nucleolar RNA (snoRNA)** A small nuclear RNA that is localized in the nucleolus.

Processing and modification of rRNA requires a class of small RNAs called **small nucleolar RNAs (snoRNAs)**. There are hundreds of snoRNAs in *S. cerevisiae*

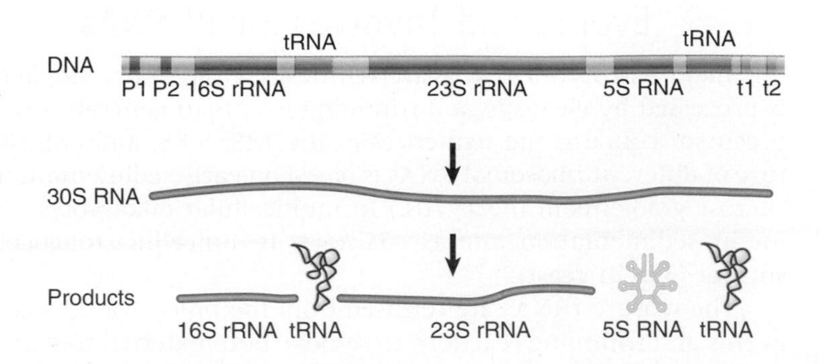

FIGURE 21.31 The *rrn* operons in *E. coli* contain genes for both rRNA and tRNA. The exact lengths of the transcripts depend on which promoters (P) and terminators (t) are used. Each RNA product must be released from the transcript by cuts on either side.

and vertebrate genomes. Some of these snoRNAs are encoded by individual genes; others are expressed from polycistrons, and many are derived from introns of their host genes. These snoRNAs themselves undergo complex processing and maturation steps. Some snoRNAs are required for cleavage of the precursor to rRNA; one example is U3 snoRNA, which is required for the first cleavage event. The U3-containing complex corresponds to the "terminal knobs" at the 5′ end of nascent rRNA transcripts, which are visible under an electron microscope. We do not know what role the snoRNA plays in cleavage. It could be required to pair with specific rRNA sequences to form a secondary structure that is recognized by an endonuclease.

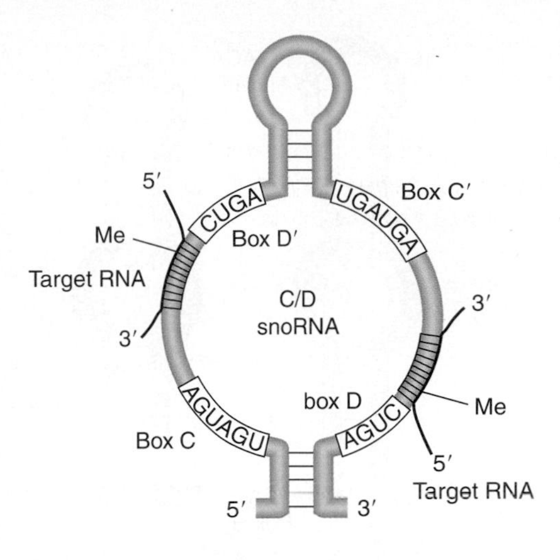

FIGURE 21.32 A snoRNA base pairs with a region of rRNA that is to be methylated.

Two groups of snoRNAs are required for the modifications that are made to bases in the rRNA. The members of each group are identified by very short conserved sequences and common features of secondary structure.

The C/D group of snoRNAs is required for adding a methyl group to the 2′ position of ribose. There are more than 100 2′-O–methyl groups at conserved locations in vertebrate rRNAs. This group takes its name from two short conserved sequence motifs called boxes C and D. Each snoRNA contains a sequence near the D box that is complementary to a region of the 18S or 28S rRNA that is methylated. Loss of a particular snoRNA prevents methylation in the rRNA region to which it is complementary.

FIGURE 21.32 shows that the snoRNA base pairs with the rRNA to create the duplex region that is recognized as a substrate for methylation. Methylation occurs within the region of complementarity at a position that is fixed five bases on the 5′ side of the D box. It is likely that each methylation event is specified by a different snoRNA; ~40 snoRNAs have been implicated in this modification. Each C+D box snoRNA is associated with three proteins Nop1 (fibrillarin in vertebrates), Nop56p, and Nop58p. The methylase(s) have not been fully characterized, although the major snoRNP protein Nop1p/fibrillarin is structurally similar to methyltransferases.

Another group of snoRNAs is involved in base modification by converting uridine to pseudouridine. There are ~50 residues in yeast rRNAs and ~100 in vertebrate rRNAs that are modified by pseudouridination. The pseudouridination reaction is shown in **FIGURE 21.33**, in which the N1 bond from uridylic acid to ribose is broken, the base is rotated, and C5 is rejoined to the sugar.

Pseudouridine formation in rRNA requires the H/ACA group of ~20 snoRNAs. They are named for the presence of an ACA triplet three nucleotides from the 3′ end and a partially conserved sequence (the H box) that lies between two stem-loop hairpin structures. Each of these snoRNAs has a sequence complementary to rRNA within

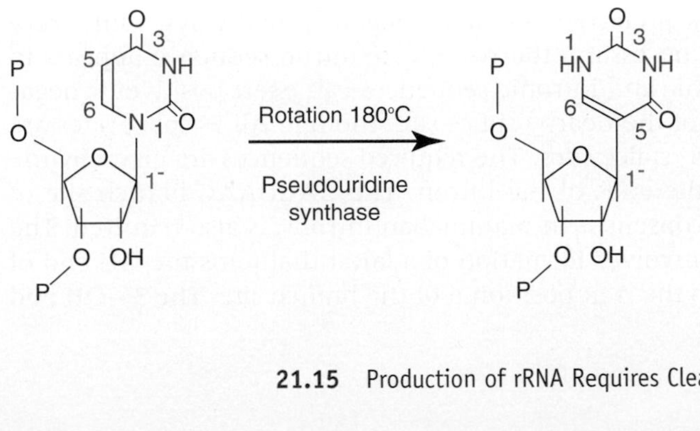

Rotation 180°C

Pseudouridine synthase

FIGURE 21.33 Uridine is converted to pseudouridine by replacing the N1-sugar bond with a C5-sugar bond and rotating the base relative to the sugar.

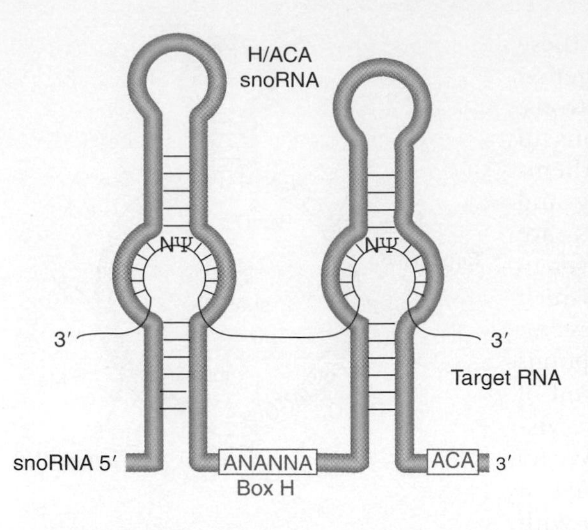

FIGURE 21.34 H/ACA snoRNAs have two short conserved sequences and two hairpin structures, each of which has regions in the stem that are complementary to rRNA. Pseudouridine is formed by converting an unpaired uridine within the complementary region of the rRNA.

the stem of each hairpin. **FIGURE 21.34** shows the structure that would be produced by pairing with the rRNA. Within each pairing region, there are two unpaired bases, one of which is a uridine that is converted to pseudouridine.

The H/ACA snoRNAs are associated with four specific nucleolar proteins, Cbf5 (dyskerin in vertebrates), Nhp2, Nop10, and Gar1. Importantly, Cbf5p/dyskerin is structurally similar to known pseudouridine synthases, and thus it likely provides the enzymatic activity in the snoRNA-guided pseudouridination reaction. Many snoRNAs are also used to guide base modifications in tRNAs as well as in snRNAs involved in pre-mRNA splicing, which are critical for their functions in prospective reactions. There are, however, a large number of snoRNAs that do not have apparent targets. These snoRNAs are called *orphan RNAs*. The existence of these orphan RNAs indicates that many biological processes may use RNA-guided mechanisms to functionally modify other expressed RNAs in a more diverse fashion than we currently understand.

KEY CONCEPTS

- The large and small rRNAs are released by cleavage from a common precursor rRNA; the 5S rRNA is separately transcribed.
- The C/D group of snoRNAs is required for modifying the 2′ position of ribose with a methyl group.
- The H/ACA group of snoRNAs is required for converting uridine to pseudouridine.
- In each case, the snoRNA base pairs with a sequence of rRNA that contains the target base to generate a typical structure that is the substrate for modification.

CONCEPT AND REASONING CHECK

What is the role of snoRNAs in the processing of rRNAs?

21.16 Summary

Splicing accomplishes the removal of introns and the joining of exons into the mature sequence of RNA. There are at least four types of reaction, as distinguished by their requirements *in vitro* and the intermediates that they generate. The systems include eukaryotic nuclear introns, group I and group II introns, and tRNA introns. Each reaction involves a change of organization within an individual RNA molecule and is, therefore, a *cis*-acting event.

Pre-mRNA splicing follows preferred but not obligatory pathways. Only very short consensus sequences are necessary; the rest of the intron sequence appears to be largely irrelevant. Both exonic and intronic sequences can exert positive or negative influence on the selection of the nearby splice site, though. All 5′ splice sites are probably equivalent, as are all 3′ splice sites. The required sequences are given by the GU-AG rule, which describes the ends of the intron. The UACUAAC branch site of yeast, or a less well conserved consensus in mammalian introns, is also required. The reaction with the 5′ splice site involves formation of a lariat that joins the GU end of the intron via a 2′–5′ linkage to the A at position 6 of the branch site. The 3′–OH end of the exon then attacks the 3′ splice site, so that the exons are ligated and the intron is released as a lariat. Lariat formation is responsible for choice of the 3′ splice site. Both reactions are transesterifications in which phosphodiester bonds are conserved. Several stages of the reaction require hydrolysis of ATP, probably to drive conformational changes in the RNA and/or protein components. Alternative splicing patterns are caused by protein factors that either facilitate the use of a new site or that block use of the default site.

Pre-mRNA splicing requires formation of a spliceosome—a large particle that assembles the consensus sequences into a reactive conformation. The spliceosome forms by the process of intron definition, involving recognition of the 5′ splice site, branch site, and 3′ splice site. If, however, introns are large, like those in vertebrates, recognition of the splice sites first follows the process of exon definition, involving the interactions across the exon between the 3′ splice site and the downstream 5′ splice site. This is then switched to paired interactions across the intron for later steps of spliceosome assembly. By either intron definition or exon definition, the initial process of splice site recognition commits the pre-mRNA substrate to the splicing pathway. The pre-mRNA complex contains U1 snRNP and a number of key protein splicing factors, including U2AF and the branch site binding factor. In multicellular eukaryotic cells, the formation of the commitment complex requires the participation of SR proteins.

The spliceosome contains the U1, U2, U4/U6, and U5 snRNPs, as well as some additional splicing factors. The U1, U2, and U5 snRNPs each contain a single snRNA and several proteins; the U4/U6 snRNP contains two snRNAs and several proteins. Some proteins are common to all snRNP particles. U1 snRNA base pairs with the 5′ splice site, U2 snRNA base pairs with the branch sequence, and U5 snRNP holds the 5′ and 3′ splice sites together via a looped sequence within the spliceosome. When U4 releases U6, the U6 snRNA base pairs with the 5′ splice site and U2, which remains base paired with the branch sequence; this may create the catalytic center for splicing. An alternative set of snRNPs provides analogous functions for splicing the U12-dependent subclass of introns. The catalytic core resembles that in group II autocatalytic introns; as a result, it is likely that the spliceosome is a giant RNA machine (like the ribosome) in which key RNA elements are at the center of the reaction.

Splicing is usually intramolecular, but *trans*-splicing (intermolecular splicing) occurs in trypanosomes and nematodes. It involves a reaction between a small SL RNA and the pre-mRNA. In nematodes there are two types of SL RNA: one is used for splicing to the 5′ end of an mRNA; the other is used for splicing to an internal site to break up the polycistronic precursor RNA. The introduction of the SL RNA to the processed mRNAs provides necessary signals for translation.

tRNA splicing involves separate endonuclease and ligase reactions. The endonuclease recognizes the secondary (or tertiary) structure of the precursor and cleaves both ends of the intron. The two half-tRNAs released by loss of the intron can be ligated by the tRNA ligase in the presence of ATP.

The termination capacity of RNA polymerase II is tightly linked to 3′ end formation of the mRNA. The sequence AAUAAA, located 11 to 30 bases upstream of the cleavage site, provides the signal for both cleavage by an endonuclease and polyadenylation by the poly(A) polymerase. This is enhanced by the complex bound on the G/U-rich element downstream from the cleavage site. Transcription is terminated when an exonuclease, which binds to the 5′ end of the nascent RNA chain created by the cleavage, catches up to RNA polymerase.

rRNA processing takes place in the nucleolus where U3 snRNA initiates a series of actions of endonucleases and exonucleases to cut and trim extra sequences in the precursor rRNA to produce individual ribosomal RNAs. Hundreds to thousands of noncoding RNAs are expressed in eukaryotic cells. In the nucleolus, two groups of such noncoding RNAs, termed snoRNAs, are responsible for pairing with rRNAs at sites that are modified. Group C/D snoRNAs identify target sites for methylation, and group H/ACA snoRNAs specify sites where uridine is converted to pseudouridine.

CHAPTER QUESTIONS

1. Which class(es) of introns have the ability to excise from the transcript autonomously, without any proteins (at least *in vitro*)?
   A. nuclear introns
   B. group I introns
   C. group II introns
   D. tRNA introns
2. The first two bases and the last two bases of nuclear introns are (almost) always:
   A. GC and AT.
   B. GU and AG.
   C. CU and AT.
   D. CG and GA.
3. Which base in the intron branch sequence is completely conserved?
   A. A
   B. G
   C. T
   D. C
4. Which two snRNPs are usually found as a single particle?
   A. U1 and U2
   B. U4 and U5
   C. U4 and U6
   D. U2 and U5
5. The U1 snRNA base pairs with:
   A. a sequence spanning the first exon-intron splicing site.
   B. the 3′ splice site of the intron.
   C. a sequence spanning the intron–second exon splicing site.
   D. the branch sequence within the intron.
6. The U2 snRNA base pairs with:
   A. a sequence spanning the first exon-intron boundary.
   B. the 3′ splice site of the intron.
   C. a sequence spanning the intron–second exon boundary.
   D. the branch sequence within the intron.
7. Removal of introns is connected to export of the mRNA from the nucleus to the cytoplasm by:
   A. binding of specific export proteins to the 5′ end of the processed mRNA.
   B. binding of specific export proteins to the first exon-exon junction.
   C. binding of specific export proteins to multiple exon-exon junctions.
   D. binding of specific export proteins to the 3′ end of the processed mRNA.
8. Which type of intron does not excise via a lariat mechanism?
   A. nuclear introns
   B. group I introns
   C. group II introns
   D. more than one of the above
9. In most cases, eukaryotic transcription actually terminates at:
   A. the mRNA cleavage site.
   B. a point upstream of the mRNA cleavage site.
   C. a point downstream of the mRNA cleavage site.
   D. a termination codon.

10. The polyadenylation signal in primary mRNA transcripts is:
   A. A-rich.
   B. G-rich.
   C. C-rich.
   D. U-rich.

KEY TERMS

A complex
alternative splicing
anti-Sm
autosplicing (self-splicing)
branch site
cap
E complex
EJC (exon junction complex)
exon definition
GU-AG rule
heterogeneous nuclear RNA (hnRNA)
hnRNP
intron definition
lariat
poly(A) polymerase (PAP)
pre-mRNA
RNA ligase
RNA splicing
SL RNA (spliced leader RNA)
small cytoplasmic RNAs (scRNA; scyrps)
small nuclear RNAs (snRNA; snurps)
small nucleolar RNA (snoRNA)
spliceosome
splicing factor
SR protein
transesterification

FURTHER READING

Black, D. (2003). Mechanisms of alternative pre-messenger RNA splicing. *Annu. Rev. Biochem.*, **72**, 291–336.

Brow, D. A. (2002). Allosteric cascade of spliceosome activation. *Annu. Rev. Gene*t. **36**, 333–360. A review of the steps in the assembly of the spliceosome and its removal of an intron.

Dreyfuss, G., Kim, V. N., and Kataoka, N. (2002). Messenger-RNA-binding proteins and the messages they carry. *Nat. Rev. Mol. Cell Biol.* **3**, 195–205. A review of the functions of proteins that bind to eukaryotic mRNAs.

Krämer, A. (1996). The structure and function of proteins involved in mammalian pre-mRNA splicing. *Annu. Rev. Biochem.* **65**, 367–409. A comprehensive review of the snRNPs and protein factors functioning in the mammalian spliceosome.

Matera, A. G., Terns, R. M., and Terns, M. P. (2007). Non-coding RNAs: lessons from the small nuclear and small nucleolar RNAs. *Nature Rev. Mol. Cell Biol.* **8**, 209–220.

Pandit, S., Wang, D., and Fu, X-D. (2008). Functional integration of transcriptional and RNA processing machineries. *Curr. Opin. Cell* **20**, 260–265.

Reed, R., and Hurt, E. (2002). A conserved mRNA export machinery coupled to pre-mRNA splicing. *Cell* **108**, 523–531. A review of the mechanism of eukaryotic mRNA transport from the nucleus to the cytoplasm and its links to pre-mRNA splicing and nonsense-mediated mRNA decay.

Shatkin, A. J., and Manley, J. L. (2000). The ends of the affair: Capping and polyadenylation. *Nature Struct. Biol.* **7**, 838–842.

Tseng, C. K., and Cheng, S. C. (2008). Both catalytic steps of nuclear pre-mRNA splicing are reversible. *Science* **320**, 1782–1784.

Wahle, E., and Keller, W. (1992). The biochemistry of 3′-end cleavage and polyadenylation of messenger RNA precursors. *Annu. Rev. Biochem.* **61**, 419–440. A review of polyadenylation of animal mRNAs.

Zhou, Z., Licklider, L. J., Gygi, S. P., and Reed, R. (2002). Comprehensive proteomic analysis of the human spliceosome. *Nature* **419**, 182–185. The identification of the ~145 proteins that make up the human spliceosomal complex, many of which have yet to be characterized. Many appear to play roles in gene expression other than RNA splicing.

22

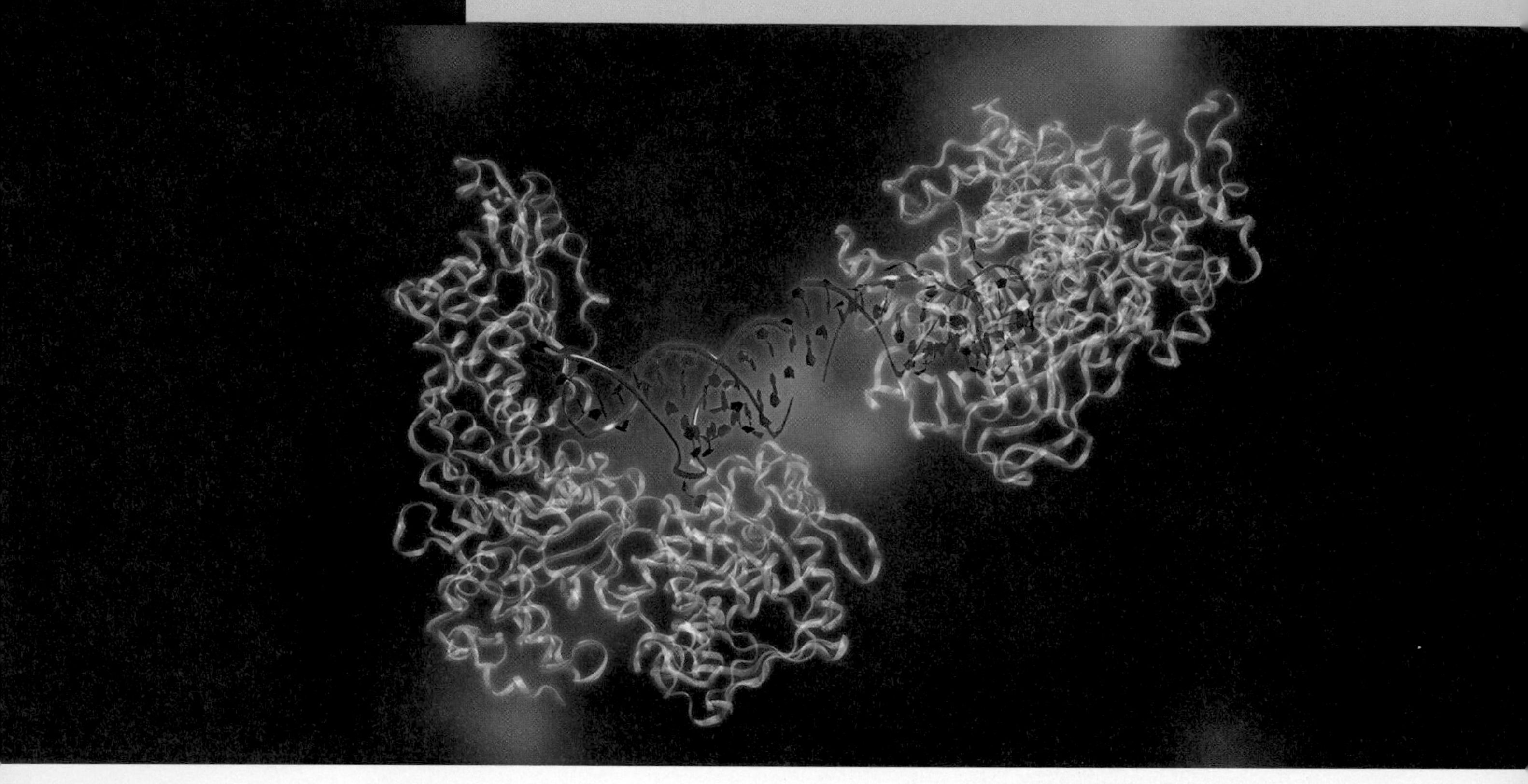

mRNA Stability and Localization

Computer artwork of aconitase (blue), in complex with ferritin mRNA (red). Aconitase is involved in the citric acid cycle but here it is performing a secondary function as an iron regulatory protein (IRP). It does this by binding to ferritin mRNA, which prevents translation into the protein product (ferritin). Ferritin acts like a sponge and helps to protect cells from the toxic effects of excess iron.

CHAPTER OUTLINE

22.1 Introduction

RNA is a central player in gene expression. It was first characterized as an intermediate in protein synthesis, but since then many other RNAs have been discovered that play structural or functional roles at other stages of gene expression. The involvement of RNA in many functions concerned with gene expression supports the general view that the entire process may have evolved in an "RNA world" in which RNA was originally the active component in maintaining and expressing genetic information. Many of these functions were subsequently assisted or taken over by proteins, with a consequent increase in versatility and probably efficiency. The focus in this chapter is messenger RNA (mRNA). The functions of other cellular RNAs are discussed in other chapters: snRNAs and snoRNAs in *Chapter 21, RNA Splicing and Processing*; tRNA and rRNA in *Chapter 24, Translation*; and miRNAs and siRNAs in *Chapter 30, Regulatory RNA*; the subset of RNAs that have retained ancestral catalytic activity are discussed in *Chapter 23, Catalytic* RNA.

Messenger RNA plays the principal role in the expression of protein-coding genes. Each mRNA molecule carries the genetic code for synthesis of a specific polypeptide during the process of translation. An mRNA carries much more information as well: it may also carry information for how frequently it will be translated, how long it is likely to survive, and where in the cell it will be translated. This information is carried in the form of RNA *cis*-elements and associated proteins. Much of this information is located in parts of the mRNA sequence that are not directly involved in encoding protein.

FIGURE 22.1 shows some of the structural features typical of mRNAs in prokaryotes and eukaryotes. Bacterial mRNA termini are not modified after transcription, so they

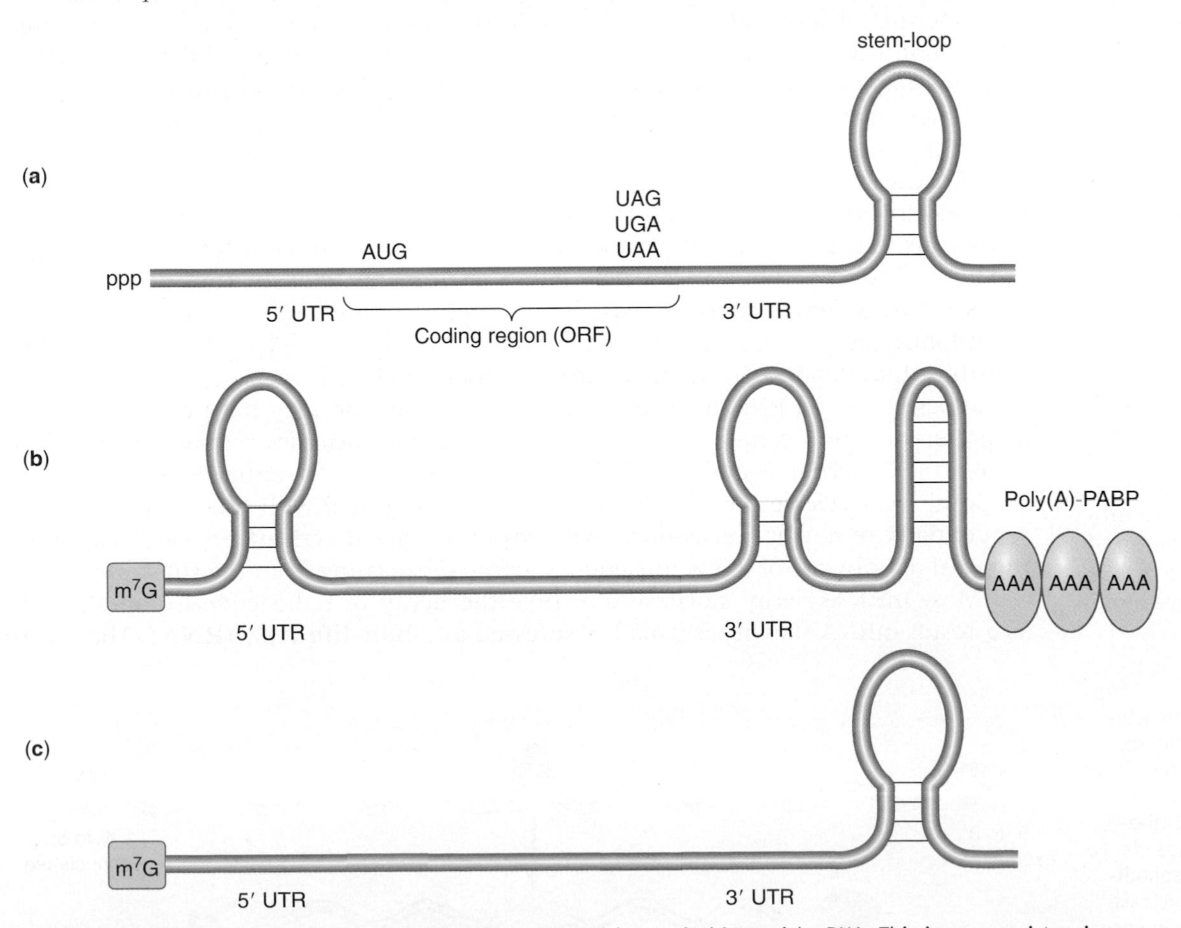

FIGURE 22.1 Features of prokaryotic and eukaryotic mRNAs. (a) A typical bacterial mRNA. This is a monocistronic mRNA, but bacterial mRNAs may also be polycistronic. Many bacterial mRNAs end in a terminal stem-loop. (b) All eukaryotic mRNAs begin with a cap (m^7G) and almost all end with a poly(A) tail. The poly(A) tail is coated with poly(A)-binding proteins (PABPs). Eukaryotic mRNAs may have one or more regions of secondary structure, typically in the 5′ and 3′ UTRs. (c) The major histone mRNAs in mammals have a 3′ terminal stem-loop in place of a poly(A) tail.

- **5′ UTR** The region in an mRNA between the start of the message and the first codon.
- **3′ untranslated region (UTR)** The region in an mRNA between the termination codon and the end of the message.
- **stem-loop** A secondary structure that appears in RNAs consisting of a base-paired region (stem) and a terminal loop of single-stranded RNA. Both are variable in size.

begin with the 5′ triphosphate nucleotide used in initiation of transcription and end with the final nucleotide added by RNA polymerase before termination. The 3′ end of many of *E. coli* mRNAs form a hairpin structure involved in intrinsic (rho-independent) transcription termination (see *Chapter 19, Prokaryotic Transcription*). Eukaryotic mRNAs are cotranscriptionally capped and polyadenylated (see *Chapter 21, RNA Splicing and Processing*). Most of the nonprotein-coding regulatory information is carried in the **5′** and **3′ untranslated regions (UTR)** of an mRNA, but some elements are present in the coding region. While all mRNAs are linear sequences of nucleotides, secondary and tertiary structures can be formed by intramolecular base-pairing. These structures can be simple, like the **stem-loop** structures illustrated in the figure, or more complex, involving branched structures or pairing of nucleotides from distant regions of the molecule. Investigation of the mechanisms by which mRNA regulatory information is deciphered and acted upon by machinery responsible for mRNA degradation, translation, and localization is an important field in molecular biology today.

22.2 Messenger RNAs Are Unstable Molecules

- **ribonuclease** An enzyme that cleaves phosphodiester linkages between RNA ribonucleotides.
- **endoribonuclease (endonuclease)** A ribonuclease that cleaves an RNA at internal site(s).
- **exoribonuclease (exonuclease)** A ribonuclease that removes terminal ribonucleotides from RNA.
- **processive (nuclease)** An enzyme that remains associated with the substrate while catalyzing the sequential removal of nucleotides.
- **distributive (nuclease)** An enzyme that catalyzes the removal of only one or a few nucleotides before dissociating from the substrate.
- **half-life ($t_{1/2}$) (RNA)** The time taken for the concentration of a given population of RNA molecules to decrease by half, in the absence of new synthesis.

Messenger RNAs are relatively unstable molecules, unlike DNA and, to a lesser extent, rRNAs and tRNAs. While it is true that the phosphodiester bonds connecting ribonucleotides are somewhat weaker than those connecting deoxyribonucleotides due to the presence of the 2′ hydroxyl group on the ribose sugar, this is not the primary reason for the instability of mRNA. Rather, cells contain a myriad of RNA degrading enzymes, called **ribonucleases**, some of which specifically target mRNA molecules.

Ribonucleases are enzymes that cleave the phosphodiester linkage connecting RNA ribonucleotides. They are diverse molecules because many different protein domains have evolved to have ribonuclease activity. The rare examples of known ribozymes (catalytic RNAs) include multiple ribonucleases, indicating the ancient origins of this important activity (see *Chapter 23, Catalytic RNA*). Ribonucleases, often just called nucleases when the RNA nature of the substrate is obvious, have many roles in a cell, including participation in DNA replication, DNA repair, processing of new transcripts (including pre-mRNAs, tRNAs, rRNAs, snRNAs, and miRNAs) and the degradation of mRNA. Ribonucleases are either **endoribonucleases (endonucleases)** or **exoribonucleases (exonucleases)**, although they are generally referred to just as endonucleases or exonucleases (though these names do not specify RNA vs. DNA substrates), as depicted in **FIGURE 22.2** (and as discussed in *Section 3.2, Nucleases*). Endonucleases cleave an RNA molecule at an internal site, and may have a requirement or preference for a certain structure or sequence. Exonucleases remove nucleotides from an RNA terminus and have a defined polarity of attack—either 5′ to 3′ or 3′ to 5′. Some exonucleases are **processive**, remaining engaged with the substrate while sequentially removing nucleotides, whereas others are **distributive**, catalyzing the removal of only one or a few nucleotides before dissociating from the substrate.

Most mRNAs decay stochastically (like the decay of radioactive isotopes), and as a result mRNA stability is usually expressed as a **half-life ($t_{1/2}$) (RNA)**. The term

FIGURE 22.2 Types of ribonucleases. Exonucleases are unidirectional. They can digest RNA either from the 5′ end or from the 3′ end, liberating individual ribonucleotides. Endonucleases cleave RNA at internal phosphosphodiester linkages. An endonuclease usually targets specific sequences and/or secondary structures.

mRNA decay is often used interchangeably with mRNA degradation. mRNA-specific stability information is encoded in *cis*-sequences (see *Section 22.7, mRNA-Specific Half-Lives Are Controlled by Sequences or Structures within the mRNA*) and is therefore characteristic of each mRNA. Different mRNAs can exhibit remarkably different stabilities, varying by 100-fold or more. In *E. coli* the typical mRNA half-life is about three minutes, but half-lives of individual mRNAs may be as short as 20 seconds and as long as 90 minutes. In budding yeast, mRNA half-lives range from 3 to 100 minutes, whereas in metazoans, half-lives range from minutes to hours and, in rare cases, even days. Abnormal mRNAs can be targeted for very rapid destruction (see *Section 22.8, Newly Synthesized RNAs Are Checked for Defects via a Nuclear Surveillance System* and *Section 22.9, Quality Control of mRNA Translation Is Performed by Cytoplasmic Surveillance Systems*). Half-life values are generally determined by some version of the method illustrated in **FIGURE 22.3**.

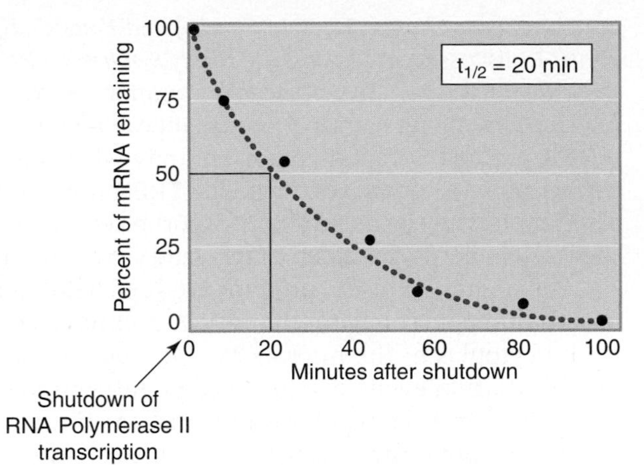

FIGURE 22.3 Method for determining mRNA half-lives. RNA polymerase II transcription is shut down, either by a drug or a temperature shift in strains with a temperature-sensitive mutation in a Pol II gene. The levels of specific mRNAs are determined by northern blot or RT-PCR at various times following shutdown. RNA degradation, once initiated, is usually so rapid that intermediates in the process are not detectible. The half-life is the time required for the mRNA to fall to one half of its initial value.

▸ **mRNA decay** mRNA degradation, assuming that the degradation process is stochastic.

The abundance of specific mRNAs in a cell is a consequence of their combined rates of synthesis (transcription and processing) and degradation. mRNA levels reach a **steady state (molecular concentration)** when these parameters remain constant. The spectrum of proteins synthesized by a cell is largely a reflection of the abundance of their mRNA templates (although differences in translational efficiency play a role). The importance of mRNA decay is highlighted by large-scale studies that have examined the relative contributions of decay rate and transcription rate to differential mRNA abundance. Decay rate predominates. The great advantage of unstable mRNAs is the ability to rapidly change the output of translation through changes in mRNA synthesis. Clearly this advantage is important enough to compensate for the seeming wastefulness of making and destroying mRNAs so quickly. Abnormal control of mRNA stability has been implicated in disease states, including cancer, chronic inflammatory responses, and coronary disease.

▸ **steady state (molecular concentration)** The concentration of a population of molecules when the rates of synthesis and degradation are equal.

KEY CONCEPTS

- mRNA instability is due to the action of ribonucleases.
- Ribonucleases differ in their substrate preference and mode of attack.
- mRNAs exhibit a wide range of half-lives.
- Differential mRNA stability is an important contributor to mRNA abundance and therefore the spectrum of proteins made in a cell.

CONCEPT AND REASONING CHECK

If the degradation rate of an mRNA is high compared to the synthesis rate of that mRNA, what is the effect on the steady state of the message? On the half-life?

22.3 Eukaryotic mRNAs Exist in the Form of mRNPs from Their Synthesis to Their Degradation

From the time pre-mRNAs are transcribed in the nucleus until their cytoplasmic destruction, eukaryotic mRNAs are associated with a changing repertoire of proteins. RNA–protein complexes are called **ribonucleoprotein particles (RNPs)**. Many of the pre-mRNA-binding proteins are involved in splicing and processing

▸ **ribonucleoprotein particles (RNPs)** A complex of RNA and proteins. "Particles" are typically used to refer to large complexes.

reactions (see *Chapter 21, RNA Splicing and Processing*), and others are involved in quality control (discussed in *Section 22.8, Newly Synthesized RNAs Are Checked for Defects via a Nuclear Surveillance System*). The nuclear maturation of an mRNA comprises multiple remodeling steps involving both the RNA sequence and its complement of proteins. The mature mRNA product is export competent only when fully processed and associated with the correct protein complexes, including TREX (for *tr*anscription *ex*port), which mediates its association with the nuclear pore export receptor. Mature mRNAs retain multiple binding sites (*cis*-elements) for different regulatory proteins, most often within their 5′ or 3′ UTRs.

While many nuclear proteins are shed before or during mRNA export to the cytoplasm, others accompany the mRNA and have cytoplasmic roles. For example, once in the cytoplasm, the nuclear cap-binding complex participates in the new mRNA's first translation event, the so-called pioneering round of translation. This first translation initiation is critical for a new mRNA; if it is found to be a defective template it will be rapidly destroyed by a surveillance system (see *Section 22.9, Quality Control of mRNA Translation Is Performed by Cytoplasmic Surveillance Systems*). An mRNA that passes its translation test will spend the rest of its existence associated with a variety of proteins that control its translation, its stability and sometimes its cellular location. The "nuclear history" of an mRNA is critical in determining its fate in the cytoplasm.

▸ **RNA-binding protein (RBP)** A protein containing one or more domains that confer an affinity for RNA, usually in an RNA sequence- or structure-specific manner.

A large number of different **RNA-binding proteins (RBPs)** are known, and many more are predicted based on genome analysis. The *S. cerevisiae* genome encodes nearly 600 different proteins predicted to bind to RNA, about one-tenth of the total gene number for this organism. Based on similar proportions, the human genome would be expected to contain more than 2000 such proteins. These estimates are based on the presence of characterized RNA-binding domains, and it is likely that additional RNA-binding domains remain to be found. The RNA targets and functions of the great majority of these RBPs are unknown, although it is considered likely that a large fraction of them interact with pre-mRNA or mRNA. This kind of analysis does not include the many proteins that do not bind RNA directly, but participate in RNA-binding complexes.

An important insight into why the number of different mRNA-binding proteins is so large has come from the finding that mRNAs are associated with distinct, but overlapping, sets of RBPs. Studies that have matched specific RBPs with their target mRNAs have revealed that those mRNAs encode proteins with shared features such as involvement in similar cellular processes or location. Thus, the repertoire of bound proteins catalogues the mRNA. For example, hundreds of yeast mRNAs are bound by one or more of six related *Puf* proteins. Puf1 and Puf2 bind mostly mRNAs encoding membrane proteins, whereas Puf3 binds mostly mRNAs encoding mitochondrial proteins, and so on. A current model, illustrated in **FIGURE 22.4**, proposes that the coordinate control

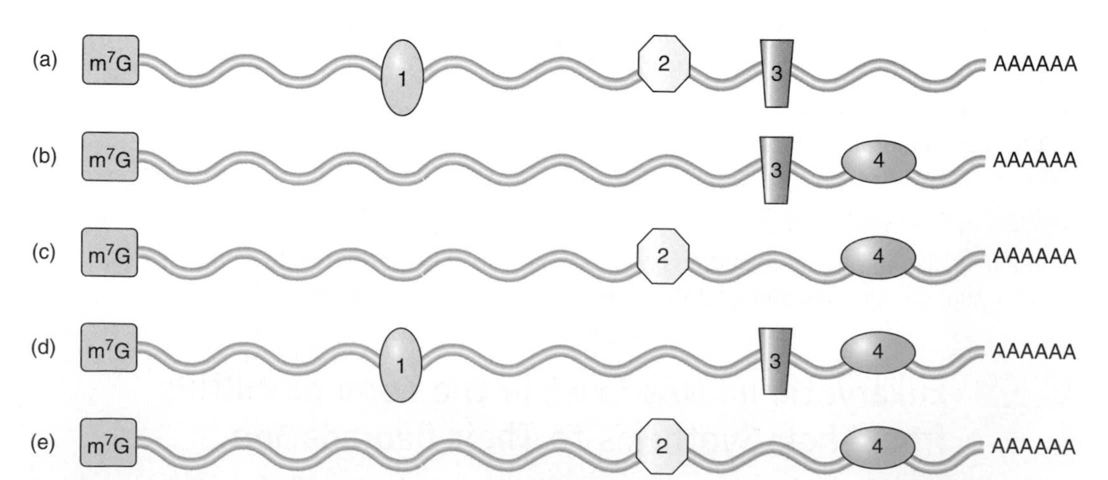

FIGURE 22.4 The concept of an RNA regulon. Eukaryotic mRNAs are bound by a variety of proteins that control their translation, localization, and stability. The subset of mRNAs that have a binding protein in common are considered part of the same regulon. In the diagram, mRNAs a and d are part of regulon 1, mRNAs (a), (c), and (e) are part of regulon 2, and so on.

of posttranscriptional processes of mRNAs is mediated by the combinatorial action of multiple RBPs, much like the coordinate control of gene transcription is mediated by the right combinations of transcription factors (see *Chapter 28, Eukaryotic Transcription Regulation*). The set of mRNAs that share a particular type of RBP has been called an **RNA regulon**.

▸ **RNA regulon** A set of mRNAs that are each bound by a particular RNA binding protein, and thus presumably subject to coordinate regulation by that protein.

KEY CONCEPTS

- mRNA associates with a changing population of proteins during its nuclear maturation and cytoplasmic life.
- Some nuclear-acquired mRNP proteins have roles in the cytoplasm.
- A very large number of RNA-binding proteins exist, most of which remain uncharacterized.
- Different mRNAs are associated with distinct, but overlapping, sets of regulatory proteins, creating RNA regulons.

CONCEPT AND REASONING CHECK

Why might it be useful for mRNAs to be continuously associated with RNA-binding proteins?

22.4 Prokaryotic mRNA Degradation Involves Multiple Enzymes

Our understanding of prokaryotic mRNA degradation comes mostly from studies of *E. coli*. So far, the general principles apply to the other bacterial species studied. In prokaryotes, mRNA degradation occurs during the process of translation. Prokaryotic ribosomes begin translation even before transcription is completed, attaching to the mRNA at an initiation site near the 5′ end and proceeding toward the 3′ end. Multiple ribosomes can initiate translation on the same mRNA sequentially, forming a **polyribosome** (or **polysome**): one mRNA with multiple ribosomes.

▸ **polyribosome (polysome)** An mRNA that is simultaneously being translated by multiple ribosomes.

E. coli mRNAs are degraded by a combination of endonuclease and 3′→5′ exonuclease activities. The major mRNA degradation pathway in *E. coli* is a multistage process illustrated in **FIGURE 22.5**. The initiating step is removal of pyrophosphate from the 5′ terminus leaving a single phosphate. The monophosphorylated form stimulates the catalytic activity of an endonuclease (RNase E), which makes an initial cut near the 5′ end of the mRNA. This cleavage leaves a 3′–OH on the upstream fragment and a 5′ monophosphate on the downstream fragment. It functionally destroys a **monocistronic mRNA**, as ribosomes can no longer initiate translation. The upstream fragment is then degraded by a 3′→5′ exonuclease (PNPase = polynucleotide phosphorylase). This two-step ribonuclease cycle is repeated along the length of the mRNA in

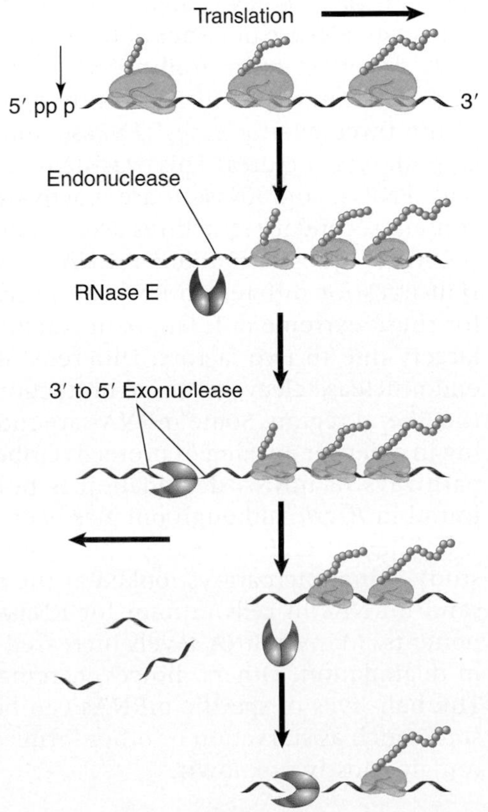

FIGURE 22.5 Degradation of bacterial mRNAs. Bacterial mRNA degradation is initiated by cleavage of the triphosphate 5′ terminus to yield a monophosphate. mRNAs are then degraded in a two-step cycle: an endonucleolytic cleavage, followed by 3′ to 5′ exonuclease digestion of the released fragment. The endonucleolytic cleavages occur in a 5′ to 3′ direction on the mRNA, following the passage of the last ribosome.

▸ **monocistronic mRNA** mRNA that codes for one polypeptide.

a 5′ to 3′ direction as more RNA gets exposed following passage of previously initiated ribosomes. This process proceeds very rapidly as the short fragments generated by RNase E can be detected only in mutant cells in which exonuclease activity is impaired.

PNPase, as well as the other known 3′→5′exonucleases in *E. coli*, are unable to progress through double-stranded regions. Thus the stem-loop structure at the 3′ end of many bacterial mRNAs protects the mRNA from direct 3′ attack. Some internal fragments generated by RNase E cleavage also have regions of secondary structure that would impede exonuclease digestion. PNPase *is*, however, able to digest through double-stranded regions if there is a stretch of single-stranded RNA at least seven to ten nucleotides long located 3′ to the stem-loop. The single-stranded sequence seems to serve as a necessary staging platform for the enzyme. Rho-independent termination leaves a single-stranded region that is too short to serve as a platform. To solve this problem a bacterial **poly(A) polymerase (PAP)** adds 10 to 40 nucleotide **poly(A)** tails to 3′ termini, making them susceptible to 3′→5′ degradation. RNA fragments terminating in particularly stable secondary structures may require repeated polyadenylation and exonuclease digestion steps. It is not known whether polyadenylation is ever the initiating step for degradation of mRNA or whether it is used only to help degrade fragments, including the 3′ terminal one. Some experiments indicate that RNase E cleavage of an mRNA may be required to activate the poly(A) polymerase. This would explain why intact mRNAs do not seem to be degraded from the 3′ end.

- **poly(A) polymerase (PAP)** The enzyme that adds the stretch of polyadenylic acid to the 3′ end of eukaryotic mRNA. It does not use a template.
- **poly(A)** A stretch of adenylic acid that is added to the 3′ end of mRNA following its synthesis.
- **degradosome** A complex of bacterial enzymes, including RNase and helicase activities,

RNase E and PNPase, along with a helicase and another accessory enzyme, form a multiprotein complex called the **degradosome**. RNase E plays dual roles in the complex. Its N-terminal domain provides the endonuclease activity, whereas its C-terminal domain provides a scaffold that holds together the other components. While RNase E and PNPase are the principal endo- and exonucleases active in mRNA degradation, others also exist, probably with more restricted roles. The role of other nucleases in mRNA degradation has been addressed by evaluating the phenotypes of mutants in each of the enzymes. For example, the inactivation of RNase E slows mRNA degradation without completely blocking it. Mutations that inactivate PNPase or either of the other two known 3′→5′ exonucleases have essentially no effect on overall mRNA stability. This reveals that any pair of the exonucleases can carry out apparently normal mRNA degradation. However, only two of the three exonucleases (PNPase and RNase R) can digest fragments with stable secondary structures. This was demonstrated in double mutant studies, in which both PNPase and RNase R are inactivated. In these mutants, mRNA fragments that contain secondary structures accumulated.

Many questions about mRNA degradation in *E. coli* remain to be answered. Half-lives for different mRNAs in *E. coli* can differ more than 100-fold. The basis for these extreme differences in stability is not fully understood but appears to be largely due to two factors. Different mRNAs exhibit a range of susceptibilities to endonuclease cleavage, some protection being conferred by secondary structure in the 5′ end region. Some mRNAs are more efficiently translated than others, resulting in a denser packing of protective ribosomes. Whether or not there are additional pathways of mRNA degradation is not known. No 5′→3′ exonuclease has been found in *E. coli*, although one has been identified in *Bacillus subtilis*. It is likely that the different endonucleases and exonucleases have distinct roles. A genomewide study using microarrays looked at the steady-state levels of more than four thousand mRNAs in cells mutant for RNase E or PNPase or other degradosome components. Many mRNA levels increased in the mutants, as expected for a decrease in degradation. Others, however, remained at the same level or even decreased. The half-lives of specific mRNAs can be altered by different cellular physiological states such as starvation or other forms of stress, and mechanisms for these changes remain mostly unknown.

KEY CONCEPTS

- Degradation of bacterial mRNAs is initiated by removal of a pyrophosphate from the 5′ terminus.
- Monophosphorylated mRNAs are degraded during translation in a two-step cycle involving endonucleolytic cleavages, followed by 3′ to 5′ digestion of the resulting fragments.
- 3′ polyadenylation can facilitate the degradation of mRNA fragments containing secondary structure.
- The main degradation enzymes work as a complex called the degradosome.

CONCEPT AND REASONING CHECK

How does the role of the poly(A) tail differ in prokaryotes and eukaryotes?

22.5 Most Eukaryotic mRNA Is Degraded via Two Deadenylation-Dependent Pathways

Eukaryotic mRNAs are protected from exonucleases by their modified ends (Figure 22.1). The 7-methyl guanosine cap protects against 5′ attack; the poly(A) tail, in association with bound proteins, protects against 3′ attack. Exceptions are the histone mRNAs in mammals, which terminate in a stem-loop structure rather than a poly(A) tail. A sequence-independent endonuclease attack—the initiating mechanism used by bacteria—is rare or absent in eukaryotes. mRNA decay has been characterized most extensively in budding yeast, although most findings apply to mammalian cells as well.

Degradation of the vast majority of mRNAs is deadenylation-dependent, i.e., degradation is initiated by breaching their protective poly(A) tail. The newly formed poly(A) tail (which is about 70–90 adenylate nucleotides in yeast and about 200 in mammals) is coated with **poly(A) binding proteins (PABP)**. The poly(A) tail is subject to gradual shortening upon entry into the cytoplasm, a process catalyzed by specific **poly(A) nucleases** (also called **deadenylases**). In both yeast and mammalian cells, the poly(A) tail is initially shortened by the PAN2/3 complex, followed by a more rapid digestion of the remaining 60 to 80 A tail by a second complex, CCR4-NOT, which contains the processive exonuclease Ccr4 and at least eight other subunits. Remarkably, similar CCR4-NOT complexes are involved in a variety of other processes in gene expression, including transcriptional activation. It is thought to be a global regulator of gene expression, integrating transcription and mRNA degradation. Other poly(A) nucleases exist in both yeast and mammalian cells, and the reason for this multiplicity is not yet clear.

▸ **poly(A) binding protein (PABP)** The protein that binds to the 3′ stretch of poly(A) on a eukaryotic mRNA.

▸ **poly(A) nuclease (deadenylase)** An exoribonuclease that is specific for digesting poly(A) tails.

Two different mRNA degradation pathways are initiated by poly(A) removal, illustrated in **FIGURE 22.6**. In the first pathway (Figure 22.6, left) digestion of the poly(A) tail down to oligo(A) length (~10 to 12 A) triggers decapping at the 5′ end of the mRNA. Decapping is catalyzed by a **decapping enzyme** complex consisting of two proteins in yeast (Dcp1 and Dcp2) and their homologs plus additional proteins in mammals. Decapping yields a 5′ monophosphorylated RNA end (the substrate for the 5′ to 3′ processive exonuclease XRN1), which rapidly digests the mRNA. In fact this digestion is so fast that intermediates could not be identified until investigators discovered that a stretch of guanosine nucleotides (poly-G) could block Xrn1 progression in yeast. As illustrated in **FIGURE 22.7**, they engineered mRNAs to contain an internal poly-G tract and found that the oligoadenylated 3′ end of the mRNAs accumulated. This result showed (1) that 5′ to 3′ exonuclease digestion was the primary route of decay, and (2) that decapping preceded complete removal of the poly(A) tail.

▸ **decapping enzyme** An enzyme that catalyzes the removal of the 7-methyl guanosine cap at the 5′ end of eukaryotic mRNAs.

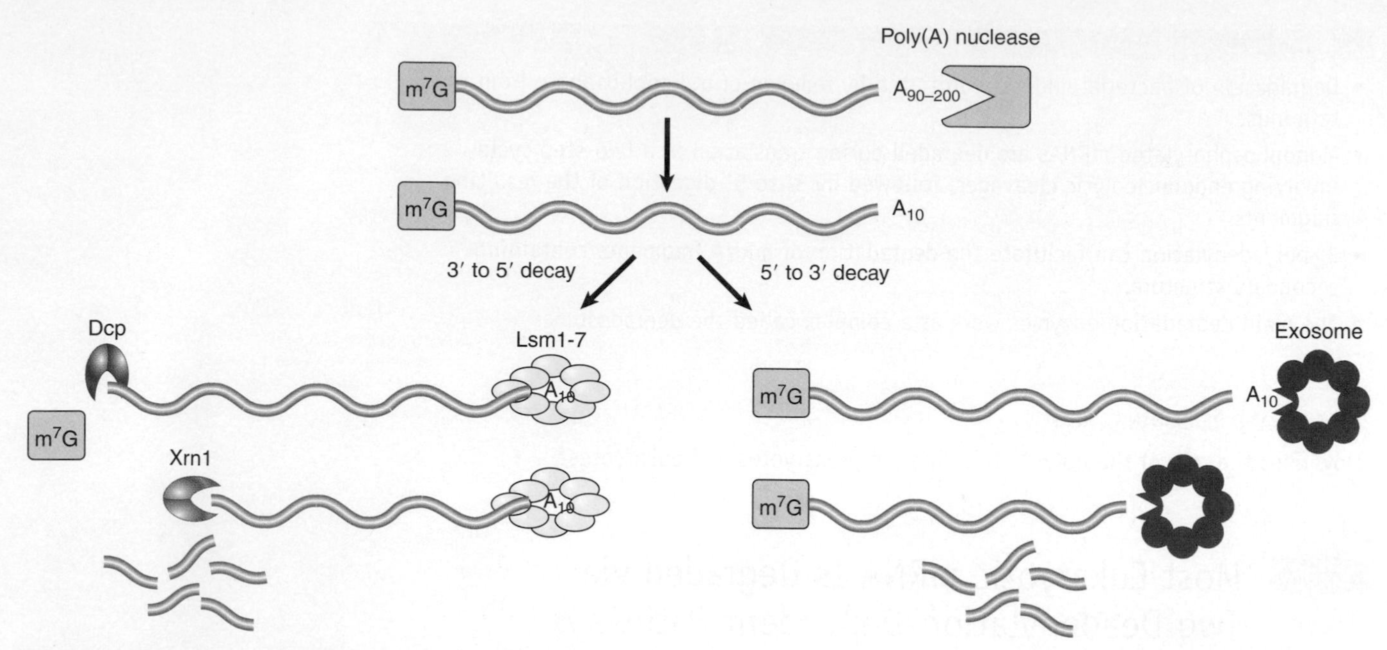

FIGURE 22.6 The major deadenylation-dependent decay pathways in eukaryotes. Two pathways are initiated by deadenylation. In both, poly(A) is shortened by a poly(A) nuclease until it reaches a length of about 10 adenylates. Then an mRNA may be degraded by the 5′ to 3′ pathway or by the 3′ to 5′ pathway. The 5′ to 3′ pathway involves decapping by Dcp and digestion by the Xrn1 exonuclease. The 3′ to 5′ pathway involves digestion by the exosome complex.

The cap is normally resistant to decapping during active translation because it is bound by the cytoplasmic cap-binding protein, a component of the eukaryotic initiation factor 4F (eIF4F) complex required for translation (described in *Chapter 24, Translation*). Thus the translation and decapping machineries compete for the cap. How does deadenylation at the 3′ end of the mRNA render the cap susceptible? Translation is known to involve a physical interaction between bound PABP at the 3′ end and the eIF4F complex at the 5′ end. Release of PABP by deadenylation is thought to destabilize the eIF4F-cap interaction, leaving the cap more frequently exposed. The mechanism is not this simple, though, because additional proteins are known to be involved the decapping event. A complex of seven related proteins, Lsm1-7, binds to the oligo(A) tract after loss of PABP and is required for decapping. Furthermore, a number of decapping enhancers have been discovered. The mechanisms by which these proteins stimulate decapping are not fully understood, although they appear to act either by recruiting/ stimulating the decapping machinery or by inhibiting translation.

In the second pathway (Figure 22.6, right), deadenylation to oligo(A) is followed by 3′ to 5′ exonuclease digestion of the body of the mRNA. This degradation

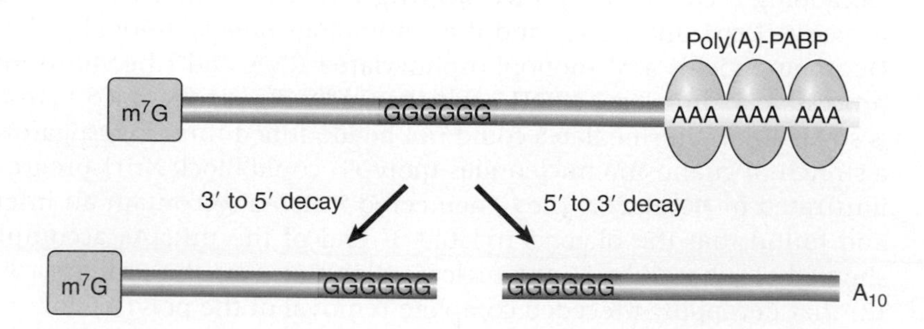

FIGURE 22.7 Use of a poly(G) sequence to determine direction of decay. A poly(G) sequence, engineered into an mRNA, will block the progression of exonucleases in yeast. The 5′ or 3′ mRNA fragment resistant to degradation accumulates in the cell and can be identified by northern blot.

step is catalyzed by the **exosome**, a ring-shaped complex consisting of a nine-subunit core with one or more additional proteins attached to its surface. A recent report showed that the exosome also has endonuclease activity, and the function of this activity in mRNA decay remains unknown. The exosome exists in similar form in Archaea and is also analogous to the bacterial degradosome in that its core subunits are structurally related to PNPase. Thus, the exosome is an ancient piece of molecular machinery. The exosome also plays an important role in the nucleus, described in *Section 22.8, Newly Synthesized RNAs Are Checked for Defects via a Nuclear Surveillance System*.

▸ **exosome** An exonuclease complex involved in nuclear processing and nuclear/cytoplasmic RNA degradation.

The relative importance of each mechanism isn't known, although in yeast, the deadenylation-dependent decapping pathway seems to predominate. The pathways are at least partially redundant. Hundreds of yeast mRNAs were examined by microarray analysis in cells in which either the 5′ to 3′ or 3′ to 5′ pathway was inactivated. In either case, only a small percentage of transcripts increased in abundance relative to wild-type cells. This finding suggests that few yeast mRNAs have a requirement for one or the other pathway. It has been proposed that these deadenylation-dependent pathways represent the default degradation pathways for all polyadenylated mRNAs, though subsets of mRNAs can be targets for other specialized pathways, described in *Section 22.6, Other Degradation Pathways Target Specific mRNAs*. Even those mRNAs that are degraded by the default pathways, however, are degraded at different mRNA-specific rates.

New studies suggest that mRNA degradation occurs within discrete particles throughout the cytoplasm, called **processing bodies (PBs)**. These structures, which are large enough to be seen with a light microscope, are clusters of nontranslating mRNPs and a variety of proteins associated with translational repression and mRNA decay, including the decapping machinery and Xrn1 exonuclease. Poly(A)-binding proteins are not generally found in PBs, suggesting that deadenylation precedes localization into these structures. Processing bodies are dynamic, increasing and decreasing in size and number, and even disappearing, under different cellular and experimental conditions that affect translation and decay. For example, release of mRNAs from polysomes by a drug that inhibits translation initiation results in a large increase in PB number and size, as does slowing degradation by partial inactivation of decay components. PBs appear to be formed by assembly of translationally repressed mRNAs and PB protein components rather than being destinations to which targeted mRNAs migrate. Not all resident mRNAs are doomed for destruction, though; some can be released for translation, but which ones and why they are freed isn't yet clear. It is not known whether all mRNA degradation normally occurs in these bodies, or even what function(s) they serve. One obvious idea is that concentrating powerful destructive enzymes in isolated locations renders mRNA degradation more safe and efficient.

▸ **processing body (PB)** A particle containing multiple mRNAs and proteins involved in mRNA degradation and translational repression, occurring in many copies in the cytoplasm of eukaryotes.

Other mRNA-containing particles related to PBs are present in specific cell types. Their similarities are based on the presence of most of the same proteins involved in translational repression and decay. **Maternal mRNA granules** are found in oocytes from a variety of organisms. These granules comprise collections of mRNAs that are held in a state of translational repression until they are activated during subsequent development. Repression is achieved by extensive deadenylation, and activation is achieved by polyadenylation. These granules may also carry mRNAs being transported to specific regions of this large cell (see *Section 22.10, Some Eukaryotic mRNAs Are Localized to Specific Regions of a Cell*). **Neuronal granules** have been identified in *Drosophila* neurons. Similar to the maternal mRNA granule, these granules function in the translational repression and transport of specific mRNAs. A fourth type of particle is called a **stress granule**. Stress granules are quite different in composition from the previous three types; however, they also contain translationally inactive mRNAs that aggregate in response to a general inhibition of translation initiation.

▸ **maternal mRNA granules** Oocyte particles containing translationally repressed mRNAs awaiting activation later in development.

▸ **neuronal granules** Particles containing translationally repressed mRNAs in transit to final cell destinations.

▸ **stress granules** Cytoplasmic particles, containing translationally inactive mRNAs, that form in response to a general inhibition of translation initiation.

KEY CONCEPTS

- The modifications at both ends of mRNA protect it against degradation by exonucleases.
- The two major mRNA decay pathways are initiated by deadenylation catalyzed by poly(A) nucleases.
- Deadenylation may be followed either by decapping and 5′ to 3′ exonuclease digestion or by 3′ to 5′ exonuclease digestion.
- The decapping enzyme competes with the translation initiation complex for 5′ cap binding.
- The exosome, which catalyzes 3′ to 5′ mRNA digestion, is a large, evolutionarily conserved complex.
- Degradation may occur within discrete cytoplasmic particles called processing bodies (PBs).
- A variety of particles containing translationally repressed mRNAs exist in different cell types.

CONCEPT AND REASONING CHECK

Summarize the two general pathways by which eukaryotic mRNA can be degraded. What do they share in common, and what is their primary difference?

22.6 Other Degradation Pathways Target Specific mRNAs

Four other pathways for mRNA degradation have been described. **FIGURE 22.8** and **FIGURE 22.9** summarize these, along with the two major pathways. These pathways are specific for subsets of mRNAs and typically involve regulated degradation events.

One pathway involves deadenylation-*in*dependent decapping, i.e., decapping proceeds in the presence of a still-long poly(A) tail. Decapping is then followed by Xrn1 digestion. Bypassing the deadenylation step requires a mechanism to recruit the decapping machinery and inhibit eIF4F binding without the help of the Lsm1-8 complex. One of the mRNAs degraded by this pathway is RPS28B mRNA, which encodes the ribosomal protein S28 and is an interesting autoregulation mechanism.

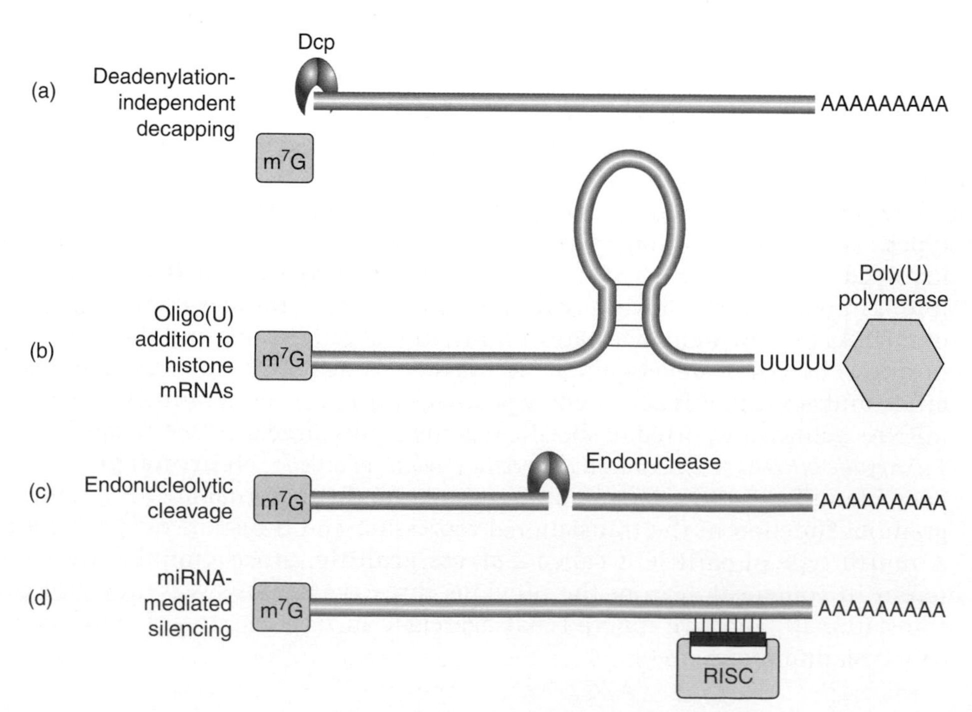

FIGURE 22.8 Other decay pathways in eukaryotic cells. The initiating event for each pathway is illustrated. (a) Some mRNAs may be decapped before deadenylation occurs. (b) Histone mRNAs receive a short poly(U) tail to become a decay substrate. (c) Degradation of some mRNAs can be initiated by a sequence-specific endonucleolytic cut. (d) Some mRNAs can be targeted for degradation or translational silencing by complementary guide miRNAs.

Pathway	Initiating event	Secondary step(s)	Substrates
Deadenylation-dependent 5′ to 3′ digestion	Deadenylation to oligo(A)	Oligo(A) binding Lsm complex Decapping 5′ to 3′ exonuclease digestion by XRN1	Probably most polyadenylated mRNAs
Deadenylation-dependent 3′ to 5′ digestion	Deadenylation to oligo(A)	3′ to 5′ exonuclease digestion by exosome	Probably most polyadenylated mRNAs
Deadenylation-independent decapping	Decapping	5′ to 3′ exonuclease digestion	Few specific mRNAs
Endonucleolytic pathway	Endonuclease cleavage	5′ to 3′ and 3′ to 5′ exonuclease digestion	Few specific mRNAs
Histone mRNA pathway	Oligouridylation	Oligo(U) binding by Lsm complex Decapping and 5′ to 3′ exonuclease digestion by XRN1 3′ to 5′ digestion by exosome	Histone mRNAs in mammals
miRNA pathway	Base-pairing with miRNA in RISC	Endonucleolytic cleavage or translational repression	Many mRNAs (extent unknown)

FIGURE 22.9 Table summarizing key elements of mRNA decay pathways in eukaryotic cells.

A stem-loop in its 3′ UTR is involved in recruiting a known decapping enhancer. The recruitment occurs only when the stem-loop is bound by S28 protein. Thus an excess of free S28 in the cell will cause the accelerated decay of its mRNA.

A second specialized pathway is used to degrade the cell-cycle regulated histone mRNAs in mammalian cells. These mRNAs are responsible for synthesis of the huge number of histone proteins needed during DNA replication. They accumulate only during S-phase and are rapidly degraded at its end. The nonpolyadenylated histone mRNAs terminate in a stem-loop structure similar to that of many bacterial mRNAs. Their mode of degradation has striking similarities to bacterial mRNA decay. A polymerase, structurally similar to the bacterial poly(A) polymerase, adds a short poly(U) tail instead of a poly(A) tail. This short tail serves as a platform for the Lsm1-7 complex and/or the exosome, activating the standard decay pathways. This mode of degradation provides an important evolutionary link between mRNA decay systems in prokaryotes and eukaryotes.

A third pathway is initiated by sequence- or structure-specific endonucleotic cleavage. The cleavage is followed by 5′ to 3′ and 3′ to 5′ digestion of the fragments, and a scavenging decapping enzyme, different from the Dcp complex, can remove the cap. Several endonucleases that cleave specific target sites in mRNAs have been identified. One interesting case is the targeted cleavage of yeast CLB2 (cyclin B2) mRNA, which occurs only at the end of mitosis. The endonuclease that catalyzes the cleavage, RNase MRP, is restricted to the nucleolus and mitochondria for most of the cell cycle where it is involved in RNA processing, but is transported to the cytoplasm in late mitosis.

The fourth pathway is the **microRNA (miRNA)** pathway, which leads directly to endonucleolytic cleavage of mRNA or to translational repression. In this case, an mRNA is targeted by the base-pairing of short (19 to 21 bp) complementary RNAs (guide miRNAs) in the context of a protein complex called RISC. The guide miRNAs are derived from transcribed miRNA genes, and are generated by cleavage from longer precursor RNAs. Thus, the destabilization of target mRNAs is controlled by regulated transcription of the miRNA genes. The details of this mechanism are described in *Chapter 30, Regulatory RNA*. The significance of this newly described pathway to total mRNA decay is not yet known but could be substantial. At least one thousand miRNAs are predicted to function in humans.

microRNA (miRNA) Very short RNAs that can regulate gene expression.

An integrated model of mRNA degradation has been proposed. This model suggests that the deadenylation-dependent decay pathways represent the default systems for degrading all polyadenylated mRNAs. The rate of deadenylation and/or other steps in degradation by these pathways can be controlled by *cis*-acting elements in each mRNA and *trans*-acting factors present in the cell. Superimposed on the default system are the mRNA decay pathways described above for targeting specific mRNAs.

KEY CONCEPTS

- Four additional degradation pathways involve regulated degradation of specific mRNAs.
- Deadenylation-independent decapping proceeds in the presence of a long poly(A) tail.
- The degradation of the nonpolyadenylated histone mRNAs is initiated by 3′ addition of a poly(U) tail.
- Degradation of some mRNAs may be initiated by sequence-specific or structure-specific endonucleolytic cleavage.
- An unknown number of mRNAs are targets for degradation or translational repression by microRNAs.

CONCEPT AND REASONING CHECK

Why would deadenylation-dependent decay serve as a default decay pathway applicable to nearly all mRNAs?

22.7 mRNA-Specific Half-Lives Are Controlled by Sequences or Structures within the mRNA

What accounts for the large range of half-lives of different mRNAs in the same cell? Specific *cis*-elements within an mRNA are known to affect its stability. The most common location for such elements is within the 3′ UTR, although they exist elsewhere. Whole genome studies have revealed many highly conserved 3′ UTR motifs, but their roles remain mostly unknown. Some are target sites for miRNA base-pairing. Others are binding sites for RBPs, some of which have known functions in stability. Rates of deadenylation can vary widely for different mRNAs, and sequences that affect this rate have been described.

▸ **destabilizing element (DE)** Any one of many different *cis* sequences, present in some mRNAs, that stimulates rapid decay of that mRNA.

▸ **AU-rich element (ARE)** A eukaryotic mRNA *cis* sequence consisting largely of A and U ribonucleotides that acts as a destabilizing element.

Destabilizing elements (DEs) have been the most widely studied. The criterion for defining a destabilizing sequence element is that its introduction into a more stable mRNA accelerates its degradation. Removal of an element from an mRNA does not necessarily stabilize it, indicating that an individual mRNA can have more than one destabilizing element. To complicate their identification further, the presence of a DE does not guarantee a short half-life under all conditions, because other sequence elements in the mRNA can modify its effectiveness.

The most well-studied type of DE is the **AU-rich element (ARE)**, found in the 3′ UTR of up to 8% of mammalian mRNAs. AREs are heterogeneous, and a number of subtypes have been characterized. One type consists of the pentamer sequence AUUUA present once or repeated multiple times in different sequence contexts. Another type does not contain AUUUA and is predominantly U-rich. A large number of ARE-binding proteins with specificity for certain ARE types and/or cell types have been identified. How do AREs work to stimulate rapid degradation? Many ARE-binding proteins have been found to interact with one or more components of the degradation machinery, including the exosome, deadenylases, and decapping enzyme, suggesting that they act by recruiting the degradation machinery. The exosome can bind some AREs directly. The AREs of a number of mRNAs have been shown to accelerate the deadenylation step of decay, although it is not likely that they all work this way. Another way they might act is by facilitating efficient engagement of the mRNA into processing bodies.

Many AU-rich DEs and other kinds of destabilizing elements have been identified in the mRNAs of budding yeast and other model organisms. For example, the

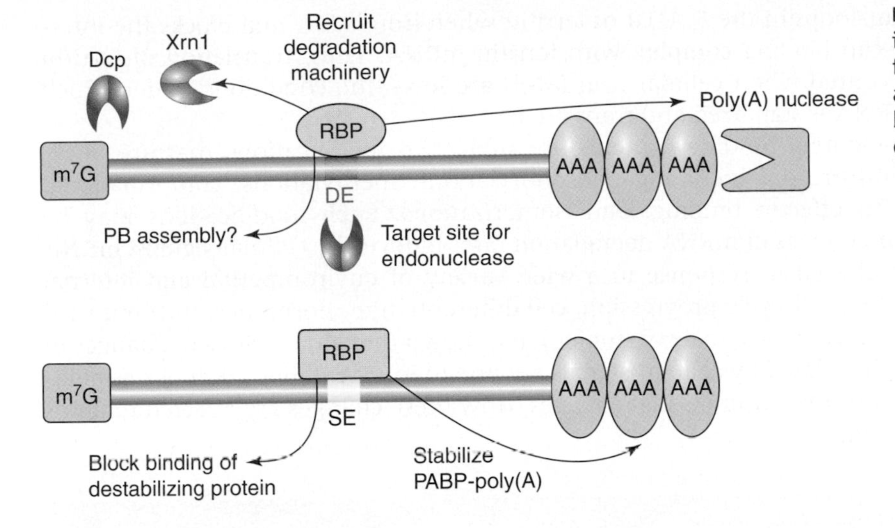

FIGURE 22.10 Mechanisms by which destabilizing elements (DEs) and stabilizing elements (SEs) function. Effects of DEs and SEs on mRNA stability are mediated primarily through the proteins that bind to them. One exception is a DE that acts as an endonuclease target site.

previously mentioned Puf proteins of yeast bind to specific UG-rich elements and accelerate the degradation of target mRNAs. In this case, the destabilizing mechanism is accelerated deadenylation by recruitment of the CCR4-NOT deadenylase. A genomics analysis of yeast 3′ UTRs has identified 53 sequence elements that correlate with the half-lives of mRNAs containing them, suggesting the number of different destabilizing elements may be large. **FIGURE 22.10** summarizes the known actions of destabilizing elements.

Stabilizing elements (SEs) have been identified in a few unusually stable mRNAs. Three mRNAs studied in mammalian cells have stabilizing pyrimidine-rich sequences in their 3′ UTRs. Proteins that bind to this element in globin mRNA have been shown to interact with PABP, suggesting they might function to protect the poly(A) tail from degradation. In some cases, an mRNA can be stabilized by inhibition of its DE. For example, certain ARE-binding proteins act to prevent the ARE from destabilizing the mRNA, presumably by blocking the ARE binding site. An example of regulated mRNA stabilization occurs for the mammalian transferrin mRNA. It is stabilized when its 3′ UTR **iron-response element (IRE)**, consisting of multiple stem-loop structures, is bound by a specific protein, as shown in **FIGURE 22.11**. The affinity of the IRE-binding protein for the IRE is altered by iron binding, exhibiting low affinity when its iron-binding site is full and high affinity when it is not. When the cellular iron concentration is low, more transferrin is needed to import iron from the bloodstream, and under these conditions the transferrin mRNA is stabilized. The IRE-binding protein stabilizes the mRNA by inhibiting the function of destabilizing sequences in the vicinity. Interestingly, the same IRE-binding protein also binds an IRE in ferritin mRNA and regulates this mRNA in a very different way. Ferritin is an iron-binding protein that sequesters excess cellular iron. The IRE-binding protein

▸ **stabilizing element (SE)** One of a variety of *cis* sequences present in some mRNAs that confers a long half-life on that mRNA.

▸ **iron-response element (IRE)** A *cis* sequence found in certain mRNAs whose stability or translation is regulated by cellular iron concentration.

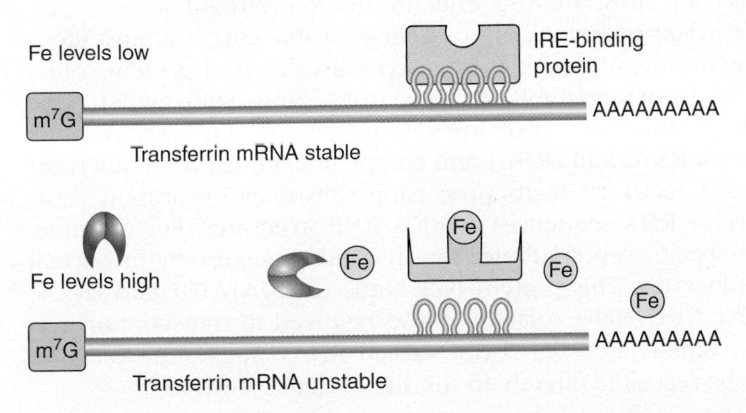

FIGURE 22.11 Regulation of transferrin mRNA stability by iron levels. The IRE in the 3′ UTR is the binding site for a protein that stabilizes the mRNA. The IRE-binding protein is sensitive to iron (Fe) levels in the cell, binding to the IRE only when iron is low.

binds IRE stem-loops in the 5′ UTR of ferritin when iron is low and blocks the interaction of the cap-binding complex with ferritin mRNA. Thus, translation of ferritin mRNA is prevented when cellular iron levels are low—the conditions under which transferrin mRNA is stabilized and translated.

Many *cis*-element-binding proteins are subject to modifications that are likely to affect their function, including phosphorylations, methylations, conformational changes due to effector binding, and isomerizations. Such modifications may be responsible for changes in mRNA degradation rates induced by cellular signals. mRNA decay can be altered in response to a wide variety of environmental and internal stimuli, including cell-cycle progression, cell differentiation, hormones, nutrient supply, and viral infection. Microarray studies have shown that almost 50% of changes in mRNA levels stimulated by cellular signals are due to mRNA stabilization or destabilization events, not to transcriptional changes. How these changes are effected remains largely unknown.

KEY CONCEPTS

- Specific *cis*-elements in an mRNA affect its rate of degradation.
- Destabilizing elements (DEs) can accelerate mRNA decay, while stabilizing elements (SEs) can reduce it.
- AU-rich elements (AREs) are common destabilizing elements in mammals and are bound by a variety of proteins.
- Some DE-binding proteins interact with components of the decay machinery and probably recruit them for degradation.
- Stabilizing elements occur on some highly stable mRNAs. mRNA degradation rates can be altered in response to a variety of signals.

CONCEPT AND REASONING CHECK

How can it be that an mRNA may be stable even when it contains a destabilizing element (DE)?

22.8 Newly Synthesized RNAs Are Checked for Defects via a Nuclear Surveillance System

All newly synthesized RNAs are subject to multiple processing steps during and after they are transcribed (see *Chapter 21, RNA Splicing and Processing*). At each step, errors may be made. While DNA errors are repaired by a variety of repair systems (see *Chapter 16, Repair Systems*), detectable errors in RNA are dealt with by destroying the defective RNA. **RNA surveillance systems** exist in both the nucleus and cytoplasm to handle different kinds of problems. Surveillance involves two kinds of activities: one to identify and tag the aberrant substrate RNA and another to destroy it.

RNA surveillance systems Systems that check RNAs (or RNPs) for errors. The system recognizes an invalid sequence or structure and triggers a response.

The destroyer is the nuclear exosome. The nuclear exosome core is almost identical to the cytoplasmic exosome, although it interacts with different protein cofactors. It removes nucleotides from targeted RNAs by 3′ to 5′ exonuclease activity. The nuclear exosome has multiple functions involving RNA processing of some noncoding RNA transcripts (snRNA, snoRNA, and rRNA) and complete degradation of aberrant transcripts. The exosome is recruited to its processing substrates by protein complexes that recognize specific RNA sequences or RNA/RNP structures. For example, Nrd1-Nab3 is a sequence-specific protein dimer that recruits the exosome to normal sn/snoRNA processing substrates. This protein pair binds to GUA[A/G] and UCUU elements, respectively. The Nrd1-Nab3 cofactor is also involved in transcription termination of these nonpolyadenylated Pol II-transcribed RNAs, suggesting that the processing exosome may be recruited directly to the site of their synthesis.

Aberrantly processed, modified, or misfolded RNAs require other protein cofactors for identification and exosome recruitment. The major nuclear complex performing this function in yeast is called **TRAMP** (an acronym for the component proteins), and it exists in at least two forms differing in the type of poly(A) polymerase present. The TRAMP complex acts in several ways to effect degradation:

1. It interacts directly with the exosome, stimulating its exonuclease activity.
2. It includes a helicase, which is probably required to unwind secondary structure and/or move RNA-binding proteins from structured RNP substrates during degradation.
3. It adds a short 3′ **oligo(A) tail** to target substrates. The oligo(A) tail is thought to make the targeted RNP a better substrate for the degradation machinery in the same way that the oligo(A) tail functions in bacteria.

▸ **TRAMP** A protein complex that identifies and polyadenylates aberrant nuclear RNAs in yeast, recruiting the nuclear exosome for degradation.

▸ **oligo(A) tail** A short poly(A) tail, generally referring to a stretch of fewer than 15 adenylates.

FIGURE 22.12 summarizes the roles of TRAMP and the exosome. It has become clear that RNA degradation in bacteria and Archaea and nuclear RNA degradation in eukaryotes are evolutionarily related processes. Their similarity suggests that the ancestral role of polyadenylation was to facilitate RNA degradation and that poly(A) was later adapted in eukaryotes for the oddly reverse function of stabilizing mRNAs in the cytoplasm.

What are the substrates for TRAMP-exosome degradation? The TRAMP complex is remarkable in that it recognizes a wide variety of aberrant RNAs synthesized by all three transcribing polymerases. It is not known how this is accomplished given that the targeted RNAs share no recognizably common features. Some researchers favor a kinetic competition model, hypothesizing that RNAs that do not get processed and assembled into final RNP form *in a timely manner* will become substrates for exosome degradation. This mechanism avoids the need to posit specific recognition of innumerable possible defects.

What kinds of abnormalities condemn pre-mRNAs to nuclear destruction? Two kinds of substrate have been identified. One type is unspliced or aberrantly spliced pre-mRNAs. Components of the spliceosome retain such transcripts either until they are degraded by the exosome or until proper splicing is completed if possible. It is thought that the kinetic competition model probably applies here, too. A pre-mRNA that is not efficiently spliced and packaged is at increased risk of being accessed by the exosome degradation machinery. The basis for recognition of aberrantly spliced pre-mRNAs is not known. The second type of pre-mRNA substrate is one that has been improperly terminated, lacking a poly(A) tail. While polyadenylation is protective in true mRNAs, it may actually be destabilizing for **cryptic unstable transcripts (CUTs)**. These nonprotein-coding RNAs (also discussed in *Section 30.3, Noncoding RNAs Can Be Used to Regulate Gene Expression*) are transcribed by RNA Pol II and do not encode recognizable proteins; however, they frequently overlap with (and may regulate) protein-coding genes. These transcripts are polyadenylated by a component of the TRAMP complex (Trf4). They are distinguished from other transcripts of unknown function by their extreme instability, normally being degraded by the TRAMP-exosome complex immediately after synthesis, possibly targeted by the Trf4-dependent polyadenylation. In

▸ **cryptic unstable transcripts (CUTs)** Nonprotein-coding RNAs transcribed by RNA Polymerase II, frequently generated from the 3′ ends of genes (resulting in antisense transcripts) and rapidly degraded after synthesis. Some CUTs have regulatory roles.

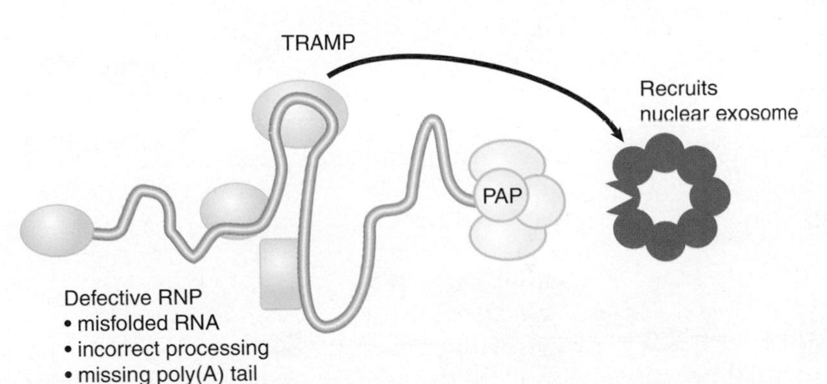

FIGURE 22.12 The role of TRAMP and the exosome in degrading aberrant nuclear RNAs. Defective RNPs are tagged by protein cofactors which then recruit the nuclear exosome. The cofactor in yeast cells is the complex TRAMP. The poly(A) polymerase (PAP, or Trf4) in TRAMP adds a short poly(A) tail to the 3′ end of the targeted RNA.

fact, the existence of these transcripts was first convincingly demonstrated in yeast strains with impaired nuclear RNA degradation. More than three-quarters of RNA Pol II transcripts may comprise noncoding RNAs and be subject to rapid degradation by the exosome! Some CUTs appear to arise from spurious transcription initiation, and the short-lived RNA products themselves typically do not appear to have a function (i.e., these RNAs do not typically act in *trans*). There are, however, examples in which there is a role for the transcription process itself in regulating nearby or overlapping coding genes (one example is described in *Section 30.3, Noncoding RNAs Can Be Used to Regulate Gene Expression*).

KEY CONCEPTS

- Aberrant nuclear RNAs are identified and destroyed by a surveillance system.
- The nuclear exosome functions both in the processing of normal substrate RNAs and in the destruction of aberrant RNAs.
- The yeast TRAMP complex recruits the exosome to aberrant RNAs and facilitates its 3′ to 5′ exonuclease activity.
- Substrates for TRAMP-exosome degradation include unspliced or aberrantly spliced pre-mRNAs and improperly terminated RNA Pol II transcripts lacking a poly(A) tail.
- The majority of RNA Pol II transcripts may be cryptic unstable transcripts (CUTs) that are rapidly destroyed in the nucleus.

CONCEPT AND REASONING CHECK

What is the apparent role of the poly(A) tail in CUTs? How does this compare to the role of the oligo(A) tail in prokaryotic mRNA decay?

22.9 Quality Control of mRNA Translation Is Performed by Cytoplasmic Surveillance Systems

Some kinds of mRNA defects can be assessed only during translation. Surveillance systems have evolved to detect three types of mRNA defects that threaten translational fidelity and to target the defective mRNAs for rapid degradation. **FIGURE 22.13** shows the

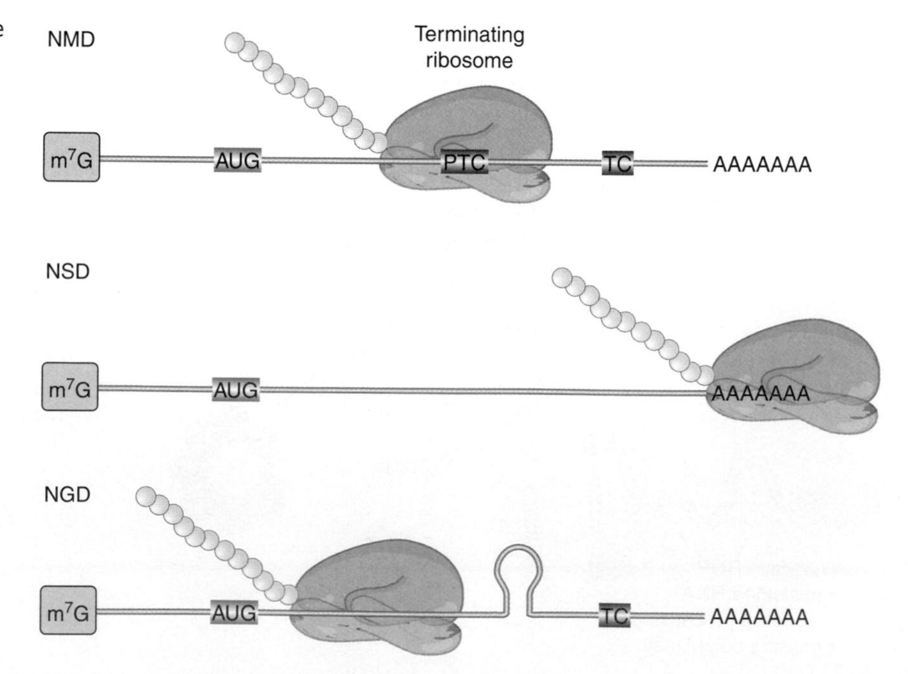

FIGURE 22.13 Substrates for cytoplasmic surveillance systems. Nonsense-mediated decay (NMD) degrades mRNAs with a premature termination codon (PTC) positioned ahead of its normal termination codon (TC). Nonstop decay (NSD) degrades mRNAs lacking an in-frame termination codon. No-go decay (NGD) degrades mRNAs having a ribosome stalled in the coding region.

substrates for each of these three systems. All three systems involve abnormal translation termination events, so it is useful to review what happens during normal termination (see *Section 24.12, Three Codons Terminate Translation and Are Recognized by Protein Factors*, for a more detailed description). When a translating ribosome reaches the termination (stop) codon, a pair of **release factors** (**RFs**; eRF1 and eRF2 in eukaryotes) enters the ribosomal A site, which is normally filled by incoming tRNAs during elongation. The release factor complex mediates the release of the completed polypeptide, followed by the mRNA, remaining tRNA, and ribosomal subunits.

Nonsense-mediated decay (NMD) targets mRNAs containing a premature termination codon (PTC). Its name comes from "nonsense mutation," which is only one way that mRNAs with a PTC can be generated. Genes without nonsense mutations can give rise to aberrant transcripts containing a PTC by (1) RNA polymerase error or (2) incomplete, incorrect, or alternative splicing. It has been estimated that almost half of alternatively spliced pre-mRNAs generate at least one form with PTC. About 30% of known disease-causing alleles probably encode an mRNA with a PTC. An mRNA with a PTC will produce C-terminal truncated polypeptides, which are considered to be particularly toxic to a cell due to their tendency to trap multiple binding partners in nonfunctional complexes. The NMD pathway has been found in all eukaryotes.

▸ **release factor (RF)** A protein required to terminate polypeptide translation to cause release of the completed polypeptide chain and the ribosome from the mRNA.

▸ **nonsense-mediated decay (NMD)** A pathway that degrades an mRNA that has a nonsense mutation prior to the last exon.

Targeting of PTC-containing mRNAs requires translation and a conserved set of protein factors. They include three **Upf proteins** (Upf1, Upf2, and Upf3) and four additional proteins (Smg1, 5, 6, and 7). Upf1 is the first NMD protein to act, binding to the terminating ribosome—specifically to its release factor complex. UPF attachment tags the mRNA for rapid decay. The specific roles of the NMD factors have not yet been defined, although phosphorylation of ribosome-bound Upf1 by Smg1 is critical. Their combined actions condemn the mRNA to the general decay machinery, and stimulate rapid deadenylation. The target mRNAs are degraded by both 5′ to 3′ and 3′ to 5′ pathways.

▸ **Upf proteins** A set of protein factors that target nonsense-mediated decay (NMD) substrates for degradation.

How are PTCs distinguished from the normal termination codon further downstream? The mechanism has been studied extensively both in yeast and in mammalian cells, where it is somewhat different; these are illustrated in **FIGURE 22.14**. The major signal that identifies a PTC in mammalian cells is the presence of a splice junction, marked by an **exon junction complex (EJC)** downstream of the premature termination codon. The majority of genes in higher eukaryotes do not have an intron interrupting the 3′ UTR, so authentic termination codons are not generally followed by a splice junction. During the **pioneer round of translation** for a normal mRNA, all EJCs occur within the coding region and are displaced by the transiting ribosome. During the pioneer round of translation for an NMD substrate, Upf2 and Upf3 proteins bind to the residual downstream EJC(s), targeting it for degradation.

▸ **exon junction complex (EJC)** A protein complex that assembles at exon-exon junctions during splicing and assists in RNA transport, localization and degradation.

▸ **pioneer round of translation** The first translation event for a newly synthesized and exported mRNA.

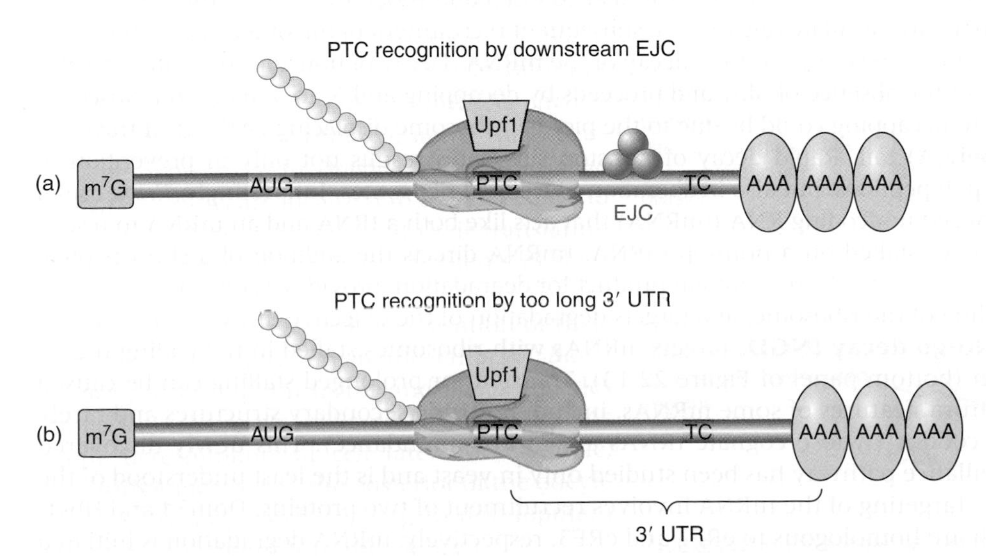

FIGURE 22.14 Two mechanisms by which a termination codon is recognized as premature. (a) In mammals, the presence of an EJC downstream of a termination codon targets the mRNA for NMD. (b) In probably all eukaryotes, an abnormally long 3′ UTR is recognized by the distance between the termination codon and the poly(A)-PABP complex. In either case, the Upf1 protein binds to the terminating ribosome to trigger decay.

2. mRNA localization also plays a role in asymmetric cell divisions; i.e., mitotic divisions that result in daughter cells that differ from one another. One way this is accomplished is by asymmetric segregation of cell fate determinants, which may be proteins and/or the mRNAs that encode them. In *Drosophila* embryos, *prospero* mRNA and its product (a transcription factor) are localized to a region of the peripheral cortex of the embryo. Later in development, oriented cell division of neuroblasts assures that only the outermost daughter cell receives *prospero*, committing it to a ganglion mother-cell fate. Asymmetric cell division is also used by budding yeast to generate a daughter cell of a different mating type than the mother cell, an event described later in this section.
3. mRNA localization in adult, differentiated cell types is a mechanism for the compartmentalization of the cell into specialized regions. Localization may be used to assure that components of multiprotein complexes are synthesized in proximity to one another and that proteins targeted to organelles or specialized areas of cells are synthesized conveniently nearby. mRNA localization is particularly important for highly polarized cells such as neurons. While most mRNAs are translated in the neuron cell body, many mRNAs are localized to its dendritic and axonal extensions. Among those is β-actin mRNA, whose product participates in dendrite and axon growth. β-actin mRNA localizes to sites of active movement in a wide variety of motile cell types. Interestingly, localization of mRNA at neuronal postsynaptic sites seems to be essential for modifications accompanying learning. In glial cells, the myelin basic protein (MBP) mRNA, which encodes a component of the myelin sheath, is localized to a specific myelin-synthesizing compartment. Plants localize mRNAs to the cortical region of cells and to regions of polar cell growth.

In some cases, mRNA localization involves transport from one cell to another. Maternal mRNPs in *Drosophila* are synthesized and assembled in surrounding nurse cells, and are transferred to the developing oocyte through cytoplasmic canals. Plants can export RNAs through plasmodesmata and transport them for long distances via the phloem vascular system. mRNAs are sometimes transported *en masse* in mRNP granules. The compositions of these granules are not yet well defined.

Three mechanisms for the localization of mRNA have been well documented:

- The mRNA is uniformly distributed but degraded at all sites except the site of translation.
- The mRNA is freely diffusible but becomes trapped at the site of translation.
- The mRNA is actively transported to a site where it is translated.

Active transport is the predominant mechanism for localization. Transport is achieved by translocation of motor proteins along cytoskeletal tracks. All three molecular motor types are exploited: dyneins and kinesins, which travel along microtubules in opposite directions, and myosins, which travel along actin fibers. This mode of localization requires at least four components: (1) *cis*-elements on the target mRNA, (2) *trans*-factors that directly or indirectly attach the mRNA to the correct motor protein, (3) *trans*-factors that repress translation, and (4) an anchoring system at the desired location.

zipcode (in RNA) Any of the number of RNA *cis* elements involved in directing cellular localization; also called localization signals.

Only a few *cis*-elements, sometimes called **zipcodes (in RNA)**, have been characterized. They are diverse, include examples of both sequence and structural RNA elements, and can occur anywhere in the mRNA, although most are in the 3′ UTR. Zipcodes have been difficult to identify, presumably because many consist of complex secondary and tertiary structures. A large number of *trans*-factors have been associated with localized mRNA transport and translational repression, some of which are highly conserved in different organisms. For example, the double-stranded RNA-binding protein *staufen* is involved in localizing mRNAs in the oocytes of *Drosophila* and *Xenopus*, as well as the nervous systems of *Drosophila*, mammals, and probably worms and zebrafish. This multitalented factor has multiple domains that can couple complexes to both actin- and microtubule-dependent transport pathways. Almost nothing is known about the fourth required component—anchoring mechanisms. Two examples of localization mechanisms are discussed below.

The localization of β-actin mRNA has been studied in cultured fibroblasts and neurons. The zipcode is a 54-nucleotide element in the 3′ UTR. Cotranscriptional binding of the zipcode element by the protein ZBP1 is required for localization, suggesting this mRNA is committed to localization before it is even processed and exported from the nucleus. Interestingly, β-actin mRNA localization is dependent on intact actin fibers in fibroblasts and intact microtubules in neurons.

Genetic analysis of *ASH1* mRNA localization in yeast has provided the most complete picture of a localization mechanism to date and is illustrated in **FIGURE 22.17**. During budding, the *ASH1* mRNA is localized to the developing bud tip, resulting in *ASH1* synthesis only in the newly formed daughter cell. Ash1 is a transcriptional repressor that disallows expression of the HO endonuclease, a protein required for mating-type switching (see *Section 15.9, Yeast Use a Specialized Recombination Mechanism to Switch Mating Type*). The result is that mating-type switching occurs only in the mother cell. The *ASH1* mRNA has four stem-loop localization elements in its coding region to which the protein She2 binds, probably in the nucleus. The protein She3 serves as an adaptor, binding both to She2 and to the myosin motor protein Myo4 (also called She1). A Puf protein, Puf6, binds to the mRNA, repressing its translation. The motor transports the Ash1 mRNP along the polarized actin fibers that lead from the mother cell to the developing bud. Additional proteins are required for proper localization and expression of the Ash1 mRNA. More than 20 yeast mRNAs use the same localization pathway.

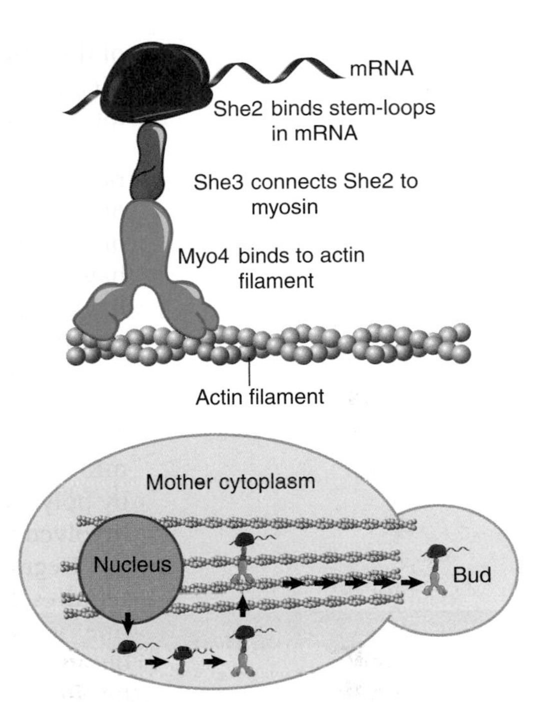

FIGURE 22.17 Localization of *ASH1* mRNA. Newly exported *ASH1* mRNA is attached to the myosin motor Myo4 via a complex with the She2 and She3 proteins. The motor transports the mRNA along actin filaments to the developing bud.

Localization mechanisms that do not involve active transport have been clearly demonstrated for only a few localized mRNAs in oocytes and early embryos. The mechanism of local entrapment of diffusible mRNAs requires the participation of previously localized anchors, which have not been identified. In *Drosophila* oocytes, diffusing *nanos* mRNA is trapped at the posterior "germ plasm," a specialized region of the cytoplasm underlying the cortex. In *Xenopus* oocytes, mRNAs localized to the vegetal pole are first trapped in a somewhat mysterious, membrane-laden structure called the mitochondrial cloud (MC), which later migrates to the vegetal pole carrying mRNAs with it. The mechanism of localized mRNA stabilization has been described for an mRNA that also localizes to the posterior pole of the *Drosophila* embryo. Early in development, the *hsp83* mRNA is uniformly distributed through the embryonic cytoplasm, but later it is degraded everywhere except at the pole. A protein called smaug is involved in destabilizing the majority of the *hsp83* mRNAs, most likely by recruiting the CCR4/NOT complex. How the pole-localized mRNAs escape is not known.

KEY CONCEPTS

- Localization of mRNAs serves diverse functions in single cells and developing embryos.
- Three mechanisms for the localization of mRNA have been documented.
- Localization requires *cis*-elements on the target mRNA and *trans*-factors to mediate the localization.
- The predominant active transport mechanism involves the directed movement of mRNPs along cytoskeletal tracks.

CONCEPT AND REASONING CHECK

After mRNAs are transported to specific locations, how could they be prevented from diffusing away?

22.11 Summary

Cellular RNAs are relatively unstable molecules due to the presence of cellular ribonucleases. Ribonucleases differ in mode of attack and are specialized for different RNA substrates. These RNA-degrading enzymes have many roles in a cell, including the decay of messenger RNA. The fact that mRNAs are short lived allows rapid adjustment

7. Most eukaryotic mRNA turnover is initiated by:
   - A. decapping.
   - B. deadenylation.
   - C. endonuclease cleavage at AU-rich regions.
   - D. no-go decay.
8. Which of the following mechanisms of mRNA localization has NOT been documented?
   - A. diffusible mRNA is trapped at a site of translation
   - B. mRNA is actively transported to a site of translation
   - C. mRNA is degraded at the site of translation but protected elsewhere
   - D. mRNA is degraded at all sites except the site of translation

KEY TERMS

3′ untranslated region (UTR)
5′ UTR
AU-rich element (ARE)
cryptic unstable transcripts (CUTs)
decapping enzyme
degradosome
destabilizing element (DE)
distributive (nuclease)
endoribonuclease (endonuclease)
exon junction complex (EJC)
exoribonuclease (exonuclease)
exosome
half-life ($t_{1/2}$) (RNA)
iron-response element (IRE)
maternal mRNA granules
microRNA (miRNA)
monocistronic mRNA
mRNA decay
neuronal granules
no-go decay (NGD)
nonsense-mediated decay (NMD)
nonstop decay (NSD)
oligo(A) tail
pioneer round of translation
poly(A)
poly(A) binding protein (PAPB)
poly(A) nuclease (deadenylase)
poly(A) polymerase (PAP)
polyribosome (polysome)
processing body (PB)
processive (nuclease)
release factor (RF)
ribonuclease
ribonucleoprotein particles (RNPs)
RNA-binding protein (RBP)
RNA regulon
RNA surveillance systems
SKI proteins
stabilizing element (SE)
steady state (molecular concentration)
stem-loop
stress granules
TRAMP
Upf proteins
zipcode (in RNA)

FURTHER READING

Du, T. G., Schmid, M., and Jansen, R. P. (2007). Why cells move messages: the biological functions of mRNA localization. *Semin. Cell Dev. Biol.* **18**, 171–177.

Filipowicz, W., Bhattacharyya, S. N., and Sonenberg, N. (2008). Mechanisms of post-transcriptional regulation by microRNAs: are the answers in sight? *Nature Rev. Genet.* **9**, 102–114.

Garneau, N. L., Wilusz, J., and Wilusz, C. J. (2007). The highways and byways of mRNA decay. *Nature Rev. Mol. Cell Biol.* **8**, 113–126.

Houseley, J., LaCava, J., and Tollervey, D. (2006). RNA-quality control by the exosome. *Nature Rev. Mol. Cell Biol.* **7**, 529–539.

Houseley, J., and Tollervey, D. (2008). The nuclear RNA surveillance machinery: the link between ncRNAs and genome structure in budding yeast? *Biochem. Biophys. Acta* **1779**, 239–246.

Houseley, J., and Tollervey, D. (2009). The many pathways of RNA degradation. *Cell* **136**, 763–776.

Keene, J. D. (2007). RNA regulons: coordination of post-transcriptional events. *Nature Reviews/Genetics* **8**, 533–543.

Martin, K. C., and Ephrussi, A. (2009). mRNA localization: gene expression in the spatial dimension. *Cell* **136**, 719–730.

Parker, R., and Sheth, U. (2007). P Bodies and the control of mRNA translation and degradation. *Mol. Cell* **25**, 635–646.

Shyu, A. B., Wilkinson, M. F., and van Hoof, A. (2008). Messenger RNA regulation: to translate or to degrade. *EMBO J.* **27**, 471–481.

Stalder, L., and Mühlemann, O. (2008). The meaning of nonsense. *Trends Cell Biol.* **18**(7), 315–321.

von Roretz, C., and Gallouzi, I. E. (2008). Decoding ARE-mediated decay: is microRNA part of the equation? *J. Cell Biol.* **181**, 189–194.

23

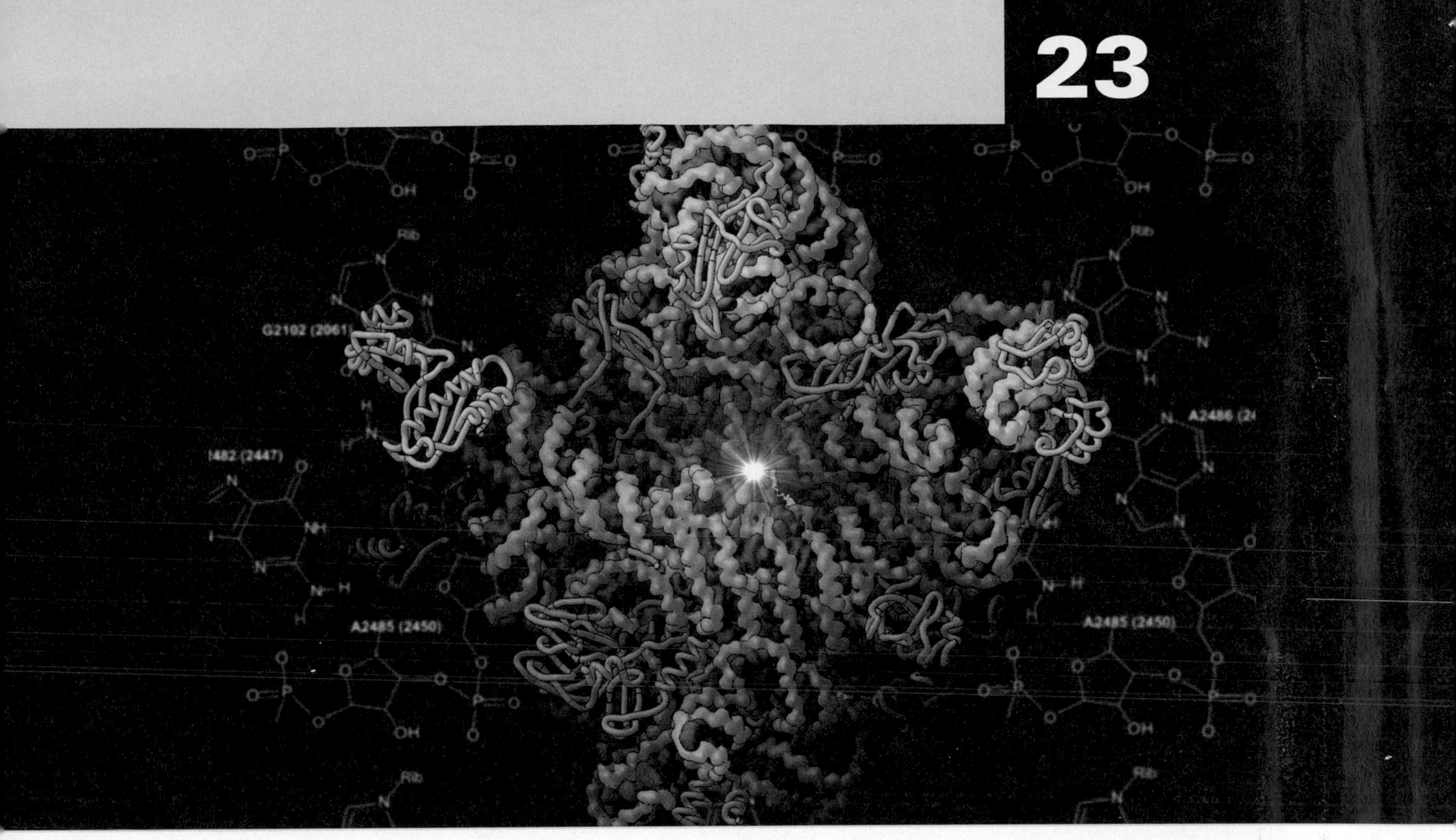

The RNA/protein architecture of the large ribosomal subunit with the active site highlighted. The background shows a schematic diagram of the peptidyl transferase active site of the ribosome. Photo courtesy of Nenad Ban, Institute for Molecular Biology & Biophysics, Zurich.

Catalytic RNA

CHAPTER OUTLINE

23.1 Introduction

The idea that only proteins could possess enzymatic activity was deeply rooted in early biochemistry. The rationale for the identification of enzymes with proteins resided in the view that only proteins, with their varied three-dimensional structures and variety of side-chain groups, had the flexibility to create the active sites that catalyze biochemical reactions. Critical studies of systems involved in RNA processing, however, have shown this view to be an oversimplification.

The first examples of RNA-based catalysis were identified in the bacterial tRNA processing enzyme, ribonuclease P (RNase P), and self-splicing group I introns in RNA from *Tetrahymena thermophilus*. For their pioneering work on RNA catalysts, Sidney Altman and Thomas Cech were awarded the 1989 Nobel Prize in Chemistry. Since the initial discovery of catalytic RNA, several other types of catalytic reactions mediated by RNA have been identified. Importantly, ribosomes, the RNA-protein complexes that manufacture peptides (see *Chapter 24, Translation*), have been identified as *ribozymes*, with RNA acting as the catalytic component and protein acting as a scaffold.

▸ **ribozyme** An RNA that has catalytic activity.

Ribozyme has become a general term used to describe an RNA with catalytic activity, and it is possible to characterize the enzymatic activity in the same way as a more conventional enzyme. Some RNA catalytic activities are directed against separate substrates (intermolecular), whereas others are intramolecular, which limits the catalytic action to a single cycle.

The enzyme RNase P is a ribonucleoprotein that contains a single RNA molecule bound to a protein. RNase P functions intermolecularly and is an example of a ribozyme that catalyzes multiple-turnover reactions. While originally identified in *E. coli*, RNase P is now known to be required for the viability of both prokaryotes and eukaryotes. The RNA possesses the ability to catalyze cleavage in a tRNA substrate, whereas the protein component plays an indirect role, probably to maintain the structure of the catalytic RNA.

The two classes of self-splicing introns, group I and group II, are good examples of ribozymes that function intramolecularly. Both group I and group II introns possess the ability to splice themselves out of their respective pre-mRNAs.

The common theme of these reactions is that the RNA can perform an intramolecular or intermolecular reaction that involves cleavage or joining of phosphodiester bonds *in vitro*. Although the specificity of the reaction and the basic catalytic activity is provided by RNA, proteins associated with the RNA may be needed in order for the reaction to occur efficiently *in vivo*.

▸ **RNA editing** A change of sequence at the level of RNA following transcription.

RNA splicing is not the only means by which changes can be introduced in the informational content of RNA. In the process of **RNA editing**, changes are introduced at individual bases, or bases are added at particular positions within an mRNA. The insertion of bases (most commonly uridine residues) occurs for several genes in the mitochondria of certain unicellular/oligocellular eukaryotes. Like splicing, RNA editing involves the breakage and reunion of bonds between nucleotides, but also requires a template for encoding the information of the new sequence.

23.2 Group I Introns Undertake Self-Splicing by Transesterification

▸ **autosplicing (self-splicing)** The ability of an intron to excise itself from an RNA by a catalytic action that depends only on the sequence of RNA in the intron.

Group I introns are found in diverse species, and more than two thousand of these introns have been identified to date. Unlike RNase P, group I introns are not essential for viability. Group I introns occur in the genes encoding rRNA in the nuclei of the unicellular/oligocellular eukaryotes *Tetrahymena thermophila* (a ciliate) and *Physarum polycephalum* (a slime mold). They are common in the genes of fungi and protists as well as occurring rarely in prokaryotes and animals. Group I introns have an intrinsic ability to splice themselves. This is called **autosplicing**, or **self-splicing**. (This property also is found in the group II introns discussed in *Section 23.6, Group II Introns May Encode Multifunction Proteins*.)

Self-splicing was discovered as a property of the transcripts of the rRNA genes in *T. thermophila*. The genes for the two major rRNAs follow the usual organization, in which both are expressed as part of a common transcription unit. The product is a 35S precursor RNA with the sequence of the small (17S) rRNA in the 5′ part and the sequence of the larger (26S) rRNA toward the 3′ end.

In some strains of *T. thermophila*, the sequence encoding the 26S rRNA is interrupted by a single, short intron. When the 35S precursor RNA is incubated *in vitro*, splicing occurs as an autonomous reaction. The intron is excised from the precursor and accumulates as a linear fragment of 400 bases, which is subsequently converted to a circular RNA. These events are summarized in **FIGURE 23.1**.

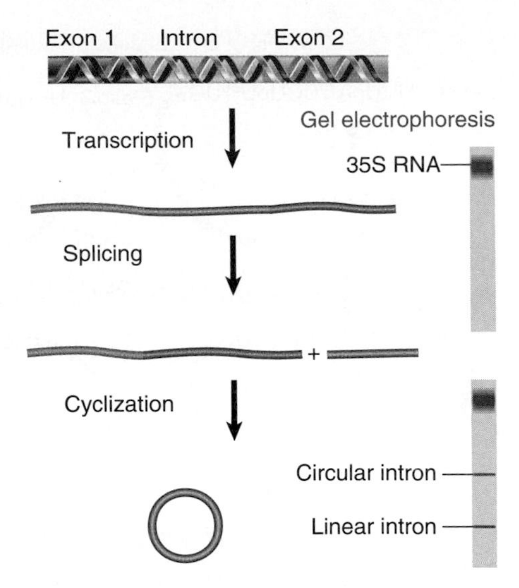

FIGURE 23.1 Splicing of the *Tetrahymena* 35S rRNA precursor can be followed by gel electrophoresis. The removal of the intron is revealed by the appearance of a rapidly moving small band. When the intron becomes circular, it electrophoreses more slowly, as seen by a higher band.

The reaction requires two metal ions and a guanosine nucleotide cofactor. No other base can be substituted for G, but a triphosphate is not needed: GTP, GDP, GMP, and guanosine itself all can be used, so there is no net energy requirement. The guanine nucleotide must have a 3′–OH group.

FIGURE 23.2 shows that three transfer reactions occur. In the first transfer, the guanosine nucleotide behaves as a cofactor that provides a free 3′–OH group that attacks the 5′ end of the intron. This reaction creates the G-intron link and generates a 3′–OH group at the end of the exon. The second transfer involves a similar chemical reaction in which the newly formed 3′–OH at the end of exon 1 attacks the second exon. The two transfers are connected; no free exons have been observed, so their ligation may occur as part of the same reaction that releases the intron. The intron is released as a linear molecule, but the third transfer reaction converts it to a circle.

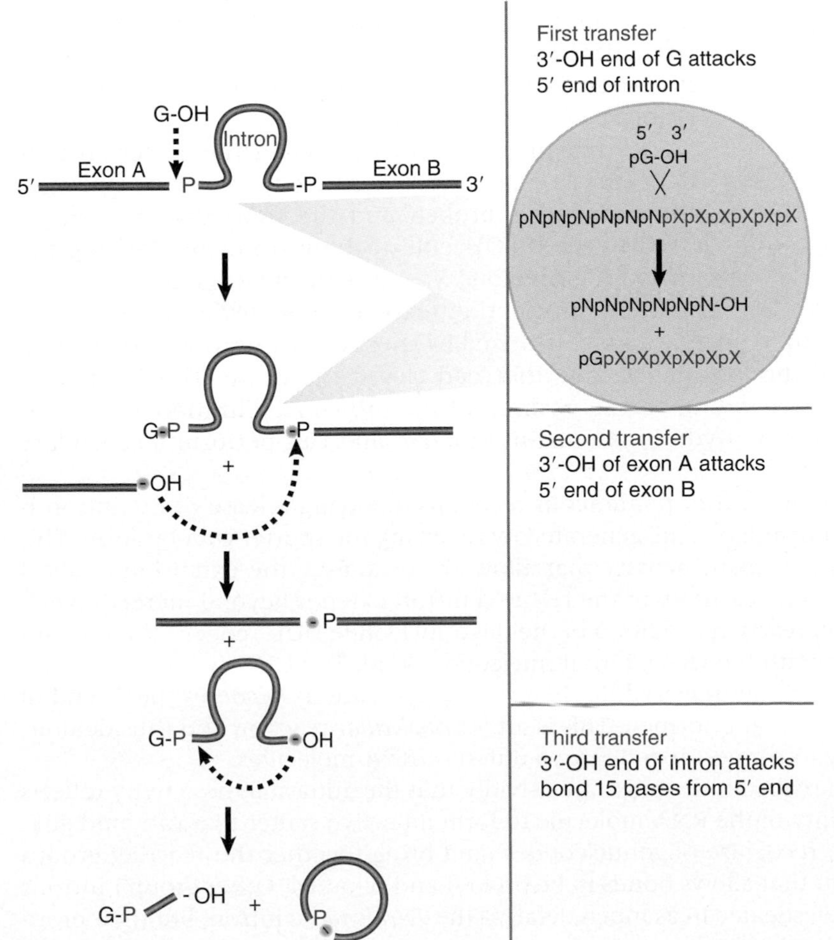

FIGURE 23.2 Self-splicing occurs by transesterification reactions in which bonds are exchanged directly. The bonds that have been generated at each stage are indicated by the blue circles.

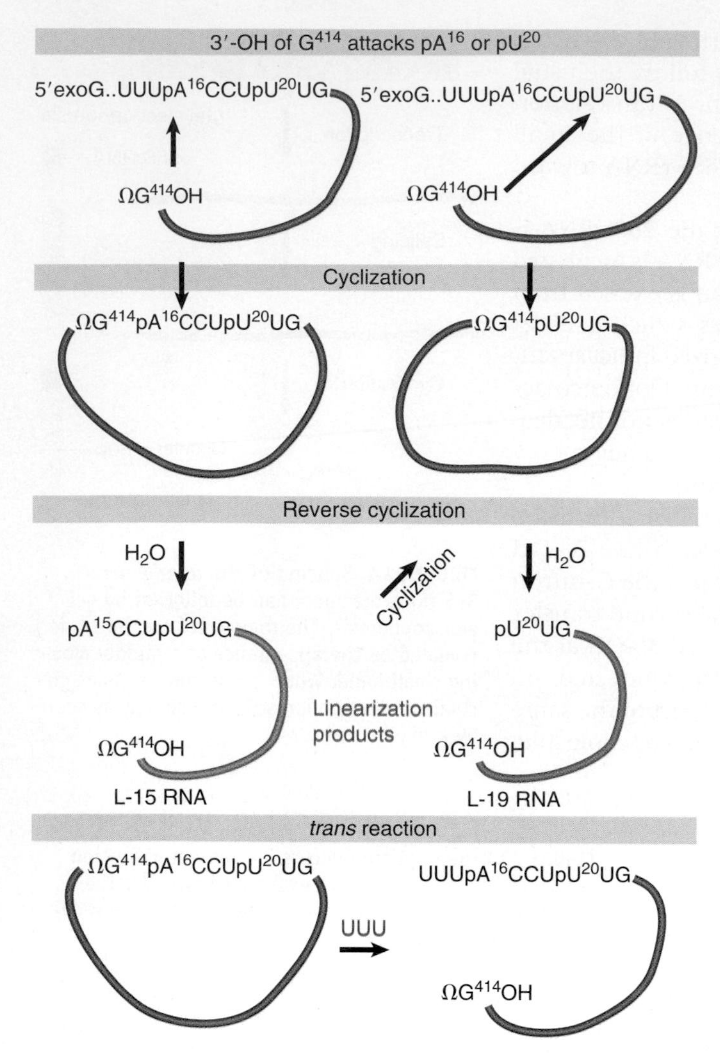

FIGURE 23.3 The excised intron can form circles by using either of two internal sites for reaction with the 5′ end and can reopen the circles by reaction with water or oligonucleotides.

Each stage of the self-splicing reaction occurs by a transesterification, in which one phosphate ester is converted directly into another without any intermediary hydrolysis. Bonds are exchanged directly and energy is conserved, so the reaction does not require input of energy from hydrolysis of ATP or GTP. Each consecutive transesterification reaction involves no net change of energy. In the cell, the concentration of GTP is high relative to that of RNA and therefore drives the reaction forward whereupon a change in secondary structure in the RNA prevents the reverse reaction. This allows the reaction to proceed to completion, instead of coming to equilibrium between spliced product and nonspliced precursors.

The ability to splice is intrinsic to the RNA and the system is able to proceed *in vitro* without addition of any protein components. The RNA forms a specific secondary/tertiary structure in which the relevant groups are brought into juxtaposition so that a guanosine nucleotide can be bound to a specific site and then the bond breakage and reunion reactions shown in Figure 23.2 can occur. Although a property of the RNA itself, the reaction is very slow *in vitro*. This is because group I intron splicing is assisted *in vivo* by proteins that serve to stabilize the RNA structure in a favorable conformation for splicing.

The ability to engage in these transfer reactions resides with the sequence of the intron, which continues to be reactive after its excision as a linear molecule. **FIGURE 23.3** summarizes catalytic activities of the excised intron from *Tetrahymena*, with residue numbers corresponding to that organism.

The intron can circularize when the 3′ terminal G (ΩG) attacks an internal position near the 5′ end. The internal bond is broken and the new 5′ end is transferred to the 3′–OH end of the intron, circularizing the intron. The previous 5′ end with the original exogenous guanosine nucleotide (exoG) is released as a linear fragment. The circularized intron can be linearized by specifically hydrolyzing the bond between ΩG and the internal residue that had closed the circle. This is called a *reverse cyclization*. Depending on the position of the primary cyclization, the linear molecule generated by hydrolysis remains reactive and can perform a secondary cyclization.

The final product of the spontaneous reactions following release of the intron is the L-19 RNA, a linear molecule generated by reversing the shorter circular form. This molecule has an enzymatic activity that allows it to catalyze the extension of short oligonucleotides. The reactivity of the released intron extends beyond merely reversing the cyclization reaction. Addition of the oligonucleotide UUU reopens the primary circle by reacting with the ΩG–internal nucleotide bond. The UUU (which resembles the 3′ end of the 15-mer released by the primary cyclization) becomes the 5′ end of the linear molecule that is formed. This is an *intermolecular* reaction and thus demonstrates the ability to connect together two different RNA molecules.

This series of reactions demonstrates vividly that the autocatalytic activity reflects a generalized ability of the RNA molecule to form an active center that can bind guanosine cofactors, recognize oligonucleotides, and bring together the reacting groups in a conformation that allows bonds to be broken and rejoined. Other group I introns have not been investigated in as much detail as the *Tetrahymena* intron, but their properties are generally similar.

KEY CONCEPTS

- The only factors required for autosplicing *in vitro* by group I introns are two metal ions and a guanosine nucleotide.
- Splicing occurs by two transesterifications, without requiring input of energy.
- The 3′-OH end of the guanine cofactor attacks the 5′ end of the intron in the first transesterification.
- The 3′-OH end generated at the end of the first exon attacks the junction between the intron and second exon in the second transesterification.
- The intron is released as a linear molecule that circularizes when its 3′-OH terminus attacks a bond at one of two internal positions.
- In *Tetrahymena*, an internal bond of the excised intron can also be attacked by other nucleotides in a *trans*-splicing reaction.

CONCEPT AND REASONING CHECK

What is the evidence that group I introns are self-splicing?

23.3 Group I Introns Form a Characteristic Secondary Structure

All group I introns can be organized into a characteristic secondary structure with nine helices (P1–P9). **FIGURE 23.4** shows a model for the secondary structure of the *Tetrahymena* intron. While structural analyses were able to elucidate the secondary structure of the group I intron, it was not until the recent determination of the crystal structure that the tertiary structure of the intron was revealed. Several crystal structures of group I introns have been solved and these confirm previous models of the secondary structure. Two of the base-paired regions are generated by pairing between conserved sequence elements that are common to group I introns. P4 is constructed from the sequences P and Q; P7 is formed from the sequences R and S. The other base-paired regions vary in sequence in individual introns. Mutational analysis identifies an intron "core" containing P3, P4, P6, and P7, which provides the minimal region that can undertake a catalytic reaction. The lengths of group I introns vary widely and the consensus sequences are located a considerable distance from the actual splice junctions.

Some of the pairing reactions are directly involved in bringing the splice junctions into a conformation that supports the enzymatic reaction. P1 includes the 3′ end of the 5′ exon. The sequence within the intron that pairs with the exon is called the internal guide sequence (IGS). The name IGS reflects the fact that originally the region immediately 3′ to the IGS sequence shown in Figure 23.4 was thought to pair with the 3′ splice junction, thus bringing the two junctions together. This interaction may occur but does not seem to be essential. A very short sequence—sometimes as short as two bases—between P7 and P9 base pairs with the sequence that immediately precedes the reactive G (ΩG, position 414 in *Tetrahymena*) at the 3′ end of the intron.

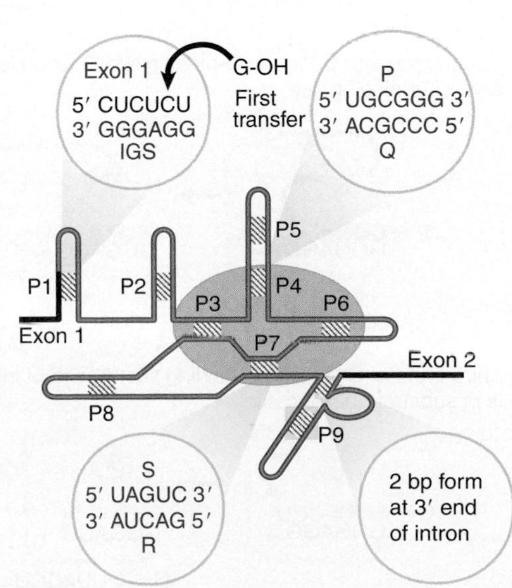

FIGURE 23.4 Group I introns have a common secondary structure that is formed by nine base-paired regions. The sequences of regions P4 and P7 are conserved, and identify the individual sequence elements P, Q, R, and S. P1 is created by pairing between the end of the left exon and the IGS of the intron; a region between P7 and P9 pairs with the 3′ end of the intron.

KEY CONCEPTS

- Group I introns form a secondary structure with nine duplex regions.
- The cores of regions P3, P4, P6, and P7 have catalytic activity.
- Regions P4 and P7 are both formed by pairing between conserved consensus sequences.
- A sequence adjacent to P7 base pairs with the sequence that contains the reactive G.

CONCEPT AND REASONING CHECK

What would be the effect of substitution mutations that weaken *P-Q* or *R-S* base pairing?

23.4 Ribozymes Have Various Catalytic Activities

The catalytic activity of group I introns was discovered by virtue of their ability to autosplice, but they also are able to undertake other catalytic reactions *in vitro*. All these reactions are based on transesterifications. We analyze these reactions in terms of their relationship to the splicing reaction itself.

The catalytic activity of a group I intron is conferred by its ability to generate particular secondary and tertiary structures that create active sites equivalent to the active sites of conventional (proteinaceous) enzymes. **FIGURE 23.5** illustrates the splicing reaction in terms of these sites (this is the same series of reactions shown previously in Figure 23.2).

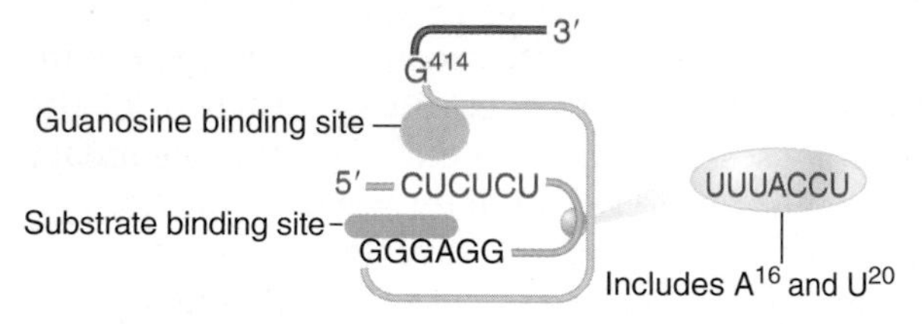

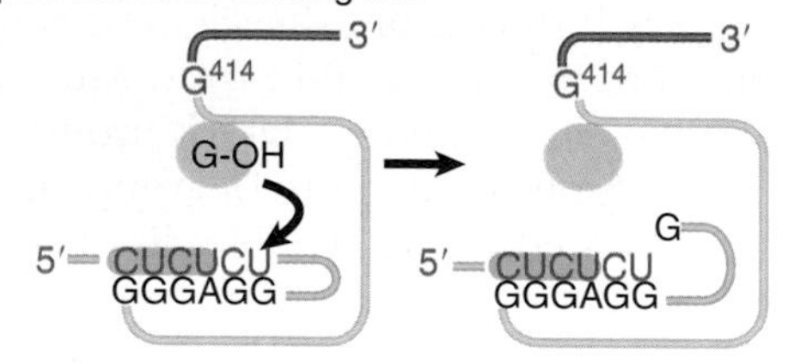

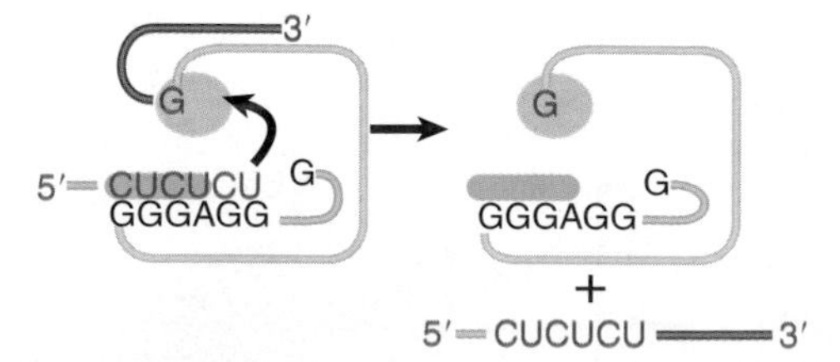

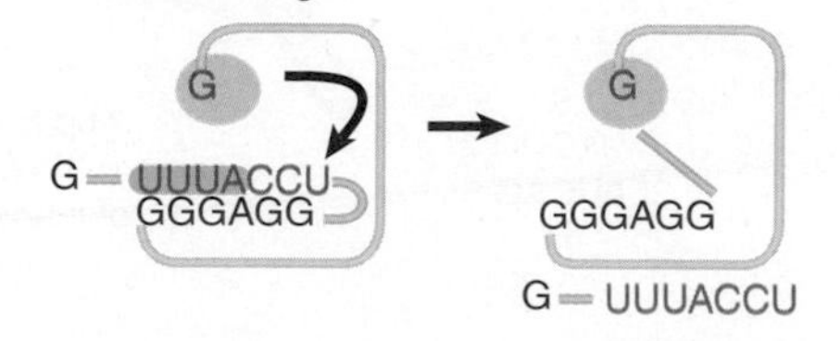

FIGURE 23.5 Excision of the group I intron in *Tetrahymena* rRNA occurs by successive reactions between the occupants of the guanosine-binding site and the substrate-binding site. The left exon is pink, and the right exon is purple.

The substrate-binding site is formed from the P1 helix, in which the 3′ end of the first intron base pairs with the IGS. A guanosine-binding site is formed by sequences in P7. This site may be occupied either by a free exogenous guanosine nucleotide (exoG) or by the ΩG residue (position 414 in *Tetrahymena*). In the first transfer reaction, the guanosine-binding site is occupied by a free guanosine nucleotide. Following release of the intron it is occupied by ΩG. The second transfer releases the joined exons. The third transfer creates the circular intron.

Binding to the substrate involves a change of conformation. Before substrate binding, the 5′ end of the IGS is close to P2 and P8; after binding, when it forms the P1 helix, it is close to conserved bases that lie between P4 and P5. The reaction is visualized by contacts that are detected in the secondary structure in **FIGURE 23.6**. In the tertiary structure, the two sites alternatively con-

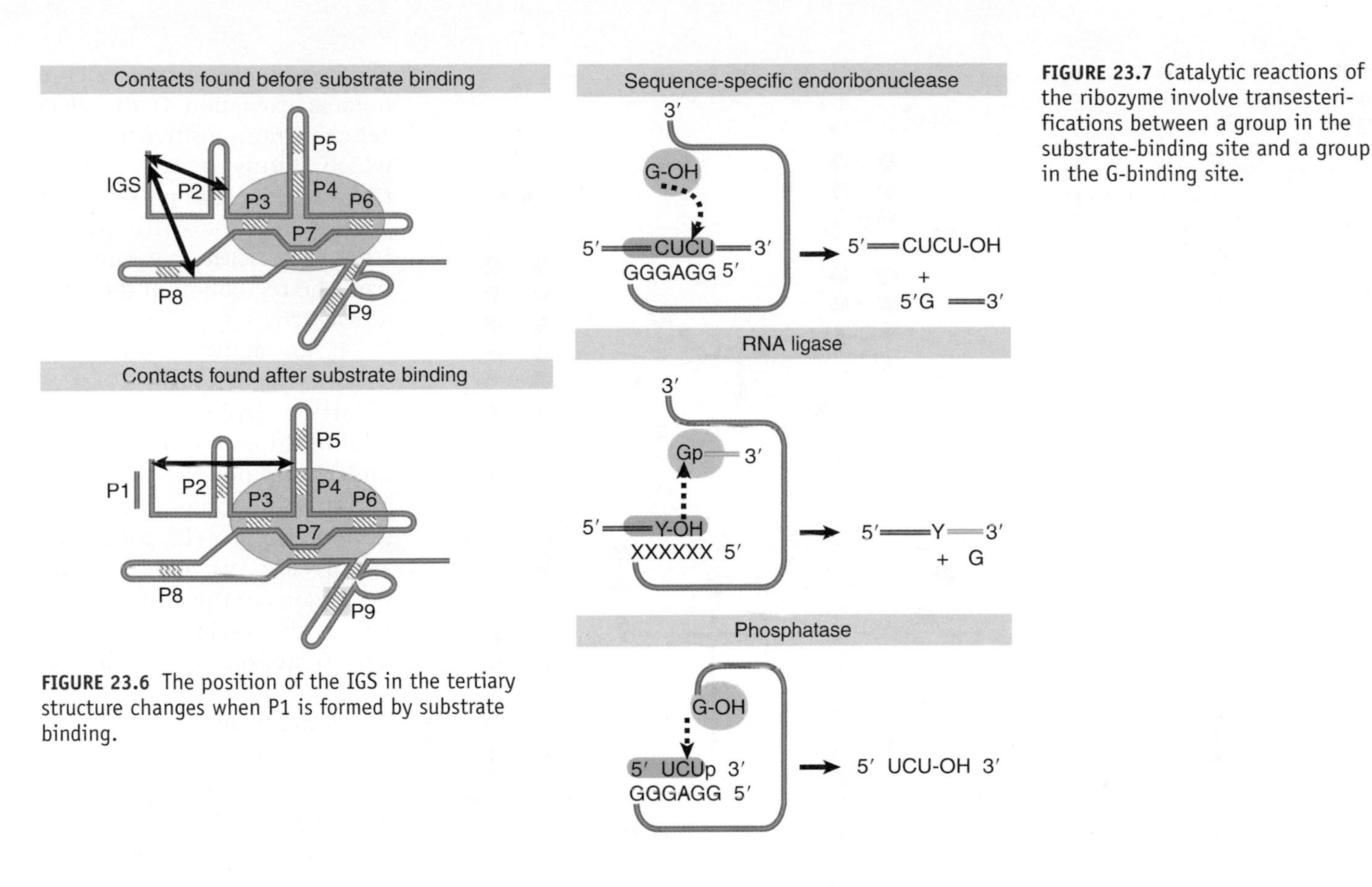

FIGURE 23.6 The position of the IGS in the tertiary structure changes when P1 is formed by substrate binding.

FIGURE 23.7 Catalytic reactions of the ribozyme involve transesterifications between a group in the substrate-binding site and a group in the G-binding site.

tacted by P1 are 37 Å apart, which implies a substantial movement in the position of P1.

Some further enzymatic reactions that *Tetrahymena* group I introns can perform are characterized in **FIGURE 23.7**. The ribozyme can function as a sequence-specific endoribonuclease by utilizing the ability of the IGS to bind complementary sequences. In this example, it binds an external substrate containing the sequence CUCU instead of binding the analogous sequence that is usually contained at the end of the 5′ exon. A guanosine-containing nucleotide is present in the G-binding site and attacks the CUCU sequence in precisely the same way that the exon is usually attacked in the first transfer reaction. This cleaves the target sequence into a 5′ molecule that resembles the 5′ exon and a 3′ molecule that bears a terminal G residue.

Enzyme	Substrate	K_M (mM)	Turnover (/min)
19-base virusoid	24-base RNA	0.0006	0.5
L-19 Intron	CCCCCC	0.04	1.7
RNase P RNA	pre-tRNA	0.00003	0.4
RNase P complete	pre-tRNA	0.00003	29
RNase T1	GpA	0.05	5,700
β galactosidase	lactose	4.0	12,500

FIGURE 23.8 Reactions catalyzed by RNA have the same features as those catalyzed by proteins, although the rate is slower. The K_M gives the concentration of substrate required for half-maximum velocity; this is an inverse measure of the affinity of the enzyme for substrate. The turnover number gives the number of substrate molecules transformed in unit time by a single catalytic site.

The reactions catalyzed by RNA can be characterized in the same way as classical enzymatic reactions in terms of Michaelis-Menten kinetics. **FIGURE 23.8** analyzes the reactions catalyzed by RNA. The K_M values for RNA-catalyzed reactions are low and, therefore, imply that the RNA can bind its substrate with high specificity. The turnover numbers for RNA catalyzed reactions, however, are low, which reflects a low catalytic rate. In effect, the RNA molecules behave in the same general manner as traditionally defined for enzymes, although they are relatively slow compared to protein catalysts (where a typical range of turnover numbers is 10^3 to 10^6 min^{-1}).

A powerful extension of the activities of ribozymes has been made with the discovery that they can be regulated by ligands (see *Section 30.3, Noncoding RNAs Can Be Used to Regulate Gene Expression*). These *cis*-acting regulatory RNA regions are called **riboswitches**. In almost all riboswitches, a conformational change determines the on

▸ **riboswitch** A catalytic RNA whose activity responds to a small ligand.

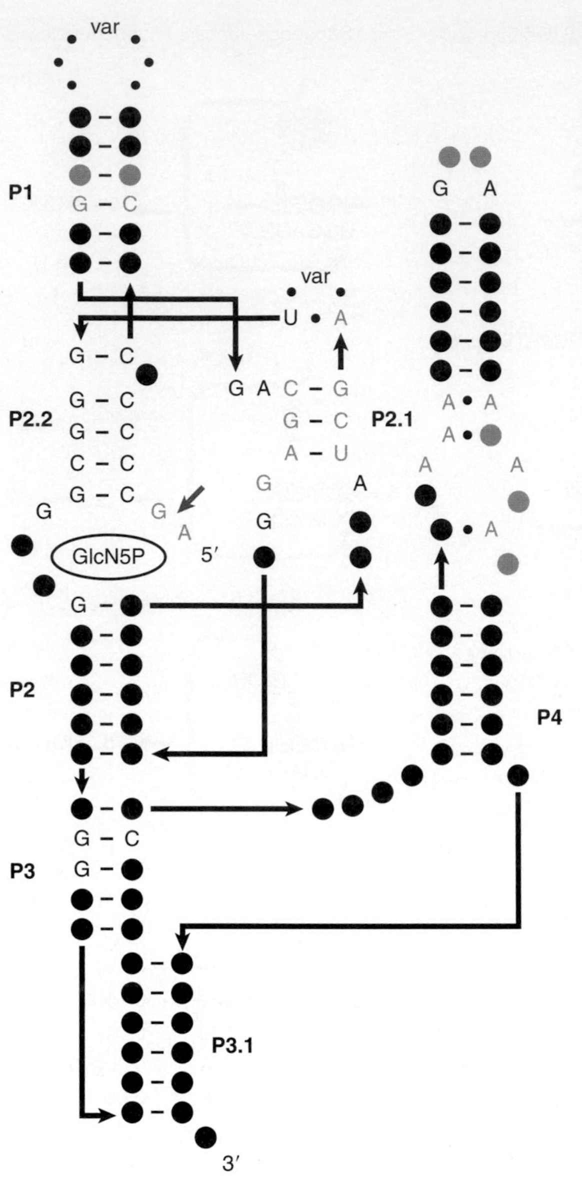

FIGURE 23.9 A ribozyme is contained within the 5′ untranslated region of the mRNA encoding the enzyme that produces glucosamine-6-phosphate. When Glc6P binds to the ribozyme, it cleaves off the 5′ end of the mRNA, thereby inactivating it and preventing further production of the enzyme.

or off state of the switch. One notable exception is the *glmS* gene in Gram-positive bacteria, which forms a self-cleaving ribozyme in the presence of glucosamine-6-phosphate (GlcN6P). **FIGURE 23.9** summarizes the regulation of the *glmS* riboswitch.

If an active center is a surface that exposes a series of active groups in a fixed relationship, it is possible to understand how RNA is capable of providing a catalytic center. In a protein, the active groups are provided by the side chains of the amino acids. The amino acid side chains have appreciable variety, including positive and negative ionic groups and hydrophobic groups. In RNA, the available moieties are more restricted, consisting primarily of the exposed groups of bases. Short regions of RNA are held in a particular secondary/tertiary conformation, providing an active surface and maintaining an environment in which bonds can be broken and formed. It seems inevitable that the interaction between the RNA catalyst and the RNA substrate will rely on base pairing to create the active environment. Divalent cations (usually Mg^{2+}) play an important role in structure, typically being present at the active site where they coordinate the positions of the various groups. Divalent metal cations also play a direct role in the endonucleolytic activity of virusoid ribozymes (see *Section 23.8, Viroids Have Catalytic Activity*).

KEY CONCEPTS

- By changing the substrate-binding site of a group I intron, it is possible to introduce alternative sequences that interact with the reactive G.
- The reactions follow classical enzyme kinetics with a low catalytic rate.

CONCEPT AND REASONING CHECK

Describe how an RNA can act as a catalytic active site like that of a protein enzyme.

23.5 Some Group I Introns Encode Endonucleases That Sponsor Mobility

Certain introns of both the group I and group II classes contain open reading frames that are translated into proteins. Expression of the proteins allows the intron (either in its original DNA form or as a DNA copy of the RNA) to be *mobile:* it is able to insert itself into a new genomic site. Introns of both groups I and II are extremely widespread, being found in both prokaryotes and eukaryotes. Group I introns migrate by DNA-mediated mechanisms, whereas group II introns migrate by RNA-mediated mechanisms.

Intron mobility was first detected by crosses in which the alleles for the relevant gene differ with regard to their possession of the intron. Polymorphisms for the presence or absence of introns are common in fungal mitochondria. This is consistent with the view that these introns originated by insertion into the gene. Some light on the process that could be involved is cast by an analysis of recombination in crosses involving the large rRNA gene of the yeast mitochondrion.

The large rRNA gene of the yeast mitochondrion has a group I intron that contains a coding sequence. The intron is present in some strains of yeast (called ω^+) but absent in others (ω^-). Progeny of genetic crosses between ω^+ and ω^- do not result in the expected genotypic ratio; the progeny are usually ω^+. If we think of the ω^+ strain as a donor and the ω^- strain as a recipient, we form the view that in $\omega^+ \times \omega^-$ crosses, a new copy of the intron is generated in the ω^- genome. As a result, the progeny are all ω^+. Mutations can occur in either parent to abolish the non-Mendelian genotypic assortment. Certain mutants show normal segregation, with equal numbers of ω^+ and ω^- progeny. When mapped, mutations in the ω^- strain occur close to the site where the intron would be inserted. Mutations in the ω^+ strain lie in the reading frame of the intron and prevent production of the protein. This suggests the model of **FIGURE 23.10**, in which the protein encoded by the intron in an ω^+ strain recognizes the site where the intron should be inserted in an ω^- strain and causes it to be preferentially inherited.

Some group I introns encode endonucleases that make them mobile. There are at least five families of homing endonuclease genes (HEGs). Two common families of HEGs are the LAGLIDADG and His-Cys Box endonucleases. These HEG-containing group I introns, however, constitute a small portion of the overall number of nuclear group I introns. While approximately 1200 nuclear group I introns have been identified, fewer than 30 of these contain HEGs.

The ω intron contains an HEG, the product of which is an endonuclease known as I-SceI. *I-SceI recognizes the ω^- gene as a target for a double-strand break.* I-SceI recognizes an 18 bp target sequence that contains the site where the intron is inserted. The target sequence is cleaved on each strand of DNA two bases to the 3′ side of the insertion site. Thus the cleavage sites are 4 bp apart and generate overhanging single strands. This type of cleavage is related to the cleavage characteristic of transposons when they migrate to new sites (see *Chapter 17, Transposable Elements and Retroviruses*). The double-strand break probably initiates a gene conversion process in which the sequence of the ω^+

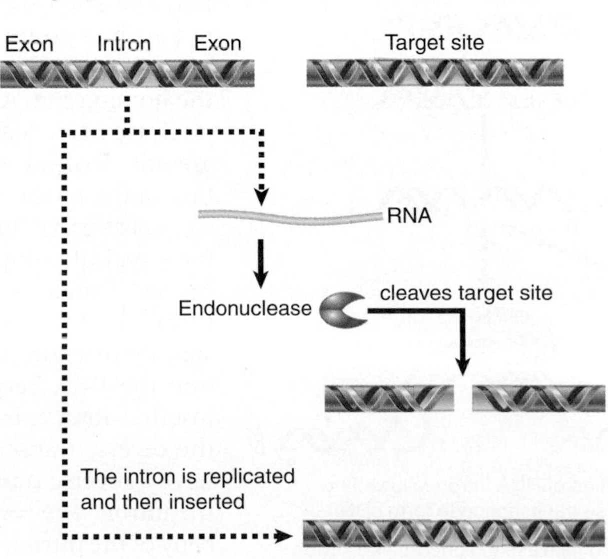

FIGURE 23.10 An intron encodes an endonuclease that makes a double-strand break in DNA. The sequence of the intron is duplicated and then inserted at the break.

gene is copied to replace the sequence of the ω^- gene. The reaction involves transposition by a duplicative mechanism and occurs solely at the level of DNA. Insertion of the intron interrupts the sequence recognized by the endonuclease, thus ensuring stability.

The variation in the endonucleases means that there is no homology between the sequences of their target sites. The target sites are among the longest and, therefore, the most specific known for any endonucleases (with a range of 14 to 40 bp). The specificity ensures that the intron perpetuates itself only by insertion into a single target site and not elsewhere in the genome. This is called **intron homing**.

▸ **intron homing** The ability of certain introns to insert themselves into a target DNA. The reaction is specific for a single target sequence.

Introns carrying sequences that encode endonucleases are found in a variety of bacteria and unicellular/oligocellular eukaryotes. These results strengthen the view that introns carrying coding sequences originated as independent elements.

KEY CONCEPTS

- Mobile introns are able to insert themselves into new sites.
- Mobile group I introns encode an endonuclease that makes a double-strand break at a target site.
- The intron transposes into the site of the double-strand break by a DNA-mediated replicative mechanism.

CONCEPT AND REASONING CHECK

How are mobile group I introns similar to transposons? How are they different?

23.6 Group II Introns May Encode Multifunction Proteins

The best characterized mobile group II introns encode a single protein in a region of the intron beyond its catalytic core. The typical protein contains an N-terminal reverse transcriptase activity, a central domain associated with an ancillary activity that assists folding of the intron into its active structure (called the *maturase*; see *Section 23.7, Some Autosplicing Introns Require Maturases*), a DNA-binding domain, and a C-terminal endonuclease domain.

The endonuclease initiates the transposition reaction and plays the same role in homing as its counterpart in a group I intron. The reverse transcriptase generates a DNA copy of the intron that is inserted at the homing site. At much lower frequency, the endonuclease also cleaves target sites that resemble, but are not identical to, the homing site, leading to insertion of the intron at new locations.

FIGURE 23.11 illustrates the transposition reaction for a typical group II intron. First, the endonuclease makes a single-strand break in the antisense strand. Cleavage of the sense strand is achieved by a reverse splicing reaction, with the RNA intron inserting itself into the DNA between the DNA exons. This newly inserted RNA intron can now act as a template for the reverse transcriptase. Almost all group II introns have a reverse transcriptase activity that is specific for the intron. The reverse transcriptase generates a DNA copy of the intron, with the end result being the insertion of the intron into the target site as a duplex DNA.

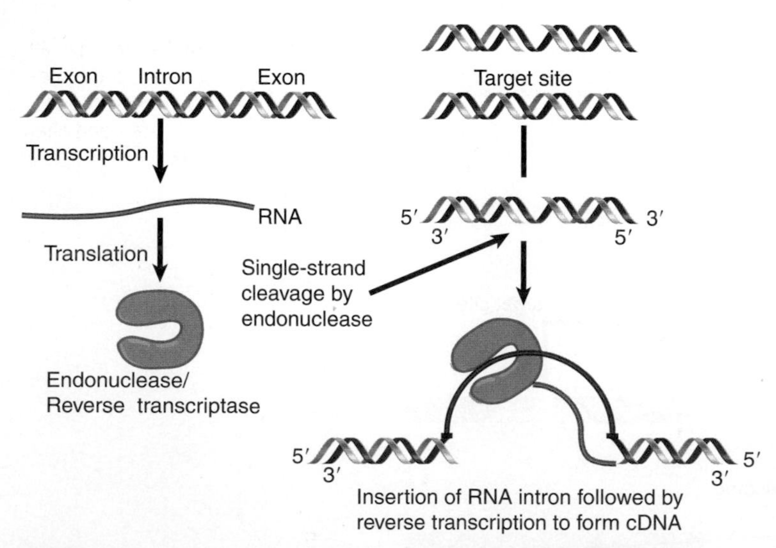

FIGURE 23.11 Reverse transcriptase/endonuclease encoded by an intron allows a copy of the RNA to be inserted into a target site.

KEY CONCEPTS

- Group II introns can autosplice *in vitro* but are usually assisted by protein activities encoded in the intron.
- A single coding frame specifies a protein with reverse transcriptase activity, maturase activity, a DNA-binding motif, and a DNA endonuclease.
- The endonuclease cleaves target DNA to allow insertion of the intron at a new site.
- The reverse transcriptase generates a DNA copy of the inserted RNA intron sequence.

CONCEPT AND REASONING CHECK

How are mobile group II introns similar to retroposons? How are they different?

23.7 Some Autosplicing Introns Require Maturases

Although group I and group II introns both have the capacity to autosplice *in vitro*, under physiological conditions they usually require assistance from proteins. Both types of intron may encode **maturase** activities that are required to assist the splicing reaction.

▸ **maturase** A protein encoded by a group I or group II intron that is needed to assist the RNA to form the active conformation that is required for self-splicing.

The maturase activity is part of the single open reading frame encoded by the intron. In the example of introns that encode homing endonucleases, the single protein product has both endonuclease and maturase activity. Mutational analysis shows that the two activities are independent. Structural analysis confirms the mutational data and shows that the endonuclease and maturase activities are provided by different active sites in the protein, each encoded by a separate domain. The coexistence of endonuclease and maturase activities in the same protein suggests a route for the evolution of the intron. **FIGURE 23.12** suggests that the intron originated in an independent autosplicing element. While Figure 23.12 depicts a group I intron, the process for group II introns is presumed to be similar. The insertion into this element of a sequence encoding an endonuclease gave it mobility. The insertion, however, might well disrupt the ability of the RNA sequence to fold into the active structure. This would create pressure for assistance from proteins that could restore folding ability. The incorporation of such a sequence into the intron would maintain its independence.

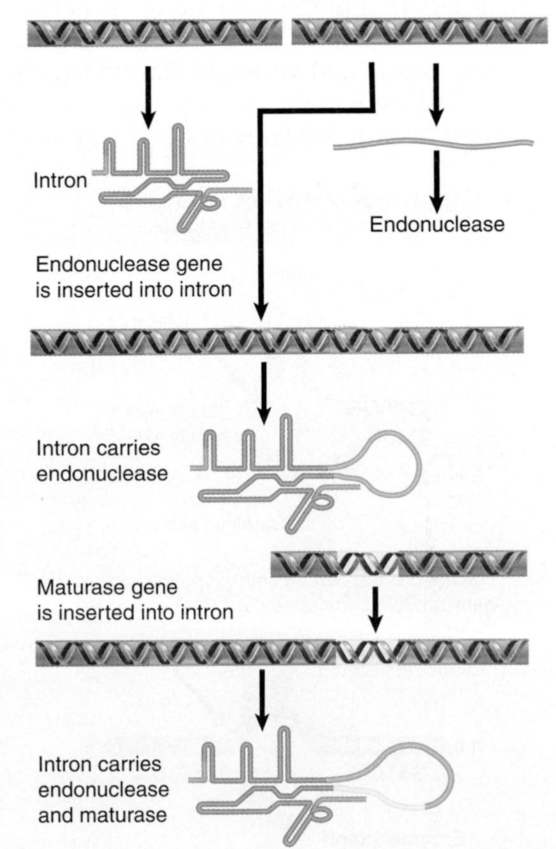

FIGURE 23.12 The intron originated as an independent sequence encoding a self-splicing RNA. The insertion of the endonuclease sequence created a homing intron that was mobile. The insertion of the maturase sequence then enhanced the ability of the intron sequences to fold into the active structure for splicing.

Some group II introns, however, do not encode maturase activity. These group II introns may use proteins (comparable to intron-encoded maturases) that are instead encoded by sequences in the host genome. This suggests a possible route for the evolution of general splicing factors. The factor may have originated as a maturase that specifically assisted the splicing of a particular intron. The coding sequence became isolated from the intron in the host genome and then it evolved to function with a wider range of substrates that the original intron sequence. The catalytic core of the intron could have evolved into an snRNA.

KEY CONCEPT

- Autosplicing introns may require maturase activities encoded within the intron to assist folding into the active catalytic structure.

CONCEPT AND REASONING CHECK

What is a maturase, and what is its role in self-splicing of introns?

23.8 Viroids Have Catalytic Activity

Another example of the ability of RNA to function as an endonuclease is provided by some small (~350 nt) plant RNAs that undertake a self-cleavage reaction. As with the case of the *Tetrahymena* group I intron, however, it is possible to engineer constructs that can function on external substrates.

These small plant RNAs fall into two general groups: viroids and virusoids. The **viroids** are infectious RNA molecules that function independently without encapsidation by any protein coat. The **virusoids** (which are sometimes called *satellite RNAs*) are similar in organization but are encapsidated by plant viruses, being packaged together with a viral genome. The virusoids cannot replicate independently because they require assistance from the virus.

- **viroid** A small infectious nucleic acid that does not have a protein coat.
- **virusoid** A small infectious nucleic acid that is encapsidated by a plant virus together with its own genome.

Viroids and virusoids both replicate via rolling circles (see Figure 14.6). The strand of RNA that is packaged into the virus is called the *plus strand*. The complementary strand, generated during replication of the RNA, is called the *minus strand*. Multimers of both plus and minus strands are found. Both types of monomer are generated by cleaving the tail of a rolling circle; circular plus-strand monomers are generated by ligating the ends of the linear monomer.

Both plus and minus strands of viroids and virusoids undergo self-cleavage *in vitro*. Some of the RNAs cleave *in vitro* under physiological conditions. Others do so only after a cycle of heating and cooling; this suggests that the isolated RNA has an inappropriate conformation but can generate an active conformation when it is denatured and renatured.

The viroids and virusoids that undergo self-cleavage form a "hammerhead" secondary structure at the cleavage site, as drawn in the upper part of **FIGURE 23.13**. Hammerhead ribozymes belong to a family of ribozymes, which include hepatitis delta virus (HDV), hairpin ribozymes, and Varkud satellite (VS) ribozyme. Functionally, HDV requires divalent metal cations to promote cleavage, while hammerhead and hairpin ribozymes do not require metal. The importance of metal for VS ribozyme cleavage is still ambiguous. All these ribozymes, however, generate a cleavage that leaves 5′-OH and 2′–3′–cyclic phosphodiester termini.

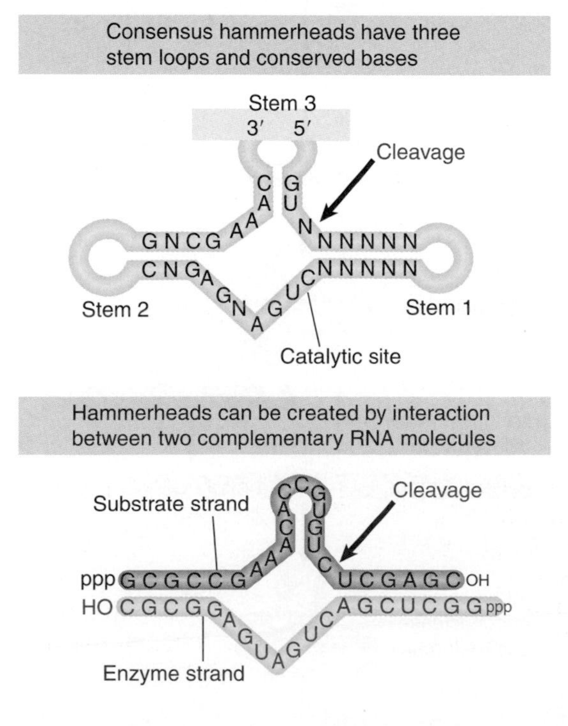

FIGURE 23.13 Self-cleavage sites of viroids and virusoids have a consensus sequence and form a hammerhead secondary structure by intramolecular pairing. Hammerheads can also be generated by pairing between a substrate strand and an "enzyme" strand.

Unlike all other ribozymes identified to date, hammerhead ribozymes and other members of the family do not require a protein component to function *in vivo* because the sequence of this structure is sufficient for cleavage. Minimally, for hammerhead ribozymes the active site is a sequence of only 58 nucleotides. The hammerhead contains three stem-loop regions whose position and size are constant and 13 conserved nucleotides, mostly in the regions connecting the center of the structure. The conserved bases and duplex stems generate an RNA with the intrinsic ability to cleave.

An active hammerhead can also be generated by pairing an RNA representing one side of the structure with an RNA representing the other side. The lower part of Figure 23.13 shows an example of a hammerhead generated by hybridizing a 19 nt molecule with a 24 nt molecule. The hybrid mimics the hammerhead structure, with the omission of loops I and III. We may regard the top (24 nt) strand of this hybrid as comprising the "substrate" and the bottom (19 nt) strand as comprising the "enzyme." When the 19 nt RNA is mixed with an excess of the 24 nt RNA, multiple copies of the 24 nt RNA are cleaved. This suggests that there is a cycle of 19 nt–24 nt pairing, cleavage, dissociation of the cleaved fragments from the 19 nt RNA, and pairing of the 19 nt RNA with a new 24 nt substrate. The 19 nt RNA is, therefore, a ribozyme with endonuclease activity. The parameters of the reaction are similar to those of other RNA-catalyzed reactions (see Figure 23.8).

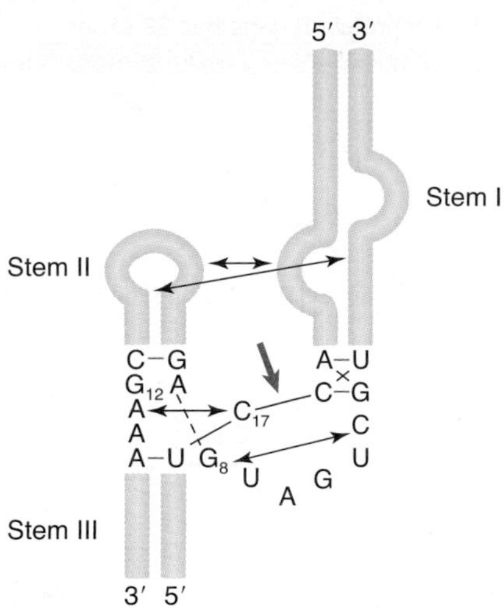

FIGURE 23.14 The hammerhead ribozyme structure is held in an active tertiary conformation by interactions between stem loops, indicated by arrows. The site of cleavage is marked with a red arrow.

Previously, the crystal structure of a minimal hammerhead ribozyme was solved. In the minimal structure, however, the architecture of the active site was such that it was unclear how catalysis could proceed. Recently, the crystal structure of the full-length hammerhead ribozyme from *Schistosoma mansoni*, a nonvirulent species, has been solved, and it gives insight into catalysis. This structure, schematically illustrated in **FIGURE 23.14**, reveals a critical tertiary interaction between a bulge in stem I and the loop of stem II. This interaction stabilizes the active site in a conformation such that G12 can deprotonate the 2′–OH of C17, the scissile bond, and create the 2′ attacking oxygen. G8, in turn, provides the hydrogen to stabilize the newly formed 5′–OH end of the 3′ cleavage product.

It is possible to design many enzyme-substrate combinations that can form minimal hammerhead structures. These structures have been used to demonstrate that introduction of the appropriate RNA molecules into a cell can allow the enzymatic reaction to occur *in vivo*. A ribozyme designed in this way essentially provides a highly specific restriction-like activity directed against an RNA target. By placing the ribozyme under control of a regulated promoter, it can be used in the same way as, for example, antisense constructs to specifically turn off expression of a target gene under defined circumstances.

KEY CONCEPTS

- Viroids and virusoids form a hammerhead structure that has a self-cleaving activity.
- Similar structures can be generated by pairing a substrate strand that is cleaved by an enzyme strand.
- When an enzyme strand is introduced into a cell, it can pair with a substrate strand target that is then cleaved.

CONCEPT AND REASONING CHECK

Describe how an engineered "hammerhead" ribozyme may be used to inhibit expression of a target gene.

23.9 RNA Editing Occurs at Individual Bases

A prime axiom of molecular biology is that the sequence of an mRNA can represent only what is encoded in the DNA. The central dogma envisaged a linear relationship in which a continuous sequence of DNA is transcribed into a sequence of mRNA that is, in turn, directly translated into polypeptide. The occurrence of interrupted genes and the removal of introns by RNA splicing introduces an additional step into the process of gene expression (see *Chapter 21, RNA Splicing and Processing*, for details). Briefly, splicing occurs at the RNA level, and it results in removal of noncoding sequences

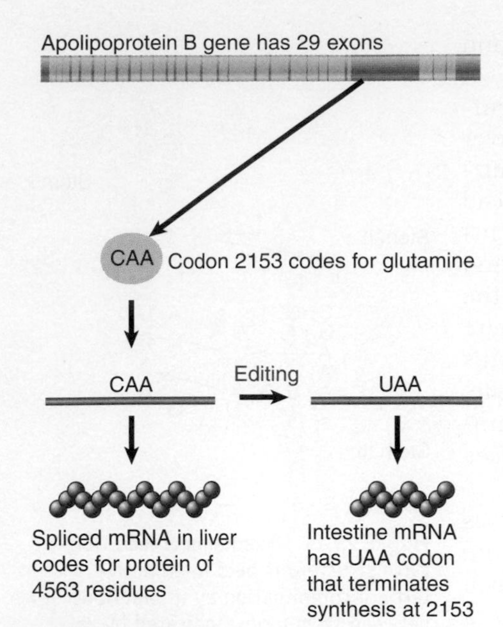

FIGURE 23.15 The sequence of the apo-B gene is the same in intestine and liver, but the sequence of the mRNA is modified by a base change that creates a termination codon in intestine.

(introns) that interrupt the coding sequences (exons) that are encoded in the DNA sequence. The process remains one of information transfer, though, in which the actual coding sequence in DNA remains unchanged.

Changes in the information encoded by DNA occur in some exceptional circumstances, most notably in the generation of new sequences encoding immunoglobulins in mammals and birds. These changes occur specifically in the somatic cells (B lymphocytes) in which immunoglobulins are synthesized (see *Chapter 18, Somatic Recombination and Hypermutation in the Immune System*). New information is generated in the DNA of an individual during the process of reconstructing an immunoglobulin gene, and information encoded in the DNA is changed by somatic mutation. The information in DNA continues to be faithfully transcribed into RNA.

RNA editing is a process in which *information is changed at the level of mRNA*. It is revealed by situations in which the coding sequence in an mRNA differs from the sequence of DNA from which it was transcribed. RNA editing occurs in two different situations, each with different causes. In mammalian cells there are cases in which a substitution occurs in an individual base in mRNA that can cause a change in the sequence of the polypeptide that is encoded. This base substitution is the result of deamination of either adenosine to become inosine or cytidine to become uridine. In trypanosome mitochondria, more widespread changes occur in transcripts of several genes when bases are systematically added or deleted.

FIGURE 23.15 summarizes the sequences of the apolipoprotein-B (*apo-B*) gene and mRNA in mammalian intestine and liver cells. The genome contains a single interrupted gene whose sequence is identical in all tissues, with a coding region of 4563 codons. This gene is transcribed into an mRNA that is translated into a protein of 512 kD representing the full coding sequence in the liver. A shorter form of the protein (~250 kDa) is synthesized in the intestine. This protein consists of the N-terminal half of the full-length protein. It is translated from an mRNA whose sequence is identical with that of liver except for a change from C to U at codon 2153. This substitution changes the codon CAA for glutamine into the ochre codon UAA for termination. Given that no alternative gene or exon is available in the genome to encode the new sequence and no change in the pattern of splicing can be discovered, we are forced to conclude that a change has been made directly in the sequence of the RNA transcript.

Another example is provided by glutamate receptors in rat brain. Editing at one position changes a glutamine codon in DNA into a codon for arginine in the mRNA. The change from glutamine to arginine affects the conductivity of the channel and, therefore, has an important effect on controlling ion flow through the neurotransmitter. At another position in the receptor, an arginine codon is converted to a glycine codon.

The events outlined for apo-B and glutamate receptors are the result of *deaminations* in which the amino group on the nucleotide ring is removed. The editing event in apo-B causes C_{2153} to be changed to U, and both changes in the glutamate receptor are from A to I (inosine). Deaminations in apolipoprotein B are catalyzed by the cytidine deaminase APOBEC (*apo*lipoprotein *B* mRNA editing *e*nzyme *c*omplex), whereas deaminations in the glutamate receptor are performed by *a*denosine *d*eaminases *a*cting on *R*NA (termed ADARs). This type of editing appears to occur largely in the nervous system. There are 16 (potential) targets for ADARs in *Drosophila melanogaster,* and all are genes involved in neurotransmission. In many cases, the editing event changes an amino acid at a functionally important position in the protein.

Enzymes that undertake deamination as such often have broad specificity—for example, the best-characterized adenosine deaminase acts on any A residues in a duplexed RNA region. Deamination of adenosine and cytidine in RNA, however, displays specificity. Editing enzymes are related to the general deaminases but have other regions or additional subunits that control their specificity. In the case of apo-B editing, the catalytic subunit of an editing complex is related to bacterial cytidine deaminase but has an additional RNA-binding region that helps to recognize the specific target site for editing. A special adenosine deaminase enzyme recognizes the target sites in the glutamate receptor RNA, and similar events occur in a serotonin receptor RNA. The complex may recognize a particular region of secondary structure in a manner analogous to tRNA-modifying enzymes or could directly recognize a nucleotide sequence. The development of an *in vitro* system for the apo-B editing event suggests that a relatively small sequence (~26 nucleotides) surrounding the editing site provides a sufficient target. **FIGURE 23.16** shows that in the case of the RNA for the glutamate receptor, GluR-B, a base-paired region that is necessary for recognition of the target site, is formed between the edited region in the exon and a complementary sequence in the downstream intron. A pattern of mispairing within the duplex region is necessary for specific recognition. Thus different editing systems may have different requirements for sequence specificity in their substrates.

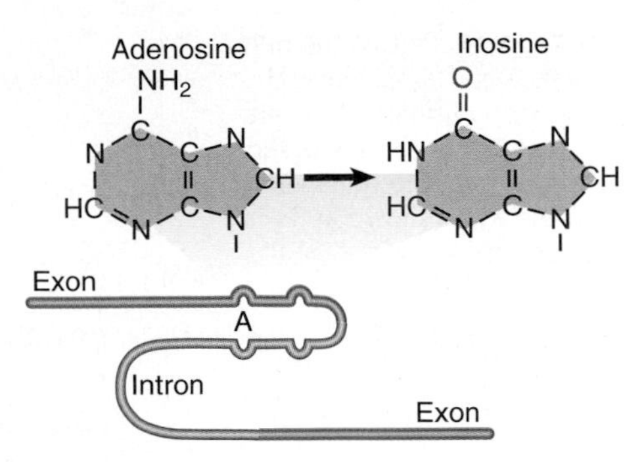

FIGURE 23.16 Editing of mRNA occurs when a deaminase acts on an adenine in an imperfectly paired RNA duplex region.

KEY CONCEPT

- Apolipoprotein-B and glutamate receptors have site-specific deaminations catalyzed by cytidine and adenosine deaminases that change the coding sequence.

CONCEPT AND REASONING CHECK

The product of a gene may be different from the predicted product based on the DNA sequence of the gene for several reasons, including RNA editing. What are the other reasons, and how can RNA editing be demonstrated to be the reason for this difference for a particular gene?

23.10 RNA Editing Can Be Directed by Guide RNAs

Another type of editing is revealed by dramatic changes in sequence in the products of several genes of trypanosome mitochondria. In the first case to be discovered, the sequence of the cytochrome oxidase subunit II protein has an internal frameshift that is not predicted based on the sequence of the *coxII* gene. The sequences of the gene and protein given in **FIGURE 23.17** are conserved in several trypanosome species, so the method of RNA editing is not unique to a single organism.

The discrepancy between the sequence of the *coxII* gene and the protein product is due to an RNA editing event. The *coxII* mRNA has an insert of an additional four nucleotides (all uridines) around the site of frameshift. The insertion establishes the proper reading frame for the protein. No second *coxII* gene carrying the frameshift sequence can be discovered; we are forced to conclude that the extra bases are inserted during or after transcription. A similar discrepancy between mRNA and genomic sequences is found in genes of the SV5 and measles paramyxoviruses, in these cases involving the addition of G residues in the mRNA.

Similar editing of RNA sequences occurs for other genes and includes deletions as well as additions of uridine. The extraordinary case of the cytochrome c oxidase III

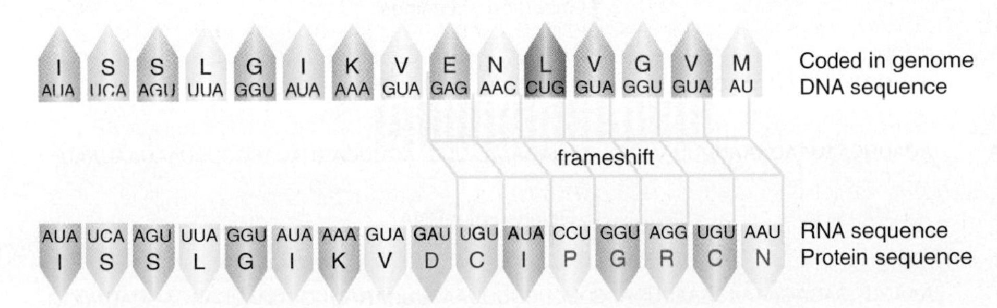

FIGURE 23.17 The mRNA for the trypanosome *coxII* gene has a frameshift relative to the DNA; the correct reading frame observed in the protein is created by the insertion of four uridines.

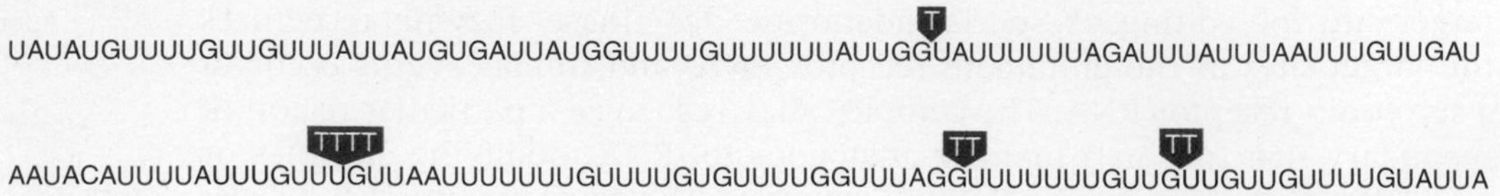

FIGURE 23.18 Part of the mRNA sequence of *T. brucei coxIII* shows many uridines that are not encoded in the DNA (shown in red) or that are removed from the RNA (shown as Ts in blue boxes).

T
UAUAUGUUUUGUUGUUUAUUAUGUGAUUAUGGUUUUGUUUUUUAUUGGUAUUUUUUAGAUUUAUUUAAUUUGUUGAU

TTTT TT TT
AAUACAUUUUAUUUGUUUGUUAAUUUUUUUGUUUUGUGUUUUGGUUUAGGUUUUUUUGUUGUUGUUGUUUUGUAUUA

(*coxIII*) gene of *Trypanosoma brucei* is summarized in **FIGURE 23.18**. More than half of the residues in the mRNA consist of uridines that are not encoded by the gene. Comparison between the genomic DNA and the mRNA shows that no stretch longer than seven nucleotides is represented in the mRNA without alteration, and runs of uridine up to seven bases long are inserted. The information for the specific insertion of uridines is provided by a *guide RNA*.

▸ **guide RNA** A small RNA whose sequence is complementary to the sequence of an RNA that has been edited. It is used as a template for changing the sequence of the preedited RNA by inserting or deleting nucleotides.

Guide RNA contains a sequence that is complementary to the correctly edited mRNA. **FIGURE 23.19** shows a model for its action in the cytochrome b gene of another trypanosome, *Leishmania*. The sequence at the top of the figure shows the original transcript, or preedited RNA. Gaps show where bases will be inserted in the editing process. Eight uridines must be inserted into this region to create the final mRNA sequence. The guide RNA is complementary to the mRNA for a significant distance, including and surrounding the edited region. Typically the complementarity is more extensive on the 3′ side of the edited region and is rather short on the 5′ side. Pairing between the guide RNA and the preedited RNA leaves gaps where unpaired A residues in the guide RNA do not find complements in the preedited RNA. The guide RNA provides a template that allows the missing U residues to be inserted at these positions in a process described below. When the reaction is completed the guide RNA separates from the mRNA, which becomes available for translation.

Specification of the final edited sequence can be quite complex. In the example of *Leishmania* cytochrome b, a lengthy stretch of the transcript is edited by the insertion of a total of 39 U residues, which appears to require two guide RNAs that act at adjacent sites. The first guide RNA pairs at the 3′-most site, and the edited sequence then becomes a substrate for further editing by the next guide RNA. The guide RNAs are encoded as independent transcription units. **FIGURE 23.20** shows a map of the relevant region of the *Leishmania* mitochondrial DNA. It includes the gene for

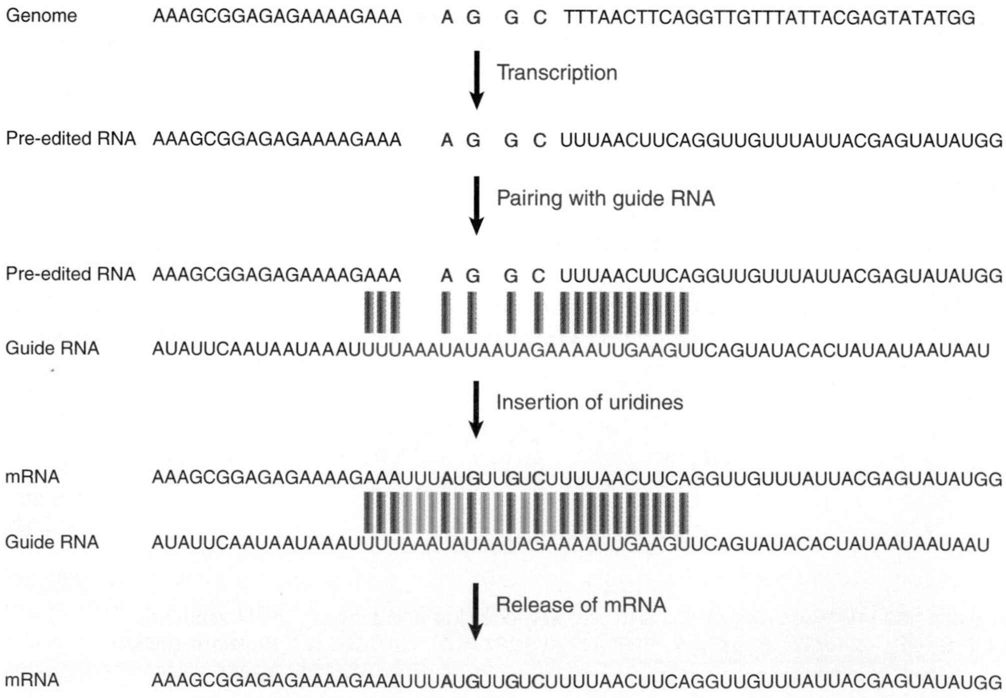

FIGURE 23.19 Preedited RNA base pairs with a guide RNA on both sides of the region to be edited. The guide RNA provides a template for the insertion of uridines. The mRNA produced by the insertions is complementary to the guide RNA.

Genome AAAGCGGAGAGAAAAGAAA A G G C TTTAACTTCAGGTTGTTTATTACGAGTATATGG

Transcription

Pre-edited RNA AAAGCGGAGAGAAAAGAAA A G G C UUUAACUUCAGGUUGUUUAUUACGAGUAUAUGG

Pairing with guide RNA

Pre-edited RNA AAAGCGGAGAGAAAAGAAA A G G C UUUAACUUCAGGUUGUUUAUUACGAGUAUAUGG

Guide RNA AUAUUCAAUAAUAAAUUUUAAAUAUAAUAGAAAAUUGAAGUUCAGUAUACACUAUAAUAAUAAU

Insertion of uridines

mRNA AAAGCGGAGAGAAAAGAAAUUUAUGUUGUCUUUUAACUUCAGGUUGUUUAUUACGAGUAUAUGG

Guide RNA AUAUUCAAUAAUAAAUUUUAAAUAUAAUAGAAAAUUGAAGUUCAGUAUACACUAUAAUAAUAAU

Release of mRNA

mRNA AAAGCGGAGAGAAAAGAAAUUUAUGUUGUCUUUUAACUUCAGGUUGUUUAUUACGAGUAUAUGG

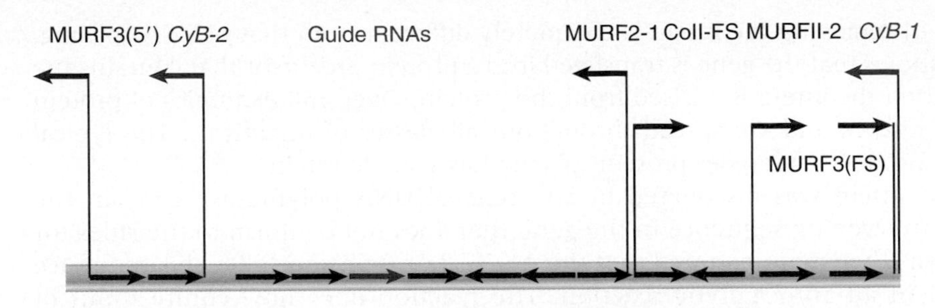

FIGURE 23.20 The *Leishmania* genome contains genes encoding preedited RNAs interspersed with units that encode the guide RNAs required to generate the correct mRNA sequences. Some genes have multiple guide RNAs. *CyB* is the gene for preedited cytochrome b, and *CyB-1* and *CyB-2* are genes for the guide RNAs involved in its editing.

cytochrome b, which encodes the preedited sequence, and two regions that specify guide RNAs. Genes for the major coding regions and for their guide RNAs are interspersed.

The characterization of intermediates that are partially edited suggests that the reaction proceeds along the preedited RNA in the 3′–5′ direction. The guide RNA determines the specificity of uridine insertions by its pairing with the preedited RNA.

Editing of uridines is catalyzed by a 20S enzyme complex called the *editosome* that is composed of about 20 proteins and contains an endonuclease, a terminal uridyltransferase (TUTase), a 3′–5′ U-specific exonuclease (exoUase), and an RNA ligase. As illustrated in **FIGURE 23.21**, the editosome binds the guide RNA and uses it to pair with the preedited mRNA. The substrate RNA is cleaved at a site that is presumably identified by the absence of pairing with the guide RNA; a uridine is inserted or deleted to base pair with the guide RNA and then the substrate RNA is ligated. Uridine triphosphate (UTP) provides the source for the uridyl residue. It is added by the TUTase activity. Deletion of U residues is mediated by an exoUase which functions in concert with a 3′ phosphatase to allow the newly edited RNA construct to religate.

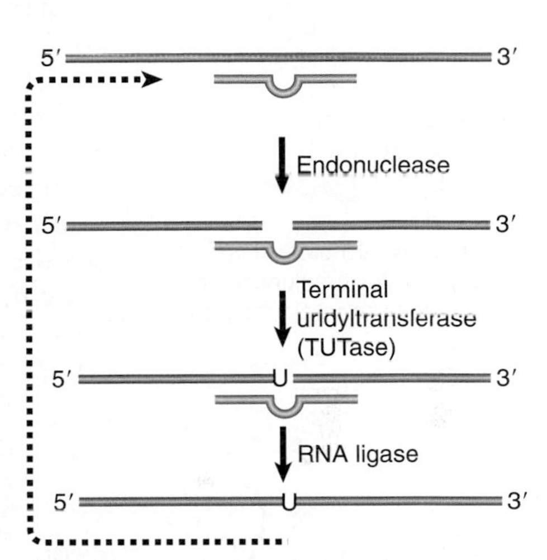

FIGURE 23.21 Addition or deletion of U residues occurs by cleavage of the RNA, removal or addition of the U, and ligation of the ends. The reactions are catalyzed by a complex of enzymes under the direction of guide RNA.

The structures of partially edited molecules suggest that the U residues are added one at a time rather than in groups. It is possible that the reaction proceeds through successive cycles in which U residues are added, tested for complementarity with the guide RNA, retained if acceptable, and removed if not so that the construction of the correct edited sequence occurs gradually. We do not know whether the same types of reaction are involved in editing reactions that add C residues.

KEY CONCEPTS

- Extensive RNA editing in trypanosome mitochondria occurs by insertions or deletions of uridine.
- The substrate RNA base pairs with a guide RNA on both sides of the region to be edited.
- The guide RNA provides the template for addition (or less often, deletion) of uridines.
- Editing is catalyzed by the editosome, a complex of endonuclease, exonuclease, terminal uridyltransferase activity, and RNA ligase.

CONCEPT AND REASONING CHECK

Describe the process by which editing of the *Leishmania* cytochrome b RNA occurs.

23.11 Protein Splicing Is Autocatalytic

Protein splicing has the same effect as RNA splicing: a sequence that is represented within the gene fails to be represented in the protein. The parts of the protein are named by analogy with RNA splicing: **exteins** are the sequences that are represented in the mature protein, and **inteins** are the sequences that are removed. The

- **protein splicing** The autocatalytic process by which an intein is removed from a protein and the exteins on either side become connected by a standard peptide bond.
- **extein** A sequence that remains in the mature protein that is produced by processing a precursor via protein splicing.
- **intein** The part that is removed from a protein that is processed by protein splicing.

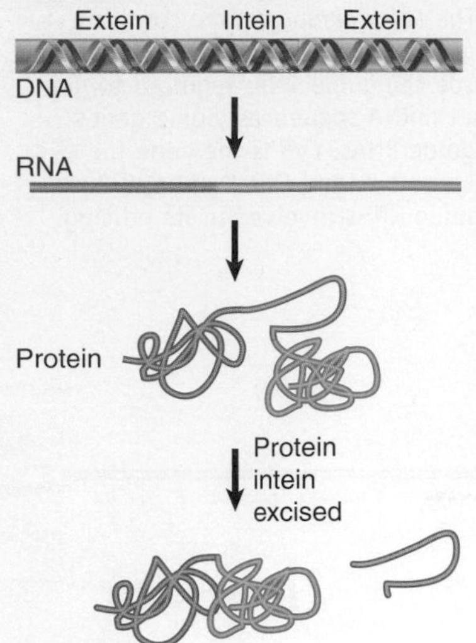

FIGURE 23.22 In protein splicing, the exteins are connected by removing the intein from the protein.

mechanism of removing the intein is completely different from that of RNA splicing. **FIGURE 23.22** shows that the gene is translated into a protein precursor that contains the intein and then the intein is excised from the protein. Over 350 examples of protein splicing are known and are spread throughout all classes of organisms. The typical gene whose product undergoes protein splicing has a single intein.

The first intein was discovered in an archaeal DNA polymerase gene in the form of an intervening sequence in the gene that does not conform to the rules for introns. It was then demonstrated that the purified protein can splice this sequence out of itself in an autocatalytic reaction. The reaction does not require input of energy and occurs through the series of bond rearrangements shown in **FIGURE 23.23**. The reaction is a function of the intein, although its efficiency can be influenced by the exteins.

The first reaction is an attack by an –OH or –SH side chain of the first amino acid in the intein on the peptide bond that connects it to the first extein. This transfers the extein from the amino-terminal group of the intein to an N-O or N-S acyl connection. This bond is then attacked by the –OH or –SH side chain of the first amino acid in the second extein. The result is to transfer extein1 to the side chain of the amino-terminal acid of extein2. Finally, the C-terminal asparagine of the intein cyclizes, and the terminal –NH of extein2 attacks the acyl bond to replace it with a conventional peptide bond. Each of these reactions can occur spontaneously at very low rates, but their occurrence in a coordinated manner that is rapid enough to achieve protein splicing requires catalysis by the intein.

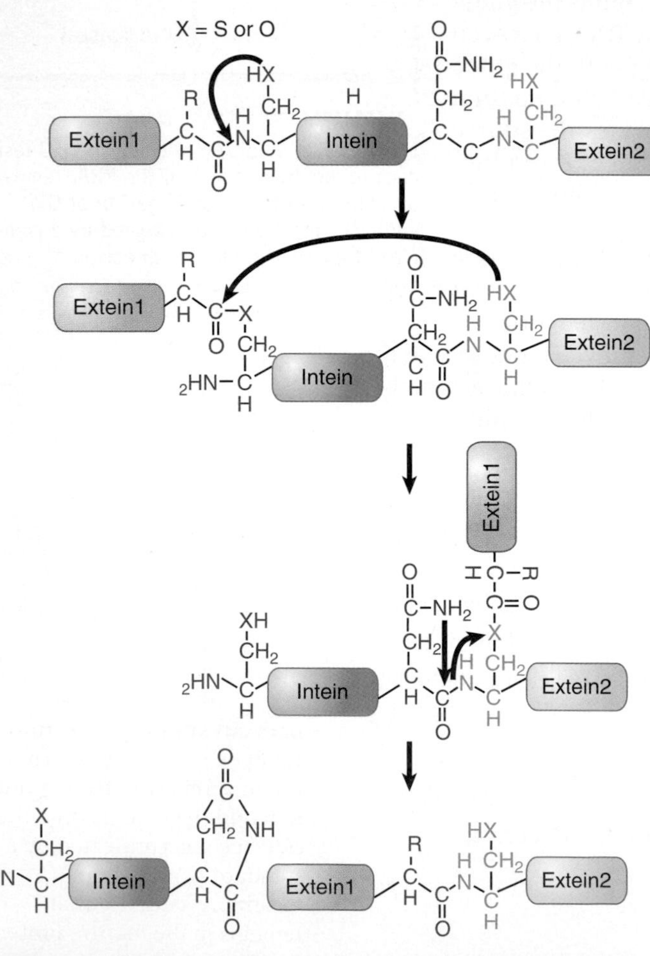

FIGURE 23.23 Bonds are rearranged through a series of transesterifications involving the -OH groups of serine or threonine or the -SH group of cysteine until the exteins are connected by a peptide bond and the intein is released with a circularized C-terminus.

Inteins have characteristic features. They are found as in-frame insertions into coding sequences. They can be recognized as such because of the existence of homologous genes that lack the insertion. They have an N-terminal serine or cysteine (to provide the –XH side chain) and a C-terminal asparagine. A typical intein has a sequence of ~150 amino acids at the N-terminal end and ~50 amino acids at the C-terminal end that are involved in catalyzing the protein splicing reaction. The sequence in the center of the intein can have other functions.

An extraordinary feature of many inteins is that they have homing endonuclease activity. A homing endonuclease cleaves a target DNA to create a site into which the DNA sequence encoding the intein can be inserted (see Figure 23.10 in *Section 23.5, Some Group I Introns Encode Endonucleases That Sponsor Mobility*). The protein splicing and homing endonuclease activities of an intein are independent.

We do not really understand the connection between the presence of both these activities in an intein, but two types of model have been suggested. One is to suppose that there was originally some sort of connection between the activities but that they have since become independent and some inteins have lost the homing endonuclease. The other is to suppose that inteins may have originated as protein splicing units, most of which (for unknown reasons) were subsequently invaded by homing endonucleases. This is consistent with the fact that homing endonucleases appear to have invaded other types of units as well, including, most notably, group I introns.

KEY CONCEPTS

- An intein has the ability to catalyze its own removal from a protein in such a way that the flanking exteins are connected.
- Protein splicing is catalyzed by the intein.
- Most inteins have two independent activities: protein splicing and a homing endonuclease.

CONCEPT AND REASONING CHECK

How can an intein be empirically distinguished from an intron?

23.12 Summary

Self-splicing is a property of two groups of introns, which are widely dispersed in unicellular/oligocellular eukaryotes, prokaryotic systems, and mitochondria. The information necessary for the reaction resides in the intron sequence, although the reaction is actually assisted by proteins *in vivo*. For both group I and group II introns, the reaction requires formation of a specific secondary/tertiary structure involving short consensus sequences. Group I intron RNA creates a structure in which the substrate sequence is held by the IGS region of the intron, and other conserved sequences generate a guanine nucleotide-binding site. It occurs by a transesterification involving a guanosine residue as cofactor. No input of energy is required. The guanosine breaks the bond at the 5′ exon–intron junction and becomes linked to the intron; the hydroxyl at the free end of the exon then attacks the 3′ exon–intron junction. The intron cyclizes and loses the guanosine and the terminal 15 bases. A series of related reactions can be catalyzed via attacks by the terminal G-OH residue of the intron on internal phosphodiester bonds. By providing appropriate substrates, it has been possible to engineer ribozymes that perform a variety of catalytic reactions, including nucleotidyl transferase activities.

Some group I and group II mitochondrial introns have open reading frames. The proteins encoded by group I introns are endonucleases that make double-stranded cleavages in target sites in DNA. The endonucleolytic cleavage initiates a gene conversion process in which the sequence of the intron itself is copied into the target site. The proteins encoded by group II introns include an endonuclease activity that initiates the transposition process and a reverse transcriptase that enables an RNA copy of the intron to be copied into the target site. These types of introns probably originated by insertion events. The proteins encoded by both groups of introns may include maturase activities that assist splicing of the intron by stabilizing the formation of the secondary/tertiary structure of the active site.

Virusoid RNAs can undertake self-cleavage at a "hammerhead" structure. Hammerhead structures can form between a substrate RNA and a ribozyme RNA, which allows cleavage to be directed at highly specific sequences. These reactions support the view that RNA can form specific active sites that have catalytic activity.

RNA editing changes the sequence of an RNA after or during its transcription. The changes are required to create a meaningful coding sequence. Substitutions of individual bases occur in mammalian systems; they take the form of deaminations in which C is converted to U or A is converted to I. A catalytic subunit related to cytidine or adenosine deaminase functions as part of a larger complex that has specificity for a particular target sequence.

Additions and deletions (most often of uridine) occur in trypanosome mitochondria and in paramyxoviruses. Extensive editing reactions occur in trypanosomes in which as many as half of the bases in an mRNA are derived from editing. The editing reaction uses a template consisting of a guide RNA that is complementary to the mRNA sequence. The reaction is catalyzed by the editosome, an enzyme complex that includes an endonuclease, exonuclease terminal uridyltransferase, and RNA ligase, using free nucleotides as the source for additions, or releasing cleaved nucleotides following deletion.

Protein splicing is an autocatalytic reaction that occurs by bond transfer reactions and input of energy is not required. The intein catalyzes its own splicing out of the flanking exteins. Many inteins have a homing endonuclease activity that is independent of the protein splicing activity.

CHAPTER QUESTIONS

1. Which class of introns has been modified to carry out catalytic activities other than the original activity?
- **A.** group I introns
- **B.** group II introns
- **C.** nuclear introns
- **D.** both group I and group II introns

2. What base is most commonly inserted into transcripts of unicellular eukaryote mitochondria by RNA editing?
- **A.** A
- **B.** U
- **C.** G
- **D.** C

3. Which class of introns is found in bacteria and rRNA genes in unicellular eukaryotes?
- **A.** group I introns
- **B.** group II introns
- **C.** nuclear introns
- **D.** both group I and group II introns

4. Group I introns, after excision from the precursor RNA:
- **A.** form lariat structures.
- **B.** form Y-shaped structures.
- **C.** form circular structures.
- **D.** remain linear.

5. As a required cofactor for group I intron splicing, a guanine nucleotide with a _________ group must be present.
- **A.** 5′ phosphate
- **B.** 3′ phosphate
- **C.** 5′–OH
- **D.** 3′–OH

6. A subset of which class(es) of introns contain(s) mobile introns that encode endonucleases that facilitate mobility of the intron to new sites?
- **A.** group I introns
- **B.** group II introns
- **C.** nuclear introns
- **D.** group I and group II introns

7. The required sequence elements needed for group II intron splicing are:
- **A.** short conserved sequences at the very ends of the intron.
- **B.** conserved sequence elements in the center of the intron.
- **C.** a characteristic secondary structure based on conserved internal inverted repeats.
- **D.** no conserved sequence elements or secondary structure.

8. Small infectious plant RNAs that function independently without encapsidation are called:
   A. viruses.
   B. viroids.
   C. virusoids.
   D. satellite RNAs.
9. Editing of apolipoprotein-B transcripts in mammalian intestine and liver involves:
   A. conversion of a U to a C.
   B. conversion of a C to a U.
   C. insertion of several U bases.
   D. deletion of several C bases.
10. RNA editing in trypanosome mitochondria involves:
   A. conversion of a U to a C.
   B. conversion of a C to a U.
   C. insertion or deletion of several U bases.
   D. insertion or deletion of several C bases.

KEY TERMS

autosplicing (self-splicing)
extein
guide RNA
intein
intron homing
maturase
protein splicing
RNA editing
riboswitch
ribozyme
viroid
virusoid

FURTHER READING

Cech, T. R. (1990). Self-splicing of group I introns. *Annu. Rev. Biochem.* **59**, 543–568. A review of the mechanism of this process.

Cochrane, J. C., and Strobel, S. A. (2008). Catalytic strategies of self-cleaving ribozymes. *Acc. Chem. Res.* **41**, 1027–1035.

Doherty, E. A., and Doudna, J. A. (2000). Ribozyme structures and mechanisms. *Annu. Rev. Biochem.* **69**, 597–615. A comparison of the structures of four ribozymes.

Haugen, P., Reeb, V., Lutzoni, F., and Bhatacharya, D. (2004). The evolution of homing endonuclease genes and group I introns in nuclear rDNA. *Mol. Biol. Evol.* **21**, 129–140.

Hoopengardner, B. (2006). Adenosine-to-inosine RNA editing: perspectives and predictions. *Mini-Rev. Med. Chem.* **6**, 1213–1216.

Lambowitz, A. M., and Zimmerly, S. (2004). Mobile group II introns. *Annu. Rev. Genet.* **38**, 1–35. A review of the molecular mechanisms for the mobility of group II introns and the evolutionary relationship between them, eukaryotic retroposons, and eukaryotic spliceosomal introns.

Liu, X.-Q. (2000). Protein-splicing intein: genetic mobility, origin, and evolution. *Annu. Rev. Genet.* **34**, 61–76. How the structure and splicing mechanism of inteins suggest the origin of these elements.

Paulus, H. (2000). Protein splicing and related forms of protein autoprocessing. *Annu. Rev. Biochem.* **69**, 447–496. A review of the process of protein splicing and a comparison to protein autoprocessing.

Saleh, L., and Perler, F. B. (2006). Protein splicing in *cis* and in *trans*. *Chem. Rec.* **6**, 183–193.

Stuart, K. D., Schnaufer, A., Ernst, N. L., and Panigrahi, A. K. (2005). Complex management: RNA editing in trypanosomes. *Trends Biochem. Sci.* **30**, 97–105.

Vicens, Q., and Cech, T. R. (2006). Atomic level architecture of group I introns revealed. *Trends Biochem. Sci.* **31**, 41–51.

Winkler, W. C., Nahvi, A., Roth, A., Collins, J. A., and Breaker, R. R. (2004). Control of gene expression by a natural metabolite-responsive ribozyme. *Nature* **428**, 281–286. A report of the discovery of a new class of ribozyme that inhibits expression of an enzyme by cleaving its mRNA. Through a negative feedback system, the ribozyme is activated by a metabolic product of the enzyme.

24

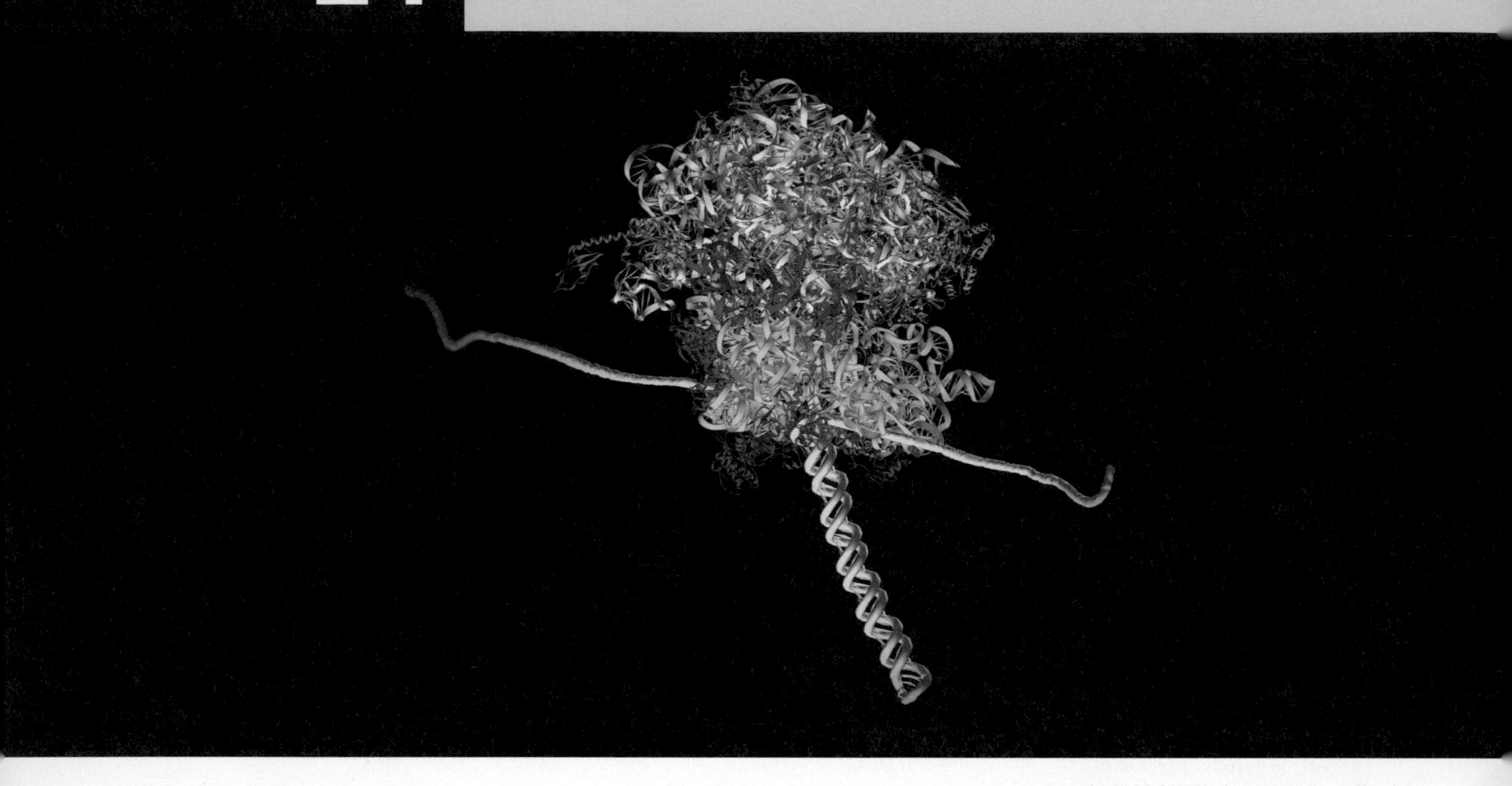

Translation

Translation of a single mRNA (in yellow) by an *E. coli* ribosome, one codon at a time. By fixing the ends of the mRNA, researchers can measure how long the ribosome spends at each codon and have learned that the ribosome often pauses after translocation cycles. Used with permission of Courtney Hodges, University of California, Berkeley, and Laura Lancaster, University of California, Santa Cruz. Photo courtesy of Carlos Bustamante, University of California, Berkeley.

CHAPTER OUTLINE

24.1 Introduction

An mRNA contains a series of codons that interact with the anticodons of aminoacyl-tRNAs so that a corresponding series of amino acids is incorporated into a polypeptide chain. The ribosome provides the environment for controlling the interaction between mRNA and aminoacyl-tRNA. The ribosome behaves like a small migrating factory that travels along the template, engaging in rapid cycles of peptide bond synthesis. tRNAs shoot in and out of the particle at an incredibly fast rate while depositing amino acids, and elongation factors cyclically associate with and dissociate from the ribosome. Together with its accessory factors, the ribosome provides the full range of activities required for all the steps of translation.

FIGURE 24.1 shows the relative dimensions of the components of the translation apparatus. The ribosome consists of two subunits that have specific roles in translation. Messenger RNA is associated with the small subunit; ~35 bases of the mRNA are bound at any time. The mRNA threads its way along the surface close to the junction of the subunits. Two tRNA molecules are active in translation at any moment, so polypeptide elongation involves reactions taking place at just two of the approximately ten codons covered by the ribosome. The two tRNAs are inserted into internal sites that stretch across the two ribosomal subunits. A third tRNA may remain on the ribosome after it has been used in translation before being recycled.

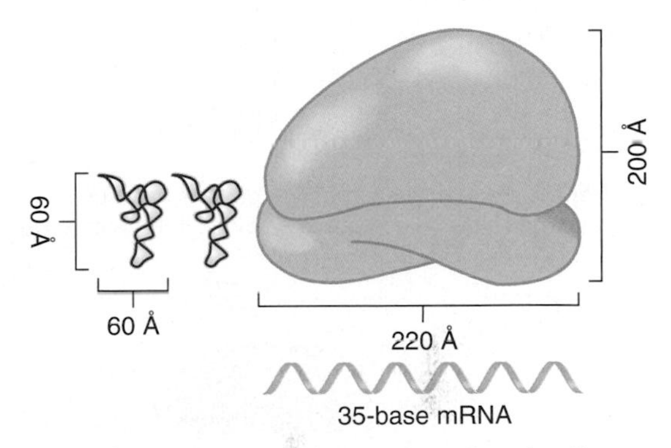

FIGURE 24.1 Size comparisons show that the ribosome is large enough to bind tRNAs and mRNA.

The basic form of the ribosome has been conserved in evolution, but there are appreciable variations in the overall size and proportions of RNA and protein in the ribosomes of bacteria, eukaryotic cytoplasm, and organelles. **FIGURE 24.2** compares the components of bacterial and mammalian ribosomes. Both are ribonucleoprotein particles that contain more RNA than protein. The ribosomal proteins are known as *r-proteins.*

Ribosomes		rRNAs	r-proteins
Bacterial (70S) mass: 2.5 MDa 66% RNA	50S	23S = 2904 bases 5S = 120 bases	31
	30S	16S = 1542 bases	21
Mammalian (80S) mass: 4.2 MDa 60% RNA	60S	28S = 4718 bases 5.8S = 160 bases 5S = 120 bases	49
	40S	18S = 1874 bases	33

FIGURE 24.2 Ribosomes are large ribonucleoprotein particles that contain more RNA than protein and dissociate into large and small subunits.

Each of the ribosome subunits contains a major rRNA and a number of small proteins. The large subunit may also contain smaller rRNA(s). In *E. coli,* the small (30S) subunit consists of the 16S rRNA and 21 r-proteins. The large (50S) subunit contains 23S rRNA, the small 5S RNA, and 31 r-proteins. With the exception of one protein present at four copies per ribosome, there is one copy of each protein. The major RNAs constitute the larger part of the mass of the bacterial ribosome. Their presence is pervasive, and most of or all the ribosomal proteins actually contact rRNA (see **FIGURE 24.3**). So the major rRNAs form what is sometimes thought of as the "backbone" of each subunit—a continuous thread whose presence dominates the structure and determines the positions of the ribosomal proteins.

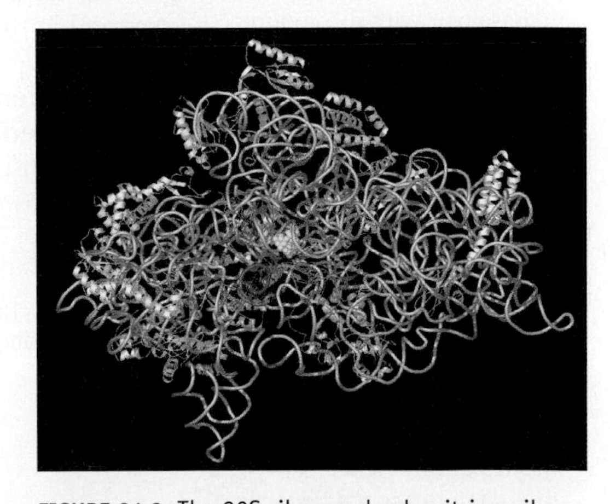

FIGURE 24.3 The 30S ribosomal subunit is a ribonucleoprotein particle. Ribosomal proteins are white and rRNA is light blue. Courtesy of Dr. Kalju Kahn.

The ribosomes in the cytosol of eukaryotes are larger than those of bacteria. The total content of both RNA and protein is greater; the major rRNA molecules are longer (called 18S and 28S rRNAs), and there are more proteins. RNA is still the predominant component by mass.

The ribosomes of mitochondria and chloroplasts are distinct from the ribosomes of the cytosol, and they take varied forms. In some cases, they are almost the size of bacterial ribosomes and have 70% RNA; in other cases, they are only 60S and have <30% RNA.

The ribosome possesses several active centers, each of which is constructed from a group of proteins associated with a region of ribosomal RNA. The active centers require the direct participation

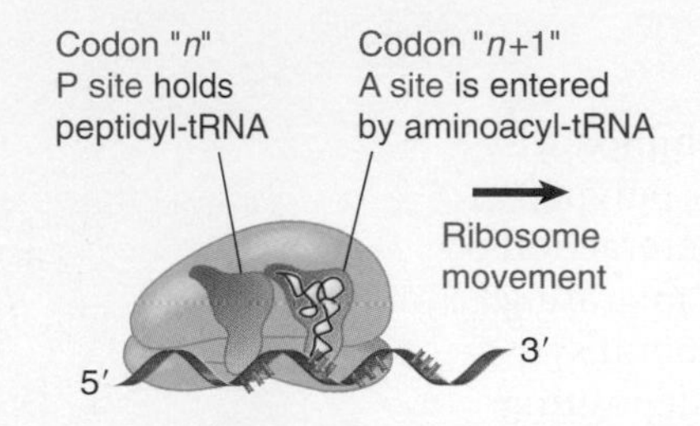

1 Before peptide bond formation peptidyl-tRNA occupies P site; aminoacyl-tRNA occupies A site

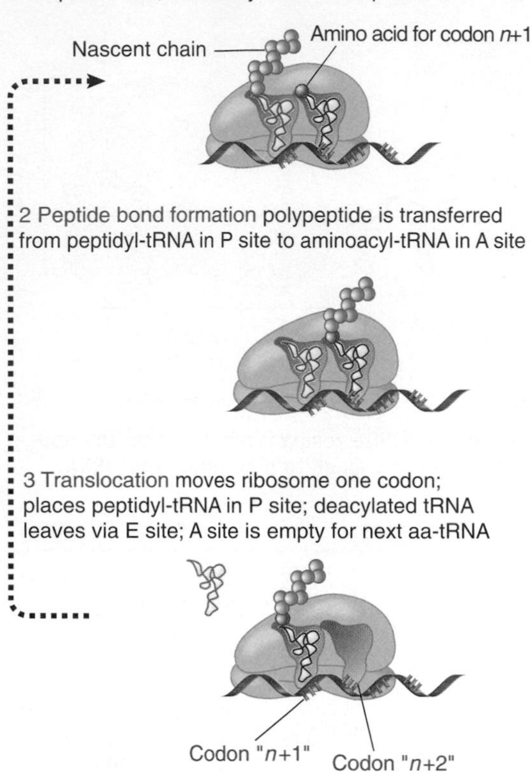

FIGURE 24.4 The ribosome has two sites for binding charged tRNA.

of rRNA in a structural or even catalytic role (where the RNA functions as a ribozyme) with proteins supporting these functions in secondary roles. Some catalytic functions require individual proteins, but none of the activities can be reproduced by isolated proteins or groups of proteins; they function only in the context of the ribosome.

In analyzing the functions of structural components of the ribosome, there are two experimental approaches. First, the effects of mutations in genes for particular ribosomal proteins or at specific bases in rRNA genes shed light on the participation of these molecules in particular reactions. Second, structural analysis, including direct modification of components of the ribosome and comparisons to identify conserved features in rRNA, identifies the physical locations of components involved in particular functions.

24.2 Translation Occurs by Initiation, Elongation, and Termination

An amino acid is brought to the ribosome by an aminoacyl-tRNA. Its addition to the growing polypeptide chain occurs by an interaction with the tRNA that brought the previous amino acid. Each of these tRNAs lies in a distinct site on the ribosome. FIGURE 24.4 shows that the two sites have different features:

- An incoming aminoacyl-tRNA binds to the **A site**. Prior to the entry of aminoacyl-tRNA, the site exposes the codon representing the next amino acid to be added to the chain.
- The codon representing the most recent amino acid to have been added to the nascent polypeptide chain lies in the **P site**. This site is occupied by **peptidyl-tRNA**, a tRNA carrying the nascent polypeptide chain.

FIGURE 24.5 shows that the aminoacyl end of the tRNA is located on the large subunit, whereas the anticodon at the other end interacts with the mRNA bound by the small subunit. So the P and A sites each extend across both ribosomal subunits.

For a ribosome to form a peptide bond, it must be in the state shown in step 1 in Figure 24.4, when peptidyl-tRNA is in the P site and aminoacyl-tRNA is in the A site. Peptide bond formation occurs when the polypeptide carried by the peptidyl-tRNA is transferred to the amino acid carried by the aminoacyl-tRNA. This reaction is catalyzed by the large subunit of the ribosome.

Transfer of the polypeptide generates the ribosome shown in step 2, in which the **deacylated tRNA**, lacking any amino acid, lies in the P site, and a new peptidyl-tRNA is in the A site. This peptidyl-tRNA is one amino acid residue longer than the peptidyl-tRNA that had been in the P site in step 1.

FIGURE 24.5 The P and A sites position the two interacting tRNAs across both ribosome subunits.

Aminoacyl ends of tRNA interact within large ribosome subunit

Anticodons are bound to adjacent triplets on mRNA in small ribosome subunit

- **A site** The site of the ribosome that an aminoacyl-tRNA enters to base pair with the codon.
- **P site** The site in the ribosome that is occupied by peptidyl-tRNA, the tRNA carrying the nascent polypeptide chain, still paired with the codon to which it bound in the A site.
- **peptidyl-tRNA** The tRNA to which the nascent polypeptide chain has been transferred following peptide bond synthesis during polypeptide translation.
- **deacylated tRNA** tRNA that has no amino acid or polypeptide chain attached because it has completed its role in translation and is ready to be released from the ribosome.

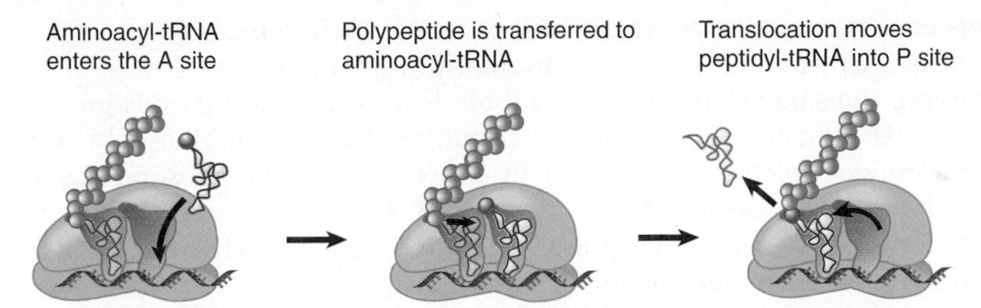

FIGURE 24.6 Aminoacyl-tRNA enters the A site, receives the polypeptide chain from peptidyl-tRNA, and is transferred into the P site for the next cycle of elongation.

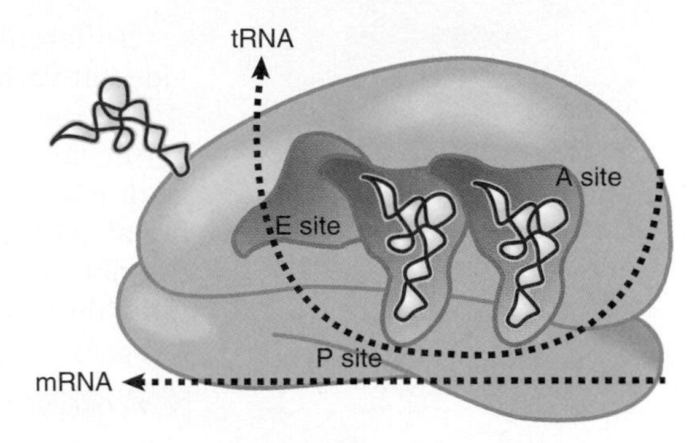

FIGURE 24.7 tRNA and mRNA move through the ribosome in the same direction.

The ribosome now moves one triplet along the messenger RNA. This stage is called **translocation**. The movement transfers the deacylated tRNA out of the P site and moves the peptidyl-tRNA into the P site (see step 3 in the figure). The next codon to be translated now lies in the A site, ready for a new aminoacyl-tRNA to enter, when the cycle will be repeated. **FIGURE 24.6** summarizes the interaction between tRNAs and the ribosome.

▸ **translocation** The movement of the ribosome one codon along mRNA after the addition of each amino acid to the polypeptide chain.

The deacylated tRNA leaves the ribosome via another tRNA-binding site, the **E site**. This site is transiently occupied by the tRNA *en route* between leaving the P site and being released from the ribosome into the cytosol. Thus the flow of tRNA is into the A site, through the P site, and out through the E site (see also Figure 24.23 in *Section 24.10*). **FIGURE 24.7** compares the movement of tRNA and mRNA, which may be thought of as a sort of ratchet in which the reaction is driven by the codon-anticodon interaction.

▸ **E site** The site of the ribosome that briefly holds deacylated tRNAs before their release.

Translation falls into the three stages shown in **FIGURE 24.8**:

- **Initiation** involves the reactions that precede formation of the peptide bond between the first two amino acids of the polypeptide. It requires the ribosome to bind to the mRNA, which forms an initiation complex that contains the first aminoacyl-tRNA. This is a relatively slow step in translation and usually determines the rate at which an mRNA is translated.
- **Elongation** includes all the reactions from synthesis of the first peptide bond to addition of the last amino acid. Amino acids are added to the chain one at a time; the addition of an amino acid is the most rapid step in translation.
- **Termination** encompasses the steps that are needed to release the completed polypeptide chain; at the same time, the ribosome dissociates from the mRNA.

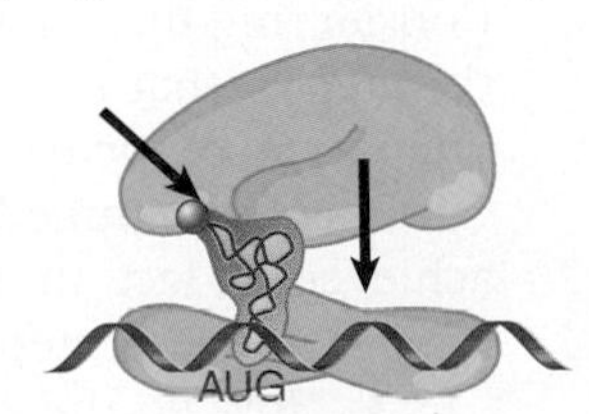

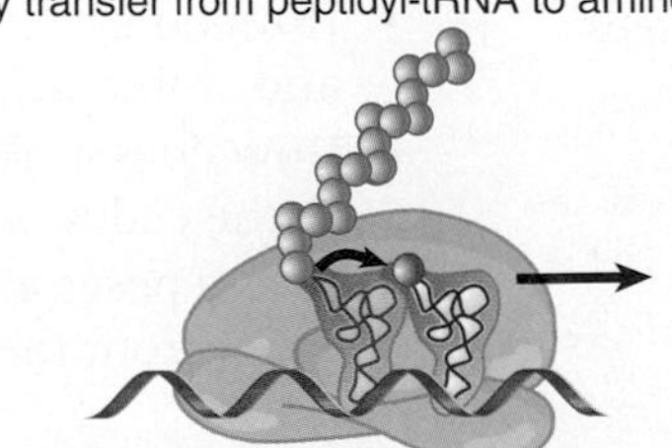

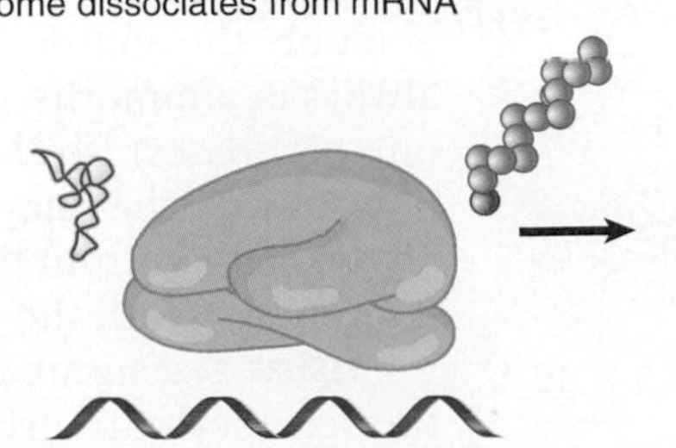

FIGURE 24.8 Translation falls into three stages.

▸ **initiation** The stages of translation up to synthesis of the first peptide bond of the polypeptide.

▸ **elongation** The stage in a macromolecular synthesis reaction (replication, transcription, or translation) in which the nucleotide or polypeptide chain is extended by the addition of individual subunits.

▸ **termination** A separate reaction that ends a macromolecular synthesis reaction (replication, transcription, or translation), by stopping the addition of subunits, and (typically) causing disassembly of the synthetic apparatus.

Different sets of accessory factors assist the ribosome at each stage. Energy is provided at various stages by the hydrolysis of guanine triphosphate (GTP).

During initiation, the small ribosomal subunit binds to mRNA and then is joined by the large subunit. During elongation, the mRNA moves through the ribosome and is translated in nucleotide triplets. (Although we usually talk about the ribosome moving along mRNA, it is more realistic to think in terms of the mRNA moving through the ribosome.) At termination, the polypeptide is released, mRNA is released, and the individual ribosomal subunits dissociate and can be used again.

KEY CONCEPTS

- The ribosome has three tRNA-binding sites.
- An aminoacyl-tRNA enters the A site.
- Peptidyl-tRNA is bound in the P site.
- Deacylated tRNA exits via the E site.
- An amino acid is added to the polypeptide chain by transferring the polypeptide from peptidyl-tRNA in the P site to aminoacyl-tRNA in the A site.

CONCEPT AND REASONING CHECK

Why would you expect the initiation stage of translation to be the slowest stage?

24.3 Special Mechanisms Control the Accuracy of Translation

We know that translation is generally accurate because of the consistency that is found when we determine the amino acid sequence of a polypeptide. There are few detailed measurements of the error rate *in vivo*, but it is generally thought to lie in the range of one error for every 10^4 to 10^5 amino acids incorporated. Considering that most polypeptides are produced in large quantities, this means that the error rate is too low to have much effect on the phenotype of the cell.

It is not immediately obvious how such a low error rate is achieved. In fact, the nature of discriminatory events is a general issue raised by several steps in gene expression:

- How do the enzymes that synthesize DNA or RNA recognize only the base complementary to the template?
- How do synthetases recognize just the corresponding tRNAs and amino acids?
- How does a ribosome recognize only the tRNA corresponding to the codon in the A site?

Each case poses a similar problem: how to distinguish one particular member from the entire set, all of which share the same general features.

Probably any substrate initially can contact the active center by a random-hit process, but then the wrong substrates are rejected and only the appropriate one is accepted. The appropriate member is always in a minority (one of 20 amino acids, one of ~30 to 50 tRNAs, one of 4 bases), so the criteria for discrimination must be strict. The point is that the enzyme must have some mechanism for increasing discrimination from the level that would be achieved merely by making contacts with the available surfaces of the substrates.

FIGURE 24.9 summarizes the error rates at the steps that can affect the accuracy of translation.

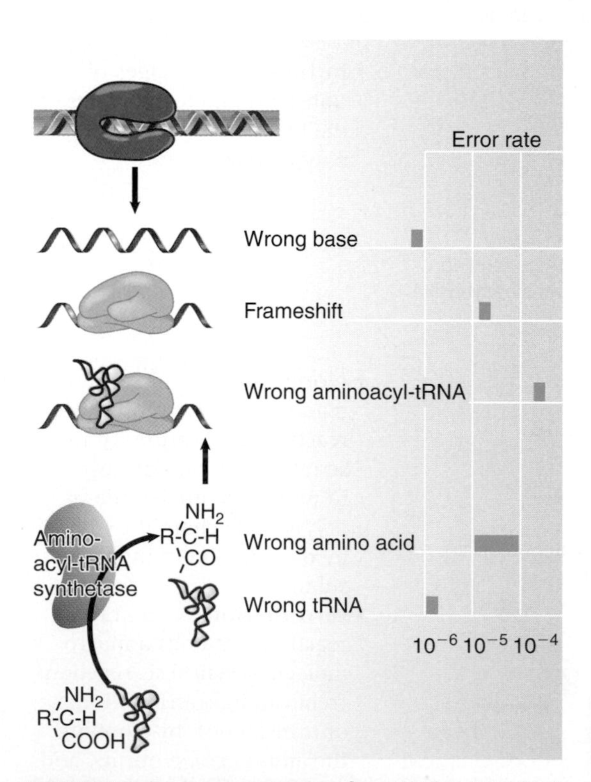

FIGURE 24.9 Errors occur at rates from 10^{-6} to 5×10^{-4} at different stages of translation.

Errors in transcribing mRNA are rare—probably less than 10^{-6}. This is an important stage for accuracy, because a single mRNA molecule is translated into many polypeptide copies. The mechanisms that ensure transcriptional accuracy are discussed in *Chapter 19, Prokaryotic Transcription*.

The ribosome can make two types of errors in translation. It may cause a frameshift by skipping a base when it reads the mRNA (or, in the reverse direction, by reading a base twice—once as the last base of one codon and then again as the first base of the next codon). These errors are rare, occurring at ~10^{-5}. Or it may allow an incorrect aminoacyl-tRNA to (mis)pair with a codon, so that the wrong amino acid is incorporated. This is probably the most common error in protein synthesis, occurring at ~5×10^{-4}. It is primarily controlled by ribosome structure and dissociation kinetics (see *Section 25.10, Synthetases Use Proofreading to Improve Accuracy*).

A tRNA synthetase can make two types of errors: It can place the wrong amino acid on its tRNA, or it can charge its amino acid with the wrong tRNA (see *Section 25.8, tRNAs Are Selectively Paired with Amino Acids by Aminoacyl-tRNA Synthetases*). The incorporation of the wrong amino acid is more common, probably because the tRNA offers a larger surface with which the enzyme can make many more contacts to ensure specificity. Aminoacyl-tRNA synthetases have specific mechanisms to correct errors before a mischarged tRNA is released (see *Section 25.10, Synthetases Use Proofreading to Improve Accuracy*).

KEY CONCEPT

- The accuracy of translation is controlled by specific mechanisms at each stage.

CONCEPT AND REASONING CHECK

Why are translation errors less serious than DNA replication errors?

24.4 Initiation in Bacteria Needs 30S Subunits and Accessory Factors

Bacterial ribosomes engaged in elongating a polypeptide chain exist as 70S particles. At termination, they are released from the mRNA as free ribosomes or ribosomal subunits. In growing bacteria, the majority of ribosomes are synthesizing polypeptides; the free pool is likely to contain ~20% of the ribosomes.

Ribosomes in the free pool can dissociate into separate subunits; this means that 70S ribosomes are in dynamic equilibrium with 30S and 50S subunits. Initiation of translation is not a function of intact ribosomes but is undertaken by the separate subunits, which reassociate during the initiation reaction. **FIGURE 24.10** summarizes the ribosomal subunit cycle during translation in bacteria.

Initiation occurs at a special sequence on mRNA called the **ribosome-binding site** (including the *Shine-Dalgarno sequence*; see *Section 24.6, mRNA Binds a 30S Subunit to Create the Binding Site for a Complex of IF-2 and fMet-tRNA$_f$*). This is a short sequence of bases that precedes the coding region (see Figure 24.15) and is complementary to a portion of the 16S rRNA (see *Section 24.15, Two rRNAs Play Active Roles in Translation*).

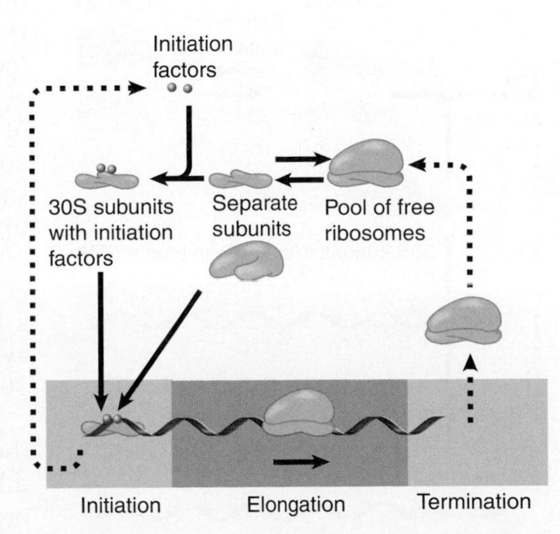

FIGURE 24.10 Initiation requires free ribosome subunits. When ribosomes are released at termination, the 30S subunits bind initiation factors and dissociate to generate free subunits. When subunits reassociate to give a functional ribosome at initiation, they release the factors.

FIGURE 24.11 Initiation factors stabilize free 30S subunits and bind initiator tRNA to the 30S-mRNA complex.

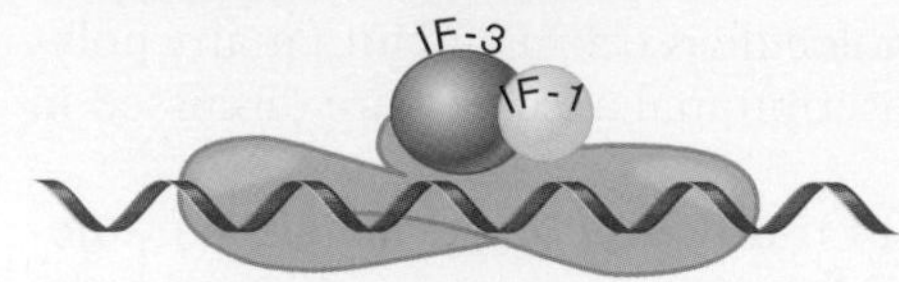

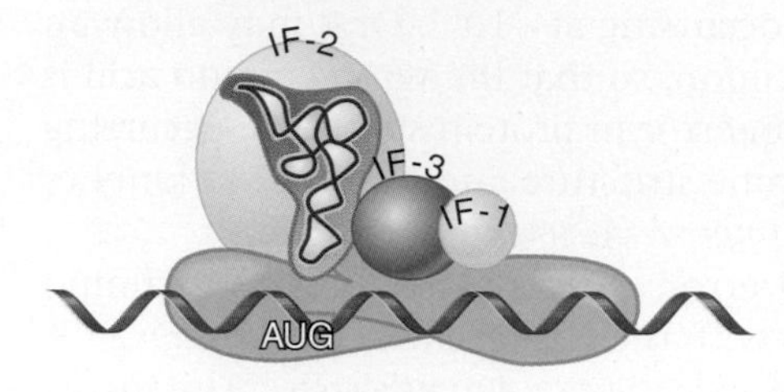

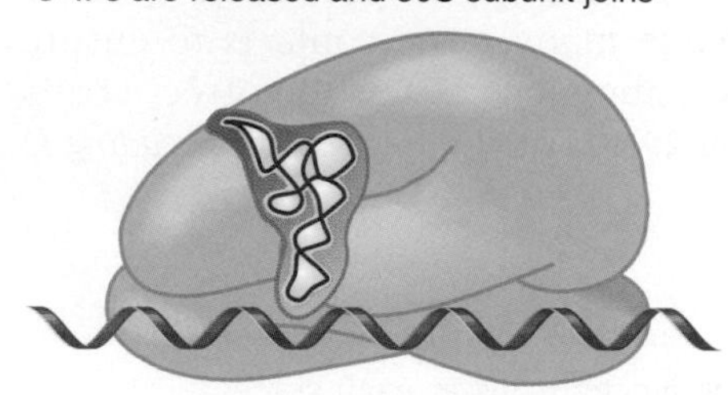

The small and large subunits associate at the ribosome-binding site to form an intact ribosome. The reaction occurs in two steps:

- Recognition of mRNA occurs when a small subunit binds to form an initiation complex at the ribosome-binding site.
- A large subunit then joins the complex to generate a complete ribosome.

Although the 30S subunit is involved in initiation, it is not sufficient by itself to undertake the reactions of binding mRNA and tRNA. It requires additional proteins called **initiation factors (IFs)**. These factors are found only on 30S subunits, and they are released when the 30S subunits associate with 50S subunits to generate 70S ribosomes. This action distinguishes initiation factors from the structural proteins of the ribosome. The initiation factors are concerned solely with formation of the initiation complex, they are absent from 70S ribosomes, and they play no part in the stages of elongation. **FIGURE 24.11** summarizes the stages of initiation.

▸ **initiation factors (IFs)** Proteins that associate with the small subunit of the ribosome specifically at the stage of initiation of polypeptide translation.

Bacteria use three initiation factors, numbered **IF-1**, **IF-2**, and **IF-3**. They are needed for both mRNA and tRNA to enter the initiation complex:

- IF-3 has multiple functions: it is needed to stabilize (free) 30S subunits and to inhibit the premature binding of the 50S subunit; it enables 30S subunits to bind to initiation sites in mRNA; and as part of the 30S-mRNA complex, it checks the accuracy of recognition of the first aminoacyl-tRNA.
- IF-2 binds a special initiator tRNA and controls its entry into the ribosome.
- IF-1 binds to 30S subunits only as a part of the complete initiation complex. It binds in the vicinity of the A site and prevents aminoacyl-tRNA from entering. Its location also may impede the 30S subunit from binding to the 50S subunit.

▸ **IF-1** A bacterial initiation factor that stabilizes the initiation complex for polypeptide translation.

▸ **IF-2** A bacterial initiation factor that binds the initiator tRNA to the initiation complex for polypeptide translation.

▸ **IF-3** A bacterial initiation factor required for 30S ribosomal subunits to bind to initiation sites in mRNA. It also prevents 30S subunits from binding to 50S ribosomal subunits.

The first function of IF-3 controls the equilibrium between ribosomal states, as shown in **FIGURE 24.12**. IF-3 binds to free 30S subunits that are released from the pool of 70S ribosomes. The presence of IF-3 prevents the 30S subunit from reassociating with a 50S subunit. IF-3 can interact directly with 16S rRNA, and there is significant overlap between the bases in 16S rRNA protected by IF-3 and those protected by binding of the 50S subunit, suggesting that it physically prevents junction of the subunits. IF-3 therefore behaves as an anti-association factor that causes a 30S subunit to remain in the pool of free subunits. The reaction between IF-3 and the 30S subunit is stoichiometric: one molecule of IF-3 binds per subunit. There is a relatively small amount of IF-3, so its availability determines the number of free 30S subunits.

FIGURE 24.12 Initiation requires 30S subunits that carry IF-3.

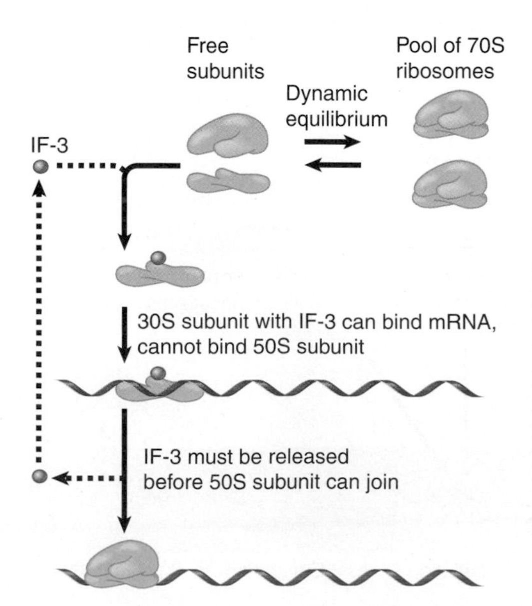

The second function of IF-3 controls the ability of 30S subunits to bind to mRNA. Small subunits must have IF-3 in order to form initiation complexes with mRNA. IF-3 must be released from the 30S-mRNA complex in order to enable the 50S subunit to join. On its release, IF-3 immediately recycles by finding another 30S subunit.

Finally, IF-3 checks the accuracy of recognition of the first aminoacyl-tRNA and helps to direct it to the P site of the 30S subunit. The former has been attributed to the C-terminal domain of IF-3. By comparison, the N-terminal domain of IF-3 is positioned to help direct the aminoacyl-tRNA into the P site of the 30S subunit by blocking the E site at the same time that IF-1 is blocking the A site.

IF-2 has a ribosome-dependent GTPase activity: it sponsors the hydrolysis of GTP in the presence of ribosomes, releasing the energy stored in the high-energy bond. The GTP is hydrolyzed when the 50S subunit joins to generate a complete ribosome. The GTP cleavage could be involved in changing the conformation of the ribosome, so that the joined subunits are converted into an active 70S ribosome.

KEY CONCEPTS

- Initiation of translation requires separate 30S and 50S ribosome subunits.
- Initiation factors (IF-1, -2, and -3), which bind to 30S subunits, are also required.
- A 30S subunit carrying initiation factors binds to an initiation site on mRNA to form an initiation complex.
- IF-3 must be released to allow 50S subunits to join the 30S-mRNA complex.

CONCEPT AND REASONING CHECK

What are the functions of the three bacterial translation initiation factors?

24.5 A Special Initiator tRNA Starts the Polypeptide Chain

Synthesis of all polypeptides starts with the same amino acid: methionine. The signal for initiating a polypeptide chain is a special initiation codon that marks the start of the reading frame. Usually the initiation codon is the triplet AUG, but in bacteria, GUG or UUG can also be used.

tRNAs recognizing the AUG codon carry methionine, and two types of tRNA can carry this amino acid. One is used for initiation, the other for recognizing AUG codons during elongation.

In bacteria, mitochondria, and chloroplasts, the initiator tRNA carries a methionine residue that has been formylated on its amino group, forming a molecule of **N-formyl-methionyl-tRNA**. The tRNA is known as **$tRNA_f^{Met}$**. The name of the aminoacyl-tRNA is usually abbreviated to fMet-$tRNA_f$.

▸ **N-formyl-methionyl-tRNA** The aminoacyl-tRNA that initiates bacterial polypeptide translation. The amino group of the methionine is formylated.

▸ **$tRNA_f^{Met}$** The special RNA used to initiate polypeptide translation in bacteria. It mostly uses AUG but can also respond to GUG and UUG.

The initiator tRNA gains its modified amino acid in a two-stage reaction. First, it is charged with the amino acid to generate Met-$tRNA_f$, and then the formylation reaction shown in **FIGURE 24.13** blocks the free NH_2 group. Although the blocked amino acid group would prevent the initiator from participating in chain elongation, it does not interfere with the ability to initiate a polypeptide.

This tRNA is used only for initiation. It recognizes the codons AUG or GUG (occasionally UUG). The codons are not recognized equally well: the extent of initiation declines by about half when AUG is replaced by GUG, and declines by about half again when UUG is employed.

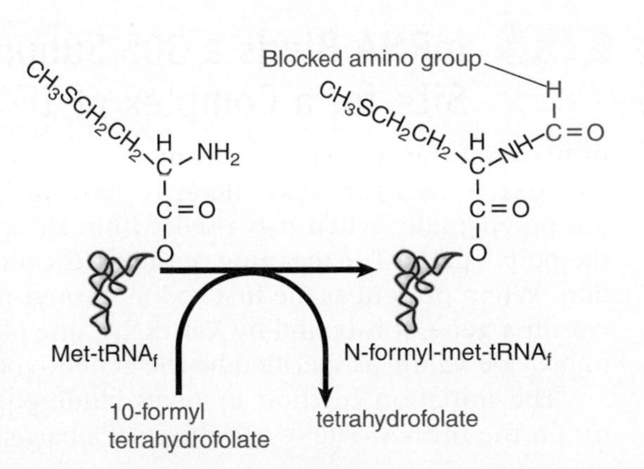

FIGURE 24.13 The initiator N-formyl-methionyl-tRNA (fMet-$tRNA_f$) is generated by formylation of methionyl-tRNA using formyl-tetrahydrofolate as cofactor.

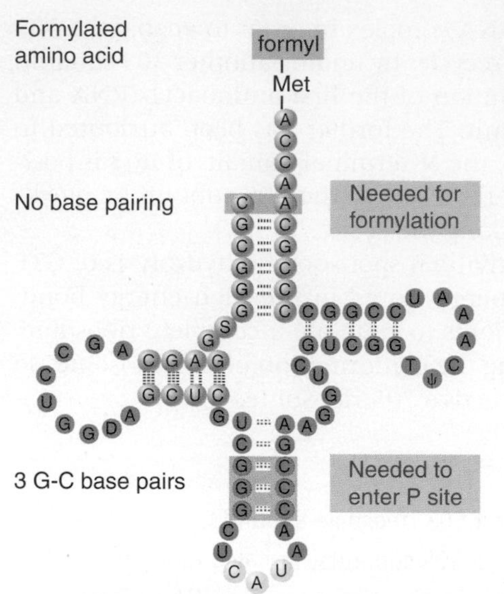

FIGURE 24.14 fMet-tRNA$_f$ has unique features that distinguish it as the initiator tRNA.

▸ **tRNA$_m^{Met}$** The bacterial tRNA that inserts methionine at internal AUG codons.

The tRNA species responsible for recognizing AUG codons in internal locations is **tRNA$_m^{Met}$**. This tRNA responds only to internal AUG codons. Its methionine cannot be formylated.

What features distinguish the fMet-tRNA$_f$ initiator and the Met-tRNA$_m$ elongator? Some characteristic features of the tRNA sequence are important, as summarized in **FIGURE 24.14**. Some of these features are needed to prevent the initiator from being used in elongation, whereas others are necessary for it to function in initiation:

- Formylation is not strictly necessary because nonformylated Met-tRNA$_f$ can function as an initiator. Formylation improves the efficiency with which the Met-tRNA$_f$ is used, though, because it is one of the features recognized by the factor IF-2 that binds the initiator tRNA.
- The bases that face one another at the last position of the stem to which the amino acid is connected are paired in all tRNAs except tRNA$_f^{Met}$. Mutations that create a base pair in this position of tRNA$_f^{Met}$ allow it to function in elongation. The absence of this pair is, therefore, important in preventing tRNA$_f^{Met}$ from being used in elongation. It is also needed for the formylation reaction.
- A series of three G-C pairs in the stem that precedes the loop containing the anticodon is unique to tRNA$_f^{Met}$. These base pairs are required to allow the fMet-tRNA$_f$ to be inserted directly into the P site.

KEY CONCEPTS

- Translation starts with a methionine amino acid usually encoded by AUG.
- In bacteria, different methionine tRNAs are involved in initiation and elongation.
- The initiator tRNA has unique structural features that distinguish it from all other tRNAs.

CONCEPT AND REASONING CHECK

What advantage is there for bacteria to have two species of tRNA that recognize the AUG codon?

24.6 mRNA Binds a 30S Subunit to Create the Binding Site for a Complex of IF-2 and fMet-tRNA$_f$

▸ **context** The fact that neighboring sequences may change the efficiency with which a codon is recognized by its aminoacyl-tRNA or is used to terminate polypeptide translation.

In bacterial translation, the meaning of the AUG and GUG codons depends on their **context**. When the AUG codon is used for initiation, a formyl-methionine begins the polypeptide; when it is used within the coding region, methionine is added to the polypeptide. The meaning of the GUG codon is even more dependent on its location. When present as the first codon, formyl-methionine is added, but when present within a gene, it is bound by Val-tRNA, one of the regular members of the tRNA set, to provide valine as specified by the genetic code.

The initiation reaction involves binding of a 30S subunit to a *ribosome-binding site* on the mRNA. The two features of a bacterial ribosome-binding site are the AUG initiation codon and a polypurine sequence preceding it by ~10 bases that corresponds to the hexamer:

5′ . . . A G G A G G . . . 3′

This polypurine stretch is known as the **Shine-Dalgarno sequence**. It is complementary to a highly conserved sequence close to the 3′ end of 16S rRNA. (The extent of complementarity differs with individual mRNAs and may extend from the four-base core sequence GAGG to a nine-base sequence extending beyond each end of the hexamer.) Written in the reverse orientation, the complementary rRNA sequence is the hexamer:

3′ . . . U C C U C C . . . 5′

▸ **Shine-Dalgarno sequence** The polypurine sequence AGGAGG centered about 10 bp before the AUG initiation codon on bacterial mRNA. It is complementary to the sequence at the 3′ end of 16S rRNA.

The Shine-Dalgarno sequence pairs with its rRNA complement during mRNA-ribosome binding. Mutations of either sequence in this reaction prevent an mRNA from being translated. The interaction is specific for bacterial ribosomes; this is a significant difference between the mechanisms of initiation for prokaryotes and eukaryotes.

FIGURE 24.15 shows how an AUG initiation codon is distinguished from an AUG codon within a coding region. When an initiation complex forms at a ribosome-binding site, the initiation codon lies within the part of the P site carried by the small subunit. The only aminoacyl-tRNA that can become part of the initiation complex is the initiator, which has the unique property of being able to enter directly into the partial P site to bind to its complementary codon.

When the large subunit joins the complex, the partial tRNA-binding sites are converted into the intact P and A sites. The initiator fMet-tRNA$_f$ occupies the P site, and the A site is available for entry of the aminoacyl-tRNA that is complementary to the second codon of the mRNA. The first peptide bond forms between the initiator and the next aminoacyl-tRNA.

Initiation occurs when an AUG (or GUG) codon lies within a ribosome-binding site because only the initiator tRNA can enter the partial P site in the 30S subunit. AUG (or GUG) codons within the message, which are encountered by a ribosome that is continuing to translate an mRNA, are not recognized by the initiator tRNA because only the regular aminoacyl-tRNAs can enter the (complete) A site in the 70S ribosome.

Accessory factors are critical in controlling the usage of aminoacyl-tRNAs. All aminoacyl-tRNAs associate with the ribosome by binding to an accessory factor. The factor used in initiation is IF-2 (see *Section 24.4, Initiation in Bacteria Needs 30S Subunits and Accessory Factors*). The accessory factor used at elongation, EF-Tu, is discussed in *Section 24.8, Elongation Factor Tu Loads Aminoacyl tRNA into the A Site.*

The initiation factor IF-2 places the initiator tRNA into the P site. By forming a complex specifically with fMet-tRNA$_f$, IF-2 ensures that only the initiator tRNA, and none of the regular aminoacyl-tRNAs, participates in the initiation reaction.

Conversely, the accessory factor that places aminoacyl-tRNAs in the A site cannot bind fMet-tRNA$_f$, which is, therefore, excluded from use during elongation.

The accuracy of initiation is also assisted by IF-3, which stabilizes binding of the initiator tRNA by recognizing correct base pairing with the second and third bases of the AUG initiation codon.

FIGURE 24.16 details the series of events by which IF-2 places the fMet-tRNA$_f$ initiator in the P site. IF-2, bound to GTP, associates with the P site of the 30S subunit. At this point, the 30S subunit carries all the initiation factors. fMet-tRNA$_f$ then binds to the IF-2 on the 30S subunit and IF-2 transfers the tRNA into the partial P site.

IF-2 has a ribosome-dependent GTPase activity that is triggered when the 50S subunit joins to generate a complete ribosome. This probably provides the energy required for conformational changes associated with the joining of the subunits.

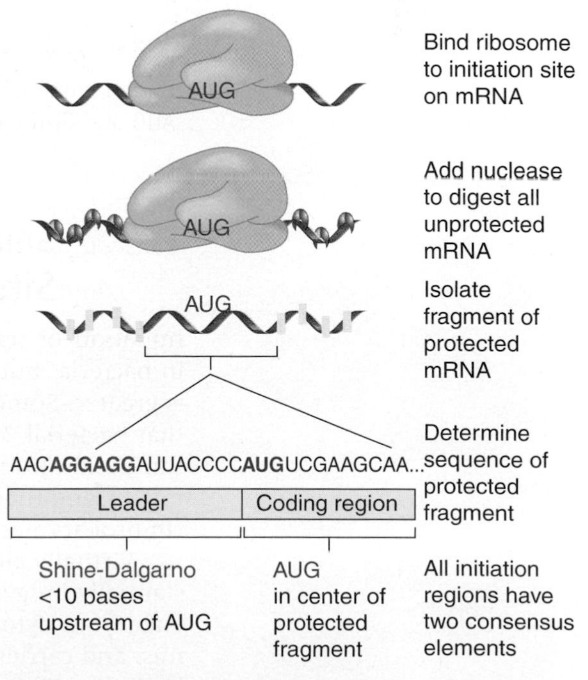

FIGURE 24.15 Ribosome-binding sites on mRNA can be recovered from initiation complexes. They include the upstream Shine-Dalgarno sequence and the initiation codon.

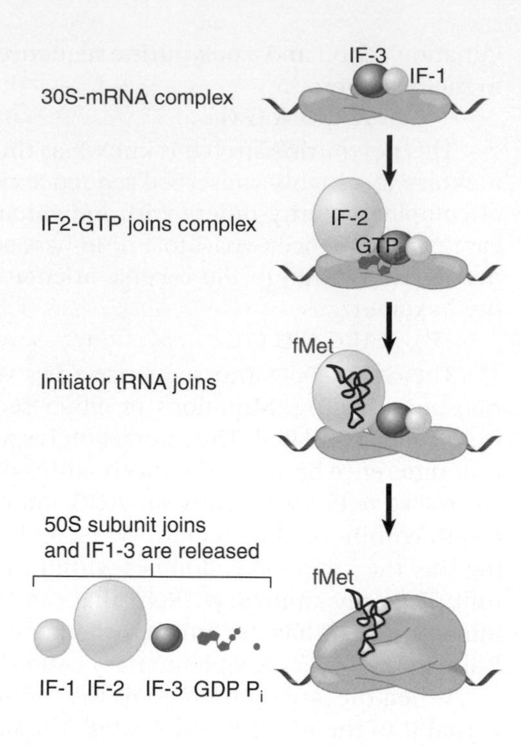

FIGURE 24.16 IF-2 is needed to bind fMet-tRNA$_f$ to the 30S-mRNA complex. After 50S binding, all IF factors are released and GTP is cleaved.

KEY CONCEPTS

- An initiation site on bacterial mRNA consists of the AUG initiation codon, preceded with a gap of ~10 bases by the Shine-Dalgarno polypurine hexamer.
- The rRNA of the 30S bacterial ribosomal subunit has a complementary sequence that base pairs with the Shine-Dalgarno sequence during initiation.
- IF-2 binds the initiator fMet-tRNA$_f$ and allows it to enter the partial P site on the 30S subunit.

CONCEPT AND REASONING CHECK

During bacterial translation, how are AUG and GUG initiation codons distinguished from AUG and GUG codons later in an ORF?

24.7 Small Subunits Scan for Initiation Sites on Eukaryotic mRNA

Initiation of translation in eukaryotic cytoplasm resembles the process that occurs in bacteria, but the order of events is different and the number of accessory factors is greater. Some of the differences in initiation are related to a difference in the way that bacterial 30S and eukaryotic 40S subunits find their binding sites for initiating translation on mRNA. In eukaryotes, small subunits first recognize the 5′ end of the mRNA and then move to the initiation site, where they are joined by large subunits. (In prokaryotes, small subunits bind directly to the initiation site.)

Virtually all eukaryotic mRNAs are monocistronic, but each mRNA usually is substantially longer than necessary just to encode its polypeptide. The average mRNA in eukaryotic cytoplasm is 1000 to 2000 bases long, has a methylated cap at the 5′ terminus, and carries 100 to 200 bases of poly(A) at the 3′ terminus. The nontranslated 5′ leader is relatively short, usually less than 100 bases. The length of the coding region is determined by the size of the polypeptide product. The nontranslated 3′ trailer is often rather long, at times reaching lengths of up to ~1000 bases.

The first feature to be recognized during translation of a eukaryotic mRNA is the methylated cap that marks the 5′ end. Messenger RNAs whose caps have been removed are not translated efficiently *in vitro*. Binding of 40S subunits to mRNA requires several initiation factors, including proteins that recognize the structure of the cap. In some mRNAs, the AUG initiation codon lies within 40 bases of the 5′ terminus of the mRNA, so that both the cap and AUG lie within the span of ribosome binding. In many mRNAs, however, the cap and AUG are farther apart; in extreme cases, they can be as much as 1000 bases away from each other. Yet the presence of the cap still is necessary for a stable complex to be formed at the initiation codon. How can the ribosome rely on two sites so far apart?

FIGURE 24.17 illustrates the "scanning" model, which supposes that the 40S subunit initially recognizes the 5′ cap and then "migrates" along the mRNA. Scanning from the 5′ end is a linear process. When 40S subunits scan the leader region, they can melt secondary structure hairpins with stabilities less than −30 kcal, but hairpins of greater stability impede or prevent migration.

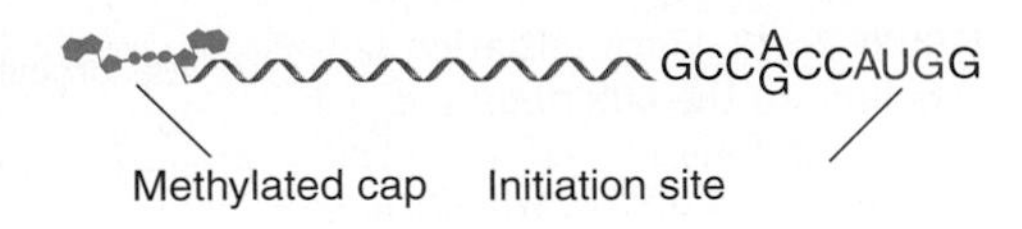

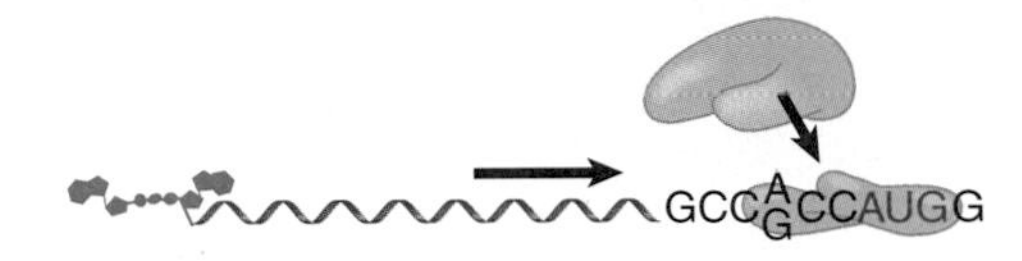

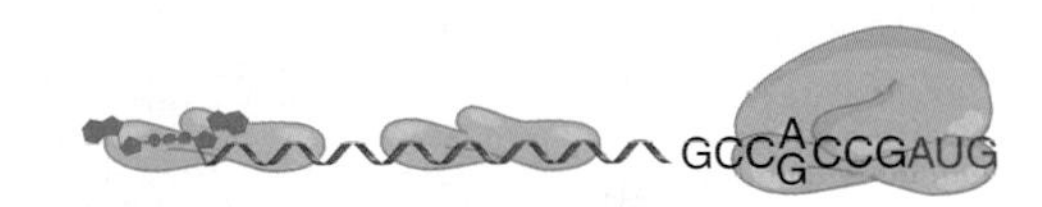

FIGURE 24.17 Eukaryotic ribosomes migrate from the 5′ end of mRNA to the ribosome binding site, which includes an AUG initiation codon.

Migration stops when the 40S subunit encounters the AUG initiation codon. Usually, although not always, the first AUG triplet sequence to be encountered will be the initiation codon. However, the AUG triplet by itself is not sufficient to halt migration; it is recognized efficiently as an initiation codon only when it is in the right context. The most important determinants of context are the bases in positions −4 and +1. An initiation codon may be recognized in the sequence NNNPuNN*AUG*G. The purine (A or G) 3 bases before the AUG codon and the G immediately following it can increase the efficiency of translation by 10×. When the leader sequence is long, further 40S subunits can recognize the 5′ end before the first has left the initiation site, creating a queue of subunits proceeding along the leader to the initiation site.

The majority of eukaryotic initiation events involve scanning from the 5′ cap, but there is an alternative means of initiation, used especially by certain viral RNAs, in which a 40S subunit associates directly with an internal site called an **IRES (*i*nternal *r*ibosome *e*ntry *s*ite)**. (This entirely bypasses any AUG codons that may be in the 5′ untranslated region.) There are few sequence homologies between known IRES elements. The most common type of IRES element includes the AUG initiation codon at its upstream boundary. The 40S subunit binds directly to it, using a subset of the same factors that are required for initiation at 5′ ends.

▸ **IRES (internal ribosome entry site)** A eukaryotic messenger RNA sequence that allows a ribosome to initiate polypeptide translation without migrating from the 5′ end.

Modification at the 5′ end occurs to almost all cellular and viral mRNAs and is essential for their translation in eukaryotic cytosol. The sole exception to this rule is provided by a few viral mRNAs (such as poliovirus) that are not capped. They use the IRES pathway. This is especially important in picornavirus infection, where it was first discovered, because the virus inhibits host translation by destroying cap structures and inhibiting the initiation factors that bind them. This prevents the translation of host mRNAs. Viral mRNAs can be translated because they use the IRES.

Eukaryotic cells have more initiation factors than bacteria: the current list includes 12 factors that are directly or indirectly required for initiation. The factors are named similarly to those in bacteria, sometimes by analogy with the bacterial factors, and are given the prefix "e" to indicate their eukaryotic origin. They act at all stages of the process, including:

- forming an initiation complex with the 5′ end of mRNA;
- forming a complex with Met-$tRNA_i$;
- binding the mRNA-factor complex to the Met-$tRNA_i$-factor complex;
- enabling the ribosome to scan mRNA from the 5′ end to the first AUG;
- detecting binding of initiator tRNA to AUG at the start site; and
- mediating joining of the 60S subunit.

FIGURE 24.18 Some initiation factors bind to the 40S ribosome subunit to form the 43S complex; others bind to mRNA. When the 43S complex binds to mRNA, it scans for the initiation codon and can be isolated as the 48S complex.

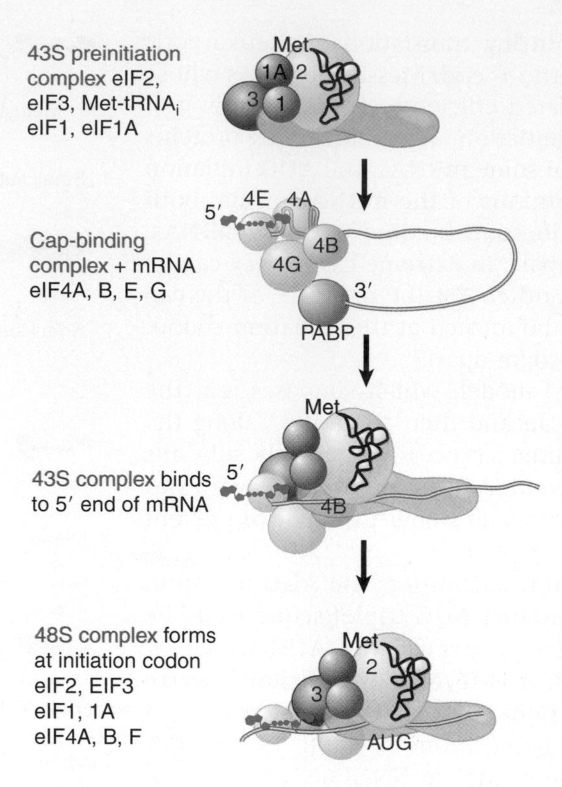

FIGURE 24.18 summarizes the stages of initiation and shows which initiation factors are involved at each stage. eIF2, together with Met-tRNA$_i$, eIF3, eIF1, and eIF1A, binds to the 40S ribosome subunit to form the 43S preinitiation complex. eIF4A, eIF4B, eIF4E, and eIF4G bind to the 5′ end of the mRNA to form the cap-binding complex. This complex associates with the 3′ end of the mRNA via eIF4G, which interacts with poly(A)-binding protein (PABP). The 43S complex binds the initiation factors at the 5′ end of the mRNA and scans for the initiation codon. It can be isolated as the 48S initiation complex.

Note the circular arrangement of the mRNA associated with the cap-binding complex in Figure 24.18 (second line), when the interaction between the PABP and eIF4G brings the 5′ and 3′ ends of the mRNA into proximity. This enhances formation of initiation complexes so that, in effect, PABP acts as an initiation factor.

KEY CONCEPTS

- Eukaryotic 40S ribosomal subunits bind to the 5′ end of mRNA and scan the mRNA until they reach an initiation site.
- A eukaryotic initiation site consists of a ten-nucleotide sequence that includes an AUG codon.
- Initiation factors are required for all stages of initiation, including binding of the initiator tRNA, 40S subunit attachment to mRNA, movement along the mRNA, and joining of the 60S subunit.
- eIF2 and eIF3 bind the initiator Met-tRNA$_i$ and GTP, and the complex binds to the 40S subunit before it associates with mRNA.

CONCEPT AND REASONING CHECK

How do eukaryotic 40S ribosomal subunits find the initiation codon of an mRNA?

24.8 Elongation Factor Tu Loads Aminoacyl-tRNA into the A Site

Once the complete ribosome is formed at the initiation codon, the stage is set for a cycle in which aminoacyl-tRNA enters the A site of a ribosome whose P site is occupied by peptidyl-tRNA. Any aminoacyl-tRNA except the initiator can enter the A site. Its entry is mediated by an **elongation factor** (**EF-Tu** in bacteria). The process is similar in eukaryotes. EF-Tu is a highly conserved protein throughout bacteria and mitochondria and is homologous to its eukaryotic counterpart.

Just like its counterpart in initiation (IF-2), EF-Tu is associated with the ribosome only during the process of aminoacyl-tRNA entry. Once the aminoacyl-tRNA is in place, EF-Tu leaves the ribosome to work again with another aminoacyl-tRNA.

- **elongation factors** Proteins that associate with ribosomes cyclically during the addition of each amino acid to the polypeptide chain.
- **EF-Tu** The elongation factor that binds aminoacyl-tRNA and places it into the A site of a bacterial ribosome.

FIGURE 24.19 depicts the role of EF-Tu in bringing aminoacyl-tRNA to the A site. The binary complex of EF-Tu-GTP binds aminoacyl-tRNA to form a ternary complex of aminoacyl-tRNA-EF-Tu-GTP. The ternary complex binds only to the A site of ribosomes whose P site is already occupied by peptidyl-tRNA. This is the critical reaction in ensuring that the aminoacyl-tRNA and peptidyl-tRNA are correctly positioned for peptide bond formation.

Aminoacyl-tRNA is loaded into the A site in two stages. First, the anticodon end binds to the A site of the 30S subunit. Then, codon-anticodon recognition triggers a change in the conformation of the ribosome. This stabilizes tRNA binding and causes EF-Tu to hydrolyze its GTP. The CCA end of the tRNA now moves into the A site on the 50S subunit. The binary complex EF-Tu-GDP is released. This form of EF-Tu is inactive and does not bind aminoacyl-tRNA effectively. Another factor, EF-Ts, mediates the regeneration of the used form, EF-Tu-GDP, into the active form, EF-Tu-GTP.

The presence of EF-Tu prevents the aminoacyl end of aminoacyl-tRNA from entering the A site on the 50S subunit (see Figure 24.23). So the release of EF-Tu-GDP is needed for the ribosome to undertake peptide bond formation. The same principle is seen at other stages of translation: one reaction must be completed properly before the next can proceed.

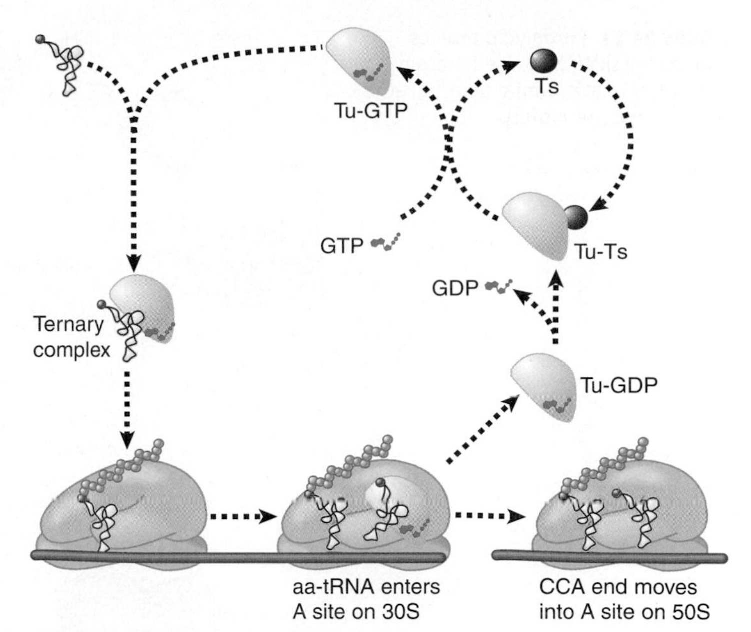

FIGURE 24.19 EF-Tu-GTP places aminoacyl-tRNA on the ribosome and then is released as EF-Tu-GDP. EF-Ts is required to mediate the replacement of GDP by GTP. The reaction consumes GTP and releases GDP. The only aminoacyl-tRNA that cannot be recognized by EF-Tu-GTP is fMet-tRNA$_f$, whose failure to bind prevents it from responding to internal AUG or GUG codons.

In eukaryotes, the factor eEF1α is responsible for bringing aminoacyl-tRNA to the ribosome, again in a reaction that involves cleavage of a high-energy bond in GTP. Like its prokaryotic homolog (EF-Tu), it is an abundant protein. After hydrolysis of GTP, the active form is regenerated by the factor eEF1βγ, a counterpart to EF-Ts.

KEY CONCEPTS

- EF-Tu is a monomeric G protein whose active form (bound to GTP) binds to aminoacyl-tRNA.
- The EF-Tu-GTP-aminoacyl-tRNA complex binds to the ribosome's A site.

CONCEPT AND REASONING CHECK

What are the roles of bacterial elongation factors EF-Tu and EF-Ts?

24.9 The Polypeptide Chain Is Transferred to Aminoacyl-tRNA

The ribosome remains in place while the polypeptide chain is elongated by transferring the polypeptide attached to the tRNA in the P site to the aminoacyl-tRNA in the A site. The reaction is shown in **FIGURE 24.20**. The activity responsible for synthesis of the peptide bond is called **peptidyl transferase**. It is a function of the large (50S or 60S) ribosomal subunit. The reaction is triggered when the aminoacyl end of the tRNA in the A site swings into a location close to the end of the peptidyl-tRNA following the release

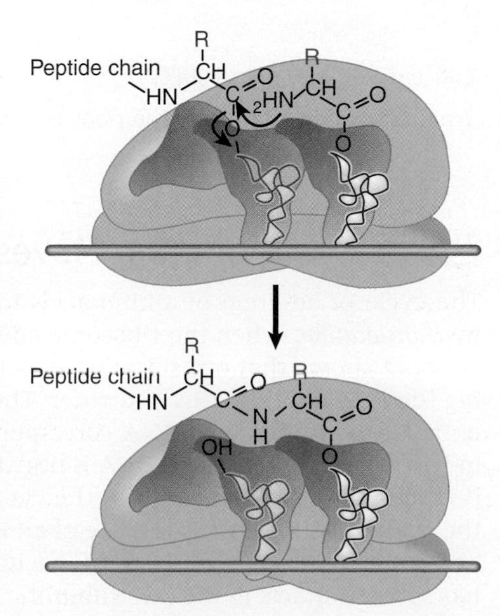

FIGURE 24.20 Peptide bond formation takes place by reaction between the polypeptide of peptidyl-tRNA in the P site and the amino acid of aminoacyl-tRNA in the A site.

▸ **peptidyl transferase** The activity of the large ribosomal subunit that synthesizes a peptide bond when an amino acid is added to a growing polypeptide chain. The actual catalytic activity is a property of the rRNA.

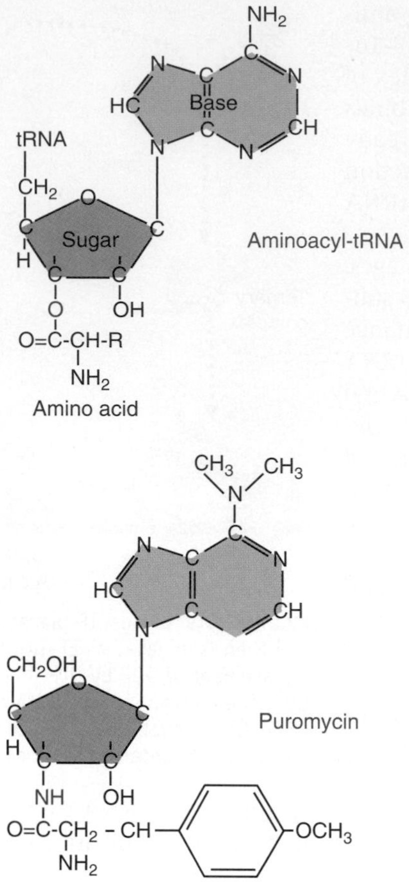

FIGURE 24.21 Puromycin mimics aminoacyl-tRNA because it resembles an aromatic amino acid linked to a sugar-base moiety.

of EF-Tu. This site has a peptidyl transferase activity that essentially ensures a rapid transfer of the peptide chain to the aminoacyl-tRNA. Both rRNA and 50S subunit proteins are necessary for this activity, but the actual act of catalysis is a property of the ribosomal RNA of the large subunit (see *Section 24.15, Two rRNAs Play Active Roles in Translation*).

The nature of the transfer reaction is revealed by the ability of the antibiotic puromycin to inhibit translation. Puromycin resembles an amino acid attached to the terminal adenosine of tRNA. **FIGURE 24.21** shows that puromycin has an N instead of the O that joins an amino acid to tRNA. The antibiotic is treated by the ribosome as though it were an incoming aminoacyl-tRNA, after which the polypeptide attached to peptidyl-tRNA is transferred to the NH_2 group of the puromycin.

The puromycin moiety is not anchored to the A site of the ribosome; as a result, the polypeptidyl-puromycin complex is released from the ribosome in the form of polypeptidyl-puromycin. This premature termination of translation is responsible for the lethal action of the antibiotic.

KEY CONCEPTS

- The 50S subunit has peptidyl transferase activity.
- The nascent polypeptide chain is transferred from peptidyl-tRNA in the P site to aminoacyl-tRNA in the A site.
- Peptide bond synthesis generates deacylated tRNA in the P site and peptidyl-tRNA in the A site.

CONCEPT AND REASONING CHECK

What catalytic molecule of the ribosome is responsible for formation of the peptide bond?

24.10 Translocation Moves the Ribosome

The cycle of addition of amino acids to the growing polypeptide chain is completed by *translocation,* when the ribosome advances three nucleotides along the mRNA. **FIGURE 24.22** shows that translocation expels the uncharged tRNA from the P site, allowing the new peptidyl-tRNA to enter. The ribosome then has an empty A site ready for entry of the aminoacyl-tRNA corresponding to the next codon. As the figure shows, in bacteria the discharged tRNA is transferred from the P site to the E site (from which it is then expelled directly into the cytosol). In eukaryotes it is expelled directly into the cytosol without the presence of an E site. The A and P sites straddle both the large and small subunits; the E site (in bacteria) is located largely on the 50S subunit, but has some contacts in the 30S subunit.

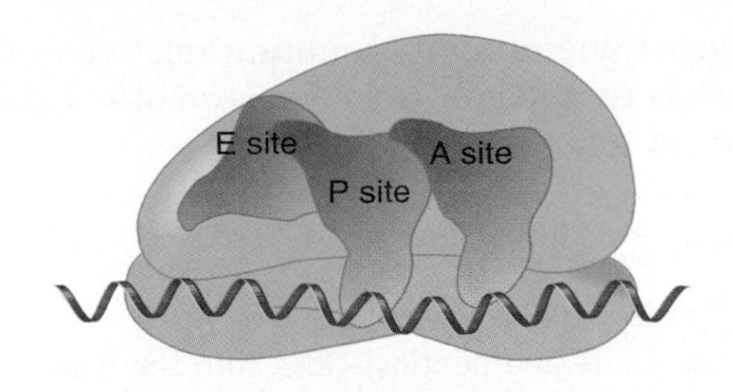

FIGURE 24.22 A bacterial ribosome has three tRNA-binding sites. Aminoacyl-tRNA enters the A site of a ribosome that has peptidyl-tRNA in the P site. Peptide bond synthesis deacylates the P site tRNA and generates peptidyl-tRNA in the A site. Translocation moves the deacylated tRNA into the E site and moves peptidyl-tRNA into the P site.

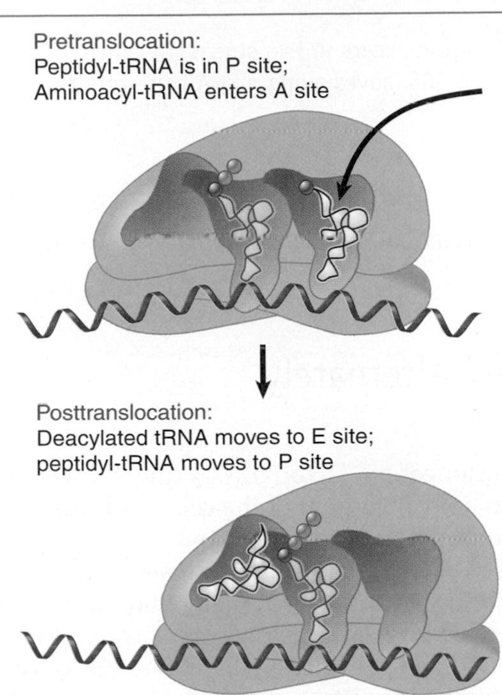

Most thinking about translocation follows the *hybrid state model,* which has translocation occurring in two stages. **FIGURE 24.23** shows that first there is a shift of the 50S subunit relative to the 30S subunit, followed by a second shift that occurs when the 30S subunit moves along mRNA to restore the original conformation. The basis for this model was the observation that the pattern of contacts that tRNA makes with the ribosome (measured by chemical footprinting) changes in two stages. When puromycin is added to a ribosome that has an aminoacylated tRNA in the P site, the contacts of tRNA on the 50S subunit change from the P site to the E site, but the contacts on the 30S subunit do not change. This suggests that the 50S subunit has moved to a posttransfer state, but the 30S subunit has not changed.

The interpretation of these results is that first the aminoacyl ends of the tRNAs (located in the 50S subunit) move into the new sites (while the anticodon ends remain bound to their anticodons in the 30S subunit). At this stage, the tRNAs are effectively bound in hybrid sites, consisting of the 50SE/30SP and the 50SP/30SA sites. Then movement is extended to the 30S subunits, so that the anticodon-codon pairing region finds itself in the right site. The most likely means of

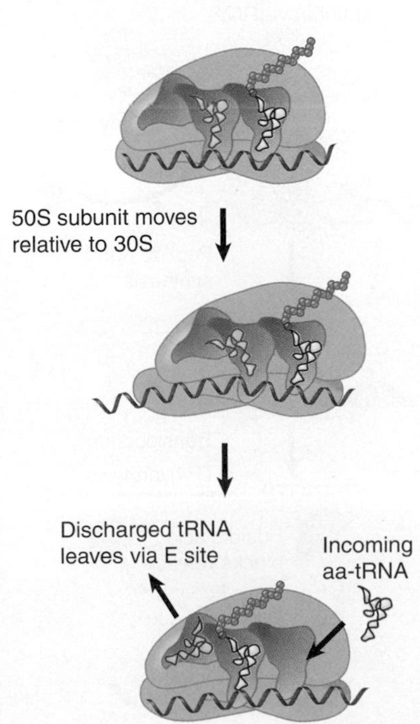

FIGURE 24.23 Models for translocation involve two stages. First, at peptide bond formation the aminoacyl end of the tRNA in the A site becomes relocated in the P site. Second, the anticodon end of the tRNA becomes relocated in the P site.

creating the hybrid state is by a movement of one ribosomal subunit relative to the other, so that translocation in effect involves two stages, with the normal structure of the ribosome being restored by the second stage.

KEY CONCEPTS

- Ribosomal translocation moves the mRNA through the ribosome by three bases.
- Translocation moves deacylated tRNA into the E site and peptidyl-tRNA into the P site and empties the A site.
- The hybrid state model proposes that translocation occurs in two stages, in which the 50S moves relative to the 30S and then the 30S moves along mRNA to restore the original conformation.

CONCEPT AND REASONING CHECK

Ribosomal translocation occurs in a way that preserves codon-anticodon pairing. Why is this important?

24.11 Elongation Factors Bind Alternately to the Ribosome

Translocation requires GTP and another elongation factor, EF-G (the eukaryotic homolog of EF-G is eEF2). This factor is a major constituent of the cell: it is present at a level of ~1 copy per ribosome (20,000 molecules per cell).

Ribosomes cannot bind EF-Tu and EF-G simultaneously, so translation follows the cycle illustrated in **FIGURE 24.24**, in which the factors are alternately bound to and released from the ribosome. Thus EF-Tu-GDP must be released before EF-G can bind; then EF-G must be released before aminoacyl-tRNA-EF-Tu-GTP can bind.

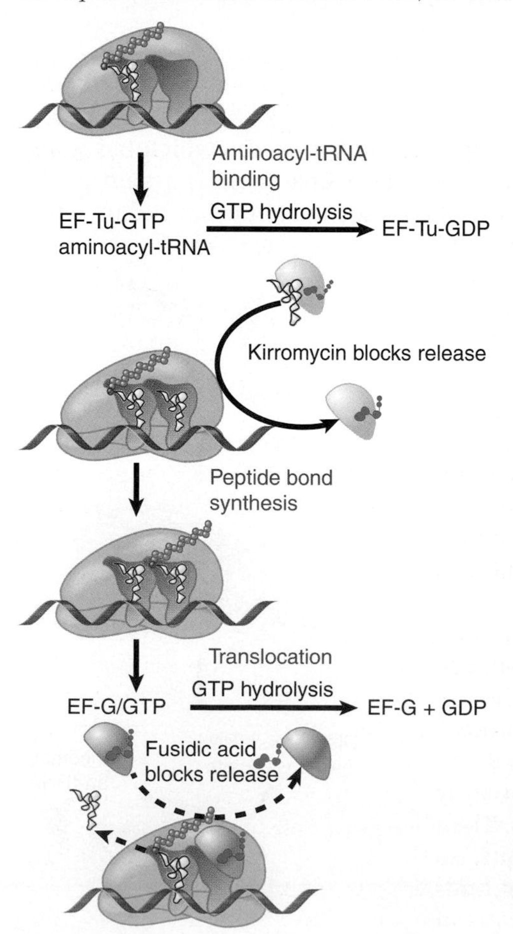

FIGURE 24.24 Binding of factors EF-Tu and EF-G alternates as ribosomes accept new aminoacyl-tRNA, form peptide bonds, and translocate.

Does the ability of each elongation factor to exclude the other rely on an allosteric effect on the overall conformation of the ribosome or on direct competition for overlapping binding sites? **FIGURE 24.25** shows an extraordinary similarity between the structures of the ternary complex of aminoacyl-tRNA-EF-Tu-GDP and EF-G. The structure of EF-G mimics the overall structure of EF-Tu bound to the amino acceptor stem of aminoacyl-tRNA. This creates the immediate assumption that they compete for the same binding site (presumably in the vicinity of the A site). The need for each factor to be released before the other can bind ensures that the events of translation proceed in an orderly manner.

Both elongation factors are monomeric GTP-binding proteins that are active when bound to GTP but inactive when bound to GDP. The triphosphate form is required for binding to the ribosome, which ensures that each factor obtains access to the ribosome only in the company of the GTP that it needs to fulfill its function.

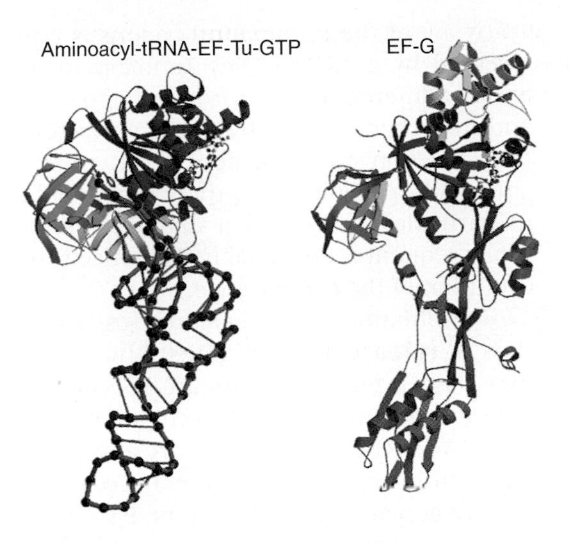

FIGURE 24.25 The structure of the ternary complex of aminoacyl-tRNA-EF-Tu-GTP (left) resembles the structure of EF-G (right). Structurally conserved domains of EF-Tu and EF-G are in red and green; the tRNA and the domain resembling it in EF-G are in purple. Photo courtesy of Poul Nissen, University of Aarhus, Denmark.

KEY CONCEPTS

- Translocation requires EF-G, whose structure resembles the aminoacyl-tRNA-EF-Tu-GTP complex.
- Binding of EF-Tu and EF-G to the ribosome is mutually exclusive.
- Translocation requires GTP hydrolysis, which triggers a change in EF-G, which in turn triggers a change in ribosome structure.

CONCEPT AND REASONING CHECK

What is the role of elongation factor EF-G in bacterial translation?

24.12 Three Codons Terminate Translation and Are Recognized by Protein Factors

Only 61 triplet codons specify amino acids. The other three triplets are termination codons (also known as nonsense codons, or **stop codons**), which end translation. They have casual names from the history of their discovery (see the accompanying *Historical Perspectives* box). The UAG triplet is called the **amber codon**, UAA is the **ochre codon**, and UGA is sometimes called the **opal codon**.

The UAG, UAA, and UGA triplet sequences are necessary and sufficient to signal the end of translation, whether it occurs naturally at the end of an ORF or is created by nonsense mutation within a coding sequence. (Sometimes the term *nonsense codon* is used to describe the termination triplets. "Nonsense" is really a term that describes the effect of a mutation in a gene rather than the meaning of the codon for translation. *Stop codon* is a better term.)

In bacterial genes, UAA is the most commonly used termination codon. UGA is used more often than UAG, although there appear to be more errors reading UGA. (An error in reading a termination codon—when an aminoacyl-tRNA improperly responds to it—results in the continuation of translation until another termination codon is encountered or the ribosome reaches the 3′ end of the mRNA, which may result in other problems. For this circumstance, bacteria have a special RNA.)

Two stages are involved in ending translation. The *termination reaction* itself involves release of the polypeptide chain from the last tRNA. The *posttermination reaction* involves release of the tRNA and mRNA and dissociation of the ribosome into its subunits.

- **stop codon** One of three triplets (UAG, UAA, or UGA) that cause polypeptide translation to terminate. They are also known historically as nonsense codons. The UAA codon is called ochre and the UAG codon is called amber, after the names of the nonsense mutations by which they were originally identified.
- **amber codon** The triplet UAG, one of the three termination codons that end polypeptide translation.
- **ochre codon** The triplet UAA, one of the three termination codons that end polypeptide translation.
- **opal codon** The triplet UGA, one of the three termination codons that end polypeptide translation. It has evolved to code for an amino acid in a small number of organisms or organelles.

FIGURE 24.26 Molecular mimicry enables the elongation factor Tu-tRNA complex, the translocation factor EF-G, and the release factors RF1/2-RF3 to bind to the same ribosomal site. RRF is the ribosome recycling factor (see Figure 24.28).

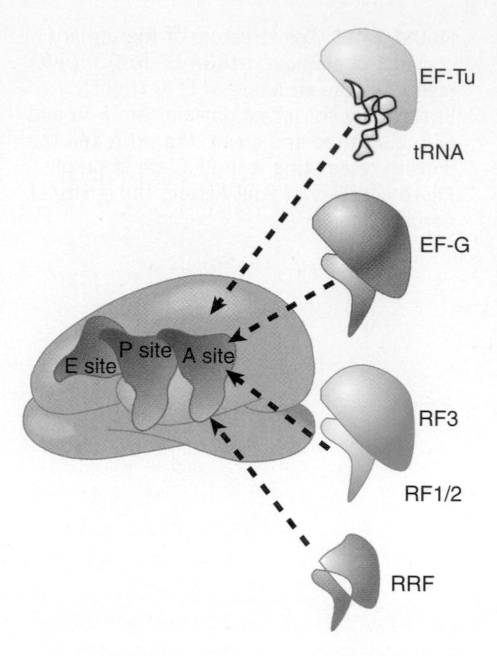

- **release factor (RF)** A protein required to terminate polypeptide translation to cause release of the completed polypeptide chain and the ribosome from mRNA.
- **RF1** The bacterial release factor that recognizes UAA and UAG as signals to terminate polypeptide translation.
- **RF2** The bacterial release factor that recognizes UAA and UGA as signals to terminate polypeptide translation.
- **RF3** A polypeptide translation termination factor related to the elongation factor EF-G. It functions to release the factors RF1 or RF2 from the ribosome when they act to terminate polypeptide translation.

None of the termination codons is represented by a tRNA. They function in an entirely different manner from other codons and are recognized directly by protein factors. (The reaction does not depend on codon-anticodon recognition, so there seems to be no particular reason why it should require a triplet sequence. Presumably this reflects the evolution of the genetic code.)

Termination codons are recognized by class 1 **release factors (RFs)**. In *E. coli*, two class 1 release factors are specific for different sequences. **RF1** recognizes UAA and UAG; **RF2** recognizes UGA and UAA. The factors act at the ribosomal A site and require polypeptidyl-tRNA in the P site. The RFs are present at much lower levels than initiation or elongation factors; there are ~600 molecules of each per cell, equivalent to one RF per ten ribosomes. In eukaryotes, there is only a single class 1 release factor, called eRF1. The efficiency with which the bacterial factors recognize their target codons is influenced by the bases on the 3′ side.

The class 1 release factors are assisted by class 2 release factors, which are not codon specific. The class 2 factors are GTP-binding proteins. In *E. coli*, the role of the class 2 factor, **RF3**, is to release the class 1 factor from the ribosome. RF3 is a GTP-binding protein that is related to the elongation factors.

RF3 resembles the GTP-binding domains of EF-Tu and EF-G, and RF1 and RF2 resemble the C-terminal domain of EF-G, which mimics tRNA. This suggests that the release factors utilize the same site that is used by the elongation factors. **FIGURE 24.26** illustrates the basic idea that these factors all have the same general shape and bind to the ribosome successively at the same site (basically the A site or a region extensively overlapping with it).

The eukaryotic class 1 release factor, eRF1, is a single protein that recognizes all three termination codons. Its sequence is unrelated to the bacterial factors. It can terminate protein synthesis *in vitro* without the class 2 factor, eRF2, although eRF2 is essential in yeast *in vivo*. The structure of eRF1 follows a familiar theme: **FIGURE 24.27** shows that it consists of three domains that mimic the structure of tRNA.

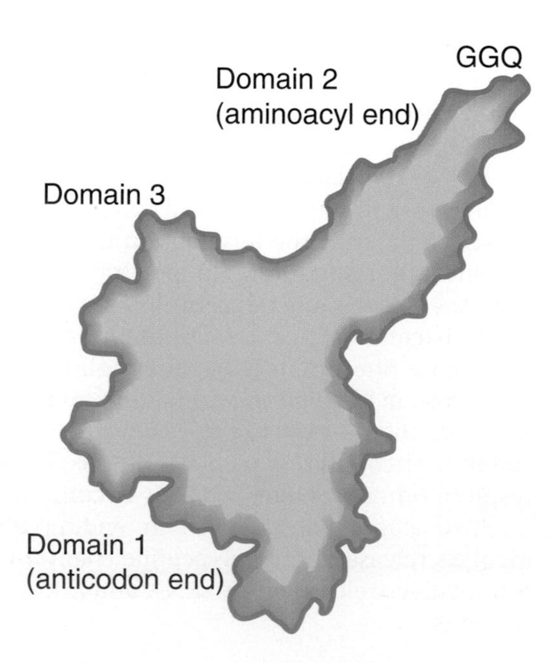

FIGURE 24.27 The eukaryotic termination factor eRF1 has a structure that mimics tRNA. The motif GGQ at the tip of domain 2 is essential for hydrolyzing the polypeptide chain from tRNA.

HISTORICAL PERSPECTIVES

The Naming of the Amber, Ochre, and Opal Codons

Amber, ochre, and opal codons are the names for the three stop codons (UAG, UAA, and UGA, respectively) used to terminate translation *in vivo*. These codons do not specify any amino acid but instead recruit release factors to terminate translation. The naming of these codons and the discovery of the nucleotide triplets that specify the stop codons date back to genetic studies of the bacteriophage T4. In the late 1950s and early 1960s, bacterial geneticists in the United States and England were isolating suppressible mutations in bacteriophage capable of growth in some bacterial strains (permissive strains) but not able to grow in others (nonpermissive strains). Furthermore, these mutations were known to result from so-called nonsense codons, nucleotide triplets that do not encode any amino acid. The presence of these nonsense mutations in coding regions interrupts the reading frame of the transcripts and results in truncated polypeptides. This premature termination of translation is suppressed when the mutant phage are grown in a permissive bacterial strain. The differences between the permissive and nonpermissive bacterial strains are mutations in suppressor genes capable of suppressing the nonsense mutations. Suppressor genes are required to generate suppressor tRNAs that have altered anticodon sequences, such that they recognize and pair with a specific nonsense codon and insert an amino acid.

The term *amber mutation* is attributed to Harris Bernstein, at the time a graduate student at Caltech in Pasadena, California. Bernstein was recruited by two other biologists at Caltech, Dick Epstein and Charley Steinberg, to settle a scientific dispute with a series of experiments designed to isolate mutations of bacteriophage T4 that were specific to one strain of bacteria. These experiments required the plating and isolation of thousands of bacteriophage plaques, and Bernstein was enticed to participate by an offer that he could name the mutations after himself. The mutations were named "amber," which is the English translation of the German name "Bernstein." Amber is a yellowish-brown gemstone that is actually fossilized tree resin. Soon it became clear that these amber mutations—and several others isolated in different labs by different methods—all shared one characteristic: they are all capable of growing in the same permissive strains. Sydney Brenner and his colleagues proposed in a paper published in 1965 that all the different suppressible mutations, including Bernstein's mutations, be referred to as amber mutations. Brenner and others also isolated additional nonsense mutations that appeared to be distinct; they were suppressible by a different set of suppressor genes. These were referred to as "ochre mutations"; ochre is a reddish-brown mineral pigment. Brenner, Stretton, and Kaplan demonstrated that the amber and ochre nonsense mutations result in the nucleotide triplets UAG and UAA, respectively. Furthermore, they postulated that these were the chain-terminating codons required for normal termination of the translation of a polypeptide. In 1967, Brenner, Barnett, Katz, and Crick demonstrated that the nucleotide triplet UGA also results from a nonsense mutation causing translation termination. In keeping with the theme of colored minerals, this nonsense codon came to be known as "opal," which is a type of silica often used as a gemstone and found in a variety of colors. These three nonsense mutations, UAG, UAA, and UGA, are now recognized as the three stop codons and are called amber, ochre, and opal codons, respectively.

The termination reaction releases the completed polypeptide but leaves a deacylated tRNA and the mRNA still associated with the ribosome. **FIGURE 24.28** shows that the dissociation of the remaining components (tRNA, mRNA, 30S, and 50S subunits) requires *ribosome recycling factor (RRF)*. RRF acts together with EF-G in a reaction that uses hydrolysis of GTP. As for the other factors involved in release, RRF has a structure that mimics tRNA, except that it lacks an equivalent for the 3′ amino acid-binding region. IF-3 is also required. RRF acts on the 50S subunit, and IF-3 acts to remove deacylated tRNA from the 30S subunit. Once the subunits have separated, IF-3 remains necessary, of course, to prevent their reassociation.

FIGURE 24.29 compares the functional and sequence homologies of the prokaryotic and eukaryotic translation factors.

FIGURE 24.28 The RF (release factor) terminates translation by releasing the polypeptide chain. The RRF (ribosome recycling factor) releases the last tRNA, and EF-G releases RRF, causing the ribosome to dissociate.

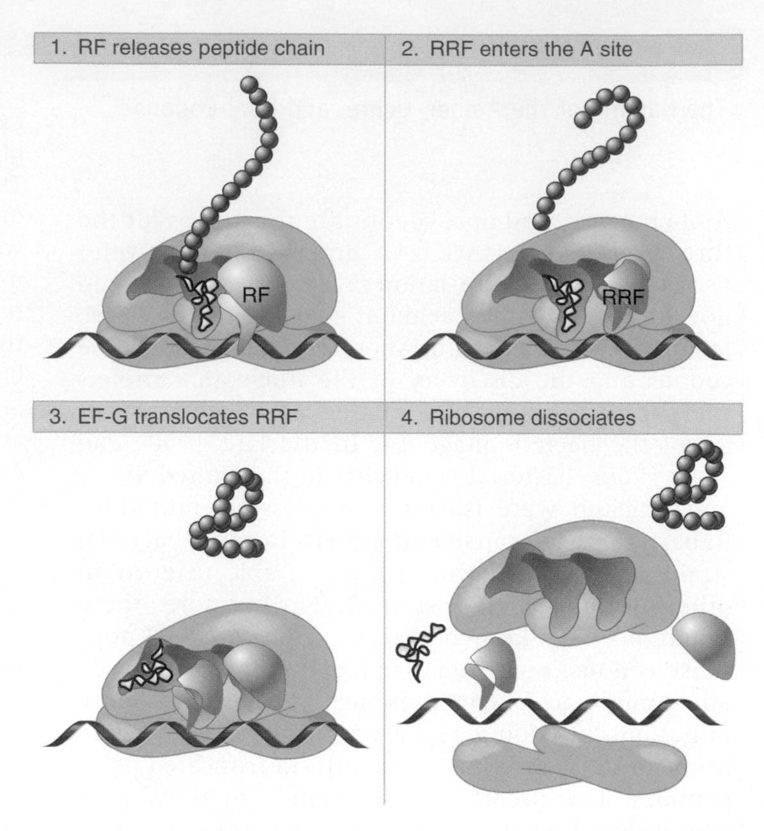

Initiation Factors			
Prokaryotic	Eukaryotic	General Function	Notes
IF-1	eIF1A	Blocks A site	eIF1A assists eIF2 in promoting Met-tRNA$_i^{Met}$ to binding to 40S; also promotes subunit dissociation
IF-2*†	eIF2, eIF3, eIF5B*	Entry of initiator tRNA	eIF2 is a GTPase eIF3 stimulates formation of the ternary complex, its binding to 40S, and binding and scanning of mRNA eIF5B is involved in initiator tRNA entry and is a GTPase
IF-3	eIF1, eIF4 complex, eIF3	Small subunit binding to mRNA	eIF4 complex functions in cap binding
Elongation Factors			
Prokaryotic	Eukaryotic	General Function	
EF-Tu†‡, EF-G†	eEF1α‡	GTP binding	
EF-Ts	eEF1β, eEF1γ	GDP exchanging	
EF-G§	eEF2§	Ribosome translocation	
Release Factors			
Prokaryotic	Eukaryotic	General Function	
RF1	eRF1	UAA/UAG recognition	
RF2	eRF1	UAA/UGA recognition	
RF3†	eRF3	Stimulation of other RF(s)	

* IF-2 and eIF5B have sequence homology.
† IF-2, EF-Tu, EF-G, and RF3 have sequence homology.
‡ EF-Tu and eEF1α have sequence homology.
§ EF-G and eEF2 have sequence homology.

FIGURE 24.29 Functional homologies of prokaryotic and eukaryotic translation factors.

KEY CONCEPTS

- The codons UAA (ochre), UAG (amber), and UGA (opal) terminate translation.
- In bacteria they are used most often with relative frequencies UAA > UGA > UAG.
- Termination codons are recognized by protein release factors, not by aminoacyl-tRNAs.
- The structures of the class 1 release factors (RF1 and RF2 in *E. coli*) resemble aminoacyl-tRNA·EF-Tu and EF-G.
- The class 1 release factors respond to specific termination codons and hydrolyze the polypeptide-tRNA linkage.
- The class 1 release factors are assisted by class 2 release factors (such as RF3) that depend on GTP.
- The mechanism is similar in bacteria (which have two types of class 1 release factors) and eukaryotes (which have only one class 1 release factor).

CONCEPT AND REASONING CHECK

Why do class 1 release factors, aminoacyl-tRNA·EF-Tu, and EF-G all have a similar three-dimensional conformation?

24.13 Ribosomal RNA Pervades Both Ribosomal Subunits

Two-thirds of the mass of the bacterial ribosome is made up of rRNA. The most penetrating approach to analyzing the secondary structure of large RNAs is to compare the sequences of corresponding rRNAs in related organisms. Those regions that are important in the secondary structure retain the ability to interact by base pairing. Thus if a base pair is required, it can form at the same relative position in each rRNA. This approach has enabled detailed models of both 16S and 23S rRNA to be constructed.

Each of the major rRNAs has a secondary structure with several discrete domains. Four general domains are formed by 16S rRNA, in which just under half of the sequence is base paired (see Figure 24.38). Six general domains are formed by 23S rRNA. The individual double-helical regions tend to be short (<8 bp). Often the duplex regions are not perfect and contain bulges of unpaired bases. Comparable models have been drawn for mitochondrial rRNAs (which are shorter and have fewer domains) and for eukaryotic cytosolic rRNAs (which are longer and have more domains). The greater length of eukaryotic rRNAs is due largely to the acquisition of sequences representing additional domains. The crystal structure of the ribosome shows that in each subunit the domains of the major rRNA fold independently and have discrete locations.

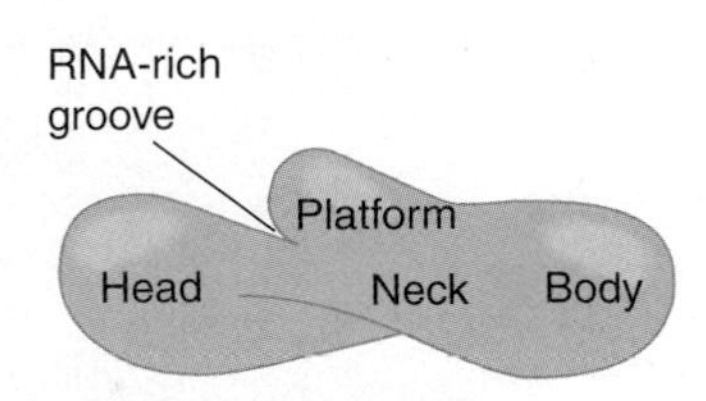

FIGURE 24.30 The 30S subunit has a head separated by a neck from the body, with a protruding platform.

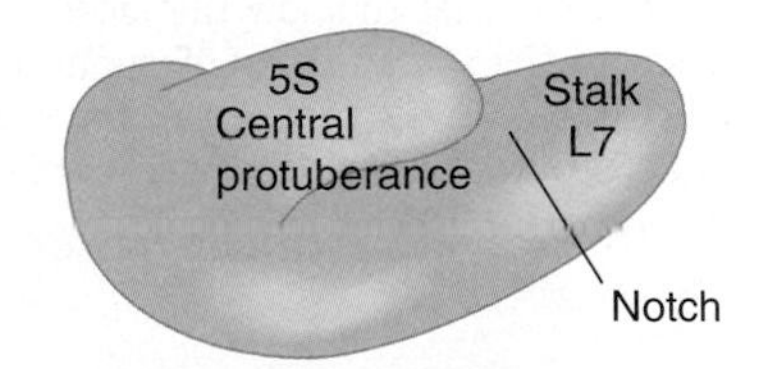

FIGURE 24.31 The 50S subunit has a central protuberance where 5S rRNA is located, separated by a notch from a stalk made of copies of the protein L7.

The 70S ribosome has an asymmetric structure. **FIGURE 24.30** shows a schematic of the structure of the 30S subunit, which is divided into four regions: the head, neck, body, and platform. **FIGURE 24.31** shows a similar representation of the 50S subunit, where two prominent features are the central protuberance (where 5S rRNA is located) and the stalk (made of multiple copies of protein L7). **FIGURE 24.32** shows that the platform of the small subunit fits into the notch of the large subunit. There is a cavity (resembling a doughnut, but not visible in Figure 24.32) between the subunits that contains some of the important sites.

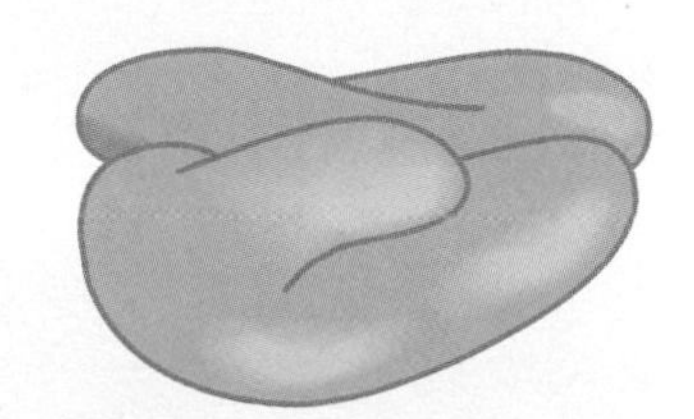

FIGURE 24.32 The platform of the 30S subunit fits into the notch of the 50S subunit to form the 70S ribosome.

The structure of the 30S subunit follows the organization of 16S rRNA, with each structural feature corresponding to a domain of the rRNA. The body is based on the 5′ domain, the platform on the central domain, and the head on the 3′ region. Figure 24.3 shows that the 30S subunit has an asymmetrical distribution of RNA and protein. One important feature is that the platform of the 30S subunit that provides the interface with the 50S subunit is composed almost entirely of RNA. At most two proteins (a small part of S7 and possibly part of S12) lie near the interface. This means that the association and

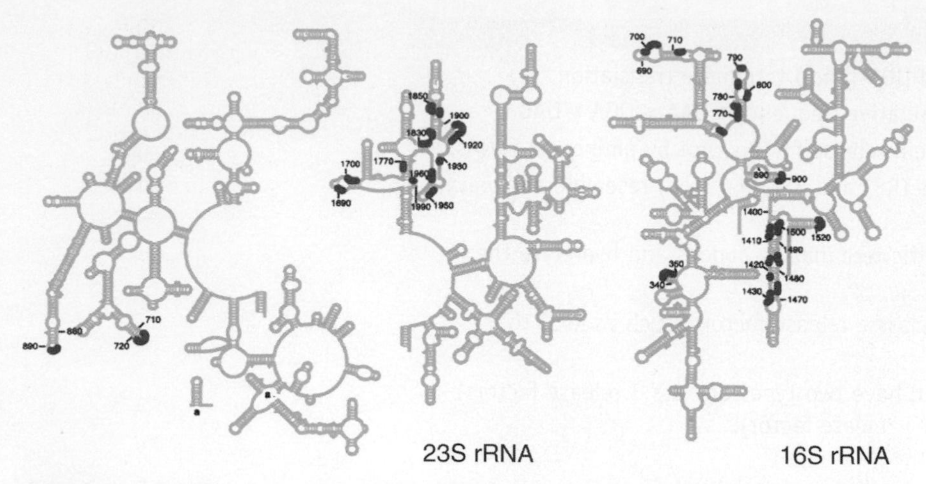

FIGURE 24.33 Contact points between the rRNAs are located in two domains of 16S rRNA and one domain of 23S rRNA. Reproduced from M. M. Yusupov, *Science* 292 (2001): 883–896.

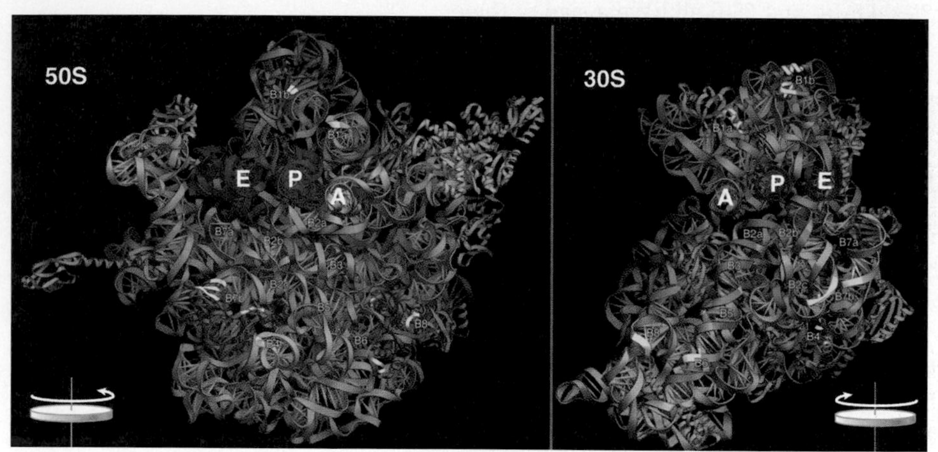

FIGURE 24.34 Contacts between the ribosomal subunits are mostly made by RNA (shown in purple). Contacts involving proteins are shown in yellow. The two subunits are rotated away from one another to show the faces where contacts are made; from a plane of contact perpendicular to the screen, the 50S subunit is rotated 90° counterclockwise, and the 30S is rotated 90° clockwise (this shows it in the reverse of the usual orientation). Photos courtesy of Harry Noller, University of California, Santa Cruz.

dissociation of ribosomal subunits must depend on interactions with the 16S rRNA. This observation supports the idea that the evolutionary origin of the ribosome may have been as a particle consisting of RNA rather than protein.

The 50S subunit has a more even distribution of components than the 30S subunit, with long rods of double-stranded RNA crisscrossing the structure. The RNA forms a mass of tightly packed helices. The exterior surface largely consists of protein, except for the peptidyl transferase center (see *Section 24.15, Two rRNAs Play Active Roles in Translation*). Almost all segments of the 23S rRNA interact with protein, but many of the proteins are relatively unstructured.

The junction of subunits in the 70S ribosome involves contacts between 16S rRNA (many in the platform region) and 23S rRNA. There are also some interactions between rRNAs of each subunit and proteins in the other and a few protein-protein contacts. **FIGURE 24.33** identifies the contact points on the rRNA structures. **FIGURE 24.34** opens out the structure (imagine the 50S subunit rotated counterclockwise and the 30S subunit rotated clockwise around the axis shown in the figure) to show the locations of the contact points on the face of each subunit.

KEY CONCEPTS

- Each rRNA has several distinct domains that fold independently.
- Virtually all ribosomal proteins are in contact with rRNA.
- Most of the contacts between ribosomal subunits are made between the 16S and 23S rRNAs.

CONCEPT AND REASONING CHECK

What evidence is there that the earliest version of the ribosome may have been composed solely of RNA?

24.14 Ribosomes Have Several Active Centers

The basic ribosomal feature to remember is that it is a cooperative structure that depends on changes in the relationships among its active sites during translation. The active sites are not small, discrete regions like the active centers of enzymes. They are large regions whose construction and activities may depend just as much on the rRNA as on the ribosomal proteins. The crystal structures of the individual subunits and bacterial ribosomes give us a good impression of the overall organization and emphasize the role of the rRNA. The most recent structure, at 3.5 Å resolution, clearly identifies the locations of the tRNAs and the functional sites. We can now account for many ribosomal functions in terms of its structure.

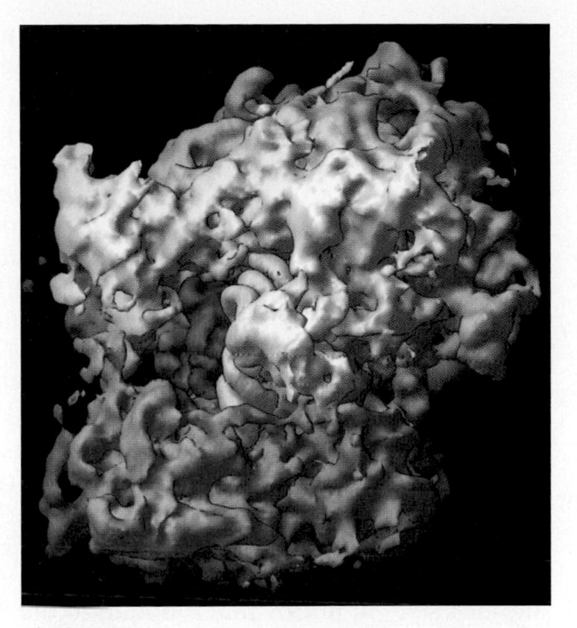

FIGURE 24.35 The 70S ribosome consists of the 50S subunit (white) and the 30S subunit (purple), with three tRNAs located superficially: yellow in the A site, blue in the P site, and green in the E site. Photo courtesy of Harry Noller, University of California, Santa Cruz.

Ribosomal functions are centered around the interactions with tRNAs. **FIGURE 24.35** shows the 70S ribosome with the positions of tRNAs in the three binding sites. The tRNAs in the A and P sites are nearly parallel to one another. All three tRNAs are aligned with their anticodon loops bound to the mRNA in the groove on the 30S subunit. The rest of each tRNA is bound to the 50S subunit. The environment surrounding each tRNA is mostly provided by rRNA. In each site, the rRNA contacts the tRNA at parts of the structure that are universally conserved.

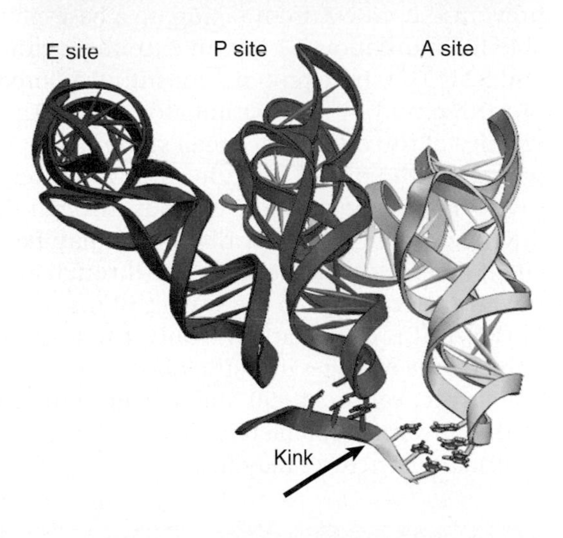

FIGURE 24.36 Three tRNAs have different orientations on the ribosome. mRNA turns between the P and A sites to allow aminoacyl-tRNAs to bind adjacent codons. Photo courtesy of Harry Noller, University of California, Santa Cruz.

It has always been a puzzle to understand how two bulky tRNAs can fit next to one another in reading adjacent codons. The crystal structure shows a 45° kink in the mRNA between the P and A sites, which allows the tRNAs to fit as shown in the expansion of **FIGURE 24.36**. The tRNAs in the P and A sites are angled at 26° relative to each other at their anticodons. The closest approach between the backbones of the tRNAs occurs at the 3′ ends, where they converge to within 5 Å (perpendicular to the plane of the page). This allows the peptide chain to be transferred from the peptidyl-tRNA in the P site to the aminoacyl-tRNA in the A site.

Translocation involves large movements in the positions of the tRNAs within the ribosome. The anticodon end of tRNA moves ~28 Å from the A site to the P site and then moves an additional 20 Å from the P site to the E site. As a result of the angle of each tRNA relative to the anticodon, the bulk of the tRNA moves much larger distances: 40 Å from the A site to the P site and 55 Å from the P site to the E site. This suggests that translocation requires a major reorganization of structure.

Much of the structure of the bacterial ribosome is occupied by its active centers. The schematic view of the ribosomal sites in **FIGURE 24.37** shows they comprise about two-thirds of the ribosomal structure. A tRNA enters the A site, is transferred by translocation into the P site, and then leaves the ribosome by the E site. The A and P sites extend across both ribosome subunits; tRNA is

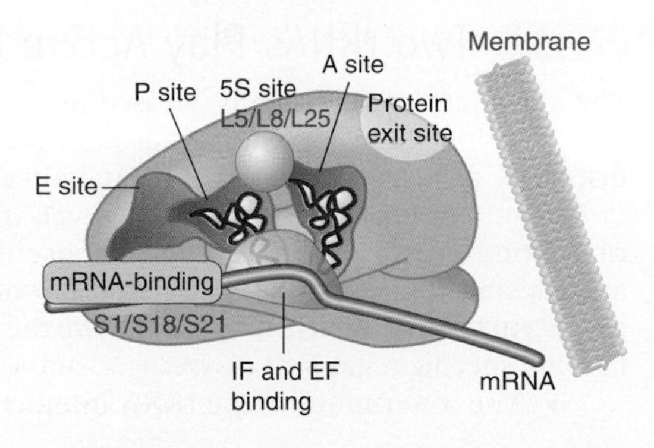

FIGURE 24.37 The ribosome has several active centers. It may be associated with a membrane. mRNA takes a turn as it passes through the A and P sites, which are angled with regard to each other. The E site lies beyond the P site. The peptidyl transferase site (not shown) stretches across the tops of the A and P sites. Part of the site bound by EF-Tu/G lies at the base of the A and P sites.

paired with mRNA in the 30S subunit, but peptide transfer takes place in the 50S subunit. The A and P sites are adjacent, enabling translocation to move the tRNA from one site into the other. The E site is located near the P site (representing a position *en route* to the surface of the 50S subunit). The peptidyl transferase center is located on the 50S subunit, close to the aminoacyl ends of the tRNAs in the A and P sites (see *Section 24.15, Two rRNAs Play Active Roles in Translation*).

All the GTP-binding proteins that function in translation (EF-Tu, EF-G, IF-2, and RF1, RF2, and RF3) bind to the same factor-binding site (sometimes called the GTPase center), which probably triggers their hydrolysis of GTP. This site is located at the base of the stalk of the large subunit, which consists of the proteins L7 and L12. (L7 is a modification of L12 and has an acetyl group on the N-terminus.) In addition to this region, the complex of protein L11 with a 58-base stretch of 23S rRNA provides the binding site for some antibiotics that affect GTPase activity. Neither of these ribosomal structures actually possesses GTPase activity, but they are both necessary for it. The role of the ribosome is to trigger GTP hydrolysis by factors bound in the factor-binding site.

Initial binding of 30S subunits to mRNA requires protein S1, which has a strong affinity for single-stranded nucleic acid. It is responsible for maintaining the single-stranded state in mRNA that is bound to the 30S subunit. This action is necessary to prevent the mRNA from taking up a base-paired conformation that would be unsuitable for translation. S1 has an extremely elongated structure and associates with S18 and S21. The three proteins constitute a domain that is involved in the initial binding of mRNA and in binding initiator tRNA. This locates the mRNA-binding site in the vicinity of the cleft of the small subunit (see Figure 24.4). The 3′ end of rRNA, which pairs with the mRNA initiation site, is located in this region.

A nascent polypeptide extends through the ribosome, away from the active sites, into the region in which ribosomes may be attached to membranes. A polypeptide chain emerges from the ribosome through an exit channel, which leads from the peptidyl transferase site to the surface of the 50S subunit. The tunnel is composed mostly of rRNA. It is quite narrow—only 1 to 2 nm wide—and is ~10 nm long. The nascent polypeptide emerges from the ribosome ~15 Å away from the peptidyl transferase site. The tunnel can hold ~50 amino acids and probably constrains the polypeptide chain so that it cannot completely fold until it leaves the exit domain, though some limited secondary structures may form.

KEY CONCEPTS

- Interactions involving rRNA are a key part of ribosome function.
- The environment of the tRNA-binding sites is largely determined by rRNA.

CONCEPT AND REASONING CHECK

What are the paths of tRNAs and the mRNA through the bacterial ribosome?

24.15 Two rRNAs Play Active Roles in Translation

The ribosome was originally viewed as a collection of proteins with various catalytic activities held together by protein-protein and RNA-protein interactions. The discovery of RNA molecules with catalytic activities (see *Chapter 21, RNA Splicing and Processing*) immediately suggests, however, that rRNA might play a more active role in ribosome function. There is now evidence that rRNA interacts with mRNA or tRNA at each stage of translation, and that the proteins are necessary to maintain the rRNA in a structure in which it can perform the catalytic functions. Several interactions involve specific regions of rRNA:

- The 3′ terminus of the rRNA interacts directly with mRNA at initiation.

- Specific regions of 16S rRNA interact directly with the anticodon regions of tRNAs in both the A site and the P site. Similarly, 23S rRNA interacts with the CCA terminus of peptidyl-tRNA in both the P site and the A site.
- Subunit interaction involves interactions between 16S and 23S rRNAs (see *Section 24.13, Ribosomal RNA Pervades Both Ribosomal Subunits*).

The functions of rRNA have been investigated by two types of approach. Structural studies show that particular regions of rRNA are located in important sites of the ribosome and that chemical modifications of these bases impede particular ribosomal functions. In addition, mutations identify bases in rRNA that are required for particular ribosomal functions. **FIGURE 24.38** summarizes the sites in 16S rRNA that have been identified by these means.

The tRNA makes contacts with the 23S rRNA in both the P and A sites. The classic criterion for demonstrating the importance of the interaction is satisfied: a mutation in tRNA can be compensated by a mutation in rRNA, so there is a required role for rRNA in both the tRNA-binding sites. Indeed, we are moving toward describing the movements of tRNA between the A and P sites in terms of making and breaking contacts with rRNA.

What is the nature of the site on the 50S subunit that provides peptidyl transferase function? A long search for ribosomal proteins that might possess the catalytic activity was unsuccessful and led to the discovery that 23S rRNA can catalyze the formation of a peptide bond between peptidyl-tRNA and aminoacyl-tRNA. But since the rRNA has the basic catalytic activity, the role of the proteins must be indirect, serving to fold the rRNA properly or to present the substrates to it. Activity is abolished by mutations in domain V, which lies in the P site. The crystal structure of an archaeal 50S subunit shows that the peptidyl transferase site basically consists of 23S

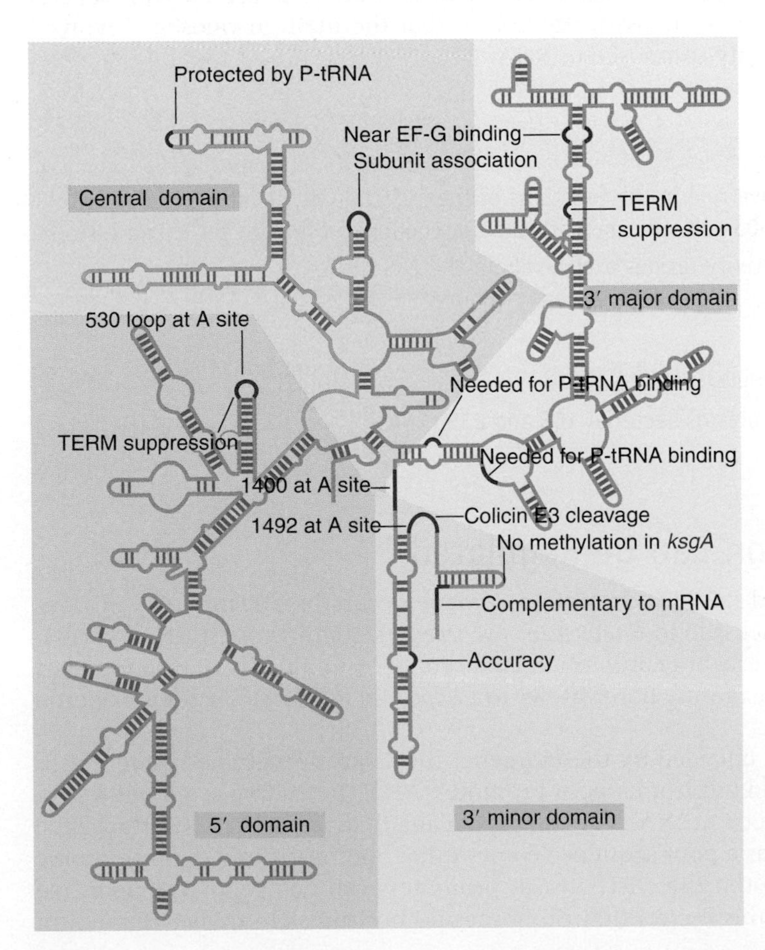

FIGURE 24.38 Some sites in 16S rRNA are protected from chemical probes when 50S subunits join 30S subunits or when aminoacyl-tRNA binds to the A site. Others are the sites of mutations that affect translation. TERM suppression sites may affect termination at some or several termination codons. The large colored blocks indicate the four domains of the rRNA.

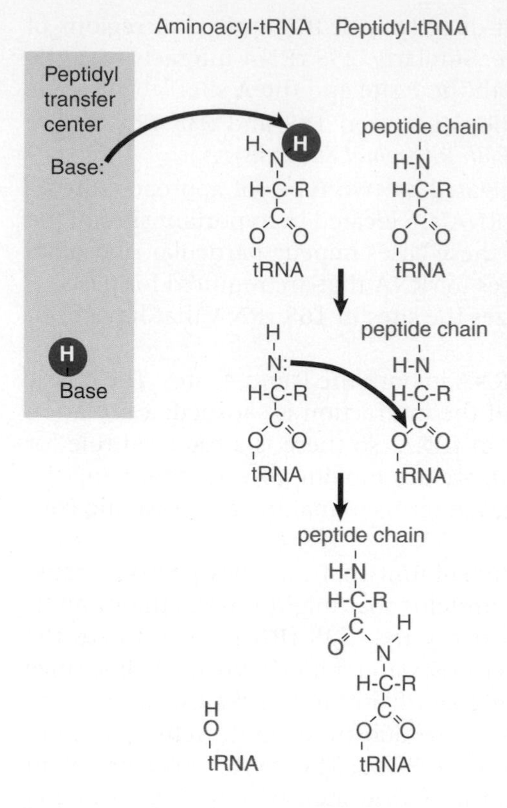

FIGURE 24.39 Peptide bond formation requires acid-base catalysis in which an H atom is transferred to a basic residue.

rRNA. There is no protein within 18 Å of the active site where the transfer reaction occurs between peptidyl-tRNA and aminoacyl-tRNA!

Peptide bond synthesis requires an attack by the amino group of one amino acid on the carboxyl group of another amino acid. Catalysis requires a basic residue to accept the hydrogen atom that is released from the amino group, as shown in **FIGURE 24.39**. If rRNA is the catalyst, it must provide this residue, but we do not know how this happens. The purine and pyrimidine bases are not basic at physiological pH. A highly conserved base (at position 2451 in *E. coli*) had been implicated but appears now neither to have the right properties nor to be crucial for peptidyl transferase activity.

The catalytic activity of isolated rRNA is quite low, and proteins that are bound to the 23S rRNA outside of the peptidyl transfer region are almost certainly required to enable the rRNA to form the proper structure *in vivo*. The idea that rRNA is the catalytic component is consistent with the results discussed in *Chapter 21, RNA Splicing and Processing*, which identify catalytic properties in RNA that are involved with several RNA processing reactions. It fits with the notion that the modern ribosome evolved from a prototype originally composed of RNA.

KEY CONCEPTS

- 16S rRNA plays an active role in the functions of the 30S subunit. It interacts directly with mRNA, with the 50S subunit, and with the anticodons of tRNAs in the P and A sites.
- Peptidyl transferase activity resides exclusively in the 23S rRNA.

CONCEPT AND REASONING CHECK

What are the functional roles of bacterial 16S and 23S rRNAs?

24.16 Translation Can Be Regulated

Control over which and how much protein is made occurs first at the level of transcription control (as discussed in *Chapter 26, The Operon*), then through RNA processing control (rare in bacteria but common in eukaryotes), and, finally, translation level control, which we will examine here. (Refer to *Chapter 26* for detail on the *lac* operon and its regulation.)

The *lac* repressor is encoded by the *lacI* gene; this is an unregulated gene that is continuously transcribed but from a poor promoter. Also, the coding region of the *lac* repressor is in a very poor mRNA. This simply means that the 5′ UTR (untranslated region) of the mRNA has a poor sequence context that does not allow rapid ribosome binding or movement onto the ORF. Just as promoters can be "good" or "poor," so can mRNAs. Together, this means that ribosomes do not translate the small amount of mRNA at the same level as the *LacZYA* polycistronic mRNA. Thus we find very little *lac* repressor in a cell—only about 10 tetramers.

A second way that translation can be modulated is by codon usage. Multiple codons exist for most of the amino acids. These codons are not decoded equally by tRNAs. Some have abundant tRNAs and some do not. An ORF consisting of codons with abundant tRNAs can be rapidly translated, whereas another ORF that contains codons with less abundant tRNAs will be translated much more slowly.

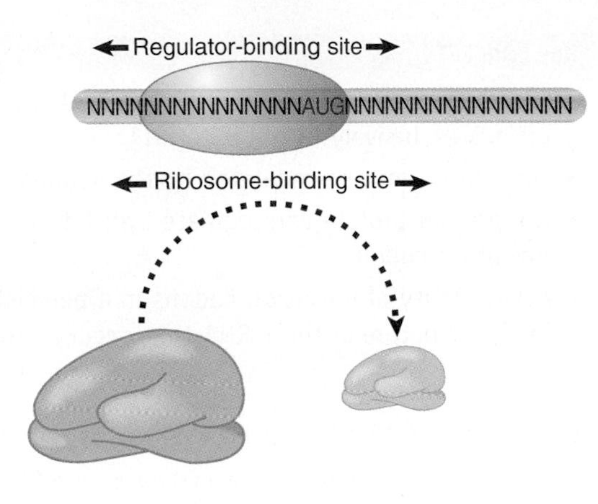

FIGURE 24.40 A regulator protein may block translation by binding to a site on mRNA that overlaps the ribosome-binding site at the initiation codon.

Additional, more active mechanisms exist for translation-level control. One mechanism for controlling gene expression at the level of translation is a parallel to the use of a repressor to prevent transcription. Translational repression occurs when a protein binds to a target region on mRNA to prevent ribosomes from recognizing the initiation region. Formally, protein-mRNA binding is equivalent to a repressor protein binding to DNA to prevent polymerase from utilizing a promoter. Polycistronic RNA allows coordinate regulation of translation, analogous to transcription repression of an operon. **FIGURE 24.40** illustrates the most common form of this interaction, in which the regulator protein binds directly to a sequence that includes the AUG initiation codon, thereby preventing the ribosome from binding.

Some examples of translational repressors and their targets are summarized in **FIGURE 24.41**. A classic example of how the product of translation can directly control the translation of its mRNA is the coat protein of the RNA phage R17; it binds to a hairpin that encompasses the ribosome-binding site in the phage mRNA. Similarly, the phage T4 RegA protein binds to a consensus sequence that includes the AUG initiation codon in several T4 early mRNAs, and T4 DNA polymerase binds to a sequence in its own mRNA that includes the Shine-Dalgarno element needed for ribosome binding.

Another form of translational control occurs when translation of one gene requires changes in secondary structure that depend on translation of an immediately preceding gene. This happens during translation of the RNA phages, whose genes always are expressed in a set order. **FIGURE 24.42** shows that the phage RNA takes up a secondary structure in which only one initiation sequence is accessible; the second cannot be recognized by ribosomes because it is base-paired with other regions of the RNA. Translation of the first gene, however, disrupts the secondary structure, allowing ribosomes to bind to the initiation site of the next gene. In this mRNA, secondary structure controls translatability.

Repressor	Target Gene	Site of Action
R17 coat protein	R17 replicase	Hairpin that includes ribosome-binding site
T4 RegA	Early T4 mRNAs	Various sequences including initiation codon
T4 DNA polymerase	T4 DNA polymerase	Shine-Dalgarno sequence
T4 p32	Gene 32	Single-stranded 5′ leader

FIGURE 24.41 Proteins that bind to sequences within the initiation regions of mRNAs may function as translational repressors.

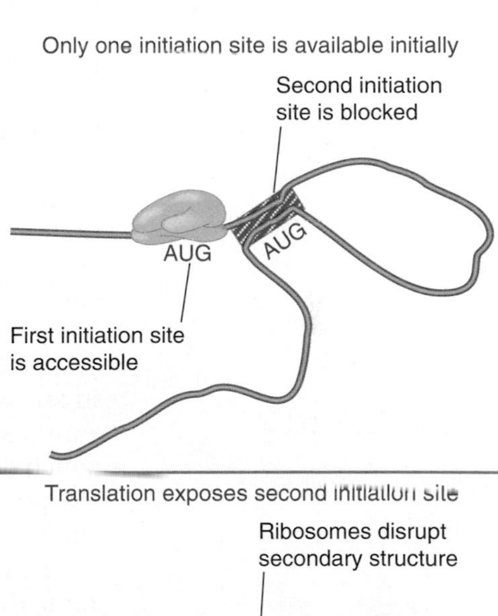

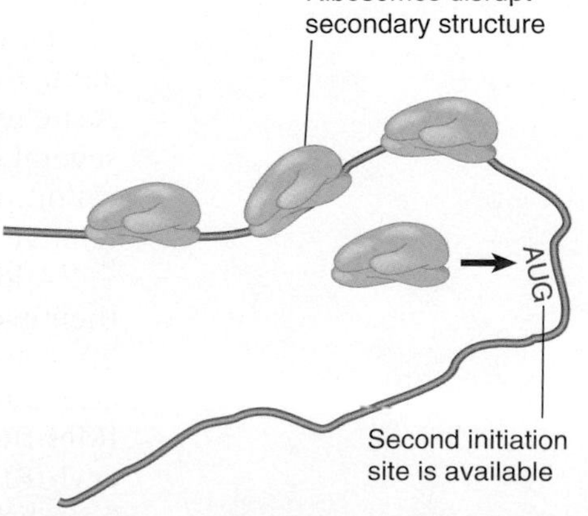

FIGURE 24.42 Secondary structure can control initiation. Only one initiation site is available in the RNA phage, but translation of the first cistron changes the conformation of the RNA so that other initiation site(s) become available.

KEY CONCEPTS

- Translation can be regulated by the 5′ UTR of the mRNA. Peptidyl transferase activity resides exclusively in the 23S rRNA.
- Translation may be regulated by the abundance of various tRNAs.
- A repressor protein can regulate translation by preventing a ribosome from binding to an initiation codon.
- Accessibility of initiation codons in a polycistronic mRNA can be controlled by changes in the structure of the mRNA that occur as the result of translation.

CONCEPT AND REASONING CHECK

What are the many parts that make up an mRNA molecule?

24.17 Summary

A codon in mRNA is recognized by an aminoacyl-tRNA, which has an anticodon complementary to the codon and carries the amino acid corresponding to the codon. A special initiator tRNA (fMet-$tRNA_f$ in prokaryotes or Met-$tRNA_i$ in eukaryotes) recognizes the AUG codon, which is used to start all coding sequences. In prokaryotes, GUG is also used. Only the termination (stop) codons, UAA, UAG, and UGA, are not recognized by aminoacyl-tRNAs.

Ribosomes are released from translation to enter a pool of free ribosomes that are in equilibrium with separate small and large subunits. Small subunits bind to mRNA and then are joined by large subunits to generate an intact ribosome that undertakes translation. Recognition of a prokaryotic initiation site involves binding of a sequence at the 3′ end of rRNA to the Shine-Dalgarno motif, which precedes the AUG (or GUG) codon in the mRNA. Recognition of a eukaryotic mRNA involves binding to the 5′ cap; the small subunit then migrates to the initiation site by scanning for AUG codons. When it recognizes an appropriate AUG codon (usually, but not always, the first it encounters), it is joined by a large subunit.

A ribosome can carry at least two aminoacyl-tRNAs simultaneously: its P site is occupied by a polypeptidyl-tRNA, which carries the polypeptide chain synthesized so far, whereas the A site is used for entry by an aminoacyl-tRNA carrying the next amino acid to be added to the chain. Bacterial ribosomes also have an E site, through which deacylated tRNA passes before it is released after being used in translation. The polypeptide chain in the P site is transferred to the aminoacyl-tRNA in the A site, creating a deacylated tRNA in the P site and a peptidyl-tRNA in the A site.

Following peptide bond synthesis, the bacterial ribosome translocates one codon along the mRNA, moving deacylated tRNA into the E site and peptidyl tRNA from the A site into the P site. Translocation is catalyzed by the elongation factor EF-G and, like several other stages of ribosome function, requires hydrolysis of GTP. During translocation, the ribosome passes through a hybrid stage in which the 50S subunit moves relative to the 30S subunit.

Additional factors are required at each stage of translation. They are defined by their cyclic association with, and dissociation from, the ribosome. Initiation factors are involved in prokaryotic initiation. IF-3 is needed for 30S subunits to bind to mRNA and also is responsible for maintaining the 30S subunit in a free form. IF-2 is needed for fMet-$tRNA_f$ to bind to the 30S subunit and is responsible for excluding other aminoacyl-tRNAs from the initiation reaction. GTP is hydrolyzed after the initiator tRNA has been bound to the initiation complex. The initiation factors must be released in order to allow a large subunit to join the initiation complex.

Eukaryotic initiation involves a greater number of protein factors. Some of them are involved in the initial binding of the 40S subunit to the capped 5′ end of the mRNA, at which point the initiator tRNA is bound by another group of factors. After this initial binding, the small subunit scans the mRNA until it recognizes the correct AUG codon. At this point, initiation factors are released and the 60S subunit joins the complex.

Prokaryotic elongation factors are involved in elongation. EF-Tu binds aminoacyl-tRNA to the 70S ribosome. GTP is hydrolyzed when EF-Tu is released, and EF-Ts is required to regenerate the active form of EF-Tu. EF-G is required for translocation. Binding of the EF-Tu and EF-G factors to ribosomes is mutually exclusive, which ensures that each step must be completed before the next can be started.

Termination occurs at any one of the three special codons, UAA, UAG, and UGA. Class 1 release factors that specifically recognize the termination codons activate the ribosome to hydrolyze the peptidyl-tRNA. A class 2 RF is required to release the class 1 RF from the ribosome. The GTP-binding factors IF-2, EF-Tu, EF-G, and RF3 all have similar structures, with the latter two mimicking the RNA-protein structure of the first two when they are bound to tRNA. They all bind to the same ribosomal site, the G-factor binding site.

Ribosomes are ribonucleoprotein particles in which a majority of the mass is provided by rRNA. The shapes of all ribosomes are generally similar, but only those of bacteria (70S) have been characterized in detail. The small (30S) subunit has a squashed shape, with a "body" containing about two-thirds of the mass divided from the "head" by a cleft. The large (50S) subunit is more spherical, with a prominent "stalk" on the right and a "central protuberance." Approximate locations of all proteins in the small subunit are known.

Each subunit contains a single major rRNA, 16S and 23S in prokaryotes. Both major rRNAs have extensive base pairing, mostly in the form of short, imperfectly paired duplex stems with single-stranded loops. Conserved features in the rRNA can be identified by comparing sequences and the secondary structures that can be drawn for rRNA of a variety of organisms. The 16S rRNA has four distinct domains; the 23S rRNA has six distinct domains. Eukaryotic rRNAs have additional domains.

The crystal structure shows that the 30S subunit has an asymmetrical distribution of RNA and protein. RNA is concentrated at the interface with the 50S subunit. The 50S subunit has a surface of protein, with long rods of double-stranded RNA crisscrossing the structure. 30S-to-50S joining involves contacts between 16S rRNA and 23S rRNA.

Each subunit has several active centers, which are concentrated in the translational domain of the ribosome where polypeptides are synthesized. Polypeptides leave the ribosome through the exit domain, which can associate with a membrane. The major active sites are the P and A sites, the E site, the EF-Tu and EF-G binding sites, peptidyl transferase, and the mRNA-binding site. Ribosome conformation may change at stages during translation; differences in the accessibility of particular regions of the major rRNAs have been detected.

The tRNAs in the A and P sites are parallel to one another. The anticodon loops are bound to mRNA in a groove on the 30S subunit. The rest of each tRNA is bound to the 50S subunit. A conformational shift of tRNA within the A site is required to bring its aminoacyl end into juxtaposition with the end of the peptidyl-tRNA in the P site. The peptidyl transferase site that links the P- and A-binding sites is a domain of the 23S rRNA, which has the peptidyl transferase catalytic activity, although proteins are probably needed to acquire the right structure.

An active role for the rRNAs in translation is indicated by mutations that affect ribosomal function, interactions with mRNA or tRNA that can be detected by chemical crosslinking, and the requirement to maintain individual base pairing interactions with the tRNA or mRNA. The 3′ terminal region of the rRNA base pairs with mRNA at initiation. Internal regions make individual contacts with the tRNAs in both the P and A sites. Ribosomal RNA is the target for some antibiotics or other agents that inhibit translation.

Gene expression may be modulated at the level of translation by the ability of an mRNA to attract a ribosome and by the abundance of specific tRNAs that recognize different codons. More active mechanisms that regulate at the level of translation are also found. Translation may be regulated by a protein that can bind to the mRNA to prevent the ribosome from binding.

CHAPTER QUESTIONS

1. How many binding sites for tRNAs are present in each bacterial ribosome?
   A. 1
   B. 2
   C. 3
   D. 4
2. In translating ribosomes, new amino acids are added to the growing peptide chain from the:
   A. P site to the A site.
   B. A site to the P site.
   C. P site to the E site.
   D. A site to the E site.
3. Which of the following is not a termination codon for translation?
   A. UGG
   B. UAG
   C. UGA
   D. UAA
4. What step in transcription/translation generally has the highest error rate?
   A. insertion of the incorrect aminoacyl tRNA into the ribosome
   B. charging of the wrong amino acid onto a tRNA
   C. insertion of the wrong amino acid into a growing protein
   D. insertion of the wrong base in mRNA during transcription
5. What is the usual start codon for translation in both prokaryotes and eukaryotes?
   A. UAG
   B. UGA
   C. AUG
   D. AGU
6. The Shine-Dalgarno sequence occurs in what location in a bacterial mRNA molecule?
   A. about 35 bases upstream of the AUG initiation codon
   B. about 10 bases upstream of the AUG initiation codon
   C. about 15 bases downstream of the AUG initiation codon
   D. about 10 bases downstream of the termination codon
7. What is the basis of interaction between the ribosome and an mRNA molecule in bacteria?
   A. specific protein-RNA interactions between ribosomal proteins and the mRNA
   B. specific base pairing between the 3′ end of the 16S rRNA in the 30S subunit and a conserved sequence in the mRNA
   C. specific base pairing between the 5′ end of the 16S rRNA in the 30S subunit and a conserved sequence in the mRNA
   D. specific interactions between ribosomal proteins and a binding protein that associates with the 5′ end of the mRNA
8. What is the modification to the initiator methionine tRNA in bacteria?
   A. methylation
   B. phosphorylation

C. acetylation
D. formylation

9. The 23S rRNA of the large subunit of bacterial ribosomes interacts with:
   A. large subunit ribosomal proteins only.
   B. the 5′ untranslated region of the mRNA.
   C. the 3′ untranslated region of the mRNA.
   D. the 3′ CCA terminus of peptidyl tRNA in the P and A sites.
10. The coat protein of the RNA phage R17 regulates its own translation by:
   A. triggering degradation of its mRNA.
   B. blocking the ribosome-binding site of its mRNA.
   C. triggering dissociation of ribosome subunits.
   D. blocking the initiation codon of its mRNA.

KEY TERMS

A site
amber codon
context
deacylated tRNA
E site
EF-Tu
elongation
elongation factors
IF-1
IF-2
IF-3
initiation
initiation factors (IFs)
IRES (internal ribosome entry site)
N-formyl-methionyl-tRNA
ochre codon
opal codon
P site
peptidyl transferase
peptidyl-tRNA
release factor (RF)
RF1
RF2
RF3
ribosome-binding site
Shine-Dalgarno sequence
stop codon
termination
translocation
$tRNA_f^{Met}$
$tRNA_m^{Met}$

FURTHER READING

Hellen, C. U., and Sarnow, P. (2001). Internal ribosome entry sites in eukaryotic mRNA molecules. *Genes Dev.* **15**, 1593–1612. A review of the known mechanisms of eukaryotic translation initiation.

Moore, P. B., and Steitz, T. A. (2003). The structural basis of large ribosomal subunit function. *Annu. Rev. Biochem.* **72**, 813–850. A review of the process that led to the detailed crystal structure of the bacterial large ribosomal subunit, the structure itself, and the functions associated with that structure.

Noller, H. F. (2005). RNA structure: reading the ribosome. *Science* **309**, 1508–1514. A review of the roles of ribosomal RNA tertiary structures in tRNA selection, peptidyl transferase activity, and other ribosomal processes.

Ramakrishnan, V. (2002). Ribosome structure and the mechanism of translation. *Cell* **108**, 557–572. A review of the known and as-yet-unknown (at the time of publication) details of bacterial protein translation, based on the recently published ribosome crystal structures.

Schuwirth, B. S., Borovinskaya, M. A., Hau, C. W., Zhang, W., Vila-Sanjurjo, A., Holton, J. M., and Doudna Cate, J. H. (2005). Structures of the bacterial ribosome at 3.5 Å resolution. *Science* **310**, 827–834. A detailed examination of the structure of the complete functional bacterial ribosome.

Selmer, M., Dunham, C. M., Murphy, IV, F. V., Weixlbaumer, A., Petry, S., Kelley, A. C., Weir, J. R., and Ramakrishnan, V. (2006). Structure of the 70S ribosome complexed with mRNA and tRNA. *Science* **313**, 1935–1942. The bacterial ribosomal structure at 2.8 Å resolution.

Stark, H., Rodnina, M. V., Wieden, H. J., van Heel, M., and Wintermeyer, W. (2000). Large-scale movement of elongation factor G and extensive conformational change of the ribosome during translocation. *Cell* **100**, 301–309. The structure of the EF-G-ribosome complex both before and after ribosome translocation, and a proposal for the mechanism of translocation.

Yonath, A. (2005). Antibiotics targeting ribosomes: resistance, selectivity, synergism and cellular regulation. *Annu. Rev. Biochem.* **74**, 649–679. A review of the structural features of the bacterial ribosome targeted by various antibiotics and the molecular bases for resistances to them.

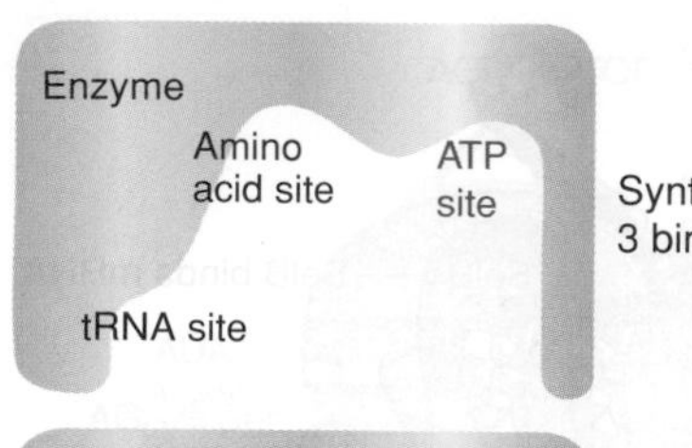

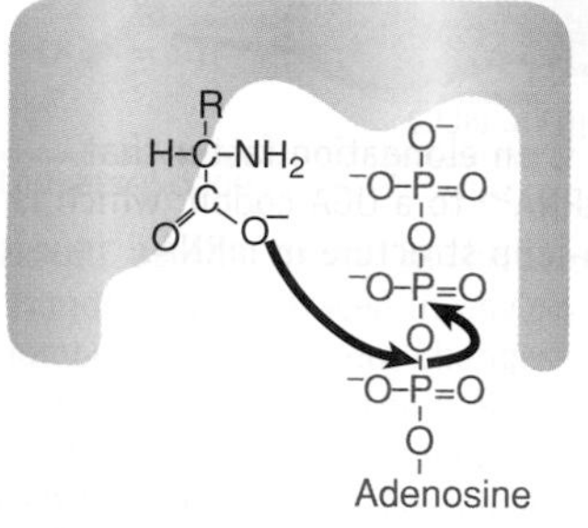

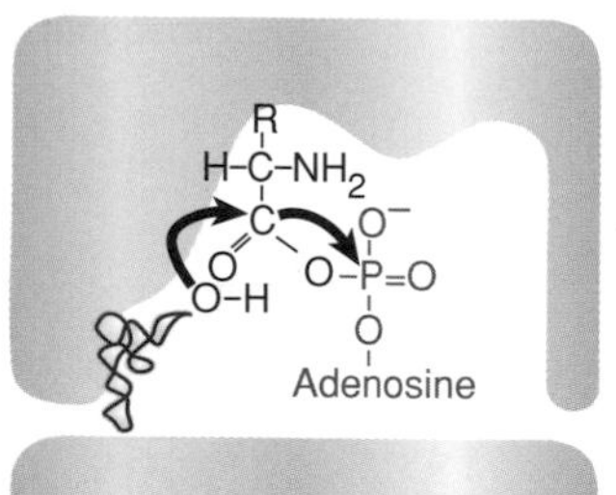

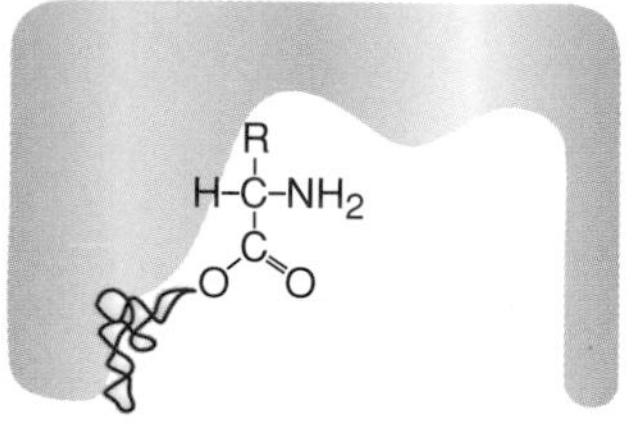

FIGURE 25.12 An aminoacyl-tRNA synthetase charges tRNA with an amino acid.

nucleic acids into the polypeptide sequence. All synthetases function by the two-step mechanism depicted in **FIGURE 25.12**:

- The amino acid first reacts with ATP to form an aminoacyl adenylate intermediate, releasing pyrophosphate. Part of the energy released in ATP hydrolysis is trapped as a high-energy mixed anhydride linkage in the adenylate.
- Next, the tRNA attacks the carbonyl carbon atom of the mixed anhydride, generating aminoacyl-tRNA with concomitant release of AMP.

Each tRNA synthetase is selective for a single amino acid among all the amino acids in the cellular pool. It also discriminates among all tRNAs in the cell. Usually, each amino acid is represented by more than one tRNA. Several tRNAs may be needed to respond to synonymous codons, and sometimes there are multiple species of tRNA that base pair with the same codon. Multiple tRNAs representing the same amino acid are called **isoaccepting tRNAs**; because they are all recognized by the same synthetase, they are also described as its **cognate tRNAs**.

A group of isoaccepting tRNAs must be charged only by the single aminoacyl-tRNA synthetase specific for their amino acid, so isoaccepting tRNAs must share some common feature(s) that enables the enzyme to distinguish them from the other tRNAs. The entire complement of tRNAs is divided into twenty isoaccepting groups, and each group is able to identify itself to its particular synthetase.

Many attempts to deduce similarities in sequence between cognate tRNAs or to induce chemical alterations that affect their charging have shown that the basis for recognition is not the same for different tRNAs, and it does not necessarily lie in some feature of primary or secondary structure alone. tRNAs are identified by their synthetases by contacts that recognize a small number of bases, typically from one to five. Three types of features are commonly used:

- Usually (but not always), at least one base of the anticodon is recognized. Sometimes all the positions of the anticodon are important.
- Often, one of the last three base pairs in the acceptor stem is recognized. An extreme case is represented by alanine tRNA, which is identified by a single unique base pair in the acceptor stem.
- The so-called discriminator base, which lies between the acceptor stem and the CCA terminus, is always invariant among isoacceptor tRNAs.

▸ **isoaccepting tRNAs** *See* **cognate tRNAs**.

▸ **cognate tRNAs** tRNAs recognized by a particular aminoacyl-tRNA synthetase. All are charged with the same amino acid.

No one of these features constitutes a unique means of distinguishing twenty sets of tRNAs or provides sufficient specificity, so it appears that recognition of tRNAs is idiosyncratic, with each following its own rules. Recognition depends on an interaction between a few points of contact in the tRNA, concentrated at the extremities, and a few amino acids constituting the active site in the protein. The relative importance of the roles played by the acceptor stem and anticodon is different for each tRNA-synthetase interaction.

KEY CONCEPTS

- Aminoacyl-tRNA synthetases are a family of enzymes that charge tRNA with an amino acid to generate aminoacyl-tRNA in a two-stage reaction that uses energy from ATP.
- Each tRNA synthetase aminoacylates all the tRNAs in an isoaccepting group, representing a particular amino acid.
- Recognition of a tRNA is based on a particular set of nucleotides that often are concentrated in the acceptor stem and anticodon loop regions of the molecule.

CONCEPT AND REASONING CHECK

What features of a tRNA allow its unique recognition by an aminoacyl-tRNA synthetase?

25.9 Aminoacyl-tRNA Synthetases Fall into Two Classes

In spite of their common function, synthetases are a very diverse group of enzymes. The individual subunits vary in molecular weight from 40 to 110 kD, and the enzymes may be monomeric, dimeric, or tetrameric. A total of 23 types of synthetases have been divided into two general classes on the basis of the structure of the domain that contains the active site. The two classes show no relationship and may have evolved independently of one another. This makes it seem possible that an early form of life with proteins that were made up of just the 11 or so amino acids encoded by one class of synthetase or the other could have existed.

A general model for synthetase-tRNA binding suggests that the enzyme binds the tRNA along the "side" of the L-shaped molecule. The same general principle applies for all synthetase-tRNA binding: the tRNA is bound principally at its two extremities, and most of the tRNA sequence is not involved in recognition by a synthetase. However, the detailed nature of the interaction is different between *class I* and *class II aminoacyl-tRNA synthetases*, as can be seen from the models of **FIGURE 25.13**, which are based on crystal structures. The two types of enzyme approach the tRNA from opposite sides, with the result that the tRNA-protein complexes look almost like mirror images of one another.

A class I synthetase (for example, $tRNA^{Gln}$ synthetase) approaches the D-loop side of the tRNA. It recognizes the minor groove of the acceptor stem at one end of the binding site and interacts with the anticodon loop at the other end. **FIGURE 25.14** is a diagrammatic representation of the crystal structure of the $tRNA^{Gln}$-synthetase complex. A revealing feature of the structure is that contacts with the enzyme change the structure of the tRNA at two important points. These can be seen by comparing the dotted and solid lines in the anticodon loop and acceptor stem.

- Bases U35 and U36 in the anticodon loop are pulled farther out of the tRNA into the enzyme.
- The end of the acceptor stem is seriously distorted, with the result that base pairing between U1 and A72 is disrupted. The single-stranded end of the stem pokes into a deep pocket in the synthetase enzyme, which also contains the binding site for ATP.

This structure shows why changes in U35, G73, or the U1-A72 base pair affect the recognition of the tRNA by its synthetase. At all these positions, hydrogen bonding occurs between the enzyme and tRNA.

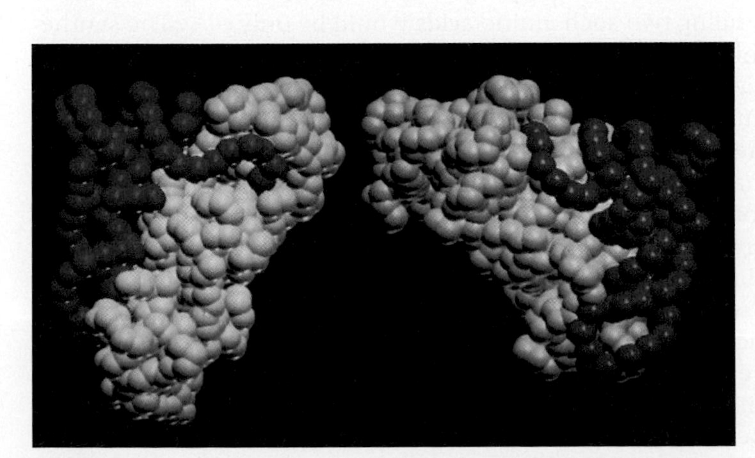

FIGURE 25.13 Crystal structures show that class I and class II aminoacyl-tRNA synthetases bind the opposite faces of their tRNA substrates. The tRNA is shown in red and the protein, in blue. Photo courtesy of Dino Moras, Institute of Genetics and Molecular and Cellular Biology.

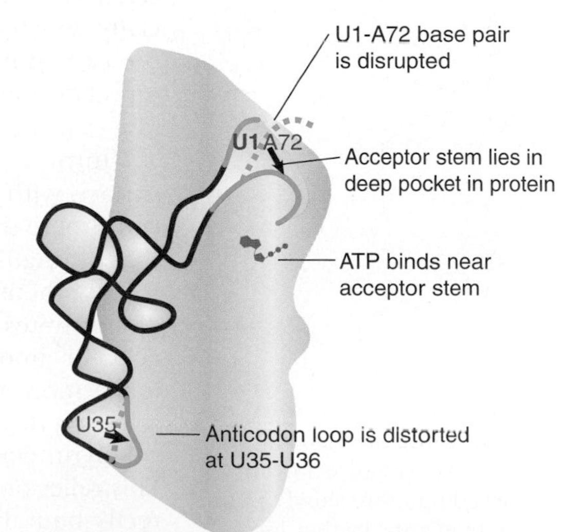

FIGURE 25.14 A class I tRNA synthetase contacts tRNA at the minor groove of the acceptor stem and at the anticodon.

FIGURE 25.15 A class II aminoacyl-tRNA synthetase contacts tRNA at the major groove of the acceptor helix and at the anticodon loop.

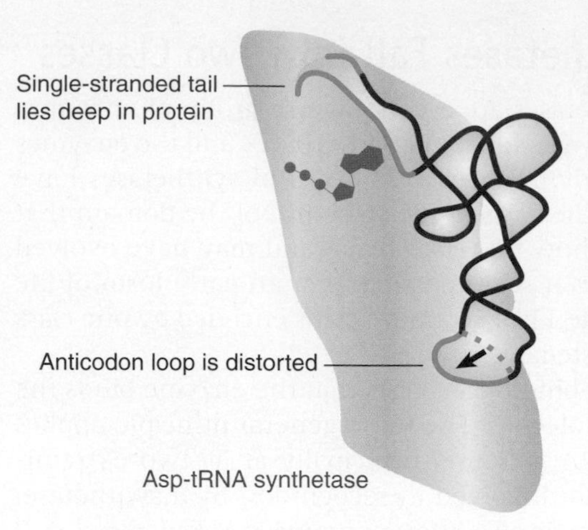

A class II enzyme (for example, tRNA^{Asp} synthetase) approaches the tRNA from the other side; it recognizes both the variable loop and the major groove of the acceptor stem, as shown in **FIGURE 25.15**. The acceptor stem remains in its regular helical conformation. ATP is probably bound near the terminal adenine. At the other end of the binding site, there is a tight contact with the anticodon loop, which has a change in conformation that allows the anticodon to be in close contact with the enzyme.

KEY CONCEPT

- Aminoacyl-tRNA synthetases are divided into class I and class II families based on mutually exclusive sets of sequence motifs and structural domains.

CONCEPT AND REASONING CHECK

What are the differences between the tRNA-enzyme interactions of class I and class II synthetases?

25.10 Synthetases Use Proofreading to Improve Accuracy

Aminoacyl-tRNA synthetases must distinguish one specific amino acid from the cellular pool of amino acids and related molecules and must also differentiate cognate tRNAs in a particular isoaccepting group (typically one to three) from the total set of tRNAs.

Many amino acids are chemically similar to one another, and all amino acids are related to the metabolic intermediates in their particular synthetic pathway. It is especially difficult to distinguish between two amino acids that differ only in the length of the carbon backbone (that is, by one —CH_2— group). Intrinsic discrimination based on relative energies of binding two such amino acids would be only ~$\frac{1}{5}$. The synthetase enzymes improve this ratio ~1000-fold.

Intrinsic discrimination between tRNAs is better, because the tRNA offers a larger surface with which to make more contacts. It is still true, however, that all tRNAs conform to the same general structure, and there may be a quite limited set of features that distinguish the cognate tRNAs from the noncognate tRNAs.

Synthetases use proofreading mechanisms to control the recognition of both types of substrates. They improve significantly on the intrinsic differences among amino acids or among tRNAs but, consistent with the intrinsic differences in each group, make more mistakes in selecting amino acids (error rates are 10^{-4} to 10^{-5}) than in selecting tRNAs (for which error rates are ~10^{-6}; see Figure 24.9).

▸ **kinetic proofreading** A proofreading mechanism that depends on incorrect events proceeding more slowly than correct events, so that incorrect events are reversed before a subunit is added to a polymeric chain.

Synthetases use **kinetic proofreading** to improve their discrimination of tRNAs. This relies on the greater intrinsic affinity of a cognate tRNA for its synthetase. A correctly bound tRNA makes contacts with the enzyme surface that allow the enzyme to aminoacylate it rapidly. An incorrectly bound tRNA makes fewer contacts, so the aminoacylation is not triggered so quickly, and this allows more time for the tRNA to dissociate from the enzyme before it is trapped by the reaction.

Specificity for amino acids varies among the synthetases. Some are highly specific for initially binding a single amino acid, whereas others can also activate amino acids closely related to the proper substrate. The analog amino acid can sometimes be converted to the adenylate form, but in none of these cases is an incorrectly activated amino acid actually used to form a stable aminoacyl-tRNA.

There are two stages at which proofreading of an incorrect aminoacyl-adenylate may occur during formation of aminoacyl-tRNA. **FIGURE 25.16** shows that both use **chemical proofreading**, in which the catalytic reaction is reversed. The extent to which one pathway or the other predominates varies with the individual synthetase. The presence of the cognate tRNA usually is needed to trigger proofreading:

- The noncognate aminoacyl-adenylate may be hydrolyzed when the cognate tRNA binds, called *pretransfer editing*. This mechanism is used predominantly by several synthetases, including those for methionine, isoleucine, and valine.
- Some synthetases use chemical proofreading at a later stage. The wrong amino acid is actually transferred to tRNA, is then recognized as incorrect by its structure in the tRNA binding site, and so is hydrolyzed and released, which is called *post-transfer editing*. The process requires a continual cycle of linkage and hydrolysis until the correct amino acid is transferred to the tRNA.

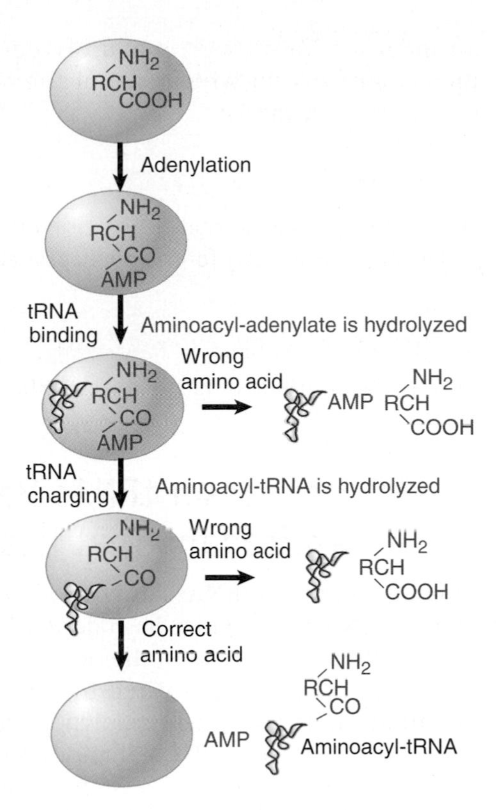

FIGURE 25.16 Proofreading by aminoacyl-tRNA synthetases may take place at the stage prior to aminoacylation (pretransfer editing), in which the noncognate aminoacyl-adenylate is hydrolyzed. Alternatively or additionally, hydrolysis of incorrectly formed aminoacyl-tRNA may occur after its synthesis (post-transfer editing).

▸ **chemical proofreading** A proofreading mechanism in which the correction event occurs after the addition of an incorrect subunit to a polymeric chain, by means of reversing the addition reaction.

A classic example in which discrimination between amino acids depends on the presence of tRNA is provided by the $tRNA^{Ile}$ synthetase of *E. coli*. The enzyme can charge valine with AMP but hydrolyzes the valyl-adenylate when $tRNA^{Ile}$ is added. The overall error rate depends on the specificities of the individual steps, as summarized in **FIGURE 25.17**. The overall error rate of 1.5×10^{-5} is less than the measured rate at which valine is substituted for isoleucine (in rabbit globin), which ranges from 2 to 5×10^{-4}. So mischarging probably provides only a small fraction of the errors that actually occur in translation.

Step	Frequency of Error
Activation of valine to Val-AMP^{Ile}	1/225
Release of Val–tRNA	1/270
Overall rate of error	1/225 × 1/270 = 1/60,000

FIGURE 25.17 The accuracy of charging $tRNA^{Ile}$ by its synthetase depends on error control at two stages.

$tRNA^{Ile}$ synthetase uses size as a basis for discrimination among amino acids in the classic double-sieve mechanism. **FIGURE 25.18** shows that it has two active sites: the synthetic (or activation) site and the editing (or hydrolytic) site. The crystal structure of the enzyme shows that the synthetic site is too small to allow leucine (a close analog of isoleucine) to enter. All amino acids larger than isoleucine are excluded from activation because they cannot enter the synthetic site. An amino acid that can enter the synthetic site is placed on tRNA. Then the enzyme tries to transfer it to the editing site. Isoleucine is safe from editing because it is too large to enter the editing site. However, valine can enter this site; as a result,

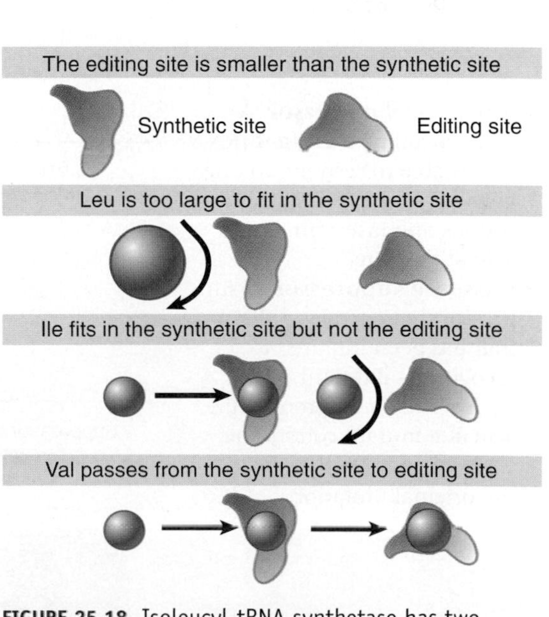

FIGURE 25.18 Isoleucyl-tRNA synthetase has two active sites. Amino acids larger than Ile cannot be activated because they do not fit in the synthetic site. Amino acids smaller than Ile are removed because they are able to enter the editing site.

an incorrect Val-tRNA^{Ile} is hydrolyzed. Essentially, the enzyme provides a double molecular sieve in which size of the amino acid is used to discriminate between closely related species.

KEY CONCEPT

- Specificity of amino acid-tRNA pairing is controlled by proofreading reactions that hydrolyze incorrectly formed aminoacyl adenylates and aminoacyl-tRNAs.

CONCEPT AND REASONING CHECK

How is the error rate of an aminoacyl-tRNA synthetase minimized?

25.11 Suppressor tRNAs Have Mutated Anticodons That Read New Codons

Isolation of mutant tRNAs has been one of the most potent tools for analyzing the ability of a tRNA to recognize its codon(s) in mRNA and for determining the effects that changes in different parts of the tRNA molecule have on codon-anticodon recognition.

Mutant tRNAs are isolated by virtue of their ability to overcome the effects of mutations in genes encoding polypeptides. In genetic terminology, a mutation that is able to overcome the effects of another mutation is called a **suppressor**.

suppressor A second mutation that compensates for or alters the effects of a primary mutation.

In tRNA suppressor systems, the primary mutation changes a codon in an mRNA so that the polypeptide product is no longer functional. The secondary suppressor mutation changes the anticodon of a tRNA so that it recognizes the mutant codon instead of (or as well as) its original target codon. The amino acid that is now inserted restores polypeptide function. The suppressors are described as **nonsense suppressors** or **missense suppressors**, depending on the nature of the original mutation.

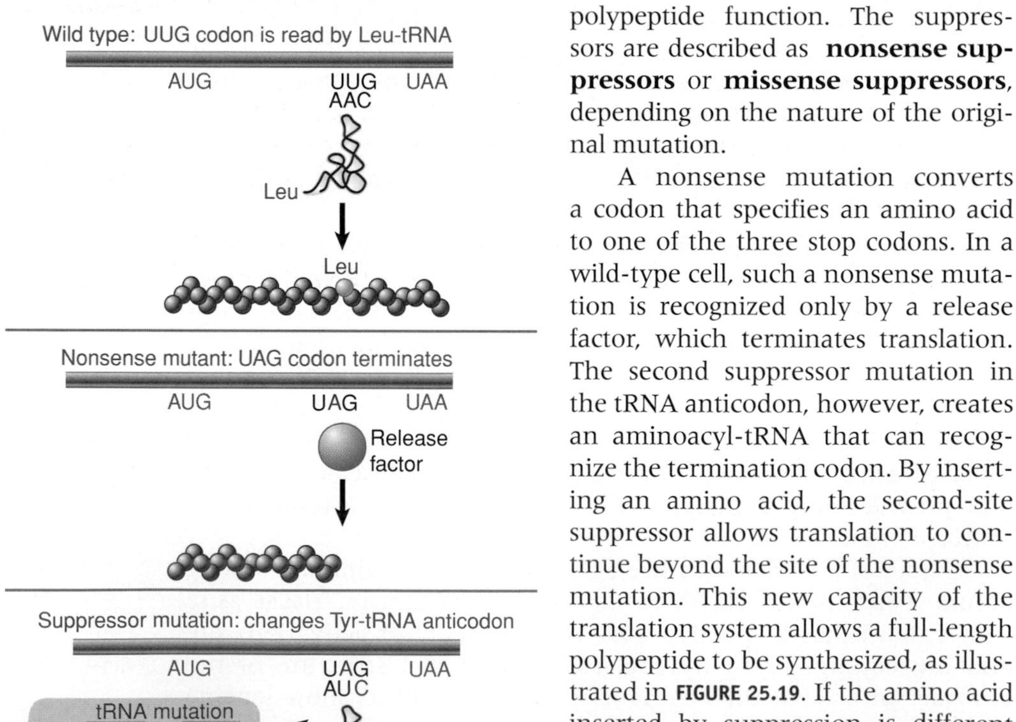

FIGURE 25.19 Nonsense mutations can be suppressed by a tRNA with a mutant anticodon, which inserts an amino acid at the mutant codon, producing a full-length polypeptide in which the original Leu residue has been replaced by Tyr.

nonsense suppressor A gene encoding a mutant tRNA that is able to respond to one or more of the termination codons and insert an amino acid at that site.

missense suppressor A suppressor that encodes a tRNA that has been mutated to recognize a different codon. By inserting a different amino acid at a mutant codon, the tRNA suppresses the effect of the original mutation.

A nonsense mutation converts a codon that specifies an amino acid to one of the three stop codons. In a wild-type cell, such a nonsense mutation is recognized only by a release factor, which terminates translation. The second suppressor mutation in the tRNA anticodon, however, creates an aminoacyl-tRNA that can recognize the termination codon. By inserting an amino acid, the second-site suppressor allows translation to continue beyond the site of the nonsense mutation. This new capacity of the translation system allows a full-length polypeptide to be synthesized, as illustrated in **FIGURE 25.19**. If the amino acid inserted by suppression is different from the amino acid that was originally present at this site in the wild-type polypeptide, the activity of the polypeptide may be altered. Each type of nonsense suppressor is specific for one of the three nonsense codons;

that is, its anticodon recognizes one of the codons UAA, UAG, or UGA.

A nonsense suppressor is isolated by its ability to recognize a mutant nonsense codon. The same triplet sequence, however, constitutes one of the normal termination signals of the cell! The mutant tRNA that suppresses the nonsense mutation must, in principle, be able to suppress natural termination at the end of any gene that uses this codon. **FIGURE 25.20** shows that this **readthrough** results in the synthesis of a longer polypeptide, with additional C-terminal sequence. The extended polypeptide will end at the next termination triplet sequence found in the reading frame. Any extensive suppression of termination is likely to be deleterious to the cell by producing extended polypeptides whose functions are thereby altered.

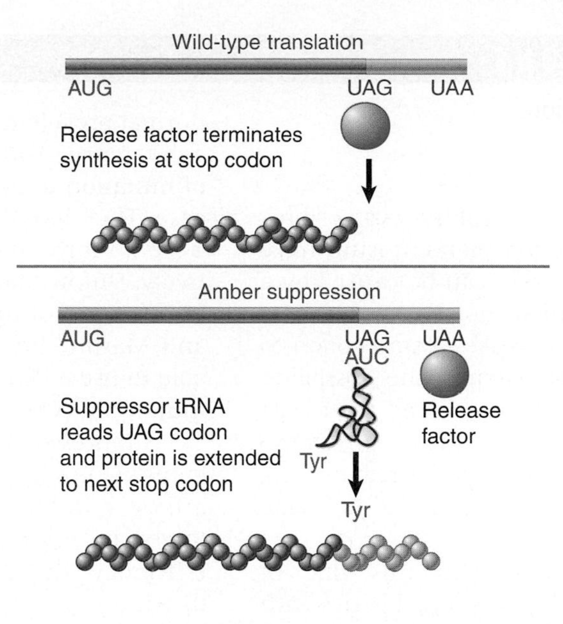

FIGURE 25.20 Nonsense suppressors also read through natural termination codons, synthesizing polypeptides that are longer than the wild type.

▸ **readthrough** This occurs at transcription or translation when RNA polymerase or the ribosome, respectively, ignores a termination signal because of a mutation of the template or the behavior of an accessory factor.

Amber codons are used relatively infrequently to terminate translation in *E. coli*; as a result, amber suppressors tend to be relatively efficient (10% to 50%). The ochre codon (UAA) is used most frequently as a natural termination signal; as a result, ochre suppressors are difficult to isolate and are always much less efficient (<10%). UGA is the least efficient of the termination codons in its natural function; it is misread by $tRNA^{Trp}$ as frequently as 1% to 3% in wild-type cells. In spite of this deficiency, however, it is used more commonly than the amber triplet UAG to terminate bacterial genes.

Missense mutations change a codon representing one amino acid into a codon representing another amino acid—one that cannot function in the polypeptide in place of the original residue. (Formally, any substitution of amino acids constitutes a missense mutation, but in practice it is detected only if it changes the activity of the polypeptide.) The mutation can be suppressed by the insertion either of the original amino acid or of some other amino acid that restores the function of the polypeptide.

FIGURE 25.21 demonstrates that missense suppression can be accomplished in the same way as nonsense suppression, by mutating the anticodon of a tRNA carrying an acceptable amino acid so that it responds to the mutant codon. So missense suppression involves a change in the meaning of the codon from one amino acid to another.

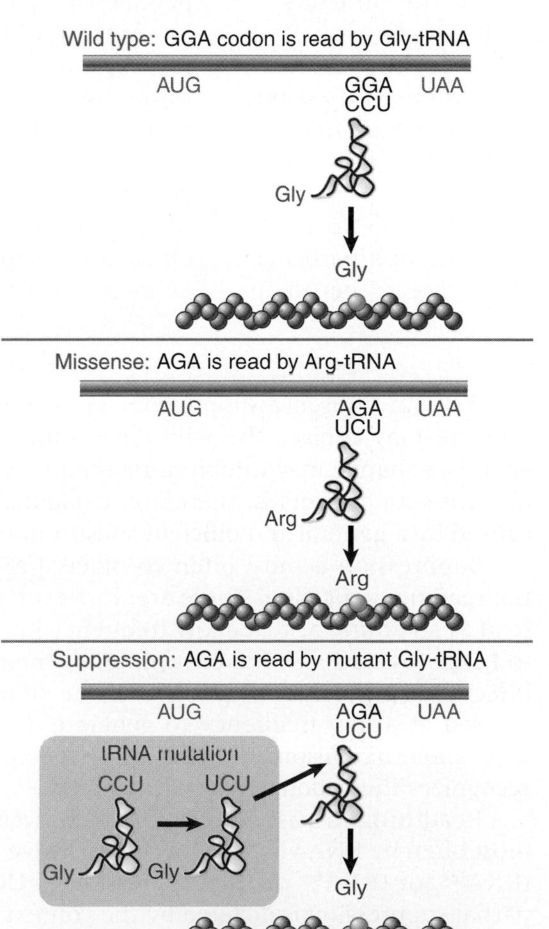

FIGURE 25.21 Missense suppression occurs when the anticodon of tRNA is mutated so that it responds to the wrong codon. The suppression is only partial because both the wild-type tRNA and the suppressor tRNA can recognize AGA.

MEDICAL APPLICATIONS

Therapies for Nonsense Mutations

What do Duchenne muscular dystrophy, cystic fibrosis, β-thalassamia, and xeroderma pigmentosum have in common? Each of these diseases can be caused by a mutation in a single gene. And when that mutation is a nonsense mutation—one that causes a stop codon to be found prematurely in the transcript—the possibility exists that a suppressor tRNA could be used to at least partially correct the defect. By introducing an amino acid at the position of the stop codon that is terminating translation, the full-length polypeptide could be made instead.

There are several advantages to this line of approach. tRNAs are encoded by very small genes with internal promoters that robustly drive expression in all cell types. Once transcribed, they have low rates of degradation. Although the delivery of genes to the target cells remains a problem for gene therapy approaches, tRNA genes are in principle very good candidates for the method.

But the very characteristics that make tRNA suppression attractive also contribute to the difficulty of carrying it out in practice. The introduction of tRNA suppressors is deleterious to cells because they have no way to distinguish between mutant nonsense codons and natural stop codons. Thus, their presence can result in readthrough of natural stop codons and the production of many abnormal polypeptides.

One way to get around this problem is to try to exploit any difference in the prevalence of a particular stop codon in mutations that cause disease versus its normal prevalence in cells. To determine if there is such a difference, Atkinson and Martin looked at 179 events of mutation to nonsense codons that cause human disease. They found that there is a ratio of approximately 1 : 2 : 3 for mutation to UAA, UAG, and UGA, respectively. Unfortunately, this ratio is similar to the proportional usage of those codons in human cells (Atkinson and Martin, 1994). Therefore, researchers will not be able to use a tRNA suppressor that preferentially affects mutant mRNAs.

Suppressor efficiency also depends on 3′ codon context. For example, UAG codons flanked 3′ by A are very inefficiently suppressed, whereas those followed by C or G are suppressed with much higher efficiency. But Atkinson and Martin found that the distribution of 3′ contexts is not significantly different in premature stop codons than in naturally occurring stop codons.

Nevertheless, as early as 1982, Temple and others were able to show that coinjection of a human tRNA suppressor gene and the poly-A^+ reticulocyte RNA from a patient with β-thalassamia into *Xenopus* oocytes resulted in the production of full-length β-globin polypeptide (Temple et al., 1982).

More recently, studies in mice have also shown some promise. Buvoli and others have shown that intramuscular injection of a plasmid containing several copies of a tRNA suppressor gene resulted in the readthrough of an ochre codon in a chloramphenicol acetyl transferase (CAT) reporter gene and expression of the active enzyme (Buvoli et al., 2000). However, the efficiency of suppression was very low, reaching 0.28% at most. In a transgenic mouse expressing the same

As with nonsense suppressors, an event that suppresses a missense mutation at one site may replace the wild-type amino acid at another site with a new amino acid. The change may inhibit normal polypeptide function. The absence of any strong missense suppressors is, therefore, explained by the damaging effects that would be caused by a general and efficient substitution of amino acids.

Suppression is most often considered in the context of a mutation that changes the reading of a codon. There are, however, some situations in which a stop codon is read as an amino acid at a low frequency in the wild-type situation. The first example to be discovered was the coat protein gene of the RNA phage Qβ. The formation of infective Qβ particles requires that the stop codon at the end of this gene be suppressed at a low frequency to generate a small proportion of coat proteins with a C-terminal extension. In effect, this stop codon is leaky. The reason is that $tRNA^{Trp}$ recognizes the codon at a low frequency.

Readthrough past stop codons also occurs in eukaryotes, where it is employed most often by RNA viruses. This may involve the suppression of UAG/UAA by $tRNA^{Tyr}$, $tRNA^{Gln}$, or $tRNA^{Leu}$ or the suppression of UGA by $tRNA^{Trp}$ or $tRNA^{Arg}$. The extent of partial suppression is dictated by the context surrounding the codon.

mutant CAT reporter gene in the heart, injection of the tRNA suppressor gene construct into the heart resulted in 1% to 2% CAT activity. The long-term effects of the treatment were not reported.

Another approach to readthrough is the use of aminoglycoside antibiotics, which bind to the decoding site of the small subunit rRNA and decrease the accuracy requirements of codon-anticodon pairing (Moazed and Noller, 1986). This results in a full-length polypeptide with a missense mutation. Gentamicin is one such drug that has been used in this manner, but it has low efficiency and toxic side effects, and it suffers from the same lack of discrimination between premature stop codons and natural ones as other readthrough approaches.

Recently, a small molecule has been identified in screens for compounds that can read through premature stop UGA codons. Called PTC124, this molecule is remarkable for its ability to discriminate between premature stop codons and naturally occurring stop codons (Welch et al., 2007). Normally, premature stop codons are recognized by the ribosome and lead to nonsense-mediated decay of the transcript. Normal termination is mechanistically different and does not lead to decay of the transcript. Presumably, PTC124 is sensitive to this difference in some way.

Finally, the success of most nonsense-suppression approaches will depend in part upon the amount of normal polypeptide required to correct a defect. Studies have suggested that to correct the muscle defect of Duchenne and Becker muscular dystrophy, the level of expression must reach at least 30% to 40% of normal polypeptide (Hoffman et al., 1988). By contrast, in canine hemophilia B, 1% of normal factor IX levels results in partial correction of the coagulation defect (Snyder et al., 1999).

Nonsense mutations in monogenic disorders seem like good candidates for new therapies because the cause of the defect is very well defined. Although intervening in cellular activities without causing unwanted side effects is always challenging, there is reason to hope that a thorough understanding of the mechanism of translation and mRNA decay will lead to effective treatments in the future.

References

Atkinson and Martin. (1994). Mutations to nonsense codons in human genetic disease: implications for gene therapy by nonsense suppressor tRNAs. *Nucleic Acids Research* **22**, 1327–1334.

Buvoli, Buvoli, and Leinwand. (2000). Suppression of nonsense mutations in cell culture and mice by multimerized suppressor tRNA genes. *Molecular and Cellular Biology* **20**, 3116–3124.

Hoffman et al. (1988). Characterization of dystrophin in muscle-biopsy specimens from patients with Duchenne's or Becker's muscular dystrophy. *New England Journal of Medicine* **318**, 1363–1368.

Moazed and Noller. (1986).Transfer RNA shields specific nucleotides in 16S ribosomal RNA from attack by chemical probes. *Cell* **47**, 985–994.

Snyder et al. (1999). Correction of hemophilia B in canine and murine models using recombinant adeno-associated viral vectors. *Nature Medicine* **5**, 64–70.

Temple et al. (1982). Construction of a functional human suppressor tRNA gene: an approach to gene therapy for β-thalassaemia. *Nature* **296**, 537–540.

Welch et al. (2007). PTC124 targets genetic disorders caused by nonsense mutations. *Nature* **447**, 87–91.

KEY CONCEPTS

- A suppressor tRNA typically has a mutation in the anticodon that changes the codons to which it responds.
- Each type of nonsense codon is suppressed by tRNAs with mutant anticodons.
- When the new anticodon corresponds to a termination codon, an amino acid is inserted and the polypeptide chain is extended beyond the termination codon. This results in nonsense suppression at a site of nonsense mutation, or in readthrough at a natural termination codon.
- Suppressor tRNAs compete with wild-type tRNAs that have the same anticodon to read the corresponding codon(s).
- Efficient suppression is deleterious because it results in readthrough past normal termination codons.
- Missense suppression occurs when the tRNA recognizes a different codon from usual, so that one amino acid is substituted for another.

CONCEPT AND REASONING CHECK

Which type of mutation is more likely to be deleterious, nonsense suppressors or missense suppressors? Why?

25.12 Recoding Changes Codon Meanings

The reading frame of a messenger RNA is usually invariant. Translation starts at an AUG codon and continues in triplets to a termination codon. Reading does not depend on the sense of the message: insertion or deletion of a base causes a frameshift mutation, in which the reading frame is changed past the site of mutation. Ribosomes and tRNAs inevitably continue reading in triplets, incorporating an entirely different series of amino acids.

There are some exceptions to the usual pattern of translation that enable a reading frame with an interruption of some sort—such as a nonsense codon or frameshift—to be translated into a full-length polypeptide. **Recoding** events are responsible for making exceptions to the usual rules.

▸ **recoding** Events that occur when the meaning of a codon or series of codons is changed from that predicted by the genetic code. It may involve altered interactions between aminoacyl-tRNA and mRNA that are influenced by the ribosome.

In one type of recoding discussed earlier in this chapter, changing the meaning of a single codon allows one amino acid to be substituted in place of another or for an amino acid to be inserted at a termination codon. **FIGURE 25.22** shows that these changes rely on the properties of an individual tRNA that responds to the codon:

- Suppression involves recognition of a codon by a (mutant) tRNA that usually would respond to a different codon (see *Section 25.11, Suppressor tRNAs Have Mutated Anticodons That Read New Codons*).
- Redefinition of the meaning of a codon occurs when an aminoacyl-tRNA is modified (see *Section 25.7, Novel Amino Acids Can Be Inserted at Certain Stop Codons*).

Changing the reading frame, another type of recoding, occurs in two types of situations:

- Frameshifting typically involves changing the reading frame when aminoacyl-tRNA slips by one base, either +1 forward or –1 backward (see *Section 25.13, Frameshifting Occurs at Slippery Sequences*). The result shown in **FIGURE 25.23** is that translation continues past a termination codon.

Suppression is caused by mutated anticodon

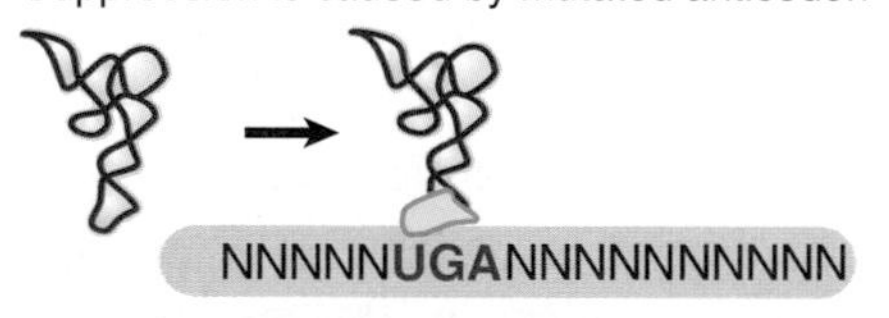

Special factor + tRNA recognizes codon

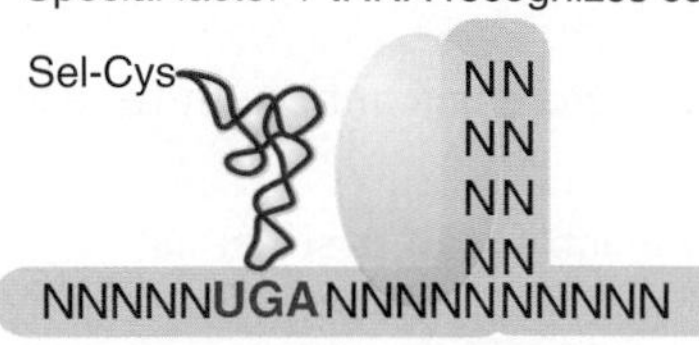

FIGURE 25.22 A mutation in an individual tRNA (usually in the anticodon) can suppress the usual meaning of that codon. In a special case, a specific tRNA is bound by an unusual elongation factor to recognize a termination codon adjacent to a hairpin loop.

–1 frameshift in HIV retrovirus

Last codon read in initial reading frame

First codon read in new reading frame

Reading without frameshift

NNNNUUUUUUAGGNNNNNNNN

Reading after frameshift

NNNNUUUUUUAGGNNNNNNNN

FIGURE 25.23 A tRNA that slips one base in pairing with a codon causes a frameshift that can suppress termination. The efficiency is usually ~5%.

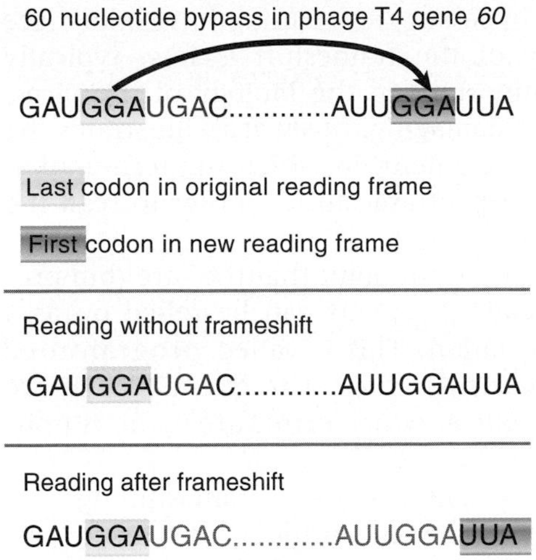

60 nucleotide bypass in phage T4 gene *60*

GAUGGAUGAC............AUUGGAUUA

Last codon in original reading frame

First codon in new reading frame

Reading without frameshift

GAUGGAUGAC............AUUGGAUUA

Reading after frameshift

GAUGGAUGAC............AUUGGAUUA

FIGURE 25.24 Bypassing occurs when the ribosome moves along mRNA so that the peptidyl-tRNA in the P site is released from pairing with its codon and then repairs with another codon farther along.

Bypassing involves a movement of the ribosome to change the codon that is paired with the peptidyl-tRNA in the P site. The sequence between the two codons fails to be represented in the resulting polypeptide. As shown in **FIGURE 25.24**, this allows translation to continue past any termination codons in the intervening region. (See *Section 25.14, Bypassing Involves Ribosome Movement,* for more detail.)

KEY CONCEPTS

- Changes in codon meaning can be caused by mutant tRNAs or by tRNAs with special properties.
- The reading frame can be changed by frameshifting or bypassing, both of which depend on properties of the mRNA.

CONCEPT AND REASONING CHECK

Although frameshifting would normally be deleterious, under what circumstances would it be advantageous?

25.13 Frameshifting Occurs at Slippery Sequences

Frameshifting is associated with specific tRNAs in two circumstances:

- Some mutant tRNA suppressors recognize a "codon" for four bases instead of the usual three bases.
- Certain "slippery" sequences allow a tRNA to move along the mRNA in the A site by one base in either the 5′ or 3′ direction.

The simplest type of external frameshift suppressor corrects the reading frame when a mutation has been caused by inserting an additional base within a stretch of identical residues. For example, a G may be inserted in a run of several contiguous G bases. The frameshift suppressor is a $tRNA^{Gly}$ that has an extra base inserted in its anticodon loop, converting the anticodon from the usual triplet sequence $CCC^{\leftarrow}$ to the quadruplet sequence $CCCC^{\leftarrow}$. The suppressor tRNA recognizes a 4-base "codon."

Situations in which frameshifting is a normal event are presented by phages and other viruses. Such events may affect the continuation or termination of translation and result from the intrinsic properties of the mRNA.

In retroviruses, translation of the first gene is terminated by a nonsense codon in phase with the reading frame. The second gene lies in a different reading frame and (in some viruses) is translated by a frameshift that changes into the second reading frame and therefore bypasses the termination codon (see Figure 25.23). The efficiency of the frameshift is low, typically ~5%. The low efficiency is important in the biology of the virus; an increase in efficiency can be damaging. **FIGURE 25.25** illustrates the similar situation of the yeast Ty element, in which the termination codon of *tya* must be bypassed by a frameshift in order to read the subsequent *tyb* gene.

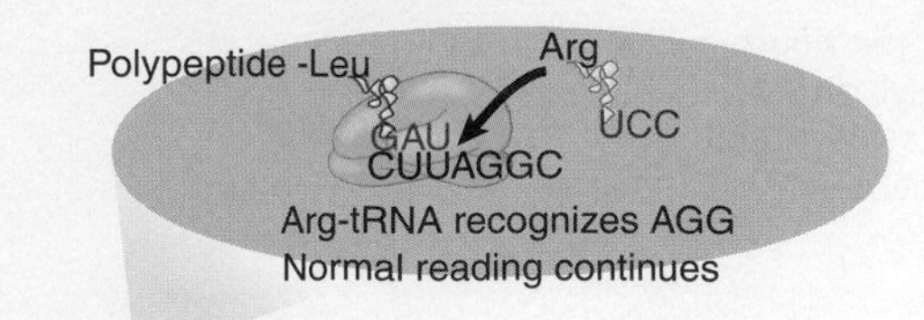

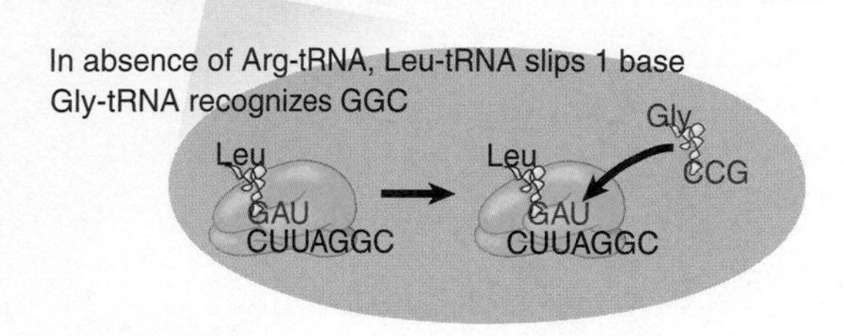

FIGURE 25.25 A +1 frameshift is required for expression of the *tyb* gene of the yeast Ty element. The shift occurs at a seven-base sequence at which two Leu codon(s) are followed by a scarce Arg codon.

Such situations make the important point that the rare (but predictable) occurrence of "misreading" events can be relied on as a necessary step in natural translation. This is called **programmed frameshifting**. It occurs at particular sites at frequencies that are 100 to 1000× greater than the rate at which errors are made at nonprogrammed sites (~3×10^{-5} per codon).

There are two common features in this type of frameshifting:

- A "slippery" sequence allows an aminoacyl-tRNA to pair with its codon and then to move +1 or −1 base to pair with an overlapping triplet sequence that can also pair with its anticodon.
- The ribosome is delayed at the frameshifting site to allow time for the aminoacyl-tRNA to rearrange its pairing. The cause of the delay can be an adjacent codon that requires a scarce aminoacyl-tRNA, a termination codon that is recognized slowly by its release factor, or a structural impediment in mRNA (for example, a "pseudoknot," a particular conformation of RNA) that impedes the ribosome.

Slippery events can involve movement in either direction; a −1 frameshift is caused when the tRNA moves backward, and a +1 frameshift is caused when it moves forward. In either case, the result is to expose an out-of-phase triplet in the A site for the next aminoacyl-tRNA. The frameshifting event occurs before peptide bond synthesis. In the most common type of case, when it is triggered by a slippery sequence in conjunction with a downstream hairpin in mRNA, the surrounding sequences influence its efficiency.

▸ **programmed frameshifting** Frameshifting that is required for expression of the polypeptide sequences encoded beyond a specific site at which a +1 or −1 frameshift occurs at some typical frequency.

The frameshifting in Figure 25.25 shows the behavior of a typical slippery sequence. The seven-nucleotide sequence CUUAGGC is usually recognized by $tRNA^{Leu}$ at CUU, followed by $tRNA^{Arg}$ at AGC. The $tRNA^{Arg}$ is scarce, though, and when its scarcity results in a delay, the $tRNA^{Leu}$ slips from the CUU codon to the overlapping UUA triplet. This causes a frameshift, because the next triplet in phase with the new pairing (GGC) is read by $tRNA^{Gly}$. Slippage usually occurs in the P site (when the $tRNA^{Leu}$ actually has become peptidyl-tRNA, carrying the nascent chain).

KEY CONCEPTS

- The reading frame may be influenced by the sequence of mRNA and the ribosomal environment.
- Slippery sequences allow a tRNA to shift by one base after it has paired with its anticodon, thereby changing the reading frame.
- Translation of some genes depends upon the regular occurrence of programmed frameshifting.

CONCEPT AND REASONING CHECK

What types of coding sequences are likely to be "slippery," allowing programmed frameshifting?

25.14 Bypassing Involves Ribosome Movement

Certain sequences trigger a bypass event, in which a ribosome stops translation, slides along mRNA with peptidyl-tRNA remaining in the P site, and then resumes translation. This is a very rare phenomenon, with only one authenticated example: in gene *60* of phage T4, the ribosome moves sixty nucleotides along the mRNA, as shown in Figure 25.24.

The key to the bypass system is that there are identical (or synonymous) codons at either end of the sequence that is skipped. They are sometimes referred to as the "take-off" and "landing" sites. Before bypass, the ribosome is positioned with a peptidyl-tRNA paired with the take-off codon in the P site, with an empty A site waiting for an aminoacyl-tRNA to enter. **FIGURE 25.26** shows that the ribosome slides along mRNA in this condition until the peptidyl-tRNA can become paired with the codon in the landing site. A remarkable feature of the system is its high efficiency, ~50%.

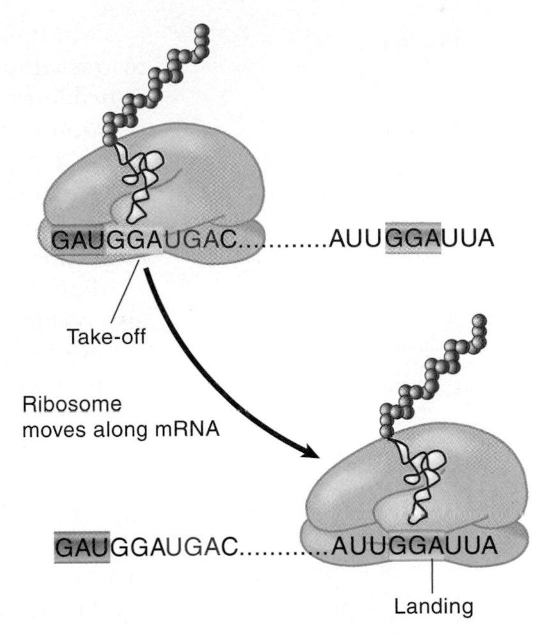

FIGURE 25.26 In bypass mode, a ribosome with its P site occupied can stop translation. It slides along mRNA to a site where peptidyl-tRNA pairs with a new codon in the P site. Then translation is resumed.

The sequence of the mRNA triggers the bypass. The important features are the two GGA codons for take-off and landing, the spacing between them, a stem-loop structure that includes the take-off codon, and the stop codon adjacent to the take-off codon.

The take-off stage requires the peptidyl-tRNA to unpair from its codon. This is followed by a movement of the mRNA that prevents it from re-pairing. Then the ribosome scans the mRNA until the peptidyl-tRNA can re-pair with the codon in the landing reaction. This is followed by the resumption of translation when aminoacyl-tRNA enters the A site in the usual way.

Like frameshifting, the bypass reaction depends on a pause by the ribosome. The probability that peptidyl-tRNA will dissociate from its codon in the P site is increased by delays in the entry of aminoacyl-tRNA into the A site. Starvation for an amino acid can trigger bypassing in bacterial genes because of the delay that occurs when there is no aminoacyl-tRNA available to enter the A site. In phage T4 gene *60*, one role of mRNA structure may be to reduce the efficiency of termination, thus creating the delay that is needed for the take-off reaction.

KEY CONCEPT

- When a ribosome encounters a GGA codon adjacent to a stop codon in a specific stem-loop structure, it moves directly to a specific GGA downstream without adding amino acids to the polypeptide.

CONCEPT AND REASONING CHECK

Explain how bypassing may allow alternate gene expression in times of bacterial starvation.

25.15 Summary

The sequence of mRNA read in triplets 5′→3′ is related by the genetic code to the amino acid sequence of a polypeptide read from N-terminus to C-terminus. Of the 64 triplets, 61 encode amino acids and 3 provide termination signals. Synonymous codons that represent the same amino acids are related, often by a difference in the third base of the codon. This third-base degeneracy, coupled with a pattern in which chemically similar amino acids tend to be encoded by related codons, minimizes the effects of mutations. The genetic code is nearly universal and must have been established very early in evolution. Variations of the code in nuclear genomes are rare, but some changes have occurred during mitochondrial evolution.

Multiple tRNAs may recognize a particular codon. The set of tRNAs responding to the various codons for each amino acid is distinctive for each organism. Codon-anticodon recognition involves wobbling at the first position of the anticodon (third position of the codon), which allows some tRNAs to recognize multiple codons. All tRNAs have modified bases, introduced by enzymes that recognize target bases in the tRNA structure. Codon-anticodon pairing is influenced by modifications of the anticodon itself and also by the context of adjacent bases, especially on the 3′ side of the anticodon.

Each amino acid is recognized by a particular aminoacyl-tRNA synthetase, which also recognizes all of the tRNAs encoding that amino acid.

Aminoacyl-tRNA synthetases vary widely but fall into two general groups according to the structure of the catalytic domain. Synthetases of each group bind the tRNA from the side, making contacts principally with the extremities of the acceptor stem and the anticodon stem-loop; the two types of synthetases bind tRNA from opposite sides. The relative importances of the acceptor stem and the anticodon region for specific recognition vary with the individual tRNA. Aminoacyl-tRNA synthetases have proofreading functions that scrutinize the aminoacyl-tRNA products and hydrolyze incorrectly joined aminoacyl-tRNAs.

Mutations may allow a tRNA to read different codons; the most common form of such mutations occurs in the anticodon itself. Alteration of its specificity may allow a tRNA to suppress a mutation in a gene encoding a polypeptide. A tRNA that recognizes a termination codon provides a nonsense suppressor, whereas a tRNA that changes the amino acid responding to a codon is a missense suppressor. Suppressors of UAG and UGA codons are more efficient than those of UAA codons, which is explained by the fact that UAA is the most commonly used natural termination codon. The efficiency of all suppressors, however, depends on the context of the individual target codon.

Frameshifts of the +1 type may be caused by aberrant tRNAs that read "codons" of four bases. Frameshifts of either +1 or −1 may be caused by slippery sequences in mRNA that allow a peptidyl-tRNA to slip from its codon to an overlapping sequence that can also pair with its anticodon. This frameshifting also requires another sequence that causes the ribosome to delay. Frameshifts determined by the mRNA sequence may be required for expression of natural genes. Bypassing occurs when a ribosome stops translation and moves along mRNA with its peptidyl-tRNA in the P site until the peptidyl-tRNA pairs with an appropriate codon; then translation resumes.

CHAPTER QUESTIONS

1. Which two amino acids are encoded by a single codon each in the universal genetic code?
   A. arginine and tryptophan
   B. cysteine and isoleucine
   C. methionine and tryptophan
   D. histidine and praline
2. Which codon position has the least meaning in determining the amino acid to be inserted during translation of an mRNA?
   A. 1
   B. 2
   C. 3
   D. All positions have equal meaning.

3. An aminoacyl-tRNA synthetase has a total of how many binding sites?
   A. 1
   B. 2
   C. 3
   D. 4
4. A suppressor tRNA has an altered:
   A. anticodon.
   B. acceptor stem.
   C. stop codon.
   D. amino acid attachment site.
5. Inosine can form base pairs with all but which one of the four usual bases in RNA?
   A. A
   B. G
   C. C
   D. U
6. How many different types of modified bases can be found in tRNA molecules?
   A. fewer than 12
   B. about 25
   C. about 36
   D. more than 70
7. What is the actual minimum number of tRNAs required for responding to the regular codons in mitochondria?
   A. 20
   B. 22
   C. 23
   D. 31
8. Isoaccepting tRNAs have which one of the following properties?
   A. They recognize two or more different triplet codons in a transcript.
   B. They recognize a single triplet codon in a transcript.
   C. They each represent the same amino acid but are recognized by different synthetases.
   D. They each represent the same amino acid and are recognized by the same synthetases.
9. A mutation that is able to overcome the effects of another mutation is commonly called a:
   A. depressor.
   B. suppressor.
   C. compensatory mutation.
   D. sense mutation.
10. A missense suppressor for a particular gene may also be a:
   A. terminator for another gene.
   B. mutator for another gene.
   C. suppressor for another gene.
   D. more than one of the above.

KEY TERMS

chemical proofreading
cognate tRNAs
isoaccepting tRNAs
kinetic proofreading
missense suppressor
nonsense suppressor
posttranscriptional modification
programmed frameshifting
readthrough
recoding
stop codon
suppressor
synonymous codons
third-base degeneracy
wobble hypothesis

FURTHER READING

Farabaugh, P. J. (1996). Programmed translational frameshifting. *Annu. Rev. Genet.* **30**, 507–528. A review of the examples of this phenomenon.

Herr, A. J., Atkins, J. F., and Gesteland, R. F. (2000). Coupling of open reading frames by translational bypassing. *Annu. Rev. Biochem.* **69**, 343–372. A review of the bypassing mechanism of take-off, scanning, and landing.

Hopper, A. K., and Phizicky, E. M. (2003). tRNA transfers to the limelight. *Genes Dev.* **17**, 162–180. A review of the processes involved in tRNA processing in yeast.

Ibba, M., and Söll, D. (2004). Aminoacyl-tRNAs: setting the limits of the genetic code. *Genes Dev.* **18**, 731–738. A brief review of the molecules involved in charging tRNAs and their activities.

Silvian, L. F., Wang, J., and Steitz, T. A. (1999). Insights into editing from an Ile-tRNA synthetase structure with $tRNA^{Ile}$ and mupirocin. *Science* **285**, 1074–1077. The 2.2 Å resolution crystal structure of a bacterial aminoacyl-tRNA synthetase and what it suggests about the editing process during the charging of a tRNA.

Srinivasan, G., James, C. M., and Krzycki, J. A. (2002). Pyrrolysine encoded by UAG in Archaea: charging of a UAG-decoding specialized tRNA. *Science* **296**, 1459–1462. A report of the amino acid pyrrolysine being encoded in the genome of a *Methanosarcina* species using the UAG codon (a termination codon in the standard genetic code). This species also produces a unique tRNA and aminoacyl-tRNA synthetase for incorporation of this amino acid in proteins.

IV

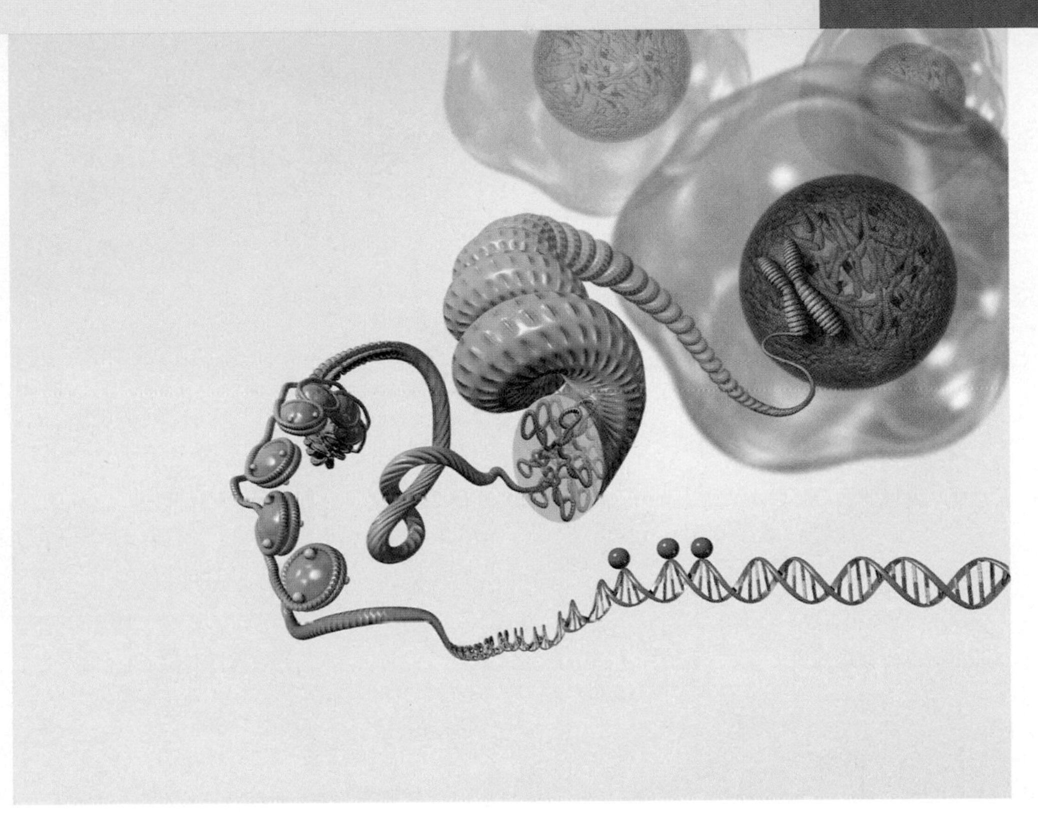

A representation of the levels of chromatin packaging from naked DNA to the chromosome. Epigenetic marks such as DNA methylation (red spheres) and histone modifications (yellow spheres) are shown.

Gene Regulation

26

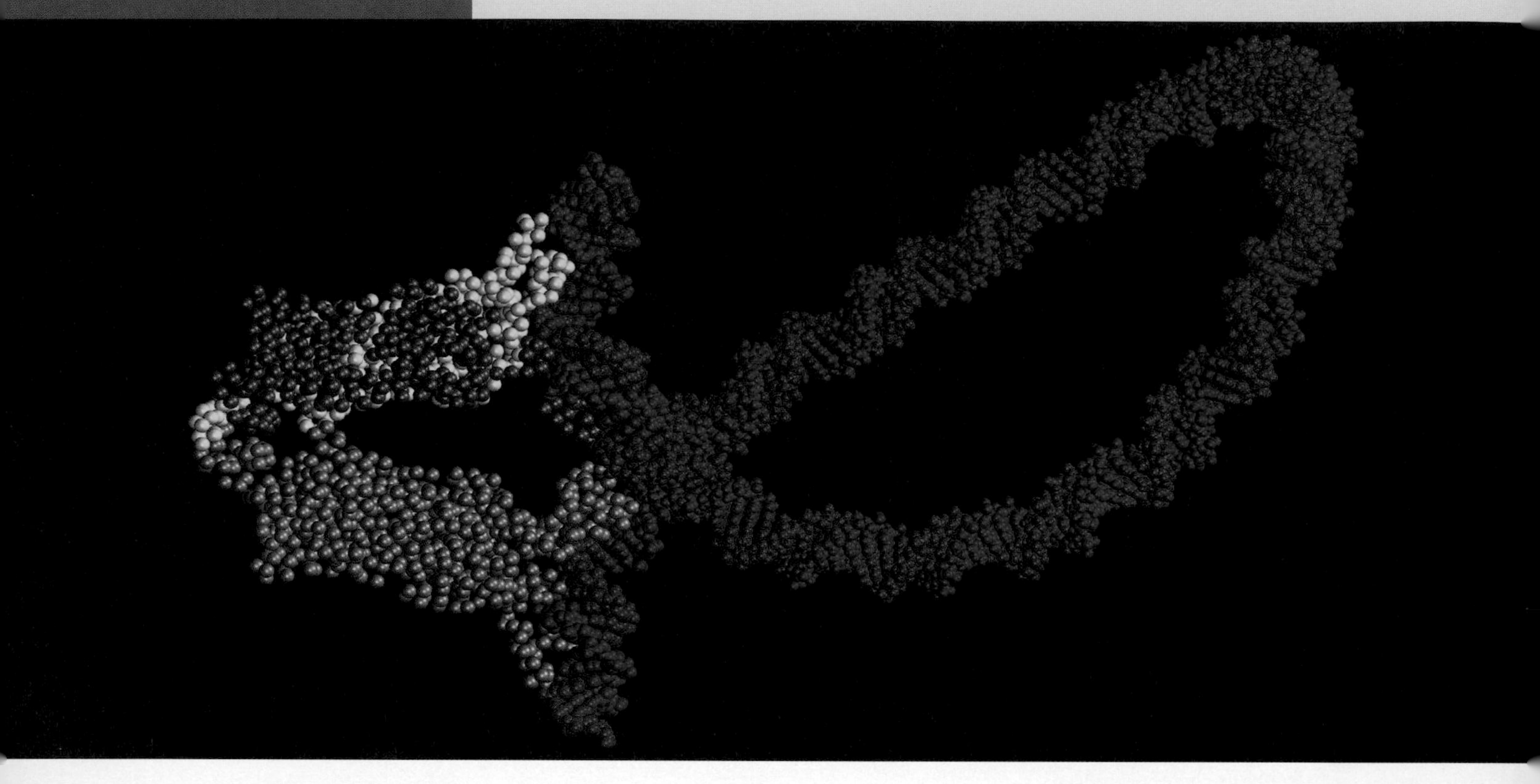

A DNA loop formed by the binding of the *lac* repressor protein to two operator sites of the *E. coli lac* operon. Photo courtesy of Noel Perkins, University of Michigan. More information at S. Goyal, *Biophys. J.* 83 (2007): 4342–4359. Graphics created with VMD (D. Humphrey, A. Dalke, and K. Schulten, "VMD—Visual Molecular Dynamics," *J. Molec. Graphics* 14 (1996): 33–38 [http://www.ks.uiuc.edu/Research/vmd/]).

The Operon

CHAPTER OUTLINE

26.1 Introduction

Gene expression can be controlled at any of several stages, which we divide broadly into transcription, processing, and translation:

- Transcription is often controlled at the stage of initiation. Transcription is not usually controlled at elongation but may be controlled at termination to determine whether RNA polymerase is allowed to proceed past a terminator to the gene(s) beyond.
- In bacteria, an mRNA is typically available for translation while it is being synthesized; this is called **coupled transcription/translation**. (In eukaryotic cells, processing of the RNA product may be regulated at the stages of modification, splicing, transport, or stability.)
- Translation in bacteria may also be directly regulated, but more commonly it is passively modulated. The coding portion, or open reading frame, of a gene can be assembled either with common codons or rare codons, which correspond to common or rare tRNAs. mRNAs containing a number of rare codons are more difficult to translate.

▸ **coupled transcription/translation** The phenomenon in bacteria where translation of the mRNA occurs simultaneously with its transcription.

The basic concept for the way transcription is controlled in bacteria is called the **operon** model and was proposed by François Jacob and Jacques Monod in 1961. They distinguished between two types of sequences in DNA: sequences that code for ***trans*-acting** products (usually proteins) and ***cis*-acting** sequences that function exclusively within the DNA. Gene activity is regulated by the specific interactions of the *trans*-acting products with the *cis*-acting sequences (see *Section 2.11, Proteins Are* trans-*Acting, but Sites on DNA Are* cis-*Acting*). In more formal terms:

- A gene is a sequence of DNA that codes for a diffusible product. *The crucial feature is that the product diffuses away from its site of synthesis to act elsewhere.* Any gene product that is free to diffuse to find its target is described as *trans*-acting.
- The description *cis*-acting applies to any sequence of DNA that functions exclusively as a DNA sequence, affecting only the DNA to which it is physically linked.

▸ **operon** A unit of bacterial gene expression and regulation, including structural genes and control elements in DNA recognized by regulator gene product(s).

▸ ***trans*-acting** A product that can function on any copy of its target DNA. This implies that it is a diffusible protein or RNA.

▸ ***cis*-acting** A site that affects the activity only of sequences on its own molecule of DNA (or RNA); this property usually implies that the site does not code for protein.

To help distinguish between the components of regulatory circuits and the genes that they regulate, we sometimes use the terms *structural gene* and *regulator gene*. A **structural gene** is simply any gene that codes for a protein (or RNA) product. Protein structural genes represent an enormous variety of structures and functions, including structural proteins, enzymes with catalytic activities, and regulatory proteins. A type of structural gene is a **regulator gene**, which simply describes a gene that codes for a protein or an RNA involved in regulating the expression of other genes.

▸ **structural gene** A gene that codes for any RNA or protein product other than a regulator.

▸ **regulator gene** A gene that codes for a product (typically protein) that controls the expression of other genes (usually at the level of transcription).

The simplest form of the regulatory model is illustrated in **FIGURE 26.1**: *a regulator gene codes for a protein that controls transcription by binding to particular site(s) on DNA.* This interaction can regulate a target gene in either a positive manner (the interaction turns the gene on) or a negative manner (the interaction turns the gene off). The sites on DNA are usually (but not exclusively) located just upstream of the target gene.

The sequences that mark the beginning and end of the transcription unit—the promoter and terminator—are examples of *cis*-acting sites. *A promoter serves to initiate transcription only of the gene or genes physically connected to it on the same stretch of DNA.* In the same way, a terminator can terminate transcription only by an RNA polymerase that has traversed the preceding gene(s). In their simplest forms, promoters and terminators are *cis*-acting elements that are recognized by RNA polymerase (although other factors also participate at each site).

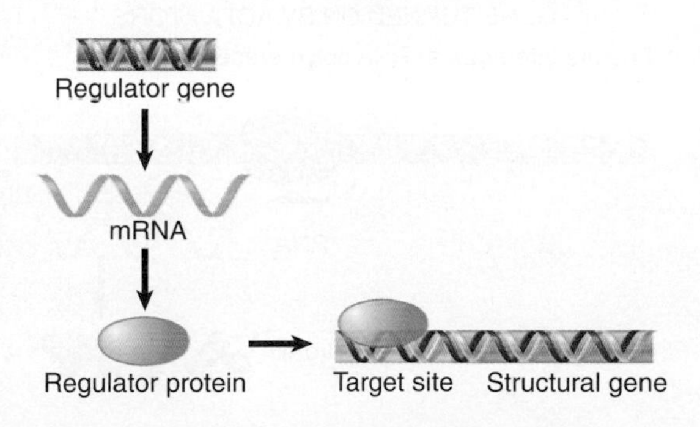

FIGURE 26.1 A regulator gene codes for a protein that acts at a target site on DNA.

Additional *cis*-acting regulatory sites are often combined with the promoter. A bacterial promoter may have one or more such sites located close by, that is, in the immediate vicinity of the startpoint. A eukaryotic promoter is likely to have a greater number of sites that are spread out over a longer distance, as we will see in *Section 28.5, Activators Interact with the Basal Apparatus*.

FIGURE 26.2 In negative control, a *trans*-acting repressor binds to the *cis*-acting operator to turn off transcription.

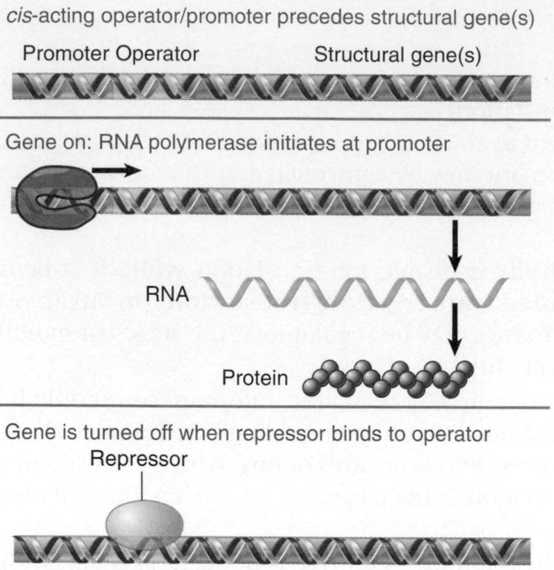

A classic mode of transcription control in bacteria is **negative control**: a repressor protein prevents a gene from being expressed. **FIGURE 26.2** shows that in the absence of the negative regulator, the gene is expressed. Close to the promoter is another *cis*-acting site called the **operator**, which is the binding site for the repressor protein. When the repressor binds to the operator, RNA polymerase is prevented from initiating transcription, and *gene expression is, therefore, turned off*. An alternative mode of control is **positive control**. This is used in bacteria (probably) with about equal frequency to negative control, and it is the most common mode of control in eukaryotes. *A transcription factor is required to assist RNA polymerase in initiating at the promoter.* **FIGURE 26.3** shows that in the absence of the positive regulator, the gene is inactive: RNA polymerase cannot by itself initiate transcription at the promoter.

‣ **negative control** A mechanism of gene regulation in which a regulator is required to turn the gene off.

‣ **operator** The site on DNA at which a repressor protein binds to prevent transcription from initiating at the adjacent promoter.

‣ **positive control** A system in which a gene is not expressed unless some action turns it on.

In addition to negative and positive control, there is another way to look at regulation of gene transcription for a gene that encodes an enzyme: whether it is controlled by the substrate or by the product of the enzyme. Bacteria need to respond swiftly to changes in their environment. Fluctuations in the supply of nutrients (such as the sugar lactose) can occur at any time, and survival depends on the ability to switch from metabolizing one substrate to another. Yet economy is important, too: a bacterium that indulges in energetically expensive ways to meet the demands of the environment is likely to be at a disadvantage. Thus a bacterium avoids synthesizing the enzymes of a pathway in the absence of the substrate but is ready to produce the enzymes if the substrate should appear. *The synthesis of enzymes in response to the appearance of a specific substrate is called* ***induction****, and the gene is an* ***inducible gene****.*

‣ **induction** The ability to synthesize certain enzymes only when their substrates are present. Applied to gene expression, it refers to switching on transcription as a result of interaction of the inducer with the regulator protein.

‣ **inducible gene** A gene that is turned on by the presence of its substrate.

The opposite of induction is **repression**, where *the* ***repressible gene*** *is controlled by the amount of the product made by the enzyme*. For example, *E. coli* synthesizes the amino acid tryptophan through the actions of an enzyme complex containing tryptophan synthetase and four other enzymes. If, however, tryptophan is provided in the medium on which the bacteria are growing, the production of the enzyme is immediately halted. This allows the bacterium to avoid devoting its resources to unnecessary synthetic activities.

FIGURE 26.3 In positive control, a *trans*-acting factor must bind to a *cis*-acting site in order for RNA polymerase to initiate transcription at the promoter.

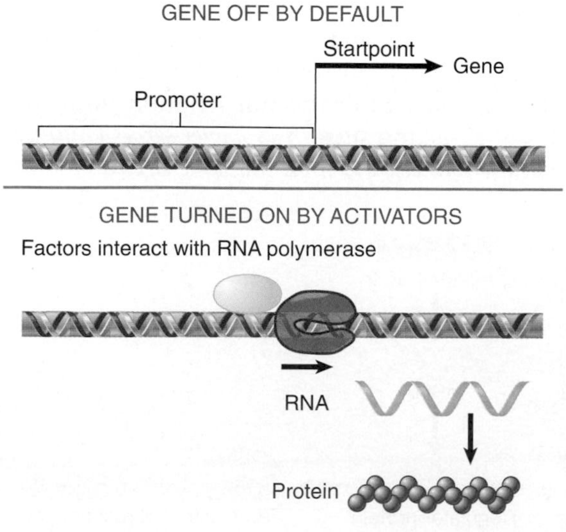

‣ **repression** The ability to prevent synthesis of certain enzymes when their products are present. More generally, it refers to inhibition of transcription (or translation) by binding of repressor protein to a specific site on DNA (or mRNA).

‣ **repressible gene** A gene that is turned off by its product.

Induction and repression represent similar phenomena. In one case the bacterium adjusts its ability to use a given substrate (such as the complex sugar lactose) for growth; in the other it adjusts its ability to synthesize a particular metabolic intermediate (such as an essential amino acid). The trigger for either type of adjustment is a small molecule that is the substrate, or related to the substrate for the enzyme, or the product of the enzyme activity, respectively. Small molecules that cause the production of enzymes that are able to metabolize them (or their analogues) are called **inducers**. Those that prevent the production of enzymes that are able to synthesize them are called **corepressors**.

These two ways of looking at regulation—negative versus positive control and inducible versus repressible control—can be combined to give four different patterns of gene regulation: **negative inducible**, **negative repressible**, **positive inducible**, **positive repressible** as shown in **FIGURE 26.4**. These enable a bacterium to perform the ultimate in inventory control of its metabolism to allow survival in rapidly changing environments.

The unifying theme is that regulatory proteins are *trans*-acting factors that recognize *cis*-acting elements (usually) upstream of the gene. The consequences of this recognition are either to activate or to repress the gene, depending on the individual type of regulatory protein. A typical feature is that the protein functions by recognizing a very short sequence in DNA, usually <10 bp in length, although the protein actually binds over a somewhat greater distance of DNA. The bacterial promoter is an example: RNA polymerase covers >70 bp of DNA at initiation, but the crucial sequences that it recognizes are the hexamers centered at −35 and −10.

A significant difference in gene organization between prokaryotes and eukaryotes is that structural genes in bacteria are organized in operons coordinately controlled

- **inducer** A small molecule that triggers gene transcription by binding to a regulator protein.
- **corepressor** A small molecule that triggers repression of transcription by binding to a regulator protein.
- **negative inducible** A control circuit in which an active repressor is inactivated by the substrate of the operon.
- **negative repressible** A control circuit in which an inactive repressor is activated by the product of the operon.
- **positive inducible** A control circuit in which an inactive positive regulator is converted into an active regulator by the substrate of the operon.
- **positive repressible** A control circuit in which an active positive regulator is inactivated by the product of the operon.

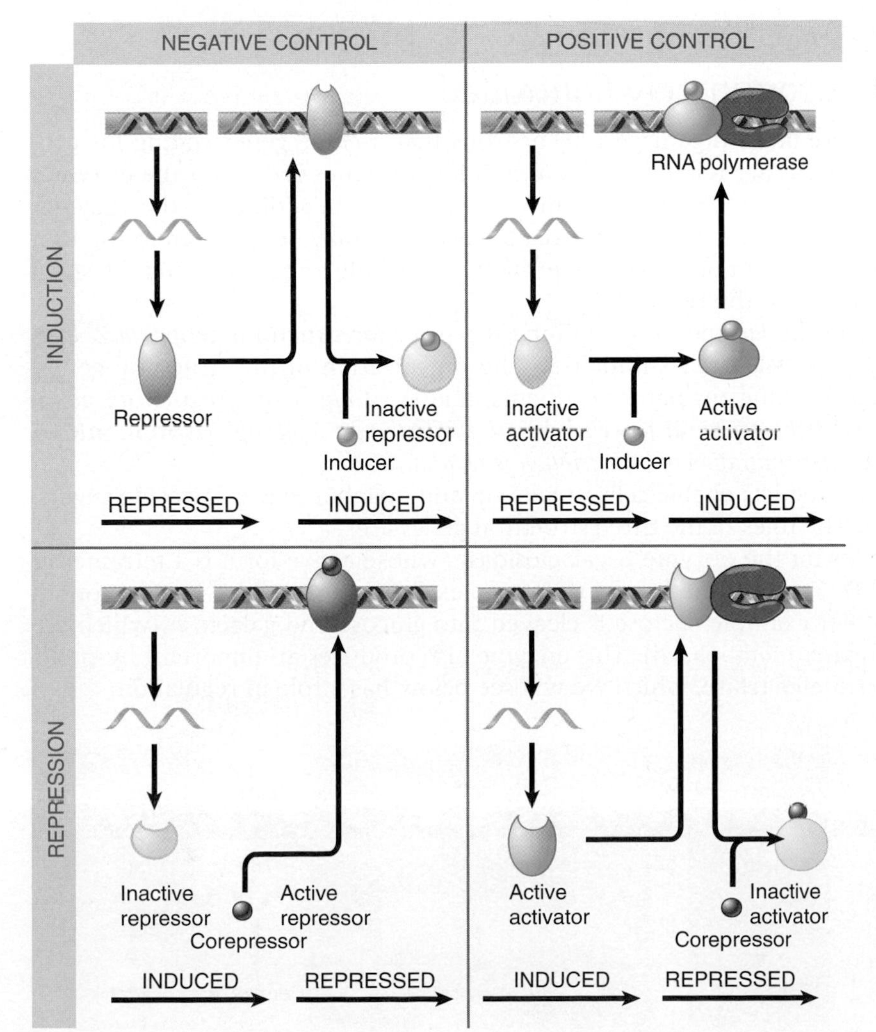

FIGURE 26.4 Regulatory circuits can be designed from all possible combinations of positive and negative control with inducible and repressible control.

by means of interactions at a single regulator, whereas genes in eukaryotes are for the most part controlled individually. As a result, in bacteria, an entire related set of genes is either transcribed or not transcribed. In this chapter, we discuss this mode of control and its use by bacteria. The means employed to coordinate control of dispersed eukaryotic genes are discussed in *Chapter 20, Eukaryotic Transcription*.

KEY CONCEPTS

- In negative regulation, a repressor protein binds to an operator to prevent a gene from being expressed.
- In positive regulation, a transcription factor is required to bind at the promoter in order to enable RNA polymerase to initiate transcription.
- In inducible regulation, the gene is regulated by the presence of its substrate.
- In repressible regulation, the gene is regulated by the product of its enzyme pathway.
- We can combine these in all four combinations: negative inducible, negative repressible, positive inducible, and positive repressible.

CONCEPT AND REASONING CHECKS

1. In a negative inducible gene system, what would be the effect of a mutation that eliminates the regulator?
2. In a positive inducible gene system, what would be the effect of a mutation that eliminates the regulator?

26.2 Structural Gene Clusters Are Coordinately Controlled

Bacterial genes are often organized into operons that include genes coding for proteins whose functions are related. It is common for the genes coding for the enzymes of a metabolic pathway to be organized into such a cluster. In addition to the enzymes actually involved in the pathway, other related activities may be included in the unit of coordinated control; for example, the protein responsible for transporting the small molecule substrate into the cell.

The cluster of the *lac* operon containing the three *lac* structural genes, *lacZ, lacY,* and *lacA,* is typical. **FIGURE 26.5** summarizes the organization of the structural genes, their associated *cis*-acting regulatory elements, and the *trans*-acting regulatory gene. *The key feature is that the structural gene cluster is transcribed into a single* ***polycistronic mRNA*** *from a promoter where initiation of transcription is regulated.*

▸ **polycistronic mRNA** mRNA that includes coding regions representing more than one gene.

The protein products enable cells to take up and metabolize β-galactoside sugars, such as lactose. The roles of the three structural genes are:

- *lacZ* codes for the enzyme β-galactosidase, whose active form is a tetramer of ~500 kD. The enzyme breaks the complex β-galactoside into its component sugars. For example, lactose is cleaved into glucose and galactose (which are then further metabolized). This enzyme also produces an important by-product, β-1,6-allolactase, which we will see below has a role in regulation.

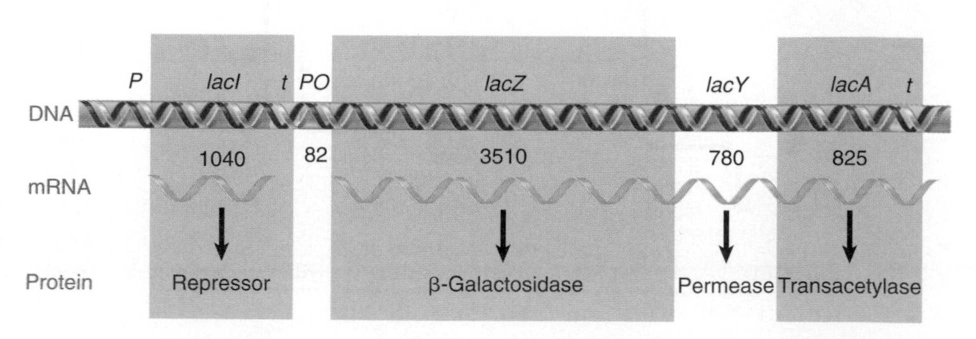

FIGURE 26.5 The *lac* operon occupies ~6000 bp of DNA. At the left the *lacI* gene has its own promoter and terminator. The end of the *lacI* region, *t*, is adjacent to the *lacZYA* promoter, *P*. Its operator, *O*, occupies the first 26 bp of the transcription unit. The long *lacZ* gene starts at base 39 and is followed by the *lacY* and *lacA* genes and a terminator, *t*.

HISTORICAL PERSPECTIVES

An Unstable Intermediate Carrying Information

Brenner and Jacob were guest investigators at the California Institute of Technology in 1960. At that time there was great interest in the mechanisms by which genes code for proteins. One possibility, which seemed reasonable at the time, was that each gene produced a different type of ribosome, differing in its RNA, which in turn produced a different type of protein. François Jacob and Jacques Monod had recently proposed an alternative, which was that the informational RNA ("messenger RNA") is actually an unstable molecule that breaks down rapidly. In this model, the ribosomes are nonspecific protein-synthesizing centers that synthesize different proteins according to specific instructions they receive from the genes through the messenger RNA. The key to the experiment is density-gradient centrifugation, which can separate macromolecules made "heavy" or "light" according to their content of ^{15}N or ^{14}N, respectively. (This technique is described in the box *Historical Perspectives* in *Chapter 12*.) The experiment is a purely biochemical proof of an issue absolutely critical for genetics—that genes code for proteins through the intermediary of a relatively short-lived messenger RNA.

> A large amount of evidence suggests that genetic information for protein structure is encoded in deoxyribonucleic acid (DNA), while the actual assembling of amino acids into proteins occurs in cytoplasmic ribonucleoprotein particles called ribosomes. The fact that proteins are not synthesized directly on genes demands the existence of an intermediate information carrier. . . . Jacob and Monod have put forward the hypothesis that ribosomes are nonspecialized structures which receive genetic information from the gene in the form of an unstable intermediate or "messenger." We present here the results of experiments on phage-infected bacteria which give direct support to this hypothesis. . . . When growing bacteria are infected with T2 bacteriophage, synthesis of DNA stops immediately, to resume 7 minutes later, while protein synthesis continues at a constant rate; in all likelihood, the protein is genetically determined by the phage. . . . Phage-infected bacteria therefore provide a situation in which the synthesis of a protein is suddenly switched from bacterial to phage control. . . . It is possible to determine experimentally [whether an unstable messenger RNA is produced] in the following way: Bacteria are grown in heavy isotopes so that all cell constituents are uniformly labeled "heavy." They are infected with phage and transferred immediately to a medium containing light isotopes so that all constituents synthesized after infection are "light." The distribution of new RNA and new protein, labeled with radioactive isotopes, is then followed by density gradient centrifugation of purified ribosomes. . . . We may summarize our findings as follows: (1) After phage infection no new ribosomes can be detected. (2) A new RNA with a relatively rapid turnover is synthesized after phage infection. This RNA, which has a base composition corresponding to that of the phage DNA, is added to pre-existing ribosomes, from which it can be detached in a cesium chloride gradient by lowering the magnesium concentration. (3) Most, if not all, protein synthesis in the infected cell occurs in pre-existing ribosomes. . . . The results also suggest that the messenger RNA may be large enough to code for long polypeptide chains. . . . It is a prediction of the messenger RNA hypothesis that the messenger RNA should be a simple copy of the gene, and its nucleotide composition should therefore correspond to that of the DNA. This appears to be the case in phage-infected cells. . . . If this turns out to be universally true, interesting implications for the coding mechanisms will be raised.

(*Source*: S. Brenner, F. Jacob, and M. Meselson. An Unstable Intermediate Carrying Information from Genes to Ribosomes for Protein Synthesis. *Nature* 190, pp. 576–581.)

- *lacY* codes for the β-galactoside permease, a 30-kD membrane-bound protein constituent of the transport system. This transports β-galactosides into the cell.
- *lacA* codes for β-galactoside transacetylase, an enzyme that transfers an acetyl group from acetyl-CoA to β-galactosides.

Mutations in either *lacZ* or *lacY* can create the *lac* genotype, in which cells cannot utilize lactose. (The genotypic description *"lac"* without a qualifier indicates loss of function.) The *lacZ* mutations abolish enzyme activity, directly preventing metabolism of lactose. The *lacY* mutants cannot take up lactose efficiently from the medium. (No defect is identifiable in *lacA* cells, which is puzzling. It is possible that the acetylation reaction gives an advantage when the bacteria grow in the presence of certain analogs of β-galactosides that cannot be metabolized, because the modification results in detoxification and excretion.)

The entire system, including structural genes and the elements that control their expression, forms a common unit of regulation called an operon. The activity of the operon is controlled by regulator gene(s) whose protein products interact with the *cis*-acting control elements.

KEY CONCEPT

- Genes coding for proteins that function in the same pathway may be located adjacent to one another and controlled as a single unit that is transcribed into a polycistronic mRNA.

CONCEPT AND REASONING CHECK

In the *lacZYA* operon, what would be the effect of a promoter mutation on *lacZ*? *lacY*?

26.3 The *lac* Operon Is Negative Inducible

We can distinguish between structural genes and regulator genes by the effects of mutations. A mutation in a structural gene deprives the cell of the particular protein for which the gene codes. A mutation in a regulator gene, however, influences the expression of all the structural genes that it controls. The consequences of a regulatory mutation reveal the type of regulation.

Transcription of the *lacZYA* genes is controlled by a regulator protein encoded by the *lacI* gene. It happens that *lacI* is located adjacent to the structural genes, but it comprises an independent transcription unit with its own promoter and terminator. In principle, *lacI* need not be located near the structural genes because it specifies a diffusible product. It can function equally well if moved elsewhere or can be carried on a separate DNA molecule (the classic test for a *trans*-acting regulator).

The *lacZYA* genes are negatively regulated: *they are transcribed unless turned off by the regulator protein*. Note that repression is not an absolute phenomenon; turing off a gene is not like turning off a lightbulb. Repression can often be a reduction in transcription by 5-fold or 100-fold. A mutation that inactivates the regulator causes the structural genes to be continually expressed, a condition called **constitutive expression**. The product of *lacI* is called the ***lac* repressor** because its function is to prevent the expression of the structural genes.

▸ **constitutive expression** Continuous expression of a given gene.

▸ ***lac* repressor** A negative gene regulator encoded by the *lacI* gene that turns off the *lac* operon.

The repressor is a tetramer of identical subunits of 38 kD each. There are ~10 tetramers in a wild-type cell. The regulator gene is not controlled; it is an unregulated gene with a poor promoter. It is transcribed into a monocistronic mRNA at a rate that appears to be governed simply by the affinity of its (poor) promoter for RNA polymerase. In addition, it is transcribed into a poor mRNA that is translated inefficiently. This is a common way to restrict the amount of protein made. In this case, the mRNA has virtually no 5′ UTR, which restricts the ability of a ribosome to start translation. These two features account for the low abundance of *lac* repressor protein in the cell.

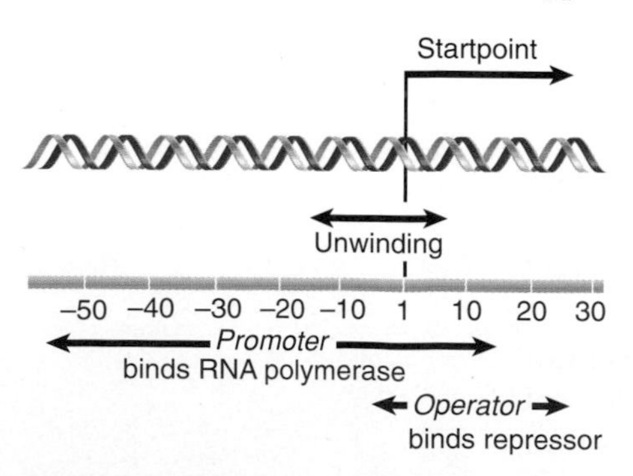

FIGURE 26.6 Repressor and RNA polymerase bind at sites that overlap around the transcription startpoint of the *lac* operon.

The repressor functions by binding to an operator (formally denoted O_{lac}) at the start of the *lacZYA* cluster. The operator lies between the promoter (P_{lac}) and the structural genes *(lacZYA)*. *When the repressor binds at the operator, it prevents RNA polymerase from initiating transcription at the promoter.* **FIGURE 26.6** expands our view of the region at the start of the *lac* structural genes. The operator extends from position −5 just upstream of the mRNA startpoint to position +21 within the transcription unit; thus it overlaps the 3′, right end of the promoter. A mutation that inactivates the operator also causes constitutive expression.

When cells of *E. coli* are grown in the absence of a β-galactoside, there is no need for β-galactosidase, and they contain very few molecules of the enzyme, about 5 per cell. When a suitable substrate is added, the enzyme activity appears very rapidly in the bacteria. Within 2 to 3 minutes some enzyme is present, and soon there are ~5000 molecules of enzyme per bacterium. (Under suitable conditions, β-galactosidase can account for 5% to 10% of the total soluble protein of the bacterium.) If the substrate is removed from the medium, the synthesis of enzyme stops as rapidly as it started.

FIGURE 26.7 summarizes the essential features of this induction. Control of transcription of the *lac* operon responds very rapidly to the inducer, as shown in the upper part of the figure. In the absence of inducer, the operon is transcribed at a very low basal level (this is an important concept; see the next section). Transcription is stimulated as soon as inducer is added; the amount of *lac* mRNA increases rapidly to an induced level that reflects a balance between synthesis and degradation of the mRNA.

The *lac* mRNA (as most mRNA is in bacteria) is extremely unstable and decays with a half-life of only ~3 minutes. This feature allows induction to be reversed rapidly by repressing transcription as soon as the inducer is removed. In a very short time all the *lac* mRNA is destroyed and enzyme synthesis ceases.

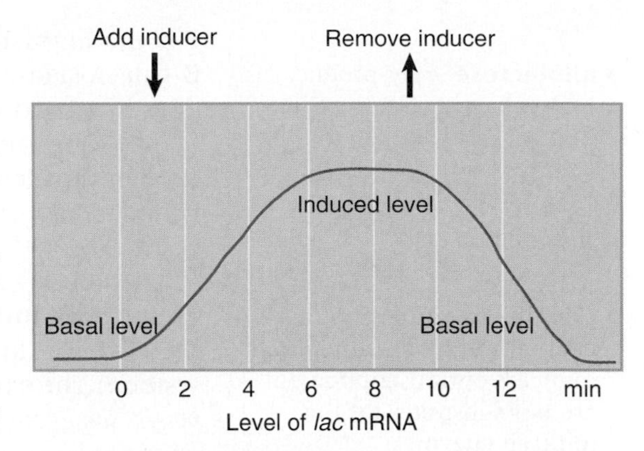

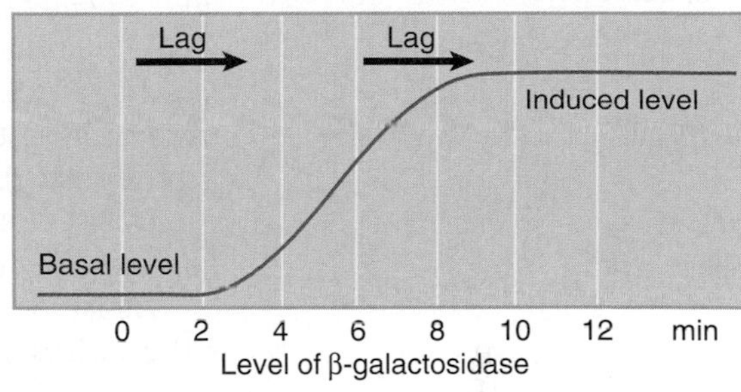

FIGURE 26.7 Addition of inducer results in rapid induction of *lac* mRNA and is followed after a short lag by synthesis of the enzymes; removal of inducer is followed by rapid cessation of synthesis.

The production of protein is followed in the lower part of the figure. Translation of the *lac* mRNA produces β-galactosidase (and the products of the other *lac* genes). There is a short lag between the appearance of *lac* mRNA and appearance of the first completed enzyme molecules (it is ~2 minutes after the rise of mRNA from basal level before protein begins to increase). There is a similar lag between reaching maximal induced levels of mRNA and protein. When inducer is removed, synthesis of enzyme ceases almost immediately (as the mRNA is degraded), but the β-galactosidase in the cell is more stable than the mRNA, so the enzyme activity remains at the induced level for longer.

KEY CONCEPTS

- Transcription of the *lacZYA* operon is controlled by a repressor protein that binds to an operator that overlaps the promoter at the start of the cluster.
- The repressor protein is a tetramer of identical subunits coded by the *lacI* gene.
- β-galactosides, the substrates of the *lac* operon, are its inducer.
- In the absence of β-galactosides, the *lac* operon is expressed only at a very low (basal) level.
- Addition of specific β-galactosides induces transcription of all three genes of the *lac* operon.
- The *lac* mRNA is extremely unstable; as a result, induction can be rapidly reversed.

CONCEPT AND REASONING CHECK

In the *lac* operon, what would be the effect of a mutation in the operator?

26.4 *lac* Repressor Is Controlled by a Small-Molecule Inducer

The ability to act as inducer or corepressor is highly specific. Only the substrate/product or a closely related molecule can serve. In most cases, the activity of the small molecule, however, does not depend on its interaction with the target enzyme.

For the *lac* system, however, the natural inducer is not lactose, but a by-product of β-galactosidase, **allolactose**. Allolactose is also a substrate of β-galactosidase, so it does not persist in the cell. Some inducers resemble the natural inducers of the *lac* operon but cannot be metabolized by the enzyme. The example *par excellence* is isopropylthiogalactoside (IPTG), one of several thiogalactosides with this property. IPTG is not recognized by β-galactosidase; even so, it is a very efficient inducer of the *lac* genes.

▸ **allolactose** A by-product of β-galactosidase; the true inducer of the *lac* operon.

Molecules that induce enzyme synthesis but are not metabolized are called **gratuitous inducers**. They are extremely useful because they remain in the cell in their original form. (A real inducer would be metabolized, interfering with study of the system.) The existence of gratuitous inducers reveals an important point. *The system must possess some component, distinct from the target enzyme, that recognizes the appropriate substrate, and its ability to recognize related potential substrates is different from that of the enzyme.*

▸ **gratuitous inducer** Inducers that resemble authentic inducers of transcription but are not substrates for the induced enzymes.

The component that responds to the inducer is the repressor protein encoded by *lacI*. Its target, the *lacZYA* structural genes, is transcribed into a single mRNA from the promoter just upstream of *lacZ*. The state of the repressor determines whether this promoter is turned off or on.

FIGURE 26.8 shows that in the absence of an inducer, the genes are not transcribed because the repressor protein is in an active form that is bound to the operator.

FIGURE 26.9 shows that when an inducer is added, the repressor is converted into either a form with lower affinity for operator or a lower affinity form that leaves the operator. Transcription then starts at the promoter and proceeds through the genes to a terminator located beyond the 3′ end of *lacA*.

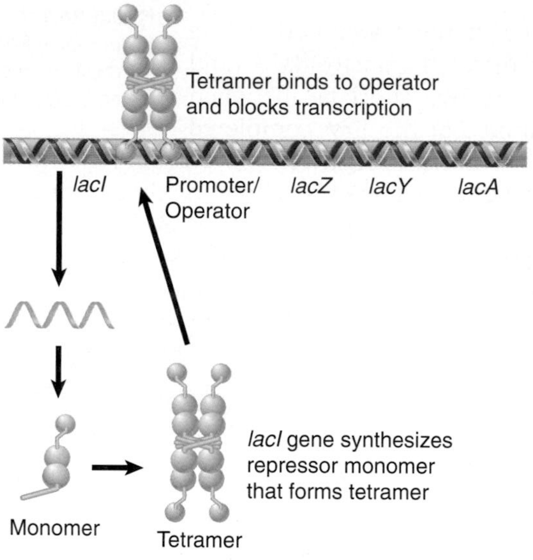

FIGURE 26.8 *lac* repressor maintains the *lac* operon in the inactive condition by binding to the operator. The shape of the repressor is represented as a series of connected domains as revealed by its crystal structure (see Figure 26.9).

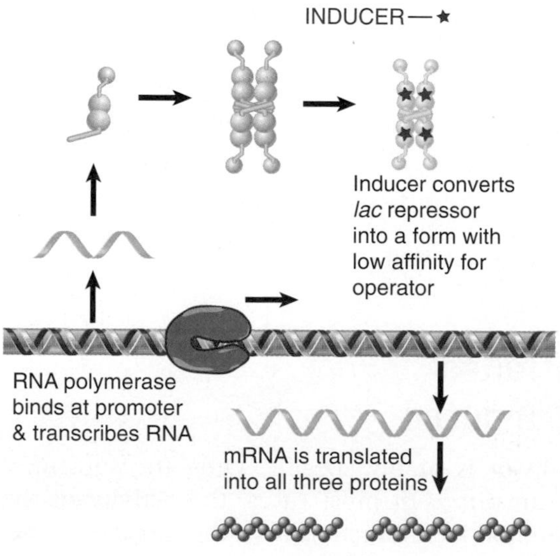

FIGURE 26.9 Addition of inducer converts repressor to a form with low affinity for the operator. This allows RNA polymerase to initiate transcription.

The crucial features of the control circuit reside in the dual properties of the repressor: it can prevent transcription, and it can recognize the small-molecule inducer. The repressor has three binding sites: one for the operator DNA, one for the inducer, and one to allow multimerization. When the inducer binds at its site, it changes the conformation of the protein in such a way as to influence the activity of the operator-binding site. The ability of one site in the protein to control the activity of another is called **allosteric control**.

▸ **allosteric control** The ability of a protein to change its conformation (and therefore activity) at one site as the result of binding a small molecule to a second site located elsewhere on the protein.

Induction accomplishes a coordinate regulation: *all the genes are expressed (or not expressed) in unison.* The mRNA is translated sequentially from its 5′ end, which explains why induction always causes the appearance of β-galactosidase, β-galactoside permease, and β-galactoside transacetylase, in that order. Translation of a common mRNA explains why the relative amounts of the three enzymes always remain the same under varying conditions of induction. Usually, the most important enzyme is first in the operon.

There are several potential paradoxes in the constitution of the operon. First, the *lac* operon contains the structural gene *(lacZ)* coding for the β-galactosidase activity needed to metabolize the sugar; it also includes the gene *(lacY)* that codes for the protein needed to transport the substrate into the cell. If the operon is in a repressed state, though, how does the inducer enter the cell to start the process of induction? The second paradox is that it requires β-galactosidase to make the inducer allolactose to induce the synthesis of β-galactosidase. That means that an operon with a mutant *lacZ* gene cannot be induced.

Two features ensure that there is always a minimal amount of the protein present in the cell—enough to start the process off. First, there is a basal level of expression of the operon: even when it is not induced, it is expressed at a residual level (0.1% of the induced level). In addition, some substrate (inducer) enters anyway via another uptake system. The basal level of β-galactosidase then converts some lactose to allolactose, leading to induction.

KEY CONCEPTS

- An inducer functions by converting the repressor protein into a form with lower operator affinity.
- *lac* repressor has two binding sites, one for the operator DNA and another for the inducer.
- *lac* repressor is inactivated by an allosteric interaction in which binding of inducer at its site changes the properties of the DNA-binding site.
- The true inducer is allolactose, not the actual substrate.

CONCEPT AND REASONING CHECK

If there are two complete copies of the *lac* operon in a cell and a mutation in one of the copies of *lacI*, what would be the effect on the cell.

26.5 *cis*-Acting Constitutive Mutations Identify the Operator

▸ **uninducible** A mutant in which the affected gene(s) cannot be expressed.

Mutations in the regulatory circuit may either abolish expression of the operon or cause constitutive expression. Mutants that cannot be expressed at all are called **uninducible**. Mutants that are continuously expressed are called *constitutive mutants*.

Components of the regulatory circuit of the operon can be identified by mutations that *affect the expression of all the structural genes and map outside them.* They fall into two classes, *cis*-acting and *trans*-acting. The promoter and the operator are identified as targets for the regulatory proteins (RNA polymerase and repressor, respectively) by *cis*-acting mutations. The locus *lacI* is identified as the gene that codes for the repressor protein by mutations that eliminate the *trans*-acting product.

The operator was originally identified by constitutive mutations, denoted O^c, whose distinctive properties provided the first evidence for *an element that functions without being represented in a diffusible product.*

The structural genes contiguous with an O^c mutation are expressed constitutively because the mutation changes the operator so that the repressor no longer recognizes it and binds to it. Thus the repressor cannot prevent RNA polymerase from initiating transcription. The operon is transcribed constitutively, as illustrated in **FIGURE 26.10**.

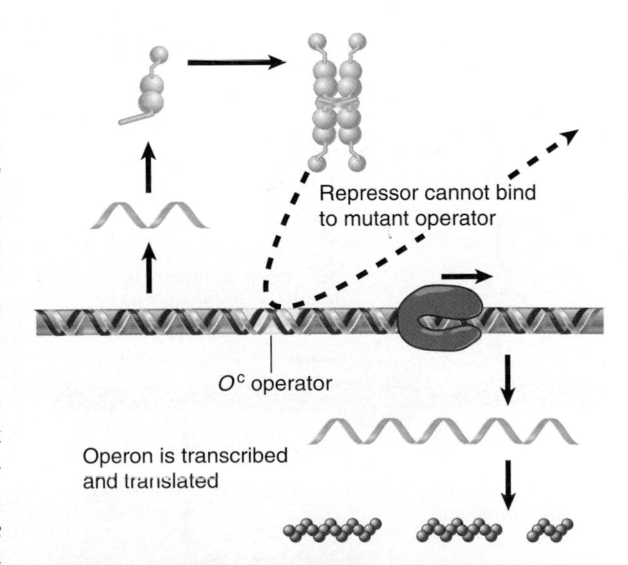

FIGURE 26.10 Operator mutations are constitutive because the operator is unable to bind repressor protein; this allows RNA polymerase to have unrestrained access to the promoter. The O^c mutations are *cis*-acting because they affect only the contiguous set of structural genes.

The operator can control only the lac genes that are adjacent to it. If a second *lac* operon is introduced into the bacterium on an independent molecule of DNA, it has its own operator. Neither operator is influenced by the other. Thus if one operon has a wild-type operator it will be repressed under the usual conditions, whereas a second operon with an O^c mutation will be expressed in its characteristic fashion. (Promoter mutations are also *cis*-acting.)

These properties define the operator as a typical *cis*-acting site, whose function depends upon recognition of its DNA sequence by some *trans*-acting factor. The operator controls the adjacent genes irrespective of the presence in the cell of other alleles of the site. A mutation in such a site—for example, the O^c mutation—is formally described as ***cis*-dominant**.

▸ ***cis*-dominant** A site or mutation that affects the properties only of its own molecule of DNA, often indicating that a site does not code for a diffusible product.

KEY CONCEPTS

- Mutations in the operator cause constitutive expression of all three *lac* structural genes.
- These mutations are *cis*-acting and affect only those genes on the contiguous stretch of DNA.

CONCEPT AND REASONING CHECK

If there are two complete copies of the *lac* operon in a cell and a mutation in only one of the operator sites controlling *lacZYA*, what would be the effect on the cell?

26.6 *trans*-Acting Mutations Identify the Regulator Gene

Two types of constitutive mutations can be distinguished genetically. O^c mutants (described in *Section 26.5*) are *cis*-dominant, whereas *lacI*⁻ mutants are recessive. This means that the introduction of a normal *lacI*⁺ gene can restore control even in the presence of a defective *lacI*⁻ gene. The *lac* repressor protein is diffusible; thus the normal *lacI* can be placed on an independent molecule of DNA. Other *lacI* mutations can cause the operon to be uninducible (unable to be turned on, denoted *lacI*^{s}, described later), similar to mutations in the promoter.

Constitutive transcription is caused by mutations of the *lacI*⁻ type, which are caused by loss of function (including deletions of the gene). When the repressor is inactive or absent, transcription of the *lac* operon can initiate at the *lac* operon promoter. **FIGURE 26.11** shows that the *lacI*⁻ mutants express the structural genes all the time (constitutively), *irrespective of whether the inducer is present or absent,* because the repressor is inactive. One important subset of *lacI*⁻ mutations (called *lacI*$^{-d}$; see the following) is localized in the DNA-binding site of the repressor. They abolish the ability to turn off the gene by damaging the site that the repressor uses to contact the operator. They are dominant mutations because they do not allow a tetramer with normal and mutant repressor subunits to bind the operator (see the following).

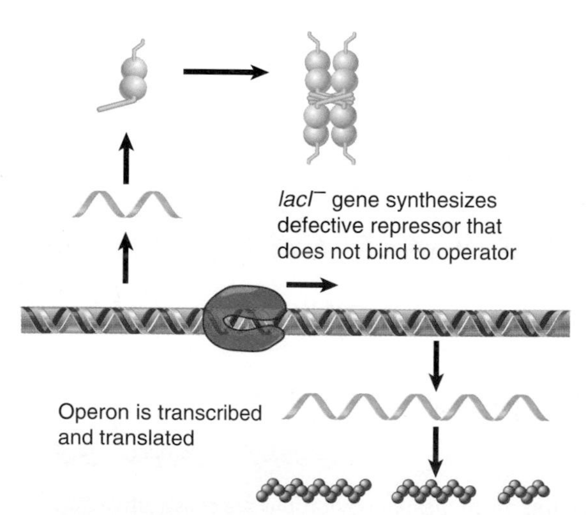

FIGURE 26.11 Mutations that inactivate the *lacI* gene cause the operon to be constitutively expressed because the mutant repressor protein cannot bind to the operator.

Uninducible mutants are caused by mutations that abolish the ability of repressor to bind the inducer. They are described as *lacI*^{s}. The repressor is "locked in" to the active form that recognizes the operator and prevents transcription. These mutations identify the inducer-binding site and other positions involved in allosteric control of the DNA binding site. The mutant repressor binds to all *lac* operators in the cell to prevent their transcription, and cannot be removed from the operator, even if wild-type protein is present.

An important feature of the repressor is that it is multimeric. Repressor subunits associate at random in the cell to form the active protein tetramer. When two different alleles of the *lacI* gene are present, the subunits made by each can associate to form a heterotetramer, whose properties differ from those of either homotetramer. This type of interaction between subunits is a characteristic feature of multimeric proteins and is described as **interallelic complementation**.

Most *lacI* mutants inactivate the repressor. Thus these genes are recessive when coexpressed with the wild-type repressor, and the *lac* operon is normally regulated. Combinations of certain repressor mutants display a form of interallelic complementation called **negative complementation**. As mentioned before, the *lacI*$^{-d}$ mutations are dominant when paired with a wild-type allele. Such mutations are called **dominant negative**. The reason for their behavior is that one mutant subunit in a tetramer can antagonize the function of the wild-type subunits, as discussed in the next section. The *lacI*$^{-d}$ mutation alone results in the production of a repressor that cannot bind the operator, and it is, therefore, constitutive like the *lacI*$^{-}$ alleles. Because the *lacI*$^{-}$ type of mutation inactivates the repressor, it is usually recessive to the wild type. However, the d notation indicates that this variant of the negative type is dominant when paired with a wild-type allele.

- **interallelic complementation** The change in the properties of a heteromultimeric protein brought about by the interaction of subunits coded by two different mutant alleles; the mixed protein may be more or less active than the protein consisting of subunits of only one or the other type.
- **negative complementation** It occurs when interallelic complementation allows a mutant subunit to suppress the activity of a wild-type subunit in a multimeric protein.
- **dominant negative** A mutation resulting in a mutant gene product that prevents the function of the wild-type gene product, causing loss or reduction of gene activity in cells containing both the mutant and wild-type alleles. The most common cause is that the gene codes for a homomultimeric protein whose function is lost if only one of the subunits is a mutant.

KEY CONCEPTS

- Mutations in the *lacI* gene are *trans*-acting and affect expression of all *lacZYA* clusters in the bacterium.
- Mutations that eliminate *lacI* function cause constitutive expression and are recessive.
- Mutations in the DNA-binding site of the repressor are constitutive because the repressor cannot bind the operator.
- Mutations in the inducer-binding site of the repressor prevent it from being inactivated and cause uninducibility.
- When mutant and wild-type subunits are present, a single *lacI*$^{-d}$ mutant subunit can inactivate a tetramer whose other subunits are wild type.
- *lacI*$^{-d}$ mutations occur in the DNA-binding site. Their effect is explained by the fact that repressor activity requires all DNA-binding sites in the tetramer to be active.

CONCEPT AND REASONING CHECK

If there are two complete copies of the *lac* operon in a cell and a mutation in one copy of *lacZ*, what would be the effect on the cell?

26.7 *lac* Repressor Is a Tetramer Made of Two Dimers

The repressor has several domains, as shown in the crystal structure illustrated in **FIGURE 26.12**. A major feature is that the DNA-binding domain is separate from the rest of the protein.

The DNA-binding domain occupies residues 1–59. It consists of two α-helices separated by a turn. This is a common DNA-binding motif known as the HTH (helix-turn-helix); the two α-helices fit into the major groove of DNA, where they make contacts with specific bases (see *Section 27.10, Lambda Repressor Uses a Helix-Turn-Helix Motif to Bind DNA*). This region is connected by

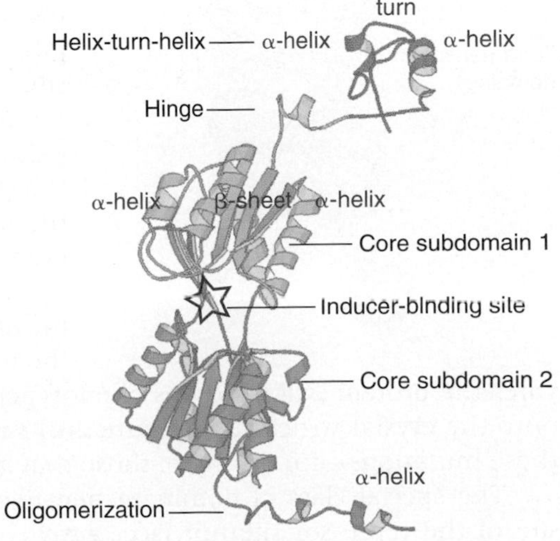

FIGURE 26.12 The structure of a monomer of *lac* repressor identifies several independent domains. Structure from Protein Data Bank 1LBG. M. Lewis et al. *Science* 271 (1996): 1247–1254. Photo courtesy of Hongli Zhan and Kathleen Matthews, Rice University.

a hinge to the main body of the protein. In the DNA-binding form of repressor, the hinge forms a small α-helix (as shown in Figure 26.12); but when the repressor is not bound to DNA, this region is disordered. The HTH and hinge together correspond to the *headpiece*.

The remainder of the protein is called the "core." The bulk of the core consists of two regions with similar structures (core domains 1 and 2). Each has a six-stranded parallel β-sheet sandwiched between two α-helices on either side. The inducer binds in a cleft between the two regions. Two monomer core domains can associate to form a dimeric version of LacI. Dimeric LacI tightly binds operator DNA because it recognizes both halves of the operator sequence, which is an inverted repeat (see the following).

The C-terminus of the monomer contains an α-helix with two leucine heptad repeats. This is the tetramerization domain. The tetramerization helices of four monomers associate to maintain the tetrameric structure. **FIGURE 26.13** shows the structure of the tetrameric core. It consists, in effect, of two dimers. The body of each dimer contains an interface between the subdomains of the two core monomers and two clefts in which two inducers bind (top). The C-terminal regions of each monomer protrude as helices. (The headpiece would join with the N-terminal regions at the top.) Together two dimers form a tetramer (center) that is held together by a C-terminal bundle of four helices.

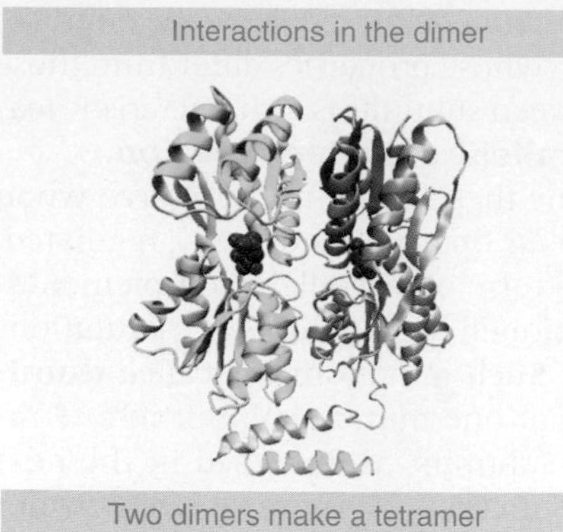

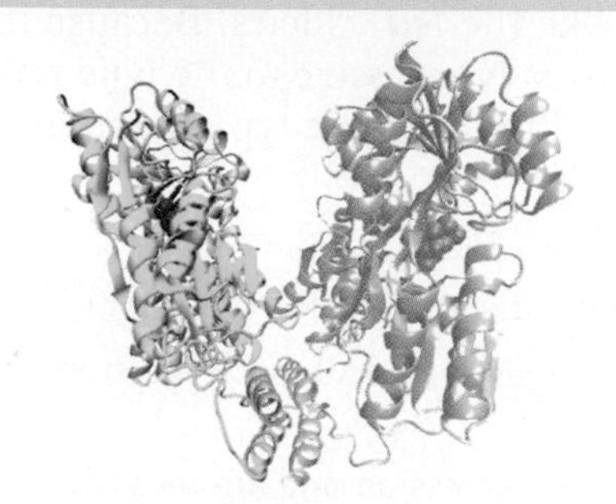

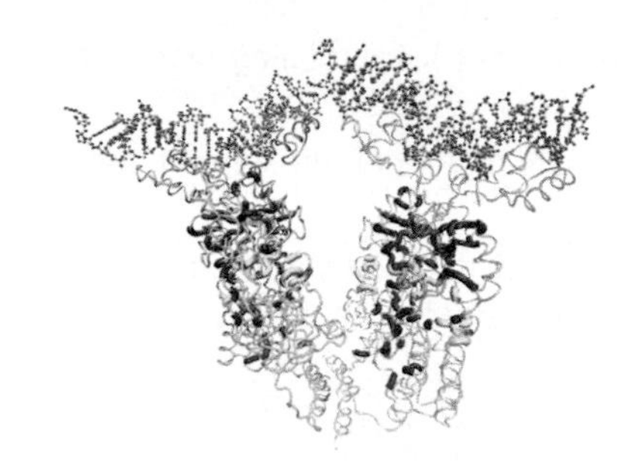

FIGURE 26.13 The crystal structure of the core region of *lac* repressor identifies the interactions between monomers in the tetramer. Each monomer is identified by a different color. Mutations are colored as: dimer interface = yellow; inducer binding = blue; oligomerization = white and purple. The protein orientation in the middle panel is rotated ~ 90° along the Z axis relative to the top panel. Photos courtesy of Benjamin Wieder and Ponzy Lu, University of Pennsylvania.

FIGURE 26.14 shows a schematic for how the monomers are organized into the tetramer. Two monomers form a dimer by means of contacts at core subdomain 2; other contacts occur between their respective tetramerization helices. The dimer has two DNA-binding domains at one end of the structure and the tetramerization helices at the other end. Two dimers then form a tetramer by interactions at the tetramerization interface. Each tetramer has four inducer-binding sites and two DNA binding sites.

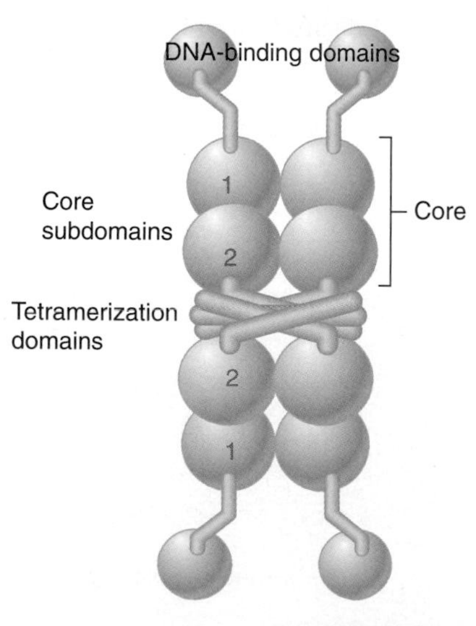

FIGURE 26.14 The repressor tetramer consists of two dimers. Dimers are held together by contacts involving core domains 1 and 2 as well as by the oligomerization helix. The dimers are linked into the tetramer by the oligomerization interface.

Mutations in the *lac* repressor identified the existence of different domains even before the structure was known. We can now explain the nature of the mutations more fully by reference to the structure, as summarized in the scheme of **FIGURE 26.15**. Recessive mutations of the *lacI*$^-$ type can occur anywhere in the bulk of the protein. Basically, any mutation that inactivates the protein will have this phenotype. The more detailed mapping of mutations onto the crystal structure in Figure 26.13 identifies specific impairments for some of these mutations—for example, those that affect oligomerization.

The special class of dominant-negative *lacI*$^{-d}$ mutations lie in the DNA-binding site of the repressor subunit (see *Section 26.6,* trans-*acting Mutations Identify Regulator*

Genes). This explains their ability to prevent mixed tetramers from binding to the operator; a reduction in the number of binding sites reduces the specific affinity for the operator. The role of the N-terminal region in specifically binding DNA is also shown by the occurrence of "tight binding" mutations in this region. These rare mutations increase the affinity of the repressor for the operator, sometimes so much that it cannot be released by inducer.

Uninducible *lacI*ˢ mutations map largely in a region of the core domain 1 extending from the inducer-binding site to the hinge. One group lies in amino acids that contact the inducer, and these mutations function by preventing binding of inducer. The remaining mutations lie at sites that must be involved in transmitting the allosteric change in conformation to the hinge when inducer binds.

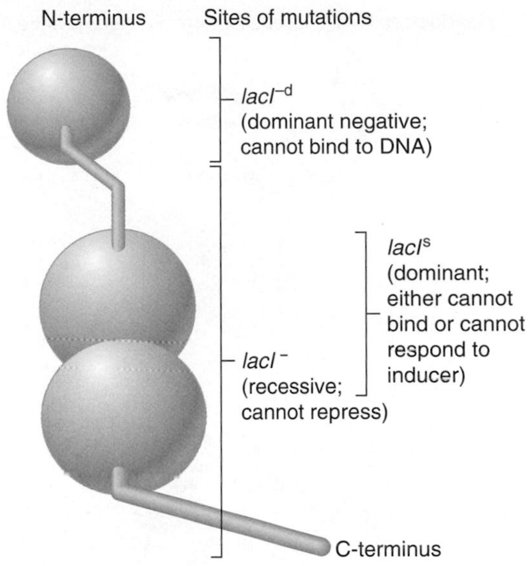

FIGURE 26.15 The locations of three types of mutations in lactose repressor are mapped on the domain structure of the protein. Recessive *lacI*⁻ mutants that cannot repress can map anywhere in the protein. Dominant negative *lacI*⁻ᵈ mutants that cannot repress map to the DNA-binding domain. Dominant *lacI*ˢ mutants that cannot induce because they do not bind inducer or cannot undergo the allosteric change map to core domain 1.

KEY CONCEPTS

- A single repressor subunit can be divided into the N-terminal DNA-binding domain, a hinge, and the core of the protein.
- The DNA-binding domain contains two short α-helical regions that bind the major groove of DNA.
- The inducer-binding site and the regions responsible for multimerization are located in the core.
- Monomers form a dimer by making contacts between core domains 1 and 2.
- Dimers form a tetramer by interactions between the oligomerization helices.
- Different types of mutations occur in different domains of the repressor protein.

CONCEPT AND REASONING CHECK

If there are two complete copies of the *lac* operon in a cell and a mutation in one of the operator sites controlling *lacZYA* as well as a second mutation in the *lacZ* gene in that operon, what would be the effect on the cell?

26.8 *lac* Repressor Binding to the Operator Is Regulated by an Allosteric Change in Conformation

How does the repressor recognize the specific sequence of operator DNA? The operator has a feature common to many recognition sites for regulator proteins: it is a type of **palindrome** known as an inverted repeat. The inverted repeats are highlighted in **FIGURE 26.16**. Each repeat can be regarded as a half site of the operator. The symmetry of

▸ **palindrome** A symmetrical sequence that reads the same forward and backward.

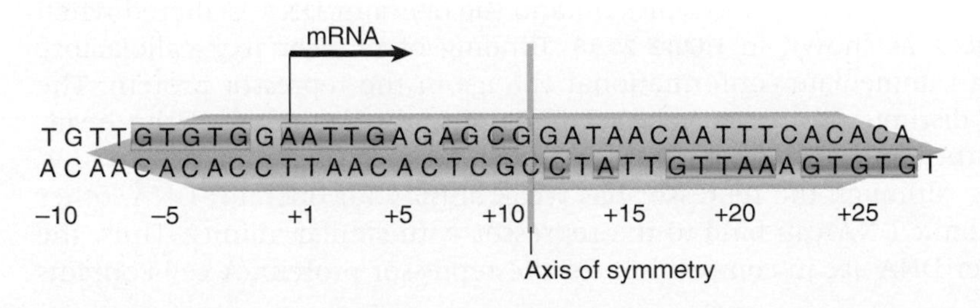

FIGURE 26.16 The *lac* operator has a symmetrical sequence. The sequence is numbered relative to the start point for transcription at +1. The pink arrows to the left and to the right identify the two dyad repeats. The green blocks indicate the positions of symmetry.

FIGURE 26.17 Inducer changes the structure of the core so that the headpieces of a repressor dimer are no longer in an orientation that permits binding to DNA bases.

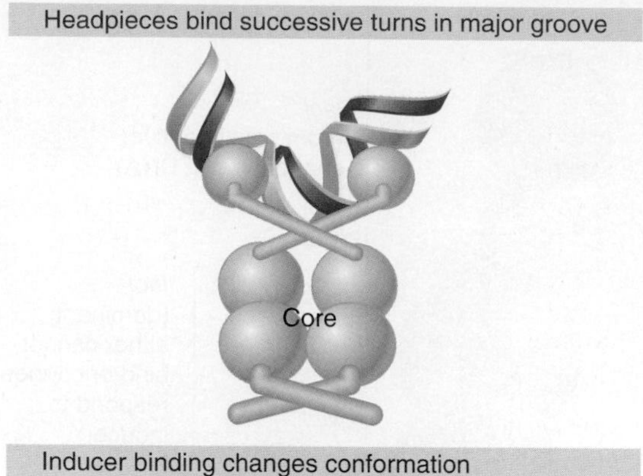

FIGURE 26.18 Does the inducer bind to the free repressor to upset an equilibrium (left) or directly to repressor bound to the operator (right)?

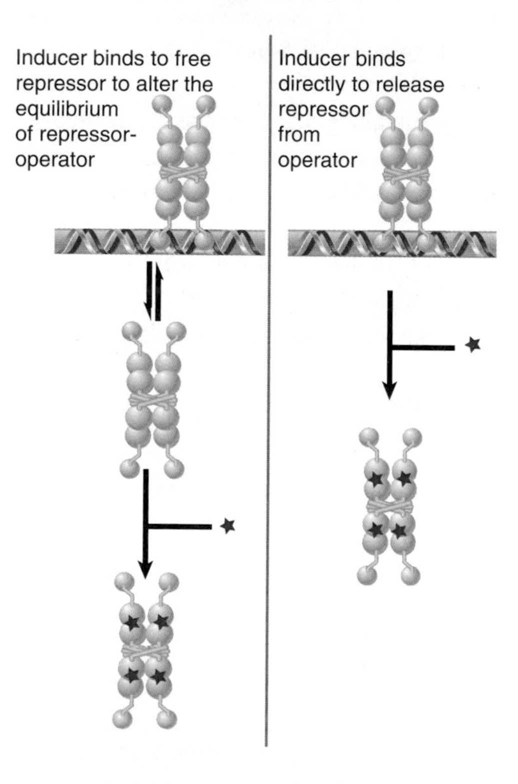

the operator matches the symmetry of the repressor protein dimer. Each DNA-binding domain of the identical subunits in a repressor can bind one half-site of the operator; two DNA-binding domains of a dimer are required to bind the full-length operator. **FIGURE 26.17** shows that the two DNA-binding domains in a dimeric unit contact DNA by inserting into successive turns of the major groove. This enormously increases affinity for the operator. Note that the *lac* operator is not a perfectly symmetrical sequence; it contains a single central base pair and the sequence of the left side binds to the repressor more strongly than the sequence of the right side.

The importance of particular bases within the operator sequence can be determined by identifying those that contact the repressor protein or in which mutations change the binding of repressor. The *lac* repressor dimer contacts the operator in such a way that each inverted repeat of the operator makes the same pattern of contacts with a repressor monomer. The region of DNA contacted by protein extends for 26 bp, and within this region are eight sites at which constitutive mutations occur. This emphasizes the same point made by promoter mutations: *A small number of essential specific contacts within a larger region can be responsible for sequence-specific association of a protein binding to DNA.*

Figure 26.17 shows another key element of repressor-operator binding: the insertion of the hinge helix into the minor groove of operator DNA, which bends the DNA by ~45°. This bend orients the major groove for HTH binding. DNA bending is commonly seen when a sequence is bound to a regulatory protein and illustrates the principle that the structure of DNA is more complicated than the canonical double helix.

The interaction of the *lac* repressor protein and the operator DNA is altered when repressor is induced as shown in **FIGURE 26.18**. Binding of inducer (e.g., allolactose or IPTG) causes an immediate conformational change in the repressor protein. The change probably disrupts the hinge helices, changing the orientation of the headpieces relative to the core, with the result that the repressor's affinity for DNA is lowered dramatically. Although the repressor has weak affinity for operator DNA, other sequences of genomic DNA can bind to the repressor with similar affinity. Thus, the operator and other DNA are in competition for the repressor protein. A cell contains much more genomic DNA than the single copy of the operator sequence; as a result, the genomic DNA "wins" the repressor protein, and the operator is vacant.

Some structural and molecular details of the induction process remain the subject of active research. The number of inducers that must be bound to a dimer (within the tetramer) in order to cause induction is under debate. The nature of the conformational change caused in *lac* repressor by binding to inducer is also not completely known.

KEY CONCEPTS

- *lac* repressor protein binds to the double-stranded DNA sequence of the operator.
- The operator is a palindromic sequence of 26 bp.
- Each inverted repeat of the operator binds to the DNA-binding site of one repressor subunit.
- Inducer binding causes a change in repressor conformation that reduces its affinity for DNA and releases it from the operator.

CONCEPT AND REASONING CHECK

How does changing the structure of a protein (like the *lac* repressor) change the way it recognizes the DNA?

26.9 *lac* Repressor Binds to Three Operators and Interacts with RNA Polymerase

The repressor dimer is sufficient to bind the entire operator sequence. Why, then, is a tetramer required to establish full repression?

Each dimer can bind an operator sequence. This enables the intact repressor to bind to two operator sites simultaneously. In fact, there are two further operator sites in the initial region of the *lac* operon. The original operator, *O1*, is located just at the start of the *lacZ* gene. It has the strongest affinity for repressor. Weaker operator sequences are located on either side; *O2* is 410 bp downstream of the startpoint in *lacZ* and *O3* is 88 bp upstream of it in *lacI*.

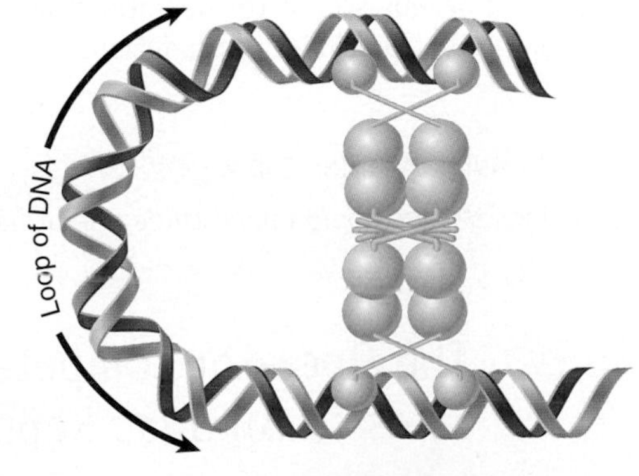

FIGURE 26.19 If both dimers in a repressor tetramer bind to DNA, the DNA between the two binding sites is held in a loop.

FIGURE 26.19 predicts what happens when a DNA-binding protein simultaneously binds to two separated sites on DNA. The DNA between the two sites forms a loop from a base where the protein has bound the two sites. The length of the loop depends on the distance between the two binding sites. When *lac* repressor binds simultaneously to *O1* and to one of the other operators, it causes the DNA between them to form a rather short loop, significantly constraining the DNA structure. A scale model for binding of tetrameric repressor to two operators is shown in **FIGURE 26.20**.

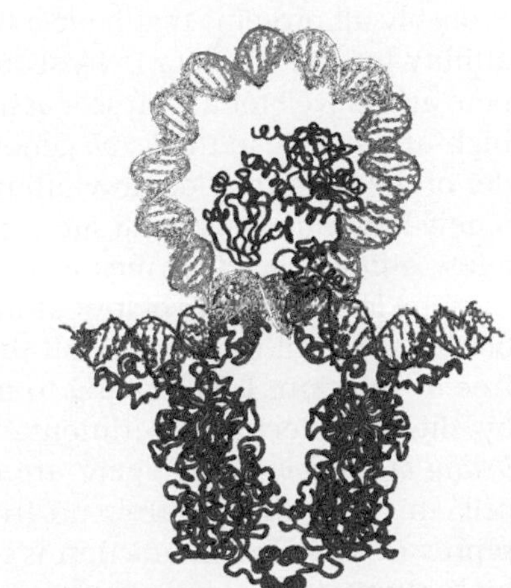

FIGURE 26.20 When a repressor tetramer binds to two operators, the stretch of DNA between them is forced into a tight loop. (The blue structure in the center of the looped DNA represents CRP, which is another regulator protein that binds in this region.) Reproduced with permission from M. Lewis et al., Science 271 (1996): cover. © 1996 AAAS. Photo courtesy of Ponzy Lu, University of Pennsylvania.

Binding at the additional operators affects the level of repression. Elimination of either the downstream operator *(O2)* or the upstream operator *(O3)* reduces the efficiency of repression by 2× to 4×. If, however, *both O2* and *O3* are eliminated, repression is reduced more than 50×. *This suggests that the ability of the repressor to bind to one of the two other operators as well as to O1 is important for establishing strong repression.* We do not know how and why this simultaneous binding increases repression. We know about the direct effects of binding of repressor to the operator *(O1)*.

We have several lines of evidence as to how binding of repressor to the *O1* operator inhibits transcription initiation by RNA polymerase. It was originally thought that repressor binding would prevent RNA polymerase from binding to the promoter. We now know that the two proteins may be bound to DNA simultaneously and that *the binding of repressor actually enhances the binding of RNA polymerase!* The bound enzytme is prevented from initiating transcription, though. The repressor in effect causes RNA polymerase to be stored at the promoter. When inducer is added, the repressor is released, and RNA polymerase can initiate transcription immediately. The overall effect of repressor is to speed up the induction process.

Does this model apply to other systems? The interaction between RNA polymerase, repressor, and the promoter/operator region is distinct in each system, because the operator does not always overlap with the same region of the promoter (see Figure 26.21). For example, in phage lambda, the operator lies in the upstream region of the promoter, and binding of repressor occludes the binding of RNA polymerase (see *Chapter 27, Phage Strategies*). Thus a bound repressor does not interact with RNA polymerase in the same way in all systems.

KEY CONCEPTS

- Each dimer in a repressor tetramer can bind an operator, so that the tetramer can bind two operators simultaneously.
- Full repression requires the repressor to bind to an additional operator downstream or upstream as well as to the operator at the *lacZ* promoter.
- Binding of repressor at the operator stimulates binding of RNA polymerase at the promoter but precludes transcription.

CONCEPT AND REASONING CHECK

Why does the *lac* operon need three operators?

26.10 The Operator Competes with Low-Affinity Sites to Bind *lac* Repressor

Probably all proteins that have a high affinity for a specific sequence also possess a low affinity for any random DNA sequence. A large number of low-affinity sites will compete just as well for a repressor as a small number of high-affinity sites. There is only one high-affinity site in the *E. coli* genome for the *lac* repressor: the *lac* operator. The remainder of the DNA provides low-affinity binding sites. Every base pair in the genome starts a new low-affinity binding site. Simply moving one base pair from the operator creates a low-affinity site! That means that there are 4.2×10^6 low-affinity sites.

The large number of low-affinity sites means that even in the absence of a specific binding site, all or virtually all the repressors are bound to DNA; very little remains free in solution. LacI binding to nonspecific genomic sites has been visualized *in vivo* by single molecule experiments. We can deduce that *all but 0.01% of repressors are bound to random DNA*. There are only about 10 molecules of repressor tetramer per cell, this says that there is no free repressor protein. Thus the critical factor of the repressor-operator interaction is the partitioning of the repressor on DNA; the single high-affinity site of the operator competes with a large number of low-affinity sites.

The efficiency of repression therefore depends on the relative affinity of the repressor for its operator compared with other random DNA sequences. The affinity must be great enough to overcome the large number of random sites. We can see how this works by comparing the equilibrium constants for *lac* repressor/operator binding with repressor/general DNA binding. **FIGURE 26.21** shows that the ratio is 10^7 for an active repressor, enough to ensure that the operator is bound by repressor 96% of the time so that transcription is effectively—but not completely—repressed. (Remember that because allolactose, not lactose, is the inducer, we always need a little β-galactosidase in the cell.) When inducer is added, the ratio is reduced to 10^4. At this level, only 3% of the operators are bound and the operon is effectively induced.

DNA	Repressor	Repressor + Inducer
Operator	2×10^{13}	2×10^{10}
Other DNA	2×10^{6}	2×10^{6}
Specificity	10^7	10^4
Operators bound	96%	3%
Operon is:	repressed	induced

FIGURE 26.21 *lac* repressor binds strongly and specifically to its operator but is released by inducer. All equilibrium constants are in M^{-1}.

The consequence of these affinities is that in an uninduced cell, one tetramer of repressor usually is bound to the operator. All, or almost all, the remaining tetramers are bound at random to other regions of DNA, as illustrated in **FIGURE 26.22**. There are likely to be very few or no repressor tetramers free within the cell.

The addition of inducer abolishes the ability of repressor to bind specifically at the operator. Those repressors bound at the operator are released and bind to random (low-affinity) sites. Thus in an induced cell, the repressor tetramers are "stored" on random DNA sites. In a noninduced cell, a tetramer is bound at the operator, whereas the remaining repressor molecules are bound to nonspecific sites. The effect of induction is, therefore, to change the distribution of repressor on DNA rather than to generate free repressor. In the same way that RNA polymerase probably moves between promoters and other DNA by swapping one sequence for another, the repressor also may directly displace one bound DNA sequence with another to move between sites.

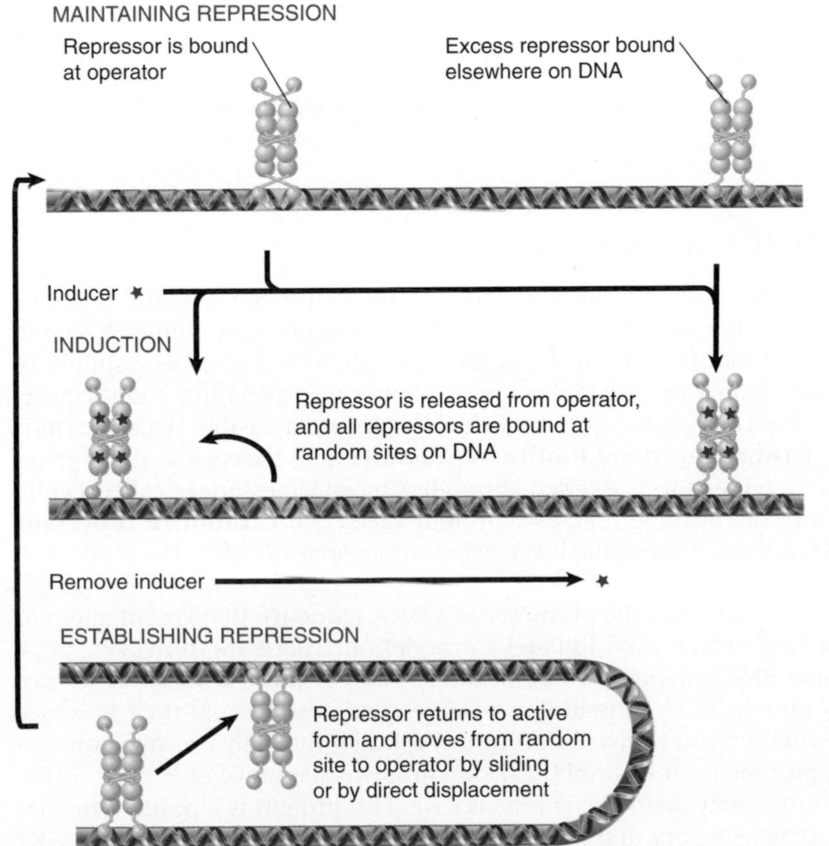

FIGURE 26.22 Virtually all the repressor in the cell is bound to DNA.

We can define the parameters that influence the ability of a regulator protein to saturate its target site by comparing the equilibrium equations for specific and nonspecific binding. As might be expected, the important parameters are:

- The size of the genome dilutes the ability of a protein to bind specific target sites (remember how large eukaryote genomes are).
- The specificity of a protein counters the effect of the mass of the DNA.
- The amount of the protein that is required increases with the total amount of DNA in the genome and decreases the specificity of DNA binding.
- The amount of the protein also must be in reasonable excess of the total number of specific target sites, so we expect regulators with many targets to be found in greater quantities than regulators with fewer targets.

KEY CONCEPTS

- Proteins that have a high affinity for a specific DNA sequence also have a low affinity for other DNA sequences.
- Every base pair in the bacterial genome is the start of a low-affinity binding site for repressor.
- The large number of low-affinity sites ensures that all repressor protein is bound to DNA.
- *lac* repressor binds to the operator by moving from a low-affinity site rather than by equilibrating from solution.
- In the absence of inducer, the operator has an affinity for repressor that is $10^7\times$ that of a low-affinity site.
- The level of 10 repressor tetramers per cell ensures that the operator is bound by repressor 96% of the time.
- Induction reduces the affinity for the operator to $10^4\times$ that of low-affinity sites, so that only 3% of operators are bound.
- Induction causes repressor to move from the operator to a low-affinity site by direct displacement.

CONCEPT AND REASONING CHECK

Why can't a mutant *lac* operon be induced without a functional *lacZ* gene?

26.11 The *lac* Operon Has a Second Layer of Control: Catabolite Repression

The *E. coli lac* operon is negative inducible. Transcription is turned on by the presence of lactose by removing the *lac* repressor. However, this operon is also under a second layer of control. It cannot be turned on by lactose if there is a sufficient supply of glucose. The rationale for this is that glucose is a better energy source than lactose, so there is no need to turn on the operon if there is glucose available. This system is part of a global network called **catabolite repression** that affects about 20 genes in *E. coli*. Catabolite repression is exerted through a second messenger called **cyclic AMP (cAMP)**, and the positive regulator protein called the **catabolite repressor protein** or **CRP** (also sometimes called *catabolite activator protein*, CAP). The *lac* operon is, therefore, under dual control.

Thus far we have dealt with the promoter as a DNA sequence that is competent to bind RNA polymerase, which then initiates transcription. There are, however, some promoters at which RNA polymerase cannot initiate transcription without assistance from an ancillary protein. Such proteins are positive regulators because their presence is necessary to switch on the transcription unit. Typically, the activator overcomes a deficiency in the promoter, for example, a poor consensus sequence at −35 or −10.

One of the most widely acting activators is CRP. This protein is a positive regulator whose presence is necessary to initiate transcription at dependent promoters. CRP

- **catabolite repression** The ability of glucose to prevent the expression of a number of genes. In bacteria this is a positive control system; in eukaryotes, it is completely different.
- **cyclic AMP (cAMP)** The coregulator of CRP, it has an internal 3′–5′ phosphodiester bond. Its concentration is inverse to the concentration of glucose.
- **catabolite repressor protein (CRP)** A positive regulator protein activated by cyclic AMP. It is needed for RNA polymerase to initiate transcription of many operons of *E. coli*.

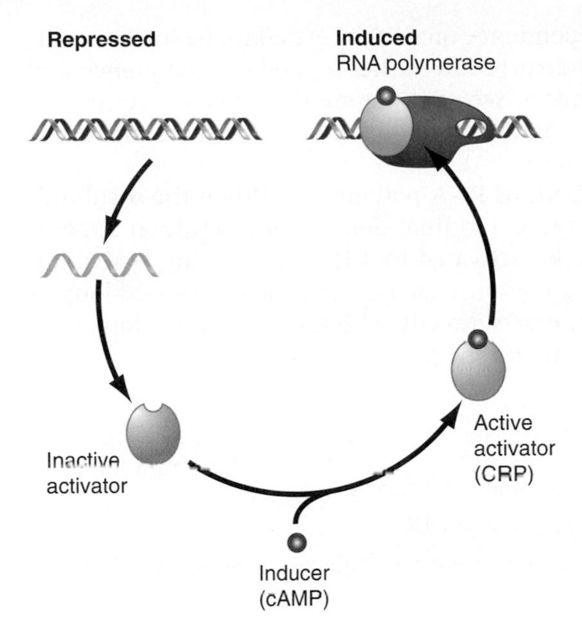

FIGURE 26.23 A small-molecule inducer, cAMP, converts an activator protein CRP to a form that binds the promoter and assists RNA polymerase in initiating transcription.

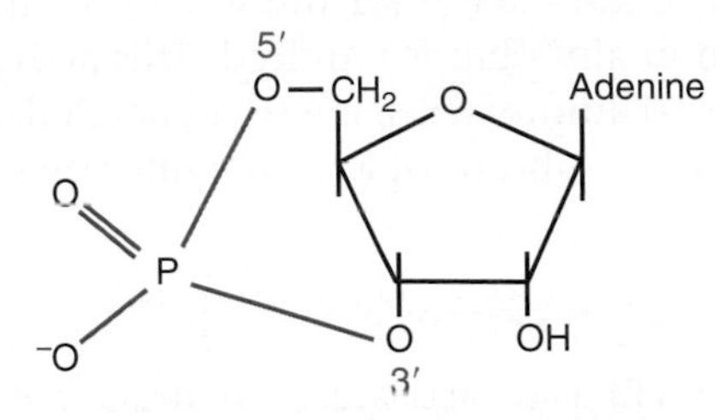

FIGURE 26.24 Cyclic AMP has a single phosphate group connected to both the 3′ and 5′ positions of the sugar ring, i.e., a single phosphodiester bond.

is active *only in the presence of cAMP*, which behaves as a classic small-molecule inducer for positive control (**FIGURE 26.23**).

cAMP is synthesized by the enzyme *adenylate cyclase*. The reaction uses ATP as substrate and introduces an internal 3′–5′ link via a phosphodiester bond, which generates the structure drawn in **FIGURE 26.24**. The level of cAMP is inversely related to the level of glucose. High levels of glucose repress adenylate cyclase as shown in **FIGURE 26.25**. Only in low levels of glucose is the enzyme active and able to synthesize cAMP. In turn, cAMP binding is required for CRP to bind to DNA and activate transcription. Thus, transcription activation only occurs when cellular glucose levels are low.

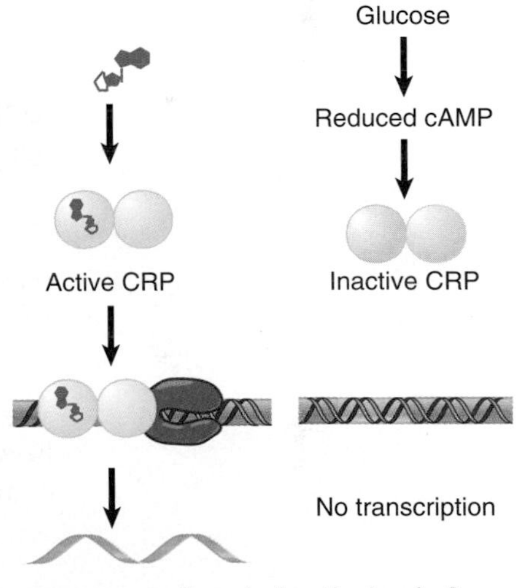

FIGURE 26.25 By reducing the level of cyclic AMP, glucose inhibits the transcription of operons that require CRP activity.

CRP is a dimer of two identical subunits of 22.5 kD, which can be activated by a single molecule of cAMP. A CRP monomer contains a DNA-binding region and a transcription-activating region. A CRP dimer binds to a site of ~22 bp at a responsive promoter. The binding sites include variations of the 5-bp consensus sequence given in **FIGURE 26.26**. Mutations preventing CRP action usually are located within the well-conserved pentamer, which appears to be the essential element in recognition. CRP binds most strongly to sites that contain two (inverted) versions of the pentamer, because this enables both subunits of the dimer to bind to the DNA.

CRP introduces a large bend when it binds DNA. In the *lac* promoter, this point lies at the center of dyad symmetry. The bend is quite severe, >90°, as illustrated in the model of **FIGURE 26.27**. There is, therefore, a dramatic change in the organization of the

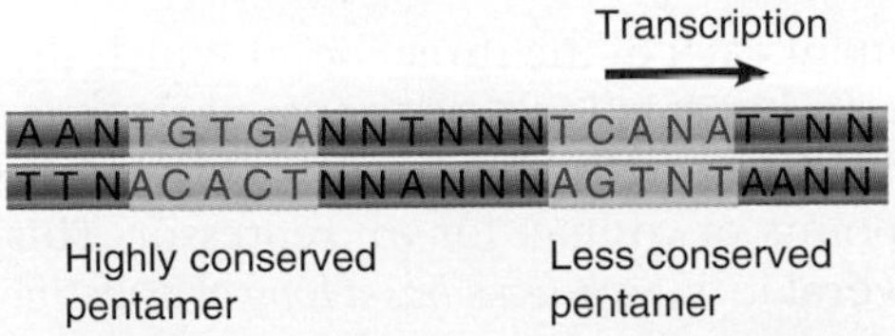

Transcription

AANTGTGANNTNNNTCANATTNN
TTNACACTNNANNNAGTNTAANN

Highly conserved pentamer | Less conserved pentamer

FIGURE 26.26 The consensus sequence for CRP contains the well-conserved pentamer TGTGA and (sometimes) an inversion of this sequence (TCANA).

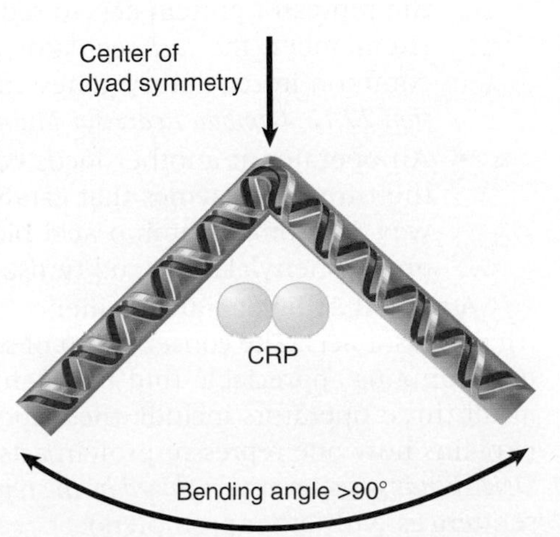

FIGURE 26.27 CRP bends DNA >90° around the center of symmetry.

DNA double helix when CRP binds. Dependence on CRP is related to the intrinsic efficiency of the promoter. No CRP-dependent promoter has a good −35 sequence and some also lack good −10 sequences. In fact, we might argue that effective control by CRP would be difficult if the promoter had effective −35 and −10 regions that interacted independently with RNA polymerase.

CRP is bound to the same face of DNA as RNA polymerase. When the α subunit of RNA polymerase has a deletion in the C-terminal domain, transcription appears normal except for the loss of ability to be activated by CRP. CRP has an "activation domain" that is required. This activating region, which consists of an exposed loop of ~10 amino acids, is a small patch that interacts directly with the C-terminal domain of the α subunit of RNA polymerase to stimulate the polymerase.

KEY CONCEPTS

- CRP is an activator protein that binds to a target sequence at a promoter.
- A dimer of CRP is activated by a single molecule of cAMP.
- cAMP is controlled by the level of glucose in the cell; a low glucose level allows cAMP to be made.
- CRP interacts with the C-terminal domain of the α subunit of RNA polymerase to activate it.

CONCEPT AND REASONING CHECK

What are the two conditions that must be met for transcription of the *lac* operon to occur?

26.12 The *trp* Operon Is a Repressible Operon with Three Transcription Units

autoregulation A site or mutation that affects the properties only of its own molecule of DNA, often indicating that a site does not code for a diffusible product.

The *lac* repressor acts only on the operator of the *lacZYA* operon. Some repressors, however, control dispersed structural genes by binding at more than one operator. An example is the *trp* repressor, which controls three unlinked sets of genes:

- An operator at the cluster of structural genes *trpEDCBA* controls coordinate synthesis of the enzymes that synthesize the amino acid tryptophan. This is an example of a *repressible operon,* one that is controlled by the product of the operon: tryptophan.
- The *trpR* regulator gene is repressed by its own product, the *trp* repressor. Thus the repressor protein acts to reduce its own synthesis: it has **autoregulation**. (Remember, the *lacI* regulator gene is unregulated.) Such circuits are quite common in regulatory genes and may be either negative or positive (see *Section 27.12, Lambda Repressor Maintains an Autoregulatory Circuit*).
- An operator at another locus controls the *aroH* gene, which codes for one of the three isoenzymes that catalyzes the initial reaction in the common pathway of aromatic amino acid biosynthesis leading to the synthesis of tryptophan, phenylalanine and tyrosine.

A related 21 bp operator sequence is present at each of the three loci at which the *trp* repressor acts. The conservation of sequence is indicated in **FIGURE 26.28**. Each operator contains appreciable (but not identical) dyad symmetry. The features conserved at all three operators include the important points of contact for *trp* repressor. This explains how one repressor protein acts on several loci: *each locus has a copy of a specific DNA-binding sequence recognized by the repressor* (just as each promoter shares consensus sequences with other promoters).

FIGURE 26.29 summarizes the variety of relationships between operators and promoters. A notable feature of the dispersed operators recognized by TrpR is their

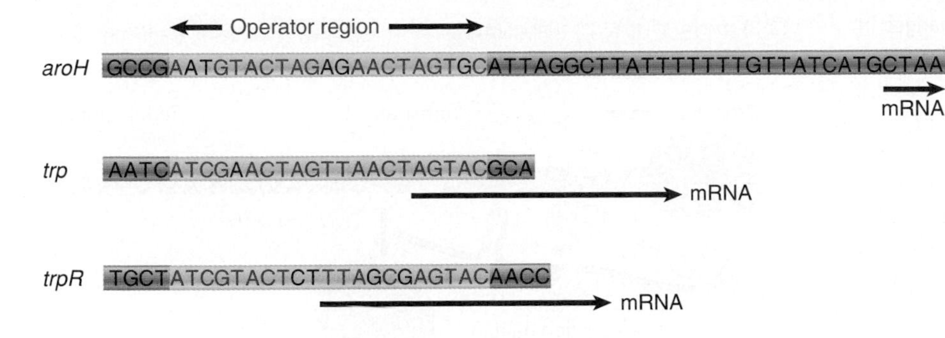

FIGURE 26.28 The *trp* repressor recognizes operators at three loci. Operators are highlighted in red; conserved bases within operators are shown in blue. The location of the startpoint and mRNA varies, as indicated by the black arrows.

presence at different locations within the promoter in each locus. In *trpR* the operator lies between positions −12 and +9, whereas in the *trp* operon it occupies positions −23 to −3. In another gene system, the *aroH* locus, it lies farther upstream, between −49 and −29. In other cases, the operator lies downstream from the promoter (as in *lac*), or apparently just upstream of the promoter (as in *gal*, for which the nature of the repressive effect is not quite clear). The ability of the repressors to act at operators whose positions are different in each target promoter suggests that there could be differences in the exact mode of repression, the common feature being that RNA polymerase is prevented from initiating transcription at the promoter.

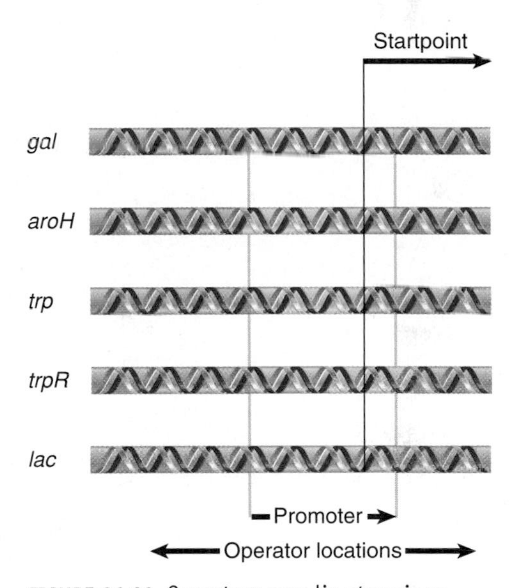

FIGURE 26.29 Operators may lie at various positions relative to the promoter.

The *trp* operon itself is under negative repressible control. This means that the *trpR* gene product, the *trp* repressor, is made as an inactive negative regulator. Repression means that the product of the *trp* operon, the amino acid tryptophan is a coregulator for the *trp* repressor. When the level of tryptophan builds up, two molecules bind to the *trp* repressor, changing its conformation to the active DNA-binding conformation and its binding to the operator. This precludes RNA polymerase binding to the overlapping promoter. Up to three *trp* repressor dimers can bind to the operator, depending on the tryptophan concentration and the concentration of repressor. The central dimer binds the tightest.

As we will see in the next section, the *trp* operon is also under dual control (like the *lac* operon earlier), but the second level of control here is quite different.

KEY CONCEPTS

- The *trp* operon is negatively controlled by the level of its product, the amino acid tryptophan.
- The amino acid tryptophan activates an inactive repressor encoded by *trpR*.
- A repressor (or activator) will act on all loci that have a copy of its target operator sequence.

CONCEPT AND REASONING CHECK

What would the effect on a cell be if the *trpR* gene were mutant?

26.13 The *trp* Operon Is Also Controlled by Attenuation

The *E. coli trp* operon, like the *lac* operon, is under two levels of control; it is under negative repressible control (described above in *Section 26.12, The trp Operon Is a Repressible Operon with Three Transcription Units*) and a complex regulatory system of **attenuation**. *Negative repressible* means that it is prevented from initiating transcription by its product, the free amino acid tryptophan. Attenuation is the second level of control. There is a region in the 5′ leader of the mRNA called the **attenuator** that contains a small ORF. Attenuation in the *E. coli trp* operon means that transcription

- **attenuation** The regulation of bacterial operons by controlling termination of transcription at a site located before the first structural gene.
- **attenuator** A terminator sequence at which attenuation occurs.

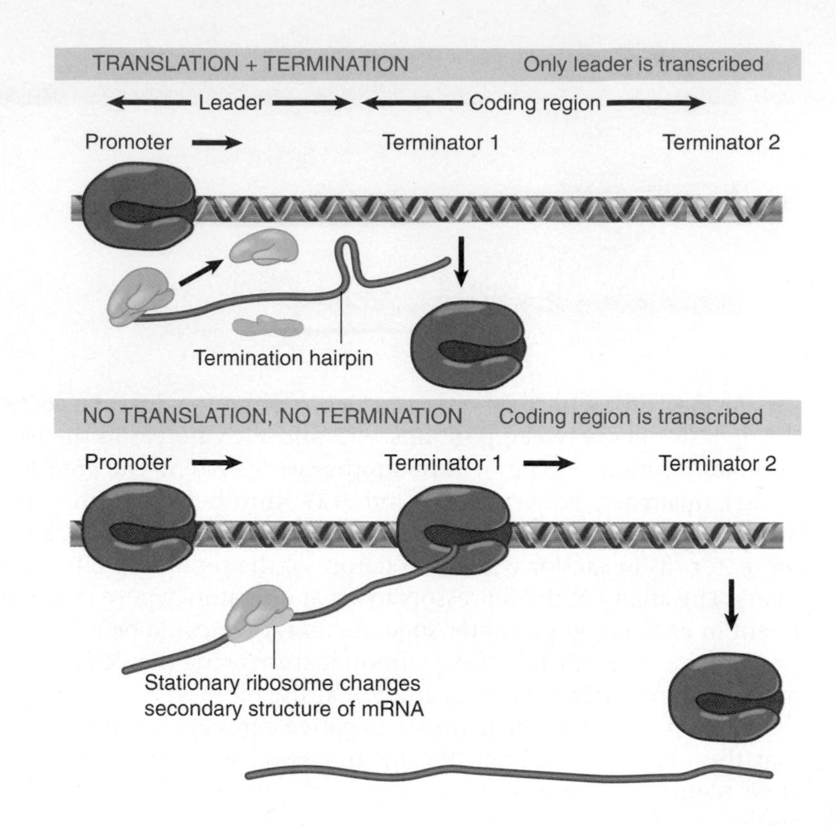

FIGURE 26.30 Termination can be controlled via changes in RNA secondary structure that are determined by ribosome movement.

termination is controlled by the rate of translation of the attenuator ORF. It is possible to establish this mode of regulation because bacteria have coupled transcription/translation. This allows *E. coli* to monitor the second pool of tryptophan, that of Trp-tRNA. High levels of Trp-tRNA will attenuate or terminate transcription, whereas if there are low levels, transcription of the *trpEDCBA* operon will be allowed to continue.

Attenuation is accomplished by changes in secondary structure of the attenuator RNA that are determined by the position of the ribosome on mRNA. **FIGURE 26.30** shows that termination requires that the ribosome can translate the attenuator. When the ribosome translates the leader region ORF, a termination hairpin forms at terminator 1. When the ribosome is slow in translating the ORF, though, the terminator hairpin does not form, and RNA polymerase can continue transcribing the coding region. *This mechanism of termination control therefore depends upon the relative levels of Trp-tRNA versus uncharged tRNA^{Trp} to influence the rate of ribosome movement in the leader region.*

Attenuation was first revealed by the observation that deleting a sequence between the operator and and the *trpE* coding region can increase the expression of the structural genes. This effect is independent of repression: both the basal and derepressed levels of transcription are increased. Thus this site influences events that occur after RNA polymerase has set out from the promoter (irrespective of the conditions prevailing at initiation).

Termination at the attenuator responds to the level of Trp-tRNA, as illustrated in **FIGURE 26.31**. In the presence of adequate amounts of Trp-tRNA, termination is efficient. With low levels of Trp-tRNA, however, RNA polymerase can continue into the structural genes.

Repression and attenuation respond in the same way to the levels of the two pools of tryptophan. When the free amino acid tryptophan is present, the operon is repressed. When tryptophan is removed, RNA polymerase has free access to the promoter and can start transcribing the operon. When Trp-tRNA is present, the operon is attenuated and transcription terminates. When the pool of tryptophan bound to its tRNA is depleted, the RNA polymerase can continue to transcribe the operon. Note the pool of free tryptophan may be low and allow transcription to begin, but if the Trp-tRNA is fully charged, transcription will terminate.

TRANSCRIPTION OF LEADER REGION

Promoter Pause Attenuator *trpE*

Polymerase initiates

Polymerase pauses

TRYPTOPHAN ABSENT: TRANSCRIPTION CONTINUES INTO OPERON

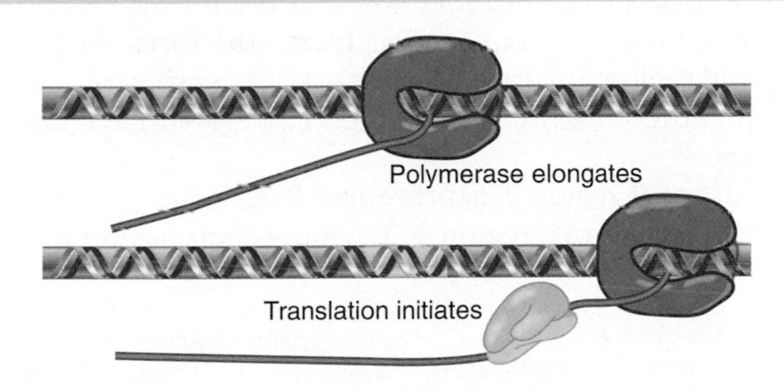

TRYPTOPHAN PRESENT: TRANSCRIPTION TERMINATES AT ATTENUATOR

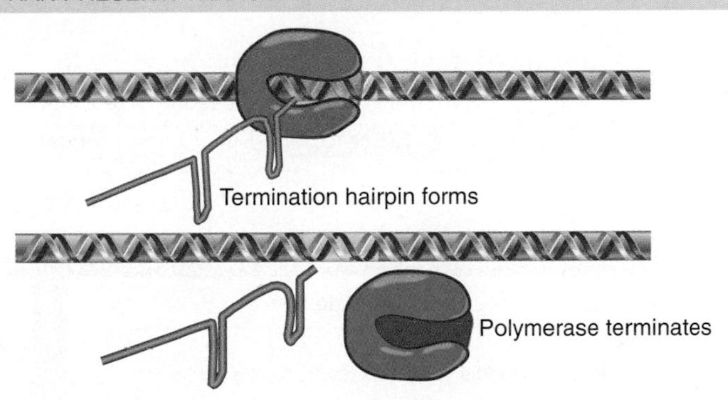

FIGURE 26.31 The *trp* operon has a short sequence coding for a leader peptide that is located between the operator and the attenuator.

Attenuation has a 10× effect on transcription. When tryptophan is present, termination is effective, and the attenuator allows only ~10% of the RNA polymerase to proceed. In the absence of tryptophan, attenuation allows virtually all the polymerases to proceed. Together with the 70× increase in initiation of transcription that results from the release of repression, this allows an ~700-fold range of regulation.

KEY CONCEPTS

- An attenuator (intrinsic terminator) is located between the promoter and the first gene of the *trp* operon.
- The absence of Trp-tRNA suppresses termination and results in a 10× increase in transcription.

CONCEPT AND REASONING CHECK

The *E. coli trp* operon is under dual control. What are the two conditions that allow transcription?

26.14 Attenuation Can be Controlled by Translation

- **leader peptide** The product that would result from translation of a short coding sequence used to regulate transcription of an operon by controlling ribosome movement.
- **ribosome stalling** The inhibition of movement that occurs when a ribosome reaches a codon for which there is no corresponding charged aminoacyl-tRNA.

How can termination of transcription at the attenuator ORF respond to the level of Trp-tRNA? The sequence of the leader region suggests a mechanism. It has a short coding sequence that could represent a **leader peptide** of fourteen amino acids. **FIGURE 26.32** shows that it contains a ribosome binding site whose AUG codon is followed by a short coding region containing two successive codons for tryptophan. When the cell has a low level of Trp-tRNA, ribosomes initiate translation of the leader peptide but slow down or stop when they reach the Trp codons. The sequence of the mRNA suggests that this **ribosome stalling** influences termination at the attenuator.

The leader sequence can be described by alternative base-paired structures. The ability of the ribosome to proceed through the leader region controls transitions between these structures. The structure determines whether the mRNA can provide the features needed for termination. **FIGURE 26.33** shows these structures. In the first, region 1 pairs with region 2 and region 3 pairs with region 4. The pairing of regions 3 and 4 generates the hairpin that precedes the U_8 sequence: *this 3–4 hairpin is the essential signal for intrinsic termination.*

A different structure is formed if region 1 is prevented from pairing with region 2. In this case, region 2 is free to pair with region 3. Region 4 then has no available

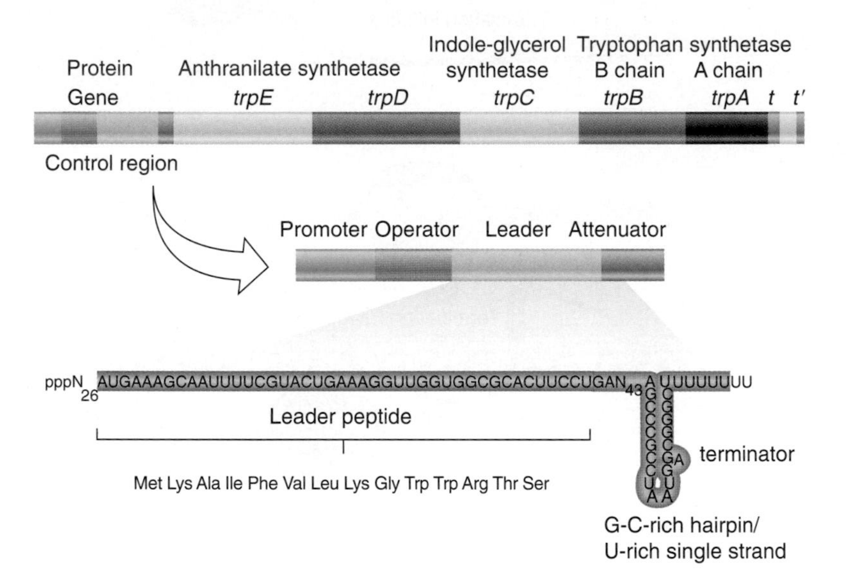

FIGURE 26.32 An attenuator controls the progression of RNA polymerase into the *trp* genes. RNA polymerase initiates at the promoter and then proceeds to position 90, where it pauses before proceeding to the attenuator at position 140. In the absence of tryptophan, the polymerase continues into the structural genes (*trpE* starts at +163). In the presence of tryptophan, there is ~90% probability of termination to release the 140 base leader RNA.

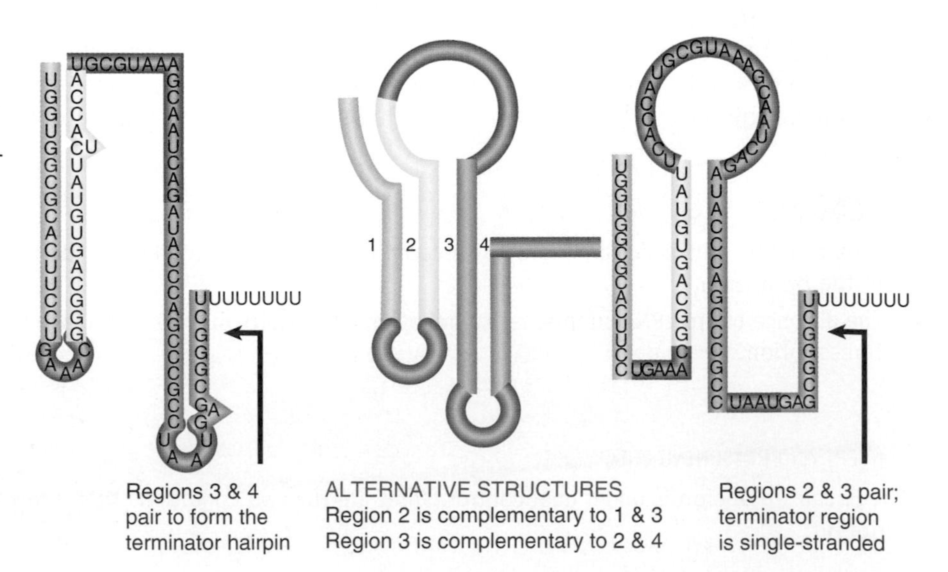

FIGURE 26.33 The *trp* leader region can exist in alternative base-paired conformations. The center shows the four regions that can base pair. Region 1 is complementary to region 2, which is complementary to region 3, which is complementary to region 4. On the left is the conformation produced when region 1 pairs with region 2 and region 3 pairs with region 4. On the right is the conformation when region 2 pairs with region 3, leaving regions 1 and 4 unpaired.

pairing partner, so it remains single-stranded. *Thus the terminator hairpin cannot be formed.*

FIGURE 26.34 shows that the position of the ribosome can determine which structure is formed in such a way that termination is attenuated only in the absence of tryptophan. The crucial feature is the position of the Trp codons in the leader peptide coding sequence.

When Trp-tRNA is abundant, ribosomes are able to synthesize the leader peptide. They continue along the leader section of the mRNA to the UGA codon, which lies between regions 1 and 2. As shown in the lower part of the figure, by progressing to this point, the ribosome extends over region 2 and prevents it from base pairing. The result is that region 3 is available to base pair with region 4, which generates the termination hairpin. Under these conditions, therefore, RNA polymerase terminates at the attenuator.

When Trp-tRNA is not abundant, ribosomes stall at the Trp codons, which are part of region 1, as shown in the upper part of the figure. Thus region 1 is sequestered within the ribosome and cannot base pair with region 2. This means that regions 2 and 3 become base paired before region 4 has been transcribed. This compels region 4 to remain in a single-stranded form. In the absence of the terminator hairpin, RNA polymerase continues transcription past the attenuator.

Control by attenuation requires a precise timing of events. For ribosome movement to determination formation of alternative secondary structure structures that control termination, translation of the leader must occur at the same time when RNA polymerase approaches the terminator site. A critical event in controlling the timing is the presence of a site that causes the RNA polymerase to pause at base 90 along the leader. The RNA polymerase paused until a ribosome translates the leader peptide. The polymerase is then released and moves off towards the attenuation site. By the time it arrives there, the secondary structure of the attenuator region has been determined.

FIGURE 26.35 summarizes the role of Trp-tRNA in controlling expression of the operon. By providing a mechanism to sense the abundance of Trp-tRNA, attenuation responds directly to the need of the cell for typtophan in protein synthesis.

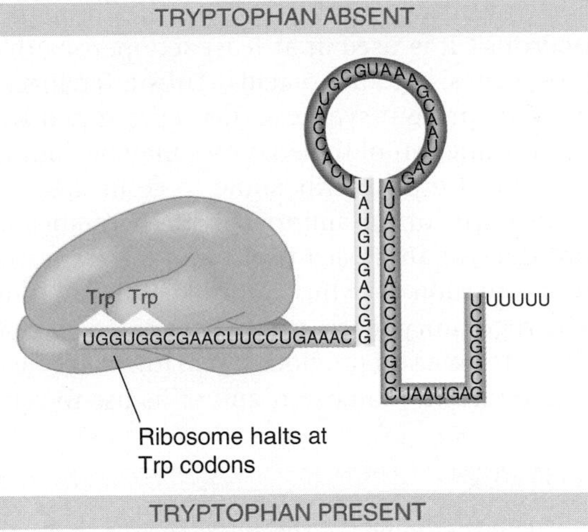

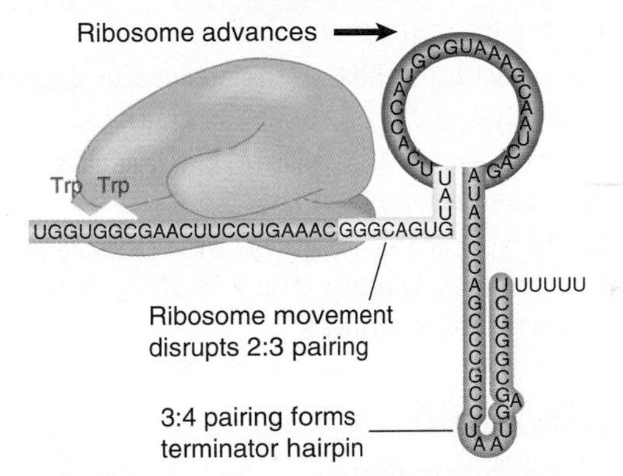

FIGURE 26.34 The alternatives for RNA polymerase at the attenuator depend on the location of the ribosome, which determines whether regions 3 and 4 can form the terminator hairpin.

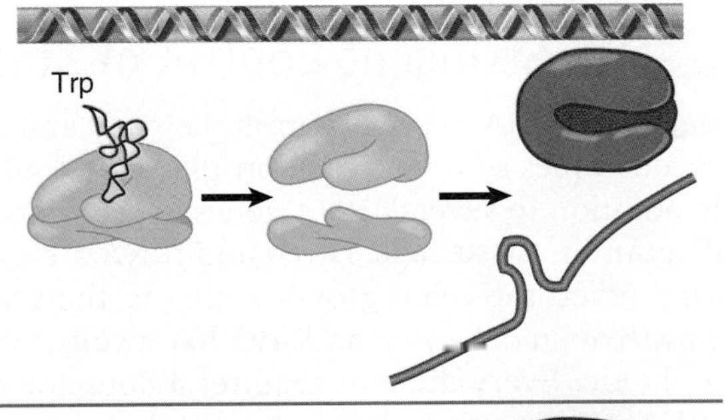

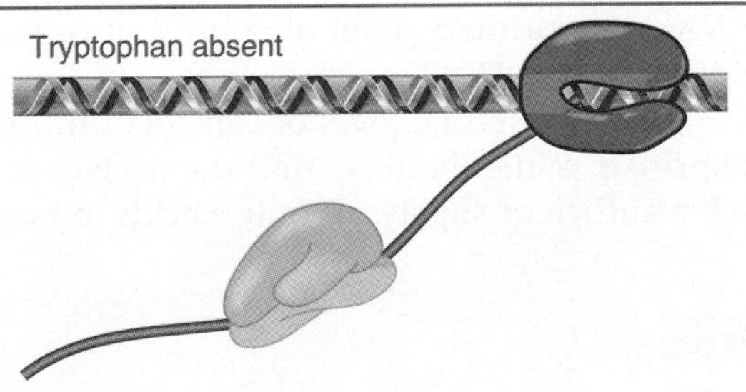

FIGURE 26.35 In the presence of Trp-tRNA, ribosomes translate the leader peptide and are released. This allows hairpin formation, so that RNA polymerase terminates. In the absence of Trp-tRNA, the ribosome is blocked, the termination hairpin cannot form, and RNA polymerase continues.

How widespread is the use of attenuation as a control mechanism for bacterial operons? It is used in at least six operons that code for enzymes concerned with the biosynthesis of amino acids. Thus a feedback from the level of the amino acid available for protein synthesis (as represented by the availability of aminoacyl-tRNA) to the production of the enzymes may be common.

The use of the ribosome to control RNA secondary structure in response to the availability of an aminoacyl-tRNA establishes an inverse relationship between the presence of aminoacy-tRNA and the transcription of the operon, which is equivalent to a situation in which aminoacy-tRNA functions as a corepressor of transcription. The regulatory mechanism is mediated by changes in the formation of duplex regions; thus attenuation provides a striking example of the importance of secondary structure in the termination event and of its use in regulation.

KEY CONCEPTS

- The leader region of the *trp* operon has a 14-codon open reading frame that includes two codons for tryptophan.
- The structure of RNA at the attenuator depends on whether this reading frame is translated.
- In the presence of Trp-tRNA, the leader is translated, and the attenuator is able to form the termination hairpin.
- In the absence of Trp-tRNA, the ribosome stalls at the tryptophan codons and an alternative secondary structure prevents formation of the termination hairpin, so that transcription continues.

CONCEPT AND REASONING CHECK

How does a hairpin structure in the mRNA cause transcription termination on the DNA?

26.15 Stringent Control of Stable RNA Transcription

Bacterial rRNA genes are multiple copy and are dispersed in the genome. *E. coli* has seven copies of a transcription unit that contains the 16S, 23S, and 5S rRNA genes, in addition to several tRNA genes in the transcribed spacer, as seen in the generalized diagram in **FIGURE 26.36**. rRNA and tRNA are stable RNAs that are required to be made only when the cell is growing, that is, the primary level of control of transcription is *growth control*. As long as *E. coli* has a sufficient supply of ATP, the cells will continue to divide. Every division requires a doubling of ribosomes and thus rRNA (as well as tRNA). The primary level of control of transcription of stable RNAs is thus the concentration of ATP.

There is a second level of control of transcription of stable RNAs called **stringent response.** When bacteria find themselves in such poor growth conditions that they lack a sufficient supply of amino acids to sustain translation, they shut down a wide

▸ **stringent response** The ability of a bacterium to shut down synthesis of tRNA and ribosomes in a poor growth medium.

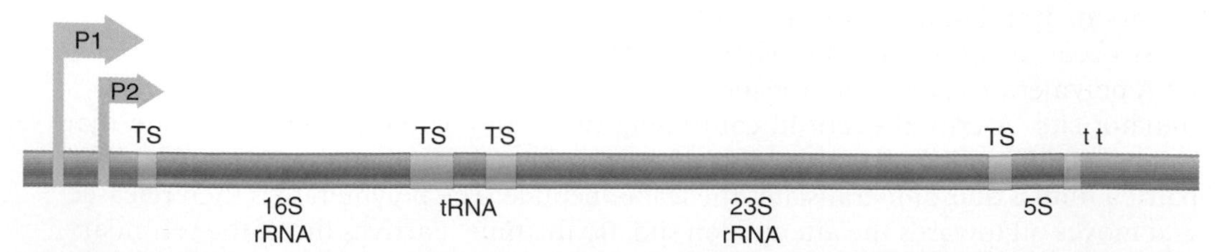

FIGURE 26.36 The *E. coli* rRNA operon structure. The two promoter, the P1 major and the P2 minor promoters are shown as arrows. Coding regions for 16S, one tRNA, 23S, and 5S rRNA are indicated as hatched boxes. Transcribed spacers (TS) are shown as open boxes. The two terminators are shown at the end of the operon.

range of activities. We can view it as a mechanism for surviving hard times: the bacterium conserves its resources by engaging in only the minimum of activities and channeling resources into the synthesis of amino acids.

The stringent response causes a massive (10–20×) reduction in the synthesis of rRNA and tRNA. This alone is sufficient to reduce the total amount of RNA synthesis to 5% to 10% of its previous level. The synthesis of certain mRNAs is reduced, leading to an overall reduction of ~3× in mRNA synthesis. The rate of protein degradation is increased. Many metabolic adjustments occur, as seen in reduced synthesis of nucleotides, carbohydrates and lipids.

The stringent response is controlled by two unusual nucleotides, **ppGpp**, guanosine tetraphosphate with diphosphates attached to both the 5′ and 3′ positions and pppGpp, guanosine pentaphosphate with a 5′ triphosphate and a 3′ diphosphate group, together denoted as *(p)ppGpp*. These nucleotides are typical small-nucleotide effectors like the second messenger cAMP (see *Section 26.11, The lac Operon Has a Second Layer of Control: Catabolite Repression*), that function by binding to target proteins to alter their activities.

▸ **ppGpp** Guanosine tetraphosphate, with diphosphate attached to both the 5′ and 3′ positions is a signaling molecule.

Deprivation of any one amino acid or a mutation that inactivates any aminoacyl-tRNA synthetase (see *Section 24.3, Special Mechanisms Control the Accuracy of Translation*) is sufficient to initiate the stringent response. The trigger that sets the entire series in motion is *the presence of uncharged tRNA in the A site of the ribosome.* Under normal conditions, of course, only aminoacyl-tRNA is placed in the A site (see *Section 24.8, Elongation Factor Tu Loads Aminoacyl-tRNA into the A Site*). But when there is not enough aminoacyl-tRNA available to respond to a particular codon, the uncharged tRNA becomes able to gain entry.

Bacterial mutants that cannot produce the stringent response are called **relaxed mutants.** The most common site of relaxed mutation lies in the gene *relA,* which codes for a protein called the **stringent factor**. This factor is associated with ribosomes, although the amount is rather low, ~1 molecule for every 200 ribosomes. So probably only a minority of ribosomes are able to produce the stringent response.

▸ **relaxed mutant** Mutants that do not display the stringent response to starvation in *E. coli* for amino acids (or other nutritional deprivations).

▸ **stringent factor** The protein RelA, which is associated with ribosomes synthesizes (p)ppGpp when an uncharged tRNA enters the A site.

The presence of uncharged tRNA in the A site blocks translation, triggering an *idling reaction* by wild type ribosomes. Provided that the A site is occupied by an uncharged tRNA specifically responding to the codon, the RelA protein catalyzes a reaction in which ATP donates a pyrophosphate group to the 3′ position of either GTP or GDP.

FIGURE 26.37 shows the pathway for synthesis of (p)ppGpp. The RelA enzyme uses GTP as substrate more frequently than GDP, so that pppGpp is the predominant product. However, pppGpp is converted to ppGpp by several enzymes. The production of ppGpp via pppGpp is the most common route, and *ppGpp is the usual effector of the stringent response*. How is ppGpp removed when conditions return to normal? A gene called *spoT* codes for an enzyme that provides the major catalyst for ppGpp degradation as shown in Figure 26.37. The activity of this enzyme causes ppGpp to be rapidly degraded with a half-life of ~20 seconds, so that the stringent response is reversed rapidly when synthesis of (p)ppGpp ceases.

ppGpp is an effector for controlling several reactions, most prominantly transcription. It activates transcription at some promoters, those involved in amino acid biosynthesis, but its major effect is to inhibit the synthesis of the stable RNA operons—rRNA (and tRNA). The unusual sequence of the major promoter of *E. coli's* rRNA genes results in a potentially unstable open complex with RNA polymerase during initiation of transcription (see *Section 19.3, The Transcription Reaction has Three Stages*) and will collapse if the ATP concentration is too low. This class of promoter also requires the activity of a transcription factor, DksA, to bind to RNA polymerase to effect the stringent response. ppGpp competes with ATP for the first nucleotide to stimulate this collapse, effectively inhibiting rRNA transcription.

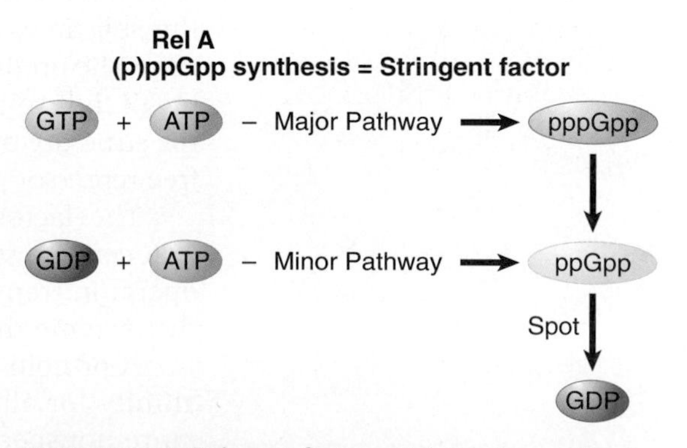

FIGURE 26.37 Stringent factor catalyzes the synthesis of pppGpp and ppGpp; ribosomal proteins can dephosphorylate pppGpp to ppGpp. ppGpp is degraded when it is no longer needed.

KEY CONCEPTS

- Poor growth conditions cause bacteria to produce the small molecule regulators (p)ppGpp.
- The trigger for the reaction is the entry of uncharged tRNA into the ribosomal A site.
- (p)ppGpp competes with ATP during formation of the open complex during transcription initiation by RNA polymerase and inhibits the reaction.

CONCEPT AND REASONING CHECK

What is the function of the stringent response?

26.16 Summary

Transcription is regulated by the interaction between *trans*-acting factors and *cis*-acting sites. A *trans*-acting factor is the product of a regulator gene. It is usually protein but also can be RNA. It diffuses in the cell, and as a result it can act on any appropriate target gene. A *cis*-acting site in DNA (or RNA) is a sequence that functions by being recognized *in situ*. It has no coding function and can regulate only those sequences with which it is physically contiguous. Bacterial genes coding for proteins whose functions are related, such as successive enzymes in a pathway, may be organized in a cluster that is transcribed into a polycistronic mRNA from a single promoter. Control of this promoter regulates expression of the entire pathway. The unit of regulation, which contains structural genes and *cis*-acting elements, is called the operon.

Initiation of transcription is regulated by interactions that occur in the vicinity of the promoter. The ability of RNA polymerase to initiate at the promoter is prevented or activated by other proteins. Genes that are active unless they are turned off by binding the regulator are said to be under negative control. Genes that are active only when the regulator is bound to them are said to be under positive control. The type of control can be determined by the dominance relationships between wild-type genes and mutants that are constitutive/derepressed (permanently on) or uninducible/super-repressed (permanently off). A second way of looking at regulation of genes that code for enzymes is whether they are controlled by the enzyme substrate (inducible) or by the enzyme product (repressible).

A repressor or activator can control multiple targets that have copies of an operator or its consensus sequence. A repressor protein prevents RNA polymerase from either binding to the promoter or activating transcription. The repressor binds to a target sequence, the operator, which is usually located around or upstream of the startpoint. Operator sequences are short and often are palindromic. The repressor is often a homomultimer whose symmetry reflects that of its target.

The ability of the repressor protein to bind to its operator is often regulated by small molecules, which provide a second level of gene regulation. In a negative inducible gene, the substrate, an inducer, prevents a repressor from binding. In a negative repressible gene, the product or corepressor activates an inactive regulator to turn off the gene. Binding of the inducer or corepressor to its site on the regulator protein produces a change in the structure of the DNA-binding site of the protein. This allosteric reaction occurs both in free repressor proteins and directly in repressor proteins already bound to DNA.

The lactose pathway in *E. coli* is controlled by negative induction. When an inducer, the substrate β-galactoside, diminishes the ability of repressor to bind to its operator, transcription and translation of the *lacZ* gene then produce β-galactosidase, the enzyme that metabolizes β-galactosides.

A protein with a high affinity for a particular target sequence in DNA has a lower affinity for all DNA. The ratio defines the specificity of the protein. There are many more nonspecific sites (any DNA sequence) than specific target sites in a genome; as a result, a DNA-binding protein such as a repressor or RNA polymerase is "stored" on DNA. (It is likely that none, or very little, is free.) The specificity for the target sequence must be great enough to counterbalance the excess of nonspecific sites over specific sites. The balance for bacterial proteins is adjusted so that the amount of protein and its specificity allow specific recognition of the target in "on" conditions but allow almost complete release of the target in "off" conditions.

Some promoters cannot be recognized by RNA polymerase or are recognized only poorly unless a specific activator protein, a positive regulator, is present. Activator proteins may also be regulated by small molecules. The CRP activator is only able to bind to target sequences when complexed with cAMP, which happens only in conditions of low glucose. All promoters that are controlled by catabolite repression have at least one copy of the CRP-binding site. Direct contact between CRP and RNA polymerase occurs through the C-terminal domain of the α subunits of RNA polymerase.

The tryptophan pathway in *E. coli* is controlled by negative repression. The corepressor tryptophan, the product of the pathway, activates the repressor protein so that it binds to the operator and prevents expression of the genes that code for the enzymes that synthesize tryptophan. The second level of control is attenuation.

Attenuation is a mechanism of regulation of gene expression used by bacteria. They are able to do so because they have coupled transcription/translation. Attenuation is the control of transcription termination by the rate of translation. The polycistronic mRNA of the *trp* operon starts with a sequence in the 5′ leader that contains an ORF with two Trp codons. The leader can fold into alternative secondary structures. One of these structures is an RNA hairpin that is an intrinsic transcription terminator. The choice of which alternate structure forms is dependent on the rate at which a ribosome can translate the ORF. The rate of translation is dependent on the ratio of $tRNA^{Trp}$ that is charged or not charged. Fully charged $tRNA^{Trp}$ will allow fast translation and will promote the formation of the intrinsic transcription terminator. Uncharged $tRNA^{Trp}$ will cause the ribosome to translate slowly and prevent the formation of the terminator hairpin, allowing the RNA polymerase to continue to transcribe the operon.

A deficiency in aminoacyl-tRNAs causes an idling reaction on the ribosome because uncharged tRNAs are brought into the A site. This leads to the synthesis of the unusual nucleotide ppGpp, a second messenger, that inhibits initiation of transcription at specific promoters such as the rRNA operons.

CHAPTER QUESTIONS

1. Gene expression at the transcription level is commonly regulated at all but which one of the following?
   A. initiation
   B. elongation
   C. termination
   D. transcript stability
2. The default level of expression of a gene under negative control is:
   A. at high level in the absence of a negative regulatory protein.
   B. at high level in the presence of a negative regulatory protein.
   C. turned off in the absence of a negative regulatory protein.
   D. turned off in the presence of a negative regulatory protein.
3. Any gene product that is free to diffuse to find its target is described as a:
   A. structural gene.
   B. regulator gene.
   C. *cis*-acting element.
   D. *trans*-acting element.
4. The *lac* mRNA is:
   A. extremely unstable with a half-life of ~1 minute.
   B. extremely unstable with a half-life of ~3 minutes.
   C. very stable with a half-life of ~15 minutes.
   D. very stable with a half-life of ~25 minutes.

5. Promoter mutations are:
   A. *trans*-dominant.
   B. *cis*-dominant.
   C. *trans*-recessive.
   D. *cis*-recessive.
6. The *trp* repressor regulates expression of its own gene. This is an example of:
   A. allosteric control.
   B. positive regulation.
   C. autogenous control.
   D. coordinate regulation.
7. When glucose levels drop in a bacterial cell:
   A. cAMP synthesis is induced.
   B. cAMP synthesis is repressed.
   C. cGMP synthesis is induced.
   D. cGMP synthesis is repressed.
8. The positive regulator of the *lac* operon is:
   A. CRP.
   B. allolactose.
   C. lactose.
   D. IPTG.
9. The *trp* repressor functions to control its own expression as an example of:
   A. induction by negative control.
   B. induction by positive control.
   C. repression by negative control.
   D. repression by positive control.
10. CRP controls expression of:
   A. the *lac* operon.
   B. many operons involved in the metabolism of sugars.
   C. the *trp* operon.
   D. many operons involved in biosynthesis of amino acids.

KEY TERMS

allolactose
allosteric control
attenuation
attenuator
autoregulation
cyclic AMP (cAMP)
catabolite repression
catabolite repressor protein
***cis*-acting**
***cis*-dominant**
constitutive expression
corepressor
coupled transcription/translation
dominant negative
gratuitous inducer
inducer
inducible gene
induction
interallelic complementation
***lac* repressor**
leader peptide
negative complementation
negative control
negative inducible
negative repressible
operon
operator
palindrome
polycistronic mRNA
positive control
positive inducible
positive repressible
ppGpp
regulator gene
relaxed mutant
repressible gene
repression
ribosome stalling
stringent factor
stringent response
structural gene
***trans*-acting**
uninducible

FURTHER READING

Elf, J., Li, G.-W., and Xie, X. S. (2007). Probing transcription factor dynamics at the single-molecule level in a living cell. *Science* **316**, 1191–1194. Using single molecules to probe the interaction between the *lac* repressor and its operator site.

Swigon, D., Coleman, B. D., and Olson, W. K. (2006). Modeling the *lac* repressor-operator assembly: the influence of DNA looping on *lac* repressor conformation. *Proc. Natl. Acad. Sci.* **103**, 9879–9884. A report on the dynamics of the assembly of *lac* repressor complexes on operator sites.

Tabaka, M., Cybutski, O., and Holyst, R. (2008). Accurate genetic switch in *E. coli*: novel mechanism of regulation corepressor. *J. Mol. Biol.* **377**, 1002–1014.

Wilson, C. J., Zahn, H., Swint-Kruse, L., and Matthews, K. S. (2007). The lactose repressor system: paradigms for regulation, allosteric behavior and protein folding. *Cell. Mol. Life Sci.* **64**, 3–16. A nice review on the *lac* system.

Yu, H., and Gertstein, M. (2006). Genomic analysis of the hierarchical structure of regulatory networks. *Proc. Natl. Acad. Sci.* **103**, 14724–14731. A review which describes how individual gene regulatory networks fit into the larger metabolic pathways.

27

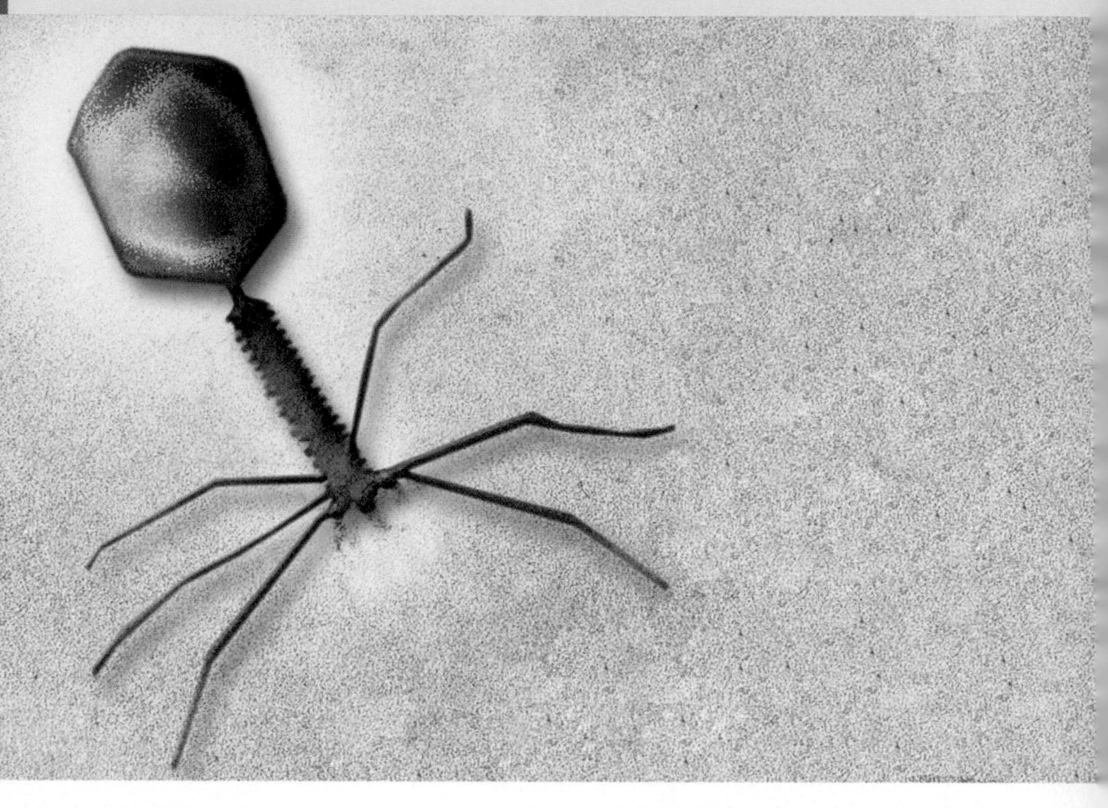

A transmission electron micrograph of a T4 phage. (Magnification: 450,000×.)

Phage Strategies

CHAPTER OUTLINE

27.1 Introduction

A virus consists of a nucleic acid genome contained in a protein coat. In order to reproduce, the virus must infect a host cell. The typical pattern of an infection is to subvert the functions of the host cell for the purpose of producing a large number of progeny viruses. Viruses that infect bacteria are generally called **bacteriophages**, often abbreviated to **phage**, or simply ϕ. Usually a phage infection kills the bacterium. The process by which a phage infects a bacterium, reproduces itself, and then kills its host is called **lytic infection**. In the typical lytic cycle, the phage DNA (or RNA) enters the host bacterium, its genes are transcribed in a set order, the phage genetic material is replicated, and the protein components of the phage particle are produced. Finally, the host bacterium is broken open (*lysed*) to release the assembled progeny particles by the process of **lysis**. For some phages, called **virulent phages**, this is their only strategy for survival.

Other phages have a dual existence. They are able to perpetuate themselves via the same sort of lytic cycle in what amounts to an open strategy for producing as many copies of the phage as rapidly as possible. They also have an alternative form of existence, though, in which the phage genome is present in the bacterium in a latent form known as a **prophage**. This form of propagation is called **lysogeny** and the infected bacteria are known as *lysogens*. Phages that follow this pathway are called **temperate phages**.

In a lysogenic bacterium, the prophage is inserted into the bacterial genome and is inherited in the same way as bacterial genes. The process by which it is converted from an independent phage genome into a prophage that is a linear part of the bacterial genome is described as **integration**. By virtue of its possession of a prophage, a lysogenic bacterium has **immunity** against infection by other phage particles of the

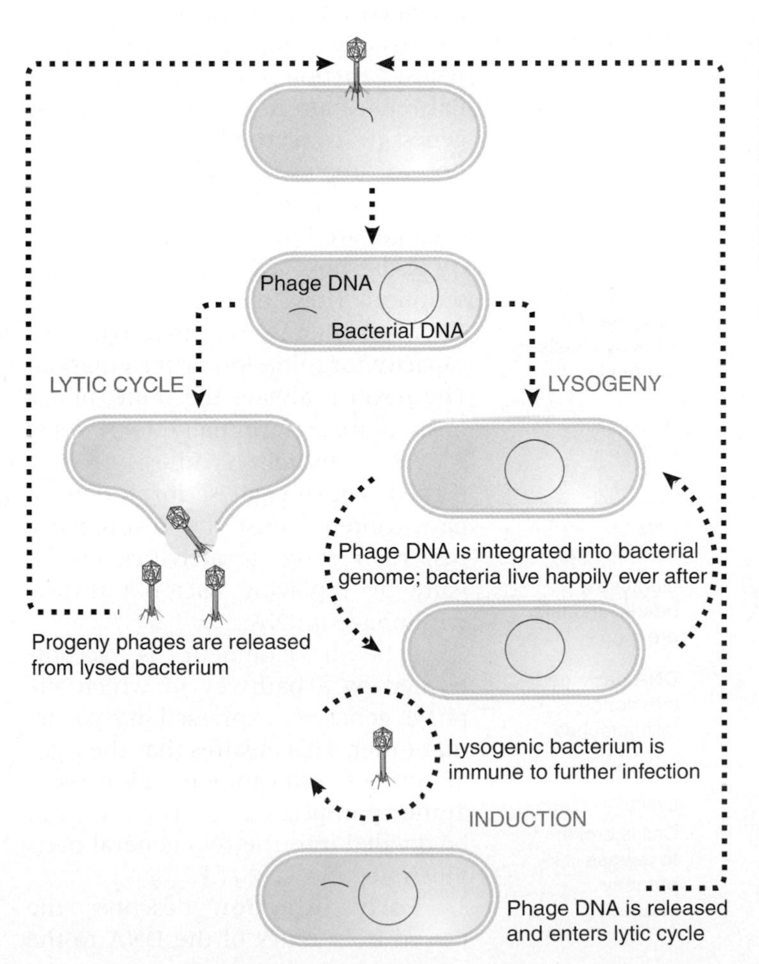

FIGURE 27.1 Lytic development involves the reproduction of phage particles with destruction of the host bacterium, but lysogenic existence allows the phage genome to be carried as part of the bacterial genetic information.

- **bacteriophage** A bacterial virus.
- **phage** An abbreviation of bacteriophage or bacterial virus.
- **lytic infection** Infection of a bacterium by a phage that ends in the destruction of the bacterium with release of progeny phage.
- **lysis** The death of bacteria at the end of a phage infective cycle when they burst open to release the progeny of an infecting phage (because phage enzymes disrupt the bacterium's cytoplasmic membrane or cell wall). The same term also applies to eukaryotic cells; for example, when infected cells are attacked by the immune system.
- **virulent phage** A bacteriophage that can only follow the lytic cycle.
- **prophage** A phage genome covalently integrated as a linear part of the bacterial chromosome.
- **lysogeny** The ability of a phage to survive in a bacterium as a stable prophage component of the bacterial genome.
- **temperate phage** A bacteriophage that can follow the lytic or lysogenic pathway.
- **integration** Insertion of a viral or another DNA sequence into a host genome as a region covalently linked on either side to the host sequences.
- **immunity** In phages, the ability of a prophage to prevent another phage of the same type from infecting a cell. In plasmids, the ability of a plasmid to prevent another of the same type from becoming established in a cell. It can also refer to the ability of certain transposons to prevent others of the same type from transposing to the same DNA molecule.

same type. Immunity is established by a single integrated prophage, so in general a bacterial genome contains only one copy of a prophage of any particular type.

There are transitions between the lysogenic and lytic modes of existence. **FIGURE 27.1** shows that when a temperate phage produced by a lytic cycle enters a new bacterial host cell, it either repeats the lytic cycle or enters the lysogenic state. The outcome depends on the conditions of infection and the genotypes of phage and bacterium.

- **induction of phage** A phage's entry into the lytic (infective) cycle as a result of destruction of the lysogenic repressor, which leads to excision of free phage DNA from the bacterial chromosome.
- **excision** Release of phage from the host chromosome as an autonomous DNA molecule.

A prophage is freed from the restrictions of lysogeny by a process called **induction of phage**. First the phage DNA is released from the bacterial chromosome by **excision**; then the free DNA proceeds through the lytic pathway.

The alternative forms in which these phages are propagated are determined by the regulation of transcription. Lysogeny is maintained by the interaction of a phage repressor with an operator. The lytic cycle requires a cascade of transcriptional controls. The transition between the two lifestyles is accomplished by the establishment of repression (lytic cycle to lysogeny) or by the relief of repression (induction of lysogen to lytic phage). These regulatory processes provide a wonderful example of how a series of relatively simple regulatory actions can be built up into complex developmental pathways.

27.2 Lytic Development Is Divided into Two Periods

Phage genomes by necessity are small. As with all viruses, they are restricted by the need to package the nucleic acid within the protein coat. This limitation dictates many of the viral strategies for reproduction. Typically, a virus takes over the apparatus of the host cell, which then replicates and expresses phage genes instead of the bacterial genes.

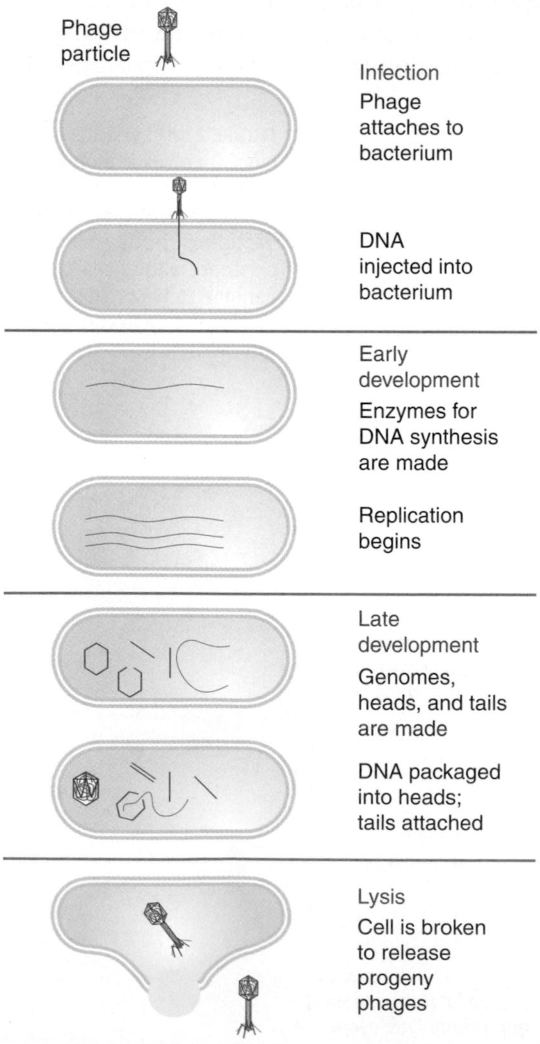

FIGURE 27.2 Lytic development takes place by producing phage genomes and protein particles that are assembled into progeny phages.

Usually the phage has genes whose function is to ensure preferential replication of phage DNA. These genes are concerned with the initiation of replication and may even include a new DNA polymerase. Changes are introduced in the capacity of the host cell to engage in transcription. They involve replacing the RNA polymerase or modifying its capacity for initiation or termination. The result is always the same: phage mRNAs are preferentially transcribed. As far as protein synthesis is concerned, the phage is, for the most part, content to use the host apparatus, redirecting its activities principally by replacing bacterial mRNA with phage mRNA.

Lytic development is accomplished by a pathway in which the phage genes are expressed in a particular order. This ensures that the right amount of each component is present at the appropriate time. The cycle can be divided into the two general parts illustrated in **FIGURE 27.2**.

- **early infection** The part of the phage lytic cycle between entry and replication of the phage DNA. During this time, the phage synthesizes the enzymes needed to replicate its DNA.

Early infection describes the period from entry of the DNA to the start of its replication. **Late infection** defines the period from the start of replication to the final step of lysing the bacterial cell to release progeny phage particles.

The early phase is devoted to the production of enzymes involved in the reproduction of DNA. These include the enzymes concerned with DNA synthesis, recombination, and, sometimes, modification. Their activities cause a *pool* of phage genomes to accumulate. In this pool, genomes are continually replicating and recombining, so that *the events of a single lytic cycle concern a population of phage genomes.*

During the late phase, the protein components of the phage particle are synthesized. Often many different proteins are needed to make up head and tail structures, so the largest part of the phage genome consists of late functions. In addition to the structural proteins, "assembly proteins" are needed to help construct the particle, although they are not incorporated into it themselves. By the time the structural components are assembling into heads and tails, replication of DNA has reached its maximum rate. The genomes then are inserted into the empty protein heads, tails are added, and the host cell is lysed to allow release of new viral particles.

▸ **late infection** The part of the phage lytic cycle from DNA replication to lysis of the cell. During this time, the DNA is replicated and structural components of the phage particle are synthesized.

KEY CONCEPTS

- A phage infective cycle is divided into the early period (before replication) and the late period (after the onset of replication).
- A phage infection generates a pool of progeny phage genomes that replicate and recombine.

CONCEPT AND REASONING CHECK

Most phages are virulent; therefore, why might it be advantageous to be temperate?

27.3 Lytic Development Is Controlled by a Cascade

The organization of the phage genetic map often reflects the sequence of lytic development. The concept of the operon is taken to somewhat of an extreme, in which the genes coding for proteins with related functions are clustered to allow their control with the maximum economy. This allows the pathway of lytic development to be controlled with a small number of regulatory switches.

The lytic cycle is under positive control, so that each group of phage genes can be expressed only when an appropriate signal is given. **FIGURE 27.3** is an overview showing that the regulatory genes function in a **cascade**, in which a gene expressed at one stage is necessary for synthesis of the genes that are expressed at the next stage.

The early part of the first stage of gene expression necessarily relies on the transcription apparatus of the host cell. In general, only a few genes are expressed at this time. Their promoters are indistinguishable from those of host genes. The name of this class of genes depends on the phage. In most cases, they are known as the **early genes**. In phage lambda, they are given the evoca-

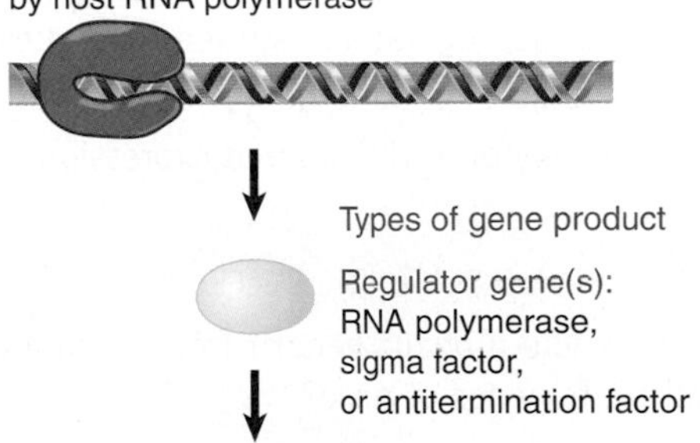

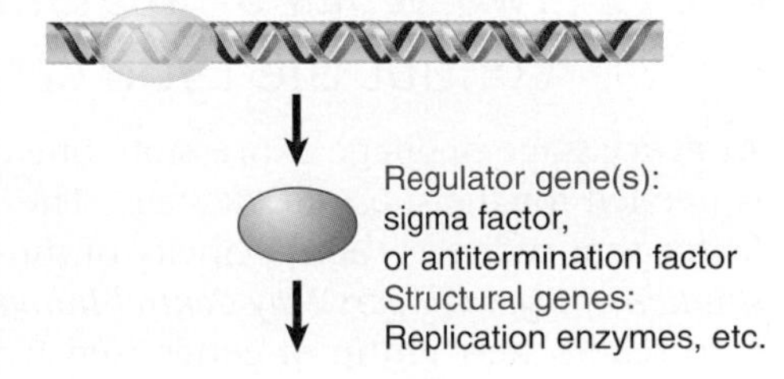

FIGURE 27.3 Phage lytic development proceeds by a regulatory cascade, in which a gene product at each stage is needed for expression of the genes at the next stage.

▸ **cascade** A sequence of events, each of which is stimulated by the previous one. In transcriptional regulation, as seen in sporulation and phage lytic development, it means that regulation is divided into stages, and at each stage, one of the genes that is expressed codes for a regulator needed to express the genes of the next stage.

▸ **early genes** Genes that are transcribed before the replication of phage DNA. They code for regulators and other proteins needed for later stages of infection.

tive description of **immediate early genes**. Irrespective of the name, they constitute only a preliminary set of genes, representing just the initial part of the early period. Sometimes they are exclusively occupied with the transition to the next period. In all cases, *one of these genes always codes for a protein, a gene regulator that is necessary for transcription of the next class of genes.*

This next class of genes in the early stage is known variously as the **delayed early** or **middle gene** group. Its expression typically starts as soon as the regulator protein coded by the early gene(s) is available. Depending on the nature of the control circuit, the initial set of early genes may or may not continue to be expressed at this stage. Often, the expression of host genes is reduced. Together the two sets of early genes account for all necessary phage functions except those needed to assemble the particle coat itself and to lyse the cell.

When the replication of phage DNA begins, it is time for the **late genes** to be expressed. Their transcription at this stage usually is arranged by embedding an additional regulator gene within the previous (delayed early or middle) set of genes. This regulator may be another antitermination factor (as in phage lambda) or it may be another sigma factor.

A lytic infection often falls into the stages described above, beginning with the early genes transcribed by host RNA polymerase (sometimes the regulators are the only products at this stage). This stage is followed by those genes transcribed under the direction of the regulator produced in the first stage (most of these genes code for enzymes needed for replication of phage DNA). The final stage consists of genes for phage components, which are transcribed under the direction of a regulator synthesized in the second stage.

The use of these successive controls, in which each set of genes contains a regulator that is necessary for expression of the next set, creates a cascade in which groups of genes are turned on (and sometimes off) at particular times. The means used to construct each phage cascade are different but the results are similar.

- **immediate early genes** Genes in phage lambda that are equivalent to the early class of other phages. They are transcribed immediately upon infection by the host RNA polymerase.
- **delayed early genes** Genes in phage lambda that are equivalent to the middle genes of other phages. They cannot be transcribed until regulator protein(s) coded by the immediate early genes have been synthesized.
- **middle genes** Phage genes that are regulated by the proteins coded by early genes. Some proteins coded by them catalyze replication of the phage DNA; others regulate the expression of a later set of genes.
- **late gene** A gene transcribed when phage DNA is being replicated. It codes for components of the phage particle.

KEY CONCEPTS

- The early genes transcribed by host RNA polymerase following infection include, or comprise, regulators required for expression of the middle set of phage genes.
- The middle group of genes includes regulators to transcribe the late genes.
- This results in the ordered expression of groups of genes during phage infection.

CONCEPT AND REASONING CHECK

Why would a phage benefit from using a cascade of gene regulators instead of transcribing all of its genes at once?

27.4 Two Types of Regulatory Events Control the Lytic Cascade

At every stage of phage expression, one or more of the active genes is a regulator that is needed for the subsequent stage. The regulator may take the form of a new sigma factor that redirects the specificity of the host RNA polymerase (see *Section 19.15, Substitution of Sigma Factors May Control Initiation*) or an antitermination factor that allows it to read a new group of genes (see *Section 19.16, Antitermination May Be a Regulated Event*). Now, let's compare the use of switching at initiation or termination to control gene expression.

FIGURE 27.4 shows that phages use two types of mechanisms for recognizing new phage promoters. One is to replace the sigma factor of the host enzyme with another factor that redirects its specificity in initiation. An alternative is to synthesize a new phage RNA polymerase. In either case, the critical feature that distinguishes the new

Holoenzyme with σ70 recognizes one set of promoters

Phage synthesizes new sigma or RNA polymerase

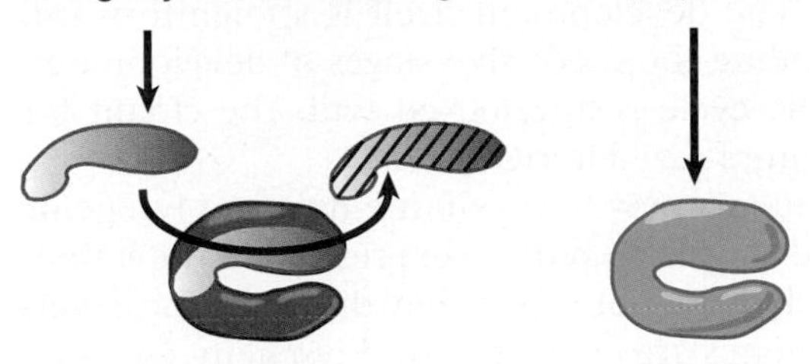

Phage sigma factor causes host enzyme to recognize new promoters

OR

Phage RNA polymerase recognizes new set of promoters

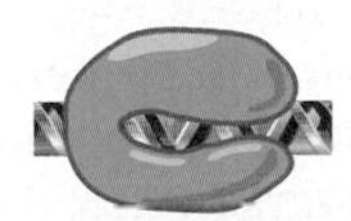

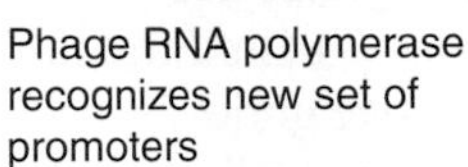

FIGURE 27.4 A phage may control transcription at initiation either by synthesizing a new sigma factor that replaces the host sigma factor or by synthesizing a new RNA polymerase.

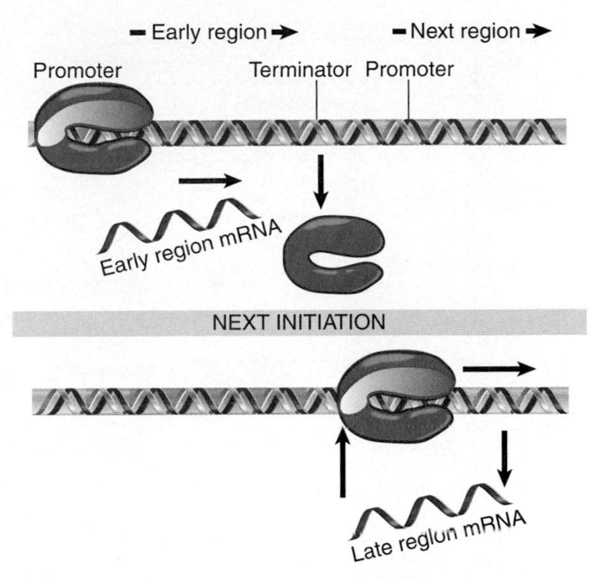

FIGURE 27.5 Control at initiation utilizes independent transcription units, each with its own promoter and terminator, which produce independent mRNAs. The transcription units need not be located near one another.

set of genes is their possession of *different promoters from those originally recognized by host RNA polymerase.* **FIGURE 27.5** shows that the two sets of transcripts are independent; as a consequence, early gene expression can cease after the new sigma factor or polymerase has been produced.

Antitermination provides an alternative mechanism for phages to control the switch from early genes to the next stage of expression. The use of antitermination depends on a particular arrangement of genes. **FIGURE 27.6** shows that the early genes lie adjacent to the genes that are to be expressed next but are separated from them by terminator sites. *If termination is prevented at these sites, the polymerase reads through into the genes on the other side.* So, in antitermination, the *same promoters* continue to be recognized by RNA polymerase. The new genes are expressed only by extending the RNA chain to form molecules that contain the early gene sequences at the 5′ end and the new gene sequences at the 3′ end. Because the two types of sequences remain linked, early gene expression inevitably continues.

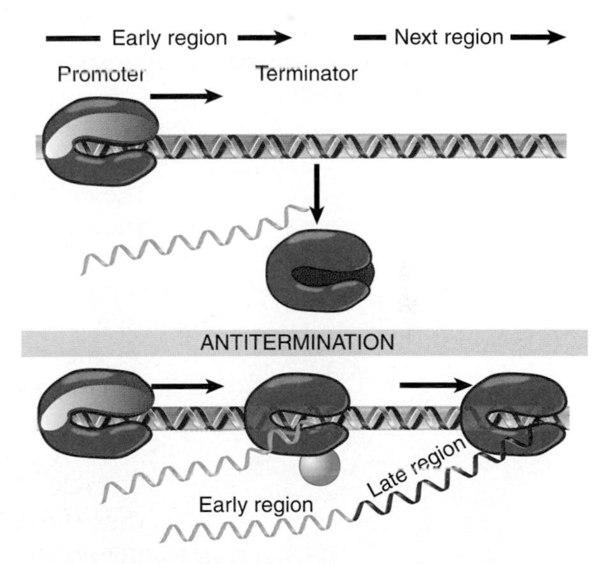

FIGURE 27.6 Control at termination requires adjacent units so that transcription can read from the first gene into the next gene. This produces a single mRNA that contains both sets of genes.

KEY CONCEPT

- Regulator proteins used in phage cascades may sponsor initiation at new (phage) promoters or cause the host polymerase to read through transcription terminators.

CONCEPT AND REASONING CHECK

What is the advantage to lambda of using an antitermination mechanism instead of a new sigma factor to progress to the delayed early genes?

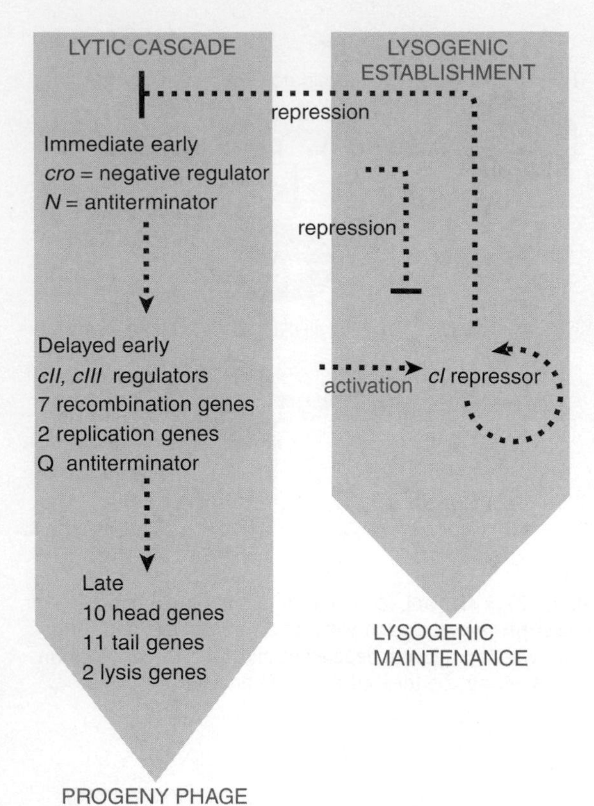

FIGURE 27.7 The lambda lytic cascade is interlocked with the circuitry for lysogeny.

27.5 Lambda Immediate Early and Delayed Early Genes Are Needed for Both Lysogeny and the Lytic Cycle

One of the most intricate cascade circuits is provided by phage lambda. Actually, the cascade for lytic development itself is straightforward, with two regulators controlling the successive stages of development. But the circuit for the lytic cycle is interlocked with the circuit for establishing lysogeny, as summarized in **FIGURE 27.7**.

When lambda DNA enters a new host cell, the lytic and lysogenic pathways start off the same way. Both require expression of the immediate early and delayed early genes, but then they diverge: lytic development follows if the late genes are expressed, and lysogeny ensues if synthesis of a gene regulator called the lambda repressor is established by turning on its gene, the *cI* gene.

Lambda has only two immediate early genes, transcribed independently by host RNA polymerase:

- The *N gene* codes for an antitermination factor whose action at the *nut* sites allows transcription to proceed into the delayed early genes (see *Section 19.16, Antitermination May Be a Regulated Event*). The *N* gene is required for both the lytic and lysogenic pathways.
- The *cro gene* has dual functions: it codes for a repressor that prevents expression of the *cI* gene coding for the lambda repressor (essentially de-repressing the late genes, a necessary action if the lytic cycle is to proceed), and it turns off expression of the immediate early genes (which are not needed later in the lytic cycle).

The delayed early genes, turned on by *N*, include two replication genes (needed for lytic infection), seven recombination genes (some involved in recombination during lytic infection, and two necessary to integrate lambda DNA into the bacterial chromosome for lysogeny), and three regulator genes. These regulator genes have opposing functions:

- The *cII-cIII* pair of regulator genes is needed to establish the synthesis of the lambda repressor.
- The Q regulator gene codes for an antitermination factor that allows host RNA polymerase to transcribe the late genes and is necessary for the lytic cycle.

Thus the delayed early genes serve two masters: some are needed for the phage to enter lysogeny, and the others are concerned with controlling the order of the lytic cycle. At this point, lambda is keeping open the option to choose either pathway.

KEY CONCEPTS

- Lambda has two immediate early genes, *N* and *cro*, which are transcribed by host RNA polymerase.
- The product of the *N* gene is required to express the delayed early genes.
- Three of the delayed early genes are regulators.
- Lysogeny requires the delayed early genes *cII-cIII*.
- The lytic cycle requires the immediate early gene *cro* and the delayed early gene *Q*.

CONCEPT AND REASONING CHECK

What would be the effect if the *N* gene were mutant?

27.6 The Lytic Cycle Depends on Antitermination by pN

To disentangle the lytic and lysogenic pathways, let's first consider just the lytic cycle. **FIGURE 27.8** gives the map of lambda phage DNA. A group of genes concerned with regulation is surrounded by genes needed for recombination and replication. The genes coding for structural components of the phage are clustered. All of the genes necessary for the lytic cycle are expressed in polycistronic transcripts from three promoters.

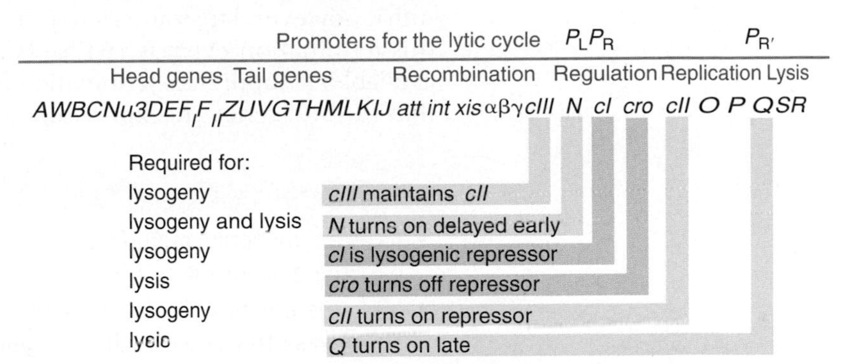

FIGURE 27.8 The lambda map shows clustering of related functions. The genome is 48,514 bp.

FIGURE 27.9 shows that the two immediate early genes, *N* and *cro*, are transcribed by host RNA polymerase. *N* is transcribed toward the left and *cro*, toward the right. Each transcript is terminated at the end of the gene. The protein pN is the regulator, the antitermination factor that allows transcription to continue into the delayed early genes by suppressing use of the terminators t_L and t_R (see *Section 19.16, Antitermination May Be a Regulated Event*). In the presence of pN, transcription continues to the left of the *N* gene into the recombination genes and to the right of the *cro* gene into the replication genes.

The map in Figure 27.8 gives the organization of the lambda DNA as it exists in the phage particle. Shortly after infection, though, the ends of the DNA join to form a circle. **FIGURE 27.10** shows

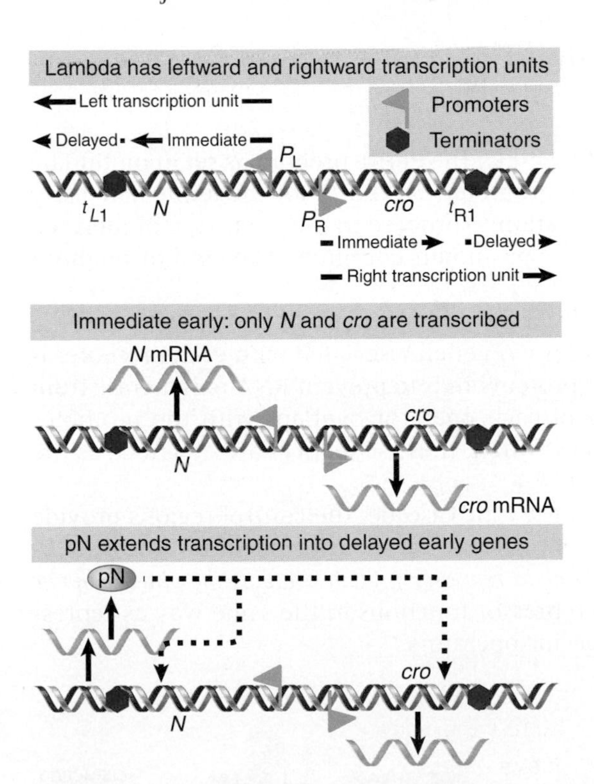

FIGURE 27.9 Phage lambda has two early transcription units. In the "leftward" unit, the "upper" strand is transcribed toward the left; in the "rightward" unit, the "lower" strand is transcribed to the right. Genes *N* and *cro* are the immediate early functions and are separated from the delayed early genes by terminators. Synthesis of N protein allows RNA polymerase to pass the terminators t_{L1} to the left and t_{R1} to the right.

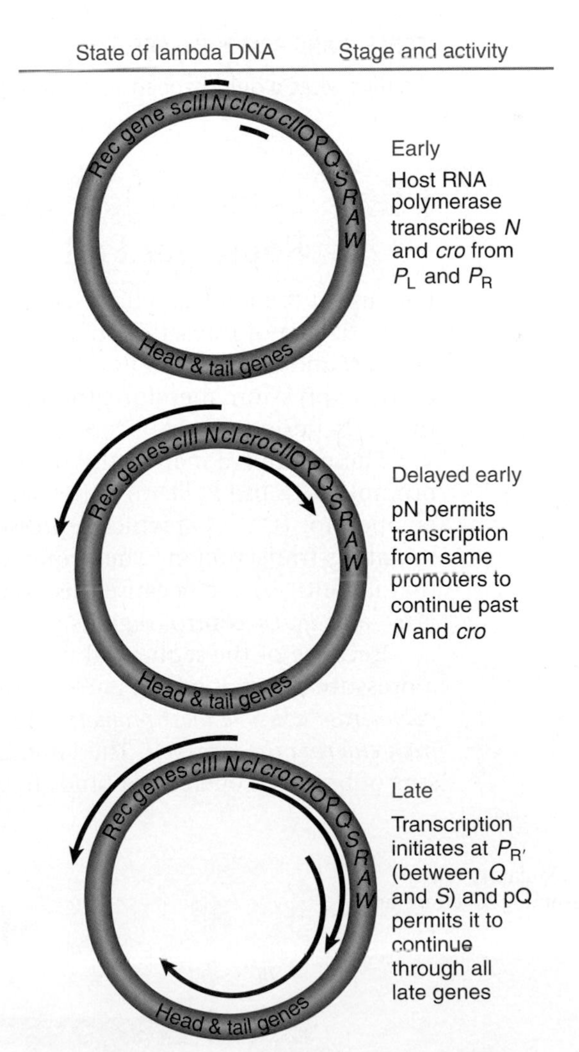

FIGURE 27.10 Lambda DNA circularizes during infection, so that the late gene cluster is intact in one transcription unit.

the true state of lambda DNA during infection. The late genes are welded into a single group, which contains the lysis genes *S-R* from the right end of the linear DNA and the head and tail genes *A-J* from the left end.

The late genes are expressed as a single transcription unit, starting from a promoter $P_{R'}$ that lies between *Q* and *S*. The late promoter is used constitutively. In the absence of the product of gene *Q* (which is the last gene in the rightward delayed early unit), however, late transcription terminates at a site t_{R3}. The transcript resulting from this termination event is 194 bases long; it is known as 6S RNA. When pQ becomes available, it suppresses termination at t_{R3} and the 6S RNA is extended, with the result that the late genes are expressed.

KEY CONCEPTS

- pN is an antitermination factor that allows RNA polymerase to continue transcription past the ends of the two immediate early genes.
- pQ is the product of a delayed early gene and is an antiterminator that allows RNA polymerase to transcribe the late genes.
- Lambda DNA circularizes after infection; as a result, the late genes form a single transcription unit.

CONCEPT AND REASONING CHECK

Predict what would happen to a cell infected by a lambda phage with a mutant *cI* gene.

27.7 Lysogeny Is Maintained by the Lambda Repressor Protein

Looking at the lambda lytic cascade, we see that the entire program is set in motion by the initiation of transcription at the two promoters P_L and P_R for the immediate early genes *N* and *cro*. Lambda uses antitermination to proceed to the next stage of (delayed early) expression; therefore, the same two promoters continue to be used throughout the early period.

The expanded map of the regulatory region drawn in **FIGURE 27.11** shows that the promoters P_L and P_R lie on either side of the *cI* gene. Associated with each promoter is an operator (O_L, O_R) at which repressor protein binds to prevent RNA polymerase from initiating transcription. The sequence of each operator overlaps with the promoter that it controls, and because this occurs so often, these sequences are described as the P_L/O_L and P_R/O_R control regions.

Because of the sequential nature of the lytic cascade, the control regions provide a pressure point at which entry to the entire cycle can be controlled. *By denying RNA polymerase access to these promoters, the lambda repressor protein prevents the phage genome from entering the lytic cycle.* The lambda repressor functions in the same way as repressors of bacterial operons: it binds to specific operators.

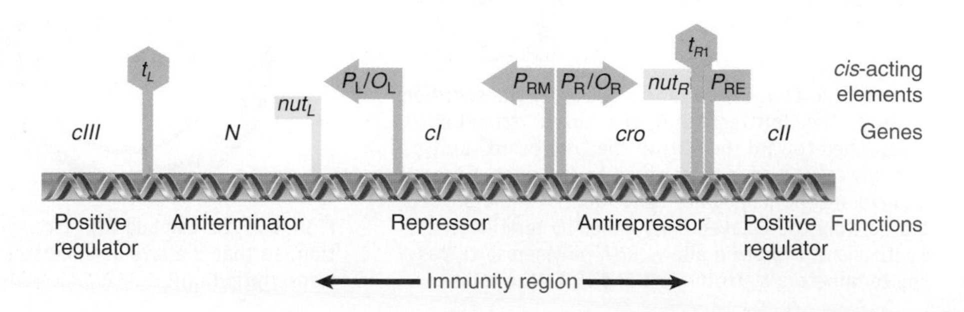

FIGURE 27.11 The lambda regulatory region contains a cluster of *trans*-acting functions and *cis*-acting elements.

The lambda repressor protein is encoded by the *cI* gene. Note in Figure 27.11 that the *cI* gene has two promoters, P_{RM} (promoter right maintenance) and P_{RE} (promoter right establishment). Mutants in this gene cannot maintain lysogeny, but always enter the lytic cycle. In the time since the original isolation of the lambda repressor protein, the characterization of the repressor protein has shown how it both maintains the lysogenic state and provides immunity for a lysogen against superinfection by new phage lambda genomes.

The lambda repressor binds independently to the two operators. Its ability to repress transcription at the associated promoters is illustrated in **FIGURE 27.12**.

At O_L the lambda repressor has the same sort of effect that we have already discussed for several other systems: it prevents RNA polymerase from initiating transcription at P_L. This stops the expression of gene *N*. P_L is used for all leftward early gene transcription, thus this action prevents expression of the entire leftward early transcription unit. So the lytic cycle is blocked before it can proceed beyond early stages.

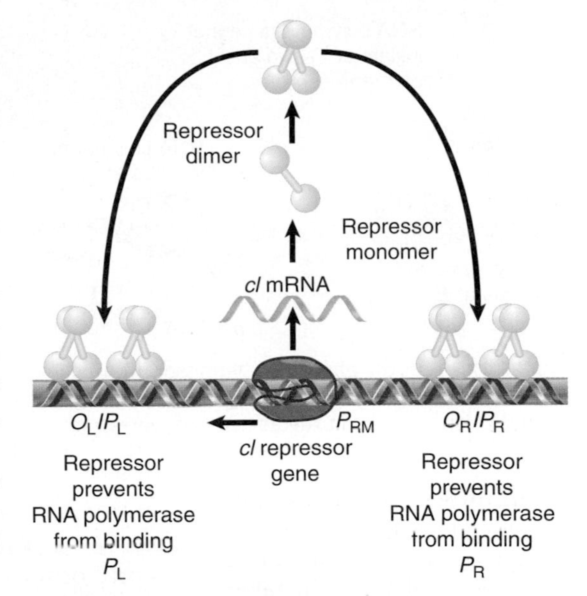

FIGURE 27.12 Repressor acts at the left operator and right operator to prevent transcription of the immediate early genes (*N* and *cro*). It also acts at the promoter P_{RM} to activate transcription by RNA polymerase of its own gene.

At O_R, repressor binding prevents the use of P_R so *cro* and the other rightward early genes cannot be expressed. The lambda repressor protein binding at O_R also stimulates transcription of *cI*, its own gene from P_{RM}.

The nature of this control circuit explains the biological features of lysogenic existence. Lysogeny is stable because the control circuit ensures that, so long as the level of lambda repressor is adequate, there is continued expression of the *cI* gene. The result is that O_L and O_R remain occupied indefinitely. By repressing the entire lytic cascade, this action maintains the prophage in its inert form.

KEY CONCEPTS

- The lambda repressor, encoded by the *cI* gene, is required to maintain lysogeny.
- The lambda repressor acts at the O_L and O_R operators to block transcription of the immediate early genes.
- The immediate early genes trigger a regulatory cascade; as a result, their repression prevents the lytic cycle from proceeding.

CONCEPT AND REASONING CHECK

Predict what would happen to a cell infected by a lambda phage with a mutant O_R.

27.8 The Lambda Repressor and Its Operators Define the Immunity Region

The presence of lambda repressor explains the phenomenon of *immunity*. If a second lambda phage DNA enters a lysogenic cell, repressor protein synthesized from the resident prophage genome will immediately bind to O_L and O_R in the new genome. This prevents the second phage from entering the lytic cycle.

The operators were originally identified as the targets for repressor action by **virulent mutations** (λ*vir*). These mutations prevent the repressor from binding at O_L or O_R, with the result that the phage inevitably proceeds into the lytic pathway when it infects a new host bacterium. Note that λ*vir* mutants can grow on lysogens because the virulent mutations in O_L and O_R allow the incoming phage to ignore the

▸ **virulent mutations** Phage mutants that are unable to establish lysogeny.

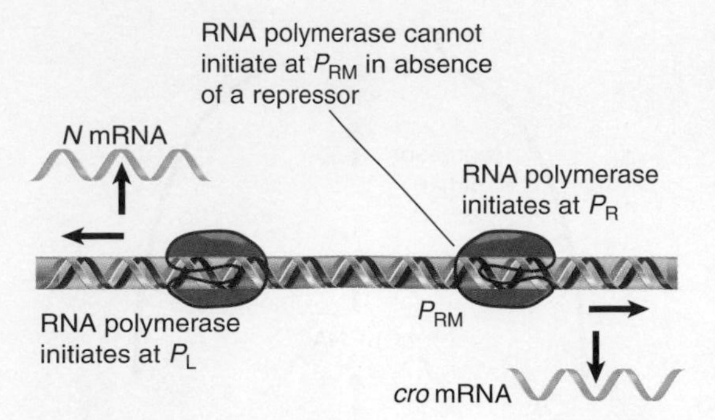

FIGURE 27.13 In the absence of repressor, RNA polymerase initiates at the left and right promoters. It cannot initiate at P_{RM} in the absence of repressor.

resident repressor and thus enter the lytic cycle. Virulent mutations in phages are the equivalent of operator-constitutive mutations in bacterial operons.

A prophage is induced to enter the lytic cycle when the lysogenic circuit is broken. This happens when the repressor is inactivated (see *Section 27.9, The DNA-Binding Form of the Lambda Repressor Is a Dimer*). The absence of repressor allows RNA polymerase to bind at P_L and P_R, starting the lytic cycle, as shown in the lower part of **FIGURE 27.13**.

The region including the left and right operators, the *cI* gene, and the *cro* gene determines the immunity of the phage. Any phage that possesses this region has the same type of immunity, because *it specifies both the repressor protein and the sites on which the repressor acts.* Accordingly, this is called the **immunity region** (as marked in Figure 27.11). Each of the four lambdoid phages ϕ80, *21*, *434*, and λ has a unique immunity region. When we say that a lysogenic phage confers immunity to any other phage of the same type, we mean more precisely that the immunity is to any other phage that has the same immunity region (irrespective of differences in other regions).

▸ **immunity region** A segment of the phage genome that enables a prophage to inhibit additional phage of the same type from infecting the bacterium. This region has a gene that encodes for the repressor, as well as the sites to which the repressor binds.

KEY CONCEPTS

- Several lambdoid phages have different immunity regions.
- A lysogenic phage confers immunity to further infection by any other phage with the same immunity region.

CONCEPT AND REASONING CHECK

Why can't another lambda phage infect a lysogenic bacterium?

27.9 The DNA-Binding Form of the Lambda Repressor Is a Dimer

The lambda repressor subunit is a polypeptide of 27 kD with the two distinct domains summarized in **FIGURE 27.14**.

- The N-terminal domain, residues 1–92, provides the operator-binding site.
- The C-terminal domain, residues 132–236, is responsible for dimerization.

Each domain can exercise its function independently of the other. The C-terminal fragment can form oligomers. The N-terminal fragment can bind the operators, although with a lower affinity than the intact lambda repressor. Thus the information for specifically contacting DNA is contained within the N-terminal domain, but the efficiency of the process is enhanced by the attachment of the C-terminal domain.

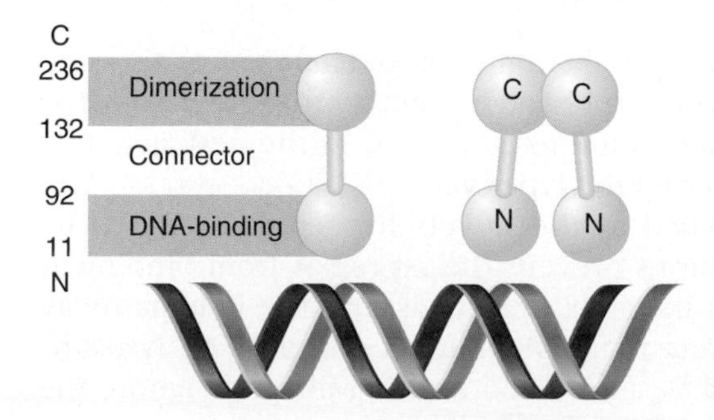

FIGURE 27.14 The N-terminal and C-terminal regions of repressor form separate domains. The C-terminal domains associate to form dimers; the N-terminal domains bind DNA.

The dimeric structure of the lambda repressor is crucial in maintaining lysogeny. The induction of a lysogenic prophage to enter the lytic cycle is caused by cleavage of the repressor subunit in the connector region, between residues 111 and 113. (This is a counterpart to the allosteric change in conformation that results when a small-molecule inducer inactivates the repressor of a bacterial operon, a capacity that the lysogenic repressor does not have.) Induction occurs under certain adverse conditions, such as exposure of lysogenic bacteria to UV irradiation, which leads to proteolytic inactivation of the repressor.

In the intact state, dimerization of the C-terminal domains ensures that when the repressor binds to DNA, its two N-terminal domains each contact DNA simultaneously. Cleavage releases the C-terminal domains from the N-terminal domains, though. As illustrated in **FIGURE 27.15**, this means that the N-terminal domains can no longer dimerize; as a result, they do not have sufficient affinity for the lambda repressor to remain bound to DNA, which allows the lytic cycle to start. Also, two dimers usually cooperate to bind at an operator, and the cleavage destabilizes this interaction. The balance between lysogeny and the lytic cycle depends on the concentration of repressor.

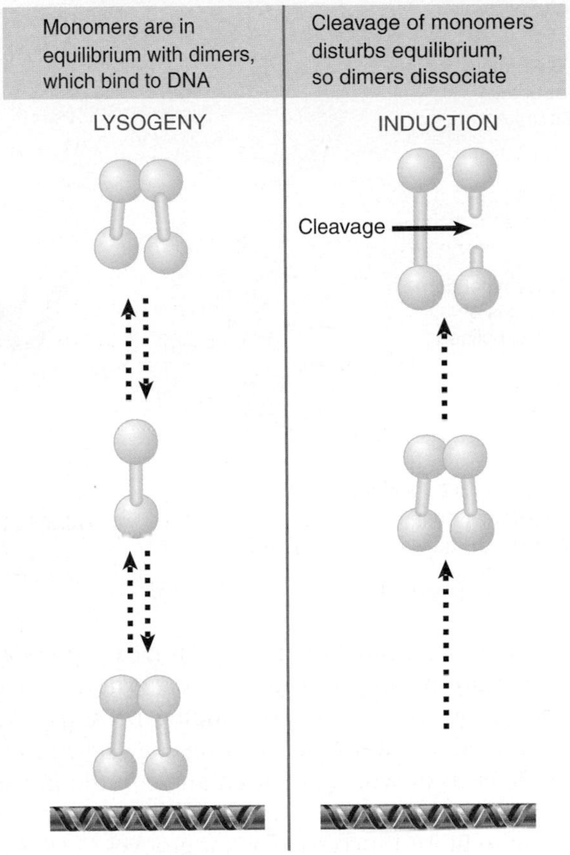

FIGURE 27.15 Repressor dimers bind to the operator. The affinity of the N-terminal domains for DNA is controlled by the dimerization of the C-terminal domains.

KEY CONCEPTS

- A repressor monomer has two distinct domains.
- The N-terminal domain contains the DNA-binding site.
- The C-terminal domain dimerizes.
- Binding to the operator requires the dimeric form so that two DNA-binding domains can contact the operator simultaneously.
- Cleavage of the repressor between the two domains reduces the affinity for the operator and induces a lytic cycle.

CONCEPT AND REASONING CHECK

What might be the advantage for a regulator to function as a dimer?

27.10 Lambda Repressor Uses a Helix-Turn-Helix Motif to Bind DNA

A repressor dimer is the unit that binds to DNA. It recognizes a sequence of 17 bp displaying partial symmetry about an axis through the central base pair. **FIGURE 27.16** shows an example of a binding site. The sequence on each side of the central base pair is sometimes called a "half-site." Each individual N-terminal region contacts a half-site. Several DNA-binding proteins that regulate bacterial transcription share a similar mode of holding DNA, in which the active domain contains two short regions of α-helix that contact DNA. (Some transcription factors in eukaryotic cells use a similar motif. See *Section 28.6, There Are Many Types of DNA-Binding Domains.*)

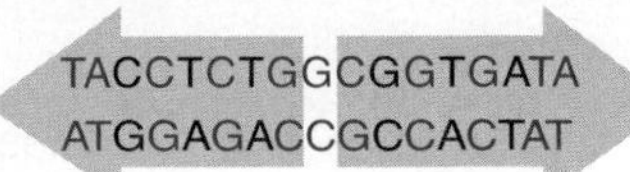

FIGURE 27.16 The operator is a 17 bp sequence with an axis of symmetry through the central base pair. Each half-site is marked by gray arrows. Base pairs that are identical in each operator half are in blue.

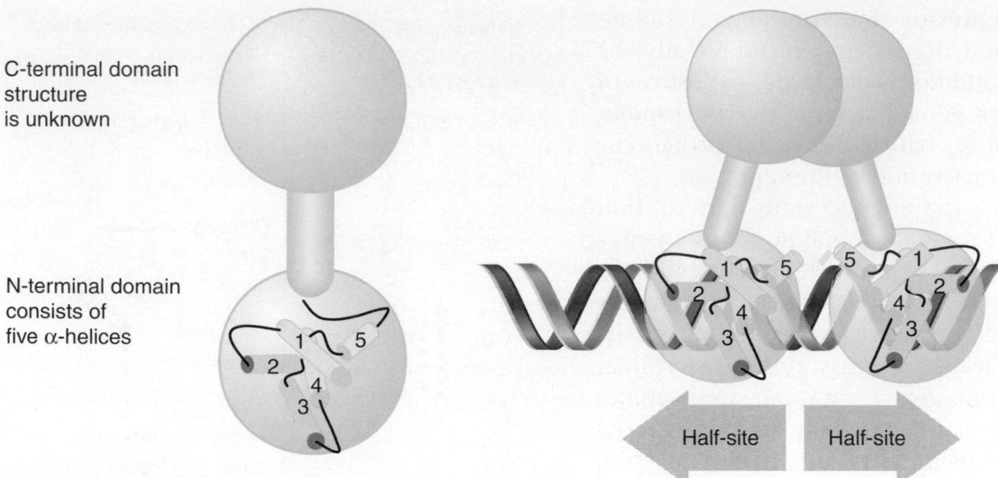

FIGURE 27.17 Lambda repressor's N-terminal domain contains five stretches of α-helix; helices 2 and 3 bind DNA.

FIGURE 27.18 In the two-helix model for DNA binding, helix-3 of each monomer lies in the wide groove on the same face of DNA, and helix-2 lies across the groove.

▸ **helix-turn-helix** The motif that describes an arrangement of two α-helices that form a site that binds to DNA, one fitting into the major groove of DNA and the other lying across it.

▸ **recognition helix** One of the two helices of the helix-turn-helix motif that makes contacts with DNA that are specific for particular bases. This determines the specificity of the DNA sequence that is bound.

The N-terminal domain of lambda repressor contains several stretches of α-helix, which are arranged as illustrated diagrammatically in **FIGURE 27.17**. Two of the helical regions are responsible for binding DNA. The **helix-turn-helix** model for contact is illustrated in **FIGURE 27.18**. Looking at a single monomer, α-helix-3 consists of nine amino acids, each of which lies at an angle to the preceding region of seven amino acids that forms α-helix-2. In the dimer, the two apposed helix-3 regions lie 34 Å apart, enabling them to fit into successive major grooves of DNA. The helix-2 regions lie at an angle that would place them across the groove. The symmetrical binding of dimer to the site means that each N-terminal domain of the dimer contacts a similar set of bases in its half-site.

Related forms of the α-helical motifs employed in the helix-turn-helix of the lambda repressor are found in several DNA-binding proteins, including catabolite repressor protein (CRP), the *lac* repressor, and several other phage repressors. By comparing the abilities of these proteins to bind DNA, we can define the roles of each helix:

- Contacts between helix 2 and helix 3 are maintained by interactions between hydrophobic amino acids.
- Contacts between helix 3 and DNA rely on hydrogen bonds between the amino acid side chains and the exposed positions of the base pairs. This helix is responsible for recognizing the specific target DNA sequence and is therefore also known as the **recognition helix**. By comparing the contact patterns summarized in **FIGURE 27.19**, we see that the lambda repressor and Cro select different sequences in the DNA as their most favored targets because they have different amino acids in the corresponding positions in helix 3.

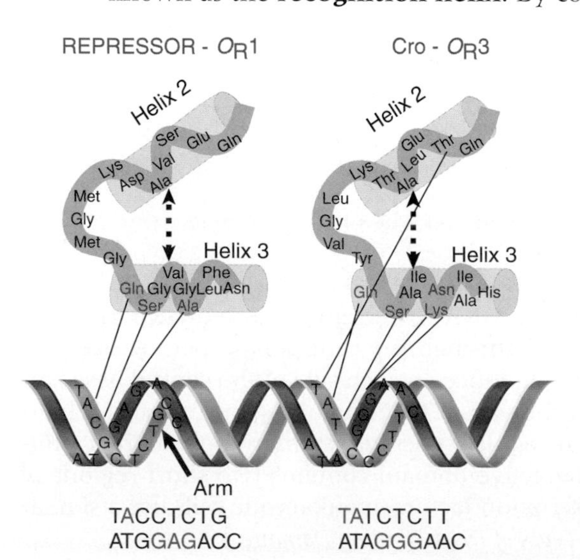

FIGURE 27.19 Two proteins that use the two-helix arrangement to contact DNA recognize lambda operators with affinities determined by the amino acid sequence of helix 3.

Contacts from helix 2 to the DNA take the form of hydrogen bonds connecting with the phosphate backbone. These interactions are necessary for binding but do not control the specificity of target recognition. In addition to these contacts, a large part of the overall energy of interaction with DNA is provided by ionic interactions with the phosphate backbone.

What happens if we manipulate the coding sequence to construct a new protein by substituting the recognition helix in one repressor with the corresponding sequence from a closely related repressor? The specificity of the hybrid protein is that of its new recognition helix. *The amino acid sequence of this short region determines the sequence specificities of the individual proteins and is able to act in conjunction with the rest of the polypeptide chain.*

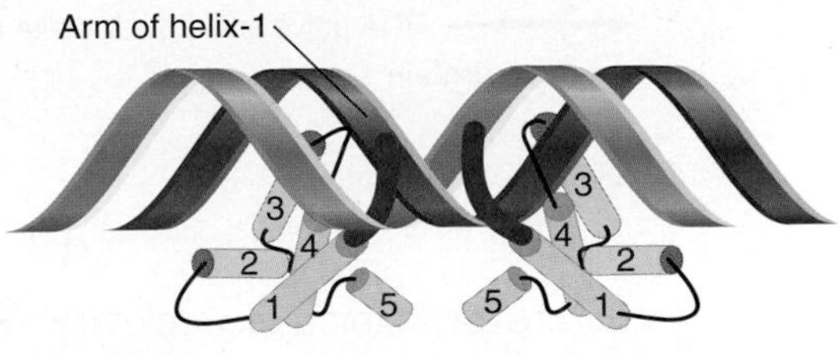

FIGURE 27.20 A view from the back shows that the bulk of the repressor contacts one face of DNA, but its N-terminal arms reach around to the other face.

The bases contacted by helix 3 lie on one face of the DNA, as can be seen from the positions indicated on the helical diagram in Figure 27.19. Repressor makes an additional contact with the other face of DNA, though. The last six N-terminal amino acids of the N-terminal domain form an "arm" extending around the back. **FIGURE 27.20** shows the view from the back. Lysine residues in the arm make contact with G residues in the major groove and also with the phosphate backbone. The interaction between the arm and DNA contributes heavily to DNA binding; the binding affinity of a mutant armless repressor is reduced by ~1000-fold.

KEY CONCEPTS

- Each DNA-binding region in the repressor contacts a half-site in the DNA.
- The DNA-binding site of the repressor includes two short α-helical regions that fit into the successive turns of the major groove of DNA.
- A DNA-binding site is a (partially) palindromic sequence of 17 bp.
- The amino acid sequence of the recognition helix makes contacts with particular bases in the operator sequence that it recognizes.

CONCEPT AND REASONING CHECK

What different functional protein domains would you expect to find in the lambda repressor?

27.11 Lambda Repressor Dimers Bind Cooperatively to the Operator

Each operator contains three repressor-binding sites. As can be seen from **FIGURE 27.21**, no two of the six individual repressor-binding sites are identical, but they all conform to a consensus sequence. The binding sites within each operator are separated by spacers of 3 to 7 bp that are rich in A-T base pairs. The sites at each operator are numbered so that O_R consists of the series of binding sites O_R1-O_R2-O_R3, whereas O_L consists of the series O_L1-O_L2-O_L3. In each case, site 1 lies closest to the startpoint for transcription in the promoter, and sites 2 and 3 lie farther upstream.

Faced with the triplication of binding sites at each operator, how does the lambda repressor decide where to start binding? At each operator, site 1 has a greater affinity (roughly tenfold) than the other sites for the lambda repressor. Thus it always binds first to O_L1 and O_R1.

Lambda repressor binds to subsequent sites within each operator in a cooperative manner. The presence of a dimer at site 1 greatly increases the affinity with which a second dimer can bind to site 2. When both sites 1 and 2 are occupied, this interaction does *not* extend farther, to site 3. At the concentrations of the lambda repressor usually found in a lysogen, both sites 1 and 2 are filled at each operator, but site 3 is not occupied.

The C-terminal domain is responsible for the cooperative interaction between dimers, as well as for the dimer formation between subunits. **FIGURE 27.22** shows that it involves both subunits of each dimer; that is, each subunit contacts its counterpart in the other dimer, forming a tetrameric structure.

A result of cooperative binding is to increase the effective affinity of repressor for the operator at physiological concentrations. This enables a lower concentration of

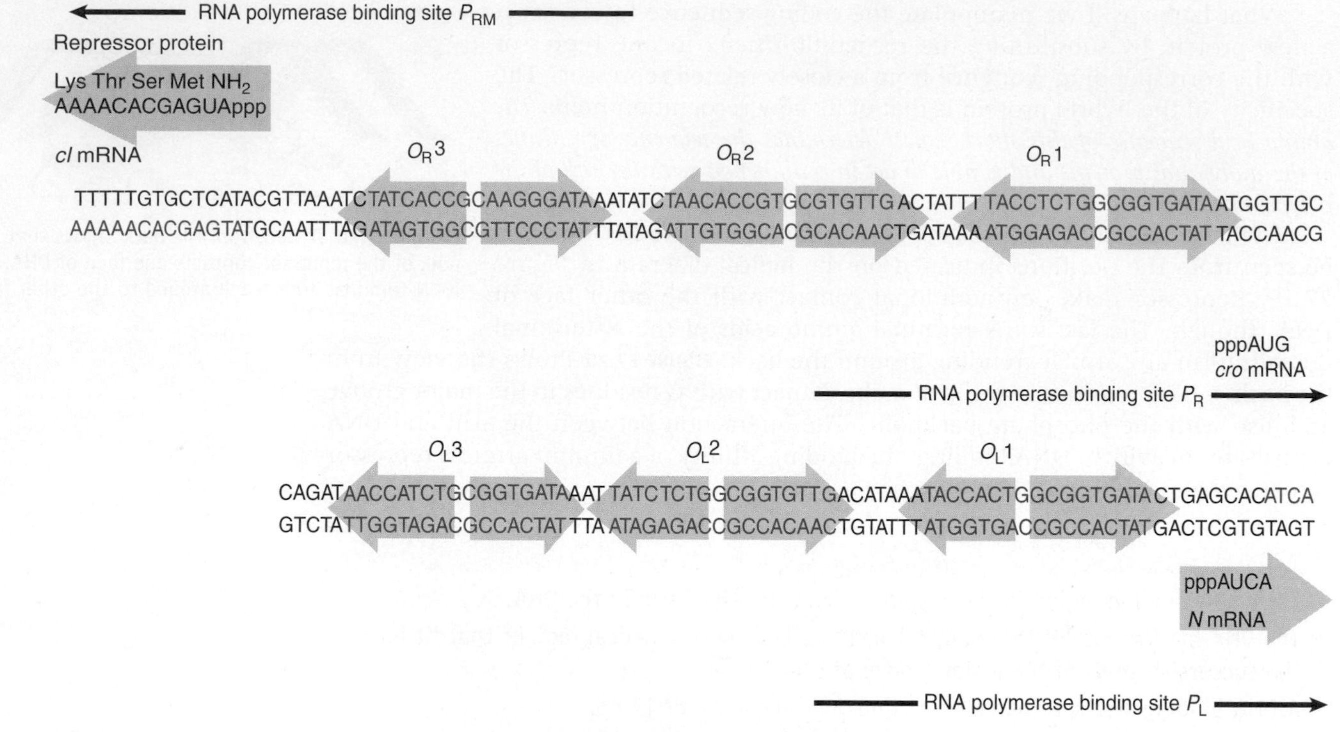

FIGURE 27.21 Each operator contains three repressor-binding sites and overlaps with the promoter at which RNA polymerase binds. The orientation of O_L has been reversed from usual to facilitate comparison with O_R.

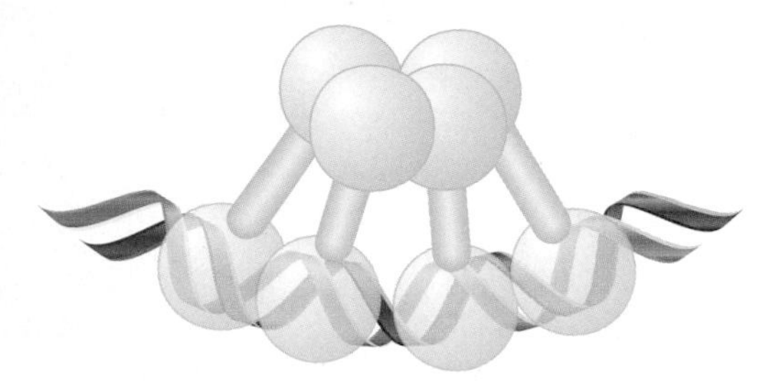

FIGURE 27.22 When two lambda repressor dimers bind cooperatively, each of the subunits of one dimer contacts a subunit in the other dimer.

repressor to achieve occupancy of the operator. This is an important consideration in a system in which release of repression has irreversible consequences. In an operon coding for metabolic enzymes, after all, failure to repress will merely allow unnecessary synthesis of enzymes. Failure to repress lambda prophage, however, will lead to induction of phage and lysis of the cell.

From the sequences shown in Figure 27.21, we see that O_L1 and O_R1 lie more or less in the center of the RNA polymerase binding sites of P_L and P_R, respectively. Occupancy of O_L1-O_L2 and O_R1-O_R2 thus physically blocks access of RNA polymerase to the corresponding promoters.

KEY CONCEPTS

- Repressor binding to one operator increases the affinity for binding a second repressor dimer to the adjacent operator.
- The affinity is 10× greater for O_L1 and O_R1 than other operators, so they are bound first.
- Cooperativity allows repressor to bind the O_L2/O_R2 sites at lower concentrations.

CONCEPT AND REASONING CHECK

Why does each operator contain three parts and not a single binding site?

27.12 Lambda Repressor Maintains an Autoregulatory Circuit

Once lysogeny has been established, the *cI* gene is transcribed from the P_{RM} promoter (see Figure 27.11) that lies to its right, close to P_R/O_R. Transcription terminates at the left end of the gene. The mRNA starts with the AUG initiation codon; because of the absence of a 5′ UTR containing a ribosome binding site, this is a very poor message that is translated inefficiently, producing only a low level of protein. Note that we have not yet described how transcription for the *cI* gene is established (see *Section 27.16, The Cro Repressor Is Needed for Lytic Infection*).

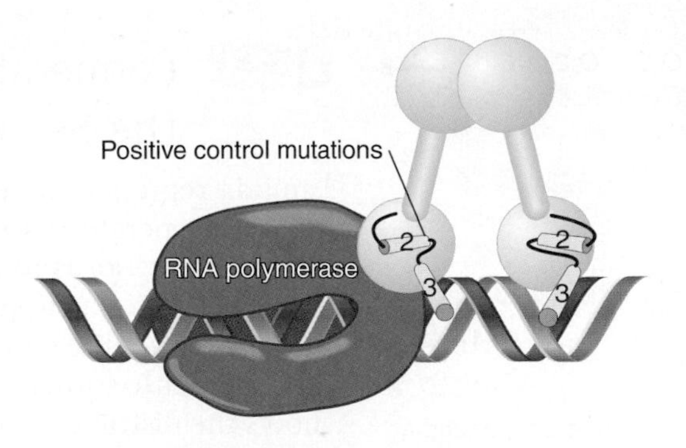

FIGURE 27.23 Positive control mutations identify a small region at helix 2 that interacts directly with RNA polymerase.

The presence of the lambda repressor at O_R has dual effects as noted earlier (*Section 27.7, Lysogeny Is Maintained by the Lambda Repressor Protein*). It blocks expression from P_R, but it assists transcription from P_{RM}. *RNA polymerase can initiate efficiently at* P_{RM} *only when the lambda repressor is bound at* O_R. The lambda repressor thus behaves as a positive regulator protein that is necessary for transcription of its own gene, *cI*. This is the definition of an autoregulatory circuit.

The RNA polymerase binding site at P_{RM} is adjacent to O_R2. This explains how the lambda repressor autoregulates its own synthesis. When two dimers are bound at O_R1-O_R2, the amino terminal domain of the dimer at O_R2 interacts with RNA polymerase. The nature of the interaction is identified by mutations in the repressor that abolish positive control because they cannot stimulate RNA polymerase to transcribe from P_{RM}. They map within a small group of amino acids, located on the outside of helix 2 or in the turn between helix 2 and helix 3. The mutations reduce the negative charge of the region; conversely, mutations that increase the negative charge enhance the activation of RNA polymerase. This suggests that the group of amino acids constitutes an "acidic patch" that functions by an electrostatic interaction with a basic region on RNA polymerase to activate it.

The location of these "positive control mutations" in the repressor is indicated in **FIGURE 27.23**. They lie at a site on repressor that is close to a phosphate group on DNA, which is also close to RNA polymerase. Thus the group of amino acids on repressor that is involved in positive control is in a position to contact the polymerase. The important principle is that *protein-protein interactions can release energy that is used to help to initiate transcription.*

The target site on RNA polymerase that the repressor contacts is in the σ^{70} subunit, which is within the region that contacts the –35 region of the promoter. The interaction between repressor and polymerase is needed for the polymerase to make the transition from a closed complex to an open complex.

This explains how low levels of repressor positively regulate its own synthesis. As long as enough repressor is available to fill O_R2, RNA polymerase will continue to transcribe the *cI* gene from P_{RM}.

KEY CONCEPTS

- The DNA-binding region of repressor at O_R2 contacts RNA polymerase and stabilizes its binding to P_{RM}.
- This is the basis for the autoregulatory control of repressor maintenance.
- Repressor binding at O_L blocks transcription of gene *N* from P_L.
- Repressor binding at O_R blocks transcription of *Cro*, but also is required for transcription of *cI*.
- Repressor binding to the operators therefore simultaneously blocks entry to the lytic cycle and promotes its own synthesis.

CONCEPT AND REASONING CHECK

How is the battle for control of which developmental pathway lambda chooses determined by *cro* and the lambda repressor?

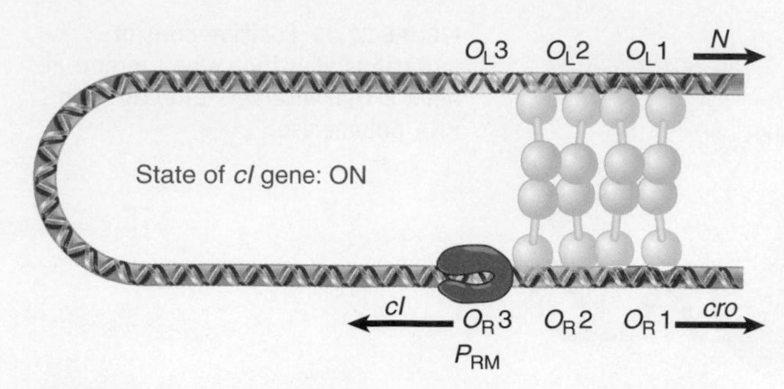

FIGURE 27.24 In the lysogenic state, the repressors bound at O_L1 and O_L2 interact with those bound at O_R1 and O_R2. RNA polymerase is bound at P_{RM} (which overlaps with O_R3) and interacts with the repressor bound at O_R2.

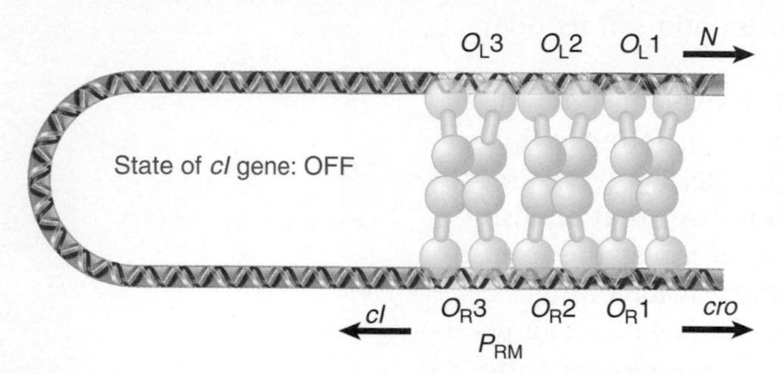

FIGURE 27.25 O_L3 and O_R3 are brought into proximity by formation of the repressor octamer, and an increase in repressor concentration allows dimers to bind at these sites and to interact.

27.13 Cooperative Interactions Increase the Sensitivity of Regulation

Lambda repressor dimers interact cooperatively at both the left and right operators, so that their normal condition when occupied by repressor is to have dimers at both the 1 and 2 binding sites. In effect, each operator has a tetramer of repressor. However, this is not the end of the story. The two dimers interact with one another to form an octamer as depicted in **FIGURE 27.24**, which shows the distribution of repressors at the operator sites that are occupied in a lysogen. Repressors are occupying O_L1, O_L2, O_R1, and O_R2, and the repressor at the last of these sites is interacting with RNA polymerase, which is initiating transcription at P_{RM}.

The interaction between the two operators has several consequences. It stabilizes repressor binding, thereby making it possible for repressor to occupy operators at lower concentrations. Binding at O_R2 stabilizes RNA polymerase binding at P_{RM}, which enables low concentrations of repressor to autogenously stimulate their own production.

The DNA between the O_L and O_R sites (that is, the gene *cI*) forms a large loop, which is held together by the repressor octamer. The octamer brings the sites O_L3 and O_R3 into proximity. As a result, two repressor dimers can bind to these sites and interact with one another, as shown in **FIGURE 27.25**. The occupation of O_R3 prevents RNA polymerase from binding to P_{RM} and, therefore, turns off expression of repressor.

This shows us how the expression of the *cI* gene becomes exquisitely sensitive to repressor concentration. At the lowest concentrations, it forms the octamer and activates RNA polymerase in a positive autogenous regulation. An increase in concentration allows binding to O_L3 and O_R3 and turns off transcription in a negative autogenous regulation. The threshold levels of repressor that are required for each of these events are reduced by the cooperative interactions, which makes the overall regulatory system much more sensitive. Any change in repressor level triggers the appropriate regulatory response to restore the lysogenic level.

Because the overall level of repressor has been reduced (about threefold from the level that would be required if there were no cooperative effects), there is less repressor that has to be eliminated when it becomes necessary to induce the phage. This increases the efficiency of induction.

KEY CONCEPTS

- Repressor dimers bound at O_L1 and O_L2 interact with dimers bound at O_R1 and O_R2 to form octamers.
- These cooperative interactions increase the sensitivity of regulation.

CONCEPT AND REASONING CHECK

What does it mean to say the *cI* gene regulates itself?

27.14 The *cII* and *cIII* Genes Are Needed to Establish Lysogeny

The control circuit for maintaining lysogeny presents a paradox. The presence of repressor protein is necessary for its own synthesis. This explains how the lysogenic condition is perpetuated. How, though, is the synthesis of repressor established in the first place?

When a lambda DNA enters a new host cell, RNA polymerase cannot transcribe *cI* because there is no repressor present to aid its binding at P_{RM}. This same absence of repressor, however, means that P_R and P_L are available. Thus the first event after lambda DNA infects a bacterium is when genes *N* and *cro* are transcribed. After this, pN allows transcription to be extended farther. This allows *cIII* (and other genes) to be transcribed on the left, whereas *cII* (and other genes) are transcribed on the right (see Figure 27.11).

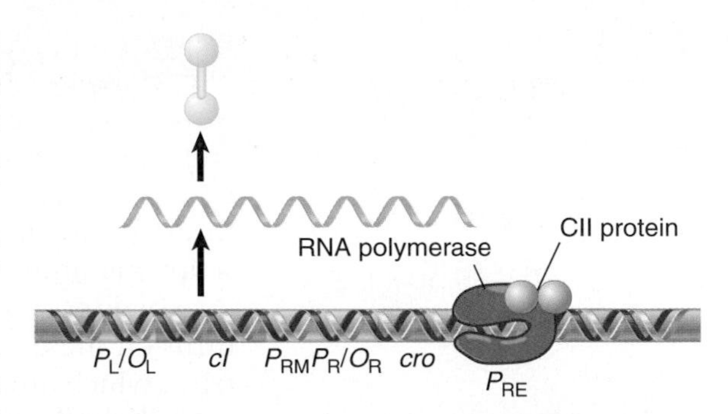

FIGURE 27.26 Repressor synthesis is established by the action of cII and RNA polymerase at P_{RE} to initiate transcription that extends from the antisense strand of *cro* through the *cI* gene.

The *cII* and *cIII* genes share with *cI* the property that mutations in them hinder lytic development. There is, however, a difference. The *cI* mutants can neither establish nor maintain lysogeny. The *cII* or *cIII* mutants have some difficulty in establishing lysogeny, but once it is established they are able to maintain it by the *cI* autoregulatory circuit.

This implicates the *cII* and *cIII* genes as positive regulators whose products are needed for an alternative system for repressor synthesis. The system is needed only to *initiate* the expression of *cI* to circumvent the inability of the autogenous circuit to engage in *de novo* synthesis. They are not needed for continued expression.

The cII protein acts directly on gene expression as a positive regulator. Between the *cro* and *cII* genes is the second *cI* promoter, called P_{RE} (P_{RE} stands for promoter right establishment). This promoter can be recognized by RNA polymerase only in the presence of cII protein, whose action is illustrated in **FIGURE 27.26**. As is the case with most promoters that require a positive regulator to function, the P_{RE} promoter has a poor fit with the consensus sequence of the TATA box and lacks a -35 sequence. This explains its dependence on the cII protein, a classic positive regulator: a protein that functions at the promoter to enable RNA polymerase to initiate transcription.

The cII protein is extremely unstable *in vivo* because it is degraded as the result of the activity of a host protein called HflA—"*hfl*" stands for *h*igh-*f*requency *l*ysogenization. The role of cIII is to protect cII against this degradation.

Transcription from P_{RE} promotes lysogeny in two ways. Its direct effect is that *cI* mRNA is translated into repressor protein. An indirect effect is that transcription proceeds through the *cro* gene in the "wrong" direction. Thus the 5′ part of the RNA corresponds to an antisense transcript of *cro;* in fact, it hybridizes to authentic *cro* mRNA, which inhibits its translation. This is important because *cro* expression is needed to enter the lytic cycle (see *Section 27.16, The Cro Repressor Is Needed for Lytic Infection*).

The *cI* coding region on the P_{RE} transcript is very efficiently translated, in contrast with the weak translation of the P_{RM} transcript noted before. In fact, repressor is synthesized approximately seven to eight times more effectively via expression from P_{RE} than from P_{RM}. This reflects the fact that the P_{RE} transcript has an efficient 5′UTR containing a strong ribosome-binding site, whereas the P_{RM} transcript has no ribosome-binding site and actually starts with the AUG initiation codon.

KEY CONCEPTS

- The delayed early gene products *cII* and *cIII* are necessary for RNA polymerase to initiate transcription at the promoter P_{RE}.
- *cII* acts directly at the promoter and *cIII* protects *cII* from degradation.
- Transcription from P_{RE} leads to synthesis of repressor and also blocks the transcription of *cro*.
- P_{RE} has atypical sequences at -10 and -35.
- RNA polymerase binds the promoter only in the presence of *cII*.
- *cII* binds to sequences close to the -35 region.

CONCEPT AND REASONING CHECK

How is the battle for control of which developmental pathway lambda chooses determined by *cII*?

27.15 Lysogeny Requires Several Events

Now we can see how lysogeny is established during an infection. **FIGURE 27.27** recapitulates the early stages and shows what happens as the result of expression of *cIII* and *cII*. *cIII* protects *cII*. The presence of *cII* allows P_{RE} to be used for transcription extending through *cI*. Lambda repressor protein is synthesized in high amounts from this transcript and immediately binds to O_L and O_R.

By directly inhibiting any further transcription from P_L and P_R, repressor binding turns off the expression of all phage genes. This halts the synthesis of *cII* and *cIII* proteins, which are unstable; they decay rapidly, with the result that P_{RE} can no longer be used. Thus the synthesis of repressor via the establishment circuit is brought to a halt.

But repressor is now present at O_R. Acting as a positive regulator, it switches on the maintenance circuit for expression from P_{RM}. Repressor continues to be synthesized, although at the lower level typical of P_{RM} function. So the establishment circuit starts off repressor synthesis at a high level; then repressor turns off all other func-

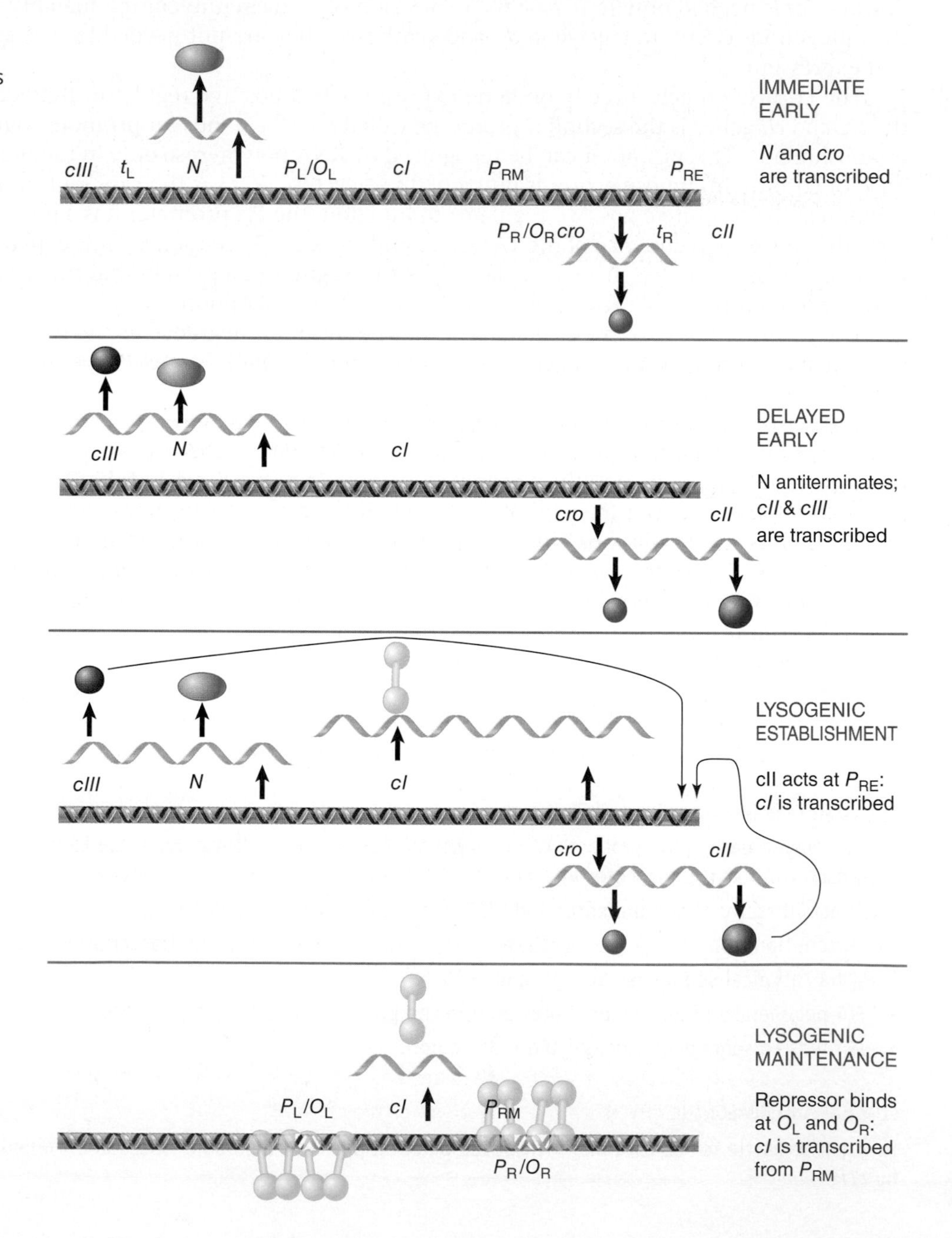

FIGURE 27.27 A cascade is needed to establish lysogeny, but then this circuit is switched off and replaced by the autogenous repressor-maintenance circuit.

ESSENTIAL IDEAS

The Mechanism of pQ Antitermination

One of the most fascinating aspects of biology is the wide variety of mechanisms that exist to regulate gene expression. Even within one approach to a problem—how to carry out transcription of genes that lie downstream of a terminator, for example—there may be several solutions. In bacteriophage λ, antitermination occurs at two points in the lytic cycle. First, the antiterminator protein pN acts to allow transcription to proceed beyond a terminator into the delayed early genes. Later on, the antiterminator protein pQ acts to allow transcription into the late genes.

Proteins often recognize sites within DNA and bind there to carry out some action. However, pN acts in a different manner. The phage DNA encodes sites called N utilization (*nut*) sites, but pN does not bind to the DNA site. Rather, it binds to the corresponding region on the nascent RNA strand (called NUT to indicate it is an RNA site). Once bound, pN recruits additional proteins called NusA, NusB, NusD, and NusE to the RNA polymerase. The resulting complex remains on the RNA polymerase and is able to proceed through the terminators t_{L1} or t_{R1} (see Figure 11.32).

The antiterminator protein pQ uses an entirely different mechanism to enable RNA polymerase to proceed through terminators downstream of the late gene promoter $P_{R'}$. pQ binds directly to the q utilization (*qut*) site on the DNA. The *qut* site is located between the promoter –10 and –35 elements of $P_{R'}$, far from the terminator elements. In addition to the *qut* site, a pause-inducing element is required for pQ-mediated antitermination to occur. The pause site is located in the initial transcribed region and resembles a promoter –10 element. When RNA polymerase pauses at this site, pQ is able to associate with the polymerase. The protein remains associated and ultimately permits transcription to occur beyond the terminator site downstream and into the late genes. No additional proteins have been found to be required for this antitermination event.

Additional research is needed to determine the details of pQ action, but it is clear that its mechanism of action is completely different from that of pN. Although the bacteriophage λ genome is only 48.5 kb, it encompasses a complexity we are still working to understand.

tions, while at the same time turning on the maintenance circuit, which functions at the low level adequate to sustain lysogeny.

We shall not at this point deal in detail with the other functions needed to establish lysogeny, but we can just briefly remark that the infecting lambda DNA must be inserted into the bacterial genome (see *Section 15.5, Specialized Enzymes Catalyze 5′ End Resection and Single-Strand Invasion*). The insertion requires the product of gene *int*, which is expressed from its own promoter P_I, at which the *cII* positive regulator also is necessary. The functions necessary for establishing the lysogenic control circuit are, therefore, under the same control as the function needed to integrate the phage DNA into the bacterial genome. Thus the establishment of lysogeny is under a control that ensures all the necessary events occur with the same timing.

Emphasizing the tricky quality of lambda's intricate cascade, we now know that *cII* promotes lysogeny in another, indirect manner. It sponsors transcription from a promoter called $P_{anti\text{-}Q}$, which is located within the *Q* gene. This transcript is an antisense version of the *Q* region, and it hybridizes with *Q* mRNA to prevent translation of Q protein, whose synthesis is essential for lytic development. Thus the same mechanisms that directly promote lysogeny by causing transcription of the *cI* repressor gene also indirectly help lysogeny by inhibiting the expression of *cro* (see above) and *Q*, the regulator genes needed for the antagonistic lytic pathway.

KEY CONCEPTS

- *cII* and *cIII* cause repressor synthesis to be established and also trigger inhibition of late gene transcription.
- Establishment of repressor turns off immediate and delayed early gene expression.
- Repressor turns on the maintenance circuit for its own synthesis.
- Lambda DNA is integrated into the bacterial genome at the final stage in establishing lysogeny.

CONCEPT AND REASONING CHECK

How does *cIII* determine the outcome of the battle between *cro* and *cI*?

27.16 The Cro Repressor Is Needed for Lytic Infection

Because lambda is a temperate virus, it has the alternatives of entering either the lysogenic pathway or the lytic pathway. Lysogeny is initiated by establishing an autoregulatory maintenance circuit that inhibits the entire lytic cascade through applying pressure at two points, P_LO_L and P_RO_R. The two pathways begin exactly the same—with the immediate early gene expression of the *N* gene and the *cro* gene (Note that the *cro* gene, required for the lytic pathway is given a headstart.), followed by the pN-directed delayed early transcription. We now face a problem. How does the phage enter the lytic cycle?

The key to the lytic cycle is the role of the *cro* gene, which codes for another repressor protein. *Cro is responsible for preventing the synthesis of the lambda repressor protein;* this action shuts off the possibility of establishing lysogeny. *Cro* mutants usually establish lysogeny rather than entering the lytic pathway, because they lack the ability to switch events away from the expression of repressor.

Cro forms a small dimer (the monomer is 9 kD) that acts within the immunity region. It has two effects:

- It prevents the synthesis of the lambda repressor via the maintenance circuit; that is, it prevents transcription via P_{RM}.
- It also inhibits the expression of early genes from both P_L and P_R.

This means that when a phage enters the lytic pathway, *Cro* has responsibility both for preventing the synthesis of the lambda repressor and subsequently for turning down the expression of the early genes once there has been enough product made.

Note that Cro achieves its function by binding to the same operators as the lambda repressor protein cI. How can two proteins have the same sites of action, yet have such opposite effects? The answer lies in the different affinities that each protein has for the individual binding sites within the operators. Let us just consider O_R, about which more is known, and where Cro exerts both its effects. The series of events is illustrated in **FIGURE 27.28**. (Note that the first two stages are identical to those of the lysogenic circuit shown in Figure 27.27.)

The affinity of Cro for O_R3 is greater than its affinity for O_R2 or O_R1. Thus it binds first to O_R3. This inhibits RNA polymerase from binding to P_{RM}. As a result, Cro's first action is to prevent the maintenance circuit for lysogeny from coming into play.

Cro then binds to O_R2 or O_R1. Its affinity for these sites is similar, and there is no cooperative effect. Its presence at either site is sufficient to prevent RNA polymerase from using P_R. This in turn stops the production of the early functions (including Cro itself). As a result of *cII*'s instability, any use of P_{RE} is brought to a halt. Thus the two actions of Cro together block *all* production of the lambda repressor.

As far as the lytic cycle is concerned, Cro turns down (although it does not completely eliminate) the expression of the early genes. Its incomplete effect is explained by its affinity for O_R1 and O_R2, which is about eight times lower than that of the lambda repressor. This effect of Cro does not occur until the early genes have become more or less superfluous, because the pQ protein is present; by this time, the phage has started late gene expression and is concentrating on the production of progeny phage particles.

Note again that in the early stages of the infection, Cro is given a head start over the lambda repressor, so it would seem that the lytic pathway is favored. Ultimately, the outcome will be determined by the concentration of the two proteins and their intrinsic DNA binding affinities.

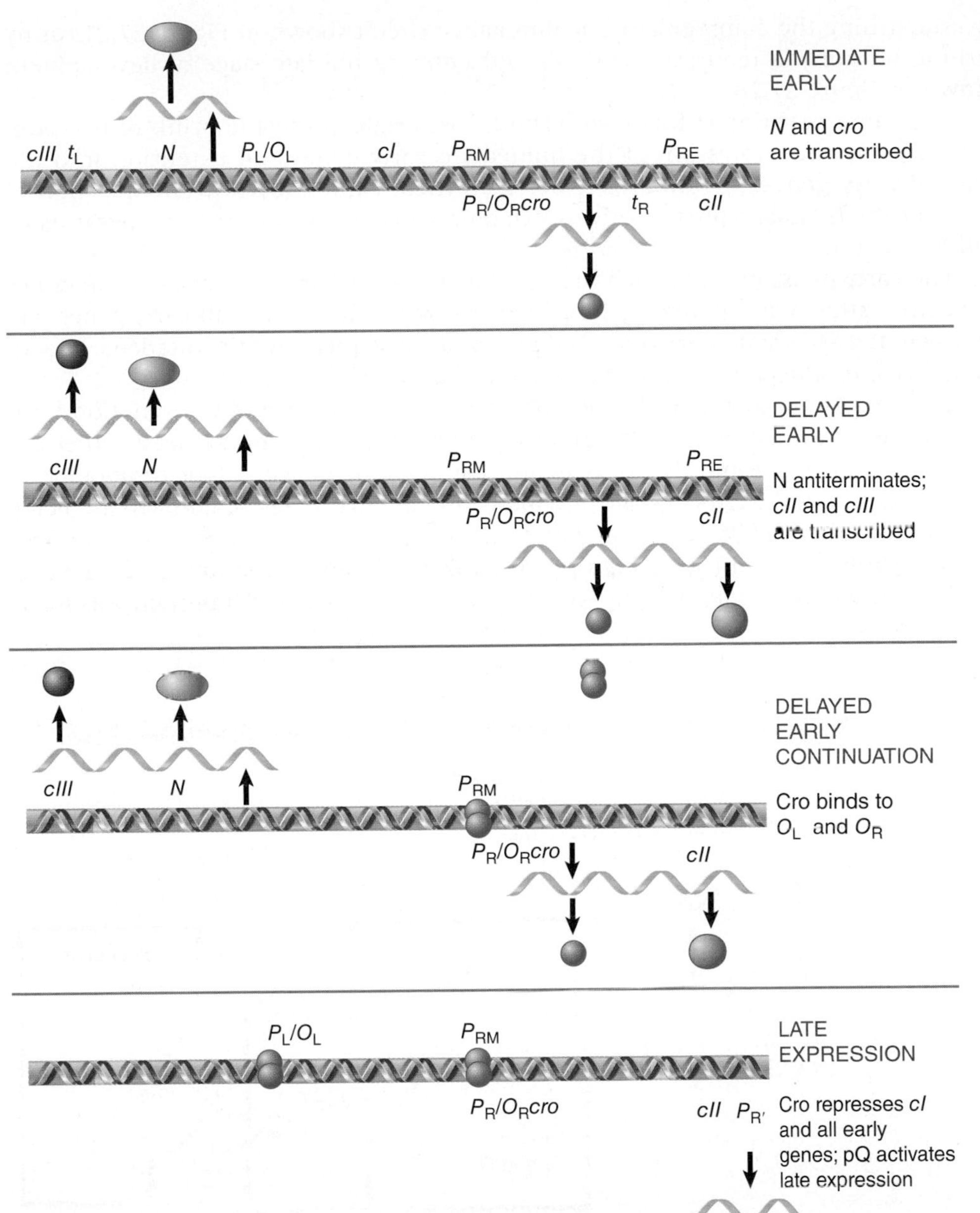

FIGURE 27.28 The lytic cascade requires Cro protein, which directly prevents repressor maintenance via P_{RM}, as well as turning off delayed early gene expression, indirectly preventing repressor establishment.

KEY CONCEPTS

- Cro binds to the same operators as the lambda repressor but with different affinities.
- When Cro binds to O_R3, it prevents RNA polymerase from binding to P_{RM} and blocks the maintenance of repressor promoter.
- When Cro binds to other operators at O_R or O_L, it prevents RNA polymerase from expressing immediate early genes, which (indirectly) blocks repressor establishment.

CONCEPT AND REASONING CHECK

If *cro* transcription begins much earlier than *cI* transcription, why doesn't *cro* always win?

27.17 What Determines the Balance Between Lysogeny and the Lytic Cycle?

The programs for the lysogenic and lytic pathways are so intimately related that it is impossible to predict the fate of an individual phage genome when it enters a new host bacterium. Will the antagonism between the lambda repressor and Cro be resolved by establishing the autoregulatory maintenance circuit shown in Figure 27.27, or by turning off lambda repressor synthesis and entering the late stage of development shown in Figure 27.28?

The same pathway is followed in both cases right up to the brink of decision. Both involve the expression of the immediate early genes and extension into the delayed early genes. The difference between them comes down to the question of whether the lambda repressor or Cro will obtain occupancy of the two operators O_L and P_L.

The early phase during which the decision is made is limited in duration in either case. No matter which pathway the phage follows, expression of all early genes will be prevented as P_L and P_R are repressed and, as a consequence of the disappearance of cII and cIII, production of repressor via P_{RE} will cease.

The critical question comes down to whether the cessation of transcription from P_{RE} is followed by activation of P_{RM} and the establishment of lysogeny, or whether P_{RM} fails to become active and the pQ regulator commits the phage to lytic development. **FIGURE 27.29** shows the critical stage at which both lambda repressor and Cro are being synthesized. This will be determined by how much lambda repressor was made. This in turn will be determined by how much cII transcription factor was made. Finally, this in turn will be—at least partly—determined by how much cIII protein was made to protect *cII*.

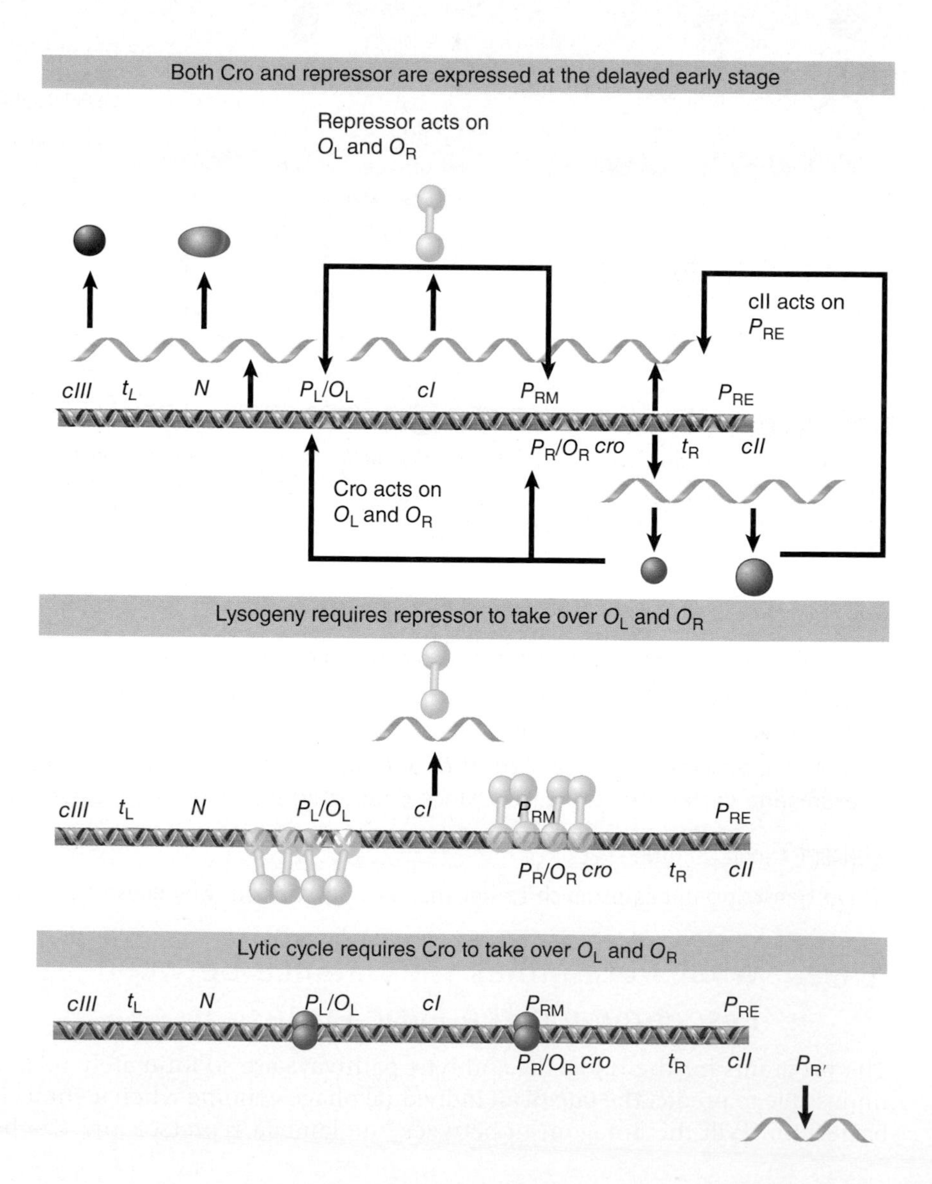

FIGURE 27.29 The critical stage in deciding between lysogeny and lysis is when delayed early genes are being expressed. If cII causes sufficient synthesis of repressor, lysogeny will result because repressor occupies the operators. Otherwise Cro occupies the operators, resulting in a lytic cycle.

The initial event in establishing lysogeny is the binding of lambda repressor at O_L1 and O_R1. Binding at the first sites is rapidly succeeded by cooperative binding of further repressor dimers at O_L2 and O_R2. This shuts off the synthesis of Cro and starts up the synthesis of lambda repressor via P_{RM}.

The initial event in entering the lytic cycle is the binding of Cro at O_R3. This stops the lysogenic-maintenance circuit from starting up at P_{RM}. Cro must then bind to O_R1 or O_R2, and to O_L1 or O_L2, to turn down early gene expression. By halting production of cII and cIII, this action leads to the cessation of lambda repressor synthesis via P_{RE}. The shutoff of lambda repressor establishment occurs when the unstable cII and cIII proteins decay.

The critical influence over the switch between lysogeny and lysis is how much cII protein is made. If cII is abundant, synthesis of repressor via the establishment promoter is effective, and, as a result, repressor gains occupancy of the operators. If cII is not abundant, lambda repressor establishment fails, and Cro binds to the operators.

The level of cII protein under any particular set of circumstances determines the outcome of an infection. Mutations that increase the stability of cII increase the frequency of lysogenization. Such mutations occur in *cII* itself or in other genes. The cause of *cII*'s instability is its susceptibility to degradation by host proteases. Its level in the cell is influenced by *cIII* as well as by host functions.

The effect of the lambda protein cIII is secondary: it helps to protect cII against degradation. The presence of cIII does not guarantee the survival of cII; however, in the absence of cIII, cII is virtually always inactivated.

Host gene products act on this pathway. Mutations in the host genes *hflA* and *hflB* increase lysogeny. The mutations stabilize cII because they inactivate host protease(s) that degrade it.

The influence of the host cell on the level of cII provides a route for the bacterium to interfere with the decision-taking process. For example, host proteases that degrade cII are activated by growth on rich medium. Thus lambda tends to lyse cells that are growing well, but is more likely to enter lysogeny on cells that are starving (and that lack components necessary for efficient lytic growth).

A different picture is seen if multiple phage infect a bacterium. Several parameters are changed. First, more cIII per bacterial cell is made to counter the amount of host protease and therefore leads to more cII being able to be made. On the other hand, in a single cell infected by multiple phage, each lambda genome will ultimately make its own decision about entering the lytic pathway or the lysogenic pathway. This is a 'noisy' decision that can be affected by minor local differences in the concentration of different molecules and proteins. The final outcome for the cell is quite different from that of a single phage infection since the status of each individual phage must be considered. Ultimately, one can imagine that a vote will be taken and for lysogeny to occur, the vote must be unanimous. Even if only one phage proceeds down the lytic pathway, cell death will occur.

KEY CONCEPTS

- The delayed early stage when both Cro and repressor are being expressed is common to lysogeny and the lytic cycle.
- The critical event is whether cII causes sufficient synthesis of repressor to overcome the action of Cro.
- When multiple phage infect a cell, all must enter the lysogenic pathway to avoid cell death.

CONCEPT AND REASONING CHECK

In the end, how does *E. coli* itself determine whether lambda will enter the lytic or lysogenic path?

27.18 Summary

Virulent phages follow a lytic life cycle, in which infection of a host bacterium is followed by production of a large number of phage particles, lysis of the cell, and release of the viruses. Temperate phages can follow the lytic pathway or the lysogenic pathway, in which the phage genome is integrated into the bacterial chromosome and is inherited in this inert, latent form like any other bacterial gene.

In general, lytic infection can be described as falling into three phases. In the first phase a small number of phage genes are transcribed by the host RNA polymerase. One or more of these genes is a regulator that controls expression of the group of genes expressed in the second phase. The pattern is repeated in the second phase, when one or more genes is a regulator needed for expression of the genes of the third phase. Genes active during the first two phases code for enzymes needed to reproduce phage DNA; genes of the final phase code for structural components of the phage particle. It is common for the very early genes to be turned off during the later phases.

In phage lambda, the genes are organized into groups whose expression is controlled by individual regulatory events. The immediate early gene *N* codes for an antiterminator that allows transcription of the leftward and rightward groups of delayed early genes from the early promoters P_R and P_L. The delayed early gene *Q* has a similar antitermination function that allows transcription of all late genes from the promoter $P_{R'}$. The lytic cycle is repressed, and the lysogenic state maintained, by expression of the *cI* gene, whose product is a repressor protein, the lambda repressor, that acts at the operators O_R and O_L to prevent use of the promoters P_R and P_L, respectively. A lysogenic phage genome expresses only the *cI* gene from its promoter, P_{RM}. Transcription from this promoter involves positive autoregulation, in which repressor bound at O_R activates RNA polymerase at P_{RM}.

Each operator consists of three binding sites for the lambda repressor. Each site is palindromic, consisting of symmetrical half-sites. Lambda repressor functions as a dimer. Each half-binding site is contacted by a repressor monomer. The N-terminal domain of repressor contains a helix-turn-helix motif that contacts DNA. Helix-3 is the recognition helix and is responsible for making specific contacts with base pairs in the operator. Helix-2 is involved in positioning helix-3; it is also involved in contacting RNA polymerase at P_{RM}. The C-terminal domain is required for dimerization. Induction is caused by cleavage between the N- and C-terminal domains, which prevents the DNA-binding regions from functioning in dimeric form, thereby reducing their affinity for DNA and making it impossible to maintain lysogeny. Lambda repressor-operator binding is cooperative, so that once one dimer has bound to the first site, a second dimer binds more readily to the adjacent site.

The helix-turn-helix motif is used by other DNA-binding proteins, including lambda Cro. Cro binds to the same operators but has a different affinity for the individual operator sites, which are determined by the sequence of helix-3. Cro binds individually to operator sites, starting with O_R3, in a noncooperative manner. It is needed for progression through the lytic cycle. Its binding to O_R3 first prevents synthesis of repressor from P_{RM}, and then its binding to O_R2 and O_R1 prevents continued expression of early genes, an effect also seen in its binding to O_L1 and O_L2.

Establishment of lambda repressor synthesis requires use of the promoter P_{RE}, which is activated by the product of the *cII* gene. The product of *cIII* is required to stabilize the *cII* product against degradation. By turning off *cII* and *cIII* expression, Cro acts to prevent lysogeny. By turning off all transcription except that of its own gene, the lambda repressor acts to prevent the lytic cycle. The choice between lysis and lysogeny depends on whether repressor or Cro gains occupancy of the operators in a particular infection. The stability of cII protein in the infected cell is a primary determinant of the outcome.

CHAPTER QUESTIONS

1. In a wild-type bacterial cell, how many copies of the lambda prophage can be present in the chromosome?
   A. none
   B. one
   C. two
   D. three or more
2. Lysogeny is maintained by the interaction of a(n) _______ with an operator.
   A. positive regulator
   B. repressor
   C. antiterminator
   D. antirepressor
3. The immediate early genes of bacteriophage lambda are:
   A. *N* and *Q*.
   B. *cI* and *Q*.
   C. *N* and *cro*.
   D. *cI* and *cro*.
4. The bacteriophage lambda Cro protein functions as a(n):
   A. antitermination factor.
   B. repressor.
   C. antirepressor.
   D. inducer.
5. What factor is needed for transcription of the phage lambda late genes?
   A. Cro protein
   B. cI protein
   C. N protein
   D. Q protein
6. What effect would an N-mutation have on phage lambda infection?
   A. no effect, infection would proceed as normal to either lytic or lysogenic pathway
   B. lytic pathway blocked; only lysogeny possible
   C. lysogenic pathway blocked; only lytic pathway possible
   D. complete abolishment of infection; neither lytic nor lysogenic pathway possible
7. What effect would a *cII*-mutation have on phage lambda infection?
   A. no effect; infection would proceed as normal to either lytic or lysogenic pathway
   B. lytic pathway blocked; only lysogeny possible
   C. lysogenic pathway blocked; only lytic pathway possible
   D. complete abolishment of infection; neither lytic nor lysogenic pathway possible
8. Transcription from the phage lambda P_{RE} promoter requires which of the following factors?
   A. cI and pN
   B. cI and cII
   C. cII and cIII
   D. pN and cII
9. Which of the following is the order of gene expression for the bacteriophage lambda lytic pathway upon infection of a wild-type host?
   A. $N \rightarrow cI \rightarrow cII \rightarrow$ late genes
   B. $N \rightarrow cro \rightarrow cII \rightarrow$ late genes
   C. $N \rightarrow cII \rightarrow Q \rightarrow$ late genes
   D. $N \rightarrow Q \rightarrow$ late genes

10. The major function of the bacteriophage Cro protein is to:
  A. prevent synthesis of the cI protein.
  B. prevent synthesis of the cII protein.
  C. prevent synthesis of the N protein.
  D. prevent synthesis of the Q protein.

KEY TERMS

bacteriophage
cascade
delayed early genes
early genes
early infection
excision
helix-turn-helix
immediate early genes
immunity
immunity region
induction of phage
integration
late gene
late infection
lysis
lysogeny
lytic infection
middle genes
phage
prophage
recognition helix
temperate phage
virulent mutations
virulent phage

FURTHER READING

Anderson, L. M., and Yang, H. (2008). DNA looping can enhance lysogenic cI transcription in phage lambda. *Proc. Natl. Acad. Sci. USA* **105**, 5827–5832.

Oppenheim, A. B., Kobiler, O., Stavans, J., Court, D. L., and Adhya, S. (2005). Switches in bacteriophage lambda development. *Annu. Rev. Gen.* **39**, 409–429.

Ptashne, M. (2004). *The Genetic Switch: Phage Lambda Revisited.* Cold Spring Harbor, NY: Cold Spring Harbor Press.

Zeng, L., Skinner, S. O., Zong, C., Skippy, J., Feiss, M., and Golding, I. (2010). Decision making at a subcellular level determines the outcome of bacteriophage infection. *Cell* **141**, 682–691.

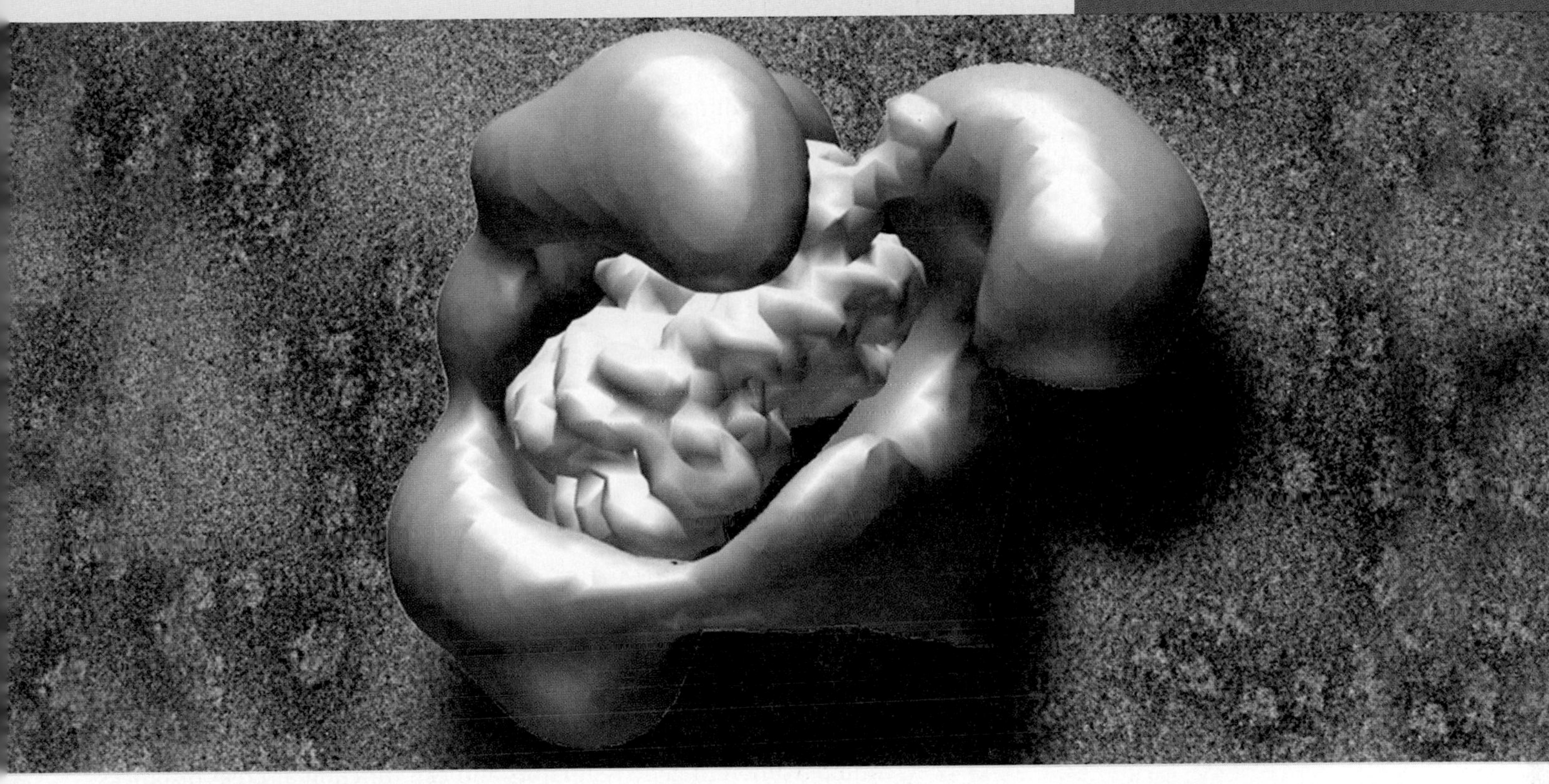

A model of the ATP-dependent chromatin remodeling complex RSC bound to a nucleosome. Photo courtesy of Andres Leschziner, Harvard University.

Eukaryotic Transcription Regulation

CHAPTER OUTLINE

28.1 Introduction

The phenotypic differences that distinguish the various kinds of cells in a higher eukaryote are largely due to differences in the expression of genes that code for proteins, that is, those transcribed by RNA polymerase II. In principle, the expression of these genes might be regulated at any one of several stages. In **FIGURE 28.1**, we can distinguish (at least) six potential control points, which form the following series:

Activation of gene structure: open chromatin
↓
Initiation of transcription and elongation
↓
Processing the transcript
↓
Transport to cytoplasm from the nucleus
↓
Translation of mRNA
↓
Degradation and turnover of mRNA

The determination of whether a gene is expressed depends on the regulatory proteins—transcription factors—that bind in and near their target gene. These binding sites are usually called enhancers (see *Section 20.9 Enhancers Contain Bidirectional Elements that Assist Initiation*). Eukaryotic enhancers differ from their bacterial counterparts in that they typically bind multiple different gene regulators and can function at great distances from the gene, including upstream, downstream, or within the gene. Ultimately, it is the combination of specific proteins binding to an enhancer (or multiple enhancers), positive and negative, that that will determine if the gene will be transcribed or not.

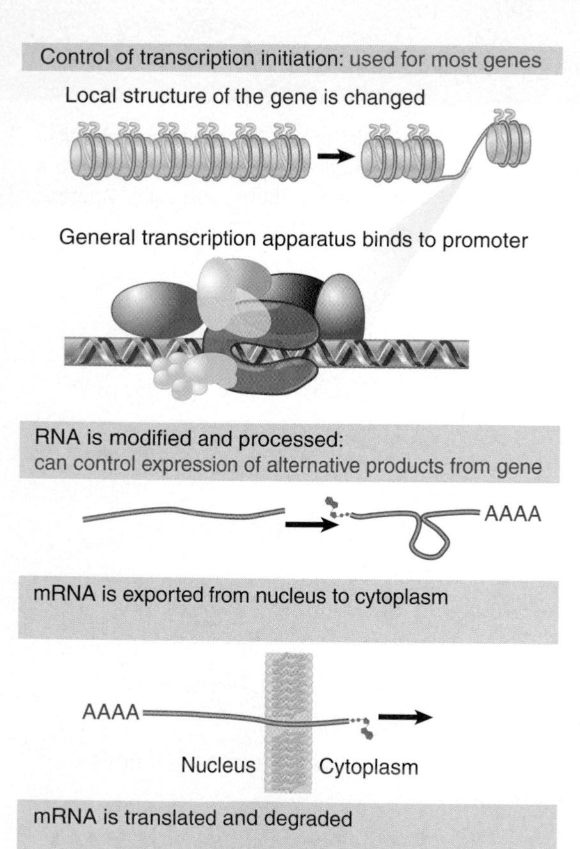

FIGURE 28.1 Gene expression is controlled principally at the initiation of transcription. Control of processing may be used to determine which form of a gene is represented in mRNA. The mRNA may be regulated during transport to the cytoplasm, during translation, and by degradation.

The first step in that decision has to do with the structure of chromatin both locally (at the promoter) and in the surrounding domain. Chromatin structure correspondingly can be regulated by individual activation events or by changes that affect a wide chromosomal region. The most localized events concern an individual target gene, where changes in nucleosomal structure and organization occur in the immediate vicinity of the promoter. Many genes have multiple promoters and/or multiple termination sites; the choice of which to use can influence how the mRNA is processed because it will change the 5′ and 3′ UTRs. More general changes may affect regions as large as a whole chromosome. Activation of a gene requires changes in the state of chromatin. The essential issue is how the transcription factors gain access to the promoter DNA.

Local chromatin structure is an integral part of controlling gene expression. Genes may exist in either of two structural conditions. Genes are found in an "active" state only in the cells in which they are expressed or potentially can be expressed. The change of structure precedes the act of transcription and indicates that the gene is "transcribable." This suggests that acquisition of the "active" structure must be the first step in gene expression. Most active genes are found in domains of euchromatin with a preferential susceptibility to nucleases, and hypersensitive sites are created at promoters before a gene is activated (see *Section 10.10, DNase Sensitivity Detects Changes in Chromatin Structure*).

There is an intimate and continuing connection between initiation of transcription and chromatin structure. Some activators of gene transcription directly modify histones; in particular, acetylation of histones is associated with gene activation. Conversely, some repressors of transcription function by deacetylating histones. Thus a reversible change in histone structure in the vicinity of the promoter is involved in the control of gene expression. These changes influence the association of histone octamers with DNA and are responsible for controlling the presence and structure of nucleosomes at specific sites. This is an important aspect of the mechanism by which a gene is maintained in an active or inactive state.

The mechanisms by which regions of chromatin are maintained in an inactive (silent) state are related to the means by which an individual promoter is repressed. The proteins involved in the formation of heterochromatin act on chromatin via the histones, and modifications of the histones are an important feature in the interaction. Once established, such changes in chromatin can persist through cell divisions, creating an **epigenetic** state in which the properties of a gene are determined by the self-perpetuating structure of chromatin. The name *epigenetic* reflects the fact that a gene may have an inherited condition (it may be active or inactive) that does not depend on its sequence (see *Chapter 29, Epigenetic Effects Are Inherited*).

▸ **epigenetic** Changes that influence the phenotype without altering the genotype. They consist of changes in the properties of a cell that are inherited but that do not represent a change in genetic information.

Once transcription begins, regulation during the elongation phase of transcription is less likely, although it does occur rarely. Attenuation as we saw in bacteria (see *Section 26.13, The* trp *Operon Is Also Controlled by Attenuation*) cannot occur in eukaryotes because of the separation of chromosomes from the cytoplasm by the nuclear membrane. The primary transcript is modified by capping at the 5′ end, and in general also is modified by polyadenylation at the 3′ end (see *Chapter 21, RNA Splicing and Processing*).

Introns must be excised from the transcripts of interrupted genes. The mature RNA must then be exported from the nucleus to the cytoplasm. Regulation of gene expression at the level of nuclear RNA processing might involve any or all of these stages, but the one for which we have most evidence concerns changes in splicing; some genes are expressed by means of alternative splicing patterns whose regulation controls the type of protein product (see *Section 21.11, Alternative Splicing Is a Rule, Rather Than an Exception, in Multicellular Eukaryotes*).

The translation of an mRNA in the cytoplasm can be specifically controlled, as can the turnover rate of the mRNA. This can also involve the localization of the mRNA to specific sites in the cell where it is expressed and/or the blocking of initiation of translation by specific protein and microRNA (miRNA) factors. Different mRNAs may have different intrinsic half-lives determined by specific sequence elements.

Regulation of tissue-specific gene transcription lies at the heart of eukaryotic development and differentiation. It is also important for control of metabolic and catabolic pathways. A regulatory transcription factor serves to provide common control of a large number of target genes, and we seek to answer two questions about this mode of regulation: How does the transcription factor identify its group of target genes? and, How is the activity of the transcription factor itself regulated in response to intrinsic or extrinsic signals?

KEY CONCEPT

- Eukaryotic gene expression is usually controlled at the level of initiation of transcription by opening the chromatin.

28.2 How Is a Gene Turned On?

Multicellular eukaryotes typically begin life through the fertilization of an egg by a sperm. In both these haploid gametes, but especially the sperm, the chromosomes are in supercondensed modified chromatin. Males of some species use positively charged polyamines like spermines and spermidines to replace the histones in sperm chromatin; others include sperm-specific histone variants. At some point after the process of fusion of the two haploid nuclei is complete in the egg, genes are then activated in a cascade of regulatory events. The general question of how a gene in closed chromatin

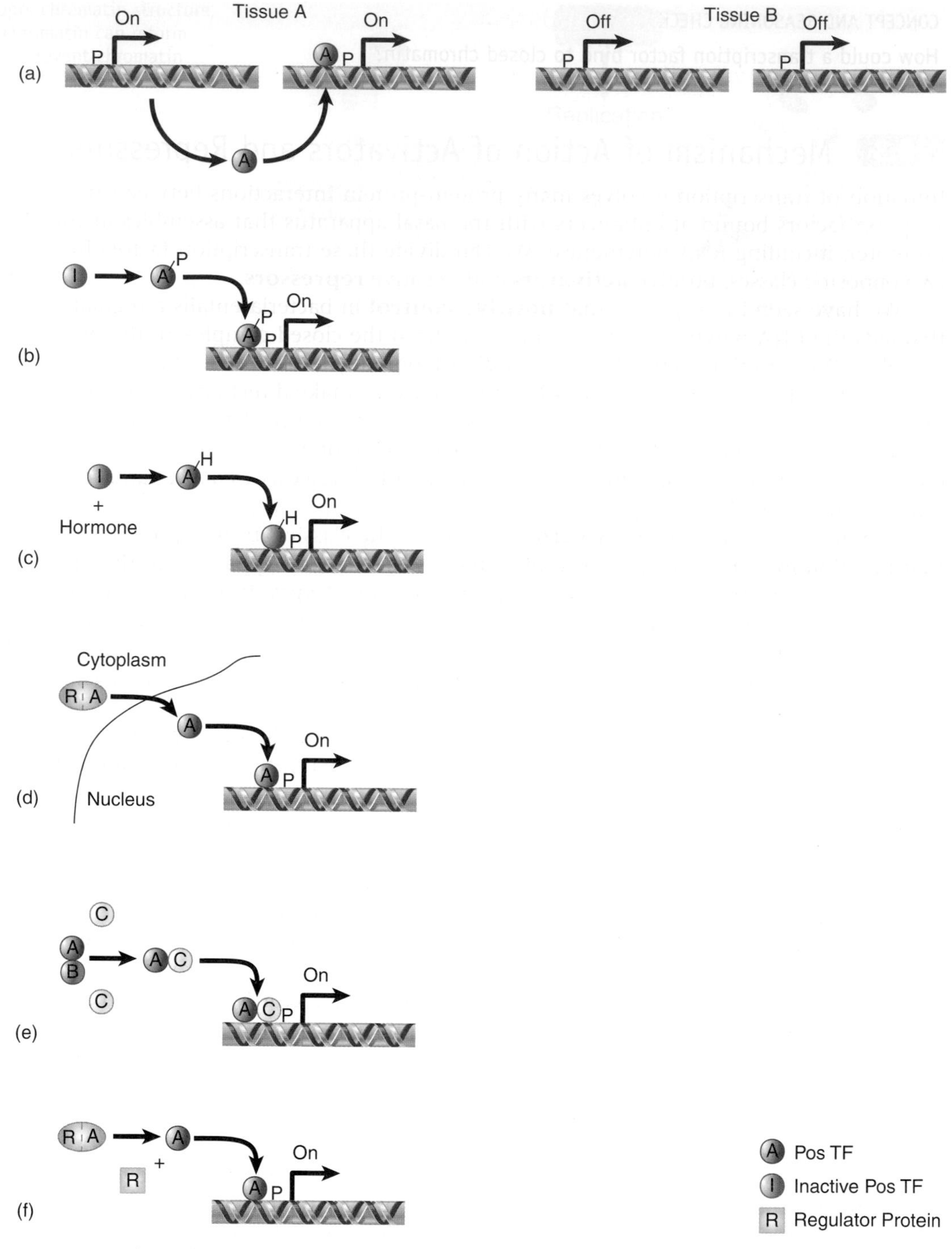

FIGURE 28.3 The activity of a positive regulatory transcription factor may be controlled by (a) the synthesis of new protein, (b) by covalent modification such as phosphorylation, (c) by ligand or hormone binding, (d) by counteracting the binding of inhibitors that sequester the regulator in the cytoplasm, (e) by the ability to select the correct binding partner for activation, and (f) by cleavage from an inactive precursor.

▸ **architectural protein** A protein that, when bound to DNA, can alter its structure, e.g., introduce a bend. They may have no other function.

The third class includes **architectural proteins**, such as Yin-Yang; these proteins function to bend the DNA, either bringing bound proteins together to facilitate forming a cooperative complex or bending the DNA the other way to prevent complex formation, as shown in **FIGURE 28.4**. Note that a strand of DNA may thus be bent in two different directions depending on whether the regulator binds to the top or to the bottom. This is a difference of one half of a turn of the helix, which is 5 bp (10.5 bp per turn).

We have seen several examples of **negative control** in bacteria, in the *lac* operon and in the *trp* operon in *Chapter 26*. Repression can occur in bacteria when the repressor prevents the RNA polymerase from converting from the closed complex to the open complex as in the *lac* operon or binds to the promoter sequence to prevent polymerase from binding as in the *trp* operon. There are many more mechanisms by which repressors act in eukaryotes, which are illustrated in **FIGURE 28.5**:

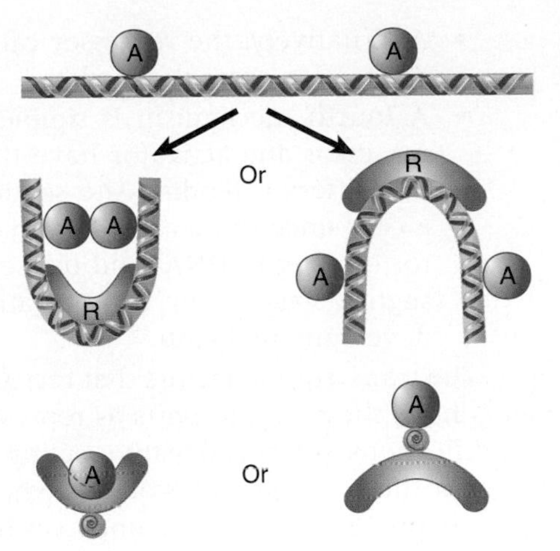

FIGURE 28.4 Architectural proteins control the structure of DNA and thus control whether bound proteins can contact each other. A = activator; R = architectural protein.

▸ **negative control** The default state of genes that are under negative control is to be expressed. A specific intervention is required to turn them off.

- One mechanism of action by which a eukaryotic repressor can prevent gene expression is to *sequester an activator* in the cytoplasm. Eukaryotic proteins are synthesized in the cytoplasm. Proteins that function in the nucleus have a domain that directs their transport through the nuclear membrane. A repressor can bind to that domain and mask it.
- Several variations of that mechanism are possible. One that takes place in the nucleus is that the repressor can bind to an activator that is already bound to an enhancer and *mask its activation domain*, preventing it from functioning (such as the Gal80 repressor; see *Section 28.13, The Yeast* GAL *Genes: A Model for Activation and Repression*).

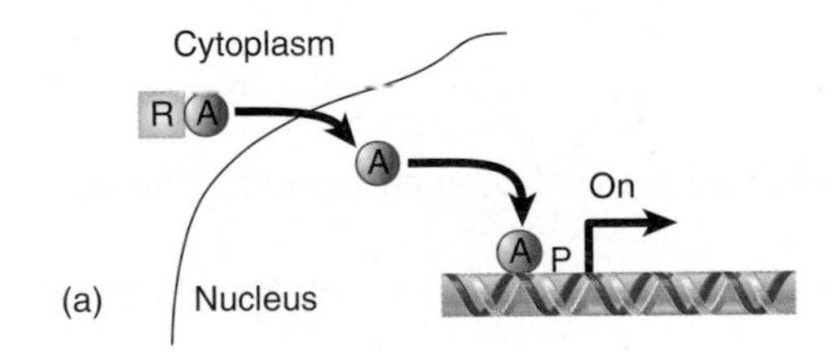

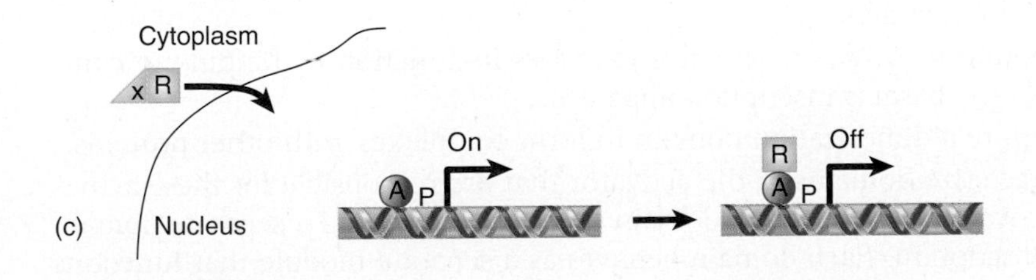

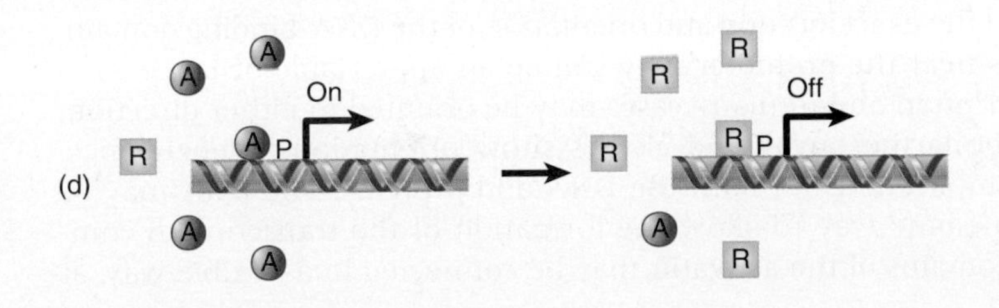

FIGURE 28.5 A repressor may control transcription by sequestering an activator in the cytoplasm, by binding an activator and masking its activation domain, by being held in the cytoplasm until it is needed, or by competition with an activator for a binding site.

- Alternatively, the repressor can be *masked and held in the cytoplasm* until it is released to enter the nucleus.
- A fourth mechanism is simple *competition for an enhancer*, where either the repressor and activator have the same binding site sequence or overlapping but different binding site sequences. This is a very versatile mechanism for a cell, since there are two variables at work here. One is the strength of factor binding to DNA, and the second variable is factor concentration. By only slightly varying the concentration of a factor, a cell can dramatically alter its developmental path.

The transcription factors that recruit the histone modifiers and chromatin remodelers have their counterparts as repressors that recruit the complexes that undo the modifications (or add negative-acting modifications) and remodeling. The same is true for the architectural proteins, where, in fact, the same protein bound to a different site prevents activator complexes from forming.

KEY CONCEPTS

- Activators determine the frequency of transcription.
- Activators work by making protein-protein contacts with the basal factors.
- Activators may work via coactivators.
- Activators are regulated in many different ways.
- Some components of the transcriptional apparatus work by changing chromatin structure.
- Repression is achieved by affecting chromatin structure or by binding to and masking activators.

CONCEPT AND REASONING CHECKS

1. How is positive control in bacteria different from positive control in eukaryotes? Why do you think it is different?
2. How is negative control in bacteria different from negative control in eukaryotes? Why do you think it is different?

28.4 Independent Domains Bind DNA and Activate Transcription

We know the most about the activator class of transcription factors. Activators require protein domains with multiple functions:

- They recognize specific DNA target sequences located in enhancers that affect a particular target gene.
- Having bound to DNA, an activator exercises its function by binding to components of the basal transcription apparatus.
- Many require a dimerization domain to form complexes with other proteins.

Can we characterize domains in the activator that are responsible for these activities? Often an activator has a separate domain that binds DNA and a separate domain that activates transcription. Each domain behaves as a separate module that functions independently when it is linked to a domain of the other type. The geometry of the overall transcription complex must allow the activating domain to contact the basal apparatus irrespective of the exact location and orientation of the DNA-binding domain.

Enhancer elements near the promoter may still be an appreciable distance from the startpoint of transcription and in many cases may be oriented in either direction. Enhancers may even be farther away and always show orientation independence. This organization has implications for both the DNA and proteins. The DNA may be looped or condensed in some way to allow the formation of the transcription complex. In addition, the domains of the activator may be connected in a flexible way, as illustrated diagrammatically in **FIGURE 28.6**. The main point here is that the DNA-binding and activating domains are independent and are connected in a way that allows the activating domain to interact with the basal apparatus irrespective of the orientation and exact location of the DNA-binding domain.

Binding to DNA is necessary for activating transcription, but there are transcription factors that function without a DNA-binding domain by virtue of protein-protein dimerization. Does activation depend on the particular DNA-binding domain? This question has been answered by making hybrid proteins that consist of the DNA-binding domain of one activator linked to the activation domain of another activator. The hybrid functions in transcription at sites dictated by its DNA-binding domain, but in a way determined by its activation domain.

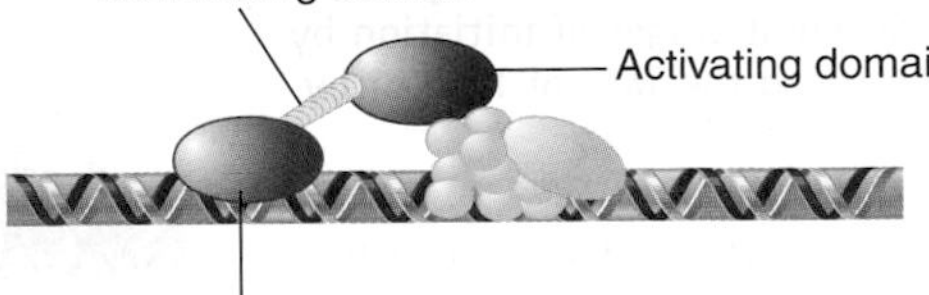

FIGURE 28.6 DNA-binding and activating functions in a transcription factor may comprise independent domains of the protein.

This result fits the modular view of transcription activators. *The function of the DNA-binding domain is to bring the activation domain to the basal apparatus at the promoter.* Precisely how or where it is bound to DNA is irrelevant, but once it is there, the activation domain can play its role. This explains why the exact locations of DNA-binding sites can vary. The ability of the two types of module to function in hybrid proteins suggests that each domain of the protein folds independently into an active structure that is not influenced by the rest of the protein.

KEY CONCEPTS

- DNA-binding and transcription-activation activities are carried by independent domains of an activator.
- The role of the DNA-binding domain is to bring the transcription-activation domain into the vicinity of the promoter.

CONCEPT AND REASONING CHECK

What would happen when one subunit of a heterodimeric activator is in a cell that contains a partner subunit with a mutation in the DNA-binding domain?

28.5 Activators Interact with the Basal Apparatus

The true activator class of transcription factors may work directly when it consists of a DNA-binding domain linked to a transcription-activating domain, as illustrated in Figure 28.4. In other cases, the activator does not itself have a transcription-activating domain (or contains only a weak activation domain but binds another protein—a coactivator—that has the transcription-activating activity. **FIGURE 28.7** shows the action of such an activator. We may regard **coactivators** as transcription factors whose specificity is conferred by the ability to bind to DNA-binding transcription factors instead of directly to DNA. A particular activator may require a specific coactivator.

▸ **coactivator** Factors required for transcription that do not bind DNA but are required for (DNA-binding) activators to interact with the basal transcription factors.

Although the protein components are organized differently, the mechanism is the same. An activator that contacts the basal apparatus directly has an activation domain covalently connected to the DNA-binding domain. When an activator works through a coactivator, the connections involve noncovalent binding between protein subunits (compare Figure 28.5 and Figure 28.6). The same interactions are responsible for activation, irrespective of whether the various domains are present in the same protein subunit or divided into multiple protein subunits. In addition, many coactivators also contain additional enzymatic activities that promote transcription activation, such as activities that modify chromatin structure (see *Section 28.9, Histone Acetylation Is Associated with Transcription Activation*).

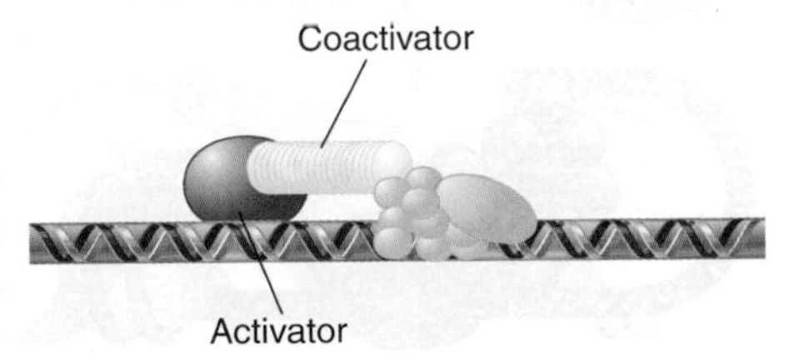

FIGURE 28.7 An activator may bind a coactivator that contacts the basal apparatus.

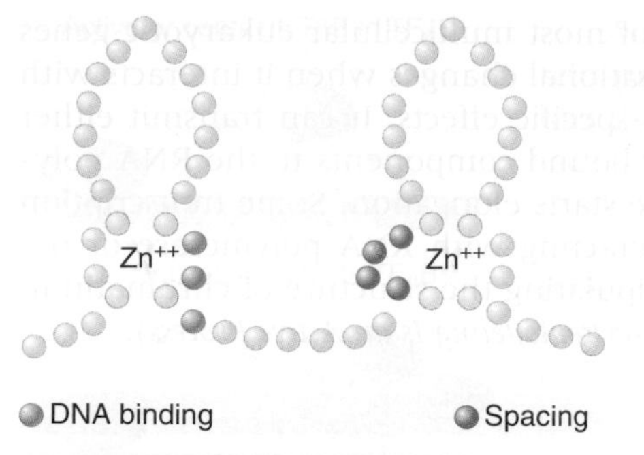

FIGURE 28.11 The first finger of a steroid receptor controls which DNA sequence is bound (positions shown in purple); the second finger controls spacing between the sequences (positions shown in blue).

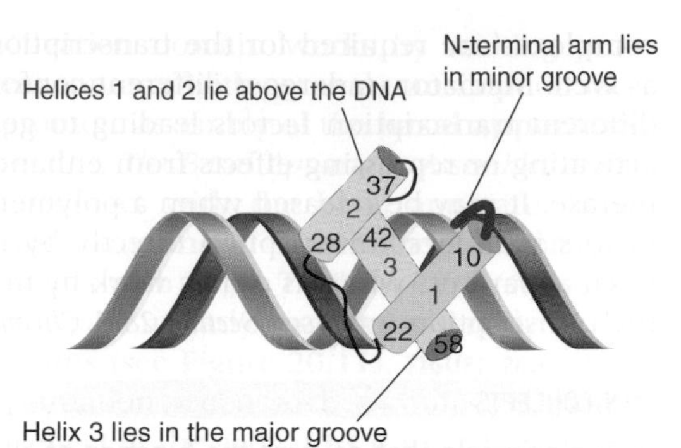

FIGURE 28.12 Helix 3 of the homeodomain binds in the major groove of DNA, with helices 1 and 2 lying outside the double helix. Helix 3 contacts both the phosphate backbone and specific bases. The N-terminal arm lies in the minor groove, and makes additional contact.

- **helix-turn-helix** The motif that describes an arrangement of two α-helices that form a site that binds to DNA, one fitting into the major groove of DNA and the other lying across it.
- **homeodomain** A class of DNA-binding motifs that contain the helix-turn-helix structure that typifies a class of transcription factors often found in developmentally regulated genes.
- **helix-loop-helix (HLH)** The motif that is responsible for dimerization of a class of transcription factors called HLH proteins. A bHLH protein has a basic DNA-binding sequence close to the dimerization motif.
- **leucine zipper** A dimerization motif that is found in a class of transcription factors.
- **bZIP (basic zipper)** A bZIP protein has a basic DNA-binding region adjacent to a leucine zipper dimerization motif.

- The **helix-turn-helix** motif was originally identified as the DNA-binding domain of phage repressors. One helix lies in the major groove of DNA and is the recognition helix; the other lies at an angle across DNA as shown in FIGURE 28.12. A related form of the motif is present in the **homeodomain**, a sequence first characterized in several proteins encoded by genes involved in developmental regulation in *Drosophila,* and by the comparable human *Hox* genes. Homeodomain proteins can be activators or repressors.
- The amphipathic **helix-loop-helix (HLH)** motif has been identified in some developmental regulators and in genes coding for eukaryotic DNA-binding proteins. Each amphipathic helix presents a face of hydrophobic residues on one side and charged residues on the other side. The length of the connecting loop varies from 12 to 28 amino acids. The motif enables proteins to dimerize, either homodimers or hetrodimers, and a basic region near this motif contacts DNA as seen in FIGURE 28.13.
- **Leucine zippers** consist of a stretch of amino acids with a leucine residue in every seventh position. The hydrophobic groups, including leucine, face one side, while the charged groups face the other side. A leucine zipper in one polypeptide interacts with a zipper in another polypeptide to form a dimer. There are rules for which zippers may dimerize. Adjacent to each zipper is a stretch of positively charged residues that is involved in binding to DNA; this is known as the **bZIP (*b*asic *zip*per)** structural motif shown in FIGURE 28.14.

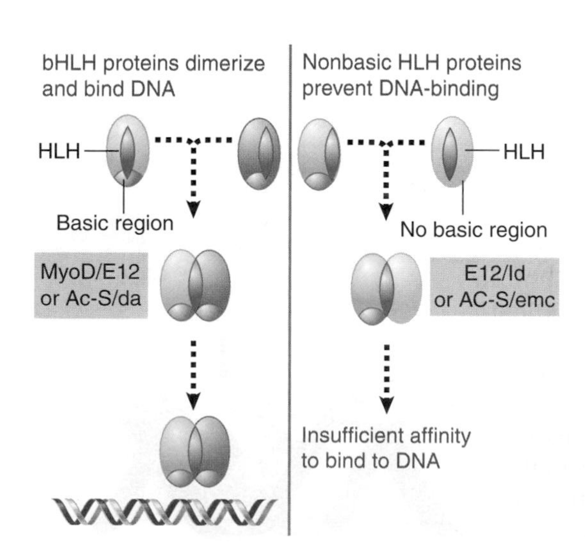

FIGURE 28.13 An HLH dimer in which both subunits are of the bHLH type can bind DNA, but a dimer in which one subunit lacks the basic region cannot bind DNA.

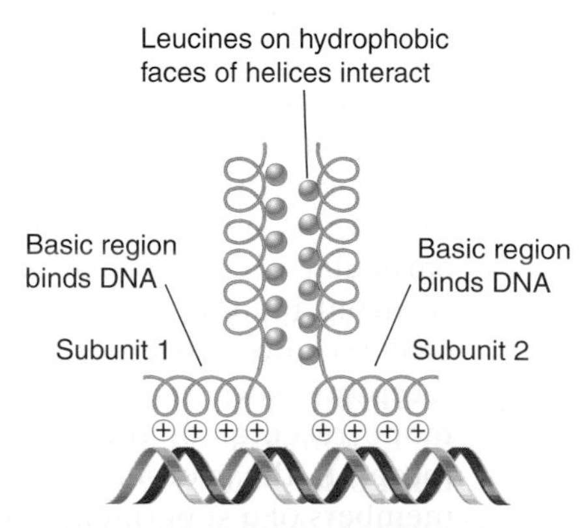

FIGURE 28.14 The basic regions of the bZIP motif are held together by the dimerization at the adjacent zipper region when the hydrophobic faces of two leucine zippers interact in parallel orientation.

KEY CONCEPTS

- Activators are classified according to the type of DNA-binding domain.
- Members of the same group have sequence variations of a specific motif that confer specificity for individual DNA target sites.

CONCEPT AND REASONING CHECK

Why do eukaryotic organisms have so many different sequence motifs for binding DNA?

28.7 Chromatin Remodeling Is an Active Process

Transcriptional activators face a challenge when trying to bind to their recognition sites in eukaryotic chromatin. **FIGURE 28.15** illustrates two general states that can exist at a eukaryotic promoter. In the inactive state, nucleosomes are present, and they prevent basal factors and RNA polymerase from binding. In the active state, the basal apparatus occupies the promoter, and histone octamers cannot bind to it. Each type of state is stable. In order to convert a promoter from the inactive state to the active state, the chromatin structure must be perturbed in order to allow binding of the basal factors.

The general process of inducing changes in chromatin structure is called **chromatin remodeling**. This consists of mechanisms for displacing histones that depend on the input of energy. Many protein-protein and protein-DNA contacts need to be disrupted to release histones from chromatin. There is no free ride: energy must be provided to disrupt these contacts. **FIGURE 28.16** illustrates the principle of a dynamic model by a factor that hydrolyzes ATP. When the histone octamer is released from DNA, other proteins (in this case transcription factors and RNA polymerase) can bind.

▸ **chromatin remodeling** The energy-dependent displacement or reorganization of nucleosomes that occurs in conjunction with activation of genes for transcription.

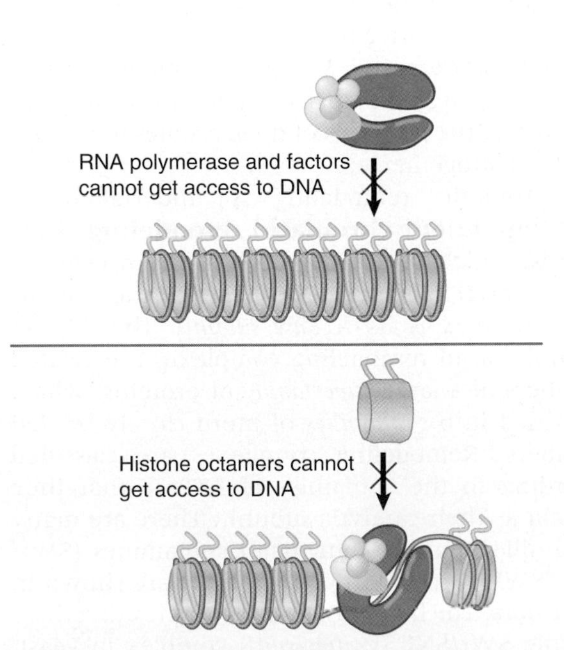

FIGURE 28.15 If nucleosomes form at a promoter, transcription factors (and RNA polymerase) cannot bind. If transcription factors (and RNA polymerase) bind to the promoter to establish a stable complex for initiation, histones are excluded.

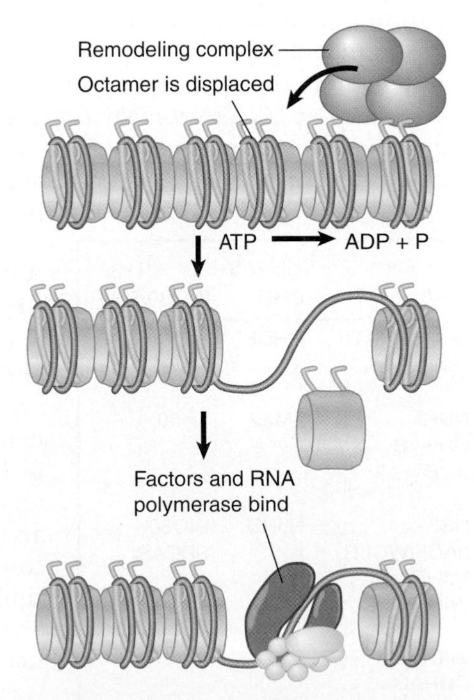

FIGURE 28.16 The dynamic model for transcription of chromatin relies upon factors that can use energy provided by hydrolysis of ATP to displace nucleosomes from specific DNA sequences.

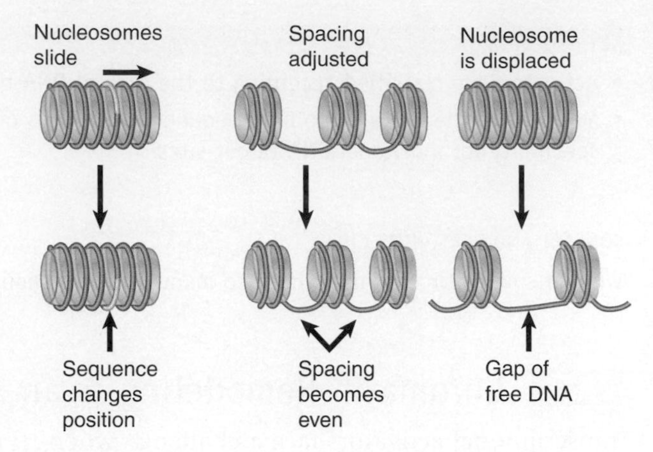

FIGURE 28.17 Remodeling complexes can cause nucleosomes to slide along DNA, can displace nucleosomes from DNA, or can reorganize the spacing between nucleosomes.

There are several alternative outcomes of chromatin remodeling, summarized in **FIGURE 28.17**:

- Histone octamers may *slide* along DNA, changing the relationship between the nucleic acid and the protein. This can alter both the rotational and the translational position of a particular sequence on the nucleosome.
- The *spacing* between histone octamers may be changed, again with the result that the positions of individual sequences are altered relative to protein.
- The most extensive change is that an octamer(s) may be *displaced entirely* from DNA to generate a nucleosome-free gap. Alternatively, one or both H2A-H2B dimers can be displaced.

A major role of chromatin remodeling is to change the organization of nucleosomes at the promoter of a gene that is to be transcribed. This is required to allow the transcription apparatus to gain access to the promoter. Remodeling can also act to prevent transcription by moving nucleosomes onto, rather than away from, essential promoter sequences. Remodeling is also required to enable other manipulations of chromatin, including repair of damaged DNA.

Remodeling often takes the form of displacing one or more histone octamers. This can result in the creation of a site that is hypersensitive to cleavage with DNase I (see *Section 10.10, DNase Sensitivity Detects Changes in Chromatin Structure*). Sometimes there are less dramatic changes in the positioning of a single nucleosome. Thus changes in chromatin structure can extend from altering the positions of nucleosomes to removing them altogether.

▸ **ATP-dependent chromatin remodeling complex** A complex of one or more proteins associated with an ATPase of the SWI2/SNF2 superfamily that uses the energy of ATP hydrolysis to alter or displace nucleosomes.

Chromatin remodeling is undertaken by **ATP-dependent chromatin remodeling complexes**, which use ATP hydrolysis to provide the energy for remodeling. The heart of the remodeling complex is its *ATPase subunit*. The ATPase subunits of all remodeling complexes are related members of a large *superfamily* of proteins, which is divided into *subfamilies* of more closely related members. Remodeling complexes are classified according to the subfamily of ATPase that they contain as their catalytic subunit. There are many subfamilies, but the four major subfamilies (SWI/SNF, ISWI, CHD, and INO80/SWR1) are shown in **FIGURE 28.18**. The first remodeling complex described was the SWI/SNF (*swi*tch/*sniff*) complex in yeast, which has homologs in all eukaryotes. The chromatin remodeling superfamily is large and diverse, and most species have multiple complexes in different subfamilies. Yeast has two SWI/SNF-related

Type of Complex	SWI/SNF	ISWI	CHD	INO80/SWRI
Yeast	SWI/SNF RSC	ISW1a, ISWb ISW2	CHDI	INO80 SWRI
Fly	dSWI/SNF (brahma)	NURF CHRAC ACF	dMI-2	Tip60
Human	hSWI/SNF	RSF hACF/WCFR hCHRAC WICH	NuRD	INO80 SRCAP
Frog		WICH CHRAC ACF	Mi-2	

FIGURE 28.18 Remodeling complexes can be classified by their ATPase subunits. This table is not exhaustive but gives some examples of each class of remodeler.

complexes and three ISWI complexes. Eight different ISWI complexes have been identified thus far in mammals. Remodeling complexes range from small heterodimeric complexes (the ATPase subunit plus a single partner) to massive complexes of 10 or more subunits. Each type of complex may undertake a different range of remodeling activities.

SWI/SNF is the prototypic remodeling complex. Its name reflects the fact that many of its subunits are encoded by genes originally identified by *swi* or *snf* mutations in *Saccharomyces cerevisiae*. (*swi* mutants cannot *swi*tch mating type, and *snf*—*s*ucrose *n*on*f*ermenting—mutants cannot use sucrose as a carbon source.) Mutations in these loci are pleiotropic (that is, they produce a range of seemingly unrelated phenotypes), and the range of defects is similar to those shown by mutants that have lost part of the carboxyl-terminal domain (CTD) of RNA polymerase II. Early hints that these genes might be linked to chromatin came from evidence that these mutations show genetic interactions with mutations in genes that code for components of chromatin: *SIN1*, which codes for a nonhistone chromatin protein, and *SIN2*, which codes for histone H3. The *SWI* and *SNF* genes are required for expression of a variety of individual loci (~120, or 2%, of *S. cerevisiae* genes require SWI/SNF for normal expression). Expression of these loci may require the SWI/SNF complex to remodel chromatin at their promoters.

SWI/SNF acts catalytically *in vitro*, and there are only ~150 complexes per yeast cell. All the genes encoding the SWI/SNF subunits are nonessential, which implies that yeast must also have other ways of remodeling chromatin. The related RSC (*r*emodels the *s*tructure of *c*hromatin) complex is more abundant and is essential. It acts at ~700 target loci.

Different subfamilies of remodeling complexes have distinct modes of remodeling, reflecting differences in their ATPase subunits as well as effects of other proteins in individual remodeling complexes. SWI/SNF complexes can remodel chromatin *in vitro* without overall loss of histones or can displace histone octamers. These reactions likely pass through the same intermediate in which the structure of the target nucleosome is altered, leading either to reformation of a (remodeled) nucleosome on the original DNA or to displacement of the histone octamer to a different DNA molecule. In contrast, the ISWI family primarily affects nucleosome positioning *without* displacing octamers, in a sliding reaction in which the octamer moves along DNA. The activity of ISWI requires the histone H4 tail as well as binding to linker DNA.

There are many contacts between DNA and a histone octamer; 14 are identified in the crystal structure. All these contacts must be broken for an octamer to be released or for it to move to a new position. How is this achieved? The ATPase subunits are distantly related to helicases (enzymes that unwind double-stranded nucleic acids), but remodeling complexes do not have any unwinding activity. Present thinking is that remodeling complexes in the SWI/NSF and ISWI classes use the hydrolysis of ATP to *twist* DNA on the nucleosomal surface. This twisting creates a mechanical force that allows a small region of DNA to be released from the surface and then repositioned. This mechanism creates transient loops of DNA on the surface of the octamer; these loops are themselves accessible to interact with other factors, or they can propagate along the nucleosome, ultimately resulting in nucleosome sliding.

Different remodeling complexes have different roles in the cell. SWI/SNF complexes are generally involved in transcriptional activation, whereas some ISWI complexes act as repressors, using their remodeling activity to slide nucleosomes *onto* promoter regions to prevent transcription. Members of the CHD (*c*hromodomain *h*elicase *D*NA-binding) family have also been implicated in repression, particularly the Mi-2/NuRD complexes, which contain both chromatin remodeling and histone deacetylase activities. Remodelers in the SWR1/INO80 class have a unique activity: in addition to their normal remodeling capabilities, some members of this class also have *histone exchange* capability, in which individual histones (usually H2A/H2B dimers) can be replaced in a nucleosome, typically with a histone variant (see *Section 10.5, Histone Variants Produce Alternative Nucleosomes*).

KEY CONCEPTS

- There are numerous chromatin remodeling complexes that use energy provided by hydrolysis of ATP.
- All remodeling complexes contain a related ATPase catalytic subunit and are grouped into subfamilies containing more closely related ATPase subunits.
- Remodeling complexes can alter, slide, or displace nucleosomes.
- Some remodeling complexes can exchange one histone for another in a nucleosome.

CONCEPT AND REASONING CHECK

How can remodeling complexes act as either activators or repressors?

28.8 Nucleosome Organization or Content May Be Changed at the Promoter

How are remodeling complexes targeted to specific sites on chromatin? They do not themselves contain subunits that bind specific DNA sequences. This suggests the model shown in **FIGURE 28.19**, in which they are recruited by activators or (sometimes) by repressors.

The interaction between transcription factors and remodeling complexes gives a key insight into their *modus operandi*. The transcription factor Swi5 activates the *HO* gene in yeast, a gene involved in mating-type switching. (Note that despite its name, Swi5 is not a member of the SWI/SNF complex.) Swi5 enters nuclei toward the end of mitosis and binds to the *HO* promoter. It then recruits SWI/SNF to the promoter. Swi5 is then released, leaving SWI/SNF at the promoter. This means that a transcription factor can activate a promoter by a "hit-and-run" mechanism, in which its function is fulfilled once the remodeling complex has bound.

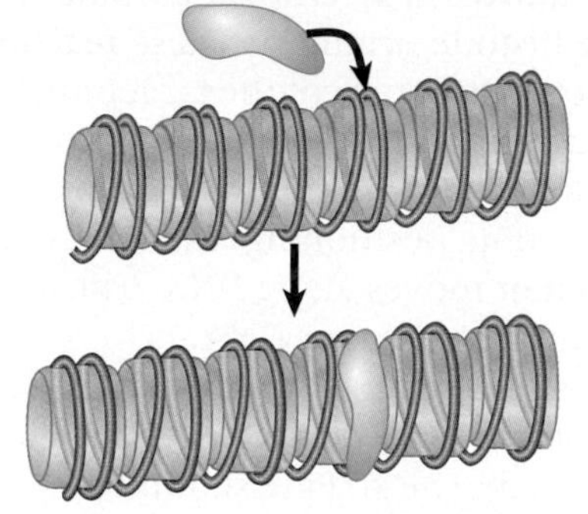

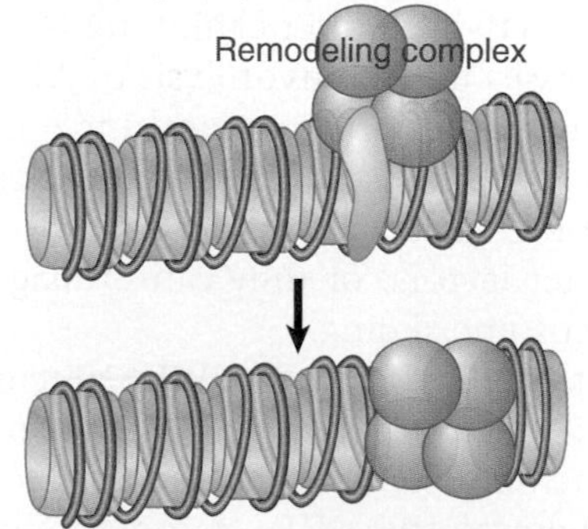

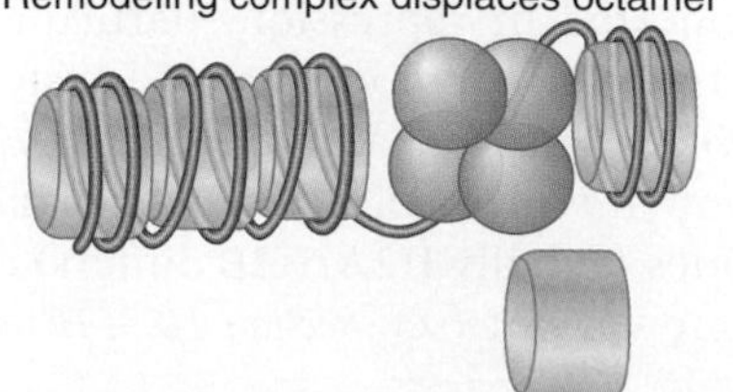

FIGURE 28.19 A remodeling complex binds to chromatin via an activator (or repressor).

The involvement of remodeling complexes in gene activation was discovered because the complexes are necessary to enable certain transcription factors to activate their target genes. One of the first examples was the GAGA factor, which activates the *Drosophila hsp70* promoter. Binding of GAGA to four $(CT)_n$-rich sites near the promoter disrupts the nucleosomes, creates a hypersensitive region, and causes the adjacent nucleosomes to be rearranged so that they occupy preferential instead of random positions. Disruption is an energy-dependent process that requires the NURF remodeling complex, a complex in the ISWI subfamily. The organization of nucleosomes is altered so as to create a boundary that determines the positions of the adjacent nucleosomes. During this process, GAGA binds to its target sites and DNA, and its presence fixes the remodeled state.

The *PHO* system in yeast was one of the first in which it was shown that a change in nucleosome organization is involved in gene activation. At the *PHO5* promoter, the bHLH activator Pho4 responds to phosphate starvation by inducing the disruption of four precisely positioned nucleosomes, as depicted in **FIGURE 28.20**. This event is independent of transcription (it occurs in a TATA⁻ mutant) and independent of replication. There are two binding sites for Pho4 (and another activator, Pho2) at the promoter. One is located between nucleosomes, which can be bound by the isolated DNA-binding domain of Pho4, and the other lies within a nucleosome, which cannot be recognized. Disruption of the nucleosome to allow DNA binding at the second site is necessary for gene activation. This action requires the presence of the transcription-activating domain and appears to involve at least two remodelers: SWI/SNF and INO80. In addition, chromatin disassembly at *PHO5* also requires a histone chaperone, Asf1, which may assist in nucleosome removal or act as a recipient of displaced histones.

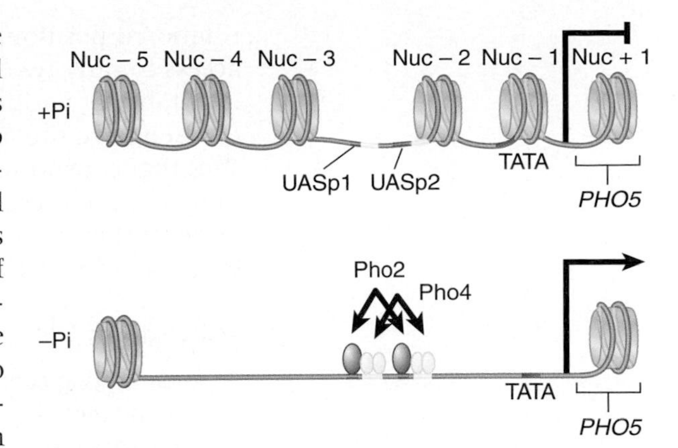

FIGURE 28.20 Nucleosomes are displaced from promoters during activation. The *PHO5* promoter contains nucleosomes positioned over the TATA box and one of the binding sites for the Pho4 and Pho2 activators. When *PHO5* is induced by phosphate starvation (-Pi), promoter nucleosomes are displaced.

A survey of nucleosome positions in a large region of the yeast genome showed that most sites that bind transcription factors are free of nucleosomes. Promoters for RNA polymerase II typically have a nucleosome-free region (NFR) ~200 bp upstream of the startpoint, which is flanked by positioned nucleosomes on either side. These positioned nucleosomes typically contain the histone variant H2AZ (called Htz1 in yeast); the deposition of H2AZ requires the SWR1 remodeling complex. This organization appears to be present in many human promoters as well. It has been suggested that H2AZ-containing nucleosomes are more easily evicted during transcription activation, thus "poising" promoters for activation; however, the actual effects of H2AZ on nucleosome stability *in vivo* are controversial.

It is not always the case, however, that nucleosomes must be excluded in order to permit initiation of transcription. Some activators can bind to DNA on a nucleosomal surface. Nucleosomes appear to be precisely positioned at some steroid hormone response elements in such a way that receptors can bind. Receptor binding may alter the interaction of DNA with histones and may even lead to exposure of new binding sites. The exact positioning of nucleosomes could be required either because the nucleosome "presents" DNA in a particular rotational phase or because there are protein-protein interactions between the activators and histones or other components of chromatin. Thus we have now moved some way from viewing chromatin exclusively as a repressive structure to considering which interactions between activators and chromatin can be required for activation.

The MMTV promoter presents an example of the need for specific nucleosomal organization. It contains an array of six partly palindromic sites, which constitute the HRE (hormone response element). Each site is bound by one dimer of hormone receptor (HR). The MMTV promoter also has a single binding site for the factor NF1, and two adjacent sites for the factor OTF. HR and NF1 cannot bind simultaneously to their sites in free DNA. **FIGURE 28.21** shows how the nucleosomal structure controls binding of the factors.

The HR protects its binding sites at the promoter when hormone is added but does not affect the micrococcal nuclease-sensitive sites that mark either side of the nucleosome. This suggests that HR is binding to the DNA on the nucleosomal surface; however, the

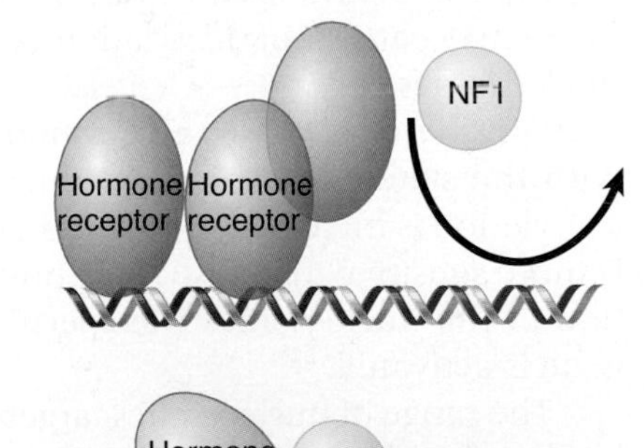

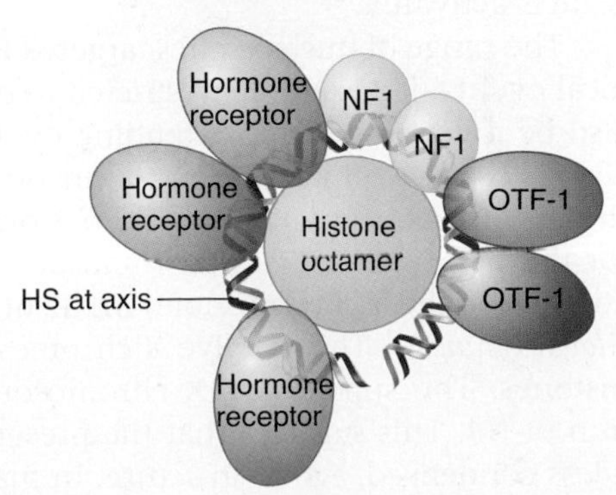

FIGURE 28.21 Hormone receptor and NF1 cannot bind simultaneously to the MMTV promoter in the form of linear DNA but can bind when the DNA is presented on a nucleosomal surface.

rotational positioning of DNA on the nucleosome prior to hormone addition allows access to only two of the four sites. Binding to the other two sites requires a change in rotational positioning on the nucleosome. This can be detected by the appearance of a sensitive site at the axis of dyad symmetry (which is in the center of the binding sites that constitute the HRE). NF1 can be detected on the nucleosome after hormone induction, so these structural changes may be necessary to allow NF1 to bind, perhaps because they expose DNA and abolish the steric hindrance by which HR blocks NF1 binding to free DNA.

KEY CONCEPTS

- A remodeling complex does not itself have specificity for any particular target site but must be recruited by a component of the transcription apparatus.
- Remodeling complexes are recruited to promoters by sequence-specific activators.
- The factor may be released once the remodeling complex has bound.
- Transcription activation often involves nucleosome displacement at the promoter.
- Promoters contain nucleosome-free regions flanked by nucleosomes containing the H2A variant H2AZ (Htz1 in yeast).
- The MMTV promoter requires a change in rotational positioning of a nucleosome to allow an activator to bind to DNA on the nucleosome.

CONCEPT AND REASONING CHECK

Why does it make sense that chromatin remodelers do not recognize promoters directly but rather are recruited by site-specific factors?

28.9 Histone Acetylation Is Associated with Transcription Activation

All the core histones are subject to multiple covalent modifications, as discussed in *Section 10.4, Nucleosomes Are Covalently Modified*. Different modifications result in different functional outcomes. One of the most extensively studied modifications (and the one characterized in the most detail) is lysine acetylation. All core histones are dynamically acetylated on lysine residues in the tails and, occasionally, within the globular core. As described in *Section 10.4*, certain patterns of acetylation are associated with newly synthesized histones that are deposited during S phase. This replication-associated pattern of acetylation is then erased after histones are incorporated into nucleosomes.

Outside of S phase, acetylation of histones in chromatin is generally correlated with the state of gene expression. The correlation was first noticed because histone acetylation is increased in a domain containing active genes, and acetylated chromatin is more sensitive to DNase I. We now know that this occurs largely because of acetylation of the nucleosomes (on specific lysines) in the vicinity of the promoter when a gene is activated.

The range of nucleosomes targeted for modification can vary. Modification can be a local event—for example, restricted to one or a few nucleosomes at a promoter. It can also be a general event, extending over large domains or even entire chromosomes. Global changes in acetylation occur on sex chromosomes. This is part of the mechanism by which the activities of genes on the X chromosome are altered to compensate for the presence of two X chromosomes in one sex but only one X chromosome (in addition to the Y chromosome) in the other sex (see *Section 29.5, X Chromosomes Undergo Global Changes*). The inactive X chromosome in female mammals has underacetylated histones. The superactive X chromosome in *Drosophila* males has increased acetylation of H4. This suggests that the presence of acetyl groups may be a prerequisite for a less condensed, active structure. In male *Drosophila*, the X chromosome is acetylated specifically at K16 of histone H4. The enzyme responsible for this acetylation is called MOF; MOF is recruited to the chromosome as part of a large protein complex. This "dosage compensation" complex is responsible for introducing general changes in the X chromosome that enable it to be more highly expressed. The increased acetylation is only one of its activities.

Acetylation, like other histone modifications, is reversible. Each direction of the reaction is catalyzed by a specific type of enzyme. Enzymes that can acetylate lysine residues in histones are called **histone acetyltransferases,** or **HATs** (these are also more generally known as **lysine (K) acetyltransferases,** or **KATs**). The acetyl groups are removed by **histone deacetylases**, or **HDACs**. There are two groups of HAT enzymes: those in group A act on histones in chromatin and are involved with the control of transcription; those in group B act on newly synthesized histones in the cytosol, and are involved with nucleosome assembly.

- **histone acetyltransferase (HAT)** An enzyme that modifies histones by addition of acetyl groups; some transcriptional coactivators have this activity. Also known as lysine (K) acetyltransferase (KAT).
- **lysine (K) acetyltransferase (KAT)** An enzyme (typically present in large complexes) that acetylates lysine residues in histones (or other proteins). Also known as histone acetyltransferase (HAT).
- **histone deacetylase (HDAC)** Enzyme that removes acetyl groups from histones; may be associated with repressors of transcription.

The breakthrough in analyzing the role of histone acetylation was provided by the characterization of the acetylating and deacetylating enzymes, and their association with other proteins that are involved in specific events of activation and repression. A basic change in our view of histone acetylation was caused by the discovery that previously identified activators of transcription turned out to also have HAT activity.

The connection was established when the catalytic subunit of a group A HAT was identified as a homolog of the yeast regulator protein Gcn5 (see the accompanying box, *Methods and Techniques*, for the history of this discovery). It then was shown that yeast Gcn5 itself has HAT activity, with histones H3 and H2B as its preferred substrates *in vivo*. Gcn5 had previously been identified as part of an adaptor complex required for the function of certain enhancers and their target promoters. It is now known that Gcn5's HAT activity is required for activation of a number of target genes.

Gcn5 was the prototypic HAT that opened the way to the identification of a large family of related acetyltransferase complexes conserved from yeast to mammals. In yeast, Gcn5 is the catalytic KAT subunit of the 1.8 MDa Spt-Ada-Gcn5-acetyltransferase (SAGA) complex, which contains several proteins that are involved in transcription. Among these proteins are several TAF_{II}s. In addition, the Taf1 subunit of $TF_{II}D$ is itself an acetyltransferase. There are some functional overlaps between $TF_{II}D$ and SAGA, most notably that yeast can survive the loss of either Taf1 or Gcn5, but cannot tolerate the deletion of both. This suggests that an acetylase activity is essential for gene expression but can be provided by either $TF_{II}D$ or SAGA. As might be expected from the size of the SAGA complex, acetylation is only one of its functions. The SAGA complex has histone H2B deubiquitylation activity (dynamic H2B ubiquitylation/deubiquitylation is also associated with transcription) and also contains subunits possessing **bromodomains** and **chromodomains**, allowing this complex to interact with acetylated and methylated histones. The bromodomain is found in a variety of proteins that interact with chromatin, including components of some chromatin remodeling complexes. Bromodomains recognize acetylated lysine, and different bromodomain-containing proteins recognize different acetylated targets. The chromodomain is a common protein motif of 60 amino acids present in a number of chromatin-associated proteins, which frequently binds to methyllysine.

- **bromodomain** A 110-amino acid domain that binds to acetylated lysines in histones.
- **chromodomain** ~60 amino acid domains that recognize can methylated lysines in histones; some chromodomains have different functions such as RNA binding.

One of the most important coactivators, p300/CREB-binding protein (CBP), is also a HAT. (Actually, p300 and CBP are different proteins, but they are so closely related that they are often referred to as a single type of activity.) p300/CBP is a coactivator that links an activator to the basal apparatus (see Figure 28.6). p300/CBP interacts with various activators, including hormone receptors, AP-1 (c-Jun and c-Fos), and MyoD. p300/CBP acetylates multiple histone targets, with a preference for the H4 tail. p300/CBP interacts with another coactivator HAT, PCAF, which is related to Gcn5 and preferentially acetylates H3 in nucleosomes. p300/CBP and PCAF form a complex that functions in transcriptional activation. In some cases yet another HAT can be involved, such as the hormone receptor coactivator ACTR, which is itself a HAT that acts on H3 and H4. One explanation for the presence of multiple HAT activities in a coactivating complex is that each HAT has a different specificity, and that multiple

METHODS AND TECHNIQUES

A Tale of Two Nuclei—*Tetrahymena thermophila*

Tetrahymena thermophila is a ciliated protist, a unicellular eukaryote that has been the subject of intense genetic studies in part because of its unique nuclear organization. Typically, the DNA in the nucleus of a eukaryotic cell has two roles: it is the stored genetic material that is transmitted to the next generation, and it is the blueprint for the gene expression of the cell. The most striking genetic feature of *Tetrahymena* is the compartmentalization of DNA into two nuclei with distinct functions: one functions as the "germline nucleus" to be used for gamete production and reproduction, and the second functions as the "somatic nucleus" to be used for transcription and gene expression. The germline nucleus is the smaller of the two and is referred to as the *micronucleus*. This nucleus is diploid and contains five pairs of chromosomes that undergo normal mitosis and meiosis. The somatic nucleus, called the *macronucleus* for its larger size, is polyploid and has an estimated 200 to 300 chromosomes. *Tetrahymena* can divide vegetatively, through mitosis of both the micronucleus and macronucleus, followed by cell division. However, under conditions of starvation, the cell undergoes meiosis and the micronucleus produces haploid gamete pronuclei. A mating pair of *Tetrahymena* cells undergoes sexual reproduction when they conjugate, and there is reciprocal exchange of the gamete pronuclei. When the two gametes fuse, they generate a diploid zygotic nucleus, which in turn undergoes mitosis to give rise to two daughter nuclei, one that becomes the new micronucleus and the second that becomes the new macronucleus. The old macronucleus is destroyed, and the two fused cells separate and continue to grow vegetatively. These life cycles are illustrated in **FIGURE B28.1**.

The newly formed macronucleus is derived from a diploid zygotic nucleus that undergoes extensive genomic reorganization and amplification. The chromosomes are fragmented in a site-specific fashion to generate approximately 200 to 300 chromosomes, ranging in size from 21 to 3000 kilobases. During the process of chromosome breakage, ~10% to 15% of the genome is eliminated in the macronucleus. In addition to this fragmentation, there is an amplification of the chromosomes resulting in approximately 50 copies of each chromosome. Telomeres are added to the newly formed chromosomes at their ends. Interestingly, these chromosomes lack centromeres so that the copies of homologous chromosomes are randomly distributed during mitosis.

The unique genomic organization of the *Tetrahymena* macronucleus and its high level of transcriptional activity have facilitated the discovery of many novel structures and mechanisms of chromosome organization and transcription, which in time have been demonstrated to be universal in all eukaryotes. For instance, telomeres and telomerase, as well as self-splicing RNAs, were first described in *Tetrahymena*. The role of chromatin remodeling, specifically the link between histone acetylation and transcription activation, was first made by the biochemical isolation of histone acetyltransferase (HAT) activity [now referred to as lysine acetyltransferase, or KAT, activity] from *Tetrahymena* macronuclei. In the mid-1990s, David Allis and his colleagues used a novel activity gel assay to isolate the first HAT from *Tetrahymena* macronuclei. Micronuclei were chosen as a likely source of HAT activity, as they are highly transcribed and enriched in histone acetylation. In these experiments, histone proteins were incorporated into a polyacrylamide gel prior to polymerization to create a "histone-containing gel." Then, crude extracts from macronuclei were subjected to electrophoresis through this histone-containing gel, separating the many different proteins. The presence of HAT activity was determined by the ability of a single polypeptide to incorporate ^{3}H [tritium]-acetate into the histones contained in the gel. Following gel electrophoresis, the gel was soaked in ^{3}H-acetate, rinsed, dried, and subjected to fluorography to detect tritiated histones. This revealed the presence of a 55-kilodalton polypeptide (p55) that was capable of specifically acetylating histone proteins. Using this assay, Allis and his colleagues were able to purify a single polypeptide with histone acetyltransferase activity. Using the peptide sequence from this polypeptide, they designed oligonucleotide primers and cloned the *Tetrahymena* p55 gene that encodes HAT activity. The sequence of this gene revealed a predicted amino acid sequence of p55 that is homologous (with 60% similarity) to the yeast Gcn5 protein, a protein known to be important for transcriptional activation in yeast. This was the first direct evidence that histone modification and the associated chromatin remodeling is a mechanism of transcriptional activation.

different acetylation events are required for activation. This enables us to redraw our picture for the action of coactivators, as shown in **FIGURE 28.22**, where RNA polymerase II is bound at a hypersensitive site and coactivators are acetylating histones on the nucleosomes in the vicinity. In fact, most HAT complexes formally function as coactivators, in that they typically do not have DNA binding activity and must be recruited to their sites of action.

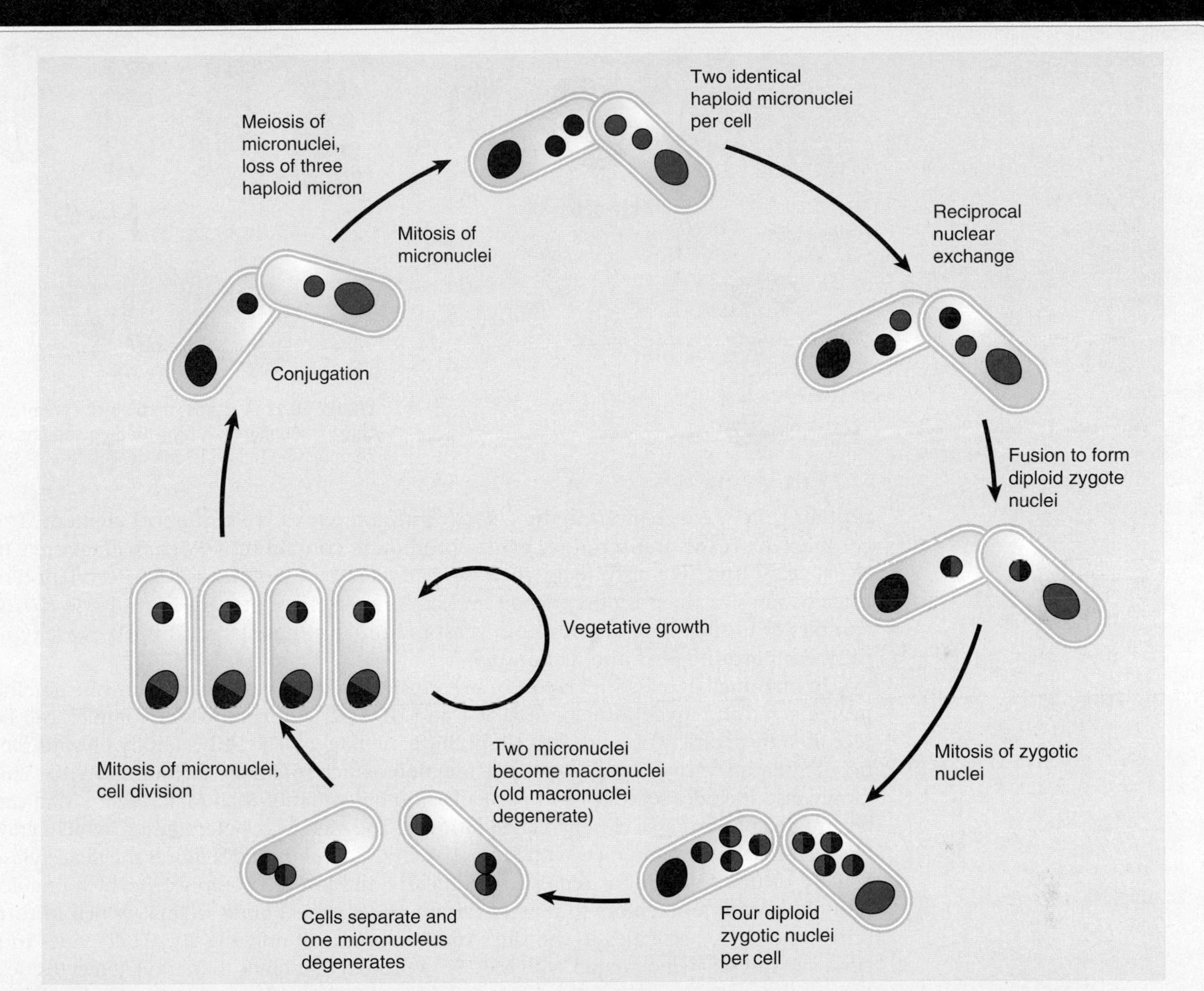

FIGURE B28.1 Life cycle of *Tetrahymena thermophila*. Vegetative cells of *T. thermophila* contain one micronucleus and one macronucleus. Conjugation begins with pairing of complementary mating types. The micronucleus undergoes meiosis to produce four haploid nuclei (three of which degenerate), followed by mitosis of the remaining haploid micronucleus. The conjugating cells then reciprocally exchange single micronuclei, which fuse to form diploid zygote nuclei. The zygote nuclei undergo two mitotic divisions to form four diploid nuclei. Two of these develop into new macronuclei (the old macronucleus degenerates), during which time chromosome fragmentation and sequence elimination occur. Cells then separate and one micronucleus degenerates. The remaining micronucleus divides mitotically and subsequent cell divisions produce four daughter cells, which can reproduce vegetatively or can enter another mating cycle.

As acetylation is linked to activation, deacetylation is linked to transcriptional repression. Where site-specific activators recruit coactivators with HAT activity, site-specific repressor proteins can recruit corepressor complexes, which often contain HDAC activity.

In yeast, mutations in *SIN3* and *RPD3* result in increased expression of a variety of genes, indicating that Sin3 and Rpd3 proteins act as repressors of transcription. Sin3 and Rpd3 are recruited to a number of genes by interacting with the DNA-binding

FIGURE 28.22 Coactivators often have HAT activities that acetylate the tails of nucleosomal histones.

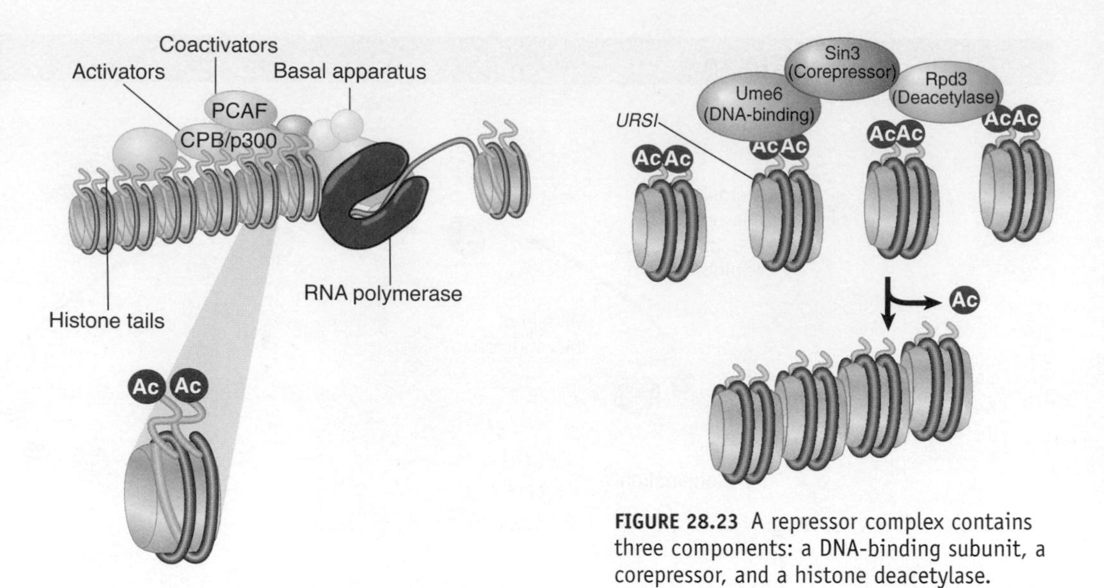

FIGURE 28.23 A repressor complex contains three components: a DNA-binding subunit, a corepressor, and a histone deacetylase.

protein Ume6, which binds to the *URS1* (upstream repressive sequence) element. The complex represses transcription at the promoters containing *URS1*, as illustrated in **FIGURE 28.23**. Rpd3 is a histone deacetylase, and its recruitment leads to deacetylation of nucleosomes at the promoter. Rpd3 and its homologs are present in multiple HDAC complexes found in eukaryotes from yeast to humans; these large complexes are typically built around Sin3 and its homologs.

In mammalian cells, Sin3 is part of a repressive complex that includes histone binding proteins and the Rpd3 homologs HDAC1 and HDAC2. This corepressor complex can be recruited by a variety of repressors to specific gene targets. The bHLH family of transcription regulators includes activators that function as heterodimers, including MyoD. This family also includes repressors, in particular the heterodimer Mad:Max, where Mad can be any one of a group of closely related proteins. The Mad:Max heterodimer (which binds to specific DNA sites) interacts with Sin3/HDAC1/2 complex, and requires the deacetylase activity of this complex for repression. Similarly, the SMRT corepressor (which enables retinoid hormone receptors to repress certain target genes) binds mSin3, which in turn brings the HDAC activities to the site. Another means of bringing HDAC activities to a DNA site can be an interaction with MeCP2, a protein that binds to methylated cytosines, a mark of transcriptional silencing (see *Section 29.6, CpG Islands Are Subject to Methylation*).

Absence of histone acetylation is also a feature of heterochromatin. This is true of both constitutive heterochromatin (typically involving regions of centromeres or telomeres) and facultative heterochromatin (regions that are inactivated in one cell although they may be active in another). Typically the N-terminal tails of histones H3 and H4 are not acetylated in heterochromatic regions (see *Section 29.3, Heterochromatin Depends on Interactions with Histones*).

KEY CONCEPTS

- Newly synthesized histones are acetylated at specific sites and then deacetylated after incorporation into nucleosomes.
- Histone acetylation is associated with activation of gene expression.
- Transcription activators are associated with histone acetylase activities in large complexes.
- Histone acetyltransferases vary in their target specificity.
- The bromodomain is found in a variety of proteins that interact with chromatin; it is used to recognize acetylated sites on histones.
- Deacetylation is associated with repression of gene activity.
- Deacetylases are present in complexes with repressor activity.

CONCEPT AND REASONING CHECK

What would be the effect on transcription of adding an HDAC inhibitor to a cell?

28.10 Methylation of Histones and Methylation of DNA Are Connected

DNA methylation (discussed in *Section 29.6, CpG Islands Are Subject to Methylation*) is generally associated with transcriptional inactivity (though there are rare exceptions), whereas histone methylation can be linked to either active or inactive regions, depending on the specific site of methylation. There are numerous sites of lysine methylation in the tail and core of histone H3 (a few of which occur only in some species), and a single lysine in the tail of H4. In addition, three arginines in H3 and one in H4 are also methylated.

Di- or trimethylation of H3K4 is associated with transcriptional activation, and trimethylated H3K4 occurs around the start sites of active genes. In contrast, H3 methylated at K9 or K27 is a feature of transcriptionally silent regions of chromatin, including heterochromatin and smaller regions that are known to not be expressed. Whole genome studies have begun to uncover general patterns of modifications linked to different transcriptional states, as shown in **FIGURE 28.24**.

Histone lysine methylation is catalyzed by histone methyltransferases (HMTs or KMTs), most of which contain a conserved region called the SET domain. Like acetylation, methylation is reversible, and two different families of demethylases have been identified: the LSD1 (lysine-specific demethylase 1, also known as KDM1) family and the Jumonji family. Different classes of enzymes demethylate arginines.

Methylated lysines (and arginines) are recognized by a number of different domains, which not only can recognize specific modified sites but also can distinguish between mono-, di-, or trimethylated lysines. The chromodomain mentioned in the previous section is a common protein motif in chromatin-associated proteins. Some chromodomain proteins are important for targeting proteins to heterochromatin by recognizing specific "silencing" modifications (such as H3 lysine 9 methylation) associated with heterochromatin. This will be discussed further in *Section 29.3, Heterochromatin Depends on Interactions with Histones* and *Section 29.4, Polycomb and Trithorax Are Antagonistic Repressors and Activators.* A number of other methyl lysine binding domains have been identified, such as the PHD (plant homeodomain), Tudor, and WD40 domains; the number of different motifs designed to recognize particular methylated sites emphasizes the importance and complexity of histone modifications.

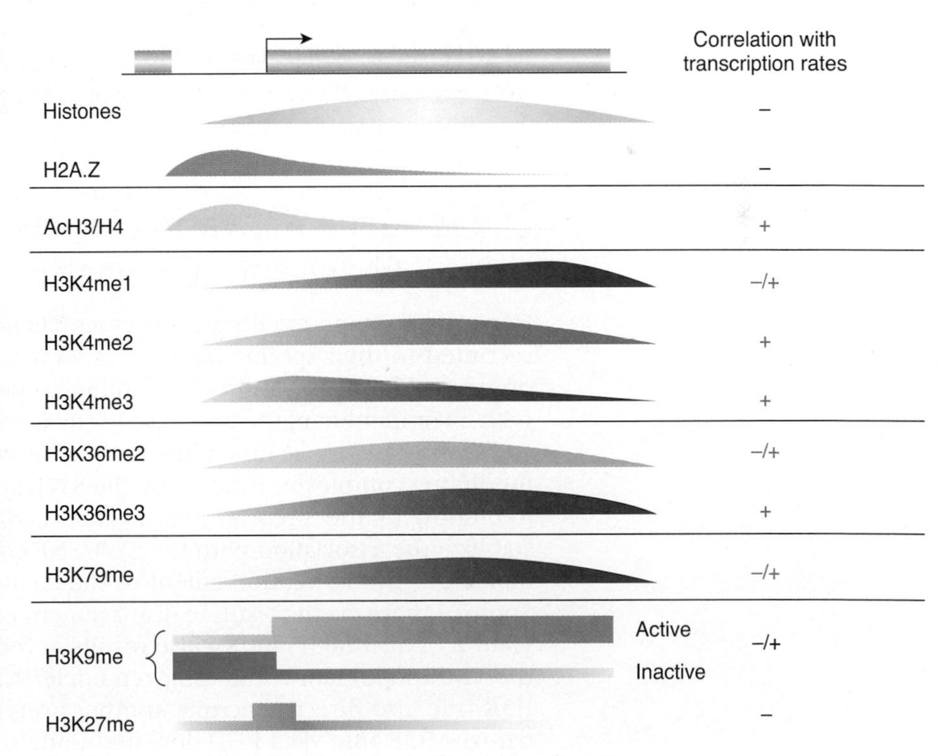

FIGURE 28.24 The distribution of histones and their modifications are mapped on an arbitrary gene relative to its promoter. The curves represent the patterns that are determined via genomewide approaches. The location of the histone variant H2A.Z is also shown. With the exception of the data on K9 and K27 methylation, most of the data are based on yeast genes. Reprinted from B. Li, M. Carey, and J. L. Workman, *Cell* 128, pp. 707–719. Copyright 2007, with permission from Elsevier (http://www.sciencedirect.com/science/journal/00928674).

In silent or heterochromatic regions, the methylation of H3 at K9 is linked to DNA methylation. The enzyme that targets this lysine is a SET-domain containing enzyme called Suv39h1. Deacetylation of H3K9 by HDACs must occur before this lysine can be methylated. H3K9 methylation then recruits a protein called HP1 (heterochromatin protein 1), which binds H3K9me via its chromodomain. HP1 then targets the activity of DNA methyltransferases (DNMTs). Most of the methylation sites in DNA are CpG islands (see *Section 29.6, CpG Islands Are Subject to Methylation*). CpG sequences in heterochromatin are usually methylated. Conversely, it is necessary for the CpG islands located in promoter regions to be unmethylated in order for a gene to be expressed.

Methylation of DNA and methylation of histones is connected in a mutually reinforcing circuit. In addition to the recruitment of DNMTs via HP1 binding to H3K4me, DNA methylation can in turn result in histone methylation. Some histone methyltransferase complexes (as well as some HDAC complexes) contain binding domains that recognize the methylated CpG doublet, so the DNA methylation reinforces the circuit by providing a target for the histone deacetylases and methyltransferases to bind. The important point is that one type of modification can be the trigger for another. These systems are widespread, as can be seen by evidence for these connections in fungi, plants, and animal cells, and for regulating transcription at promoters used by both RNA polymerases I and II, as well as maintaining heterochromatin in an inert state.

KEY CONCEPTS

- Methylation of both DNA and specific sites on histones is a feature of inactive chromatin.
- The SET domain is part of the catalytic site of protein methyltransferases.
- Chromodomains and other conserved domains can bind to specific methylation sites on histones.
- The two types of methylation event are connected.

CONCEPT AND REASONING CHECK

HP1 recognizes H3 methylated on lysine 9, but NOT H3 methylated on lysine 4. Why is this specificity critical?

28.11 Gene Activation Involves Multiple Changes to Chromatin

How are histone-modifying enzymes such as acetyltransferases (or deacetylases) recruited to their specific targets? As we have seen with remodeling complexes, the process is likely to be indirect. A sequence-specific activator (or repressor) may interact with a component of the acetylase (or deacetylase) complex to recruit it to a promoter.

There can also be direct interactions between remodeling complexes and histone-modifying complexes. Binding by the SWI/SNF remodeling complex may lead in turn to binding by the SAGA acetyltransferase complex. Acetylation of histones may then stabilize the association with the SWI/SNF complex (which contains bromodomains), making a mutual reinforcement of the changes in the components at the promoter. Some of these events result in displacement of nucleosomes from the promoter. Methylation of histone H3 on K4 also results in recruitment of numerous factors, including the chromodomain-containing remodeler Chd1, which also associates with SAGA. H3K4me also directly recruits another acetyltransferase complex, NuA3, which recognizes H3K4me via a PHD domain in one of its subunits. These are just a few of the interactions that occur during transcription activation in yeast; similar complex networks of interactions also facilitate transcription in multicellular eukaryotes.

We can connect all of the events at the promoter into the series summarized in **FIGURE 28.25**. The initiating event is binding of a sequence-specific component, which is either able to find its target DNA sequence in the context of chromatin or which binds to a site in a nucleosome-free region. This activator recruits remodeling

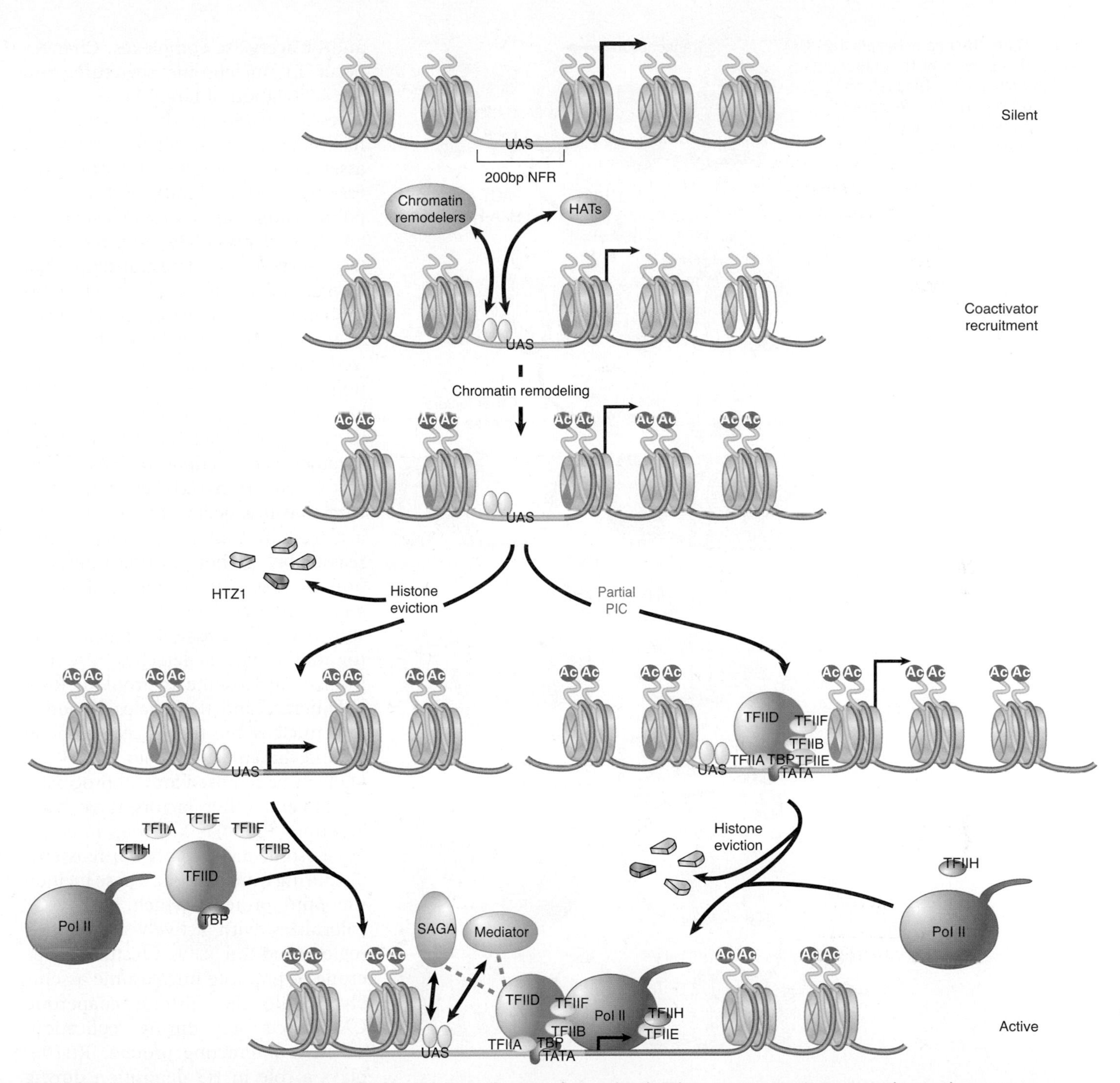

FIGURE 28.25 Htz1-containing nucleosomes flank a 200 bp nucleosome-free region (NFR) at a promoter. Upon targeting to the upstream-activation sequence (UAS), activators recruit various coactivators (such as Swi/Snf or SAGA). This recruitment further increases the binding of activators, particularly for those bound within nucleosomal regions. More importantly, histones are acetylated at promoter-proximal regions, and these nucleosomes become much more mobile. In one model (left), a combination of acetylation and chromatin remodeling directly results in the loss of Htz1-containing nucleosome, thereby exposing the entire core promoter to the GTFs and Pol II. SAGA and mediator then facilitate PIC formation through direct interactions. In the other model (right), which represents the remodeled state, partial PICs could be assembled at the core promoter without loss of Htz1. It is the binding of Pol II and $TF_{II}H$ that leads to the displacement of Htz1-containing nucleosomes and the full assembly of PIC. Reprinted from B. Li, M. Carey, and J. L. Workman, *Cell* 128, pp. 707–719. Copyright 2007, with permission from Elsevier [http://www.sciencedirect.com/science/journal/00928674].

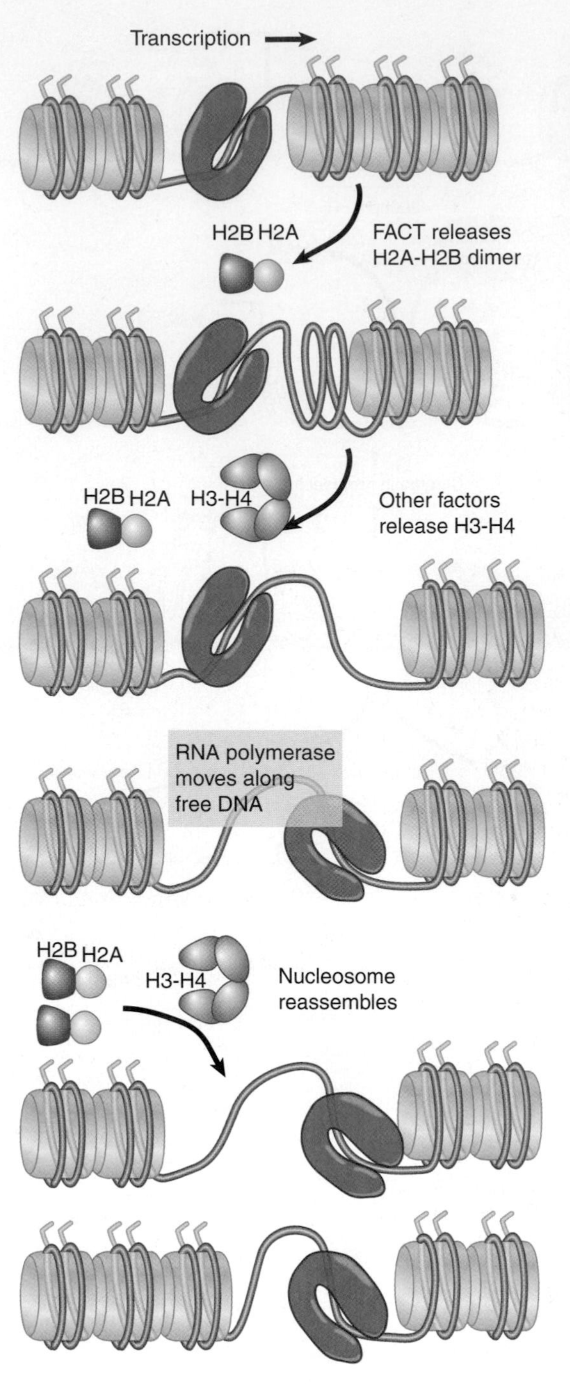

FIGURE 28.26 Histone octamers are disassembled ahead of the transcribing polymerase. They reform following transcription. Release of H2A-H2B dimers probably initiates the disassembly process.

and/or acetylase complexes. Changes occur in nucleosome structure, and the acetylation of target histones provides a covalent mark that the locus has been activated. Initiation complex assembly follows (after any other necessary activators bind), and at some point histones are typically displaced.

A further set of dynamic modifications serve to facilitate transcriptional elongation, and to "reset" the chromatin behind the elongating polymerase. Several factors have been characterized that are critical during transcription elongation. The first of these to be identified is a conserved heterodimeric factor called FACT (*fa*cilitates *c*hromatin *t*ranscription). FACT can act as an H2A-H2B chaperone *in vitro*, and it is believed that it may be involved in H2A-H2B displacement, reassembly, or both during transcription *in vivo*. The activities of FACT and similar factors suggest the model shown in **FIGURE 28.26**, in which FACT (or another factor) detaches H2A-H2B from a nucleosome in front of RNA polymerase and then helps to add it to a nucleosome that is reassembling behind the enzyme. Other factors are likely to be required for this process.

Several other factors have been identified that play key roles in either nucleosome displacement or reassembly during transcription. These include the Spt6 protein, which like FACT colocalizes with actively transcribed regions and can act as a histone chaperone to promote nucleosome assembly. While the histone chaperone CAF-1 acts only during replication, a CAF-1-interacting protein, Rtt106, plays a role in H3 deposition during transcription. The important outcome of these activities is that even at highly transcribed genes, no free DNA is generated but instead is rapidly reformed into chromatin. This is important to prevent "cryptic transcription," which is transcription of noncoding RNA from sites that may resemble promoters but are not normally accessible to the transcription apparatus.

KEY CONCEPTS

- Remodeling complexes can facilitate binding of acetyltransferase complexes, and vice versa.
- Histone methylation can also recruit chromatin modifying complexes.
- Different modifications and complexes facilitate transcription elongation.

CONCEPT AND REASONING CHECK

How does acetylation facilitate chromatin remodeling?

28.12 Histone Phosphorylation Affects Chromatin Structure

All histones can be phosphorylated *in vivo* in different contexts. Histones are phosphorylated in three circumstances:

- cyclically during the cell cycle,
- in association with chromatin remodeling during transcription, and
- during DNA repair.

It has been known for a long time that the linker histone H1 is phosphorylated at mitosis, and more recently it was discovered that H1 is an extremely good substrate for the Cdc2 kinase that controls cell division. This led to speculation that the phosphorylation might be connected with the condensation of chromatin, but so far no direct effect of this phosphorylation event has been demonstrated, and we do not know whether it plays a role in cell division. In *Tetrahymena*, it is possible to delete all the genes for H1 without significantly affecting the overall properties of chromatin (though there are some local affects on gene expression).

Phosphorylation of serine 10 of histone H3 is linked to transcriptional activation (where it promotes acetylation of K14 in the same tail), as well to chromosome condensation and mitotic progression. In *Drosophila melanogaster*, loss of a kinase that phosphorylates histone H3S10 (JIL-1) has devastating effects on chromatin structure. **FIGURE 28.27** compares the usual extended structure of the polytene chromosome (upper photograph) with the structure that is found in a null mutant that has no JIL-1 kinase (lower photograph). The absence of JIL-1 is lethal, but the chromosomes can be visualized in the larvae before they die.

This suggests that H3 phosphorylation is required to generate the more extended chromosome structure of euchromatic regions. Evidence supporting the idea that JIL-1 acts directly on chromatin is that it associates with the complex of proteins that binds to the X chromosome to increase its gene expression in males (see *Section 29.5, X Chromosomes Undergo Global Changes*), and JIL-1-dependent H3 serine 10 phosphorylation antagonizes methylation of H3 on lysine 9, a heterochromatin mark. This is consistent with a role for JIL-1 in promoting an active chromatin conformation. It is not clear how this role of H3 phosphorylation is related to the requirement for H3 phosphorylation to initiate chromosome condensation in at least some species (including mammals and the ciliate *Tetrahymena*).

This leaves us with somewhat conflicting impressions of the roles of histone phosphorylation. Where it is important in the cell cycle, it is likely to be as a signal for condensation. Its effect in transcription and repair appears to be the opposite, where it contributes to open chromatin structures compatible with transcription activation and repair processes. (Histone phosphorylation during repair is discussed in *Section 10.5, Histone Variants Produce Alternative Nucleosomes,* and in *Chapter 16, Repair Systems*.)

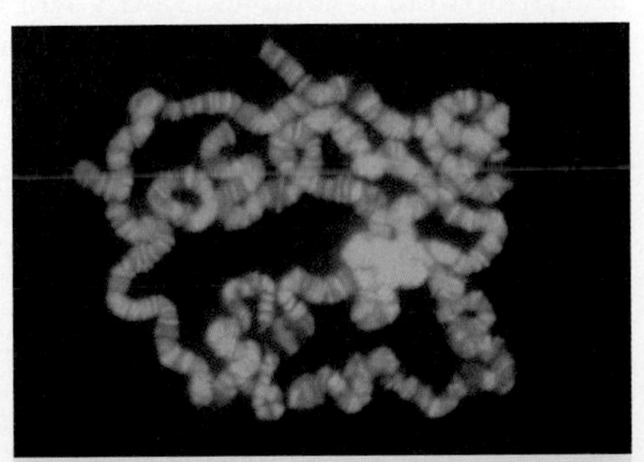

Loss of JIL-1 causes condensation

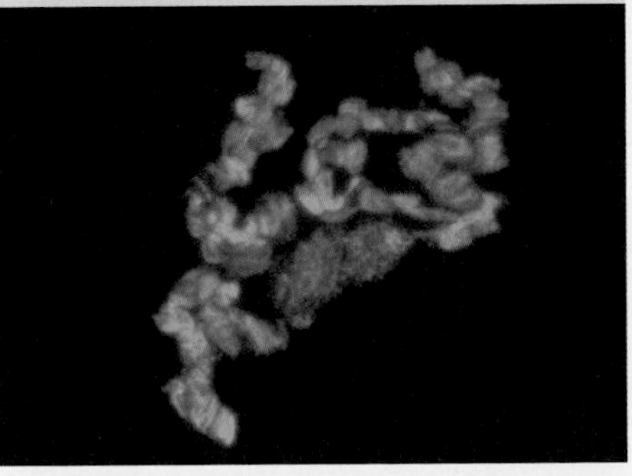

FIGURE 28.27 Flies that have no JIL-1 kinase have abnormal polytene chromosomes that are condensed (bottom panel) instead of extended. Photos courtesy of Jorgen Johansen and Kristen M. Johansen, Iowa State University.

It is possible, of course, that phosphorylation of different histones, or even of different amino acid residues in one histone, has opposite effects on chromatin structure, comparable to the ways in which histone methylation can have either repressive or activating effects on transcription.

KEY CONCEPT

- Histone phosphorylation is linked to transcription, repair, chromosome condensation, and cell cycle progression.

CONCEPT AND REASONING CHECK

What is the likely effect of JIL-1 mutation on transcription in polytene chromosomes?

28.13 The Yeast *GAL* Genes: A Model for Activation and Repression

Yeast, like bacteria, need to be able to rapidly respond to their environment (see *Section 26.3, The* lac *Operon Is Negative Inducible*). In the yeast *Saccharomyces cerevisiae* the *GAL* genes serve a similar function to the *lac* operon in *E. coli*. In an emergency, when there is little or no glucose as an energy source and only galactose (or in *E. coli*, lactose) is available, then the cell will survive because it can catabolize the alternate sugar to generate ATP. The *GAL* system in *S. cerevisiae* has been a model system to investigate gene regulation in eukaryotes for many years. We will focus on two of the genes, *GAL1* and *GAL10*, shown in **FIGURE 28.28**. Like most eukaryotic genes, the *GAL* genes are monocistronic. This pair of genes are divergently transcribed and regulated from a central control region called the **UAS (*u*pstream *a*ctivating *s*equence)**, which is similar to an enhancer. Like the *lac* operon in *E. coli*, the *GAL* genes are induced by their substrate, galactose. For the same reason as in *E. coli*, the *GAL* genes are also under a second level of control: catabolite repression. They cannot be activated by the substrate galactose when there is a sufficient supply of glucose, the preferred energy source.

▸ **UAS (upstream activating sequence)** The equivalent in yeast of the enhancer in higher eukaryotes and is bound by transcriptional regulatory proteins.

The *GAL* genes are under five different levels of control. The first level of control is chromatin structure. Mutations in any of the subunits of SWI/SNF and in the acetyltransferase complex SAGA will result in reduced expression of the *GAL* genes. Second, in the UAS there are both general enhancer and Mig1 repressor binding sites (not shown here). The third level is through a noncoding antisense RNA transcript

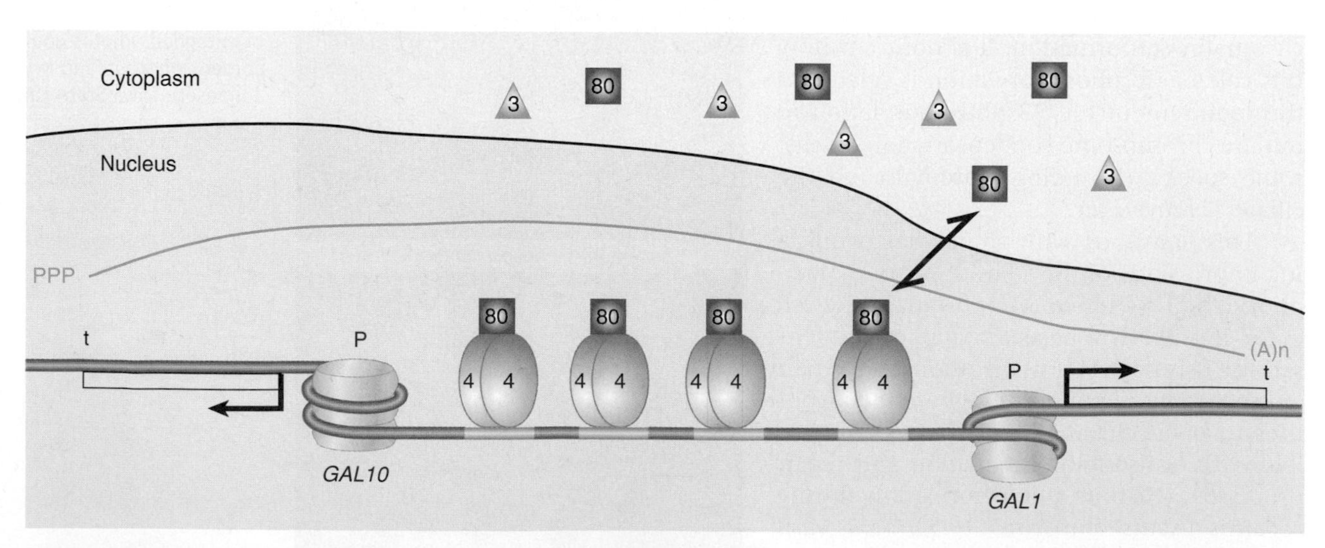

FIGURE 28.28 The yeast *GAL1/GAL10* locus highlighting the UAS and showing the Gal4, Gal80, and Gal3 regulatory proteins. Nucleosomes are positioned at the promoters when the genes are not being transcribed. The noncoding transcript antisense to *GAL10* is also shown.

that assists in maintaining repressed chromatin over the open reading frames. The fourth level is the *GAL* gene-specific, galactose induction mechanism. The fifth level is catabolite (glucose) repression.

GAL1 is an unusual gene in that it lacks the typical nucleosome-free region present at the start sites of most yeast genes. Instead, the start site is contained in a well-positioned nucleosome, whereas the ~170 bp UAS region is held in a nucleosome-free state, which may be partly dependent on the chromatin remodeler SWI/SNF. This DNA region has an unusual base composition, short-phased AT repeats every 10 base pairs, which causes the DNA to bend. Nucleosomes containing the histone variant H2AZ (Htz in yeast) are positioned over the promoters of both *GAL1* and *GAL10*, presumably aided in their positioning by the bent DNA.

GAL10 is also an unusual gene in that it has a cryptic promoter in open chromatin at its 3′ end. This promoter transcribes a noncoding RNA (a cryptic unstable transcript, or CUT, see *Section 30.3, Noncoding RNAs Can Be Used to Regulate Gene Expression*) that is antisense to *GAL10* and extends through and includes *GAL1*. Transcription is very inefficient and the RNA abundance is extremely low (less than one copy per cell) due in part to rapid degradation. Under repressed conditions this promoter is stimulated by the Reb1 transcription factor (usually thought to be an RNA polymerase I transcription factor). The noncoding transcript represses transcription of the *GAL1/10* pair of genes by recruiting a methyltransferase leading to H3K4 and H3K36 methylation, which in turn leads to the recruitment of HDAC to deacetylated the chromatin, leading to repressed chromatin.

The *GAL* genes are ultimately controlled by the positive regulator Gal4, which binds as a dimer to four binding sites in the UAS as shown in Figure 28.28 and **FIGURE 28.29**. Gal4 in turn is regulated by Gal80, a negative regulator that binds to Gal4 and masks its activation domain, preventing it from activating transcription. This is the normal state for the *GAL* genes: turned off and waiting to be induced. Gal80 normally shuttles back and forth between the cytoplasm and the nucleus, reentering the nucleus because of a nuclear localization domain. Gal80 in turn is regulated in the cytoplasm by its negative regulator Gal3, which is itself controlled by the inducer galactose.

Gal3, when it is activated by a phosphorylated derivative of galactose in conjunction with NADP, has the ability to bind to Gal80. When it does, Gal3 masks the nuclear localization signal of Gal80, preventing it from shuttling back into the nucleus. Gal3 is thus a negative regulator of a negative regulator, which makes it a positive regulator of Gal4. This depletes the nuclear level of Gal80, unmasking Gal4 and allowing activation of the genes. NADP is thought to be a "second messenger" metabolic sensor.

Unmasked Gal4 is now able to begin the process of turning on the *GAL1/10* genes through direct contact with a number of proteins at the promoter. During induction, Reb1 no longer binds to the cryptic promoter in *GAL10*, shutting of the cryptic antisense transcript. Gal4 recruits an H2B histone ubiquitylation factor (Rad6), which then stimulates histone methylation of histone H3. Next, the SAGA acetyltransferase complex both deubiquitylates H2B and acetylates histone H3, ultimately resulting in the eviction of the poised nucleosomes from the two promoters. The removal is facilitated by the remodeler SWI/SNF and histone chaperones Hsp90/70. This allows the

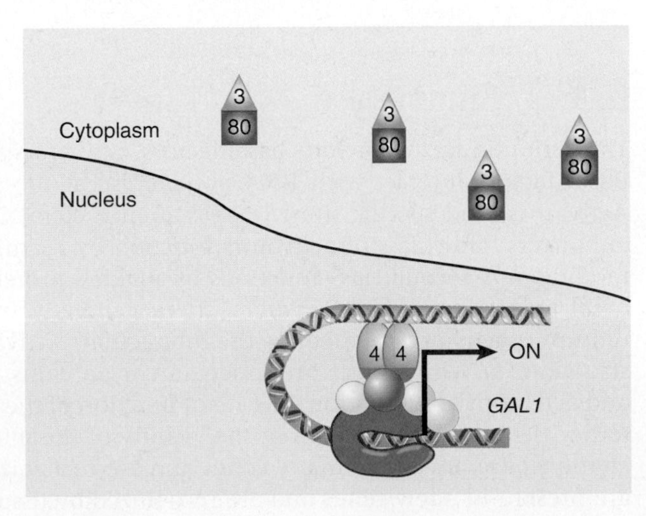

FIGURE 28.29 The yeast *GAL1* gene as it is being activated. Gal3 is holding Gal80 in the cytoplasm, allowing Gal4 to recruit the transcription machinery and activate transcription.

recruitment of TBP/$TF_{II}D$, which then recruits RNA polymerase II and the coactivator complex Mediator. The elongation control factor $TF_{II}S$ is also recruited.

During the elongation phase of transcription, nucleosomes over the open reading frames are disrupted (see *Section 20.8, Initiation Is Followed by Promoter Clearance and Elongation*). In order to prevent spurious transcription from cryptic promoters on either strand, histone octamers must reform as RNA polymerase II passes. A number of histone chaperones and the FACT complex play a role in the dynamics of octamer disassembly and assembly during elongation.

Although catabolite repression in eukaryotes is used for the same purpose as in *E. coli* (which uses cAMP as a positive coregulator), it has a completely different mechanism. Glucose is a preferred sugar source compared to galactose. If the cell has both sugars, it will preferentially use the best source, glucose, and repress the genes for galactose utilization. Glucose repression of the yeast *GAL* genes is multifaceted. The glucose dependent switch is the protein kinase Snf1.

Glucose repression inactivates Snf1, which allows Mig1 to be active and interact with the Cyc8-Tup1 corepressor, known to recruit histone deacetylases to reestablish the pattern of repressed nucleosomes. Transcription of several other genes, including *GAL3* and the galactose permease transporter, are also similarly repressed. Overcoming repression requires only low glucose levels. This activates SNF1, which inactivates Mig1 by phosphorylation. Mig1 phosphorylation disrupts its interaction with the Cyc8-Tup1 corepressor and also leads to its export from the nucleus, relieving glucose repression.

KEY CONCEPTS

- *GAL1/10* genes are positively regulated by the activator Gal4.
- *GAL1/10* genes are negatively regulated by a noncoding RNA synthesized from a cryptic promoter that controls chromatin structure.
- Gal4 is negatively regulated by Gal80, which shuttles between the nucleus and the cytoplasm.
- Gal80 is negatively regulated in the cytoplasm by Gal3, which is activated by the inducer, galactose.
- Activated Gal4 recruits the machinery necessary to alter the chromatin and recruit RNA polymerase.
- Catabolite repression is mediated by a glucose dependent protein kinase, Snf1.

CONCEPT AND REASONING CHECK

How can Gal3, a negative regulator, positively regulate *GAL1* and *GAL10*?

28.14 Summary

Transcription factors include basal factors, activators and repressors, and coactivators. Basal factors interact with RNA polymerase at the startpoint within the promoter. Activators bind specific short DNA sequence elements located near promoters or in enhancers. One class of activators function by recruiting chromatin remodelers and modifiers. A second class functions by making protein-protein interactions with the basal apparatus. Some activators interact directly with the basal apparatus; others require coactivators to mediate the interaction. Activators often have a modular construction, in which there are independent domains responsible for binding to DNA and activating transcription. The main function of the DNA-binding domain may be to tether the activating domain in the vicinity of the initiation complex. Some response elements are present in many genes and are recognized by ubiquitous factors; others are present in a few genes and are recognized by tissue-specific factors.

Near the promoters for RNA polymerase II are a variety of short *cis*-acting elements, each of which is recognized by a *trans*-acting factor. The *cis*-acting elements can be located upstream of the TATA box and may be present in either orientation and at a variety of distances with regard to the startpoint, downstream within an intron, or at the end of the gene. These elements are recognized by activators or repressors that interact with the basal transcription complex to determine the efficiency with which the promoter is used. Some activators interact directly with components of the basal apparatus; others interact via intermediaries called coactivators. The targets in the basal apparatus are the TAFs of $TF_{II}D$, or $TF_{II}B$ or $TF_{II}A$. The interaction stimulates assembly of the basal apparatus.

Several groups of transcription factors have been identified by sequence homology. The homeodomain is a 60–amino acid sequence that regulates development in insects, worm, and humans. It is related to the prokaryotic helix-turn-helix motif and is the DNA-binding motif for these transcription factors.

Another motif involved in DNA binding is the zinc finger, which is found in proteins that bind DNA or RNA (or sometimes both). A zinc finger has cysteine and histidine residues that bind zinc. One type of finger is found in multiple repeats in some transcription factors; another is found in single or double repeats in others.

The leucine zipper contains a stretch of amino acids rich in leucine that are involved in dimerization of transcription factors. An adjacent basic region is responsible for binding to DNA in the bZIP transcription factors.

Steroid receptors were the first members identified of a group of transcription factors in which the protein is activated by binding a small hydrophobic hormone. The activated factor becomes localized in the nucleus and binds to its specific response element, where it activates transcription. The DNA-binding domain has zinc fingers.

HLH (helix-loop-helix) proteins have amphipathic helices that are responsible for dimerization, which are adjacent to basic regions that bind to DNA. bHLH proteins have a basic region that binds to DNA and fall into two groups: ubiquitously expressed and tissue specific. An active protein is usually a heterodimer between two subunits, one from each group. When a dimer has one subunit that does not have the basic region, it fails to bind DNA, so such subunits can prevent gene expression. Combinatorial associations of subunits form regulatory networks.

Many transcription factors function as dimers, and it is common for there to be multiple members of a family that form homodimers and heterodimers. This creates the potential for complex combinations to govern gene expression. In some cases, a family includes inhibitory members, whose participation in dimer formation prevents the partner from activating transcription.

Genes whose control regions are organized in nucleosomes usually are not expressed. In the absence of specific regulatory proteins, promoters and other regulatory regions are organized by histone octamers into a state in which they cannot be activated. This may explain the need for nucleosomes to be precisely positioned in the vicinity of a promoter, so that essential regulatory sites are appropriately exposed. Some transcription factors have the capacity to recognize DNA on the nucleosomal surface, and a particular positioning of DNA may be required for initiation of transcription.

Chromatin remodeling complexes have the ability to slide or displace histone octamers by a mechanism that involves hydrolysis of ATP. Remodeling complexes range from small to extremely large and are classified according to the type of the ATPase subunit. Common types are SWI/SNF, ISWI, CHD, and SWR1/INO80. A typical form of this chromatin remodeling is to displace one or more histone octamers from specific sequences of DNA, creating a boundary that results in the precise or preferential positioning of adjacent nucleosomes. Chromatin remodeling may also involve changes in the positions of nucleosomes, sometimes involving sliding of histone octamers along DNA.

Extensive covalent modifications occur on histone tails, all of which are reversible. Acetylation of histones occurs at both replication and transcription and facilitates formation of a less compact chromatin structure. Some coactivators, which connect transcription factors to the basal apparatus, have histone acetylase activity. Conversely, repressors may be associated with deacetylases. The modifying enzymes are usually specific for particular amino acids in particular histones. Some histone modifications may be exclusive or synergistic with others. Conserved domains (such as bromodomains and chromodomains) allow proteins to bind to specifically modified histones.

Large activating (or repressing) complexes often contain several activities that undertake different modifications of chromatin. Some common motifs found in proteins that modify chromatin are the chromodomain (which binds methylated lysine), the bromodomain (which targets acetylated lysine), and the SET domain (which is part of the active sites of histone methyltransferases).

CHAPTER QUESTIONS

1. Eukaryotic gene expression is controlled principally at:
A. initiation of transcription.
B. elongation of transcription.
C. initiation of translation.
D. RNA processing.

2. True activators are transcription factors that bind to:
A. other proteins to enhance transcription.
B. promoters.
C. enhancers.
D. promoters and enhancers.

3. Which of the following has a DNA-binding domain important in genes involved in developmental regulation in *Drosophila?*
A. homeodomains
B. steroid receptors
C. leucine zippers
D. zinc fingers

4. Leucine zippers have a stretch of amino acids with a leucine residue in every:
A. 5th position.
B. 7th position.
C. 10th position.
D. 12th position.

5. The leucine zipper is a(n):
A. positively charged helix that binds as a monomer in the major groove.
B. negatively charged helix that binds as a monomer in the major groove.
C. amphipathic helix that dimerizes, with a basic region that binds DNA.
D. amphipathic helix that dimerizes, with an acidic region that binds DNA.

6. Changes in chromatin independent of DNA sequence that persist through multiple cell divisions are called:
A. epigenetic effects.
B. remodeling effects.
C. translational effects.
D. prions.

7. Promoters of protein-coding genes in eukaryotic chromatin have a nucleosome-free region of about:
A. 75 bp upstream of the startpoint.
B. 75 bp downstream of the startpoint.
C. 200 bp upstream of the startpoint.
D. 200 bp with the startpoint about in the middle.

8. Methylation of histones occurs most often on which of the following pairs of amino acids?
  A. lysine and asparagine
  B. asparagine and glutamine
  C. lysine and arginine
  D. lysine and glutamine

9. Bromodomains bind:
  A. methylated lysine.
  B. methylated arginine.
  C. acetylated lysine.
  D. phosphorylated serine.

10. Which histone variant is associated with nucleosomes that flank nucleosome-free regions in promoters?
  A. H2AX
  B. H2AZ
  C. macroH2A
  D. H3.3

KEY TERMS

activator
antirepressor
architectural protein
ATP-dependent chromatin remodeling complex
basal apparatus
boundary element
bromodomain
bZIP (basic zipper)
chromatin remodeling
chromodomain
coactivator
epigenetic
helix-loop-helix (HLH)
helix-turn-helix
histone acetyltransferase (HAT)
histone deacetylase (HDAC)
homeodomain
insulator
leucine zipper
lysine (K) acetyltransferase (KAT)
mediator
negative control
positive control
repressor
steroid receptor
true activator
UAS (upstream activating sequence)
zinc finger

FURTHER READING

Cairns, B. (2005). Chromatin remodeling complexes: strength in diversity, precision through specialization. *Curr. Opin. Gen. Devel.* **15**, 185–190. A review of the diverse family of chromatin remodelers and their roles in transcription and other functions.

Houseley, J., Rubbi, L., Grunstein, M., Tollervey, D., and Vogelauer, M. (2008). A ncRNA modulates histone modification and mRNA induction in the yeast *GAL* gene cluster. *Mol. Cell*, **32**, 685–695.

Lee, K. K., and Workman, J. L. (2007). Histone acetyltransferase complexes: one size doesn't fit all. *Nature Revs. Molec. Cell. Biol.* **8**, 284–295. A review of a variety of HAT (KAT) complexes and their roles in multiple DNA transactions, including transcription, chromosome decondensation, and others.

Li, B., Carey, M., and Workman, J. L. (2007). The role of chromatin during transcription. *Cell* **128**, 707–719. A review of many aspects of chromatin modification and remodeling and their relationship to transcriptional activation.

Peng, G., and Hopper, J. E. (2002). Gene activation by interaction of an inhibitor with a cytoplasmic signaling protein. *Proc. Natl. Acad. Sci. USA*. **99**, 8548–8553. A report describing the *GAL* gene system regulation.

Ruthenburg, A. J., Li, H., Patel, D. J., and Allis, C. D. (2007). Multivalent engagement of chromatin modifications by linked binding modules. *Nature Revs. Molec. Cell. Biol.* **8**, 983–994. A review of histone modifications and their recognition, with an emphasis on models for recognition of combinations of histone modifications.

Schnitzler, G. R. (2008). Control of nucleosome positions by DNA sequence and remodeling machines. *Cell Biochem. Biophys.* **51**, 67–80. A review of the interplay between intrinsic DNA sequence and chromatin remodeling in the establishment of activating vs. repressing nucleosome positions.

Science. (2008). **319**, 1781–1799. A collection of reviews on eukaryotic gene regulation.

29

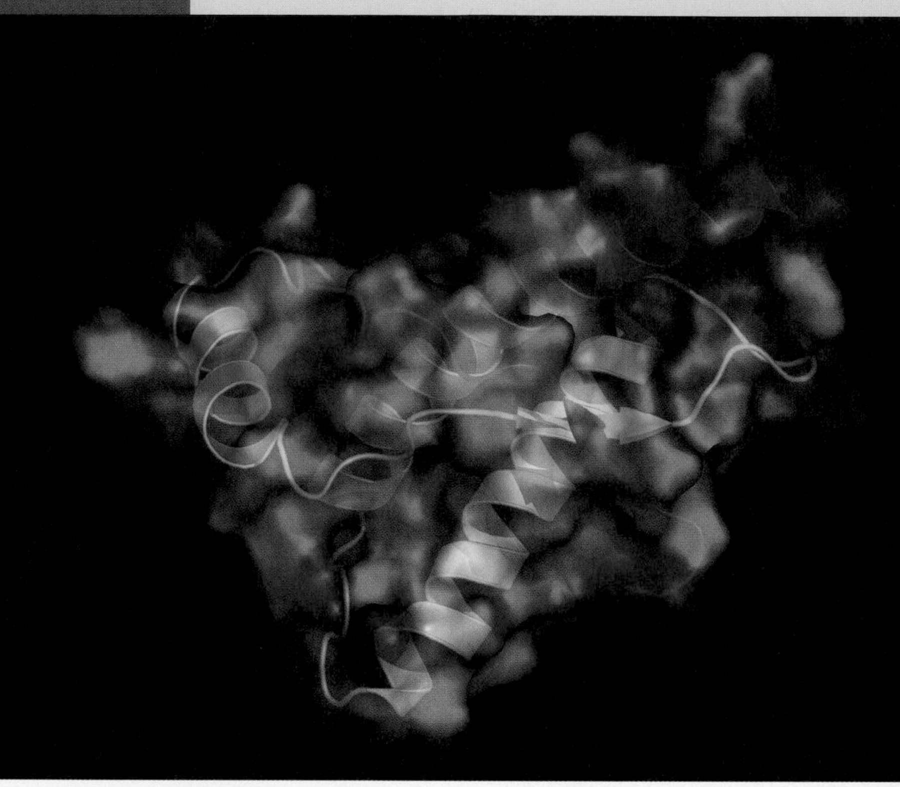

A prion protein in cellular (noninfectious) form. The inner ribbon diagram shows the secondary structure motifs of the protein backbone. The transparent overlay depicts the actual protein surface, with the charged surfaces color-coded in blue (negative) and red (positive). Photo courtesy of Martin Stumpe, Max-Planck-Institute for Biophysical Chemistry, Germany (http://www.martinstumpe.com).

Epigenetic Effects Are Inherited

CHAPTER OUTLINE

29.1 Introduction

Epigenetic inheritance describes the inheritance of different functional states, which may have different phenotypic consequences, without any change in the *sequence* of DNA. This means that two individuals with the *same* DNA sequence at a locus that can exhibit different epigenetic patterns may show *different* phenotypes. The basic cause of this phenomenon is the existence of a self-perpetuating structure in one of the individuals that does not depend on DNA sequence. Several different types of structure have the ability to sustain epigenetic effects:

- A covalent modification of DNA (such as methylation of a base).
- A proteinaceous structure that assembles on DNA.
- A protein aggregate that controls the conformation of new subunits as they are synthesized.

▸ **epigenetic** Changes that influence the phenotype without altering the genotype. They consist of changes in the properties of a cell that are inherited, but that do not represent a change in genetic information.

In each case the epigenetic state results from a difference in function (typically inactivation) that is determined by the structure.

In the case of DNA methylation, a DNA sequence methylated in its control region may fail to be transcribed, whereas the unmethylated sequence will be expressed. **FIGURE 29.1** shows how this situation is inherited. One allele has a sequence that is methylated on both strands of DNA, whereas the other allele has an unmethylated sequence. Replication of the methylated allele creates hemimethylated daughters that are restored to the methylated state by a constitutively active DNA **methyltransferase** enzyme, sometimes called a "maintenance methylase." Replication does not affect the state of the unmethylated allele. If the state of methylation affects transcription, the two alleles differ in their state of gene expression, even though their sequences are identical. As we will see, DNA methylation, histone modifications, and heterochromatin assembly can all mutually reinforce an epigenetically silenced state, contributing to stability of the silenced state over multiple cell divisions.

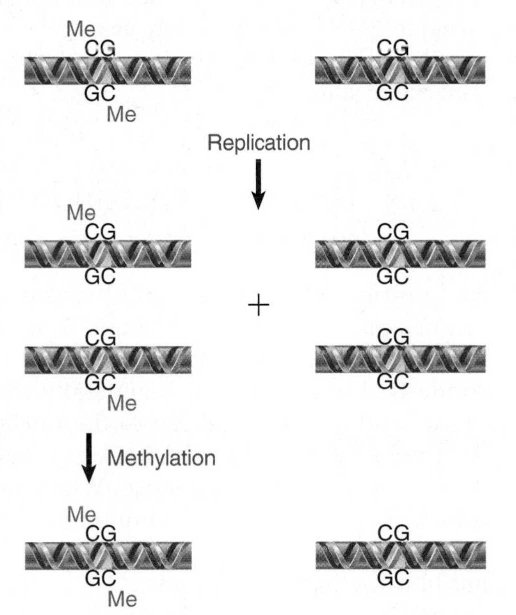

FIGURE 29.1 Replication of a methylated site produces hemimethylated DNA, in which only the parental strand is methylated. A perpetuation methylase recognizes hemimethylated sites and adds a methyl group to the base on the daughter strand. This restores the original situation, in which the site is methylated on both strands. An unmethylated site remains unmethylated after replication.

▸ **methyltransferase** An enzyme that adds a methyl group to a substrate, which can be a small molecule, a protein, or a nucleic acid.

Self-perpetuating structures that assemble on DNA usually have a repressive effect by forming heterochromatic regions that prevent the expression of genes within them. Their perpetuation depends on the ability of proteins in a heterochromatic region to remain bound to those regions after replication and then to recruit more protein subunits to sustain the complex. If individual subunits are distributed at random to each daughter duplex at replication, the two daughters will continue to be marked by the protein, although its density will be reduced to half of the level before replication. **FIGURE 29.2** shows that the existence of epigenetic effects forces us to take the view that a protein responsible for such a situation must have some sort of self-templating or self-assembling capacity to restore the original complex.

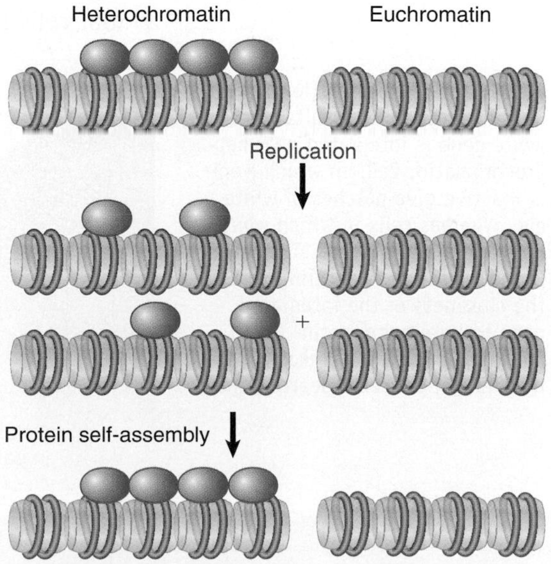

FIGURE 29.2 Heterochromatin is created by proteins that associate with histones. Perpetuation through division requires that the proteins associate with each daughter duplex and then recruit new subunits to reassemble the repressive complexes.

It can be the state of protein modification, rather than the presence of the protein *per se*, that is responsible for an epigenetic effect. Usually the tails of histones H3 and H4 are not acetylated in constitutive heterochromatin (and instead are methylated at specific sites). If heterochromatin is inappropriately acetylated, though, silenced genes may become active. The effect may be perpetuated through mitosis and meiosis, which suggests that an epigenetic effect has been created by changing the state of histone acetylation.

▸ **prion** A proteinaceous infectious agent that behaves as an inheritable trait, although it contains no nucleic acid. Examples are PrP^{Sc}, the agent of scrapie in sheep and bovine spongiform encephalopathy, and [*PSI*+], which confers an inherited state in yeast.

Independent protein aggregates that cause epigenetic effects (called **prions**) work by sequestering the protein in a form in which its normal function cannot be displayed. Once the protein aggregate has formed, it forces newly synthesized protein subunits to join it in the inactive conformation.

KEY CONCEPT

- Epigenetic effects can result from modification of a nucleic acid after it has been synthesized or by the perpetuation of protein structures.

CONCEPT AND REASONING CHECK

Histones present in nucleosomes prior to replication are randomly distributed to both daughter duplexes immediately behind the replication fork and are mixed together with newly synthesized histones. How might this contribute to the maintenance of a particular epigenetic state?

29.2 Heterochromatin Propagates from a Nucleation Event

An interphase nucleus contains both euchromatin and heterochromatin. The condensation state of heterochromatin is close to that of mitotic chromosomes. Heterochromatin is distinct from euchromatin in a number of ways. Heterochromatin remains condensed in interphase, is generally transcriptionally repressed, replicates late in S phase, and may be localized to the nuclear periphery. Centromeric heterochromatin typically consists of satellite DNAs; however, the formation of heterochromatin is not rigorously defined by sequence. When a gene is transferred, either by a chromosomal translocation or by transfection and integration, into a position adjacent to heterochromatin, it may become inactive as the result of its new location, implying that it has become heterochromatic.

▸ **position effect variegation (PEV)** Silencing of gene expression that occurs as the result of proximity to heterochromatin.

Such inactivation is the result of an epigenetic effect. It may differ between individual cells in an animal and results in the phenomenon of **position effect variegation (PEV)**, in which genetically identical cells have different phenotypes (PEV was introduced in the *Chapter 10* box, *Methods and Techniques*). This has been well characterized in *Drosophila*. **FIGURE 29.3** shows an example of position effect variegation in the fly eye. Some of the regions in the eye lack color, whereas others are red. This occurs because the *white* gene (required to develop red pigment) is inactivated by adjacent heterochromatin in some cells but remains active in others.

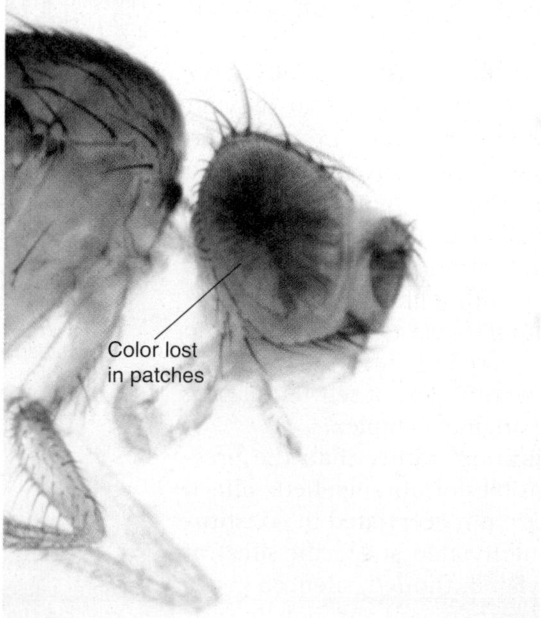

FIGURE 29.3 Position effect variegation in eye color results when the *white* gene is integrated near heterochromatin. Cells in which *white* is inactive give patches of white eye, whereas cells in which *white* is active give red patches. The severity of the effect is determined by the closeness of the integrated gene to heterochromatin. Photo courtesy of Steven Henikoff, Fred Hutchinson Cancer Research Center.

The explanation for this effect is shown in **FIGURE 29.4**. Inactivation spreads from heterochromatin into the adjacent region for a variable distance. In some cells it goes far enough to inactivate a nearby gene, but in others it does not. This takes

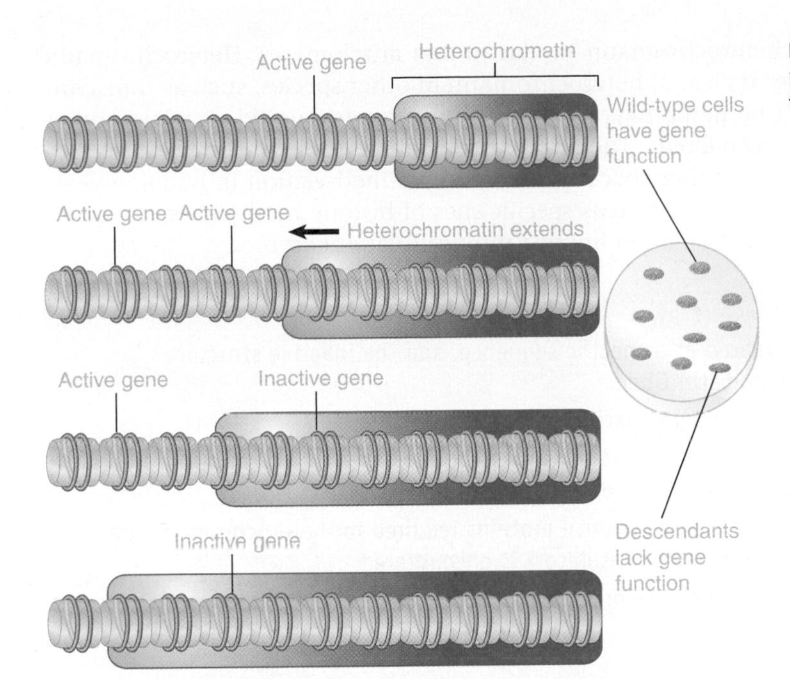

FIGURE 29.4 Extension of heterochromatin inactivates genes. The probability that a gene will be inactivated depends on its distance from the heterochromatin region.

place at a certain point in embryonic development, and after that point the state of the gene is stably inherited by the progeny cells. Cells descended from an ancestor in which the gene was inactivated form patches corresponding to the phenotype of loss-of-function (in the case of *white*, absence of color).

The closer a gene lies to heterochromatin, the higher the probability that it will be inactivated. This results from the fact that the formation of heterochromatin is a two-stage process: a *nucleation* event occurs at a specific sequence (triggered by binding of a protein that recognizes this sequence), and then the inactive structure *propagates* along the chromatin fiber. The distance for which the inactive structure extends is not precisely determined and may be stochastic, being influenced by parameters such as the quantities of limiting protein components. One factor that may affect the spreading process is the activation of promoters in the region; an active promoter may inhibit spreading. In addition, insulators can protect a transcriptionally active region by preventing heterochromatin spreading (see *Section 10.12, Insulators Define Independent Domains*).

Genes that are closer to heterochromatin are more likely to be inactivated, and will, therefore, be inactive in a greater proportion of cells. This model predicts that the boundaries of a heterochromatic region might be terminated by exhausting the supply of one of the proteins that is required. In fact, many proteins that regulate or contribute to heterochromatin formation were originally identified in *Drosophila* using screens for dosage-dependent effects of these proteins on heterochromatin spreading.

The effect of **telomeric silencing** in yeast is analogous to position effect variegation in *Drosophila;* genes translocated to a telomeric location show the same sort of variable loss of activity. This results from a spreading effect that propagates from the telomeres. In this case, the binding of the Rap1 protein to telomeric repeats triggers the nucleation event, which results in the recruitment of heterochromatin proteins, as described in *Section 29.3, Heterochromatin Depends on Interactions with Histones*.

▸ **telomeric silencing** The repression of gene activity that occurs in the vicinity of a telomere.

In addition to the telomeres, there are two other sites at which heterochromatin is nucleated in yeast. Yeast mating type is determined by the activity of a single active locus (*MAT*), but the genome contains two other copies of the mating type sequences (*HML* and *HMR*), which are maintained in an inactive form (this system was introduced in *Section 15.9, Yeast Use a Specialized Recombination Mechanism to Switch Mating Type*). The silent loci *HML* and *HMR* nucleate heterochromatin via binding of several proteins (rather than the single protein, Rap1, required to initiate the process at telomeres), which then lead to propagation of heterochromatin similar to that at telomeres. Heterochromatin in yeast exhibits features typical of heterochromatin in other species, such as transcriptional inactivity and self-perpetuating protein structures superimposed on nucleosomes (which are generally deacetylated). The only notable difference between yeast heterochromatin and that of most other species is that histone methylation in budding yeast is not associated with silencing, whereas specific sites of histone methylation are a key feature of heterochromatin formation in most multicellular eukaryotes.

KEY CONCEPTS

- Heterochromatin is nucleated at a specific sequence, and the inactive structure propagates along the chromatin fiber.
- Genes within regions of heterochromatin are inactivated.
- The length of an inactive region can vary from cell to cell; as a result, inactivation of genes in this vicinity causes position effect variegation.
- Heterochromatin spreading continues until proteins required for heterochromatin are depleted or an obstacle (such as an insulator) is encountered.
- Similar spreading effects occur at telomeres and at the silent mating type loci in yeast.

CONCEPT AND REASONING CHECK

Why is a transcriptionally active gene likely to prevent heterochromatin spreading?

29.3 Heterochromatin Depends on Interactions with Histones

Inactivation of chromatin occurs by the addition of proteins to the nucleosomal fiber. The inactivation may be due to a variety of effects, including the condensation of chromatin to make it inaccessible to the apparatus needed for gene expression, the addition of proteins that directly block access to regulatory sites, or the presence of proteins that directly inhibit transcription.

Two systems that have been characterized at the molecular level involve HP1 in mammals and the SIR complex in yeast. Although many of the proteins involved in each system are not evolutionarily related, the general mechanism of reaction is similar: the points of contact in chromatin are the N-terminal tails of the histones.

Our insights into the molecular mechanisms for regulating the formation of heterochromatin originated with mutants that affect position effect variegation. Some 30 genes have been identified in *Drosophila*. They are named systematically as *Su(var)* for genes whose products act to *su*ppress *var*iegation and *E(var)* for genes whose products *e*nhance *var*iegation. Remember that the genes were named for the behavior of the *mutant* loci; thus *Su(var)* mutations lie in genes whose products are needed for the formation of heterochromatin. They include enzymes that act on chromatin, such as histone deacetylases, and proteins that are localized to heterochromatin. In contrast, *E(var)* mutations lie in genes whose products are needed to activate gene expression. They include members of the SWI/SNF complex (see *Section 28.7, Chromatin Remodeling Is an Active Process*).

HP1 (*h*eterochromatin *p*rotein 1) is one of the most important Su(var) proteins. It was originally identified as a protein that is localized to heterochromatin by staining polytene chromosomes with an antibody directed against the protein. It was later shown to be the product of the gene *Su(var)2-5*. Its homolog in the fission yeast *Schizosaccaromyces pombe* is encoded by the *SWI6* gene. HP1 is now called HP1α because two related proteins, HP1β and HP1γ, have since been found.

HP1 contains a chromodomain near the N-terminus and another domain that is related to it, called the chromo-shadow domain, at the C-terminus. The HP1 chromodomain binds to histone H3 that is trimethylated at lysine 9 (H3K9me3).

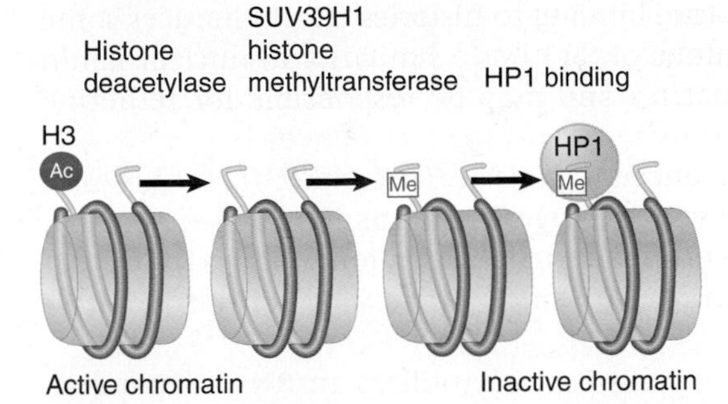

FIGURE 29.5 SUV39H1 is a histone methyltransferase that acts on K9 of histone H3. HP1 binds to the methylated histone.

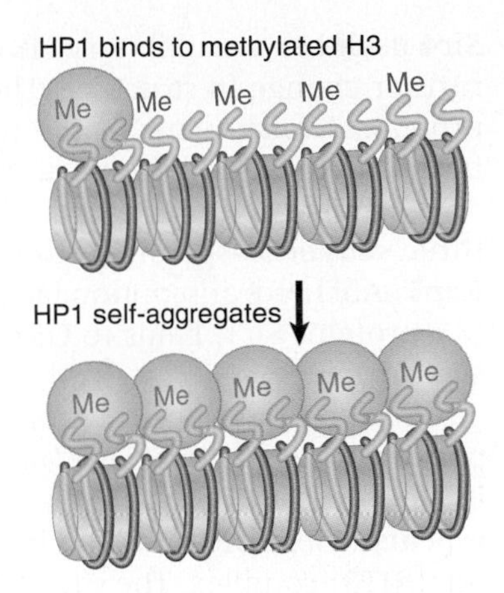

FIGURE 29.6 Binding of HP1 to methylated histone H3 forms a trigger for silencing because further molecules of HP1 aggregate on the nucleosome chain.

Mutation of a deacetylase that acts on the H3 N-terminus prevents the methylation at K9. This leads to a model for initiating formation of heterochromatin shown in **FIGURE 29.5**. First the deacetylase acts to remove acetyl groups from the H3 tail, which then allows the SUV39H1 methyltransferase (also known as KMT1A) to methylate H3K9 to create the signal to which HP1 will bind. This is a trigger for forming inactive chromatin. **FIGURE 29.6** shows that the inactive region may then be extended by the ability of further HP1 molecules to interact with one another.

Heterochromatin formation at telomeres and silent mating-type loci in yeast relies on an overlapping set of genes, known as *silent information regulators* (*SIR* genes). Mutations in *SIR2*, *SIR3*, or *SIR4* cause *HML* and *HMR* to become activated and also relieve the inactivation of genes that have been integrated near telomeric heterochromatin. The products of these loci therefore function to maintain the inactive state of both types of heterochromatin.

FIGURE 29.7 shows a model for actions of these proteins. At telomeres, the process is initiated by a sequence-specific DNA-binding protein, Rap1, which binds to the $C_{1-3}A$ repeats at the telomeres. The proteins Sir3 and Sir4 interact with Rap1 and also with one another (they may function as a heteromultimer). Sir3/Sir4 interact with the N-terminal tails of the histones H3 and H4, with a preference for unacetylated tails. Another SIR protein, Sir2, is a deacetylase, and its activity is necessary to maintain binding of the Sir3/4 complex to chromatin.

Rap1 has the crucial role of identifying the DNA sequences at which heterochromatin forms. It recruits Sir3/Sir4, and they interact directly with the histones H3/H4. Once Sir3/Sir4 have bound to histones H3/H4, the complex may polymerize further and spread along the chromatin fiber. This may inactivate the region, either because coating with Sir3/

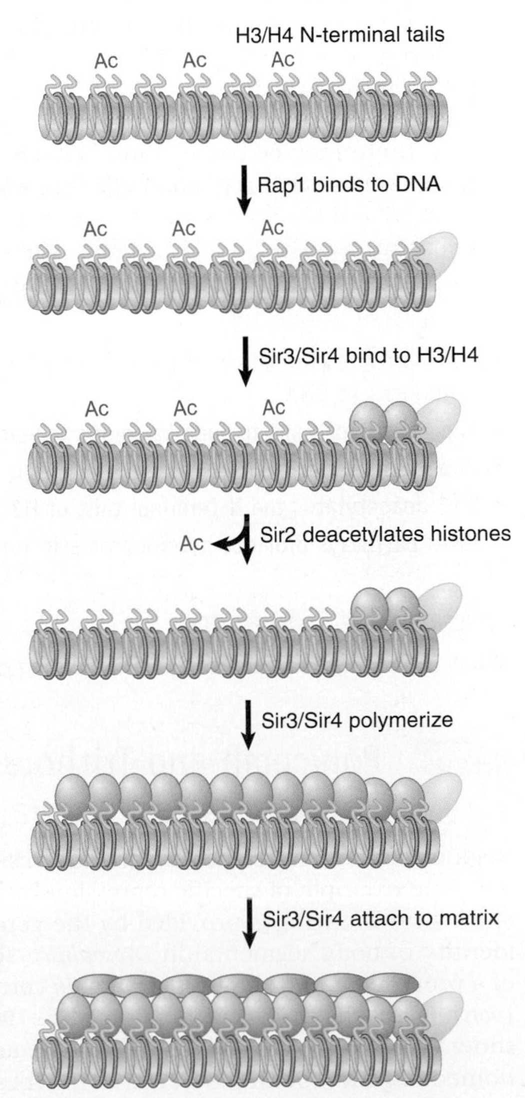

FIGURE 29.7 Formation of heterochromatin is initiated when Rap1 binds to DNA. Sir3/4 bind to Rap1 and also to histones H3/H4. Sir2 deacetylates histones. The SIR complex polymerizes along chromatin and may connect telomeres to the nuclear matrix.

Sir4 itself has an inhibitory effect or because binding to histones H3/H4 induces some further change in structure. The C-terminus of Sir3 has a similarity to nuclear lamin proteins (constituents of the nuclear matrix) and may be responsible for tethering heterochromatin to the nuclear periphery.

A similar series of events forms the silenced regions at *HMR* and *HML*. In this case, three sequence-specific factors are involved in triggering formation of the complex: Rap1, Abf1 (a transcription factor), and ORC (the *o*rigin *r*eplication *c*omplex). Another SIR protein, Sir1, binds to ORC and then recruits Sir2, -3, and -4 to form the repressive structure.

Formation of heterochromatin in the yeast *S. pombe* utilizes an RNAi-dependent pathway (see *Section 30.6, How Does RNA Interference Work?*). This pathway is initiated by the production of siRNA molecules resulting from transcription of centromeric repeats. These siRNAs result in formation of RNA-induced transcriptional gene silencing (RITS) complex. The siRNA components are responsible for localizing the complex at centromeres. The siRNA complex promotes methylation of histone H3K9 by the Clr4 methyltransferase (also known as KMT1, a homolog of *Drosophila Su(Var)3-9*). H3K9 methylation recruits the *S. pombe* homolog of HP1, Swi6.

How does a silencing complex repress chromatin activity? It could condense chromatin so that regulator proteins cannot find their targets. The simplest case would be to suppose that the presence of a silencing complex is mutually incompatible with the presence of transcription factors and RNA polymerase. The cause could be that silencing complexes block remodeling (and thus indirectly prevent factors from binding) or that they directly obscure the binding sites on DNA for the transcription factors. The situation may not be this simple, though, because transcription factors and RNA polymerase can be found at promoters in silenced chromatin. This could mean that the silencing complex prevents the factors from working rather than from binding as such. In fact, there may be competition between gene activators and the repressing effects of chromatin, so that activation of a promoter inhibits spread of the silencing complex.

KEY CONCEPTS

- HP1 is the key protein in forming mammalian heterochromatin, and acts by binding to methylated histone H3.
- Rap1 initiates formation of heterochromatin in yeast by binding to specific target sequences in DNA.
- The targets of Rap1 include telomeric repeats and silencers at *HML* and *HMR*.
- Rap1 recruits Sir3/Sir4, which interact with the N-terminal tails of H3 and H4.
- Sir2 deacetylates the N-terminal tails of H3 and H4 and promotes spreading of Sir3/Sir4.
- RNAi pathways promote heterochromatin formation at centromeres.

CONCEPT AND REASONING CHECK

What is the role of sequence-specific binding proteins in heterochromatin formation?

29.4 Polycomb and Trithorax Are Antagonistic Repressors and Activators

Regions of constitutive heterochromatin, such as at telomeres and centromeres, provide one example of specific repression by chromatin structure. Another insight into chromatin silencing is provided by the genetics of homeotic genes (which affect the identity of body segments) in *Drosophila*. These studies have led to the identification of a protein complex that may *maintain* certain genes in a repressed state. The gene *Pc* (*polycomb*) is the prototype for a class of ~15 loci called the *Pc* group (*Pc-G*). *Pc* mutants show transformations of cell type that are equivalent to gain-of-function mutations in homeotic genes, because in *Pc* mutants these genes are expressed in tissues in which

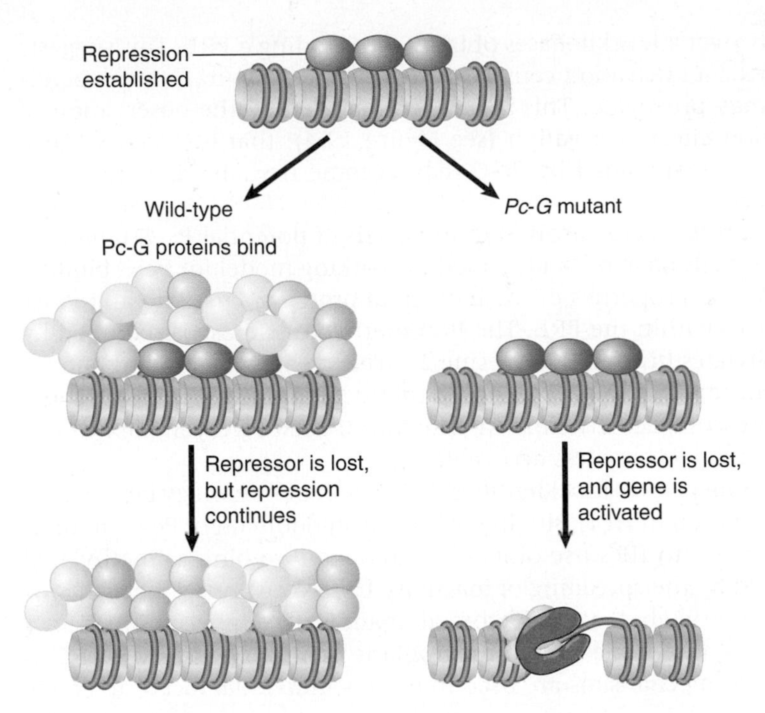

FIGURE 29.8 Pc-G proteins do not initiate repression but are responsible for maintaining it.

they are normally repressed. This implicates *Pc* in negatively regulating transcription. Mutations in various genes in the *Pc* group generally have the same result of derepressing homeotic genes, which suggests the possibility that the group of proteins has some common regulatory role.

The Pc-G proteins are not conventional repressors. They are not responsible for determining the *initial* pattern of expression of the genes on which they act. In the absence of Pc-G proteins, these genes are initially repressed as usual, but later in development the repression is lost without Pc-G group functions. This suggests that *the Pc-G proteins in some way recognize the state of repression when it is established, and they then act to perpetuate it through cell division of the daughter cells.* In other words, Pc-G proteins are necessary for the *maintenance* phase of repression, but not for the initial *establishment* phase. **FIGURE 29.8** shows a model in which Pc-G proteins bind in conjunction with a repressor, but the Pc-G proteins remain bound after the repressor is no longer available. This is necessary to maintain repression, so that if Pc-G proteins are absent, the gene becomes activated.

The Pc proteins function in large complexes, known as PRCs (*P*olycomb-*r*epressive *c*omplexes). These complexes contain proteins with a variety of functions and enzymatic activities, such as histone deacetylase, methyltransferase, and ubiquitin ligase activities. For example, the PRC1 complex contains Pc itself, several other Pc-G proteins, and five general transcription factors, while the Esc-E(z) complex contains the Pc-G protein Esc, the E(z) methyltransferase, a histone binding protein, and a histone deacetylase. Pc itself has a chromodomain that binds to methylated H3. These properties reveal a connection between chromatin remodeling and repression by PRCs.

A region of DNA that is sufficient to enable the response to the *Pc-G* genes is called a PRE (*p*olycomb *r*esponse *e*lement). It can be defined operationally by the property that it maintains repression in its vicinity throughout development. The assay for a PRE is to insert it close to a reporter gene that is controlled by an enhancer that is repressed in early development and then to determine whether the reporter becomes expressed subsequently in the descendants. An effective PRE will prevent such reexpression.

The PRE is a complex structure that measures ~10 kb. Several proteins, including Pho, Phol and GAGA factor (GAF), with DNA-binding activity for sites within the PRE have been identified, but there could be others. When a locus is repressed by Pc-G, however, the Pc-G proteins occupy a much larger region of DNA than the PRE itself. Pc is found locally over a few kilobases of DNA surrounding a PRE. This suggests that the PRE may provide a nucleation center, from which a structural state depending on Pc-G proteins may propagate. This model is supported by the observation of effects related to position effect variegation (see Figure 29.4); that is, a gene near a locus whose repression is maintained by Pc-G may become heritably inactivated in some cells but not others.

The effects of Pc-G proteins are vast in that hundreds of potential Pc-G targets in plants, insects, and mammals have been identified. A working model for Pc-G binding at a PRE is suggested by the properties of the individual proteins. First Pho and Pho1 bind to specific sequences within the PRE. The E(z) methyltransferase is recruited by Pho/Pho1; it then methylates histone H3 at lysine 27. This creates the binding site for the PRC, because the chromodomain of Pc binds to methylated H3K27. The Polycomb complex induces a more-compact structure in chromatin; each PRC complex causes about three nucleosomes to become less accessible.

In fact, the chromodomain was first identified as a region of homology between Pc and the heterochromatin protein HP1. Binding of the chromodomain of Pc to methylated K27 of H3 is analogous to HP's use of its chromodomain to bind to methylated K9. Variegation is caused by the spreading of inactivity from constitutive heterochromatin, and, as a result, it is likely that the chromodomain is used by Pc and HP1 in a similar way to induce the formation of heterochromatic or inactive structures. This model implies that similar mechansims are used to repress individual loci or to create heterochromatin.

The *trithorax* group (*trxG*) of proteins have the opposite effect to the Pc-G proteins: they act to maintain genes in an active state. There may be some similarities in the actions of the two groups: mutations in some loci prevent both Pc-G and trxG from functioning, suggesting that they could rely on common components. trxG proteins are quite diverse; some comprise subunits of chromatin remodeling enzymes such as SWI/SNF, whereas others also possess important histone modification activities, such as histone **demethylases**, which could oppose the activities of Pc-G proteins. The trxG proteins act by making chromatin continuously accessible to transcription factors. Although PcG and trxG proteins promote opposite outcomes, they bind to the same PREs and thus are directly competing for control of homeotic gene expression.

‣ **demethylase** A name for an enzyme that removes a methyl group, typically from DNA, RNA, or protein.

KEY CONCEPTS

- Polycomb group proteins (Pc-G) perpetuate a state of repression through cell divisions.
- The PRE is a DNA sequence that is required for the action of Pc-G.
- The PRE provides a nucleation center from which Pc-G proteins propagate an inactive structure.
- Trithorax group proteins antagonize the actions of the Pc-G.

CONCEPT AND REASONING CHECK

Mutations in *Pc* result in derepression of homeotic genes. What would be the likely effect on expression of these genes of mutations in *brahma*, a trxG gene that is the ATPase subunit of *Drosophila* SWI/SNF? What about a *Pc brahma* double mutant?

29.5 X Chromosomes Undergo Global Changes

For species with chromosomal sex determination, the sex of the individual presents an interesting problem for gene regulation, because of the variation in the number of X chromosomes. If X-linked genes were expressed equally well in each sex, females would have twice as much of each product as males. The importance of avoiding this situation is shown by the existence of **dosage compensation**, which equalizes the level of expression of X-linked genes in the two sexes. Mechanisms used in different species are summarized in **FIGURE 29.9**:

‣ **dosage compensation** Mechanisms employed to compensate for the discrepancy between the presence of two X chromosomes in one sex but only one X chromosome in the other sex.

	Mammals	Flies	Worms
	Inactivate one ♀ X	Double expression ♂ X	Halve expression two ♀ X
X X			
X Y			

FIGURE 29.9 Different means of dosage compensation are used to equalize X-chromosome expression in male and female.

- In mammals, one of the two female X chromosomes is inactivated. The result is that females have only one active X chromosome, which is the same situation found in males. The active X chromosome of females and the single X chromosome of males are expressed at the same level.
- In *Drosophila*, the expression of the single male X chromosome is doubled relative to the expression of each female X chromosome.
- In *Caenorhabditis elegans*, the expression of each female (really a hermaphrodite) X chromosome is halved relative to the expression of the single male X chromosome.

The common feature in all these mechanisms of dosage compensation is that *the entire chromosome is the target for regulation*. A global change occurs that quantitatively affects all of the promoters on the chromosome. We know most about the inactivation of the X chromosome in mammalian females, where the entire chromosome becomes heterochromatic.

The twin properties of heterochromatin are its condensed state and its associated transcriptional inactivity. Heterochromatin can be divided into two types:

- **Constitutive heterochromatin** contains specific sequences that have no coding function. In general these include satellite DNAs, often found at the centromeres, and telomeric repeats. These regions are invariably heterochromatic because of their intrinsic sequence.
- **Facultative heterochromatin** takes the form of chromosome segments or entire chromosomes that are inactive in one cell lineage, although they can be expressed in other lineages. The example *par excellence* is the mammalian X chromosome. The inactive X chromosome in females is perpetuated in a heterochromatic state, whereas the active X chromosome is euchromatic. Thus *identical DNA sequences are involved in both states*. Once the inactive state has been established, it is inherited by descendant cells. This is an example of epigenetic inheritance, because it does not depend on the DNA sequence.

▸ **constitutive heterochromatin** The inert state of permanently nonexpressed sequences, usually satellite DNA.

▸ **facultative heterochromatin** The inert state of sequences that also exist in active copies, for example, one mammalian X chromosome in females.

▸ **single-X hypothesis** The theory that describes the inactivation of one X chromosome in female mammals.

Our basic view of the situation of the female mammalian X chromosomes was formed by the **single-X hypothesis** in 1961. Female mice that are heterozygous for X-linked coat-color mutations have a variegated phenotype in which some areas of the coat are wild type but others are mutant. FIGURE 29.10 shows that this can be explained *if one of the two X chromosomes is inactivated at random in each cell of a small precursor population*. Cells in which the X chromosome carrying the wild-type gene is inactivated give rise to progeny that express only the mutant allele on the active chromosome. Cells derived from a precursor where the other chromosome was inactivated have an active wild-type gene. In the case of coat color, cells descended from a particular precursor stay together and thus form a patch of the same color, creating the pattern of

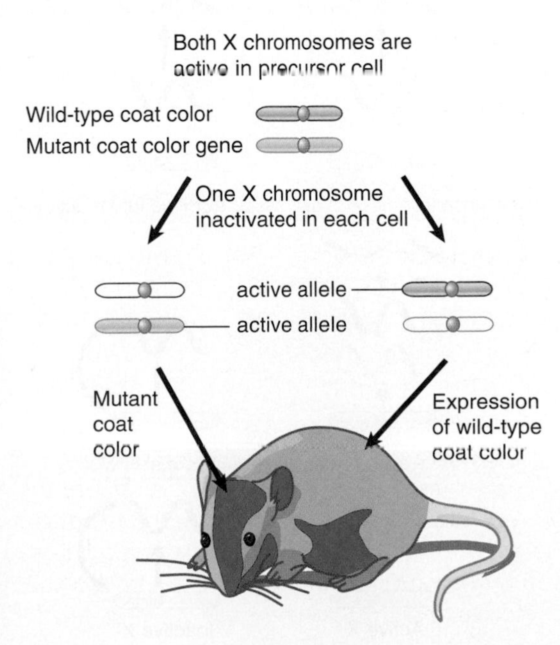

FIGURE 29.10 X-linked variegation is caused by the random inactivation of one X chromosome in each precursor cell. Cells in which the + allele is on the active chromosome have a wild-type phenotype; cells in which the – allele is on the active chromosome have mutant phenotype.

visible variegation. In other cases, individual cells in a population will express one or the other of X-linked alleles; for example, in heterozygotes for the X-linked locus *G6PD*, any particular red-blood cell will express only one of the two allelic forms.

▸ **$n - 1$ rule** The rule that states that only one X chromosome is active in female mammalian cells; any others are inactivated.

Inactivation of the X chromosome in females is governed by the **$n - 1$ rule**: However many X chromosomes are present, all but one will be inactivated. In normal females there are, of course, two X chromosomes, but in rare cases where nondisjunction has generated a 3X or greater genotype, only one X chromosome remains active. This suggests a general model in which a specific event is limited to one X chromosome and protects it from an inactivation mechanism that applies to all the others.

A single locus on the X chromosome is sufficient for inactivation. When a translocation occurs between the X chromosome and an autosome, this locus is present on only one of the reciprocal products, and only that product can be inactivated. By comparing different translocations, it was possible to map this locus, which is called the *Xic* (*X-i*nactivation *c*enter). A cloned region of 450 kb contains all the properties of the *Xic*. When this sequence is inserted as a transgene onto an autosome, the autosome can become subject to inactivation.

Xic is a *cis*-acting locus that contains the information necessary to both count X chromosomes and inactivate all copies but one. Pairing of *Xic* loci on the two X chromosomes has been implicated in the mechanism for the random choice of X inactivation. Inactivation spreads from *Xic* along the entire X chromosome. When *Xic* is present on an X chromosome–autosome translocation, inactivation spreads into the autosomal regions (although the effect is not always complete).

Xic is a complex genetic locus that expresses several long noncoding RNAs (ncRNAs). The most important of these is a gene called *Xist* (*X i*nactive *s*pecific *t*ranscript), which is stably expressed only on the *inactive* X chromosome. The behavior of this gene is effectively the opposite from all other loci on the chromosome, which are turned off. Deletion of *Xist* prevents an X chromosome from being inactivated. It does not, however, interfere with the counting mechanism (because other X chromosomes can be inactivated). Thus we can distinguish two features of *Xic*: the unidentified element(s) required for counting and the *Xist* gene required for inactivation.

The $n - 1$ rule suggests that stabilization of *Xist* RNA is the "default" and that some blocking mechanism prevents stabilization at one X chromosome (which will be the active X). This means that, although *Xic* is necessary and sufficient for a chromosome to be *inactivated*, the products of other loci are necessary for the establishment of the *active* X chromosome.

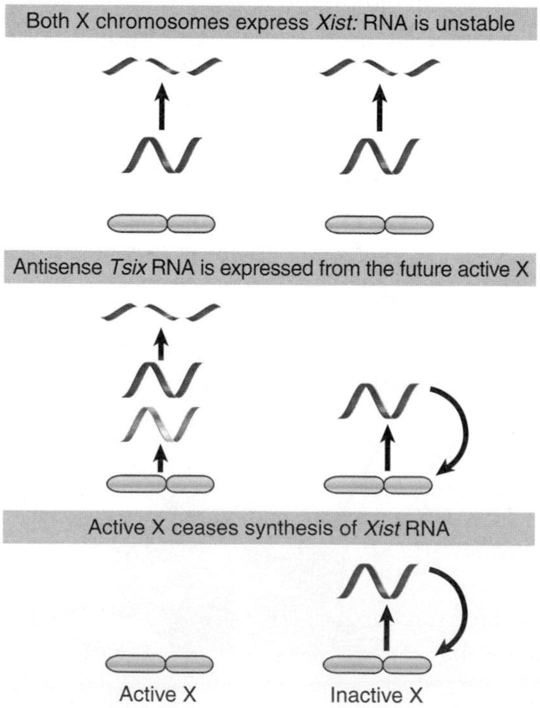

FIGURE 29.11 X-inactivation involves stabilization of *Xist* RNA, which coats the inactive chromosome.

The *Xist* transcript is regulated in a negative manner by *Tsix*, its antisense partner. Lack of *Tsix* expression on the future inactive X chromosome permits *Xist* to become upregulated and stabilized, and persistence of *Tsix* on the future active X prevents *Xist* upregulation.

FIGURE 29.11 illustrates the role of *Xist* and *Tsix* RNA in X-inactivation. *Xist* codes for a noncoding RNA that lacks open reading frames. The *Xist* RNA "coats" the X chromosome from which it is synthesized; this suggests that it has a structural role. Prior to X-inactivation, it is synthesized by both female X chromosomes.

Following inactivation, the RNA is found only on the inactive X chromosome. The transcription rate remains the same before and after inactivation, so the transition depends on posttranscriptional events.

Accumulation of *Xist* on the future inactive X results in exclusion of transcription machinery (such as RNA polymerase II), and leads to the recruitment of Polycomb repressor complexes (PRC1 and PRC2), which trigger a series of chromosomewide histone modifications (H2AK119 ubiquitination, H3K27 methylation, and H4K20 methylation, and H4 deacetylation). Late in the process, an inactive X-specific histone variant, macroH2A, is incorporated into the chromatin, and promoter DNA is methylated. These changes are summarized in **FIGURE 29.12**. (The repressive effects of promoter methylation are discussed in the following sections.) At this point, the heterochromatic state of the inactive X is stable, and *Xist* is not required to maintain the silent state of the chromosome. Despite the extensive silencing of genes on the inactive X, ~5% of genes are still transcribed from the inactive X chromosome. It is important to note that genes on both X chromosomes are active prior to X inactivation, which occurs at ~1000 cell stage of development. X "counting" occurs during these early stages, and the function of the two X's is necessary; otherwise, there would be no detectable effects of the XO genotype (Turner Syndrome).

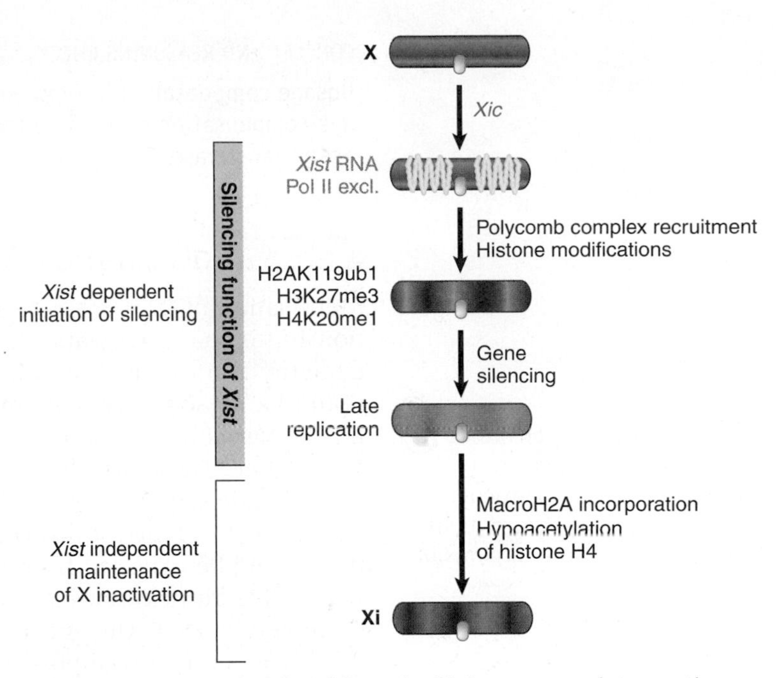

FIGURE 29.12 *Xist* RNA produced from the *Xic* locus accumulates on the future inactive X (Xi). This excludes transcription machinery, such as RNA polymerase II (Pol II). Polycomb group complexes are recruited to the *Xist*-covered chromosome and establish chromosomewide histone modifications. Histone macroH2A becomes enriched on the Xi and promoters of genes on the Xi are methylated. In this phase X inactivation is irreversible and *Xist* is not required for maintenance of the silent state.

Global changes also occur in other types of dosage compensation. In *Drosophila*, a large ribonucleoprotein complex, MSL, is found only in males, where it localizes on the X chromosome. This complex contains two noncoding RNAs, which appear to be needed for localization to the male X (perhaps analogous to the localization of *Xist* to the inactive mammalian X), and a histone acetyltransferase that acetylates histone H4 on K16 throughout the male X. The net result of the action of this complex is the twofold increase in transcription of all genes on the male X. In *C. elegans*, females/hermaphrodites use mechanisms to globally condense the two X chromosomes in order to reduce expression of each by half compared to the single male X.

KEY CONCEPTS

- One of the two female X chromosomes is inactivated at random in each cell during embryogenesis of placental mammals.
- In exceptional cases where there are >2 X chromosomes, all but one are inactivated.
- The *Xic* (X-inactivation center) is a *cis*-acting region on the X chromosome that is necessary and sufficient to ensure that only one X chromosome remains active.
- *Xic* includes the *Xist* gene, which codes for an RNA that is found only on inactive X chromosomes.
- *Xist* recruits Polycomb complexes, which modify histones on the inactive X.
- The mechanism that is responsible for preventing *Xist* RNA from accumulating on the active chromosome is unknown.

CONCEPT AND REASONING CHECK

Dosage compensation in *Drosophila* also depends on noncoding RNAs, which recruit a dosage compensation complex to the single X chromosome in males. This complex contains an acetyltransferase. Explain how this promotes dosage compensation in *Drosophila*.

29.6 CpG Islands Are Subject to Methylation

▸ **CpG islands** Stretches of 1–2 kb in mammalian genomes that are enriched in CpG dinucleotides; frequently found in promoter regions of genes.

Methylation of DNA is a critical parameter that affects transcription. The typical relationship is that methylation in the vicinity of a promoter inhibits transcription and demethylation is required for gene expression. The reality is a little more complex than that, as some areas are methylated in active genes. Critical regions for methylation in vertebrates usually occur in **CpG islands**, which are found in the 5′ region of the gene. These islands are detected by the presence of an increased density of the dinucleotide sequence, CpG (CpG = 5′–CG-3′).

The CpG dinucleotide occurs in vertebrate DNA at only ~20% of the frequency that would be expected from the proportion of G-C base pairs. This may be because when CpG dinucleotides are methylated on C, spontaneous deamination of methyl-C converts it to T. This introduces a mutation that can result in the loss of the CpG dinucleotide, as it is impossible for the damage repair system to determine if the T or the G should be replaced (although some repair systems are biased to remove the T in this context). In certain regions, however, the density of CpG doublets reaches the predicted value; in fact, it is increased by 10× relative to the rest of the genome. The CpG doublets in these regions can be methylated or unmethylated, which affects expression of genes in the area. In some cases, CpG islands begin just upstream of a promoter and extend downstream into the transcribed region before petering out. An example is shown in **FIGURE 29.13**.

These CpG-rich islands have an average G-C content of ~60%, compared with the 40% average in bulk DNA. They take the form of stretches of DNA typically 1 to 2 kb long. There are ~45,000 such islands in the human genome. Some of the islands are present in repeated Alu elements and may just be the consequence of their high G-G content. The human genome sequence confirms that, excluding these, there are ~29,000 islands. There are fewer in the mouse genome; ~15,500. About 10,000 of the islands in both species appear to reside in a context of sequences that are conserved between the species, consistent with a regulatory significance of these islands. From 2% to 7% of the cytosines of animal cell DNA are methylated (the value varies with the species).

All the housekeeping genes that are constitutively expressed have CpG islands; this accounts for about half of the islands. The remaining islands occur at the promoters of tissue-regulated genes; about half these genes have islands. In these cases, the presence

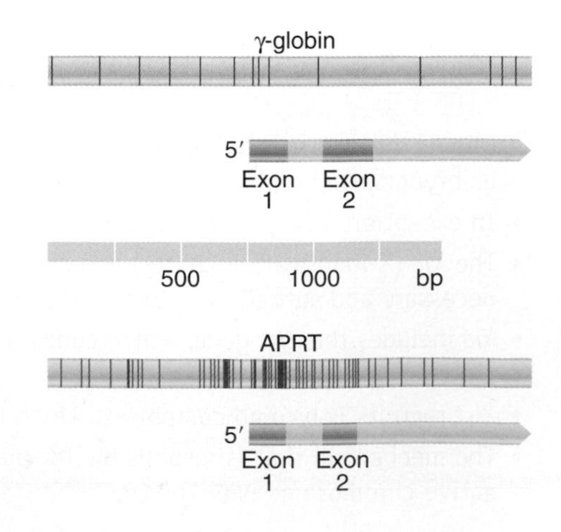

FIGURE 29.13 The typical density of CpG doublets in mammalian DNA is ~1/100 bp, as seen for a γ-globin gene (top). In a CpG island, the density is increased to >10 CpGs/100 bp. The island in the constitutively expressed APRT gene starts ~100 bp upstream of the promoter and extends ~400 bp into the gene. Each vertical line represents a CpG doublet.

of an unmethylated CpG island may be necessary for transcription, but it is not sufficient; all the other necessary events, such as binding of tissue-specific activators, must also occur. In general, however, methylation of promoter CpG islands is frequently correlated with transcriptional silencing. One dramatic example of this is the methylation of promoters along the inactive X chromosome in female mammals.

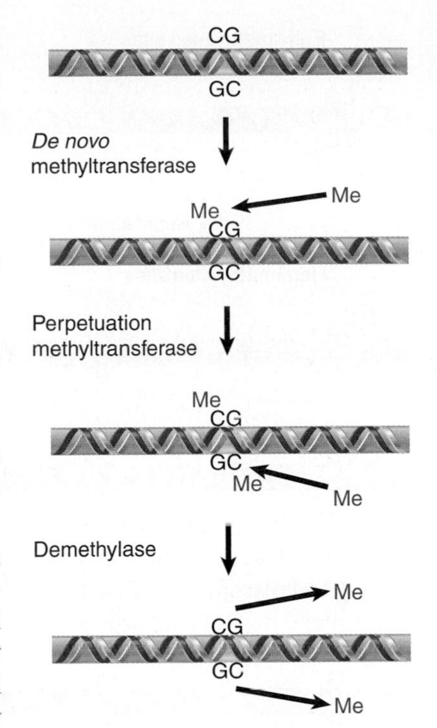

FIGURE 29.14 The state of DNA methylation is controlled by three types of enzyme. *De novo* and perpetuation methyltransferases are responsible for methylating DNA, whereas demethylases remove the methyl groups.

Methylation of a CpG island can affect transcription. One of two mechanisms can be involved:

- Methylation of a binding site for some factor may prevent that factor from binding.
- Methylation may cause specific repressors to bind to the DNA, which promotes the formation of a region of heterochromatin.

Repression may be caused by classes of protein that bind to methylated CpG sequences. The protein MeCP1 requires the presence of several methyl groups to bind to DNA, whereas MeCP2 and a family of related proteins can bind to a single methylated CpG base pair (although some of this family may also recognize unmethylated DNA). This explains why a methylation-free zone is required for initiation of transcription. The absence of methyl groups is associated with gene expression. It is important to note that DNA methylation in not used for controlling gene expression in every species. In the case of *Drosophila melanogaster* (and other Dipteran insects), there are very low levels of methylation of DNA, and neither the yeast *S. cerevisiae* nor the nematode *C. elegans* have any DNA methylation. The other differences between inactive and active chromatin appear to be the same as in species that display methylation. Thus, in these organisms, any role that methylation has in vertebrates is replaced by some other mechanism.

Methylation occurs by the action of DNA methyltransferases. There are two types of DNA methyltransferase, whose actions are distinguished by the state of the methylated DNA, as shown in FIGURE 29.14. A **maintenance** (or "perpetuation") **methyltransferase** acts constitutively *only on* **hemimethylated sites** to convert them to **fully methylated** sites. Its existence means that any methylated site is perpetuated after replication, as replication results in the conversion of fully methylated sites to hemimethylated (the daughter strand is unmethylated), as shown in FIGURE 29.15. Maintenance methylation is virtually 100% efficient; this figure shows that if maintenance methylation fails to act at a hemimethylated site, all methylation will be lost on one copy after the next round of replication. There is one maintenance methyltransferase (Dnmt1) in mouse, and it is essential: mouse embryos in which its gene has been disrupted do not survive past early embryogenesis. The absence of Dnmt1 causes widespread demethylation at promoters, and we assume this is lethal because of the uncontrolled gene expression.

- **maintenance methyltransferase** Adds a methyl group to a target site that is already hemimethylated.
- **hemimethylated site** A palindromic sequence that is methylated on only one strand of DNA.
- **fully methylated** A site that is a palindromic sequence that is methylated on both strands of DNA.

To modify DNA at a new position requires the action of a ***de novo* methyltransferase**, which recognizes DNA by virtue of a specific sequence or is recruited to a specific site in DNA. It acts *only* on unmethylated DNA to add a methyl group to one strand. There are two *de novo* methyltransferases (Dnmt3A and Dnmt3B) in mouse; they have different target sites, and both are essential for development. Mutations in Dnmt3B prevent methylation of satellite DNA, which causes centromere instability at the cellular level. Mutations in the corresponding human gene cause a disease called ICF (immunodeficiency, centromere instability, facial anomalies). The importance of methylation is emphasized by another human disease, Rett syndrome, which is caused by mutation of the gene for the protein MeCP2 that binds methylated CpG sequences. Patients with Rett syndrome exhibit autismlike symptoms, which appear

- ***de novo* methyltransferase** Adds a methyl group to an unmethylated target sequence on DNA.

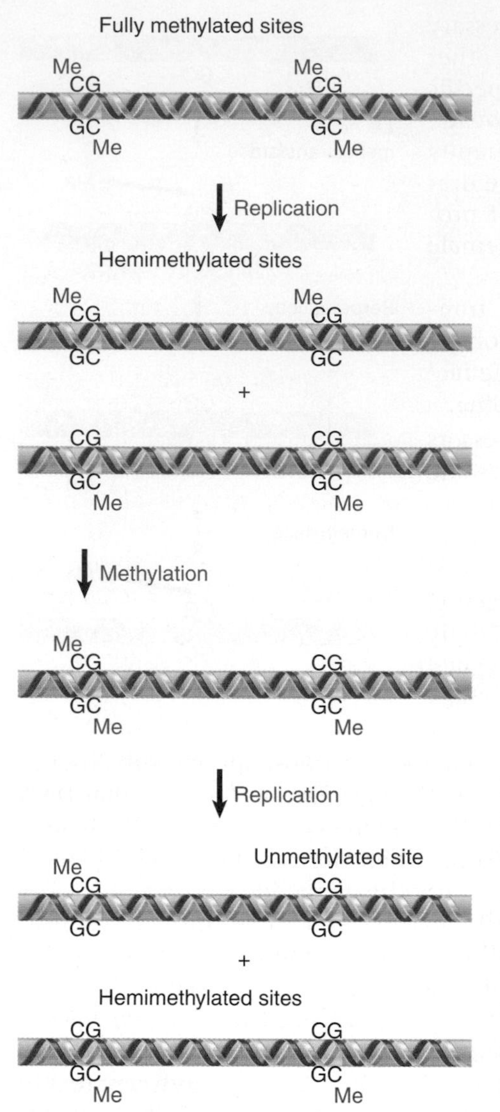

FIGURE 29.15 Replication converts fully methylated sites to hemimethylated sites. For the methylation state to be maintained following replication, a perpetuation methyltransferase must recognize the hemimethylated substrate and methylate the daughter strands.

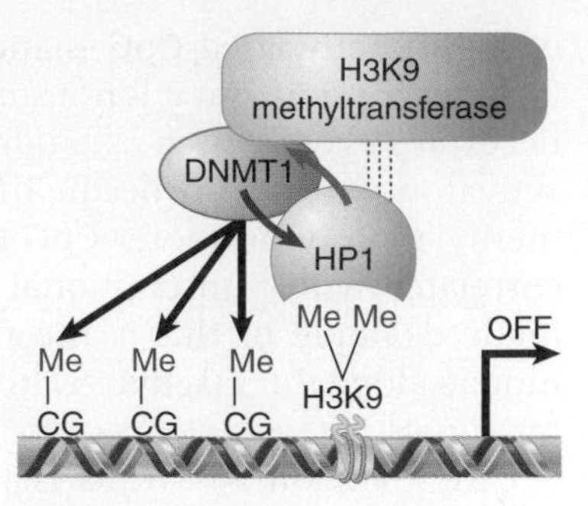

FIGURE 29.16 Mammalian HP1 is recruited to regions where lysine 9 of histone H3 (H3K9) has been methylated by a histone methyltransferase. HP1 then binds to DNMT1 and potentiates its DNA methyltransferase activity (blue arrow), thereby enhancing cytosine methylation (meCG) on nearby DNA. DNMT1 could in turn assist HP1 loading onto chromatin (red arrow). Furthermore, association of DNMT1 with the histone methyltransferase could allow a positive feedback loop to stabilize inactive chromatin.

to be the result of a failure of normal gene silencing in the brain; they also have RNA processing defects.

DNA methylation is closely linked to histone modifications associated with heterochromatin formation; in fact, DNA methylation and heterochromatin are mutually reinforcing, as depicted in **FIGURE 29.16** (also see *Section 28.10, Methylation of Histones and Methylation of DNA Are Connected*). Recall that HP1 is recruited to regions in which histone H3 has been methylated at lysine 9, a modification involved in heterochromatin formation. It turns out that HP1 can interact with Dnmt1, which can promote DNA methylation in the vicinity of HP1 binding. Furthermore, Dnmt1 can directly interact with the methyltransferase responsible for H3K9 methylation, creating a positive feedback loop to ensure continued DNA and histone methylation. These interactions (and other similar networks of interactions) contribute to the stability of epigenetic states, allowing a heterochromatin region to be maintained through many cell divisions.

KEY CONCEPTS

- Demethylation at the 5′ end of the gene is necessary for transcription.
- CpG islands surround the promoters of some genes.
- There are ~45,000 CpG islands in the human genome; ~29,000 exist outside Alu elements.
- Methylation of a CpG island prevents activation of a promoter within it.
- Repression is caused by proteins that bind to methylated CpG doublets.
- Replication converts a fully methylated site to a hemimethylated site.
- Hemimethylated sites are converted to fully methylated sites by a maintenance methyltransferase.
- DNA and histone methylation are mutually reinforcing.

CONCEPT AND REASONING CHECK

Some methyl-DNA binding complexes also contain histone deacetylases. Explain how this would also reinforce the stability of a silenced region.

29.7 DNA Methylation Is Responsible for Imprinting

The pattern of methylation of germ cells is established in each sex during gametogenesis by a two-stage process. First, the existing pattern is erased by a genomewide demethylation, and then the pattern specific for each sex is imposed during meiosis.

All allelic differences are lost when primordial germ cells develop in the embryo; irrespective of sex, the previous patterns of methylation are erased, and a typical gene is then unmethylated. In males, the pattern develops in two stages. The methylation pattern that is characteristic of mature sperm is established in the spermatocyte, but further changes are made in this pattern after fertilization. In females, the maternal pattern is imposed during oogenesis, when oocytes mature through meiosis after birth.

Systematic changes occur in early embryogenesis. Some sites will continue to be methylated, whereas others will be specifically unmethylated in cells in which a gene is expressed. From the pattern of changes, we may infer that individual sequence-specific demethylation events occur during somatic development of the organism as particular genes are activated.

The specific pattern of methyl groups in germ cells is responsible for the phenomenon of **imprinting**, which describes a difference in behavior between the alleles inherited from each parent. The expression of certain genes in mouse embryos depends upon the sex of the parent from which they were inherited. For example, the allele coding for IGF-II (insulinlike growth factor II) that is inherited from the father is expressed, but the allele that is inherited from the mother is not expressed. The *IGF-II* gene in oocytes is methylated, but the *IGF-II* gene in sperm is not methylated, so the two alleles behave differently in the zygote. The dependence on sex is reversed (that is, the maternal copy is expressed) for IGF-IIR, a receptor that causes the rapid turnover of IGF-II. The allele that is silenced is referred to as the *imprinted* allele.

imprinting A change in a gene that occurs during passage through the sperm or egg, with the result that the paternal and maternal alleles have different properties in the very early embryo. This is caused by methylation of DNA.

This sex-specific mode of inheritance requires that the pattern of methylation be established specifically during each gametogenesis. The fate of a hypothetical locus in a mouse is illustrated in **FIGURE 29.17**. In the early embryo, the paternal allele is unmethylated and expressed, and the maternal allele is methylated and silent. What happens when this mouse itself forms gametes? If it is a male, the allele contributed to the sperm must be unmethylated, irrespective of whether or not it was originally methylated. Thus when the maternal allele finds itself in a sperm, it must be demethylated. If the mouse is a female, the allele contributed to the egg must be methylated; if it was originally the paternal allele, methyl groups must be added.

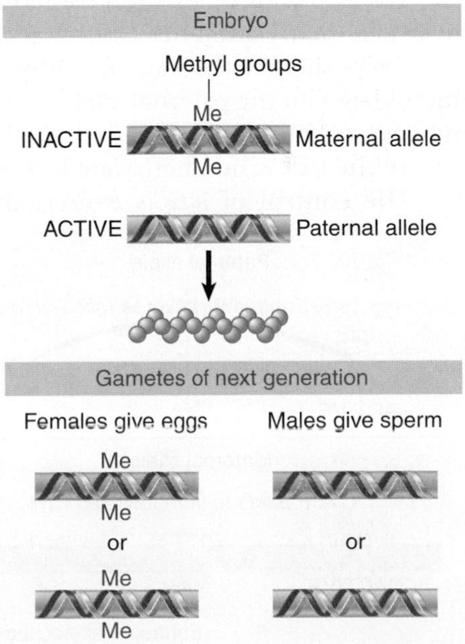

FIGURE 29.17 The typical pattern for imprinting is that a methylated locus is inactive. If this is the maternal allele, only the paternal allele is active and will, therefore, be essential to produce a functional product of the locus. The methylation pattern is reset when gametes are formed, so that all sperm have the paternal type and all oocytes have the maternal type.

The consequence of imprinting is that an embryo is *hemizygous* for any imprinted gene. Thus in the case of a heterozygous cross where the allele of one parent has an inactivating mutation, the embryo will survive if the wild-type allele comes from the parent in which this allele is active but will die

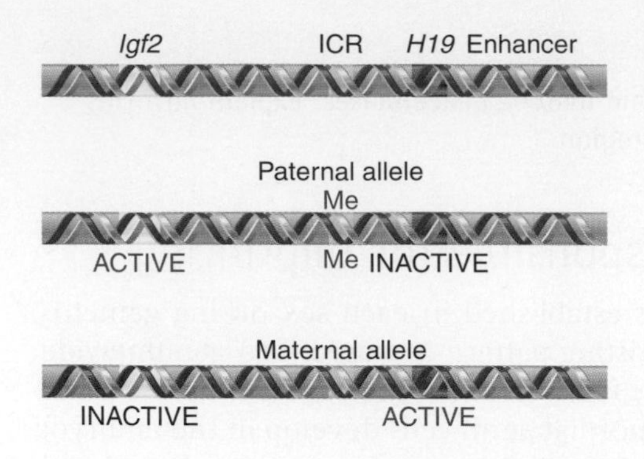

FIGURE 29.18 *ICR* is methylated on the paternal allele, where *Igf2* is active and *H19* is inactive. *ICR* is unmethylated on the maternal allele, where *Igf2* is inactive and *H19* is active.

if the wild-type allele is the imprinted (silenced) allele (assuming the gene is essential). Imprinted alleles exhibit classic epigenetic inheritance (as opposed to Mendelian inheritance): although the paternal and maternal alleles can have identical sequences, they display different properties, depending on which parent provided them. These properties are inherited through meiosis and the subsequent somatic mitoses.

Imprinted genes are estimated to comprise 1% to 2% of the mammalian transcriptome, and they are sometimes clustered. More than half of the ~25 known imprinted genes in mouse are contained in two particular regions, each containing both maternally and paternally expressed genes. This suggests the possibility that imprinting mechanisms may function over long distances. Some insights into this possibility come from deletions in the human population that cause the Prader-Willi and Angelman diseases. Most cases of these neurodevelopmental disorders are caused by the same 4 Mb deletion on chromosome 15, but the syndromes are different, depending on which parent contributed the deletion. The reason is that the deleted region includes at least one gene that is paternally imprinted and at least one that is maternally imprinted (in fact there are several imprinted genes in the region, but the contributions of each to the diseases are not fully understood). In addition to these genes, this region also contains an "imprint center" that acts at a distance to control sex-specific imprinting of genes in the surrounding area, and small deletions in this imprint center can also cause Prader-Willi or Angelman's. The basic effect of these deletions is to prevent a father from resetting the paternal mode to a chromosome inherited from his mother. The result is that these genes remain in maternal mode, so that the paternal as well as maternal alleles are silent in the offspring, resulting in Prader-Willi. The inverse effect is found in Angelman's syndrome.

Imprinting is determined by the state of methylation of a *cis*-acting site near a target gene or genes. These regulatory sites are known as *d*ifferentially *m*ethylated *d*omains (DMDs) or *i*mprinting *c*ontrol *r*egions (ICRs). Deletion of these sites removes imprinting, and the target loci then behave the same in both maternal and paternal genomes.

The behavior of a region containing two genes, *Igf2* and *H19*, illustrates the ways in which methylation can control gene activity. **FIGURE 29.18** shows that these two genes react oppositely to the state of methylation at the ICR located between them. The ICR is methylated on the paternal allele. *H19* shows the typical response of inactivation. Note, however, that *Igf2* is expressed. The reverse situation is found on a maternal allele, where the ICR is not methylated. *H19* now becomes expressed, but *Igf2* is inactivated.

The control of *Igf2* is exercised by an insulator function of the ICR (Insulators were discussed in *Section 10.12, Insulators Define Independent Domains*). **FIGURE 29.19** shows that when the ICR is unmethylated, it binds the protein CTCF. This creates a functional insulator that blocks an enhancer from activating the *Igf2* promoter. This is an unusual effect in which methylation indirectly activates a gene by blocking an insulator. The regulation of *H19* shows the more usual mechanism of control in which methylation creates an inactive imprinted state.

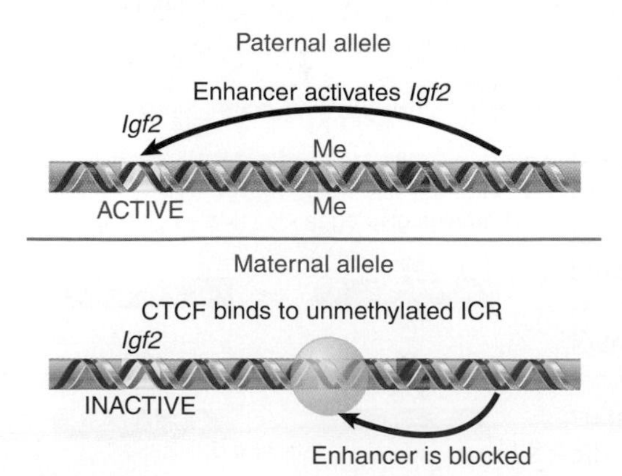

FIGURE 29.19 The *ICR* is an insulator that prevents an enhancer from activating *Igf2*. The insulator functions only when it binds CTCF to unmethylated DNA.

KEY CONCEPTS

- Paternal and maternal alleles may have different patterns of methylation at fertilization.
- Methylation is usually associated with inactivation of the gene.
- When genes are differentially imprinted, survival of the embryo may require that the functional allele be provided by the parent with the unmethylated allele.
- Survival of heterozygotes for imprinted genes is different, depending on the direction of the cross.
- Imprinted genes occur in clusters and may depend on a local control site where *de novo* methylation occurs unless specifically prevented.
- Imprinted genes are controlled by methylation of *cis*-acting sites.

CONCEPT AND REASONING CHECK

In placental mammals, X-inactivation is random, but in marsupials, the paternal X chromosome is always inactivated, analogous to imprinting an entire chromosome. What does this mean for the expression of X-linked traits in marsupials?

29.8 Yeast Prions Show Unusual Inheritance

One of the clearest cases of the dependence of epigenetic inheritance on the condition of a protein is provided by the behavior of proteinaceous infectious agents, or prions. They have been characterized in two circumstances: by genetic effects in yeast and as the causative agents of neurological diseases in mammals, including humans. A striking epigenetic effect is found in yeast, where two different states can be inherited that map to a single genetic locus, *although the sequence of the gene is the same in both states*. A number of prions have been identified in yeast, one of the best characterized is [*PSI*+]. The two different states are [*psi*$^-$] (noninfectious) and [*PSI*$^+$] (prion). A switch in condition occurs at a low frequency as the result of a spontaneous transition between the states.

The [*psi*] genotype maps to the locus *SUP35*, which codes for a translation termination factor. **FIGURE 29.20** summarizes the effects of the Sup35 protein in yeast. In wild-type cells, which are characterized as [*psi*$^-$], the gene is active, and Sup35 protein terminates protein synthesis. In cells of the mutant [*PSI*$^+$] type, the factor does not function, which causes a failure to terminate protein synthesis properly.

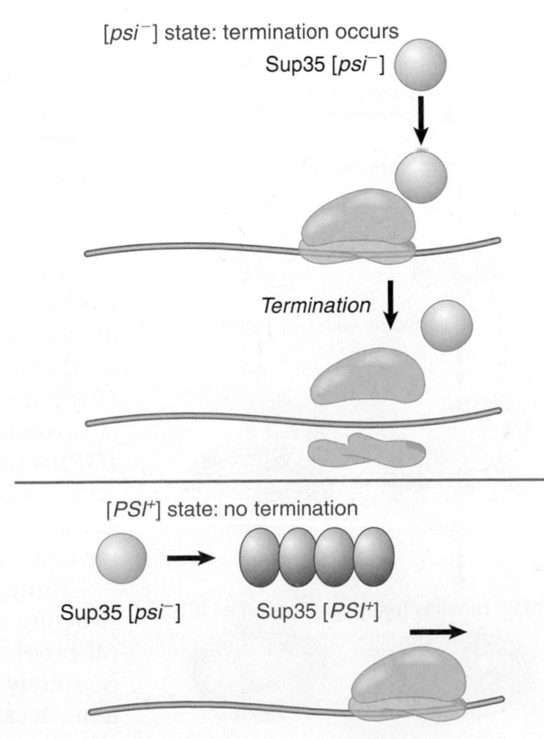

FIGURE 29.20 The state of the Sup35 protein determines whether termination of translation occurs.

[*PSI*$^+$] strains have unusual genetic properties. When a [*psi*$^-$] strain is crossed with a [*PSI*$^+$] strain, *all the progeny are* [*PSI*$^+$]. This is a pattern of inheritance that would be expected of an extrachromosomal agent, but the [*PSI*$^+$] trait cannot be mapped to any such nucleic acid. The [*PSI*$^+$] trait is *metastable*, which means that although it is inherited by most progeny, it is lost at a higher rate than is consistent with mutation. Similar behavior is shown also by the locus *URE2*, which

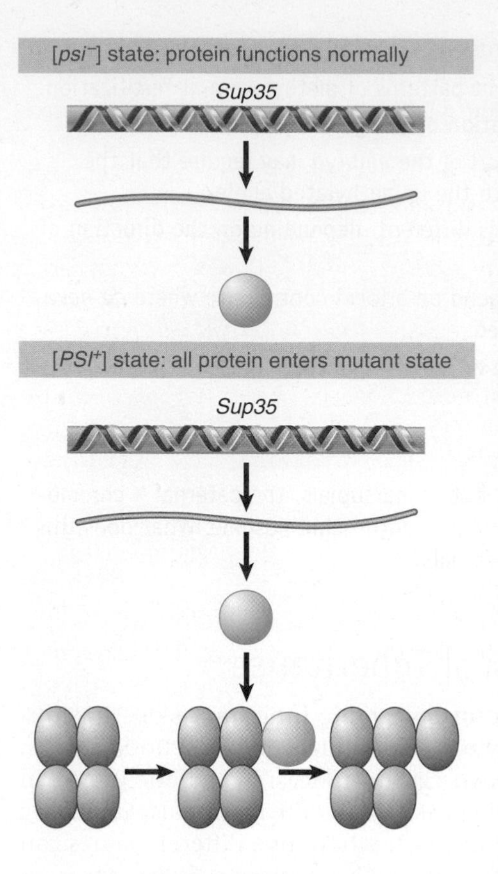

FIGURE 29.21 Newly synthesized Sup35 protein is converted into the [*PSI*⁺] state by the presence of preexisting [*PSI*⁺] protein.

codes for a protein required for nitrogen-mediated repression of certain catabolic enzymes. When a yeast strain is converted into an alternative state, called [*URE3*], the Ure2 protein is no longer functional.

The [*PSI*⁺] state is determined by the conformation of the Sup35 protein. In a wild-type [*psi*⁻] cell, the protein displays its normal function. In a [*PSI*⁺] cell, though, the protein is present in an alternative conformation in which its normal function has been lost. To explain the unilateral dominance of [*PSI*⁺] over [*psi*⁻] in genetic crosses, we must suppose that *the presence of protein in the* [*PSI*⁺] *state causes all the protein in the cell to enter this state*. This requires an interaction between the [*PSI*⁺] protein and newly synthesized protein, which probably reflects the generation of an oligomeric state in which the [*PSI*⁺] protein has a nucleating role, as illustrated in **FIGURE 29.21**.

A feature common to both the Sup35 and Ure2 proteins is that each consists of two domains that function independently. The C-terminal domain is sufficient for the activity of the protein. The N-terminal domain is sufficient for formation of the structures that make the protein inactive. Thus yeast in which the N-terminal domain of Sup35 has been deleted cannot acquire the [*PSI*⁺] state, and the presence of a [*PSI*⁺] N-terminal domain is sufficient to maintain Sup35 protein in the [*PSI*⁺] condition. The critical feature of the N-terminal domain is that it is rich in glutamine and asparagine residues.

Loss of function in the [*PSI*⁺] state is due to the sequestration of the protein in an oligomeric complex. Sup35 protein in [*PSI*⁺] cells is clustered in discrete foci, whereas the protein in [*psi*⁻] cells is diffused in the cytosol. Sup35 protein from [*PSI*⁺] cells forms **amyloid fibers** *in vitro*; these have a characteristic high content of β-sheet structures, which are believed provide a surface at the ends of the filaments that "template" further assembly of the variant structure.

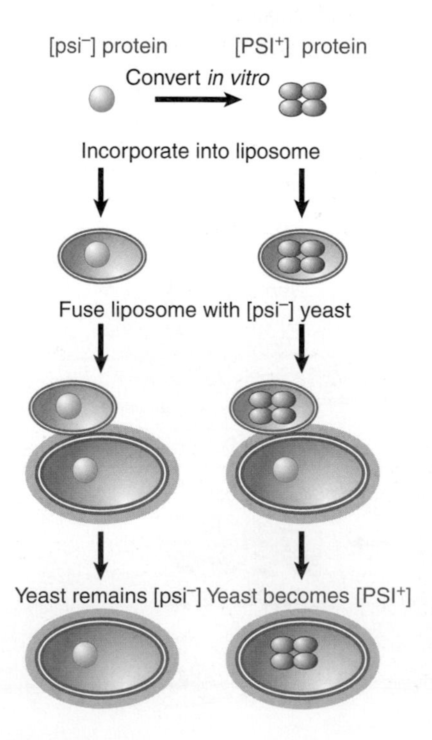

FIGURE 29.22 Purified protein can convert the [*psi*⁻] state of yeast to [*PSI*⁺].

▸ **amyloid fibers** Insoluble fibrous protein polymers with a cross β-sheet structure, generated by prions or other dysfunctional protein aggregations (such as in Alzheimers).

The involvement of protein conformation (rather than covalent modification) is suggested by the effects of conditions that affect protein structure. Denaturing treatments cause loss of the [*PSI*⁺] state. In particular, the chaperone Hsp104 is involved in inheritance of [*PSI*⁺]. Deletion of *HSP104* prevents maintenance of the [*PSI*⁺] state, and it is believed that Hsp104 actually disaggregates amyloid filaments to create growth points for new filaments.

Using the ability of Sup35 to form the inactive structure *in vitro*, it is possible to provide biochemical proof that the inheritance of the [*PSI+*] state is entirely dependent on the structure of the protein. **FIGURE 29.22** illustrates a striking experiment in which the protein was converted to the inactive form *in vitro*, put into liposomes (where, in effect, the protein is surrounded by an artificial membrane), and then introduced directly into cells by fusing the liposomes with [*psi*$^-$] yeast. The yeast cells were converted to [*PSI*$^+$]!

The ability of yeast to form the [*PSI*$^+$] prion state depends on the genetic background. The yeast must be [*PIN*$^+$] in order for the [*PSI*$^+$] state to form. The [*PIN*$^+$] condition itself is an epigenetic state: the prion form of the protein Rnq1. Rnq1 and other prion-forming proteins share the characteristic of Sup35 of having Gln/Asn-rich domains. Overexpression of these domains in yeast stimulates formation of the [*PSI*$^+$] state. This suggests that there is a common model for the formation of the prion state that involves aggregation of the Gln/Asn domains into self-propagating amyloid structure.

How does the presence of one Gln/Asn protein influence the formation of prions by another? We know that the formation of Sup35 prions is specific to Sup35 protein; that is, it does not occur by cross-aggregation with other proteins. This suggests that the yeast cell may contain soluble proteins that antagonize prion formation. These proteins are not specific for any one prion. As a result, the introduction of any Gln/Asn domain protein that interacts with these proteins will reduce the concentration. This will allow other Gln/Asn proteins to aggregate more easily.

Prions have recently been linked to chromatin remodeling factors. Swi1 is a subunit of the SWI/SNF chromatin remodeling complex (see *Section 28.7, Chromatin Remodeling Is an Active Process*), and this protein can become a prion. Swi1 aggregates in [SWI$^+$] cells but not in nonprion cells and is dominantly and cytoplasmically inherited. This suggests that inheritance through proteins can impact chromatin remodeling and potentially affect gene expression throughout the genome.

KEY CONCEPTS

- The Sup35 protein in its wild-type soluble form [*psi*−] is a termination factor for translation.
- It can also exist in an alternative form of oligomeric aggregates [*PSI*+], in which it is not active in protein synthesis.
- The presence of the oligomeric form causes newly synthesized protein to acquire the inactive structure.
- Conversion between the two forms is influenced by chaperones.
- A number of other prion-forming proteins, such as Swi1, have been identified in yeast; all share Gln/Asn-rich domains.

CONCEPT AND REASONING CHECK

Hsp104 promotes prion formation, but *overexpression* of Hsp104 interferes with prion inheritance. Why might this be?

29.9 Prions Cause Diseases in Mammals

Prion diseases have been found in sheep, elk, cows, and humans. The basic phenotype is an ataxia—a neurodegenerative disorder that is manifested by an inability to remain upright. The name of the disease in sheep, **scrapie**, reflects the phenotype: The sheep rub against walls in order to stay upright. Scrapie can be perpetuated by inoculating sheep with tissue extracts from infected animals. The human disease **kuru** was discovered in New Guinea, where it appeared to be perpetuated by cannibalism, in particular the eating of brains. Related diseases in Western populations with a pattern of genetic transmission include Gerstmann-Straussler syndrome and the related Creutzfeldt-Jakob disease (CJD), which occurs sporadically. Most recently, a disease resembling CJD appears to have been transmitted by consumption of meat from cows suffering from "mad cow" disease.

▸ **scrapie** A disease caused by an infective agent made of protein (a prion).

▸ **kuru** A human neurological disease caused by prions. It may be caused by eating infected brains.

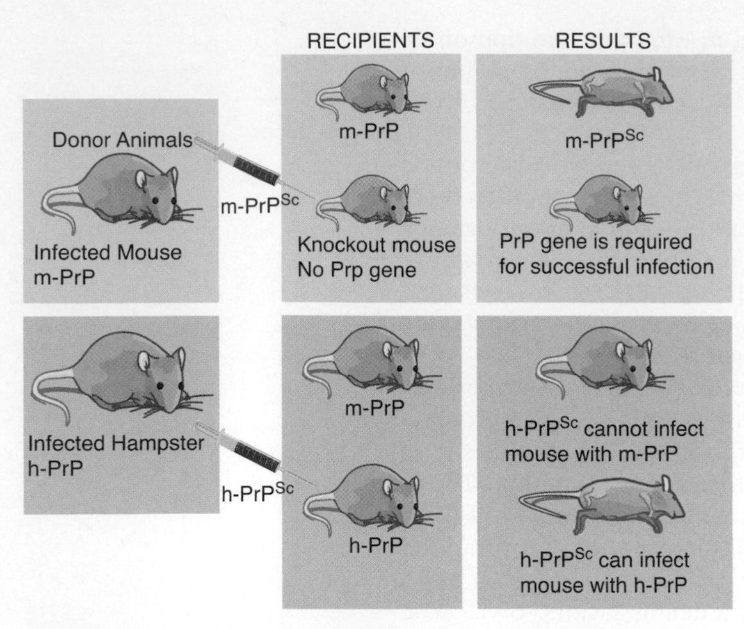

FIGURE 29.23 A PrP^{Sc} protein can infect only an animal that has the same type of endogenous PrP^{C} protein.

When tissue from scrapie-infected sheep is inoculated into mice, the disease occurs in a period ranging from 75 to 150 days. The active component is a protease-resistant protein. The protein is encoded by a gene that is normally expressed in the brain. The form of the protein in normal brain, called PrP^{C}, is sensitive to proteases. Its conversion to the resistant form, called PrP^{Sc}, is associated with occurrence of the disease; which causes neurotoxicity and eventual death. As with the yeast prions, this is a case of epigenetic inheritance in which there is no change in genetic information (because normal and diseased cells have the same *PrP* gene sequence), but the PrP^{Sc} form of the protein is the infectious agent (whereas PrP^{C} is harmless). The PrP^{Sc} form has a high content of β sheets, which form an amyloid fibrillous structure that is absent from the PrP^{C} form. The basis of the difference between the PrP^{Sc} and PrP^{C} froms appears to lie with a change of conformation rather than with any covalent alteration. Both proteins are glycosylated and linked to the membrane by a GPI-linkage.

The assay for infectivity in mice allows the dependence on protein sequence to be tested. **FIGURE 29.23** illustrates the results of some critical experiments. In the normal situation, PrP^{Sc} protein extracted from an infected mouse will induce disease (and ultimately kill) when it is injected into a recipient mouse. If the *PrP* gene is deleted, a mouse becomes resistant to infection. This experiment demonstrates two things. First, the endogenous protein is necessary for an infection, because it provides the raw material that is converted into the infectious agent. Second, the cause of disease is not the removal of the PrP^{C} form of the protein, because a mouse with no PrP^{C} survives normally: the disease is caused by a gain-of-function in PrP^{Sc}. If the PrP gene is altered to prevent the GPI linkage from occurring, mice infected with PrP^{Sc} do not develop disease, which suggests that the gain of function involves an altered signaling function for which the GPI-linkage is required.

The existence of species barriers allows hybrid proteins to be constructed to delineate the features required for infectivity. The original preparations of scrapie were perpetuated in several types of animal, but these cannot always be transferred readily. For example, mice are resistant to infection from prions of hamsters. This means that hamster-PrP^{Sc} cannot convert mouse-PrP^{C} to PrP^{Sc}. The situation changes, though, if the mouse *PrP* gene is replaced by a hamster *PrP* gene. A mouse with a hamster *PrP* gene is sensitive to infection by hamster PrP^{Sc}. This suggests that the conversion of cellular PrP^{C} protein into the Sc state requires that the PrP^{Sc} and PrP^{C} proteins have matched sequences.

There are different "strains" of PrP^{Sc}, which are distinguished by characteristic incubation periods upon inoculation into mice. This implies that the protein is not restricted solely to alternative states of PrP^{C} and PrP^{Sc}, but rather that there may be multiple Sc states. These differences must depend on some self-propagating property of the protein other than its sequence. If conformation is the feature that distinguishes PrP^{Sc} from PrP^{C}, then there must be multiple conformations, each of which has a self-templating property when it converts PrP^{C}.

The probability of conversion from PrP^{C} to PrP^{Sc} is affected by the sequence of PrP. Gerstmann-Straussler syndrome in humans is caused by a single amino acid change in PrP. This is inherited as a dominant trait. If the same change is made in the mouse PrP gene, mice develop the disease. This suggests that the mutant protein has an increased probability of spontaneous conversion into the Sc state. Similarly, the sequence of the PrP gene determines the susceptibility of sheep to develop the disease spontaneously; the combination of amino acids at three positions (codons 136, 154, and 171) determines susceptibility.

The prion offers an extreme case of epigenetic inheritance, in which the infectious agent is a protein that can adopt multiple conformations, each of which has a self-templating property. This property is likely to involve the state of aggregation of the protein.

KEY CONCEPTS

- The protein responsible for scrapie exists in two forms: the wild-type noninfectious form PrP^{C}, which is susceptible to proteases, and the disease-causing form PrP^{Sc}, which is resistant to proteases.
- The neurological disease can be transmitted to mice by injecting the purified PrP^{Sc} protein into mice.
- The recipient mouse must have a copy of the PrP gene coding for the mouse protein.
- The PrP^{Sc} protein can perpetuate itself by causing the newly synthesized PrP protein to take up the PrP^{Sc} form instead of the PrP^{C} form.
- Multiple strains of PrP^{Sc} may have different conformations of the protein.

CONCEPT AND REASONING CHECK

Would injection of a [*PSI*+] prion from yeast induce PrP^{Sc} in mice? Why or why not?

29.10 Summary

The formation of heterochromatin occurs by proteins that bind to specific chromosomal regions (such as telomeres) and that interact with histones. The formation of an inactive structure may propagate along the chromatin thread from an initiation center. Similar events occur in silencing of the inactive yeast mating-type loci. Repressive structures that are required to maintain the inactive states of particular genes are formed by the Pc-G protein complex in *Drosophila*. They share with heterochromatin the property of propagating from an initiation center.

Formation of heterochromatin may be initiated at certain sites and then propagated for a distance that is not precisely determined. When a heterochromatic state has been established, it is inherited through subsequent cell divisions. This gives rise to a pattern of epigenetic inheritance, in which two identical sequences of DNA may be associated with different protein structures, and therefore have different abilities to be expressed. This explains the occurrence of position effect variegation in *Drosophila*.

Modification of histone tails is a trigger for chromatin reorganization. Acetylation is generally associated with gene activation. Histone acetylases are found in activating complexes, whereas histone deacetylases are found in inactivating complexes. Histone methylation at specific sites in the histone N-termini is associated with gene inactivation. Some histone modifications may be exclusive or synergistic with others.

Inactive chromatin at yeast telomeres and silent mating-type loci appears to have a common cause and involves the interaction of silent information regulator (SIR) proteins with the N-terminal tails of histones H3 and H4. Formation of the inactive complex may be initiated by binding of one protein to a specific sequence of DNA; the other components may then polymerize in a cooperative manner along the chromosome.

Inactivation of one X chromosome in female placental mammals occurs at random. The *Xic* locus is necessary and sufficient to count the number of X chromosomes. The $n - 1$ rule ensures that all X chromosomes but one are inactivated. *Xic* contains the gene *Xist*, which codes for an RNA that is expressed only on the inactive X chromosome. Stabilization of *Xist* RNA is the mechanism by which the inactive X chromosome is distinguished; it is then inactivated by the activities of Polycomb complexes, heterochromatin formation, and DNA methylation.

Methylation of DNA is inherited epigenetically. Replication of DNA creates hemimethylated products, and a maintenance methylase restores the fully methylated state. Some methylation events depend on parental origin. Sperm and eggs contain specific and different patterns of methylation, with the result that paternal and maternal alleles are differently expressed in the embryo. This is responsible for imprinting, in which the unmethylated allele inherited from one parent is essential because it is the only active allele; the allele inherited from the other parent is silent. Patterns of methylation are reset during gamete formation in every generation.

Prions are proteinaceous infectious agents that are responsible for the disease of scrapie in sheep and for related diseases in human beings. The infectious agent is a variant of a normal cellular protein. The PrP^{Sc} form has an altered conformation that is self-templating: The normal PrP^{C} form does not usually take up this conformation, but does so in the presence of PrP^{Sc}. A similar effect is responsible for inheritance of the [*PSI*] element in yeast.

CHAPTER QUESTIONS

List five characteristics of heterochromatin.

1. ______________________

2. ______________________

3. ______________________

4. ______________________

5. ______________________

Explain briefly the characteristics of constitutive heterochromatin and facultative heterochromatin.

6. Constitutive ______________________

7. Facultative ______________________

Dosage compensation equalizes the level of expression of X-linked genes in the two sexes to avoid double expression of X chromosome-encoded genes in females. List three ways this has been shown to occur (in different organisms).

8. ______________________

9. ______________________

10. ______________________

What are two general mechanisms by which epigenetic effects can be inherited, or self-perpetuated?

11. ______________________

12. ______________________

List two animal diseases known to be caused by prions.

13. ______________________ **14.** ______________________

15. Hemimethylated DNA is methylated on:

A. only the newly synthesized DNA strand.
B. only the parental DNA strand.
C. only some scattered methylation sites on both DNA strands.
D. only methylated sites associated with heterochromatin.

16. Position effect variegation describes:

A. phenotypically identical cells with different genetic elements.
B. phenotypically identical cells with similar genetic elements.
C. genetically identical cells with similar phenotypes.
D. genetically identical cells with different phenotypes.

17. The HP1 protein plays a key role in formation of heterochromatin in mammals by:

A. binding to methylated histone H3.
B. binding to methylated histone H5.
C. binding to acetylated histone H1.
D. binding to acetylated histone H2B.

18. A *de novo* methylase recognizes:
   A. non-methylated sites and adds a methyl group to the base on one strand.
   B. hemimethylated sites and adds a methyl group to the base on one strand.
   C. methylated sites and adds methyl groups to additional bases on both strands.
   D. methylated sites and adds methyl groups to additional bases on the parental strand.
19. Prion diseases in animals cause:
   A. muscular degeneration.
   B. neurodegenerative disorder.
   C. digestive system disorder.
   D. blindness.

KEY TERMS

amyloid fibers
constitutive heterochromatin
CpG islands
***de novo* methyltransferase**
demethylase
dosage compensation
epigenetic
facultative heterochromatin
fully methylated
hemimethylated site
imprinting
kuru
maintenance methyltransferase
methyltransferase
***n* − 1 rule**
position effect variegation (PEV)
prion
proteinaceous infectious agent (prion)
scrapie
single-X hypothesis
telomeric silencing

FURTHER READING

Horn, P. J., and Peterson, C. L. (2006). Heterochromatin assembly: a new twist on an old model. *Chromo. Res.* **14**, 83–94. A review of heterochromatin assembly and maintenance focused on data obtained from studies of fission yeast.

Horsthemke, B., and Buiting, K. (2008). Genomic imprinting and imprinting defects in humans. *Adv. Gen.* **61**, 225–246. A review of imprinting in placental mammals and defects in imprinting that lead to human disease.

Kim, J. K., Samaranayake, M., and Pradhan, S. (2009). Epigenetic mechanisms in mammals. *Cell. Mol. Life Sci.* **66**, 596–612. A comprehensive review of mammalian epigenetics, which includes discussion of DNA methylation, histone modification, insulators, X inactivation, imprinting, and developmental epigenetics (e.g., PcG and trxG).

Payer, B., and Lee, J. T. (2008). X chromosome dosage compensation: how mammals keep the balance. *Ann. Rev. Gen.* **42**, 733–772. A review of the mechanism and evolution of X chromosome inactivation.

Schwartz, Y. B., and Pirotta, V. (2008) Polycomb complexes and epigenetic states. *Curr. Opin. Cell Biol.* **20**, 266–273. A review of Polycomb Group (PcG) complexes and the regulation of PcG target genes.

Shkundina, I. S., and Ter-Avanesyan, M. D. (2007) Prions. *Biochemistry (Moscow)* **72**, 1519–1536. A review of prions focusing on lessons learned from the [*PSI*+] prion of yeast.

30

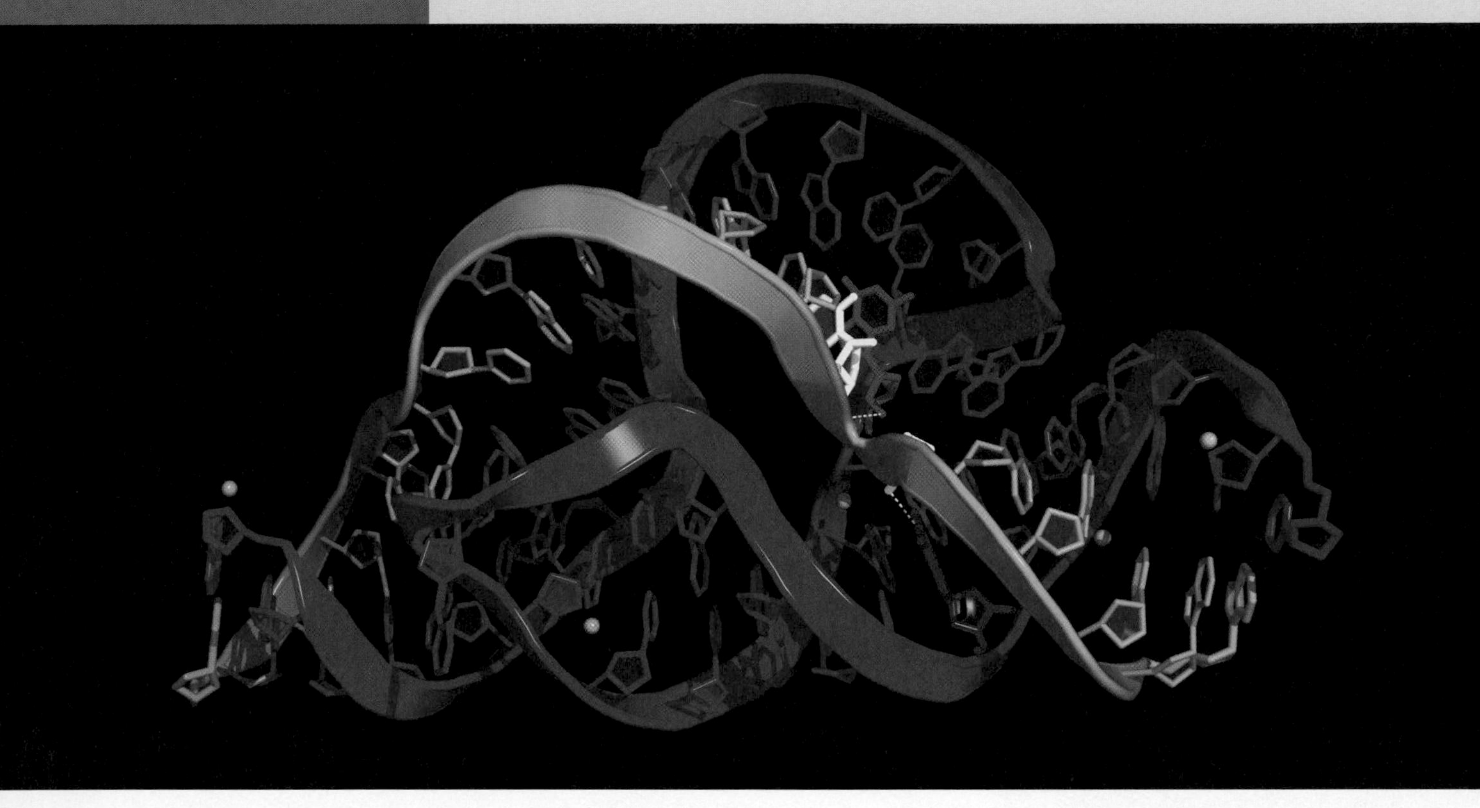

A hammerhead ribozyme that catalyzes a self-cleaving reaction. Many plant viroids are hammerhead RNAs. Reproduced from M. Martick et al., Solvent Structure and Hammerhead Ribozyme Catalysis. *Chem. Bio.* **15**, pp. 332–342. Copyright 2008, with permission from Elsevier (http://www.sciencedirect.com/science/journal/10745521). Photo courtesy of William Scott, University of California, Santa Cruz.

Regulatory RNA

CHAPTER OUTLINE

30.1 Introduction

The basic principle of regulation is that gene expression is controlled by a regulator that interacts with a specific sequence or structure in either DNA or mRNA at some stage prior to the synthesis of protein. The stage of expression that is controlled can be transcription, when the target for regulation is DNA, or it can be at translation, when the target for regulation is RNA. Control during transcription can be at initiation, elongation, or termination. The regulator can be a protein or an RNA. "Controlled" can mean that the regulator turns off (represses) or turns on (activates) the target. Expression of many genes can be coordinately controlled by a single regulator gene on the principle that each target contains a copy of the sequence or structure that the regulator recognizes. Regulators may themselves be regulated, most typically in response to small molecules whose supply responds to environmental conditions. Regulators may be controlled by other regulators to make complex circuits.

Let's compare the ways that different types of regulators work.

Many protein regulators work on the principle of allosteric changes. The protein has two binding sites: one for a nucleic acid target, the other for a small molecule. Binding of the small molecule to its site changes the conformation in such a way as to alter the affinity of the other site for the nucleic acid. The way in which this happens is known in detail for the *lac* repressor in *E. coli* (see *Chapter 26, The Operon*). Protein regulators are often multimeric, with a symmetrical organization that allows two subunits to contact a palindromic target on DNA. This can generate cooperative binding effects that create a more sensitive response to regulation.

Regulation via RNA may use changes in secondary structure base pairing as the guiding principle. The ability of an RNA to shift between different conformations with regulatory consequences is the nucleic acid's alternative to the allosteric changes of protein conformation. The changes in structure may result from either intramolecular or intermolecular interactions.

The most common role for intramolecular changes is for an RNA molecule to assume alternative secondary structures by utilizing different schemes for base pairing. The properties of the alternative conformations may be different. Changes in secondary structure of an mRNA can result in a change in its ability to be translated. Secondary structure also is used to regulate the termination of transcription, when the alternative structures differ in whether they permit termination or not (as we saw with attenuation in *Chapter 26, The Operon*).

In intermolecular interactions, an RNA regulator recognizes its target by the familiar principle of complementary base pairing. **FIGURE 30.1** shows that the regulator is sometimes a small RNA molecule with extensive secondary structure, but with a single-stranded region(s) that is complementary to a single-stranded region in its target. The formation of a double-stranded region between regulator and target can have two types of consequence:

- Formation of the double-stranded structure may itself be sufficient for regulatory purposes. In some cases, a protein(s) can bind only to the single-stranded form of the target sequence and is, therefore, prevented from acting by duplex formation. In other cases, the duplex region becomes a target for binding, for example, by nucleases that degrade the RNA and therefore prevent its expression.
- Duplex formation may be important because it sequesters a region of the target RNA that would otherwise participate in some alternative secondary structure.

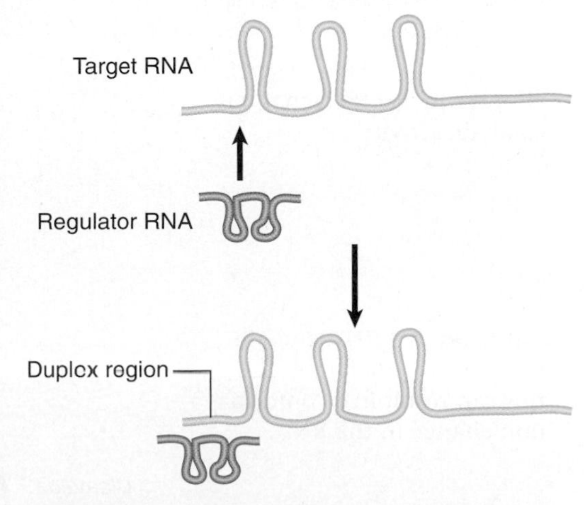

FIGURE 30.1 A regulator RNA is a small RNA with a single-stranded region that can pair with a single-stranded region in a target RNA.

We once thought that RNA was merely structural: mRNA carried the blueprint for the synthesis of proteins, rRNA was the structural component of the ribosome, and tRNA shuttled amino acids to the ribosome. We now see a vast new RNA world where RNAs have numerous functions, where mRNA can regulate its own translation (see *Section 26.13*,

The trp *Operon Is Controlled by Attenuation*), where rRNA catalyzes peptide bond formation and with tRNA participates in the mechanism of fidelity of translation (see *Section 24.3, Special Mechanisms Control the Accuracy of Translation*).

This new RNA world extends far beyond the three major RNA types just described to include dozens of different new RNAs. These RNAs can function as guide RNAs, or splicing cofactors. In addition, there is a large and very heterogenous class of RNAs with regulatory function, to be described shortly. We have not yet uncovered all the mysteries of the RNA world.

KEY CONCEPT

- RNA functions as a regulator by forming a region of secondary structure (either inter- or intramolecular) that changes the properties of a target sequence.

CONCEPT AND REASONING CHECK

How could mRNA sequences that are not coding be important in the mRNA?

30.2 A Riboswitch Can Control Expression of an mRNA

As we have seen in *Section 26.13, The* trp *Operon Is Also Controlled by Attenuation,* and in *Section 26.15, Translation Can Be Regulated,* an mRNA is more than simply an open reading frame. We have seen in bacteria that regions in the 5′ UTR (5′ *un*translated *r*egion) contain elements that, due to coupled transcription/translation, can control transcription termination. We have also seen that the 5′ UTR sequence itself can determine if a mRNA is a "good" message, which supports a high level of translation, or a "poor" message, which does not. What we will see next is another type of element that can be present in the 5′ UTR—one that can control expression of the mRNA with a different mechanism called a **riboswitch**.

▸ **riboswitch** A catalytic RNA whose activity responds to a metabolite product or another small ligand.

A *riboswitch* is an RNA element that can assume alternate base pairing configurations with alternate outcomes. **FIGURE 30.2** summarizes one type of riboswitch in the regulation of a system that produces the metabolite GlcN6P (Glucosamine-6-phosphate). The gene *glmS* codes for an enzyme that synthesizes GlcN6P from fructose-6-phosphate and glutamine. GlcN6P is a fundamental intermediate in bacterial cell wall biosynthesis. The mRNA contains a long 5′ UTR before the coding region of the mRNA. Within the 5′ UTR is a **ribozyme**—a sequence of RNA that has catalytic activity (see *Section 23.4, Ribozymes Have Various Catalytic Activities*). In this case, the catalytic activity is an endonuclease that cleaves its own RNA. It is activated by binding of the metabolite product, GlcN6P, to the **aptamer** region of the ribozyme. The aptamer is the RNA domain that binds the metabolite. The consequence is that accumulation of GlcN6P activates the ribozyme, which cleaves the mRNA, which, in turn, prevents further translation. This is a parallel

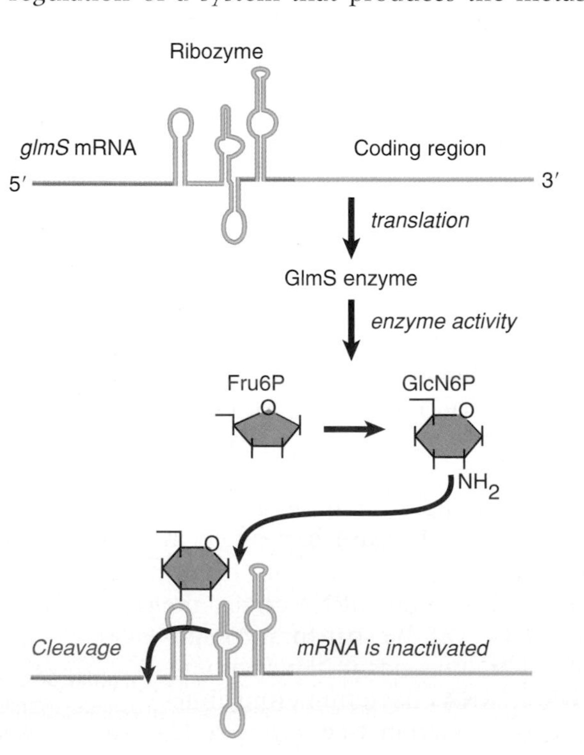

FIGURE 30.2 The 5′ untranslated region of the mRNA for the enzyme that synthesizes GlcN6P contains a ribozyme that is activated by the metabolic product. The ribozyme inactivates the mRNA by cleaving it.

▸ **ribozyme** An RNA that has catalytic activity.

▸ **aptamer** An RNA domain that binds a small molecule; this can result in a conformation change in the RNA.

to allosteric control of a repressor protein by the end product of a metabolic pathway. There are numerous examples of such riboswitches in bacteria.

Not all riboswitches encode a ribozyme that controls the mRNA stability. Other riboswitches have alternative configurations of the RNA that allow or prevent expression of the mRNA by affecting ribosome binding or transcription termination. Riboswitches are found more commonly in bacteria than in eukaryotes. There are multiple classes of riboswitches.

An interesting eukaryotic riboswitch has been described in the fungus *Neurospora* as a control mechanism for alternative splicing. The gene *NMT1* (involved with vitamin B1 synthesis) produces an mRNA precursor with a single intron that has two splice donor sites. Alternate use of these two sites can produce a functional or nonfunctional message, depending on the concentration of a vitamin B1 metabolite, TPP (*t*hiamine *p*yrophos*p*hate). Thus product concentration controls product formation, a form of repressible control. The selection of the splice site is controlled by a riboswitch in the intron. At a low concentration of TPP, the proximal splice donor site is chosen and the distal splice donor site is blocked by the riboswitch, as seen in **FIGURE 30.3**. This splice site produces a functional mRNA. At high concentration, TPP binds the riboswich to alter its configuration and prevents blocking the distal splice donor site to allow the alternate splice, which produces a nonfunctional mRNA.

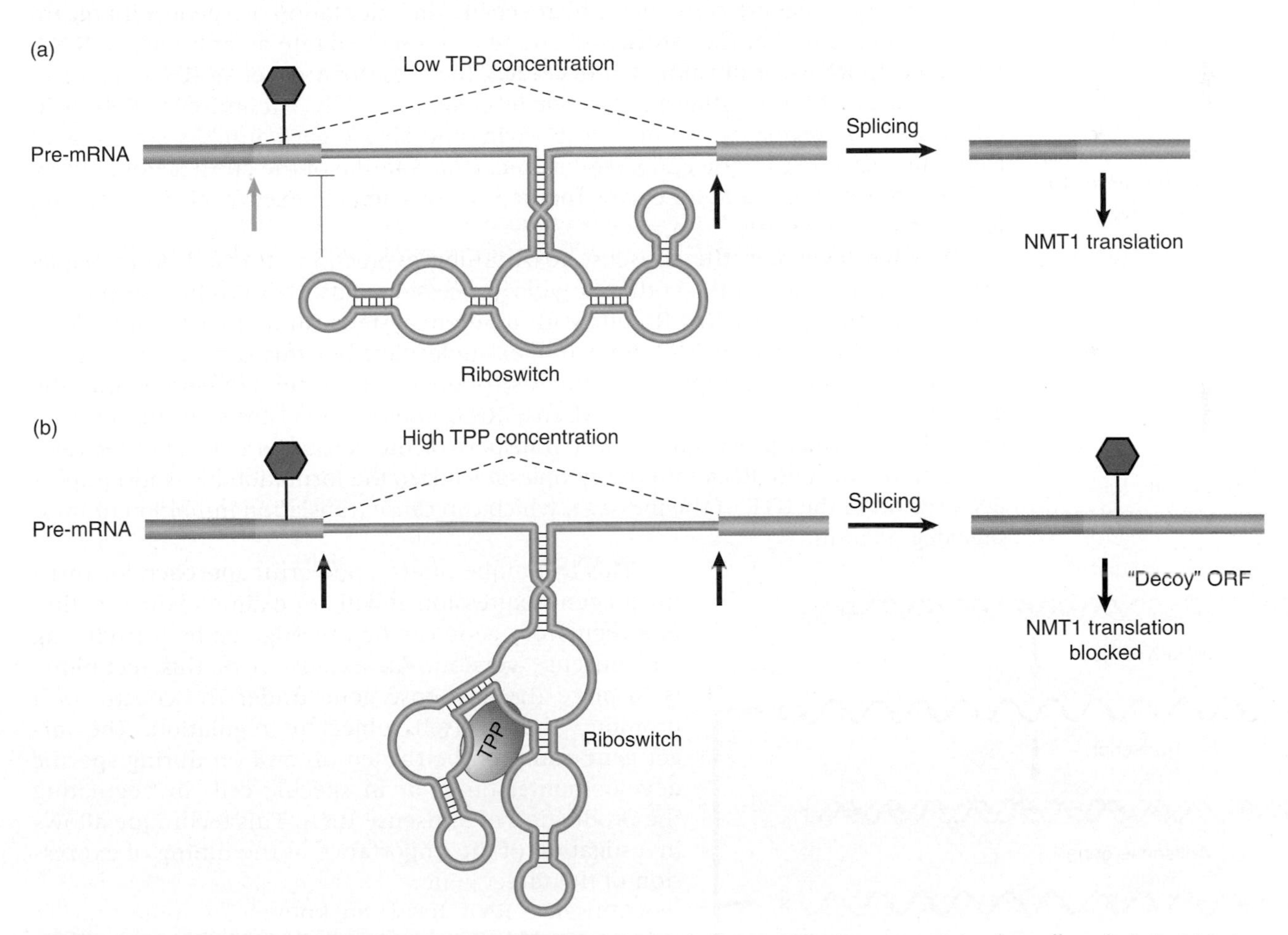

FIGURE 30.3 The expression of the *NMT1* gene is regulated at the level of pre-mRNA alternate splicing by a riboswitch that binds to TPP. (a) At low concentration of TPP, the TPP-binding aptamer of the riboswitch base pairs with sequences surrounding a splice site (red blocking line) in a nearby noncoding sequence and prevents its selection by the splice machinery. A distal splice site (green arrow) is selected, however, resulting in a shorter functional mRNA containing an open reading frame. (b) At high TPP levels, the aptamer undergoes a conformational rearrangement so that the region that was previously bound to the nearby splice site is now used to bind to TPP. This ultimately generates a longer but nonproductive splice variant that contains a short decoy region with a stop codon (indicated by the red stop sign) preventing gene expression. Reprinted by permission from Macmillan Publishers Ltd: B. J. Blencowe and M. Khanna, *Nature* **447**, pp. 391–393, copyright 2007.

KEY CONCEPTS

- A riboswitch is an RNA whose activity is controlled by the metabolite product or another small ligand. (A ligand is any molecule that binds to another.)
- A riboswitch may be a ribozyme.

CONCEPT AND REASONING CHECK

An mRNA consists of how many different elements?

30.3 Noncoding RNAs Can Be Used to Regulate Gene Expression

Base pairing offers a powerful means for one RNA to control the activity of another. There are many cases in both prokaryotes and eukaryotes where a (usually rather short) single-stranded RNA base pairs with a complementary region of a mRNA, and as a result it prevents expression of the mRNA. One of the earliest illustrations of this effect was provided by an artificial situation in which **antisense genes** were introduced into eukaryotic cells.

- **antisense gene** A gene that has a complementary sequence to an RNA that is its target.
- **ncRNA (noncoding RNA)** A noncoding RNA that does not contain an open reading frame.
- **antisense RNA** RNA that has a complementary sequence to an RNA that is its target.

Antisense genes are constructed by reversing the orientation of a gene with regard to its promoter, so that the "antisense" strand is transcribed into an antisense **ncRNA (*noncoding RNA*)**, as illustrated in FIGURE 30.4. Synthesis of **antisense RNA** can inactivate a target RNA in either prokaryotic or eukaryotic cells. An antisense RNA is in effect an RNA regulator. An antisense thymidine kinase gene inhibits synthesis of thymidine kinase from the endogenous gene. Quantitation of the effect is not entirely reliable, but it seems that an excess (perhaps a considerable excess) of the antisense RNA may be necessary.

At what level does the antisense RNA inhibit expression? It could, in principle, prevent transcription of the authentic gene, processing of its RNA product, or translation of the messenger RNA. Results with different systems show that the inhibition depends on formation of RNA-RNA duplex molecules, but this can occur either in the nucleus or in the cytoplasm. In the case of an antisense gene integrated into the genome in a cultured cell, sense-antisense RNA duplexes may form in the nucleus, preventing normal processing and/or transport of the sense RNA. In another case, injection of antisense RNA into the cytoplasm leads to the formation of a short duplex RNA region in the UTR of the message, which can cause translation inhibition or message degradation.

- **nested gene** A gene located within the intron of another gene.

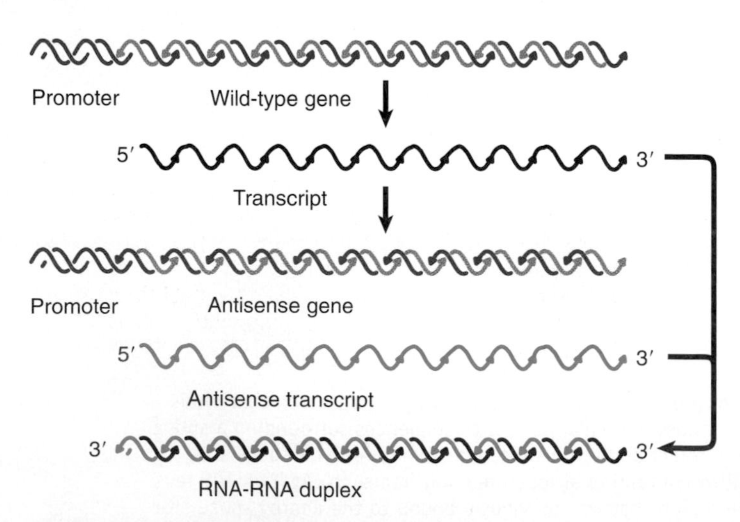

FIGURE 30.4 Antisense RNA can be generated by reversing the orientation of a gene with respect to its promoter and can anneal with the wild-type transcript to form duplex DNA.

This technique offers a powerful approach for turning off gene expression at will; for example, the function of a regulatory gene can be investigated by introducing an antisense version. An extension of this technique is to place the antisense gene under the control of a promoter that is itself subject to regulation. The target gene can then be turned off and on during specific developmental times or in specific cells by regulating the production of antisense RNA. This technique allows investigation of the importance of the timing of expression of the target gene.

Antisense RNA has been known for some time in eukaryotes. The first genome sequencing projects demonstrated that **nested genes** (genes located within the introns of other genes) are widespread. They are more common than was first thought, comprising as much as 5% to 10% of genes. If the nested gene is transcribed from the opposite strand, then antisense RNA is produced. This head-to-head arrangement of a nested gene will also lead to **transcriptional interference (TI)** because both genes cannot be simultaneously transcribed.

▸ **transcription interference (TI)** The phenomenon in which transcription from one promoter interferes directly with transcription from a second, linked promoter.

Transcriptional interference is emerging as a significant mechanism of transcriptional regulation, and it can actually occur both when an interfering RNA is produced in an antisense orientation, as described before, or even in the sense orientation. For example, the yeast *SER3* gene (involved in serine biosynthesis) is normally repressed in the presence of serine and turned on in its absence (a repressible gene). It turns out that under serine-rich, repressive conditions, a noncoding RNA is expressed from the intergenic region upstream of the *SER3* promoter and is transcribed from the same strand as *SER3*. This RNA (named *SER3 regulatory gene*, or *SRG1*) does not encode a protein, but its high expression serves to disrupt transcription initiation at the *SER3* promoter. *SRG1* is induced by serine, so in this case the end product of the biosynthetic pathway regulates *SER3* by causing transcriptional interference by a sense transcript at the *SER3* promoter. It is important to note that in transcriptional interference, it can be transcription *per se*, rather than the RNA product, that is responsible for the regulatory effect.

Recent experiments using both whole genome tiling arrays (probing not just the genes but the entire genome) and massive whole-cell RNA sequencing have shown that the vast majority of the eukaryotic genome is transcribed. This includes gene regions, of course, but surprisingly also includes both the coding and noncoding strands. The estimate is that as much as 70% of human genes produce antisense RNA. This pattern varies with cell type and is presumably regulated. Also transcribed are intergenic regions and simple sequence repeat heterochromatin regions, both centromeres and telomeres, previously assumed to house no information. Transcripts from both the coding (sense) and noncoding (antisense) strands can result in noncoding RNAs with regulatory properties.

A direct role for antisense RNA in transcription control has been demonstrated. For example, in the yeast *S. cerevisiae,* the gene *PHO84* is regulated in part by a class of noncoding RNAs called **cryptic unstable transcripts**, or **CUTs**. As shown in **FIGURE 30.5**, in addition to the promoter at the 5′ end of the gene, there is another promoter (which is unregulated) on the opposite strand. Transcription from this promoter on the opposite strand produces an antisense RNA. Under normal conditions this RNA is degraded by the TRAMP and exosome complex (see *Section 22.8, Newly Synthesized RNAs are Checked for Defects via a Nuclear Surveillance System*) as it is produced. In the absence of degradation or in aging cells, the antisense RNA persists. This antisense RNA or CUT, recruits histone deacetylase enzymes that remove acetate groups from histones, thereby causing the chomatin over the gene to be remodeled and condensed so that the gene can no longer be transcribed (see *Section 28.9, Histone Acetylation is Associated with Transcription Activation*). This is gene-specific remodeling directed by the antisense RNA and does not extend to the neighboring genes.

▸ **cryptic unstable transcript (CUT)** Nonprotein-coding RNAs frequently generated by promoters located at the 3′ end of a gene resulting antisense transcripts.

Since this discovery, similar examples of ncRNAs that result in alterations of local chromatin structure have been described, such as a long ncRNA transcribed from the *GAL1-10* locus (see *Section 28.13, The Yeast* GAL *Genes: A Model for Activation*

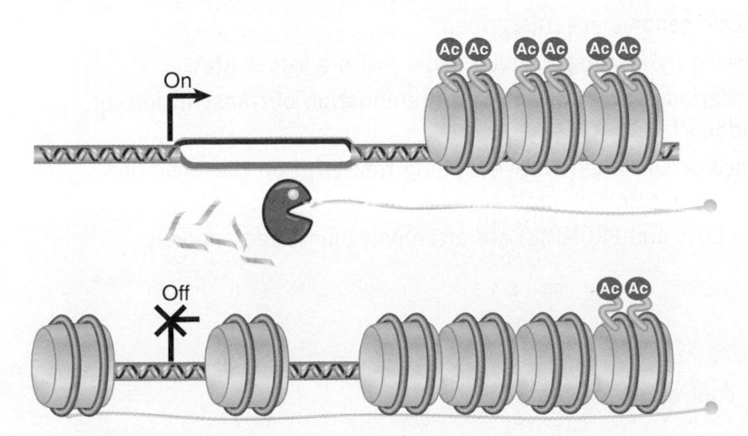

FIGURE 30.5 *PHO84* antisense RNA stabilization is paralleled by histone deacetylase recruitment, histone deacetylation, and *PHO84* transcription repression. In wild-type cells, the RNA is rapidly degraded. In aging cells, antisense transcripts are stabilized and recruit the histone deacetylase to repress transcription. Adapted from J. Camblong et al., *Cell* **131** (2007): 706–717.

and Repression) that also results in histone deacetylation (as well as methylation) to promote *GAL* gene repression. ncRNAs also prevent Ty retrotransposition through changes in chromatin structure in *trans*; this is reminiscent of the role of piRNAs in *Drosophila* (discussed in *Section 30.5, MicroRNAs Are Widespread Regulators in Eukaryotes*). One of the better understood examples of a ncRNA system is that of *XIST* transcripts used in X chromosome inactivation, described in *Section 29.5, X Chromosomes Undergo Global Changes*.

▸ **lincRNA** Large noncoding RNAs originating from chromosomal regions in between classical genes.

A different class of ncRNA is **lincRNA**, or *l*arge *i*ntergenic *n*on*c*oding RNA, transcribed from regions between genes. A number of these lincRNAs have been implicated in diseases such as cancer by causing misregulation. One example in humans is HOTAIR (*HO*X *t*ranscript *a*nt*i*sense *R*NA), a transcript that, when misregulated in adult cells, causes cancer cell metastasis as the region becomes more embryonic-like by silencing supressor genes. HOTAIR originates from the homeotic gene region *HOXC*, a region containing a set of genes that give identity to cells and tissues. These genes are important in multicellular eukaryotes during early development; misregulation of the gene can lead to tissues assuming the incorrect cell fate. The normal function of HOTAIR is to establish the chromatin conditions that are appropriate for each gene in each cell—that is, open chromatin in cells that should express the gene and closed chromatin in cells that should not. HOTAIR carries out this function by acting as an RNA scaffold to assemble the Polycomb Repressive Complex 2 (PCR2) to silence those genes that should be turned off. Note that this chromatin remodeling is one that requires transcription from the region to repress transcription.

▸ **PROMPT** Promoter upstream transcripts, short RNAs produced from both strands of DNA from active promoters.

This phenomenon may be quite widespread. In human HeLa cells, when a component of the RNA degradation machinery is disabled, vast amounts of upstream transcripts called **PROMPTs** (*prom*oter u*p*stream *t*ranscripts) are observed from active promoters. Like CUTs in yeast, this RNA is polyadenylated and very unstable. It can occur in both directions from the promoter and may be related to the fact that open chromatin is available.

▸ **paraspeckles** Small, irregularly shaped ribonucleoprotein bodies found in interchromatin nuclear locations, typically 10–20 per nucleus.

An architectural role has been proposed for another class of ncRNAs, as a scaffold for **paraspeckles**, small nuclear ribonucleoprotein bodies that are believed to be processing centers for mRNA maturation and export control from the nucleus. These are found only in mammalian cells. Typical human nuclei have 10 to 20 paraspeckles. The RNA NEAT1 (*n*uclear *e*nriched *a*utosomal *t*ranscript) is a large 4 Kb, abundant polyadenylated RNA found only in the nucleus and is the major RNA in paraspeckles.

These mechanisms offer powerful approaches for turning off genes at will. It is not, however necessarily a one-way street where a regulatory RNA is produced and simply turns off expression of a message. This system can also be balanced by the production of a counter protein that can bind to and interfere with the RNA. Thus dynamic systems can exist that can change over time according to demands placed on the cell.

KEY CONCEPTS

- Vast tracts of the eukaryotic genome are transcribed.
- A regulator RNA can function by forming a duplex region with a target RNA.
- The duplex may block initiation of translation, cause termination of transcription, or create a target for an endonuclease.
- Transcriptional interference occurs when an overlapping transcript on the same or opposite strand prevents transcrption of another gene.
- Noncoding RNAs (such as CUTs and PROMPTs) are often polyadenylated and very unstable.

CONCEPT AND REASONING CHECK

How can an antisense RNA control gene expression?

30.4 Bacteria Contain Regulator sRNAs

Bacteria contain many genes (up to hundreds) that code for regulator RNAs. These are short RNA molecules ranging from about 50 nucleotides to about 200$^+$ nucleotides that are collectively known as **sRNAs**. Some of the sRNAs are general regulators that affect many target genes; others are specific for a single transcript. These sRNAs typically function as imperfect (meaning that only small regions within the sRNA are complementary to the target) antisense RNA.

▸ **sRNA** A small bacterial RNA that functions as a regulator of gene expression.

At what level does the antisense RNA inhibit expression? It could, in principle, (1) prevent transcription of the gene, (2) affect processing of its RNA product, (3) affect translation of the messenger, or (4) affect stability of the RNA. Results with different systems show that inhibition depends on formation of RNA-RNA duplex molecules.

Base pairing offers a powerful means for one RNA to control the activity of another, as we have already seen. There are many cases in both prokaryotes and eukaryotes where a (usually rather short) single-stranded RNA base pairs with a complementary region of an mRNA, and, as a result, it prevents expression of the mRNA.

Oxidative stress in *E. coli* provides an interesting example of a general control system in which an sRNA is the regulator. When exposed to reactive oxygen species, bacteria respond by inducing antioxidant defense genes. Hydrogen peroxide activates the transcription activator OxyR, which controls the expression of several inducible genes. One of these genes is *oxyS*, which codes for a small RNA.

FIGURE 30.6 shows two salient features of the control of *oxyS* expression. In a wild-type bacterium under normal conditions, it is not expressed. The pair of gels on the left side of the figure show that it is expressed at high levels in a mutant bacterium with a constitutively active *oxyR* gene. This identifies *oxyS* as a target for activation by *oxyR*. The pair of gels on the right side of the figure show that *oxyS* RNA is transcribed within one minute of exposure to hydrogen peroxide.

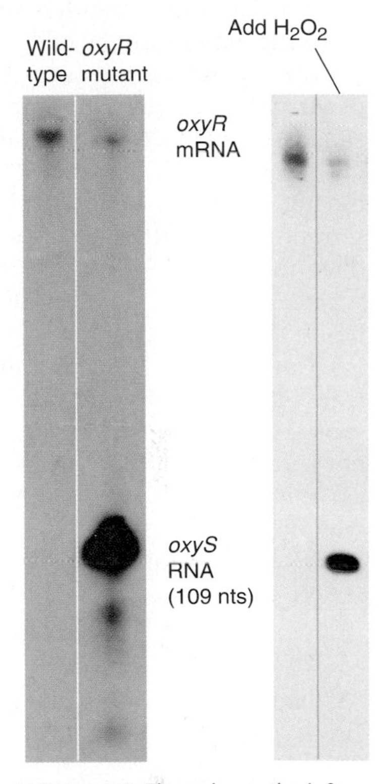

FIGURE 30.6 The gels on the left show that *oxyS* RNA is induced in an *oxyR* constitutive mutant. The gels on the right show that *oxyS* RNA is induced within one minute of adding hydrogen peroxide to a wild-type culture. Reproduced from S. Altuvia et al., A small stable RNA . . . , *Cell* **90**, pp. 44–53. Copyright 1997, with permission from Elsevier (http://www.sciencedirect.com/science/journal/00928674). Photo courtesy of Gisela Storz, National Institutes of Health.

The *oxyS* RNA is a short ncRNA of 109 nucleotides. It is a *trans*-acting regulator that affects gene expression at the level of translation. It has >10 target mRNAs; at some of them, it activates expression, and at others it represses expression. **FIGURE 30.7** shows the mechanism of repression of one target, the *flhA* mRNA. Three stem-loop double-stranded RNA structures protrude in the secondary structure of *oxyS* mRNA, and the loop closest to the 3′ terminus is complementary to a sequence just preceding the initiation codon of *flhA* mRNA. Base pairing between *oxyS* RNA and *flhA* RNA prevents the ribosome from binding to the initiation codon and therefore represses translation. There is also a second pairing interaction that involves a sequence within the coding region of *flhA*.

Another target for *oxyS* is *rpoS*, the gene coding for an alternative sigma factor (which activates the general stress response). By inhibiting production of the sigma factor, *oxyS* ensures that the specific response to oxidative stress does not trigger the response that is appropriate for other stress conditions. The *rpoS* gene is also regulated by two other sRNAs (*dsrA* and *rprA*), which activate it. These three sRNAs appear to be global regulators that coordinate responses to various environmental conditions.

The actions of all three sRNAs are assisted by an RNA-binding protein called Hfq. The Hfq protein was originally identified as a bacterial host factor needed for

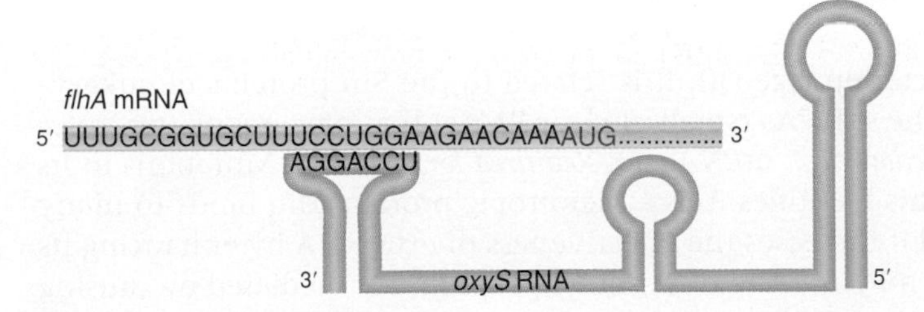

FIGURE 30.7 *oxyS* RNA inhibits translation of *flhA* mRNA by base pairing with a sequence just upstream of the AUG initiation codon.

MEDICAL APPLICATIONS

Artificial Antisense Genes Can Be Used to Turn Off Viruses and Cancer Genes

The therapeutic application of antisense technology is a turning point in molecular biology and medicine. Artificial or laboratory synthesized antisense DNA or RNA oligonucleotides can be designed and synthesized to target and consequently turn off, or *silence*, the function of virtually any gene unique to cells or viruses. Short, single-stranded DNA oligonucleotides bind between the groove of double-stranded DNA gene sequences, producing a triple helix structure that prevents its transcription into mRNA (**FIGURE B30.1**). Short RNA antisense oligonucleotides target the mRNA transcripts (sense strand) of viral or cellular genes, preventing their translation (**FIGURE B30.2**).

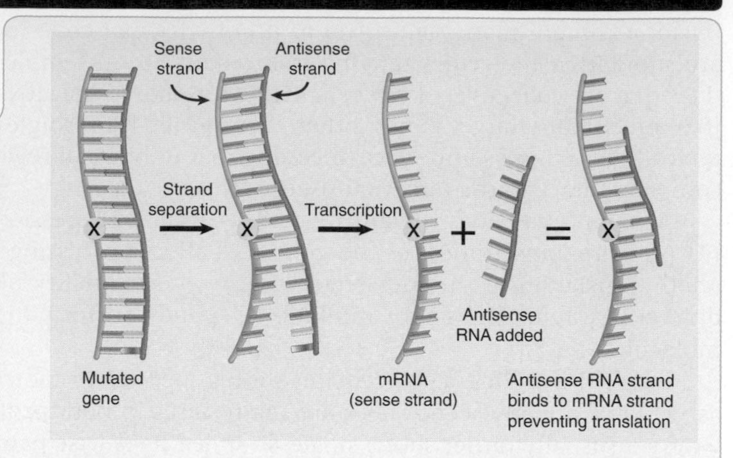

FIGURE B30.2 Preventing translation using antisense technology. Reproduced from *New Approaches to Gene Therapy* from the Genetics Science Learning Center. Used under license from the University of Utah [http://learn.genetics.utah.edu].

Compared to DNA antisense oligonucleotides, RNAi is less toxic to cells and much more potent. RNA antisense technology takes advantage of a natural phenomenon called *RNA interference*, or RNAi (also referred to as *gene silencing*). The RNA interference mechanism was originally discovered in the worm *Caenorhabditis elegans* but is known to play a pivotal role in regulating gene expression in a wide variety of organisms, including plants and humans. It is also considered a form of antiviral defense, especially in plants. When laboratory-made small interfering dsRNAs of approximately 21–23 bp in length are injected into uninfected or infected mammalian cells, the molecules are recognized by the enzymatic machinery of RNAi. Through the RNAi pathway, the end result is homology-dependent inhibition of expression of a target mRNA. Exploiting this native gene-silencing pathway is a way to regulate gene expression. Applications of RNAi have the potential to give rise to a cornucopia of drugs that silence disease-causing genes or disarm viral pathogens.

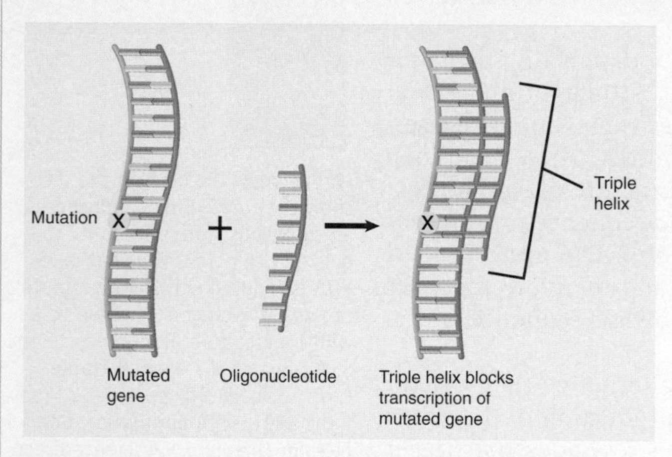

FIGURE B30.1 Preventing transcription of a mutated gene using triple-helix-forming oligonucleotides. Reproduced from *New Approaches to Gene Therapy* from the Genetics Science Learning Center. Used under license from the University of Utah (http://learn.genetics.utah.edu).

Theoretically, viral and bacterial infections should be amenable to antisense-mediated silencing. In addition, diseases caused by dominant mutations should also be amenable. The early phases of clinical trials using RNAi therapeutics are in progress to treat certain viral infections and cancers. Known and emerging viruses for which there are no effective vaccines or antiviral drugs available are an increasingly serious threat to public health. Drug designers of RNAi-based antivirals have prioritized their efforts toward treating viral infections caused by cytomegaloviruses, HIV-1,

replication of the RNA bacteriophage Qβ. It is related to the Sm proteins of eukaryotes that bind to many of the snRNAs (small nuclear RNAs) that have regulatory roles in gene expression (see *Section 21.5, snRNAs Are Required for Splicing*). Mutations in its gene have many effects; this identifies it as a pleiotropic protein. Hfq binds to many of the sRNAs of *E. coli*, and it increases the effectiveness of *oxyS* RNA by enhancing its ability to bind to its target mRNAs. The effect of Hfq is probably mediated by causing respiratory syncytial viruses (RSV), hepatitis B and C viruses, and human papilloma viruses. Critical to the design of RNAi drugs is ensuring that the antisense molecules target conserved sequences of the viral RNA genome or mRNAs that are essential viral factors. This approach may allow drug development to keep pace with viruses that mutate rapidly.

HIV-1 was the first virus targeted by RNAi treatment because the life cycle and gene expression pattern of the virus was well understood. Synthetic antisense RNAs were used to target HIV-1-encoded RNAs such as the TAR element, or *tat, rev, gag, env, vif, nef,* and reverse transcriptase mRNAs. The high mutation rate of HIV represents a substantial challenge. For this reason, an alternative approach is being tried. This targets conserved cellular cofactors, such as CCR5, required for HIV infection. This is possible because CCR5 is nonessential for normal immune system function, but central to HIV entry into cells. HIV-1 binds to the CD4 antigen and CCR5 coreceptor present on macrophages. Binding to the CCR5 coreceptor causes a conformational change in HIV's gp41 protein, allowing the virus to fuse with the plasma membrane of the host cell and subsequent entry. People living with a genetic variation of the CCR5 gene are resistant to HIV infection and still maintain a healthy immune system. CCR5 plays a role in the inflammatory response to infection; however, when this gene is deleted or not functional in the host, other cellular chemokines compensate for this loss of function.

Hepatitis C virus (HCV) infects an estimated 3% of the world's population. It is a major cause of chronic hepatitis. About 80% of HCV-infected people suffer from liver cirrhosis and hepatocellular cancer. It is the leading cause of liver transplantation in the United States. The only therapy available to treat patients is a combination of interferon and the drug ribavirin. The majority of HCV-infected individuals do not respond to this therapy because it often produces very toxic side effects in patients. At the time of this writing, RNAi studies *in vitro* were able to show that Huh-7.5 hepatoma cells containing persistently replicating HCV replicons could be "cured." To date, experimental RNAi directed at HCV RNA targets have not moved past Phase II clinical trials.

RNAi may also be used to silence errant genes containing dominant mutations that cause certain cancers. Even though there have been recent advances in surgery, radiotherapy, and chemotherapy, the prognosis and survival rate for patients with brain tumors remains poor. New therapies that specifically target and eradicate tumor cells are needed to improve the life expectancy of these patients. The human epidermal growth factor receptor (EGFR) gene represents a potential target for treating cancers in general. In normal cells, epidermal growth factor (EGF) binds to EGFR, causing the induction of cell proliferation or differentiation in mammalian cells. In cancer cells, the EGFR gene is overexpressed or mutated, causing the induction of uncontrolled cell growth and a malignant phenotype (solid tumors). Studies are under way that use RNAi treatment to target EGFR in an experimental human brain tumor in a SCID (severe combined immunodeficiency) mouse model.

There are still a number of hurdles and safety concerns before RNAi can be successfully applied as a therapeutic. The major hurdle of RNAi therapies is the delivery of these macromolecules to the desired cell type, tissue, or organ. RNA molecules do not readily cross membranes because of their negative charges and sizes. Liposome-based carriers, nanoparticles or other delivery schemes may eventually overcome this challenge. Other hurdles include stability and specificity of RNAi. Despite these obstacles, RNAi is a powerful tool to study gene function. RNAi discovery enables the development of a new class of human therapeutics in the pipeline that has the potential to treat a wide array of important diseases.

References

Aagaard and Rossi. (2007). RNAi Therapeutics: Principles, Prospects and Challenges. *Adv. Drug Deliv. Rev.* **59**(2–3): 75–86.

Boado. (2005). RNA Interference and Nonviral Targeted Gene Therapy of Experimental Brain Cancer. *NeuroRX* **25**: 139–150.

Haasnoot et al. (2007). RNA Interference Against Viruses: Strike and Counterstrike. *Nature Biotech.* **25**: 1435–1443.

a small change in the secondary structure of *oxyS* RNA that improves the exposure of the single-stranded sequences that pair with the target mRNAs.

We are just beginning to realize the vast potential that small RNAs possess in controlling so much of the life cycle of an organism. A system of bacterial defense against foreign invaders, both viruses and certain plasmids, in the very well-known bacterium *E. coli K12* provides an example of just how much we have yet to learn. This

▸ **CRISPR** Clusters of regularly interspaced repeats in prokaryotes that are transcribed and processed into short RNAs that function by RNA interference as an immune system against phage and plasmid infection.

system is based upon sets of clusters of short palindromic repeats called **CRISPRs** (*c*lusters of *r*egularly *i*nterspersed *s*hort *p*alindromic *r*epeats) separated by hypervariable spacer sequences derived from captured phage and plasmids. These are widespread in both eubacteria and archaea. These hypervariable spacer sequences are used to provide the host bacterium with resistance to further phage and plasmid infection, as seen in **FIGURE 30.8**.

The CRISPR defense system requires transcription of the repeat-spacer array from a leader sequence (acting as a promoter) and is used in conjunction with an RNA processing system encoded by eight genes, called *cas* (*C*RISPR-*as*sociated) genes in *E. coli,* usually located adjacent to each CRISPR locus. These genes code for a variety of polymerases, nucleases, helicases and RNA-binding proteins. A multimeric complex of Cas

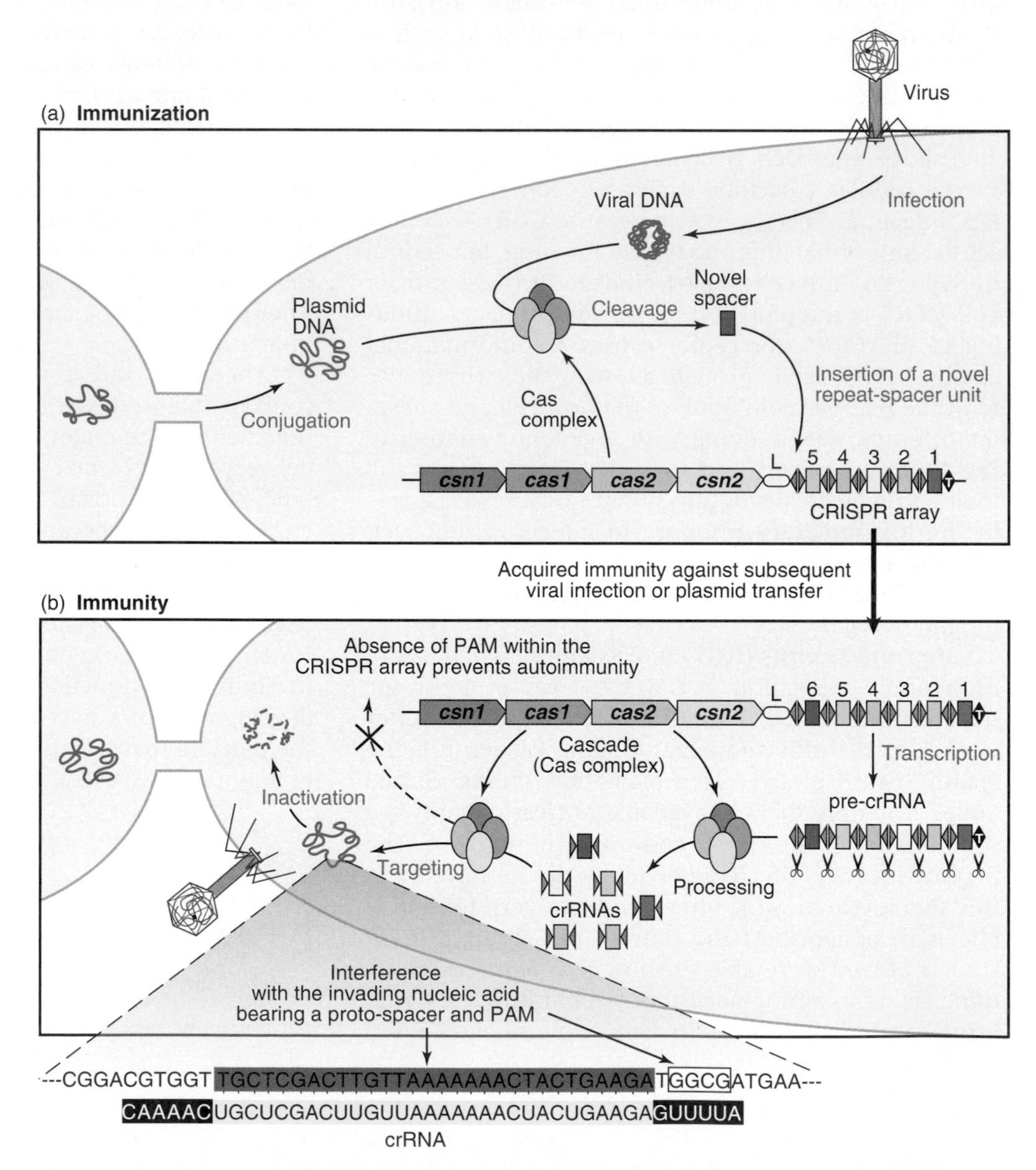

FIGURE 30.8 Overview of the CRISPR/Cas mechanism of action. (a) Immunization process: after insertion of exogenous DNA from viruses or plasmids, a Cas complex recognizes foreign DNA and integrates a novel repeat-spacer unit at the leader end of the CRISPR locus. (b) Immunity process: the CRISPR repeat-spacer array is transcribed into a pre-crRNA that is processed into mature crRNAs, which are subsequently used as a guide by a Cas complex to interfere with the corresponding invading nucleic acid. Repeats are represented as diamonds, spacers as rectangles, and the CRISPR leader is labeled L. Reproduced from P. Horvath and R. Barrangou, *Science* **327** (2010): 167–170. Reprinted with permission from AAAS.

proteins can be identified and is called Cascade (*C*RISPR-*as*sociated *c*omplex for *a*ntiviral *de*fense). The CRISPR region is transcribed into a long RNA, pre-crRNA, which is processed into short CRISPR RNAs of about 57 nucleotides containing a spacer flanked by two partial repeats. The model proposed is that these RNAs, complementary to phage DNA, are used as guides for the Cas interference machinery. The complex will base pair with either the virus genome or its RNA to prevent expression of the phage genes and ultimately degradation.

This is an actively evolving system. Virus infection leads to the incorporation of new spacers into a CRISPR locus. That virus can only then successfully infect that bacterium if it undergoes mutation of its sequence element.

KEY CONCEPTS

- Bacterial regulator RNAs are called sRNAs.
- Several of the sRNAs are bound by the protein Hfq, which increases their effectiveness.
- The *oxyS* sRNA activates or represses expression of >10 loci at the posttranscriptional level.
- Tandem repeats can be transcribed into powerful antiviral RNAs.

CONCEPT AND REASONING CHECK

How could an sRNA activate translation?

30.5 MicroRNAs Are Widespread Regulators in Eukaryotes

Eukaryotes, like bacteria, use RNAs to regulate gene expression. Noncoding RNAs are used to control gene expression in the nucleus at the level of DNA; in many cases the expression and function of these RNAs are inextricably linked to chromatin structure. Transcription of tandemly repeated simple sequence satellite DNA is required for the very formation of heterochromatin itself (see *Chapter 28, Eukaryotic Transcription Regulation*, and *Chapter 29, Epigenetic Effects are Inherited*). We will focus here mainly on control in the cytoplasm at the level of mRNA. As we will see, the eukaryotic mechanisms, while related to the bacterial mechanisms, are very different.

Attenuation is not possible in eukaryotes as it is in bacteria because the nuclear membrane separates the processes of transcription and translation. Given that eukaryotic mRNA is so much more stable than bacterial mRNA, with an average half-life of hours as opposed to minutes, much more translation-level control is necessary in eukaryotes, both at the level of translation initiation and mRNA stability control itself (see *Chapter 22, mRNA Stability and Localization*).

There are numerous classes of small noncoding RNAs in eukaryotes. We have already seen some of these, such as the different classes of guide RNAs that are involved in RNA splicing, editing, and modification (see *Chapter 21, RNA Splicing and Processing*, and *Chapter 23, Catalytic RNA*).

Very small RNAs, or microRNAs (**miRNAs**), are gene-expression regulators found in most, if not all, eukaryotes. These bear some resemblance to their bacterial sRNA counterparts, but as we will see, they are smaller and their mechanism of action is different. The human genome has an estimated 1000 genes that code for miRNAs that participate in **RNA interference (RNAi)**, half from the introns of coding genes, and about half from large ncRNAs. Even more interesting, miRNAs can originate from pseudogenes, supposedly inactive genelike regions that were once thought to have no function. RNA interference is a general mechanism to repress gene expression, usually (but not always) at the level of translation. These miRNAs go by a number of names and are sometimes called **stRNA**, or *s*hort *t*emporal RNA (because they are involved in development). Some miRNAs have also

- **miRNA** Very short RNAs that may regulate gene expression.
- **RNA interference (RNAi)** A process by which short 21 to 23 nucleotide antisense RNAs, derived from longer double-stranded RNAs, can modulate expression of mRNA by translation inhibition or degradation.
- **stRNA** Short temporal RNA, a form of miRNA in eukaryotes that modulates mRNA expression during development.

been shown to affect transcription initiation by binding to the gene's promoter. It is estimated that hundreds of miRNAs control thousands of mRNAs. It may be that a large fraction of mRNAs are at some point targeted by miRNAs at all stages of development. Each miRNA may have hundreds of target mRNAs and each mRNA may be controlled by multiple miRNAs.

- **piRNA** Piwi RNA, a special form of miRNA found in germ cells.
- **siRNA** Short interfering RNA, a miRNA that prevents gene expression.

Piwi-associated RNAs, **piRNA**, are a special class of miRNA found in germ cells. Another type of very small RNA is **siRNA** (*s*hort *i*nterfering *RNA*), which is typically produced during a virus infection. Both piRNA and siRNA can also be used to control the expression of transposable elements. These classes are summerized in **FIGURE 30.9**.

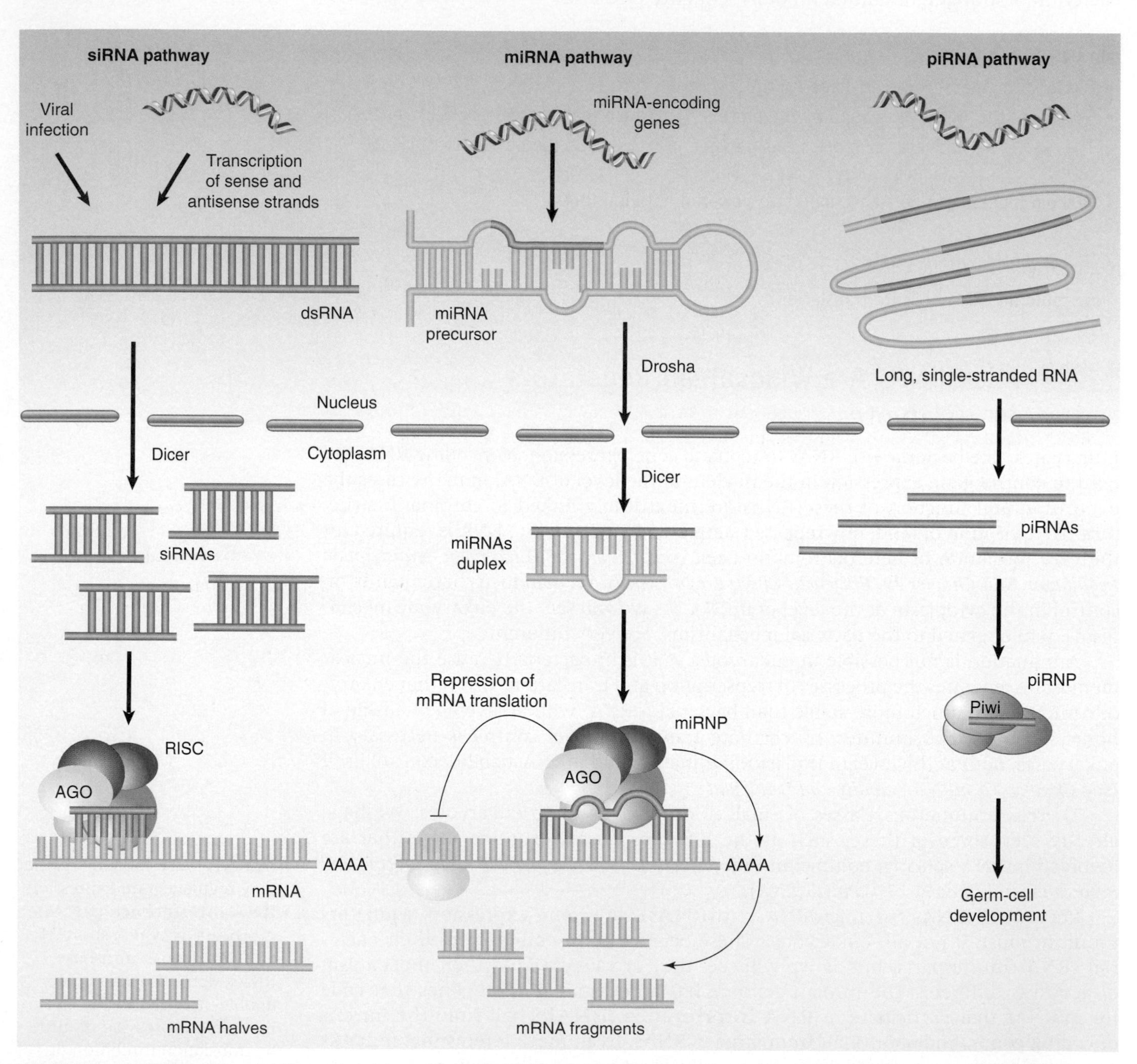

FIGURE 30.9 Small RNAs are generally produced by processing of longer precursors. Three separate pathways exist for processing siRNAs, miRNAs and piRNAs. Reprinted by permission from Macmillan Publishers Ltd: H Großhans and W. Filipowicz, *Nature* **451**, pp. 414–416, copyright 2008.

These RNAs have multiple origins and multiple mechanisms of synthesis and processing. Most are produced as larger precursor RNAs that are processed and cleaved to the correct size and then delivered to their target.

The miRNAs used in RNAi are produced as large RNA primary transcripts called pri-miRNA that are self complementary and can automatically fold into a double-stranded hairpin structure, usually with some imperfect base pairing. The pri-miRNA is processed in a two-step reaction. The first step is catalyzed by **Drosha**, an RNase III superfamily member endonuclease, in the nucleus. Drosha reduces the pri-miRNA to about a 70 bp precursor fragment, pre-miRNA. This cleavage determines the 5′ and 3′ ends of the precursor. After export from the nucleus to the cytoplasm, the second step is catalyzed by **Dicer** to produce a short double-stranded ~22 base pair segment with short, ~2 nucleotide single-stranded ends. Dicer has an N-terminal helicase activity, which enables it to unwind the double-stranded region, and two nuclease domains that are also related to the bacterial RNase III. Related enzymes are nearly universal in eukaryotes.

▸ **Drosha** An endonuclease that processes double-stranded primary RNAs into short, ~70 bp precursors for Dicer processing.

▸ **Dicer** An endonuclease that processes double stranded precursor RNA to 21 to 23 nucleotide molecules.

These short double-stranded RNA fragments are delivered to, or loaded onto, a complex called **RISC** (*R*NA-*i*nduced *s*ilencing *c*omplex). Proteins in the Argonaute (Ago) family are components of this complex and are required for the final processing to a single strand, to be delivered to the 3′ UTR of its target mRNA. Humans have eight Ago family members, *Drosophila* has five. These proteins have an ancient origin and are found in bacteria, archaea, and eukaryotes (this system is absent in the yeast *Saccharyomyces cerevisiae* but is present in some of its close relatives). RISC has endonuclease activity that cleaves the passenger strand, the one that will not be used, in the duplex miRNA.

▸ **RISC** RNA-induced silencing complex, a ribonucleoprotein particle composed of a short single-stranded siRNA and a nuclease that may cleave mRNAs complementary to the siRNA. It receives siRNA from Dicer and delivers it to the mRNA.

The degree of base pairing and the sequence of the ends (determined by Dicer cleavage) of the duplex dictate which of the multiple Ago family members picks up the RNA duplex and which strand is selected as the passenger strand to be degraded, as shown in **FIGURE 30.10**. The RISC complex is now in a position to use the mature miRNA to guide it to its target mRNA.

A germline subset of miRNA is the more recently discovered Piwi-interacting RNA, or piRNA (P-element *i*nduced wimpy testis). In *Drosophila*, these are sometimes called **rasiRNAs**, for *r*epeat-*a*ssociated *siRNA*s. These are named piRNA because they interact with a different subfamily member of the Ago class proteins, known as Piwi (also called Hiwi in humans). Piwi-class proteins are found only in metazoan organisms (muticellular eukaryotes). In addition, the piRNAs are somewhat longer than miRNAs, ranging from 24 to 31 nucleotides. piRNAs are found in giant tandem clusters; there can be tens of thousands of copies. The processing pathway has not yet been completely determined.

▸ **rasiRNA** A germline subset of miRNA transcribed from transposable elements and other repeated elements that is used to silence transposable elements.

The function of piRNA is also different than miRNAs. Their primary function is in the nucleus, to repress the expression of transposable elements, preserve genome integrity, and control chromatin structure (see *Chapter 17, Transposable Elements and Retroviruses*, and *Chapter 28, Eukaryotic Transcription Regulation*). Only a small fraction of the piRNAs are complementary to transposable elements. Most map to single-copy DNA, both genes and intergenic regions. In *Drosophila*, it is maternally inherited piRNAs that provide protection against transposon activation to the female from P element–mediated hybrid dysgenesis (see *Section 17.6, Transposition of P Elements Cause Hybrid Dysgenesis*).

siRNAs have a different origin. These are frequently derived from viral infections, which typically transcribe both genomic strands to produce complememtary double-stranded RNAs. These large double-stranded RNAs are processed by Dicer in a manner similar to that of the miRNAs described above and are delivered to RISC. sRNAs are also derived from transcription of transposable elements and are used to silence them. This process can be amplified in plants and in *C. elegans* by an RNA-dependent RNA polymerase.

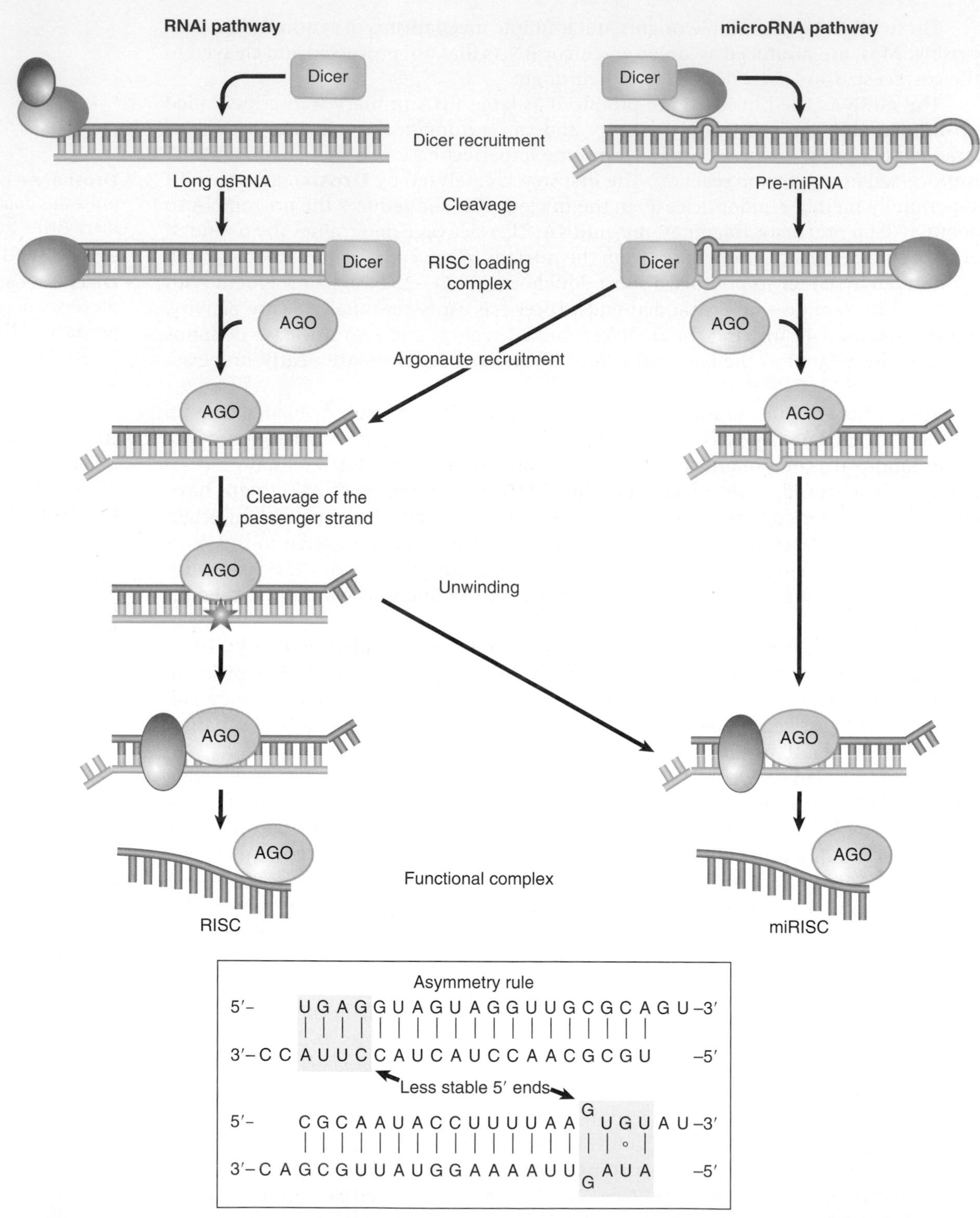

FIGURE 30.10 Assembly of the Argonaut-small RNA complex. Inside the cell, a double-stranded RNA duplex is bound by a recognition complex that contains a Dicer-family member and a dsRNA-binding protein (blue). In *Drosophila melanogaster,* a double-stranded RNA binding protein forms a RISC complex with Dicer (right panel), whereas in the RNAi pathway (left panel) Dicer is important for recruiting the Ago protein. Once Ago is associated with the RNA duplex, the passenger, blue, strand is cleaved. RNA strand separation and incorporation into the Ago protein are guided by the strength of base pairing at the 5′ ends of the duplex; this is known as the asymmetry rule. In this example the easiest 5′ end to unwind is highlighted in yellow. Once unwound, the siRNA or miRNA will associate in a complex with Ago to form the RISC complex. The degree of complementarity between the two strands of the duplex can define how miRNAs are sorted into different Ago proteins. The purple oval represents an unidentified "unwindase" protein. The star represents an endonuclease event. Reprinted by permission from Macmillan Publishers Ltd: G. Hutvagner and M. J. Simard, *Nat. Rev. Mol. Cell Biol.* **9**, pp. 22–32, copyright 2008.

KEY CONCEPTS

- Eukaryotic genomes code for many short (~22 base) RNA molecules called microRNAs.
- piRNAs regulate gene expression in germ cells and act to silence transposable elements.
- siRNAs are complementary to viruses and transposable elements.

CONCEPT AND REASONING CHECK

Predict the consequence of a mutation that alters a miRNA.

30.6 How Does RNA Interference Work?

RISC is the complex that carries out translational control, guided to its mRNA target in the cytoplasm by the associated miRNA. There are two primary mechanisms used to control mRNA expression: degradation of the mRNA or inhibition of translation of the mRNA. Plants use RNAi primarily for mRNA degradation, whereas animals primarily use translation inhibition. Both groups, however, do use both systems. The choice is primarily determined by the degree of base pairing between the miRNA and the mRNA. The higher the degree of base pairing, the more likely that target mRNA will be degraded.

This is an essential mechanism for fine-tuned control of translation in eukaryotes. As noted earlier, eukaryotic mRNA is much more stable than bacterial mRNA; because degradation of some mRNAs is stochastic, cells must be able to tightly control which mRNAs will be translated into protein and for how long. During development, it is especially critical to ensure rapid and complete turnover of key mRNAs, as we will see shortly.

RISC uses the miRNA as a guide to scan RNAs for small regions of homology. These regions are usually found in an AU-rich region in the 3′ UTR of mRNAs. A given mRNA may have multiple different target sites and thus respond to different miRNAs. In binding to its target site on the mRNA, the 5′ end of the miRNA from nucleotide 2 to 8 is the most important—the *seed sequence*. These should have perfect base pairing.

Once binding has occurred, there are several different possible outcomes, as shown in **FIGURE 30.11**, ranging from various mechanisms of inhibiting translation to degradation of the message. RISC can interfere with translation that is already underway from a ribosome by blocking translation elongation (Figure 30.11a) or by inducing proteolysis of the nascent polypeptide being produced (Figure 30.11b).

RISC can also inhibit translation initiation in multiple ways, presumably by virtue of the fact that the central domain of the Ago polypeptide has homology to the cap-binding initiation factor eIF4E (see *Section 24.7, Small Subunits Scan for Initiation Sites on Eukaryotic mRNA*). RISC can bind to the cap and inhibit eIF4E from joining (Figure 30.11c) or prevent the large 60S ribosomal subunit from joining (Figure 30.11d). RISC can also prevent the circularization of the mRNA by preventing cap binding to the polyA tail (Figure 30.11e). One way in which RISC can promote mRNA degradation is by promoting deadenylation and subsequent decapping of the message (Figure 30.11f). RISC can also indirectly facilitate mRNA degradation by targeting the mRNA to existing degradation pathways. RISC mediates the sequestration of mRNAs to processing centers called P bodies (cytoplasmic processing bodies). These are sites where mRNA can both be stored for future use and where decapped mRNA is degraded.

Although translation repression is the most common outcome (that we currently know about) for miRNA action, miRNAs can also lead to translation activation. The 3′ UTR of tumor necrosis factor-α (TNF-α) contains a regulatory element called an ARE (*A*U-*r*ich *e*lement). These are common elements that are usually involved in translation repression. In this case, however, the ARE is involved in activation of translation after serum starvation. This activation has now been shown to require RISC and its

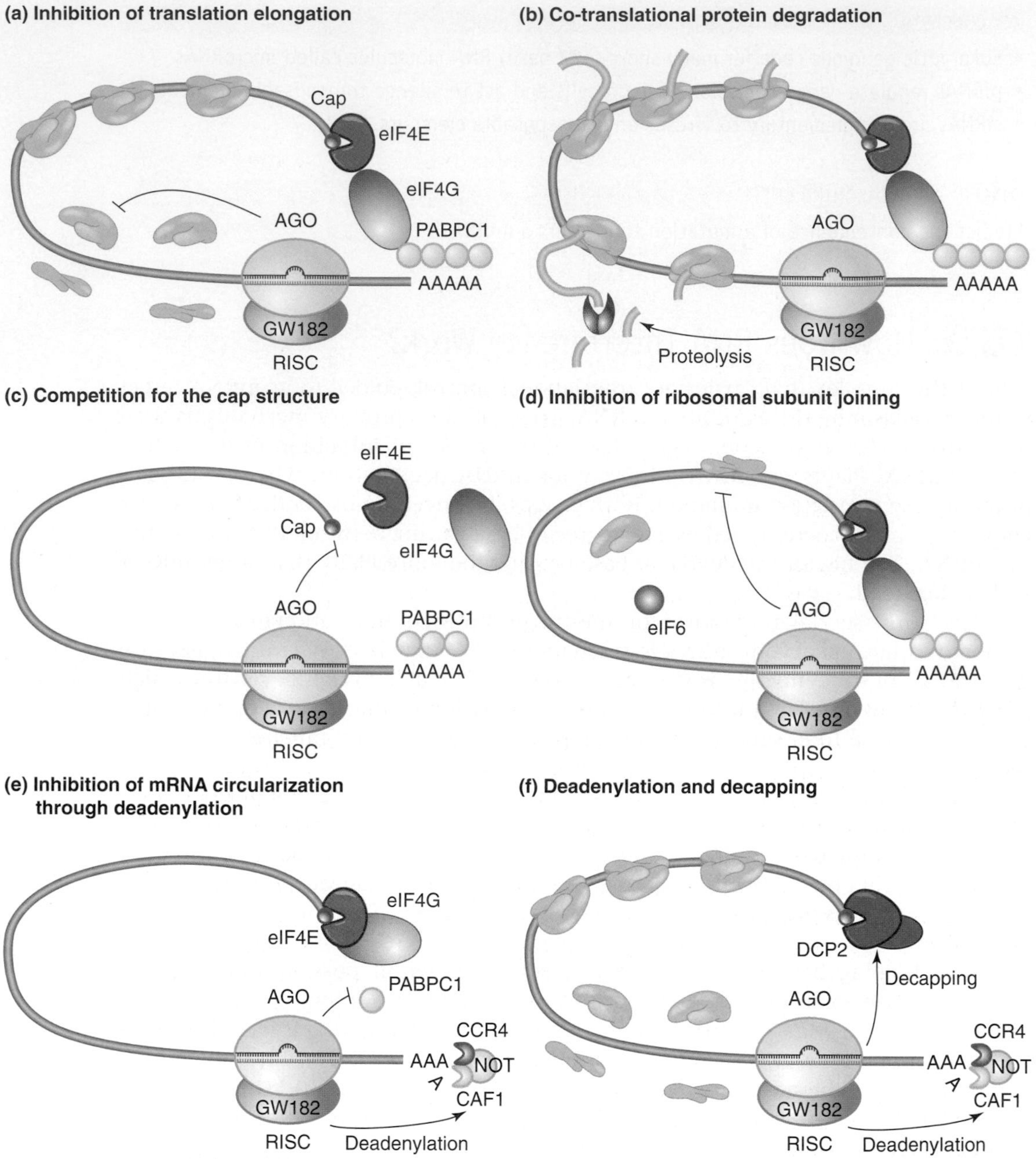

(a) Inhibition of translation elongation

(b) Co-translational protein degradation

(c) Competition for the cap structure

(d) Inhibition of ribosomal subunit joining

(e) Inhibition of mRNA circularization through deadenylation

(f) Deadenylation and decapping

FIGURE 30.11 Mechanism of miRNA-mediated gene silencing. (a) Postinitiation mechanisms. miRNAs (red) repress translation of target mRNAs by blocking translation elongation or by promoting premature dissociation of ribosomes. (b) Cotranslational protein degradation. This model proposes that translation is not inhibited, but rather the nascent polypeptide chain is degraded cotranslationally. (c–e) Initiation mechanisms; miRNAs interfere with a very early step of translation. (c) Ago proteins compete with eIF4E for binding to the cap structure (red dot). (d) Ago proteins recruit eIF6, which prevents the large ribosomal subunit from joining the small subunit. (e) Ago proteins prevent the formation of the closed loop mRNA configuration by an ill-defined mechanism that includes deadenylation. (f) miRNA-mediated mRNA decay. miRNAs trigger deadenylation and subsequent decapping of the mRNA target. Proteins required for this process are shown including components of the major deadenylation complex (CAF1, CCR4, and the NOT complex), the decapping enzyme DCP2, and several decapping activators (dark blue circles). (Note that mRNA decay could be an independent mechanism of silencing, or a consequence of translational repression, irrespective of whether repression occurs at the initiation or postinitiation levels of translation.) RISC is shown as a minimal complex including an Ago protein (blue) and the subunit GW182 (yellow). Reprinted from A. Eulalio, E. Huntzinger, and E. Izaurralde, Getting to the root of miRNA . . . , *Cell* **132,** pp.9–14. Copyright 2008, with permission from Elsevier (http://www.sciencedirect.com/science/journal/00928674).

miRNA in a complex with the fragile X-related protein FXR1, an RNA binding protein. The question of how the RISC complex is converted from its normal repression mode to activation hinges on the exact makeup of the complex. Different protein partners in the complex will elicit different responses. Serum starvation leads to the recruitment of FXR1, which alters RISC action, perhaps because RISC is communicating between the 3′ UTR and the mRNA cap, where translation is controlled.

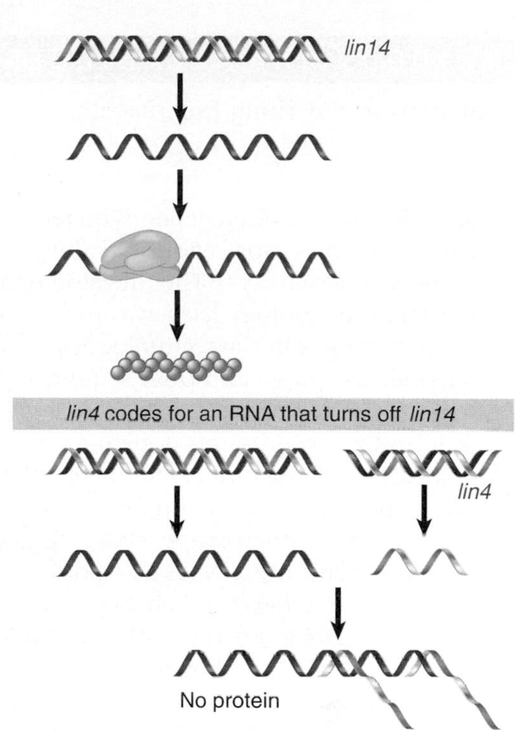

FIGURE 30.12 *lin4* RNA regulates expression of lin14 by binding to the 3′ nontranslated region of *lin14* mRNA.

One of the earliest known examples of RNAi in animals was discovered in the nematode *Caenorhabditis elegans* as the result of the interaction between the regulator gene *lin4* (lineage) and its target gene, *lin14*. **FIGURE 30.12** illustrates the behavior of this regulatory system. The *lin14* target gene produces an mRNA that regulates larval development. Lin14 is a critical protein for specifying the timing of mitotic divisions in a special group of cells. Expression of *lin14* is controlled by *lin4*, which codes for a miRNA. The *lin4* transcripts are complementary to a ten-base sequence that is imperfectly repeated seven times in the 3′ UTR of the *lin14* mRNA.

As we described for bacterial sRNA, there can be a dynamic interplay between different elements that can modulate the ultimate outcome. There are multiple mechanisms to control the reaction between RISC and its target mRNA. Proteins can bind to mRNA target sequences to prevent their utilization by RISC, and the 3′ UTR of the mRNA itself may have alternate base-pairing structures that can influence the ability of RISC to identify and target a binding site. miRNA precursors can be edited by ADAR, an adenosine deaminase editing enzyme, which converts A to I and disrupts A:U base pairing. This can result in either activation or inactivation of a miRNA. *C. elegans* and some viruses express a ncRNA, which can interfere with Dicer and alter the mRNA profile in a cell. Even more interesting is that some genes have alternate poly(A) cleavage sites and are able to produce alternate versions of the mRNA, with different 3′ UTRs and thus with different target sequences for different miRNAs.

RNAi has become a powerful technique for ablating the expression of a specific target gene. The technique, however, has been limited in mammalian cells, which have the more generalized response to dsRNA of shutting down protein synthesis and degrading mRNA. **FIGURE 30.13** shows that this happens because of two reactions. The dsRNA activates the enzyme PKR, which inactivates the translation initiation factor eIF2α by phosphorylating it. It also activates 2′5′ oligoadenylate synthetase,

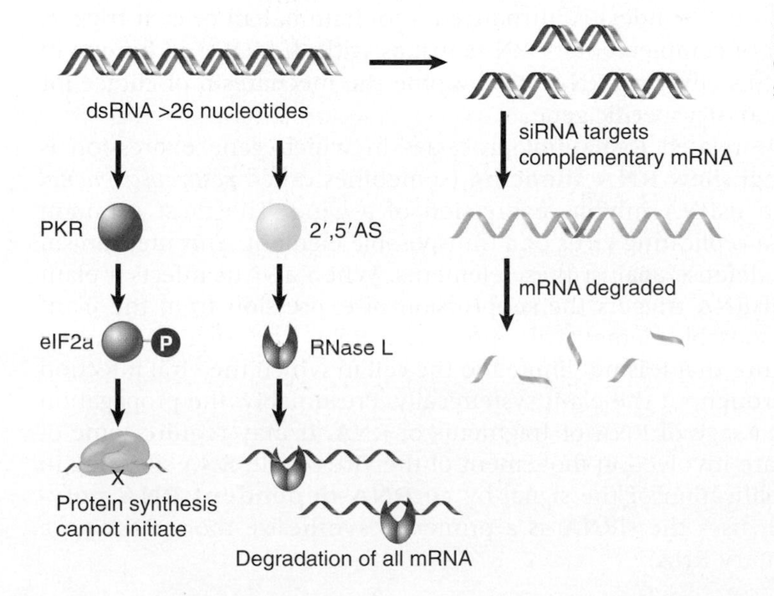

FIGURE 30.13 Long dsRNA inhibits protein synthesis and triggers degradation of all mRNA in mammalian cells, as well as having sequence-specific effects.

METHODS AND TECHNIQUES

Microarrays and Tiling Experiments

Scientific progress often depends on technological breakthroughs, such as those witnessed after cloning and PCR, and now microarrays. The development of genomic microarray technology itself was made possible because of breakthroughs in the technology of inexpensive oligonucleotide synthesis and DNA sequencing technologies. Microarray experiments are simply classical nucleic acid hybridization experiments done on a massive scale.

A DNA microarray, sometimes called "DNA on a chip," is basically a very simple device. A support platform can be a microscope slide. Target nucleic acid sequences, typically DNA, are spotted onto the slide. The earliest microarrays had been made by manually spotting the DNA. By the 1990s, automated machines were developed that allowed precise and high-density arrays to be made. Schematics to build these devices circulated on the Internet, allowing many laboratories to assemble their own devices.

The array density can range from 10 to 1000 spots per square millimeter. Commercially produced arrays have become very popular for a number of reasons, which are discussed shortly. Standardized arrays for many model organisms are readily available from numerous vendors, and custom arrays for individual projects can also be synthesized.

Design of the target sequence probes to be spotted is a complex task. The DNA can be genomic PCR fragments, PCR fragments of cloned cDNA (*c*omplementary DNA), copies of mRNA from known genes, or oligonucleotides, typically of 25 to 60 bp. For example, one can make an array containing a PCR product from every gene in an organism. For a typical higher eukaryote with 20,000 genes, that would require 40,000 PCR primers. The products have to be designed so that they are each unique sequences that do not have self-complementarity. The melting temperatures (i.e., the GC contents) have to be within a certain range for mass production and use. The same is true if oligonucleotides are used as the target sequence. Computer programs are available to assist in target design. An array set to query exon usage and alternative splicing may be ten times larger.

Once the decision on target sequences has been made, the next issue is the RNA to be tested. In a typical experiment, mRNAs are labeled with fluorescent dyes. The mRNA pool must first be amplified and then labeled. Amplification can be achieved by the synthesis of cDNA or RT-PCR (reverse transcriptase-PCR) copies. Experiments of this nature can ask a number of different questions, including how gene expression changes during development, or with altered metabolic challenges such as starvation. It can also compare the gene expression profile in a normal cell to that of a cancer cell.

In these types of experiments, where two cell types are being compared, the amplified mRNAs each have a different color of fluorescent tag attached, usually red and green. The two pools of amplified tagged RNAs are

whose product activates RNase L, which degrades all RNAs in the cell. It turns out, however, that these reactions require dsRNA that is longer than 26 nucleotides. If shorter dsRNA (21 to 23 nucleotides) is introduced into mammalian cells, it triggers the specific degradation of complementary RNAs just as with the RNAi technique in worms and flies. With this advance, RNAi has become the mechanism of choice for turning off the expression of a specific gene.

RNA interference is related to natural processes in which gene expression is silenced. Plants and fungi show **RNA silencing** (sometimes called *posttranscriptional gene silencing*), in which dsRNA inhibits expression of a gene. The most common sources of the RNA are a replicating virus or a transposable element. This mechanism may have evolved as a defense against these elements. When a virus infects a plant cell, the formation of dsRNA triggers the suppression of expression from the plant genome. Similarly, transposable elements also produce dsRNA. RNA silencing has the further remarkable feature that it is not limited to the cell in which the viral infection occurs: it can spread throughout the plant systemically. Presumably, the propagation of the signal involves passage of RNA or fragments of RNA. It may require some of the same features that are involved in movement of the virus itself. RNA silencing in plants involves an amplification of the signal by an **RNA-dependent RNA polymerase** (RDRP), which uses the siRNA as a primer to synthesize more RNA on a template of complementary RNA.

▸ **RNA silencing** The ability of an RNA, especially ncRNA, to alter chromatin structure in order to prevent gene transcription.

▸ **RNA-dependent RNA polymerase** An enzyme that catalyzes the synthesis of RNA from an RNA template.

then mixed and applied to the microarray. Classical hybridization kinetics dictate that the most abundant RNA will hybridize to the probe more readily than the least abundant RNA.

The third step in the process is data acquisition and interpretation. Special scanners need to be employed in order to read each dot on the microarray at different fluorescent wavelengths. As might be expected, a massive data set can come from a single experiment. Bioinformatic software packages are required to sort and organize the data into a usable format. In the experiments described above, this entails the grouping together of genes that are expressed together. The choice of red and green labels results in data that range from a red spot indicating that the red labeled pool of RNA predominated in the mixture, to a yellow spot which indicates that both pools are equally represented, to a green spot indicating that the second pool predominates. A black spot indicates that the gene was not expressed in either pool (**FIGURE B30.3**).

A genomic tiling array is an extension of the expression microarray in that the target covers the entire genome, chromosome, or region of interest on both strands. The oligonucleotides can be of different lengths and spacing to obtain uniform coverage. The same issues of target probe design apply here as in the expression arrays described above. Repeat and self-complementary sequences must be avoided, and the melting temperatures must be within a critical range. The question asked here is also somewhat different. Instead of querying known genes at different times, now the question becomes, Is this region ever expressed?

For further reading see *Annual Review of Biochemistry* **74**, 53–82, 2005.

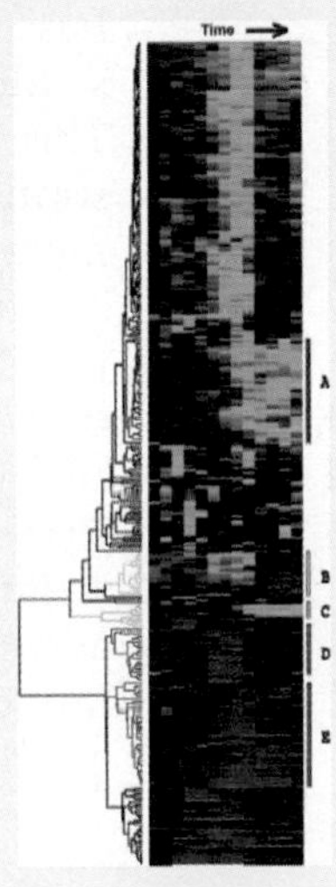

FIGURE B30.3 An example of analysis of gene expression in which similarly expressed genes have been grouped together. This experiment identified genes expressed over time during serum stimulation in human fibroblast cells. Green represents genes whose expression is reduced compared to unstimulated cells, while red indicates genes whose expression increases after stimulation. Groups A–E are genes with related functions that also show similar expression patterns. Reproduced from M. B. Eisen et al., *Proc. Natl. Acad. Sci. USA* **95** (1998): 14863–14868. Copyright 1998 National Academy of Sciences, USA. Photo courtesy of Michael Eisen, University of California, Berkeley.

KEY CONCEPTS

- MicroRNAs regulate gene expression by base pairing with complementary sequences in target mRNAs.
- RNA interference triggers degradation or translation inhibition of mRNAs complementary to miRNA or siRNA.
- dsRNA may cause silencing of host genes.

CONCEPT AND REASONING CHECK

How can a miRNA that binds to the mRNA 3′ UTR affect functions like translation initiation at the 5′ end?

30.7 Heterochromatin Formation Requires MicroRNAs

As we saw in the last chapter (see *Section 29.3, Heterochromatin Depends on Interaction with Histones*), heterochromatin is one of the major substructures that can be seen in chromosomes. It is visually different when stained because it is more condensed than euchromatin. It is late replicating and contains few genes. The underlying sequence of constitutive heterochromatin is different from euchromatin in that it consists primarily of simple sequence satellite DNA organized in giant blocks. Small islands of unique sequence DNA containing genes are found scattered within heterochromatin. The simple sequence regions had long been thought to be transcriptionally silent. We now understand that virtually the entire genome is transcribed, including the simple sequence satellite DNA that is often found surrounding centromeres. In fact, transcripts from these sequences are used to organize the heterochromatin structure and repress its transcription.

The centromic heterochromatin of the fission yeast *Schizosaccharomyces pombe* has been a model for understanding heterochromatin formation. The outer region sequences of the heterochromatin are transcribed into ncRNAs by RNA polymerase II. This transcript is copied by an RNA-dependent RNA polymerase to give a double-stranded RNA which is processed into siRNA. Plants use a variation of the RNA polymerase called RNA polymerase IVb/V to amplify the ncRNA signal.

In a manner similar to what we saw in *Section 30.6, How Does RNA Interference Work?*, the RNA is processed by Dicer. The complex to which the fragments are delivered is called **RITS (*R*NA-*i*nduced *t*ranscriptional *s*ilencing)**. RITS contains an Argonaut subunit, Ago1. RITS and RDRP are in a complex together. Again, as we saw before, RITS uses the siRNA as a targeting mechanism to begin the process of repressing transcription. This entails the recruitment of factors to begin chromatin modification, such as a histone H3K9 methyltransferase, as seen in **FIGURE 30.14**. An analogous system is found in *Drosophila*, where rasiRNAs (*r*epeat *a*ssociated *s*mall *i*nterfering *RNAs*) are targeted to the alternative RISC complex containing Piwi, Aubergine, and Ago3 proteins.

▸ **RITS (RNA-induced transcriptional silencing** A form of RNAi in which siRNAs are used to downregulate transcription by modification and remodeling chromatin.

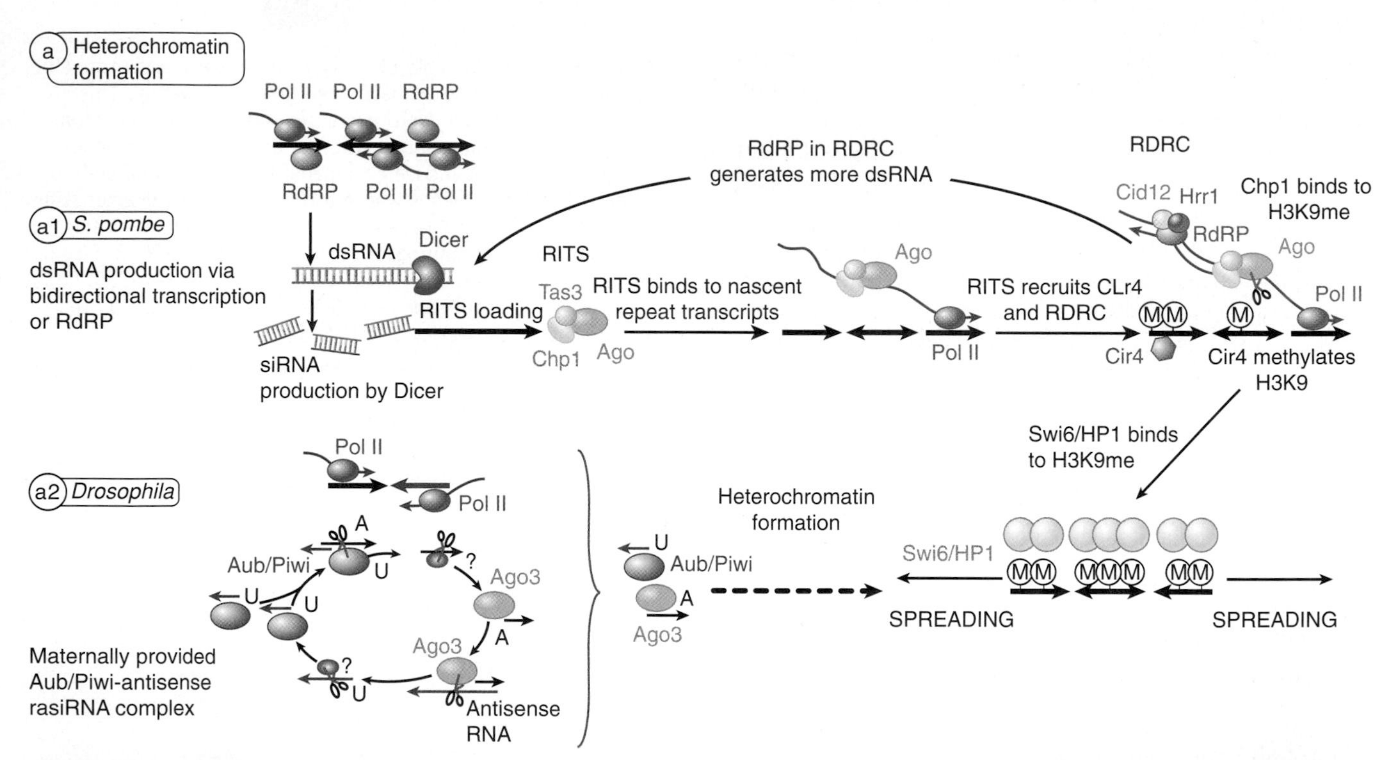

FIGURE 30.14 (a) Heterochromatin formation: (a1) In *Schizosaccharomyces pombe*, DNA repeats produce double-stranded RNAs through bidirectional transcription or RNA-dependent RNA synthesis. dsRNAs are cut into siRNAs that are loaded into a RITS complex. RITS finds the DNA repeats through siRNA base pairing with the nascent transcript and recruits the RdRP complex and a histone methyltransferase that methylates histone H3 at lysine 9 (H3K9me). RdRP uses the Ago-cut RNA as template to synthesize more dsRNA, which will in turn be cut into siRNAs to reinforce heterochromatin formation. The RITS complex binds to H3K9me, resulting in stable interaction of RITS and heterochromatic Swi6, an HP1 homolog, leading to spreading of heterochromatin. (a2) In *Drosophila* rasiRNAs are produced in a Dicer independent mechanism. The alternate RISC complex containing Piwi and Aubergine associates with antisense-strand derived rasiRNA with a preference for a U at the 5′ end, whereas Ago associates with sense-strand rasiRNA with a preference for an A. An unknown nuclease (?) generates the sense rasiRNAs that associate with Ago. In turn, Ago-sense siRNA binds to antisense RNA and generates more antisense rasiRNA. The resulting rasiRNA complexes initiate heterochromatin formation (dotted arrow line). As in yeast above, H3K9me binds to HP1 protein, leading to the spread of heterochromatin. A similar mechanism has been reported in mammals. Reprinted from Y. Bei, S. Pressman, and R. Carthew, Snapshot: Small RNA-mediated . . . , *Cell* **130**, pp. 756–756e1. Copyright 2007, with permission from Elsevier (http://www.sciencedirect.com/science/journal/00928674).

KEY CONCEPT

- MicroRNAs can promote heterochromatin formation.

CONCEPT AND REASONING CHECK

How can the cell tell the difference between euchromatin and heterochromatin?

30.8 Summary

Gene expression can be regulated positively by factors that activate a gene or negatively by factors that repress a gene. The first and most common level of control is at the initiation of transcription, but elongation and termination of transcription may also be controlled. Translation may be controlled by regulators that interact with mRNA. The regulatory products may be proteins, which often are controlled by allosteric interactions in response to the environment, or RNAs, which function by base pairing with the target RNA to change its secondary structure or interfere with its function. Small metabolites can also bind to RNA aptamer domains and affect an alteration in secondary structure, as seen in riboswitches. Regulatory networks can be created by linking regulators so that the production or activity of one regulator is controlled by another.

ncRNAs such as antisense RNA are used in bacterial and eukaryotic cells as a powerful system to regulate gene expression. This regulation can be direct, at the level of interference with an RNA polymerase, or indirect, by affecting the chromatin configuration of the gene. Antisense transcripts can also function in the cytoplasm by giving rise to a host of small regulatory RNAs.

Small regulator RNAs are found in both bacteria and eukaryotes. *E. coli* has ~80 sRNA species. The *oxyS* sRNA controls about ten target loci at the posttranscriptional level; some of them are repressed, whereas others are activated. Repression is caused when the sRNA binds to a target mRNA to form a duplex region that includes the ribosome-binding site. MicroRNAs are ~22 bases long and are produced in most eukaryotes by Dicer cleavage of a longer transcript, which is then delivered to RISC for delivery to its target mRNA. They function by base pairing with target mRNAs to form duplex regions that are susceptible to cleavage by endonucleases or inhibition of translation. These are dynamic systems, which themselves are controlled by both accessory protein and enzymes and by other RNAs. The technique of RNA interference is becoming the method of choice for inactivating eukaryotic genes. It uses the introduction of short dsRNA sequences with one strand complementary to the target RNA, and it works by inducing degradation of the targets. This may be related to a natural defense system in plants called RNA silencing.

CHAPTER QUESTIONS

1. At what level does antisense RNA function to inhibit gene expression?
   A. It may prevent transcription of the original gene in the cell.
   B. It may affect processing of the RNA product of the gene.
   C. It may alter translation of the mRNA.
   D. All of the above.
2. The *lin*4 microRNA in *C. elegans* regulates expression of the *lin*14 gene by:
   A. binding to complementary sequences in an early exon of *lin*14 mRNA.
   B. binding to complementary sequences in the first intron of *lin*14 mRNA.
   C. binding to complementary sequences in the 5′ untranslated region of *lin*14 mRNA.
   D. binding to complementary sequences in the 3′ untranslated region of *lin*14 mRNA.

3. Dicer cleaves double-stranded RNA molecules or regions to generate oligonucleotides of:
   A. 18–20 bases.
   B. 21–23 bases.
   C. 24–26 bases.
   D. 27–29 bases.

KEY TERMS

antisense gene
antisense RNA
aptamer
CRISPR
cryptic unstable transcripts (CUT)
Dicer
Drosha
lincRNA
miRNA
ncRNA
nested genes
paraspeckles
piRNA
PROMPT
rasiRNA
riboswitch
ribozyme
RITS (RNA-induced transcriptional silencing)
RNA-dependent RNA polymerase
RNA interference (RNAi)
RNA silencing
RISC
siRNA
sRNA
stRNA
transcription interference (TI)

FURTHER READING

A collection of current news and a number of reviews about microRNAs and other regulatory RNAs in the RNA World can be found in *Science* **319**, 1781–1799, 2008.

Horvath, P., and Barrangou, R. (2010). CRISPR/Cas, the immune system of bacteria and archaea. *Science* **327**, 167–170.

Proudfoot, N., and Gullerova, M. (2007). Gene silencing cuts both ways. *Cell* **131**, 649–651. An interesting article about how noncoding RNA can control gene expression.

Tijsterman, M., Ketting, R. E., and Plasterk, R. H. (2002). The genetics of RNA silencing. *Annu. Rev. Genet*. **36**, 485–490.

Walters, L. S., and Storz, G. (2009). Regulatory RNAs in bacteria. *Cell* **136**, 615–628.

Glossary

10 nm fiber—A linear array of nucleosomes, generated by unfolding from the natural condition of chromatin.

3′ UTR—The region in an mRNA between the termination codon and the end of the message.

2R hypothesis—The proposal that the vertebrate genome is a result of two rounds of polyploidization.

30 nm fiber—A coil of nucleosomes. It is the basic level of organization of nucleosomes in chromatin.

3′ UTR—The untranslated sequence downstream from the coding region of an mRNA.

-35 box—The consensus sequence centered about 35 bp before the start point of a bacterial gene. It is involved in initial recognition by RNA polymerase.

5′ UTR—In mRNA, the untranslated sequence at the 5′ end that precedes the initiation codon.

5′ end resection—The generation of 3′ overhanging single-stranded regions that occurs via exonucleolytic digestion of the 5′ ends at a double strand break.

abortive initiation—A process in which RNA polymerase starts transcription but terminates before it has left the promoter. It then reinitiates. Several cycles may occur before the elongation stage begins.

abundance—The average number of mRNA molecules (or other specific molecules) per cell.

abundant mRNA—Consists of a small number of individual species, each present in a large number of copies per cell.

acceptor arm—A short duplex on tRNA that terminates in the CCA sequence to which an amino acid is linked.

***Ac* element**—Activator element; an autonomous transposable element in maize.

acentric fragment—A fragment of a chromosome (generated by breakage) that lacks a centromere and is lost at cell division.

A complex—The second splicing complex, formed by the binding of U2 snRNP to the E complex.

acridines—Mutagens that act on DNA to cause the insertion or deletion of a single base pair. They were useful in defining the triplet nature of the genetic code.

activator—A protein that stimulates the expression of a gene, typically by acting at a promoter to stimulate RNA polymerase.

adaptive (acquired) immunity—The response mediated by lymphocytes that are activated by their specific interaction with antigen.

allele—One of several alternative forms of a gene occupying a given locus on a chromosome.

allelic exclusion—The expression in any particular lymphocyte of only one allele coding for the expressed immunoglobulin. This is caused by feedback from the first immunoglobulin allele to be expressed that prevents activation of a copy on the other chromosome.

allolactose—A by-product of β-galactosidase; the true inducer of the *lac* operon.

allopolyploidy—Having more than two complete sets of chromosomes derived from two different species.

allosteric control—The ability of a protein to change its conformation (and, therefore, activity) at one site as the result of binding a small molecule to a second site located elsewhere on the protein.

alternative splicing—The production of different RNA products from a single product by changes in the usage of splicing junctions.

Alu domain—The parts of the 7S RNA of the SRP that are related to Alu RNA.

amber codon—The triplet UAG, one of the three termination codons that end polypeptide translation.

aminoacyl-tRNA—A tRNA linked to an amino acid. The COOH group of the amino acid is linked to the 3′- or 2′-OH group of the terminal base of the tRNA.

aminoacyl-tRNA synthetases—Enzymes responsible for covalently linking amino acids to the 2′- or 3′-OH position of tRNA.

amplicon—The precise, primer-to-primer double-stranded nucleic acid product of a PCR or RT-PCR reaction.

amyloid fibers—Insoluble fibrous protein polymers with a cross β-sheet structure, generated by prions or other dysfunctional protein aggregations (such as in Alzheimer's disease).

annealing—The renaturation of a duplex structure from single strands that were obtained by denaturing duplex DNA.

antibody (immunoglobulin)—A protein that is produced by B lymphocytes and that binds a particular antigen. They are synthesized in membrane-bound and secreted forms. Those produced during an immune response recruit effector functions to help neutralize and eliminate the pathogen.

anticodon—A trinucleotide sequence in tRNA that is complementary to the codon in mRNA and enables the tRNA to place the appropriate amino acid in response to the codon.

anticodon arm—A stem-loop structure in tRNA that exposes the anticodon triplet at one end.

antigen—A molecule that can bind specifically to an antigen receptor, such as an antibody.

antiparallel—Strands of the double helix are organized in opposite orientation, so that the 5′ end of one strand is aligned with the 3′ end of the other strand.

antirepressor—A positive regulator that functions in opening chromatin.

antisense gene—A gene that has a complementary sequence to an RNA that is its target.

antisense RNA—RNA that has a complementary sequence to an RNA that is its target.

antisense template strand—The DNA strand that is complementary to the sense strand and acts as the template for synthesis of mRNA.

anti-Sm—An autoimmune antiserum that defines the Sm domain that is common to a group of proteins found in snRNPs that are involved in RNA splicing.

antitermination—A mechanism of transcriptional control in which termination is prevented at a specific terminator site, allowing RNA polymerase to read into the genes beyond it.

antitermination protein—Protein that allows RNA polymerase to transcribe through certain terminator sites.

anucleate cell—Bacteria that lack a nucleoid but are of similar shape to wild-type bacteria.

apoptosis—The capacity of a cell to respond to a stimulus by initiating a signal transduction pathway that leads to its death through the activation of a characteristic set of reactions.

aptamer—An RNA domain that binds a small molecule; this can result in a conformation change in the RNA.

architectural protein—A protein that, when bound to DNA, can alter its structure, e.g., introduce a bend. They may have no other function.

arm—One of the four (or in some cases five) stem-loop structures that make up the secondary structure of tRNA.

ARS—An origin for replication in yeast. The common feature among different examples of these sequences is a conserved 11 bp sequence called the A domain.

A site—The site of the ribosome that an aminoacyl-tRNA enters to base pair with the codon.

assembly factors—Proteins that are required for formation of a macromolecular structure but are not themselves part of that structure.

ATP-dependent chromatin remodeling complex—A complex of one or more proteins associated with an ATPase of the SWI2/SNF2 superfamily that uses the energy of ATP hydrolysis to alter, reposition, or displace nucleosomes.

***att* sites**—The loci on a lambda phage and the bacterial chromosome at which recombination integrates the phage into, or excises it from, the bacterial chromosome.

attenuation—The regulation of bacterial operons by controlling termination of transcription at a site located before the first structural gene.

attenuator—A terminator sequence at which attenuation occurs.

AU-rich element (ARE)—A eukaryotic mRNA *cis* sequence consisting largely of A and U ribonucleotides that acts as a destabilizing element.

autoimmune disease—A pathological condition in which the immune response is directed against self antigens.

autonomous controlling element—An active transposon in maize with the ability to transpose.

autopolyploidy—Having more than two complete sets of chromosomes derived from the same species.

autoradiography—A method of capturing an image of radioactive materials on film or nuclear emulsion.

autoregulation—A site or mutation that affects the properties only of its own molecule of DNA, often indicating that a site does not code for a diffusible product.

autosplicing (self-splicing)—The ability of an intron to excise itself from an RNA by a catalytic action that depends only on the sequence of RNA in the intron.

axial element—A proteinaceous structure around which the chromosomes condense at the start of synapsis.

back mutation—A mutation that reverses the effect of a mutation that had inactivated a gene; thus, it restores the original sequence or function of the gene product.

bacteriophage—A bacterial virus.

Bam islands—A series of short, repeated sequences found in the nontranscribed spacer of *Xenopus* rDNA genes.

bands—Structures visible on polytene chromosomes as dense regions that contain the majority of DNA; they include active genes. Also, chromosomal bands revealed by DNA stains.

basal apparatus—The complex of transcription factors that assembles at the promoter before RNA polymerase is bound.

basal transcription factors—Transcription factors required by RNA polymerase II to form the initiation complex at all RNA polymerase II promoters. Factors are identified as $TF_{II}X$, where X is a letter.

B cell—A lymphocyte that produces antibodies. Development occurs primarily in bone marrow.

B cell receptor—The antigen receptor on B lymphocytes.

bidirectional replication—A system in which an origin generates two replication forks that proceed away from the origin in opposite directions.

bivalent—The structure containing all four chromatids (two representing each homolog) at the start of meiosis.

boundary element—A DNA sequence element bound by proteins that prevent the spread of open or closed chromatin.

branch migration—The ability of a DNA strand partially paired with its complement in a duplex to extend its pairing by displacing the resident strand with which it is homologous.

branch site—A short sequence just before the end of an intron at which the lariat intermediate is formed in splicing by joining the 5′ nucleotide of the intron to the 2′ position of an adenosine.

breakage and reunion—The mode of genetic recombination in which two DNA duplex molecules are broken at corresponding points and then rejoined crosswise (involving formation of a length of heteroduplex DNA around the site of joining).

bromodomain—A 110–amino acid domain that binds to acetylated lysines in histones.

bZIP (basic zipper)—A bZIP (basic zipper) protein has a basic DNA-binding region adjacent to a leucine zipper dimerization motif.

cap—The structure at the 5′ end of eukaryotic mRNA, which is introduced after transcription by linking the terminal phosphate of 5′ GTP to the terminal base of the mRNA.

cap 0—A cap at the 5′ end of mRNA that has only a methyl group on 7-guanine.

cap 1—A cap at the 5′ end of mRNA that has methyl groups on the terminal 7-guanine and the 2′-O position of the next base.

cap 2—A cap that has three methyl groups (7-guanine, 2′-O position of next base, and N6 adenine) at the 5′ end of mRNA.

capsid—The external protein coat of a virus particle.

carboxy terminal domain (CTD)—The domain of the largest subunit of eukaryotic RNA polymerase II that is phosphorylated at initiation and is involved in coordinating several activities with transcription.

cascade—A sequence of events, each of which is stimulated by the previous one. In transcriptional regulation, as seen in sporulation and phage lytic development, it means that regulation is divided into stages, and at each stage, one of the genes that is expressed codes for a regulator needed to express the genes of the next stage.

catabolite regulation—The ability of glucose to prevent expression of a number of genes. In bacteria this is a positive control system; in eukaryotes, it is completely different.

catabolic repressor protein (CRP)—A positive regulator protein activated by cyclic AMP. It is needed for RNA polymerase to initiate transcription of many operons of *E. coli*.

catenate—To link together two circular molecules, as in a chain.

cDNA—A single-stranded DNA complementary to an RNA and synthesized from it by reverse transcription *in vitro*.

cell-mediated response—The immune response that is mediated primarily by T lymphocytes. It is defined based on immunity that cannot be transferred from one organism to another by serum antibody.

central dogma—The paradigm that information cannot be transferred from protein to protein or protein to nucleic acid but can be transferred between nucleic acids and from nucleic acid to protein.

central element—A structure that lies in the middle of the synaptonemal complex, along which the lateral elements of homologous chromosomes align. It is formed from Zip proteins.

centromere—A constricted region of a chromosome that includes the site of attachment (the kinetochore) to the mitotic or meiotic spindle. It may consist of unique DNA sequences or highly repetitive sequences and contains proteins not found anywhere else in the chromosome.

C genes—Gene segments that code for the constant regions of immunoglobulin protein chains.

checkpoint—A biochemical control mechanism that prevents the cell from progressing from one stage to the next unless specific goals and requirements have been met.

chemical proofreading—A proofreading mechanism in which the correction event occurs after the addition of an incorrect subunit to a polymeric chain, by means of reversing the addition reaction.

chiasma (pl. chiasmata)—A site at which two homologous chromosomes have exchanged material during meiosis.

chromatin—The complex of nuclear DNA and associated its proteins.

chromatin immunoprecipitation (ChIP)—A method for detecting *in vivo* protein-DNA interactions that entails isolating proteins with an antibody and identifying DNA sequences that are associated with these proteins.

chromatin remodeling—The energy-dependent displacement or reorganization of nucleosomes that occurs in conjunction with activation of genes for transcription.

chromocenter—An aggregate of heterochromatin from different chromosomes.

chromodomain—~60 amino acid domains that recognize methylated lysines in histones; some chromodomains have different functions such as RNA binding.

chromomeres—Densely staining granules visible in chromosomes under certain conditions, especially early in meiosis, when a chromosome may appear to consist of a series of chromomeres.

chromosomal walk—A technique for locating a gene by using the most closely linked markers as a probe for a genetic library.

chromosome—A discrete unit of the genome carrying many genes. Each consists of a very long molecule of duplex DNA and an approximately equal mass of proteins. It is visible as a morphological entity only during cell division.

chromosome pairing—The coupling of the homologous chromosomes at the start of meiosis.

chromosome scaffold—A proteinaceous structure in the shape of a sister chromatid pair, generated when chromosomes are depleted of histones.

***cis*-acting**—A site that affects the activity only of sequences on its own molecule of DNA (or RNA); this property usually implies that the site does not code for protein.

***cis*-dominant**—A site or mutation that affects the properties only of its own molecule of DNA, often indicating that a site does not code for a diffusible product.

cistron—The genetic unit defined by the complementation test; it is equivalent to a gene and includes all noncomplementing alleles.

clamp—A protein complex that forms a circle around the DNA; by connecting to DNA polymerase, it ensures that the enzyme action is processive.

clamp loader—A five-subunit protein complex that is responsible for loading the β clamp onto DNA at the replication fork.

class switching—A change in Ig gene organization in which the C region of the heavy chain is changed, but the V region remains the same.

clonal expansion—The production of numerous daughter cells all arising from a single cell.

cloning—Propagation of a DNA sequence by incorporating it into a hybrid construct that can be replicated in a host cell.

cloning vectors—DNA (often derived from a plasmid or a bacteriophage genome) that can be used to propagate an incorporated DNA sequence in a host cell; vectors contain selectable markers and replication origins to allow identification and maintenance of the vector in the host.

closed (blocked) reading frame—A reading frame that cannot be translated into protein because of the occurrence of termination codons.

closed complex—The stage of initiation of transcription before RNA polymerase causes the two strands of DNA to separate to form the transcription bubble.

cloverleaf—The structure of tRNA drawn in two dimensions, forming four distinct arm-loops.

coactivator—Factors required for transcription that do not bind DNA but are required for (DNA-binding) activators to interact with the basal transcription factors.

coding end—The free DNA end of coding sequences produced during recombination of immunoglobulin and T cell receptor genes. Coding ends are at the termini of the cleaved V and (D)J coding regions. Their subsequent joining yields a coding joint.

coding joint—The DNA junction created by the joining of two coding ends during V(D)J recombination.

coding region—A part of a gene that encodes for a polypeptide sequence.

coding (sense) strand—The DNA strand that has the same sequence as the mRNA and is related by the genetic code to the polypeptide sequence that it represents.

codon—A triplet of nucleotides that codes for an amino acid or a termination signal.

codon bias—A higher usage of one codon in genes to encode amino acids for which there are several synonymous codons.

codon usage—A description of the relative abundance of tRNAs for each codon.

cognate tRNAs—tRNAs recognized by a particular aminoacyl-tRNA synthetase. All are charged with the same amino acid.

cointegrate—A structure that is produced by fusion of two replicons, one originally possessing a transposon and the other lacking it; the cointegrate has copies of the transposon present at both junctions of the replicons, oriented as direct repeats.

colinearity—The relationship that describes the 1:1 correspondence of a sequence of triplet nucleotides to a sequence of amino acids.

compatibility group—A group of plasmids that contains members unable to coexist in the same bacterial cell.

complement—A set of ~20 proteins that function through a cascade of proteolytic actions to lyse infected target cells, or to attract macrophages.

complementary—Base pairs that match up in the pairing reactions in double helical nucleic acids (A with T in DNA or with U in RNA, and C with G).

complementation test—A test that determines whether two mutations are alleles of the same gene. It is accomplished by crossing two different recessive mutations that have the same phenotype and determining whether the wild-type phenotype can be produced. If so, the mutations are said to complement each other and are probably not mutations in the same gene.

complex mRNA—See **scarce mRNA**.

composite elements—Transposable elements consisting of two IS elements (can be the same or different) and the DNA sequences between the IS elements; the non-IS sequences often include gene(s) conferring antibiotic resistance.

concerted evolution (coincidental evolution)—The ability of two or more related genes to evolve together as though constituting a single locus.

conditional lethal—A mutation that is lethal under one set of conditions but not lethal under a second set of conditions, such as temperature.

conjugation—A process in which two cells come in contact and transfer genetic material. In bacteria, DNA is transferred from a donor to a recipient cell. In protozoa, DNA passes from each cell to the other.

consensus sequence—An idealized sequence in which each position represents the base most often found when many actual sequences are compared.

conserved sequences—Sequences in which many examples of a particular nucleic acid or protein are compared and the same individual bases or amino acids are always found at particular locations.

constant region (C region)—The part of an immunoglobulin, or T cell receptor, that varies least in amino acid sequence between different molecules. They are en coded by C gene segments. The heavy chain regions identify the type of immunoglobulin and recruits effector functions.

constitutive expression—Continuous expression of a gene.

constitutive gene—See **housekeeping genes**.

constitutive heterochromatin—The inert state of darkly stained regions, usually satellite DNA.

context—The fact that neighboring sequences may change the efficiency with which, for example, a codon is recognized by its aminoacyl-tRNA or is used to terminate polypeptide translation.

controlling elements—Transposable units in maize originally identified solely by their genetic properties. They may be autonomous (able to transpose independently) or nonautonomous (able to transpose only in the presence of an autonomous element).

copy number—The number of copies of a plasmid that is maintained in a bacterium relative to the number of copies of the origin of the bacterial chromosome.

core enzyme—The complex of RNA polymerase subunits needed for elongation. It does not include additional subunits or factors that may be needed for initiation or termination.

core histone—One of the four types of histone (H2A, H2B, H3, and H4 and their variants) found in the core particle derived from the nucleosome. (This excludes linker histones.)

core promoter—The shortest sequence at which an RNA polymerase can initiate transcription (typically at a much lower level than that displayed by a promoter containing additional elements). For RNA polymerase II it is the minimal sequence at which the basal transcription apparatus can assemble, and it often includes one or more of three common controller elements: the Inr, the TATA box, and the DPE. It is typically ~40 bp long.

core sequence—The segment of DNA that is common to the attachment sites on both the phage lambda and bacterial genomes. It is the location of the recombination event that allows phage lambda to integrate.

corepressor—A small molecule that triggers repression of transcription by binding to a regulator protein.

cosmid—Cloning vector derived from a bacterial plasmid by incorporating the *cos* sites of phage lambda, which make the plasmid DNA a substrate for the lambda packaging system.

coupled transcription/translation—The process in bacteria where a message is simultaneously being translated while it is being transcribed.

CpG islands—Stretches of 1–2 kb in mammalian genomes that are enriched in CpG dinucleotides; frequently found in promoter regions of genes.

CRISPR—Clusters of regularly interspaced repeats in prokaryotes that are transcribed and processed into short RNAs that function by RNA interference as an immune system against phage and plasmid infection.

crossover fixation—A possible consequence of unequal crossing over that allows a mutation in one member of a tandem cluster to spread through the whole cluster (or to be eliminated).

crown gall disease—A tumor that can be induced in many plants by infection with the bacterium *Agrobacterium tumefaciens.*

cryptic satellite—A satellite DNA sequence not identified as such by a separate peak on a density gradient; that is, it remains present in main band DNA.

cryptic unstable transcripts (CUTs)—Nonprotein-coding RNAs transcribed by RNA Polymerase II, frequently generated from the 3′ ends of genes (resulting in antisense transcripts) and rapidly degraded after synthesis. Some CUTs have regulatory roles.

ctDNA (cpDNA)—Chloroplast DNA.

C-terminal domain—The domain of RNA polymerase that is involved in stimulating transcription by contact with regulatory proteins. It also has roles in RNA capping, splicing, and polyadenylation.

C-value—The total amount of DNA in the genome (per haploid set of chromosomes).

C-value paradox—The lack of relationship between the DNA content (C-value) of an organism and its coding potential.

cyclic AMP (cAMP)—The coregulator of CRP, it has an internal 3′–5′ phosphodiester bond. Its concentration is inverse to the concentration of glucose.

cyclin-dependent kinases—A family of kinases that are inactive unless bound to a cyclin molecule. Most CDKs participate in cell cycle control.

cyclins—Proteins that bind and help activate cyclin-dependent kinases. Cyclin concentration varies throughout the cell cycle and their periodic availability plays an important role in regulating cell cycle progression.

cytotoxic T cell—A T lymphocyte (usually CD8+) that can be stimulated to kill cells containing intracellular pathogens, such as viruses.

cytotype—A cytoplasmic condition that affects P element activity. The effect of cytotype is due to the presence or absence of a repressor of transposition, which is provided by the mother to the egg.

D arm—The arm of tRNA that has a high content of the base dihydrouridine.

deacylated tRNA—tRNA that has no amino acid or polypeptide chain attached because it has completed its role in protein synthesis and is ready to be released from the ribosome.

decapping enzyme—An enzyme that catalyzes the removal of the 7-methyl guanosine cap at the 5′ end of eukaryotic mRNAs.

degradosome—A complex of bacterial enzymes, including RNase and helicase activities, that is involved in degrading mRNA.

delayed early genes—Genes in phage lambda that are equivalent to the middle genes of other phages. They cannot be transcribed until regulator protein(s) coded by the immediate early genes have been synthesized.

deletion—The removal of a sequence of DNA, the regions on either side being joined together except in the case of a terminal deletion at the end of a chromosome.

demethylase—A casual name for an enzyme that removes a methyl group, typically from DNA, RNA, or protein.

denaturation—A molecule's conversion from the physiological conformation to some other (inactive) conformation. In DNA, this involves the separation of the two strands due to breaking of hydrogen bonds between bases.

***de novo* methyltransferase**—Adds a methyl group to an unmethylated target sequence on DNA.

destabilizing element—Any one of many different *cis* sequences, present in some mRNAs, that stimulates rapid decay of that mRNA.

Dicer—An endonuclease that processes double-stranded RNA to 21 to 23 nucleotide molecules.

dideoxynucleotide (ddNTP)—A chain-terminating nucleotide that lacks a 3′-OH group and, therefore, is not a substrate for DNA polymerization; used in DNA sequencing.

direct repeats—Identical (or closely related) sequences present in two or more copies in the same orientation in the same molecule of DNA.

distributive (nuclease)—An enzyme that catalyzes the removal of only one or a few nucleotides before dissociating from the substrate.

divergence—The corrected percent difference in nucleotide sequence between two related DNA sequences or in amino acid sequences between two polypeptides.

D loop (displacement loop)—1. A region of the animal mitochondrial DNA molecule that is variable in size and sequence and contains the origin of replication. 2. The loop of displaced DNA generated by strand invasion and extension during homologous recombination.

DNA fingerprinting—Analysis of the differences between individuals in the fragments generated by using restriction enzymes to cut regions that contain short repeated sequences or by PCR. The lengths of the repeated regions are unique to every individual, and, as a result, the presence of a particular subset in any two individuals can be used to define their common inheritance (e.g., a parent-child relationship).

DNA ligase—The enzyme that makes a bond between an adjacent 3′–OH and 5′–phosphate end where there is a nick in one strand of duplex DNA.

DNA polymerase—An enzyme that synthesizes a daughter strand(s) of DNA (under direction from a DNA template). Any particular enzyme may be involved in repair or replication (or both).

DNA repair—The removal and replacement of damaged DNA by the correct sequence.

DNase—An enzyme that degrades DNA.

domain—In reference to a chromosome, it refers either to a discrete structural entity defined as a region within which supercoiling is independent of other regions or to an extensive region including an expressed gene that has heightened sensitivity to degradation by the enzyme DNase I. In a protein, it is a discrete continuous part of the amino acid sequence that can be equated with a particular function.

dominant negative—A mutation that results in a mutant gene product that prevents the function of the wild-type gene product, causing loss or reduction of gene activity in cells containing both the mutant and wild-type alleles. The most common cause is that the gene codes for a homomultimeric protein whose function is lost if only one of the subunits is a mutant.

dosage compensation—Mechanisms employed to compensate for the discrepancy between the presence of two X chromosomes in one sex but only one X chromosome in the other sex.

double-strand breaks (DSB)—Breaks that occur when both strands of a DNA duplex are cleaved at the same site. Genetic recombination is initiated by such breaks. The cell also has repair systems that act on breaks that are created at other times.

doubling time—The period (usually measured in minutes) that it takes for a bacterial cell to reproduce.

down mutation—A mutation in a promoter that decreases the rate of transcription.

downstream—Sequences proceeding farther in the direction of expression within the transcription unit.

downstream promoter element (DPE)—A common component of RNA polymerase II promoters that do not contain a TATA box.

Drosha—An endonuclease that processes double-stranded primary RNAs into short, ~70 bp precursors for Dicer processing.

D segment—An additional sequence that is found between the V and J regions of an immunoglobulin heavy chain.

***Ds* element**—Dissociation element; a non-autonomous transposable element in maize, related to the autonomous Activator (*Ac*) element.

early genes—Genes that are transcribed before the replication of phage DNA. They code for regulators and other proteins needed for later stages of infection.

early infection—The part of the phage lytic cycle between entry and replication of the phage DNA. During this time, the phage synthesizes the enzymes needed to replicate its DNA.

E complex—The first (early) complex to form at a splice site, consisting of U1 snRNP bound at the splice site together with factor ASF/SF2, U2AF bound at the branch site, and the bridging protein SF1/BBP.

EF-Tu—The elongation factor that binds aminoacyl-tRNA and places it into the A site of a bacterial ribosome.

EJC (exon junction complex)—A protein complex that assembles at exon-exon junctions during splicing and assists in RNA transport, localization, and degradation.

elongation—The stage in a macromolecular synthesis reaction (replication, transcription, or translation) in which the nucleotide or polypeptide chain is extended by the addition of individual subunits.

elongation factors—Proteins that associate with ribosomes cyclically during the addition of each amino acid to the polypeptide chain.

endonuclease—An enzyme that cleaves bonds within a nucleic acid chain; it may be specific for RNA or for single-stranded or double-stranded DNA.

endoribonuclease (endonuclease)—A ribonuclease that cleaves an RNA at internal site(s).

enhancer—A *cis*-acting sequence that increases the utilization of (most) eukaryotic promoters and can function in either orientation and in any location (upstream or downstream) relative to the promoter.

epigenetic—Changes that influence the phenotype without altering the genotype. They consist of changes in the properties of a cell that are inherited that do not represent a change in genetic information.

episome—A plasmid able to integrate into bacterial DNA.

epitope tag—A polypeptide that has been added to a protein that allows its identification by an antibody.

equilibrium density-gradient centrifugation—A gradient method used to separate macromolecules on the basis of differences in their density. For DNA, it is prepared from a heavy soluble compound such as CsCl.

error-prone polymerase—A DNA polymerase that incorporates noncomplementary bases into the daughter strand.

error-prone synthesis—A repair process in which noncomplementary bases are incorporated into the daughter strand.

E site—The site of the ribosome that briefly holds deacylated tRNAs before their release.

euchromatin—The form of chromatin that comprises most of the genome in the interphase nucleus, which is less tightly condensed than heterochromatin, and contains most of the active or potentially active single-copy genes.

excision—1. Release of phage or episome or other sequence from the host chromosome as an autonomous DNA molecule. 2. The step in an excision-repair system that consists of removing a single-stranded stretch of DNA by the action of a 5′ to 3′ exonuclease.

excision repair—A type of repair system in which one strand of DNA is directly excised and then replaced by resynthesis using the complementary strand as template.

exon—Any segment of an interrupted gene that is represented in the mature RNA product.

exon definition—The process in which a pair of splicing sites are recognized by interactions involving the 5′ site of the intron and the 5′ site of the next intron downstream.

exon junction complex (EJC)—A protein complex that assembles at exon-exon junctions during splicing and assists in RNA transport, localization and degradation.

exon shuffling—The hypothesis that genes have evolved by the recombination of various exons coding for functional protein domains.

exon trapping—Inserting a genomic fragment into a vector whose function depends on the provision of splicing junctions by the fragment.

exonuclease—An enzyme that cleaves nucleotides one at a time from the end of a polynucleotide chain; it may be specific for either the 5′ or 3′ end of DNA or RNA.

exoribonuclease—A ribonuclease that removes terminal ribonucleotides from RNA.

exosome—A complex of several exonucleases involved in degrading mRNA.

expressed sequence tag (EST)—A short sequenced fragment of a cDNA sequence that can be used to identify an actively expressed gene.

expression vector—A cloning vector that allows the expression, either translation or just transcription, of the insert.

extein—A sequence that remains in the mature protein that is produced by processing a precursor via protein splicing.

extra arm—The arm of tRNA that shows length variation among different tRNAs.

extranuclear genes—Genes that reside outside the nucleus in organelles such as mitochondria and chloroplasts.

facultative heterochromatin—The inert state of sequences that also exist in active copies, for example, one mammalian X chromosome in females.

fixation—The process by which a new allele replaces the allele that was previously predominant in a population.

fluorescence resonant energy transfer (FRET)—A process whereby the emission from an excited fluorophore is captured and reemitted at a longer wavelength by a nearby second fluorophore whose excitation spectrum matches the emission frequency of the first fluorophore.

forward mutation—A mutation that inactivates a functional gene.

F plasmid—An episome that can be free or integrated in *E. coli* and that can sponsor conjugation in either form.

frameshift—A mutation caused by deletions or insertions that are not a multiple of three base pairs. They change the frame in which triplets are translated into polypeptide.

fully methylated—A site that is a palindromic sequence that is methylated on both strands of DNA.

gain-of-function mutation—A mutation that causes an increase in the normal gene activity. It sometimes represents acquisition of certain abnormal properties. It is often, but not always, dominant.

G-bands—Bands generated on eukaryotic chromosomes by staining techniques that appear as a series of lateral striations. They are used for karyotyping (identifying chromosomes and chromosomal regions by the banding pattern).

gene cluster—A group of adjacent genes that are identical or related.

gene conversion—The alteration of one strand of a heteroduplex DNA to make it complementary with the other strand at any position(s) where there were mispaired bases, or the complete replacement of genetic material at one locus by a homologous sequence.

gene expression—The process by which the information in a sequence of DNA in a gene is used to produce an RNA or polypeptide, involving transcription and (for polypeptides) translation.

gene family—A set of genes within a genome that encode related or identical proteins or RNAs. The members originated from duplication of an ancestral gene followed by accumulation of changes in sequence between the copies. Most often the members are related but not identical.

gene redundancy—The concept that two or more genes may fulfill the same function, so that no single one of them is essential.

genetic code—The correspondence between triplets in DNA (or RNA) and amino acids in polypeptide.

genetic drift—The chance fluctuation (without selective pressure) of the frequencies of alleles in a population.

genetic hitchhiking—The change in frequency of an allele due to its association (functional association or physical linkage) with an allele at another locus that is also changing in frequency.

genetic map—See **linkage map**.

genetic recombination—A process by which separate DNA molecules are joined into a single molecule, due to such processes as crossing over or transposition.

genome—The complete set of sequences in the genetic material of an organism. It includes the sequence of each chromosome plus any DNA in organelles.

glycosylase—A repair enzyme that removes damaged bases by cleaving the bond between the base and the sugar.

gratuitous inducer—Inducers that resemble authentic inducers of transcription but are not substrates for the induced enzymes.

growing point—See **replication fork**.

GU-AG rule—The rule that describes the presence of these constant dinucleotides at the first two and last two positions of introns of nuclear genes.

guide RNA—A small RNA whose sequence is complementary to the sequence of an RNA that has been edited. It is used as a template for changing the sequence of the preedited RNA by inserting or deleting nucleotides.

gyrase—Enzyme that introduces negative supercoils into DNA.

hairpin—An RNA sequence that can fold back on itself forming double-stranded RNA.

half-life ($t_{1/2}$) (RNA)—The time taken for the concentration of a given population of RNA molecules to decrease by half, in the absence of new synthesis.

haplotype—The particular combination of alleles in a defined region of some chromosome—a small portion of the genotype. Originally used to described combinations of (MHC) alleles, it now may be used to describe major-histocompatibility couples.

Hb anti-Lepore—A fusion gene produced by unequal crossing over that has the N-terminal part of β globin and the C-terminal part of δ globin.

Hb Kenya—A fusion gene produced by unequal crossing over between the between Aγ and β globin genes.

Hb Lepore—An unusual globin protein that results from unequal crossing over between the β and δ genes. The genes become fused together to produce a single β-like chain that consists of the N-terminal sequence of δ joined to the C-terminal sequence of β.

HbH (hemoglobin) disease—A condition in which there is a disproportionate amount of the abnormal tetramer β_4 relative to the amount of normal hemoglobin ($\alpha_2\beta_2$).

heavy chain—The larger of the two types of subunits that make up an antibody tetramer. Each antibody contains two heavy chains. The N-terminus of the heavy chain forms part of the antigen recognition site, whereas the C-terminus determines the antibody subclass.

helicase—An enzyme that uses energy provided by ATP hydrolysis to separate the strands of a nucleic acid duplex.

helix-loop-helix (HLH)—The motif that is responsible for dimerization of a class of transcription factors called HLH proteins. A bHLH protein has a basic DNA-binding sequence close to the dimerization motif.

helix-turn-helix—The motif that describes an arrangement of two α-helices that form a site that binds to DNA, one fitting into the major groove of DNA and other lying across it.

helper T (T_H) cell—A T lymphocyte that activates macrophages and stimulates B cell proliferation and antibody production. They usually express cell surface CD4 but not CD8.

helper virus—A virus that provides functions absent from a defective virus, enabling the latter to complete the infective cycle during a mixed infection with the helper virus.

hemimethylated DNA—DNA that is methylated on one strand of a target sequence that has a cytosine on each strand.

heterochromatin—Regions of the genome that are highly condensed, are not transcribed, and are late replicating. It is divided into two types: constitutive and facultative.

heteroduplex DNA—DNA that is generated by base pairing between complementary single strands derived from the different parental duplex molecules; it occurs during genetic recombination.

heterogeneous nuclear RNA (hnRNA)—RNA that constitutes transcripts of nuclear genes made primarily by RNA polymerase II; it has a wide size distribution and variable stability.

heteromultimer—A molecular complex (such as a protein) composed of different subunits.

heteroplasmy—Having more than one mitochondrial allelic variant in a cell.

Hfr—A bacterium that has an integrated F plasmid within its chromosome. Hfr stands for high-frequency recombination, referring to the fact that chromosomal genes are transferred from an Hfr cell to an F^- cell much more frequently than from an F^+ cell.

histone acetyltransferase (HAT)—An enzyme that modifies histones by addition of acetyl groups; some transcriptional coactivators have this activity. Also known as lysine acetyltransferase (KAT).

histone code—The hypothesis that combinations of specific modifications on specific histone residues act cooperatively to define chromatin function.

histone deacetylase (HDAC)—Enzyme that removes acetyl groups from histones; typically associated with repressors of transcription.

histone fold—A motif found in all four core histones in which three α-helices are connected by two loops.

histone octamer—The complex of two copies of each of the four different core histones (H2A, H2B, H3, and H4); DNA wraps around the histone octamer to form the nucleosome.

histone tails—Flexible amino- or carboxy-terminal regions of the core histones that extend beyond the surface of the nucleosome; histone tails are sites of extensive post-translational modification.

histone variant—Any of a number of histones closely related to one of the core histones (H2A, H2B, H3 or H4) that can assemble into a nucleosome in the place of the related core histone; many histone variants have specialized functions or localization. There are also numerous linker histone variants.

histones—Conserved DNA-binding proteins that form the basic subunit of chromatin in eukaryotes. H2A, H2B, H3, and H4 form an octameric core around which DNA coils to form a nucleosome. Linker histones are external to the nucleosome.

hnRNP—The ribonucleoprotein form of hnRNA (heterogeneous nuclear RNA), in which the hnRNA is complexed with proteins. Pre-mRNAs are not exported until processing is complete; thus, they are found only in the nucleus.

Holliday junction—An intermediate structure in homologous recombination in which the two duplexes of DNA are connected by the genetic material exchanged between two of the four strands, one from each duplex.

holoenzyme—1. The DNA polymerase complex that is competent to initiate replication. 2. The RNA polymerase form that is competent to initiate transcription. It consists of the five subunits of the core enzyme ($\alpha_2\beta\beta'\omega$) and σ factor.

homeodomain—A class of DNA-binding motifs that contain the helix-turn-helix structure that typifies a class of transcription factors often found in developmentally regulated genes.

homologous genes (homologs)—Related genes in the same species, such as alleles on homologous chromosomes or multiple genes in the same genome sharing common ancestry.

homologous recombination—Recombination involving a reciprocal exchange of sequences of DNA, e.g., between two chromosomes that carry the same genetic loci.

homomultimer—A molecular complex (such as a protein) in which the subunits are identical.

horizontal transfer—The transfer of DNA from one cell to another by a process other than cell division, such as bacterial conjugation.

hotspots—A site in the genome at which the frequency of mutation (or recombination) is very much increased, usually by at least an order of magnitude relative to neighboring sites.

housekeeping gene (constitutive gene)—A gene that is (theoretically) expressed in all cells because it provides basic functions needed for sustenance of all cell types.

humoral response—An immune response that is mediated primarily by antibodies. It is defined as immunity that can be transferred from one organism to another by serum antibody.

hybrid dysgenesis—The inability of certain strains of *D. melanogaster* to interbreed, because the hybrids are sterile (although otherwise they may be phenotypically normal).

hybridization—The pairing of complementary RNA and DNA strands to give an RNA-DNA hybrid.

hydrops fetalis—A fatal disease resulting from the absence of the hemoglobin α gene.

hypersensitive site—A short region of chromatin detected by its extreme sensitivity to cleavage by DNase I and other nucleases; it comprises an area from which nucleosomes are excluded.

IF-1—A bacterial initiation factor that stabilizes the initiation complex for polypeptide translation.

IF-2—A bacterial initiation factor that binds the initiator tRNA to the initiation complex for polypeptide translation.

IF-3—A bacterial initiation factor required for 30S ribosomal subunits to bind to initiation sites in mRNA. It also prevents 30S subunits from binding to 50S ribosomal subunits.

immediate early genes—Genes in phage lambda that are equivalent to the early class of other phages. They are transcribed immediately upon infection by the host RNA polymerase.

immune response—An organism's reaction, mediated by components of the immune system, to an antigen.

immunity—In phages, the ability of a prophage to prevent another phage of the same type from infecting a cell. In plasmids, the ability of a plasmid to prevent another of the same type from becoming established in a cell. It can also refer to the ability of certain transposons to prevent others of the same type from transposing to the same DNA molecule.

immunity region—A segment of the phage genome that enables a prophage to inhibit additional phage of the same type from infecting the bacterium. This region has a gene that codes for the repressor, as well as the sites to which the repressor binds.

immunoglobulin (antibody)—A protein that is produced by B cells and that binds to a particular antigen. They are synthesized in membrane-bound and secreted forms. Those produced during an immune response recruit effector functions to help neutralize and eliminate the pathogen.

imprecise excision—Excision that occurs when a transposon removes itself from the original insertion site but leaves behind some of its sequence.

imprinting—A change in a gene that occurs during passage through the sperm or egg with the result that the paternal and maternal alleles have different properties in the very early embryo. This is caused by methylation of DNA.

incision—A step in a mismatch excision-repair system in which an endonuclease recognizes the damaged area in the DNA and isolates it by cutting the DNA strand on both sides of the damage.

indirect end labeling—A technique for examining the organization of DNA by making a cut at a specific site and identifying all fragments containing the sequence adjacent to one side of the cut; it reveals the distance from the cut to the next break(s) in DNA.

induced mutations—Mutations that result from the action of a mutagen. The mutagen may act directly on the bases in DNA, or it may act indirectly to trigger a pathway that leads to a change in DNA sequence.

inducer—A small molecule that triggers gene transcription by binding to a regulator protein.

inducible gene—A gene that is turned on by the presence of its substrate.

induction—The ability to synthesize certain enzymes only when their substrates are present. Applied to gene expression, it refers to switching on transcription as a result of interaction of the inducer with the regulator protein.

induction of phage—A phage's entry into the lytic (infective) cycle as a result of destruction of the lysogenic repressor, which leads to excision of free phage DNA from the bacterial chromosome.

initiation—The stages of transcription up to synthesis of the first bond in RNA. These include binding of RNA polymerase to the promoter (this is sometimes referred to as preinitiation) and melting a short region of DNA into single strands.

initiation codon—A special codon (usually AUG) used to start synthesis of a polypeptide.

initiation factors (IFs)—Proteins that associate with the small subunit of the ribosome specifically at the stage of initiation of polypeptide translation.

initiator (Inr)—The sequence of a pol II promoter between −3 and +5 and has the general sequence Py_2CAPy_5. It is the simplest possible pol II promoter.

innate immunity—An immune response that depends on recognition of predefined patterns in pathogens.

insert—A fragment of DNA that is to be cloned to a vector.

insertion sequence (IS)—A small bacterial transposon that carries only the genes needed for its own transposition.

***in situ* hybridization**—Hybridization performed by denaturing the DNA of cells squashed on a microscope slide so that reaction is possible with an added single-stranded RNA or DNA; the added preparation is labeled and its hybridization is followed by autoradiography or fluorescence. *In situ* hybridization can also be performed in intact tissues.

insulator—A sequence that prevents an activating or inactivating effect passing from one side to the other.

integrase—An enzyme that is responsible for a site-specific recombination that inserts one molecule of DNA into another.

integration—Insertion of a viral or another DNA sequence into a host genome as a region covalently linked on either side to the host sequences.

intein—The part that is removed from a protein that is processed by protein splicing.

interactome—The complete set of protein complexes/protein-protein interactions present in a cell, tissue, or organism.

interallelic complementation—The change in the properties of a heteromultimeric protein brought about by the interaction of subunits coded by two different mutant alleles; the mixed protein may be more or less active than the protein consisting of subunits only of one or the other type.

interbands—The gene-rich regions of chromosomes that lie between the bands in Giemsa-stained chromosomes. More generally, interbands refer to any region between identified bands (G bands, bands of polytene chromosomes, etc.).

intercistronic region—In a polycistronic mRNA, the distance between the termination codon of one cistron and the initiation codon of the next cistron.

interrupted gene—A gene in which the coding sequence is not continuous due to the presence of introns.

intron—A segment of DNA that is transcribed but later removed from within the transcript by splicing together the sequences (exons) on either side of it.

intron definition—The process in which a pair of splicing sites are recognized by interactions involving only the 5′ site and the branchpoint/3′ site.

intron homing—The ability of certain introns to insert themselves into a target DNA. The reaction is specific for a single target sequence.

"introns early" model—The hypothesis that the earliest genes contained introns and some genes subsequently lost them.

"introns late" model—The hypothesis that the earliest genes did not contain introns and that introns were subsequently added to some genes.

invariant sequences—Base positions in tRNA that have the same nucleotide in virtually all (>95%) tRNAs.

inverted-terminal repeats—The short related or identical sequences present in reverse orientation at the ends of some transposons.

IRES (internal ribosome entry site)—A eukaryotic messenger RNA sequence that allows a ribosome to initiate polypeptide translation without migrating from the 5′ end.

iron-response element (IRE)— A *cis* sequence found in certain mRNAs whose stability or translation is regulated by cellular iron concentration.

isoaccepting tRNAs—See **cognate tRNAs**.

joint molecule—A pair of DNA duplexes that are connected together through a reciprocal exchange of genetic material.

J segments—Coding sequences in the immunoglobulin and T cell receptor loci. They are between the variable (V) and constant (C) gene segments.

kinetic proofreading—A proofreading mechanism that depends on incorrect events proceeding more slowly than correct events, so that incorrect events are reversed before a subunit is added to a polymeric chain.

kinetochore—A small organelle associated with the surface of the centromere that attaches a chromosome to the microtubules of the mitotic spindle. Each mitotic chromosome contains two "sisters" that are positioned on opposite sides of its centromere and face in opposite directions.

knockdown—A genetically engineered organism that has had a gene downregulated by introducing a silencing vector to reduce the expression (usually translation) of the gene.

knock-in— A genetically engineered organism that has had a gene sequence replaced by a different sequence.

knockout—A genetically engineered organism that has had a gene disabled by a targeted mutation.

kuru—A human neurological disease caused by prions. It may be caused by eating infected brains.

***lac* repressor**—A negative gene regulator encoded by the *lacI* gene that turns off the *lac* operon.

lagging strand—The strand of DNA that must grow overall in the 3′ to 5′ direction and is synthesized discontinuously in the form of short fragments (5′–3′) that are later connected covalently.

large subunit—The subunit of the ribosome (50S in bacteria, 60S in eukaryotes) that has the peptidyl transferase active site that synthesizes the peptide bond.

lariat—An intermediate in RNA splicing in which a circular structure with a tail is created by a 5′ to 2′ bond.

late gene—A gene transcribed when phage DNA is being replicated. It codes for components of the phage particle.

late infection—The part of the phage lytic cycle from DNA replication to lysis of the cell. During this time, the DNA is replicated and structural components of the phage particle are synthesized.

lateral element—A structure in the synaptonemal complex that forms when a pair of sister chromatids condenses on to an axial element.

leader (5′ UTR)—In mRNA, the untranslated sequence at the 5′ end that precedes the initiation codon.

leader peptide—The product that would result from translation of a short coding sequence used to regulate transcription of an operon by controlling ribosome movement.

leading strand—The strand of DNA that is synthesized continuously in the 5′ to 3′ direction.

lesion bypass—Replication by an error prone DNA polymerase on a template that contains a damaged base. The polymerase can incorporate a noncomplementary base into the daughter strand.

leucine-rich region—A motif found in the extracellular domains of some surface receptor proteins in animal and plant cells.

leucine zipper (leu zipper)—A dimerization motif that is found in a class of transcription factors.

licensing factor—A factor located in the nucleus and necessary for replication; it is inactivated or destroyed after one round of replication. New factors must be provided for further rounds of replication to occur.

ligate—To covalently link two ends of nucleic acid chains; they may be two ends of one chain or two ends of different chains, either DNA or RNA.

light chain—The smaller of the two types of subunits that make up an antibody tetramer. Each antibody contains two light chains. The N-terminus of the light chain forms part of the antigen recognition site.

lincRNA—Large intergenic noncoding RNAs originating from chromosomal regions in between classical genes.

linkage—The tendency of genes to be inherited together as a result of their location on the same chromosome; measured by percent recombination between loci.

linkage disequilibrium—A nonrandom association between alleles at different loci, whether due to linkage or some other cause, such as selection on a specific multilocus combination of alleles.

linkage map—A map of the positions of loci or other genetic markers on a chromosome obtained by measuring recombination frequencies between markers.

linker DNA—Nonnucleosomal DNA present between nucleosomes.

linker histones—A family of histones (such as histone H1) that are not components of the nucleosome core; linker histone bind nucleosomes and/or linker DNA and promote 30 nm fiber formation.

locus—The position on a chromosome at which the gene for a particular trait resides; it may be occupied by any one of the alleles for the gene.

locus control region (LCR)—The region that is required for the expression of several genes in a domain.

long interspersed elements (LINEs)—Long interspersed elements; a major class of retrotransposons that occupy ~21% of the human genome.

long-terminal repeat (LTR)—The sequence that is repeated at each end of the provirus (integrated retroviral sequence).

loop—A single-stranded region at the end of a hairpin in RNA (or single-stranded DNA); it corresponds to the sequence between inverted repeats in duplex DNA.

loss-of-function mutation—A mutation that eliminates or reduces the activity of a gene. It is often, but not always, recessive.

luxury gene—A gene coding for a specialized function, usually synthesized in large amounts in particular cell types.

lyase—A repair enzyme (usually also a glycosylase) that opens the sugar ring at the site of a damaged base.

lysis—The death of bacteria at the end of a phage infective cycle when they burst open to release the progeny of an infecting phage (because phage enzymes disrupt the bacterium's cytoplasmic membrane or cell wall). The same term also applies to eukaryotic cells; for example, when infected cells are attacked by the immune system.

lysogenic/lysogeny—The ability of a phage to survive in a bacterium as a stable prophage component of the bacterial genome.

lytic infection—Infection of a bacterium by a phage that ends in the destruction of the bacterium with release of progeny phage.

maintenance methyltransferase—Enzyme that adds a methyl group to a target site that is already hemimethylated.

major groove—A fissure running the length of the DNA double helix that is 22 Å across.

major histocompatibility complex (MHC)—A chromosomal region containing genes that are involved in the immune response. The genes encode proteins for antigen presentation, cytokines, and complement, as well as other functions. It is highly polymorphic. Its genes and proteins are divided into three classes.

maternal inheritance—The preferential survival in the progeny of genetic markers provided by the female parent.

maternal mRNA granules—Oocyte particles containing translationally repressed mRNAs awaiting activation later in development.

mating-type cassette—A locus containing the genes required for mating type in yeast (either **a** or α), which can either be active or inactive (silent) copies of the locus. Mating type is changed when an active cassette of one type is replaced by a silent cassette of the other type.

matrix attachment region (MAR)—A region of DNA that attaches to the nuclear matrix. It is also known as a scaffold attachment site (SAR).

maturase—A protein coded by a group I or group II intron that is needed to assist the RNA to form the active conformation that is required for self-splicing.

mature transcript—A modified RNA transcript. Modification may include the removal of intron sequences and alterations to the 5′ and 3′ ends.

mediator—A large protein complex associated with yeast RNA polymerase II. It contains factors that are necessary for transcription from many or most promoters.

melting temperature—The midpoint of the temperature range over which the strands of DNA separate.

messenger RNA (mRNA)—The intermediate that represents one strand of a gene coding for polypeptide. Its coding region is related to the polypeptide sequence by the triplet genetic code.

metaphase (or mitotic) scaffold—A proteinaceous structure in the shape of a sister chromatid pair, generated when chromosomes are depleted of histones.

methyltransferase—An enzyme that adds a methyl group to a substrate, which can be a small molecule, a protein, or a nucleic acid.

microarray—An arrayed series of thousands of tiny DNA oligonucleotide samples imprinted on a small chip. mRNAs can be hybridized to microarrays to assess the amount and level of gene expression.

micrococcal nuclease (MNase)—An endonuclease that cleaves DNA; in chromatin, DNA is cleaved preferentially between nucleosomes.

microRNA (miRNA)—Very short RNAs that can regulate gene expression.

microsatellite—DNAs consisting of repetitions of extremely short (typically <10 bp) units.

microtubule organizing center (MTOC)—A region from which microtubules emanate. In animal cells the centrosome is the major microtubule organizing center.

middle genes—Phage genes that are regulated by the proteins coded by early genes. Some proteins coded by them catalyze replication of the phage DNA; others regulate the expression of a later set of genes.

minicell—An anucleate bacterial (*E. coli*) cell produced by a division that generates a cytoplasm without a nucleus.

minisatellite—DNA consisting of many copies of a short repeating sequence. The length of the repeating unit is measured in tens of base pairs. The number of repeats varies between individual genomes.

minor groove—A fissure running the length of the DNA double helix that is 12 Å across.

minus-strand DNA—The single-stranded DNA sequence that is complementary to the viral RNA genome of a plus strand virus.

miRNA—Very short RNAs that may regulate gene expression.

mismatch repair—Repair that corrects recently inserted bases that do not pair properly. The process preferentially corrects the sequence of the daughter strand by distinguishing the daughter strand and parental strand; in prokaryotes this is on the basis of their states of methylation.

missense suppressor—A suppressor that codes for a tRNA that has been mutated to recognize a different codon. By inserting a different amino acid at a mutant codon, the tRNA suppresses the effect of the original mutation.

modification—All changes made to the nucleotides of DNA or RNA after their initial incorporation into the polynucleotide chain.

molecular clock—An approximately constant rate of evolution that occurs in DNA sequences, such as by the genetic drift of neutral mutations.

monocistronic mRNA—mRNA that codes for one polypeptide.

mRNA decay—mRNA degradation, assuming that the degradation process is stochastic.

mtDNA—mitochondrial DNA.

multicopy replication control—Occurs when the control system allows the plasmid to exist in more than one copy per individual bacterial cell.

multiforked chromosome—A bacterial chromosome that has more than one set of replication forks, because a second initiation has occurred before the first cycle of replication has been completed.

multiple cloning site—An artificial DNA sequence in a cloning vector containing multiple restriction endonuclease sites for cloning.

mutagens—Substances that increase the rate of mutation by inducing changes in DNA sequence, directly or indirectly.

mutation hotspot—A site in the genome at which the frequency of mutation (or recombination) is very much increased, usually by at least an order of magnitude relative to neighboring sites.

mutator—A mutation or a mutated gene that increases the basal level of mutation. Such genes often code for proteins that are involved in repairing damaged DNA.

n – 1 rule—The rule that states that only one X chromosome is active in female mammalian cells; any others are inactivated.

nascent polypeptide—A protein that has not yet completed its synthesis; the polypeptide chain is still attached to the ribosome via a tRNA.

nascent RNA—An RNA chain that is still being synthesized, so that its 3′ end is paired with DNA where RNA polymerase is elongating.

ncRNA (noncoding RNA)—A noncoding RNA that does not contain an open reading frame.

negative complementation—Occurring when interallelic complementation allows a mutant subunit to suppress the activity of a wild-type subunit in a multimeric protein.

negative control—A method of gene regulation in which a regulator is required to turn the gene off.

negative inducible—A control circuit in which an active repressor is inactivated by the substrate of the operon.

negative repressible—A control circuit in which an inactive repressor is activated by the product of the operon.

nested genes—A gene located within an intron of another gene.

neuronal granules—Particles containing translationally repressed mRNAs in transit to final cell destinations.

neutral mutation—A mutation that has no significant effect on evolutionary fitness and usually has no effect on the phenotype.

neutral substitutions—Mutations that cause changes in amino acids of the protein product but that do not affect the protein's activity.

N-formyl-methionyl-tRNA—The aminoacyl-tRNA that initiates bacterial polypeptide translation. The amino group of the methionine is formylated.

N nucleotide—A short nontemplated sequence that is added randomly by the enzyme at coding joints during rearrangement of immunoglobulin and T cell receptor genes. They augment the diversity of antigen receptors.

no-go decay (NGD)—A pathway that rapidly degrades an mRNA with ribosomes stalled in its coding region.

nonallelic genes—Two (or more) copies of the same gene that are present at different locations in the genome (contrasted with alleles, which are copies of the same gene derived from different parents and present at the same location on the homologous chromosomes).

nonautonomous controlling element—A transposon in maize that encodes a non-functional transposase; it can transpose only in the presence of a *trans*-acting autonomous member of the same family.

nonhistone—Any structural protein found in a chromosome except one of the histones.

nonhomologous end joining (NHEJ)—The pathway that ligates blunt ends. It is used to repair DNA double-strand breaks and in certain recombination pathways (such as immunoglobulin recombination).

non-Mendelian inheritance—A pattern of inheritance that does not follow that expected by Mendelian principles (each parent contributing a single allele to offspring). Extranuclear genes show a non-Mendelian inheritance pattern.

nonprocessed pseudogenes—Nonfunctional gene copies that are formed by gene duplication and mutational inactivation.

nonproductive rearrangement—Occurs as a result of the recombination of V, (D), J gene segments if the rearranged gene segments are not in the correct reading frame. It occurs when nucleotide addition or subtraction disrupts the reading frame or when a functional protein is not produced.

nonrepetitive DNA—DNA that is unique (present only once) in a genome.

nonreplicative transposition—The movement of a transposon that leaves a donor site (usually generating a double-strand break) and moves to a new site.

nonsense-mediated decay (NMD)—A pathway that degrades an mRNA that has a nonsense mutation prior to the last exon.

nonsense suppressor—A gene coding for a mutant tRNA that is able to respond to one or more of the termination codons and insert an amino acid at that site.

nonstop delay (NSD)—A pathway that rapidly degrades an mRNA that lacks an in-frame termination codon.

nonsynonymous mutation—A mutation in a coding region that alters the amino acid sequence of the polypeptide product.

nontranscribed spacer—The region between transcription units in a tandem gene cluster.

nuclear pore—Pore in the nuclear membrane that allows transport of RNA, proteins, and other molecules in and out of the nucleus.

nuclease—An enzyme that can break a phosphodiester bond.

nucleation center—A duplex hairpin in TMV (tobacco mosaic virus) in which assembly of coat protein with RNA is initiated.

nucleoid—The structure in a prokaryotic cell that contains the genome. The DNA is bound to proteins and is not enclosed by a membrane.

nucleolar organizer—The region of a chromosome carrying genes coding for rRNA.

nucleolus—A discrete region of the nucleus where ribosomes are produced.

nucleoside—A molecule consisting of a purine or pyrimidine base linked to the 1′ carbon of a pentose sugar.

nucleosome—The basic structural subunit of chromatin, consisting of ~200 bp of DNA and an octamer of histone proteins.

nucleosome positioning—The placement of nucleosomes at defined sequences of DNA instead of at random locations with regard to sequence.

nucleotide—A molecule consisting of a purine or pyrimidine base linked to the 1′ carbon of a pentose sugar and a phosphate group linked to either the 5′ or 3′ carbon of the sugar.

null mutation—A mutation that completely eliminates the function of a gene.

nut—An acronym for N utilization site, the sequence of DNA that is recognized by the N antitermination factor.

ochre codon—The triplet UAA, one of the three termination codons that end polypeptide translation.

Okazaki fragment—Short stretches of 1000 to 2000 bases produced during discontinuous replication; they are later joined into a covalently intact strand.

oligo (A) tail—A short poly(A) tail, generally referring to a stretch of fewer than 15 adenylates

oncogene—A gene that when altered may cause cancer, typically by a dominant mechanism.

one gene : one enzyme hypothesis—Beadle and Tatum's hypothesis that a gene is responsible for the production of a single enzyme.

one gene : one polypeptide hypothesis—A modified version of the not generally correct one gene : one enzyme hypothesis; the hypothesis that a gene is responsible for the production of a single polypeptide.

opal codon—The triplet UGA, one of the three termination codons that end polypeptide translation. It has evolved to code for an amino acid in a small number of organisms or organelles.

open complex—The stage of initiation of transcription when RNA polymerase causes the two strands of DNA to separate to form the transcription bubble.

open reading frame (ORF)—A sequence of DNA consisting of triplets that can be translated into amino acids starting with an initiation codon and ending with a termination codon.

operator—The site on DNA at which a repressor protein binds to prevent transcription from initiating at the adjacent promoter.

operon—A unit of bacterial gene expression and regulation, including structural genes and control elements in DNA recognized by regulator gene product(s).

opine—A derivative of arginine that is synthesized by plant cells infected with crown gall disease.

ORC—Origin recognition complex, a multiprotein complex found in eukaryotes that binds to the replication origin (the ARS in yeast) and remains associated with it throughout the cell cycle.

origin—A sequence of DNA at which replication is initiated.

orthologous genes (orthologs)—Related genes in different species.

overlapping gene—A gene in which part of the sequence is found within part of the sequence of another gene.

overwound—B-form DNA that has fewer than 10.4 base pairs per turn of the helix.

packing ratio—The ratio of the length of DNA to the unit length of the fiber containing it.

palindrome—A symmetrical sequence that reads the same forward and backward.

paralogous genes (paralogs)—Genes that share a common ancestry due to gene duplication.

paraspeckles—Small, irregularly shaped ribonucleoprotein bodies found in interchromatin nuclear locations, typically 10–20 per nucleus.

patch recombinant—DNA that results from a Holliday junction being resolved by cutting the exchanged strands. The duplex is largely unchanged, except for a DNA sequence on one strand that came from the homologous chromosome.

pathogenicity islands—DNA segments that are present in pathogenic bacterial genomes but absent in their nonpathogenic relatives.

P element—A type of transposon in *D. melanogaster*.

peptidyl transferase—The activity of the large ribosomal subunit that synthesizes a peptide bond when an amino acid is added to a growing polypeptide chain. The actual catalytic activity is a property of the rRNA.

peptidyl-tRNA—The tRNA to which the nascent polypeptide chain has been transferred following peptide bond synthesis during polypeptide translation.

phage—An abbreviation of bacteriophage or bacterial virus.

phosphatase—An enzyme that breaks a phosphomonoester bond, cleaving a terminal phosphate.

photoreactivation—A repair mechanism that uses a white light-dependent enzyme to split cyclobutane pyrimidine dimers formed by ultraviolet light.

pilin—The subunit that is polymerized into the pilus in bacteria.

pilus—A surface appendage on a bacterium that allows the bacterium to attach to other bacterial cells. It appears as a short, thin, flexible rod. During conjugation, it is used to transfer DNA from one bacterium to another.

pioneer round of translation—The first translation event for a newly synthesized and exported mRNA.

piRNA—Piwi RNA, a special form of miRNA found in germ cells.

plasmid—Circular, extrachromosomal DNA. It is autonomous and can replicate itself.

plus-strand DNA—The strand of the duplex sequence representing a retrovirus that has the same sequence as that of the RNA.

plus-strand virus—A virus with a single-stranded nucleic acid genome whose sequence directly codes for the protein products.

P nucleotide—A short palindromic (inverted repeat) sequence that is generated during rearrangement of immunoglobulin and T cell receptor V, (D), and J gene segments. They are generated at coding joints when RAG proteins cleave the hairpin ends generated during rearrangement.

point mutation—A change in the sequence of DNA involving a single base pair.

poly(A)—A stretch of ~200 bases of adenylic acid that is added to the 3′ end of mRNA following its synthesis.

poly(A) binding protein (PABP)—The protein that binds to the 3′ stretch of poly(A) on a eukaryotic mRNA.

poly(A) nuclease (deadenylase)—An exoribonuclease that is specific for digesting poly(A) tails.

poly(A) polymerase (PAP)—The enzyme that adds the stretch of polyadenylic acid to the 3′ end of eukaryotic mRNA. It does not use a template.

poly(A)+ mRNA—mRNA that has a 3′ terminal stretch of poly(A).

polycistronic mRNA—mRNA that includes coding regions representing more than one gene.

polymerase chain reaction (PCR)—A process for the amplification of a defined nucleic acid section through repeated thermal cycles of denaturation, annealing, and polymerase extension.

polymorphism—The simultaneous occurrence in the population of alleles showing variations at a given position.

polynucleotide—A chain of nucleotides, such as DNA or RNA.

polyploidization—The process by which a viable organismal lineage gains additional chromosome sets, often by hybridization and chromosome doubling.

polyribosome (polysome)—An mRNA that is simultaneously being translated by several ribosomes.

polytene chromosomes—Chromosomes that are generated by successive replications of a chromosome set without separation of the replicas.

position effect variegation (PEV)—Silencing of gene expression that occurs as the result of proximity to heterochromatin.

positive control—A system in which a gene is not expressed unless some action turns it on.

positive inducible—A control circuit in which an inactive positive regulator is converted into an active regulator by the substrate of the operon.

positive repressible—A control circuit in which an active positive regulator is inactivated by the product of the operon.

postreplication complex—A protein-DNA complex in *S. cerevisiae* that consists of the ORC complex bound to the origin.

posttranscriptional modification—All changes made to the nucleotides of RNA after their initial incorporation into the polynucleotide chain.

posttranslational translocation—The movement of a protein across a membrane after the synthesis of the protein is completed and it has been released from the ribosome.

ppGpp—Guanosine tetraphosphate, with diphosphate groups attached to both the 5′ and 3′ positions, a signaling molecule.

pppGpp—Guanosine pentaphosphate, with a triphosphate group attached to the 5′ position and a diphosphate group attached to the 3′ position, is an alarmone.

precise excision—The removal of a transposon plus one of the duplicated target sequences from the chromosome. Such an event can restore function at the site where the transposon inserted.

preinitiation complex—The assembly of transcription factors at the promoter before RNA polymerase binds in eukaryotic transcription.

pre-mRNA—The nuclear transcript that is processed by modification and splicing to produce a mature mRNA.

prereplication complex—A protein-DNA complex at the origin in *S. cerevisiae* that is required for DNA replication. The complex contains the ORC complex, Cdc6, and the MCM proteins.

presynaptic filaments—Single-stranded DNA bound in a helical nucleoprotein filament with a strand transfer protein such as Rad51 or RecA.

primary transcript—The original unmodified RNA product corresponding to a transcription unit.

primase—A type of RNA polymerase that synthesizes short segments of RNA that will be used as primers for DNA replication.

primer—A short sequence (often of RNA) that is paired with one strand of DNA and provides a free 3′-OH end at which a DNA polymerase starts synthesis of a deoxyribonucleotide chain.

prion—A proteinaceous infectious agent that behaves as an inheritable trait, although it contains no nucleic acid. One example is PrP^{Sc}, the agent of scrapie in sheep and bovine spongiform encephalopathy.

probe—A labeled nucleic acid used to identify a complementary sequence.

processed pseudogenes—Nonfunctional gene copies that are framed by reverse transcription of mRNA and insertion of a duplex copy into the genome.

processing body (PB)—A particle containing multiple mRNAs and proteins involved in mRNA degradation and translational repression, occurring in many copies in the cytoplasm of eukaryotes.

processive (nuclease)—An enzyme that remains associated with the substrate while catalyzing the sequential removal of nucleotides.

processivity—The ability of an enzyme to perform multiple catalytic cycles with a single template instead of dissociating after each cycle.

productive rearrangement—Occurs as a result of the recombination of V, (D), and J gene segments if all the rearranged gene segments are in the correct reading frame.

programmed frameshifting—Frameshifting that is required for expression of the polypeptide sequences encoded beyond a specific site at which a +1 or –1 frameshift occurs at some typical frequency.

promoter—A region of DNA where RNA polymerase binds to initiate transcription.

PROMPT—Promoter upstream transcripts, short RNAs produced from both strands of DNA from active promoters.

proofreading—A mechanism for correcting errors in DNA synthesis that involves scrutiny of individual units after they have been added to the chain.

prophage—A phage genome covalently integrated as a linear part of the bacterial chromosome.

proteinaceous infectious agent (prion)—A proteinaceous agent that behaves as an inheritable trait, although it contains no nucleic acid. Examples are PrPSc, the agent of scrapie in sheep and bovine spongiform encephalopathy, and Psi, which confers an inherited state in yeast.

protein splicing—The autocatalytic process by which an intein is removed from a protein and the exteins on either side become connected by a standard peptide bond.

proteome—The complete set of proteins that is expressed by the entire genome. Sometimes the term is used to describe the complement of proteins expressed by a cell at any one time.

provirus—A duplex sequence of DNA integrated into a eukaryotic genome that represents the sequence of the RNA genome of a retrovirus.

pseudogenes—Inactive but stable components of the genome derived by mutation of an ancestral active gene. Usually they are inactive because of mutations that block transcription or translation or both.

P site—The site in the ribosome that is occupied by peptidyl-tRNA, the tRNA carrying the nascent polypeptide chain, still paired with the codon to which it bound in the A site.

puff—An expansion of a band of a polytene chromosome associated with the synthesis of RNA at some locus in the band.

purine—A double-ringed nitrogenous base, such as adenine or guanine.

pyrimidine—A single-ringed nitrogenous base, such as cytosine, thymine, or uracil.

pyrimidine dimer—A dimer that forms when ultraviolet irradiation generates a covalent link directly between two adjacent pyrimidine bases in DNA. It blocks DNA replication and transcription.

quantitative PCR (qPCR)—A PCR reaction used to amplify and simultaneously quantify an amplicon.

rDNA—Genes encoding ribosomal RNA (rRNA).

reading frame—One of three possible ways of reading a nucleotide sequence. Each divides the sequence into a series of successive triplets.

readthrough—Occurring at transcription or translation when RNA polymerase or the ribosome, respectively, ignores a termination signal because of a mutation of the template or the behavior of an accessory factor.

real-time PCR—A PCR technique with continuous monitoring of product formation as synthesis proceeds, usually through flourometric methods.

***rec* mutations**—Mutations of *E. coli* that cannot undertake general recombination.

recoding—Events that occur when the meaning of a codon or series of codons is changed from that predicted by the genetic code. It may involve altered interactions between aminoacyl-tRNA and mRNA that are influenced by the ribosome.

recognition helix—One of the two helices of the helix-turn-helix motif that makes contacts with DNA that are specific for particular bases. This determines the specificity of the DNA sequence that is bound.

recombinant DNA—An artificial DNA molecule created by joining two (or more) DNA molecules from different sources.

recombinant joint—The point at which two recombining molecules of duplex DNA are connected (the edge of the heteroduplex region).

recombinase—Enzyme that catalyzes site-specific recombination.

recombination nodules (nodes)—Dense objects present on the synaptonemal complex; they may represent protein complexes involved in crossing over.

recombination-repair—A mode of filling a gap in one strand of duplex DNA by retrieving a homologous single strand from another duplex.

recombination signal sequences (RSS)—The combinations of heptamer-spacer-nonamer consensus sequence that are required for accurate Ig locus recombination.

regulator gene—A gene that codes for a product (typically protein) that controls the expression of other genes (usually at the level of transcription).

relaxase—An enzyme that cuts one strand of DNA and binds to the free 5′ end.

relaxed mutant—Mutants that do not display the stringent response to starvation in *E. coli* for amino acids (or other nutritional deprivation).

release factor (RF)—A protein required to terminate polypeptide translation to cause release of the completed polypeptide chain and the ribosome from mRNA.

renaturation—The reassociation of denatured complementary single strands of a DNA double helix.

repetitive DNA—DNA that is present in many (related or identical) copies in a genome.

replication bubble (replication eye)—A region in which DNA has been replicated within a longer, unreplicated region.

replication-defective virus—A virus that cannot perpetuate an infective cycle because some of the necessary genes are absent (replaced by host DNA in a transducing virus) or mutated.

replication fork—The point at which strands of parental duplex DNA are separated so that replication can proceed. A complex of proteins, including DNA polymerase, is found there.

replicative transposition—The movement of a transposon by a mechanism in which first it is replicated, and then one copy is transferred to a new site.

replicon—A unit of the genome in which DNA is replicated. Each contains an origin for initiation of replication.

replisome—The multiprotein structure that assembles at the bacterial replication fork to undertake synthesis of DNA. It contains DNA polymerase and other enzymes.

reporter gene—A sequence that is attached to another gene, which codes for a peptide that is easily identified or measured.

repressible gene—A gene that is turned off by its product.

repression—The ability to prevent synthesis of certain enzymes when their products are present. More generally, it refers to inhibition of transcription (or translation) by binding of repressor protein to a specific site on DNA (or mRNA).

repressor—A protein that inhibits expression of a gene. It may act to prevent transcription by binding to an enhancer or silencer.

resolution—A process occurring by a homologous recombination reaction between the two copies of the transposon in a cointegrate. The reaction generates the donor and target replicons, each with a copy of the transposon.

resolvase—The enzyme activity involved in site-specific recombination between two copies of a transposon that has been duplicated.

restriction endonuclease—An enzyme that recognizes specific short sequences of DNA and cleaves the duplex (sometimes at the target site, sometimes elsewhere, depending on type).

restriction fragment length polymorphism (RFLP)—Inherited differences in target sites for restriction enzymes (for example, caused by base changes in the target site) that result in differences in the lengths of the fragments produced by cleavage with the relevant restriction enzyme. They are used for genetic mapping to link the genome directly to a conventional genetic marker.

restriction map—A linear array of restriction sites on DNA, determined by cleaving the DNA with various restriction endonucleases or by scanning a known sequence for restriction sites.

restriction point—The point during G1 at which a cell becomes committed to division. (In yeast this point is known as START.)

retroposon (non-LTR retrotransposon)—A transposon that mobilizes via an RNA intermediate, similar to an LTR retrotransposon, but that lacks LTRs and uses a distinct transposition mechanism.

retrotransposon (LTR retrotransposon)—A transposon that mobilizes via an RNA form; the DNA element is transcribed into RNA and then reverse transcribed into DNA, which is inserted at a new site in the genome. It does not have an infective (viral) form. This name typically refers to retroelements that contain retrovirus-like LTRs and resemble retroviruses.

retrovirus—An RNA virus with the ability to convert its sequence into DNA by reverse transcription.

reverse gyrase—Enzyme that introduces positive supercoils into DNA.

reverse transcriptase—An enzyme that uses single-stranded RNA as a template to synthesize a complementary DNA strand.

reverse transcription—Synthesis of DNA on a template of RNA. It is accomplished by the enzyme reverse transcriptase.

reverse transcription polymerase chain reaction (RT-PCR)—A technique for the detection and quantification of expression of a gene by reverse transcription and amplification of RNAs from a cell sample.

revertants—Reversions of a mutant cell or organism to the wild-type phenotype.

RF1—The bacterial release factor that recognizes UAA and UAG as signals to terminate polypeptide translation.

RF2—The bacterial release factor that recognizes UAA and UGA as signals to terminate polypeptide translation.

RF3—A polypeptide translation termination factor related to the elongation factor EF-G. It functions to release the factors RF1 or RF2 from the ribosome when they act to terminate polypeptide translation.

Rho-dependent termination—Transcriptional termination by bacterial RNA polymerase in the presence of the Rho factor.

Rho factor—A protein involved in assisting *E. coli* RNA polymerase to terminate transcription at certain terminators (called Rho-dependent terminators).

ribonuclease—An enzyme that cleaves phosphodiester linkages between RNA ribonucleotides.

ribonucleoprotein particles (RNPs)—A complex of RNA and proteins. "Particles" are typically used to refer to large complexes.

ribosomal RNAs (rRNAs)—A major component of the ribosome.

ribosome—A large assembly of RNA and proteins that synthesizes polypeptides under direction from an mRNA template.

ribosome-binding site—A sequence on bacterial mRNA that includes an initiation codon that is bound by a 30S subunit in the initiation phase of polypeptide translation.

ribosome stalling—The inhibition of movement that occurs when a ribosome reaches a codon for which there is no corresponding charged aminoacyl-tRNA.

riboswitch—A catalytic RNA whose activity responds to a metabolite product or another small ligand.

ribozyme—An RNA that has catalytic activity.

RISC—RNA-induced silencing complex, a ribonucleoprotein particle composed of a short single-stranded siRNA and a nuclease that cleaves mRNAs complementary to the siRNA. It receives siRNA from Dicer and delivers it to the mRNA.

risiDNA—A germline subset of miRNA transcribed from transposable elements and other repeated elements that is used to silence transposable elements.

RITS (RNA-induced transcriptional silencing)—A form of RNAi in which siRNAs are used to downregulate transcription by modification and remodeling chromatin.

RNA-binding protein (RPB)—A protein containing one or more domains that confer an affinity for RNA, usually in an RNA sequence- or structure-specific manner.

RNA-dependent RNA polymerase—An enzyme that catalyzes the synthesis of RNA from an RNA template.

RNA editing—A change of sequence at the level of RNA following transcription.

RNA interference (RNAi)—A process by which short 21 to 23 nucleotide antisense RNAs, derived from longer double-stranded RNAs, can modulate expression of mRNA by translation inhibition or degradation.

RNA ligase—An enzyme that functions in tRNA splicing to make a phosphodiester bond between the two exon sequences that are generated by cleavage of the intron.

RNA polymerase—An enzyme that synthesizes RNA using a DNA template (formally described as DNA-dependent RNA polymerases).

RNA processing—Modifications to RNA transcripts of genes. This may include alterations to the 3′ and 5′ ends and the removal of introns.

RNA regulation—A set of mRNAs that are each bound by a particular RNA binding protein, and thus presumably subject to coordinate regulation by that protein.

RNA silencing—The ability of an RNA, especially ncRNA, to alter chromatin structure in order to prevent gene transcription.

RNA splicing—The process of excising introns from RNA and connecting the exons into a continuous mRNA.

RNA surveillance systems—Systems that check RNAs (or RNPs) for errors. The system recognizes an invalid sequence or structure and triggers a response.

RNase—An enzyme that degrades RNA.

rolling circle—A mode of replication in which a replication fork proceeds around a circular template for an indefinite number of revolutions; the DNA strand newly synthesized in each revolution displaces the strand synthesized in the previous revolution, giving a tail containing a linear series of sequences complementary to the circular template strand.

rotational positioning—The location of the histone octamer relative to turns of the double helix, which determines which face of DNA is exposed on the nucleosome surface.

R segments—The sequences that are repeated at the ends of a retroviral RNA. They are called R-U5 and U3-R.

rut—An acronym for rho utilization site, the sequence of RNA that is recognized by the rho termination factor.

satellite DNA—DNA that consists of many tandem repeats (identical or related) of a short basic repeating unit.

scarce mRNA (complex mRNA)—mRNA that consists of a large number of individual mRNA species, each present in very few copies per cell. This accounts for most of the sequence complexity in RNA.

scrapie—A disease caused by an infective agent made of protein (a prion).

second-site reversion—A second mutation suppressing the effect of a first mutation.

selfish DNA—DNA sequences that do not contribute to the phenotype of the organism but have self-perpetuation within the genome as their primary function.

semiconservative replication—DNA replication accomplished by separation of the strands of a parental duplex, with each strand then acting as a template for synthesis of a complementary strand.

semidiscontinuous replication—The mode of replication in which one new strand is synthesized continuously while the other is synthesized discontinuously.

septal ring—A complex of several proteins coded by *fts* genes of *E. coli* that forms at the midpoint of the cell. It gives rise to the septum at cell division. The first of the proteins to be incorporated is FtsZ, which gave rise to the original name of the Z-ring.

septum—The structure that forms in the center of a dividing bacterium, providing the site at which the daughter bacteria will separate. The same term is used to describe the cell wall that forms between plant cells at the end of mitosis.

sequence context—The sequence surrounding a consensus sequence. It may modulate the activity of the consensus sequence.

Shine-Dalgarno sequence—The polypurine sequence AGGAGG centered about 10 bp before the AUG initiation codon on bacterial mRNA. It is complementary to the sequence at the 3′ end of 16S rRNA.

short interspersed elements (SINEs)—Short interspersed elements; a major class of short (<500 bp) nonautonomous retrotransposons that occupy ~13% of the human genome (see **retrotransposon**).

shuttle vector—A cloning vector that can replicate in two different species.

sigma factor—The subunit of bacterial RNA polymerase needed for initiation; it is the major influence on selection of promoters.

signal end—The free DNA end of excised sequences produced during recombination of immunoglobulin and T cell receptor genes. Signal ends lie at the termini of the cleaved fragment containing the recombination signal sequences. Their subsequent joining yields a signal joint.

signal transduction pathway—The process by which a stimulus or cellular state is sensed by and transmitted to pathways within the cell.

silencer—A short sequence of DNA that can inactivate expression of a gene in its vicinity.

silent mutation—A mutation that does not change the sequence of a polypeptide because it produces synonymous codons.

simple sequence DNA—Short repeating units of DNA sequence.

single-copy replication control—A control system in which there is only one copy of a replicon per unit bacterium. The bacterial chromosome and some plasmids have this type of regulation.

single nucleotide polymorphism (SNP)—A polymorphism (variation in sequence between individuals) caused by a change in a single nucleotide. This accounts for most of the genetic variation between individuals.

single-strand binding protein (SSB)—The protein that attaches to single-stranded DNA, thereby preventing the DNA from forming a duplex.

single-strand exchange—A reaction in which one of the strands of a duplex of DNA leaves its former partner and instead pairs with the complementary strand in another molecule, displacing its homologue in the second duplex.

single-strand invasion—The process in which a single strand of DNA displaces its homologous strand in a duplex.

single-X hypothesis—The theory that describes the inactivation of one X chromosome in female mammals.

siRNA—Short interfering RNA, a miRNA that prevents gene expression.

sister chromatid—Each of two identical copies of a replicated chromosome; this term is used as long as the two copies remain linked at the centromere. Sister chromatids separate during anaphase in mitosis or anaphase II in meiosis.

site-specific recombination—Recombination that occurs between two specific sequences, as in phage integration/excision or resolution of cointegrated structures during transposition.

SKI proteins— A set of protein factors that target nonstop decay (NSD) substrates for degradation.

SL RNA (spliced leader RNA)—A small RNA that donates an exon in the *trans*-splicing reaction of trypanosomes and nematodes.

small cytoplasmic RNAs (scRNA; scyrps)—RNAs that are present in the cytoplasm (and sometimes are also found in the nucleus). Scyrps are the ribonucleoprotein particles that include an associated snRNA and its protein partners.

small nuclear RNA (snRNA; snurps)—Small RNA species confined to the nucleus; several of them are involved in splicing or other RNA-processing reactions. Snurps are the ribonucleoprotein particles that include a specific snRNA and its protein partners.

small nucleolar RNA (snoRNA)—A small nuclear RNA that is localized in the nucleolus.

small subunit—The subunit of the ribosome (30S in bacteria, 40S in eukaryotes) that binds the mRNA.

somatic hypermutation (SMH)—The introduction of somatic mutations in a rearranged immunoglobulin gene. The mutations can change the sequence of the corresponding antibody, especially in its antigen-binding site.

somatic mutation—A mutation occurring in a somatic cell, therefore affecting only its daughter cells; it is not inherited by descendants of the organism.

somatic recombination—Recombination that occurs in nongerm cells (i.e., it does not occur during meiosis); most commonly used to refer to the process of joining a V gene to a C gene in a lymphocyte to generate an immunoglobulin or T cell receptor.

Southern blotting—A process for the transfer of DNA bands separated by gel electrophoresis from the gel matrix to a solid support membrane for subsequent probing and detection.

S phase—The restricted part of the eukaryotic cell cycle during which synthesis of DNA occurs.

spindle—A structure made up of microtubules that guides the movements of the chromosomes during mitosis.

splice recombinant—DNA that results from a Holliday junction being resolved by cutting the nonexchanged strands. Both strands of DNA before the exchange point come from one chromosome; the DNA after the exchange point comes from the homologous chromosome.

spliceosome—A complex formed by snRNPs and additional protein factors that is required for RNA splicing.

splicing—The process of excising introns from RNA and connecting the exons into a continuous mRNA.

splicing factor—A protein component of the spliceosome that is not part of one of the snRNPs.

spontaneous mutations—Mutations that occur in the absence of any added reagent to increase the mutation rate as the result of errors in replication (or other events involved in the reproduction of DNA) or by random changes to the chemical structure of bases.

sporulation—The generation of a spore by a bacterium (by morphological conversion) or by a yeast (as the product of meiosis).

S region—A sequence involved in immunoglobulin class switching. They consist of repetitive sequences at the 5′ ends of gene segments encoding the heavy-chain constant regions.

SR protein—A protein that has a variable length of a Ser-Arg-rich region and is involved in splicing.

sRNA—A small bacterial RNA that functions as a regulator of gene expression.

stabilizing element (SE)—One of a variety of *ci* sequences present in some mRNAs that confers a long half-life on that mRNA.

start point—The position on DNA corresponding to the first base incorporated into RNA.

steady state (molecular concentration)—The concentration of a population of molecules when the rates of synthesis and degradation are constant.

stem—The base-paired segment of a hairpin structure in RNA.

stem-loop—A secondary structure that appears in RNAs consisting of a base-paired region (stem) and a terminal loop of single-stranded RNA. Both are variable in size.

steroid receptor—Transcription factor that is activated by binding of a steroid ligand.

stop codon—One of three triplets (UAG, UAA, or UGA) that cause polypeptide translation to terminate. They are also known historically as nonsense codons. The UAA codon is called ochre and the UAG codon is called amber, after the names of the nonsense mutations by which they were originally identified.

strand displacement—A mode of replication of some viruses in which a new DNA strand grows by displacing the previous (homologous) strand of the duplex.

stress granules—Cytoplasmic particles, containing translationally inactive mRNAs, that form in response to a general inhibition of translation initiation.

stringency—A measure of the exactness of complementarity required between two nucleic acid strands to allow them to hybridize. Stringency is related to buffer ionic strength and reaction temperature.

stringent factor—The protein RelA, which is associated with ribosomes. It synthesizes ppGpp and pppGpp when an uncharged tRNA enters the A site.

stringent response—The ability of a bacterium to shut down synthesis of tRNA and ribosomes in a poor growth medium.

stRNA—Short temporal RNA, a form of miRNA in eukaryotes that modulates mRNA expression during development.

structural gene—A gene that encodes any RNA or polypeptide product other than a regulator.

subclone—The process of breaking a cloned fragment into smaller fragments for further cloning.

supercoiling—The coiling of a closed duplex DNA in space so that it crosses over its own axis.

superfamily—A set of genes all related by presumed descent from a common ancestor but now showing considerable variation.

suppression mutation—A second event eliminates the effects of a mutation without reversing the original change in DNA.

suppressor—A second mutation that compensates for or alters the effects of a primary mutation.

synapsis—The association of the two pairs of sister chromatids (representing homologous chromosomes) that occurs at the start of meiosis; the resulting structure is called a bivalent.

synaptonemal complex—The protein structure that forms between synapsed homologous chromosomes that is believed to be necessary for recombination to occur.

synonymous codons—Codons that have the same meaning (specifying the same amino acid or specifying termination of translation) in the genetic code.

synonymous mutation—A mutation in a coding region that does not alter the amino acid sequence of the polypeptide product.

synteny—A relationship between chromosomal regions of different species where homologous genes occur in the same order.

synthetic genetic array analysis (SGA)—An automated technique in budding yeast whereby a mutant is crossed to an array of approximately 5000 deletion mutants to determine if the mutations interact to cause a synthetic lethal phenotype.

synthetic lethality—An event that occurs when two mutations that by themselves are viable cause lethality when combined.

TAFs—The subunits of $TF_{III}B$ that assist TBP in binding to DNA. They also provide points of contact for other components of the transcription apparatus.

TATA-binding protein (TBP)—The subunit of transcription factor $TF_{II}D$ that binds to the TATA box in the promoter and is positioned at the promoters that do not contain a TATA box by other factors. Also present in SLI and $TF_{III}B$.

TATA box—A conserved A-T-rich octamer found about 25 bp before the start point of each eukaryotic RNA polymerase II transcription unit; it is involved in positioning the enzyme for correct initiation.

TATA-less promoter—A promoter lacking a TATA box in the sequence upstream of its start point.

T cell receptor (TCR)—The antigen receptor on T lymphocytes. It is clonally expressed and binds to a complex of MHC class I or class II protein and antigen-derived peptide.

T cells—Lymphocytes of the T (thymic) lineage; they may be subdivided into several functional types. They carry T cell receptors and are involved in the cell-mediated immune response.

T-DNA—The segment of the Ti plasmid of *Agrobacterium tumefaciens* that is transferred to the plant cell nucleus during infection. It carries genes that transform the plant cell.

telomerase—The ribonucleoprotein enzyme that creates repeating units of one strand at the telomere by adding individual bases to the DNA 3′ end, as directed by an RNA sequence in the RNA component of the enzyme.

telomere—The natural end of a chromosome; the DNA sequence consists of a simple repeating unit with a protruding single-stranded end.

telomeric silencing—The repression of gene activity that occurs in the vicinity of a telomere.

temperate phage—A phage that can enter a lysogenic cycle within the host (can become a prophage integrated into the host genome) or enter the lytic cycle.

template strand—The DNA strand that is copied by the polymerase.

terminal protein—A protein that allows replication of a linear phage genome to start at the very end. It attaches to the 5′ end of the genome through a covalent bond, is associated with a DNA polymerase, and contains a cytosine residue that serves as a primer.

terminase—An enzyme cleaves multimers of a viral genome and then uses hydrolysis of ATP to provide the energy to translocate the DNA into an empty viral capsid starting with the cleaved end.

termination—A separate reaction that ends a macromolecular synthesis reaction (replication, transcription, or translation) by stopping the addition of subunits and (typically) causing disassembly of the synthetic apparatus.

termination codon—One of the three codons (UAA, UAG, UGA) that signal the termination of translation of a polypeptide. They are also known as stop codons.

terminator—A sequence of DNA that causes RNA polymerase to terminate transcription.

ternary complex—The complex in initiation of transcription that consists of RNA polymerase and DNA as well as a dinucleotide that represents the first two bases in the RNA product.

$TF_{II}D$—The transcription factor that binds to the TATA sequence or Inr of promoters for RNA polymerase II. It consists of TBP (TATA binding protein) and the TAF subunits that bind to TBP.

thalassemia—A disease of red blood cells resulting from lack of either α or β globin.

third-base degeneracy—The lesser effect on codon meaning of the nucleotide present in the third (3′) codon position.

threshold cycle (C_T)—The thermocycle number in a real-time PCR or RT-PCR reaction at which the product signal rises above a specified cutoff value to indicate amplicon production is occurring.

Ti plasmid—An episome of the bacterium *Agrobacterium tumefaciens* that carries the genes responsible for the induction of crown gall disease in infected plants.

tight binding—The binding of RNA polymerase to DNA in the formation of an open complex (when the strands of DNA have separated).

tiling array—An array of immobilized nucleic acid sequences, which together represent the entire genome of an organism. The shorter each array sequence is, the larger the total required number of spots is, but the greater the genetic resolution of the array.

T_m—The theoretical melting temperature of a duplex nucleic acid segment into separate strands. T_m is dependent on parameters that include sequence composition, duplex length, and buffer ionic strength.

Tn—Followed by a number, it denotes bacterial transposons carrying markers that are not related to their function, e.g., drug resistance.

topoisomerase—An enzyme that changes the number of times the two strands in a closed DNA molecule cross each other. It does this by cutting the DNA, passing DNA through the break, and resealing the DNA.

trailer (3′ UTR)—An untranslated sequence at the 3′ end of an mRNA following the termination codon.

TRAMP—A protein complex that identifies and polyadenylates aberrant nuclear RNAs in yeast, recruiting the nuclear exosome for degradation.

***trans*-acting**—A product that can function on any copy of its target DNA. This implies that it is a diffusible protein or RNA.

***trans*-acting sequence**—DNA sequence encoding for a product that can function on any copy of its target DNA. This implies that it is a diffusible protein or RNA.

transcription—Synthesis of RNA from a DNA template.

transcription interference (TI)—The phenomenon in which transcription from one promoter interferes directly with transcription from a second, linked promoter.

transcription unit—The sequence between sites of initiation and termination by RNA polymerase; it may include more than one gene.

transcriptome—The complete set of RNAs present in a cell, tissue, or organism. Its complexity is due mostly to mRNAs, but it also includes noncoding RNAs.

transducing virus—A virus that carries part of the host genome in place of part of its own sequence. The best known examples are retroviruses in eukaryotes and DNA phages in *E. coli*.

transesterification—A reaction that breaks and makes chemical bonds in a coordinated transfer so that no energy is required.

transfection—In eukaryotic cells, the acquisition of new genetic material by incorporation of added DNA.

transfer region—A segment on the F plasmid that is required for bacterial conjugation.

transfer RNA (tRNA)—The intermediate in protein synthesis that interprets the genetic code. Each molecule can be linked to an amino acid. It has an anticodon sequence that is complementary to a triplet codon representing the amino acid.

transformation—In bacteria, the acquisition of new genetic material by incorporation of added DNA.

transforming principle—DNA that is taken up by a bacterium and whose expression then changes the properties of the recipient cell.

transgenic—An organism created by introducing DNA prepared in test tubes into the germline. The DNA may be inserted into the genome or exist in an extrachromosomal structure.

transition—A mutation in which one pyrimidine is replaced by the other, or in which one purine is replaced by the other.

translation—Synthesis of protein from an mRNA template.

translational positioning—The location of a histone octamer at successive turns of the double helix, which determines which sequences are located in linker regions.

translocation—1. The movement of the ribosome one codon along mRNA after the addition of each amino acid to the polypeptide chain. 2. The reciprocal or nonreciprocal exchange of chromosomal material between nonhomologous chromosomes.

transposable element—See **transposon**.

transposase—The enzyme activity involved in insertion of transposon at a new site.

transposition—Movement of mobile genetic elements from one location in the genome to another.

transposon—A DNA sequence able to insert itself (or a copy of itself) at a new location in the genome without having any sequence relationship with the target locus.

transversion—A mutation in which a purine is replaced by a pyrimidine, or vice versa.

$\text{tRNA}_\text{f}^\text{Met}$—The special RNA used to initiate polypeptide translation in bacteria. It mostly uses AUG but can also respond to GUG and CUG.

$\text{tRNA}_\text{m}^\text{Met}$—The bacterial tRNA that inserts methionine at internal AUG codons.

true activator—A positive transcription faction that functions by making contact, direct or indirect, with the basal apparatus to activate transcription.

true reversion—A mutation that restores the original sequence of the DNA.

tumor suppressor—Proteins that usually act by blocking cell proliferation or promoting cell death. Cancer may result when a tumor-suppressor gene is inactivated by a loss-of-function mutation.

Ty—It stands for transposon yeast, the first transposable element to be identified in yeast.

type I topoisomerase—An enzyme that changes the topology of DNA by nicking and resealing one strand of DNA.

type II topoisomerase—An enzyme that changes the topology of DNA by nicking and resealing both strands of DNA.

UAS (upstream activating sequence)—The equivalent in yeast of the enhancer in higher eukaryotes and is bound by transcriptional regulatory proteins.

U5—The repeated sequence at the 5′ end of a retroviral RNA.

underwound—B-form DNA that has greater than 10.4 base pairs per turn of the helix.

unequal crossing over (nonreciprocal recombination)—The results of an error in pairing and crossing over in which nonequivalent sites are involved in a recombination event. It produces one recombinant with a deletion of material and one with a duplication.

unidentified reading frame (URF)—An open reading frame with an as yet undetermined function.

uninducible—A mutant in which the affected gene(s) cannot be expressed.

Upf proteins—A set of protein factors that target nonsense-mediated decay (NMD) substrates for degradation.

U3—The repeated sequence at the 3′ end of a retroviral RNA.

up mutation—A mutation in a promoter that increases the rate of transcription.

upstream—Sequences in the opposite direction from expression.

upstream activating sequence (UAS)—The equivalent in yeast of the enhancer in higher eukaryotes; a UAS cannot function downstream of the promoter.

variable number tandem repeat (VNTR)—Very short repeated sequences, including microsatellites and minisatellites.

variable region (V region)—An antigen-binding site of an immunoglobulin, or T cell receptor, molecule. They are composed of the variable domains of the component chains. They are coded by V gene segments and vary extensively among antigen receptors as the result of multiple, different genomic copies and of changes introduced during synthesis.

variegation—It is produced by a change in genotype during somatic development.

vector—A plasmid or phage chromosome that is used to perpetuate a cloned DNA segment.

vegetative phase—The period of normal growth and division of a bacterium. For a bacterium that can sporulate, this contrasts with the sporulation phase, when spores are being formed.

V gene—A sequence coding for the major part of the variable (N-terminal) region of an immunoglobulin chain.

viroid—A small infectious nucleic acid that does not have a protein coat.

virulent mutations—Phage mutants that are unable to establish lysogeny

virulent phage—A bacteriophage that can only follow the lytic cycle.

virusoid—A small infectious nucleic acid that is encapsidated by a plant virus together with its own genome.

wobble hypothesis—The ability of a tRNA to recognize more than one codon by unusual (non-G-C, non-A-T) pairing with the third base of a codon.

xeroderma pigmentosum (XP)—A disease caused by mutation in one of the XP genes that results in hypersensitivity to sunlight (particularly ultraviolet light), skin disorders, and cancer predisposition.

yeast artificial chromosome (YAC)— A cloning vector used to clone very large DNA fragments, up to 3000 kb in size, containing yeast telomeres, a centromere, and a replication origin so that it can propagate in yeast cells.

zinc finger—A DNA-binding motif that typifies a class of transcription factor that contains one or more zinc ions to help stabilize the proteins.

zipcode (in RNA)—Any of the number of RNA *cis* elements involved in directing cellular localization; also called localization signals.

zoo blot—The use of Southern blotting to test the ability of a DNA probe from one species to hybridize with the DNA from the genomes of a variety of other species.

Z-ring—See septal ring.**risiDNA**—A germline subset of miRNA transcribed from transposable elements and other repeated elements that is used to silence transposable elements.

Appendix

Answers to Even-Numbered End-of-Chapter Questions

Chapter 1

2. D; **4.** A; **6.** A; **8.** D; **10.** A

Chapter 2

2. B; **4.** B; **6.** C; **8.** C; **10.** C

Chapter 3

2. C; **4.** A; **6.** E, B, G, F, D, C, A

Chapter 4

2. C; **4.** A; **6.** B; **8.** B; **10.** D

Chapter 5

2. B; **4.** D; **6.** C; **8.** C; **10.** B

Chapter 6

2. C; **4.** C; **6.** B; **8.** D; **10.** B

Chapter 7

2. A; **4.** A; **6.** C; **8.** B; **10.** A

Chapter 8

2. B; **4.** A; **6.** A; **8.** C; **10.** A

Chapter 9

2. filaments; **4.** spherical or icosohedron; **6.** it often consists of multiple repeats that are not transcribed; **8.** it replicates late in S phase and has a reduced incidence of genetic recombination; **10.** C; **12.** B; **14.** A; **16.** A

Chapter 10

2. prevents; **4.** promoter; **6.** repressive or silencing; **8.** B; **10.** C; **12.** B

Chapter 11

2. C; **4.** A; **6.** D; **8.** A

Chapter 12

2. A; **4.** A; **6.** B; **8.** C

Chapter 13

2. B; **4.** D; **6.** D; **8.** B; **10.** A

Chapter 14

2. B; **4.** D; **6.** C; **8.** D; **10.** B

Chapter 15

2. bacterial protein IHF, phage protein Int, *attB* sequence in the bacterial chromosome, *attP* sequence in the phage genome; **4.** A; **6.** C; **8.** D; **10.** C

Chapter 16

2. removes DNA between nicks around the damaged site; **4.** helicases; **6.** seals nick in DNA to reestablish a continuous phosphodiester backbone; **8.** lyases; **10.** A; **12.** C; **14.** D

Chapter 17

2. C; **4.** A; **6.** any of the following: matrix protein, capsid protein, nucleocapsid protein; **8.** any of the following: protease, reverse transcriptase, integrase; **10.** surface protein (spikes allow virion to interact with host); **12.** C; **14.** A; **16.** B; **18.** D

Chapter 18

2. A; **4.** B, **6.** D; **8.** heptamer; **10.** signal or noncoding region

Chapter 19

2. C; **4.** A; **6.** D; **8.** A; **10.** B

Chapter 20

2. B, C, C; **4.** B; **6.** A; **8.** B; **10.** C

Chapter 21

2. B; **4.** C; **6.** D; **8.** B; **10.** A

Chapter 22

2. A; **4.** B; **6.** C; **8.** D

Chapter 23

2. B; **4.** C; **6.** A; **8.** B; **10.** C

Chapter 24

2. B; **4.** A; **6.** B; **8.** D; **10.** B

Chapter 25

2. C; **4.** A; **6.** D; **8.** A; **10.** B

Chapter 26

2. A; **4.** B; **6.** C; **8.** A; **10.** B

Chapter 27

2. B; **4.** B; **6.** D; **8.** C; **10.** A

Chapter 28

2. C; **4.** B; **6.** A; **8.** C; **10.** B

Chapter 29

2. it remains condensed in interphase; **4.** it replicates late in S phase; **6.** constitutive heterochromatin contains specific DNA sequences that have no coding function and includes (in some species) satellite DNAs located at centromeres; **8.** in mammals, one of the two female X chromosomes is inactivated; **10.** in *C. elegans*, gene expression from each female X chromosome is only half relative to expression from the single male X chromosome; **12.** a self-perpetuating state or conformation of a protein may be established and maintained; **14.** Scrapie, Gerstmann-Staussler syndrome or Creuzfeldt-Jakob disease [CJD]

Chapter 30

2. D

Index

B

C

E